RABIES

FOURTH EDITION

RABIES
Scientific Basis of the Disease and Its Management

FOURTH EDITION

Edited by

ANTHONY R. FOOKS

Director of a World Health Organization Communicable Disease Surveillance and Response Collaborating Centre for the Characterization of Rabies and Rabies-Related Viruses/Head of an OIE (World Organisation for Animal Health) Reference Laboratory for Rabies, The Animal and Plant Health Agency, Addlestone, Surrey, United Kingdom

ALAN C. JACKSON

Professor of Medicine (Neurology), University of Manitoba, Winnipeg, MB, Canada

Academic Press is an imprint of Elsevier
125 London Wall, London EC2Y 5AS, United Kingdom
525 B Street, Suite 1650, San Diego, CA 92101, United States
50 Hampshire Street, 5th Floor, Cambridge, MA 02139, United States
The Boulevard, Langford Lane, Kidlington, Oxford OX5 1GB, United Kingdom

© 2020 Elsevier Inc. All rights reserved.
Exception to the above: Chapter 4: 2020 Published by Elsevier Inc.

No part of this publication may be reproduced or transmitted in any form or by any means, electronic or mechanical, including photocopying, recording, or any information storage and retrieval system, without permission in writing from the publisher. Details on how to seek permission, further information about the Publisher's permissions policies and our arrangements with organizations such as the Copyright Clearance Center and the Copyright Licensing Agency, can be found at our website: www.elsevier.com/permissions.

This book and the individual contributions contained in it are protected under copyright by the Publisher (other than as may be noted herein).

Notices
Knowledge and best practice in this field are constantly changing. As new research and experience broaden our understanding, changes in research methods, professional practices, or medical treatment may become necessary.

Practitioners and researchers must always rely on their own experience and knowledge in evaluating and using any information, methods, compounds, or experiments described herein. In using such information or methods they should be mindful of their own safety and the safety of others, including parties for whom they have a professional responsibility.

To the fullest extent of the law, neither the Publisher nor the authors, contributors, or editors, assume any liability for any injury and/or damage to persons or property as a matter of products liability, negligence or otherwise, or from any use or operation of any methods, products, instructions, or ideas contained in the material herein.

Library of Congress Cataloging-in-Publication Data
A catalog record for this book is available from the Library of Congress

British Library Cataloguing-in-Publication Data
A catalogue record for this book is available from the British Library

ISBN 978-0-12-818705-0

For information on all Academic Press publications
visit our website at https://www.elsevier.com/books-and-journals

Front Cover Credits in Rabies, 4th edition (Fooks and Jackson, eds)

Images were adapted with permission from Scott et al. in Journal of Virology 82:513–521, 2008; doi: 10.1128/JVI.01677-07 Copyright © 2008 (in upper row on the left and right) and from Jackson et al. in Journal of Virology 84:4697–4705, 2010; doi: 10.1128/JVI.02654-09 Copyright ©2010 (in lower row in the middle), American Society for Microbiology.

Publisher: Andre Gerhard Wolff
Acquisitions Editor: Kattie Washington
Editorial Project Manager: Samantha Allard
Production Project Manager: Selvaraj Raviraj
Cover Designer: Victoria Pearson

Typeset by SPi Global, India

Contents

Contributors ix
Foreword xi
Preface xvii

1. A history of rabies—The foundation for global canine rabies elimination

Charles E. Rupprecht, Conrad M. Freuling, Reeta S. Mani, Carlos Palacios, Claude T. Sabeta, and Michael Ward

1.1 Introduction 1
1.2 Asia 2
1.3 Europe 7
1.4 Africa 11
1.5 South America, Central America, and the Caribbean 16
1.6 North America 23
1.7 Australia 25
References 28

2. Rabies virus

William H. Wunner and Karl-Klaus Conzelmann

2.1 Introduction 43
2.2 Rabies virus architecture 44
2.3 Genome and RNP structures 56
2.4 Life cycle of rabies virus infection 62
References 70
Further reading 81

3. Evolution of rabies virus

Daniel G. Streicker and Roman Biek

3.1 Introduction 83
3.2 Microevolutionary dynamics of rabies in stable host virus associations 85
3.3 Macroevolutionary dynamics of rabies 87
3.4 Reconciling strains of evidence for the role of adaptive evolution in cross-species transmission 91
3.5 Applications of evolutionary data for rabies prevention and control—genetics as a tag on transmission 94
3.6 Conclusions 95
References 96

4. Epidemiology

Ryan MacLaren Wallace and Jesse Blanton

4.1 Introduction 103
4.2 Global rabies epidemiology 104
4.3 RABV transmission 114
4.4 Surveillance general 117
4.5 Human rabies: Public health measures 129
4.6 Animal rabies: Epidemiology and control 130
References 133

5. Molecular epidemiology

Susan A. Nadin-Davis

5.1 Introduction 143
5.2 Key aspects of *Lyssavirus* biology 143
5.3 Methods of viral typing 145
5.4 Lyssavirus taxonomy 153
5.5 Emerging trends 174
5.6 Concluding remarks 177
Acknowledgments 178
References 178

6. Rabies in terrestrial animals

Thomas Müller and Conrad M. Freuling

6.1 Introduction 195
6.2 Lyssavirus infections in reservoir species 198

6.3 Spillover (dead-end) hosts for terrestrial rabies   212
Acknowledgments   217
References   217

7. Bat rabies

Ashley C. Banyard, April Davis, Amy T. Gilbert, and Wanda Markotter

7.1 Introduction   231
7.2 Bat rabies in the New World   233
7.3 Bat rabies in the Old World   239
7.4 African lyssaviruses   243
7.5 Asian and Australian lyssaviruses   247
7.6 Experimental studies with lyssaviruses in bats   248
7.7 Important knowledge gaps and challenges to lyssavirus research   253
7.8 Future prospects for controlling lyssaviruses in bats   260
7.9 Conclusions   262
Acknowledgments   262
References   263

8. Human disease

Alan C. Jackson

8.1 Introduction   277
8.2 Exposures, incubation period, and prodromal symptoms   277
8.3 Clinical forms of disease   282
8.4 Investigations   287
8.5 Differential diagnosis   290
8.6 Rabies due to other Lyssavirus species   291
8.7 Conclusions   295
References   296

9. Pathogenesis

Alan C. Jackson

9.1 Introduction   303
9.2 Virus entry into the nervous system   303
9.3 RABV receptors   307
9.4 Spread to the CNS   310
9.5 Spread within the CNS   312
9.6 Spread from the CNS   313
9.7 Animal models of RABV neurovirulence   314

9.8 Structural damage caused by RABV infection in the CNS   315
9.9 Brain dysfunction in rabies   323
9.10 Recovery from rabies and chronic RABV infection   332
9.11 Conclusions   335
References   335

10. Pathology

John P. Rossiter and Alan C. Jackson

10.1 Introduction   347
10.2 Macroscopic findings   348
10.3 Pathology in the central nervous system   348
10.4 Pathology in the peripheral nervous system   364
10.5 Pathology involving the inoculation site, eye, and extraneural organs   368
10.6 Summary and conclusions   370
References   371

11. Immunology

Monique Lafon

11.1 Introduction   379
11.2 RABV innate immune response   381
11.3 RABV adaptive immune response   386
11.4 RABV infection triggers a CNS-mediated immune unresponsiveness   390
11.5 Paradoxical role of IFN in RABV virulence   390
11.6 Conclusions   392
References   392

12. Laboratory diagnosis of rabies

Lorraine M. McElhinney, Denise A. Marston, Megan Golding, and Susan A. Nadin-Davis

12.1 Laboratory-based rabies diagnostic testing   401
12.2 History of rabies diagnostic tests   410
12.3 Detection of viral antigen   411
12.4 Molecular methods of viral RNA detection   416
12.5 Detection of live virus   433
12.6 Antemortem diagnosis of rabies   434
12.7 Conclusions   436
Acknowledgments   437
References   437

13. Measures of rabies immunity
Susan M. Moore and Chandra R. Gordon

13.1 Introduction   445
13.2 Rabies serology methods   449
13.3 Assay selection   459
13.4 Assuring quality results   461
13.5 Defining "adequate" or "minimum" response to rabies vaccination   466
13.6 Regulatory compliance   470
13.7 Conclusions   474
References   475

14. Human and animal vaccines
Thirumeni Nagarajan and Hildegund C.J. Ertl

14.1 Introduction   481
14.2 History of rabies vaccines   482
14.3 Pre-exposure prophylaxis   482
14.4 Post-exposure prophylaxis   483
14.5 Rabies virus strains for vaccine production   483
14.6 Production of rabies vaccines   485
14.7 Cell substrates for rabies virus propagation   487
14.8 Primary cells   488
14.9 Diploid cells   488
14.10 Continuous cell lines   489
14.11 Production systems   490
14.12 Viral inactivation   491
14.13 Downstream processing   491
14.14 RABV purification   492
14.15 Formulation   493
14.16 Potency testing   494
14.17 Safety issues and mitigation strategies   495
14.18 Oral rabies vaccination   496
14.19 Wildlife   497
14.20 Domestic animals   498
14.21 Monitoring oral rabies vaccination   499
14.22 Future directions   499
References   500

15. Next generation of rabies vaccines
Hildegund C.J. Ertl

15.1 Introduction   509
15.2 Current rabies vaccine regimens   510
15.3 Incidence and risk for rabies and vaccine failures   511
15.4 Correlates of protection   511
15.5 Novel vaccines to rabies   512
15.6 The "ideal" rabies vaccine   520
15.7 Summary   521
References   522

16. Public health management of humans at risk
Deborah J. Briggs and Susan M. Moore

16.1 Introduction   527
16.2 Obstacles to preventing rabies in humans   529
16.3 Exposure   529
16.4 Vaccination protocols to prevent rabies   531
16.5 Special populations   536
16.6 Adverse reactions to cell culture vaccines   537
16.7 Interchangeability and coadministration of vaccines   537
16.8 Educational initiatives   538
16.9 Reducing vaccination regimens   539
16.10 Conclusions and future   539
References   540
Further reading   545

17. Therapy of human rabies
Alan C. Jackson

17.1 Human cases with recovery from rabies   551
17.2 Future prospects for the aggressive management of rabies in humans   557
17.3 Importance of palliation in the management of rabies   561
References   561

18. Dog rabies and its control
Darryn L. Knobel, Katie Hampson, Tiziana Lembo, Sarah Cleaveland, and Alicia Davis

18.1 Introduction   567
18.2 The influence of epidemiological and socioeconomic frameworks on dog rabies control   570
18.3 Practical aspects of dog rabies control   581
18.4 Conclusions   592
References   593

19. Rabies control in wild carnivores
Amy T. Gilbert and Richard B. Chipman

19.1 Introduction 605
19.2 Historical aspects, milestones, and epizootiology 606
19.3 Conceptual foundations and principles 616
19.4 Structure and operation of control programs 617
19.5 Bait and vaccine principles, research and developments 634
19.6 Conclusions 638
Acknowledgments 638
References 638

20. Modeling canine rabies virus transmission dynamics
Malavika Rajeev, C. Jessica E. Metcalf, and Katie Hampson

20.1 Introduction 655
20.2 Existing modeling studies 660
20.3 The gap between models and data 663
20.4 Conclusions 665
Data and code availability 666
References 666

21. Strategies for the elimination of dog-mediated human rabies by 2030
Terrence P. Scott, Andre Coetzer and Louis H. Nel

21.1 Introduction 671
21.2 Sustainability requires government ownership and commitment 672
21.3 International guidelines/recommendations 678
21.4 Networks 679
21.5 Monitoring and surveillance 680
21.6 Conclusions 684
References 684

22. Future developments and challenges
Anthony R. Fooks and Alan C. Jackson

22.1 Introduction 689
22.2 Pathogenesis 690
22.3 Therapy of human rabies 690
22.4 Epidemiology 691
22.5 Prevention of human rabies 692
22.6 Control of animal rabies 694
22.7 Health impact and economic burden of canine rabies 695
22.8 Conclusions 695
References 696

Index 699

Contributors

Ashley C. Banyard Animal and Plant Health Agency (Weybridge), Addlestone, Surrey, United Kingdom

Roman Biek Institute of Biodiversity, Animal Health and Comparative Medicine, College of Medical Veterinary and Life Sciences, University of Glasgow, Glasgow, Scotland

Jesse Blanton United States Centers for Disease Control and Prevention, Poxvirus and Rabies Branch, Atlanta, GA, United States

Deborah J. Briggs Department of Diagnostic Medicine, College of Veterinary Medicine, Kansas State University, Manhattan, KS, United States

Richard B. Chipman United States Department of Agriculture, Animal and Plant Health Inspection Service, Wildlife Services, National Rabies Management Program, Concord, NH, United States

Sarah Cleaveland Institute of Biodiversity, Animal Health and Comparative Medicine, College of Medical, Veterinary and Life Sciences, University of Glasgow, Glasgow, United Kingdom

Andre Coetzer Department of Biochemistry, Genetics and Microbiology, Faculty of Natural and Agricultural Sciences, University of Pretoria, Pretoria, Gauteng, South Africa; Global Alliance for Rabies Control, Manhattan, KS, United States

Karl-Klaus Conzelmann Max von Pettenkofer Institute Virologie, Medical Faculty & Gene Center, Ludwig-Maximilians University of Munich, Munich, Germany

Alicia Davis School of Social and Political Sciences/Institute of Health and Wellbeing, University of Glasgow, Glasgow, United Kingdom

April Davis Rabies Laboratory, Wadsworth Center New York State Department of Health, Slingerlands, NY, United States

Hildegund C.J. Ertl The Wistar Institute, Philadelphia, PA, United States

Anthony R. Fooks Director of a World Health Organization Communicable Disease Surveillance and Response Collaborating Centre for the Characterization of Rabies and Rabies-Related Viruses/Head of an OIE (World Organisation for Animal Health) Reference Laboratory for Rabies, The Animal and Plant Health Agency, Addlestone, Surrey, United Kingdom

Conrad M. Freuling Institute of Molecular Virology and Cell Biology, Friedrich-Loeffler-Institut, Federal Research Institute for Animal Health, Greifswald-Insel Riems, Germany

Amy T. Gilbert United States Department of Agriculture, Animal and Plant Health Inspection Service, Wildlife Services, National Wildlife Research Center, Fort Collins, CO, United States

Megan Golding Animal & Plant Health Agency (Weybridge), Addlestone, Surrey, United Kingdom

Chandra R. Gordon Rabies Laboratory/Veterinary Diagnostic Laboratory, College of Veterinary Medicine, Kansas State University, Manhattan, KS, United States

Katie Hampson Institute of Biodiversity, Animal Health and Comparative Medicine, College of Medical, Veterinary and Life Sciences, University of Glasgow, Glasgow, United Kingdom

Alan C. Jackson Professor of Medicine (Neurology), University of Manitoba, Winnipeg, MB, Canada

Darryn L. Knobel Department of Biomedical Sciences, Ross University School of Veterinary Medicine, Basseterre, St Kitts and Nevis

Monique Lafon Institut Pasteur, Paris, France

Tiziana Lembo Institute of Biodiversity, Animal Health and Comparative Medicine, College of Medical, Veterinary and Life Sciences, University of Glasgow, Glasgow, United Kingdom

Reeta S. Mani Department of Neurovirology, National Institute of Mental Health & Neurosciences, Bangalore, India

Wanda Markotter Centre for Viral Zoonoses, Department of Medical Virology, Faculty of Health Sciences, University of Pretoria, Pretoria, South Africa

Denise A. Marston Animal & Plant Health Agency (Weybridge), Addlestone, Surrey, United Kingdom

Lorraine M. McElhinney Animal & Plant Health Agency (Weybridge), Addlestone, Surrey, United Kingdom

C. Jessica E. Metcalf Department of Ecology and Evolutionary Biology, Princeton University, Princeton, NJ, United States

Susan M. Moore Rabies Laboratory/Veterinary Diagnostic Laboratory; Department of Diagnostic Medicine, College of Veterinary Medicine, Kansas State University, Manhattan, KS, United States

Thomas Müller Institute of Molecular Virology and Cell Biology, Friedrich-Loeffler-Institut, Federal Research Institute for Animal Health, Greifswald-Insel Riems, Germany

Susan A. Nadin-Davis Canadian Food Inspection Agency, Ottawa Laboratory-Fallowfield, Ottawa, ON, Canada

Thirumeni Nagarajan R&D center, Vaccines Division, Biological E Limited, Hyderabad, India

Louis H. Nel Department of Biochemistry, Genetics and Microbiology, Faculty of Natural and Agricultural Sciences, University of Pretoria, Pretoria, Gauteng, South Africa; Global Alliance for Rabies Control, Manhattan, KS, United States

Carlos Palacios Pablo Cassara Foundation, Buenos Aires, Argentina

Malavika Rajeev Department of Ecology and Evolutionary Biology, Princeton University, Princeton, NJ, United States

John P. Rossiter Department of Pathology and Molecular Medicine, Queen's University and Kingston Health Sciences Centre, Kingston, ON, Canada

Charles E. Rupprecht LYSSA LLC, Cumming, GA, United States

Claude T. Sabeta Onderstepoort Veterinary Institute, Pretoria, South Africa

Terrence P. Scott Department of Biochemistry, Genetics and Microbiology, Faculty of Natural and Agricultural Sciences, University of Pretoria, Pretoria, Gauteng, South Africa; Global Alliance for Rabies Control, Manhattan, KS, United States

Daniel G. Streicker Institute of Biodiversity, Animal Health and Comparative Medicine, College of Medical Veterinary and Life Sciences, University of Glasgow; MRC-University of Glasgow Centre for Virus Research, Glasgow, Scotland

Ryan MacLaren Wallace United States Centers for Disease Control and Prevention, Poxvirus and Rabies Branch, Atlanta, GA, United States

Michael Ward The University of Sydney, Camden, Australia

William H. Wunner The Wistar Institute, Philadelphia, PA, United States

Foreword

I give great credit to Alan Jackson for having had the vision to develop the first edition of this, our "encyclopedia of rabies," and for the perseverance to continue the project, now through four editions. As each edition has had to stand as a complete information source at the time of its publication, this has meant reviewing every chapter every time, whether the chapter subject has been relatively quiet or rip-roaring. I wonder if Alan had any notion at the beginning that there could be no end. Now, eighteen years later (2002–20), it seems more important than ever that the series be continued—all of the foundational information has become the stuff supporting rabies prevention, control and clinical activities, and now the ambitious goal to eliminate dog-mediated human rabies, globally, by 2030. Perhaps the plan is more than ambitious but, indeed, so was oral vaccination of wildlife when it was first attempted. It is a worthy enterprise—grounded in a perspective of "science-in-support-of-public-health." And, I see it as a fitting goal, given that rabies control is the signature historic emblem of veterinary medicine's role in public health and comparative medicine—in many countries it has been the rabies control program in the "veterinary unit/branch/division" of the public health department that has led to all the other ways such units now contribute to public health.

Along the way, Alan has had pillars in rabies research and public health practice as co-editors and chapter authors; now, we all wish Anthony Fooks a "welcome aboard" and a bon voyage.

I am interested in all matters rabies and have spent some time with each chapter; as I categorize my thoughts I can imagine an informal social gathering of all the editors and chapter authors as they air out thoughts, notions, reflections, ruminations, and crystal ball gazing that might follow upon consideration of the formal contents of the book per se. My own thoughts might extend from:

- *The history of rabies*. The history of medical/veterinary virology cannot be told without paying homage to Louis Pasteur and his colleagues, and their influence on the many seminal discoveries of others made in following years. Since I believe that a history lesson should be part of every scholar's introduction to the discipline, the bolder the history chapter, the better! And, the history lesson should always retell some of the grand stories that might hook the next generation of rabies scholars.
- *The rabies viruses (the lyssaviruses), their molecular biology, evolution, and phylogeography*. Some might say that the molecular structure/function of the viruses is now an established subject, with the discoveries of the past 40 years now fully appreciated and applied, but there have been so many nuanced advances in the past few years affecting practical downstream subjects, such as diagnostics, vaccinology, etc., that certainly there is much more to come. I wonder how others see the future of rabies molecular biology—I recall Niels Bohr's rejoinder: "Prediction is very difficult, especially if it's about the future."

- *The disease, rabies, its clinical presentation, pathology, and pathogenesis.* Is it not true that there are still too many mysteries in rabies pathogenesis/pathophysiology to ease off on this area of research? The dominance in recent years of in vitro technologies, the choice in most rabies research programs of laboratory-adapted strains of virus, the ethical reluctance to employ experimental animals, especially dogs and other reservoir host animals, and the scarcity of individuals trained in all necessary (tertiary[a]) subdisciplines of pathology make me wonder about future progress.
- *The host, its innate and immune responses.* Recent research findings about the host response to rabies infection are intriguing, but I wonder whether there are an appropriate number of immunologists in the rabies enterprise. When a research field is as complex as has been proven to be the case with rabies immunology, should there not be more research funding made available so as to draw in more of the diverse expertise called for?
- *Vaccination, of humans, domestic animals, and wildlife.* As has been stated clearly and forcefully in each edition of this book, the major impediment in advancing human and animal vaccination rates is cost. The authors of pertinent chapters have over the years explained the situation, yet the problem persists—economists can explain the various public and private business models underpinning the near status quo, but they are not likely the people who will foment change. The goal of global elimination of canine-mediated human rabies requires cost cutting via technological advances, but really applying the "miracle of genetic engineering" seems further off than if we were thinking of human hepatitis vaccines. Failed public understanding of human vaccinology, the failing that has given rise to human and veterinary antivaccination activist movements, is likely also involved.
- *Diagnostics.* As always, rabies control/prevention programs start with the need to show that in quite a few countries this is an important disease—this concerns the never-ending need for public education to drive "the public will." This matter is nicely covered in this edition. One key is more quantal knowledge of incidence/prevalence/etc., all translated into lay language—I think knowledge stemming from diagnostics data is central here. Almost separately from the place of diagnostic data in the population-based perspective is the need for reliable diagnostics in support of clinical medical care—there must be no failure to rapidly answer the question, "was the raccoon that bit my child rabid?" Focus here may not be so much on the technology of diagnosis (which is excellent in most countries), but its availability as the budgets of local health departments are compromised.
- *The natural history and ecology of the lyssaviruses and the epidemiology of the disease rabies.* In the past few years, there has been an explosion of interest in the ecology of the pathogens (*ecology*, defined by some as the science dealing with the

---

[a] I use the word *tertiary* here, thinking that our education as scientists progresses through levels, some of which are terminal for most students, from a primary level (general biological science, organismal, cellular and molecular biology, general virology, etc.), to a secondary level (professional and specialty education/training/experience, graduate and postgraduate academic education/training/experience, clinical training/experience), to a tertiary level (where expertise and experience in the interactions or interfaces between multiple scholarly disciplines, which characterize so much of modern research, must be assimilated).

intersection of the pathogen, the host, and its environment). As covered so well in this book, the wonderfully complex natural history of rabies and the other lyssaviruses provides many generalizing lessons for ecologists from outside the world of real rabies expertise. I rebel when ecologists overstep their expertise and generalize beyond the facts. Often this leading edge of ecology has devolved to computer modeling and predicting the time/place of next disease episode, epidemic or pandemic, all for public media, social media, and particular funding streams. There is often too little linkage to the science of epidemiology and the other infectious disease disciplines, and their roots in on-the-ground investigation and lifelong experience with practical intervention actions. I am thinking of the great "rabies-ologists" of several generations who have advanced our understanding to the level where it now stands. They have provided a grand foundation, from which it seems reasonable to set the goal of globally eliminating dog-mediated human rabies.

At least in the United States, rabies has been dealt with at the interface between public health and veterinary public health agencies and institutions, and as everywhere else its research base has extended from diverse sciences: virology, immunology, pathology, animal biology, wildlife biology, geography, epidemiology, social sciences, economics, government, and clinical medical and veterinary specialties—and ecology. Surely, this will continue into the future. In this regard, I have come to appreciate Joshua Lederberg's *Convergence Model* (Fig. F.1) more and more, again seeing necessary new expertise/experience sets that are being called for to attack: (1) the emergence of the virus, rabies virus, in new niches; (2) the establishment of new rabies transmission patterns; (3) the spread or reintroduction of the virus from poorly understood reservoir hosts; (4) the risk to humans and domestic animals of rabies in still poorly understood reservoirs such as bats; and (5) the human societal factors that in some circumstances seem to favor the continued success of the virus. And on and on. I am still thinking of how the genius of Joshua Lederberg would point to new ways to attack rabies viruses and rabies the disease.

- *The global elimination of canine-mediated human rabies*. I am a student of Frank Fenner, the great renaissance-man, scholar, and virologist. As chairman of the expert group that certified to the WHO World Health Assembly that the global eradication of smallpox had been accomplished, Frank Fenner considered the terms used to be quite important. He favored reserving the term [virus disease] *elimination* for regional success (measles/mump/rubella, foot-and-mouth disease), and the term *eradication* for global success (smallpox, rinderpest, soon, wishfully, polio). I think this choice of terms has been useful, as canine-mediated rabies elimination activities come to be followed by the international public health community and the public at large. So, with my bow to Frank Fenner and his terminology, I return to the heart of this edition, the global elimination canine-mediated human rabies.

The tone taken in this volume is that the goal has been firmly established—the crusade is underway; the lessons from smallpox and rinderpest eradication have been taken; needed tools are in hand; the public and its leaders, globally, are being enrolled and the public health priority for action promulgated; the enterprise is being organized with leadership,

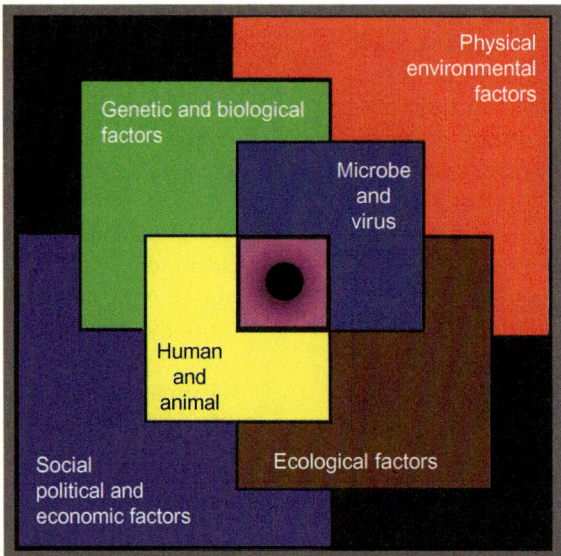

FIG. F.1 Joshua Lederberg's *Convergence Model*. At the center of the model is a *box* representing the ultimate convergence of all the factors leading to the emergence, establishment, spread, and continued success of the infectious agent/disease, in the case at hand, rabies viruses/rabies disease. The *boxes* present the targets for disease prevention, control, and elimination. The interior of the *central box* is presented as a gradient, the *pink outer edges* representing what is known about the factors involved, and the *black center* representing unknown interacting factors (the ol' "*black box*")—action programs always have to start without full knowledge—such is the history and reality of the public health sciences. Interlocking with the *center box* are the two main players—the individual host and the infectious agent, the rabies viruses. This interaction is influenced by interlocking, converging determinant "factors" affecting emergence, progression, spread, consequences, etc. More and more we are learning of complex genetic and biological factors, physical environmental factors, macroecological factors (such as climate), and social, societal, political, and economic factors of crucial importance. The *Convergence Model* can help us broaden our view as we try to think of the interactions between the boxes, even if thinking about all of them at once, as Joshua Lederberg envisioned, may be beyond our reach. *Modified from Smolinski, M. S., Hamburg, M. A., & Lederberg, J. (Eds.). (2003). Microbial threats to health: Emergence, detection and response. Committee on Emerging Microbial Threats to Health in the 21st Century. Board on Global Health, Institute of Medicine (now National Academy of Medicine), U.S. National Academy of Sciences. National Academies Press.*

managerial, financial, technical, logistical, operational, and other resources identified; international agencies (WHO, OIE, FAO, GARC, and national ministries) have been brought on board; and implementation is set to follow a detailed plan—the Global Strategic Plan (GSP). The leaders of this enterprise have understood that there must be consistent advocacy and a unified sense of optimism. These leaders have written about all this extensively and they have written the appropriate chapters in this edition (the title/subtitle of this edition matches this theme—"Rabies: the Scientific Basis of the Disease and its Management;" Chapter 1, "A History of Rabies—The Foundation for Global Canine Rabies Elimination," etc.). One of the most forceful lessons from these leaders is that sustainability is the core requirement for success, and that this must be country-centric, with governments taking ownership of national programs and their costs (of course, with international funding and advocacy support). The line between what is already in the pipeline vs what is still in the future may be a bit vague [e.g., "... Moving forward, political will, international interest and public backing must be garnered to build international pressure and gather resources to facilitate efforts..." (from Chapter 21)], but this is not the place for an armchair critique. Nevertheless, perhaps one notion is acceptable—since science and public health communities exist in silos that often need to be actively opened: as it is clear that all of the leaders of this plan are supportive of its feasibility and merits, might it not be appropriate to establish a hypercritical group of public health experts, a "devil's advocacy group," to hear from all naysayers and their counterarguments and thereby gain a step toward dealing with them before they appear in the harsh halls of public funding?

There is still much to be done in this most interesting, intriguing, and important min-world of medical science/veterinary science/public health science, the world of rabies and its prevention and control. I applaud Alan Jackson, Anthony Fooks, and their colleagues in "the rabies club" for carrying on Pasteur's legacy.

Frederick A. Murphy

Preface

Rabies is one of the oldest diseases known to mankind and yet still causes untold misery and many thousands of human lives lost every year, enduring as one of the most feared viral zoonotic diseases of the 21st century. Rabies remains one of the foremost neglected diseases. This volume *Rabies: Scientific Basis of the Disease and its Management* (4th Edition) provides clinicians, veterinarians, scientists, policy makers, and stakeholders involved in rabies control with an up-to-date review of information known about the disease and also the rabies situation worldwide from a "One Health" perspective. This edition provides a global strategic vision for the elimination of rabies in animal reservoirs and explains how this strategy will result in the concomitant reduction in cases of human rabies. Substantial progress has already been made in tackling the worldwide rabies problem in certain geographic regions, resulting in reduction of the rabies global death rate. The methods to accomplish this are now very well established. Unfortunately, setbacks have occurred in resource-limited regions where management of rabies is overwhelmed by other priorities. The disease has a case-fatality rate approaching 100% and threatens tens of thousands of people worldwide, especially in Asia and Africa, with endemic dog rabies being the largest risk factor for human disease. Dogs are the most important reservoir for rabies and account for greater than 99% of the human cases of rabies. The reduction of dog-mediated rabies will, therefore, result in a concomitant decrease in human rabies cases. Rabies remains a neglected disease of poverty and impoverished people and children carry a disproportional share of the burden of rabies, especially in resource-limited regions where individuals live on less than 1 USD a day. The many thousands of human fatalities that occur each year is an estimated figure because confirmation of human rabies still carries taboos in some cultures and, therefore, is underreported. A multifaceted approach for human rabies elimination is badly needed involving government support, advocacy, education, and preimmunization of "at risk" human populations. The high cost of rabies biologicals, including rabies vaccine and rabies immune globulin, is a barrier for effective human rabies prevention and remains an important future challenge. Neural-tissue-derived rabies vaccines, which are associated with high complication rates, are disappearing and, fortunately, only a few countries still continue to use them.

Effective therapy of human rabies is an important challenge for the future. The number of survivors has recently increased greatly, particularly in India, without using any specific therapy other than supportive critical care. Unfortunately, many of the survivors have severe neurological sequelae. An improved understanding of rabies pathogenesis would be useful in the development of novel therapies for human rabies. New approaches need to be taken for the therapy of human rabies rather than repeating failed protocols from the past.

There needs to be a much more concerted effort to control both dog rabies and rabies in wildlife hosts using mass vaccination campaigns. Dog-mediated rabies control is the highest priority in preventing human rabies cases. This will require the commitment of governments of many countries with the allocation of the necessary resources. However, it remains to be seen whether they will actually act upon the empirical evidence and provide the necessary financial support. Mass vaccination campaigns using parenteral vaccines, and advances in oral vaccines for wildlife, have facilitated the elimination of rabies in terrestrial carnivores in rabies-enzootic countries worldwide.

Rabies control is a dynamic field, and this volume will be an invaluable reference source. Momentum is clearly developing for commitments from countries for global elimination of dog-mediated human rabies cases by 2030. The next decade will be a very exciting time in the rabies field. The vision for a world free of dog-mediated human rabies is aligned with the Sustainable Development Goals to combat poverty, hunger, and disease. The eradication endgame for rabies poses a diverse set of challenges in the 21st century, in reaching the goal that dog-mediated human rabies is eliminated worldwide by 2030. This in itself would be an unprecedented achievement and the elimination of dog-mediated human rabies would be the first zoonotic viral disease to be globally eradicated.

This volume describes developments and progress in many areas of the rabies field since the Third edition was published in 2013, including in the areas of epidemiology, molecular epidemiology, prevention and therapy of human rabies, and control of wildlife and dog rabies. This work is truly a multidisciplinary effort by authors in many different fields. We are confident that this edition will give a diverse audience including physicians, veterinarians, virologists, immunologists, wildlife biologists, and students in different fields a much better understanding of the disease. We would like to thank Elsevier for allowing us to serve as the editors of this Fourth edition. We thank all of the authors for sharing their expertise in their insightful and up-to-date chapters. All of the contributors have done a superb job, and we very much appreciate their efforts for their excellent work. Finally, special thanks to Frederick Murphy for writing the Foreword, not only in this Fourth edition, but for writing excellent Forewords in all preceding editions.

Anthony R. Fooks
The Animal and Plant Health Agency,
Addlestone, Surrey, United Kingdom

Alan C. Jackson
University of Manitoba, Winnipeg,
MB, Canada

CHAPTER

1

A history of rabies—The foundation for global canine rabies elimination

Charles E. Rupprecht[a], Conrad M. Freuling[b], Reeta S. Mani[c], Carlos Palacios[d], Claude T. Sabeta[e], Michael Ward[f]

[a]LYSSA LLC, Cumming, GA, United States [b]Institute of Molecular Virology and Cell Biology, Friedrich-Loeffler-Institut, Federal Research Institute for Animal Health, Greifswald-Insel Riems, Germany [c]Department of Neurovirology, National Institute of Mental Health & Neurosciences, Bangalore, India [d]Pablo Cassara Foundation, Buenos Aires, Argentina [e]Onderstepoort Veterinary Institute, Pretoria, South Africa [f]The University of Sydney, Camden, Australia

1.1 Introduction

Rabies, an acute, progressive encephalitis caused by a lyssavirus, is ancient, with countless historical events occurring "before the common era" (BCE). Most treatises on the topic begin with a notation of rabies as one of the oldest infectious diseases.

The objective of this chapter is a spatiotemporal summary of seemingly important occurrences on rabies and, in context, an attempt at reanalysis of critical past events related to a more current understanding of the disease, as a backdrop to individual themes within this current edition, as well as a subtext for the programmatic global elimination of human rabies via dogs (GEHRD), as envisioned by international organizations (Abela-Ridder, de Balogh, Kessels, Dieuzy-Labaye, & Torres, 2018). Reinterpretations about rabies origins are based in part upon collective logic, inference, evidence, and parsimony (Rupprecht, Kuzmin, & Meslin, 2017). Many prehistoric events had obvious consequences for rabies (and other neglected tropical diseases (NTDs)), as interpreted today. Physicochemical actions, geological mechanisms, organic evolution, ecological dynamics and biological processes, at a molecular basis to a community level, have all shaped collectively what we call "rabies"—a tale of viruses and mammals; domestication and migrations; dogs and people. With the advent of recorded history, superstitions and parroted expert opinions alone gave way to the seeds of distinctive observations of nature, which budded during the Renaissance and flowered throughout the 18th–19th centuries, from philosophy to eventual acceptance. Details on pre-20th century luminaries, facts and figures may be found in earlier editions of this book.

Clearly, canine rabies was recognized and controlled before the 20th century. With an improved 21st century understanding, sensitive and specific diagnostics and highly safe and efficacious biologics, human rabies is preventable and canine rabies can be eliminated, in an even more effective, economical, and ethical manner, at a broad scale (Cleaveland et al., 2018).

1.2 Asia

1.2.1 A seat of canine domestication?

Rabies appears in Asian accounts nearly as long as there are records (Table 1.1). Based on such accounts and molecular evidence, the seeds of canine rabies may have originated in Asia (Bourhy et al., 2008; Meng et al., 2011; Nadin-Davis, Turner, Paul, Madhusudana, & Wandeler, 2007). Dogs were the first human companion species and the only large carnivore

TABLE 1.1 Major historical rabies events in Asia.

Period	Event	Reference
~1930 BCE	Eshnunna code—record of the causal link between the bite from a dog and human death	Roth, Hoffner, & Michalowski, 1997
600–1000 BCE	"Hydrophobia" and associated fatality well recognized in Susruta Samhita, a classic text of Ayurveda	Bhishagratna, 1911
~56 BCE	Early mention of rabies in China	Wang & Huang, 2001
1891 CE	Persons exposed to animal bites receive PEP at the Pasteur Institute in Saigon (Vietnam), the first in Asia, Africa, or Latin America	Tarantola, 2017
1911 CE	Semple develops the inactivated (phenolized) NTV at the Pasteur Institute of India at Kasauli—the most common rabies vaccine to be used in the world until 2000	Chakrabarti, 2010
1912 CE	Princess Banlusirisan (Princess Pao) of Thailand dies of rabies, which results in establishment of the Pasteur Institute in Thailand "Paturasapha" on April 26, 1913 in her honor (now known as the Queen Saovabha Memorial Institute, Bangkok)	Garden, 2009
1921 CE	First major national program to vaccinate dogs started in Japan (in Nagasaki and Tokyo) using phenolized vaccine	Baer, 2007
1932 CE (6th July)	"The Bombay Riot over Dogs"—The Parsi community in Bombay took up arms to protect their dogs against a regulation issued in 1813, which mandated that Indian pariah dogs be killed every year to control the menace of stray and rabid dogs in British-ruled India	Shah, 2018
1953 CE	Last case of rabies in Singapore	Yang, Cho, & Kim, 2018

TABLE 1.1 Major historical rabies events in Asia—cont'd

Period	Event	Reference
1954 CE	Reported human case after bat bite in India	Veeraraghavan, 1954
1956 CE	Last case of rabies in Japan	Yang et al., 2018
1961 CE	Elimination of canine RABV transmission in Taiwan	Yang et al., 2018
1970 CE	The most convincing evidence supporting efficacy of antirabies vaccines until 1970—from exhaustive reports of the Pasteur Institute of Southern India, Coonoor	Veeraraghavan, 1970
1970 CE	The first WHO regional seminar on veterinary public health held in Mukteshwar (India) during April 1970	Schwabe, 1971
1974–76 CE	First clinical trials of HDCV (WI-38) in human volunteers in Asia (Iran and India)	Bahmanyar, 1974
1975–76 CE	"Iran Wolf Study"—First use of HDCV (along with rabies antiserum) for PEP in 45 persons severely bitten by confirmed rabid wolves (2) and dogs (6) between June 1975 and January 1976 in Iran with remarkable success	Bahmanyar, Fayaz, Nour-Salehi, Mohammadi, & Koprowski, 1976
1980 CE	Report of a case in a *Pteropus* bat in northern India	Pal et al., 1980
1980 CE	An adjuvanted vaccine with Beijing strain of RABV grown on primary Syrian hamster kidney cell line (PHKCV) released for human use in China	Lin, 1990
1991 CE	Novel bat lyssavirus Aravan (and later Khujand) identified in central Asia	Kuzmin, Botvinkin, Rybin, & Baialiev, 1992
1993 CE	Recognition of rabies mediated by raccoon dogs in the Republic of Korea, with ORV occurring by 2000, the last human case in 2004 and no further animal cases since 2013	Yang et al., 2018
1997 CE	Outbreak of canine rabies on Flores Island, Indonesia	Windiyaningsih, Wilde, Meslin, Suroso, & Widarso, 2004
2007 CE	Peak of the current wave in China of ~3303 human cases	Tu, Feng, & Wang, 2018
2008 CE	Recognition of a major outbreak on the island of Bali, Indonesia	Susilawathi et al., 2012
2012 CE	Isolation of Irkut virus from a bat in China	Liu, Zhang, Zhao, Zhang, & Hu, 2013
2013 CE	Finding of RABV in ferret badgers in Taiwan	CDC, 2014
2014 CE	Serological evidence of lyssavirus activity among bats in northern India	Mani et al., 2017
2014 CE	Novel lyssavirus isolated from brains of flying foxes (*Pteropus medius*) in Sri Lanka, designated as Gannoruwa bat lyssavirus	Gunawardena et al., 2016
2015 CE	Loss of "rabies-free" status in Malaysia due to canine rabies	Yang et al., 2018
2016 CE	Isolation of new bat lyssavirus in Taiwan	Hu et al., 2018

ever domesticated. When, where, and how domestication occurred remain controversial (Botigué et al., 2017; Frantz et al., 2016; Ollivier et al., 2018; Wang et al., 2016). Evidence of dog-like canids in the fossil records from Siberia date to ~30,000 years ago (Ovodov et al., 2011). Early dog domestications in the Near East are estimated to ~15,000 BCE (Dayan, 1994). The temple of Gobekli-Tepe in south-east Turkey, dated to ~12,000 BCE, depicted dogs in stone carvings. The Natufian Grave (~12,000 BCE), in Ain Mallaha, Israel, contains an elderly man buried with a puppy (Bevan, 2018; Davis & Valla, 1978; Mark, 2019). Collared dog figurines appear at Harappa and other Indus civilization sites (~3300–1300 BCE). In the mature Harappan phase (~2600–1900 BCE), canine remains are found at almost all localities.

1.2.2 Dogs, deaths, and divinities

Mesopotamians recognized risks posed by dog bites and resorted to incantations and enchanted water to treat rabies (Yuhong, 2001). The cuniform tablets of Eshnunna, excavated at Tell Harmal (Shaduppum), Baghdad, Iraq, during 1945–47 (Roth et al., 1997), refer to Sumerian laws, indicating a possible causal link between the bite of a dog and human death (A iv 20–23), recognized ~4000 years ago (Roth et al., 1997; Tarantola, 2017). Whether such passages relate to rabies per se, or simply death from dog bite alone, remains open to interpretation (Wilkinson, 1977). Biblical, Talmudic, and Islamic narratives offer similar concerns.

A dog bite was considered as dangerous as the venom of a snake or scorpion and a disease outbreak was considered an ominous sign of divine anger. Mesopotamians attributed rabies as one of the destructive powers of the gods. The rabid dog was reflected as a disease demon, under the control of Ea and Asarlubi, healing gods (Yuhong, 2001). The dog of Gula, the Babylonian goddess of healing, espoused power of the divine against the terror that dogs evoked (Mark, 2017; Yuhong, 2001). People could appease the goddess by worship at her temples, where dogs roamed freely. The famous Nimrud Dogs, ceramic and bronze statuettes found in the 12th century BCE (in modern Northern Iraq), are among the best-known examples of amulets dedicated to Gula, for those blessed by her benevolence (Mark, 2017). Dogs appear as companions of the Innana (Ishtar), the Sumerian goddess of love and war, who traveled with seven prized dogs in collar and leash, as recorded in *The Epic of Gilgamesh*, dated to ~2150–1400 BCE (Mark, 2019). In the Vedic period of India (~1500–500 BCE), Yama, the God of Death, is described as attended by two four-eyed, brindled watch-dogs, constant companions and emissaries of death: "… *broad-nosed and brown, the messengers of Yama, greedy of lives, wander among people* …" (Macdonell, 1900). Since the 19th century, *Hadakai Maa*, a goddess believed to control mad-dogs and prevent hydrophobia, was worshipped in Western India (Monier-Williams, 1891). Notably, the word "rabies" is derived from Sanskrit *"rabhas"* (to do violence), providing inspiration for related terms, as Indo-Aryans dispersed westward.

1.2.3 The disease and perceived remedies

Though rabies is recognized in ancient texts, efforts at treatment were mostly faith-based and magical. In China, rabies was described ~56 BCE (Wang & Huang, 2001). Ko Hung's handy therapy for emergencies (~ the year 300) prescribed sucking blood from a bite and subsequent moxibustion, and to "kill the dog responsible for the bite, remove its brain, and rub it on the wound. After this, there will be no recurrence." (Kerr, Needham, & Wood, 2004). Such beliefs on sucking "viruses or slimy poisons" from a wound still have disastrous consequences today (Zhao, Zhang, Cheng, & Zhou, 2019).

The Persian Avesta, the Sacred Book of Zoroastrianism (~7th century BCE), advocated avoiding bites altogether for primary rabies prevention. One quote appears in the *Susruta Samhita*, a classic text of Ayurveda, the Indian system of traditional medicine (~6th–10th century BCE), recognized the neurological nature of infection: "... *if the patient becomes exceedingly frightened at the sight or mention of the very name of water, he should be understood to have been afflicted with Jala-trasa (water-scare) and be deemed to have been doomed...*" (Bhishagratna, 1911). Bleeding and cauterization of the wounds with clarified butter was recommended. Mixtures with the milky exudate of the Arka plant (*Calotropis* spp.) and a compound of white Punarnava (*Boerhavia diffusa*) and Dhuttura (*Datura metel*) were also prescribed as an errhine (i.e., to induce a nasal discharge).

1.2.4 Postcolonial occurrence

Rabies was a longstanding problem throughout ancient Asia, but details during European engagement are sketchy. Canine rabies in Turkey was described as early as 1586. During 1852, a rabid wolf attacked >100 people in the town of Adalia in Anatolia, several of whom subsequently "went mad" and died (Nadin-Davis & Bingham, 2004). Rabies in Japan was reported during 1732 in Nagasaki, which spread rapidly and caused multiple human deaths (Tojinbara, 2002). Rabies was noted in Hong-Kong during 1857, when an English bloodhound became rabid, with at least one person dying of hydrophobia (Steele & Fernandez, 1991). An outbreak occurred in China during 1860 and a man died of rabies near Canton. During 1867, many English dogs in Shanghai were infected, suggesting rabies was translocated via infected European dogs (Nadin-Davis & Bingham, 2004; Steele & Fernandez, 1991).

1.2.5 Semple's antirabic vaccine

No progress in postexposure prophylaxis (PEP) occurred until Pasteur's celebrated success during the late 19th century, rapidly replicated around the globe (Steele & Fernandez, 1991). For example, during 1900, the Pasteur Institute of India at Kasauli, in the foothills of the Himalayas, was founded under David Semple, a British officer in the Indian Medical Service. Semple focused on passive immunity initially. During 1903–04, he treated ~200 patients with antirabic serum, as a prelude to Pasteurian vaccination. Patient outcomes were unclear, which led him to focus efforts on a more conventional antirabic vaccine. Breaking from Pasteurian practices of attenuated viruses, Semple, a protégé of British physician and typhoid vaccine developer Almroth Wright, developed a carbolized (phenolized) inactivated rabies virus (RABV) vaccine during 1911. Initially, the vaccine was derived from rabbit brains, but later from infected sheep and goats. The earliest vaccine used in India was Pasteur's dried spinal cord, replaced during 1911 by the original Semple vaccine (1% carbolized rabbit brain), followed by experimental vaccines ranging from 1% to 5% from 1923. From the 1930s, a standardized Semple vaccine (5% carbolized sheep brain) was used exclusively (Chakrabarti, 2010; Cheema, 2015). The finished Semple vaccine contained ~5%–10% of sheep or goat brain tissue, 0.25%–0.5% phenol, and 1:10,000 thiomersal, as preservative. The vaccine was stored at

2–5°C, not frozen. Given extremely high usage in the region, Semple's product remained the most commonly used rabies vaccine in the world, until ~2000. Production in India was discontinued only during 2004. Though effective, it had a significant risk of neurological complications and was replaced gradually by vaccines using duck embryos (late 1950s) and cell-culture vaccine (CCV) techniques (~1970s).

1.2.6 Additional progress throughout the 20th century

The promising immunogenicity of the human diploid cell vaccine (HDCV) was tested under severe field conditions during 1976 on wolf bite victims in Iran (Bahmanyar et al., 1976). Given the high case fatalities after rabid wolf bites, spectacular results in surviving patients paved the way for licensing of this first modern CCV. Thereafter, primary chick embryo cell vaccine (PCECV), after rigorous worldwide clinical trials, became the first CCV produced on an industrial scale in Asia (i.e., India) during 1990. In addition, purified Vero cell rabies vaccine (PVRV) was available commercially in the region since 1986. Considering concerns over basic safety and efficacy, nerve-tissue vaccines (NTV) were abandoned gradually. By 1992, Thailand was the first Asian country to use only CCV.

Introduction of CCV also improved upon the tedious, painful PEP regimens with NTV to be shortened. Studies on dose- and cost-sparing intradermal (ID) regimens (Madhusudana, Saha, Sood, & Saxena, 1988; Nicholson, Prestage, Cole, Turner, & Bauer, 1981; Warrell et al., 1983), were employed in Thailand and other Asian countries, such as Bangladesh, India, and Sri Lanka. Today, no Asian countries use NTV, discontinued in Bangladesh, Myanmar, and Pakistan since 2011, 2013, and 2015, respectively (World Health Organization, 2018).

With establishment of Pasteur Institutes throughout Asia to decentralize antirabic vaccination at the beginning of the 20th century, and annual surveys initiated during 1959 by the Veterinary Public Health Unit of the World Health Organization (WHO), a systematic documentation of regional rabies problems was possible. Individuals exposed to animal bites received PEP at the Pasteur Institute in Saigon as early as 1891, the first in Asia, Africa, or Latin America. The founding WHO regional seminar on veterinary public health was held in Mukteswar (India) during April 1970, with participants from India, Indonesia, Mongolia, Nepal, Sri Lanka, and Thailand (Ahuja, Tripathi, Saha, & Saxena, 1985; Steele & Fernandez, 1991; Tarantola, 2017).

Though dogs were recognized as major RABV reservoirs throughout Asia, several wildlife species, notably wolves (in Iran), jackals (in Afghanistan, Pakistan, Nepal, and India), and mongooses (in India), were also considered significant vectors (Baltazard & Ghodssi, 1954; Greval, 1932; Johnson, 1819).

Regarding animal control, the first major national program to vaccinate dogs was begun in Japan (in Nagasaki and Tokyo) using phenolized vaccine during 1921 (Baer, 2007). Through public education, canine population management, strict regulations, and vaccination, several areas were considered canine "rabies-free" between 1959 and 1968, including Hong Kong, Japan, Malaysia, Singapore, and Taiwan (Steele & Fernandez, 1991).

1.2.7 Current status

Most human deaths, and PEP costs, are highest in Asia, with annual estimates >US$1.5 billion. India alone accounts for about 60% of Asian rabies fatalities and 35% of deaths globally (Hampson et al., 2015; World Health Organization, 2018). Others with a high incidence include Afghanistan, Bangladesh, Cambodia, China, Indonesia, Myanmar, Nepal, Pakistan, the Philippines, and Vietnam. Notwithstanding historical burden, several countries, such as Sri Lanka and Thailand, have demonstrated a significant decline in human rabies cases (Hampson et al., 2015).

The focus on mass dog vaccination is the most pragmatic and cost-effective approach to canine rabies elimination (Cleaveland & Hampson, 2017). Though mass dog vaccination poses a challenge in multiple countries with large populations of free-ranging dogs, several success stories appear (Gibson et al., 2015; Harischandra, Gunesekera, Janakan, Gongal, & Abela-Ridder, 2016). Albeit shown in smaller areas, this goal is achievable, leading to substantial health and economic benefits. Oral RABV vaccines, extensively tested for efficacy and safety in dogs, should be used as an adjunct to parenteral immunization (Cliquet et al., 2018). Given the considerable historical burden, incorporation of enhanced laboratory-based surveillance, canine vaccination, modern human PEP, innovative financing models (Hampson et al., 2015), and a strong political commitment are required to achieve the formidable goal of the GEHRD in Asia (Yang et al., 2018).

1.3 Europe

1.3.1 Early incursions

How and when rabies arose in Europe are unknown. Based on viral evolutionary data and coevolution with primary reservoirs among the Chiroptera (Aiewsakun & Katzourakis, 2015), by inference, rabies was in European bat populations much earlier than other mammals. Bat rabies went unnoticed until the early 20th century, deemed of negligible medical importance, compared to canine rabies (Rupprecht et al., 2017).

Besides bats, other reservoirs reside in the Carnivora. The earliest arrival of rabies in Europe may either have been be a result of wildlife-mediated rabies (i.e., by foxes), or by domesticated dogs. As for the latter, earliest fossil records provide the presence of dogs in Europe in the Upper Paleolithic (Benecke, 1987). Although Africa, Australia, and the Americas are excluded as localities, origins, divergence, and timing of pre-Neolithic domestication in Eurasia remain debatable (Botigué et al., 2017; Frantz et al., 2016; Thalmann & Perri, 2019).

Mass human migrations, invasions, trade route travels and tribal dispersals, accompanied by dogs, resulted in disease introduction and spread, as exemplified by modern dog RABV (Troupin et al., 2016). Unfortunately, there is no definitive record of RABV linked to dog and human phylogenies and thus temporal or directional movements. One possible introduction could have occurred after a massive migration took place from the Eurasian steppes to Central Europe (Haak et al., 2015). Indeed, rabies was present in ancient Eurasian cultures, such as Greece, Rome, Egypt, and the Levant in the West, to China and Siberia in the East (Baer, 2007;

Tarantola, 2017). However, at least within Europe, accounts of wildlife rabies were reported only since early medieval times (Blancou, 2004).

1.3.2 Recognition and control measures

Multiple European scientists contributed to a better understanding of rabies pathobiology and management, somewhat overshadowed by Pasteur's accomplishments during the late 19th century (Table 1.2). Nevertheless, effective animal rabies control began before vaccination. Since human PEP was yet to be discovered, actions focused on RABV vectors. Control included canine confinement and quarantine (Blancou, 1994; Fleming, 1872). Regulations

TABLE 1.2 Selected occurrence of rabies recognition, prevention, and control in Europe.

Date	Event	Reference
~500–300 BCE	Earliest references to rabies by pre-Christian authors	Neville, 2004
~1000 CE	"Laws of Wales" authorized any suspect dog to be killed, provided that clinical signs of rabies observed	Wilkinson, 1988
~1793 CE	Plan for eliminating rabies from the British Isles proposed, consisting of universal quarantine and ban of importations	Wilkinson, 1988
~1890 CE	Dog-mediated rabies effectively controlled in several European countries, through the establishment of regulations for management and their enforcement, for example, containing and muzzling dogs and the elimination of strays	Blancou, 2004
1903 CE	Last indigenous human RABV case in Great Britain	Fooks, Roberts, Lynch, Hersteinsson, & Runolfsson, 2004
1903 CE	Detection of intracytoplasmic inclusion bodies within the brains of rabid animals, as utility in diagnosis	Negri, 1903
1903 CE	Recognition of infection via a filterable agent	Remlinger, 1903
~1940 CE	Establishment and spread across Europe of fox-mediated rabies, presumably originating in areas of present-day Kaliningrad oblast	Wandeler, 2004
~1950 CE	Broad-scale application of parenteral vaccination of dogs with substantial progress in the elimination of dog-mediated rabies	Blancou, 2004
1968 CE	The front wave of the fox-rabies epizootic reached France	Aubert et al., 2004
1978 CE	First field trials of oral rabies vaccination (ORV) in Switzerland, followed by, for example,	Blancou, Andral, Aubert, & Artois, 1982; Schneider, Wachendörfer, Schmittdiel, & Cox,

TABLE 1.2 Selected occurrence of rabies recognition, prevention, and control in Europe—cont'd

Date	Event	Reference
	Germany, France, and other countries to develop a strategic approach, including machine-made baits and aerial distribution	1983; Steck, Wandeler, Bichsel, Capt, & Schneider, 1982
1990 CE	Field trials were turned into rabies control programs by subsidy of the European Union	Müller & Freuling, 2018
1994 CE	Generation of infectious virus from cloned cDNA	Schnell, Mebatsion, & Conzelmann, 1994
2002 CE	Divergent lyssavirus isolated from the brain of a male common bent-winged bat (*Miniopterus schreibersi*), in the western Caucasus Mountains, termed West Caucasian bat virus	Botvinkin et al., 2003
2008 CE	Rabies reemergence among foxes in Italy, followed by ORV	De Benedictis et al., 2008
2012 CE	Fox rabies reached Greece, followed by ORV	Tsiodras et al., 2013
2019 CE	Due to the implementation of ORV, RABV is virtually absent in the EU and countries of the Western Balkans	WHO Rabies Bulletin Europe database, 2019

existed for registration and keeping of dogs, and elimination of strays locally (e.g., Utrecht, Netherlands) during 1446, within a Dutch province (e.g., Friesland) during 1714, and throughout a country (e.g., Prussia) during 1787. Such conventions led to elimination of dog-mediated rabies from Denmark, Norway, and Sweden by 1826 and England by 1902 (Blancou, 2004; Fooks et al., 2004). Which measure was most important remained elusive. Ultimately, destruction of all suspect animals was drastic, as documented for British kennels during the 18th century (Fleming, 1872). Similarly, effects of muzzling were hotly debated, alongside concerns on animal welfare.

Despite success in human PEP post-Pasteur, only later did substantial technical progress achieved during the first half of the 20th century lead to development of more safe, affordable, and efficacious inactivated animal rabies vaccines (Steele & Fernandez, 1991). During the 1930–80s, application of mass (sometimes compulsory) vaccination to dogs, together with registration, ownership tax collection, and movement restrictions, became a cornerstone in canine rabies control at a European level, resulting in a declining burden within a few decades (Müller et al., 2012).

1.3.3 Wildlife rabies appreciation

Despite success in canine rabies management, a major epizootic of fox rabies followed during the 1940s. Sylvatic rabies began in a focus south of Kaliningrad during World War II, perhaps by sustained spillover from dogs (Taylor, 1976). Shifts toward fox-mediated rabies were also documented in the former Soviet Union and molecular characterization revealed closely related RABV lineages in the Asian part of Russia, questioning this "canine-origin" hypothesis (Kuzmin et al., 2004). Regardless, prevailing conditions favored RABV perpetuation in red foxes (Wandeler, 2004), similar to a situation in Turkey, where dog-mediated RABV successfully crossed species barriers (Marston et al., 2017; Vos et al., 2009).

Red foxes spread rabies across Europe within a few decades (Taylor, 1976). By the mid-1970s, large parts of Central and Western Europe were affected (Wandeler, 2004). Traditional measures to control urban rabies failed to stop vulpine rabies. Thus, shifts from urban to sylvatic rabies posed new challenges for control, requiring substantial changes in policies (Wandeler, Capt, Kappeler, & Hauser, 1988). Efforts included intensive culling, killing of cubs at dens, poisoning, gassing, and hormonal sterilization. Attempts to interrupt RABV transmission within fox populations had variable success at a local level. In fact, it was exceedingly difficult to reduce densities below levels at which social networks ceased to operate and transmission was interrupted at $R^0 < 1$ (Aubert, 1992; Wandeler, 2004). Besides ethical considerations, disruption of fox social systems increased contact rates and disease incidence, and were deemed counterproductive (Aubert, 1992).

1.3.4 Oral vaccination applications

Throughout the 1970s, research showed that foxes could be immunized against rabies orally, opening new avenues for control, due to the concept of oral rabies vaccination (ORV) of wildlife (Baer, Abelseth, & Debbie, 1971). During 1978, the first ORV field trial was conducted in Switzerland (Steck et al., 1982), followed by efforts in other countries (e.g., Germany, France, and Belgium). Thanks to such pioneering attempts and later financial support from the European Union (EU), ORV was recognized as a breakthrough for fox rabies control. For approved national programs, 50% (since 2010, 75%) of the costs for vaccine baits, distribution and associated laboratory-based surveillance and follow-up investigations were reimbursed by the EU. The EU also provided resources for ORV in neighboring non-EU countries (Demetriou & Moynagh, 2011). Since implementation of ORV, continuous decreases in rabies were reported throughout Europe: from 18,778 in 1990 to 9137 in 2006 and 3641 in 2018, reflecting >80% reduction in the number of reported cases over the past 28 years. Notably, as of 2018, 99% of all rabies cases are eastward, in the Russian Federation, Ukraine, Turkey, Moldova, and Georgia (World Health Organization, 2019). The EU may be free from carnivore rabies by 2020 (European Union, 2017). Thereafter, a permanent *cordon sanitaire* is needed (Freuling, Selhorst, Batza, & Müller, 2008).

The success of wildlife ORV is not conceivable without setbacks (Müller et al., 2015; Müller & Freuling, 2018). For instance, there were several border incursions: France-Switzerland (1990); France-Belgium-Germany (1993); Italy-Slovenia-Austria (1993); Germany-Poland-Czech Republic (1995); Italy-Slovenia (2008); Greece-Former Yugoslavian Republic of Macedonia (FYROM), etc. (Freuling et al., 2013; Mulatti et al., 2013; Tasioudi et al., 2014).

1.3.5 Reintroductions

Rabies (re)introduction by domestic animals remains a threat from imported rabid animals (Johnson, Freuling, Horton, Müller, & Fooks, 2011). The legislative framework of the "Pet Travel Scheme" (Regulation (EU) No 576/2013) requires animal identification, vaccination, and sometimes serological tests. When these are met, risk of reintroduction into the EU is generally low (EFSA, 2006; Goddard et al., 2012). However, importation of rabid animals from endemic countries through failure of border controls, ignorance of rules, or active subversion remains a threat to the rabies-free status of EU member states. During 2001–13, 22 rabid animals were brought into the EU (Cliquet, Picard-Meyer, & Robardet, 2014; Johnson et al., 2011), associated with costly public health responses (Gautret, Ribadeau-Dumas, Parola, Brouqui, & Bourhy, 2011). With increasing mobility, both legal and illegal importation of animals continue, requiring member states to maintain vigilance, strengthen public education about animal movements, and implement measures based on heightened surveillance to mitigate risks of pathogen introduction (Fevre, Bronsvoort, Hamilton, & Cleaveland, 2006).

1.3.6 Europe as rabies-free?!

Perhaps in the truest sense, ultimate European rabies elimination is much discussed—but for sure eradication is not realistic (Rupprecht et al., 2008). Diverse bat-associated lyssaviruses, including European bat lyssavirus types 1 and 2 (EBLV-1, -2) as well as Bokeloh Bat Lyssavirus (BBLV) (Freuling et al., 2011; Schatz et al., 2013), Lleida bat lyssavirus (LLBLV) (Arechiga Ceballos et al., 2013), and Kotalahti bat lyssavirus (KBLV) (Nokireki, Tammiranta, Kokkonen, Kantala, & Gadd, 2018), among others (Table 1.2), preclude true rabies freedom and raise concerns that carnivores could act as a vector for such viruses in case of sustained intraspecies spillover, as experienced with insectivorous bat RABV variants in foxes, raccoons, and skunks in North America (Daoust, Wandeler, & Casey, 1996; Kuzmin et al., 2012; Leslie et al., 2006). However, studies showed that susceptibility of mesocarnivores to EBLVs is low, resulting in abortive infection, suggesting a negligible risk of sustained EBLV spillover from bats to European carnivores (Cliquet et al., 2009; Vos et al., 2004). While studies with other lyssaviruses are lacking, inference of viral characteristics based upon their phylogenetic relationships also supports assumptions that sustained spillovers are unlikely with non-RABV lyssaviruses (Marston et al., 2018). Thus, given the prior burden in Europe and the progress post-World War II in human rabies prevention, canine rabies elimination, wildlife rabies control, laboratory-based surveillance, lyssavirus discovery, and major epidemiological insights, recent history is a dramatic demonstration of program success at a continental level.

1.4 Africa

1.4.1 Indigenous roots or regional translocations?

As throughout Eurasia, rabies in Africa has a long and interesting history (Table 1.3), but somewhat lost to antiquity. More than 2000 years later, the disease remains largely uncontrolled throughout a continent often considered to be its birthplace. Rabies may have dispersed to parts of Africa via Eurasia (Taylor, Latham, & Woolhouse, 2001). Unfortunately, the African history is poorly understood.

Rabies was present in Northern Africa for hundreds of years (Baer, 1991; Nadin-Davis & Bingham, 2004; Nel & Rupprecht, 2007). Such occurrences would be associated with canine

TABLE 1.3 Key rabies events within Africa.

Year	Event	Remarks	References
~1842 CE	Maleke, a chief of the Bakwains, who formerly lived on the hill Litubaruba (now Molepolole), died after the bite of a rabid dog	Reported to David Livingstone, at Litubaruba, during his travels in southern Africa, but he could not authenticate the claim	Fleming, 1872
~1851 CE	Rabies reported as common in Algeria	Apparently well known among many local Arab tribes, well before 1830 and European colonization	Fleming, 1872
~1862 CE	Reports of outbreaks in Ethiopia	Described by Sir Samuel Baker, while exploring the tributaries of the Nile, including the practice of forcing suspect dogs through a fire, believed as a curative practice	Fleming, 1872
1893 CE	Outbreak of rabies in the Eastern Cape province of South Africa	Outbreak limited to domestic animals	Snyman, 1940; Swanepoel et al., 1993
1950s CE	Waves of canine rabies spread from Namibia through Botswana into the northern Transvaal	Additional fronts penetrated south from Mozambique into Swaziland and northern Natal by 1961	King, Meredith, & Thomson, 1993
1951 CE	Mass vaccinations began in Zimbabwe with the modified-live low egg passage (LEP) live vaccine	Improved efficacy	Adamson, 1954
1956 CE	Isolation of a new rhabdovirus from the brain of a straw-colored fruit bats (*Eidolon helvum*) in Nigeria	Later identified as the first nonrabies lyssavirus, Lagos bat virus	Boulger & Porterfield, 1958
1968 CE	Isolation of a new rhabdovirus originally from pooled lung, liver, spleen, kidney, and heart of shrews (*Crocidura* sp.) collected in and near Ibadan, Nigeria and also in 2 children	Later identified as a second new lyssavirus, Mokola virus	Kemp, Moore, Causey, Odelola, & Fabiyi, 1972
1970 CE	Virus isolated from the brain of a South African man bitten on the lip by a bat	Later identified as a new lyssavirus, Duvenhage virus	Meredith, Prossouw, & Koch, 1971
1977–85 CE	Epidemic spread of rabies in the kudu antelope (*Tragelaphus strepsiceros*) in the central ranching areas of Namibia	Spread of rabies within this species was believed to be through social grooming, after infection by jackals	Hübschle, 1988; Swanepoel et al., 1993
1981–82 CE	Mokola virus outbreak occurred in Bulawayo (Zimbabwe) involving 6 cats and a dog	One of the cats and the dog had previously been vaccinated against rabies, suggesting that not all lyssaviruses can be prevented by current rabies vaccines	Foggin, 1988
1991 CE	Recognition of rabies in Ethiopian wolves	Example of the impact the disease has in conservation biology of highly endangered species (including African wild dogs)	Sillero-Zubiri, King, & Macdonald, 1996

TABLE 1.3 Key rabies events within Africa—cont'd

Year	Event	Remarks	References
2008 CE	Creation of the Africa Rabies Expert Bureau (AfroREB) in Grand-Bassam, Côte-d'Ivoire	Representative of several regional NGOs throughout the continent, including the Pan African Rabies Control Network (Paracon), the Rabies in West Africa Group (RIWA), the Southern and Eastern African Rabies Group (SEARG), etc.	Dodet et al., 2008
2009 CE	New lyssavirus isolated from the brain of a dead Commerson's leaf-nosed bat (*Hipposideros commersoni*), found in a cave in a coastal region of Kenya	Later called Shimoni bat virus	Kuzmin et al., 2010
2009 CE	New lyssavirus isolated from the brain of an African civet (*Civettictis civetta*), with clinical rabies in the Serengeti National Park of Tanzania	Subsequently named Ikoma lyssavirus	Horton et al., 2014
2010 CE	Serological reactivity detected among bats on south western Indian Ocean islands	Demonstration of broader lyssavirus activity in bat populations of the region	Melade et al., 2016

RABV from the Middle East and Eurasia. This record is somewhat confusing, in contrast to Eurasia and the Levant, considering domestic dog breeds appear in Egyptian artwork ~5000 years ago and rabies seems to have been known in ancient Egypt (Rollinson, 1956). In Sudan, rabies appeared during ~1904, present continuously since at least 1925 (Harbi, 1976). Ethiopia also had a long history, as illustrated by their pharmacopoeias and traditional treatments (Fekadu, 1982). Some Ethiopian RABV, that seemed to have reduced virulence and a suggested tendency to produce a "carrier status," were studied at Institut Pasteur (France) prior to the 20th century (Fekadu, 1972). Blancou (2004) advocated that rabies was present in northern Africa long before the 20th century, suggesting the disease disappeared periodically or became rare. According to Ghobashy (1986), Prunel reported four definite cases of hydrophobia in Alexandria during ~1850–57, and subsequently the disease became common both there and in Cairo (Curasson, 1942; Nadin-Davis & Bingham, 2004). Few reliable human deaths are accounted in Egypt at the beginning of the 20th century, but canine rabies was rampant during the early 1900s.

In West Africa, there were unconfirmed reports in the early 20th century, more frequently observed during the ~1920–30s. Rabies was confirmed in Mali, Senegal, and Nigeria between 1906 and 1912, but later in Ghana, between 1918 and 1955 (Belcher, Wurapa, & Atuora, 1976; Tomori & David-West, 1985). In the Democratic Republic of Congo (DRC), rabies was reported by 1933. Today, dogs and jackals found between Angola and the DRC contribute to rabies spread, particularly in the western portion of the country (Belcher et al., 1976).

Within East Africa, rabies was seen during ~1912 (in Kenya), whereas neighboring Tanzania and Uganda only reported cases during 1936 and 1945, respectively (Hudson, 1944;

Macharia, Ombacho, Kasiiti, Mbugua, & Gacheru, 2003). This area has continued as an enzootic area to date. Rabies was also confirmed at Tororo during 1946, at the Kenya-Uganda border. Apparently, the disease was recognized clinically in Tanganyika during 1932 in the Musoma-Loliando area. The initial report of rabies in Kenya was in South Nyanza during 1902, but the first confirmed case was during 1912, in a dog that fought with a jackal on the Nairobi outskirts (Hudson, 1944).

Explorers, such as Barrow and Neele (1801) and Livingstone (1857), commented on the apparent "complete absence" of rabies during their travels. However, reports from the 18th century indicated its presence in southern Africa (Neitz & Marais, 1932). Cluver (1927) noted that Thurnberg referred to rabies in South Africa during 1772, and that further outbreaks were reported during 1828, 1857, 1861, and 1932 (Neitz & Thomas, 1933). In addition, local folklore maintained that the bites of ground squirrels in Nigeria (McMillan & Boulger, 1960) and genets in South Africa (Neitz & Marais, 1932) caused a fatal disease, resembling rabies in humans. Hudson (1944) records that rabies was known to South Kavirondo inhabitants (Foggin, 1988; Swanepoel et al., 1993), but no confirmation was made until an outbreak at Port Elizabeth during 1893 (sometimes cited as 1892), which was traced to an Airedale dog imported from the United Kingdom. This outbreak was confined to dogs, cats, and a smaller number of cases in ruminants.

During 1956, 28 sub-Saharan countries recorded 1309 cases in dogs (Libeau, 1960), compared to 1985 with 2126 cases (WHO/Rabies/87.198). In southwest Africa (now called Namibia) during 1926, several local inhabitants died at a mission hospital in Ovamboland with symptoms of hydrophobia and a history of having been bitten by suspect dogs. Further investigations indicated that rabies existed there for the prior 20 years, but only during 1938 did a dog (that showed signs of rabies in Okavango) test positive at Onderstepoort, South Africa.

1.4.2 Wildlife involvement

Rabies became established in dogs after European colonization. In southern Africa, black-backed jackals, *Canis mesomelas*, and bat-eared foxes, *Otocyon megalotis*, maintained a canid rabies biotype (Bingham, 1999; Bingham, Foggin, Wandeler, & Hill, 1999a, 1999b; Foggin, 1988; Sabeta, Bingham, & Nel, 2003; Sabeta, Mansfield, McElhinney, Fooks, & Nel, 2007; Snyman, 1940; Swanepoel et al., 1993; Zulu, Sabeta, & Nel, 2009). Distinct RABV occurred in yellow mongoose, *Cynictis penicillata*, found on the Highveld plateau of South Africa and in Botswana, and in the slender mongoose, *Galerella sanguinea*, in Zimbabwe (Snyman, 1940). Fitzsimons, as reported in Snyman (1940), made the following observations about spotted genet (*Genetta felina*) during ~1919: "... saliva of this animal apparently has some poisonous property, but this has not been satisfactorily demonstrated...". A description of African wildlife cannot be complete without a mention of "kudu rabies," a unique phenomenon in this farmed antelope species in central Namibia, speculated by some of kudu-to-kudu transmission (Hassel et al., 2018; Hikufe et al., 2019). Locally, rabies in kudu occurs in a magnitude not reported widely outside of Namibia (Hübschle, 1988; Swanepoel et al., 1993).

1.4.3 Molecular insights

Molecular studies of canine RABV retrieved from repositories in Africa revealed branching patterns supporting four clades (Bourhy et al., 2008). These included Africa 1 and 2 clades, closely related to the Eurasian RABV lineage, grouped into a larger "cosmopolitan" clade, together with an Africa 4 clade, identified in Egypt. In contrast, an Africa 3 clade is restricted to southern Africa, adapted to wild carnivores, especially yellow mongoose, *Cynictis penicillata*, supporting an epidemiological cycle distinct from dog RABV (Snyman, 1940). The Africa 1 clade has a wide geographical distribution and consists of subclades Africa 1a and 1b (Bourhy et al., 2008), with the former predominantly circulating in Northern and Western Africa, while the latter circulates in Southern and Eastern Africa. Rabies in most East African countries is poorly resolved and comparatively few RABV sequences are available in the public domain. Unfortunately, most African laboratories lack reliable diagnosis and typing of lyssaviruses, except for a few reference laboratories. Recent development of more field-focused tests may significantly improve laboratory-based surveillance by decentralizing diagnostic testing (World Health Organization, 2018).

1.4.4 Control measures

Rabies outbreaks were generally controlled through culling, imposing strict dog laws, and restricting dog movements (Bingham, 1999). Other measures included "tie up" orders operating within a 20-mile radius of an infected focus, compulsory reporting of suspicious signs and deaths, improvement of diagnostic services, and reduction of wildlife (e.g., jackals). This "tie up" of dogs was extremely unpopular. Customarily, dogs were let loose at night to forage for food and to protect villages against wild animals. Hence, illicit movements of dogs were impossible to prevent.

Mass vaccinations began during 1951, with the modified-live low egg passage (LEP) vaccine (Adamson, 1954). The LEP product was subsequently replaced by inactivated CCV during the early 1990s, used either at central point locations or house to house. Although the focus was on dog rabies, per se, wildlife rabies, particularly jackal rabies, caused major problems, given the relative ease of transmission across canid species (Bingham, 1999; Sabeta et al., 2003; Zulu et al., 2009). As in Asia, research was conducted on oral vaccination of dogs, but was never implemented (Cliquet et al., 2018).

1.4.5 Other lyssaviruses

The first nonrabies lyssaviruses were described from Nigeria during the 1950s (Table 1.3). Several others were characterized into the 21st century (Markotter & Coertse, 2018; Rupprecht, Kuzmin, Yale, Nagarajan, & Meslin, 2019). Significantly, rabies related to these diverse lyssaviruses was believed to have evolved from Africa, but recent literature disputes this idea (Hayman, Fooks, Marston, & Garcia, 2016). Such nonrabies lyssaviruses (e.g., Duvenhage, Lagos bat virus, Mokola, Shimoni and Ikoma) are believed to be comparatively rare and perhaps insignificant, compared to the over-riding public health hazard of dog rabies. As with wildlife rabies in more developed countries, epidemiological implications of lyssaviruses in bats and other mammals may only be appreciated if the GEHRD is achieved.

1.4.6 Toward 2030?

Today, Africa faces the same challenges in rabies monitoring, prevention, and control as other lesser developed regions, namely widespread poverty, limited public health infrastructure, high dog: human densities, and less-than-ideal laboratory-based surveillance (Rupprecht & Salahuddin, 2019). Progress observed to date at several sites, including Chad (Zinsstag et al., 2017), Malawi (Gibson et al., 2016), South Africa (LeRoux et al., 2018), and Tanzania (Mpolya et al., 2017), benefited tremendously from local and international partnerships and underscore the realization that canine rabies prevention is possible. Such critical lessons learnt from these public-private demonstration projects must be replicated, and more importantly sustained, throughout Africa, if the GEHRD is to become a reality by 2030 (Sabeta & Ngoepe, 2018).

1.5 South America, Central America, and the Caribbean

1.5.1 Colonization impacts

The Americas were the last regions to receive humans, dogs, and enzootic canine RABV. In pre-Columbian cultures, there was no evidence of human or canine rabies. After European colonization, Spanish documents only provide hints about vampire bats (Hughes, Orciari, & Rupprecht, 2005; Steele & Fernandez, 1991; Vos et al., 2011). Such notations are in agreement with later phylogenetic reconstructions about bat RABV (Kuzmina et al., 2013; Troupin et al., 2016). Relatively low probability of RABV spillover from vampire bats to humans or other animals initially may have limited a direct linkage to rabies, which was already well known by European colonists from the Old World. Additionally, other "new" diseases were severely affecting indigenous New World populations (Berlinguer, 1992; Esparza, 2017; Nunn & Qian, 2010; O'Fallon & Fehren-Schmitz, 2011). Under such conditions, human rabies may have been overlooked.

1.5.2 Centuries absent-canine rabies?

South America was apparently free of canine rabies before, and remained so for almost three centuries after, European colonization. This canine "rabies-free" status in the pre-Columbian period is supported by limitations of host switching and apparent low spillover events from circulating bat RABV variants to dogs over ~14,000 years (Badrane & Tordo, 2001; Troupin et al., 2016; Vos et al., 2011). Moreover, complementary evidence supports ideas that spread of domestic dogs (*Canis lupus familiaris*) throughout South America occurred later (not before ~3000 BCE) than in North and Central America. Such dog populations were directly associated with the first complex societies of Ecuador, Bolivia, and Peru, but quite limited in extent beyond the Andes, absent from large parts of the continent (e.g., Amazonia, the Gran Chaco, and much of the Southern Cone) at the period of European contact (Guedes Milheira, Loponte, García Esponda, Acosta, & Ulguim, 2017; Mitchell, 2017; Prates, Prevosti, & Berón, 2010). This situation agrees with an essential control facet of canine rabies: maintaining a lower, less connected naïve free-ranging dog population, to minimize RABV spread. Local canid populations were fragmented and presented a more restricted diversity between indigenous and introduced dogs (NíLeathlobhair et al., 2018; Stahl, 2013; van Asch et al., 2013).

Besides connectivity, transportation and pathogenesis issues limited emergence opportunities. During the earliest European colonization period, the typical trans-Atlantic passage was ~5–10 weeks, restricting RABV introduction. Considering an incubation period of ~3–8 weeks, affected animals were likely discarded if ill during a long voyage (Márquez Ruiz, 2008). Another factor influencing dissemination was coastal concentration. Europeans landed at continental margins during the 1500s, not reaching interiors until the 1700s (Roberts, 1989). Estimations suggest the most recent common ancestor of canine RABV dates to between 1308 and 1510 (Troupin et al., 2016). Introduction of a Cosmopolitan clade of dog-maintained RABV lineage into the Western Hemisphere, based on genomic data, began between 1687 and 1773, coincident with the apogee of colonization, with no canine RABV members outside the "Cosmopolitan Group" (Badrane & Tordo, 2001; Bourhy et al., 2008; Smith & Seidel, 1993; Troupin et al., 2016).

1.5.3 The "spillover" phenomenon

Once established, colonization spread rabies throughout the continents, introducing RABV into new areas and novel hosts. Such phenomena were possible due to primary characteristics of RABV, namely, an ability to adapt to new hosts and variable incubation periods. Moreover, the enormous social and ecological changes following European colonization created a high urban population growth rate, together with introduced dog breeds, which completely replaced native dogs (NíLeathlobhair et al., 2018), increasing the probabilities of wildlife infection of dog-maintained RABV, most likely in the 18th century, with clear examples thereafter (Table 1.4) among native carnivores (Kobayashi et al., 2011; Nadin-Davis & Bingham, 2004; Troupin et al., 2016).

Transmission by hematophagous and nonhematophagous bats was only noted by Carini during 1911 and Haupt and Rehaag during 1921 (Steele & Fernandez, 1991)—now widely reported across South America (Johnson, Aréchiga-Ceballos, & Aguilar-Setien, 2014; Tobergte & Curtis, 2013). Typically, detection of affected taxa with bat-associated RABV variants, such as dogs and cats, occurs in localities where canine-maintained RABV were reduced or mass vaccination of domestic animals ceased (Escobar et al., 2015).

1.5.4 From initial evidence to the successful vaccinations

Eventually, canine and human rabies were reported throughout Central (Fleming, 1872; Smith, Orciari, Yager, Seidel, & Warner, 1992; Steele & Fernandez, 1991) and later in South America, where dog rabies was undocumented between 1552 and 1802 (de Azara, 1802; López de Gómara, 1922). Rabies was described in Peru during 1803 (Unanue, 1806), Argentina and Uruguay after British invasions during 1806–07 (Barcat, 2011; Steele & Fernandez, 1991), in Chile during 1835 (Darwin, 1845), and Brazil during 1844 (Sigaud, 1844), as described (Table 1.4).

TABLE 1.4 Highlights of notable rabies-related events in the Americas.

Year	Event	Reference
1552 CE	Canine rabies reportedly unknown in Perú	López de Gómara, 1922
1709 CE	Rabies reported in Mexico	de Esteyneffer, 1712
1741–78 CE	Human rabies reported throughout Central America	Fleming, 1872
1753–62 CE	Rabies described in archives of colonial U.S.	Johnson, 1952
1768 CE	Rabies identified in New England (Boston area) among dogs and foxes	Ravenel, 1901
1786 CE	Rabies spread from the northern to the southern U.S. states	Kerr & Stimson, 1909
1802 CE	Rabies unknown in either Paraguay or Rio de la Plata	de Azara, 1802
1803 CE	First report in South America, with "canine madness" described in Peru	Unanue, 1806
1806–07 CE	Canine rabies entered by the River Plate into Uruguay and Argentina	Johnson, 1952; Barcat, 2011
1819 CE	First recorded Canadian human death (Governor-General), by fox	Tabel, Corner, Webster, & Casey, 1974
1835 CE	Canine rabies is common throughout South America, including Brazil	Sigaud, 1844
1835 CE	Rabies described in several valleys from Chile	Darwin, 1845
1839 CE	First Canadian human death from dog rabies, in Quebec	Mitchell, 1967
1860s CE	Skunk rabies recognized throughout the northern North American prairies to the present	Ma et al., 2018
1886 CE	First human vaccination within Latin America, in Argentina	Amasino, Garbi, & Amasino, 2002; Esparza, 2017
1888 CE	Pasteur vaccine use in Mexico	Rodriguez de Roma, 2008
1907–17 CE	Large canine rabies outbreak in southern Ontario	Tabel et al., 1974
1908 CE	Distribution of antirabic dried, glycerinated cord materials for human vaccination by the U.S. Hygienic Laboratory to state boards of health	Anderson, 1914
1909 CE	Major outbreak reported among coyotes in western U.S.	Anonymous, 1909
1920 CE	PEP administered to thousands of humans and hundreds of domestic animals, using a modification of the Hogyes dilution method (including 22 naturally exposed dogs, all of whom survived) in the U.S.	Sellers, 1923a
1921 CE	Discontinuation of national distribution of animal cord preserved in glycerin for human vaccination, forcing states to produce or purchase commercial products in the U.S.	Sellers, 1923b
1922 CE	Reports of canine madness in the Canadian Arctic	Elton, 1931
1923 CE	Introduction of canine rabies vaccination in the U.S.	Eichhorn & Lyon, 1922

TABLE 1.4 Highlights of notable rabies-related events in the Americas—cont'd

Year	Event	Reference
1925 CE	Laboratory demonstration that virus passes within axons and produces degenerative changes	Goodpasture, 1925
1927 CE	Use of Sellers stain for enhanced detection of Negri bodies in U.S.	Young & Sellers, 1927
1930 CE	Single dose of inactivated vaccine shown for dogs in U.S.	Schoening, 1930
1935 CE	Description of a mouse protection test in U.S.	Webster & Dawson, 1935
1936 CE	Propagation of RABV in murine tissue culture in U.S.	Webster & Clow, 1936, 1937
1939 CE	Laboratory susceptibility of chicks and chick embryos to RABV	Dawson, 1939
1940s CE	Sporadic raccoon rabies described in Florida	Scatterday, Schneider, Jennings, & Lewis, 1960
1940 CE	Use of UV for inactivation in U.S.	Hodes, Webster, & Lavin, 1940
1940 CE	Test for vaccine potency	Habel, 1940
1942 CE	Use of chemical inactivation by phenol and chloroform for canine vaccination in U.S.	Johnson & Leach, 1942
1946 CE	Creation of a national control program by the U.S. Public Health Service, Communicable Disease Center	Held, Tierkel, & Steele, 1967
1947 CE	Extensive fox rabies outbreak in Canada, from the Arctic to southern provinces	Tabel et al., 1974
1948 CE	Avianized chick rabies virus vaccine	Koprowski & Cox, 1948
1948 CE	Mass dog vaccination and proof of concept in Shelby county, U.S.	Tierkel, Graves, Tuggle, & Wadley, 1950
1953 CE	Bat rabies described in Florida	Sacatterday & Galton, 1954
1955 CE	Major raccoon rabies epizootic recognized in southern Florida and potential reservoir threat recognized for future spread	Kappus, Bigler, McLean, & Trevino, 1970
1956 CE	Initiation of duck embryo vaccination	Peck, Powell, & Culbertson, 1956
1957 CE	Wildlife cases exceed canine rabies reports in U.S.	Held et al., 1967
1957 CE	First rabid bat diagnosed in British Columbia	Avery & Tailyour, 1960
1958 CE	Non-nervous tissue propagation of RABV	Kissling, 1958
1958 CE	Application of the direct fluorescent antibody test, U.S.	Goldwasser & Kissling, 1958
1962 CE	Observation of elongated rod-like particles in infected murine neurons by use of thin-section electron microscopy	Matsumoto, 1962
1962 CE	Transmission of RABV by the nonbite route	Constantine, 1962
1964 CE	Infection of the human diploid cell WI38, Wistar Institute	Wiktor, Fernandes, & Koprowski, 1964
1964 CE	Adaptation of the ERA strain for use as a veterinary vaccine	Abelseth, 1964

Continued

TABLE 1.4 Highlights of notable rabies-related events in the Americas—cont'd

Year	Event	Reference
1965 CE	Experimental demonstration of viral spread in the peripheral and central nervous systems via neuronal route	Johnson, 1965
1967 CE	Reproducible plaquing system for RABV in cell culture	Sedwick & Wiktor, 1967
1970 CE	Nonfatal rabies in a vaccinated Ohio child, Matthew Winkler	Hattwick, Weis, Stechschulte, Baer, & Gregg, 1972
1971 CE	Preparation and clinical testing of human rabies immune globulin	Cabasso, Loofbourow, Roby, & Anuskiewicz, 1971
1971 CE	Oral vaccination proof of concept in U.S.	Baer et al., 1971
1973 CE	Evidence of centrifugal peripheral neural spread of viral infection to multiple organs after CNS infection	Murphy, Harrison, Winn, & Bauer, 1973
1973 CE	Design of the RFFIT	Smith, Yager, & Baer, 1973
1973 CE	Determination of the viral excretion period in dogs relative to clinical signs	Vaughn Jr., Gerhardt, & Newell, 1965
1973 CE	Airborne transmission to a laboratory worker, N.Y.	Winkler, Fashinell, Leffingwell, Howard, & Conomy, 1973
1977 CE	Report of the Mid-Atlantic raccoon rabies epizootic	Jenkins & Winkler, 1987
1978 CE	Creation of monoclonal antibodies (MAbs)	Wiktor & Koprowski, 1978
1978 CE	Transmission by corneal transplant in U.S.	Houff et al., 1979
1982 CE	Identification of acetylcholine as a viral receptor	Lentz, Burrage, Smith, Crick, & Tignor, 1982
1983 CE	Cloning of mature full-length CVS glycoprotein gene into plasmids for direct expression	Yelverton, Norton, Obijeski, & Goeddel, 1983
1983 CE	Characterization of an antigenic determinant of the viral glycoprotein correlating with pathogenicity	Dietzschold et al., 1983
1984 CE	Construction of the vaccinia-rabies glycoprotein (V-RG) vaccine	Wiktor et al., 1984
1985 CE	First oral rabies vaccination field trials of foxes in Canada	Charlton et al., 1986
1988 CE	Major coyote rabies outbreak in south Texas	Clark et al., 1994
1989 CE	Proof-of-concept use of MAbs for experimental PEP	Schumacher et al., 1989
1989 CE	Construction of a recombinant human adenovirus-rabies vaccine in Ontario	Prevec, Campbell, Christie, Belbeck, & Graham, 1990
1990s CE	Initiation of oral rabies vaccine use with V-RG in Canada and the U.S.	Maki et al., 2017
1990 CE	Beginning of the Rabies in the Americas (RITA) conference	http://www.rabiesintheamericas.org/home
1990 CE	Mass parenteral canine vaccination programs in Mexico	Lucas et al., 2008

TABLE 1.4 Highlights of notable rabies-related events in the Americas—cont'd

Year	Event	Reference
1993 CE	Elimination of rabies from red foxes in eastern Ontario by ORV	MacInnes et al., 2001
1994 CE	Partial recovery in a Mexican child	Alvarez et al., 1994
2004 CE	Last canine RABV case reported in U.S.	Blanton, Hanlon, & Rupprecht, 2007
2004 CE	Rabies survival, without prior vaccination, in a Wisconsin teenager, Jeanie Giese	Willoughby et al., 2005
2005 CE	Last reported human case from canine RABV in Mexico	Blanton, Krebs, Hanlon, & Rupprecht, 2006
2006 CE	Field distribution of Onrab in Canada	Rosatte et al., 2009
2007 CE	Designation of World Rabies Day (WRD)	Briggs & Hanlon, 2007
2008 CE	Initiation of a continent-wide North American Rabies Management Program (NARMP)	Slate et al., 2009
2012 CE	Successful oral rabies vaccine developed using transgenic maize in Mexico	Loza-Rubio et al., 2012
2013 CE	Contraceptive use of Gonacon in Mexican dogs	Vargas-Pino et al., 2013
2016 CE	Last reported case of canine RABV in Mexico	Ma et al., 2018

Introductions of European dogs occurred unabated. Free-ranging dogs became a health concern, facilitating RABV spread and a need for human rabies prevention (Badrane & Tordo, 2001; Bourhy et al., 2008; Nadin-Davis & Bingham, 2004). After Pasteur's breakthroughs, during the following year, on September 4th, 1886 in Argentina, the physician Desiderio Davel reproduced NTV production, vaccinating two children, exposed to a rabid dog (Barcat, 2011; Esparza, 2017), as iconic examples of human PEP in South America (Table 1.4).

1.5.5 Regional efforts: Large-scale vaccination and laboratory-based surveillance

After the original Pasteur Institute was created during 1887, thereafter in South America, during the early 1900s, several institution-initiated activities began in Brazil, Chile, and Argentina (Anonymous, 1941; Suárez, 1940; Teixeira, Sandoval, & Takaoka, 2005).

Concomitantly, during 1902, an International Sanitary Bureau was born, constituting during 1958 the *"Pan American Health Organization"* (PAHO). This institution's fundamental purpose was promoting and coordinating efforts throughout the Americas to combat disease, including establishment of networks to investigate regional problems and promoting local solutions (Alleyne, 2002; Saldaña, 2006; Victora, 2002). Within South America, classical NTV were used until the 1950s. To reduce adverse effects, Fuenzalida and Palacios (1955) developed an inactivated suckling mouse brain (SMB) NTV in Chile, used for human PEP and canine rabies control (Fuenzalida & Palacios, 1955; Held et al., 1972; Schneider & Santos-Burgoa, 1994). This SMB vaccine was licensed in Chile, Uruguay, Argentina, Peru, Brazil,

Venezuela, Mexico, Bolivia, Ecuador, and Guatemala between 1960 and 1968. The SMB became the most commonly used vaccine in the region, was safer and reduced the number of NTV doses needed (Schneider & Santos-Burgoa, 1994). Thereafter, SMB products were replaced by human CCV, which were also used in dog vaccination campaigns.

Long before the GEHRD plan, an opportunity arose during 1983, at the Inter-American Meeting at the Ministerial Level on Health and Agriculture (RIMSA). The Latin American Countries (LAC) made formal political commitments to eliminate canine rabies from urban areas (e.g., >100,000 persons). This commitment was ratified at the Meeting of the Directors of National Programs for Rabies in the Americas (REDRIPA), strengthening collaborative efforts (Freire de Carvalho et al., 2018).

With coordination by PAHO, South American countries, following a Regional Program of Elimination of Human Rabies, applied mass dog vaccination across a wide range of settings (Clavijo et al., 2013). Concomitantly, improvement in laboratory-based surveillance, between reference laboratories (mainly from the Ministries of Agriculture and Health), provided countries with tools to process large numbers of samples, in Argentina, Brazil, Colombia, and Peru. These data were shared regularly by reporting to the Regional Epidemiological Surveillance System for Rabies (SIRVERA), maintained by the Veterinary Public Health unit of PAHO (Freire de Carvalho et al., 2018; Vigilato et al., 2013). Progressive improvement of diagnostic infrastructure in major cities, even with a high variation from country to country, led some countries to establish an extensive network of national laboratories for diagnosis, maintaining essential epidemiological surveillance for the detection of foci and monitoring of areas already free (Clavijo et al., 2013). Sustained efforts over the past 4 decades produced a significant impact upon dog rabies. Support provided by PAHO (e.g., by SIRVERA, REDIPRA, etc.) was extremely useful, reducing canine rabies cases by 98% and human cases by 95% (Freire de Carvalho et al., 2018).

1.5.6 Setbacks

Despite progress, success was not homogeneous throughout Central, South America, and the Caribbean, due in part to socioeconomic differences. Canine rabies remains endemic in Bolivia, Haiti, and zones in which sporadic dog-transmitted rabies occurs, such as in Perú, Venezuela and northeastern Brazil. Cases concentrate on the periphery of large cities, neglected communities (geographically remote, with large free-roaming dog populations) and border areas. Gaps exist due to misinformation, limited availability of quality health services, cultural disparities, poverty, scarce accessibility to biologics and prophylaxis (Clavijo et al., 2013; Vigilato, Clavijo, et al., 2013). In many areas, surveillance is deficient (Seetahal et al., 2018; Taylor, Hampson, Fahrion, Abela-Ridder, & Nel, 2017). Culling is ineffective and reducing turnover maintains population immunity at high levels between campaigns (Castillo-Neyra, Levy, & Naquira, 2016; Cleaveland & Hampson, 2017). Regional cooperation among countries will ensure elimination of canine rabies in remaining foci (Del Rio Vilas et al., 2017; Freire de Carvalho et al., 2018; Seetahal et al., 2018). Such setbacks are expected and manageable. Hormone-based vaccine-induced contraception methods may be promising choices for stabilizing uncontrolled populations. Oral dog vaccination could be useful when populations are difficult to reach, as demonstrated by examples in Africa, Asia, Europe, and

North America (Vargas-Pino et al., 2013; Wu, Franka, Svoboda, Pohl, & Rupprecht, 2009). New thermostable vaccines can overcome cold-chain constraints (Lankester et al., 2016). Ultimately, responsible pet ownership is essential to maintain healthy dog populations.

1.5.7 The emerging challenges of bat rabies

The reduction of human infections by dogs revealed long-term RABV circulation in previously under-reported areas, reservoirs and vectors, such as bats. This challenge demands development of a common strategy for preventing human rabies transmitted by bats, especially in remote areas in Amazonia (e.g., Peru, Ecuador, and Brazil), with an increasing number of human cases. Bat rabies presents a complex dynamic often dependent on human activities, such as livestock production or environmental disturbances (BottoNuñez, Becker, & Plowright, 2019). An increased number of rabies cases, associated with bat-related RABV variants, has occurred in wild mesocarnivores and nonhuman primates over the past 15 years (Kotait et al., 2019). Anthropogenic factors drive bat-derived RABV within a high diversity of taxa within the Neotropics (Kobayashi et al., 2007; Streicker et al., 2012).

1.5.8 Future perspectives

As elsewhere, the primary factors determining canine rabies elimination in this region are maintenance of mass canine vaccination, prompt and proper PEP, greater awareness about resurgence risks, enhanced laboratory-based surveillance, and responsible pet ownership. Sustainability of a healthy animal population is one of the basics of the "One World-One Health" concept (Beran & Frith, 1988). Unfortunately, a decrease in "dog-maintained" RABV in the region may create a false sense of security. Clearly, the risk of dog-derived RABV re-emergence is widely present, demanding a continuous surveillance network and elimination of "hot spots": at the borders of Bolivia; in portions of Central America; and in the Caribbean (Seetahal et al., 2018). Success requires continuous advocacy, champions and economic support to definitively eliminate dog-mediated rabies by 2030, as this region forms the basic proof of concept for the developing world and cannot be ignored.

1.6 North America

The major negative impacts of colonization associated with introduced diseases upon indigenous peoples throughout North America are identical to historical accounts elsewhere. Unlike the situation in Eurasia and North Africa, but united with Central and South America and the Caribbean (and certainly with Australia), true canine rabies was a new phenomenon in North America (Table 1.4). Although likely present earlier in the form of wildlife rabies, no suspect dog rabies cases were described prior to the 18th century (Baer, 2007; Rupprecht et al., 2017). Thereafter, canine rabies was reported throughout the region, as a byproduct of European introduction. One of the first reports of a canine rabies outbreak within the New World occurred in Mexico (Carrada Bravo, 1978). Although reports of rabies in dogs occurred throughout North America, Canada does not appear to have been as widely impacted, compared to Mexico and the USA (Fehlner-Gardiner, 2018).

As in the Old World, myths predominated during the 18th–19th century, such as in the use of "madstones" in rural folk medicine (Fig. 1.1). Although bites were most common by rabid dogs, wildlife were also a major concern, especially exposures from rabid coyotes and wolves. As in Eurasia and Africa during the prevaccination era, in response to outbreaks among mesocarnivores, tens of thousands of furbearers were killed annually during the late 19th–20th centuries (Records, 1932).

During the early part of the 20th century, European contributions to laboratory diagnosis and human PEP were incorporated rapidly (Hoenig, Jackson, & Dickinson, 2018; Watson, 1913). These included formation of Pasteur Institutes for local production and distribution of biologics by mail order (Fig. 1.2). After World War II, the elimination of canine RABV transmission provided the epidemiological luxury to document viral variants among multiple wildlife taxa, including bats, foxes, raccoons, and skunks. Significantly, Mexico represented the northern-most extent of vampire bat rabies, approximately 150 km to the border with the USA (Hayes & Piaggio, 2018; Rupprecht et al., 2018).

As summarized in Table 1.4, most of the notable 20th century events ascribed to North America relate to characterization of a diversity of RABV variants (especially associated with bats and as aligned with South America, the sole region with RABV as the only lyssavirus documented to date) and major technological advancements, such as the birth of additional diagnostic tests, potency assays, contributions to pathobiology, modern human CCV and biologics, the conception of ORV for wildlife, development of monoclonal antibodies, recombinant virus expression systems, global collaborations via the NARMP, RITA, WRD, and unprecedented enhanced laboratory-based surveillance, in excess of 100,000 submissions per year (Ma et al., 2018). Well in advance of the GEHRD by 2030, Canada, Mexico, and the USA all achieved this goal, serving as a New World role model (Fehlner-Gardiner, 2018; Rupprecht et al., 2018).

FIG. 1.1 In folk culture, "madstones" were believed to be curative after animal bites via the absorption of poisons, toxins, and other substances. *Photo taken by Charles Rupprecht of samples courtesy of the National Museum of American History, Smithsonian Institution, Washington, DC.*

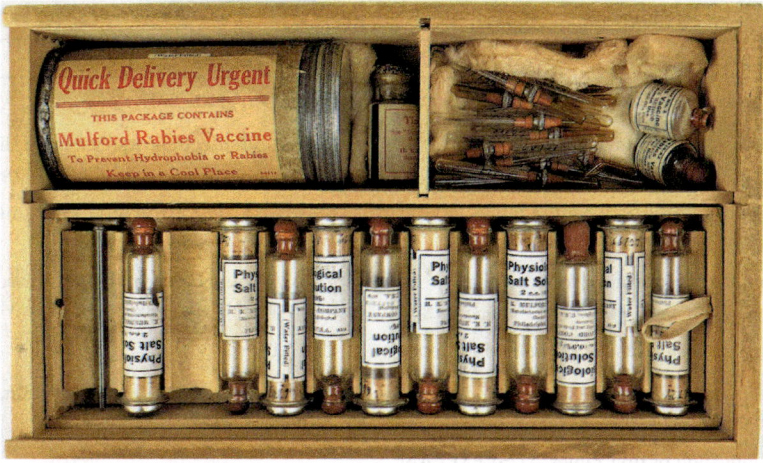

FIG. 1.2 Example of a mail-order human nerve tissue vaccine used for rabies prophylaxis in the USA during the early part of the 20th century. *Samples courtesy of the National Museum of American History, Smithsonian Institution, Washington, DC.*

1.7 Australia

1.7.1 RABV

Significantly, RABV infection was never documented in Australian native fauna (Table 1.5). Prior to European colonization, presumed maintenance of RABV - if an incursion had occurred - would be via native dingos (*Canis lupus dingo*). Domestic dogs (*Canis lupus familiaris*) were introduced during European colonization. Puppies, as well as Governor Phillip's greyhounds, were reportedly aboard the First Fleet (https://firstfleetfellowship.org.au/library/first-fleetlist-livestock-provisions-plants-seeds/). The first (and only) reported rabies outbreak occurred in Tasmania during the summer of 1866–67, in and near Hobart, as reviewed by Robertson (1932) and Crowther (1946), based on reporting in the local newspaper, *The Hobart Town Mercury*, between February 18 and April 4, 1867. An official enquiry was held by the City Council Health Committee, and Robertson (1932), concluded that the outbreak was rabies, whereas Crowther (1946) favored traumatic tetanus. More recently, Pullar and McIntosh (1954) reviewed the evidence and concluded that "there is no reason whatsoever to doubt the official diagnosis [of rabies] made at the time." The outbreak consisted of four incidents affecting up to five dogs and three pigs, and five people that were bitten (one death) by dogs. In the second incident, on January 17, 1867, a dog was observed to have "totally changed its usual aspect and demeanour" became vicious and quarrelsome, and developed pica. On 19th January, it attacked a boy, causing an inch-long tear in his lower lip. Two days later, the dog died. On 12th February, the boy began showing signs of rabies (e.g., fever, listlessness, headaches, hallucinations, and hydrophobia) and died on 17th February. No inquest was held nor autopsy performed. However, within a week, the City Council appointed two additional constables to collect all stray dogs (Pullar & McIntosh, 1954). No further cases were reported. Regarding the source, ships from all parts of the world frequently

TABLE 1.5 Inter-related historical events on rabies in Australia.

Year	Event	Reference
~2000 BCE	Dingoes introduced to Australia	Oskarsson et al., 2012
1788 CE	Domesic dogs introduced to Australia	https://firstfleetfellowship.org.au/library/first-fleetlist-livestock-provisions-plants-seeds/
1866–67 CE	Suspect outbreak of canine rabies in Hobart, Tasmania—one human death	Robertson, 1932; Crowther, 1946; Pullar & McIntosh, 1954
1871 CE	Red foxes introduced to Australia	https://www.feralscan.org.au/foxscan/pagecontent.aspx?page=fox_historyandbiology
1987 CE	First imported case of human rabies reported	CDC, 1988
1990 CE	Second imported case of human rabies reported	Bek, Smith, Levy, Sullivan, & Rubin, 1992; Grattan-Smith et al., 1992
1991 CE	Publication of first rabies AUSVETPLAN	
1996 CE	ABLV first identified in a black flying fox (*Pteropus alecto*), NSW	Hooper et al., 1997
1996 CE	First human case of ABLV (QLD)	Allworth, Murray, & Morgan, 1996; Samaratunga, Searle, & Hudson, 1998
1996 CE	Publication of second rabies AUSVETPLAN	
1997 CE	Report of first identification of ABLV from archival samples collected in 1995	Bunn & Garner, 1997
1998 CE	Second human case of ABLV (QLD)	Hanna et al., 2000
2009 CE	Publication of third AUSVETPLAN	
2012 CE	Risk prioritization of northern Australia rabies risk	Cookson, Sergeant, & Martin, 2012
2013 CE	Third human case of ABLV (QLD)	Francis et al., 2014
2013 CE	First exposure of a dog to ABLV reported	NSW Department of Primary Industries, 2013
2014 CE	First detection of infection of horses with ABLV	Annand & Reid, 2014
2015 CE	Model of RABV spread in domestic dogs in northern Australia	Durr & Ward, 2015
2016 CE	Free-roaming domestic and wild dog model	Sparkes et al., 2016
2017 CE	Model of RABV spread in wild dogs in northern Australia	Johnstone-Robertson, Fleming, Ward, & Davis, 2017
2017 CE	Risk assessment of RABV incursion in northern Australia	Hudson, Brookes, & Ward, 2017

NSW, *New South Wales*; QLD, *Queensland*.

called at Hobart during the 1860s. In the virtual absence of quarantine restrictions, it was assumed a dog incubating RABV was introduced. Pullar and McIntosh (1954) estimated that such an introduction probably occurred ~August 1866. No further RABV outbreaks were recorded (Geering, 1992). Two confirmed reports of rabies in humans in Australia occurred (i.e., 1987 and 1990), but the disease was considered contracted abroad, in India (CDC, 1988), and North Vietnam (Bek et al., 1992; Grattan-Smith et al., 1992), respectively.

During the 1990s, the threat of dog rabies was largely discounted, although the presence of dingoes, domestic dogs, and their hybrids in northern Australia was recognized as a risk (Newsome & Catling, 1992). Forman (1993) highlighted the potential of dogs to become infected and maintain both sylvatic (involving introduced red foxes) and urban rabies cycle within Australia. Susceptibility of Australian native fauna, including carnivorous quolls (*Dasyurus* spp.), is unknown. Due to presumed low abundance and limited distributions, native carnivorous marsupials have not been considered a high risk for RABV transmission (Garner, 1992). Australia has disease preparedness plans (AUSVETPLAN) to address incursions of exotic diseases, such as rabies (Animal Health Australia, 2019). The first edition was published during 1991. An exercise to test fox rabies preparedness and contingency planning was conducted during 1990 (O'Brien & Berry, 1992). The most current rabies AUSVETPLAN (version 4) is under review.

Northern Australia is recognized as a risk area for a RABV incursion. Rabies spread in eastern Indonesia prompted a re-evaluation of this risk. In particular, the Bali rabies outbreak during 2008 caused great concern and prompted a new research effort focused on northern Australia. The risk of a RABV incursion in northern Australia was assessed via expert opinion during 2012 (Cookson et al., 2012). Risks included the illegal importations of rabid animals via boats (continuing pre-European settlement Australian-South East Asian crosscultural traditions), unauthorized fishing vessels, and itinerant yachts. Sparkes et al. (2015) summarized current incursion scenarios through the Top End of the Northern Territory and via Papua New Guinea and the Torres Strait to Cape York Peninsula.

The first simulation model of rabies spread in northern Australia was published during 2015 (Durr & Ward, 2015). This model was parameterized based on GPS studies of roaming dogs in the Indigenous communities of the Northern Peninsula Area (NPA), Queensland (Durr & Ward, 2014). This was followed during 2016 by Sparkes et al., who focused on the Top End of the Northern Territory. The first model of rabies spread in wild dog populations was published during 2017 (Johnstone-Robertson et al., 2017), also focusing on the NPA. The first formal risk assessment of a RABV incursion in northern Australia (e.g., NPA) was published the same year (Hudson et al., 2017).

Considering the dual spread of humans and dogs throughout the world over tens of thousands of years, it is indeed intriguing that canine rabies never became established on the continent. In the case of the presumed incursion of rabies in Hobart during the 1860s, Pullar and McIntosh (1954) concluded that good fortune, prompt recognition, sound advice and immediate action eliminated further opportunity to spread. During the latter 19th century, most colonial travel and trade occurred with Europe (especially with the British Isles) and travel typically consisted of a journey of >100 days. The likely low numbers of dogs being transported, and extended journey time, would have made risks negligible. The new Australian government formed at independence (1901) was aware of rabies, and the introduction of strict quarantine laws mitigated this risk further during the 20th century, when travel times were rapidly reduced.

Recently, the focus on RABV incursions has shifted to northern Australia due to the proximity of rabies endemic islands in eastern Indonesia, existing cultural ties, a large population of free-roaming domestic dogs in Indigenous communities, and the presence of a wild dog/dingo population. Risk analyses suggest there is a nonnegligible risk of canine rabies incursion (Hudson et al., 2017). Increased surveillance of dogs on boats is necessary. Also, the need for incursion planning and response policy has prompted the development of rabies spread models (Durr & Ward, 2015; Johnstone-Robertson et al., 2017; Sparkes et al., 2016).

1.7.2 ABLV

Australian Bat Lyssavirus (ABLV) was identified in a juvenile black flying fox (*Pteropus alecto*) collected in Ballina, northern New South Wales, during May 1996 (Hooper et al., 1997). Testing of archival material identified a positive black flying fox sampled in Townsville during January 1995 (Bunn & Garner, 1997). The first human case was reported during 1996 (Allworth et al., 1996; Samaratunga et al., 1998). A second and third human case was reported during 1998 and 2013, respectively (Francis et al., 2014; Hanna et al., 2000). Exposure of a dog to ABLV was reported during 2013 (NSW Department of Primary Industries, 2013) and infection of two horses during 2014 (Annand & Reid, 2014). Limited molecular analysis infers that ABLV may have been endemic to Australia well before European colonization (Warrilow, Smith, Harrower, & Smith, 2002).

Serological evidence suggests a wide geographical distribution of ABLV in bats throughout Australia (Field, 2018; Prada, Boyd, Baker, Jackson, & O'Dea, 2019). To date, ABLV-infected bats were reported from coastal areas around Australia. Inland, ABLV was found among bats in Victoria, New South Wales, Queensland, and the Northern Territory (Bunn & Garner, 1997; Field & Ross, 1999). ABLV infection has been commonly detected in four species of flying-fox in Australia - the black flying-fox (*Pteropus alecto*), little red flying-fox (*Pteropus scapulatus*), grey-headed flying-fox (*Pteropus poliocephalus*), and spectacled flying-fox (*Pteropus conspicillatus*), although at an extremely low prevalence of <1% (Field, 2018). ABLV has also been detected in one species of microbat (e.g., *Saccolaimus flaviventris*, the yellow-bellied sheath-tailed bat), at a higher prevalence (Field, 2018). Preexposure vaccination with CCV is recommended for people whose occupation or recreational activities place them at increased risk of being exposed to bats (https://www.health.nsw.gov.au/Infectious/controlguideline/Pages/rabies.aspx#3; accessed 17 February 2019).

As the only continent to have rabies via ABLV, but significantly not RABV since the 19th century, Australia represents a major example of the ability to achieve and maintain such an historical status, in support of the concept of the GEHRD, at a major scale. Such an achievement serves as a model for Asia, which would reduce the risk significantly for potential canine rabies translocation to Australia and elsewhere.

References

Abela-Ridder, B., de Balogh, K., Kessels, J. A., Dieuzy-Labaye, I., & Torres, G. (2018). Global rabies control: The role of international organisations and the Global Strategic Plan to eliminate dog-mediated human rabies. *Revue Scientifique et Technique, 37*(2), 741–749.

References

Abelseth, M. K. (1964). An attenuated rabies vaccine for domestic animals produced in tissue culture. *Canadian Veterinary Journal*, 5(11), 279–286.

Adamson, J. S. (1954). Ecology of rabies in Southern Rhodesia. *Bulletin of the World Health Organization*, 10(5), 753–759.

Ahuja, S., Tripathi, K. K., Saha, S. M., & Saxena, S. N. (1985). Epidemiology of rabies in India. In E. Kuwert, C. Mérieux, H. Koprowski, & K. Bogel (Eds.), *Rabies in the tropics* (pp. 571–582). Berlin: Springer.

Aiewsakun, P., & Katzourakis, A. (2015). Endogenous viruses: Connecting recent and ancient viral evolution. *Virology*, 479-480, 26–37. https://doi.org/10.1016/j.virol.2015.02.011.

Alleyne, G. A. O. (2002). The Pan American Health Organization's first 100 years: Reflections of the Director. *American Journal of Public Health*, 92(12), 1890–1894.

Allworth, A., Murray, K., & Morgan, J. (1996). A human case of encephalitis due to a lyssavirus recently identified in fruit bats. *Communicable Diseases Intelligence*, 20(24), 504.

Alvarez, L., Fajardo, R., Lopez, E., Pedroza, R., Hemachudha, T., Kamolvarin, N., ... Baer, G. M. (1994). Partial recovery from rabies in a nine-year-old boy. *Pediatric Infectious Disease Journal*, 13(12), 1154–1155. https://doi.org/10.1097/00006454-199412000-00020.

Amasino, C. F., Garbi, C. J., & Amasino, M. F. (2002). The urban rabies in the Province of Buenos Aires, Argentina: Origin-evolution-present time. *Analecta Veterinaria*, 22(1), 17–31.

Anderson, J. F. (1914). The anti-rabic service of the hygienic laboratory for the past 5 years. *American Journal of Public Health*, 4, 322–326.

Animal Health Australia (2019). *AUSVETPLAN manuals and documents*. https://animalhealthaustralia.com.au/our-publications/ausvetplan-manuals-and-documents.

Annand, E. J., & Reid, P. A. (2014). Clinical review of two fatal equine cases of infection with the insectivorous bat strain of Australian bat lyssavirus. *Australian Veterinary Journal*, 92(9), 324–332. https://doi.org/10.1111/avj.12227.

Anonymous (1909). Monthly Bulletin. California State Board of Health. In *November*. .

Anonymous. (1941). Scientific institutions in Latin America. Municipal Rabies Institute of Buenos Aires. *Boletin de la Oficina Sanitaria Panamericana*, 20(10), 1017–1018. Retrieved from http://iris.paho.org/xmlui/handle/123456789/13320.

Arechiga Ceballos, N., Moron, S. V., Berciano, J. M., Nicolas, O., Lopez, C. A., Juste, J., ... Echevarria, J. E. (2013). Novel lyssavirus in bat, Spain. *Emerging Infectious Diseases*, 19(5), 793–795. https://doi.org/10.3201/eid1905.121071.

Aubert, M. (1992). Epidemiology of fox rabies. In K. Bögel, F. X. Meslin, & M. Kaplan (Eds.), *Wildlife rabies control* (pp. 9–18): Kent Wells Medical Ltd.

Aubert, M., Cliquet, F., Smak, J. A., Brochier, B., Schon, J., & Kappeler, A. (2004). Rabies in France, The Netherlands, Belgium, Luxembourg and Switzerland. In A. A. King, A. R. Fooks, M. Aubert, & A. I. Wandeler (Eds.), *Historical perspective of rabies in Europe and the Mediterranean Basin* (pp. 129–145). Paris: OIE.

Avery, R. J., & Tailyour, J. M. (1960). The isolation of the rabies virus from insectivorous bats in British Columbia. *Canadian Journal of Comparative Medicine and Veterinary Science*, 24(5), 143–146.

Badrane, H., & Tordo, N. (2001). Host switching in Lyssavirus history from the Chiroptera to the Carnivora orders. *Journal of Virology*, 75(17), 8096–8104.

Baer, G. M. (1991). *The natural history of rabies* (2nd ed.). Boca Raton: CRC Press.

Baer, G. M. (2007). The history of rabies. In A. C. Jackson & W. H. Wunner (Eds.), *Rabies: Scientific basis of the disease and its management* (2nd ed., pp. 1–22). Amsterdam: Elsevier Academic Press.

Baer, G. M., Abelseth, M. K., & Debbie, J. G. (1971). Oral vaccination of foxes against rabies. *American Journal of Epidemiology*, 93(6), 487–490.

Bahmanyar, M. (1974). Results of antibody profiles in man vaccinated with the HDCV vaccine with various schedules. In R. H. Regamey, W. Hennessen, R. Lang, F. T. Perkins, & R. Triau (Eds.), *Vol. 21. Symposia series in immunobiological standardization* (pp. 231–239). Basel.

Bahmanyar, M., Fayaz, A., Nour-Salehi, S., Mohammadi, M., & Koprowski, H. (1976). Successful protection of humans exposed to rabies infection. Postexposure treatment with the new human diploid cell rabies vaccine and antirabies serum. *Journal of the American Medical Association*, 236(24), 2751–2754.

Baltazard, M., & Ghodssi, M. (1954). Prevention of human rabies; treatment of persons bitten by rabid wolves in Iran. *Bulletin of the World Health Organization*, 10(5), 797–803.

Barcat, J. A. (2011). Rabies in the river plate. *Medicina (B Aires)*, 71(1), 91–93.

Barrow, J., & Neele, S. J. (1801). *An account of travels into the interior of southern Africa, in the years 1797 and 1798*. London: Printed by A. Strahan. for T. Cadell Jun. and W. Davies, in the Strand. Retrieved from https://www.biodiversitylibrary.org/bibliography/101967#/summary.

Bek, M. D., Smith, W. T., Levy, M. H., Sullivan, E., & Rubin, G. L. (1992). Rabies case in New South Wales, 1990: Public health aspects. *Medical Journal of Australia, 156*(9), 596–597. 600.

Belcher, D. W., Wurapa, F. K., & Atuora, D. O. (1976). Endemic rabies in Ghana. Epidemiology and control measures. *American Journal of Tropical Medicine and Hygiene, 25*(5), 724–729. https://doi.org/10.4269/ajtmh.1976.25.724.

Benecke, N. (1987). Studies on early dog remains from Northern Europe. *Journal of Archaeological Science, 14*(1), 31–49. https://doi.org/10.1016/S0305-4403(87)80004-3.

Beran, G. W., & Frith, M. (1988). Domestic animal rabies control: An overview. *Reviews of Infectious Diseases, 10*(Suppl. 4), S672–S677.

Berlinguer, G. (1992). The interchange of disease and health between the old and new worlds. *American Journal of Public Health, 82*(10), 1407–1413.

Bevan, R. (2018). Turkey's Göbekli Tepe: Is this the world's first architecture? In *The Art Newspaper*. Retrieved from https://www.theartnewspaper.com/news/is-this-the-world-s-first-architecture.

Bhishagratna, K. K. L. (1911). *Chapter VI. In An English translation of the Sushruta Samhita: With a full and comprehensive introduction, additional texts, different readings, notes, comparative views, index, glossary and plates. Vol. II: Nidána-Sthána, S'árira-Sthána, Chikitsitasthána and Kalapa-Sthána* (pp. 728–736). Calcutta.

Bingham, J. (1999). *The control of rabies in jackals in Zimbabwe*. PhD Harare, Zimbabwe: University of Zimbabwe.

Bingham, J., Foggin, C. M., Wandeler, A. I., & Hill, F. W. (1999a). The epidemiology of rabies in Zimbabwe. 1. Rabies in dogs (*Canis familiaris*). *Onderstepoort Journal of Veterinary Research, 66*(1), 1–10.

Bingham, J., Foggin, C. M., Wandeler, A. I., & Hill, F. W. (1999b). The epidemiology of rabies in Zimbabwe. 2. Rabies in jackals (*Canis adustus* and *Canis mesomelas*). *Onderstepoort Journal of Veterinary Research, 66*(1), 11–23.

Blancou, J. (1994). Early methods for the surveillance and control of rabies in animals. *Revue Scientifique et Technique, 13*(2), 361–372.

Blancou, J. (2004). Rabies in Europe and the Mediterranean Basin: From antiquity to the 19th century. In A. A. King, A. R. Fooks, M. Aubert, & A. I. Wandeler (Eds.), *Historical perspective of rabies in Europe and the Mediterranean Basin* (pp. 15–24). Paris: OIE.

Blancou, J., Andral, L., Aubert, M., & Artois, M. (1982). Oral vaccination of foxes against rabies. Results of trials in France. *Bulletin de l'Academie Veterinaire de France, 55*(3), 351–359.

Blanton, J. D., Hanlon, C. A., & Rupprecht, C. E. (2007). Rabies surveillance in the United States during 2006. *Journal of the American Veterinary Medical Association, 231*(4), 540–556.

Blanton, J. D., Krebs, J. W., Hanlon, C. A., & Rupprecht, C. E. (2006). Rabies surveillance in the United States during 2005. *Journal of the American Veterinary Medical Association, 229*(12), 1897–1911.

Botigué, L. R., Song, S., Scheu, A., Gopalan, S., Pendleton, A. L., Oetjens, M., … Veeramah, K. R. (2017). Ancient European dog genomes reveal continuity since the Early Neolithic. *Nature Communications, 8*. 16082 https://doi.org/10.1038/ncomms16082.

BottoNuñez, G., Becker, D. J., & Plowright, R. K. (2019). The emergence of vampire bat rabies in Uruguay within a historical context. *Epidemiology and Infection, 147*, e180. https://doi.org/10.1017/S0950268819000682.

Botvinkin, A. D., Poleschuk, E. M., Kuzmin, I., Borisova, T. I., Gazaryan, S. V., Yager, P., & Rupprecht, C. E. (2003). Novel lyssaviruses isolated from bats in Russia. *Emerging Infectious Diseases, 9*(12), 1623–1625.

Boulger, L. R., & Porterfield, J. S. (1958). Isolation of a virus from Nigerian fruit bats. *Transactions of the Royal Society of Tropical Medicine and Hygiene, 52*(5), 421–424. https://doi.org/10.1016/0035-9203(58)90127-5.

Bourhy, H., Reynes, J. M., Dunham, E. J., Dacheux, L., Larrous, F., Huong, V. T. Q., … Holmes, E. C. (2008). The origin and phylogeography of dog rabies virus. *Journal of General Virology, 89*(11), 2673–2681.

Briggs, D., & Hanlon, C. A. (2007). World rabies day: Focusing attention on a neglected disease. *Veterinary Record, 161*(9), 288–289.

Bunn, C., & Garner, G. (1997). Update on surveillance for Australian bat lyssavirus. *Australasian Epidemiologist, 4*(3), 27–30.

Cabasso, V. J., Loofbourow, J. C., Roby, R. E., & Anuskiewicz, W. (1971). Rabies immune globulin of human origin: Preparation and dosage determination in non-exposed volunteer subjects. *Bulletin of the World Health Organization, 45*, 303–315.

Carrada Bravo, T. (1978). Investigacion documental de la primeraepidemia de rabiaregistradaen la Republica Mexicana en 1709. *Salud Publica Mex, 20*(6), 705–716.

Castillo-Neyra, R., Levy, M. Z., & Naquira, C. (2016). Efecto del sacrificio de perrosvagabundosen el control de la rabiacanina. *Revista Peruana de Medicina Experimental y Salud Publica, 33*(4), 772–779. https://doi.org/10.17843/rpmesp.2016.334.2564.

References

Centres for Disease Control (CDC). (1988). Imported human rabies—Australia, 1987. *MMWR. Morbidity and Mortality Weekly Report*, 37(22), 351–353.

Centres for Disease Control (CDC). (2014). Notes from the field: Wildlife rabies on an island free from canine rabies for 52 years—Taiwan, 2013. *MMWR. Morbidity and Mortality Weekly Report*, 63(8), 178.

Chakrabarti, P. (2010). "Living vs. dead:" The Pasteurian paradigm and imperial vaccine research. *Bulletin of the History of Medicine*, 84(3), 387–423.

Charlton, K. M., Webster, W. A., Casey, G. A., Rhodes, A. J., Macinnes, C. D., & Lawson, K. F. (1986). Recent advances in rabies diagnosis and research. *Canadian Veterinary Journal*, 27(2), 85–89.

Cheema, G. (2015). Historical notes: Rabies, anti-rabic vaccine and the raj. *Indian Journal of History of Science*, 50(3), 514–520. https://doi.org/10.16943/ijhs/2015/v50i4/48320.

Clark, K. A., Neill, S. U., Smith, J. S., Wilson, P. J., Whadford, V. W., & McKirahan, G. W. (1994). Epizootic canine rabies transmitted by coyotes in south Texas. *Journal of the American Veterinary Medical Association*, 204(4), 536–540.

Clavijo, A., Del Rio Vilas, V. J., Mayen, F. L., Yadon, Z. E., Beloto, A. J., Vigilato, M. A. N., … Cosivi, O. (2013). Gains and future road map for the elimination of dog-transmitted rabies in the Americas. *American Journal of Tropical Medicine and Hygiene*, 89(6), 1040–1042. https://doi.org/10.4269/ajtmh.13-0229.

Cleaveland, S., & Hampson, K. (2017). Rabies elimination research: Juxtaposing optimism, pragmatism and realism. *Proceedings of the Royal Society B: Biological Sciences*, 284(1869) https://doi.org/10.1098/rspb.2017.1880.

Cleaveland, S., Thumbi, S. M., Sambo, M., Lugelo, A., Lushasi, K., Hampson, K., & Lankester, F. (2018). Proof of concept of mass dog vaccination for the control and elimination of canine rabies. *Revue Scientifique et Technique*, 37(2), 559–568. https://doi.org/10.20506/rst.37.2.2824.

Cliquet, F., Guiot, A. L., Aubert, M., Robardet, E., Rupprecht, C. E., & Meslin, F. X. (2018). Oralvaccination of dogs: A well-studied and undervalued tool for achieving human and dog rabies elimination. *Veterinary Research*, 49(1), 61.

Cliquet, F., Picard-Meyer, E., Barrat, J., Brookes, S. M., Healy, D. M., Wasniewski, M., … Fooks, A. R. (2009). Experimental infection of foxes with European Bat Lyssaviruses type-1 and 2. *BMC Veterinary Research*, 5, 19.

Cliquet, F., Picard-Meyer, E., & Robardet, E. (2014). Rabies in Europe: What are the risks? *Expert Review of Anti-Infective Therapy*, 12(8), 905–908. https://doi.org/10.1586/14787210.2014.921570.

Cluver, E. (1927). Rabies in South Africa. *Journal of the Medical Association of South Africa*, 1(5), 247–253.

Constantine, D. G. (1962). Rabies transmission by Nonbite Route. *Public Health Reports*, 77(4), 287–289.

Cookson, B., Sergeant, E. S. G., & Martin, P. A. J. (2012). Risk-based prioritisation of surveillance for exotic animal diseases in northern Australia. In *Proceedings of the 13th International Symposium on Veterinary Epidemiology and Economics (ISVEE), Belgium, Netherlands* (p. 379).

Crowther, W. E. (1946). A case of so-called hydrophobia; a matter of diagnosis. *Medical Journal of Australia*, 1, 69–72.

Curasson, G. (1942). *Traité de pathologie exolique vétérinaire et comparée. Tome I: Maladies à ultra-virus* (2nd. ed.). Paris: Vigot Frères.

Daoust, P. Y., Wandeler, A. I., & Casey, G. A. (1996). Cluster of rabies cases of probable bat origin among red foxes in Prince Edward Island, Canada. *Journal of Wildlife Diseases*, 32(2), 403–406.

Darwin, C. (1845). In J. Hubert & J. Gil (Eds.), *Viaje de un naturalistaalrededor del mundo* (1st ed. 1945). Buenos Aires: Librería El Ateneo. Retrieved from https://archive.org/details/rabieshydrophobi00flem.

Davis, S. J. M., & Valla, F. R. (1978). Evidence for domestication of the dog 12,000 years ago in the Natufian of Israel. *Nature*, 276(5688), 608–610.

Dawson, J. R. (1939). Infection of chicks and chick embryos with rabies. *Science*, 89(2309), 300–301. https://doi.org/10.1126/science.89.2309.300-a.

Dayan, T. (1994). Early domesticated dogs of the near east. *Journal of Archaeological Science*, 21(5), 633–640.

de Azara, F. (1802). De los perros. In *Apuntamientos para la Historia Natural de los quadrúpedos del Paragüay y Rio de la Plata* (pp. 277–291). Madrid: Imprenta de la Viuda de Ibarra.

De Benedictis, P., Gallo, T., Iob, A., Coassin, R., Squecco, G., Ferri, G., … Mutinelli, F. (2008). Emergence of fox rabies in north-eastern Italy. *Euro Surveillance*, 13(45). pii: 19033.

de Esteyneffer, J. (1712). Libro II - Capitulo XXVII: De las Heridas o Mordeduras Ponzoñosas. In *Florilegio medicinal de todas las enfermedades* (pp. 382–386).

Del Rio Vilas, V. J., Freire de Carvalho, M. J., Vigilato, M. A. N., Rocha, F., Vokaty, A., Pompei, J. A., … Cosivi, O. (2017). Tribulations of the Last Mile: Sides from a regional program. *Frontiers in Veterinary Science*, 4(4).

Demetriou, P., & Moynagh, J. (2011). The European Union strategy for external cooperation with neighbouring countries on rabies control. *Rabies Bulletin Europe*, 35(1), 5–7.

Dietzschold, B., Wunner, W. H., Wiktor, T. J., Lopes, M., Lafon, M., Smith, C. L., & Koprowski, H. (1983). Characterization of an antigenic determinant of the glycoprotein that correlates with pathogenicity of rabies virus. *Proceedings of the National Academy of Sciences of the United States of America*, 80(1), 70–74.

Dodet, B., Adjogoua, E. V., Aguemon, A. R., Amadou, O. H., Atipo, A. L., Baba, B. A., ... Wateba, M. I. (2008). Fighting rabies in Africa: The Africa Rabies Expert Bureau (AfroREB). *Vaccine*, 26(50), 6295–6298.

Durr, S., & Ward, M. P. (2014). Roaming behaviour and home range estimation of domestic dogs in Aboriginal and Torres Strait Islander communities in northern Australia using four different methods. *Preventive Veterinary Medicine*, 117(2), 340–357. https://doi.org/10.1016/j.prevetmed.2014.07.008.

Durr, S., & Ward, M. P. (2015). Development of a novel rabies simulation model for application in a non-endemic environment. *PLOS Neglected Tropical Diseases*, 9(6), e0003876. https://doi.org/10.1371/journal.pntd.0003876.

EFSA. (2006). Assessment of the risk of rabies introduction into the UK, Ireland, Sweden, Malta, as a consequence of abandoning the serological test measuring protective antibodies to rabies. *EFSA Journal*, 436, 1–54.

Eichhorn, A., & Lyon, B. M. (1922). Prophylactic vaccination of dogs against rabies. *Journal of the American Medical Association*, LXI(14), 38.

Elton, C. (1931). Epidemics among sledge dogs in the Canadian arctic and their relation to disease in the arctic fox. *Canadian Journal of Research*, 5(6), 673–692. https://doi.org/10.1139/cjr31-106.

Escobar, L. E., Restif, O., Yung, V., Favi, M., Pons, D. J., & Medina-Vogel, G. (2015). Spatial and temporal trends of bat-borne rabies in Chile. *Epidemiology and Infection*, 143(7), 1486–1494. https://doi.org/10.1017/S095026881400226X.

Esparza, J. (2017). Viral epidemics in Latin America from the sixteenth to the nineteenth centuries and the early days of virology in the region. In J. E. Ludert, F. H. Pujol, & J. Arbiza (Eds.), *Human virology in Latin America: From biology to control* (pp. 3–16). Cham: Springer International Publishing.

European Union. (2017). *Rabies eradication in the EU*. ISBN: 978-92-79-43519-5. https://doi.org/10.2772/58274.

Fehlner-Gardiner, C. (2018). Rabies control in North America—Past, present and future. *Revue Scientifique et Technique*, 37(2), 421–437. https://doi.org/10.20506/rst.37.2.2812.

Fekadu, M. (1972). Atypical rabies in dogs in Ethiopia. *Ethiopian Medical Journal*, 10, 79–86.

Fekadu, M. (1982). Rabies in Ethiopia. *American Journal of Epidemiology*, 115(2), 266–273.

Fevre, E. M., Bronsvoort, B. M., Hamilton, K. A., & Cleaveland, S. (2006). Animal movements and the spread of infectious diseases. *Trends in Microbiology*, 14(3), 125–131. https://doi.org/10.1016/j.tim.2006.01.004.

Field, H. E. (2018). Evidence of Australian bat lyssavirus infection in diverse Australian bat taxa. *Zoonoses and Public Health*, 65, 742–748. https://doi.org/10.1111/zph.12480.

Field, H. E., & Ross, A. D. (1999). Emerging viral diseases of bats. In *Wildlife in Australia—Healthcare and management; proceedings 327, 13–17 September 1999, Dubbo, Australia* (pp. 501–513). Sydney: Post-Graduate Foundation in Veterinary Science.

Fleming, G. (1872). *Rabies and hydrophobia: Their history, nature, causes, symptoms, and prevention* (1st ed.). London: Chapman and Hall.

Foggin, C. M. (1988). *Rabies and rabies-related viruses in Zimbabwe: Historical, virological and ecological aspects, virological and ecological aspects*. PhD thesis University of Zimbabwe.

Fooks, A. R., Roberts, D. H., Lynch, M., Hersteinsson, P., & Runolfsson, H. (2004). Rabies in the United Kingdom, Ireland and Iceland. In A. A. King, A. R. Fooks, M. Aubert, & A. I. Wandeler (Eds.), *Historical perspective of rabies in Europe and the Mediterranean Basin* (pp. 25–32). Paris: OIE.

Forman, A. J. (1993). The threat of rabies introduction and establishment in Australia. *Australian Veterinary Journal*, 70(3), 81–83. https://doi.org/10.1111/j.1751-0813.1993.tb03281.x.

Francis, J. R., Nourse, C., Vaska, V. L., Calvert, S., Northill, J. A., McCall, B., & Mattke, A. C. (2014). Australian Bat Lyssavirus in a child: The first reported case. *Pediatrics*, 133(4), e1063–e1067. https://doi.org/10.1542/peds.2013-1782.

Frantz, L. A. F., Mullin, V. E., Pionnier-Capitan, M., Lebrasseur, O., Ollivier, M., Perri, A., ... Larson, G. (2016). Genomic and archaeological evidence suggest a dual origin of domestic dogs. *Science*, 352(6290), 1228–1231. https://doi.org/10.1126/science.aaf3161.

Freire de Carvalho, M., Vigilato, M. A. N. N., Pompei, J. A., Rocha, F., Vokaty, A., Molina-Flores, B., ... Del Rio Vilas, V. J. (2018). Rabies in the Americas: 1998–2014. *PLOS Neglected Tropical Diseases*, 12(3), e0006271. https://doi.org/10.1371/journal.pntd.0006271.

Freuling, C. M., Beer, M., Conraths, F. J., Finke, S., Hoffmann, B., Keller, B., ... Müller, T. (2011). Novel lyssavirus in Natterer's bat, Germany. *Emerging Infectious Diseases*, 17(8), 1519–1522. https://doi.org/10.3201/eid1708.110201.

Freuling, C. M., Hampson, K., Selhorst, T., Schroder, R., Meslin, F. X., Mettenleiter, T. C., & Müller, T. (2013). The elimination of fox rabies from Europe: Determinants of success and lessons for the future. *Philosophical Transactions of the Royal Society B, Biological Sciences, 368*(1623), 20120142. https://doi.org/10.1098/rstb.2012.0142.

Freuling, C. M., Selhorst, T., Batza, H. J., & Müller, T. (2008). The financial challenge of keeping a large region rabies-free—The EU example. *Developments in Biologicals, 131,* 273–282.

Fuenzalida, E., & Palacios, R. (1955). Un métodomejoradoen la preparación de la vacunaantirrábica. *Boletín del Instituto Bacteriológico de Chile, 8*(1–4), 3–10.

Garden, D. (2009). Rabies research in Thailand. *Asian Biomedicine, 3*(2), 213–220.

Garner, M. G. (1992). World rabies picture—Implications for Australia. In P. O'Brien & G. Berry (Eds.), *Wildlife rabies contingency planning in Australia: National wildlife rabies workshop, 12–16 March 1990.* Australian Govt. Pub. Service: Canberra.

Gautret, P., Ribadeau-Dumas, F., Parola, P., Brouqui, P., & Bourhy, H. (2011). Risk for rabies importation from North Africa. *Emerging Infectious Diseases, 17*(12), 2187–2193.

Geering, W. (1992). Rabies: An overview of the disease. In P. O'Brien & G. Berry (Eds.), *Wildlife rabies contingency planning in Australia: National wildlife rabies workshop, 12–16 March 1990.* Australian Govt. Pub. Service: Canberra.

Ghobashy, H. M. M. (1986). Rabies control in Egypt. In *MZCP Seminar on planning and management of national rabies control programmes, Tunis, 20–21 October 1986* (pp. 109–112): WHO.

Gibson, A. D., Handel, I. G., Shervell, K., Roux, T., Mayer, D., Muyila, S., … Gamble, L. (2016). The vaccination of 35,000 dogs in 20 working days using combined static point and door-to-door methods in Blantyre, Malawi. *PLOS Neglected Tropical Diseases, 10*(7), e0004824. https://doi.org/10.1371/journal.pntd.0004824.

Gibson, A. D., Ohal, P., Shervell, K., Handel, I. G., Bronsvoort, B. M., Mellanby, R. J., & Gamble, L. (2015). Vaccinate-assess-move method of mass canine rabies vaccination utilising mobile technology data collection in Ranchi, India. *BMC Infectious Diseases, 15*(1), 589. https://doi.org/10.1186/s12879-015-1320-2.

Goddard, A. D., Donaldson, N. M., Horton, D. L., Kosmider, R., Kelly, L. A., Sayers, A. R., … Snary, E. L. (2012). A quantitative release assessment for the noncommercial movement of companion animals: Risk of rabies reintroduction to the United Kingdom. *Risk Analysis, 32*(10), 1769–1783.

Goldwasser, R. A., & Kissling, R. E. (1958). Fluorescent antibody staining of street and fixed rabies virus antigens. *Proceedings of the Society for Experimental Biology and Medicine, 98*(2), 219–223.

Goodpasture, E. W. (1925). A study of rabies, with reference to a neural transmission of the virus in rabbits, and the structure and significance of Negri bodies. *American Journal of Pathology, 1*(6), 547–582.543.

Grattan-Smith, P. J., O'Regan, W. J., Ellis, P. S., O'Flaherty, S. J., McIntyre, P. B., & Barnes, C. J. (1992). Rabies. A second Australian case, with a long incubation period. *Medical Journal of Australia, 156*(9), 651–654.

Greval, S. D. S. (1932). Rabies in mongoose. *Indian Medical Gazette, 67*(8), 451–453.

Guedes Milheira, R., Loponte, D. M., García Esponda, C., Acosta, A., & Ulguim, P. (2017). The first record of a pre-columbian domestic dog (*Canis lupus familiaris*) in Brazil. *International Journal of Osteoarchaeology, 27*(3), 488–494. https://doi.org/10.1002/oa.2546.

Gunawardena, P. S., Marston, D. A., Ellis, R. J., Wise, E. L., Karawita, A. C., Breed, A. C., … Fooks, A. R. (2016). Lyssavirus in Indian flying foxes, Sri Lanka. *Emerging Infectious Diseases, 22*(8), 1456–1459. https://doi.org/10.3201/eid2208.151986.

Haak, W., Lazaridis, I., Patterson, N., Rohland, N., Mallick, S., Llamas, B., … Reich, D. (2015). Massive migration from the steppe was a source for Indo-European languages in Europe. *Nature, 522*(7555), 207–211. https://doi.org/10.1038/nature14317.

Habel, K. (1940). Evaluation of a mouse test for the standardization of the immunizing power of anti-rabies vaccines. *Public Health Reports, 55*(33), 1473–1487. https://doi.org/10.2307/4583406.

Hampson, K., Coudeville, L., Lembo, T., Sambo, M., Kieffer, A., Attlan, M., … Prevention on behalf of the Global Alliance for Rabies Control Partners for Rabies (2015). Estimating the global burden of endemic canine rabies. *PLOS Neglected Tropical Diseases, 9*(4), e0003709.

Hanna, J. N., Carney, I. K., Smith, G. A., Tannenberg, A. E., Deverill, J. E., Botha, J. A., … Searle, J. W. (2000). Australian bat lyssavirus infection: A second human case, with a long incubation period. *Medical Journal of Australia, 172*(12), 597–599.

Harbi, M. S. (1976). The incidence of rabies in animals in the Sudan. *Bulletin of Animal Health and Production in Africa, 24*(1), 43–46.

Harischandra, P. L., Gunesekera, A., Janakan, N., Gongal, G., & Abela-Ridder, B. (2016). Sri Lanka takes action towards a target of zero rabies death by 2020. *WHO South-East Asia Journal of Public Health, 5*(2), 113–116.

Hassel, R., Vos, A., Clausen, P., Moore, S., van der Westhuizen, J., Khaiseb, S., ... Müller, T. (2018). Experimental screening studies on rabies virus transmission and oral rabies vaccination of the Greater Kudu (*Tragelaphus strepsiceros*). *Scientific Reports, 8*(1), 16599. https://doi.org/10.1038/s41598-018-34985-5.

Hattwick, M. A., Weis, T. T., Stechschulte, C. J., Baer, G. M., & Gregg, M. B. (1972). Recovery from rabies. A case report. *Annals of Internal Medicine, 76*(6), 931–942. https://doi.org/10.7326/0003-4819-76-6-931.

Hayes, M. A., & Piaggio, A. J. (2018). Assessing the potential impacts of a changing climate on the distribution of a rabies virus vector. *PLoS ONE, 13*(2), e0192887. https://doi.org/10.1371/journal.pone.0192887.

Hayman, D. T., Fooks, A. R., Marston, D. A., & Garcia, R. J. (2016). The global phylogeography of lyssaviruses—Challenging the 'Out of Africa' hypothesis. *PLOS Neglected Tropical Diseases, 10*(12), e0005266. https://doi.org/10.1371/journal.pntd.0005266.

Held, J. R., Fuenzalida, E., López Adaros, H., Arrossi, J. C., Poles, N. O., & Scivetti, A. (1972). Inmunizaciónhumana con vacunaantirrábica de cerebro de ratonlactante. *Boletin de la Oficina Sanitaria Panamericana, 72*(1), 565–575.

Held, J. R., Tierkel, E. S., & Steele, J. H. (1967). Rabies in man and animals in the United States, 1946–65. *Public Health Reports, 82*(11), 1009–1018.

Hikufe, E. H., Freuling, C. M., Athingo, R., Shilongo, A., Ndevaetela, E. -E., Helao, M., ... Maseke, A. (2019). Ecology and epidemiology of rabies in humans, domestic animals and wildlife in Namibia, 2011–2017. *PLOS Neglected Tropical Diseases, 13*(4), e0007355.

Hodes, H. L., Webster, L. T., & Lavin, G. I. (1940). The use of ultraviolet light in preparing a non-virulent antirabeis vaccine. *Journal of Experimental Medicine, 72*(4), 437–444. https://doi.org/10.1084/jem.72.4.437.

Hoenig, L. J., Jackson, A. C., & Dickinson, G. M. (2018). The early use of Pasteur's rabies vaccine in the United States. *Vaccine, 36*(30), 4578–4581.

Hooper, P. T., Lunt, R. A., Gould, A. R., Samaratunga, H., Hyatt, A. D., Gleeson, L. J., ... Murray, P. K. (1997). A new lyssavirus—The first endemic rabies-related virus recognized in Australia. *Bulletin de l'Institut Pasteur, 95*(4), 209–218. https://doi.org/10.1016/S0020-2452(97)83529-5.

Horton, D. L., Banyard, A. C., Marston, D. A., Wise, E., Selden, D., Nunez, A., ... Fooks, A. R. (2014). Antigenic and genetic characterization of a divergent African virus, Ikoma lyssavirus. *Journal of General Virology, 95*(Pt 5), 1025–1032. https://doi.org/10.1099/vir.0.061952-0.

Houff, S. A., Burton, R. C., Wilson, R. W., Henson, T. E., London, W. T., Baer, G. M., ... Sever, J. L. (1979). Human-to-human transmission of rabies virus by corneal transplant. *New England Journal of Medicine, 300*(11), 603–604. https://doi.org/10.1056/nejm197903153001105.

Hu, S. C., Hsu, C. L., Lee, M. S., Tu, Y. C., Chang, J. C., Wu, C. H., ... Hsu, W. C. (2018). Lyssavirus in Japanese Pipistrelle, Taiwan. *Emerging Infectious Diseases, 24*(4), 782–785. https://doi.org/10.3201/eid2404.171696.

Hübschle, O. J. (1988). Rabies in the kudu antelope (*Tragelaphus strepsiceros*). *Reviews of Infectious Diseases, 10*(Suppl. 4), 629–633.

Hudson, J. R. (1944). A short note on the history of rabies in Kenya. *East African Medical Journal, 21*, 322–327.

Hudson, E. G., Brookes, V. J., & Ward, M. P. (2017). Assessing the risk of a Canine rabies incursion in Northern Australia. *Frontiers in Veterinary Science, 4*, 141. https://doi.org/10.3389/fvets.2017.00141.

Hughes, G. J., Orciari, L. A., & Rupprecht, C. E. (2005). Evolutionary timescale of rabies virus adaptation to North American bats inferred from the substitution rate of the nucleoprotein gene. *Journal of General Virology, 86*(5), 1467–1474.

Jenkins, S. R., & Winkler, W. G. (1987). Descriptive epidemiology from an epizootic of raccoon rabies in the middle Atlantic States, 1982–1983. *American Journal of Epidemiology, 126*(3), 429–437.

Johnson, D. E. (1819). Observations on rabies Contagiosa. *Medico-Chirurgical Journal, 1*(4), 494–498.

Johnson, H. N. (1952). Rabies. In T. M. Rivers (Ed.), *Viral and rickettsial infections of man* (pp. 267–299). Philadelphia: Lippincott.

Johnson, R. T. (1965). Experimental rabies. Studies of cellular vulnerability and pathogenesis using fluorescent antibody staining. *Journal of Neuropathology and Experimental Neurology, 24*(4), 662–674.

Johnson, N., Aréchiga-Ceballos, N., & Aguilar-Setien, A. (2014). Vampire bat rabies: Ecology, epidemiology and control. *Viruses, 6*(5), 1911–1928. https://doi.org/10.3390/v6051911.

Johnson, N., Freuling, C., Horton, D., Müller, T., & Fooks, A. R. (2011). Imported rabies, European Union and Switzerland, 2001–2010. *Emerging Infectious Diseases, 17*(4), 753–754.

Johnson, H. N., & Leach, C. N. (1942). Studies on the single injection method of canine rabies vaccination. *American Journal of Public Health and the Nation's Health*, *32*(2), 176–180. https://doi.org/10.2105/ajph.32.2.176.

Johnstone-Robertson, S. P., Fleming, P. J. S., Ward, M. P., & Davis, S. A. (2017). Predicted spatial spread of canine rabies in Australia. *PLOS Neglected Tropical Diseases*, *11*(1), e0005312. https://doi.org/10.1371/journal.pntd.0005312.

Kappus, K. D., Bigler, W. J., McLean, R. G., & Trevino, H. A. (1970). The raccoon an emerging rabies host. *Journal of Wildlife Diseases*, *6*(4), 507–509.

Kemp, G. E., Moore, D. L., Causey, O. R., Odelola, A., & Fabiyi, A. (1972). Mokola virus—Further studies on Iban 27377, a new rabies-related etiologic agent of zoonosis in Nigeria. *American Journal of Tropical Medicine and Hygiene*, *21*(3), 356–359.

Kerr, R., Needham, J., & Wood, N. (2004). Science and civilisation in China: Chemistry and chemical technology. *Ceramic technology*, Retrieved from, https://books.google.co.in/books?hl=en&lr=&id=mabcHwmAD5oC&oi=fnd&pg=PR22&dq=15.%09Joseph+Needham.+Science+and+Civilization+in+China,+Vol+6%3B+Part+6+(Medicine).+Cambridge+University+Press,+United+Kingdom,+2004&ots=5Td5IsaPK3&sig=coAF3km4XwNPrsryMqUNolWn9p4.

Kerr, J. W., & Stimson, A. H. (1909). The prevalence of rabies in the United States. *Journal of the American Medical Association*, *LIII*(13), 989–994.

King, A. A., Meredith, C. D., & Thomson, G. R. (1993). Canid and viverrid rabies viruses in South Africa. *Onderstepoort Journal of Veterinary Research*, *60*(4), 295–299.

Kissling, R. E. (1958). Growth of rabies virus in non-nervous tissue culture. *Proceedings of the Society for Experimental Biology and Medicine*, *98*(2), 223–225.

Kobayashi, Y., Sato, G., Kato, M., Itou, T., Cunha, E. M. S., Silva, V. M., … Sakai, T. (2007). Genetic diversity of bat rabies viruses in Brazil. *Archives of Virology*, *152*(11), 1995–2004.

Kobayashi, Y., Suzuki, Y., Itou, T., Ito, F. H., Sakai, T., & Gojobori, T. (2011). Evolutionary history of dog rabies in Brazil. *Journal of General Virology*, *92*(1), 85–90.

Koprowski, H., & Cox, H. R. (1948). Studies on chick embryo adapted rabies virus. I. Culture Characteristics and Pathogenicity. *Journal of Immunology*, *60*(4), 533–554.

Kotait, I., Oliveira, R. d. N., Carrieri, M. L., Castilho, J. G., Macedo, C. I., Pereira, P. M. C., … E, C. (2019). Non-human primates as a reservoir for rabies virus in Brazil. *Zoonoses and Public Health*, *66*(1), 47–59. https://doi.org/10.1111/zph.12527.

Kuzmin, I. V., Botvinkin, A. D., McElhinney, L. M., Smith, J. S., Orciari, L. A., Hughes, G. J., … Rupprecht, C. E. (2004). Molecular epidemiology of terrestrial rabies in the former Soviet Union. *Journal of Wildlife Diseases*, *40*(4), 617–631.

Kuzmin, V. I., Botvinkin, A. D., Rybin, S. N., & Baialiev, A. B. (1992). A lyssavirus with an unusual antigenic structure isolated from a bat in southern Kyrgyzstan. *Voprosyvirusologii*, *37*(5–6), 256–259.

Kuzmin, I. V., Mayer, A. E., Niezgoda, M., Markotter, W., Agwanda, B., Breiman, R. F., & Rupprecht, C. E. (2010). Shimoni bat virus, a new representative of the Lyssavirus genus. *Virus Research*, *149*(2), 197–210.

Kuzmin, I. V., Shi, M., Orciari, L. A., Yager, P. A., Velasco-Villa, A., Kuzmina, N. A., … Rupprecht, C. E. (2012). Molecular inferences suggest multiple host shifts of rabies viruses from bats to mesocarnivores in Arizona during 2001–2009. *PLOS Pathogens*, *8*(6), e1002786.

Kuzmina, N. A., Kuzmin, V. I., Ellison, J. A., Taylor, S. T., Bergman, D. L., Dew, B., & Rupprecht, C. E. (2013). A reassessment of the evolutionary timescale of bat rabies viruses based upon glycoprotein gene sequences. *Virus Genes*, *47*(2), 305–310. https://doi.org/10.1007/s11262-013-0952-9.

Lankester, F. J., Wouters, P. A. W. M., Czupryna, A., Palmer, G. H., Mzimbiri, I., Cleaveland, S., … Sonnemans, D. G. P. (2016). Thermotolerance of an inactivated rabies vaccine for dogs. *Vaccine*, *34*(46), 5504–5511. https://doi.org/10.1016/j.vaccine.2016.10.015.

Lentz, T. L., Burrage, T. G., Smith, A. L., Crick, J., & Tignor, G. H. (1982). Is the acetylcholine receptor a rabies virus receptor? *Science*, *215*(4529), 182–184. https://doi.org/10.1126/science.7053569.

LeRoux, K., Stewart, D., Perrett, K. D., Nel, L. H., Kessels, J. A., & Abela-Ridder, B. (2018). Rabies control in Kwa Zulu-Natal, South Africa. *Bulletin of the World Health Organization*, *96*(5), 360–365. https://doi.org/10.2471/blt.17.194886.

Leslie, M. J., Messenger, S., Rohde, R. E., Smith, J., Cheshier, R., Hanlon, C., & Rupprecht, C. E. (2006). Bat-associated rabies virus in skunks. *Emerging Infectious Diseases*, *12*(8), 1274–1277.

Libeau, L. (1960). Enquete sur les cas de rage en Afrique. *Bulletin for the Epizootiological Diseases of Africa*, *8*, 289–294.

Lin, F. T. (1990). The protective effect of the large-scale use of PHKC rabies vaccine in humans in China. *Bulletin of the World Health Organization*, *68*(4), 449–454.

Liu, Y., Zhang, S., Zhao, J., Zhang, F., & Hu, R. (2013). Isolation of Irkut virus from a *Murina leucogaster* bat in China. *PLOS Neglected Tropical Diseases, 7*(3), e2097. https://doi.org/10.1371/journal.pntd.0002097.

Livingstone, D. (1857). *Missionary travels and researches in South Africa.* London: John Murray.

López de Gómara, F. (1922). Cosasnotables que hay y que no hay en el Perú. In *Vol. Tomo II. Historia general de las Indias 1511–1564* (pp. 194–196). Calpe: Madrid.

Loza-Rubio, E., Rojas-Anaya, E., Lopez, J., Olivera-Flores, M. T., Gomez-Lim, M., & Tapia-Perez, G. (2012). Induction of a protective immune response to rabies virus in sheep after oral immunization with transgenic maize, expressing the rabies virus glycoprotein. *Vaccine, 30*(37), 5551–5556.

Lucas, C. H., Pino, F. V., Baer, G., Morales, P. K., Cedillo, V. G., Blanco, M. A., & Avila, M. H. (2008). Rabies control in Mexico. *Developments in Biologicals, 131*, 167–175.

Ma, X., Monroe, B. P., Cleaton, J. M., Orciari, L. A., Li, Y., Kirby, J. D., … Blanton, J. D. (2018). Rabies surveillance in the United States during 2017. *Journal of the American Veterinary Medical Association, 253*(12), 1555–1568.

Macdonell, A. A. (1900). Philosophy of Rigveda. In *A history of Sanskrit literature* (pp. 116–138). New York: D. Appleton.

Macharia, M. J., Ombacho, K. M., Kasiiti, L. J., Mbugua, H. C. W., & Gacheru, S. G. (2003). Status of rabies in Kenya covering 5 years (1998–2002). In *Proceedings of the Seventh Southern and Eastern African Rabies Group: World Health Organization Meeting. Ezulwini, Swaziland: 12–15 May 2003* (pp. 31–38).

MacInnes, C. D., Smith, S. M., Tinline, R. R., Ayers, N. R., Bachmann, P., Ball, D. G., … Voigt, D. R. (2001). Elimination of rabies from red foxes in eastern Ontario. *Journal of Wildlife Diseases, 37*(1), 119–132.

Madhusudana, S. N., Saha, S. M., Sood, M., & Saxena, S. N. (1988). Multisite intradermal vaccination using tissue culture vaccine as an economical prophylactic regimen against rabies. *Indian Journal of Medical Research, 87*, 1–4.

Maki, J., Guiot, A. L., Aubert, M., Brochier, B., Cliquet, F., Hanlon, C. A., … Lankau, E. W. (2017). Oral vaccination of wildlife using a vaccinia-rabies-glycoprotein recombinant virus vaccine (RABORAL V-RG((R))): A global review. *Veterinary Research, 48*(1), 57. https://doi.org/10.1186/s13567-017-0459-9.

Mani, R. S., Dovih, D. P., Ashwini, M. A., Chattopadhyay, B., Harsha, P. K., Garg, K. M., … Madhusudana, S. N. (2017). Serological evidence of lyssavirus infection among bats in Nagaland, a North-Eastern state in India. *Epidemiology and Infection, 145*(8), 1635–1641.

Mark, J. J. (2017). *Gula. In Ancient history encyclopedia.* https://www.ancient.eu/Gula/.

Mark, J. J. (2019). *Dogs in the ancient world. In Ancient history encyclopedia.* https://www.ancient.eu/.

Markotter, W., & Coertse, J. (2018). Bat lyssaviruses. *Revue Scientifique et Technique, 37*(2), 385–400. https://doi.org/10.20506/rst.37.2.2809.

Márquez Ruiz, M. Á. J. (2008). Interchange of pathogens between the old and new world. Revista de l'Academia de Ciencies Veterinaries de Catalunya *(Curs 2007–2008)*, 16–21.

Marston, D. A., Banyard, A. C., McElhinney, L. M., Freuling, C. M., Finke, S., de Lamballerie, X., … Fooks, A. R. (2018). The lyssavirus host-specificity conundrum—Rabies virus—The exception not the rule. *Current Opinion in Virology, 28*, 68–73. https://doi.org/10.1016/j.coviro.2017.11.007.

Marston, D. A., Horton, D. L., Nunez, J., Ellis, R. J., Orton, R. J., Johnson, N., … Fooks, A. R. (2017). Genetic analysis of a rabies virus host shift event reveals within-host viral dynamics in a new host. *Virus Evolution, 3*(2). https://doi.org/10.1093/ve/vex038. vex038.

Matsumoto, S. (1962). Electron microscopy of nerve cells infected with street rabies virus. *Virology, 17*(1), 198–202.

McMillan, B., & Boulger, L. R. (1960). The susceptibility of the ground-squirrel *Xerus* (*Euxerus*) erythropus Geoffroy, 1803, to rabies street virus and its potentiality as a reservoir of rabies in Northern Nigeria. *Annals of Tropical Medicine and Parasitology, 54*(2), 165–171. https://doi.org/10.1080/00034983.1960.11685972.

Melade, J., McCulloch, S., Ramasindrazana, B., Lagadec, E., Turpin, M., Pascalis, H., … Dellagi, K. (2016). Serological evidence of lyssaviruses among bats on southwestern Indian Ocean islands. *PLoS ONE, 11*(8), e0160553. https://doi.org/10.1371/journal.pone.0160553.

Meng, S., Sun, Y., Wu, X., Tang, J., Xu, G., Lei, Y., … Rupprecht, C. E. (2011). Evolutionary dynamics of rabies viruses highlights the importance of China rabies transmission in Asia. *Virology, 410*(2), 403–409.

Meredith, C. D., Prossouw, A. P., & Koch, H. P. (1971). An unusual case of human rabies thought to be of chiropteran origin. *South African Medical Journal, 45*(28), 767–769.

Mitchell, C. A. (1967). Rabies in Quebec City. Case report 1839. *Medical Services Journal, Canada, 23*(5), 809–812.

Mitchell, P. (2017). Disease: A Hitherto unexplored constraint on the spread of dogs (*Canis lupus familiaris*) in pre-Columbian South America. *Journal of World Prehistory, 30*(4), 301–349. https://doi.org/10.1007/s10963-017-9111-x.

Monier-Williams, M. (1891). Tutelary and Village deities. In *Brāhmanism and Hindūism: Or, religious thought and life in India, as based on the Veda a. other sacred books of the Hindūs* (pp. 209–229). London: J. Murray.

Mpolya, E. A., Lembo, T., Lushasi, K., Mancy, R., Mbunda, E. M., Makungu, S., ... Hampson, K. (2017). Toward elimination of dog-mediated human rabies: Experiences from implementing a large-scale demonstration project in Southern Tanzania. *Frontiers in Veterinary Science, 4*, 21.

Mulatti, P., Bonfanti, L., Patregnani, T., Lorenzetto, M., Ferre, N., Gagliazzo, L., ... Marangon, S. (2013). 2008–2011 sylvatic rabies epidemic in Italy: Challenges and experiences. *Pathogens and Global Health, 107*(7), 346–353. https://doi.org/10.1179/2047772413Z.000000000175.

Müller, T., Demetriou, P., Moynagh, J., Cliquet, F., Fooks, A. R., Conraths, F. J., ... Freuling, C. M. (2012). Rabies elimination in Europe—A success story. In A. R. Fooks & T. Müller (Eds.), *Rabies control—Towards sustainable prevention at the source, compendium of the OIE global conference on rabies control, Incheon-Seoul, 7–9 September 2011, Republic of Korea* (pp. 31–44). Paris: OIE.

Müller, T., & Freuling, C. M. (2018). Rabies control in Europe: An overview of past, current and future strategies. *Revue Scientifique et Technique, 37*(2), 409–419. https://doi.org/10.20506/rst.37.2.2811.

Müller, T., Freuling, C. M., Wysocki, P., Roumiantzeff, M., Freney, J., Mettenleiter, T. C., & Vos, A. (2015). Terrestrial rabies control in the European Union: Historical achievements and challenges ahead. *Veterinary Journal, 203*(1), 10–17. https://doi.org/10.1016/j.tvjl.2014.10.026.

Murphy, F. A., Harrison, A. K., Winn, W. C., & Bauer, S. P. (1973). Comparative pathogenesis of rabies and rabies-like viruses—Infection of central nervous-system and centrifugal spread of virus to peripheral tissues. *Laboratory Investigation, 29*(1), 1–16.

Nadin-Davis, S. A., & Bingham, J. (2004). Europe as a source of rabies for the rest of the world. In A. A. King, A. R. Fooks, M. Aubert, & A. I. Wandeler (Eds.), *Historical perspective of rabies in Europe and the Mediterranean Basin* (pp. 259–280). Paris: OIE.

Nadin-Davis, S. A., Turner, G., Paul, J. P., Madhusudana, S. N., & Wandeler, A. I. (2007). Emergence of Arctic-like rabies lineage in India. *Emerging Infectious Diseases, 13*(1), 111–116.

Negri, A. (1903). Beitragzum Studium der Aetiologie der Tollwuth. *Zeitschriftfür Hygiene und Infektionskrankheiten, 43*(1), 507–528. https://doi.org/10.1007/BF02217551.

Neitz, W. O., & Marais, I. P. (1932). Rabies as it occurs in the Union of South Africa. In Union of South Africa, Department of Agriculture (Ed.), *Eighteenth Report of the Director of Veterinary Services and Animal Industry, Onderstepoort, Pretoria* (pp. 71–89). The Government Printer: Pretoria.

Neitz, W. O., & Thomas, A. D. (1933). Rabies in South Africa. Occurrence and distribution of cases during 1932. *Onderstepoort Journal of Veterinary Science and Animal Industry, 1*(1), 51–56.

Nel, L. H., & Rupprecht, C. E. (2007). Emergence of lyssaviruses in the old world: The case of Africa. In J. E. Childs, J. S. Mackenzie, & J. A. Richt (Eds.), *Wildlife and Emerging Zoonotic Diseases: The Biology, Circumstances and Consequences of Cross-Species Transmission* (pp. 161–193). Berlin: Springer-Verlag Berlin Heidelberg.

Neville, J. (2004). Rabies in the ancient world. In A. A. King, A. R. Fooks, M. Aubert, & A. I. Wandeler (Eds.), *Historical perspective of rabies in Europe and the Mediterranean Basin* (pp. 1–13). Paris: OIE.

New South Wales Department of Primary Industries. (2013). A dog tests antibody positive for lyssavirus. *Animal Health Surveillance Quarterly, 3*, 3–4.

Newsome, A., & Catling, P. (1992). Host range and its implications for wildlife rabies in Australia. In P. O'Brien & G. Berry (Eds.), *Wildlife rabies contingency planning in Australia: National Wildlife Rabies Workshop, 12–16 March 1990*. Canberra: Australian Government Publishing Service.

Nicholson, K. G., Prestage, H., Cole, P. J., Turner, G. S., & Bauer, S. P. (1981). Multisite intradermal antirabies vaccination: Immune responses in man and protection of rabbits against death from street virus by postexposure administration of human diploid-cell-strain rabies vaccine. *Lancet, 318*(8252), 915–918.

NíLeathlobhair, M., Perri, A. R., Irving-Pease, E. K., Witt, K. E., Linderholm, A., Haile, J., ... Frantz, L. A. F. (2018). The evolutionary history of dogs in the Americas. *Science, 361*(6397), 81–85. https://doi.org/10.1126/science.aao4776.

Nokireki, T., Tammiranta, N., Kokkonen, U. -M., Kantala, T., & Gadd, T. (2018). Tentative novel lyssavirus in a bat in Finland. *Transboundary and Emerging Diseases, 65*(3), 593–596. https://doi.org/10.1111/tbed.12833.

Nunn, N., & Qian, N. (2010). The Columbian exchange: A history of disease, food, and ideas. *Journal of Economic Perspectives, 24*(2), 163–188. https://doi.org/10.1257/jep.24.2.163.

O'Brien, P., & Berry, G. E. (1992). *Wildlife rabies contingency planning in Australia: Bureau of rural resources proceedings No. 11*. Canberra: Australian Government Publishing Service.

O'Fallon, B. D., & Fehren-Schmitz, L. (2011). Native Americans experienced a strong population bottleneck coincident with European contact. *Proceedings of the National Academy of Sciences, 108*(51), 20444–20448. https://doi.org/10.1073/pnas.1112563108.

Ollivier, M., Tresset, A., Frantz, L. A. F., Brehard, S., Balasescu, A., Mashkour, M., … Vigne, J. D. (2018). Dogs accompanied humans during the Neolithic expansion into Europe. *Biology Letters, 14*(10). pii: 20180286.

Oskarsson, M. C. R., Klütsch, C. F. C., Boonyaprakob, U., Wilton, A., Tanabe, Y., & Savolainen, P. (2012). Mitochondrial DNA data indicate an introduction through Mainland Southeast Asia for Australian dingoes and Polynesian domestic dogs. *Proceedings of the Royal Society B: Biological Sciences, 279*(1730), 967–974. https://doi.org/10.1098/rspb.2011.1395.

Ovodov, N. D., Crockford, S. J., Kuzmin, V. Y., Higham, T. F. G., Hodgins, G. W. L., & van der Plicht, J. (2011). A 33,000-year-old incipient dog from the Altai Mountains of Siberia: Evidence of the earliest domestication disrupted by the last glacial maximum. *PLoS ONE, 6*(7), e22821. https://doi.org/10.1371/journal.pone.0022821.

Pal, S. R., Arora, B., Chhuttani, P. N., Broor, S., Choudhury, S., Joshi, R. M., & Ray, S. D. (1980). Rabies virus infection of a flying fox bat, *Pteropus policephalus* in Chandigarh, Northern India. *Tropical and Geographical Medicine, 32*(3), 265–267.

Peck, F. B., Powell, H. M., & Culbertson, C. G. (1956). Duck embryo rabies vaccine: Study of fixed virus vaccine grown in embryonating duck eggs and killed with beta-propiolactone. *Journal of the American Medical Association, 162*, 1373–1376.

Prada, D., Boyd, V., Baker, M., Jackson, B., & O'Dea, M. (2019). Insights into Australian bat lyssavirus in insectivorous bats of Western Australia. *Tropical Medicine and Infectious Disease, 4*(1), 46. https://doi.org/10.3390/tropicalmed4010046.

Prates, L., Prevosti, F. J., & Berón, M. (2010). First records of prehispanic dogs in Southern South America (Pampa-Patagonia, Argentina). *Current Anthropology, 51*(2), 273–280. https://doi.org/10.1086/650166.

Prevec, L., Campbell, J. B., Christie, B. S., Belbeck, L., & Graham, F. L. (1990). A recombinant human adenovirus vaccine against rabies. *Journal of Infectious Diseases, 161*(1), 27–30. https://doi.org/10.1093/infdis/161.1.27.

Pullar, E. M., & McIntosh, K. S. (1954). The relation of Australia to the world rabies problem. *Australian Veterinary Journal, 30*(11), 326–336. https://doi.org/10.1111/j.1751-0813.1954.tb05387.x.

Ravenel, M. P. (1901). Rabies. In *Bulletin 79*: Department of Agriculture of Pennsylvania.

Records, E. (1932). Rabies—Its history in Nevada. *California and Western Medicine, 37*(2), 90–94.

Remlinger, P. (1903). Le passage du virus rabique à travers les filtres. *Annales de L'Institut Pasteur, 101*, 765–774.

Roberts, L. (1989). Disease and death in the New World. *Science, 246*(4935), 1245–1247.

Robertson, W. A. N. (1932). Milestones in the pastoral age of Australia. *Australiasian Association for the Advancement of Science, 31*, 295–325.

Rodriguez de Roma, A. (2008). La "cienciapausteriana" a través de la vacunaantirrábica: el casoméxicano. *Dynamis: Acta Hispanicaad Medicinae ScientiarumqueHistoriamIllustrandam, 16*, 291–316.

Rollinson, D. H. L. (1956). Problems of rabies control in Africa. *Bulletin Epizootic Diseases in Africa, 4*, 7–16.

Rosatte, R. C., Donovan, D., Davies, J. C., Allan, M., Bachmann, P., Stevenson, B., … Lawson, K. (2009). Aerial distribution of ONRAB baits as a tactic to control rabies in raccoons and striped skunks in Ontario, Canada. *Journal of Wildlife Diseases, 45*(2), 363–374. https://doi.org/10.7589/0090-3558-45.2.363.

Roth, M. T., Hoffner, H. A., & Michalowski, P. (1997). *Law collections from Mesopotamia and Asia minor.* Atlanta: Scholars Press.

Rupprecht, C. E., Bannazadeh Baghi, H., Del Rio Vilas, V. J., Gibson, A. D., Lohr, F., Meslin, F. X., … Gamble, L. (2018). Historical, current and expected future occurrence of rabies in enzootic regions. *Revue Scientifique et Technique, 37*(2), 729–739. https://doi.org/10.20506/rst.37.2.2836.

Rupprecht, C. E., Barrett, J., Briggs, D., Cliquet, F., Fooks, A. R., Lumlertdacha, B., … Wandeler, A. I. (2008). Can rabies be eradicated? *Developments in biologicals, 131*, 95–121.

Rupprecht, C., Kuzmin, I., & Meslin, F. (2017). Lyssaviruses and rabies: Current conundrums, concerns, contradictions and controversies. *F1000Research, 6*, 184. https://doi.org/10.12688/f1000research.10416.1.

Rupprecht, C. E., Kuzmin, I. V., Yale, G., Nagarajan, T., & Meslin, F. X. (2019). Priorities in applied research to ensure programmatic success in the global elimination of canine rabies. *Vaccine, 37*(Suppl. 1), A77–A84. https://doi.org/10.1016/j.vaccine.2019.01.015.

Rupprecht, C. E., & Salahuddin, N. (2019). Current status of human rabies prevention: Remaining barriers to global biologics accessibility and disease elimination. *Expert Review of Vaccines, 18*(6), 629–640. https://doi.org/10.1080/14760584.2019.1627205.

Sabeta, C. T., Bingham, J., & Nel, L. H. (2003). Molecular epidemiology of canid rabies in Zimbabwe and South Africa. *Virus Research, 91*(2), 203–211. https://doi.org/10.1016/s0168-1702(02)00272-1.

Sabeta, C. T., Mansfield, K. L., McElhinney, L. M., Fooks, A. R., & Nel, L. H. (2007). Molecular epidemiology of rabies in bat-eared foxes (*Otocyon megalotis*) in South Africa. *Virus Research, 129*(1), 1–10. https://doi.org/10.1016/j.virusres.2007.04.024.

Sabeta, C., & Ngoepe, E. C. (2018). Controlling dog rabies in Africa: Successes, failures and prospects for the future. *Revue Scientifique et Technique, 37*(2), 439–449. https://doi.org/10.20506/rst.37.2.2813.

Sacatterday, J. E., & Galton, M. M. (1954). Bat rabies in Florida. *Veterinary Medicine, 149*, 133–135.

Saldaña, J. J. (2006). *Science in Latin America: A history*. Austin: University of Texas Press.

Samaratunga, H., Searle, J. W., & Hudson, N. (1998). Non-rabies Lyssavirus human encephalitis from fruit bats: Australian bat Lyssavirus (pteropid Lyssavirus) infection. *Neuropathology and Applied Neurobiology, 24*(4), 331–335. https://doi.org/10.1046/j.1365-2990.1998.00129.x.

Scatterday, J. E., Schneider, N. J., Jennings, W. L., & Lewis, A. L. (1960). Sporadic animal rabies in Florida. *Public Health Reports, 75*, 945–953.

Schatz, J., Fooks, A. R., McElhinney, L., Horton, D., Echevarria, J., Vazquez-Moron, S., … Freuling, C. M. (2013). Bat rabies surveillance in Europe. *Zoonoses and Public Health, 60*(1), 22–34. https://doi.org/10.1111/zph.12002.

Schneider, M. C., & Santos-Burgoa, C. (1994). Tratamiento contra la rabiahumana: un poco de suhistoria. *Revista de Saúde Pública, 28*(6), 454–463.

Schneider, L. G., Wachendörfer, G., Schmittdiel, E., & Cox, J. H. (1983). Ein Feldversuchzuroralen Immunisierung von Füchsengegen Tollwut in der Bundesrepublik Deutschland. II. Planung, Durchführung und Auswertung des Feldversuches. *Tierärztliche Umschau, 38*, 476–480.

Schnell, M. J., Mebatsion, T., & Conzelmann, K. K. (1994). Infectious rabies viruses from cloned cDNA. *Embo Journal, 13*(18), 4195–4203.

Schoening, H. W. (1930). Experimental studies with killed canine rabies vaccine. *Journal of the American Medical Association, LXXVI*(29), 25.

Schumacher, C. L., Dietzschold, B., Ertl, H. C., Niu, H. S., Rupprecht, C. E., & Koprowski, H. (1989). Use of mouse anti-rabies monoclonal antibodies in postexposure treatment of rabies. *Journal of Clinical Investigation, 84*(3), 971–975. https://doi.org/10.1172/jci114260.

Schwabe, C. W. (1971). *Report on the First WHO regional seminar on veterinary public health held in Mukteswar, India, 8–18 April 1970* [46 p.]. WHO Regional Office for South-East Asia: New Delhi

Sedwick, W. D., & Wiktor, T. J. (1967). Reproducible plaquing system for rabies, lymphocytic choriomeningitis, k and other ribonucleic acid viruses in BHK-21-13S agarose suspensions. *Journal of Virology, 1*(6), 1224–1226.

Seetahal, J., Vokaty, A., Vigilato, M., Carrington, C., Pradel, J., Louison, B., … Rupprecht, C. (2018). Rabies in the Caribbean: A situational analysis and historic review. *Tropical Medicine and Infectious Disease, 3*(3), 89. https://doi.org/10.3390/tropicalmed3030089.

Sellers, T. F. (1923a). A simple modification of hogyes dilution method of preparing antirabic treatment: Report on 3080 treatments. *American Journal of Public Health, 13*(10), 813–815. https://doi.org/10.2105/ajph.13.10.813.

Sellers, T. F. (1923b). Status of rabies in the United States in 1921. *American Journal of Public Health, 13*(9), 742–747. https://doi.org/10.2105/ajph.13.9.742.

Shah, A. (2018). *Bombay's riot over dogs*. https://www.livehistoryindia.com/snapshort-histories/2018/05/09/bombays-riot-over-dogs.

Sigaud, J. F. X. (1844). Maladies Generales et Speziales. De la rage. In *Du climat et des maladies du Brésiloustistiquemédicale de cet empire* (pp. 424–425). Paris: Fortin.

Sillero-Zubiri, C., King, A. A., & Macdonald, D. W. (1996). Rabies and mortality in Ethiopian wolves (*Canis simensis*). *Journal of Wildlife Diseases, 32*(1), 80–86. https://doi.org/10.7589/0090-3558-32.1.80.

Slate, D., Algeo, T. P., Nelson, K. M., Chipman, R. B., Donovan, D., Blanton, J. D., … Rupprecht, C. E. (2009). Oral rabies vaccination in North America: Opportunities, complexities, and challenges. *PLOS Neglected Tropical Diseases, 3*(12), e549. https://doi.org/10.1371/journal.pntd.0000549.

Smith, J. S., Orciari, L. A., Yager, P. A., Seidel, H. D., & Warner, C. K. (1992). Epidemiologic and historical relationships among 87 rabies virus isolates as determined by limited sequence analysis. *Journal of Infectious Diseases, 166*(2), 296–307.

Smith, J. S., & Seidel, H. D. (1993). Rabies: A new look at an old disease. *Progress in Medical Virology, 40*, 82–106.

Smith, J. S., Yager, P. A., & Baer, G. M. (1973). A rapid reproducible test for determining rabies neutralizing antibody. *Bulletin of the World Health Organization, 48*(5), 535–541.

Snyman, P. S. (1940). The study and control of the vectors of rabies in South Africa. *Onderstepoort Journal of Veterinary Research*, *15*(1–2), 9–140.

Sparkes, J., Fleming, P. J. S., Ballard, G., Scott-Orr, H., Durr, S., & Ward, M. P. (2015). Canine rabies in Australia: A review of preparedness and research needs. *Zoonoses and Public Health*, *62*(4), 237–253. https://doi.org/10.1111/zph.12142.

Sparkes, J., McLeod, S., Ballard, G., Fleming, P. J. S., Kortner, G., & Brown, W. Y. (2016). Rabies disease dynamics in naive dog populations in Australia. *Preventive Veterinary Medicine*, *131*, 127–136. https://doi.org/10.1016/j.prevetmed.2016.07.015.

Stahl, P. W. (2013). Early dogs and endemic South American canids of the Spanish main. *Journal of Anthropological Research*, *69*(4), 515–533. https://doi.org/10.3998/jar.0521004.0069.405.

Steck, F., Wandeler, A. I., Bichsel, P., Capt, S., & Schneider, L. G. (1982). Oral immunisation of foxes against rabies. *Zentralblatt für Veterinärmedizin Reihe B*, *29*, 372–396.

Steele, J. H., & Fernandez, P. J. (1991). History of rabies and global aspects. In G. M. Baer (Ed.), *The natural history of rabies* (2nd ed., pp. 1–24). Boca Raton: CRC Press.

Streicker, D. G., Recuenco, S., Valderrama, W., Benavides, J. G., Vargas, I., Pacheco, V., ... Altizer, S. (2012). Ecological and anthropogenic drivers of rabies exposure in vampire bats: Implications for transmission and control. *Proceedings of the Royal Society B: Biological Sciences*, *279*(1742), 3384–3392. https://doi.org/10.1098/rspb.2012.0538.

Suárez, E. H. (1940). Scientific institutions in Latin America: The Bacteriological Institute of Chile. *Boletin de la Oficina Sanitaria Panamericana*, *19*(10), 997–1000.

Susilawathi, N. M., Darwinata, A. E., Dwija, I. B., Budayanti, N. S., Wirasandhi, G. A., Subrata, K., ... Mahardika, G. N. (2012). Epidemiological and clinical features of human rabies cases in Bali 2008–2010. *BMC Infectious Diseases*, *12*. 81. https://doi.org/10.1186/1471-2334-12-81.

Swanepoel, R., Barnard, B. J. H., Meredith, C. D., Bishop, G. C., Bruchner, G. K., Foggin, C. M., & Hübschle, O. J. (1993). Rabies in southern Africa. *Onderstepoort Journal of Veterinary Research*, *60*(323–346).

Tabel, H., Corner, A. H., Webster, W. A., & Casey, C. A. (1974). History and epizootiology of rabies in Canada. *Canadian Veterinary Journal*, *15*(10), 271–281.

Tarantola, A. (2017). Four thousand years of concepts relating to rabies in animals and humans, its prevention and its cure. *Tropical Medicine and Infectious Disease*, *2*(2). https://doi.org/10.3390/tropicalmed2020005.

Tasioudi, K. E., Iliadou, P., Agianniotaki, E. I., Robardet, E., Liandris, E., Doudounakis, S., ... Mangana-Vougiouka, O. (2014). Recurrence of animal rabies, Greece, 2012. *Emerging Infectious Diseases*, *20*(2), 326–328. https://doi.org/10.3201/eid2002.130473.

Taylor, D. (1976). Rabies: Epizootic aspects; diagnosis; vaccines; notes for guidance; official policy—Epizootic aspects. *Veterinary Record*, *99*, 157–160.

Taylor, L. H., Hampson, K., Fahrion, A., Abela-Ridder, B., & Nel, L. H. (2017). Difficulties in estimating the human burden of canine rabies. *Acta Tropica*, *165*, 133–140. https://doi.org/10.1016/j.actatropica.2015.12.007.

Taylor, L. H., Latham, S. M., & Woolhouse, M. E. (2001). Risk factors for human disease emergence. *Philosophical Transactions of the Royal Society B, Biological Sciences*, *356*(1411), 983–989. https://doi.org/10.1098/rstb.2001.0888.

Teixeira, L. A., Sandoval, M. R. C., & Takaoka, N. Y. (2005). Instituto Pasteur de São Paulo: Cemanos de combate à raiva. *História, Ciências, Saúde-Manguinhos*, *11*(3), 751–766.

Thalmann, O., & Perri, A. R. (2019). Paleogenomic inferences of dog domestication. In C. Lindqvist & O. P. Rajora (Eds.), *Paleogenomics: Genome-Scale Analysis of Ancient DNA* (pp. 273–306). Cham: Springer International Publishing.

Tierkel, E. S., Graves, L. M., Tuggle, H. G., & Wadley, S. L. (1950). Effective control of an outbreak of rabies in memphis and Shelby County, Tennessee. *American Journal of Public Health and the Nations Health*, *40*(9), 1084–1088. https://doi.org/10.2105/AJPH.40.9.1084.

Tobergte, D. R., & Curtis, S. (2013). Bat rabies and other lyssavirus infections. *Journal of Chemical Information and Modeling*, *53*(9), 1689–1699. https://doi.org/10.1017/CBO9781107415324.004.

Tojinbara, K. (2002). History of epidemics and prevention of rabies in the dogs in Japan. *Japanese Journal of Veterinary History*, *39*, 14–30.

Tomori, O., & David-West, K. B. (1985). Epidemiology of rabies in Nigeria. In E. Kuwert, C. Mérieux, H. Koprowski, & K. Bogel (Eds.), *Rabies in the Tropics* (pp. 485–490). Berlin: Springer.

Troupin, C., Dacheux, L., Tanguy, M., Sabeta, C., Blanc, H., Bouchier, C., ... Bourhy, H. (2016). Large-scale phylogenomic analysis reveals the complex evolutionary history of rabies virus in multiple carnivore hosts. *PLOS Pathogens*, *12*(12), e1006041. https://doi.org/10.1371/journal.ppat.1006041.

Tsiodras, S., Dougas, G., Baka, A., Billinis, C., Doudounakis, S., Balaska, A., … Kremastinou, J. (2013). Re-emergence of animal rabies in northern Greece and subsequent human exposure, October 2012–March 2013. *Euro Surveillance*, *18*(18), 20474.

Tu, C., Feng, Y., & Wang, Y. (2018). Animal rabies in the People's Republic of China. *Revue Scientifique et Technique*, *37*(2), 519–528. https://doi.org/10.20506/rst.37.2.2820.

Unanue, J. H. (1806). Influencia del climasobre los animales. In *Observacionessobre el clima de Lima, y sus influenciasen los seresorganizados, en especial el hombre* (pp. 66–71). Lima: En la Imprenta Real de los Huérfanos.

van Asch, B., Zhang, A. b., Oskarsson, M. C. R., Klutsch, C. F. C., Amorim, A., Savolainen, P., … Oskarsson, M. C. R. (2013). Pre-Columbian origins of native American dog breeds, with only limited replacement by European dogs, confirmed by mtDNA analysis. *Proceedings of the Royal Society B: Biological Sciences*, *280*(1766), 20131142. https://doi.org/10.1098/rspb.2013.1142.

Vargas-Pino, F., Gutiérrez-Cedillo, V., Canales-Vargas, E. J., Gress-Ortega, L. R., Miller, L. A., Rupprecht, C. E., … Slate, D. (2013). Concomitant administration of GonaCon™ and rabies vaccine in female dogs (*Canis familiaris*) in Mexico. *Vaccine*, *31*(40), 4442–4447. https://doi.org/10.1016/j.vaccine.2013.06.061.

Vaughn, J. B., Jr., Gerhardt, P., & Newell, K. W. (1965). Excretion of street rabies virus in the saliva of dogs. *Journal of the American Medical Association*, *193*, 363–368. https://doi.org/10.1001/jama.1965.03090050039010.

Veeraraghavan, N. (1954). A Case of Hydrophobia Following Bat Bite. In S. Report (Ed.), *Pasteur Institute of Southern India*. Coonoor: Madras Diocesan Press.

Veeraraghavan, N. (1970). *Annual report: Director Pasteur Institute of Southern India.* Coonoor: Diocesan Pr.

Victora, C. G. (2002). 100 years of PAHO: A personal testimony. *American Journal of Public Health*, *92*(12), 1887–1888.

Vigilato, M. A. N., Clavijo, A., Knobl, T., Silva, H. M. T., Cosivi, O., Schneider, M. C., … Espinal, M. A. (2013). Progress towards eliminating canine rabies: Policies and perspectives from Latin America and the Caribbean. *Philosophical Transactions of the Royal Society B, Biological Sciences*, *368*(1623), 20120143. https://doi.org/10.1098/rstb.2012.0143.

Vos, A., Freuling, C., Eskiizmirliler, S., Un, H., Aylan, O., Johnson, N., … Askaroglu, H. (2009). Rabies in foxes, Aegean region, Turkey. *Emerging Infectious Diseases*, *15*(10), 1620–1622.

Vos, A., Müller, T., Neubert, L., Zurbriggen, A., Botteron, C., Pöhle, D., … Jackson, A. C. (2004). Rabies in red foxes (*Vulpes vulpes*) experimentally infected with European bat lyssavirus type 1. *Journal of Veterinary Medicine Series B*, *51*(7), 327–332.

Vos, A., Nunan, C., Bolles, D., Müller, T., Fooks, A. R., Tordo, N., & Baer, G. M. (2011). The occurrence of rabies in pre-Columbian Central America: An historical search. *Epidemiology and Infection*, *139*(10), 1445–1452. https://doi.org/10.1017/S0950268811001440.

Wandeler, A. (2004). Epidemiology and ecology of fox rabies in Europe. In A. A. King, A. R. Fooks, M. Aubert, & A. I. Wandeler (Eds.), *Historical perspective of rabies in Europe and the Mediterranean Basin* (pp. 201–214). Paris: OIE.

Wandeler, A. I., Capt, S., Kappeler, A., & Hauser, R. (1988). Oral immunization of wildlife against rabies: Concept and first field experiments. *Reviews of Infectious Diseases*, *10*(Suppl. 4), S649–S653. https://doi.org/10.1093/clinids/10.Supplement_4.S649.

Wang, X. J., & Huang, J. T. (2001). Epidemiology. In Y. X. Yu (Ed.), *Rabies and rabies vaccine* (pp. 127–144). Beijing: Chinese Medicine Technology Press.

Wang, G. -D., Zhai, W., Yang, H. -C., Wang, L., Zhong, L., Liu, Y. -H., … Zhang, Y. -P. (2016). Out of southern East Asia: The natural history of domestic dogs across the world. *Cell Research*, *26*(1), 21–33. https://doi.org/10.1038/cr.2015.147.

Warrell, M. J., Suntharasamai, P., Sinhaseni, A., Phanfung, R., Vincent-Falquet, J. C., Bunnag, D., … Harinasuta, T. (1983). An economical regimen of human diploid cell strain anti-rabies vaccine for post-exposure prophylaxis. *Lancet*, *322*(8345), 301–304.

Warrilow, D., Smith, I. L., Harrower, B., & Smith, G. A. (2002). Sequence analysis of an isolate from a fatal human infection of Australian bat lyssavirus. *Virology*, *297*(1), 109–119. https://doi.org/10.1006/viro.2002.1417.

Watson, E. M. (1913). The Negri bodies in rabies. *Journal of Experimental Medicine*, *17*(1), 29–42. https://doi.org/10.1084/jem.17.1.29.

Webster, L. T., & Clow, A. D. (1936). Propagation of rabies virus in tissue culture and the successful use of culture virus as an antirabic vaccine. *Science*, *84*(2187), 487–488. https://doi.org/10.1126/science.84.2187.487.

Webster, L. T., & Clow, A. D. (1937). Propagation of rabies virus in tissue culture. *Journal of Experimental Medicine*, *66*(1), 125–131. https://doi.org/10.1084/jem.66.1.125.

Webster, L. T., & Dawson, J. R. (1935). Early diagnosis of rabies by mouse inoculation. Measurement of humoral immunity to rabies by mouse protection test. *Proceedings of the Society for Experimental Biology and Medicine*, *32*(4), 570–573. https://doi.org/10.3181/00379727-32-7767P.

Wiktor, T. J., Fernandes, M. V., & Koprowski, H. (1964). Cultivation of rabies virus in human diploid cell strain WI-38. *Journal of immunology, 93*, 353–366.

Wiktor, T. J., & Koprowski, H. (1978). Monoclonal antibodies against rabies virus produced by somatic cell hybridization: Detection of antigenic variants. *Proceedings of the National Academy of Sciences of the United States of America, 75*(8), 3938–3942. https://doi.org/10.1073/pnas.75.8.3938.

Wiktor, T. J., Macfarlan, R. I., Reagan, K. J., Dietzschold, B., Curtis, P. J., Wunner, W. H., & Mackett, M. (1984). Protection from rabies by a vaccinia virus recombinant containing the rabies virus glycoprotein gene. *Proceedings of the National Academy of Sciences of the United States of America, 81*(22), 7194–7198. https://doi.org/10.1073/pnas.81.22.7194.

Wilkinson, L. (1977). The development of the virus concept as reflected in corpora of studies on individual pathogens. 4. Rabies—Two millennia of ideas and conjecture on the aetiology of a virus disease. *Medical History, 21*(1), 15–31.

Wilkinson, L. (1988). Understanding the nature of rabies: An historical perspective. In J. B. Campbell & K. M. Charlton (Eds.), *Rabies* (pp. 1–23). Boston, MA: Springer US.

Willoughby, R. E., Jr., Tieves, K. S., Hoffman, G. M., Ghanayem, N. S., Amlie-Lefond, C. M., Schwabe, M. J., … Rupprecht, C. E. (2005). Survival after treatment of rabies with induction of coma. *New England Journal of Medicine, 352*(24), 2508–2514. https://doi.org/10.1056/NEJMoa050382.

Windiyaningsih, C., Wilde, H., Meslin, F. X., Suroso, T., & Widarso, H. S. (2004). The rabies epidemic on flores island, Indonesia (1998–2003). *Journal of the Medical Association of Thailand, 84*(11), 1389–1393.

Winkler, W. G., Fashinell, T. R., Leffingwell, L., Howard, P., & Conomy, P. (1973). Airborne rabies transmission in a laboratory worker. *Journal of the American Medical Association, 226*(10), 1219–1221.

World Health Organization. (2018). WHO expert consultation on rabies: Third report. In *Vol. 1012. WHO technical report series*. Geneva: WHO.

World Health Organization. (2019). *Rabies bulletin Europe database*. https://www.who-rabies-bulletin.org.

Wu, X., Franka, R., Svoboda, P., Pohl, J., & Rupprecht, C. E. (2009). Development of combined vaccines for rabies and immunocontraception. *Vaccine, 27*(51), 7202–7209.

Yang, D. K., Cho, I. S., & Kim, H. H. (2018). Strategies for controlling dog-mediated human rabies in Asia: Using 'One Health' principles to assess control programmes for rabies. *Revue Scientifique et Technique, 37*(2), 473–481. https://doi.org/10.20506/rst.37.2.2816.

Yelverton, E., Norton, S., Obijeski, J. F., & Goeddel, D. V. (1983). Rabies virus glycoprotein analogs: Biosynthesis in *Escherichia coli*. *Science, 219*(4585), 614–620. https://doi.org/10.1126/science.6297004.

Young, C. C., & Sellers, T. F. (1927). A new method for staining Negri bodies of rabies. *American Journal of Public Health, 17*(10), 1080–1081. https://doi.org/10.2105/ajph.17.10.1080.

Yuhong, W. (2001). Rabies and Rabid Dogs in Sumerian and Akkadian Literature. *Journal of the American Oriental Society, 121*(1), 32–43.

Zhao, H., Zhang, J., Cheng, C., & Zhou, Y. H. (2019). Rabies acquired through mucosal exposure, China, 2013. *Emerging Infectious Diseases, 25*(5), 1028–1029. https://doi.org/10.3201/eid2505.181413.

Zinsstag, J., Lechenne, M., Laager, M., Mindekem, R., Naissengar, S., Oussiguere, A., … Chitnis, N. (2017). Vaccination of dogs in an African city interrupts rabies transmission and reduces human exposure. *Science Translational Medicine, 9*(421). https://doi.org/10.1126/scitranslmed.aaf6984.

Zulu, G. C., Sabeta, C. T., & Nel, L. H. (2009). Molecular epidemiology of rabies: Focus on domestic dogs (*Canis familiaris*) and black-backed jackals (*Canis mesomelas*) from northern South Africa. *Virus Research, 140*(1–2), 71–78. https://doi.org/10.1016/j.virusres.2008.11.004.

CHAPTER 2

Rabies virus

William H. Wunner[a], Karl-Klaus Conzelmann[b]

[a]The Wistar Institute, Philadelphia, PA, United States [b]Max von Pettenkofer Institute Virologie, Medical Faculty & Gene Center, Ludwig-Maximilians University of Munich, Munich, Germany

2.1 Introduction

2.1.1 Lyssaviruses

Rabies virus (RABV) is the prototype virus of the genus *Lyssavirus* (from the Greek *lyssa* meaning "rage") in the family *Rhabdoviridae* (from the Greek *rhabdos* meaning "rod") of the order *Mononegavirales* (MNV). RABV, the causative agent of classic rabies in animals and humans, is a highly neurotropic virus in the mammalian host invariably causing a fatal encephalomyelitis once the infection is established and has reached the brain. RABV is distributed worldwide among specific mammalian reservoir hosts comprising various carnivore and bat species. Fifteen other lyssaviruses, which share certain morphological and structural characteristics with RABV, have been identified. Only 6 of the 16 currently recognized lyssaviruses within the *Lyssavirus* genus have caused a rabies-like encephalomyelitis in humans. Of note, of the 16 lyssaviruses recognized only RABV has multiple host reservoirs, while the other lyssaviruses are exclusively associated with bat reservoirs (Marston et al., 2018). The 16 lyssaviruses segregate into 2 phylogroups based on phylogenetic analyses. Phylogroup I includes the classic (prototype) RABV, Duvenhage virus (DUVV), European bat lyssavirus, type 1 (EBLV-1), and type 2 (EBLV-2), Australian bat lyssavirus (ABLV), Aravan virus (ARAV), Khujand virus (KHUV), Irkut virus (IRKV), Bokeloh bat lyssavirus (BBLV), Ikoma lyssavirus (IKOV), Lleida bat lyssavirus (LLEBV), and Gannoruwa bat lyssavirus (GBLV), the newest species to be characterized and considered an independent species with phylogroup I (Gunawardena et al., 2016; Hanlon et al., 2005; Kuzmin, Niezgoda, et al., 2008). All phylogroup I lyssaviruses are transmitted by bats; only RABV is adapted to and spread by carnivores as their reservoir host. The evolutionary relationship between bat transmitted and carnivore transmitted lyssaviruses is not well understood (Rupprecht, Kuzmin, & Meslin, 2017).

Phylogroup II includes LBV, Mokola virus (MOKV), and Shimoni bat virus (SHIBV), the newest species to be characterized and considered an independent species within phylogroup II (Kuzmin et al., 2010). West Caucasian bat virus (WCBV), which does not crossreact serologically with any members of the two phylogroups, could tentatively belong to a third phylogroup (Freuling et al., 2011) (See Chapters 4 and 5). Phylogenetic analyses suggest that all lyssaviruses have originated from a precursor bat virus (Rupprecht et al., 2017).

Lyssavirus species share many of the biologic and physicochemical features that are associated with other viruses of the *Rhabdoviridae* family. These include the bullet-shaped virus morphology, helical nucleocapsid (NC) or ribonucleoprotein (RNP) core, and general organization of the viral RNA (vRNA) genome and structural proteins. In contrast to all other rhabdoviruses, however, lyssaviruses are not transmitted by insect vectors and have adapted to direct transmission. The five structural proteins of the lyssavirus particle (virion) include a nucleocapsid protein (N), phosphoprotein (P), matrix protein (M), glycoprotein (G), and RNA-dependent RNA polymerase or large protein (L). These lyssavirus proteins generally share many of the biologic functions that the same viral proteins have in other rhabdoviruses. Some of the structural proteins of lyssaviruses, on the other hand, can differ dramatically in their antigenic properties and in their post-translational modifications to convey different, often specific properties that distinguish lyssaviruses from other rhabdoviruses. Lyssaviruses, like other rhabdoviruses, also use similar mechanisms to enter susceptible cells (albeit, they may use different receptors) express and replicate their genome RNA, assemble and release mature progeny virions from the plasma membrane or internal membranes of infected cells.

2.2 Rabies virus architecture

2.2.1 Virus structure and composition

Lyssaviruses, like other rhabdoviruses, consist mainly of protein (67%–74%), lipid (20%–26%), carbohydrate (3%), and RNA (2%–3%), as integral components (percent of total mass) of their structure. The viral RNA genome of approximately 12 kilobases (kb) forms the backbone of the tightly coiled helical RNP (RNA plus protein) core, which extends along the longitudinal axis of the bullet-shaped virus particle. Included in the RNP core are the N, P, and L protein components, which are surrounded by the viral membrane proteins, M and G, and a mixture of lipoprotein components derived from the cell membrane that form the outer envelope or "membrane matrix" of the virion (Fig. 2.1). The vRNA genome is single-stranded and nonsegmented and has a negative-sense (minus-strand) polarity. This implies that the genomic RNA is not infectious.

The five viral genes of the vRNA, which encode the structural proteins of the virus (Fig. 2.2), are arranged in a strictly conserved order (3'-N-P-M-G-L-5') and are flanked by short terminal regulatory sequences (see Section 2.3.1).

2.2.2 Morphology and core structure of standard and defective virus particles

RABV particles are best described as bacilliform, rod or bullet-shaped, with one conical end (rounded) (hemispherical) and the other flattened (planar), the morphological hallmark

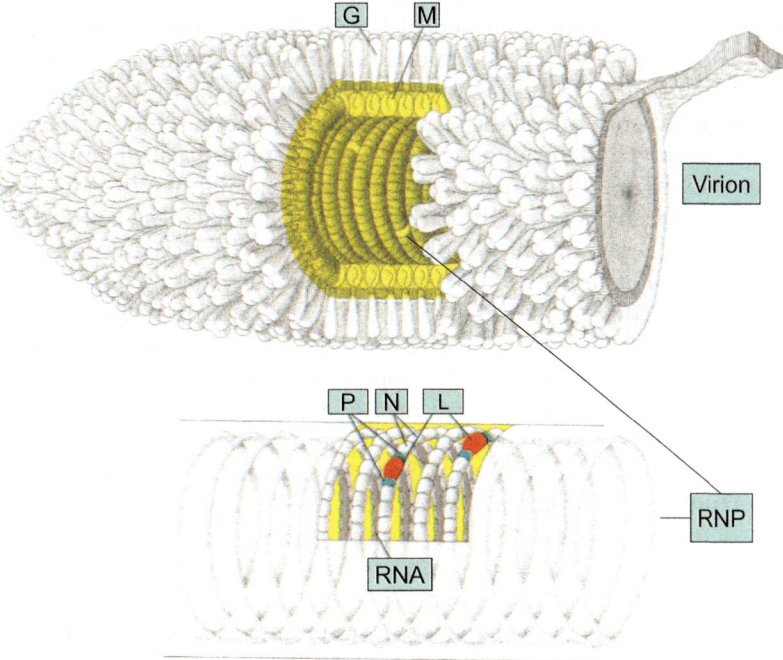

FIG. 2.1 Schematic representation of the rabies virion. The drawing shows the internal ribonucleoprotein (RNP) core consisting of the single-strand, negative-sense genome RNA encapsidated with nucleocapsid protein (N), the virion-associated RNA polymerase (L) and polymerase cofactor phosphoprotein (P). The RNP core in association with the matrix protein (M) is condensed into the typical bullet-shape particle that is characteristic of rhabdoviruses. A lipid bilayer envelope (or membrane) in which the surface trimeric glycoprotein (G) spikes are anchored surrounds the RNP-M structure. The membrane "tail" depicted in the drawing represents the trailing piece of envelope that is frequently observed in the electron microscope attached to the virus as it buds from the plasma membrane of the infected cell. *Reproduced from Wunner, W.H., Larson, J.K., Dietzschold, B., & Smith, C.L. (1988). Review of Infectious Diseases, 10 (Suppl. 4), S771–S784, with permission.*

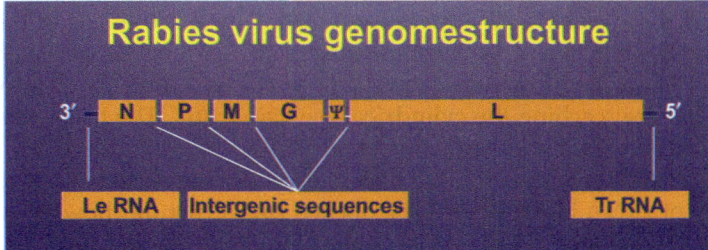

FIG. 2.2 Organization of the rabies virus genome. The nucleoprotein (N), phosphoprotein (P), matrix protein (M), glycoprotein (G), and large RNA-polymerase protein (L) genes are separated by intergenic di- and penta-nucleotide sequences and the long pseudogene (ψ) sequence and are flanked by the leader (Le) RNA and trailer (Tr) RNA sequences at the 3′ and 5′ ends, respectively.

of all rhabdoviruses (Fig. 2.1). At the core of all standard and defective virions is a ribbon of tightly coiled yet flexible, left-handed helical RNP that has a periodicity of approximately 7.5 nm per turn. The length of the tightly coiled RNP core in the standard-size infectious virion measures approximately 165 × 50 nm. During virus assembly, the RNP core is associated with M to form the "skeleton" structure of the virus (Mebatsion, Weiland, & Conzelmann, 1999). As virus particles mature and bud through the cellular membrane, the skeleton structure acquires a lipid bilayer envelope (7.5–10 nm thick) that surrounds the mature virion. Located on the external surface of the viral envelope are the surface projections that measure 8.3–10 nm in length. Each projection or spike contains three molecules (a trimer) of the viral G (Gaudin, Ruigrok, Tuffereau, Knossow, & Flamand, 1992). These have been described when viewed in the electron microscope (EM) as the "short spikes extending outward with the appearance of hollow knobs at their distal ends" (Murphy & Harrison, 1979). It is estimated that the height of the "hollow knobs" or "heads" of the spike is about 4.8 nm; the rest of the spike is made up of the thin ∼3.5 nm-"stalk" on which the head rests (Gaudin et al., 1992).

The average length of standard-size, infectious rabies virions measures 180 nm (130–250 nm) and the average diameter is 75 nm (60–110 nm) (Davies, Englert, Sharpless, & Cabasso, 1963; Hummeler, Koprowski, & Wiktor, 1967). Cryo-EM of RABV indicates an average length of 198 nm (range: 183–222 nm) and an average diameter of 86 nm (range: 77–95 nm) (Riedel et al., 2019) or 81 nm and a length of 188 nm but showed this for only 33% of the total particle number (Guichard et al., 2011).

The average outer diameter of the helix was 67 nm (range: 56–74 nm) and its average inner diameter 51 nm (range: 43–57 nm) (Riedel et al., 2019). The helical turns of the RNP trunk deviate by 44 degrees from the central axis of the virion and are 71 Å apart. The RABV RNP helix is left-handed and based on the localization of N-proteins in the electron density map, the 3′ end of the viral RNA is located at the conical end of the RNP, which comprises up to seven helical turns. As compared to vesicular stomatitis virus (VSV) (Ge et al., 2010), the higher-level organization of M and N proteins within the RNP complex differs in several ways. The RABV helical turns deviate by 19 degrees less from the central axis of the virus and are interspaced by an additional 20 Å when compared to VSV. This increase in distance between neighboring turns is reflected by differences in the molecular interactions between turns. In contrast to the VSV, where M seems to form a mesh around the N-RNP, there is no evidence of an interaction between M molecules of neighboring turns in the RABV RNP.

Cryo-electronmicroscopy more recently revealed that the outermost lipid bilayer leaflet of the bullet-shaped rhabdovirus, VSV, measures 700 Å. The exact length of the virion varies, 1960 ± 80 Å, from the conical end, which comprises ∼25% of the total length, through the cylindrical (helical) trunk, comprising ∼75%. The conical end contains approximately seven turns of a spiral before reaching the cylindrical (helical) trunk. The trunk of a typical virion contains approximately 29 spiral turns (Ge et al., 2010). Docking of the "ring" structure of N and RNA (for details, see Section 2.3.6) into the cryo-EM structure of the RNP establishes the directionality of vRNA in the virion. The docked ring structure shows that the 3′ end is at the conical tip of the bullet and the 5′ end is at the base of the trunk. Cryo-EM tomography of RABV confirms the canonical bullet-shaped morphology of the lyssavirus with a diameter of 86 nm (range: 77–85 nm) (Riedel et al., 2019).

Beside standard-size bullet-shaped rabies virions, other shorter (truncated), often cone-shaped "defective" virions are sometimes coproduced, particularly in cell culture.

They contain RNA genomes, which are typically shorter than full-size genomes as a result of genome truncations or internal sequence deletions from the genome (Conzelmann, Cox, & Thiel, 1991; Lazzarini, Keene, & Schubert, 1981; Marriott & Dimmock, 2010).

Some defective rabies virions, like defective virions of other rhabdoviruses and other RNA viruses, replicate at the expense of standard "helper" virus. Defective virions can grow to become the dominant particle type(s) in infected cells and do so by interfering with production of standard infectious virus. These are therefore known as defective-interfering (DI) virions (Clark, Parks, & Wunner, 1981; Wiktor, Dietzschold, Leamnson, & Koprowski, 1977). DI virions of RABV are readily generated in standard cell cultures infected with laboratory-adapted (fixed) strains of RABV (Clark et al., 1981; Wunner & Clark, 1980). They have not been described in rabies virus infections in vivo, although their role in controlling production of infectious virions in vivo has been suggested (Marriott & Dimmock, 2010).

2.2.3 Viral proteins

The three viral proteins located in the viral RNP core (Fig. 2.1) are the N, the noncatalytic polymerase-associated P, and the catalytic L component of the virion-associated RNA polymerase. All three proteins are involved in the RNA polymerase activity of the virion. Both the N and P are phosphorylated in RABV, unlike in other rhabdoviruses, including VSV, in which only the P is phosphorylated (Gupta, Blondel, Choudhary, & Banerjee, 2000; Sokol et al., 1974). The most abundant protein in the RNP core is N (1325 or 1800 copies) followed by P (691 or 950 copies) and L (25 or 72 copies) (Flamand, Raux, Gaudin, & Ruigrok, 1993; Madore & England, 1977). The stoichiometric relationship that emerges from these independent estimates, however, indicates that the N:P ratio in the RNP complex is 2:1 per virion, which is supported by recent structural analyses. The proteins of the RNP core and the association of N with P and L will be discussed in more detail later (see Section 2.3.2).

The remaining two structural proteins of the RABV, G and M, are associated with the lipid-bilayer envelope that surrounds the RNP core. The M is a small-size protein that lines the viral envelope. A part of the M is attached at one end to the viral envelope and the rest sticks into the interior of the virion, where it interacts with N to connect between the envelope and RNP core (Guichard et al., 2011). Interestingly, the multifunctional M associates with both the RNP and the viral G, collaborating with G in infected cells, to produce progeny virions in the budding process at the cell membrane (Mebatsion et al., 1999; Nakahara et al., 1999).

The G is the only glycosylated protein. It produces the trimeric spike-like projections on the surface of the viral envelope (Gaudin et al., 1992). The viral G molecule is glycosylated with branched-chain oligosaccharides, which account for 10%–12% of the total mass of the protein. The number of G and M molecules per virion has been estimated to be 1205 and 1148 (Flamand et al., 1993) and 1800 and 1547 (Madore & England, 1977), respectively. This calculates to approximately 450 trimeric G spikes distributed on the outer surface of each virion. The rabies virion envelope contains other host-derived minor protein components such as actin and heat shock proteins of the Hsp70 type, CD44 and CD99-related glycoprotein (VAP21) similar to other negative-strand RNA viruses (Lahaye, Vidy, Fouquet, & Blondel, 2012; Naito & Matsumoto, 1978; Sagara et al., 1998; Sagara & Kawai, 1992; Sagara, Tsukita, Yonemura, & Kawai, 1995). It is possible that the molecular chaperones, such as the heat shock protein calnexin, that associate with the viral proteins during synthesis are incorporated into virions after binding to and assisting in G folding (Gaudin, 1997). In a similar manner, cytoskeleton proteins normally expressed on the host cell surface may be incorporated into virions as a consequence of their proximal location and function in virus budding (Riedel et al., 2019; Sagara et al., 1995, 1998). Cellular kinases that activate the transcriptional function of P in RABV may also be packaged into rabies virions (Gupta et al., 2000).

2.2.3.1 Nucleoprotein (N)

The RABV N Pasteur virus (PV) strain contains 450 amino acids, one of which, the serine residue at position 389 (S389), is phosphorylated, and has a molecular weight of ~57 kDa. The N in RABV appears to be phosphorylated by a cellular casein kinase II (Anzai et al., 1997; Gupta et al., 2000; Wu, Lei, & Fu, 2003). The amino acid sequence of N is the most conserved of the viral proteins among the lyssaviruses (Marston et al., 2007; Warrilow, Smith, Harrower, & Smith, 2002). Despite the overall conserved nature of the N, there is a relatively high degree of genetic diversity within short segments of the N gene between the genotypes (Bourhy et al., 1999; Bourhy, Kissi, & Tordo, 1993; Conzelmann, Cox, Schneider, & Thiel, 1990; Johnson, McElhinney, Smith, Lowings, & Fooks, 2002; Kissi, Tordo, & Bourhy, 1995; Kuzmin, Hughes, Botvinkin, Orciari, & Rupprecht, 2005). An important reason for the high level of amino acid sequence conservation within specific regions in N is that it must retain certain key functions that are dependent on specific protein-RNA genome interactions (see Section 2.3.2). On the other hand, the noted amino acid differences provide unique, genotype-specific epitopes on the N that define antigenic relationships between virus strains within and between genotypes on the basis of their reactivity patterns (antigenicity) with a panel of anti-N monoclonal antibodies (MAbs) (Dietzschold, Lafon, et al., 1987; Flamand, Wiktor, & Koprowski, 1980a, 1980b; Smith, 1989). The exploitation of the qualitative diversity in the N gene at the nucleotide level has also led to an extensive analysis of phylogenetic relationships of lyssaviruses and suggested quantitative criteria for lyssavirus species definition using the polymerase chain reaction (PCR) and nucleotide sequencing technologies (Kuzmin et al., 2005) (see Chapters 4 and 12).

The binding site on N for its interaction with vRNA has been localized to both the C-terminal and N-terminal regions of N with amino acid residues 149, 161, 168, 225, 235, 237, 290, 298, 323, 352, and 434 being involved in direct RNA binding (Kouznetzoff, Buckle, & Tordo, 1998). After binding to vRNA, the N undergoes sufficient conformational change to acquire a number of conformation-dependent epitopes, one of which enables Ser389 of N to be phosphorylated (Anzai et al., 1997; Dietzschold, Lafon, et al., 1987; Kawai et al., 1999; Toriumi & Kawai, 2004). Phosphorylation of N Ser389 stabilizes the interaction between N and P in the rabies viral RNP complex (Toriumi & Kawai, 2004). It has also been suggested that the phosphorylation of N during vRNA or complementary viral RNA (cRNA) encapsidation is important to regulate vRNA transcription and replication (Liu, Yang, Wu, & Fu, 2004; Wu, Gong, Foley, Schnell, & Fu, 2002; Yang, Koprowski, Dietzschold, & Fu, 1999).

The N is the second most extensively analyzed of the RABV proteins (after the G) with respect to its antigenic and immunogenic structure and function. The immunological interest in N stems from the observation that the RNP of RABV induces protective immunity against a peripheral challenge of lethal rabies virus in animals (Dietzschold, Wang, et al., 1987; Tollis, Dietzschold, Volia, & Koprowski, 1991). Three linear epitopes (antibody binding sites) on N were mapped to amino acids 358 to 367 (antigenic site I) and three linear epitopes (antigenic site IV) were mapped to two independent regions, amino acids 359 to 366 and 375 to 383 (Goto et al., 2000; Minamoto et al., 1994). Although the stretch of amino acids between residues 359 and 366 is shared by the two independent antigenic sites I and IV, the MAbs that recognize epitopes within these sites do not compete with each other for binding to the N antigen. Thus, it would appear that the respective epitopes are detected on different forms of N, one that represents N that is diffusely distributed in the cytoplasm and the other that is associated with viral RNPs (Goto et al., 2000; Jiang et al., 2010). The fact that the N associated with cytoplasmic inclusion bodies (IBs) or Negri bodies (NBs) (Lahaye et al., 2009) may represent N in its mature form is suggested by a MAb specific for antigenic site II, which only recognizes a conformation-specific epitope on the IB-associated N antigen. These and the conformation-dependent epitopes present in antigenic sites II and III and also at the phosphorylation site (serine 389) of N-RNA, and others yet to be mapped for which MAbs are available, provide valuable diagnostic tools (Kawai et al., 1999; Lahaye et al., 2009; Minamoto et al., 1994).

The N is also a major target for T helper (Th) cells that crossreact among rabies and rabies-related viruses (Celis, Karr, Dietzschold, Wunner, & Koprowski, 1988; Celis, Ou, Dietzschold, & Koprowski, 1988; Ertl et al., 1989). Several Th cell epitopes in the RABV N were identified and mapped using a series of overlapping synthetic peptides corresponding to N sequences of approximately 15 amino acids in length (Ertl et al., 1989).

Additionally, the RABV N functions as an exogenous superantigen (Lafon et al., 1992). It is perhaps the only viral superantigen that has been identified in humans. Some of the properties and responses found not only in humans but also in mice that are attributable to the RABV N in the role of superantigen include (1) its potent activation of peripheral blood lymphocytes in human vaccinees, (2) its ability to produce a more rapid and heightened VNA response upon injection of inactivated rabies vaccines, (3) its induction of early T-cell activation steps and expansion and mobilization of CD4+ Vb8 T cells to trigger and support production of VNA, and (4) its ability to bind to HLA class II antigens expressed on the surface of cells (Lafon et al., 1992).

2.2.3.2 Phosphoprotein (P)

The RABV P (PV strain) contains 297 amino acids (38-kDa) and is the least conserved of the five RABV proteins. The diversity of P sequences is also greatest (43%–97%) among certain pairings of lyssaviruses with two regions of greatest variability between residues 52–78 and 155–178 (Marston et al., 2007). The P is a multifunctional and multifaceted protein. It interacts with N to form N-P complexes and acts as a chaperone for newly synthesized N preventing its polymerization (self-assembly) and nonspecific binding to cellular RNA (Mavrakis et al., 2003) and specifically directs N encapsidation of the vRNA (Chenik, Chebli, Gaudin, & Blondel, 1994; Fu, Zheng, Wunner, Koprowski, & Dietzschold, 1994; Gigant, Iseni, Gaudin, Knossow, & Blondel, 2000). As a subunit of the RNA polymerase (P-L) complex, the P plays a pivotal role as a noncatalytic cofactor in transcription and replication of the viral genome. The P stabilizes the L and places the P-L complex on the RNA template, which the L protein alone is unable to do (Chenik et al., 1994; Chenik, Schnell, Conzelmann, & Blondel, 1998; Fu et al., 1994).

The P of lyssaviruses, like the P in other negative-strand viruses, exists in a variety of phosphorylated forms. Two prominent forms of P are present in both rabies virions and in virus-infected cells. One is a major hypophosphorylated 37-kDa form and the other is a minor hyperphosphorylated 40-kDa form (Toriumi & Kawai, 2004). RABV P is phosphorylated in the N-terminal portion by two distinct types of protein kinases, one of which is a unique heparin-sensitive protein kinase (Gupta et al., 2000; Takamatsu et al., 1998). This unique 71-kDa kinase, designated RABV protein kinase (RABV-PK), phosphorylates recombinant P (36 kDa, expressed in *E. coli*) at S63 and S64 (sequence of the challenge virus standard [CVS] strain) and nascent P (37 kDa) expressed in baby hamster kidney (BHK)-21 cells infected with the high egg passage (HEP)-Flury strain of RABV. In both cases, hyperphosphorylation alters the mobility of P (36 kDa and 37 kDa) in sodium dodecyl sulfate-polyacrylamide gel electrophoresis (SDS-PAGE) causing it to migrate more slowly, as a protein of 40 kDa (Toriumi & Kawai, 2004). The other phosphorylating enzyme is protein kinase C, which has several isomers (PKCα, β, χ, and δ). In contrast to the RABV-PK, phosphorylation of P by the PKC isoforms, dominated by PKCγ activity, did not alter the migration of P in SDS-PAGE (Gupta et al., 2000). Upon analyzing the PK activity in rabies virions for the presence of these two types of enzymes, it was concluded the RABV-PK is selectively packaged in mature rabies virions along with a smaller amount of the predominant PKCγ isoform as the rabies virion-associated PKs.

RABV P interacts with the nascent soluble N (N°) and L via domains of P that are specific for each of these two proteins (Castel et al., 2009; Chenik et al., 1998). At least two independent N-binding sites have been found on P. One binds nascent RABV P to nascent N° to maintain N° in a competent form for RNA encapsidation and the other binds to the C-terminal part of N in the RNA-N complex (Chenik et al., 1994; Fu et al., 1994; Schoehn, Iseni, Mavrakis, Blondel, & Ruigrok, 2001). The N-RNA site is located within the C-terminal 30 amino acids (between amino acids 267 and 297) of P and N° binding is mediated by the N-terminal portion of the protein between amino acids 69 and 177 (Chenik et al., 1994). Both sites interact with N in a manner that is mutually independent (Fu et al., 1994). The interaction involving the N-terminal binding site (first 40 amino acids) of P requires that the P interact with N° soon after the two proteins are synthesized in vivo (Castel et al., 2009; Mavrakis et al., 2006), but that it may compete with endogenous RNA (Albertini, Ruigrok, & Blondel, 2011). The P is able to form elongated dimers (Gerard et al., 2007; Mavrakis et al., 2003).

In the formation of progeny RNP, the P is required to also bind L, the large catalytic subunit of the viral RNA polymerase, to produce a virus-encoded RNA polymerase complex that is fully active. The P subunit has a major binding site for L within the first 19 amino acids of P (Chenik et al., 1998), in the N-terminal unfolded region of P (Castel et al., 2009; Gao, Greenfield, Cleverley, & Lenard, 1996; Gerard et al., 2007; Gigant et al., 2000; Spadafora, Canter, Jackson, & Perrault, 1996). Oligomerization of RABV P does not require phosphorylation nor is the N-terminal domain (first 52 amino acids) necessary for oligomerization or binding to N-RNA [cRNA and vRNA] template (Gigant et al., 2000; Mavrakis et al., 2003). This is in contrast to the P of VSV, which requires phosphorylation for oligomer formation, to be fully active and available for binding both to L and to the RNA template (Albertini et al., 2011; Ding, Green, Lu, & Luo, 2006; Gao et al., 1996; Gerard et al., 2007).

A particular feature of the RABV P protein is its ability to interact with multiple host cell proteins (and to modulate cellular pathways). The 10-kDa cytoplasmic dynein light chain (LC8), which is involved in the intracellular transport of organelles, was found to interact strongly with the P of RABV and MOKV (Jacob, Badrane, Ceccaldi, & Tordo, 2000;

Raux, Flamand, & Blondel, 2000). The P domain that interacts with dynein LC8 was mapped to the N-terminal half of the P, between amino acids 138 and 172 in the P. Binding of P to LC8 seems not to be involved in axonal transport of RABV, since deletion of the LC8 binding domain did not change RABV entry into the central nervous system (CNS) (Rasalingham, Rosssiter, Mebatsion, & Jackson, 2005). Deletion of the LC8 binding domain however significantly suppressed viral transcription and replication in the CNS resulting in loss of viral infectivity in the CNS (Tan, Preuss, Willliams, & Schnell, 2007). In addition, the P protein of several RABV strains (dimerization domain), but not of other lyssaviruses, was shown to bind to focal adhesion kinase (FAK), which positively regulates viral protein synthesis (Fouquet et al., 2015). RABV P protein (with critical positions at 162 and 166) was also shown to interact with mitochondrial complex I and to increase levels of reactive oxygen species (Kammouni et al., 2015; Kammouni, Wood, & Jackson, 2017).

A well-established and critical role of RABV P protein is counteracting the host innate immune response, in particular the type I/III interferon system. These IFNs are produced in virus-infected cells after sensing of pathogen- or danger-associated molecular patterns (PAMPs or DAMPs), mostly represented by viral nucleic acids. In auto- and paracrine fashion, the secreted IFNs can, via JAK/STAT signaling, stimulate the transcription of hundreds of antiviral and/or immune stimulatory genes, resulting in a powerful antiviral state. The RABV P is able to paralyze the signaling cascades leading to transcriptional activation of interferon genes as well as interferon signaling pathways, thereby limiting the establishment of an antiviral host response.

Specifically, RABV P can effectively counteract the activation of latent interferon regulatory factors (IRF-3), the initial key transcription factor for IFN expression (Honda & Taniguchi, 2006). P interferes with the phosphorylation of serine 386 at the C terminus of the IRF-3 protein, which allows dimers of IRF-3 to form and be recruited to the IFN-β enhancer (Brzózka, Finke, & Conzelmann, 2005). In addition, P blocks IFN-mediated JAK/STAT signaling, by sequestering phosphorylated STAT1, -2, and -3 molecules in the cytoplasm (Blondel, Maarifi, Nisole, & Chelbi-Alix, 2015; Brzózka, Finke, & Conzelmann, 2006; Vidy, Chelbi-Alix, & Blondel, 2005).

2.2.3.3 Virion-associated RNA polymerase or large protein (L)

The RABV L (PV strain) contains 2142 amino acids (2127 amino acids in SAD-B19 strain) (244 kDa) and is encoded in the fifth gene, which comprises more than half (54%) of the coding potential of the RABV genome. The L is the catalytic component of the polymerase complex, which along with the noncatalytic cofactor P is responsible for the majority of enzymatic activities involved in viral RNA transcription and replication, including an RNA-dependent RNA polymerase (RdRP), an unconventional set of mRNA capping enzymes—a GDP polyribonucleotidyltransferase (PRNTase) and a dual-specificity mRNA cap methylase (MTase) (Morin, Liang, Gardner, Ross, & Whelan, 2017; Ogino & Banerjee, 2007; Ogino, Ito, Sugiyama, & Ogino, 2016). Comparisons of L sequences from different negative-strand RNA viruses have initially helped to map the functionally homologous and unique sequences in attempts to locate the ascribed enzyme activities. As in all negative-strand RNA viruses, the RABV virion-associated viral RNA polymerase plays a unique role at the start of infection by initiating the primary transcription of the genome RNA after the RNP core is released into the cytoplasm of the infected cell. The enzymatic steps of transcription include initiation and elongation of the leader RNA and mRNA transcripts as well as the cotranscriptional modifications of the mRNAs that include 5′-capping, methylation, and 3′-polyadenylation. Comparisons of L sequences from different negative-strand RNA viruses have helped to map the functionally homologous and unique sequences in attempts to locate the ascribed enzyme activities (Barik, Rud, Luk, Banerjee, & Kang, 1990; Poch, Blumberg, Bougueleret, & Tordo, 1990; Tordo, Poch, Ermine, Keith, & Rougeon, 1988). One of the main features of L that comes out of the sequence comparison is the number of clusters of conserved amino acid residues that appear to be purposefully aligned in blocks (or boxes), I–VI, along the protein (Lij, Rahmeh, Morelli, & Whelan, 2008; Ogino, Yadav, & Banerjee, 2010; Poch et al., 1990). Within these boxes, some residues form strongly conserved domains, with a high proportion of amino acids either strictly or conservatively maintained in identical positions, while other domains are more variable, consistent with the multifunctional nature of L (Banerjee & Chattopadhyay, 1990; Poch et al., 1990; Tordo et al., 1988). One of the blocks (box III), the catalytic domain, in the central part of the RABV L, between residues 530 and 1177, contains four motifs, A, B, C, and D, that represent regions of highest similarity (Poch, Sauvaget, Delarue, & Tordo, 1989; Tordo et al., 1988). These motifs, which are thought to constitute the "polymerase module" of L, maintain the same linear arrangement and location in all viral RNA-dependent RNA and DNA polymerases (Barik et al., 1990; Delarue, Poch, Tordo, Moras, & Argos, 1990; Poch et al., 1990). Among the conserved sequences in these four motifs is the triamino acid catalytic center sequence GDN (glycine, aspartic acid, and asparagine) in motif C, which is extensively conserved in all nonsegmented negative-strand RNA viruses (Poch et al., 1989). At least two other sequences between amino acid residues 754 to 778 and 1332 to 1351 in the VSV L have been identified as consensus sites for binding and utilization of ATP, similar to those identified in cellular kinases (Barik et al., 1990; Canter, Jackson, & Perrault, 1993). Three essential activities encoded by L are involved in the binding and utilization of ATP. These are (1) the transcriptional activity that requires binding to substrate ribonucleoside triphosphates (rNTPs), (2) polyadenylation, and (3) protein kinase activity for specific phosphorylation of the P in transcriptional activation (Banerjee & Chattopadhyay, 1990; Sanchez, De, & Banerjee, 1985).

2.2.3.4 Matrix protein (M)

The RABV M (PV strain) is the smallest of the virion proteins with 202 amino acids (25 kDa) (Bourhy et al., 1993; Conzelmann et al., 1990; Gould et al., 1998; Hiramatsu, Mannen, Mifune, Nishizono, & Takita-Sonoda, 1993; Tordo, Poch, Ermine, & Keith, 1986) and can form dimers (54 kDa; 10%–20% of total M in the cell) that form a strong association with G (Nakahara et al., 2003). M binds to and condenses the nascent NC core into a tightly coiled, helical RNP-M protein complex, forming a string-like structure between the NC core turns and producing the bullet-shaped "skeleton" structure of the virion. Approximately 1200 to 1500 copies of M molecules bind to rabies virus RNP core. At the same time, M mediates binding of the viral core structure to the host membrane where it initiates virus budding (Mebatsion et al., 1999). A proline-rich motif (PPEY) in the M located at residues 35–38 within the highly conserved 14-amino acid sequence near the N-terminus of the RABV M was the first motif to be associated with virus budding from the cell membrane surface (Harty et al., 2001; Harty, Paragas, Sudol, & Palese, 1999). Other proline-rich (PPxY, PT/SAP, YxxL, and FPIV) motifs and related core sequences are found in the M of VSV, and other viruses and are referred to as a late-budding domains (L-domains) (references in Irie, Licata, McGettigan, Schnell, & Harty, 2004). These domains can contribute to late steps of virus-budding processes by interacting with components of the endosomal sorting complex required for transport (ESCRT) system (Schoneberg, Lee, Iwasa, & Hurley, 2017). Although the exocytotic release of progeny virions requires the RABV M in the RNP-M "skeleton" complex, the efficiency of virus budding is greatly enhanced by the interaction of the RNP-M skeleton complex with the envelope G (Mebatsion et al., 1999; Mebatsion, Konig, & Conzelmann, 1996). Increased virion production as a result of interaction of the cytoplasmic domain of the transmembrane spike G and the viral RNP-M core in RABV assembly suggests that a concerted action of both core and spike proteins supports efficient recovery of virions (Mebatsion et al., 1999). Notably, the interaction of M-RNPs with the cytoplasmic domain of G does not need to be optimal, as the G proteins of related viruses can substitute for the homologous G in budding virions (Albisetti et al., 2017; Mebatsion, Schnell, & Conzelmann, 1995; Morimoto, Foley, McGettigan, Schnell, & Dietzschold, 2000).

2.2.3.5 Glycoprotein (G)

The RABV G (all species) is translated from a G-mRNA transcript that encodes 524 amino acids, which includes an N-terminal 19-amino acid signal peptide (SP) domain. The SP provides a membrane insertion signal, which transports the nascent protein into the rough endoplasmic reticulum (ER)-Golgi-plasma membrane secretion pathway before it is cleaved from the N-terminus of the G molecule in the Golgi apparatus. (Conzelmann et al., 1990; Tordo et al., 1988). The mature G (without the SP) of all RABV strains for which the amino acid sequence as been determined has 505-amino acids ($\sim$65 kDa), and that of MOKV has 503 amino acids (Benmansour et al., 1992; Bourhy et al., 1993). The G is a type I membrane glycoprotein with an N-terminal ectodomain (ED), which extends outward on the plasma membrane and surface of mature virus particles, a 22-amino acid transmembrane (TM) (anchoring) domain and a 44-amino acid long C-terminal domain (CD) locating in the cytoplasm. The resulting transmembrane G is organized into trimers (three monomers of 65 kDa each) in the Golgi apparatus, which later form the G spikes embedded in the plasma membrane and on the virion surface (Gaudin et al., 1992; Whitt, Buonocore, Prehaud, & Rose, 1991). The G spikes in the viral envelope extend 8.3 nm from the virus surface and represent the major surface protein of the virion. The CD (last 44 amino acids) of the G extends inward from the plasma membrane into the cytoplasm of the infected cell where it interacts with M of the skeleton particle to complete the virion assembly. The ED of G (residues 1 to 439 of the mature RABV G) in each spike is the business end of the molecule for a variety of functional virus interactions. It is responsible for interaction with cellular binding sites (receptors), and, therefore, is important in viral pathogenesis by targeting the appropriate cells for infection (Sissoeff, Mousli, England, & Tuffereau, 2005). It is also responsible for low pH-induced fusion of the viral envelope with plasma and endosomal membranes in the cell in the early phase of the RABV life cycle (Albertini, Baquero, Ferlin, & Gaudin, 2012; Gaudin, Ruigrok, Knossow, & Flamand, 1993). And it is critical for the induction of a host humoral immune response to RABV infection and as the target of viral neutralizing antibodies (VNAs) as well as for virus specific helper and cytotoxic T cells (Celis, Karr, et al., 1988; Macfarlan, Dietzschold, & Koprowski, 1986).

Appropriate glycosylation of RABV G is important for its proper expression and function. The oligosaccharides that are associated with the G are linked to asparagine (N, in single letter code) residues in the tripeptide sequence (sequon) asparagine-X-serine (NXS) or asparagine-X-threonine (NXT), where "X" is any amino acid other than proline (Shakineshleman, Remaley, Eshleman, Wunner, & Spitalnik, 1992). RABV G molecules can have one to four sequons or N-linked carbohydrate (N-glycan) sites per molecule depending on the virus strain. Even if an N-glycan site is present, it may be inefficiently glycosylated or not occupied at all depending on the amino acid composition of the sequon (Dietzschold, Wiktor, Wunner, & Varrichio, 1983; Kasturi, Eshleman, Wunner, & Shakineshleman, 1995; Shakineshleman, Wunner, & Spitalnik, 1993; Wunner, Dietzschold, Smith, Lafon, & Golub, 1985). A virus can be heterogeneous with regard to the number of sites that are occupied with N-glycans (macroheterogeneity) and with regard to the types of glycan structures (microheterogeneity) present at each site (Wojczyk et al., 2005). Clearly, various factors can affect the glycosylation of individual sequons in G, some influence efficiency of N-glycosylation and others influence processing of N-glycans into a variety of oligosaccharide structures. The sugar residues of the N-glycan are presynthesized in the cell and transferred from a lipid precursor by the enzyme oligosaccharyltransferase as a precursor core-oligosaccharide unit ($Glc_3Man_9GlcNAc_2$) to the specific sequons of the nascent G molecule. Typically, the transfer occurs as the protein, which is synthesized on the membrane-bound ribosomes (rough ER), begins to fold cotranslationally, and during translocation to the cytoplasmic ER membrane. After the precursor core-oligosaccharide is transferred, the high-mannose triglucosylated oligosaccharide is trimmed and processed in the lumen of the rough ER and Golgi stacks to form the "complex type" N-linked monoglucosylated oligosaccharide of the mature G molecule (see references in Gaudin, 1997). The molecular chaperone calnexin recognizes the partially trimmed monoglucosylated glycan and binds to it, assisting the G to fold correctly and completely in order to achieve its biological activity, stability, and antigenicity (Gaudin, 1997; Okazaki, Ohno, Takase, Ochiai, & Saito, 2000; Shakineshleman et al., 1992). This dependence of the RABV G on the molecular interaction with calnexin explains why it is critical that at least one of the asparagine residues, that is, N319 in the RABV G, is conserved in all virus species. N319-glycosylation is essential for correct and complete folding of the nascent RABV G and required for subsequent transport to the cell surface (Wojczyk et al., 2005).

The RABV G is also modified by the addition of palmitic acid (referred to as fatty acid acylation or palmitoylation) at cysteine 461, located in the intra-CD on the C-terminal side of its TM region (Gaudin, Tuffereau, Benmansour, & Flamand, 1991). Although the functional significance of palmitoylation is not entirely clear, it is presumed to have a stabilizing effect on the trimeric G spike anchored in the membrane. Palmitoylation may also play a role in the virus budding process by facilitating the interaction between the CD "tail" of the G and the M in the RNP-M complex at the cell membrane.

Another important function of the RABV G that is critical in establishing the phenotype of the virus is its role in defining the pathogenicity and determining the neuroinvasive pathway of the virus (Ito, Takayama, Yamada, Sugiyama, & Minamoto, 2001; Kucera, Dolivo, Coulon, & Flamand, 1985; Yan, Mohankumar, Dietzschold, Schnell, & Fu, 2002). While the neurotropism of a particular RABV strain is primarily a function and major defining characteristic of its G, it is relevant to note that other viral components and attributes are also important in altering the viral pathogenesis of rabies (Faber et al., 2004; Morimoto et al., 2000). Whether or not the virus will cause a lethal infection (follow a specific pathway to induce a fatal disease) is determined first by the interaction of the G spike with a specific receptor on neuronal cells in vivo (Lafon, 2005; Sissoeff et al., 2005). An attempt to map the p75NTR receptor-binding site on RABV G has suggested that the receptor binds to a region of the G (within residues 318–352) on both sides of antigenic site III and site "a" that is not neutralized by anti-G antibody (Langevin & Tuffereau, 2002). Lentiviral vectors pseudotyped with RABV G have demonstrated that the G not only allows entry into the nervous system, upon infection of neurons at distal connected sites within the nervous system, it facilitates and enhances the retrograde axonal transport of the virus to the CNS (Mazarakis et al., 2001). Second, in many fixed RABV strains, the pathogenic or virulence phenotype of the virus correlates with a single amino acid at a specific position in the RABV G. For example, arginine (R) 333 or lysine (K) 333 of wild-type (virulent) G determines the virulence phenotype or neuroinvasive pattern of RABV in the CNS. Virus variants that substitute glutamine (Q), isoleucine (I), glycine (G), methionine (M), or serine (S) for R333 in the rabies virus G express a phenotype that is either less pathogenic or totally avirulent compared to the parental wild-type virus in adult immunocompetent mice (Dietzschold, Wunner, et al., 1983; Mebatsion, 2001; Seif, Coulon, Rollin, & Flamand, 1985; Tao et al., 2010; Tuffereau et al., 1989). While these particular mutations in the G correlate with reduced or abolished neuroinvasiveness, they do not impair the ability of the virus to multiply in cell culture. The same substitutions (e.g., Q333 for R333 or R333Q) in the G can, however, affect the rate of virus spread from cell to cell (Dietzschold et al., 1985; Faber et al., 2005) and the ability to infect motor neurons in vivo and in vitro (Coulon, Ternaux, Flamand, & Tuffereau, 1998). They can modify the host range spectrum of the virus and determine the choice of neuronal pathways the virus uses to reach the CNS (Etessami et al., 2000; Kucera et al., 1985) as well as the distribution of virus to different areas of the brain (Yan et al., 2002). The R333Q substitution is also associated with a greater ability of the virus to induce apoptosis in neuronal cells (Tao et al., 2010). In field isolates, the basic 333 residues seem not to have this importance for pathogenicity (Sato et al., 2009). Still, other amino acid substitutions at other positions on the RABV G appear to confer or influence viral pathogenicity. For example, amino acid substitutions located between positions 34 and 42 and at positions 198 and 200 (related to epitopes specific to antigenic site II) of the CVS strain G reduced the pathogenicity of the CVS strain when inoculated intramuscularly in adult mice (Prehaud, Coulon, Lafay, Thiers, & Flamand, 1988). One or more amino acids between residues 164 and 303 in the G of the pathogenic parental Nishigahara Ni-CE strain compared with the avirulent Ni-CE variant RC-HL, also appear to define the lethality that is characteristic of this virus since R333 is present in the G of both strains (Takayama-Ito, Ito, Yamada, Minamoto, & Sugiyama, 2004).

Finally, the RABV G is a potent immunogen with major importance immunologically for the induction of the host immune response against virus infection. Because it is such an important antigen for the immune system to mount a response against, it is probably the most extensively studied RABV protein in terms of structure in relation to its immunogenicity and antigenicity for VNAs, i.e., as a target for VNAs. The G induces VNAs that recognize both conformational and linear epitope (antibody binding) sites and stimulates helper as well as cytotoxic T-cell activity.

2.2.4 Viral lipids and carbohydrate

A lipoprotein bilayer forms the viral envelope (or membrane matrix) surrounding the helical RNP core. The lipids, which constitute 20%–26% of the viral lipoprotein envelope, are derived entirely from the host cell and depending on where the virus buds through the cellular membrane, concentrations of certain lipids may be higher in the viral envelope than is represented in the rest of the plasma membrane. In general, the RABV membrane contains a mixture of lipids, including phospholipids (mainly sphingomyelin, phosphatidylethanolamine, and phosphatidylcholine), neutral lipids (mainly triglycerides and cholesterol), and glycolipids (reviewed in Wunner, 1991).

2.3 Genome and RNP structures

One of the characteristic features of MNV like RABV is the stable helical (supercoiled) RNP complex in which the genome RNA is tightly encapsidated by N (N-vRNA) and associated with P and L. RNP structure and function is not only important for progeny virus morphogenesis it is directly interrelated with the typical mode of gene expression. This involves sequential transcription of nonencapsidated subgenomic mRNAs from genome RNPs as well as replication of full-length antigenome and progeny genome RNPs. Release of the RNP from infecting virions into the cytoplasm of cells and dissociation of M condensing the RNP results in transition of the RNP from the supercoiled state to a relaxed structure, in which the N-vRNA complex can serve as a template for RNA synthesis (Iseni, Barge, Baudin, Blondel, & Ruigrok, 1998). In fact, exclusively N-vRNA (and N-cRNA) rather than naked vRNA is a suitable template for the viral polymerase, both for synthesis of subgenomic nonencapsidated mRNAs (transcription) and of full-length progeny as well as novel DI RNPs (replication) (Albertini et al., 2011).

Details of the molecular structure of the helical N-RNA complexes of RABV and VSV were revealed by X-ray diffraction of crystals obtained from ring structures of N-vRNA complexes containing 11 or 10 N molecules (Albertini et al., 2006; Green, Zhang, Wertz, & Luo, 2006). The RNA in these structures is tightly sequestered in a cavity at the interface between the N- and C-terminal lobes of the N, which appear to clamp down onto the bound RNA. The characteristics of RNP folding, RNA binding, and assembly are highly conserved in VSV and RABV, despite their lack of significant homology in amino acid sequences of N (Luo, Green, Zhang, Tsao, & Qiu, 2007). The complete occlusion of the vRNA by N explains the excellent protection of the vRNA against high salt, attack by RNases, and silencing by siRNAs (Albertini et al., 2006, 2007; Albertini, Schoehn, Weissenhorn, & Ruigrok, 2008). The tight N-vRNA clamping appears to open exclusively and transiently during genome transcription and replication to specifically give access for the polymerase complex (L/P or L-P/N) to the vRNA template. This requires conformational changes in N caused by the polymerase and/or P alone (Albertini et al., 2008). Easiest access to the N-vRNA would be at the 3′-end of the genome and antigenome. Indeed, exclusive entry at or close to the 3′-terminus of the genome and (obligatory) sequential transcription of the downstream succession of genes is the characteristic feature of all MNV viruses.

2.3.1 Genome primary structure

The single-strand negative-sense RNA genomes (vRNAs) of lyssaviruses represent the prototypical MNV genomes. They are about 12 kb in length and have unmodified 3'-hydroxyl (3'-OH) and 5'-triphosphate (5'-PPP) ends. They consist of a succession of five genes in the strictly conserved order 3'-N-P-M-G-L-5', flanked by short 3'- and 5'- terminal regulatory extragenic regions known as Le and Tr regions, respectively (Fig. 2.3).

The 3'-terminal Le region can be regarded as the genomic promoter (GP) that directs the transcription of monocistronic mRNAs, as well as the synthesis of full-length antigenome RNA (cRNA). The 3'-end of the cRNA (i.e., the complement of the genomic Tr region) functions as the antigenome promoter (AGP), but unlike the GP, only directs synthesis of full-length N-encapsidated cRNAs, and not of unencapsidated RNAs. GP and AGP share a similar composition with a high A and U content, and in RABV the 11 3'-terminal nucleotides are identical (3'-UGCGAAUUGUU...5') (Tordo et al., 1988; Tordo, Poch, Ermine, & Keith, 1986; Tordo, Poch, Ermine, Keith, & Rougeon, 1986). Such terminal complementarity is a common feature of MNV genomes and provides strong evidence that these terminal sequences are important for promoter function (Tordo et al., 1988; Whelan, Barr, & Wertz, 2004).

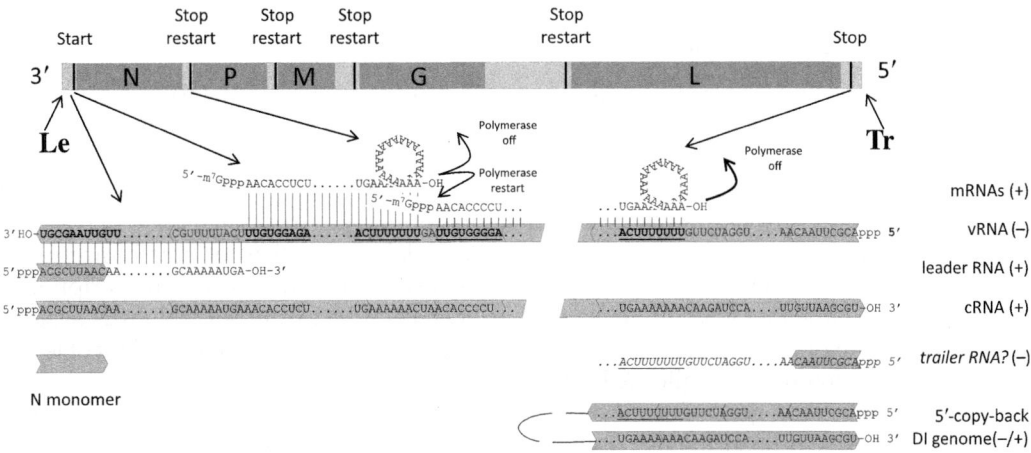

FIG. 2.3 RABV genome transcription and replication products. On top, the 12-kb genome (vRNA) of RABV is schematically shown. Localization of the five open reading frames, nucleoprotein (N), phosphoprotein (P), matrix protein (M), glycoprotein (G), and RNA polymerase or large protein (L) (gray) and transcription signals (black bars) are indicated. In the lower part, details of the vRNA sequence are shown, including terminal sequences and transcription signals (underlined) of the N and L genes. Transcription of the 5'-capped mRNAs is obligatory sequential and starts at the 3' proximal N start signal. At the stop/polyadenylation signal, the polymerase may resume transcription of the downstream gene at the nearby start signal or dissociate from the template, which leads to the transcription gradient. The intergenic regions between the stop/polyadenylation and restart signals are probably not transcribed. Replication involves concurrent encapsidation of RNA by N. Incomplete encapsidation of the growing leader RNA might lead to abortive replication at the leader/N gene border or a switch to transcription. Whether a corresponding regulatory product of the cRNA (trailer RNA) exists in RABV is not clear. Defective interfering RNAs of RABV comprise the cRNA 5'-end and an exact complementary 3'-end, which is generated by backcopying of the nascent positive strand by the polymerase.

2.3.2 Lyssavirus genes

Most of the genome (>99%) comprises the protein-encoding genes. They are transcribed by the viral polymerase to yield mRNAs that can be efficiently translated by the host cell machinery. The mRNAs have a typical 5′-terminal 7-methylguanylate cap structure (m^7G) facilitating translation initiation, which in all lyssaviruses invariantly start with the sequence 5′-m^7G-AACAYYNCU, where Y=C or U, N=any nucleotide (nt). The mRNA sequences also have very short 5′-noncoding regions in the range of only 20–30 nts, and longer 3′-noncoding regions in the range of 100–500 nts. A particularly long 3′-noncoding region is observed in the G mRNA of RABV corresponding to the region in the genome that was initially described as a pseudogene (Ψ).

Differences in the 3′-noncoding region of genes mostly account for the observed differences in length of lyssavirus genomes. Among the lyssaviruses, RABVs seem to have the shortest genomes, which are in the range of 11,923–11,932 nts. The genome lengths of other lyssavirus species seem to increase somewhat, though the overall variation remains very low. The longest lyssavirus genome reported is that of WCBV and still it comprises only 12,178 nts (Kuzmin, Wu, Tordo, & Rupprecht, 2008). For more details on the genetic variability of lyssavirus genomes and proteins, refer to Banyard, Hayman, Johnson, McElhinney, and Fooks (2011) and Nadin-Davis and Real (2011); see also Chapter 7.

2.3.3 Transcription of genes

According to the widely accepted stop/start model of transcription of MNV viruses, RABV transcription initiates exclusively at or close to the 3′ terminal GP (Fig. 2.3), and the genes are transcribed in an obligatory sequential manner (Flamand & Delagneau, 1978; Holloway & Obijeski, 1980). The observed synthesis of an abundant 58 nt-long 5′-PPP- and nonpolyadenylated leader RNA from the 3′-terminal Le region of rhabdoviruses (Colonno & Banerjee, 1978; Leppert, Rittenhouse, Perrault, Summers, & Kolakofsky, 1979) originally argued in favor of a model in which leader RNA synthesis is absolutely required for transcription of the downstream protein-encoding genes. However, recent data from several virus systems argue as well in favor of transcription initiation mechanisms, which are independent of leader RNA synthesis (Noton, Tremaglio, & Fearns, 2019).

The protein-encoding genes downstream of the Le region are separated by conserved gene border signals, which direct the activity of the polymerase during vRNA transcription, but which must be ignored during vRNA replication. They specify the end of the upstream mRNA by a stop/polyadenylation (poly(A)) signal and the beginning of the downstream mRNA (start signal). Typically, stop and start signals are separated by short, most likely nontranscribed intergenic spacer sequences (IGSs). While in VSV these IGSs invariably comprise 2 nucleotides, lyssaviruses mostly have IGSs of 2 (at the N/P gene border), 5 (at P/M and M/G borders), and 19–28 nts (at G/L borders), though some variation (16 nts) is observed at the M/G border in some lyssaviruses.

The stop/poly(A) signals are characterized by a run of mostly 7 U residues (in RABV 3′-…ACUUUUUU…-5′) on which the polymerase repeatedly stutters back and forth to synthesize a 50–150 nt-long poly(A) tail (Barr, Whelan, & Wertz, 1997). Without leaving the N-vRNA template, the polymerase is thought to release the newly generated polyadenylated mRNA and to initiate transcription of the downstream gene at the conserved restart signal

(3′…UUGURRNGA…-5′) leaving the border signals between sequential genes untranscribed. As was shown for VSV, restart involves the addition of a typical 5′-m^{7}G cap (Both, Moyer, & Banerjee, 1975; Ogino, 2014) by the L protein through an unconventional capping mechanism (Lij et al., 2008; Ogino & Banerjee, 2007). The stop/poly(A) signal in the L gene is not followed by a restart signal, which means the downstream Tr sequence, comprising about 70 nts of the lyssavirus genome RNA, is not transcribed.

As a consequence of the exclusive entry of the polymerase at the 3′ end of the genome and eventual dissociation of the polymerase at every gene junction, upstream (promoter proximal) mRNAs are produced more abundantly than downstream mRNAs. This type of transcriptional attenuation is a very simple and efficient method to achieve differential expression levels of gene products. The conserved gene order of MNV genomes therefore seems to reflect the relative protein amounts needed for completion of the viral life cycle.

The steepness of the mRNA transcript gradient depends on the rate of dissociation of the polymerase from the template at the gene borders. For RABV, there is evidence that reinitiation is influenced by the length of the intergenic (IG) region, i.e., the distance between stop/poly (A) and restart signal. A conserved feature of lyssaviruses is the elongated IGS of the G/L gene border, which was interpreted to reflect a need for downregulation of L expression. Indeed, the short 2 or 5 nt-long IGSs of the N/P-, P/M-, and M/G-gene borders directed high-level expression of reporter genes from artificial bicistronic RABV minigenomes, whereas the 24-nt IGS from the G/L border was poorly active in mediating expression of the downstream reporter gene (Finke, Mueller-Waldeck, & Conzelmann, 2003). Moreover, replacement of the long G/L IGS with the 2 nts of the N/P IG region in recombinant RABV resulted in more abundant transcription of the L mRNA. This was correlated with higher accumulation of L and an overall increase in vRNA synthesis (Finke, Cox, & Conzelmann, 2000). The resulting viruses showed pronounced cytopathic effects in vitro, a phenotype predicted to be detrimental for viruses like RABV, which rely on a stealth strategy for pathogenesis. Thus, attenuating L expression through transcriptional attenuation by a long IG region between the G and L genes seems to be of advantage for RABV and other lyssaviruses.

Another observation strongly supporting the need for specifically limiting the expression of the RABV polymerase is the presence of extra stop signal or stop signal-like sequences in the noncoding region downstream of the G open reading frame (ORF) in several virus strains, including Evelyn-Rokotniki-Abelseth (ERA) and PV strains. These signals comprise 5 or 6 U residues and can lead to transcription stop and polyadenylation in a certain percentage of G mRNAs, such that in addition to the standard long G mRNAs, shorter G mRNAs are generated (Conzelmann et al., 1990; Morimoto, Ohkubo, & Kawai, 1989; Tordo, Poch, Ermine, Keith, et al., 1986). It is assumed that a polymerase complex that terminates transcription at the upstream signal is not able to reinitiate at the start signal of the L gene, because it cannot cope with an "intergenic region" of several hundred nucleotides. Therefore, only those polymerase complexes terminating at the downstream G/L border with the 24 nt-long IGS are supposed to contribute to L mRNA transcription. Rather than representing the remnants of a former gene, as proposed initially (Tordo, Poch, Ermine, Keith, et al., 1986), the extra transcription stop signals between G and L genes may serve the goal of limiting L mRNA transcription. The noncoding region of the G gene is not critical for virus growth and has been used preferably for substitution with single or multiple extra genes (Finke & Conzelmann, 2005; Gomme, Wanjalla, Wirblich, & Schnell, 2011; Schnell, Mebatsion, & Conzelmann, 1994).

The simple and modular organization of rhabdovirus genomes, where (i) cistrons are defined by short gene border signals and (ii) gene expression primarily depends on the position of the gene, has been exploited in recent years by numerous reverse genetics approaches. These involved addition of extra single or multiple genes and rearrangement of gene order (Conzelmann, 2004). Indeed, the artificial shifting of viral genes to other positions in the genome has become a versatile tool to study the impact of gene dosage on the rhabdovirus life cycle and pathogenesis (Ball, Pringle, Flanagan, Perepelitsa, & Wertz, 1999; Brzózka et al., 2005; Novella, Ball, & Wertz, 2004). For RABV, for example, moving the P gene from the second to the most downstream fifth position has revealed a requirement for high protein levels in order to successfully counteract the cellular antiviral interferon response (Brzózka et al., 2006). Moving the RABV M gene to the most promoter-proximal position has revealed its contribution to the regulation of viral transcription (Finke et al., 2003), and insertion of multiple G genes has been used to enhance antigen expression for vaccine purposes (Faber et al., 2009).

2.3.4 Replication of RABV N-vRNA genomes

Replication of genome N-vRNA complexes involves the synthesis of complementary full-length antigenome N-cRNA. Consequently, the viral polymerase here acts as an RNP-dependent RNP polymerase (a replicase). Like N-vRNA, the nascent N-cRNA is encapsidated concurrently with elongation. The coupling of cRNA elongation and N-encapsidation also dictates that replication of full-length N-cRNA is assured. In addition, transcriptional signal sequences must be ignored by the replicase. In contrast to the transcriptase, which is composed of L-P heteromers, the replicase is considered to be composed of L associated with P-N heteromers in which the P chaperones N for specific encapsidation of nascent cRNA (Chenik et al., 1994; Fu et al., 1994; Schoehn et al., 2001).

N-cRNA synthesis starts at the 3′-end of the N-vRNA genome by primer-independent synthesis to produce the 5′-end of the antigenome (i.e., producing a sequence corresponding to that of the short leader RNA) (Fig. 2.3). The finding of N-encapsidated leader RNA in rhabdovirus-infected cells (Blumberg, Giorgi, & Kolakofsky, 1983) indicated that encapsidation of nascent antigenome sequences is initiated by a replicase, or that the Le RNA sequence comprises a specific N-packaging signal that further directs the polymerase to the replicase mode. As revealed by X-ray analyses, every RABV N protomer covers 9–10 nts of RNA, and, as suggested by studies revealing the importance of flush-end N phasing in other MNV viruses, should be formed in a way that the first 9–10 nts are covered by the first N protomer. In the presence of high levels of N-P (or L-P-N) heteromers, synthesis of encapsidated cRNA (or vRNA) is occurring in a processive mode until the 5′ end of the template is reached. The resulting antigenome N-cRNA has a 3′-end identical in the first 11 nts to that of the genome vRNA and can efficiently direct the synthesis of abundant amounts of genome N-vRNAs, provided that N is available (Fig. 2.3).

Notably, the anti-genome promoter (AGP) of rhabdoviruses are much more active in directing replication than the genome promoter (GP), such that the overwhelming amount of full-length N-RNAs generated in infected cells comprises negative-strand vRNAs, which are useful for further transcription, replication, and assembly of virus progeny (Finke & Conzelmann, 1997).

Specifically, vRNAs make up 98% and 90% of full-length RNAs in RABV- and VSV- infected cells, respectively (Finke & Conzelmann, 1997, 1999; Whelan et al., 2004). In contrast, recombinant RABV in which both genome and antigenome RNAs have identical promoters derived from the AGP produce equal amounts of genome and antigenome N-RNAs. The viral AGP therefore seems to effectively outcompete the GP for replicase activity. This is most impressively illustrated by DI particles (Clark et al., 1981; Wiktor et al., 1977; Wunner & Clark, 1980). The genomes of VSV DI particles, which most efficiently interfere with the replication and propagation of VSV, belong to the so-called 5′ copy-back type (Marriott & Dimmock, 2010) (see Fig. 2.3). These are generated during RNA synthesis initiating from the viral AGP by dissociation of the polymerase from the template N-cRNA and backcopying of the nascent N-cRNA, and positive- and negative-strand RNAs have an identical 3′ terminal AGP promoter. In addition to directly interfering with replication of the helper virus, DI RNAs of MNV are also readily recognized by pattern receptors and stimulate an antiviral immune response (Pfaller, Donohue, Nersisyan, Brodsky, & Cattaneo, 2018; Sanchez-Aparicio et al., 2017).

2.3.5 Regulation of transcription and replication

The mechanisms and regulation of genome transcription and replication involved in the postulated switch between the two modes of RNA synthesis are still insufficiently known. Transcription is obviously required throughout infection, initially for providing the structural proteins for replication like N and P, and later for providing the proteins, M and G, for assembly of progeny virions. In fact, transcription remains the major mode of RNA synthesis throughout infection, even in the presence of high amounts of N and P.

The term "primary transcription" has been used to describe transcription from the incoming genome RNPs, when the amount of N produced is insufficient to support replication. According to the conventional view, transcription involves the synthesis and release of a positive strand 5′-PPP leader RNA and reinitiation of transcription at the N gene start signal. A switch from the transcription mode to the replication mode is thought to be caused by the presence of sufficient N-P or L-P-N complexes. For leader-dependent transcription, an intriguing model was proposed (Vidal & Kolakofsky, 1989) in which complete encapsidation of the nascent Le RNA prevents the polymerase from recognizing the transcription stop/start signal at the Le region-N gene junction. In this case, the polymerase would not release the leader RNA from the "leader RNA/N-mRNA" transcript and continue RNA encapsidation in the replication mode. In case of insufficient N-P availability, naked or incompletely encapsidated leader RNA would not prevent the polymerase from switching to the transcription mode. Once in the transcription mode and downstream of the Le RNA/N gene junction, the polymerase is thought not to be able convert back to the replicase mode. In this model, the synthesis of leader RNA might be regarded as abortive replication.

The alternative models of leader-independent transcription suggest the initiation of transcription by a transcriptase form of the polymerase. They suggest that the transcriptase either enters at the 3′ end of the genome and scans the GP without synthesizing a leader RNA until it encounters the start signal for N mRNA synthesis or that the transcriptase (in contrast to the replicase) can enter directly at the N start signal (for recent review see Noton et al., 2019). Any of these models would yield unencapsidated mRNAs and is compatible with the stop/restart model of MNV transcription. Whether the mechanisms are mutually exclusive, or may coexist, remains to be determined.

Interestingly, rather than the concentration of N alone, it is the level of M that seems to influence the mode of RABV RNA synthesis. Early in vitro work suggested that purified M inhibits transcription of RABV (Ito, Nishizono, Mannen, Hiramatsu, & Mifune, 1996). Reverse genetics experiments involving M gene-shifting in recombinant RABV or complementation of minigenomes with M confirmed an inhibitory effect on transcription while replication remained unaffected (Finke et al., 2003). Moreover, specific mutations introduced in the M resulted in viruses with a "high transcription" phenotype, which produced much more mRNAs relative to genome templates than the parental viruses (Finke & Conzelmann, 2003). The molecular mechanisms behind such models of transcription regulation are not yet resolved but may involve association of M with the template N-vRNA or different types of the polymerase complex.

2.3.6 Structural aspects of RABV RNA synthesis

Obviously, any RNA synthesis of RABV requires elaborate interplay of the proteins involved, namely, N, P, and L and their interaction with the vRNA genome. A better understanding of RABV RNA synthesis requires insight into the structure and dynamics of these protein interactions. Great progress has been achieved regarding RABV RNA synthesis by resolving the atomic structure of the RABV N-RNA complex, as template for RNA synthesis, but atomic details of how this template is used for RNA synthesis are limited so far. The N-vRNA structure was determined from short N-RNA rings that formed after expression of recombinant N^o and encapsidation of cellular RNA in cells expressing N^o in the absence of other RABV proteins (reviewed in Albertini et al., 2011). The N^o forms two domains with a positively charged cleft in which 9 nts of the RNA are buried. Terminal extensions of the protein extend to neighboring protomers, thereby stabilizing the N-RNA complex. Binding of P to N^o (see Section 2.2.3.2) probably shields the N-RNA cleft and the terminal N extensions to prevent unspecific RNA binding and aggregation of N^o. Simultaneous binding of P to L via its N-terminal domain and to N-RNA via the C-terminal domain is probably involved in bringing the L in contact with the N-RNA template. The contact to N-RNA is probably mediated via the C-terminal lobe of N^o (Schoehn et al., 2001), in a way that P covers two N^o-protomers, resulting in a 1:2 ratio of P:N in the artificial rings (Ribeiro et al., 2008, 2009). In the infected cell, binding of P to N-vRNA is enhanced by phosphorylation of N at Ser389 (Toriumi & Kawai, 2004). It is assumed that several N protomers might dissociate locally to provide sufficient space for the large RNA polymerase (L) molecule to bind.

2.4 Life cycle of rabies virus infection

The sequence of events in the RABV life cycle, that is, replication in vitro and in vivo (in cell culture or animal) can be roughly divided into three phases. The first or early phase includes virus attachment to receptors on susceptible host cells, entry via direct virus fusion externally with the plasma membrane or internally with endosomal membranes of the cell, and uncoating of virus particles and liberation of the helical RNP in the cytoplasm. The second or middle phase includes transcription and replication of the viral genome and viral protein synthesis, and the third or late phase includes virus assembly and egress from the infected cell. The early phase of the RABV life cycle has been studied in many different cell culture systems. These include neuronal and non-neuronal cell lines and primary, dissociated cell cultures derived from dissected pieces of nervous tissue. One caveat that overshadows the use of experimental cell culture systems is that the cells appear to behave differently ex vivo in their susceptibility for RABV infection compared with their susceptibility to infection in vivo. That is, once cells are removed from their in vivo environment, particularly neuronal cells, they can lose their natural control over susceptibility or resistance to RABV infection. Nevertheless, many studies using in vitro cell culture systems describe how virus enters the host cell by direct fusion with the plasma and endosomal membranes (Iwasaki, Wiktor, & Koprowsk, 1973; Le Blanc et al., 2005) or by receptor-mediated endocytosis (Hummeler et al., 1967; Iwasaki et al., 1973; Lycke & Tsiang, 1987; Piccinotti & Whelan, 2016; Superti, Derer, & Tsiang, 1984; Tsiang, Delaporte, Ambroise, Derer, & Koenig, 1986). No in vitro system has yet provided a detailed explanation of how RABV enters muscle cells in vivo to support the experimental infections in hamster (Murphy & Bauer, 1974; Murphy, Bauer, Harrison, & Winn, 1973) and skunk (Charlton & Casey, 1979) that show virus replication in striated muscle cells near the site of inoculation.

2.4.1 Early-phase events: Role of the RABV receptor, endocytosis, and G-mediated fusion

RABV infection starts with virus attachment to the surface of a target cell and penetration into the cell through an endosomal transport pathway (Le Blanc et al., 2005; Piccinotti & Whelan, 2016). Most likely, the virus attaches itself to a "receptor" molecule or cellular receptor unit (CRU) on the cell surface that leads to or permits direct virus entry into susceptible cells in culture (in vitro) or specific target cells at the site of inoculation (in vivo) (Fig. 2.4). Studies using various cell culture systems have implicated various lipids, gangliosides, carbohydrate, and protein of the plasma membrane in RABV binding to cells in culture (Broughan & Wunner, 1995; Conti et al., 1988; Conti, Superti, & Tsiang, 1986; Perrin, Portnoi, & Sureau, 1982; Superti et al., 1986; Superti, Derer, & Tsiang, 1984; Wunner, Reagan, & Koprowski, 1984), but none have proven to be "specific" receptors. Others have focused on specific cell receptor units (CRUs) in vivo that appear to correlate with the defined neurotropism of the virus (Lafon, 2005); see also Chapter 9. One of these CRUs is the nicotinic acetylcholine receptor (nAChR) found at neuromuscular junctions, where RABV can also be found colocalized in situ (Lewis, Fu, & Lentz, 2000). Not all cell lines infected with RABV in vitro, however, express the nAChR (Reagan & Wunner, 1985; Tsiang, 1993; Tsiang et al., 1986) and some neuronal cells infected with rabies virus in vivo may not express the nAChR (Kucera et al., 1985; Lafay et al., 1991; Tsiang et al., 1986). Two other possibilities are the neural cell adhesion molecule (NCAM) CD56 on the cell surface of RABV-susceptible cell lines (Thoulouze et al., 1998) and the low-affinity p75 neurotrophin receptor (p75NTR), a nerve growth factor (Langevin & Tuffereau, 2002; Tuffereau, Benejean, Blondel, Kieffer, & Flamand,

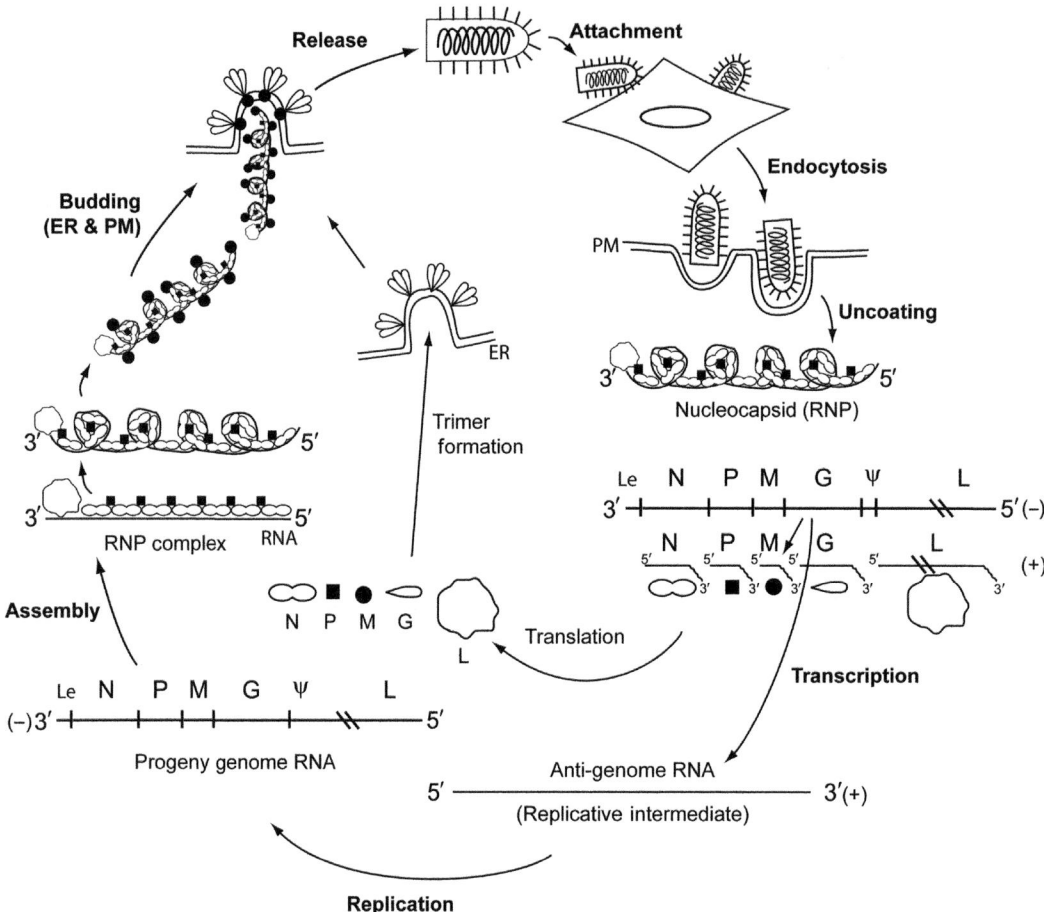

FIG. 2.4 Rabies virus life cycle in the cell. Virus enters the cell following attachment through coated pits (viropexis) or via cell surface receptors, mediated by the viral glycoprotein (G) fusing with the cellular membrane (endocytosis). After internalization, the viral G mediates low pH-dependent fusion with the endosomal membrane and the virus is uncoated, releasing the helical nucleocapsid (NC) of the ribonucleoprotein (RNP) core. The five structural genes (N, P, M, G, and L) of the genome RNA (vRNA) in the NC are transcribed into five positive (+) strand monocistronic messenger (m)RNAs and a full-length + strand (anti-genome) replicative intermediate RNA (cRNA). The antigenome cRNA serves as the template for replication of progeny genome (− strand) vRNA. The proteins (N, P, M, and L) are synthesized from their respective mRNAs on membrane-free ribosomes in the cytoplasm and the G is synthesized from the G-mRNA on membrane-bound ribosomes (rough endoplasmic reticulum). Some of the N-P molecular complexes produce cytoplasmic inclusion bodies (Negri bodies) in vivo and some N-P complexes encapsidate the + strand and − strand viral RNAs. After progeny genome vRNA is encapsidated by N-P protein complex, and L is incorporated to form progeny RNP (both full-length standard and shorter defective) structures, the M binds to the RNP and condenses the RNP into the "skeleton" structures. The skeleton structures interact with the trimeric G structures anchored in the plasma membrane and assemble into virus particles (virions) that bud from the plasma membrane of the infected cell into adjacent extracellular or interstitial space.

1998). More recently, mGluR2 has been added to the list of CRUs (Wang et al., 2018). In any case, the lack of the proteins described does not prevent RABV entry, suggesting the possibility that the virus is not limited to choosing only one type of receptor in order to initiate the virus life cycle in the infected animal.

After targeted binding of virus to its receptor(s) on host cells, virus is internalized by endocytosis RABV, like VSV, and may also enter the cell through coated pits and uncoated vesicles (viropexis or pinocytosis)(Piccinotti & Whelan, 2016; Tsiang, Derer, & Taxi, 1983). As part of the internalization process, fusion between the viral and endosomal membranes is activated in the acidic environment (pH 6.3–6.5) within the endosomal compartment. At the threshold pH for fusion activation, a series of specific and discrete conformational changes in RABV G takes place whereby it assumes at least three structurally distinct "conformational" states; for review see Albertini et al. (2012).

Prior to binding to the cellular receptor, the G on the virion surface is in its "native" state. After the virus attaches to the receptor and is internalized, the G is "activated" to a hydrophobic state, which enables it to interact with the hydrophobic endosomal membrane. Upon entering the endosomal compartment and low pH environment of the cellular compartment, the fusion capacity of the G is activated via a major structural change in the G that exposes the fusion domain, which interacts with the target cell membrane. By further rearrangements in the G, the viral and cell membrane are brought into close vicinity such that (via hemifusion) a fusion pore may develop. For review, see Albertini et al. (2012). The functional state of the RABV G in which it acquires fusogenic activity is correlated with at least one specific conformational epitope. This epitope, which appears to be formed by combining two separate regions, the neurotoxin-like region (residues 189–214) of RABV G and the conformational antigenic site III (residues 330–340) is abrogated when the G is exposed to acidic conditions (Kankanamge, Irie, Mannen, Tochikura, & Kawai, 2003; Sakai et al., 2004). The transition from virus interaction to generating the fusion pore is a high-cost energy step and depends on the integrity and correct folding of the G trimers directly involved in the fusion process. It has been shown that more than one trimer of G is required to build a competent fusion site (Gaudin et al., 1993).

After low-pH fusion, the G assumes a reversible "fusion-inactive" conformation, which makes the G monomer appear longer than the "native" conformation and assume selective antigenic distinctions (Gaudin et al., 1993; Kankanamge et al., 2003). The fusion-inactivated G, which is no longer relevant to the fusion process is highly sensitive to cellular proteases and appears to be in a dynamic equilibrium with the "native" G that is regulated by lowering and raising the pH (Gaudin, Raux, Flamand, & Ruigrok, 1996; Gaudin, Tuffereau, Segretain, Knossow, & Flamand, 1991). Interestingly, the fusion-inactive conformation serves the G in another capacity. During nascent viral protein synthesis, the G assumes an "inactive state"-like conformation, protecting the G post-translationally from fusing with the acid nature of Golgi vesicles, while it is transported through the Golgi stacks to the cell surface. At the cell surface, the G acquires its "native" conformation and structure (Gaudin et al., 1999; Gaudin, Ruigrok, & Brunner, 1995; Gaudin, Tuffereau, Durrer, Flamand, & Ruigrok, 1995). The MAbs that recognize specific low pH-sensitive conformational epitopes of the G can identify certain acid-induced conformational changes, as well as detect the various stages of nascent G monomer folding and its association with molecular chaperones like BiP and calnexin (Gaudin, 1997; Kankanamge et al., 2003; Maillard & Gaudin, 2002).

2.4.2 Middle-phase events: Transcription, replication, and nascent protein synthesis

2.4.2.1 Viral RNP release, initiation of genome RNA transcription, replication, and protein synthesis

In the second phase of the RABV life cycle, vRNA genome transcription is initiated in the cytoplasm of the infected cell after the tightly coiled transcriptionally "frozen" RNP core is released from endosomal vesicles (Fig. 2.4). With the dissociation of M from the RNP during the "uncoating" process, the tightly coiled RNP structure relaxes to form a loosely coiled helix (Iseni et al., 1998), conceivably to facilitate the ensuing vRNA transcription and replication events in the cell as described in detail in Section 2.3.3. The incoming RNP serves as a template for RNA synthesis by approximately 50 RNP-associated RNA polymerase (L-P) complexes. The location of these complexes on the RNP is not known. It is assumed that the polymerase either initiates transcription at the 3′ end of the genome RNA, or it resumes transcription at the next downstream internal mRNA start site on the viral genome, close to where the polymerase complex was "frozen" in place during progeny virus assembly in a previously infected cell. This part of the replication process has been termed "primary transcription" (see Section 2.3.5). According to the stop/restart model of transcription (see Section 2.3.3), 5′-capped and polyadenylated monocistronic mRNA transcripts are sequentially produced from the genes and eventually are translated into one of the viral proteins (Flamand & Delagneau, 1978; Holloway & Obijeski, 1980). Re-entry of polymerase released after transcription of the most downstream L gene, or of newly produced polymerase, leads to the typical transcript gradient in which the mRNAs of 3′ proximal genes are transcribed more abundantly than those of distal genes.

The proteins of the virus are synthesized from the viral mRNAs using the protein synthesis machinery of the host cell. The G-mRNA is translated on membrane-bound polyribosomes (polysomes) and inserted cotranslationally into the lumen of the ER where disulfide bond formation occurs and the molecular chaperones are available to assist in the folding of G monomers before the molecule is transported out of the ER (Gaudin, 1997). While in the lumen of the ER, the G monomers undergo modification at specific asparagine (N) residues by core glycosylation and N-glycan processing (see Section 2.2.3.5) and form homotrimers (Gaudin et al., 1992; Whitt et al., 1991). The final processing of the N-linked carbohydrate side chains takes place in the Golgi apparatus of the intracellular membrane network. The other four viral mRNAs (N-, P-, M-, and L-mRNA) are translated on "free" polysomes in the cytoplasm. As the M accumulates in the cytoplasm of the infected cell, it is likely to specifically interact with the eukaryotic translation initiation factor, eIF3h, and inhibit translation of cellular (host) mRNAs that have a Kozak-like 5′-UTR and seize control of the host translational machinery in infected cells for the translation of viral mRNAs (Komarova et al., 2007). Accumulation of other viral proteins, specifically N and P, allows replication to initiate, which involves N-encapsidation-dependent synthesis of full-length antigenome N-cRNAs. These serve as templates for amplification of genome N-vRNAs, which can serve in (secondary) transcription and protein expression.

Accumulation of proteins leads to formation of cytoplasmic IBs, which increase in size over time (Lahaye et al., 2009; Nikolic et al., 2017). These IBs are known as NBs in neurons (Kristensson, Dastur, Manghani, Tsiang, & Bentivoglio, 1996), which are readily detectable by staining, and have been used as diagnostic markers for RABV infection. The IBs are viral factories with properties of liquid organelles. Their assembly is driven by intrinsically ordered domains and the dimerization domain of the P protein and association with N protein (Nikolic et al., 2017). The IBs contain the entire replication machinery (including N, P, and L proteins, vRNA, cRNA, and mRNAs, as well as M protein, and a couple of cellular proteins, including HSP70 and FAK, which interact with N and P proteins, respectively) (Fouquet et al., 2015; Lahaye et al., 2009; Pollin, Granzow, Kollner, Conzelmann, & Finke, 2013; Sagara & Kawai, 1992). Live-cell imaging indicates that viral nucleocapsids are ejected from IBs and transported along microtubules to form either new virions or secondary viral factories (Nikolic et al., 2017).

Viral mRNAs are always much more abundant than genome and antigenome RNPs, suggesting that transcription of mRNAs remains the major mode of RNA synthesis throughout infection. While initially the proteins produced are required for vRNA replication, at later stages they are needed in abundant amounts for assembly of progeny virus. Genome N-vRNA templates, which must serve both processes, are produced at higher levels than antigenome cRNAs. Quantification with strand-specific probes revealed a 50-fold excess of genomic vRNAs over antigenome cRNAs. This biased replication could be attributed to a stronger activity of the GP, which successfully competes with the AGP for polymerase (Finke & Conzelmann, 1997).

2.4.3 Late phase: Assembly and budding of progeny virus

The process of virus assembly (RABV morphogenesis) actually begins already in the middle phase of the life cycle with encapsidation of vRNA and formation of vRNA-N-P, i.e., RNP, which requires accumulation of sufficient pools of viral N, P, and L to support RNP formation (Iseni et al., 1998; Liu et al., 2004; Mavrakis et al., 2003). RABV morphogenesis seems to be associated with the formation of the IBs (NBs) as sites of virus assembly commonly found particularly in neurons in brain tissue, as well as in tissue culture (Hummeler et al., 1967; Hummeler, Tomassini, Sokol, Kuwert, & Koprowski, 1968; Matsumoto, 1962; Matsumoto & Miyamoto, 1966; Nikolic et al., 2017); see Chapter 10. Assembly of progeny RNP and virions continues into the late phase of the life cycle as long as the cells remain metabolically active.

The M is the next viral protein to associate with RNP complexes. From the time M enters the virus assembly pathway as a soluble protein in the cytoplasm, it is involved in all steps that lead to virus budding. M is a multifunctional protein and it plays several key roles in the dynamics of the formation of progeny virus. First, the association of M with newly formed transcriptionally active RNP changes the balance between viral RNA transcription and replication by inhibiting transcription and stimulating replication (Finke et al., 2003; Ito et al., 1996). M imparts this differential effect in RNA synthesis on encapsidated vRNA either by binding to the vRNA+N template or the L+P polymerase in the complex (Finke et al., 2003). It is conceivable that the different regulatory functions of M in RNA synthesis may also be provided by the different conformational forms of the monomer M, Mα, and Mβ (see Section 2.2.3.4). The M will then localize the RNP coil at the cellular membrane or internal membranes where the glycosylated trimeric G is concentrated and where the M is able to interact with G (Mebatsion et al., 1999). Finally, the M alters the structure of the RNP by condensing it into a tightly coiled "skeleton-like" form in which the polymerase activity is "frozen." In this function of M, which may involve M as a dimer, M initiates virus budding at cellular membranes where they become enveloped by the cellular membrane and enter the virus budding process (Mebatsion et al., 1996, 1999; Nakahara et al., 2003). In the mature rabies virion that buds from the cell membrane, the M lies between the lipid bilayer envelope (formed by interaction with the host cell membrane) and the helical RNP that it covers (Mebatsion et al., 1999). Thus, the M covering and condensing the helical RNP is thought to play an important role in virion morphogenesis, i.e., giving the particle its bullet-shape morphology. If the M is missing from RABV particles that contain G spikes on the surface, a morphological variation in budded particles is observed suggesting that the particles contain uncondensed RNP. M-deficient RABV also causes increased cell-cell fusion and enhanced cell death, in contrast to the relatively benign cytopathic effect that is observed with wild-type virus. While assembly and budding of bullet-shaped particles can occur in the absence of the transmembrane G spikes that are normally associated with infectious particles (Mebatsion et al., 1996, 1999; Robison & Whitt, 2000), more significantly, virus budding is much less efficient in the absence of M.

In the final stages of RABV assembly, the mature virions acquire their lipid bilayer envelope as the assembled skeleton (RNP+M) structure buds through the host cell plasma membrane. Mature virions that bud through the plasma membrane into the extracellular space are frequently observed in extraneural tissue cells in vivo and in a variety of in vitro tissue culture systems (Davies et al., 1963; Hummeler et al., 1967; Iwasaki et al., 1973; Matsumot & Kawai, 1969; Matsumoto, Schneider, Kawai, & Yonezawa, 1974; Tsiang et al., 1983). However, virions can also mature intracellularly by budding through the cytoplasmic ER or Golgi apparatus as is often observed in infected neuronal cells of brain (Matsumoto, 1975; Matsumoto et al., 1974). RABV budding at internal membranes was also observed after intergenotypic replacement of lyssavirus matrix proteins, which demonstrates the central role of lyssavirus M proteins in virus assembly (Finke, Granzow, Hurst, Pollin, & Mettenleiter, 2010).

If virion budding occurs at a site in the cell membrane (or cytoplasmic ER) where the glycosylated trimeric RABV transmembrane G is also targeted, then infectious virions will be produced bearing the G molecules arranged as trimeric spike-like structures tightly packed and anchored in the viral envelope (Whitt et al., 1991). The C-terminal tail of the G molecule is then free to interact with the M of the RNP+M skeleton structure. Interaction of G with M is essential for stabilization of the G in trimers on the virion surface and for efficient budding of RABV (Mebatsion et al., 1996, 1999). This does not exclude the possibility that skeleton structures may bud from membrane regions where no G exists (Mebatsion et al., 1996, 1999; Robison & Whitt, 2000). In this case, budding would be inefficient, producing low levels, and the bullet-shaped particles produced would be spikeless (free of G) and noninfectious (Mebatsion et al., 1996). If skeletons bud through ER or Golgi membranes, they bud into the lumen of vesicles produced from these membranes and may be secreted from the cell through the normal secretory pathway.

2.4.4 Cell-to-cell spread (transport) of progeny virus

As most studies on RABV cell biology, biochemistry, and reverse genetics have been performed in nonpolarized cell culture systems, the results of these studies do not necessarily reflect the natural behavior of wt RABV or "street" isolates from animals. In fact, RABV street isolates require several passages for adaptation. The changes required to yield so-called fixed RABV, or RABV strains, are poorly defined and may affect receptor usage, replication, and intracellular transport and budding (Dietzschold, Li, Faber, & Schnell, 2008).

Wild-type RABV is prone to quickly enter the immune privileged nervous system and has developed multiple traits contributing to this "stealth" behavior (Schnell, McGettigan, Wirblich, & Papaneri, 2010). These include neurotropism, effective axonal transport mechanisms, tools to dampen innate and adaptive immune responses and to prevent premature neuronal damage, as well as ways to interfere with migration of immune cells through the blood-brain barrier.

In any case, successful infection of cells and spread in vitro and in vivo requires the G protein (Etessami et al., 2000), and the G protein determines the phenotypic differences of the viruses with respect to neuroinvasiveness and virulence as discussed further in Section 2.2.3.5.

RABV, via its G protein, can infect many cells in vitro and in vivo, including muscle cells or, after direct injection into the brain, astrocytes, oligodendrocytes, and neurons. This is probably reflecting the broad receptor usage described earlier. Infection of astrocytes, the most abundant glial cells in the CNS, can lead to abortive or productive infection dependent on the strain used (Pfefferkorn et al., 2016; Tian et al., 2017). While non-neuronal cells apparently can release progeny virus, probably as described earlier, the situation in neurons is different. Once in a neuron, RABV seems to be transmitted exclusively to presynaptic neurons, but the molecular mechanisms behind this specificity are not understood. Budding of cell-free virus from the postsynaptic membranes of infected cells and particle uptake by a presynaptic axon terminal may be involved, as suggested from early electron microscopy examinations in vivo (Charlton & Casey, 1979). The selective presence at presynaptic nerve terminals of the described or yet unknown RABV receptors, is not sufficient as a rational explanation for the specificity observed. Additional factors, like preferential release of virus at postsynaptic membrane sites, possibly contribute, though in cell lines RABV budding is not restricted to specialized membranes.

The exclusive retrograde trans-synaptic transmission of RABV between neurons has been utilized for some time by using replication competent RABV as tracer for neuronal connections (Astic, Saucier, Coulon, Lafay, & Flamand, 1993; Ugolini, 1995). However, in this approach, the differentiation of 1st-, 2nd-, and further order neurons is difficult, due to stochastic rather than simultaneous transmission over synapses. The more recent development of genetically engineered "monosynaptic" RABV tracers, which allows interpretations to unambiguously identify 1st-order presynaptic neurons from a genetically defined starter neuron (Wickersham et al., 2007), has now been established as a gold standard in neuronal connection tracing. The system employs G-gene deficient RABV (ΔG RABV), which are pseudotyped with an avian sarcoma leucosis virus envelope protein (ASLV-EnvA) to allow selective entry into neurons, which express the avian receptor TVA (tumor virus A). If these "starter cells" express in addition the RABV G protein, the virus will be able to cross one synapse and then halt transmission because of the lack of the G gene (Callaway & Luo, 2015; Ghanem & Conzelmann, 2016). This system is highly specific as, almost exclusively, neurons are infected from the starter cell, and, notably, specific hippocampal astrocytes, which are known to form so-called tripartite synapses with neurons, further illustrating the exclusive trans-synaptic transmission of RABV from a neuron.

Cell-free RABV infects neurons predominantly at axon ends and, within intra-axonal retrograde transport vesicles carried along microtubules, can travel long distances within a neuron before membrane fusion and RNP release occurs (Ceccaldi, Gillet, & Tsiang, 1989; Coulon et al., 1989; Gillet, Derer, & Tsiang, 1986; Klingen, Conzelmann, & Finke, 2008; Kucera et al., 1985; Lafay et al., 1991) (see Chapter 9). An important role could be attributed to the interaction of the virus with p75NTR, one of the reported RABV receptors. Notably, axonal transport of p75NTR-positive vesicles was shown to be significantly faster when loaded with RABV particles, suggesting RABV not only hijacks the axonal transport machinery but can also manipulate it to get faster access to the CNS (Gluska et al., 2014).

After acidification of axonal transport vesicles, membrane fusion at low pH and release of RNPs, the virus replicates in NBs, which are mostly located at the neuronal cell body. However, the late steps of virus transport in neurons, including transport of new viruses or of individual virus components to dendrites in the anterograde direction, and release at postsynaptic areas, are completely obscure. Further research on this late phase is essential to gain insight into the mechanisms of spread and to devise ways for therapeutic intervention.

It is well established that the spread of cell-free virus particles can be limited by the presence, in vitro and in vivo, of VNAs that block virus attachment to cellular receptors, fusion and subsequent virus entry into a susceptible cell (Dietzschold et al., 1985; Flamand et al., 1993). However, in the case of direct cell-to-cell spread, RABV spreads despite a continuous presence of serum VNA and is able to survive, at least, the humoral arm of the immune protection mechanisms during infection (Dietzschold et al., 1985). Rabies post-exposure prophylaxis with VNA has proven reliable if applied soon after exposure, possibly by capturing (neutralizing) most of the free viruses before they have entered neurons. Whether RABV-infected neurons can be destroyed by the cellular arm of the immune system, is questionable, perhaps unknown, but cannot be excluded in view of successful PEP.

References

Albertini, A. A., Baquero, E., Ferlin, A., & Gaudin, Y. (2012). Molecular and cellular aspects of rhabdovirus entry. *Viruses, 4*(1), 117–139. https://doi.org/10.3390/v4010117.

Albertini, A. A., Wernimont, A. K., Muziol, T., Ravelli, R. B., Clapier, C. R., Schoehn, G., ... Ruigrok, R. W. H. (2006). Crystal structure of the rabies virus nucleoprotein-RNA complex. *Science, 313*(5785), 360–363.

Albertini, A. A. V., Clapier, C. R., Wernimont, A. K., Schoehn, G., Weissenhorn, W., & Ruigrok, R. W. H. (2007). Isolation and crystallization of a unique size category of recombinant rabies virus nucleoprotein-RNA rings. *Journal of Structural Biology, 158*(1), 129–133. https://doi.org/10.1016/J.Jsb.2006.10.011.

Albertini, A. A. V., Ruigrok, R. W. H., & Blondel, D. (2011). Rabies virus transcription and replication. *Advances in Virus Research: Research Advances in Rabies, 79*, 1–22. https://doi.org/10.1016/B978-0-12-387040-7.00001-9.

Albertini, A. A. V., Schoehn, G., Weissenhorn, W., & Ruigrok, R. W. H. (2008). Structural aspects of rabies virus replication. *Cellular and Molecular Life Sciences, 65*(2), 282–294. https://doi.org/10.1007/S00018-007-7298-1.

Albisetti, G. W., Ghanem, A., Foster, E., Conzelmann, K. K., Zeilhofer, H. U., & Wildner, H. (2017). Identification of two classes of somatosensory neurons that display resistance to retrograde infection by rabies virus. *The Journal of Neuroscience, 37*(43), 10358–10371. https://doi.org/10.1523/jneurosci.1277-17.2017.

Anzai, J., Takamatsu, F., Takeuchi, K., Kohno, T., Morimoto, K., Goto, H., ... Kawai, A. (1997). Identification of a phosphatase-sensitive epitope of rabies virus nucleoprotein which is recognized by a monoclonal antibody 5-2-26. *Microbiology and Immunology, 41*(3), 229–240.

Astic, L., Saucier, D., Coulon, P., Lafay, F., & Flamand, A. (1993). The CVS strain of rabies virus as transneuronal tracer in the olfactory system of mice. *Brain Research, 619*(1-2), 146–156.

References

Ball, L. A., Pringle, C. R., Flanagan, B., Perepelitsa, V. P., & Wertz, G. W. (1999). Phenotypic consequences of rearranging the P, M, and G genes of vesicular stomatitis virus. *Journal of Virology, 73*(6), 4705–4712.

Banerjee, A. K., & Chattopadhyay, D. (1990). Structure and function of the RNA polymerase of vesicular stomatitis virus. *Advances in Virus Research, 38*, 99–124.

Banyard, A. C., Hayman, D., Johnson, N., McElhinney, L., & Fooks, A. R. (2011). Bats and lyssaviruses. *Advances in Virus Research: Research Advances in Rabies, 79*, 239–289. https://doi.org/10.1016/B978-0-12-387040-7.00012-3.

Barik, S., Rud, E. W., Luk, D., Banerjee, A. K., & Kang, C. Y. (1990). Nucleotide sequence analysis of the L gene of vesicular stomatitis virus (New Jersey serotype): identification of conserved domains in L proteins of nonsegmented negative-strand RNA viruses. *Virology, 175*(1), 332–337.

Barr, J. N., Whelan, S. P. J., & Wertz, G. W. (1997). cis-Acting signals involved in termination of vesicular stomatitis virus mRNA synthesis include the conserved AUAC and the U7 signal for polyadenylation. *Journal of Virology, 71*(11), 8718–8725.

Benmansour, A., Brahimi, M., Tuffereau, C., Coulon, P., Lafay, F., & Flamand, A. (1992). Rapid sequence evolution of street rabies glycoprotein is related to the highly heterogeneous nature of the viral population. *Virology, 187*(1), 33–45.

Blondel, D., Maarifi, G., Nisole, S., & Chelbi-Alix, M. K. (2015). Resistance to rhabdoviridae infection and subversion of antiviral responses. *Viruses, 7*(7), 3675–3702. https://doi.org/10.3390/v7072794.

Blumberg, B. M., Giorgi, C., & Kolakofsky, D. (1983). N-Protein of vesicular stomatitis-virus selectively encapsidates leader RNA in vitro. *Cell, 32*(2), 559–567.

Both, G. W., Moyer, S. A., & Banerjee, A. K. (1975). Translation and identification of the mRNA species synthesized in vitro by the virion-associated RNA polymerase of vesicular stomatitis virus. *Proceedings of the National Academy of Sciences of the United States of America, 72*(1), 274–278.

Bourhy, H., Kissi, B., Audry, L., Smreczak, M., Sadkowska-Todys, M., Kulonen, K., ... Holmes, E. C. (1999). Ecology and evolution of rabies virus in Europe. *Journal of General Virology, 80*(Pt 10), 2545–2557.

Bourhy, H., Kissi, B., & Tordo, N. (1993). Molecular diversity of the Lyssavirus genus. *Virology, 194*(1), 70–81. S0042-6822(83)71236-5 [pii]. https://doi.org/10.1006/viro.1993.1236.

Broughan, J. H., & Wunner, W. H. (1995). Characterization of protein involvement in rabies virus binding to Bhk-21-cells. *Archives of Virology, 140*(1), 75–93.

Brzózka, K., Finke, S., & Conzelmann, K. K. (2005). Identification of the rabies virus alpha/beta interferon antagonist: Phosphoprotein P interferes with phosphorylation of interferon regulatory factor 3. *Journal of Virology, 79*(12), 7673–7681. https://doi.org/10.1128/JVI.79.12.7673-7681.2005.

Brzózka, K., Finke, S., & Conzelmann, K. K. (2006). Inhibition of interferon signaling by rabies virus phosphoprotein P: Activation-dependent binding of STAT1 and STAT2. *Journal of Virology, 80*(6), 2675–2683.

Callaway, E. M., & Luo, L. (2015). Monosynaptic circuit tracing with glycoprotein-deleted rabies viruses. *The Journal of Neuroscience, 35*(24), 8979–8985. https://doi.org/10.1523/jneurosci.0409-15.2015.

Canter, D. M., Jackson, R. L., & Perrault, J. (1993). Faithful and efficient in vitro reconstitution of vesicular stomatitis virus transcription using plasmid-encoded L and P proteins. *Virology, 194*(2), 518–529. S0042-6822(83)71290-0 [pii] https://doi.org/10.1006/viro.1993.1290.

Castel, G., Chteoui, M., Caignard, G., Prehaud, C., Mehouas, S., Real, E., ... Tordo, N. (2009). Peptides that mimic the amino-terminal end of the rabies virus phosphoprotein have antiviral activity. *Journal of Virology, 83*(20), 10808–10820. https://doi.org/10.1128/Jvi.00977-09.

Ceccaldi, P. E., Gillet, J. P., & Tsiang, H. (1989). Inhibition of the transport of rabies virus in the central nervous system. *Journal of Neuropathology and Experimental Neurology, 48*(6), 620–630.

Celis, E., Karr, R. W., Dietzschold, B., Wunner, W. H., & Koprowski, H. (1988). Genetic restriction and fine specificity of human T cell clones reactive with rabies virus. *Journal of Immunology, 141*(8), 2721–2728.

Celis, E., Ou, D., Dietzschold, B., & Koprowski, H. (1988). Recognition of rabies and rabies-related viruses by T cells derived from human vaccine recipients. *Journal of Virology, 62*(9), 3128–3134.

Charlton, K. M., & Casey, G. A. (1979). Experimental rabies in skunks: Immunofluorescence light and electron microscopic studies. *Laboratory Investigation, 41*(1), 36–44.

Chenik, M., Chebli, K., Gaudin, Y., & Blondel, D. (1994). In-vivo interaction of rabies virus phosphoprotein (P) and nucleoprotein (N)—Existence of 2 N-binding sites on P-protein. *Journal of General Virology, 75*, 2889–2896.

Chenik, M., Schnell, M., Conzelmann, K. K., & Blondel, D. (1998). Mapping the interacting domains between the rabies virus polymerase and phosphoprotein. *Journal of Virology, 72*(3), 1925–1930.

Clark, H. F., Parks, N. F., & Wunner, W. H. (1981). Defective interfering particles of fixed rabies viruses: Lack of correlation with attenuation or auto-interference in mice. *Journal of General Virology, 52*(Pt 2), 245–258.

Colonno, R. J., & Banerjee, A. K. (1978). Complete nucleotide-sequence of leader RNA synthesized in vitro by vesicular stomatitis-virus. *Cell, 15*(1), 93–101.

Conti, C., Hauttecoeur, B., Morelec, M. J., Bizzini, B., Orsi, N., & Tsiang, H. (1988). Inhibition of rabies virus-infection by a soluble membrane-fraction from the rat central nervous-system. *Archives of Virology, 98*(1-2), 73–86.

Conti, C., Superti, F., & Tsiang, H. (1986). Membrane carbohydrate requirement for rabies virus binding to chicken-embryo related cells. *Intervirology, 26*(3), 164–168.

Conzelmann, K. K. (2004). Reverse genetics of mononegavirales. *Current Topics in Microbiology and Immunology, 283*, 1–41.

Conzelmann, K. K., Cox, J. H., Schneider, L. G., & Thiel, H. J. (1990). Molecular cloning and complete nucleotide sequence of the attenuated rabies virus SAD B19. *Virology, 175*(2), 485–499.

Conzelmann, K. K., Cox, J. H., & Thiel, H. J. (1991). An L (polymerase)-deficient rabies virus defective interfering particle RNA is replicated and transcribed by heterologous helper virus L-proteins. *Virology, 184*(2), 655–663.

Coulon, P., Derbin, C., Kucera, P., Lafay, F., Prehaud, C., & Flamand, A. (1989). Invasion of the peripheral nervous systems of adult mice by the CVS strain of rabies virus and its avirulent derivative AvO1. *Journal of Virology, 63*(8), 3550–3554.

Coulon, P., Ternaux, J. P., Flamand, A., & Tuffereau, C. (1998). An avirulent mutant of rabies virus is unable to infect motoneurons in vivo and in vitro. *Journal of Virology, 72*(1), 273–278.

Davies, M. C., Englert, M. E., Sharpless, G. R., & Cabasso, V. J. (1963). The electron microscopy of rabies virus in cultures of chicken embryo tissues. *Virology, 21*, 642–651.

Delarue, M., Poch, O., Tordo, N., Moras, D., & Argos, P. (1990). An attempt to unify the structure of polymerases. *Protein Engineering, 3*(6), 461–467.

Dietzschold, B., Lafon, M., Wang, H., Otvos, L., Jr., Celis, E., Wunner, W. H., & Koprowski, H. (1987). Localization and immunological characterization of antigenic domains of the rabies virus internal N and NS proteins. *Virus Research, 8*(2), 103–125. https://doi.org/10.1016/0168-1702(87)90023-2.

Dietzschold, B., Li, J., Faber, M., & Schnell, M. (2008). Concepts in the pathogenesis of rabies. *Future Virology, 3*(5), 481–490.

Dietzschold, B., Wang, H. H., Rupprecht, C. E., Celis, E., Tollis, M., Ertl, H., … Koprowski, H. (1987). Induction of protective immunity against rabies by immunization with rabies virus ribonucleoprotein. *Proceedings of the National Academy of Sciences of the United States of America, 84*(24), 9165–9169.

Dietzschold, B., Wiktor, T. J., Trojanowski, J. Q., Macfarlan, R. I., Wunner, W. H., Torresanjel, M. J., & Koprowski, H. (1985). Differences in cell-to-cell spread of pathogenic and apathogenic rabies virus in vivo and in vitro. *Journal of Virology, 56*(1), 12–18.

Dietzschold, B., Wiktor, T. J., Wunner, W. H., & Varrichio, A. (1983). Chemical and immunological analysis of the rabies soluble glycoprotein. *Virology, 124*(2), 330–337.

Dietzschold, B., Wunner, W. H., Wiktor, T. J., Lopes, A. D., Lafon, M., Smith, C. L., & Koprowski, H. (1983). Characterization of an antigenic determinant of the glycoprotein that correlates with pathogenicity of rabies virus. *Proceedings of the National Academy of Sciences of the United States of America, 80*(1), 70–74.

Ding, H. T., Green, T. J., Lu, S. Y., & Luo, M. (2006). Crystal structure of the oligomerization domain of the phosphoprotein of vesicular stomatitis virus. *Journal of Virology, 80*(6), 2808–2814. https://doi.org/10.1128/Jvi.80.6.2808-2814.2006.

Ertl, H. C., Dietzschold, B., Gore, M., Otvos, L., Jr., Larson, J. K., Wunner, W. H., & Koprowski, H. (1989). Induction of rabies virus-specific T-helper cells by synthetic peptides that carry dominant T-helper cell epitopes of the viral ribonucleoprotein. *Journal of Virology, 63*(7), 2885–2892.

Etessami, R., Conzelmann, K. K., Fadai-Ghotbi, B., Natelson, B., Tsiang, H., & Ceccaldi, P. E. (2000). Spread and pathogenic characteristics of a G-deficient rabies virus recombinant: An in vitro and in vivo study. *Journal of General Virology, 81*, 2147–2153.

Faber, M., Faber, M. L., Papaneri, A., Bette, M., Weihe, E., Dietzschold, B., & Schnell, M. J. (2005). A single amino acid change in rabies virus glycoprotein increases virus spread and enhances virus pathogenicity. *Journal of Virology, 79*(22), 14141–14148. https://doi.org/10.1128/JVI.79.22.14141-14148.2005.

Faber, M., Li, J., Kean, R. B., Hooper, D. C., Alugupalli, K. R., & Dietzschold, B. (2009). Effective preexposure and postexposure prophylaxis of rabies with a highly attenuated recombinant rabies virus. *Proceedings of the National Academy of Sciences of the United States of America, 106*(27), 11300–11305. https://doi.org/10.1073/pnas.0905640106.

Faber, M., Pulmanausahakul, R., Nagao, K., Prosniak, M., Rice, A. B., Koprowski, H., ... Dietzschold, B. (2004). Identification of viral genomic elements responsible for rabies virus neuroinvasiveness. *Proceedings of the National Academy of Sciences of the United States of America, 101*(46), 16328–16332. https://doi.org/10.1073/pnas.0407289101.

Finke, S., & Conzelmann, K. K. (1997). Ambisense gene expression from recombinant rabies virus: Random packaging of positive- and negative-strand ribonucleoprotein complexes into rabies virions. *Journal of Virology, 71*(10), 7281–7288.

Finke, S., & Conzelmann, K. K. (1999). Virus promoters determine interference by defective RNAs: Selective amplification of mini-RNA vectors and rescue from cDNA by a 3′ copy-back ambisense rabies virus. *Journal of Virology, 73*(5), 3818–3825.

Finke, S., & Conzelmann, K. K. (2005). Recombinant rhabdoviruses: Vectors for vaccine development and gene therapy. *Current Topics in Microbiology and Immunology, 292*, 165–200.

Finke, S., & Conzelmann, K. M. (2003). Dissociation of rabies virus matrix protein functions in regulation of viral RNA synthesis and virus assembly. *Journal of Virology, 77*(22), 12074–12082. https://doi.org/10.1128/Jvi.77.22.12074-12082.2003.

Finke, S., Cox, J. H., & Conzelmann, K. K. (2000). Differential transcription attenuation of rabies virus genes by intergenic regions: Generation of recombinant viruses overexpressing the polymerase gene. *Journal of Virology, 74*(16), 7261–7269.

Finke, S., Granzow, H., Hurst, J., Pollin, R., & Mettenleiter, T. C. (2010). Intergenotypic replacement of lyssavirus matrix proteins demonstrates the role of lyssavirus M proteins in intracellular virus accumulation. *Journal of Virology, 84*(4), 1816–1827.

Finke, S., Mueller-Waldeck, R., & Conzelmann, K. K. (2003). Rabies virus matrix protein regulates the balance of virus transcription and replication. *Journal of General Virology, 84*, 1613–1621. https://doi.org/10.1099/Vir.0.19128-0.

Flamand, A., & Delagneau, J. F. (1978). Transcriptional mapping of rabies virus in vivo. *Journal of Virology, 28*(2), 518–523.

Flamand, A., Raux, H., Gaudin, Y., & Ruigrok, R. W. (1993). Mechanisms of rabies virus neutralization. *Virology, 194*(1), 302–313. S0042-6822(83)71261-4 [pii]. https://doi.org/10.1006/viro.1993.1261.

Flamand, A., Wiktor, T. J., & Koprowski, H. (1980a). Use of hybridoma monoclonal-antibodies in the detection of antigenic differences between rabies and rabies-related virus proteins. 1. The nucleocapsid protein. *Journal of General Virology, 48*, 97–104.

Flamand, A., Wiktor, T. J., & Koprowski, H. (1980b). Use of hybridoma monoclonal-antibodies in the detection of antigenic differences between rabies and rabies-related virus proteins. 2. The glycoprotein. *Journal of General Virology, 48*, 105–109.

Fouquet, B., Nikolic, J., Larrous, F., Bourhy, H., Wirblich, C., Lagaudriere-Gesbert, C., & Blondel, D. (2015). Focal adhesion kinase is involved in rabies virus infection through its interaction with viral phosphoprotein P. *Journal of Virology, 89*(3), 1640–1651. https://doi.org/10.1128/jvi.02602-14.

Freuling, C. M., Beer, M., Conraths, F. J., Finke, S., Hoffmann, B., Keller, B., ... Muller, T. (2011). Novel Lyssavirus in Natterer's Bat, Germany. *Emerging Infectious Diseases, 17*(8), 1519–1522. https://doi.org/10.3201/Eid1708.110201.

Fu, Z. F., Zheng, Y. M., Wunner, W. H., Koprowski, H., & Dietzschold, B. (1994). Both the N-terminal and the C-terminal domains of the nominal phosphoprotein of rabies virus are involved in binding to the nucleoprotein. *Virology, 200*(2), 590–597.

Gao, Y., Greenfield, N. J., Cleverley, D. Z., & Lenard, J. (1996). The transcriptional form of the phosphoprotein of vesicular stomatitis virus is a trimer: Structure and stability. *Biochemistry, 35*(46), 14569–14573.

Gaudin, Y. (1997). Folding of rabies virus glycoprotein: Epitope acquisition and interaction with endoplasmic reticulum chaperones. *Journal of Virology, 71*(5), 3742–3750.

Gaudin, Y., Moreira, S., Benejean, J., Blondel, D., Flamand, A., & Tuffereau, C. (1999). Soluble ectodomain of rabies virus glycoprotein expressed in eukaryotic cells folds in a monomeric conformation that is antigenically distinct from the native state of the complete, membrane-anchored glycoprotein. *Journal of General Virology, 80*, 1647–1656.

Gaudin, Y., Raux, H., Flamand, A., & Ruigrok, R. W. H. (1996). Identification of amino acids controlling the low-pH-induced conformational change of rabies virus glycoprotein. *Journal of Virology, 70*(11), 7371–7378.

Gaudin, Y., Ruigrok, R. W. H., & Brunner, J. (1995). Low-pH induced conformational changes in viral fusion proteins—Implications for the fusion mechanism. *Journal of General Virology, 76*, 1541–1556.

Gaudin, Y., Ruigrok, R. W. H., Knossow, M., & Flamand, A. (1993). Low-pH conformational changes of rabies virus glycoprotein and their role in membrane fusion. *Journal of Virology, 67*(3), 1365–1372.

Gaudin, Y., Ruigrok, R. W. H., Tuffereau, C., Knossow, M., & Flamand, A. (1992). Rabies virus glycoprotein is a trimer. *Virology, 187*(2), 627–632.

Gaudin, Y., Tuffereau, C., Benmansour, A., & Flamand, A. (1991). Fatty acylation of rabies virus proteins. *Virology, 184*(1), 441–444.

Gaudin, Y., Tuffereau, C., Durrer, P., Flamand, A., & Ruigrok, R. W. H. (1995). Biological function of the low-pH, fusion inactive conformation of rabies virus glycoprotein (G)—G is transported in a fusion inactive state-like conformation. *Journal of Virology, 69*(9), 5528–5534.

Gaudin, Y., Tuffereau, C., Segretain, D., Knossow, M., & Flamand, A. (1991). Reversible conformational changes and fusion activity of rabies virus glycoprotein. *Journal of Virology, 65*(9), 4853–4859.

Ge, P., Tsao, J., Schein, S., Green, T. J., Luo, M., & Zhou, Z. H. (2010). Cryo-EM model of the bullet-shaped vesicular stomatitis virus. *Science, 327*(5966), 689–693. https://doi.org/10.1126/Science.1181766.

Gerard, F. C. A., Ribeiro, E. D., Albertini, A. A. V., Gutsche, I., Zaccai, G., Ruigrok, R. W. H., & Jamin, M. (2007). Unphosphorylated Rhabdoviridae phosphoproteins form elongated dimers in solution. *Biochemistry, 46*(36), 10328–10338. https://doi.org/10.1021/Bi7007799.

Ghanem, A., & Conzelmann, K. K. (2016). G gene-deficient single-round rabies viruses for neuronal circuit analysis. *Virus Research, 216*, 41–54. https://doi.org/10.1016/j.virusres.2015.05.023.

Gigant, B., Iseni, F., Gaudin, Y., Knossow, M., & Blondel, D. (2000). Neither phosphorylation nor the amino-terminal part of rabies virus phosphoprotein is required for its oligomerization. *Journal of General Virology, 81*, 1757–1761.

Gillet, J. P., Derer, P., & Tsiang, H. (1986). Axonal transport of rabies virus in the central nervous system of the rat. *Journal of Neuropathology and Experimental Neurology, 45*(6), 619–634.

Gluska, S., Zahavi, E. E., Chein, M., Gradus, T., Bauer, A., Finke, S., & Perlson, E. (2014). Rabies virus hijacks and accelerates the p75NTR retrograde axonal transport machinery. *PLoS Pathogens, 10*(8) e1004348. https://doi.org/10.1371/journal.ppat.1004348.

Gomme, E. A., Wanjalla, C. N., Wirblich, C., & Schnell, M. J. (2011). Rabies virus as a research tool and viral vaccine vector. *Advances in Virus Research: Research Advances in Rabies, 79*, 139–164. https://doi.org/10.1016/B978-0-12-387040-7.00009-3.

Goto, H., Minamoto, N., Ito, H., Ito, N., Sugiyama, M., Kinjo, T., & Kawai, A. (2000). Mapping of epitopes and structural analysis of antigenic sites in the nucleoprotein of rabies virus. *The Journal of General Virology, 81*(Pt 1), 119–127.

Gould, A. R., Hyatt, A. D., Lunt, R., Kattenbelt, J. A., Hengstberger, S., & Blacksell, S. D. (1998). Characterisation of a novel lyssavirus isolated from Pteropid bats in Australia. *Virus Research, 54*(2), 165–187. https://doi.org/10.1016/S0168-1702(98)00025-2.

Green, T. J., Zhang, X., Wertz, G. W., & Luo, M. (2006). Structure of the vesicular stomatitis virus nucleoprotein-RNA complex. *Science, 313*(5785), 357–360. https://doi.org/10.1126/Science.1126953.

Guichard, P., Krell, T., Chevalier, M., Vaysse, C., Adam, O., Ronzon, F., & Marco, S. (2011). Three dimensional morphology of rabies virus studied by cryo-electron tomography. *Journal of Structural Biology, 176*(1), 32–40. https://doi.org/10.1016/J.Jsb.2011.07.003.

Gunawardena, P. S., Marston, D. A., Ellis, R. J., Wise, E. L., Karawita, A. C., Breed, A. C., … Fooks, A. R. (2016). Lyssavirus in Indian flying foxes, Sri Lanka. *Emerging Infectious Diseases, 22*(8), 1456–1459. https://doi.org/10.3201/eid2208.151986.

Gupta, A. K., Blondel, D., Choudhary, S., & Banerjee, A. K. (2000). The phosphoprotein of rabies virus is phosphorylated by a unique cellular protein kinase and specific isomers of protein kinase C. *Journal of Virology, 74*(1), 91–98.

Hanlon, C. A., Kuzmin, I. V., Blanton, J. D., Weldon, W. C., Manangan, J. S., & Rupprecht, C. E. (2005). Efficacy of rabies biologics against new lyssaviruses from Eurasia. *Virus Research, 111*(1), 44–54. S0168-1702(05)00073-0 [pii]. https://doi.org/10.1016/j.virusres.2005.03.009.

Harty, R. N., Brown, M. E., McGettigan, J. P., Wang, G., Jayakar, H. R., Huibregtse, J. M., … Schnell, M. J. (2001). Rhabdoviruses and the cellular ubiquitin-proteasome system: A budding interaction. *Journal of Virology, 75*(22), 10623–10629. https://doi.org/10.1128/JVI.75.22.10623-10629.2001.

Harty, R. N., Paragas, J., Sudol, M., & Palese, P. (1999). A proline-rich motif within the matrix protein of vesicular stomatitis virus and rabies virus interacts with WW domains of cellular proteins: Implications for viral budding. *Journal of Virology, 73*(4), 2921–2929.

Hiramatsu, K., Mannen, K., Mifune, K., Nishizono, A., & Takita-Sonoda, Y. (1993). Comparative sequence analysis of the M gene among rabies virus strains and its expression by recombinant vaccinia virus. *Virus Genes, 7*(1), 83–88.

Holloway, B. P., & Obijeski, J. F. (1980). Rabies virus-induced RNA synthesis in Bhk-21 cells. *Journal of General Virology*, 49, 181–195.

Honda, K., & Taniguchi, T. (2006). IRFs: Master regulators of signalling by Toll-like receptors and cytosolic pattern-recognition receptors. *Nature Reviews Immunology*, 6(9), 644–658. https://doi.org/10.1038/Nri1900.

Hummeler, K., Koprowski, H., & Wiktor, T. J. (1967). Structure and development of rabies virus in tissue culture. *Journal of Virology*, 1(1), 152–170.

Hummeler, K., Tomassini, N., Sokol, F., Kuwert, E., & Koprowski, H. (1968). Morphology of the nucleoprotein component of rabies virus. *Journal of Virology*, 2(10), 1191–1199.

Irie, T., Licata, J. M., McGettigan, J. P., Schnell, M. J., & Harty, R. N. (2004). Budding of PPxY-containing rhabdoviruses is not dependent on host proteins TGS101 and VPS4A. *Journal of Virology*, 78(6), 2657–2665.

Iseni, F., Barge, A., Baudin, F., Blondel, D., & Ruigrok, R. W. (1998). Characterization of rabies virus nucleocapsids and recombinant nucleocapsid-like structures. *The Journal of General Virology*, 79(Pt 12), 2909–2919.

Ito, N., Takayama, M., Yamada, K., Sugiyama, M., & Minamoto, N. (2001). Rescue of rabies virus from cloned cDNA and identification of the pathogenicity-related gene: Glycoprotein gene is associated with virulence for adult mice. *Journal of Virology*, 75(19), 9121–9128.

Ito, Y., Nishizono, A., Mannen, K., Hiramatsu, K., & Mifune, K. (1996). Rabies virus M protein expressed in Escherichia coli and its regulatory role in virion-associated transcriptase activity. *Archives of Virology*, 141(3–4), 671–683.

Iwasaki, Y., Wiktor, T. J., & Koprowsk, H. (1973). Early events of rabies virus replication in tissue cultures—Electron microscopic study. *Laboratory Investigation*, 28(2), 142–148.

Jacob, Y., Badrane, H., Ceccaldi, P. E., & Tordo, N. (2000). Cytoplasmic dynein LC8 interacts with lyssavirus phosphoprotein. *Journal of Virology*, 74(21), 10217–10222.

Jiang, Y., Luo, Y. H., Michel, F., Hogan, R. J., He, Y., & Fu, Z. F. (2010). Characterization of conformation-specific monoclonal antibodies against rabies virus nucleoprotein. *Archives of Virology*, 155(8), 1187–1192. https://doi.org/10.1007/S00705-010-0709-X.

Johnson, N., McElhinney, L. M., Smith, J., Lowings, P., & Fooks, A. R. (2002). Phylogenetic comparison of the genus Lyssavirus using distal coding sequences of the glycoprotein and nucleoprotein genes. *Archives of Virology*, 147(11), 2111–2123. https://doi.org/10.1007/S00705-002-0877-4.

Kammouni, W., Wood, H., & Jackson, A. C. (2017). Serine residues at positions 162 and 166 of the rabies virus phosphoprotein are critical for the induction of oxidative stress in rabies virus infection. *Journal of Neurovirology*, 23(3), 358–368. https://doi.org/10.1007/s13365-016-0506-8.

Kammouni, W., Wood, H., Saleh, A., Appolinario, C. M., Fernyhough, P., & Jackson, A. C. (2015). Rabies virus phosphoprotein interacts with mitochondrial Complex I and induces mitochondrial dysfunction and oxidative stress. *Journal of Neurovirology*, 21(4), 370–382. https://doi.org/10.1007/s13365-015-0320-8.

Kankanamge, P. J., Irie, T., Mannen, K., Tochikura, T. S., & Kawai, A. (2003). Mapping of the low pH-sensitive conformational epitope of rabies virus glycoprotein recognized by a monoclonal antibody #1-30-44. *Microbiology and Immunology*, 47(7), 507–519.

Kasturi, L., Eshleman, J. R., Wunner, W. H., & Shakineshleman, S. H. (1995). The hydroxy amino-acid in an Asn-X-Ser/Thr sequon can influence N-linked core glycosylation efficiency and the level of expression of a cell-surface glycoprotein. *Journal of Biological Chemistry*, 270(24), 14756–14761.

Kawai, A., Toriumi, H., Tochikura, T. S., Takahashi, T., Honda, Y., & Morimoto, K. (1999). Nucleocapsid formation and/or subsequent conformational change of rabies virus nucleoprotein (N) is a prerequisite step for acquiring the phosphatase-sensitive epitope of monoclonal antibody 5-2-26. *Virology*, 263(2), 395–407. https://doi.org/10.1006/viro.1999.996 S0042-6822(99)99962-2 [pii].

Kissi, B., Tordo, N., & Bourhy, H. (1995). Genetic polymorphism in the rabies virus nucleoprotein gene. *Virology*, 209(2), 526–537. S0042-6822(85)71285-8 [pii]. https://doi.org/10.1006/viro.1995.1285.

Klingen, Y., Conzelmann, K. K., & Finke, S. (2008). Double-labeled rabies virus: Live tracking of enveloped virus transport. *Journal of Virology*, 82(1), 237–245.

Komarova, A. V., Real, E., Borman, A. M., Brocard, M., England, P., Tordo, N., … Jacob, Y. (2007). Rabies virus matrix protein interplay with eIF3, new insights into rabies virus pathogenesis. *Nucleic Acids Research*, 35(5), 1522–1532. https://doi.org/10.1093/nar/gkl1127.

Kouznetzoff, A., Buckle, M., & Tordo, N. (1998). Identification of a region of the rabies virus N protein involved in direct binding to the viral RNA. *The Journal of General Virology*, 79(Pt 5), 1005–1013.

Kristensson, K., Dastur, D. K., Manghani, D. K., Tsiang, H., & Bentivoglio, M. (1996). Rabies: Interactions between neurons and viruses. A review of the history of Negri inclusion bodies. *Neuropathology and Applied Neurobiology, 22*(3), 179–187.

Kucera, P., Dolivo, M., Coulon, P., & Flamand, A. (1985). Pathways of the early propagation of virulent and avirulent rabies strains from the eye to the brain. *Journal of Virology, 55*(1), 158–162.

Kuzmin, I. V., Hughes, G. J., Botvinkin, A. D., Orciari, L. A., & Rupprecht, C. E. (2005). Phylogenetic relationships of Irkut and West Caucasian bat viruses within the Lyssavirus genus and suggested quantitative criteria based on the N gene sequence for lyssavirus genotype definition. *Virus Research, 111*(1), 28–43. S0168-1702(05)00072-9 [pii] https://doi.org/10.1016/j.virusres.2005.03.008.

Kuzmin, I. V., Mayer, A. E., Niezgoda, M., Markotter, W., Agwanda, B., Breiman, R. F., & Rupprecht, C. E. (2010). Shimoni bat virus, a new representative of the Lyssavirus genus. *Virus Research, 149*(2), 197–210. S0168-1702 (10)00044-4 [pii]. https://doi.org/10.1016/j.virusres.2010.01.018.

Kuzmin, I. V., Niezgoda, M., Franka, R., Agwanda, B., Markotter, W., Beagley, J. C., … Rupprecht, C. E. (2008). Possible emergence of West Caucasian bat virus in Africa. *Emerging Infectious Diseases, 14*(12), 1887–1889.

Kuzmin, I. V., Wu, X., Tordo, N., & Rupprecht, C. E. (2008). Complete genomes of Aravan, Khujand, Irkut and West Caucasian bat viruses, with special attention to the polymerase gene and non-coding regions. *Virus Research, 136* (1–2), 81–90. S0168-1702(08)00170-6 [pii]. https://doi.org/10.1016/j.virusres.2008.04.021.

Lafay, F., Coulon, P., Astic, L., Saucier, D., Riche, D., Holley, A., & Flamand, A. (1991). Spread of the CVS strain of rabies virus and of the avirulent mutant AvO1 along the olfactory pathways of the mouse after intranasal inoculation. *Virology, 183*(1), 320–330.

Lafon, M. (2005). Rabies virus receptors. *Journal of Neurovirology, 11*(1), 82–87. https://doi.org/10.1080/13550280 590900427.

Lafon, M., Lafage, M., Martinez-Arends, A., Ramirez, R., Vuillier, F., Charron, D., … Scott-Algara, D. (1992). Evidence for a viral superantigen in humans. *Nature, 358*(6386), 507–510. https://doi.org/10.1038/358507a0.

Lahaye, X., Vidy, A., Fouquet, B., & Blondel, D. (2012). Hsp70 protein positively regulates rabies virus infection. *Journal of Virology*. https://doi.org/10.1128/JVI.06501-11.

Lahaye, X., Vidy, A., Pomier, C., Obiang, L., Harper, F., Gaudin, Y., & Blondel, D. (2009). Functional characterization of negri bodies (NBs) in rabies virus-infected cells: Evidence that NBs are sites of viral transcription and replication. *Journal of Virology, 83*(16), 7948–7958. https://doi.org/10.1128/Jvi.00554-09.

Langevin, C., & Tuffereau, C. (2002). Mutations conferring resistance to neutralization by a soluble form of the neurotrophin receptor (p75NTR) map outside of the known antigenic sites of the rabies virus glycoprotein. *Journal of Virology, 76*(21), 10756–10765. https://doi.org/10.1128/Jvi.76.21.10756-10765.2002.

Lazzarini, R. A., Keene, J. D., & Schubert, M. (1981). The origins of defective interfering particles of the negative-strand RNA viruses. *Cell, 26*(2 Pt 2), 145–154.

Le Blanc, I., Luyet, P. P., Pons, V., Ferguson, C., Emans, N., Petiot, A., … Gruenberg, J. (2005). Endosome-to-cytosol transport of viral nucleocapsids. *Nature Cell Biology, 7*(7), 653–664. https://doi.org/10.1038/Ncb1269.

Leppert, M., Rittenhouse, L., Perrault, J., Summers, D. F., & Kolakofsky, D. (1979). Plus and minus strand leader RNAs in negative strand virus-infected cells. *Cell, 18*(3), 735–747.

Lewis, P., Fu, Y. G., & Lentz, T. L. (2000). Rabies virus entry at the neuromuscular junction in nerve-muscle cocultures. *Muscle & Nerve, 23*(5), 720–730.

Lij, J., Rahmeh, A., Morelli, M., & Whelan, S. P. J. (2008). A conserved motif in region V of the large polymerase proteins of nonsegmented negative-sense RNA viruses that is essential for mRNA capping. *Journal of Virology, 82*(2), 775–784. https://doi.org/10.1128/Jvi.02107-07.

Liu, P., Yang, J., Wu, X., & Fu, Z. F. (2004). Interactions amongst rabies virus nucleoprotein, phosphoprotein and genomic RNA in virus-infected and transfected cells. *The Journal of General Virology, 85*(Pt 12), 3725–3734. https://doi.org/10.1099/vir.0.80325-0.

Luo, M., Green, T. J., Zhang, X., Tsao, J., & Qiu, S. H. (2007). Conserved characteristics of the rhabdovirus nucleoprotein. *Virus Research, 129*(2), 246–251. https://doi.org/10.1016/J.Virusres.2007.07.011.

Lycke, E., & Tsiang, H. (1987). Rabies virus-infection of cultured rat sensory neurons. *Journal of Virology, 61*(9), 2733–2741.

Macfarlan, R. I., Dietzschold, B., & Koprowski, H. (1986). Stimulation of cytotoxic T-lymphocyte responses by rabies virus glycoprotein and identification of an immunodominant domain. *Molecular Immunology, 23*(7), 733–741.

Madore, H. P., & England, J. M. (1977). Rabies virus protein synthesis in infected BHK-21 cells. *Journal of Virology, 22*(1), 102–112.

Maillard, A. P., & Gaudin, Y. (2002). Rabies virus glycoprotein can fold in two alternative, antigenically distinct conformations depending on membrane-anchor type. *Journal of General Virology, 83*, 1465–1476.

Marriott, A. C., & Dimmock, N. J. (2010). Defective interfering viruses and their potential as antiviral agents. *Reviews in Medical Virology, 20*(1), 51–62. https://doi.org/10.1002/Rmv.641.

Marston, D. A., Banyard, A. C., McElhinney, L. M., Freuling, C. M., Finke, S., de Lamballerie, ... Fooks, A. R. (2018). The lyssavirus host-specificity conundrum—rabies virus—the exception not the rule. *Current Opinion in Virology, 28*, 68–73.

Marston, D. A., McElhinney, L. M., Johnson, N., Muller, T., Conzelmann, K. K., Tordo, N., & Fooks, A. R. (2007). Comparative analysis of the full genome sequence of European bat lyssavirus type 1 and type 2 with other lyssaviruses and evidence for a conserved transcription termination and polyadenylation motif in the G-L 3' non-translated region. *The Journal of General Virology, 88*(Pt 4), 1302–1314. https://doi.org/10.1099/vir.0.82692-0.

Matsumot, S., & Kawai, A. (1969). Comparative studies on development of rabies virus in different host cells. *Virology, 39*(3), 449–459.

Matsumoto, S. (1962). Electron microscopy of nerve cells infected with street rabies virus. *Virology, 17*, 198–202.

Matsumoto, S. (1975). Electron microscopy of central nervous system infection. In G. M. Baer (Ed.), *The natural history of rabies* (pp. 217–233). New York: Academic Press.

Matsumoto, S., & Miyamoto, K. (1966). Electron-microscopic studies on rabies virus multiplication and the nature of the Negri body. *Symposium Series in Immunobiology Standardization, 1*, 45–54.

Matsumoto, S., Schneider, L. G., Kawai, A., & Yonezawa, T. (1974). Further studies on the replication of rabies and rabies-like viruses in organized cultures of mammalian neural tissues. *Journal of Virology, 14*(4), 981–996.

Mavrakis, M., Iseni, F., Mazza, C., Schoehn, G., Ebel, C., Gentzel, M., ... Ruigrok, R. W. H. (2003). Isolation and characterisation of the rabies virus N degrees-P complex produced in insect cells. *Virology, 305*(2), 406–414. https://doi.org/10.1006/Viro.2002.1748.

Mavrakis, M., Mehouas, S., Real, E., Iseni, F., Blondel, D., Tordo, N., & Ruigrok, R. W. H. (2006). Rabies virus chaperone: Identification of the phosphoprotein peptide that keeps nucleoprotein soluble and free from non-specific RNA. *Virology, 349*(2), 422–429. https://doi.org/10.1016/J.Virol.2006.01.030.

Mazarakis, N. D., Azzouz, M., Rohll, J. B., Ellard, F. M., Wilkes, F. J., Olsen, A. L., ... Mitrophanous, K. A. (2001). Rabies virus glycoprotein pseudotyping of lentiviral vectors enables retrograde axonal transport and access to the nervous system after peripheral delivery. *Human Molecular Genetics, 10*(19), 2109–2121.

Mebatsion, T. (2001). Extensive attenuation of rabies virus by simultaneously modifying the dynein light chain binding site in the P protein and replacing Arg333 in the G protein. *Journal of Virology, 75*(23), 11496–11502.

Mebatsion, T., Konig, M., & Conzelmann, K. K. (1996). Budding of rabies virus particles in the absence of the spike glycoprotein. *Cell, 84*(6), 941–951. S0092-8674(00)81072-7 [pii].

Mebatsion, T., Schnell, M. J., & Conzelmann, K. K. (1995). Mokola virus glycoprotein and chimeric proteins can replace rabies virus glycoprotein in the rescue of infectious defective rabies virus particles. *Journal of Virology, 69*(3), 1444–1451.

Mebatsion, T., Weiland, F., & Conzelmann, K. K. (1999). Matrix protein of rabies virus is responsible for the assembly and budding of bullet-shaped particles and interacts with the transmembrane spike glycoprotein G. *Journal of Virology, 73*(1), 242–250.

Minamoto, N., Tanaka, H., Hishida, M., Goto, H., Ito, H., Naruse, S., ... Mifune, K. (1994). Linear and conformation-dependent antigenic sites on the nucleoprotein of rabies virus. *Microbiology and Immunology, 38*(6), 449–455.

Morimoto, K., Foley, H. D., McGettigan, J. P., Schnell, M. J., & Dietzschold, B. (2000). Reinvestigation of the role of the rabies virus glycoprotein in viral pathogenesis using a reverse genetics approach. *Journal of Neurovirology, 6*(5), 373–381.

Morimoto, K., Ohkubo, A., & Kawai, A. (1989). Structure and transcription of the glycoprotein gene of attenuated HEP-Flury strain of rabies virus. *Virology, 173*(2), 465–477.

Morin, B., Liang, B., Gardner, E., Ross, R. A., & Whelan, S. P. J. (2017). An in vitro RNA synthesis assay for rabies virus defines ribonucleoprotein interactions critical for polymerase activity. *Journal of Virology, 91*(1). https://doi.org/10.1128/jvi.01508-16.

Murphy, F. A., & Bauer, S. P. (1974). Early street rabies virus-infection in striated-muscle and later progression to central nervous-system. *Intervirology, 3*(4), 256–268.

Murphy, F. A., Bauer, S. P., Harrison, A. K., & Winn, W. C. (1973). Comparative pathogenesis of rabies and rabies-like viruses—Viral infection and transit from inoculation site to central nervous system. *Laboratory Investigation, 28*(3), 361–376.

Murphy, F. A., & Harrison, A. K. (1979). Electron microscopy of the rhabdoviruses of animals. In D. H. L. Bishop (Ed.), *Rhabdoviruses* (pp. 65–106). Boca Raton: CRC Press.

Nadin-Davis, S. A., & Real, L. A. (2011). Molecular phylogenetics of the Lyssaviruses—Insights from a coalescent approach. *Advances in Virus Research: Research Advances in Rabies, 79,* 203–238. https://doi.org/10.1016/B978-0-12-387040-7.00011-1.

Naito, S., & Matsumoto, S. (1978). Identification of cellular actin within the rabies virus. *Virology, 91*(1), 151–163.

Nakahara, K., Ohnuma, H., Sugita, S., Yasuoka, K., Nakahara, T., Tochikura, T. S., & Kawai, A. (1999). Intracellular behavior of rabies virus matrix protein (M) is determined by the viral glycoprotein (G). *Microbiology and Immunology, 43*(3), 259–270.

Nakahara, T., Toriumi, H., Irie, T., Takahashi, T., Ameyama, S., Mizukoshi, M., & Kawai, A. (2003). Characterization of a slow-migrating component of the rabies virus matrix protein strongly associated with the viral glycoprotein. *Microbiology and Immunology, 47*(12), 977–988.

Nikolic, J., Le Bars, R., Lama, Z., Scrima, N., Lagaudriere-Gesbert, C., Gaudin, Y., & Blondel, D. (2017). Negri bodies are viral factories with properties of liquid organelles. *Nature Communications, 8*(1)58. https://doi.org/10.1038/s41467-017-00102-9.

Noton, S. L., Tremaglio, C. Z., & Fearns, R. (2019). Killing two birds with one stone: How the respiratory syncytial virus polymerase initiates transcription and replication. *PLoS Pathogens, 15*(2) e1007548. https://doi.org/10.1371/journal.ppat.1007548.

Novella, I. S., Ball, L. A., & Wertz, G. W. (2004). Fitness analyses of vesicular stomatitis strains with rearranged genomes reveal replicative disadvantages. *Journal of Virology, 78*(18), 9837–9841. https://doi.org/10.1128/Jvi.78.18.9837-9841.2004.

Ogino, M., Ito, N., Sugiyama, M., & Ogino, T. (2016). The rabies virus L protein catalyzes mRNA capping with GDP polyribonucleotidyltransferase activity. *Viruses, 8*(5). https://doi.org/10.3390/v8050144.

Ogino, T. (2014). Capping of vesicular stomatitis virus pre-mRNA is required for accurate selection of transcription stop-start sites and virus propagation. *Nucleic Acids Research, 42*(19), 12112–12125. https://doi.org/10.1093/nar/gku901.

Ogino, T., & Banerjee, A. K. (2007). Unconventional mechanism of mRNA capping by the RNA-dependent RNA polymerase of vesicular stomatitis virus. *Molecular Cell, 25*(1), 85–97. https://doi.org/10.1016/J.Molcel.2006.11.013.

Ogino, T., Yadav, S. P., & Banerjee, A. K. (2010). Histidine-mediated RNA transfer to GDP for unique mRNA capping by vesicular stomatitis virus RNA polymerase. *Proceedings of the National Academy of Sciences of the United States of America, 107*(8), 3463–3468. https://doi.org/10.1073/pnas.0913083107.

Okazaki, Y., Ohno, H., Takase, K., Ochiai, T., & Saito, T. (2000). Cell surface expression of calnexin, a molecular chaperone in the endoplasmic reticulum. *The Journal of Biological Chemistry, 275*(46), 35751–35758. https://doi.org/10.1074/jbc.M007476200.

Perrin, P., Portnoi, D., & Sureau, P. (1982). Étude de l'adsorption et de la pénétration du virus rabique; interactions avec les cellules BHK21 et des membrane artificielles. *Annales de l'Institut Pasteur/Virologie, 133E,* 403–422.

Pfaller, C. K., Donohue, R. C., Nersisyan, S., Brodsky, L., & Cattaneo, R. (2018). Extensive editing of cellular and viral double-stranded RNA structures accounts for innate immunity suppression and the proviral activity of ADAR1p150. *PLoS Biology, 16*(11) e2006577. https://doi.org/10.1371/journal.pbio.2006577.

Pfefferkorn, C., Kallfass, C., Lienenklaus, S., Spanier, J., Kalinke, U., Rieder, M., … Staeheli, P. (2016). Abortively infected astrocytes appear to represent the main source of interferon beta in the virus-infected brain. *Journal of Virology, 90*(4), 2031–2038. https://doi.org/10.1128/jvi.02979-15.

Piccinotti, S., & Whelan, S. P. (2016). Rabies internalizes into primary peripheral neurons via clathrin coated pits and requires fusion at the cell body. *PLoS Pathogens, 12*(7) e1005753. https://doi.org/10.1371/journal.ppat.1005753.

Poch, O., Blumberg, B. M., Bougueleret, L., & Tordo, N. (1990). Sequence comparison of five polymerases (L proteins) of unsegmented negative-strand RNA viruses: Theoretical assignment of functional domains. *The Journal of General Virology, 71*(Pt 5), 1153–1162.

Poch, O., Sauvaget, I., Delarue, M., & Tordo, N. (1989). Identification of four conserved motifs among the RNA-dependent polymerase encoding elements. *The EMBO Journal, 8*(12), 3867–3874.

Pollin, R., Granzow, H., Kollner, B., Conzelmann, K. K., & Finke, S. (2013). Membrane and inclusion body targeting of lyssavirus matrix proteins. *Cellular Microbiology, 15*(2), 200–212.

Prehaud, C., Coulon, P., Lafay, F., Thiers, C., & Flamand, A. (1988). Antigenic site II of the rabies virus glycoprotein—Structure and role in viral virulence. *Journal of Virology, 62*(1), 1–7.

Rasalingham, P., Rosssiter, J. F., Mebatsion, T., & Jackson, A. C. (2005). Comparative pathogenesis of the SAD-L16 strain of rabies virus and a mutant modifying the dynein light chain binding site of the rabies virus phosphoprotein in young mice. *Virus Research*, *111*(1), 55–60. 50168-1702(05)00074-2 [pii]. https://doi.org/10.1026/j.virusres.2005.03.010.

Raux, H., Flamand, A., & Blondel, D. (2000). Interaction of the rabies virus P protein with the LC8 dynein light chain. *Journal of Virology*, *74*(21), 10212–10216.

Reagan, K. J., & Wunner, W. H. (1985). Rabies virus interaction with various cell-lines is independent of the acetylcholine-receptor—Brief report. *Archives of Virology*, *84*(3-4), 277–282.

Ribeiro, E. A., Favier, A., Gerard, F. C. A., Leyrat, C., Brutscher, B., Blondel, D., ... Jamin, M. (2008). Solution structure of the C-terminal nucleoprotein-RNA binding domain of the vesicular stomatitis virus phosphoprotein. *Journal of Molecular Biology*, *382*(2), 525–538. https://doi.org/10.1016/J.Jmb.2008.07.028.

Ribeiro, E. D., Leyrat, C., Gerard, F. C. A., Albertini, A. A. V., Falk, C., Ruigrok, R. W. H., & Jamin, M. (2009). Binding of rabies virus polymerase cofactor to recombinant circular nucleoprotein-RNA complexes. *Journal of Molecular Biology*, *394*(3), 558–575. https://doi.org/10.1016/J.Jmb.2009.09.042.

Riedel, C., Vasishtan, D., Pražák, V., Ghanem, A., Conzelmann, K. -K., & Rumenapf, T. (2019). Cryo EM structure of the rabies virus ribonucleoprotein complex. *Scientific Reports*, *9*, 0639. https://doi.org/10.1038/s41598-019-46126-7.

Robison, C. S., & Whitt, M. A. (2000). The membrane-proximal stem region of vesicular stomatitis virus G protein confers efficient virus assembly. *Journal of Virology*, *74*(5), 2239–2246.

Rupprecht, C., Kuzmin, I., & Meslin, F. (2017). Lyssaviruses and rabies: Current conundrums, concerns, contradictions and controversies. *F1000Res*, *6*, 184. https://doi.org/10.12688/f1000research.10416.1.

Sagara, J., & Kawai, A. (1992). Identification of heat shock protein 70 in the rabies virion. *Virology*, *190*(2), 845–848.

Sagara, J., Tochikura, T. S., Tanaka, H., Baba, Y., Tsukita, S., & Kawai, A. (1998). The 21-kDa polypeptide (VAP21) in the rabies virion is a CD99-related host cell protein. *Microbiology and Immunology*, *42*(4), 289–297.

Sagara, J., Tsukita, S., Yonemura, S., & Kawai, A. (1995). Cellular actin-binding ezrin-radixin-moesin (ERM) family proteins are incorporated into the rabies virion and closely associated with viral envelope proteins in the cell. *Virology*, *206*(1), 485–494. S0042-6822(95)80064-6 [pii].

Sakai, M., Kankanamge, P. J., Shoji, J., Kawata, S., Tochikura, T. S., & Kawai, A. (2004). Studies on the conditions required for structural and functional maturation of rabies virus glycoprotein (G) in G cDNA-transfected cells. *Microbiology and Immunology*, *48*(11), 853–864.

Sanchez, A., De, B. P., & Banerjee, A. K. (1985). In vitro phosphorylation of NS protein by the L protein of vesicular stomatitis virus. *The Journal of General Virology*, *66*(Pt 5), 1025–1036.

Sanchez-Aparicio, M. T., Garcin, D., Rice, C. M., Kolakofsky, D., Garcia-Sastre, A., & Baum, A. (2017). Loss of Sendai virus C protein leads to accumulation of RIG-I immunostimulatory defective interfering RNA. *The Journal of General Virology*, *98*(6), 1282–1293. https://doi.org/10.1099/jgv.0.000815.

Sato, G., Kobayashi, Y., Motizuki, N., Hirano, S., Itou, T., Cunha, E. M., ... Sakai, T. (2009). A unique substitution at position 333 on the glycoprotein of rabies virus street strains isolated from non-hematophagous bats in Brazil. *Virus Genes*, *38*(1), 74–79. https://doi.org/10.1007/s11262-008-0290-5.

Schnell, M. J., McGettigan, J. P., Wirblich, C., & Papaneri, A. (2010). The cell biology of rabies virus: Using stealth to reach the brain. *Nature Reviews. Microbiology*, *8*(1), 51–61.

Schnell, M. J., Mebatsion, T., & Conzelmann, K. K. (1994). Infectious rabies viruses from cloned cDNA. *The EMBO Journal*, *13*(18), 4195–4203.

Schoehn, G., Iseni, F., Mavrakis, M., Blondel, D., & Ruigrok, R. W. (2001). Structure of recombinant rabies virus nucleoprotein-RNA complex and identification of the phosphoprotein binding site. *Journal of Virology*, *75*(1), 490–498. https://doi.org/10.1128/JVI.75.1.490-498.2001.

Schoneberg, J., Lee, I. H., Iwasa, J. H., & Hurley, J. H. (2017). Reverse-topology membrane scission by the ESCRT proteins. *Nature Reviews. Molecular Cell Biology*, *18*(1), 5–17. https://doi.org/10.1038/nrm.2016.121.

Seif, I., Coulon, P., Rollin, P. E., & Flamand, A. (1985). Rabies virulence—Effect on pathogenicity and sequence characterization of rabies virus mutations affecting antigenic site III of the glycoprotein. *Journal of Virology*, *53*(3), 926–934.

Shakineshleman, S. H., Remaley, A. T., Eshleman, J. R., Wunner, W. H., & Spitalnik, S. L. (1992). N-linked glycosylation of rabies virus glycoprotein—Individual sequons differ in their glycosylation efficiencies and influence on cell-surface expression. *Journal of Biological Chemistry*, *267*(15), 10690–10698.

Shakineshleman, S. H., Wunner, W. H., & Spitalnik, S. L. (1993). Efficiency of N-linked core glycosylation at asparagine-319 of rabies virus glycoprotein is altered by deletions C-terminal to the glycosylation sequon. *Biochemistry*, *32*(36), 9465–9472.

Sissoeff, L., Mousli, M., England, P., & Tuffereau, C. (2005). Stable trimerization of recombinant rabies virus glycoprotein ectodomain is required for interaction with the p75(NTR) receptor. *Journal of General Virology*, 86, 2543–2552. https://doi.org/10.1099/Vir.0.81063-0.

Smith, J. S. (1989). Rabies virus epitopic variation: Use in ecologic studies. *Advances in Virus Research*, 36, 215–253.

Sokol, F., Clark, H. F., Wiktor, T. J., McFalls, M. L., Bishop, D. H., & Obijeski, J. F. (1974). Structural phosphoproteins associated with ten rhabdoviruses. *The Journal of General Virology*, 24(3), 433–445.

Spadafora, D., Canter, D. M., Jackson, R. L., & Perrault, J. (1996). Constitutive phosphorylation of the vesicular stomatitis virus P protein modulates polymerase complex formation but is not essential for transcription or replication. *Journal of Virology*, 70(7), 4538–4548.

Superti, F., Derer, M., & Tsiang, H. (1984). Mechanism of rabies virus entry into CER cells. *Journal of General Virology*, 65, 781–789.

Superti, F., Hauttecoeur, B., Morelec, M. J., Goldoni, P., Bizzini, B., & Tsiang, H. (1986). Involvement of gangliosides in rabies virus infection. *Journal of General Virology*, 67, 47–56.

Takamatsu, F., Asakawa, N., Morimoto, K., Takeuchi, K., Eriguchi, Y., Toriumi, H., & Kawai, A. (1998). Studies on the rabies virus RNA polymerase: 2. Possible relationships between the two forms of the non-catalytic subunit (P protein). *Microbiology and Immunology*, 42(11), 761–771.

Takayama-Ito, M., Ito, N., Yamada, K., Minamoto, N., & Sugiyama, M. (2004). Region at amino acids 164 to 303 of the rabies virus glycoprotein plays an important role in pathogenicity for adult mice. *Journal of Neurovirology*, 10(2), 131–135. https://doi.org/10.1080/13550280490279799.

Tan, G. S., Preuss, M. A., Willliams, J. C., & Schnell, M. J. (2007). The dynein light chain 8 binding motif of rabies virus phosphoprotein promotes efficient viral transcription. *Proceedings of the National Academy of Sciences of the United States of America*, 104(17), 7229–7234. 0701397104 [pii]. https://doi.org/10.1073/pnas.0701397104.

Tao, L., Ge, J., Wang, X., Zhai, H., Hua, T., Zhao, B., ... Bu, Z. (2010). Molecular basis of neurovirulence of flury rabies virus vaccine strains: Importance of the polymerase and the glycoprotein R333Q mutation. *Journal of Virology*, 84(17), 8926–8936. https://doi.org/10.1128/JVI.00787-10.

Thoulouze, M. I., Lafage, M., Schachner, M., Hartmann, U., Cremer, H., & Lafon, M. (1998). The neural cell adhesion molecule is a receptor for rabies virus. *Journal of Virology*, 72(9), 7181–7190.

Tian, B., Zhou, M., Yang, Y., Yu, L., Luo, Z., Tian, D., ... Zhao, L. (2017). Lab-attenuated rabies virus causes abortive infection and induces cytokine expression in astrocytes by activating mitochondrial antiviral-signaling protein signaling pathway. *Frontiers in Immunology*, 8, 2011. https://doi.org/10.3389/fimmu.2017.02011.

Tollis, M., Dietzschold, B., Volia, C. B., & Koprowski, H. (1991). Immunization of monkeys with rabies ribonucleoprotein (RNP) confers protective immunity against rabies. *Vaccine*, 9(2), 134–136.

Tordo, N., Poch, O., Ermine, A., & Keith, G. (1986). Primary structure of leader RNA and nucleoprotein genes of the rabies genome: Segmented homology with VSV. *Nucleic Acids Research*, 14(6), 2671–2683.

Tordo, N., Poch, O., Ermine, A., Keith, G., & Rougeon, F. (1986). Walking along the rabies genome: Is the large G-L intergenic region a remnant gene? *Proceedings of the National Academy of Sciences of the United States of America*, 83(11), 3914–3918.

Tordo, N., Poch, O., Ermine, A., Keith, G., & Rougeon, F. (1988). Completion of the rabies virus genome sequence determination: Highly conserved domains among the L (polymerase) proteins of unsegmented negative-strand RNA viruses. *Virology*, 165(2), 565–576.

Toriumi, H., & Kawai, A. (2004). Association of rabies virus nominal phosphoprotein (P) with viral nucleocapsid (NC) is enhanced by phosphorylation of the viral nucleoprotein (N). *Microbiology and Immunology*, 48(5), 399–409.

Tsiang, H. (1993). Pathophysiology of rabies virus-infection of the nervous system. *Advances in Virus Research*, 42, 375–412.

Tsiang, H., Delaporte, S., Ambroise, D. J., Derer, M., & Koenig, J. (1986). Infection of cultured rat myotubes and neurons from the spinal-cord by rabies virus. *Journal of Neuropathology and Experimental Neurology*, 45(1), 28–42.

Tsiang, H., Derer, M., & Taxi, J. (1983). An in vivo and in vitro study of rabies virus infection of the rat superior cervical ganglia. *Archives of Virology*, 76(3), 231–243.

Tuffereau, C., Benejean, J., Blondel, D., Kieffer, B., & Flamand, A. (1998). Low-affinity nerve-growth factor receptor (P75NTR) can serve as a receptor for rabies virus. *The EMBO Journal*, 17(24), 7250–7259. https://doi.org/10.1093/emboj/17.24.7250.

Tuffereau, C., Leblois, H., Benejean, J., Coulon, P., Lafay, F., & Flamand, A. (1989). Arginine or lysine in position-333 of ERA and CVS glycoprotein is necessary for rabies virulence in adult mice. *Virology*, 172(1), 206–212.

Ugolini, G. (1995). Specificity of rabies virus as a transneuronal tracer of motor networks: Transfer from hypoglossal motoneurons to connected second-order and higher order central nervous system cell groups. *Journal of Comparative Neurology, 356*(3), 457–480.

Vidal, S., & Kolakofsky, D. (1989). Modified model for the switch from Sendai virus transcription to replication. *Journal of Virology, 63*(5), 1951–1958.

Vidy, A., Chelbi-Alix, M., & Blondel, D. (2005). Rabies virus P protein interacts with STAT1 and inhibits interferon signal transduction pathways. *Journal of Virology, 79*(22), 14411–14420.

Wang, J., Wang, Z., Liu, R., Shuai, L., Wang, X., Luo, J., … Bu, Z. (2018). Metabotropic glutamate receptor subtype 2 is a cellular receptor for rabies virus. *PLoS Pathogens, 14*(7) e1007189. https://doi.org/10.1371/journal.ppat.1007189.

Warrilow, D., Smith, I. L., Harrower, B., & Smith, G. A. (2002). Sequence analysis of an isolate from a fatal human infection of Australian bat lyssavirus. *Virology, 297*(1), 109–119 S0042682202914170 [pii].

Whelan, S. P. J., Barr, J. N., & Wertz, G. W. (2004). Transcription and replication of nonsegmented negative-strand RNA viruses. *Current Topics in Microbiology and Immunology, 283*, 61–119.

Whitt, M. A., Buonocore, L., Prehaud, C., & Rose, J. K. (1991). Membrane-fusion activity, oligomerization, and assembly of the rabies virus glycoprotein. *Virology, 185*(2), 681–688.

Wickersham, I. R., Lyon, D. C., Barnard, R. J., Mori, T., Finke, S., Conzelmann, K. K., … Callaway, E. M. (2007). Monosynaptic restriction of transsynaptic tracing from single, genetically targeted neurons. *Neuron, 53*(5), 639–647. S0896-6273(07)00078-5 [pii]. https://doi.org/10.1016/j.neuron.2007.01.033.

Wiktor, T. J., Dietzschold, B., Leamnson, R. N., & Koprowski, H. (1977). Induction and biological properties of defective interfering particles of rabies virus. *Journal of Virology, 21*(2), 626–635.

Wojczyk, B. S., Takahashi, N., Levy, M. T., Andrews, D. W., Abrams, W. R., Wunner, W. H., & Spitalnik, S. L. (2005). N-glycosylation at one rabies virus glycoprotein sequon influences N-glycan processing at a distant sequon on the same molecule. *Glycobiology, 15*(6), 655–666. https://doi.org/10.1093/Glycob/Cwi046.

Wu, X., Lei, X., & Fu, Z. F. (2003). Rabies virus nucleoprotein is phosphorylated by cellular casein kinase II. *Biochemical and Biophysical Research Communications, 304*(2), 333–338. S0006291X03005941 [pii].

Wu, X. F., Gong, X. M., Foley, H. D., Schnell, M. J., & Fu, Z. F. (2002). Both viral transcription and replication are reduced when the rabies virus nucleoprotein is not phosphorylated. *Journal of Virology, 76*(9), 4153–4161. https://doi.org/10.1128/Jvi.76.9.4153-4161.2002.

Wunner, W. H. (1991). The chemical composition and molecular structure of rabies viruses. In G. M. Baer (Ed.), *The natural history of rabies* (pp. 31–67). (2nd ed.). Boca Raton: CRC Press.

Wunner, W. H., & Clark, H. F. (1980). Regeneration of DI particles of virulent and attenuated rabies virus: Genome characterization and lack of correlation with virulence phenotype. *The Journal of General Virology, 51*(Pt 1), 69–81.

Wunner, W. H., Dietzschold, B., Smith, C. L., Lafon, M., & Golub, E. (1985). Antigenic variants of CVS rabies virus with altered glycosylation sites. *Virology, 140*(1), 1–12.

Wunner, W. H., Reagan, K. J., & Koprowski, H. (1984). Characterization of saturable binding sites for rabies virus. *Journal of Virology, 50*(3), 691–697.

Yan, X. Z., Mohankumar, P. S., Dietzschold, B., Schnell, M. J., & Fu, Z. F. (2002). The rabies virus glycoprotein determines the distribution of different rabies virus strains in the brain. *Journal of Neurovirology, 8*(4), 345–352. https://doi.org/10.1080/13550280290100707.

Yang, J., Koprowski, H., Dietzschold, B., & Fu, Z. F. (1999). Phosphorylation of rabies virus nucleoprotein regulates viral RNA transcription and replication by modulating leader RNA encapsidation. *Journal of Virology, 73*(2), 1661–1664.

Further reading

Superti, E., Seganti, L., Tsiang, H., & Orsi, N. (1984). Role of phospholipids in rhabdovirus attachment to CER cells. Brief report. *Archives of Virology, 81*(3–4), 321–828.

CHAPTER 3

Evolution of rabies virus

Daniel G. Streicker[a,b], *Roman Biek*[a]

[a]Institute of Biodiversity, Animal Health and Comparative Medicine, College of Medical Veterinary and Life Sciences, University of Glasgow, Glasgow, Scotland [b]MRC-University of Glasgow Centre for Virus Research, Glasgow, Scotland

3.1 Introduction

Rabies virus (RABV) thrives in a realm of extremes and paradoxes that challenge our view of how viruses evolve and sustain transmission in natural and novel hosts. RABV is universally lethal yet able to maintain itself without driving populations of its generally long-lived and slow-reproducing bat and carnivore hosts extinct. RABV has the biological capacity to infect any mammalian species and cross-species transmission events are readily observed in nature, yet diverse viral variants rely on single host species for their independent, long-term perpetuation. Finally, despite apparent host specificity over ecological timescales, the evolutionary history of RABV is dominated by host shifts both within and between bats and carnivores (Fig. 3.1). This potential to establish transmission cycles in novel host species remains largely unpredictable. RABV achieves each of these seemingly improbable feats with a nonsegmented, single-stranded genome of approximately 12,000 nucleotides encoding only five genes (N, P, M, G, and L). Moreover, like many other RNA viruses, the molecular evolution of RABV is inherently constrained. The absence of a proofreading mechanism in the virus-encoded RNA polymerase leads to high rates of spontaneous mutation (Drake, 1993). While these mutations potentially provide a diverse population of genetic variants upon which selection might act (sometimes referred to as a "quasispecies" or "mutant spectrum"), most mutations will be deleterious and pleiotropic. Thus, despite constant generation of novel genetic diversity, pathways for adaptive evolution are likely to be highly constrained in RABV (Holmes, Woelk, Kassis, & Bourhy, 2002). How has a virus with such an extreme life history, small genome, and constrained evolution become a global threat to human and animal health?

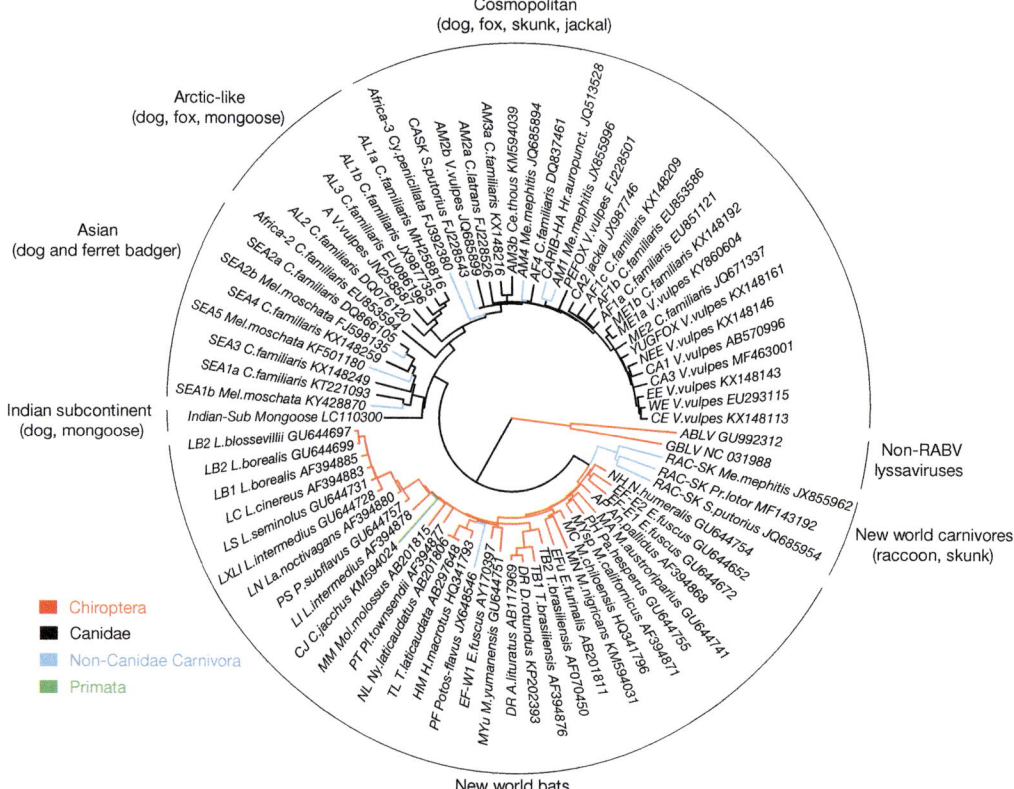

FIG. 3.1 Macroevolutionary dynamics of rabies virus. Maximum likelihood phylogenetic tree of the nucleoprotein gene showing relationships among major RABV lineages. Lineage names follow Troupin et al. (2016), with additional annotations following Streicker et al. (2010) and Velasco-Villa et al. (2017). Branch colors indicate the suspected reservoir association. Host genera are abbreviated as follows: A=*Artibeus*; An=*Antrozous*; C=*Canis*; Ca=*Callithrix*; Ce=*Cerdocyon*; D=*Desmodus*; H=*Histiotus*; Hr=*Herpestes*; L=*Lasiurus*; La=*Lasionycteris*; M=*Myotis*; Me=*Mephitis*; Mel=*Melogale*; Mol=*Molossus*; Ny=*Nyctinomops*; N=*Nycticeius*; P=*Perimyotis*; Pa=*Parastrellus*; Pt=*Plecotus*; S=*Spilogale*; T=*Tadarida*; V=*Vulpes*.

This chapter aims to synthesize the evolutionary repertoire that allowed rabies to spread globally and establish transmission cycles in diverse reservoir hosts. We review both microevolutionary changes at the level of the viral genome and macroevolutionary patterns that explain the origins of new viral variants and discuss the ultimate evolutionary origins of the Lyssaviruses in general and RABV in particular. We then discuss several evolutionary mysteries surrounding RABV: how to reconcile the lack of strong evidence for adaptive evolution using traditional genetic measures of adaptation with the apparent requirements for adaptive evolution implied by host specificity in nature? What are the adaptive barriers that RABV must overcome to establish in new species? Finally, we discuss how the evolution of RABV can inform prevention and control of RABV in wild and domestic animal reservoirs.

3.2 Microevolutionary dynamics of rabies in stable host virus associations

Mutations provide the necessary raw material for any evolutionary change to occur. Like other ssRNA viruses, RABV lacks a proof-reading mechanism in its RNA polymerase, resulting in a high spontaneous mutation rate, causing about one random mutation per genome replication on average (Drake & Holland, 1999; Duffy, Shackelton, & Holmes, 2008). Across the genome, most of these mutations will be deleterious and result in nonviable virus, especially those resulting in nonsynonymous changes in the amino acid sequence. However, a small proportion of variants is maintained, resulting in some accumulation of viral diversity within an infected host. That these variants might exhibit phenotypic differences, such as tissue tropism, has been suggested for some time (Morimoto et al., 1998, 1996). More recently, high-throughput sequencing methods have confirmed substantial variation below the consensus level during serial passage in the same host environment. However, these experiments have failed to provide clear evidence for the virus adapting to specific tissues (Bonnaud et al., 2019) and the significance of within-host variation for maintaining RABV transmission within the same host species remains unclear.

Transmission from one host individual to the next generally results in a reduction in viral diversity due to the combined effect of genetic bottlenecks and negative selection causing the removal of less fit variants. The vast majority of amino acid positions in the RABV genome are under strong purifying selection, as indicated by a severe bias toward synonymous over nonsynonymous changes (Holmes et al., 2002; Troupin et al., 2016). There is little variation among the five RABV genes in this respect, though there might be a slightly higher tolerance toward amino acid changes in the P and G genes (Troupin et al., 2016). Importantly, there is a striking absence of sites under positive selection in the RABV genome, at least while the virus remains associated with the same host species. This is probably explained by the way virus and host interact, in particular the lack of any host factors, such as immune modulation or resistance, that would be able to change the outcome of infection. A constant host environment and the absence of selective pressures changing dynamically are expected to favor conserved protein structures and to negate the need for adaptive change. Where synonymous or nonsynonymous substitutions in the RABV genome become fixed over time, resulting in measurable evolutionary change, they are most likely to represent selectively neutral changes.

The RABV genome evolves on the order of 10^{-4} substitutions per site per year, with little systematic variation among protein coding genes. The noncoding region between the glycoprotein and polymerase (G-L) evolves faster; up to 1.68×10^{-3} substitutions per site per year (Bourhy et al., 2008; Davis, Rambaut, Bourhy, & Holmes, 2007; Troupin et al., 2016). The increasing number of phylogenetic studies using molecular clock methods to quantify evolutionary rates has revealed that RABV evolves at different tempos in different maintenance host species (Fig. 3.2). For example, mongoose-associated RABV strains evolved slower than sympatric canid-associated strains in southern Africa, and RABV evolved nearly four-fold faster in ferret badgers compared to in domestic dogs (Davis et al., 2007; Troupin et al., 2016). Intriguingly, some of the fastest-evolving RABVs are carnivore viruses, while the slowest-evolving viruses are in bats, although there is substantial overlap at intermediate evolutionary rates (Fig. 3.2). In carnivore-associated strains, the underlying drivers of this variation have yet to be robustly explored but could arise from differences in host ecology and life history, sampling biases associated with the temporal or geographic span of datasets,

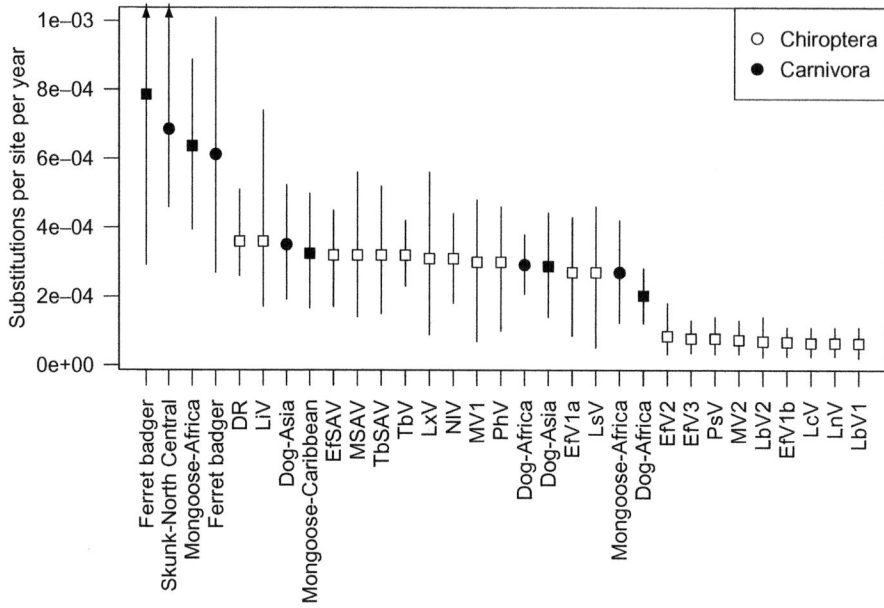

FIG. 3.2   Rates of evolution observed across host-associated viral lineages for the nucleoprotein gene (squares) and glycoprotein gene (circles) in carnivores and bats. Clock rates are from Streicker, Lemey, et al. (2012), Troupin et al. (2016), and Pepin et al. (2017), which contain details on each variant and methods used to estimate clock rates. Bars represent 95% highest posterior density intervals; triangles indicate intervals that were truncated for ease of graphical presentation.

or time-varying rates of evolution with faster evolution following recent host shifts. The larger number of reservoir host-associated viral lineages in bats has enabled more comprehensive analyses of rate variation which lend support to the idea that variation in host ecology and life history alter rates of viral evolution. In particular, viruses associated with tropical and subtropical bat species showed nearly four-fold increases in the tempo of evolution in relative to temperate bats, presumably a consequence of year-round versus seasonally constrained transmission as a result of bat hibernation in the temperate zone (Streicker, Lemey, Velasco-Villa, & Rupprecht, 2012). This seasonal pause in transmission was supported by experimental infections and mathematical modeling, highlighting the value of evolutionary inference to reveal aspects of rabies biology (Davis et al., 2016; George et al., 2011). Further insights on the complex interplay between host ecology and viral evolution may come from comparisons of viruses circulating in widespread host species that occur in different ecological conditions.

Homologous recombination, representing an alternative pathway for introducing genetic novelty and rapid evolutionary change, was long considered absent in RABV. However, several independent studies from the past decade suggest that recombination is biologically possible in RABV and other lyssaviruses though it appears to be extremely rare (Deviatkin & Lukashev, 2018; Ding, Xu, Sun, He, & He, 2017; Liu, Liu, Liu, Zhai, & Xie, 2011). Of the small number of RABV recombination instances reported, most have only been seen in a single sequence. These cases thus provide no indication that the recombinant virus had been transmitted and been able to propagate itself in the host population, possibly because these recombinants experience some fitness costs. A notable exception is a putative recombination event detected within a RABV clade associated with New World carnivores (Ding et al., 2017). The clade consists of several lineages maintained by skunk and raccoon hosts. The glycoprotein carried by the viruses within these lineages appears to be the result of a historical recombination event between an American bat-associated RABV variant and a cosmopolitan strain of RABV that is mainly associated with domestic dogs and which was introduced to the New World during the post-Columbian era. Ding et al. propose that it is the acquisition of the head domain of the glycoprotein from the cosmopolitan RABV strain that facilitated the virus host shift from bat to skunk and raccoon hosts. Recent analyses have also revealed indications for historic recombination events in other RABV strains and between bat lyssaviruses (Deviatkin & Lukashev, 2018), but the evolutionary significance of these events remains unclear. While these results demonstrate the absence of fundamental molecular barriers preventing recombination in RABV and other lyssaviruses, the necessary conditions for this will only be met on very rare occasions. Specifically, the short acute phase of a few days and low population level prevalence make coinfection of the same host individual with two different viruses that are actively replicating a highly unlikely event. To be transmitted, the recombinant virus would also need to reach the salivary glands before the host succumbs to the first infecting virus, making the transmission of recombinants even more improbable. But despite its apparent rarity, recombination has the potential to cause a fundamental shift in the evolutionary trajectory of RABV. More suspected cases are likely to be reported as the amount of published genome data increases and will allow to examine RABV recombination, as well as its molecular mechanisms and epidemiological significance, with further scrutiny.

3.3 Macroevolutionary dynamics of rabies

Numerous host shifts within and among bat and carnivore species are apparent both in deep branches of the RABV phylogeny and in more contemporary events (Fig. 3.1). Host shifts are important as they establish novel animal reservoirs that create new threats to human and animal health (Leslie et al., 2006; Randall et al., 2004; Rupprecht, Smith, Makonnen Fekadu, & Childs, 1995; Smith, Orciari, & Yager, 1995; Zhang et al., 2009). As such, identifying which hosts are most susceptible to establish novel transmission cycles and from which current reservoir these are most likely to emerge is a key evolutionary challenge. Among bats, host shifts are more likely to occur among closely related species; however, RABVs perpetuated by divergent bat families (Vespertillionidae, Mollossidae, and Phyllostomidae) illustrate that host shifts have also occurred over large evolutionary distances (Streicker et al., 2010; Velasco-Villa et al., 2017). Host shifts among carnivores have involved a diverse set of mesocarnivore families including Canidae (coyote, fox, dog); Herpestidae (mongoose), Mephitidae (skunk), Mustelidae (ferret badger), and Procyonidae (raccoon) (Fig. 3.1). However, for reasons that remain unclear, in both carnivores and bats, taxonomic families supporting transmission contain numerous species without evidence of maintaining RABV transmission cycles. One possibility is that ecological or life history traits such as litter size, geographic range area, or population size predispose certain species to maintain RABV. Biological traits such as having sufficient bite force and tooth morphology to pierce the skin of conspecifics may also be prerequisites. Trait-based analyses that use advanced tools like machine learning, as carried out for rodent hosts of zoonoses or mosquito vectors of arboviruses could be a fruitful way forward to proactively identify high-risk species pairs of cross-species transmission (Evans, Dallas, Han, Murdock, & Drake, 2017; Han, Schmidt, Bowden, & Drake, 2015). Alternatively, RABV may not yet have spread to its full potential host range and many species that are currently RABV-free may eventually become new reservoirs. Moreover, establishment of RABV in new host species may allow the virus to explore novel phenotypic and evolutionary space that facilitates further host shifts (Fisher, Streicker, & Schnell, 2018; Mollentze, Biek, & Streicker, 2014). This "snowball effect" hypothesis predicts an expansion in the host range of RABV over time. Host shifts that span large evolutionary distances between donor and recipient species might be particularly important as they would be expected to precipitate additional host shifts among more closely related species. Possible examples of this may be the establishment of RABV in the common vampire bat (*Desmodus rotundus*), which appears to have precipitated one or more host shifts to fruit-eating *Artibeus* bats in the same Phyllostomidae family of neotropical bats (Calderón et al., 2019; Fahl et al., 2012; Pawan, 1948; Shoji et al., 2004) and the spread of RABV among American mesocarnivores following a presumed host shift from bats (Ding et al., 2017).

The association of the vast majority of Lyssaviruses with Old World bats (i.e., Aravan lyssavirus, Bokeloh bat lyssavirus, Duvenhage lyssavirus, European bat lyssaviruses 1 and 2, Khujand lyssavirus, Irkut lyssavirus, Kotalahti bat lyssavirus, Lagos bat lyssavirus, Lleida bat lyssavirus, Shimoni bat lyssavirus, Taiwan bat lyssavirus, West Caucasian bat lyssavirus), relative to the low species diversity found in the Americas (Rabies lyssavirus) and Oceania (Australian bat lyssavirus and Gannoruwa bat lyssavirus) implies a Chiropteran origin of the genus which phylogeographic reconstructions suggest arose in the Palearctic regions of Europe, northern Asia, or North Africa (Hayman, Fooks, Marston, & Garcia-R, 2016; Rupprecht, Kuzmin, & Meslin, 2017). Lyssavirus circulation in the New World therefore likely arose more recently following one or more introductions from the Old World. The ultimate host and geographic origins of RABV remain an unsolved riddle despite decades of research. The closest relatives of RABV are Gannoruwa bat lyssavirus from Sri Lanka and Australian bat lyssavirus, consistent with an Oceanic origin of RABV (Gunawardena et al., 2016). However, among bats RABV only circulates in the New World, and as such, is absent from its putative Old World or Oceanic bat origin. How RABV became established in Old World carnivores and American bats is therefore unclear. Our best current understanding is that RABV originated in Old World bats and shifted to Old World carnivores, which spread the virus globally (Fig. 3.3). However, this would have required selective extinction of RABV from Old World bats amid conditions that favored the perpetuation of RABV in New World bats, of other Old World bat lyssaviruses that predated RABV, and of RABV in Old World carnivores (Kuzmin & Tordo, 2012).

This scenario further requires explaining how RABV reached New World bats. If RABV is ancient, this could have occurred near the origin of Chiroptera (~64MYA) when Australia and South America were connected via a forested Antarctica (Teeling et al., 2005). Rare dispersal of bats to the Americas also occurred during the Eocene (Africa to South America) and during the Pleistocene (Palaearctic to Nearctic) (Lim, 2009). Molecular clock-based phylogenetic estimates have not supported an ancient origin of RABV in New World bats (Hayman et al., 2016; Holmes et al., 2002; Hughes, Orciari, & Rupprecht, 2005), but such analyses are inevitably biased by effects of sampling timescales on inferences of molecular evolutionary rates and should be revisited as new approaches become possible (Duchêne, Holmes, & Ho, 2014; Membrebe, Suchard, Rambaut, Baele, & Lemey, 2019).

RABV in indigenous American carnivores (raccoon/skunk clade) seems most likely to have arisen from New World bats. Some evidence suggests this host shift was mediated by a recombination event with Old World carnivore RABVs when these viruses were introduced in the post-Columbian era (Ding et al., 2017; Velasco-Villa et al., 2017). This scenario is further supported by phylogenetic estimates that indigenous American carnivore RABV originated within the last 400 years (Kuzmina et al., 2013). However, for reasons that remain unclear, American bat RABVs appear only slightly older and were not discovered until 1911 in Trinidad and 1953 in North America (Carini, 1911; Hughes et al., 2005; Pawan, 1936; Scatterday, 1954; Streicker, Altizer, Velasco-Villa, & Rupprecht, 2012). Geographic and host range expansions of both American bat and carnivore RABVs also suggest that neither are at an epidemiological equilibrium, though it is difficult to conclude whether this reflects recent viral introduction to the Americas, effects of environmental change on reservoir distributions or interspecific interactions, or improved surveillance (Benavides, Valderrama, & Streicker, 2016; Biek, Henderson, Waller, Rupprecht, & Real, 2007; Johnson, Aréchiga-Ceballos, & Aguilar-Setien, 2014; Kuzmina et al., 2013; Pepin et al., 2017; Streicker et al., 2016). The accelerating discovery of novel RABV and non-RABV lyssaviruses through field studies may fill evolutionary gaps that ultimately resolve the conundrum of RABV origins (Aréchiga Ceballos et al., 2013; Arechiga-Ceballos et al., 2010; Hu et al., 2018; Nokireki, Tammiranta, Kokkonen, Kantala, & Gadd, 2018). Increasing potential for using metagenomics to sequence degraded viruses from formalin-fixed collections or fortuitous discoveries of preserved tissues may similarly yield new insights (Ng et al., 2014; Xiao, Halbur, & Opriessnig, 2012).

Whereas multiple historic host shifts are evident from the RABV phylogeny, their timing, location, and the circumstances under which they occurred remain largely uncertain.

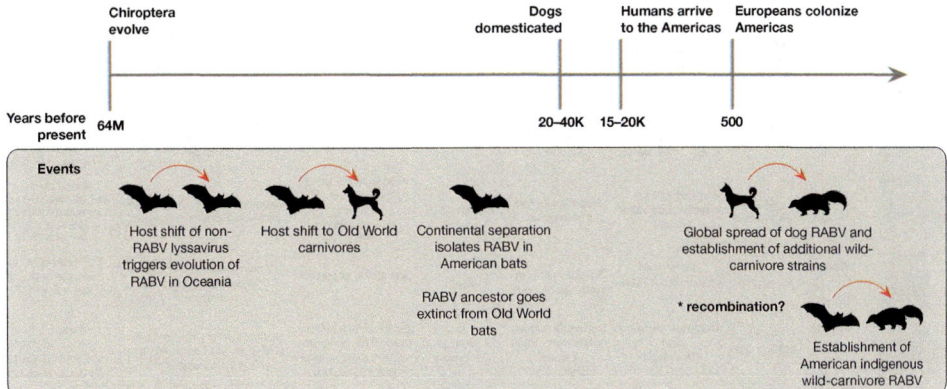

FIG. 3.3 Timeline of potential events in the evolutionary history of RABV. Figure shows the hypothesized order of key events, but timescales are approximate and some events could have happened over long time intervals. *Animal silhouettes from https://creazilla.com.*

90   3. Evolution of rabies virus

This includes lack of precise information about the host species that donated the virus and the one that received it, as subsequent host shifts and lineage extinctions might have resulted in extant RABV lineages providing an incomplete picture of historic processes (Fig. 3.4). However, there are numerous examples of host shifts that occurred in contemporary times, some of which have been documented in considerable detail (Fig. 3.5). These cases can provide

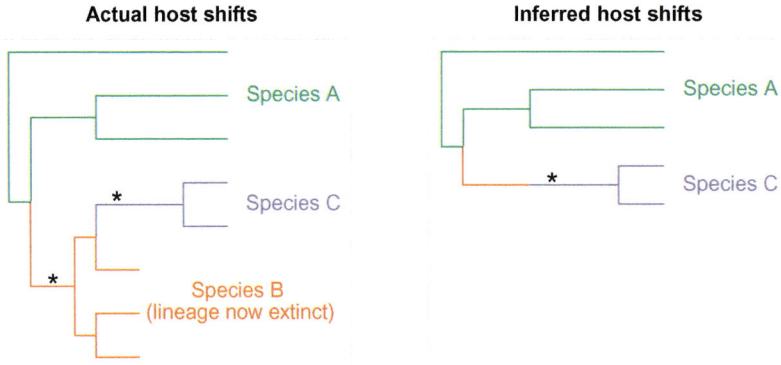

FIG. 3.4 Historic host shifts might not be accurately inferred from contemporary data. Left panel: Two host shifts occurred (represented by asterisks), with virus jumping from species A to species B and from species B to species C. Subsequent to the second host shift, the virus lineage in species B became extinct. As a consequence, virus sequence data from the two extant lineages in species A and C provide an incomplete picture by indicating that the virus jumped directly from species A to C (right panel). In addition, all genetic changes that occurred after the host shift from species A to species B (shown in orange) will also be incorrectly attributed to the wrongly inferred host shift from species A to species C.

Recipient host(s)	Source host	Introductions/Outcome	Location	Timeline	Evolutionary adaptation	Key references
Formosan ferret badger (*Melogale moschata*)	Domestic dog (*Canis familiaris*)	Multiple/Sustained transmission	Taiwan, China	Detection in 2012, Molecular evidence of circulation for 91-113 years before	Parallel evolution observed in two sites in two independent host shifts	Zhao et al. (2014) Chiou et al. (2014) Lin et al. (2016) Troupin et al. (2016)
Small Indian mongoose (*Herpestes auropunctatus*)	Domestic dog (*Canis familiaris*)	Multiple/Sustained transmission	Caribbean (Cuba, Grenada, Puerto Rico)	Multiple host shifts between 1870s and early 1900s	Not investigated	Everard and Everard (1992) Zieger et al. (2014) Seetahal et al. (2018)
Striped skunk (*Mephitis mephitis*) Gray fox (*Urocyon cinereoargenteus*)	Big brown bat (*Eptesicus fuscus*)	Multiple/Sustained transmission, but ultimately controlled by vaccination	Arizona, USA	Distinct outbreaks in 2001 (twice) and 2009	Positive selection on M and L, but considered false positives One candidate site (242G) linked to bat to carnivore host shifts	Leslie et al. (2006) Kuzmin et al. (2012)
Greater kudu (*Tragelaphus strepsiceros*)	Black-backed jackal* (*Canis mesomelas*)	Multiple?/Sustained transmission, but controversial	Namibia	1975 to present	3 sites in (34G, 112G, 191G) have Kudu-specific amino acid changes	Barnard et al. (1982) Mansfield et al. (2006) Scott et al. (2013) Hassel et al. (2018)
Raccoon (*Procyon lotor*)	Unidentified insectivorous bat or skunk	Single/Sustained transmission	Eastern USA	Host shift pre 1950s Epizootic spread following translocation in 1970s	Recombination between bat and carnivores strains suggested	Nettles, Shaddock, Keith Sikes and Reyes (1979) Rupprecht and Smith (1994) Biek et al. (2007) Ding et al. (2017)
White tufted marmoset (*Callithrix jacchus*)	Unidentified (insectivorous?) bat	Single/Sustained transmission likely	Brazil	Late 1980s to present	69 marmoset-specific amino acid changes across the genome	Favoretto et al. (2001) Favoretto et al. (2013) Kotait et al. (2018)
Fruit bats (*Artibeus spp.*)	Common vampire bat (*Desmodus rotundus*)	Unclear/Sustained or recurrent cross-species transmission?	Brazil, Colombia	Evidence of infection from 1920s onwards, existence or timing of host shift unclear	Distinct clusters in amino acid phylogeny of G separate fruit-eating from vampire bats	Pawan (1948) Shoji et al. (2004) Fahl et al. (2012) Calderón et al. (2019)

* Potentially domestic dog

FIG. 3.5 Examples of well-studied and putative host shifts to diverse orders of mammals. *Images from Wikimedia commons (https://commons.wikimedia.org)*.

useful insights into the factors involved when RABV emerges in a novel host, including an opportunity to better understand the role of evolution in this process. We define host shifts here as cross-species transmission events that were followed by subsequent transmission in the novel host, even if the virus did not persist in the novel host in the long run.

Interestingly, several reports of putatively novel RABV transmission cycles involve noncanonical hosts, including in South American primates (Condori-Condori, Streicker, Cabezas-Sanchez, & Velasco-Villa, 2013; Favoretto et al., 2013; Favoretto, de Mattos, Morais, Araujo, & de Mattos, 2001; Kotait et al., 2018) and an African artiodactyl (Barnard, Hassel, Geyer, & De Koker, 1982; Hassel et al., 2018; Mansfield et al., 2006; Scott et al., 2013). These apparent host shifts have questioned the longstanding perception that RABV is exclusively maintained by bats and carnivores (Fig. 3.5). Sustained transmission is supported by large numbers of cases over multiple years in these species and their viruses clustering together on the RABV phylogeny. Still, it is often difficult to discern whether these are true host shifts, recurrent spillover from poorly sampled species, or some combination of the two. This is exacerbated where there is uncertainty about the origins of these new variants because no closely related viruses in other species have been detected. Such cases prevent assessment of potential adaptive evolution since it is difficult to evaluate whether observed amino acid changes occurred in the novel host or the unidentified reservoir (Fig. 3.4). For example, the marmoset RABV encountered in South America occurs on a long branch within a larger clade of insectivorous bats, making it unlikely that all of the divergence from known bat viruses occurred in marmosets (Kotait et al., 2018). An emergent pattern from available data is that successful introductions of RABV into a novel host species often occurred more than once from the same source reservoir host species (Fig. 3.5). This suggests that these emergence events might not be simply due to chance but also involve deterministic features (e.g., pre-adaptations in the viral lineages involved). It also creates the intriguing opportunity to test for parallel evolutionary changes in the virus during multiple introductions, which can provide a strong indication that such changes are adaptive (Gutierrez, Escalera-Zamudio, & Pybus, 2019).

3.4 Reconciling strains of evidence for the role of adaptive evolution in cross-species transmission

Evolutionary specialization of viruses to their host may preadapt them to establish transmission cycles in alternative hosts or restrict their host range if adaptations to one host species are disadvantageous in another (Holmes, 2009; Longdon, Brockhurst, Russell, Welch, & Jiggins, 2014). Cell-culture passaging, serial infections in laboratory animals, and naturally occurring cross-species emergence events have revealed how the genetic background of RABV influences the outcome of cross-species transmission at different scales. Morimoto et al. (1998) showed that passaging standard neurotropic virus on baby hamster kidney (BHK) cells selected for amino acid changes in the G protein that reduced neurotropism, increased replication in non-neuronal cells, and lowered virulence in adult mice. However, the extent and genomic location of evolutionary changes also seem to depend on the viral genetic background and context of the novel host environment, making dynamics less predictable.

For example, a fox RABV passaged on BHK cells, mice, dogs, and cats maintained a relatively stable consensus sequence, but underwent dramatic changes in mutation frequency in N, P, and G genes along with host-dependent mutations, suggesting potential for rapid positive selection (Kissi et al., 1999). Similarly, Bonnaud et al. (2019) showed elevated mutation rates in foxes compared to dogs following reciprocal cross-infections and passaging. This led to positively selected changes across the genome, several of which could be linked to natural RABVs host shifts.

Experimental studies have also produced compelling evidence for phenotypic changes and host specialization. For example, raccoons inoculated with a raccoon-derived RABV strain died acutely or showed severe signs, whereas signs were more subtle among raccoons inoculated with a strain isolated from dogs (Hamir, Moser, & Rupprecht, 1996). Interestingly, the evolutionary pathways allowing the virus to adapt to new species might involve directional biases. Dog-adapted RABV was able to easily infect red fox cell lines and to establish in experimentally infected foxes. In contrast, the fox-adapted strain struggled to cause infection in the dog host environment both in vitro and in vivo (Bonnaud et al., 2019). Dose dependency could be a potential explanation for such asymmetries. When experimentally infecting skunks and foxes with a fox-derived RABV strain, Sikes (1962) found viral titers in fox saliva were generally not high enough to cause infections in skunks, while skunks experimentally infected with the same virus produced titers that were able to kill foxes. Importantly, these differences appear to be attributable to virus effects rather than variation in susceptibility among host species. For example, striped skunks succumbed to infection with their own RABV variant during experiment infection (Hill Jr & Beran, 1992; Hill Jr, Smith, Beran, & Beard, 1993) but failed to develop signs in the aforementioned experiment involving a virus from foxes (Sikes, 1962).

Based on these laboratory findings and the frequency of host shifts in RABV evolutionary history, it may be surprising that large-scale studies across RABV have rarely detected sites under positive selection (Holmes et al., 2002; Troupin et al., 2016). Generation of viral diversity does not appear to be a limiting factor: Marston et al. (2017) found that bursts in subpopulation viral heterogeneity are detectable at the population level during the early stages of natural host shifts, providing ample raw material for adaptive evolution. Yet despite this, it has not been possible to identify specific sites in the RABV genome that are necessarily involved in multiple host shifts. One explanation could be that the methods used had little power to detect selection dependent on viral genetic background or the virus-host combination involved, rather than codons that underwent nonsynonymous changes each time the virus established in a new host. In support of this possibility, Streicker, Altizer, et al. (2012) found episodic selection on a diverse set of codons in G and L genes following host shifts among bats, but the number and identity of sites under selection were inconsistent across host shifts, confirming that molecular signatures of host adaptation are highly idiosyncratic and would be missed by efforts to find repeated selection on the same sites.

Detailed investigations of outbreaks initiated by cross-species transmission have supported the context dependency on the evolutionary dynamics of host shifts that arises from the interaction between viral genetic background and host environment. Two independent host jumps of RABV from dogs into ferret badgers were found to be associated with the same two amino changes in both cases, Leu-N374-Ser and Lys-L200-Arg (Troupin et al., 2016). The former is a particularly strong candidate for adaptive change, given that this position is otherwise highly conserved in RABVs evolutionary history and that the change from Leu to Ser is nonconservative. However, it is not clear from the current data whether these variants were circulating at low frequency within dogs prior to the host shift or whether they arose subsequent to the virus crossing over into ferret badgers. In another host shift involving multiple outbreaks of bat RABVs in carnivores in the southwest USA, Kuzmin et al. (2012) identified that bat viruses with T242 rather than S242 in the G ectodomain—a site previously implicated in pathogenesis (Takayama-Ito, Ito, Yamada, Sugiyama, & Minamoto, 2006)—seemed disproportionately likely to transmit among carnivores, but found little evidence for post-host shift adaptation. Although it should be noted that circulation in novel carnivores was rapidly extinguished by outbreak responses, making it unclear whether positive selection would have been detected if transmission had continued. A deep sequencing study of an outbreak of skunk rabies in gray foxes similarly pointed to selection of preadapted variants (Borucki et al., 2013). Few amino acids changed during transmission among foxes, and genotypes associated with the outbreak in foxes were detected as rare variants in the viral population in earlier samples from skunks. The emerging consensus is that RABV establishment in novel hosts requires species-specific viral genotypes, which can either be pre-existing ancestral states, arise through enrichment of pre-existing viral subpopulations that already existed in the donor host or, perhaps more rarely, evolve de novo in the novel host. This may explain the geography of where host shifts have and have not occurred in broadly distributed host species. For example, raccoons are occasionally infected by RABV variants throughout the USA (Wallace et al., 2014), but only a single viral variant has established in raccoons in the east coast. Similarly, bats maintain rabies throughout the USA, but outbreaks of bat RABV in skunks are restricted to a single viral variant, which circulates in the southwestern bats (Kuzmin et al., 2012).

It is clear from the infection experiments cited here that heterologous transfer of virus to a novel host might result in disease and host death before transmission could occur. In general, longer incubation times in RABV have been found to be associated with more widespread dissemination of the virus and higher titers in the salivary gland (Baer & Bales, 1967; Fekadu, Shaddock, & Baer, 1982). In contrast, shorter incubation times tend to result in lower titers or even host death prior to the virus reaching the salivary glands (Baer & Bales, 1967; Fekadu et al., 1982; Sikes, 1962). This suggests that the virus faces a tradeoff between efficient and rapid replication and prolonging the infectious period to maximize the opportunity for transmission. This kind of virulence-transmission tradeoff, which has received widespread attention in the infectious disease literature (Ewald, 1983; Galvani, 2003), would be expected to lead to continuous selection for optimal replication and virulence (i.e., time to host death) even after the virus has been able to establish in a novel host. In fact, temporal or spatial changes in the ecological conditions affecting host populations, such as fluctuations in densities and contact rates, could continue to alter selection pressures in this respect. Consistent with this hypothesis, Aubert, Blancou, Barrat, Artois, and Barrat (1991) found that an isolate of red fox RABV taken early during an outbreak caused transmission intervals of more variable lengths in experimental infections compared to an isolate collected from the same population ten years later. The mechanisms driving this phenotypic variation among RABV isolates, as well as their genetic determinants, are not yet understood.

Although most studies of the evolutionary underpinnings and phenotypic consequences of RABV adaptation have focused on nonsynonymous substitutions, fully resolving how

RABV adapts to new host species may further require considering less canonical forms of viral adaptation. For example, synonymous substitutions are increasingly appreciated to regulate translation and influence mRNA stability and secondary structure, forming a functional basis for their involvement in adaptive evolution (Cuevas, Domingo-Calap, & Sanjuán, 2012; Novella, Zárate, Metzgar, & Ebendick-Corpus, 2004). A related possibility is that adaptation may involve synonymous genome-wide biases in dinucleotide, codon, or codon-pair use; a pattern which is broadly associated with animal reservoir and arthropod vector associations across a variety of RNA viruses (Babayan, Orton, & Streicker, 2018; Coleman et al., 2008; Martínez, Jordan-Paiz, Franco, & Nevot, 2016; Plotkin & Kudla, 2011; Shen et al., 2015). Synonymous changes that modulate replication rate in the novel host could favor RABV arrival in the salivary glands before overt host morbidity and death, enabling transmission to new hosts. Finally, the diversity of viral and host-origin components within the bolus of natural rabies exposures has been almost entirely overlooked. In influenza, host proteins embedded in the viral envelope have significant impact on viral replication (Hutchinson et al., 2014); though in RABV, this effect could only be expected to influence the earliest events during exposure.

3.5 Applications of evolutionary data for rabies prevention and control—genetics as a tag on transmission

The remarkable volume and quality of data collected through public health and veterinary surveillance of RABV has facilitated the integration of evolutionary data into rabies management. A salient epidemiological feature of bat and carnivore rabies is epizootic waves of infection emanating from point source introductions (Benavides et al., 2016; Biek et al., 2007; Pepin et al., 2017). These invasions can be readily identified and characterized using recently developed phylogeographic tools (Dellicour, Rose, & Pybus, 2016; Lemey, Rambaut, Welch, & Suchard, 2010). Particularly when constrained by geographic features such as mountain ranges or rivers, as observed for vampire bat and raccoon rabies respectively, spread can be predictable (Benavides et al., 2016; Real et al., 2005; Russell, Smith, Waller, Childs, & Real, 2004; Smith, Lucey, Waller, Childs, & Real, 2002; Streicker et al., 2016). This potentially allows for evidence-based design of the location and width of vaccine corridors targeting either natural reservoirs or spillover hosts such as humans or domestic animals. Even in RABV-endemic areas, molecular tools are increasingly useful to identify the determinants of viral persistence and dispersal, which ultimately could inform management, for example by directing vaccination campaigns to crucial areas for viral maintenance or mitigating human-mediated viral spread, respectively (Bourhy et al., 2016; Brunker et al., 2018; Brunker, Hampson, Horton, & Biek, 2012; Streicker, González, Luconi, Barrientos, & Leon, 2019; Talbi et al., 2010).

The need for rapid and precise characterization of RABV variants arises especially for new incursions into rabies-free areas. Many parts of Canada near the USA border, for example, have experienced outbreaks of raccoon rabies over the past decades through introductions across a vaccination corridor that is meant to contain the endemic area on the USA side.

Trewby, Nadin-Davis, Real, and Biek (2017) used RABV genomic data from both sides of the border to identify potential source populations for such incursions and to better understand the processes causing them. They found that the majority of outbreaks were caused by viruses circulating locally near the vaccination zone and short-distance dispersal. In some cases, this involved repeated virus movement across the border, pointing to more systematic breaches of the vaccine corridor. Similarly, genetic source attribution plays an important role in situations where cases are observed after elimination has been achieved, such as after widespread dog vaccination campaigns (Brunker et al., 2015). Whereas sequence data from single genes can be sufficient to broadly characterize emerging strains and to establish their relationship to previously observed variants, full genome data might be required to achieve sufficient phylogenetic resolution for finer placement. Thanks to high-throughput sequencing technologies becoming more affordable and more portable, such data can increasingly be generated in situ, which is expected to become an increasingly relevant aspect of RABV elimination and control programs (Brunker, Nadin-Davis, & Biek, 2018). Strategic planning based on model forecasts remains relatively uncommon but is an emerging area where evolutionary data can plausibly enhance existing RABV control.

3.6 Conclusions

Lyssaviruses that cause rabies are globally distributed, but the richness of species found in Eurasia and Africa (16 established or putative species), relative to the low diversity found elsewhere in the world (three species) together with phylogenetic evidence point to an Old World origin of the genus, with later spread to Oceania and the Americas. Although details on the geographic and species origins of RABV remain speculative, its subsequent evolutionary history has been dominated by host shifts, which can be observed both on historical and contemporary time scales. The case for host adaptation in cross-species emergence is clear from phenotypic and molecular studies, but the molecular mechanisms and predictability remain elusive. A key challenge is to bridge the gap between tractable experimental passaging studies in vitro and in vivo and the real-world complexities of transmission in natural settings. For example, experimental infections are a necessary but imperfect proxy for natural transmission through bite. The molecular changes associated with host shifts currently appear idiosyncratic. However, as an ever-increasing diversity of host shifts is being documented, patterns are beginning to emerge, particularly from parallel host shifts between the same donor and recipient host species. The revolution in sequencing capabilities is further now allowing interrogation of these host shifts at the genomic scale, including noncanonical forms of adaptation by synonymous changes. A logical next step may be to use reverse genetics to explore the consequences of hypothesized adaptive changes on infection, pathogenesis, and transmission via controlled infections. In nature, continued surveillance and characterization of RABV variants and non-RABV lyssaviruses will help clarify the origins of viral diversity and identify missing links associated with contemporary host shifts. Along with targeted field studies, these data will provide insights into emerging rabies reservoirs in nontraditional host groups. Although prediction seems far away, it increasingly appears possible to identify the necessary conditions for future RABV host shifts and the role of evolution therein.

References

Aréchiga Ceballos, N., Vázquez Morón, S., Berciano, J. M., Nicolás, O., Aznar López, C., Juste, J., & Echevarría, J. E. (2013). Novel lyssavirus in bat, Spain. *Emerging Infectious Diseases, 19*(5), 793–795. https://doi.org/10.3201/eid1905.121071.

Arechiga-Ceballos, N., Velasco-Villa, A., Shi, M., Flores-Chavez, S., Barron, B., Cuevas-Dominguez, E., & Aguilar-Setien, A. (2010). New rabies virus variant found during an epizootic in white-nosed coatis from the Yucatan Peninsula. *Epidemiology and Infection, 138*(11), 1586–1589.

Aubert, M. F., Blancou, J., Barrat, J., Artois, M., & Barrat, M. J. (1991). Transmissibility and pathogenicity in the red fox of two rabies viruses isolated at a 10 year interval. *Annales de Recherches Veterinaires. Annals of Veterinary Research, 22*(1), 77–93.

Babayan, S. A., Orton, R. J., & Streicker, D. G. (2018). Predicting reservoir hosts and arthropod vectors from evolutionary signatures in RNA virus genomes. *Science, 362*(6414), 577–580. https://doi.org/10.1126/science.aap9072.

Baer, G. M., & Bales, G. L. (1967). Experimental rabies infection in the Mexican freetail bat. *Journal of Infectious Diseases, 117*(1), 82–90.

Barnard, B. J. H., Hassel, R. H., Geyer, H. J., & De Koker, W. C. (1982). Non-bite transmission of rabies in kudu (Tragelaphus strepsiceros). *The Onderstepoort Journal of Veterinary Research, 49*(4), 191–192.

Benavides, J. A., Valderrama, W., & Streicker, D. G. (2016). Spatial expansions and travelling waves of rabies in vampire bats. *Proceedings of the Royal Society B: Biological Sciences, 283*(1832), 1–9. https://doi.org/10.1098/rspb.2016.0328.

Biek, R., Henderson, J. C., Waller, L. A., Rupprecht, C. E., & Real, L. A. (2007). A high-resolution genetic signature of demographic and spatial expansion in epizootic rabies virus. *Proceedings of the National Academy of Sciences of the United States of America, 104*(19), 7993–7998.

Bonnaud, E. M., Troupin, C., Dacheux, L., Holmes, E. C., Monchatre-Leroy, E., Tanguy, M., & Bourhy, H. (2019). Comparison of intra- and inter-host genetic diversity in rabies virus during experimental cross-species transmission. *PLoS Pathogens, 15*(6), e1007799. https://doi.org/10.1371/journal.ppat.1007799.

Borucki, M. K., Chen-Harris, H., Lao, V., Vanier, G., Wadford, D. A., Messenger, S., & Allen, J. E. (2013). Ultra-deep sequencing of intra-host rabies virus populations during cross-species transmission. *PLoS Neglected Tropical Diseases. 7*(11),e2555. https://doi.org/10.1371/journal.pntd.0002555.

Bourhy, H., Nakouné, E., Hall, M., Nouvellet, P., Lepelletier, A., Talbi, C., & Rambaut, A. (2016). Revealing the microscale signature of endemic zoonotic disease transmission in an African urban setting. *PLoS Pathogens, 12*(4), 1–15. https://doi.org/10.1371/journal.ppat.1005525.

Bourhy, H., Reynes, J. -M., Dunham, E. J., Dacheux, L., Larrous, F., Huong, V. T. Q., & Holmes, E. C. (2008). The origin and phylogeography of dog rabies virus. *The Journal of General Virology, 89*(Pt 11), 2673–2681. https://doi.org/10.1099/vir.0.2008/003913-0.

Brunker, K., Hampson, K., Horton, D. L., & Biek, R. (2012). Integrating the landscape epidemiology and genetics of RNA viruses: rabies in domestic dogs as a model. *Parasitology*, 1–15. https://doi.org/10.1017/S003118201200090X.

Brunker, K., Lemey, P., Marston, D. A., Fooks, A. R., Lugelo, A., Ngeleja, C., & Biek, R. (2018). Landscape attributes governing local transmission of an endemic zoonosis: Rabies virus in domestic dogs. *Molecular Ecology, 27*(3), 773–788. https://doi.org/10.1111/mec.14470.

Brunker, K., Marston, D. A., Horton, D. L., Cleaveland, S., Fooks, A. R., Kazwala, R., & Hampson, K. (2015). Elucidating the phylodynamics of endemic rabies virus in eastern Africa using whole-genome sequencing. *Virus Evolution, 1*(1), 1–11. https://doi.org/10.1093/ve/vev011.

Brunker, K., Nadin-Davis, S., & Biek, R. (2018). Genomic sequencing, evolution and molecular epidemiology of rabies virus. *Revue Scientifique et Technique (International Office of Epizootics), 37*(2), 401–408.

Calderón, A., Guzmán, C., Mattar, S., Rodríguez, V., Acosta, A., & Martínez, C. (2019). Frugivorous bats in the Colombian Caribbean region are reservoirs of the rabies virus. *Annals of Clinical Microbiology and Antimicrobials 18*(1)11. https://doi.org/10.1186/s12941-019-0308-y.

Carini, A. (1911). About one large epizootic of rabies. *Annales de l'Institut Pasteur, 25*, 843–846.

Chiou, H. Y., Hsieh, C. H., Jeng, C. R., Chan, F. T., Wang, H. Y., & Pang, V. F. (2014). Molecular characterization of cryptically circulating rabies virus from ferret badgers, Taiwan. *Emerging Infectious Diseases, 20*(5), 790–798. https://doi.org/10.3201/eid2005.131389.

Coleman, J. R., Papamichail, D., Skiena, S., Futcher, B., Wimmer, E., & Mueller, S. (2008). Virus attenuation by genome-scale changes in codon pair bias. *Science, 320*(5884), 1784–1787. https://doi.org/10.1126/science.1155761.

Condori-Condori, R. E., Streicker, D. G., Cabezas-Sanchez, C., & Velasco-Villa, A. (2013). Enzootic and epizootic rabies associated with vampire bats, Peru. *Emerging Infectious Diseases, 19*(9), 1463–1469. https://doi.org/10.3201/eid1909.130083.

Cuevas, J. M., Domingo-Calap, P., & Sanjuán, R. (2012). The fitness effects of synonymous mutations in DNA and RNA viruses. *Molecular Biology and Evolution, 29*(1), 17–20. https://doi.org/10.1093/molbev/msr179.

Davis, A. D., Morgan, S. M. D., Dupuis, M., Poulliott, C. E., Jarvis, J. A., Franchini, R., ... Rudd, R. J. (2016). Overwintering of rabies virus in silver haired bats (Lasionycteris noctivagans). *PLoS One, 11*(5), e0155542. https://doi.org/10.1371/journal.pone.0155542.

Davis, P. L., Rambaut, A., Bourhy, H., & Holmes, E. C. (2007). The evolutionary dynamics of canid and mongoose rabies virus in southern Africa. *Archives of Virology, 152*(7), 1251–1258.

Dellicour, S., Rose, R., & Pybus, O. G. (2016). Explaining the geographic spread of emerging viruses: a new framework for comparing viral genetic information and environmental landscape data. *BMC Bioinformatics, 17*(82), 1–12. https://doi.org/10.1186/s12859-016-0924-x.

Deviatkin, A. A., & Lukashev, A. N. (2018). Recombination in the rabies virus and other lyssaviruses. *Infection, Genetics and Evolution, 60*, 97–102. https://doi.org/10.1016/J.MEEGID.2018.02.026.

Ding, N. Z., Xu, D. S., Sun, Y. Y., He, H. B., & He, C. Q. (2017). A permanent host shift of rabies virus from Chiroptera to Carnivora associated with recombination. *Scientific Reports, 7*(1), 1–9. https://doi.org/10.1038/s41598-017-00395-2.

Drake, J. W. (1993). Rates of spontaneous mutation among RNA viruses. *Proceedings of the National Academy of Sciences of the United States of America, 90*(9), 4171–4175.

Drake, J. W., & Holland, J. J. (1999). Mutation rates among RNA viruses. *Proceedings of the National Academy of Sciences of the United States of America, 96*(24), 13910–13913.

Duchêne, S., Holmes, E. C., & Ho, S. Y. W. (2014). Analyses of evolutionary dynamics in viruses are hindered by a time-dependent bias in rate estimates. *Proceedings of the Royal Society B: Biological Sciences, 281*(1786). 20140732. https://doi.org/10.1098/rspb.2014.0732.

Duffy, S., Shackelton, L. A., & Holmes, E. C. (2008). Rates of evolutionary change in viruses: patterns and determinants. *Nature Reviews Genetics, 9*(4), 267–276. https://doi.org/10.1038/nrg2323.

Evans, M. V., Dallas, T. A., Han, B. A., Murdock, C. C., & Drake, J. M. (2017). Data-driven identification of potential zika virus vectors. *eLife, 6*, 1–38. https://doi.org/10.7554/eLife.22053.

Everard, C. O. R., & Everard, J. D. (1992). Mongoose rabies in the Caribbean. *Annals of the New York Academy of Sciences, 653*(1), 356–366. https://doi.org/10.1111/j.1749-6632.1992.tb19662.x.

Ewald, P. W. (1983). Host-parasite relations, vectors, and the evolution of disease severity. *Annual Review of Ecology and Systematics, 14*(1), 465–485.

Fahl, W. O., Carnieli, P., Jr., Castilho, J. G., Carrieri, M. L., Kotait, I., Iamamoto, K., ... Brandão, P. E. (2012). Desmodus rotundus and Artibeus spp. bats might present distinct rabies virus lineages. *The Brazilian Journal of Infectious Diseases, 16*(6), 545–551.

Favoretto, S., de Mattos, C., de Mattos, C., Campos, C., Sacramento, D., & Durigon, E. (2013). The emergence of wildlife species as a source of human rabies infection in Brazil. *Epidemiology and Infection, 141*(7), 1552–1561. https://doi.org/10.1017/S0950268813000198.

Favoretto, S. R., de Mattos, C. C., Morais, N. B., Araujo, F. A. A., & de Mattos, C. A. (2001). Rabies in marmosets (Callithrix jacchus), Ceara, Brazil. *Emerging Infectious Diseases, 7*(6), 1062–1065.

Fekadu, M., Shaddock, J. H., & Baer, G. M. (1982). Excretion of rabies virus in the saliva of dogs. *Journal of Infectious Diseases, 145*(5), 715–719.

Fisher, C. R., Streicker, D. G., & Schnell, M. J. (2018). The spread and evolution of rabies virus: Conquering new frontiers. *Nature Reviews Microbiology, 16*(4), 241–255. https://doi.org/10.1038/nrmicro.2018.11.

Galvani, A. P. (2003). Epidemiology meets evolutionary ecology. *Trends in Ecology & Evolution, 18*(3), 132–139.

George, D. B., Webb, C. T., Farnsworth, M. L., O'Shea, T. J., Bowen, R. A., Smith, D. L., ... Rupprecht, C. E. (2011). Host and viral ecology determine bat rabies seasonality and maintenance. *Proceedings of the National Academy of Sciences, 108*(25), 10208–10213. https://doi.org/10.1073/pnas.1010875108.

Gunawardena, P. S., Marston, D. A., Ellis, R. J., Wise, E. L., Karawita, A. C., Breed, A. C., ... Fooks, A. R. (2016). Lyssavirus in Indian Flying Foxes, Sri Lanka. *Emerging Infectious Diseases*, 22(8), 1456–1459. https://doi.org/10.3201/eid2208.151986.

Gutierrez, B., Escalera-Zamudio, M., & Pybus, O. G. (2019). Parallel molecular evolution and adaptation in viruses. *Current Opinion in Virology*, 34, 90–96.

Hamir, A. N., Moser, G., & Rupprecht, C. E. (1996). Clinicopathologic variation in raccoons infected with different street rabies virus isolates. *Journal of Veterinary Diagnostic Investigation*, 8(1), 31–37.

Han, B. A., Schmidt, J. P., Bowden, S. E., & Drake, J. M. (2015). Rodent reservoirs of future zoonotic diseases. *Proceedings of the National Academy of Sciences*, 112(22), 201501598. https://doi.org/10.1073/pnas.1501598112.

Hassel, R., Vos, A., Clausen, P., Moore, S., van der Westhuizen, J., Khaiseb, S., ... Müller, T. (2018). Experimental screening studies on rabies virus transmission and oral rabies vaccination of the Greater Kudu (Tragelaphus strepsiceros). *Scientific Reports*. 8(1), 16599. https://doi.org/10.1038/s41598-018-34985-5.

Hayman, D. T. S., Fooks, A. R., Marston, D. A., & Garcia-R, J. C. (2016). The Global Phylogeography of Lyssaviruses—Challenging the "Out of Africa" Hypothesis. *PLoS Neglected Tropical Diseases*. 10(12), e0005266. https://doi.org/10.1371/JOURNAL.PNTD.0005266.

Hill, R. E., Jr., & Beran, G. W. (1992). Experimental inoculation of raccoons (Procyon lotor) with rabies virus of skunk origin. *Journal of Wildlife Diseases*, 28(1), 51–56.

Hill, R. E., Jr., Smith, K. E., Beran, G. W., & Beard, P. D. (1993). Further studies on the susceptibility of raccoons (Procyon lotor) to a rabies virus of skunk origin and comparative susceptibility of striped skunks (Mephitis mephitis). *Journal of Wildlife Diseases*, 29(3), 475–477.

Holmes, E. C. (2009). *The evolution and emergence of RNA viruses*. USA: Oxford University Press.

Holmes, E. C., Woelk, C. H., Kassis, R., & Bourhy, H. (2002). Genetic constraints and the adaptive evolution of rabies virus in nature. *Virology*, 292(2), 247–257.

Hu, S. -C., Hsu, C. -L., Lee, M. -S., Tu, Y. -C., Chang, J. -C., Wu, C. -H., ... Cheng, M. -C. (2018). Lyssavirus in Japanese Pipistrelle, Taiwan. *Emerging Infectious Diseases*, 24(4), 782.

Hughes, G. J., Orciari, L. A., & Rupprecht, C. E. (2005). Evolutionary timescale of rabies virus adaptation to North American bats inferred from the substitution rate of the nucleoprotein gene. *The Journal of General Virology*, 86(Pt 5), 1467–1474. https://doi.org/10.1099/vir.0.80710-0.

Hutchinson, E. C., Charles, P. D., Hester, S. S., Thomas, B., Trudgian, D., Martínez-Alonso, M., & Fodor, E. (2014). Conserved and host-specific features of influenza virion architecture. *Nature Communications*, 5(May), 4816. https://doi.org/10.1038/ncomms5816.

Johnson, N., Aréchiga-Ceballos, N., & Aguilar-Setien, A. (2014). Vampire bat rabies: ecology, epidemiology and control. *Viruses*, 6(5), 1911–1928. https://doi.org/10.3390/v6051911.

Kissi, B., Badrane, H., Audry, L., Lavenu, A., Tordo, N., Brahimi, M., & Bourhy, H. (1999). Dynamics of rabies virus quasispecies during serial passages in heterologous hosts. *Journal of General Virology*, 80, 2041–2050.

Kotait, I., De Novaes, R., Maria, O., Carrieri, L., Castilho, J. G., & Isabel, C. (2018). Non-human primates as a reservoir for rabies virus in Brazil. *Zoonoses and Public Health*, (June), 1–13. https://doi.org/10.1111/zph.12527.

Kuzmin, I. V., Shi, M., Orciari, L. A., Yager, P. A., Velasco-Villa, A., Kuzmina, N. A., ... Rupprecht, C. E. (2012). Molecular inferences suggest multiple host shifts of rabies viruses from bats to mesocarnivores in Arizona during 2001–2009. *PLoS Pathogens*. 8(6),e1002786. https://doi.org/10.1371/journal.ppat.1002786.

Kuzmin, I. V., & Tordo, N. (2012). Genus Lyssavirus. *Rhabdoviruses: Molecular Taxonomy, Evolution, Genomics, Ecology, Host-Vector Interactions, Cytopathology and Control*, 37–58.

Kuzmina, N. A., Lemey, P., Kuzmin, I. V., Mayes, B. C., Ellison, J. A., Orciari, L. A., ... Rupprecht, C. E. (2013). The phylogeography and spatiotemporal spread of south-central skunk rabies virus. *PLoS One*, 8(12), 1–11. https://doi.org/10.1371/journal.pone.0082348.

Lemey, P., Rambaut, A., Welch, J. J., & Suchard, M. A. (2010). Phylogeography takes a relaxed random walk in continuous space and time. *Molecular Biology and Evolution*, 27(8), 1877–1885. https://doi.org/10.1093/molbev/msq067.

Leslie, M. J., Messenger, S., Rohde, R. E., Smith, J., Cheshier, R., Hanlon, C., & Rupprecht, C. E. (2006). Bat-associated rabies virus in skunks. *Emerging Infectious Diseases*, 12(8), 1274–1277.

Lim, B. K. (2009). Review of the origins and biogeography of bats in South America. *Chiroptera Neotropical*, 15(1), 391–410.

Lin, Y. -C., Chu, P. -Y., Chang, M. -Y., Hsiao, K. -L., Lin, J. -H., & Liu, H. -F. (2016). Spatial temporal dynamics and molecular evolution of re-emerging rabies virus in Taiwan. *International Journal of Molecular Sciences*, 17(3), 392.

Liu, W., Liu, Y., Liu, J., Zhai, J., & Xie, Y. (2011). Evidence for inter-and intra-clade recombinations in rabies virus. *Infection, Genetics and Evolution, 11*(8), 1906–1912.

Longdon, B., Brockhurst, M. A., Russell, C. A., Welch, J. J., & Jiggins, F. M. (2014). The evolution and genetics of virus host shifts. *PLoS Pathogens, 10*(11). https://doi.org/10.1371/journal.ppat.1004395.

Mansfield, K., McElhinney, L., Hübschle, O., Mettler, F., Sabeta, C., Nel, L. H., & Fooks, A. R. (2006). A molecular epidemiological study of rabies epizootics in kudu (Tragelaphus strepsiceros) in Namibia. *BMC Veterinary Research, 2*, 2. https://doi.org/10.1186/1746-6148-2-2.

Marston, D. A., Horton, D. L., Nunez, J., Ellis, R. J., Orton, R. J., Johnson, N., ... Fırat, M. (2017). Genetic analysis of a rabies virus host shift event reveals within-host viral dynamics in a new host. *Virus Evolution, 3*(2)vex038.

Martínez, M. A., Jordan-Paiz, A., Franco, S., & Nevot, M. (2016). Synonymous virus genome recoding as a tool to impact viral fitness. *Trends in Microbiology, 24*(2), 134–147. https://doi.org/10.1016/j.tim.2015.11.002.

Membrebe, J. V., Suchard, M. A., Rambaut, A., Baele, G., & Lemey, P. (2019). Bayesian inference of evolutionary histories under time-dependent substitution rates. *Molecular Biology and Evolution, 36*(8), 1793–1803. https://doi.org/10.1093/molbev/msz094.

Mollentze, N., Biek, R., & Streicker, D. G. (2014). The role of viral evolution in rabies host shifts and emergence. *Current Opinion in Virology, 8*, 68–72. https://doi.org/10.1016/j.coviro.2014.07.004.

Morimoto, K., Hooper, D. C., Carbaugh, H., Fu, Z. F., Koprowski, H., & Dietzschold, B. (1998). Rabies virus quasispecies: implications for pathogenesis. *Proceedings of the National Academy of Sciences of the United States of America, 95*(6), 3152–3156.

Morimoto, K., Patel, M., Corisdeo, S., Hooper, D. C., Fu, Z. F., Rupprecht, C. E., ... Dietzschold, B. (1996). Characterization of a unique variant of bat rabies virus responsible for newly emerging human cases in North America. *Proceedings of the National Academy of Sciences of the United States of America, 93*(11), 5653–5658.

Nettles, V. F., Shaddock, J. H., Keith Sikes, R., & Reyes, C. R. (1979). Rabies in translocated raccoons. *American Journal of Public Health, 69*(6), 601–602. https://doi.org/10.2105/AJPH.69.6.601.

Ng, T. F. F., Chen, L. -F., Zhou, Y., Shapiro, B., Stiller, M., Heintzman, P. D., ... Delwart, E. (2014). Preservation of viral genomes in 700-y-old caribou feces from a subarctic ice patch. *Proceedings of the National Academy of Sciences, 111*(47), 16842–16847. https://doi.org/10.1073/pnas.1410429111.

Nokireki, T., Tammiranta, N., Kokkonen, U., Kantala, T., & Gadd, T. (2018). Tentative novel lyssavirus in a bat in Finland. *Transboundary and Emerging Diseases, 65*(3), 593–596.

Novella, I. S., Zárate, S., Metzgar, D., & Ebendick-Corpus, B. E. (2004). Positive selection of synonymous mutations in vesicular Stomatitis Virus. *Journal of Molecular Biology, 342*(5), 1415–1421. https://doi.org/10.1016/J.JMB.2004.08.003.

Pawan, J. L. (1936). The transmission of paralytic rabies in Trinidad by the vampire bat (Desmodus rotundus). *Annals of Tropical Medicine and Parasitology, 30*, 101.

Pawan, J. L. (1948). Fruit-eating bats and paralytic rabies in Trinidad. *Annals of Tropical Medicine and Parasitology, 42*(2), 173–177.

Pepin, K. M., Davis, A. J., Streicker, D. G., Fischer, J. W., VerCauteren, K. C., & Gilbert, A. T. (2017). Predicting spatial spread of rabies in skunk populations using surveillance data reported by the public. *PLoS Neglected Tropical Diseases, 11*(7), e0005822. https://doi.org/10.1371/journal.pntd.0005822.

Plotkin, J. B., & Kudla, G. (2011). Synonymous but not the same: the causes and consequences of codon bias. *Nature Reviews Genetics, 12*(1), 32–42. https://doi.org/10.1038/nrg2899.

Randall, D. A., Williams, S. D., Kuzmin, I. V., Rupprecht, C. E., Tallents, L. A., Tefera, Z., ... Laurenson, M. K. (2004). Rabies in endangered Ethiopian wolves. *Emerging Infectious Diseases, 10*(12), 2214–2217.

Real, L. A., Henderson, J. C., Biek, R., Snaman, J., Jack, T. L., Childs, J. E., ... Nadin-Davis, S. A. (2005). Unifying the spatial population dynamics and molecular evolution of epidemic rabies virus. *Proceedings of the National Academy of Sciences of the United States of America, 102*(34), 12107–12111.

Rupprecht, C. E., & Smith, J. S. (1994). Raccoon rabies: The re-emergence of an epizootic in a densely populated area. *Seminars in Virology, 5*(2), 155–164. https://doi.org/10.1006/smvy.1994.1016.

Rupprecht, C., Kuzmin, I., & Meslin, F. (2017). Lyssaviruses and rabies: current conundrums, concerns, contradictions and controversies. *F1000Research, 6*(0), 184. https://doi.org/10.12688/f1000research.10416.1.

Rupprecht, C. E., Smith, J. S., Makonnen Fekadu, M. S., & Childs, J. E. (1995). The ascension of wildlife rabies: A cause for public health concern or intervention? *Emerging Infectious Diseases, 1*(4), 107–114. https://doi.org/10.3201/eid0104.950401.

Russell, C. A., Smith, D. L., Waller, L. A., Childs, J. E., & Real, L. A. (2004). A priori prediction of disease invasion dynamics in a novel environment. *Proceedings of the Royal Society B: Biological Sciences, 271*(1534), 21–25. https://doi.org/10.1098/rspb.2003.2559.

Scatterday, J. E. (1954). Bat rabies in Florida. *Journal of the American Veterinary Medical Association, 124*(923), 125.

Scott, T. P., Fischer, M., Khaiseb, S., Freuling, C., Höper, D., Hoffmann, B., ... Nel, L. H. (2013). Complete genome and molecular epidemiological data infer the maintenance of rabies among kudu (Tragelaphus strepsiceros) in Namibia. *PLoS One, 8*(3), e58739.

Seetahal, J., Vokaty, A., Vigilato, M., Carrington, C., Pradel, J., Louison, B., ... Rupprecht, C. E. (2018). Rabies in the Caribbean: a situational analysis and historic review. *Tropical Medicine and Infectious Disease, 3*(3), 89. https://doi.org/10.3390/tropicalmed3030089.

Shen, S. H., Stauft, C. B., Gorbatsevych, O., Song, Y., Ward, C. B., Yurovsky, A., ... Wimmer, E. (2015). Large-scale recoding of an arbovirus genome to rebalance its insect versus mammalian preference. *Proceedings of the National Academy of Sciences of the United States of America, 112*(15), 4749–4754. https://doi.org/10.1073/pnas.1502864112.

Shoji, Y., Kobayashi, Y., Sato, G., Itou, T., Miura, Y., Mikami, T., ... Sakai, T. (2004). Genetic characterization of rabies viruses isolated from frugivorous bat (Artibeus spp.) in Brazil. *Journal of Veterinary Medical Science, 66*(10), 1271–1273.

Sikes, R. K. (1962). Pathogenesis of rabies in wildlife. I. Comparative effect of varying doses of rabies virus inoculated into foxes and skunks. *American Journal of Veterinary Research, 23*, 1041.

Smith, D. L., Lucey, B., Waller, L. A., Childs, J. E., & Real, L. A. (2002). Predicting the spatial dynamics of rabies epidemics on heterogeneous landscapes. *Proceedings of the National Academy of Sciences of the United States of America, 99*(6), 3668–3672. https://doi.org/10.1073/pnas.042400799.

Smith, J. S., Orciari, L. A., & Yager, P. A. (1995). Molecular epidemiology of rabies in the united states. *Seminars in Virology, 6*(6), 387–400. https://doi.org/10.1016/S1044-5773(05)80016-2.

Streicker, D. G., Altizer, S. M., Velasco-Villa, A., & Rupprecht, C. E. (2012). Variable evolutionary routes to host establishment across repeated rabies virus host shifts among bats. *Proceedings of the National Academy of Sciences, 109*(48), 19715–19720. https://doi.org/10.1073/pnas.1203456109.

Streicker, D. G., González, S. L. F., Luconi, G., Barrientos, R. G., & Leon, B. (2019). Phylodynamics reveals extinction-recolonization dynamics underpin apparently endemic vampire bat rabies in Costa Rica. *Proceedings of the Royal Society B: Biological Sciences*, https://doi.org/10.1098/rspb.2019.1527.

Streicker, D. G., Lemey, P., Velasco-Villa, A., & Rupprecht, C. E. (2012). Rates of viral evolution are linked to host geography in bat rabies. *PLoS Pathogens, 8*(5), e1002720. https://doi.org/10.1371/journal.ppat.1002720.

Streicker, D. G., Turmelle, A. S., Vonhof, M. J., Kuzmin, I. V., McCracken, G. F., & Rupprecht, C. E. (2010). Host phylogeny constrains cross-species emergence and establishment of rabies virus in bats. *Science, 329*(5992), 676–679. https://doi.org/10.1126/science.1188836.

Streicker, D. G., Winternitz, J. C., Satterfield, D. A., Condori-Condori, R. E., Broos, A., Tello, C., & Valderrama, W. (2016). Host–pathogen evolutionary signatures reveal dynamics and future invasions of vampire bat rabies. *Proceedings of the National Academy of Sciences, 113*(39), 10926–10931. https://doi.org/10.1073/pnas.1606587113.

Takayama-Ito, M., Ito, N., Yamada, K., Sugiyama, M., & Minamoto, N. (2006). Multiple amino acids in the glycoprotein of rabies virus are responsible for pathogenicity in adult mice. *Virus Research, 115*(2), 169–175.

Talbi, C., Lemey, P., Suchard, M. A., Abdelatif, E., Elharrak, M., Jalal, N., & Bourhy, H. (2010). Phylodynamics and Human-mediated dispersal of a zoonotic virus. *PLoS Pathogens, 6*(10), 10. https://doi.org/10.1371/journal.ppat.1001166.

Teeling, E. C., Springer, M. S., Madsen, O., Bates, P., O'brien, S. J., & Murphy, W. J. (2005). A molecular phylogeny for bats illuminates biogeography and the fossil record. *Science, 307*(5709), 580–584. https://doi.org/10.1126/science.1105113.

Trewby, H., Nadin-Davis, S. A., Real, L. A., & Biek, R. (2017). Processes underlying rabies virus incursions across US–Canada border as revealed by whole-genome phylogeography. *Emerging Infectious Diseases, 23*(9), 1454–1461. https://doi.org/10.3201/eid2309.170325.

Troupin, C., Dacheux, L., Tanguy, M., Sabeta, C., Holmes, E. C., Bouchier, C., & Vignuzzi, M. (2016). Large-scale phylogenomic analysis reveals the complex evolutionary history of rabies virus in multiple carnivore hosts. *PLoS Pathogens*, 1–20. https://doi.org/10.1371/journal.ppat.1006041.

Velasco-Villa, A., Mauldin, M. R., Shi, M., Escobar, L. E., Gallardo-Romero, N. F., Damon, I., & Emerson, G. (2017). The history of rabies in the Western Hemisphere. *Antiviral Research*, https://doi.org/10.1016/j.antiviral.2017.03.013.

Wallace, R. M., Gilbert, A., Slate, D., Chipman, R., Singh, A., Wedd, C., & Blanton, J. D. (2014). Right place, wrong species: A 20-year review of rabies virus cross species transmission among terrestrial mammals in the United States. *PLoS One*, *9*(10), e107539. https://doi.org/10.1371/journal.pone.0107539.

Xiao, C. -T., Halbur, P. G., & Opriessnig, T. (2012). Complete genome sequence of a newly identified porcine astrovirus genotype 3 strain US-MO123. *Journal of Virology*, *86*(23), 13126.

Zhang, S., Tang, Q., Wu, X., Liu, Y., Zhang, F., Rupprecht, C. E., & Hu, R. (2009). Rabies in ferret badgers, Southeastern China. *Emerging Infectious Diseases*, *15*(6), 946–949. https://doi.org/10.3201/eid1506.081485.

Zhao, J., Liu, Y., Zhang, S., Zhang, F., Wang, Y., Mi, L., ... Hu, R. (2014). Molecular characterization of three ferret badger (*Melogale moschata*) rabies virus isolates from Jiangxi province, China. *Archives of Virology*, *159*(8), 2059–2067. https://doi.org/10.1007/s00705-014-2044-0.

Zieger, U., Marston, D. A., Sharma, R., Chikweto, A., Tiwari, K., Sayyid, M., ... Horton, D. L. (2014). The phylogeography of rabies in Grenada, West Indies, and implications for control. *PLoS Neglected Tropical Diseases*, *8*(10), e3251. https://doi.org/10.1371/journal.pntd.0003251.

CHAPTER 4

Epidemiology

Ryan MacLaren Wallace, Jesse Blanton

United States Centers for Disease Control and Prevention, Poxvirus and Rabies Branch, Atlanta, GA, United States

4.1 Introduction

Epidemiology is the study of the occurrence and distribution of disease over time (Giesecke, 2017). Often, the objective in epidemiology is to develop or improve interventions that can reduce the burden of disease. Epidemiology is primarily a denominator science focused on characterizing populations rather than drawing conclusions based on individual case reports. The practice of epidemiology is critical in the practice of public health because it is a framework for defining risk and inferring causality. Take, for example, a common scenario experienced around the world, where a canine rabies vaccination campaign comes to a small, underserved village once a year. During a particular campaign, a participant's dog unexpectedly dies in the days after receiving its rabies vaccination. The dog's owner observes a temporal association between the events, assumes causality, and convinces the rest of the community to refuse to participate in the vaccination campaign the following year. These commonplace scenarios can significantly disrupt efforts to control and eliminate disease transmission. Epidemiology provides a framework for investigating trends and designing experiments to evaluate the mechanisms and validity of proposed associations. In many developing countries, up to one-third of dogs die each year from a multitude of health conditions (Kitala et al., 2001; Schildecker et al., 2017; Wallace, Mehal, et al., 2017). Many dog deaths that occur after vaccination events result from the handling of the animals during transport to the clinic, or other underlying health conditions, rather than any fault of the vaccine itself. Having this information and incorporating it into community engagement efforts can serve a critical role in a successful intervention campaign. It is through the practice of epidemiology that we can better understand when an event in a population is expected or unexpected, and it is through this understanding that we can help foster communities to support impactful health interventions.

Rabies is the disease caused by viruses in the genus *Lyssavirus* (Family: *Rhabdoviridae*). The epidemiology of rabies is complex. Lyssaviruses can infect any mammal and are zoonotic pathogens, those that can be transmitted from animals to humans. The virus is maintained by numerous reservoir species, each with distinct geographic boundaries and transmission dynamics. Rabies epidemiology must consider these unique transmission cycles, involving both humans and animals, that this group of viruses has adapted to allow them to move within populations through at least three epidemiologic patterns: intrareservoir transmission, spillover events, and host-shift events. These features of RABV, and other viruses of the *Lyssavirus* genus, result in an intricate and ever-changing landscape of animal and human disease.

Rabies presents as an acute encephalitis in affected humans and animals, characterized by the development of neurologic abnormalities that range from aggressive hyperactivity to paralytic syndromes (Jackson, 2014), typically death occurs within 14 days of symptom onset if no intensive care is provided. Rabies is caused by infection with any one of the 16 currently recognized viruses in the genus *Lyssavirus* (Evans, Horton, Easton, Fooks, & Banyard, 2012; Fooks et al., 2014). The vast majority of annual human rabies deaths (>59,000 per year) are attributed to RABV (Hampson et al., 2015), whereas there are only 16 known human deaths from all other members of the *Lyssavirus* genus combined (WHO TRS).

Successful interventions to control rabies have existed since the 18th century. Yet, over 200 years later, rabies continues to kill more people than any other zoonotic disease (Fooks et al., 2014; Hampson et al., 2015). Rabies primarily affects people residing in low- and middle-income countries and has been deemed a neglected disease by the World Health Organization. Numerous economic, political, and logistical factors have contributed to rabies persistence around the world, and a lack of adequate disease surveillance capacity in many endemic countries is thought to heavily contribute to stalled enthusiasm for disease elimination (Taylor, Hampson, Fahrion, Abela-Ridder, & Nel, 2017). Prior studies have shown that rabies deaths are severely underdetected, and in some settings less than 1% of suspected human rabies deaths are confirmed by modern diagnostic methods (Cleaveland, Fevre, Kaare, & Coleman, 2002; Suraweera et al., 2012). However, an increase in global disease awareness since the mid-2000s has introduced more resources for the study and prevention of this complex disease. In just the past decade (2010–19), five new species of the *Lyssavirus* genus were classified (Walker et al., 2018). Additionally, as wildlife disease surveillance improves, RABV is being detected in places where it had previously been considered absent. There are constant natural and anthropomorphic changes in rabies epidemiology that will continue in the future.

4.2 Global rabies epidemiology

Lyssaviruses are globally distributed, affecting human and animal populations on all continents with the exception of Antarctica (Fig. 4.1). With the exception of RABV, Lyssaviruses are almost exclusively associated with transmission cycles in bats throughout Europe, Asia, Africa, and Australia. In contrast, RABV transmission cycles have been documented in numerous terrestrial carnivore species, most notably in domestic dogs, on a near-global scale. Furthermore, RABV transmission associated with bats is restricted to the Western Hemisphere. Globally, an

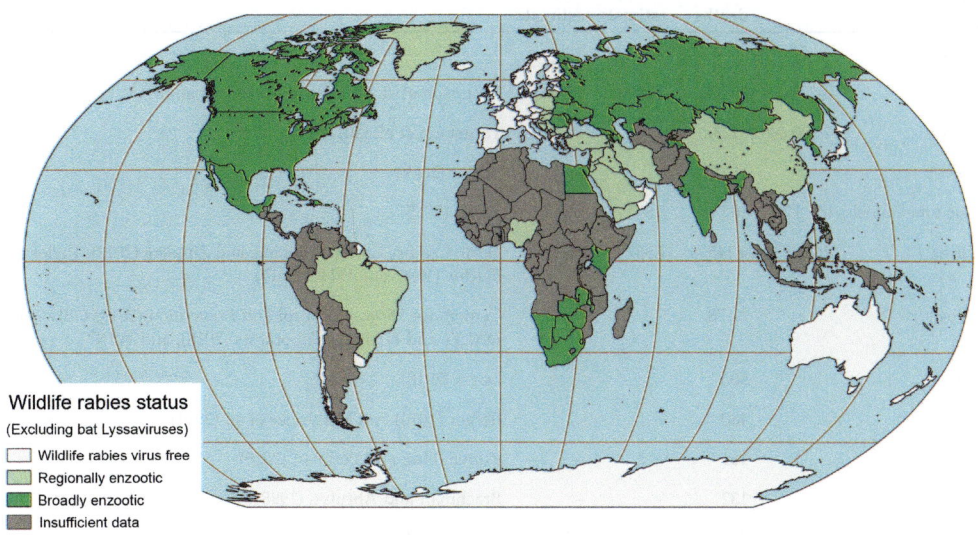

FIG. 4.1   Distribution of RABV maintained by wildlife.

estimated 59,000 people die from rabies each year (range: 25,000–159,000), resulting in over 3.7 million disability adjusted life years (DALYs) (Hampson et al., 2015). In terms of burden of disease and lethality, rabies is the deadliest of all zoonotic diseases (Table 4.1). Substantial economic losses are realized from rabies since human deaths alone cost the global economy USD $7.8 billion each year with an estimated USD $650 million in additional losses from livestock and animal vaccination. Despite the divergence of Lyssaviruses and reservoir species globally, over 98% of the deaths and economic losses are due to infections with RABV transmitted from dog bites (Fig. 4.2).

4.2.1 Asia

Asia, including the Indian subcontinent, accounts for over half of the global human rabies burden with over 37,000 estimated deaths occurring each year (0.8/100,000 persons) (Hampson et al., 2015). Of the 30 countries that make up the Asian continent, only 4 are considered rabies-free: Japan, Maldives, Republic of Korea, and Singapore. Aside from these four rabies-free countries, canine rabies continues to cause the vast majority of human and animal rabies deaths throughout Asia. Considerable variability in the level of canine rabies control practices exists between countries in this continent. Dog vaccination coverage ranges from less than 1% in countries such as Cambodia, Myanmar, and North Korea to well over 50% in Sri Lanka, Malaysia, and Thailand. Similarly, rates of human rabies deaths also vary drastically in the region. India has the highest number of human rabies cases in the world (per capita death rate 1.7 per 100,000). Myanmar and Afghanistan have human rabies death rates of 9 and 6 per 100,000 residents, respectively, whereas numerous Eurasian countries experience less than 1 human rabies death per 1000,000 residents.

TABLE 4.1 Global human deaths attributed to zoonotic pathogens as listed by the World Health Organization.

Pathogen	DALYs	Human deaths	Lower estimate	Upper estimate	References
Zoonotic pathogens					
Rabies	3,700,000	59,000	25,000	159,000	Hampson et al. (2015) and Murray et al. (2012)
Leptospirosis	2,900,000	51,500	–	–	Torgerson et al. (2015)
Lassa fever	–	45,000	–	–	Senior (2009)
Zoonotic tuberculosis	607,775	10,545	7,894	14,472	WHO (2015)
Brucellosis	264,073	4,145	1,557	95,894	Dean, Crump, Greter, Schelling, and Zinsstag (2012), Taylor and Perdue (1989) and WHO (2015)
Ebola virus	–	2,178	–	–	Centers for Disease Control and Prevention (2019) GBD 2017 DALYs and HALE Collaborators (2018), and WHO
Botulism	–	500	–	–	Davis (2018)
Anthrax	–	480	–	–	Berger (2016) and Hendricks et al. (2014)
MERS	–	161	–	–	Zumla, Hui, and Perlman (2015)
Plague	–	132	–	–	Brachman and Abrutyn (1998)
SARS	–	48	–	774	WHO (2003)
Rift valley fever	–	48	4	91	Labeaud, Bashir, and King (2011)
Zoonotic influenza	–	32	2	47	WHO (2018)
Marburg virus	–	24	–	–	Christou (2011) and Green (2012)
Streptococcus suis	133,021	13	–	–	Huong et al. (2019) and Wertheim, Nghia, Taylor, and Schultsz (2009)
Variant CJD	–	7	–	–	WHO (n.d.)
Zoonotic foodborne pathogens					
Salmonella	3,895,547	57,000	–	–	Healy and Bruce (2019), Majowicz et al. (2010), and WHO (2015)
Campylobacter	3,733,822	37,604	27,738	55,101	Kaakoush, Castano-Rodriguez, Mitchell, and Man (2015) and Ternhag, Torner, Svensson, Giesecke, and Ekdahl (2005)
Trematode	1,875,000	7,158	–	–	Keiser and Utzinger (2009), Murray et al. (2012), and Torgerson et al. (2015)
E. coli (zoonotic)	26,827	269	111	814	Majowicz et al. (2014) and Wasteson (2001)
Zoonotic parasitic pathogens					
Leishmaniosis	3,317,000	30,000	20,000	40,000	Alvar et al. (2012) and Murray et al. (2012)
Echinococcosis	1,666,434	18,000	11,900	28,200	Torgerson et al. (2014) and Torgerson, Keller, Magnotta, and Ragland (2010)
Cysticercosis	503,000	462	–	–	Murray et al. (2012)
Zoonotic vectorborne pathogens					
Dengue virus	825,000	20,000	–	–	Fredericks and Fernandez-Sesma (2014) and Murray et al. (2012)
Zika virus	–	13,087	–	–	GBD 2017 DALYs and HALE Collaborators (2018), Honein, Cetron, and Meaney-Delman (2019), and Sarmiento-Ospina, Vasquez-Serna, Jimenez-Canizales, Villamil-Gomez, and Rodriguez-Morales (2016)
Chagas disease	546,000	12,500	–	–	Moncayo and Silveira (2009) and Murray et al. (2012)
Japanese encephalitis	–	10,000	–	–	Diagana, Preux, and Dumas (2007)
Chikungunya	–	1,400	–	–	Manimunda, Mavalankar, Bandyopadhyay, and Sugunan (2011) and Wahid, Ali, Rafique, and Idrees (2017)
Crimean Congo fever	–	30	–	–	Ince et al. (2014)

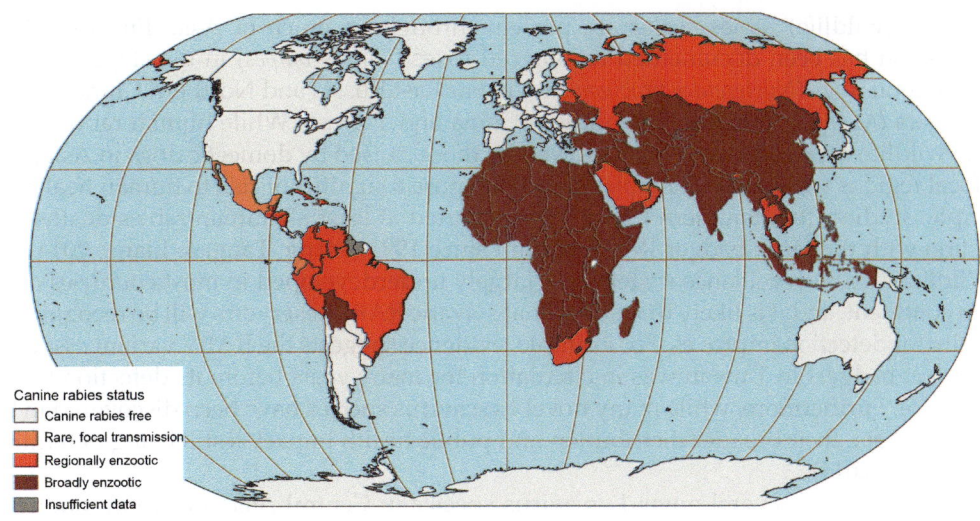

FIG. 4.2 Distribution of RABV maintained by domestic dogs.

Rabies control activities on the Asian continent highlight both promise for numerous countries' prospects for rabies elimination, as well as challenges in the remaining areas with Rabies virus circulation. Significant improvements have been made across Asia to increase access to biologics for human PEP. These improvements are most obvious in urban centers, and therefore the majority of human deaths tend to now be clustered in rural parts of these countries where access to healthcare facilities and PEP can still be limited. Use of nerve tissue vaccines have been discontinued across Asia, while access to high-quality tissue culture vaccines has improved, but has resulted in high costs across the region, estimated at $1.5 billion annually for PEP (Hampson et al., 2015). This high cost for PEP illustrates the importance of integrated canine rabies control programs to reduce the local transmission of RABV and subsequent human exposure (Undurraga et al., 2017).

In 2017, the Republic of Korea self-declared national elimination of enzootic transmission of canine rabies (Yang et al., 2017). The country is one of very few to have achieved this accomplishment in the past decade (Blanton, Hanlon, & Rupprecht, 2007). Counter to the success in the Republic of Korea, Malaysia experienced a canine rabies outbreak in 2015; the first in this country in nearly two decades. The outbreak was located on the southern border with Indonesia and porous borders as well as low dog-vaccination coverage in Malaysia were contributing factors. The extent of the outbreak was originally underestimated, and between 2015 and 2019, at least 19 human rabies deaths were reported to health authorities (citation: http://bernama.com/en/news.php?id=1745751). This event highlights the importance of regional rabies elimination strategies and the fragility of rabies freedom when the threat of reintroduction from endemic bordering areas exists (Jeon et al., 2019).

Sylvatic (wildlife) rabies cycles are present throughout much of Asia. Enzootic RABV transmission has been documented among red foxes (*Vulpes vulpes*) in the steppe zones of Asia, raccoon dogs (*Nyctereutes procyonoides*) in Far East Russia and Northern China, and ferret badgers (*Melogale moschata*) in Southern China and Taiwan. While human rabies deaths due to wildlife exposures are overshadowed by those caused by domestic dogs in Asia, there are focal regions where wildlife rabies reservoirs pose a significant risk to human health. For example, studies from Southern China have reported at least 87 human rabies deaths from infection with the ferret badger RABV variant since 1994 (Wang, Tang, & Liang, 2014).

Wildlife rabies surveillance systems are largely underdeveloped in most countries on the Asian continent, and it is likely that additional sylvatic RABV reservoirs will be recognized as capacity for detection improves. For example, evidence suggests the RABV variant associated with ferret badgers in Taiwan was in circulation for many years before its detection (Chang et al., 2016). Furthermore, while many novel Lyssavirus species have been discovered in bats in Central Asia, their broader distribution and public health impact is still being determined (Fooks et al., 2014).

The discovery of several novel Lyssavirus species in Central Asia, including one of the most divergent species (West Caucasian Bat Lyssavirus), suggests a high degree of diversity in Asia. The ongoing findings in this region challenge some of the previous "out of Africa" theories regarding the evolution of Lyssaviruses overall (Fisher, Streicker, & Schnell, 2018). However, the majority of these discoveries have been restricted to the Central Asia region around Kyrgyzstan and have included few isolates from sampled bats (WHO TRS). No human deaths have been reported among the five Lyssaviruses discovered in the region since the early 1990s. Despite limited surveillance in Eastern Asia, serological surveillance in bats confirmed the presence of Lyssavirus in the region, and in 2014, a new Lyssavirus (*Gannoruwa bat lyssavirus*) was isolated in Sri Lanka (Gunawardena et al., 2016).

4.2.2 Africa

An estimated 21,500 people die from rabies each year in the African continent (1.8/100,000 persons) (Hampson et al., 2015). Dogs remain the primary global rabies virus reservoir—they cause nearly all human rabies deaths and canine rabies is endemic in every country on the African continent. However, other sylvatic cycles for RABV have been described in jackals (*Canis adustus* and *Canis mesomelas*), bat-eared foxes (*Otocyon megalotis*), and mongooses (*Herpestidae family*) (Sabeta, Mansfield, McElhinney, Fooks, & Nel, 2007; Zulu, Sabeta, & Nel, 2009). The kudu has been proposed as a reservoir for RABV in Namibia, which, if true, would be the only known herbivore Lyssavirus reservoir. The proposed mechanism of transmission among kudu is salivary contamination of Acacia trees transmitted via oral mucosal abrasions during communal feeding (Scott et al., 2013). However, recent studies have cast doubt on theories of sustained transmission and instead lend more evidence for continuous cross-species transmission from jackals to kudu (Hassel et al., 2018; Hikufe et al., 2019).

While RABV transmission has the most significant public health impact in Africa, at least four other Lyssavirus species are known to circulate in insectivorous and frugivorous bat species (Fooks et al., 2014). Among these, the *Duvenhage lyssavirus* (DUVV) and *Mokola lyssavirus* (MOKV) are the only species associated with human rabies cases (WHO TRS). The MOKV was first identified in a shrew in Nigeria in the late 1960s and has been periodically isolated from cats and other small carnivores since that time. In the early 1970s, MOKV was associated with two human rabies cases in Nigeria. The presentation in both of these cases was atypical and only one died. While apparently widely distributed throughout sub-Saharan Africa, the reservoir for MOKV remains unknown, though shrews have been proposed as a possible reservoir species (Kgaladi, Nel, & Markotter, 2013; Kgaladi, Wright, et al., 2013). The DUVV was first isolated from a South African man in 1970. Since that time, two additional human rabies deaths due to DUVV infection have been reported: one from South Africa and one from the Netherlands, though the exposure in the second case took place in Kenya. Insectivorous bats in the genus *Miniopterus* are a likely reservoir of DUVV in sub-Saharan Africa. While no human rabies cases are associated with the *Lagos bat lyssavirus* (LBV), it is well documented in frugivorous African bats, particularly *Eidolon helvum* and *Epomophorus* spp. Cases of LBV spillover to other animal species have been reported. More recently, *Shimoni bat lyssavirus* (SHBV) and *Ikoma lyssavirus* (IKOV) have been identified from Kenya and Tanzania, respectively. Only a single isolate has been isolated for these viruses, so the reservoir remains unknown (particularly for IKOV, which was isolated from a likely spillover infection to an African civet (*Civettictis civetta*)) (WHO TRS).

4.2.3 Middle East

Over 220 human rabies deaths are estimated to occur each year in the countries that comprise the Middle Eastern region, yet very few are diagnosed or reported to international health agencies (Bengoumi, Mansouri, Ghram, & Merot, 2018). While the Kingdom of Bahrain, Qatar, and United Arab Emirates are generally considered to be rabies-free, geographic considerations have typically been cited as the basis for their status rather than comprehensive epidemiologic or surveillance data. Advancements in the elimination of rabies in Israel have been reported, primarily through the implementation of mandatory dog vaccination legislation as well as the implementation of a wildlife oral rabies vaccination program (Yakobson et al., 2006, 2017). Despite early successful efforts of these programs, canine rabies re-emerged in Northern Israel in 2010 and, like other countries in the region, Israel remains endemic for canine and sylvatic rabies (David, Bellaiche, & Yakobson, 2010; David, Dveres, Yakobson, & Davidson, 2009).

Sylvatic rabies transmission cycles occur throughout the Middle East in terrestrial mammals, including the red fox and golden jackal (Bengoumi et al., 2018). Rabies cases in wolves, livestock, and cats are frequently reported, but phylogenetic evidence supports the notion that these are instances of spillover from the aforementioned reservoir species. To date, there have been no reports of nonrabies lyssaviruses in the Middle East. However, limited rabies surveillance, particularly among bats, precludes any definitive conclusion that other Lyssavirus species circulate in this region.

4.2.4 Europe

Successful control and elimination of canine rabies in Europe occurred from the mid-1800s to the mid-1900s. Control efforts began with targeted dog control programs integrating canine vaccination as reliable vaccines became available. However, as canine rabies was largely eliminated in Europe (with the exception of small pockets in Belarus, Georgia, Russia, and Ukraine), the sylvatic cycles in red foxes (*Vulpes vulpes*) and raccoon dogs (*Nyctereutes procyonoides*) were recognized as an ongoing public health issue. Control of rabies in these wildlife species was limited until the introduction of a live attenuated oral rabies vaccine in Europe. Oral rabies vaccination (ORV) programs began in Switzerland in 1978 and have been successful in eliminating the RABV from most of Europe.

Rabies elimination in Europe is perhaps the greatest achievement of any region in the world in terms of zoonotic disease control; intensive efforts successfully eliminated RABV variants in both domestic dogs and wildlife. However, this rabies-free status is continuously tested by incursions from rabies-endemic countries. Since 2000, there have been 25 recorded events of rabid dogs imported into rabies-free Europe (Euro CDC). The majority of these incursions have resulted from dogs originating in Morocco ($n=14$), but other countries responsible for importation of rabid dogs include Algeria (2015), Bosnia-Herzegovina (2010), Afghanistan (2009), Croatia (2008), Gambia (2008), Sri Lanka (2008), India (2007), North Africa (2003), Estonia (2003), Azerbaijan (2002), and Nepal (2001). Many of these animals were imported illegally with falsified or incorrect vaccination records. While nearly all events were recognized by public health officials before secondary transmission events could occur, a rabid dog imported into France in 2008 was not recognized before infecting a local dog, leading to a secondary transmission event. As a result, OIE revoked the rabies-free status of France for a 2-year period while local officials instituted enhanced surveillance and control efforts to ensure that no additional cases occurred.

Despite the successful control of RABVs, the discovery of new bat Lyssaviruses presents an ongoing risk of rabies in Europe. *European bat lyssaviruses 1* and *2* (EBLV1 and EBLV2) were isolated in the late 1970s and have been associated with four human rabies cases in Ukraine, Russia, Finland, and Scotland. More recently, the *Bokeloh bat lyssavirus* (BBLV) was isolated in France, Germany, and Poland (Smreczak, 2018), and *Lleida bat lyssavirus* (LLEBV), one of the more divergent phylogroup III Lyssaviruses, was discovered in Spain and France (Picard-Meyer et al., 2019). The public health importance of the two most recently discovered species in Europe has not yet been determined.

4.2.5 North America

Unlike other parts of the world, RABV is the only known Lyssavirus in the Western Hemisphere. However, the diversity of wildlife reservoirs in North America, many with overlapping geographic ranges, makes animal rabies control a challenge (Fig. 4.3). Though canine rabies was controlled in Canada by the early 1950s, a national strategy and mass

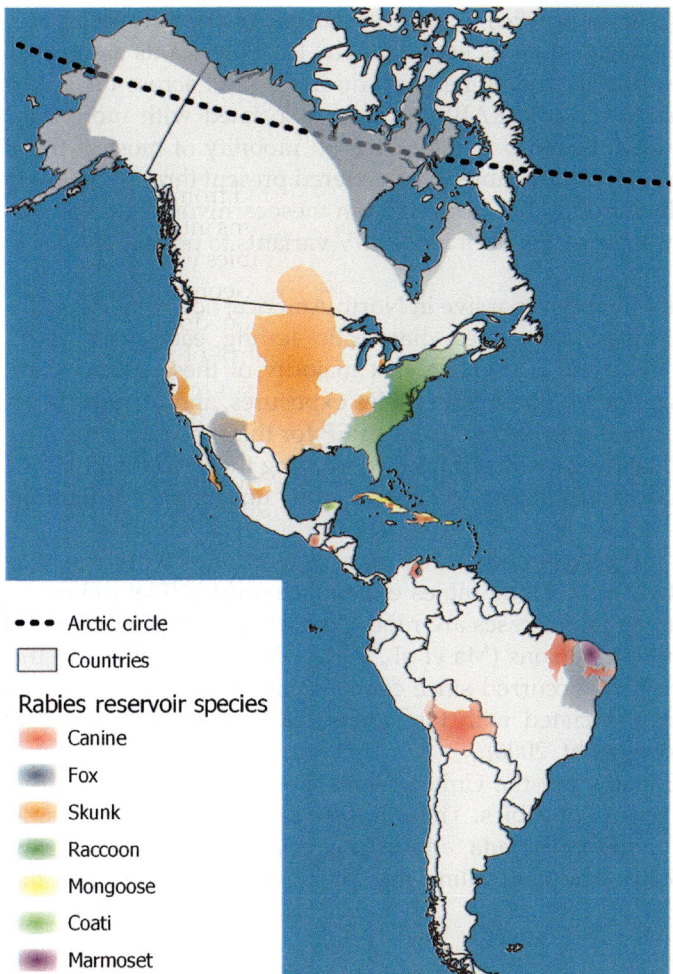

FIG. 4.3 Distribution of terrestrial wildlife and dog RABV variants in the Western Hemisphere.

vaccination program for dogs was necessary in the United States to ensure control across more than 3,000 counties and 39,000 municipalities (Rosatte, 1988; Tabel, Corner, Webster, & Casey, 1974). With these methods, canine rabies was successfully eliminated from the United States by the late 1970s (Steele, 1951; Steele & Tierkel, 1949; Velasco-Villa et al., 2008). In Mexico, instituting similar mass vaccination programs has nearly eliminated canine rabies (Fehlner-Gardiner, 2018; Ma et al., 2018). However, all three countries continuously face the challenge of importation of rabid dogs from endemic countries, including crossborder movement of rabid dogs from Guatemala (Hercules et al., 2018; McQuiston et al., 2008; Sinclair et al., 2015).

In all three North American countries, the public health importance of controlling the sylvatic rabies transmission cycle became increasingly apparent as canine rabies was controlled. Multiple carnivorous hosts continue to maintain the RABV, including coatis (*Nasua narica*), coyotes (*Canis latrans*), Arctic foxes (*Alopex lagopus*), gray foxes (*Urocyon cinereoargenteus*), raccoons (*Procyon lotor*), and striped skunks (*Mephitis mephitis*). The RABV variants associated with these species are typically geographically defined and do not extend across the entire home range of a species. In contrast, multiple RABV variants associated with more than 30 species of bats have been identified. Because of the increased mobility of most of these bat species compared to other reservoir species, rabies is considered present throughout their home ranges (Ma et al., 2018). Spillover of RABV variants from mesocarnivore reservoirs to other species is fairly common; however, spillover of bat RABV variants to nonbat species is observed infrequently.

The majority of animal rabies surveillance is passive in North America; however, typically 125,000–150,000 animals are submitted for rabies diagnostic testing each year (>2.5 submissions/10,000 population) (Ma et al., 2018). The vast majority of these submissions are part of an integrated bite management approach to rabies exposures, and diagnostic results govern clinical decisions regarding the need for rabies PEP for the exposed. Typically, less than 10% of these animals will be diagnosed as rabid, precluding the need for PEP that would otherwise be administered if laboratory diagnosis was unavailable (Manning et al., 2008).

Cases of human rabies reported in North America are primarily associated with exposure to a bat or represent cases imported from canine rabies endemic countries (Pieracci et al., 2019). In the United States, 1–3 human rabies cases are reported each year, representing an incidence of less than 1 per 100 million persons (Ma et al., 2018). Similar reductions in the burden of human rabies in Mexico have occurred since canine rabies has been controlled, with the last case of human rabies associated with the canine RABV variant reported in 2005 (Blanton, Krebs, Hanlon, & Rupprecht, 2006).

Oral vaccination programs in Canada and the United States are used to control, and in some cases eliminate, sylvatic reservoirs of rabies. These efforts played an important role in eliminating a cluster of red fox rabies in Canada, eliminating coyote and the Texas gray fox RABV variant in the United States, and controlling the spread of raccoon rabies in the United States and Canada.

4.2.6 Caribbean islands

Of the five countries that continue to report dog-mediated human rabies deaths in the Western Hemisphere, three are Caribbean nations: Cuba, Dominican Republic, and Haiti. Of the 182 estimated annual human rabies deaths in the Western Hemisphere, 130 occur in Haiti, 7 in Dominican Republic, and 1 in Cuba (Hampson et al., 2015; Undurraga et al., 2017). A canine rabies control program implemented in Haiti in 2013, and scaled to national levels in 2018, has been credited with reducing human rabies deaths by more than twofold through improved postexposure patient counseling as well as improved access to human rabies vaccines (Etheart et al., 2017; Wallace, Etheart, et al., 2017).

The small Indian mongoose (*Herpestes auropunctatus*) was introduced into many Caribbean islands during the second half of the 19th century to control rodent populations and, therefore, to improve sugar-cane-farming practices (Everard & Everard, 1992). In at least four separate host-shift events, the local canine RABV variant became established within the local mongoose populations in Cuba, Dominican Republic, Grenada, and Puerto Rico (Nadin-Davis, Velez, Malaga, & Wandeler, 2008). While evidence is lacking to support that a mongoose RABV variant is present in Haiti, limited wildlife surveillance efforts and the country's proximity to Dominican Republic suggest that it is cryptically circulating. Several human rabies deaths due to mongoose rabies are regularly reported from the Caribbean each year, and there are no proven methods to eliminate mongooses or mongoose rabies (Styczynski et al., 2017).

RABVs in bats have been reported in several Caribbean islands. This includes, most notably, the enzootic transmission of RABV in vampire bats in Trinidad and sporadic reports of rabies in insectivorous bats in Cuba and Dominican Republic (Seetahal et al., 2018). Limited wildlife rabies surveillance efforts have been reported from most Caribbean islands, precluding any conclusion as to the presence or absence of bat-associated RABVs there. Nonrabies lyssaviruses have never been reported in the Western Hemisphere and are not hypothesized to be found on any Caribbean islands.

4.2.7 South America

Considerable progress has been made toward the elimination of canine rabies in South America. However, Bolivia remains a significant hot spot with occasional crossborder movement of cases into Brazil and Peru. In addition, canine rabies cases are continually reported from El Salvador, Guatemala, Honduras, Nicaragua, and Venezuela (Freire de Carvalho et al., 2018; Seetahal et al., 2018). Several distinct wildlife reservoirs have been identified across the region, including the marmoset (*Callithrix jacchus*) and the crab-eating fox (*Cerdocyon thous*). Surveillance of wildlife for rabies is, however, generally inadequate to allow major epidemiological inferences (Seetahal et al., 2018).

As canine rabies has come under control in most South American countries, Vampire bat rabies has become an increasingly major public health problem. The common vampire bat (*Desmodus rotundus*) is the primary species associated with transmission to humans and domestic animals in many parts of this region, especially in remote areas of the Amazon rainforest, where these bats commonly feed on humans. Vampire bat-transmitted bovine paralytic rabies also has a significant economic effect on the livestock industry. Ongoing evaluations of childhood rabies vaccination programs in Peru have suggested these programs may be cost effective in settings where there is a high exposure rate from vampire bats and access to medical care is limited (e.g., the Amazon) (Gilbert et al., 2012). Alternative efforts focused on controlling vampire bat populations (e.g., anticoagulant pastes) have been largely ineffective and frequently have negative impacts on other bat species.

4.2.8 Oceania

The Oceania regions is largely considered to be free of Lyssaviruses, in particular: Melanesia, Micronesia, Polynesia, and New Zealand. In 1996, the *Australian bat Lyssavirus* (ABLV) was detected in Australia and has subsequently been detected in all species of frugivorous mega bats and an insectivorous bat species in Australia. Furthermore, three human deaths due to ABLV have been reported, as well as spillover infection into horses.

4.3 RABV transmission

4.3.1 Typical routes of transmission

Rabies is most commonly transmitted through the bite of an infected animal, after virus-laden saliva is inoculated into an innervated area of the body (Fooks et al., 2014). Virus spreads within peripheral nerves to the central nervous system. Upon reaching the brain, the virus proliferates within neuronal cells, disrupting their routine function but not directly resulting in neuronal death. RABV then spreads centrifugally back through the peripheral nerves. It is at this time that unique behavioral and neurologic abnormalities develop in the victim that favor virus-laden saliva inoculation into new hosts, such as aggression leading to bites.

4.3.2 Atypical routes of transmission

Rabies cases due to nonbite exposures are rare, but documented cases have occurred due to ingestion of infected animals, contamination of oral mucosa, and organ and tissue transplantation (Afshar, 1979). Nonbite exposures with the greatest risk appear to be among organ or tissue recipients from a RABV-infected donor (Manning et al., 2008). Prior to 2004 reports of iatrogenic rabies cases were rare, with only eight cases of rabies following cornea transplant from a rabid donor (Ross et al., 2015). However, in 2004, a cluster of four human rabies cases were reported in the United States following solid organ and vascular tissue transplants from the same rabies infected donor (Centers for Disease Control and Prevention, 2004). Additional occurrences have been reported in Germany, Kuwait, and China (Chen et al., 2017; Ross et al., 2015; Saeed & Al-Mousawi, 2017; Vora et al., 2013).

Rabies transmission via ingestion of experimentally infected animals has been established; however, human cases resulting from consumption of meat from a rabid animals are extremely rare (Correa-Giron, Allen, & Sulkin, 1970; Ekanem et al., 2013; Wertheim et al., 2009). Human rabies cases from consumption of rabid animals have been reported; however, the risk from killing and preparing meat from a rabid animal is likely higher and in some reported cases cannot be excluded as the possible route of transmission (Nguyen et al., 2011; Odeh, Umoh, & Dzikwi, 2013). Furthermore, consumption of animal products, such as milk, is not considered a risk of rabies transmission. Few case studies are reported for humans and animals suggesting transmission from the rabid mother to a nursing infant; however, these studies did not exclude contact with saliva from the mothers (Afshar, 1979).

In addition to direct mucosal exposures, possible cases of droplet or aerosol exposure have been described. Most notably, two cases of aerosol exposure have been attributed to laboratory exposures and two cases to possible airborne exposures in researchers working in caves containing millions of free-tailed bats (*Tadarida brasiliensis*) in the Southwest United States (Gibbons, 2002). However, alternative infection routes cannot be discounted in these cases. Similar airborne incidents have not occurred in approximately 25 years, presumably due in part to improved personal protective equipment use by persons working in these environments.

4.3.3 Incubation period

The incubation period for rabies is highly variable, with reports of symptoms developing within days of exposure to rare cases of multiyear incubation periods (Manning et al., 2008). Typical incubation periods are in the range of 3 weeks to 3 months after the exposure. Factors that influence the incubation period include the amount of virus inoculated into the body, proximity of the exposure to innervated tissues, and proximity of the exposure to the central nervous system. Studies from Iran found that persons exposed on locations closer to the head were more likely to develop rabies and developed disease faster than persons bitten at anatomical locations more distant (Baltazard & Ghodssi, 1954). Long incubation periods have also been reported, often associated with nonbite or loosely characterized exposures. Confirmation of the date of exposure with cases of long incubation periods, such as one report in India suggesting a 25-year incubation period, can be difficult to confirm (Shankar et al., 2012). However, reliable clinical, epidemiologic, and molecular data were available to confirm one case with an 8-year incubation period. Both were attributed to nonbite exposures (Boland et al., 2014; Shankar et al., 2012). Exposure to highly innervated anatomical locations favors successful infection and exposure to locations closer to the brain result in shorter incubation periods.

4.3.4 Viral shedding

Excretion of RABV in saliva is the primary route of transmission, and has been well studied among several domestic animals, particularly cats, dogs, and ferrets (Niezgoda, Briggs, Shaddock, & Rupprecht, 1998; Vaughn Jr., Gerhardt, & Newell, 1965; Vaughn, Gerhardt, & Paterson, 1963). In these seminal studies, virus excretion was not observed more than 3 days before onset of signs in animals. The onset of viral shedding is variable among animals, and rabies was not detected in saliva at all in nearly half of the animals in the experiments (with the exception of cats which had a high rate of RABV RNA detection in saliva). Clinical diagnosis of human rabies cases in the United States has similarly detected RABV in saliva in approximately 43% of cases between 5 and 28 days after the onset of symptoms (Petersen & Rupprecht, 2011). Despite reports of longer viral excretion periods, particularly associated with an Ethiopian canine RABV variant, a conservative 10-day observation period has been widely utilized to rule out possible rabies exposures over the past three to four decades (Fekadu, Chandler, & Harrison, 1982; Fekadu, Shaddock, & Baer, 1982; National Association of State Public Health Veterinarians et al., 2016). No human rabies cases have been reported as a failure of this observation period.

4.3.5 Enzootic cycles

Lyssaviruses are maintained within species-specific reservoir populations and typically across defined geographic regions. Enzootic cycles occur mostly through intraspecies transmission; however, cross-species transmission does occur and is typically related to species ecology and local resource competition (Borchering, Liu, Steinhaus, Gardner, & Kuang, 2012; Fisher et al., 2018; Ma et al., 2018; Streicker, Lemey, Velasco-Villa, & Rupprecht, 2012; Streicker et al., 2010).

4.3.6 Cross-species transmission

Cross-species interactions resulting in RABV transmission are commonly reported around the world (Wallace et al., 2014). Transmission typically occurs between different reservoir and nonreservoir bat species; or from a mesocarnivore reservoir to a nonbat species (Fig. 4.4). Transmission from a bat reservoir to a nonbat species is less frequently reported. Overall, there appears to be variability in the frequency of cross-species transmission; likely due to the particular RABV variant, reservoir species, and local ecology (Wallace et al., 2014). In the United States, the raccoon RABV variant is associated with a high rate of spillover (nearly one rabid nonraccoon for every reported rabid raccoon) (Ma et al., 2018). Multiple factors may be involved to allow raccoons to act as a sort of "superspreader," or the raccoon RABV variant may not be exclusively maintained by raccoons (skunks may be involved). Resolving these difficult transmission cycles is critical for planning ongoing ORV programs in the region. In contrast to the raccoon, cross-species transmission from bats to nonbat species is rare. A study evaluating rabid skunks in Texas (the primary reservoir in the state) over a 13-year period found only a single case of a bat RABV variant among nearly 5000 animals tested (Oertli, Wilson, Hunt, Sidwa, & Rohde, 2009).

4.3.7 Host shift events

While host shift events, the adaptation of a RABV variant and subsequent maintenance in a new reservoir species, were once considered to be rare, they have been observed multiple times over the past 80 years. Several mechanisms for host shift events have been proposed.

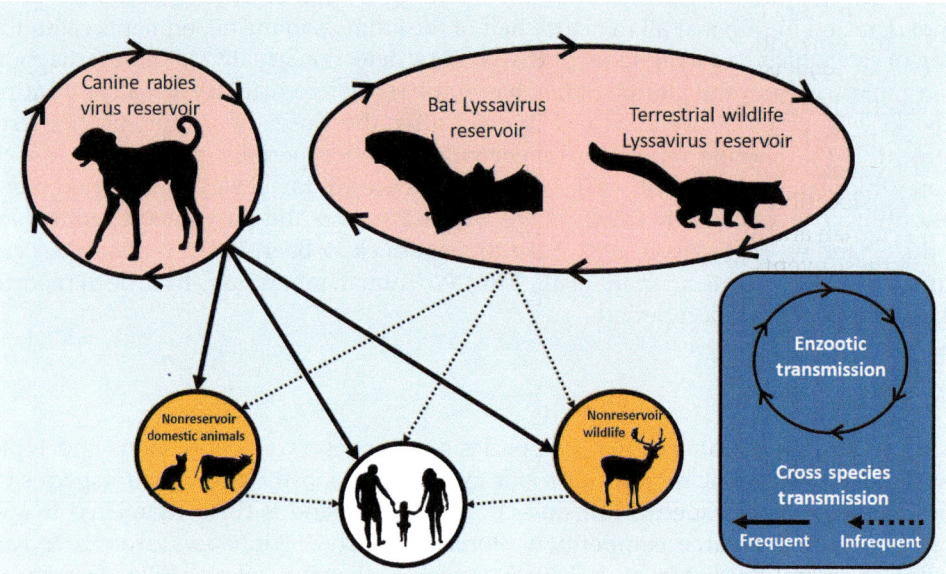

FIG. 4.4 Enzootic and cross-species transmission pathways for Lyssavirus species.

These include preadaptation viral mutations, postadaptation viral mutations, and host adaptations (Kuzmin et al., 2012; Smith & Seidel, 1993; Streicker et al., 2010). Preadaptation viral mutations occur when genetic changes to the virus occur in the reservoir host, prior to inoculation into a nonreservoir species; the genetic changes altering the pathogenesis of the virus to allow favorable replication in the new species. Postadaptation mutations occur when genetic changes to the virus occur after inoculation from the reservoir species into a nonreservoir. As little viral replication occurs at the site of infection, this mechanism for host shift events is less likely. Host adaptations may occur through either genetic or ecological factors.

Regardless of the mechanisms leading to host shift events, the epidemiologic impact when host shifts occur can be catastrophic. In the Western Hemisphere, independent cycles of RABV transmission occur in at least 20 species of bats, each the result of a host shift event (Kobayashi et al., 2007; Kuzmina et al., 2013; Streicker et al., 2010). North America is endemic for no fewer than eight terrestrial RABV variants, each the result of a host shift event from either bats or canines (Velasco-Villa et al., 2008).

4.3.8 Translocation events

Epidemiologic changes can also be impacted by translocation of RABV-infected animals into naïve populations. Translocations can be human-mediated or natural events. Human-mediated translocations have been associated with large-scale epidemiologic changes in the Western Hemisphere. Prior to 1977, the raccoon RABV variant was only known to be present in the Southeastern United States (Florida and Georgia). Human-mediated translocation of a rabies-infected raccoon from Florida into a raccoon-rabies-free mid-Atlanta state initiated the largest epizootic in recorded history (Centers for Disease Control and Prevention, 2000; Jenkins & Winkler, 1987; Wyatt, Barker, Bennett, & Hanlon, 1999). Over the following two decades, the epizootic front expanded in all directions by approximately 18–24 miles each year (Moore, 1999; Wilson et al., 1997). By 1999, the virus had crossed the St. Lawrence River into Canada. Today, the raccoon RABV variant is considered endemic to the entire eastern United States, where it accounts for nearly 75% of all terrestrial rabies cases in the country and led to a sixfold increase in human rabies exposures (Pieracci et al., 2019; Wyatt et al., 1999). Translocation events, be they natural or human-mediated, can have tremendous impacts for human and animal health. Areas prone to RABV translocations should maintain vigilance for these events through surveillance capacity, community education, and enforcement of legislation designed to prevent the intentional movement of unvaccinated animals.

4.4 Surveillance general

4.4.1 Principles of rabies surveillance

Epidemiology is dependent upon observable events, or outcomes, upon which to study. While there are many methods for obtaining disease occurrence data, systematic and routine reporting of animal and human rabies deaths through a defined surveillance system is the gold standard. Disease surveillance systems can be used to identify rabies outbreaks, monitor trends, and evaluate interventions such as animal vaccination programs or improvements in access to human rabies biologics (Franka & Wallace, 2018; World Health Organization, 2013). However, not all surveillance systems produce reliable information. The World Organization for Animal Health has established standards for surveillance systems used to declare an area free from disease. Most notably, these standards include:

(a) The disease is notifiable (see later)
(b) An ongoing system of effective disease surveillance has been in operation for at least 2 years without any case detection, with a minimum requirement being an ongoing early detection program to ensure investigation and reporting of rabies-suspect animals.
(c) Postdisease elimination, infrastructure is maintained to enable detection of disease reincursion and any disease presence is reported to national and international agencies.
(d) Regulatory measures for the prevention of rabies are implemented, particularly restrictions on the importation of high-risk animals from rabies-endemic countries.

Notifiability is a foundational component of an effective disease surveillance system and implies that health authorities have established policies requiring the reporting of a disease by health care professionals, have established a standard case definition of what should be reported, and have established a mechanism to routinely inform health care professionals of these requirements. While many countries have made human or animal rabies reportable, many have not developed or published case definitions. To help facilitate standard reporting, the Word Health Organization has published recommended case definitions for human and animal rabies surveillance (Table 4.2) (WHO TRS).

4.4.2 Global limitations in surveillance capacity

The cycle of neglect is often described in public health as a lack of resources to detect cases of a disease, which leads to a lack of community support to control the disease, which results in a lack of political support to provide adequate resources to detect the disease. The outcome of this cycle is complacency in ineffective health interventions and persistence of the disease burden in the community. This is the unfortunate situation for many rabies-endemic countries. A global study conducted in 2013 among 91 rabies-endemic countries found that 41% had ineffective surveillance systems to detect human rabies cases (Taylor, Knopf, & Partners for Rabies, 2015). Most commonly cited reasons for ineffective surveillance included underreporting from stakeholders, lack of follow-up on case reports, lack of diagnostics to confirm cases, poor funding for surveillance activities, and lack of legislation to enforce reporting requirements.

While there is no global repository for documenting human rabies deaths, several regional initiatives have developed reporting platforms to allow countries to report human rabies deaths. Countries have agreed to participate in regional data sharing platforms covering the Americas (El Sistema de Información Regional para la Vigilancia Epidemiológica de la Rabia (SIRVERA)), Europe (Rabies Bulletin Europe), and Africa (Rabies Epidemiology Bulletin). Inclusion of Asia and the Middle East in similar surveillance reporting platforms is underway, as well as mechanisms to consolidate key metrics from these regional platforms into

TABLE 4.2 Rabies case definitions as recommended by the World Health Organization.

Suspect case	An animal that presents with any of the following signs • Hypersalivation • Paralysis • Lethargy • Unprovoked abnormal aggression (biting two or more people or animals, and/or inanimate objects) • Abnormal vocalization • Diurnal activity of nocturnal species	• Notify appropriate local authorities of suspect rabid animal • Collect primary animal history if available (i.e., ownership status, vaccination status, prior exposures, date of onset for signs, etc.) • Collect CNS samples for laboratory diagnosis if available
Probable case	A suspect animal with a history of a bite by another suspect/Probable/Confirmed animal. AND/OR A suspect animal that is killed, died, or disappears within 4–5 days of observing illness.	• Systematically record secondary information and link to primary history • Notify appropriate authorities of probable animal rabies cases according to national protocols
Confirmed case	A suspect or probable animal confirmed using a standard diagnostic test as defined by WHO or OIE (see this chapter) or OIE manual (OIE, 2016).[a]	• Notify appropriate authorities for follow-up of any human or animal exposures • Systematically record laboratory diagnostic results and link with case record
Not a case	A suspect or probable animal in which rabies is ruled out by laboratory diagnosis	• Systematically record laboratory diagnostic results and link with primary history

[a] *If other diagnostic tests are used, depending on the sensitivity and specificity of those tests, a confirmation of results by a validated secondary test may be needed (particularly for negative results).*

global data repositories. Despite these efforts, comparison of reported surveillance data for country-reported human rabies cases to estimated numbers from various models shows a large degree of systematic underdetection across countries (Table 4.3); many countries have estimated case detection rates below 5%. Case detection for animal rabies cases is equally low, estimated to be less than 10% even in countries with well-performing rabies surveillance systems (Townsend et al., 2013).

The laboratory requirements to achieve confirmation of human and animal deaths from rabies are an important factor limiting adequate surveillance (Table 4.4). Confirmatory rabies diagnosis is most often performed on postmortem brain tissue, which requires at least partial autopsy/necropsy to obtain appropriate specimens (Rupprecht, Fooks, & Abela-Ridder, 2019). Many cultures have an aversion to autopsy's or otherwise manipulating a deceased body, limiting the ability to collect human samples for testing. Qualified medical personnel to collect a brain tissue sample also limits sample collection in many rabies-endemic countries. Further limiting case detection are the complicated diagnostic methods that are required; current rabies diagnostics must be conducted in a biosafety level two or greater facility and are based on immunohistochemical, fluorescence, and PCR assays. Reliable bedside diagnostic assays are not yet available.

TABLE 4.3 Underdetection of human rabies cases for selected canine-rabies endemic countries.

Country	Average reported human rabies deaths per year	Estimated human rabies deaths per year (Hampson et al., 2015)	Surveillance case detection rate (%)
Haiti	10	130	7.7
Cuba	1	1	100
Dominican Republic	1	7	14.3
Ethiopia	15	2771	0.5
South Africa	11	42	26.2
Madagascar	10	197	5.1
Ivory Coast	22	569	3.9
Ghana	9	112	8.0
Sierra Leone	14	301	4.7
Liberia	1	226	0.4
Nigeria	3	1637	0.1
Kenya	372	523	71.1
Mozambique	49	1326	3.7
Uganda	5	64	7.8
Cambodia (Ly et al., 2009)	7	446	1.6
Vietnam (CDC, 2019)	91	360	25.3
Thailand	8	62	12.9
Myanmar	168	4552	3.7
China (Zhou et al., 2016)	924	6002	15.4

Asia-other: Gongal and Wright (2011).
Caribbean countries: Seetahal et al. (2018).
African country human rabies deaths reported from: https://rabiesalliance.org/networks/paracon/bulletin (Accessed 22 August 2019).

4.4.3 Rationale for implementing rabies surveillance systems

Rabies surveillance programs are implemented for numerous reasons, often associated with the stage of rabies control in the program area. Franka et al. proposed five stages of rabies control and their associated surveillance activities (Franka & Wallace, 2018). These stages are ordered from least to most capacity for rabies control and include "Proof of Burden," "Human Rabies Prevention," "Monitoring Control Measures," "Verification of Elimination," and "Post-elimination." The type of surveillance activities and the related diagnostic assays that are acceptable for the activity are dependent on these stages and the rationale for conducting surveillance.

TABLE 4.4 Comparison of rabies assays and utility under different surveillance programs.

Assay		Stage 1: Proof of burden	Stage 2: Human rabies prevention	Stage 3: Monitoring of control measures	Stage 4: Verification of rabies elimination	Stage 5: Postelimination	Sample required	Notes
Molecular	RT-PCR	+++	+++	+++	+++	+++	Brain Skin biopsy Saliva	• Primary test • Recognized by OIE and WHO[a]
	Nested RT-PCR	+++	+++	+++	+++	+++	Brain Skin biopsy Saliva	• Primary test • Recognized by OIE and WHO[a]
	RTq-PCR	+++	+++	+++	+++	+++	Brain Skin biopsy Saliva	• Primary test • Recognized by OIE and WHO[a]
Antigen	Direct Fluorescent Antigen (DFA) test	+++	+++	+++	+++	+++	Brain Skin biopsy	• Primary test • Recognized by OIE and WHO[a]
	Direct Rapid Immunohistochemistry Test (DRIT)	+++	+++	+++	+++	+++	Brain Skin biopsy	• Primary test • Recognized by OIE and WHO[a]
	Indirect Rapid Immunohistochemistry Test (IRIT)	++	−	+	−	−	Brain Skin biopsy	Can also be used for antigenic typing studies

Continued

TABLE 4.4 Comparison of rabies assays and utility under different surveillance programs—cont'd

Assay		Stage 1: Proof of burden	Stage 2: Human rabies prevention	Stage 3: Monitoring of control measures	Stage 4: Verification of rabies elimination	Stage 5: Postelimination	Sample required	Notes
	Immunochromatographic test (Lateral Flow device—LFD)	++	−	+	−	−	Brain Skin biopsy	• Potentially low-cost • Transportable • Variable sensitivity and specificity
Antibody	RFFIT/FAVN	++	−	+++	−	−	Serum CSF	
	IFA	++	−	++	−	−	Serum	
	Indirect ELISA	++	−	++	−	−	Serum	
	Competitive ELISA	++	−	++	−	−	Serum	

[a] Recognized by OIE and WHO as test for population freedom from infection, for eradication policies, for confirmation of clinical cases and prevalence of infection (Black, Lovings, Smith, Heaton, & McElhinney, 2002; David, 2012; Dupuis, Brunt, Appler, Davis, & Rudd, 2015; Dyer et al., 2013; Faye et al., 2017; Fischer et al., 2013; Koch, Berger, & Kammann, 1974; Lee, Hutchinson, & Ziegler, 1977; Markotter et al., 2015; Nadin-Davis, Sheen, & Wandeler, 2009; Wacharapluesadee et al., 2011; Wadhwa et al., 2017).
−: not recommended, +: low recommendation, useful for confirmatory testing, ++: moderate recommendation, useful for primary or confirmatory testing, +++: high recommendation, useful for primary testing.
From Wallace, R. M., Blanton, J. D., Muller, T., Freuling, C. (2018). The role of diagnostics in surveillance.

Stage 1: Proof of Burden. In many countries, the burden and impact of rabies within animals and even human populations is not known, which contributes to the cycle of neglect and lack of action to support control activities. In these early stages of rabies surveillance and epidemiologic capacity, there is often limited diagnostic capacity or technical expertise to appropriately investigate rabies cases. Epidemiologic activities for programs in this stage should focus on capacity building and improving the understanding of the burden and transmission dynamics. Targeted sampling of suspected rabid animals (see case definition above) and development of laboratory methods consisting of assays that have acceptable sensitivity and specificity for lyssaviruses should be pursued. An example of the benefits of establishing such capacity can be found in the recent events in Taiwan. After elimination of rabies from dog populations in 1961, surveillance for rabies waned. In 2012, a wildlife disease surveillance program was created to determine presence of zoonotic and emerging pathogens (Chang et al., 2016; Chiou et al., 2014). The program was a collaboration between wildlife rehabilitation centers and government researches in which sick and found-dead animals were sampled and tested for a range of pathogens. In 2012, a spate of neurologic and found-dead ferret badgers were treated at the participating clinics; early testing ruled out common causes before rabies was considered. The animals were subsequently found to have died from RABV infection. Further investigation led to the revelation that RABV had been cryptically circulating in the ferret badger population for decades, likely as a result of a host-shift event from the locally endemic canine RABV.

Serological surveys, which often target healthy (nonrabies suspect) animals, may be applicable under certain settings as screening tools for presence of Lyssavirus circulation, particularly in relation to bat-associated lyssaviruses. In 2016, researchers published results from a serological survey of bats in seven southwestern Indian Ocean islands (Melade et al., 2016). Lyssaviruses were not known to be present on these islands, and the long distance separating them from mainland Africa provided a plausible expectation that the islands were Lyssavirus-free. However, as a testament to the widespread presence of rabies and nonrabies Lyssaviruses, bats from each of the seven islands were found to have serologic evidence for three nonrabies Lyssaviruses: LBV (7.3% sero-positive), DUVV (18.0% sero-positive), and EBLV-1 (5.9% sero-positive).

Molecular characterization of lyssaviruses should be undertaken to better describe the epidemiology of rabies and to characterize transmission dynamics. In 2017, as part of a laboratory-capacity-building effort, researchers in Lesotho, which is surrounded by South Africa, sequenced 21 samples from dogs and livestock (Coetzer et al., 2017). Results of virus characterization showed that there were at least three independent transmission cycles occurring in Lesotho, each associated with crossborder infections.

Rabies programs in early development stages may benefit form a range of epidemiologic activities. As seen in these examples, establishing surveillance and diagnostic capacity is a foundational step for a strong surveillance and epidemiology program. Activities ranging from active surveillance for sick animals, to serologic monitoring in a target species, through to characterization of viruses can all provide important information to describe disease presence and inform control strategies.

Stage 2: Human Rabies Prevention. After establishing the epidemiology of rabies in a program area, surveillance efforts should be designed to prevent human disease; these systems are often referred to as Passive Public Health Surveillance Systems. Passive surveillance operates on the premise that timely testing of suspected rabid animals can inform treatment decisions for persons with an exposure (Manning et al., 2008). These activities can prevent rabies deaths while also increasing program cost effectiveness through risk-based provision of rabies biologics (Etheart et al., 2017; Rysava et al., 2018; Undurraga et al., 2017; WHO Rabies Modelling Consortium, 2019). When test results are used to advocate for, or against, human prophylactic immunization, only WHO- and OIE-designated primary rabies diagnostic tests should be considered and laboratories should comply with international recommendations (WHO TRS).

Passive surveillance systems often have a high case detection rate, as the sampling frame is relegated to animals with clinical signs consistent with neurologic infections or otherwise acting abnormally (Franka & Wallace, 2018; Ma et al., 2018). The United States operates one of the world's most extensive rabies passive surveillance systems, testing over 90,000 animals each year. Among these, approximately 20,000 bats are tested, primarily after they have exposed a person or domestic animal, and among these 20,000 approximately 7% are found to be infected with the RABV. A study conducted in Washington State, United States, in 2017 highlighted the bias created through passive rabies surveillance, reporting that abnormally acting bats submitted to public health for testing had four-times greater odds of being infected with RABV compared to apparently healthy bats that were also tested by public health (21.5% positivity rates compared to 6.5% positivity rate, respectively) (Bonwitt, Oltean, Lang, Kelly, & Goldoft, 2018). In contrast, randomized sampling efforts for RABV detection in North American bats have yielded significantly lower case detection rates of 0%–0.3% (Bowen et al., 2013; O'Shea, Bowen, Stanley, Shankar, & Rupprecht, 2014).

Passive public health surveillance systems are designed from the perspective of an individual's health and providing information that can be used to inform healthcare treatment decisions such as whether or not to initiate rabies postexposure prophylaxis. Despite the biased epidemiologic information obtained from these systems, they are highly cost-effective endeavors for the judicious use of human rabies biologics. The aforementioned passive surveillance program in the United States is credited with preventing upwards of 90,000 unnecessary rabies postexposure prophylaxis treatments each year; a potential cost savings of over $600 million USD each year (Ma et al., 2018; Pieracci et al., 2019). A canine rabies passive surveillance program enacted in Haiti reported similar benefits, showing a twofold reduction in the cost per human rabies death averted under the passive surveillance system and a similar evaluation in the Philippines showed that up to half of the rabies biologics provided each year could be avoided through implementation of passive public health surveillance systems (Rysava et al., 2018; Undurraga et al., 2017). Perhaps most supportive for the global acceptance of passive public health surveillance as a core component of a rabies control program are the findings from the World Health Organization's rabies modeling consortium, which reported in 2018 that global adoption of risk-based provision of rabies biologics (i.e., passive public health surveillance) could prevent nearly 500,000 human rabies deaths between 2020 and 2035 (WHO Rabies Modelling Consortium, 2019).

Stage 3: Monitoring and Assessment of Control Measures. Globally, passive public health surveillance provides the vast majority of epidemiologic data by which we are able to monitor trends in rabies. However, in addition to the biased results that are prone in passive surveillance data, these data are limited to geographical areas where both people and animals reside and where infrastructure exists to recognize exposures, collect samples, and conduct diagnostic testing. The OIE recommends that surveillance systems are capable of conducting surveillance that is representative of the target species and their geographic ranges; something that may not be possible with passive surveillance systems. In such cases, active rabies surveillance programs can be used to inform the epidemiologic situation where passive surveillance has provided insufficient epidemiologic data.

Active rabies surveillance is based on identifying rabies suspected animals through targeted sampling of abnormally acting, found-dead, or roadkill animals. The United States Department of Agriculture operates an extensive active surveillance program to inform the progress of a large-scale wildlife rabies control program. Annually, over 5000 animals are actively surveilled to determine if vaccination programs are reducing the burden of disease and monitor for atypical epidemiologic events. Detection rates for active surveillance programs are often much lower than passive surveillance, resulting in higher costs per case-detected. Kirby et al. summarized case detection rates by method of active surveillance in the United States between 2005 and 2016 and found an overall rate of detection of 2.1%, nearly threefold lower than detection rates from the comparable passive surveillance system (Kirby et al., 2017) (OIE). The study authors also found vast differences in detection rates based on the type of surveillance activity conducted; strange-acting animals had detection rates similar to passive surveillance systems (5.7% positive), however road-kill, surveillance-trapping, and nuisance animals all had very low rates of detection (0.7%, 0.0%, and 0.2% rates of case detection, respectively). While surveillance efforts are increased for active surveillance, these results are less biased than passive surveillance data and provide epidemiologic information in settings that passive surveillance does not.

Stage 4: Verification of Rabies Elimination. Official recognition of freedom from animal diseases is regulated by the World Organization for Animal Health (OIE). Currently, only seven animal disease have official control programs by which a formal declaration of disease-freedom is assigned: foot and mouth disease, rinderpest, bovine spongiform encephalopathy, african horse sickness, peste des petits ruminants, classical swine fever, and contagious bovine pleuropneumonia. Official disease status declarations undergo thorough review of epidemiologic data by a panel of international experts. All other diseases, including rabies, are governed on a self-declared disease status. Self-declared statuses undergo less rigorous review processes, but are encouraged to adhere to OIE guidance.

The self-declaration of rabies freedom can have profound consequences for animal trade and human vaccination policies. According to international standards, verification of rabies freedom should be considered if rabies is notifiable, effective systems of rabies surveillance are in place, all regulatory measures for prevention and control are implemented, and that there have been no indigenous rabies cases for at least 2 years. Epidemiologic evidence to support elimination should be geographically representative of the declared area. Only internationally recognized rabies diagnostic assays should be considered when making a determination of rabies freedom and surveillance efforts should be focused on animals satisfying the suspect and/or probable case definition. Testing of healthy animals is of limited value for surveillance purposes (likely to give negative results); hence, intended bias in sampling for suspect animals is beneficial in this regard. Combinations of passive and active surveillance efforts may be required to support a declaration of rabies freedom.

Two events have occurred since 2010 that underscore the importance of epidemiologic monitoring for rabies to establish a rabies-free status, and to ensure that the status is maintained. From 1961 until 2013, Taiwan had a self-declared rabies-free status after canine rabies was eliminated from the island. However, in 2013, it was discovered that a unique sylvatic transmission cycle had been present the entire time in the Taiwanese ferret-badger. Retrospective assessment of surveillance activities conducted during the period of self-declared rabies-free status highlighted several inconsistencies with OIE recommendations for adequate surveillance. Most notably, surveillance efforts targeted apparently healthy dogs and cats that were euthanized at animal shelters. This population of animals was not representative geographically nor did they satisfy the rabies suspect or probable case definitions. It was only after institution of a passive wildlife disease surveillance system, focusing on testing of clinically ill animals, was the cryptically circulating RABV identified.

The second event to underscore the importance of adequate surveillance was observed in Malaysia in 2015. Previously self-declared as rabies-free, passive surveillance systems detected rabies cases in dogs in villages near the border with Indonesia. Despite rapid implementation of control measures, the virus had already infected numerous villages resulting in the deaths of at least 18 people. Verification of rabies freedom should not be taken lightly, and the self-declaration process should be interpreted carefully by the international community.

Stage 5: Post-elimination. After rabies elimination has been achieved, the intensity of surveillance should consider the risk for possible reintroduction. Appropriate border and trade security policies should be in place. Given the large degree of international trade and travel, most countries should consider continuation of surveillance activities, albeit at a reduced intensity. Any confirmed cases should undergo additional virus characterization to determine if the event is due to cryptic transmission cycles or due to incursion of a nonnative virus variant.

4.4.4 Critical components of functional surveillance systems

Key components for effective and useful rabies surveillance systems are: (a) adequate and clear reporting regulations (policies and laws for notification); (b) laboratory diagnostic systems and infrastructure that meet the needs of the country; (c) integrated/coordinated surveillance systems (paper based or electronic, clear cases definitions, standardized indicators, zero reporting and minimum requirements); (d) resources; and (e) regional approach to surveillance (in addition to national or subnational) (Franka & Wallace, 2018). Rabies surveillance systems often suffer from a lack of coordination, in part because the responsible authority is not always identified. Rabies is a disease of animals, most commonly domestic dogs and wildlife; therefore, the animal health sector may not view control as their responsibility since the economic impact on livestock is relatively minor. Alternatively, the human health sector may not have the resources or authority necessary to control the disease in the animal reservoir population. Departmental and legal separation of human and animal health sectors as well as absence of requirements for disease notification and reporting (or lack of enforcement thereof) in many rabies-endemic countries creates a barrier for effective and responsive rabies surveillance (Taylor et al., 2015).

Multiple international rabies reporting system have been implemented with various degrees of success. WHO's Rabnet platform (no longer active) (World Health Organization, 1998), OIE's World Animal Health Information Database (WAHIS) Interface (http://www.oie.int/wahis_2/public/wahid.php/Wahidhome/Home) as well as PAHO's epidemiologic surveillance system—SIEPI (Latin America) have experienced challenges in adequate engagement from countries to report rabies cases. WHO's Rabies Bulletin Europe (https://www.who-rabies-bulletin.org/site-page/what-rabies), which serves as a surveillance database for 44 European countries (including Russian Federation), is the most effective database providing both numerator (positive cases) and denominator (number of tested animals) data (Freuling, Kloss, Schroder, Kliemt, & Muller, 2012).

To assess and address the disconnect between estimated, reported, and true rabies data in Africa, the Pan-African Rabies Control Network (PARACON) is currently developing and implementing a regional rabies-specific disease surveillance bulletin based on the District Health Information System 2 platform (Scott, Coetzer, Fahrion, & Nel, 2017). This surveillance bulletin allows participating countries to analyze submitted data, but also facilitates the sharing of submitted data with approved international authorities, reducing redundancy and reporting fatigue. Providing a data repository that has epidemiologic benefits to the submitting countries may soon prove to be a more productive means for sustained data sharing, rather than one-way reporting channels to international health authorities.

4.4.5 Laboratory diagnostic methods for effective surveillance system

Laboratory diagnosis of rabies, by detection of lyssavirus antigen or nucleic acid, is a prerequisite for effective rabies surveillance (Banyard, Horton, Freuling, Muller, & Fooks, 2013). Many laboratory diagnostic techniques for rabies have been assessed and evaluated, but only very few have shown adequate sensitivity and specificity to be considered primary or confirmatory tests by OIE and WHO (Table 4.5) (OIE, 2016). The diagnostic assay(s) chosen should complement the goal of the surveillance activities being undertaken. Surveillance efforts focused on establishing a rabies burden may sacrifice sensitivity to facilitate more rapid and field-based activities. However, any surveillance activities that may impact human treatment decisions or for which results will be used to make a declaration of rabies-freedom should be supported by internationally recognized primary rabies diagnostic assays.

Point-of-care-testing with simple, reliable, and easy-to-operate devices would be an ideal approach to provide rabies test results for immediate decision making for administration of prophylaxis following a bite exposure as well as for surveillance purposes. Such testing would eliminate the challenges associated with the need for preservation of samples and their transportation to laboratory facility. New, field-ready PCR technologies intended to be used as point-of-care are currently being assessed and implemented for detection of various pathogens (Biava et al., 2018). Lateral flow devices (LFD), based on immunochromatographic reaction, have been extensively assessed for potential point-of-care-testing assays (Lechenne et al., 2016). Although very promising, evaluations of LFDs have shown a high number of false-negative results and significant batch-to-batch variability, rendering these tests incompatible with surveillance programs that impact human treatment decisions or that are attempting to confirm a rabies-free status. It has been recommended that further development and validation for such devices is needed, before their licensure and routine use in the field.

4.4.6 Future of integrated surveillance approaches

Rabies is a highly fatal viral zoonosis that has no reliably effective treatment after the onset of symptoms. To balance the lethality of the virus with expensive vaccines that are often in

TABLE 4.5 Risk matrix for rabies exposure, considering the type of exposure and the status of the offending dog.

Rabies exposure risk matrix[a]

Exposure consideration	Probability of death based on level of exposure (%)[b]	Information collected at time of bite						Quarantine or testing	
		Dog symptomatic	Dog dead at follow-up	Dog bite was not provoked	Stray dog	Dog bit multiple people	Dog not vaccinated	Dog healthy and available for quarantine	Dog healthy 10 Days postbite	Tested negative
Bite to head/neck	45.0	High	High	High	High	High	High	Low	No risk	No risk
Multiple severe bite wounds	27.5	High	High	High	High	Moderate	Moderate	Low	No risk	No risk
Bites to young children	27.5	High	High	High	High	Moderate	Moderate	Low	No risk	No risk
Bites to extremities	5.0	High	Moderate	Moderate	Moderate	Moderate	Low	Low	No risk	No risk
Minor bites or scratches (no break in skin)	1.0	Moderate	Moderate	Moderate	Moderate	Moderate	Low	Low	No risk	No risk
Probability the dog has rabies (%)[c]		62.2	39.7	15.0	13.9	10.6	4.7	0.08	0	0

[a] Risk calculated by the product of the probability the dog had rabies and the probability that if rabid, the level of exposure would result in death.
[b] Babes, V. (1912). Traité de la rage. Paris: Ballière, http://gallica.bnf.fr/ark:/12148/bpt6k5462676f.
[c] Medley, A. M., Millien, M. F., Blanton, J. D., Ma, X., Augustin, P., Crowdis, K., & Wallace, R. M. (2017). Retrospective cohort study to assess the risk of rabies in biting dogs, 2013–2015, Republic of Haiti. Tropical Medicine and Infectious Disease, 2, 14.

From WHO. (2018). WHO Expert Consultation on Rabies: Third Report. In WHO technical report series No 1012 (Vol. 183, pp. 183). Geneva: WHO.

short supply, human and animal health sectors can institute a One Health-based system termed Integrated Bite Case Management (IBCM) to improve the efficiency of vaccine usage. IBCM encompasses the investigation of exposures from rabies suspected animals, which is complemented by healthcare counseling (Etheart et al., 2017). The results of IBCM investigations are intended to deliver a patient-tailored risk-assessment and healthcare instructions. Since rabies diagnostic assays are equivalent for humans and animals, laboratory systems for rabies may also implement a One Health approach by accepting samples submitted from either sector. IBCM programs should facilitate near-real-time information sharing between sectors at the subnational and national levels to ensure that rabies control and elimination programs channel resources according to the situation in both humans and animals (Fig. 4.3) (WHO TRS).

In Haiti, an independent, laboratory-based surveillance system, comprising active community bite investigation and passive animal rabies investigation, has proven successful in real-time monitoring and assessment of rabies burden (Wallace et al., 2015). In certain instances, where the emphasis is only on human rabies prevention and control and where surveillance of domestic animals or wildlife exceeds public health surveillance needs, targeted surveillance platforms are being developed in parallel. In Canada, RageDB is a spatiotemporal, web-based database application that focuses on storage, analysis, and communication regarding raccoon rabies variant surveillance data between public health, agriculture, and wildlife agencies (Rees, Gendron, Lelievre, Cote, & Belanger, 2011). In the United States, a GIS-based, centralized, rabies surveillance database and internet mapping application was developed and piloted in eight states as a new tool for the rapid real-time mapping and dissemination of data on animal rabies cases in relation to unaffected, enzootic, and baited areas where current interventions are underway (Blanton, Manangan, et al., 2006). Integration of surveillance systems and human-animal-vector approach should be emphasized, sought, and planned for over the development and dissemination of incompatible, temporary, or individual disease/species surveillance systems.

4.5 Human rabies: Public health measures

The development of postexposure prophylaxis, particularly using modern high-quality vaccines, represented the greatest advancement in preventing human rabies cases. While distribution systems for childhood and other vaccines is well established in most countries, distribution of rabies vaccines has not routinely taken advantage of these networks (Li et al., 2018). This increases inefficiencies in distribution and likely reduces overall accessibility in many countries and adds to the already high costs associated with rabies PEP. One of the difficulties in integrating rabies vaccines with other national distribution networks is its use as a postexposure administration as opposed to broader mass vaccination programs that follow routine schedules. However, in some cases, the administration networks remain the same.

Alternative approaches to reducing human rabies have been to integrate rabies vaccination with childhood immunization schedules in countries where canine rabies is endemic. Most studies of this approach to human rabies prevention have reported administration to infants and children to be safe and immunogenic. However, in most canine rabies-endemic setting, childhood immunization against rabies would not be cost effective and ensuring access to PEP is preferred (Kessels, Tarantola, Salahuddin, Blumberg, & Knopf, 2019). In contrast, a study evaluating mass rabies vaccination in hard-to-reach populations in the Amazon found such programs were cost effective due to the high exposure rate from vampire bats and the low accessibility to healthcare in these regions (Gilbert et al., 2012).

Overall human rabies prevention will focus on recognition of a possible rabies exposure and timely administration of rabies PEP. Ultimately, successful canine rabies control programs remain the most cost-effective way to prevent human rabies exposure through the control and elimination of canine rabies. The general public should be better informed about avoiding direct contact with wildlife in general and with abnormally behaving and sick animals in particular. Any person bitten by a wild or domestic animal, particularly in areas where wildlife rabies is endemic, should seek medical attention. However, careful consideration of the exposure scenario and the subsequent risk of rabies is important to reduce overutilization of rabies vaccine for low-risk exposures (Table 4.5). Such overuse of biologics increases the cost to individuals and national rabies control programs and other supply restricted systems can result in vaccine not being available to persons with high-risk exposures. Translocation of wildlife for any purpose except conservation should be banned or strongly discouraged. Movement of unvaccinated domestic pets should similarly be restricted.

4.6 Animal rabies: Epidemiology and control

4.6.1 Bats

Lyssaviruses have been detected in bats throughout the world, displaying distinct species and geographical epidemiologic patterns (Fooks et al., 2017). Bats have been identified as reservoir species for all Lyssavirus species except MOKV and IKOV, for which the true primary host is yet to be identified. Bat-associated nonrabies lyssaviruses are found in Europe, Africa, Asia, and Australia, whereas RABV is only found in bats in the Western Hemisphere. RABV is present in at least 30 species of bats indigenous to North America. Spillover of RABV infections from bats into domestic mammals has been reported to occur at a rate of approximately 12 spillover events per 1000 domestic animal rabies cases, based on passive surveillance data in the United States (Ma et al., 2018). Spillover of RABV from bats into terrestrial wildlife is documented less frequently through passive surveillance, at a rate of 6 spillover events per 1000 terrestrial wildlife rabies cases. Reverse spillover of Lyssavirus species from terrestrial carnivores into bats is thought to be an extremely rare event (Matsumoto et al., 2017). Spillover infections from bats into humans are a significant public health concern in North America, where 48 of the 54 human RABV infections acquired in the United States between 1984 and 2014 were attributable to variants of the RABV associated with bats (Pieracci et al., 2019). Vampire bat-transmitted RABV is a major concern for livestock health in Central and South America; vampire bat-transmitted human rabies deaths occur with the same relative frequency as insectivorous bat-transmitted human rabies deaths reported in North America (Freire de Carvalho et al., 2018).

Controlling RABV infections in bats is challenging. Support for investigating control measures in bats is countered by the high relative burden of disease due to domestic dogs. Further complicating control efforts are the complex epidemiologic cycles in bats and diverse species distribution. Parenteral vaccination is impossible and bats are unlikely to be a candidate for oral vaccination due to safety concerns for environmentally distributed modified-live vaccines. Culling of bats has not proven effective at controlling RABV transmission, and in some events has been reported to be counter-productive as culling might increase the proportion of the colony that is susceptible to infection and dispersal of colony survivors can move the virus to other colonies. Vampire bat-transmitted rabies in livestock can be controlled by incorporating routine rabies vaccination into herd health management plans. Preventive immunization of populations living in highly enzootic areas with limited access to anti-rabies biologicals should be considered.

4.6.2 Terrestrial carnivores

Terrestrial carnivores have been identified as reservoirs for RABV on numerous continents. In Africa, Asia, and Europe terrestrial RABV reservoirs are the result of host shift events from domestic dogs. In the Western Hemisphere, terrestrial RABV reservoirs have been the result of host shifts from both bats and domestic dogs. In the United States, rabies is primarily a disease of terrestrial wildlife. This was not always the case, but as canine rabies was brought under control in the 1950s, cases of rabies in wildlife species quickly outnumbered those of dogs. Before canine rabies was eliminated in the 1970s, it made the jump into several wildlife species, including skunks, foxes, and mongoose. Detection of cryptically circulating wildlife terrestrial rabies reservoirs has been reported in other countries as well and is consistently associated with local host shift events from the endemic canine RABV variant into terrestrial mammals. Human rabies deaths due to terrestrial mammals are rarely reported; in the United States only 6 of 54 human rabies deaths since 1984 have been associated with transmission from terrestrial wildlife. Control of terrestrial wildlife rabies has been successful in numerous countries in Europe as well as Canada through the implementation of oral rabies vaccination programs.

4.6.3 Domestic dogs

Domestic dogs are responsible for over 98% of the global human rabies burden (WHO TRS). Over 120 countries have enzootic transmission of the canine RABV variant, placing over 3 billion people at risk for infection. Canine rabies is maintained in susceptible dog populations, with a relatively low viral reproductive ratio of 1.1–1.8 (Borse et al., 2018; Zinsstag et al., 2017). Outbreaks in urban centers have reported higher reproductive ratios, but enzootic transmission occurs, overall, at a relatively low rate (Kadowaki, Hampson, Tojinbara, Yamada, & Makita, 2018).

Dog vaccination remains the most cost-effective means of rabies control and elimination (Elser, Hatch, Taylor, Nel, & Shwiff, 2018). Accessibility of dogs for vaccination is a key factor in both the success and cost effectiveness of a dog vaccination program (Wallace et al., 2019). Highly accessible dogs, such as those that can be presented on leash by an owner, are good targets for fixed-point, parenteral vaccination strategies. However, these types of dogs may play relatively little role in enzootic rabies transmission if they are primarily confined and have limited opportunities to interact freely with other dogs (Taylor, Wallace, et al., 2017). Less accessible dogs, such as community dogs or owned, free-roaming dogs, play a critical role in RABV transmission in many communities. While some of these dogs may be accessible through fixed-point, parenteral vaccination, alternative methods might be indicated. Door-to-door vaccination and capture-vaccinate-release have been described in literature as highly effective means of reaching less accessible dog populations (Cleaton et al., 2018; Mazeri et al., 2019; Mazeri et al., 2018; Smith et al., 2017). However, these methods are often more costly and time consuming compared to fixed-point vaccination methods. Oral rabies vaccination of inaccessible dogs is continuing to be evaluated in terms of cost effectiveness and role (Head et al., 2019; Wallace et al., 2019). Numerous international organizations have recognized the importance of investigating appropriate applications of oral rabies vaccination of dogs, but to date there are no examples of large-scale application of oral vaccines for dogs.

Evaluation of dog vaccination programs is a critical component of a rabies control program. Postvaccination assessments should be conducted as often as possible, but the frequency is dependent upon the vaccination methods being practiced. The end goal of a vaccination program is not to achieve 70% coverage, rather it is to reduce and eventually eliminate rabies cases. Therefore, the gold-standard evaluation of a vaccination program is documentation of a reduction of cases, postvaccination. In settings where surveillance capacity is robust and can detect significant reductions in cases across representative geographic and species boundaries, then minimal postvaccination evaluation is necessary (LeRoux et al., 2018).

Robust surveillance systems capable of evaluating canine rabies vaccination programs are implemented in very few canine rabies-endemic countries (Taylor, Hampson, et al., 2017). When surveillance is not reliable to evaluate vaccination efforts, postvaccination evaluations must be conducted. Numerous methods have been described to conduct these evaluations. Most commonly cited are household surveys and field surveys. Household surveys capture owner-reported attendance at vaccination campaigns. This method is applicable in communities with majority-owned dogs, but is prone to error if owners misrepresent their attendance or in communities with a large proportion of community-owned dogs. Field surveys, sometimes referred to as mark-recapture surveys, rely on field teams to canvas communities and document dogs with, and without, a mark of vaccination. This method is applicable in communities with numerous free-roaming dogs and provides a more accurate reflection of the vaccination coverage among the truly susceptible dog population.

Variations in the timing of postvaccination evaluations have also been described in the literature. Traditional vaccination methods often rely upon evaluation after the vaccination program is complete. Results can be compiled and vaccination programs are revised in the following campaign-year. More recently, a method of vaccinate-assess-move has been described in which smartphone technologies are used to capture near real-time vaccination activities (Gibson et al., 2015). Vaccination records are uploaded to a central coordinator who can assess coverages and coordinate vaccination activities on a day-to-day basis. Once geographic, numeric, and community factors are satisfied, vaccination teams are directed to new communities. The benefit of vaccinate-assess-move programs is that revisions to strategies and improvements in vaccination coverage can be achieved during the same campaign-year (Gibson et al., 2018).

Canine rabies vaccination programs should strive to reach as many dogs as possible, but when resources are limited (which is the reality for most canine rabies-endemic countries), programs should devise strategies to reach the highest possible proportion of susceptible dogs (Wallace et al., 2019). There is no single standard program that will be effective in every setting, and mixed-methods approaches will likely be necessary across any heterogeneous, large-scale dog vaccination program. Postvaccination success is best determined through documented decline in cases; however, evaluations should be conducted when such capacity is not present. Vaccinate-assess-move programs, reliant on real-time electronic data collection, have shown great promise at increasing dog vaccination coverages in low-infrastructure settings.

4.6.4 Wildlife species of special concern

Despite extensive surveillance efforts to detect RABV in rodent populations in areas endemic for rabies across the world, only rare exceptional instances of infections have been reported primarily in large-bodied rodents such as groundhogs and beavers (Fitzpatrick, Dyer, Blanton, Kuzmin, & Rupprecht, 2014). Rodents are thought not to have a role in the epidemiology and transmission of RABV. Furthermore, there are no documented human rabies deaths due to a rodent bite. Only in extremely rare situations should a rodent bite result in a recommendation for PEP, such as when the rodent is confirmed rabies-positive (WHO, 2018).

Spillover of RABV into populations of endangered species can have detrimental effects. Rabies outbreaks have been reported in the remaining population of the endangered Ethiopian wolf (*Canis simensis*), African wild dogs (*Lycaon pictus*) in southern Africa, and Blanford fox (*Vulpus cana*) in Israel. Control of enzootic transmission in the local dog population is the most effective means at preventing these spillover infections. Programs have also successfully implemented oral vaccines to prevent rabies cases in these endangered species.

References

Afshar, A. (1979). A review of non-bite transmission of rabies virus infection. *The British Veterinary Journal*, *135*(2), 142–148.

Alvar, J., Velez, I. D., Bern, C., Herrero, M., Desjeux, P., Cano, J., … WHO Leishmaniasis Control Team. (2012). Leishmaniasis worldwide and global estimates of its incidence. *PLoS One*, *7*(5), e35671. https://doi.org/10.1371/journal.pone.0035671.

Baltazard, M., & Ghodssi, M. (1954). Prevention of human rabies; treatment of persons bitten by rabid wolves in Iran. *Bulletin of the World Health Organization*, *10*(5), 797–803.

Banyard, A. C., Horton, D. L., Freuling, C., Muller, T., & Fooks, A. R. (2013). Control and prevention of canine rabies: The need for building laboratory-based surveillance capacity. *Antiviral Research*, *98*(3), 357–364. https://doi.org/10.1016/j.antiviral.2013.04.004.

Bengoumi, M., Mansouri, R., Ghram, B., & Merot, J. (2018). Rabies in North Africa and the Middle East: Current situation, strategies and outlook. *Revue Scientifique et Technique*, *37*(2), 497–510. https://doi.org/10.20506/rst.37.2.2818.

Berger, S. (2016). *Anthrax: Global status*. GIDEON Informatics, Inc.

Biava, M., Colavita, F., Marzorati, A., Russo, D., Pirola, D., Cocci, A., … Di Caro, A. (2018). Evaluation of a rapid and sensitive RT-qPCR assay for the detection of Ebola Virus. *Journal of Virological Methods*, 252, 70–74. https://doi.org/10.1016/j.jviromet.2017.11.009.

Black, E. M., Lowings, J. P., Smith, J., Heaton, P. R., & McElhinney, L. M. (2002). A rapid RT-PCR method to differentiate six established genotypes of rabies and rabies-related viruses using TaqMan technology. *Journal of Virological Methods*, 105(1), 25–35.

Blanton, J. D., Hanlon, C. A., & Rupprecht, C. E. (2007). Rabies surveillance in the United States during 2006. *Journal of the American Veterinary Medical Association*, 231(4), 540–556. https://doi.org/10.2460/javma.231.4.540.

Blanton, J. D., Krebs, J. W., Hanlon, C. A., & Rupprecht, C. E. (2006). Rabies surveillance in the United States during 2005. *Journal of the American Veterinary Medical Association*, 229(12), 1897–1911. https://doi.org/10.2460/javma.229.12.1897.

Blanton, J. D., Manangan, A., Manangan, J., Hanlon, C. A., Slate, D., & Rupprecht, C. E. (2006). Development of a GIS-based, real-time Internet mapping tool for rabies surveillance. *International Journal of Health Geographics*, 5, 47. https://doi.org/10.1186/1476-072X-5-47.

Boland, T. A., McGuone, D., Jindal, J., Rocha, M., Cumming, M., Rupprecht, C. E., … Rosenthal, E. S. (2014). Phylogenetic and epidemiologic evidence of multiyear incubation in human rabies. *Annals of Neurology*, 75(1), 155–160. https://doi.org/10.1002/ana.24016.

Bonwitt, J., Oltean, H., Lang, M., Kelly, R. M., & Goldoft, M. (2018). Bat rabies in Washington State: Temporal-spatial trends and risk factors for zoonotic transmission (2000-2017). *PLoS One*, 13(10), e0205069. https://doi.org/10.1371/journal.pone.0205069.

Borchering, R. K., Liu, H., Steinhaus, M. C., Gardner, C. L., & Kuang, Y. (2012). A simple spatiotemporal rabies model for skunk and bat interaction in northeast Texas. *Journal of Theoretical Biology*, 314, 16–22. https://doi.org/10.1016/j.jtbi.2012.08.033.

Borse, R. H., Atkins, C. Y., Gambhir, M., Undurraga, E. A., Blanton, J. D., Kahn, E. B., … Meltzer, M. I. (2018). Cost-effectiveness of dog rabies vaccination programs in East Africa. *PLoS Neglected Tropical Diseases*, 12(5), e0006490. https://doi.org/10.1371/journal.pntd.0006490.

Bowen, R. A., O'Shea, T. J., Shankar, V., Neubaum, M. A., Neubaum, D. J., & Rupprecht, C. E. (2013). Prevalence of neutralizing antibodies to rabies virus in serum of seven species of insectivorous bats from Colorado and New Mexico, United States. *Journal of Wildlife Diseases*, 49(2), 367–374. https://doi.org/10.7589/2012-05-124.

Brachman, P. S., & Abrutyn, E. (Eds.), (1998). *Bacterial infections of humans: Epidemiology and control* (3rd ed.). Springer Science & Business Media, LLC.

CDC. (2019). *Rabies in Vietnam*. Retrieved from https://www.cdc.gov/worldrabiesday/vietnam.html.

Centers for Disease Control and Prevention. (2000). Update: Raccoon rabies epizootic—United States and Canada, 1999. *MMWR. Morbidity and Mortality Weekly Report*, 49(2), 31–35.

Centers for Disease Control and Prevention. (2004). Investigation of rabies infections in organ donor and transplant recipients—Alabama, Arkansas, Oklahoma, and Texas, 2004. *MMWR. Morbidity and Mortality Weekly Report*, 53(26), 586–589.

Centers for Disease Control and Prevention. (2019). *2014–2016 Ebola outbreak in West Africa*. Retrieved from https://www.cdc.gov/vhf/ebola/history/2014-2016-outbreak/index.html.

Chang, S. S., Tsai, H. J., Chang, F. Y., Lee, T. S., Huang, K. C., Fang, K. Y., … Fei, C. Y. (2016). Government response to the discovery of a rabies virus reservoir species on a previously designated rabies-free island, Taiwan, 1999-2014. *Zoonoses and Public Health*, 63(5), 396–402. https://doi.org/10.1111/zph.12240.

Chen, S., Zhang, H., Luo, M., Chen, J., Yao, D., Chen, F., … Chen, T. (2017). Rabies virus transmission in solid organ transplantation, China, 2015-2016. *Emerging Infectious Diseases*, 23(9), 1600–1602. https://doi.org/10.3201/eid2309.161704.

Chiou, H. Y., Hsieh, C. H., Jeng, C. R., Chan, F. T., Wang, H. Y., & Pang, V. F. (2014). Molecular characterization of cryptically circulating rabies virus from ferret badgers, Taiwan. *Emerging Infectious Diseases*, 20(5), 790–798. https://doi.org/10.3201/eid2005.131389.

Christou, L. (2011). The global burden of bacterial and viral zoonotic infections. *Clinical Microbiology and Infection*, 17(3), 326–330. https://doi.org/10.1111/j.1469-0691.2010.03441.x.

Cleaton, J. M., Wallace, R. M., Crowdis, K., Gibson, A., Monroe, B., Ludder, F., … King, A. (2018). Impact of community-delivered SMS alerts on dog-owner participation during a mass rabies vaccination campaign, Haiti 2017. *Vaccine*, 36(17), 2321–2325. https://doi.org/10.1016/j.vaccine.2018.03.017.

Cleaveland, S., Fevre, E. M., Kaare, M., & Coleman, P. G. (2002). Estimating human rabies mortality in the United Republic of Tanzania from dog bite injuries. *Bulletin of the World Health Organization*, 80(4), 304–310.

Coetzer, A., Coertse, J., Makalo, M. J., Molomo, M., Markotter, W., & Nel, L. H. (2017). Epidemiology of rabies in Lesotho: The importance of routine surveillance and virus characterization. *Tropical Medicine and Infectious Disease, 2*(3). https://doi.org/10.3390/tropicalmed2030030.

Correa-Giron, E. P., Allen, R., & Sulkin, S. E. (1970). The infectivity and pathogenesis of rabiesvirus administered orally. *American Journal of Epidemiology, 91*(2), 203–215. https://doi.org/10.1093/oxfordjournals.aje.a121129.

David, D. (2012). Role of the RT-PCR method in ante-mortem & post-mortem rabies diagnosis. *The Indian Journal of Medical Research, 135*(6), 809–811.

David, D., Bellaiche, M., & Yakobson, B. A. (2010). Rabies in two vaccinated dogs in Israel. *The Veterinary Record, 167*(23), 907–908. https://doi.org/10.1136/vr.c3614.

David, D., Dveres, N., Yakobson, B. A., & Davidson, I. (2009). Emergence of dog rabies in the northern region of Israel. *Epidemiology and Infection, 137*(4), 544–548. https://doi.org/10.1017/S0950268808001180.

Davis, C. P. (2018). Botulism. Retrieved from https://www.medicinenet.com/botulism/article.htm#botulism_facts.

Dean, A. S., Crump, L., Greter, H., Schelling, E., & Zinsstag, J. (2012). Global burden of human brucellosis: A systematic review of disease frequency. *PLoS Neglected Tropical Diseases, 6*(10), e1865. https://doi.org/10.1371/journal.pntd.0001865.

Diagana, M., Preux, P. M., & Dumas, M. (2007). Japanese encephalitis revisited. *Journal of the Neurological Sciences, 262*(1–2), 165–170. https://doi.org/10.1016/j.jns.2007.06.041.

Dupuis, M., Brunt, S., Appler, K., Davis, A., & Rudd, R. (2015). Comparison of automated quantitative reverse transcription-PCR and direct fluorescent-antibody detection for routine rabies diagnosis in the United States. *Journal of Clinical Microbiology, 53*(9), 2983–2989. https://doi.org/10.1128/JCM.01227-15.

Dyer, J. L., Niezgoda, M., Orciari, L. A., Yager, P. A., Ellison, J. A., & Rupprecht, C. E. (2013). Evaluation of an indirect rapid immunohistochemistry test for the differentiation of rabies virus variants. *Journal of Virological Methods, 190*(1-2), 29–33. https://doi.org/10.1016/j.jviromet.2013.03.009.

Ekanem, E. E., Eyong, K. I., Philip-Ephraim, E. E., Eyong, M. E., Adams, E. B., & Asindi, A. A. (2013). Stray dog trade fuelled by dog meat consumption as a risk factor for rabies infection in Calabar, southern Nigeria. *African Health Sciences, 13*(4), 1170–1173. https://doi.org/10.4314/ahs.v13i4.44.

Elser, J. L., Hatch, B. G., Taylor, L. H., Nel, L. H., & Shwiff, S. A. (2018). Towards canine rabies elimination: Economic comparisons of three project sites. *Transboundary and Emerging Diseases, 65*(1), 135–145. https://doi.org/10.1111/tbed.12637.

Etheart, M. D., Kligerman, M., Augustin, P. D., Blanton, J. D., Monroe, B., Fleurinord, L., … Wallace, R. M. (2017). Effect of counselling on health-care-seeking behaviours and rabies vaccination adherence after dog bites in Haiti, 2014-15: A retrospective follow-up survey. *The Lancet Global Health, 5*(10), e1017–e1025. https://doi.org/10.1016/S2214-109X(17)30321-2.

Evans, J. S., Horton, D. L., Easton, A. J., Fooks, A. R., & Banyard, A. C. (2012). Rabies virus vaccines: Is there a need for a pan-lyssavirus vaccine? *Vaccine, 30*(52), 7447–7454. https://doi.org/10.1016/j.vaccine.2012.10.015.

Everard, C. O., & Everard, J. D. (1992). Mongoose rabies in the Caribbean. *Annals of the New York Academy of Sciences, 653*, 356–366. https://doi.org/10.1111/j.1749-6632.1992.tb19662.x.

Faye, M., Dacheux, L., Weidmann, M., Diop, S. A., Loucoubar, C., Bourhy, H., … Faye, O. (2017). Development and validation of sensitive real-time RT-PCR assay for broad detection of rabies virus. *Journal of Virological Methods, 243*, 120–130. https://doi.org/10.1016/j.jviromet.2016.12.019.

Fehlner-Gardiner, C. (2018). Rabies control in North America—Past, present and future. *Revue Scientifique et Technique, 37*(2), 421–437. https://doi.org/10.20506/rst.37.2.2812.

Fekadu, M., Chandler, F. W., & Harrison, A. K. (1982). Pathogenesis of rabies in dogs inoculated with an Ethiopian rabies virus strain. Immunofluorescence, histologic and ultrastructural studies of the central nervous system. *Archives of Virology, 71*(2), 109–126.

Fekadu, M., Shaddock, J. H., & Baer, G. M. (1982). Excretion of rabies virus in the saliva of dogs. *The Journal of Infectious Diseases, 145*(5), 715–719. https://doi.org/10.1093/infdis/145.2.715.

Fischer, M., Wernike, K., Freuling, C. M., Muller, T., Aylan, O., Brochier, B., … Hoffmann, B. (2013). A step forward in molecular diagnostics of lyssaviruses—Results of a ring trial among European laboratories. *PLoS One, 8*(3), e58372. https://doi.org/10.1371/journal.pone.0058372.

Fisher, C. R., Streicker, D. G., & Schnell, M. J. (2018). The spread and evolution of rabies virus: Conquering new frontiers. *Nature Reviews. Microbiology, 16*(4), 241–255. https://doi.org/10.1038/nrmicro.2018.11.

Fitzpatrick, J. L., Dyer, J. L., Blanton, J. D., Kuzmin, I. V., & Rupprecht, C. E. (2014). Rabies in rodents and lagomorphs in the United States, 1995-2010. *Journal of the American Veterinary Medical Association, 245*(3), 333–337. https://doi.org/10.2460/javma.245.3.333.

Fooks, A. R., Banyard, A. C., Horton, D. L., Johnson, N., McElhinney, L. M., & Jackson, A. C. (2014). Current status of rabies and prospects for elimination. *Lancet, 384*(9951), 1389–1399. https://doi.org/10.1016/S0140-6736(13)62707-5.

Fooks, A. R., Cliquet, F., Finke, S., Freuling, C., Hemachudha, T., Mani, R. S., … Banyard, A. C. (2017). Rabies. *Nature Reviews. Disease Primers, 3*, 17091. https://doi.org/10.1038/nrdp.2017.91.

Franka, R., & Wallace, R. (2018). Rabies diagnosis and surveillance in animals in the era of rabies elimination. *Revue Scientifique et Technique, 37*(2), 359–370. https://doi.org/10.20506/rst.37.2.2807.

Fredericks, A. C., & Fernandez-Sesma, A. (2014). The burden of dengue and chikungunya worldwide: Implications for the southern United States and California. *Annals of Global Health, 80*(6), 466–475. https://doi.org/10.1016/j.aogh.2015.02.006.

Freire de Carvalho, M., Vigilato, M. A. N., Pompei, J. A., Rocha, F., Vokaty, A., Molina-Flores, B., … Del Rio Vilas, V. J. (2018). Rabies in the Americas: 1998-2014. *PLoS Neglected Tropical Diseases, 12*(3), e0006271. https://doi.org/10.1371/journal.pntd.0006271.

Freuling, C. M., Kloss, D., Schroder, R., Kliemt, A., & Muller, T. (2012). The WHO Rabies Bulletin Europe: A key source of information on rabies and a pivotal tool for surveillance and epidemiology. *Revue Scientifique et Technique, 31*(3), 799–807.

GBD 2017 DALYs, & HALE Collaborators. (2018). Global, regional, and national disability-adjusted life-years (DALYs) for 359 diseases and injuries and healthy life expectancy (HALE) for 195 countries and territories, 1990–2017: A systematic analysis for the Global Burden of Disease Study 2017. *Lancet, 392*(10159), 1859–1922. https://doi.org/10.1016/S0140-6736(18)32335-3.

Gibbons, R. V. (2002). Cryptogenic rabies, bats, and the question of aerosol transmission. *Annals of Emergency Medicine, 39*(5), 528–536. https://doi.org/10.1067/mem.2002.121521.

Gibson, A. D., Mazeri, S., Lohr, F., Mayer, D., Burdon Bailey, J. L., Wallace, R. M., … Gamble, L. (2018). One million dog vaccinations recorded on mHealth innovation used to direct teams in numerous rabies control campaigns. *PLoS One, 13*(7), e0200942. https://doi.org/10.1371/journal.pone.0200942.

Gibson, A. D., Ohal, P., Shervell, K., Handel, I. G., Bronsvoort, B. M., Mellanby, R. J., & Gamble, L. (2015). Vaccinate-assess-move method of mass canine rabies vaccination utilising mobile technology data collection in Ranchi, India. *BMC Infectious Diseases, 15*, 589. https://doi.org/10.1186/s12879-015-1320-2.

Giesecke, J. (2017). *Modern infectious disease epidemiology* (3rd ed.). Boca Raton, FL: CRC Press.

Gilbert, A. T., Petersen, B. W., Recuenco, S., Niezgoda, M., Gomez, J., Laguna-Torres, V. A., & Rupprecht, C. (2012). Evidence of rabies virus exposure among humans in the Peruvian Amazon. *The American Journal of Tropical Medicine and Hygiene, 87*(2), 206–215. https://doi.org/10.4269/ajtmh.2012.11-0689.

Gongal, G., & Wright, A. E. (2011). Human rabies in the WHO Southeast Asia Region: Forward steps for elimination. *Advances in Preventive Medicine, 2011*. 383870. https://doi.org/10.4061/2011/383870.

Green, A. (2012). Uganda battles Marburg fever outbreak. *Lancet, 380*(9855), 1726. https://doi.org/10.1016/s0140-6736(12)61973-4.

Gunawardena, P. S., Marston, D. A., Ellis, R. J., Wise, E. L., Karawita, A. C., Breed, A. C., … Fooks, A. R. (2016). Lyssavirus in Indian flying foxes, Sri Lanka. *Emerging Infectious Diseases, 22*(8), 1456–1459. https://doi.org/10.3201/eid2208.151986.

Hampson, K., Coudeville, L., Lembo, T., Sambo, M., Kieffer, A., Attlan, M., … Global Alliance for Rabies Control Partners for Rabies Prevention. (2015). Estimating the global burden of endemic canine rabies. *PLoS Neglected Tropical Diseases, 9*(4), e0003709. https://doi.org/10.1371/journal.pntd.0003709.

Hassel, R., Vos, A., Clausen, P., Moore, S., van der Westhuizen, J., Khaiseb, S., … Muller, T. (2018). Experimental screening studies on rabies virus transmission and oral rabies vaccination of the Greater Kudu (*Tragelaphus strepsiceros*). *Scientific Reports, 8*(1), 16599. https://doi.org/10.1038/s41598-018-34985-5.

Head, J. R., Vos, A., Blanton, J., Muller, T., Chipman, R., Pieracci, E. G., … Wallace, R. (2019). Environmental distribution of certain modified live-virus vaccines with a high safety profile presents a low-risk, high-reward to control zoonotic diseases. *Scientific Reports, 9*(1), 6783. https://doi.org/10.1038/s41598-019-42714-9.

Healy, J. M., & Bruce, B. B. (2019). Travel-related infectious diseases. In *Salmonellosis (Nontyphoidal)*. Retrieved from https://wwwnc.cdc.gov/travel/yellowbook/2020/travel-related-infectious-diseases/salmonellosis-nontyphoidal.

Hendricks, K. A., Wright, M. E., Shadomy, S. V., Bradley, J. S., Morrow, M. G., Pavia, A. T., … Workgroup on Anthrax Clinical Guidelines. (2014). Centers for disease control and prevention expert panel meetings on prevention and treatment of anthrax in adults. *Emerging Infectious Diseases, 20*(2). https://doi.org/10.3201/eid2002.130687.

Hercules, Y., Bryant, N. J., Wallace, R. M., Nelson, R., Palumbo, G., Williams, J. N., ... Brown, C. (2018). Rabies in a dog imported from Egypt—Connecticut, 2017. *MMWR. Morbidity and Mortality Weekly Report, 67*(50), 1388–1391. https://doi.org/10.15585/mmwr.mm6750a3.

Hikufe, E. H., Freuling, C. M., Athingo, R., Shilongo, A., Ndevaetela, E. E., Helao, M., ... Maseke, A. (2019). Ecology and epidemiology of rabies in humans, domestic animals and wildlife in Namibia, 2011-2017. *PLoS Neglected Tropical Diseases, 13*(4), e0007355. https://doi.org/10.1371/journal.pntd.0007355.

Honein, M. A., Cetron, M. S., & Meaney-Delman, D. (2019). Endemic Zika virus transmission: implications for travellers. *The Lancet Infectious Diseases, 19*(4), 349–351. https://doi.org/10.1016/S1473-3099(18)30793-X.

Huong, V. T. L., Turner, H. C., Kinh, N. V., Thai, P. Q., Hoa, N. T., Horby, P., ... Wertheim, H. F. L. (2019). Burden of disease and economic impact of human *Streptococcus suis* infection in Viet Nam. *Transactions of the Royal Society of Tropical Medicine and Hygiene, 113*(6), 341–350. https://doi.org/10.1093/trstmh/trz004.

Ince, Y., Yasa, C., Metin, M., Sonmez, M., Meram, E., Benkli, B., & Ergonul, O. (2014). Crimean-Congo hemorrhagic fever infections reported by ProMED. *The International Journal of Infectious Diseases, 26*, 44–46. https://doi.org/10.1016/j.ijid.2014.04.005.

Jackson, A. C. (2014). Rabies. *Handbook of Clinical Neurology, 123*, 601–618. https://doi.org/10.1016/B978-0-444-53488-0.00029-8.

Jenkins, S. R., & Winkler, W. G. (1987). Descriptive epidemiology from an epizootic of raccoon rabies in the Middle Atlantic States, 1982-1983. *American Journal of Epidemiology, 126*(3), 429–437. https://doi.org/10.1093/oxfordjournals.aje.a114674.

Jeon, S., Cleaton, J., Meltzer, M. I., Kahn, E. B., Pieracci, E. G., Blanton, J. D., & Wallace, R. (2019). Determining the post-elimination level of vaccination needed to prevent re-establishment of dog rabies. *PLoS Neglected Tropical Diseases, 13*(12); e0007869. https://doi.org/10.1371/journal.pntd.0007869.

Kaakoush, N. O., Castano-Rodriguez, N., Mitchell, H. M., & Man, S. M. (2015). Global epidemiology of campylobacter infection. *Clinical Microbiology Reviews, 28*(3), 687–720. https://doi.org/10.1128/CMR.00006-15.

Kadowaki, H., Hampson, K., Tojinbara, K., Yamada, A., & Makita, K. (2018). The risk of rabies spread in Japan: A mathematical modelling assessment. *Epidemiology and Infection, 146*(10), 1245–1252. https://doi.org/10.1017/S0950268818001267.

Keiser, J., & Utzinger, J. (2009). Food-borne trematodiases. *Clinical Microbiology Review, 22*(3), 466–483.

Kessels, J., Tarantola, A., Salahuddin, N., Blumberg, L., & Knopf, L. (2019). Rabies post-exposure prophylaxis: A systematic review on abridged vaccination schedules and the effect of changing administration routes during a single course. *Vaccine*. https://doi.org/10.1016/j.vaccine.2019.01.041.

Kgaladi, J., Nel, L. H., & Markotter, W. (2013). Comparison of pathogenic domains of rabies and African rabies-related lyssaviruses and pathogenicity observed in mice. *The Onderstepoort Journal of Veterinary Research, 80*(1), 511. https://doi.org/10.4102/ojvr.v80i1.511.

Kgaladi, J., Wright, N., Coertse, J., Markotter, W., Marston, D., Fooks, A. R., ... Nel, L. H. (2013). Diversity and epidemiology of Mokola virus. *PLoS Neglected Tropical Diseases, 7*(10), e2511. https://doi.org/10.1371/journal.pntd.0002511.

Kirby, J. D., Chipman, R. B., Nelson, K. M., Rupprecht, C. E., Blanton, J. D., Algeo, T. P., & Slate, D. (2017). Enhanced rabies surveillance to support effective oral rabies vaccination of raccoons in the eastern United States. *Tropical Medicine and Infectious Disease. 2*(3). https://doi.org/10.3390/tropicalmed2030034.

Kitala, P., McDermott, J., Kyule, M., Gathuma, J., Perry, B., & Wandeler, A. (2001). Dog ecology and demography information to support the planning of rabies control in Machakos District, Kenya. *Acta Tropica, 78*(3), 217–230.

Kobayashi, Y., Sato, G., Kato, M., Itou, T., Cunha, E. M., Silva, M. V., ... Sakai, T. (2007). Genetic diversity of bat rabies viruses in Brazil. *Archives of Virology, 152*(11), 1995–2004. https://doi.org/10.1007/s00705-007-1033-y.

Koch, I., Berger, J., & Kammann, M. (1974). Rabies antibody determination with the neutralization test (N-test) and the indirect fluorescence antibody test (IFA-test) in comparison (author's transl). *Zentralblatt fur Bakteriologie, Parasitenkunde, Infektionskrankheiten und Hygiene. Erste Abteilung Originale. Reihe A: Medizinische Mikrobiologie und Parasitologie, 226*(3), 291–297.

Kuzmin, I. V., Shi, M., Orciari, L. A., Yager, P. A., Velasco-Villa, A., Kuzmina, N. A., ... Rupprecht, C. E. (2012). Molecular inferences suggest multiple host shifts of rabies viruses from bats to mesocarnivores in Arizona during 2001-2009. *PLoS Pathogens, 8*(6), e1002786. https://doi.org/10.1371/journal.ppat.1002786.

Kuzmina, N. A., Kuzmin, I. V., Ellison, J. A., Taylor, S. T., Bergman, D. L., Dew, B., & Rupprecht, C. E. (2013). A reassessment of the evolutionary timescale of bat rabies viruses based upon glycoprotein gene sequences. *Virus Genes, 47*(2), 305–310. https://doi.org/10.1007/s11262-013-0952-9.

Labeaud, A. D., Bashir, F., & King, C. H. (2011). Measuring the burden of arboviral diseases: The spectrum of morbidity and mortality from four prevalent infections. *Population Health Metrics*, *9*(1), 1. https://doi.org/10.1186/1478-7954-9-1.

Lechenne, M., Naissengar, K., Lepelletier, A., Alfaroukh, I. O., Bourhy, H., Zinsstag, J., & Dacheux, L. (2016). Validation of a rapid rabies diagnostic tool for field surveillance in developing countries. *PLoS Neglected Tropical Diseases*, *10*(10), e0005010. https://doi.org/10.1371/journal.pntd.0005010.

Lee, T. K., Hutchinson, H. D., & Ziegler, D. W. (1977). Comparison of rabies humoral antibody titers in rabbits and humans by indirect radioimmunoassay, rapid-fluorescent-focus-inhibition technique, and indirect fluorescent-antibody assay. *Journal of Clinical Microbiology*, *5*(3), 320–325.

LeRoux, K., Stewart, D., Perrett, K. D., Nel, L. H., Kessels, J. A., & Abela-Ridder, B. (2018). Rabies control in KwaZulu-Natal, South Africa. *Bulletin of the World Health Organization*, *96*(5), 360–365. https://doi.org/10.2471/BLT.17.194886.

Li, A. J., Sreenivasan, N., Siddiqi, U. R., Tahmina, S., Penjor, K., Sovann, L., ... Hyde, T. B. (2018). Descriptive assessment of rabies post-exposure prophylaxis procurement, distribution, monitoring, and reporting in four Asian countries: Bangladesh, Bhutan, Cambodia, and Sri Lanka, 2017-2018. *Vaccine*. https://doi.org/10.1016/j.vaccine.2018.10.011.

Ly, S., Buchy, P., Heng, N. Y., Ong, S., Chhor, N., Bourhy, H., & Vong, S. (2009). Rabies situation in Cambodia. *PLoS Neglected Tropical Diseases*, *3*(9), e511. https://doi.org/10.1371/journal.pntd.0000511.

Ma, X., Monroe, B. P., Cleaton, J. M., Orciari, L. A., Li, Y., Kirby, J. D., ... Blanton, J. D. (2018). Rabies surveillance in the United States during 2017. *Journal of the American Veterinary Medical Association*, *253*(12), 1555–1568. https://doi.org/10.2460/javma.253.12.1555.

Majowicz, S. E., Musto, J., Scallan, E., Angulo, F. J., Kirk, M., O'Brien, S. J. ... International Collaboration on Enteric Disease 'Burden of Illness, S. (2010). The global burden of nontyphoidal *Salmonella* gastroenteritis. *Clinical Infectious Diseases*, *50*(6), 882–889. https://doi.org/10.1086/650733.

Majowicz, S. E., Scallan, E., Jones-Bitton, A., Sargeant, J. M., Stapleton, J., Angulo, F. J., ... Kirk, M. D. (2014). Global incidence of human Shiga toxin-producing *Escherichia coli* infections and deaths: A systematic review and knowledge synthesis. *Foodborne Pathogens and Disease*, *11*(6), 447–455. https://doi.org/10.1089/fpd.2013.1704.

Manimunda, S. P., Mavalankar, D., Bandyopadhyay, T., & Sugunan, A. P. (2011). Chikungunya epidemic-related mortality. *Epidemiology & Infection*, *139*(9), 1410–1412. https://doi.org/10.1017/S0950268810002542.

Manning, S. E., Rupprecht, C. E., Fishbein, D., Hanlon, C. A., Lumlertdacha, B., Guerra, M., ... Advisory Committee on Immunization Practices Centers for Disease Control and Prevention (CDC). (2008). Human rabies prevention—United States, 2008: Recommendations of the Advisory Committee on Immunization Practices. *MMWR - Recommendations and Reports*, *57*(RR-3), 1–28.

Markotter, W., Coertse, J., le Roux, K., Peens, J., Weyer, J., Blumberg, L., & Nel, L. H. (2015). Utility of forensic detection of rabies virus in decomposed exhumed dog carcasses. *Journal of the South African Veterinary Association*, *86*(1), 1220. https://doi.org/10.4102/jsava.v86i1.1220.

Matsumoto, T., Nanayakkara, S., Perera, D., Ushijima, S., Wimalaratne, O., Nishizono, A., & Ahmed, K. (2017). Terrestrial animal-derived rabies virus in a juvenile Indian flying fox in Sri Lanka. *Japanese Journal of Infectious Diseases*, *70*(6), 693–695. https://doi.org/10.7883/yoken.JJID.2017.249.

Mazeri, S., Gibson, A. D., de Clare Bronsvoort, B. M., Handel, I. G., Lohr, F., Bailey, J. B., ... Mellanby, R. J. (2019). Sociodemographic factors which predict low private rabies vaccination coverage in dogs in Blantyre, Malawi. *The Veterinary Record*, *184*(9), 281. https://doi.org/10.1136/vr.105000.

Mazeri, S., Gibson, A. D., Meunier, N., Bronsvoort, B. M. D., Handel, I. G., Mellanby, R. J., & Gamble, L. (2018). Barriers of attendance to dog rabies static point vaccination clinics in Blantyre, Malawi. *PLoS Neglected Tropical Diseases*, *12*(1), e0006159. https://doi.org/10.1371/journal.pntd.0006159.

McQuiston, J. H., Wilson, T., Harris, S., Bacon, R. M., Shapiro, S., Trevino, I., ... Marano, N. (2008). Importation of dogs into the United States: Risks from rabies and other zoonotic diseases. *Zoonoses and Public Health*, *55*(8–10), 421–426.

Melade, J., McCulloch, S., Ramasindrazana, B., Lagadec, E., Turpin, M., Pascalis, H., ... Dellagi, K. (2016). Serological evidence of lyssaviruses among bats on southwestern Indian Ocean Islands. *PLoS One*, *11*(8), e0160553. https://doi.org/10.1371/journal.pone.0160553.

Moncayo, A., & Silveira, A. C. (2009). Current epidemiological trends for Chagas disease in Latin America and future challenges in epidemiology, surveillance and health policy. *Memórias do Instituto Oswaldo Cruz*, *104*(Suppl. 1), 17–30. https://doi.org/10.1590/s0074-02762009000900005.

Moore, D. A. (1999). Spatial diffusion of raccoon rabies in Pennsylvania, USA. *Preventive Veterinary Medicine*, *40*(1), 19–32.

Murray, C. J., Vos, T., Lozano, R., Naghavi, M., Flaxman, A. D., Michaud, C., ... Memish, Z. A. (2012). Disability-adjusted life years (DALYs) for 291 diseases and injuries in 21 regions, 1990-2010: A systematic analysis for the Global Burden of Disease Study 2010. *Lancet, 380*(9859), 2197–2223. https://doi.org/10.1016/S0140-6736(12)61689-4.

Nadin-Davis, S. A., Sheen, M., & Wandeler, A. I. (2009). Development of real-time reverse transcriptase polymerase chain reaction methods for human rabies diagnosis. *Journal of Medical Virology, 81*(8), 1484–1497. https://doi.org/10.1002/jmv.21547.

Nadin-Davis, S. A., Velez, J., Malaga, C., & Wandeler, A. I. (2008). A molecular epidemiological study of rabies in Puerto Rico. *Virus Research, 131*(1), 8–15. https://doi.org/10.1016/j.virusres.2007.08.002.

National Association of State Public Health Veterinarians, Compendium of Animal Rabies Prevention and Control Committee, Brown, C. M., Slavinski, S., Ettestad, P., ... Sorhage, F. E. (2016). Compendium of Animal Rabies Prevention and Control, 2016. *Journal of the American Veterinary Medical Association, 248*(5), 505–517. https://doi.org/10.2460/javma.248.5.505.

Nguyen, A. K., Nguyen, D. V., Ngo, G. C., Nguyen, T. T., Inoue, S., Yamada, A., ... Nguyen, H. T. (2011). Molecular epidemiology of rabies virus in Vietnam (2006-2009). *Japanese Journal of Infectious Diseases, 64*(5), 391–396.

Niezgoda, M., Briggs, D. J., Shaddock, J., & Rupprecht, C. E. (1998). Viral excretion in domestic ferrets (*Mustela putorius furo*) inoculated with a raccoon rabies isolate. *American Journal of Veterinary Research, 59*(12), 1629–1632.

Odeh, L. E., Umoh, J. U., & Dzikwi, A. A. (2013). Assessment of risk of possible exposure to rabies among processors and consumers of dog meat in Zaria and Kafanchan, Kaduna state, Nigeria. *Global Journal of Health Science, 6*(1), 142–153. https://doi.org/10.5539/gjhs.v6n1p142.

Oertli, E. H., Wilson, P. J., Hunt, P. R., Sidwa, T. J., & Rohde, R. E. (2009). Epidemiology of rabies in skunks in Texas. *Journal of the American Veterinary Medical Association, 234*(5), 616–620. https://doi.org/10.2460/javma.234.5.616.

OIE. (2016). Rabies (infection with rabies virus). In *Vol. 2. Manual of diagnostic tests and vaccines for terrestrial animals*. Paris: OIE (chapter 2.1.17).

O'Shea, T. J., Bowen, R. A., Stanley, T. R., Shankar, V., & Rupprecht, C. E. (2014). Variability in seroprevalence of rabies virus neutralizing antibodies and associated factors in a Colorado population of big brown bats (*Eptesicus fuscus*). *PLoS One, 9*(1), e86261. https://doi.org/10.1371/journal.pone.0086261.

Petersen, B., & Rupprecht, C. (2011). Human rabies epidemiology and diagnosis. In S. Tkachev (Ed.), *Non-flavivirus encephalitis*. Intech.

Picard-Meyer, E., Beven, V., Hirchaud, E., Guillaume, C., Larcher, G., Robardet, E., ... Cliquet, F. (2019). Lleida Bat Lyssavirus isolation in *Miniopterus schreibersii* in France. *Zoonoses and Public Health, 66*(2), 254–258. https://doi.org/10.1111/zph.12535.

Pieracci, E. G., Pearson, C. M., Wallace, R. M., Blanton, J. D., Whitehouse, E. R., Ma, X., ... Olson, V. (2019). Vital signs: Trends in human rabies deaths and exposures—United States, 1938-2018. *MMWR. Morbidity and Mortality Weekly Report, 68*(23), 524–528. https://doi.org/10.15585/mmwr.mm6823e1.

Rees, E. E., Gendron, B., Lelievre, F., Cote, N., & Belanger, D. (2011). Advancements in web-database applications for rabies surveillance. *International Journal of Health Geographics, 10*, 48. https://doi.org/10.1186/1476-072X-10-48.

Rosatte, R. C. (1988). Rabies in Canada—History, epidemiology and control. *The Canadian Veterinary Journal, 29*(4), 362–365.

Ross, R. S., Wolters, B., Hoffmann, B., Geue, L., Viazov, S., Gruner, N., ... Muller, T. (2015). Instructive even after a decade: Complete results of initial virological diagnostics and re-evaluation of molecular data in the German rabies virus "outbreak" caused by transplantations. *International Journal of Medical Microbiology, 305*(7), 636–643. https://doi.org/10.1016/j.ijmm.2015.08.013.

Rupprecht, C., Fooks, A. R., & Abela-Ridder, B. (Eds.), (2019). In *Vol. 1. Laboratory techniques in rabies* (5th ed.). Geneva: WHO.

Rysava, K., Miranda, M. E., Zapatos, R., Lapiz, S., Rances, P., Miranda, L. M., ... Hampson, K. (2018). On the path to rabies elimination: The need for risk assessments to improve administration of post-exposure prophylaxis. *Vaccine*. https://doi.org/10.1016/j.vaccine.2018.11.066.

Sabeta, C. T., Mansfield, K. L., McElhinney, L. M., Fooks, A. R., & Nel, L. H. (2007). Molecular epidemiology of rabies in bat-eared foxes (*Otocyon megalotis*) in South Africa. *Virus Research, 129*(1), 1–10. https://doi.org/10.1016/j.virusres.2007.04.024.

Saeed, B., & Al-Mousawi, M. (2017). Rabies acquired through kidney transplantation in a child: A case report. *Experimental and Clinical Transplantation, 15*(3), 355–357. https://doi.org/10.6002/ect.2017.0046.

Sarmiento-Ospina, A., Vasquez-Serna, H., Jimenez-Canizales, C. E., Villamil-Gomez, W. E., & Rodriguez-Morales, A. J. (2016). Zika virus associated deaths in Colombia. *The Lancet Infectious Diseases*, *16*(5), 523–524. https://doi.org/10.1016/S1473-3099(16)30006-8.

Schildecker, S., Millien, M., Blanton, J. D., Boone, J., Emery, A., Ludder, F., … Wallace, R. M. (2017). Dog ecology and barriers to canine rabies control in the Republic of Haiti, 2014-2015. *Transboundary and Emerging Diseases*, *64*(5), 1433–1442. https://doi.org/10.1111/tbed.12531.

Scott, T. P., Coetzer, A., Fahrion, A. S., & Nel, L. H. (2017). Addressing the disconnect between the estimated, reported, and true rabies data: The development of a regional African rabies bulletin. *Frontiers in Veterinary Science*, *4*, 18. https://doi.org/10.3389/fvets.2017.00018.

Scott, T. P., Fischer, M., Khaiseb, S., Freuling, C., Hoper, D., Hoffmann, B., … Nel, L. H. (2013). Complete genome and molecular epidemiological data infer the maintenance of rabies among kudu (*Tragelaphus strepsiceros*) in Namibia. *PLoS One*, *8*(3), e58739. https://doi.org/10.1371/journal.pone.0058739.

Seetahal, J. F. R., Vokaty, A., Vigilato, M. A. N., Carrington, C. V. F., Pradel, J., Louison, B., … Rupprecht, C. E. (2018). Rabies in the Caribbean: A situational analysis and historic review. *Tropical Medicine and Infectious Disease*. *3*(3) https://doi.org/10.3390/tropicalmed3030089.

Senior, K. (2009). Lassa fever: Current and future control options. *The Lancet Infectious Diseases*, *9*(9), 532. https://doi.org/10.1016/S1473-3099(09)70217-8.

Shankar, S. K., Mahadevan, A., Sapico, S. D., Ghodkirekar, M. S., Pinto, R. G., & Madhusudana, S. N. (2012). Rabies viral encephalitis with proable 25 year incubation period!. *Annals of Indian Academy of Neurology*, *15*(3), 221–223. https://doi.org/10.4103/0972-2327.99728.

Sinclair, J. R., Wallace, R. M., Gruszynski, K., Freeman, M. B., Campbell, C., Semple, S., … Murphy, J. (2015). Rabies in a dog imported from Egypt with a falsified rabies vaccination certificate—Virginia, 2015. *MMWR Morbidity and Mortality Weekly Report*, *64*(49), 1359–1362. https://doi.org/10.15585/mmwr.mm6449a2.

Smith, T. G., Millien, M., Vos, A., Fracciterne, F. A., Crowdis, K., Chirodea, C., … Wallace, R. (2017). Evaluation of immune responses in dogs to oral rabies vaccine under field conditions. *Vaccine*. https://doi.org/10.1016/j.vaccine.2017.09.096.

Smith, J. S., & Seidel, H. D. (1993). Rabies: A new look at an old disease. *Progress in Medical Virology*, *40*, 82–106.

Smreczak, M., Orłowska, A., Marzec, A., Trębas, P., Müller, T., Freuling, C. M., & Żmudziński, J. F. (2018). Bokeloh bat lyssavirus isolation in a Natterer's bat. *Poland*. https://doi.org/10.1111/zph.12519.

Steele, J. H. (1951). Rabies. *Veterinary Medicine*, *46*(8), 327.

Steele, J. H., & Tierkel, E. S. (1949). Rabies problems and control. *Public Health Reports*, *64*(25), 785–796.

Streicker, D. G., Lemey, P., Velasco-Villa, A., & Rupprecht, C. E. (2012). Rates of viral evolution are linked to host geography in bat rabies. *PLoS Pathogens*, *8*(5), e1002720. https://doi.org/10.1371/journal.ppat.1002720.

Streicker, D. G., Turmelle, A. S., Vonhof, M. J., Kuzmin, I. V., McCracken, G. F., & Rupprecht, C. E. (2010). Host phylogeny constrains cross-species emergence and establishment of rabies virus in bats. *Science*, *329*(5992), 676–679. https://doi.org/10.1126/science.1188836.

Styczynski, A., Tran, C., Dirlikov, E., Zapata, M. R., Ryff, K., Petersen, B., … Garcia, B. R. (2017). Human rabies—Puerto Rico, 2015. *MMWR. Morbidity and Mortality Weekly Report*, *65*(52), 1474–1476. https://doi.org/10.15585/mmwr.mm6552a4.

Suraweera, W., Morris, S. K., Kumar, R., Warrell, D. A., Warrell, M. J., & Jha, P. Million Death Study Collaborators. (2012). Deaths from symptomatically identifiable furious rabies in India: A nationally representative mortality survey. *PLoS Neglected Tropical Diseases*, *6*(10), e1847. https://doi.org/10.1371/journal.pntd.0001847.

Tabel, H., Corner, A. H., Webster, W. A., & Casey, C. A. (1974). History and epizootiology of rabies in Canada. *The Canadian Veterinary Journal*, *15*(10), 271–281.

Taylor, J. P., & Perdue, J. N. (1989). The changing epidemiology of human brucellosis in Texas, 1977–1986. *American Journal of Epidemiology*, *130*(1), 160–165. https://doi.org/10.1093/oxfordjournals.aje.a115308.

Taylor, L. H., Hampson, K., Fahrion, A., Abela-Ridder, B., & Nel, L. H. (2017). Difficulties in estimating the human burden of canine rabies. *Acta Tropica*, *165*, 133–140. https://doi.org/10.1016/j.actatropica.2015.12.007.

Taylor, L. H., Knopf, L., & Partners for Rabies Prevention. (2015). Surveillance of human rabies by national authorities—A global survey. *Zoonoses Public Health*, *62*(7), 543–552. https://doi.org/10.1111/zph.12183.

Taylor, L. H., Wallace, R. M., Balaram, D., Lindenmayer, J. M., Eckery, D. C., Mutonono-Watkiss, B., … Nel, L. H. (2017). The role of dog population management in rabies elimination—A review of current approaches and future opportunities. *Frontiers in Veterinary Science*, *4*, 109. https://doi.org/10.3389/fvets.2017.00109.

Ternhag, A., Torner, A., Svensson, A., Giesecke, J., & Ekdahl, K. (2005). Mortality following Campylobacter infection: A registry-based linkage study. *BMC Infectious Diseases*, *5*, 70. https://doi.org/10.1186/1471-2334-5-70.

Torgerson, P. R., de Silva, N. R., Fevre, E. M., Kasuga, F., Rokni, M. B., Zhou, X. N., ... Stein, C. (2014). The global burden of foodborne parasitic diseases: an update. *Trends in Parasitology*, *30*(1), 20–26. https://doi.org/10.1016/j.pt.2013.11.002.

Torgerson, P. R., Hagan, J. E., Costa, F., Calcagno, J., Kane, M., Martinez-Silveira, M. S., ... Abela-Ridder, B. (2015). Global burden of leptospirosis: Estimated in terms of disability adjusted life years. *PLoS Neglected Tropical Diseases*, *9*(10), e0004122. https://doi.org/10.1371/journal.pntd.0004122.

Torgerson, P. R., Keller, K., Magnotta, M., & Ragland, N. (2010). The global burden of alveolar echinococcosis. *PLoS Neglected Tropical Diseases*, *4*(6), e722. https://doi.org/10.1371/journal.pntd.0000722.

Townsend, S. E., Lembo, T., Cleaveland, S., Meslin, F. X., Miranda, M. E., Putra, A. A., ... Hampson, K. (2013). Surveillance guidelines for disease elimination: A case study of canine rabies. *Comparative Immunology, Microbiology and Infectious Diseases*, *36*(3), 249–261. https://doi.org/10.1016/j.cimid.2012.10.008.

Undurraga, E. A., Meltzer, M. I., Tran, C. H., Atkins, C. Y., Etheart, M. D., Millien, M. F., ... Wallace, R. M. (2017). Cost-effectiveness evaluation of a novel integrated bite case management program for the control of human rabies, Haiti 2014-2015. *The American Journal of Tropical Medicine and Hygiene*, *96*(6), 1307–1317. https://doi.org/10.4269/ajtmh.16-0785.

Vaughn, J. B., Jr., Gerhardt, P., & Newell, K. W. (1965). Excretion of street rabies virus in the saliva of dogs. *JAMA*, *193*, 363–368. https://doi.org/10.1001/jama.1965.03090050039010.

Vaughn, J. B., Gerhardt, P., & Paterson, J. C. (1963). Excretion of street rabies virus in saliva of cats. *JAMA*, *184*, 705–708. https://doi.org/10.1001/jama.1963.73700220001013.

Velasco-Villa, A., Reeder, S. A., Orciari, L. A., Yager, P. A., Franka, R., Blanton, J. D., ... Rupprecht, C. E. (2008). Enzootic rabies elimination from dogs and reemergence in wild terrestrial carnivores, United States. *Emerging Infectious Diseases*, *14*(12), 1849–1854. https://doi.org/10.3201/eid1412.080876.

Vora, N. M., Basavaraju, S. V., Feldman, K. A., Paddock, C. D., Orciari, L., Gitterman, S., ... Transplant-Associated Rabies Virus Transmission Investigation Team. (2013). Raccoon rabies virus variant transmission through solid organ transplantation. *JAMA*, *310*(4), 398–407. https://doi.org/10.1001/jama.2013.7986.

Wacharapluesadee, S., Phumesin, P., Supavonwong, P., Khawplod, P., Intarut, N., & Hemachudha, T. (2011). Comparative detection of rabies RNA by NASBA, real-time PCR and conventional PCR. *Journal of Virological Methods*, *175*(2), 278–282. https://doi.org/10.1016/j.jviromet.2011.05.007.

Wadhwa, A., Wilkins, K., Gao, J., Condori Condori, R. E., Gigante, C. M., Zhao, H., ... Li, Y. (2017). A pan-lyssavirus Taqman real-time RT-PCR assay for the detection of highly variable rabies virus and other lyssaviruses. *PLoS Neglected Tropical Diseases*, *11*(1), e0005258. https://doi.org/10.1371/journal.pntd.0005258.

Wahid, B., Ali, A., Rafique, S., & Idrees, M. (2017). Global expansion of chikungunya virus: Mapping the 64-year history. *International Journal of Infectious Diseases*, *58*, 69–76. https://doi.org/10.1016/j.ijid.2017.03.006.

Walker, P. J., Blasdell, K. R., Calisher, C. H., Dietzgen, R. G., Kondo, H., Kurath, G., ... Ictv Report Consortium. (2018). ICTV virus taxonomy profile: Rhabdoviridae. *The Journal of General Virology*, *99*(4), 447–448. https://doi.org/10.1099/jgv.0.001020.

Wallace, R., Etheart, M., Ludder, F., Augustin, P., Fenelon, N., Franka, R., ... Millien, M. (2017). The health impact of rabies in Haiti and recent developments on the path toward elimination, 2010-2015. *The American Journal of Tropical Medicine and Hygiene*, *97*(4_Suppl), 76–83. https://doi.org/10.4269/ajtmh.16-0647.

Wallace, R. M., Gilbert, A., Slate, D., Chipman, R., Singh, A., Cassie, W., & Blanton, J. D. (2014). Right place, wrong species: A 20-year review of rabies virus cross species transmission among terrestrial mammals in the United States. *PLoS One*, *9*(10), e107539. https://doi.org/10.1371/journal.pone.0107539.

Wallace, R. M., Mehal, J., Nakazawa, Y., Recuenco, S., Bakamutumaho, B., Osinubi, M., ... Wamala, J. (2017). The impact of poverty on dog ownership and access to canine rabies vaccination: Results from a knowledge, attitudes and practices survey, Uganda 2013. *Infectious Diseases of Poverty*, *6*(1), 97. https://doi.org/10.1186/s40249-017-0306-2.

Wallace, R. M., Reses, H., Franka, R., Dilius, P., Fenelon, N., Orciari, L., ... Millien, M. (2015). Establishment of a high canine rabies burden in Haiti through the implementation of a novel surveillance program [corrected]. *PLoS Neglected Tropical Diseases*, *9*(11), e0004245. https://doi.org/10.1371/journal.pntd.0004245.

Wallace, R., Undurraga, E., Gibson, A., Boone, J., Pieracci, E., Gamble, L., & Blanton, J. (2019). Estimating the effectiveness of vaccine programs in dog populations. *Epidemiology and Infection*, *147*, e247.

Wang, L., Tang, Q., & Liang, G. (2014). Rabies and rabies virus in wildlife in mainland China, 1990-2013. *International Journal of Infectious Diseases*, 25, 122–129. https://doi.org/10.1016/j.ijid.2014.04.016.

Wasteson, Y. (2001). Zoonotic *Escherichia coli*. *Acta Veterinaria Scandinavica*, 95, 79–84.

Wertheim, H. F., Nghia, H. D., Taylor, W., & Schultsz, C. (2009). *Streptococcus suis*: An emerging human pathogen. *Clinical Infectious Diseases*, 48(5), 617–625. https://doi.org/10.1086/596763.

Wertheim, H. F., Nguyen, T. Q., Nguyen, K. A., de Jong, M. D., Taylor, W. R., Le, T. V., … Nguyen, H. D. (2009). Furious rabies after an atypical exposure. *PLoS Medicine*, 6(3), e44. https://doi.org/10.1371/journal.pmed.1000044.

WHO. (n.d.). Transmissible spongiform encephalopathies (TSE). Retrieved from https://www.who.int/bloodproducts/tse/en/.

WHO. (2003). Summary of probable SARS cases with onset of illness from 1 November 2002 to 31 July 2003. Retrieved from https://www.who.int/csr/sars/country/table2004_04_21/en/.

WHO. (2015). *WHO estimates of the global burden of foodborne diseases*. Retrieved from https://apps.who.int/iris/bitstream/handle/10665/199350/9789241565165_eng.pdf?sequence=1%20.

WHO. (2018). Cumulative number of confirmed human cases for avian influenza A(H5N1) reported to WHO, 2003–2018. World Health Organization.

WHO. (2018). Ebola situation reports: Democratic Republic of the Congo (Archive). Retrieved from https://www.who.int/ebola/situation-reports/drc-2018/en/.

WHO. (2018). WHO Expert Consultation on Rabies. Third Report *WHO technical report series No 1012. Vol. 2018.* (p. 183) Geneva: WHO.

WHO Rabies Modelling Consortium. (2019). The potential effect of improved provision of rabies post-exposure prophylaxis in Gavi-eligible countries: A modelling study. *The Lancet Infectious Diseases*, 19(1), 102–111. https://doi.org/10.1016/S1473-3099(18)30512-7.

Wilson, M. L., Bretsky, P. M., Cooper, G. H., Jr., Egbertson, S. H., Van Kruiningen, H. J., & Cartter, M. L. (1997). Emergence of raccoon rabies in Connecticut, 1991-1994: Spatial and temporal characteristics of animal infection and human contact. *The American Journal of Tropical Medicine and Hygiene*, 57(4), 457–463. https://doi.org/10.4269/ajtmh.1997.57.457.

World Health Organization. (1998). RABNET-strengthening international surveillance of human and animal rabies. *Weekly Epidemiological Record*, 73(33), 254–256.

World Health Organization. (2013). WHO Expert Consultation on Rabies. Second report *World Health Organ Tech Rep Ser (982)* (pp. 1–139). back cover.

Wyatt, J. D., Barker, W. H., Bennett, N. M., & Hanlon, C. A. (1999). Human rabies postexposure prophylaxis during a raccoon rabies epizootic in New York, 1993 and 1994. *Emerging Infectious Diseases*, 5(3), 415–423. https://doi.org/10.3201/eid0503.990312.

Yakobson, B. A., King, R., Amir, S., Devers, N., Sheichat, N., Rutenberg, D., … David, D. (2006). Rabies vaccination programme for red foxes (*Vulpes vulpes*) and golden jackals (*Canis aureus*) in Israel (1999-2004). *Developmental Biology (Basel)*, 125, 133–140.

Yakobson, B., Taylor, N., Dveres, N., Rotblat, S., Spero, Z., Lankau, E. W., & Maki, J. (2017). Impact of rabies vaccination history on attainment of an adequate antibody titre among dogs tested for international travel certification, Israel—2010-2014. *Zoonoses and Public Health*, 64(4), 281–289. https://doi.org/10.1111/zph.12309.

Yang, D. K., Kim, H. H., Lee, K. K., Yoo, J. Y., Seomun, H., & Cho, I. S. (2017). Mass vaccination has led to the elimination of rabies since 2014 in South Korea. *Clinical and Experimental Vaccine Research*, 6(2), 111–119. https://doi.org/10.7774/cevr.2017.6.2.111.

Zhou, H., Vong, S., Liu, K., Li, Y., Mu, D., Wang, L., … Yu, H. (2016). Human rabies in China, 1960-2014: A descriptive epidemiological study. *PLoS Neglected Tropical Diseases*, 10(8), e0004874. https://doi.org/10.1371/journal.pntd.0004874.

Zinsstag, J., Lechenne, M., Laager, M., Mindekem, R., Naissengar, S., Oussiguere, A., … Chitnis, N. (2017). Vaccination of dogs in an African city interrupts rabies transmission and reduces human exposure. *Science Translational Medicine*. 9(421) https://doi.org/10.1126/scitranslmed.aaf6984.

Zulu, G. C., Sabeta, C. T., & Nel, L. H. (2009). Molecular epidemiology of rabies: Focus on domestic dogs (*Canis familiaris*) and black-backed jackals (*Canis mesomelas*) from northern South Africa. *Virus Research*, 140(1–2), 71–78. https://doi.org/10.1016/j.virusres.2008.11.004.

Zumla, A., Hui, D. S., & Perlman, S. (2015). Middle East respiratory syndrome. *Lancet*, 386(9997), 995–1007. https://doi.org/10.1016/S0140-6736(15)60454-8.

CHAPTER 5

Molecular epidemiology

Susan A. Nadin-Davis

Canadian Food Inspection Agency, Ottawa Laboratory-Fallowfield, Ottawa, ON, Canada

5.1 Introduction

The discipline of molecular epidemiology, in which patterns of disease transmission are followed using selected markers that distinguish different populations of the disease-causing agent, is now a well-established central theme in the study of infectious diseases. Knowledge generated by such an approach improves understanding of the mechanisms that operate on viral emergence and spread and contributes to better disease control. Over the last few decades, many technological innovations, particularly in the development of new tools that are used for genetic characterization of viruses, have enabled the development and application of increasingly sophisticated and sensitive viral detection and typing methods. This chapter will summarize the knowledge gained by applying such methods to rabies and the rabies-related viruses that constitute the *Lyssavirus* genus.

5.2 Key aspects of *Lyssavirus* biology

Several biological characteristics of lyssaviruses make them especially amenable to molecular epidemiological investigation. As for all members of the *Mononegavirales*, the lyssaviruses use error-prone RNA polymerases for replication and the resulting infidelity in genome copying ensures that any population of these viruses does not exist as a discrete entity but as a collection of genetic variants that are distributed around a central consensus sequence, a phenomenon frequently referred to as a "quasispecies" (Lauring & Andino, 2010), though use of the more general term "population sequence heterogeneity" may be a more accurate description of the genetic structure of viral populations (Nadin-Davis & Real, 2011). While the high level of mutability of the lyssaviruses could in principle allow them to overcome barriers to their spread, both in the individual host and between species, in practice their genetic diversity is limited by the process of purifying selection, which removes transient deleterious mutations and effectively places substantial limitations on viral mutant fixation (Bourhy et al., 2008; Holmes, Woelk, Kassis, & Bourhy, 2002; Troupin et al., 2016). Accordingly, the observed diversity of a particular viral strain or variant is often limited and changes slowly with time due to acquisition of many neutral mutations. Sometimes, a virus population undergoes a severe bottleneck and the subpopulation that successfully emerges establishes a founder lineage (Marston, Horton, et al., 2017). If the consensus sequence of this founder lineage differs from that of its precursor, even by just a small number of mutations, these differences can be the basis of molecular epidemiological investigation (Marston et al., 2018).

Each lyssavirus is always associated with and maintained by a particular mammalian host reservoir (Hanlon, 2013). It is presumed that this association depends upon both the speed with which the virus can reach the CNS and propagate prior to the host mounting an immune response to the virus and the relative efficiency by which the virus can be transmitted between conspecifics (Fooks et al., 2017). It thus follows that the geographical spread of a particular viral population is determined both by the overall range of the animal species itself and aspects of the host animal's biology that determines the extent of interaction between different subpopulations of the host. The maintenance of this virus-host relationship is assumed to involve significant coadaptation by both parties but despite intensive scrutiny, the biological or molecular mechanisms that contribute to this association are not fully understood. Long-term maintenance of such a virus-host relationship will isolate this virus population from others and lead via genetic drift to the emergence of a virus subpopulation with distinct distinguishing markers (Marston, Horton, et al., 2017).

A "spill-over event" (cross-species transmission) occurs when a rabies virus variant that is normally maintained in a reservoir host successfully infects a different species but fails to adapt to the new host. In most cases, sustained transmission of the virus within the second species does not occur and this results in a "dead-end" infection for the virus. However, rarely such an event can initiate a new virus-host relationship in which sustained propagation and intra-species transmission of the virus within the new host occurs. Such a "host shift" will usually be associated with the emergence of a distinct viral population and hence a new variant. Accordingly, in this chapter, a viral variant is defined as a viral population that is maintained within a particular host reservoir in a geographically defined area and which is clearly genetically distinguished from other sympatric or allopatric viral populations.

All lyssavirus genomes studied (Delmas et al., 2008; Freuling et al., 2011; Kuzmin, Wu, Tordo, & Rupprecht, 2008; Markotter, Kuzmin, Rupprecht, & Nel, 2008; Marston, Ellis, et al., 2017; Marston, Horton, et al., 2012; Marston et al., 2007; Troupin et al., 2016) have an extremely similar organization in which most of the genome comprises five genes, N, P, M, G, and L in 3′ to 5′ order along its length. Consequently, lyssavirus mutation almost always occurs via single-base substitutions, necessary to preserve the reading frame of encoded proteins, or through small insertions/deletions in noncoding regions. Within protein-coding regions, most mutations are third base synonymous changes, although first and second base changes are also observed; the less common second base changes are always nonsynonymous. Observation of a mutation through consensus sequencing requires not only that a base change be incorporated into a newly synthesized viral genome, but also that the change becomes predominant within the viral population, even if only within a single individual. While the initial mutation event occurs by chance at a constant rate along the genome, as dictated by the error rate of the viral RNA polymerase, only those mutations that either confer a fitness advantage to the virus or neutral changes which are retained by chance become fixed in the population long enough to be observed. Different regions of the genome vary in the extent of mutation that can be tolerated depending upon whether the function of the encoded protein depends upon specific amino acid residues or more general biochemical characteristics. Consequently, levels of nucleotide similarity observed between lyssaviruses vary significant along the length of the genome. Functional constraints result in highest conservation within the N and L genes, the P and G genes are more variable while the M gene exhibits intermediate variability. Even within each gene distinct domains exhibiting variable levels of conservation are evident; for example, the most variable P gene is recognized as having a highly modular structure due to inclusion of multiple functionally distinct domains (Leyrat et al., 2011), while the limited conservation of the transmembrane and cytoplasmic domains of the glycoprotein is reflected in high variability for the G gene over this region (see Chapter 2).

While most intergenic regions are very short, there is a long, highly variable noncoding G-L intergenic region of unknown function. The reason for its retention is thus unclear, but the presence of mutational hot spots along its length suggests that some mutational constraints operate on parts of this sequence, suggesting that some G-L region sequences may have regulatory functions yet to be identified (Szanto, Nadin-Davis, Rosatte, & White, 2011).

5.3 Methods of viral typing

Epidemiological investigation relies on the ability to differentiate those viruses that have a common origin from those that emerged independently, thus gaining insight into the origins of an outbreak and its subsequent spread. Methods of viral typing are thus a key component to this process. Historically, differentiation of such variants has been achieved by antigenic typing, which relies on the discriminatory ability of a panel of monoclonal antibodies (MAbs). The process involves application of each Mab separately to sections or smears of rabies-positive brain material and evaluation of its binding to the viral target using indirect fluorescent antibody methods. The reactivity of each MAb is scored as positive or negative depending on whether it binds to a specific viral epitope, the nature of which can depend on both the primary amino acid sequence of the targeted protein and its secondary or tertiary structure. Highly discriminatory antigenic typing panels employ multiple MAbs targeting several epitopes associated with distinct viral variants.

For laboratories in which robust Mab panels have already been established, this method of viral typing remains a cost-effective tool for the rapid identification of viral variants (Fehlner-Gardiner et al., 2008). However, several factors limit the more widespread application of this approach. In the Americas, the complexity of the viral variants associated with different bat species has often challenged typing efforts employing a CDC panel developed for Latin American countries (Velasco-Villa et al., 2006). Generation of additional novel MAbs with the appropriate discriminatory properties can overcome this problem but requires the generation and screening of a large number of Mab-secreting hybridomas, a process which is time consuming and costly. As a result, recent efforts to further develop antigenic typing panels for use in different geographical areas have been limited (Chaves et al., 2015). At the same time, genetic methods of typing, which have proven to be more sensitive to small differences between viral isolates, have increased in popularity due to advances in nucleic acid amplification and sequencing methods and it is this approach that will be the focus of this review.

5.3.1 Genetic characterization of lyssaviruses

In the early 1990s, the development of nucleic acid amplification technologies, and especially the reverse-transcription polymerase chain reaction (RT-PCR) technique that generates dsDNA copies of portions of the viral genome for characterization, established an entirely new approach to rabies virus characterization (Smith, Orciari, Yager, Seidel, & Warner, 1992). The later inclusion into these assays of sequence–specific probes or DNA-intercalating dyes, that generate fluorescent signals as amplicon is produced, facilitates real-time lyssavirus detection and such assays are now increasingly used for rabies diagnosis (see Chapter 12). While newer methods of viral sequencing do not require targeted amplification, many studies still employ this approach due to its relative ease and manageable cost.

The choice of lyssavirus gene or region to be targeted for characterization is often dependent on the reason for the study. If differentiation of viruses circulating in distinct reservoirs is the main goal, then characterization of relatively short regions from any gene may be sufficient. Where the study is more focused on better understanding of the spread of a specific variant, targeting of less conserved genomic regions such as the P gene or the G-L intergenic region may be more useful. Ultimately characterization of the complete viral genome yields the highest possible resolution, and technologies supporting such detailed analyses are further addressed later.

Once an amplicon has been generated using RT-PCR, several different techniques can be used for its characterization. In the past, such methods have included Restriction Fragment Length Polymorphism Analysis, Heteroduplex Mobility Assays, as well as the use of target-specific reagents to selectively amplify particular viral variants by RT-PCR (Nadin-Davis, 2013). A different viral typing approach employed variant-specific probes based on P gene sequences for in situ hybridization to formalin-fixed brain tissue to discriminate many Canadian variants (Nadin-Davis, Sheen, & Wandeler, 2003). More recent approaches include pyrosequencing technology, in which a very short segment of the genome is sequenced (De Benedictis et al., 2011), and amplification-based assays, which employ variant specific reagents in reverse transcription loop-mediated isothermal amplification (RT-LAMP) (Saitou et al., 2010) or real-time RT-PCR (Nadin-Davis, 2019) assays.

While some of these methods were appropriate for the available technology of the period, the development of efficient Sanger sequencing methods using fluorescent dyes and capillary electrophoresis made direct nucleotide sequencing of PCR products readily achievable. As a result, over the last 35 years, a large body of lyssavirus sequence data has been generated, much of which is available in publicly accessible databases such as NCBI and EMBL, thereby providing a comprehensive source of reference material for viral comparisons.

5.3.2 Application of high-throughput sequencing (HTS) methods

While whole lyssavirus genome sequences could be determined using the traditional methods just described (Delmas et al., 2008), this process was time consuming due to the need to amplify many overlapping amplicons and then sequence them using traditional Sanger sequencing methods in which each individual reaction typically generates a single sequence read between 600 and 900 bases. The development of commercial HTS platforms in the first decade of the 21st century, beginning with the Roche 454 system and followed by Illumina and Ion Torrent technologies (Glenn, 2011), initiated another revolution in the approaches used to study viral molecular epidemiology. These HTS platforms generate thousands of shorter sequence reads from pooled indexed samples and then, through physical and bioinformatics means, separate these reads according to the sample source (Mardis, 2008). The large amount of data generated by these platforms results in thousands of overlapping reads that can be aligned and assembled to generate whole viral genomes using a fraction of the technical manpower needed for more conventional approaches. Even newer platforms, such as the Oxford Nanopore MinION sequencer which can generate much longer read lengths, promise to further revolutionize viral genomics (Kilianski et al., 2016).

However, depending on the goals of specific investigations, the application of these new technologies to lyssavirus epidemiology can involve rather different sequencing strategies (Brunker, Nadin-Davis, & Biek, 2018). The principal differences between these methods lie in whether the RNA recovered from infected brain tissue is used for a shotgun approach (Marston et al., 2013), which may or may not involve removal of some host nucleic acids but avoids viral amplification, or whether overlapping viral amplicons generated by RT-PCR are employed for sequencing (Nadin-Davis, Colville, Trewby, Biek, & Real, 2017). Both approaches have their pros and cons (Brunker, Nadin-Davis, & Biek, 2018). The shotgun approach avoids any bias that might be introduced by RT-PCR, which depends upon use of primers appropriately matched with their target sequence for successful amplification. This becomes especially important when attempting to sequence novel lyssaviruses that may exhibit significant sequence diversity from known members of the genus. However, depending upon the sample's level of infection, this approach can result in limited levels of coverage along the length of the viral genome if too many samples are combined in the sequencing run; limited pooling of samples in this way can increase costs significantly. Approaches involving viral genome amplification ensure that high coverage along the length of the genome is obtained even when large numbers of samples are pooled together for the sequencing run thus minimizing the cost per sample. These latter methods have been successfully applied to molecular epidemiological studies of single viral variants at relatively local or regional levels (Nadin-Davis, Colville, et al., 2017; Trewby, Nadin-Davis, Real, & Biek, 2017) but could become more complicated with respect to primer design if amplification of multiple variants is desired.

The application of HTS to lyssavirus studies around the world is reflected in the increasing number of whole viral genomes being submitted to public sequence databases (Brunker, Nadin-Davis, & Biek, 2018). Regardless of the methods employed for sequencing, the primary goal is usually to compile the consensus sequence of an isolate, i.e., a sequence that represents the most common nucleotide base at each position along the genome. In some situations, however, the extent of viral diversity within an isolate is of interest and traditionally this has been achieved by insertion of amplicons into a plasmid vector followed by sequencing of multiple clones. With the development of highly parallel sequencing methods, direct interrogation of the genetic diversity of samples is now possible and several applications of this capability are described later.

5.3.3 Principles of phylogenetic analysis

Phylogenetic analysis of sequence data has become an essential tool for molecular epidemiological analysis of RNA viruses (Grubaugh et al., 2019; Lam, Hon, & Tang, 2010) and lyssaviruses are no exception in this regard. Using consensus sequence data, similar tools are applied for viral typing and phylogenetic studies regardless of sequence length and method of data acquisition. However, the greater resolution afforded by more extensive data such as complete genomes vastly improves the interpretive capability of the analysis. Once nucleotide sequence data have been collected over a predetermined sequence window, these data are analyzed to explore the interrelationships between the samples thus represented. The first step involves aligning the sequence data such that each position can be directly compared for all samples. This is followed by an interrogation of the data to generate a phylogenetic tree, a diagram of hierarchical branches that depicts the evolutionary relationships between all samples. Those samples that form a discrete cluster on one branch of the tree are said to form a clade; where there is strong support for this cluster (see later) the samples are said to form a monophyletic clade, indicating that all members originated from a common precursor. However, the term "clade" is often applied loosely and all trees contain clades within clades, a structure that represents continued viral evolution and subtype emergence. A taxon is normally defined as a species or group of species that clearly identifies a specific group of organisms; this term is sometimes used to refer to a group of specimens that form a monophyletic clade. Terms that describe the groupings of a phylogenetic tree, such as phylogroup, genotype, clade, cluster, group, type, and lineage can be, and often are, used interchangeably. However, for clarity when describing a phylogenetic study, it is helpful to assign specific designations to distinct levels of sample association and apply these designations consistently. Furthermore, the use of descriptive clade/group designations and, wherever possible, reference to designations described previously, facilitates comparison of samples between studies.

The sequence database employed in the analysis usually includes one or more reference sequences to provide context and allow interpretation of the tree. While the sequence window targeted for such a study will be determined in part by its goal, in general it has been observed that similar conclusions on the overall epidemiological relationships of a group of lyssaviruses are obtained regardless of the target sequence employed. Thus, studies on the relationships between representatives of all the recognized lyssavirus genotypes exhibit similar results irrespective of the use of a specific gene, a portion of a gene, or the whole viral genome (Wu, Franka, Velasco-Villa, & Rupprecht, 2007), an observation consistent with the notion that members of this genus rarely, if ever, undergo recombination events. However, any analysis requires enough information to generate phylogenies having sufficiently robust statistical support. While short target sequences (200–300 bases) can yield phylogenetic predictions, the longer a sequence window, the greater the likelihood of finding differences between samples; as genetic variation increases, the number of informative characters available to a phylogenetic analysis increases and this improves the chances of obtaining well-supported phylogenetic trees (Nadin-Davis, Colville, et al., 2017).

To identify and type a virus sample, it must be compared over the same portion of the genome with viruses that are representative of that geographical region, and thus the choice of target sequence will depend on the availability of sequence data previously generated for reference isolates. Indeed, although the N gene is relatively conserved, it has often been targeted successfully to differentiate sympatric viral variants and the large amount of N gene sequence data that has been deposited into publicly accessible databases facilitates comparison with new isolates. For studies that seek to monitor variation within a closely related viral population, targeting of more variable sequence is preferred and the longer the sequence window, the more informative the analysis.

Phylogenetic trees are generated from sequence data by computer programs that employ a variety of different algorithms and methods for tree reconstruction (Choudhuri, 2014). While there are a variety of packages available for this purpose, one of the most commonly used is the user-friendly software MEGA (Molecular Evolutionary Genetic Analysis), the latest version of which (MEGA-X) has recently been released (Kumar, Stecher, Li, Knyaz, & Tamura, 2018). This software incorporates a variety of algorithms and methods for phylogenetic tree construction. The popular distance-based methods, such as neighbor joining (NJ) or unweighted pair-group method with arithmetic means (UPGMA), consider the overall genetic distance between all pairs of sequences to generate a distance matrix, which is then used as the basis for phylogenetic construction. Alternative algorithms, including those employing character-based maximum parsimony (MP) and maximum likelihood (ML) methods, consider individual substitutions to determine all possible tree constructions supported by the data and then identify the optimal tree by a comparative process. MP identifies the optimal tree by selecting the minimal number of evolutionary steps required to explain the data. ML identifies the optimal tree as that most likely to have occurred according to an assumed evolutionary model, several of which have been described; a program to interrogate the data for the best fitting model is incorporated into the MEGA package.

The choice of method to be employed for an analysis will often depend on the study's purpose. Distance methods are frequently preferred from a practical standpoint due to their relatively rapid execution and their ability to identify groups or clades as efficiently as other more computationally intensive methods. For many epidemiological studies, the association of an isolate within a clade, rather than its precise position within this clade, normally identifies the variant responsible for that case thereby providing the key information sought. Thus, for many analyses, distance methods are sufficiently predictive. However, when the data are to be analyzed to explore mechanisms of viral evolution, use of other algorithms such as ML is considered more appropriate. Fig. 5.1 illustrates trees generated for representative members of the *Lyssavirus* genus using several of these methods.

It is apparent that when sufficiently informative sequence data are employed, all methods generate very similar trees. Indeed, the analysis of a sequence dataset using multiple methods of phylogenetic reconstruction can be most helpful in supporting the predicted relationships between a group of viruses and thereby enhancing the overall predictive strength of the study. Another important measure of the robustness of a phylogenetic tree is the use of nonparametric bootstrap analysis. Most phylogeny software can incorporate this statistical method into their analyses, and all molecular epidemiological studies should be encouraged to include this statistic. The method is valuable because even relatively small datasets generate multiple possible branching patterns or trees and statistical methods must be employed to predict the most likely branching pattern referred to as the consensus tree. In nonparametric

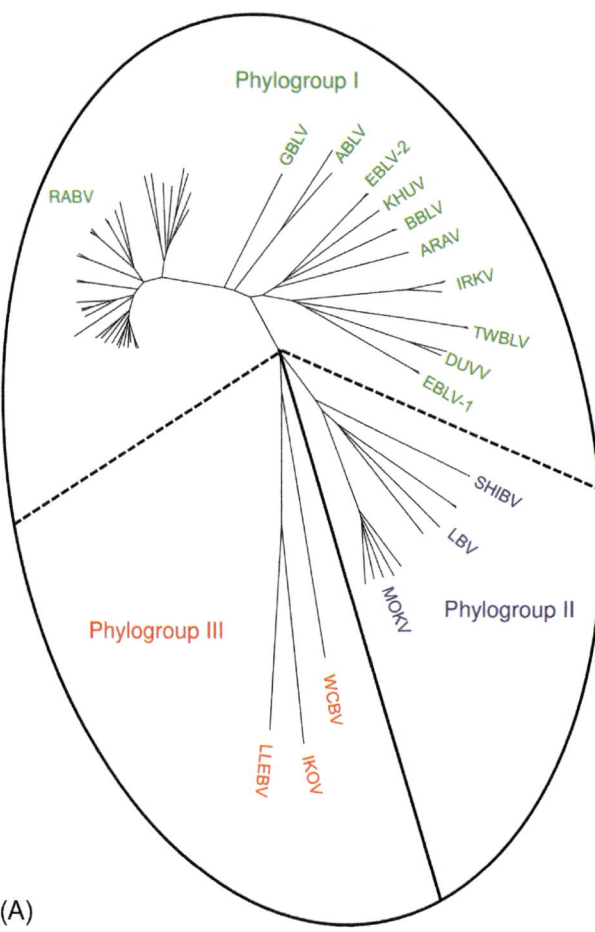

FIG. 5.1 (A) Radial NJ tree generated from whole genome sequences of 70 representative lyssaviruses. The NCBI accession numbers for all samples employed are indicated in B and C; the KBLV sample is not included in this tree due to lack of complete genome sequence. The division of the genus into three phylogroups is illustrated.

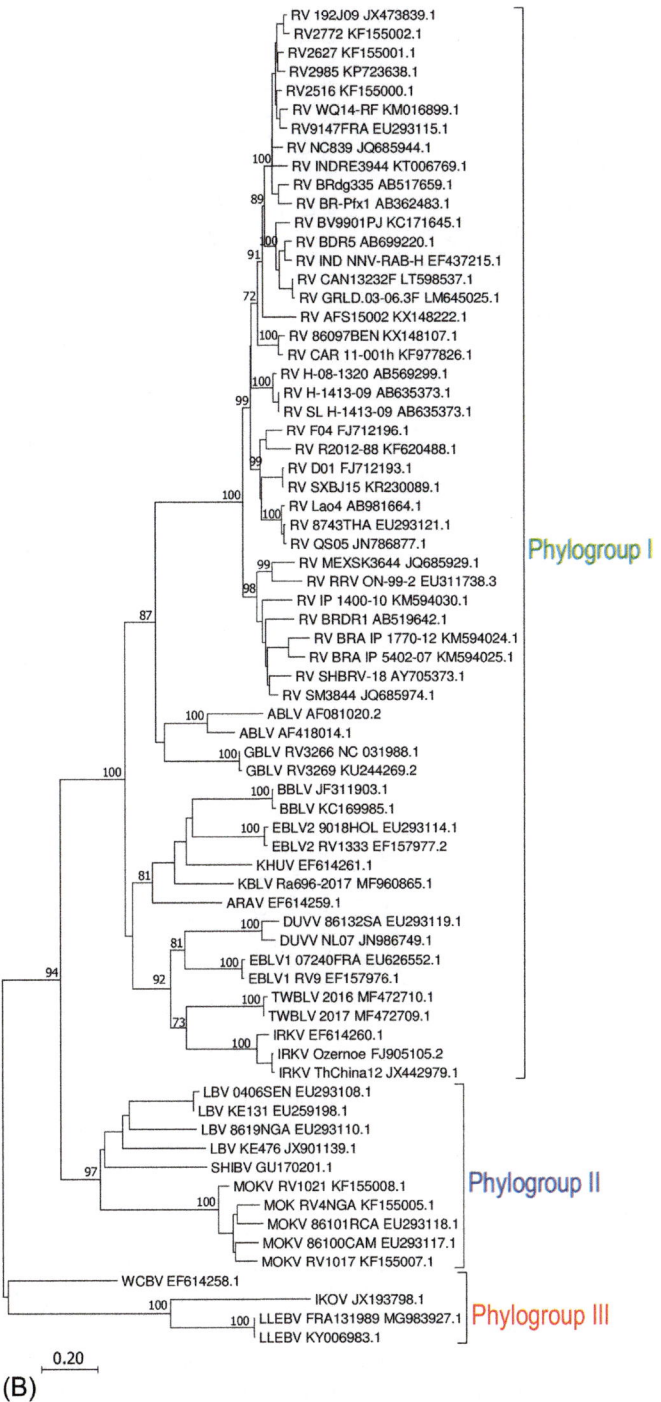

FIG.5.1, CONT'D (B) An ML tree generated from the complete N gene-coding sequence (1350bp) of 71 representative lyssaviruses, including the KBLV sample. Values at major nodes indicate bootstrap values as a percentage. Branch lengths indicate distances between taxa according to the scale at bottom. *(Continued)*

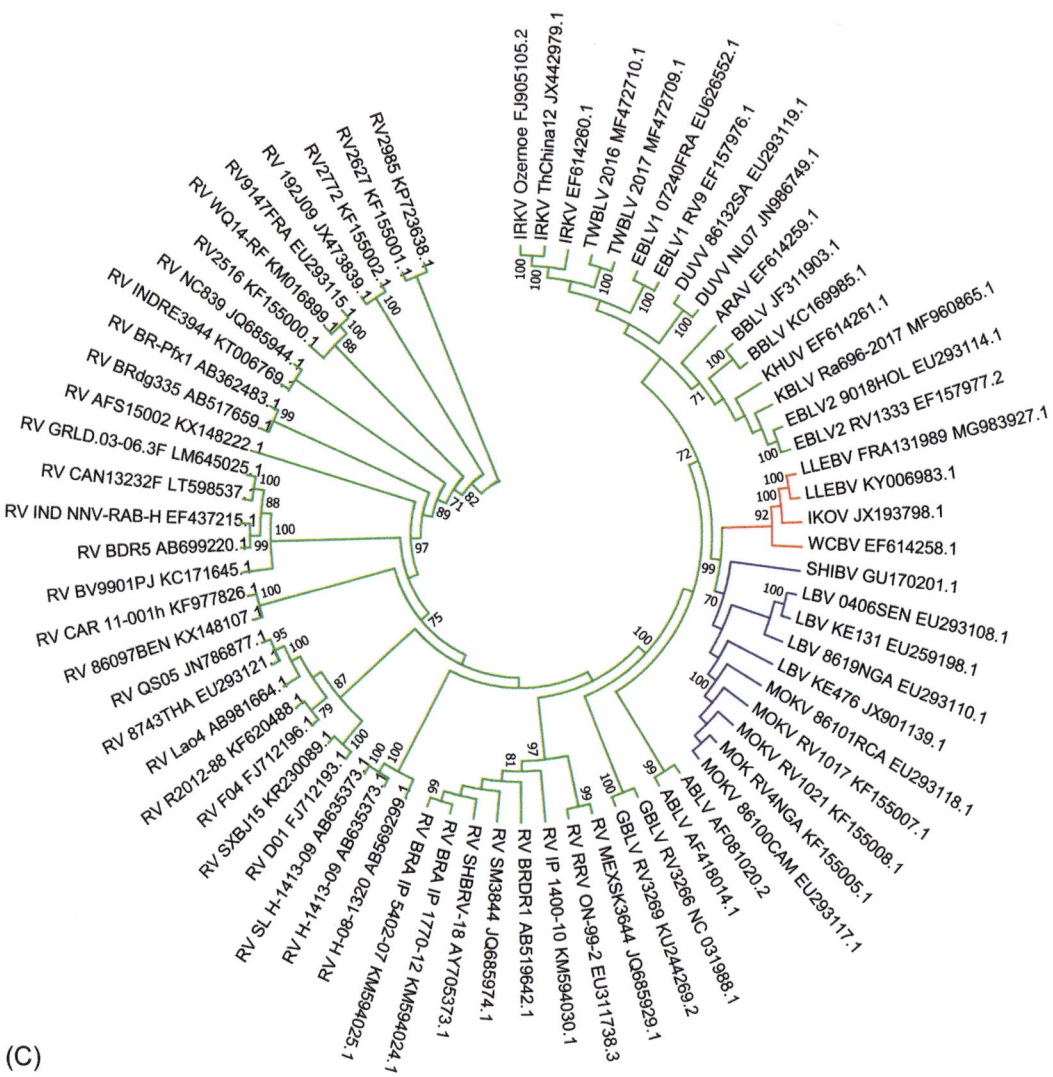

FIG. 5.1, CONT'D (C) A circular MP tree generated from the same sequence data set as the ML tree shown in panel B. Bootstrap data are indicated for nodes with values ≥ 70%. Note that MP trees indicate the overall topology of the phylogeny without indicating distances between taxa. Species abbreviations are as indicated in Table 5.1 except that RABV is shortened to RV.

bootstrap analysis, the nucleotide sequence data are resampled randomly with replacement thereby generating pseudoreplicates of the original data; the number of replicates is set by the operator, usually between 100 and 1000. A smaller number of replicates is generally employed when applying computationally intensive programs such as MP, while an NJ analysis can normally readily incorporate 1000 replicates. Upon analysis by one of these algorithms, the proportion of times that each clade occurs within all trees is calculated and this value is considered as a measure of support for that grouping. In the case of RNA viruses, bootstrap values >90% are generally regarded as providing strong support for a clade, while values >70% are often considered significant (Bauldauf, 2003). Clades having bootstrap values below this level must be interpreted with caution.

Another useful tool for tree construction is the inclusion of an outgroup, a sequence representative of a taxon that is known or assumed to be less closely related to all other taxa of the analysis than these taxa are to each other. Thus, an outgroup helps to define the base of the tree and the direction of change as viruses emerge. An unexpected association of an unknown sample with an outgroup implies that the original assumptions regarding the origins of the unknown were incorrect and a different sample set should be employed for an alternate phylogenetic prediction.

Developments in coalescent theory and the application of these principles to develop new phylogenetic analysis programmes such as the BEAST (Bayesian Evolutionary Analysis Sampling Trees) package (Bouckaert et al., 2014; Drummond, Suchard, Xie, & Rambaut, 2012) have provided epidemiologists with alternative methods of tree generation that are being applied to lyssavirus studies (Nadin-Davis & Real, 2011). In such Bayesian statistical methods, posterior probabilities are employed in place of bootstrap values as an indication of the likelihood that certain taxa group into the same clade. These tools are frequently used to infer estimates of the viral nucleotide substitution rate, which can then be used to apply time scale estimates to phylogenetic trees. Furthermore, these tools also permit the exploration of viral population dynamics, thus providing insights into the demographics of viral outbreaks (Grubaugh et al., 2019) and a better understanding of viral evolution (see Chapter 3). Demonstration of convergence of duplicate runs using this approach is an important indicator that the method has reached an appropriate conclusion.

Phylogenetic trees can also be constructed with protein translation products of coding regions using approaches similar to those applied to nucleotide sequence data. Comparison of trees generated using nucleotide and amino acid sequence data of the lyssavirus N locus shows rather similar topologies, although the former tend to have higher bootstrap values. This observation may simply reflect the fact that nucleotide sequences, with three times more datapoints than amino acid sequences, can include synonymous mutations and thus contain greater informative diversity.

One important aspect of any molecular epidemiological study is the selection of samples to be included for comparison. Ideally, an appropriately representative selection of samples, as required to answer the questions posed by the phylogenetic study, should be included but compliance with this criterion may not be simple. Most collections of rabies viruses are drawn from passive surveillance systems that can introduce significant bias. Areas of low human population density, lack of human contact with rabies reservoir species, or an inadequate infrastructure for laboratory confirmation of suspect cases all limit the numbers of samples available for study. The first two limitations have precluded collection of many samples from northern temperate and arctic regions, while the latter issue has severely hampered studies of the nature of the lyssaviruses circulating in many parts of Africa. Thus, while collections from such areas are incomplete, phylogenetic analysis provides the best available means of exploring the epidemiology of the disease in these regions.

5.4 Lyssavirus taxonomy

5.4.1 Lyssavirus species

As detailed in Table 5.1, all lyssaviruses are currently divided into 16 different species, with rabies lyssavirus (RABV) assigned as the type species of the genus by the International Committee on Taxonomy of Viruses (Amarasinghe et al., 2019).

TABLE 5.1 Species currently assigned to the *Lyssavirus* genus.

Species	Distribution	Reservoir
Rabies lyssavirus (RABV)	Worldwide with the exception of Antarctica and some islands. Some regions traditionally affected (e.g., western Europe) have recently eradicated the disease and attained rabies-free status	*Carnivora*: many species of domestic and wild canids, foxes, mongooses, skunks, and the raccoon, *Chiroptera*: (Americas only) several species of insectivorous, vampire and frugivorous bats
Aravan lyssavirus (ARAV)	Only single case reported from Kyrgyzstan	*Microchiroptera*: *Myotis blythi*[a]
Australian bat lyssavirus (ABLV)	Australia and possibly areas of SE Asia	*Megachiroptera*: Pteropid species *Microchiroptera*: the insectivorous yellow-bellied sheathtail bat (*Saccolaimus flavicentris*)
Bokeloh bat lyssavirus (BBLV)	Several cases recovered from Germany, France, and Poland	Natterer's bat (*Myotis nattererii*)
Duvenhage virus (DUVV)	African nations, including Guinea, South Africa, Zimbabwe	*Microchiroptera*: Single cases assigned to *Miniopterus schreibersii*, *Nycteris gambiensis*, and *N. thebaica*
European bat lyssavirus 1 (EBLV-1)	Europe, including Denmark, France, Germany, The Netherlands, Poland, Russia, Spain, Ukraine	*Microchiroptera*: *Eptesicus serotinus*
European bat lyssavirus 2 (EBLV-2)	Several countries of western Europe particularly The Netherlands, Switzerland, Finland, and the UK	*Microchiroptera*: *Myotis* species, especially *M. dasycneme* and *M. daubentonii*
Gannoruwa bat lyssavirus (GBLV)	Four cases recovered from flying foxes in Sri Lanka	*Pteropus* sp.
Ikoma virus (IKOV)	Single case in Tanzania	Isolated from an African civet (*Civettictis civetta*) but bat reservoir possible[a]
Irkut virus (IRKV)	One case from Irkutsk province, Russia. A closely related virus (Ozernoe Virus) was recovered in Primorye Territory, Russia; One case from China	*Microchiroptera*: Irkut virus was recovered from a bat (*Murina leucogaster*),[a] while Ozernoe virus was from a human case after exposure to an unidentified bat
Khujand virus (KHUV)	Single case from Tajikistan	*Microchiroptera*: single case from a bat (*Myotis daubentonii*)[a]
Lagos bat virus (LBV)	Several African countries, including Central African Republic, Ethiopia, Nigeria, Senegal, South Africa, Zimbabwe	*Megachiroptera*: Fruit bats, including *Eidolon helvum*, *Rousettus aegyptiacus* and *Epomophorus wahlbergi*
Lleida bat lyssavirus (LLEBV)	Two cases (one each in Spain and France)	Isolated from *Miniopterus schreibersii*[a]
Mokola virus (MOKV)	Several African countries, including Cameroon, Central African Republic, Ethiopia, Nigeria, South Africa, Zimbabwe	Reservoir unknown—single cases in shrews (*Crocidura* sp.) and a small rodent (*Lopyhromys sikapusi*) with most reported cases in domestic cats and dogs

TABLE 5.1 Species currently assigned to the *Lyssavirus* genus—cont'd

Species	Distribution	Reservoir
Shimoni bat virus (SHIBV)	Single case from a bat in Kenya	Commerson's leaf-nosed bat (*Hipposideros commersoni*)[a]
West Caucasian bat virus (WCBV)	Single case from the Krasnodar region of Russia	Isolated from *Miniopteris schreibersii*[a]
Isolates awaiting classification:		
Kotalahti bat lyssavirus	Single case from Finland	Isolated from *Myotis brandtii*[a]
Taiwan bat lyssavirus (TWBLV)	Two cases in Taiwan	Isolated from *Pipistrellus abramus*

[a]*Reservoir host to be confirmed.*

While this classification has in the past been supported by serological, immunological, and epidemiological information (Badrane, Bahloul, Perrin, & Tordo, 2001), it is increasingly based on sequence data. A nucleotide identity value of 80%–82% for the N gene is used as an appropriate cut-off for discriminating between species (Kuzmin, Hughes, Botvinkin, Orciari, & Rupprecht, 2005). In general, members of the same species have identity values between 82% and 100%, values within the 82%–95% range often define distinct lineages within the species and all members of a particular variant usually exhibit >95% identity.

Efforts to organize these species into higher-order groupings based upon genetic, immunologic, and pathologic characteristics of certain members now suggest that the genus can be organized into at least three distinct phylogroups (Badrane et al., 2001; Kuzmin et al., 2005). As illustrated in Fig. 5.1, phylogroup 1 comprises RABV, ABLV, ARAV, BBLV, DUVV, GBLV, IRKV, KHUV, and both EBLVs, phylogroup 2 constitutes the distinctive African viruses LBV, SHIBV, and MOKV and phylogroup 3 is represented by IKOV, LLEBV, and WCBV. However, as novel isolates continue to be identified the diversity of known lyssaviruses will increase and this organization of the genus may have to be revisited in the future (Fooks, 2004).

This differentiation does however have functional significance since current vaccines are efficacious against all phylogroup 1 viruses but have limited to no efficacy against viruses of the other phylogroups (Hanlon et al., 2005).

5.4.2 Molecular epidemiology of rabies-related lyssaviruses

With the exception of the type species, rabies lyssavirus, the numbers of isolates collected for many lyssavirus species remain limited and hence our knowledge of their diversity and range is incomplete. As surveillance intensifies, especially in bat populations, this situation is slowly improving. Serological studies in many areas suggest the broader circulation of several lyssaviruses, especially in Asia and Africa, and, as described later, there have been several recent reports of the isolation of recognized species in new geographical areas. Our knowledge of lyssavirus diversity, distribution, and molecular epidemiology is summarized here with the understanding that the known complexity of the genus will most certainly increase significantly in the near future.

5.4.2.1 Australian bat lyssavirus

Since the identification of a rabies-like virus circulating in Australian fruit bats in 1996 (Fraser et al., 1996) and its subsequent designation as Australian bat lyssavirus (ABLV) (Hooper et al., 1997), many isolates of this species have been characterized (Guyatt et al., 2003). ABLVs are readily differentiated into two biotypes believed to have diverged several centuries ago, making the recent identification of this species all the more remarkable (Field, 2018). One biotype, associated with *Pteropus* (flying fox) species, has been recovered from all four *Pteropus* species that inhabit the Australian mainland, while the other biotype has been recovered from insectivorous bats, especially the yellow-bellied sheathtail bat, *Saccolaimus flaviventris* (Gould et al., 1998; Gould, Kattenbelt, Gumley, & Lunt, 2002). ABLVs have been the etiologic agents for three human rabies cases (Francis et al., 2014; Moore, Jansen, Graham, Smith, & Craig, 2010) and have also caused disease in domestic animals (Annand & Reid, 2014). Surveillance, to detect either viral antigen in brain tissue or antibodies in sera, suggests that although ABLV prevalence rates are low, both viral biotypes circulate widely in several Australian bat taxa consistent with the long-term presence of the virus on the continent (Field, 2018; Moore et al., 2010). Serological evidence suggests that ABLVs, or closely related viruses, may circulate in bat populations of the Philippines (Arguin et al., 2002).

5.4.2.2 European bat lyssaviruses

The association of lyssaviruses with European insectivorous bats was brought into focus in 1985 with the death of a bat researcher from rabies. Due to the public health concerns regarding these viruses, greater surveillance efforts since then have identified >1000 cases of rabid bats in Western Europe. As a result, the known distribution of these viruses is being extended and new bat lyssaviruses have been recognized in recent years. Five lyssaviruses are currently known to circulate in European bats: two of these species, European bat lyssavirus type 1 (EBLV-1) and type 2 (EBLV-2), have been recognized for several decades while the other three, Bokeloh bat lyssavirus (BBLV), Lleida bat lyssavirus (LLEBV), and the West Caucasian bat lyssavirus (WCBV), have been recognized only recently.

EBLV-1, by far the most frequently isolated lyssavirus in Europe and responsible for >97% of all bat rabies cases, is maintained in *Eptesicus serotinus* bats and is widely distributed throughout many European countries (McElhinney et al., 2013). EBLV-1 can be divided into two genetic lineages, 1a and 1b, that exhibit differences in their evolution and dispersal patterns and which may represent two independent introductions of this biotype into Europe. The relatively homogeneous EBLV-1a lineage occurs predominantly in northern countries of Germany, Denmark, the Netherlands, and Poland in areas comprising flat, low-lying landscape with a mix of urban and rural areas and some forest, landscape which is ideal habitat for *Eptesicus serotinus*, and which likely facilitated rapid east-west spread of EBLV-1a across Northern Europe (Müller et al., 2007). This subtype is the most prevalent EBLV lineage in western France (Picard-Meyer et al., 2014) from where it appears to have spread southwards into northern Spain (Mingo-Casas et al., 2018).

In contrast, the more divergent EBLV-1b lineage appears to exhibit a north-south axis of spread and occurs in the Netherlands (Van der Poel et al., 2005), Germany (Johnson et al., 2007), and in northern France, with some overlap in range with EBLV-1a in central France (Picard-Meyer et al., 2014). In Spain, a collection of distinct EBLV-1b variants have been isolated from two bat species, *E. isabellinus* and *E. serotinus*, the former apparently being the main viral reservoir (Vázquez-Morón, Juste, Ibáñez, Berciano, & Echevarría, 2011) with possible interspecific transmission to *E. serotinus* (Mingo-Casas et al., 2018).

Due to challenges in studying the presence of EBLVs in many protected bat species in Western Europe, a number of studies have approached this by inferring exposure through detection of neutralizing antibodies in sera collected from these animals. Presence of such antibodies is taken as an indication of the circulation of these viruses in these populations.

Bat species seropositive for antibodies against EBLV-1 in Italy and Croatia included two members of the *Myotis* genus, *M. myotis* and *M. blythii*, while additional species included *Tadarida teniotis* in Italy and *Miniopteris schreibersii* in Croatia, species not considered EBLV-1 reservoirs in other parts of Europe, but limitations in acquiring specimens of *E. serotinus* precluded investigation of this species in these studies (Leopardi et al., 2018; Šimić et al., 2018). Such sero-surveillance studies have also suggested the circulation of lyssaviruses closely related to EBLV-1 in bats collected from sites in Morocco and Algeria (Leopardi et al., 2018; Serra-Cobo et al., 2018; Šimić et al., 2018), thereby supporting the speculation that EBLVs were originally introduced into Western Europe from Africa by migrating bats. This hypothesis has heightened interest in the role of other bat species, which are more likely to have been responsible for this spread compared to the more sedentary *E. serotinus*. Spillover of EBLV-1 to *Myotis* species has been observed in Spain (Serra-Cobo, Amengual, Abellán, & Bourhy, 2002) and there is increasing serological evidence from several countries (Picard-Meyer et al., 2011; Serra-Cobo et al., 2013) for circulation of EBLV-1 in *Miniopteris schreibersii*, a migratory species that could be a potential carrier by which the virus was introduced into Europe and subsequently spread. Modeling studies have proposed that persistence of the virus could be achieved in the absence of *E. serotinus* through interactions between *Myotis* species and *Miniopteris schreibersii* (Colombi et al., 2019). This raises the possibility that given the differences in roosting patterns and anthropomorphic interactions of many bat species our current knowledge of their relative roles in maintaining these viruses may be biased and may change significantly with further study.

EBLV-2, first isolated in 1985 from a Swiss bat biologist working on bats in Finland, has been confirmed as the etiological agent in a total of 34 cases in bats and is suspected in a few others (McElhinney et al., 2018). It has been recovered from two bat species, *M. daubentonii*, which is widely distributed across western Europe, and *M. dasycneme*, which has a more restricted range on the European mainland (McElhinney et al., 2013). EBLV-2 has been identified in many European countries, including the Netherlands, Switzerland, Denmark, Germany, and the United Kingdom (McElhinney et al., 2013) where passive surveillance was initiated following a case of EBLV-2-mediated human rabies in Scotland (Fooks et al., 2003). In the United Kingdom, 13 cases of EBLV-2 have been recorded in *M. daubentonii* suggesting the virus is indigenous to the country (Wise et al., 2017) and recent isolations from this species have been reported in Finland (Nokireki et al., 2017) and Norway (Moldal et al., 2017). The relatively small number of isolates, in addition to some confusion between laboratories in the archiving and sharing of these isolates, has complicated their phylogenetic analysis and brought into question earlier suggestions that these viruses segregate into two distinct lineages, 2a and 2b (McElhinney et al., 2013). More recent analysis using whole genome sequencing data shows that these viruses exhibit regional localization according to country of origin and undergo relatively low levels of nucleotide substitution. Dutch isolates, which all originated from *M. dasycneme*, form a separate clade, while the UK isolates segregate into multiple regional groups distinct from other European groups (McElhinney et al., 2013, 2018). Genetic studies on bat populations suggest that entry of EBLV-2 into the UK has been facilitated by regular movement of *M. daubentonii* between mainland Europe and the UK; in contrast, *E. serotinus* populations are more fragmented thereby limiting opportunities for incursion of EBLV-1 into the country (Smith et al., 2011).

While bats are the established reservoirs for EBLVs, spillover infection of EBLV-1 to terrestrial species has occasionally been reported (McElhinney et al., 2013). While experimental studies have shown that EBLVs can cause clinical features of rabies in red and silver foxes (Picard-Meyer et al., 2008; Vos et al., 2004), the likelihood that these viruses could adapt and become capable of sustained transmission in a terrestrial host appears to be negligible.

Bokeloh bat lyssavirus (BBLV) was first isolated from a rabid Natterer's bat (*Myotis nattererii*) in Germany in 2010 (Freuling et al., 2011). Additional isolates have subsequently been reported from Germany, France, and Poland; all were recovered from *M. nattererii* except for one recovered from a *Pipistrellus pipistrellus* bat in Germany (Eggerbauer et al., 2017; Picard-Meyer, Servat, et al., 2013; Smreczak et al., 2018). All nine BBLV sequences identified form a monophyletic clade distinguishing these viruses from all other European bat lyssaviruses (Eggerbauer et al., 2017; Smreczak et al., 2018). Interestingly, this clade diverges into two distinct clusters, though the basis for this separation remains unclear as there appears to be no correlation with host subspecies or spatial distribution.

Another novel lyssavirus, first recovered in Spain from a *Miniopterus schreibersii* bat, was designated Lleida bat lyssavirus (LLEBV). Initial studies limited to N gene analysis suggested that LLEBV was a distinct species most closely related to the more divergent lyssaviruses IKOV and WCBV (Aréchiga Ceballos et al., 2013), a conclusion subsequently confirmed using whole genome analysis (Banyard et al., 2018; Marston, Ellis, et al., 2017). A second isolation of LLEBV in France further supported the diverse nature of this virus, its classification as a distinct species, and the probable role of *Miniopterus schreibersii* as a reservoir host for this virus (Picard-Meyer et al., 2019).

Only one isolation of the West Caucasian bat lyssavirus (WCBV), from a *Miniopteris schreibersii* specimen recovered in Russia, has been reported but the highly diverse nature of its genome sequence has clearly supported its classification as a distinct species (Kuzmin et al., 2005; Kuzmin, Wu, et al., 2008). While the true range of this species remains unclear, the identification of neutralizing antibodies against WCBV in *Miniopterus* bats of Kenya has fueled the suggestion that this species also circulates in Africa (Kuzmin, Niezgoda, et al., 2008b).

A recently described, but yet unclassified, lyssavirus, Kotalahti bat lyssavirus recovered from a *Myotis* bat in Finland, may represent another species based on its distinct phylogenetic placement (Nokireki, Tammiranta, Kokkonen, Kantala, & Gadd, 2018).

5.4.2.3 *Asian bat lyssaviruses*

A small number of viral isolates from across the Asian continent define several lyssavirus species. This includes Aravan virus (ARAV), Irkut virus (IRKV), Khujand virus (KHUV), and Gannoruwa bat lyssavirus (GBLV) (Gunawardena et al., 2016; Kuzmin et al., 2003, 2005). Only a single isolation of each of ARAV and KHUV is yet reported and further isolates are needed to establish the true reservoirs of these lyssaviruses and their geographical ranges. A total of three IRKV isolates have been described: the original case recovered from a *Murina leucogaster* bat in Irkutsk province in Russia (Kuzmin et al., 2005), an isolate designated Ozernoe virus (Leonova et al., 2009) was recovered from a fatal human case after exposure to a bat, and another isolate recovered from a greater tube-nosed bat (*Murina leucogaster*) in China (Liu, Zhang, Zhao, Zhang, & Hu, 2013). This suggests that this species circulates over an extensive region of Asia and serological evidence supports the circulation of these or closely related viruses in certain bat populations of Thailand (Lumlertdacha et al., 2005), Cambodia (Reynes et al., 2004), and Bangladesh (Kuzmin et al., 2006).

Four isolates of GBLV, recovered from flying foxes in Sri Lanka, form a tight but separate lyssavirus clade (Gunawardena et al., 2016). It remains to be seen if this species circulates beyond this island nation. Two lyssavirus isolates from the Japanese pipistrelle (*Pipistrellus abramus*) recovered in Taiwan and designated Taiwan bat lyssavirus (TWBLV) have been sequenced along their entire genomes; based on their nucleotide identity and clustering between KHUV and EBLV-1, it is proposed that they represent yet another species (Hu et al., 2018), a request yet to be considered by the ICTV.

5.4.2.4 Lyssaviruses of Africa

Duvenhage lyssavirus

Beginning in 1970, there have to date been only five reports of the isolation of Duvenhage virus (DUVV) (Van Eeden, Markotter, & Nel, 2011; van Thiel et al., 2009). The first isolate was in a fatal human case from South Africa, the second from an insectivorous bat in South Africa, initially identified as *Miniopterus schreibersii*, but now designated as *Miniopterus natalensis*, the third from another species of insectivorous bat, *Nycteris thebaica*, captured in Zimbabwe, and the fourth was again from a human in South Africa (Paweska et al., 2006). The last report was of a human case diagnosed in the Netherlands after exposure of the patient to a bat in Kenya (van Thiel et al., 2009). While review of all DUVV sequence data might suggest a larger number of isolations, the assignment of individual isolates with different designations by different groups has confused the issue significantly. The four DUVV viruses recovered from southern Africa are highly homogeneous, possibly reflecting the limited geographical range over which they were retrieved, while the isolate from Kenya is genetically the most diverse (van Thiel et al., 2009) and may represent a second lineage of this species (Van Eeden et al., 2011). This specimen clearly indicates that our knowledge of the extent of DUVV diversity and geographical range will increase substantially with additional isolations across Africa.

Lagos bat lyssavirus

Since the first Lagos Bat Lyssavirus (LBV) isolate was recovered from the brain of a Nigerian fruit bat (*Eidolon helvum*) in 1956, and later identified as a rabies-related virus, just 16 LBV isolations have been documented (Markotter et al., 2008). LBV has a wide range across much of the African continent having been identified in Nigeria, Ghana, the Central African Republic, Guinea and Senegal, South Africa, Ethiopia, Zimbabwe, and Kenya (Kuzmin, Niezgoda, et al., 2008a; Markotter et al., 2008), while a single isolation made in Europe originated from a bat imported from Africa (Picard-Meyer et al., 2004). Although one LBV isolate was from an insectivorous bat, *Nycteris gambiensis*, these viruses are usually recovered from several species of frugivorous bats. Based on viral isolations, *Epomophorus wahlbergi* is the likely LBV reservoir species in South Africa (Markotter et al., 2008), while serological studies suggest reservoir roles for *Eidolon helvum* in Ghana and Nigeria (Dzikwi et al., 2010; Hayman et al., 2008; Wright et al., 2010), for both *Eidolon helvum* and *Rousettus aegyptiacus* in Kenya (Kuzmin, Niezgoda, et al., 2008a) and for *Eidolon dupreanum* in Madagascar (Jean-Marc Reynes et al., 2011). The virus has also been recovered from dogs, several cats, and a mongoose (*Atilax paludinosus*) presumably due to spillover infections, but it has never been reported in humans. Despite the limited sampling, significant antigenic and genetic variation occurs within this species (Kuzmin, Niezgoda, et al., 2008a; Markotter et al., 2008). It appears that the greater the geographical separation between samples, the greater their genetic diversity and four distinct lineages (A-D) of LBV have been described. Due to its high divergence from the other lineages, it was suggested that LBV lineage A be considered a separate species, designated Dakar bat lyssavirus (DBLV), for which the isolate from Dakar, Senegal, would represent the type virus (Markotter et al., 2008).

Shimoni bat lyssavirus

The Shimoni bat virus (SHIBV) isolate, recovered from the brain of a Commerson's leaf-nosed bat (*Hipposideros commersoni*) in Kenya (Kuzmin et al., 2010), is an outlier to the LBV clade and now recognized as a distinct species. Serological studies support the role of the Commerson's leaf-nosed bat as a reservoir for this virus (Kuzmin et al., 2011).

Mokola lyssavirus

Since the original isolation of Mokola Lyssavirus (MOKV) from shrews in Nigeria, this viral species has been recovered from human rabies cases in Nigeria and from other species in several African countries, including Cameroon, Zimbabwe, the Central African Republic, Ethiopia, and South Africa (Nel, Jacobs, Jaftha, von Teichman, & Bingham, 2000). All South African isolates have been recovered with the help of enhanced antigenic tools for screening all rabies-positive specimens (Sabeta et al., 2010; Sabeta, Markotter, et al., 2007). However, in the last 40 years, this virus has been detected in just 24 cases in host species that included shrews, cats, and dogs and a single isolation from a rodent (Kgaladi et al., 2013). While the high proportion of cases in cats suggests a bat or rodent reservoir for MOKV, the actual reservoir host is currently unknown. The most recent comprehensive analysis of the diversity of Mokola virus, which examined just 18 isolates recovered from several countries, has shown the existence of regional groupings with distinct phylogenetic clades circulating in different parts of the continent (Kgaladi et al., 2013). The lack of surveillance infrastructure and diagnostic capability in most African nations precludes a systematic assessment of the prevalence of this virus, but it clearly circulates across much of Africa.

Ikoma lyssavirus

A single isolate of Ikoma lyssavirus (IKOV), recovered from an African civet in Tanzania, is another member of phylogroup III (Marston, Ellis, et al., 2012; Marston, Horton, et al., 2012). It is unknown if the civet, or some other mammalian species, most probably a Chiropteran species, is the reservoir host and how extensively this virus circulates across Africa.

5.4.2.5 Future issues

The recent spate of isolations of novel lyssaviruses, even in Western Europe where surveillance of bats for diseases of public health importance is relatively comprehensive, underscores the limitations in our knowledge of the diversity of this genus. Continued surveillance in all corners of the world will undoubtedly expose the existence and circulation of other lyssavirus species. For example, there is now serological evidence that lyssaviruses related to DUVV and LBV circulate in bats on several southwestern Indian Ocean islands (Mélade et al., 2016). The isolation and characterization of such diverse lyssaviruses is of notable interest not least since it is known that current rabies prophylactics lack the ability to protect against members of phylogroups II and III, thus highlighting the need for alternative treatments against infection with these agents (Evans et al., 2018).

5.4.3 Molecular epidemiology of rabies lyssaviruses

One of the most comprehensive phylogenetic analyses of RABVs circulating globally identified eight major lineages (Troupin et al., 2016), including six canid-related lineages, estimated to have emerged between 500 and 700 years ago, and two lineages of the Americas previously referred to collectively as the American Indigenous lineage (Smith et al., 1992). The trees illustrated in Fig. 5.2 depict all these lineages and provide a framework for the

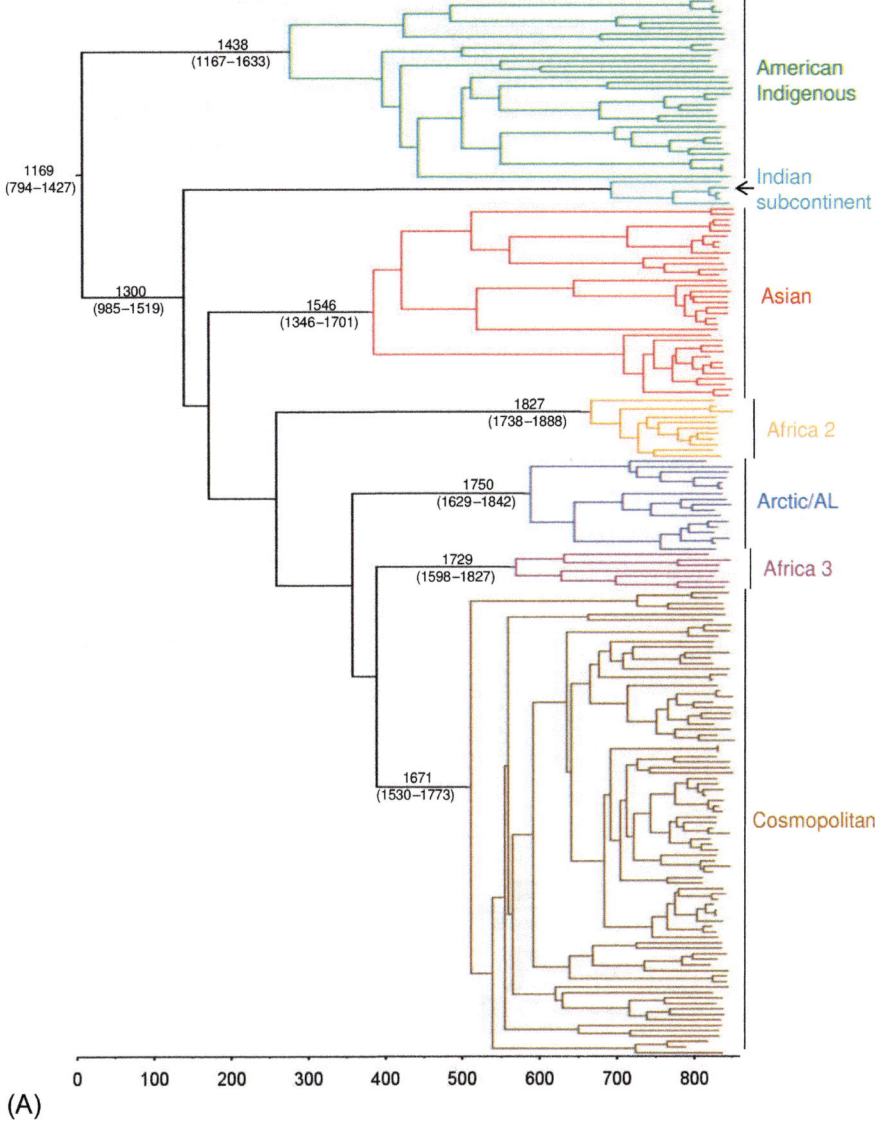

FIG. 5.2 Phylogenetic analysis of complete N gene sequences of representative RABVs. (A) BEAST analysis of 195 RABVs using a relaxed molecular clock with the GTR+G+I substitution model. Values at nodes indicate the estimated year of emergence of the clade and range based on the 95% highest posterior density (HPD) values in brackets. A time scale is shown at bottom. All posterior values for the seven lineages, as identified to the right, were at the maximum value of 1. All samples used in this tree are included in panels B–D with the exception of 5 that lacked date information.

(Continued)

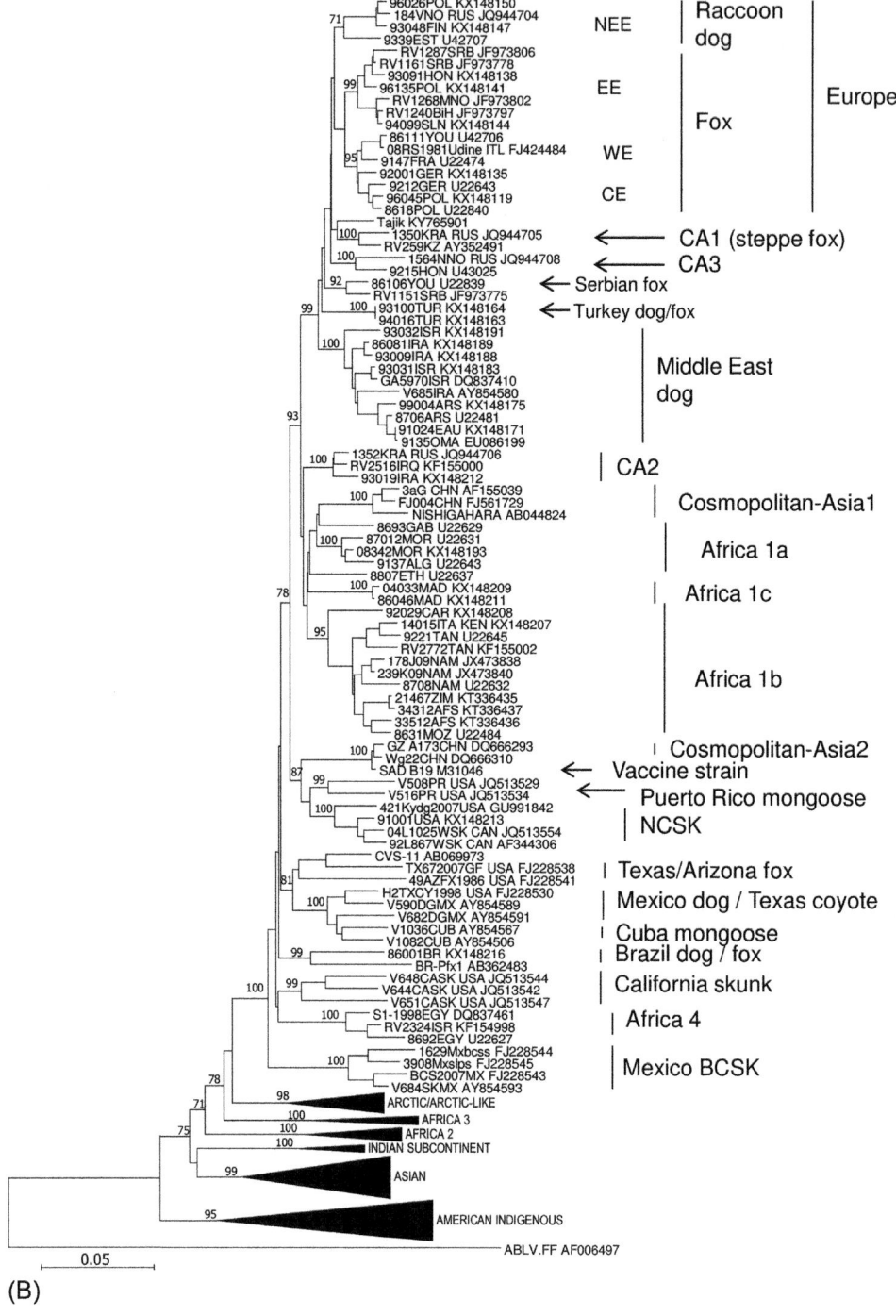

FIG. 5.2, CONT'D (B–D) These show the same NJ tree, generated from the N gene of 200 RABVs, with different clades collapsed in each panel so as to illustrate details of each lineage, as described in the text. Each sample is identified by its designation and NCBI Accession number. Bootstrap support, expressed as a percentage, is shown for most major nodes and branch lengths represent genetic distance according to the scale at the bottom of each panel. The N gene of an ABLV isolate was used as an outgroup. (B) detail of cosmopolitan lineage;

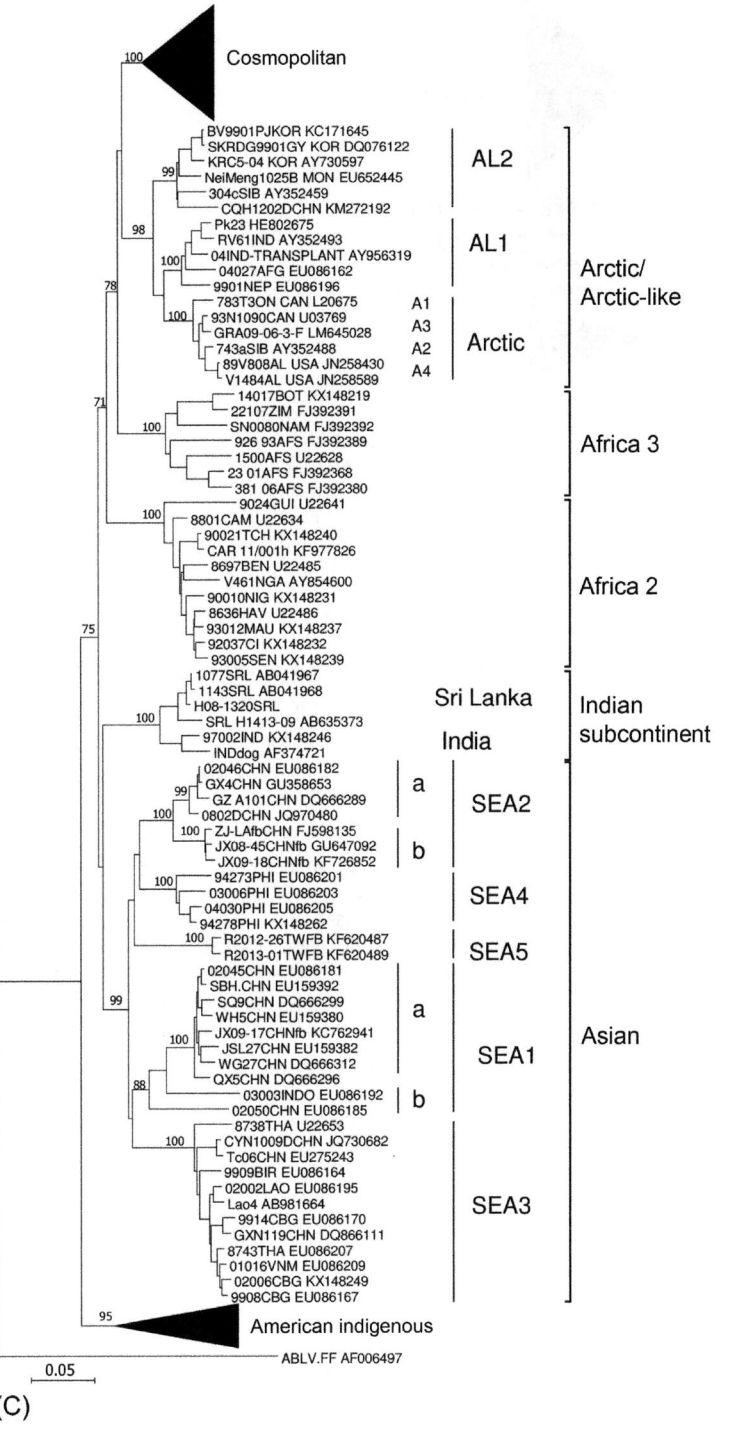

FIG.5.2, CONT'D (C) details of Africa 2, Africa 3, Arctic/Arctic-like, Asian and Indian subcontinent lineages;
(Continued)

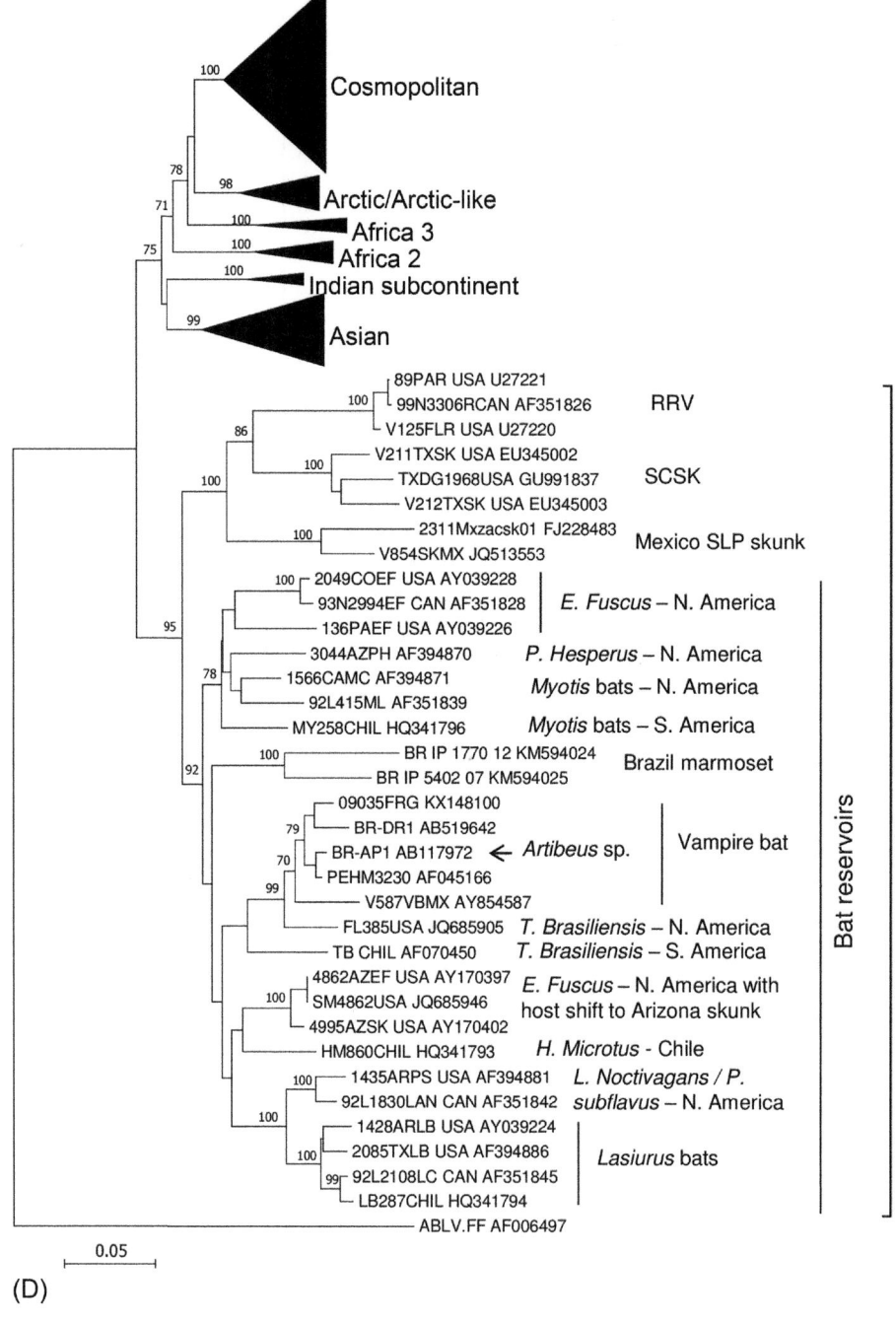

FIG.5.2, CONT'D (D) detail of American indigenous lineage.

following description of the global status of RABV molecular epidemiology for which space allows description of just major themes.

5.4.3.1 Cosmopolitan lineage

The geographically widely distributed cosmopolitan lineage is thought to have originated in Europe approximately 300 years ago. Based on the genetic relatedness of all members of this lineage, despite their geographical separation and association with multiple host species, it is widely accepted that this lineage was distributed to many parts of the world during colonial activities and subsequent mass human migrations. Due to the often lengthy incubation period of rabies, it was possible for infected animals, probably dogs, to accompany humans in their travels to several countries on different continents where viral transmission could initiate new maintenance cycles, either in dogs or other indigenous mammalian populations (Nadin-Davis & Bingham, 2004). Within this lineage, several distinct clades are recognized.

European dog rabies, well documented since the Middle Ages, was replaced in the early 20th century, following a host-switch event, by an outbreak in red foxes, which spread westwards from Eastern Europe following World War II. Genetic studies of viruses from this European fox clade identified several geographically restricted variants circulating in western (WE), central (CE), eastern (EE), and north-eastern (NEE) regions of Europe (Bourhy et al., 1999). Oral vaccination programs in several countries of western Europe have successfully eradicated the WE fox rabies variant (Cliquet & Aubert, 2004) and as several eastern European countries seek improved economic ties with the European Union, increased rabies surveillance and control in these countries is yielding an improved understanding of the disease situation (Müller et al., 2015). As a result, the persistence of several of these variants in foxes of Eastern Europe has been recognized. The WE variant persists in Bosnia-Herzegovina, Montenegro, and Slovenia (McElhinney et al., 2011), from where it entered Italy in 2008 (De Benedictis et al., 2009), while in the eastern Balkan states the EE variant predominates (McElhinney et al., 2011; Rihtarič, Hostnik, Grom, & Toplak, 2011); both types have been documented in Croatia (Lojkić et al., 2012). Many different variants, including EE and NEE as well as Russian variants, are known to have circulated in Romania, centrally located within the continent (Turcitu et al., 2010); southwards spread of several of these variants into Serbia and Bulgaria has been documented (Robardet et al., 2013) while a distinct type referred to as the Serbian fox variant has also been described (McElhinney et al., 2011). The EE variant has spread into the Republic of Macedonia (Picard-Meyer, Mrenoshki, et al., 2013) and subsequently into Greece (Tasioudi et al., 2014) in recent years. A fox rabies outbreak in Hungary in 2013–2014 was also linked to viruses spreading from Romania (Hornyák, Juhász, Forró, Kecskeméti, & Bányai, 2018). Additional variants of the cosmopolitan lineage, including possibly older now extinct dog-associated viruses, have also been observed in the Balkan region (McElhinney et al., 2011).

The NEE and CE variants continue to circulate in fox populations in Poland (Orłowska & Żmudziński, 2014) and the Ukraine (Picard-Meyer et al., 2012), while the NEE variant became established in raccoon dog populations in the three Baltic countries of Estonia, Latvia, and Lithuania (Bourhy et al., 1999; Metlin, Rybakov, Gruzdev, Neuvonen, & Huovilainen, 2007) following the introduction of this wild canid into Europe from east Asia (Kauhala & Kowalczyk, 2011). Efforts to eradicate this enzootic by ORV have met with some success (Robardet et al., 2016). In Russia, several variants of the cosmopolitan lineage occur

(Kuzmin et al., 2004); the most prevalent steppe RABV (CA1), reported in red and corsac foxes, circulates widely in central Russia and in many other countries situated in the steppe ecological zones having spread eastwards from its likely origins in the western part of the country. Another variant identified in the Tver region of Russia is similar to viruses from central Europe (CA3), cases of the NEE variant have been reported in Western Russia and in southern regions viruses related to those of the Middle East have been recovered (Deviatkin et al., 2017).

Except for a few studies, the rabies situation in the Middle East generally remains poorly documented due to a culture of neglect and ongoing conflict in several areas (Bannazadeh Baghi, Alinezhad, Kuzmin, & Rupprecht, 2018). Viruses closely related to those in Europe, designated as the Middle-Eastern variant, circulate in several countries and have been divided into four distinct types (Horton et al., 2015), which exhibit a complex pattern of regional spread. One of these appears to be restricted mostly to Turkey; the variants found here are distinct from those in the rest of Europe, likely due to the country's separation from the Balkan states to the west by the Balkan Mountains, which may pose a significant barrier to disease spread. Two distinct viral types, circulating in western and eastern regions of the country, are spread primarily by dogs, although a focus of fox-associated viruses has been recognized in a region of western Turkey, likely representing a host shift for that virus (Johnson et al., 2009). In Israel, jackals and foxes have served as reservoirs for several types of the Middle-Eastern variant (David et al., 2007) and despite control efforts the situation continues to emerge in the north of the country due to incursions of dog-associated variants phylogenetically closely related to Turkish isolates (David, Dveres, Yakobson, & Davidson, 2009). A mixture of variants (see CA2 and CA3) circulate in several countries such as Jordan, Iran, and Iraq and areas of the Caucasus central plains, including Azerbaijan and Georgia, while just one discrete recently emerged variant circulates in large sections of the Arabian Peninsula (Horton et al., 2015). While the Middle Eastern variant viruses have spread as far as the eastern border of Iran, viruses further east are generally represented by the Arctic-like group; indeed just two viruses of the Arctic-like group have been recovered from eastern border areas of Iran (Nadin-Davis, Simani, Armstrong, Fayaz, & Wandeler, 2003) and the area corresponding to 60 degrees longitude appears to clearly divide these two lineages (Horton et al., 2015).

In Asia, the steppe fox virus of the cosmopolitan lineage has spread as far east as the Russian-Mongolian border (Adelshin et al., 2015) and into Mongolia (Boldbaatar et al., 2010), Inner Mongolia, and the northwestern provinces of China (Feng et al., 2014). Although most cases of rabies in China are due to distinct Asian lineages, the cosmopolitan lineage has also been reported in dogs in several other provinces of China (Meng et al., 2011; Tao et al., 2013) where it is unclear if these cases are the result of historical introduction from other countries or the consequence of poorly inactivated vaccines.

The cosmopolitan lineage is also represented by many variants in the western hemisphere, including much of the American continent (Velasco-Villa, Mauldin, et al., 2017). Within the Caribbean, distinct viral types have been independently introduced into dogs on the island of Hispaniola, comprising both Haiti and the Dominican Republic, and into mongoose populations in Puerto Rico, Cuba, and Grenada (Nadin-Davis et al., 2006; Nadin-Davis, Velez, Malaga, & Wandeler, 2008; Seetahal et al., 2018; Zieger et al., 2014). In continental North America, skunks are associated with several distinct cosmopolitan variants, including the north central skunk strain (NCSK) that circulates in north central USA and western Canada, the California skunk (CASK) variant and the Mexican Baja California skunk (BCSK) variant (Davis, Nadin-Davis, Moore, & Hanlon, 2013). Two distinct viral variants have been associated with foxes in Texas and Arizona; while no fox variant cases have been recorded in Texas in recent years such variants have been identified in northern Mexico (Nadin-Davis & Loza-Rubio, 2006). Across Mexico, dog variants, which had circulated extensively in the past, have been virtually eliminated (Velasco-Villa, Escobar, et al., 2017) apart from small foci of cases in the southeastern states of Chiapas and Yucatan. Interestingly, some Yucatan cases were due to atypical viral variants, which clustered as an outlier to several North American skunk variants, suggesting the existence of an unidentified sylvatic reservoir in the region (Garcés-Ayala et al., 2017).

Throughout Central and South America, the cosmopolitan lineage has been represented by dog-associated variants established in the area since colonization by Spanish and English explorers (Velasco-Villa, Mauldin, et al., 2017). In Brazil, it was estimated that the currently circulating dog-associated viral variants were introduced in the late 19th or early 20th centuries during a time of high levels of immigration from Europe (Kobayashi et al., 2011). While extensive dog vaccination programs across much of Latin America have been very successful in reducing both dog and human cases of the disease (Velasco-Villa, Escobar, et al., 2017), it has become apparent that distinct viral variants are maintained in wildlife reservoirs (Favoretto et al., 2013). In northeastern Brazil, foxes harbor variants closely related to those which have circulated in dogs (Bernardi et al., 2005; Carnieli et al., 2009; Cordeiro et al., 2016); foxes have also been identified as a potential reservoir in Colombia (Páez, Saad, Nůňez, & Bóshell, 2005) and some of the recent cases of dog rabies reported in several countries, including Paraguay, Argentina, and Bolivia, may be due to spillover from wildlife reservoirs such as the crab-eating fox (Amarilla et al., 2018). These sylvatic reservoirs are garnering increased attention due to their ability to transmit the disease to both domestic animals and humans and potentially undermine dog rabies control efforts (Rocha, de Oliveira, Heinemann, & Gonçalves, 2017).

Another clade contained within the cosmopolitan lineage, Africa 1, is widely distributed throughout the continent and is currently divided into three groups 1a, 1b, and 1c (Troupin et al., 2016). Africa 1a has been identified in most countries of northern and eastern Africa with detailed studies performed on samples from Ethiopia (Johnson et al., 2010), Sudan (Marston et al., 2009), and Tunisia (Amouri et al., 2011). Canid isolates constituting Africa 1b have been recovered mostly from dogs in the Central African Republic, Kenya, Mozambique (Coetzer et al., 2017), Namibia, Tanzania (Brunker et al., 2015), Zaire, Zambia (Muleya et al., 2012), Botswana (Johnson, Letshwenyo, Baipoledi, Thobokwe, & Fooks, 2004), Zimbabwe (Coetzer, Gwenhure, Makaya, Markotter, & Nel, 2019), and South Africa (Sabeta, Bingham, & Nel, 2003). Members of both 1a and 1b groups were identified in Uganda, which may define a border zone between these types (Hirano, Itou, Shibuya, Kashiwazaki, & Sakai, 2010). The distinct Africa 1c group has been recovered only from the island of Madagascar (Troupin et al., 2016). This clear geographical separation of the Africa 1 subtypes has been complicated, however, with the apparent recent spread of Africa 1 viruses into west African countries such as Nigeria (Talbi et al., 2009) and Ghana (Hayman et al., 2011), where the Africa 2 lineage was thought to circulate exclusively (see later). Indeed, a recent examination of the situation in Cameroon found a mixture of Africa 1a, Africa 1b, and Africa 2 RABVs (Sadeuh-Mba, Momo, Besong, Loul, & Njouom, 2017).

While dogs are the major viral reservoir, in some jurisdictions certain wildlife species can maintain particular viral variants. In southern Africa, epizootics associated with bat-eared foxes (Sabeta, Mansfield, McElhinney, Fooks, & Nel, 2007) and jackals (Zulu, Sabeta, & Nel, 2009) have been described as well as an unusual herbivore host, the kudu antelope, for which there is now evidence that a jackal-associated variant underwent a host shift into the kudu population to initiate an independent transmission cycle (Mansfield, McElhinney, et al., 2006; Scott et al., 2013). Of some concern in Africa is the threat that rabies, including transmission of the disease to endangered wildlife species from domestic dogs, could have devastating effects on such populations; ongoing surveillance and in some cases vaccination of vulnerable wildlife species has been proposed to counter such threats (Hassel et al., 2018; Johnson et al., 2010; Sabeta et al., 2018).

An outlying clade of the cosmopolitan lineage, designated Africa 4 and represented by a small number of isolates recovered from Egypt and Israel, appears to be maintained by dog populations of the region (David et al., 2007; Troupin et al., 2016).

5.4.3.2 Africa 2

The Africa 2 lineage comprises a group of dog-associated viruses that now circulates widely across sub-Saharan western and central Africa (Troupin et al., 2016). It has been estimated that this lineage emerged within the last 200 years and has slowly spread east to west across the continent generating multiple regionally restricted variants in the process (Talbi et al., 2009). Detailed studies on the viruses circulating within specific countries, e.g., Burkina Faso (De Benedictis et al., 2010), Mali (Traoré et al., 2016), and Ghana (Hayman et al., 2011) have been described while the phylogeny and phylogeography of dog-associated Africa 1 and Africa 2 RABV groups across the continent have been reviewed (Talbi & Bourhy, 2011).

5.4.3.3 Africa 3

The distinct Africa 3 lineage consists of a group of viruses known to circulate in mongooses in South Africa, Botswana, and Zimbabwe (Johnson, Letshwenyo, Baipoledi, Thobokwe, & Fooks, 2004; Nel et al., 2005; Troupin et al., 2016), although its true range may extend to other parts of Africa (Nel et al., 2005). Although the yellow mongoose is the principal host for these viruses in South Africa, they are often also reported in various viverrid species (e.g., civets in Zimbabwe) that may contribute to virus maintenance (Sabeta et al., 2008). High levels of genetic heterogeneity and strong geographical partitioning of variants of this lineage have been described (Nel et al., 2005). It has been suggested that the mongoose and canid biotypes exhibit distinct evolutionary dynamics (Davis, Rambaut, Bourhy, & Holmes, 2007), a claim further supported by evidence that the Africa 3 lineage, together with the viruses associated with Asian ferret badgers, evolves 2–4 times more quickly than dog-associated variants, possible indicative of host adaptation by these viruses (Troupin et al., 2016).

5.4.3.4 Arctic/Arctic-like

This lineage, originally designated simply as the Arctic lineage, was historically considered to be limited to red and arctic fox populations of arctic and temperate regions of North America (Canada and Alaska, USA), Greenland and Russia, especially Siberia, with the occasional incursion into the remote Svalbard archipelago of Norway, probably by animals venturing over pack ice from Russia (Kuzmin et al., 2004; Mansfield, Racloz, et al., 2006; Mørk, Bohlin, Fuglei, Åsbakk, & Tryland, 2011; Mørk & Prestrud, 2004). The Arctic-like (AL) designation was added in recognition of its extensive geographical range into many areas south of the temperate zone (Kuzmin, Hughes, Botvinkin, Gribencha, & Rupprecht, 2008; Nadin-Davis, Turner, Paul, Madhusudana, & Wandeler, 2007) and it now appears that the lineage emerged somewhere within Asia before spreading northwards and westwards to yield the true Arctic clade, which probably emerged in the first half of the 20th century (Kuzmin, Hughes, et al., 2008; Nadin-Davis, Sheen, & Wandeler, 2012). Currently the entire lineage is divided into three main groupings, which have been designated somewhat differently by different investigators.

The true arctic clade is further divided into four distinct types, A-1 to A-4, that exhibit distinct temporal and spatial distributions (Kuzmin, Hughes, et al., 2008; Nadin-Davis, Sheen, & Wandeler, 2012). A1 has been recovered from southern parts of Canada only; A2 has been found in Alaska as well as northern parts of the European region of Russia and Franz Josef land (Deviatkin et al., 2017), A3 ranges across northern latitudes of North America and Greenland (Hanke et al., 2016), while A4 is restricted to the state of Alaska and segregates into two distinct clades based on geographical location (Hueffer & Murphy, 2018).

Another group, often referred to as arctic-like 1 (AL1), comprises isolates from Afghanistan, Pakistan, Iran, Nepal, Bhutan, Mongolia, Bangladesh, and most of India and further subtyping of this group into geographically localized clades has been described (Jamil et al., 2012; Kuzmin, Hughes, et al., 2008; Nadin-Davis, Sheen, & Wandeler, 2012; Pant et al., 2013; Reddy et al., 2018; Tenzin et al., 2010; Troupin et al., 2016). The last group, often designated as arctic-like 2 (AL2), circulates primarily in South Korea (Yang et al., 2011), Mongolia (Boldbaatar et al., 2010), and neighboring parts of Siberia and Russia (Deviatkin et al., 2017); this type was also present in Japan in the 1950s prior to rabies eradication from this country (Hatakeyama, Sadamasu, & Kai, 2011). All Arctic-like viruses circulate widely in dogs, but wild canids probably also help to maintain these viral populations; for example, raccoon dogs act as a viral reservoir in Korea and foxes have been proposed as a reservoir in India (Madhusudana, Mani, Ashwin, & Desai, 2013). In recent years, outbreaks of rabies due to the Arctic-like type have occurred in Inner Mongolia (Shao et al., 2010) as well as other regions of northern China (Li et al., 2016; Tao et al., 2015; Tao, Li, Guo, Yan, & Zhu, 2019) raising concerns about further expansion of this epizootic to other parts of the country.

5.4.3.5 Indian subcontinent

This distinctive lineage is found only in Sri Lanka (Nanayakkara, Smith, & Rupprecht, 2003) and limited parts of southern India (Reddy et al., 2014), probably due to human-mediated transfer of infected animals between these two areas. The dog is the main host with frequent spillover to other domestic animals, but wildlife reservoirs may also maintain distinct variants of this lineage in Sri Lanka (Matsumoto et al., 2011).

5.4.3.6 Asian

The Asian lineage, estimated to have emerged approximately 450 years ago, circulates across much of mainland and south-east Asia. Many studies of this lineage have been reported in recent years, but unfortunately no standard nomenclature has been adopted and only when careful comparisons are made is it evident that the main conclusions of all

TABLE 5.2 Various designations assigned to RABV lineages recovered in Asia.

Group designation (Troupin et al., 2016)	Group designation (Gong et al., 2010)	Group designation (Miao et al., 2018)	Description
SEA1a	Asian 1c	CHINA V	Chongquing/Ningxia region of China
SEA1b	Asian 1a Asian 1b	CHINA I	Southern and Eastern China Indonesia
SEA2a	Asian 2a1	CHINA II	Southern China
SEA2b	Asian 2a2	CHINA II	Southern China
SEA3	Asian 2c	CHINA VI	Indochina (Thailand, Laos, Cambodia, Vietnam, Burma) Yunnan dog
SEA4	Asian 2b		Philippines
SEA5			Taiwan ferret badger
Cosmopolitan		CHINA III	
Arctic-like		CHINA IV	

studies are indeed quite similar (Gong et al., 2010; Meng et al., 2011; Ming et al., 2010; Zhang et al., 2009). Table 5.2 indicates the correspondence of the different groupings most often employed and the following narrative refers to the typings described by Troupin et al (Troupin et al., 2016) in which the lineage is subdivided into five main branches designated SEA1 to SEA5.

In China, most viral isolates fall into two main groups, SEA1 and SEA2, each of which has been further subdivided (Zhang et al., 2009). SEA1b viruses make up a large percentage of the cases identified throughout the eastern provinces of China with regional variants emerging as the disease spreads (Gong et al., 2010), while the SEA1a subgroup is represented by smaller numbers of cases from localized areas of central China. SEA2, comprising subgroups 2a and 2b, is localized mostly to the southern provinces where it overlaps in range with SEA1. Detailed phylogenetic analyses of Chinese samples have identified a complex mix of regionally localized viral types as well as cocirculation of multiple types within an individual province, observations that support both local spread and frequent long-distance movement of viruses as a result of human-mediated transportation of diseased animals. In particular, the southern provinces of Guangxi and Yunnan have been identified as hot spots of rabies activity; their role as major trading hubs appears to have facilitated spread of Asia variants within China as well as providing a source of infection for other southeast Asian countries (Ming et al., 2010; Yao et al., 2015; Zhang et al., 2014).

While dogs and other domestic animals still comprise a large percentage of animal rabies cases in China, the ferret badger appears to constitute an emerging wildlife rabies reservoir (Miao et al., 2018). First identified in Jiangxi province as a RABV with several unique coding characteristics (Zhang et al., 2010; Zhao et al., 2014), isolates have now been recovered from ferret badgers in the neighboring Zhejiang province (Zhao, Zhao, Liu, Jiang, & Yang, 2019) and they cluster to both major viral clades circulating in the country, i.e., SEA1 and SEA2. It has been suggested that recent host shift events from the dog reservoir have initiated this new wildlife threat, which is becoming an increasingly significant public health problem (Miao et al., 2018; Zhao et al., 2019). Interestingly, the same species of ferret badger (*Melogale moschata*) has recently been identified as a rabies reservoir in Taiwan, an island previously considered for many decades to be rabies free (Chiou et al., 2014). Several cases have been identified since the first reported case in 2013 and phylogenetic studies place the Taiwan virus as a separate cluster (SEA5) within the Asian lineage (Miao et al., 2018). The Taiwan ferret badger virus has diverged into two types geographically separated east-west by a mountain range (Tsai et al., 2016) and it is believed to have existed on the island undetected for many decades (Lan et al., 2017; Lin et al., 2016).

There is now strong evidence that rabies spread from China to several other countries throughout the region during periods of extensive movement of Chinese migrants (Gong et al., 2010; Meng et al., 2011). Dog-associated viruses from Indonesia form a separate cluster within the SEA1b group (Troupin et al., 2016). Phylogeographic analysis has indicated that the virus was first introduced into Java from where it radiated outwards to several other islands of the archipelago where isolation has resulted in the emergence of a number of distinct types (Dibia et al., 2015; Susetya et al., 2008). The SEA4 group, probably introduced from China around the beginning of the 20th century, has been responsible for dog rabies throughout the Philippines (Meng et al., 2011) with subsequent division into three major clades circulating in distinct island groups following island to island spread (Saito et al., 2013; Tohma et al., 2014). Vaccination programs have had limited success in many parts of the country due to low vaccination coverage and continued opportunities for introduction of the disease into islands previously rabies-free (Tohma et al., 2016).

Samples from most other countries of south-east Asia, including Thailand, Malaysia, Vietnam, Cambodia, Myanmar (Burma), and Laos, have been grouped into the SEA3 group (Meng et al., 2011) with regional diversification in many areas. However, detailed analysis undertaken on the circulating viruses for some of these countries suggests that ongoing crossborder spread of the disease may be complicating this epidemiology. Studies of rabies viruses collected in Vietnam indicate that viruses from the southern part of the country cluster paraphyletically within the SEA2 group and that Philippine isolates (SEA4) form an internal clade to the Vietnamese samples, while in northern regions viruses of both SEA1 and SEA2 have been recovered (Nguyen et al., 2011). This suggests both a complex association of RABVs between the Philippines and Vietnam and ongoing spread of viruses across the northern Vietnamese/southern Chinese border. A study of viruses collected from two well-separated regions of Laos defined one large temporally evolving clade and two other separate clades strongly suggestive of multiple incursions from neighboring countries (Ahmed et al., 2015).

5.4.3.7 *American indigenous*

The American indigenous lineage, isolated only on the American continent, comprises viruses with a very distinct evolutionary origin compared to other RABVs. These viruses are found predominantly in many species of *Chiroptera* while a few distinctive strains are harbored by mesocarnivores, likely as a result of several independent host switch events from bat reservoirs. The latter include the south central skunk (SCSK) variant associated with skunks of the south central USA, another skunk-associated variant often referred to as the SLP skunk type found in the Sinaloa-Durango region of Mexico, and the North American raccoon rabies variant (RRV). In addition, a distinct RABV variant of this lineage is apparently associated with the white-tufted ear marmoset (*Callithrix jacchus*), a small diurnal primate indigenous to northeastern Brazil. Frequent human contact with this primate has resulted in several human deaths from rabies, a situation that represents an emerging public health issue (Favoretto, de Mattos, Morais, Alves Araújo, & de Mattos, 2001; Kotait et al., 2019). This primate-associated variant, which is most closely related to viruses harbored by *Lasiurus* and *Lasionycteris* bats of the Americas, appears to have evolved into multiple distinct types over several centuries but has only recently been recognized perhaps due to expanding human demographics in the area.

The RRV, first reported in Florida in the 1940s, was inadvertently introduced, through transportation of infected animals, into the mid-Atlantic region in the 1970s from where it spread rapidly throughout the eastern seaboard of the USA (Winkler & Jenkins, 1991). It spread northwards as far as the USA-Canada border and since 1999 has entered southern Canada on several occasions (Trewby et al., 2017). Extensive control efforts using primarily oral rabies vaccination (ORV) strategies have been undertaken in both countries with varying success (Rosatte et al., 2009; Slate et al., 2005). In light of the huge public and animal health issues raised by this particular variant due to the urban nature of its host, extensive sequence characterization of collections of RRV have been undertaken. In general agreement with the surveillance data, time-scaled phylogenetic studies have suggested that this variant emerged in the mid-1940s (Szanto et al., 2011) and was introduced into Virginia in 1973 with emergence of a number of distinct types as the epizootic progressed northwards (Biek, Henderson, Waller, Rupprecht, & Real, 2007). High-resolution phylogeographic studies have illustrated the impact of topographical features on variant evolution and dispersion and identified mechanisms of crossborder spread between Canada and the USA involving natural animal-to-animal transmission as well as probable anthropomorphic intervention (Nadin-Davis et al., 2018; Nadin-Davis, Buchanan, Nituch, & Fehlner-Gardiner, 2020; Trewby et al., 2017).

The SCSK variant probably emerged in Texas in the mid 19th century (Kuzmina et al., 2013) before spreading to many central US states and as far north as South Dakota (Davis et al., 2013). A comparison of the genetic diversity of this variant with that of the NCSK variant, with which its range overlaps in the states of Kansas, Nebraska, and South Dakota, has revealed some interesting differences in their molecular epidemiology and evolution (Barton, Gregory, Davis, Hanlon, & Wisely, 2010). Some RABVs from Mexican skunk specimens also fall into the American indigenous lineage but lie on branches distinct from those of the SCSK variant suggesting that the virus has been introduced into this host on multiple occasions (Davis et al., 2013; Velasco-Villa et al., 2005). Indeed, in Flagstaff, Arizona, a big brown bat RABV variant was maintained in skunks for several years, demonstrating the opportunity for emergence of new virus:host associations in this species (Kuzmin et al., 2012).

The role of vampire bats in harboring RABV in the Americas was first recognized on the island of Trinidad following outbreaks of rabies in cattle early in the 20th century (Seetahal et al., 2013). Indeed it is apparent that the introduction of large-scale cattle ranching activities over extensive areas of South America has facilitated substantial increases in populations of the common vampire bat (*Desmodus rotundus*) resulting in the spread of a RABV variant harbored by this species over much of its range (Johnson, Aréchiga-Ceballos, & Aguilar-Setien, 2014).

Currently, vampire bat rabies is distributed as far south as Argentina up through South and Central America into much of Mexico; although the vampire bat is the reservoir host, most reported cases are in domestic livestock species. Three antigenic variants, AgV3, AgV5, and AgV11, identified by the CDC MAb panel are associated with vampire bats, but all of these samples form a monophyletic clade with multiple regionally associated branches (Condori-Condori, Streicker, Cabezas-Sanchez, & Velasco-Villa, 2013; Ellison et al., 2014; Guarino et al., 2013; Macedo et al., 2010; Nadin-Davis & Loza-Rubio, 2006; Streicker et al., 2016; Torres et al., 2014; Velasco-Villa, Mauldin, et al., 2017). In Brazil, the AgV3 variant commonly found in *Desmodus rotundus* has also been detected in the hairy-legged vampire bat (*Diphylla ecaudata*) possibly as a result of interspecific transmission (Castilho, Carnieli Jr., et al., 2010). In both Brazil and Colombia, the isolation of vampire bat variants from frugivorous bats (*Artibeus* spp.) has suggested that these hosts support an independent transmission cycle (Calderón et al., 2019; Mochizuki et al., 2011). Beyond the huge economic and animal health costs, vampire bat rabies has become a serious public health issue in Amazonian regions causing several human rabies cases in Brazil (Barbosa et al., 2008) and Ecuador (Castilho, Carnieli, et al., 2010).

The association of RABVs with insectivorous bats of North America was first recorded in the 1950s (see Chapter 7), and subsequent surveillance and viral typing of isolates from across the region have identified many distinct variants that are associated with particular bat species. The species most frequently recorded as rabies positive include the big brown bat (*Eptesicus fuscus*), which harbors several distinct variants (Nadin-Davis, Feng, Mousse, Wandeler, & Aris-Brosou, 2010; Neubaum et al., 2008; Shankar et al., 2005), members of the *Lasiurus* genus, including hoary (*L. cinereus*), red (*L. borealis*), and northern yellow (*L. intermedius*) bats, each of which harbor distinct but closely related variants (Streicker et al., 2010), the silver-haired bat (*Lasionycteris noctivagans*), which together with the tricolored bat (*Perimyotis subflavus*), harbors a variant frequently implicated in human rabies cases (Messenger, Smith, & Rupprecht, 2002), and in the USA and Mexico, the Brazilian free-tailed bat (*Tadarida brasiliensis*) (Shankar et al., 2005; Streicker et al., 2010; Velasco-Villa, Mauldin, et al., 2017). Additional viral variants that are associated with more rarely reported species, for example the western pipistrelle (*Pipistrellus hesperus*) in the USA (Franka et al., 2006), have been documented. While bats of the *Myotis* genus are infrequently reported with rabies, several North American species harbor distinct viral types (Nadin-Davis, Alnabelseya, & Knowles, 2017; Streicker et al., 2010). Moreover, additional viral variants for which the reservoirs are yet to be determined may be harbored by other bat species (Aréchiga-Ceballos et al., 2010; Nadin-Davis & Loza-Rubio, 2006; Velasco-Villa et al., 2008). While control of bat rabies through human-mediated efforts remains problematic due to the aerial life style and feeding habits of these hosts, the decline in populations of certain bat species, attributable to the recent white-nose syndrome outbreak (Blehert, 2012), may have a notable impact on bat rabies epidemiology in eastern North America.

As information on the viruses associated with insectivorous bats of North America grew, increased surveillance in South America has identified large numbers of rabid insectivorous bats some of which represent reservoir species (Escobar, Peterson, Favi, Yung, & Medina-Vogel, 2015). In the most southern parts of the continent, in Chile and Argentina, variants associated with small numbers of specimens of the *Myotis*, *Eptesicus*, *Histiotus*, and *Lasiurus* genera have been recovered, but the Brazilian free-tailed bat, *Tadarida brasiliensis*, is the principal bat rabies reservoir (Escobar et al., 2013; Pinero et al., 2012). On the basis of limited numbers of isolates in most cases, many of these RABV variants appear to be widely disseminated throughout much of Latin America (Escobar et al., 2015). In Brazil, viral variants associated with *Eptesicus furinalis*, *Molossus molossus*, and species of the genera *Histiotus*, *Nyctinomops*, *Myotis*, and *Lasiurus* have been identified (Albas et al., 2011; Bernardi et al., 2005; Kobayashi et al., 2007; Menozzi, de Novaes Oliveira, Paiz, Richini-Pereira, & Langoni, 2017; Oliveira et al., 2010; Queiroz et al., 2012). In addition, single reports of rabies in other species have appeared (Castilho et al., 2008), but the reservoir role of these species requires further study since interspecific transmission and/or difficulties in bat speciation can confound the identification of true reservoir hosts.

5.5 Emerging trends

5.5.1 The importance of viral typing

Supplementation of rabies surveillance data with viral typing and molecular epidemiology provides for better understanding of the association of a particular viral variant with a specific host. This becomes especially important when new lyssavirus species are identified since identification of the reservoir host can impact risk assessments following a potential exposure. In addition, beyond increased fundamental knowledge of the range and spread of a particular variant, this information is of critical importance to control programs to ensure appropriate targeting of the disease reservoir. Viral typing also provides an important means of identifying incursions of variants into new areas; for example, dogs imported from countries where rabies is enzootic have occasionally brought the disease to western countries (Hercules et al., 2018; Ribadeau-Dumas et al., 2016) and timely intervention is needed to prevent establishment of a new disease focus and reversal of past gains in disease elimination (Velasco-Villa, Escobar, et al., 2017). Viral sequence analysis can also reveal unexpected complexity of outbreak situations, as for recent cases in Tibet where multiple viral lineages are responsible for reemergence of the disease, thus indicating the need for a multipronged approach to control (Tao, Li, Wang, et al., 2019).

5.5.2 Understanding the factors impacting viral spread

Many rabies studies increasingly employ a combination of tools, including surveillance, viral phylogeny linked to spatial data and landscape topography (phylogeography), host ecology and population genetics, and epidemiological modeling to better understand viral spread and hence infer the best approaches for control (Dellicour et al., 2017; Dellicour, Rose, & Pybus, 2016). Examples of such investigations range from very localized studies in an African urban setting (Coetzer et al., 2019) to broader-scale studies of a country or region (Brunker et al., 2015; Cleaveland et al., 2014; Tian et al., 2018). It is apparent that in many parts of Africa while geopolitical boundaries often limit rabies virus spread, human-mediated dispersal of the disease occurs often and is responsible for an increasingly complex pattern of viral admixture at both national and local levels (Brunker et al., 2015; Hayman et al., 2011;

Talbi et al., 2010). Moreover, knowledge of topographical barriers to rabies spread can be beneficial in maximizing the cost effectiveness of control campaigns (Brunker, Lemey, et al., 2018).

5.5.3 Role of the host

When speciation of diseased animals is in question, either due to submission of partial or badly damaged carcasses, this can usually be addressed by sequencing of certain host loci; portions of the mitochondrial genome, such as the cytochrome b (Carnieli Jr, Castilho, Oliveira, Brandão, & Batista, 2016) or cytochrome oxidase subunit I (Nadin-Davis, Guerrero, Knowles, & Feng, 2012) genes are frequent targets. Indeed, due to their small size and subtle morphological differences, this approach is especially useful for confirming bat reservoir species, submissions of which may not reach the laboratory in good condition (Nadin-Davis, Alnabelseya, & Knowles, 2017; Streicker et al., 2010) and such studies have helped to provide some clues as to the evolutionary paths by which so many distinct viral types have emerged in American chiropteran hosts. Distinct viral variants are associated with Brazilian free-tailed bats in the southern and northern hemispheres (Velasco-Villa, Mauldin, et al., 2017) and a similar situation exists for variants of *Myotis* bats (Nadin-Davis, Alnabelseya, & Knowles, 2017) indicating that multiple introductions of the virus into members of these genera have occurred. It has been postulated that migratory species such as members of the *Lasiurus* genus may be central to the dispersion of viral variants to other genera followed by independent evolution of these variants into distinct types (Faria, Suchard, Rambaut, Streicker, & Lemey, 2013; Streicker, Altizer, Velasco-Villa, & Rupprecht, 2012). Host genetics and feeding habits appear to impact interspecific transmission patterns of many insectivorous bat-associated rabies viruses of North America (Streicker et al., 2010). Geographic factors have almost certainly also had an impact; for example, the Rocky Mountain range in western Canada appears to have been a significant barrier to the spread of variants associated with *Eptesicus fuscus* and *Myotis* species (Nadin-Davis et al., 2010; Nadin-Davis, Alnabelseya, & Knowles, 2017).

Sequencing of mitochondrial loci or other variable loci such as microsatellites can also be used to study the population structure of the host; correlation of the association of viral subtypes with host subpopulations can provide insight into mechanisms of disease persistence and spread. Such studies have again been especially useful in examining patterns of rabies spread in bat populations due to their aerial lifestyle (Harris et al., 2008; Smith et al., 2011). Genetic studies of vampire bats and their viral variants in South America suggest that viral spread occurs predominantly through dispersal of males, while climactic conditions and topographical features such as mountain ranges, which restrict host movements, can result in the establishment of regionally localized viral variants (Kobayashi et al., 2008; Streicker et al., 2016; Torres et al., 2014).

Similar studies on North American raccoons suggested that in the host's northern range, rivers impede host movement and disease spread while further south other topographical features such as valleys and ridges appeared to have minimal effect (Cullingham et al., 2008; Cullingham, Kyle, Pond, Rees, & White, 2009; Root et al., 2009). Analysis of skunk microsatellite loci showed limited host population substructure in the USA, thereby allowing overlap of the ranges of both the NCSK and SCSK variants (Barton et al., 2010). In Europe, studies on red fox populations have identified a correlation between distinct subpopulations of this host and the spread of two different foci of rabies into Italy (Zecchin et al., 2019). In Alaska, with the help of microsatellite analysis to explore the relative importance of two fox hosts in maintenance of the arctic RABV variant, it was concluded that the arctic fox is probably the primary maintenance host, while the red fox serves mainly as a spillover species in this environment (Goldsmith et al., 2016). An elegant study of arctic fox rabies in Greenland employed HTS techniques to sequence both the virus and selected host loci together (Hanke et al., 2016), thus circumventing the need for separate sequencing efforts, an approach that may become more widely used in the future.

5.5.4 What controls host shift events?

It remains noteworthy that, of all the lyssaviruses, RABV appears to be the only member with the capability of undergoing relatively frequent host shifts (Marston et al., 2018). However, the molecular mechanisms associated with such host shifts remain unclear. It has often been assumed that specific amino acid residues in one or more viral proteins dictate the virus' fitness for maintenance in a particular host, and that the codons for such residues would be expected to be subject to positive selection. While a study on bat-associated RABVs identified some residues which appeared to be under positive selection (Streicker et al., 2010), examination of consensus sequence datasets involving host shifts between nonflying mammals have consistently failed to identify residues under positive selection (Kuzmin et al., 2012; Marston, Horton, et al., 2017). However, a rare mutation at the subconsensus level may have facilitated a skunk to fox host shift by the CASK variant (Borucki et al., 2013). It has been proposed that preadaptation of RABVs to many carnivore hosts may facilitate host shifts (Kuzmin et al., 2012; Marston et al., 2018), while other factors such as host ecology and interspecies contact rates have a significant impact on such events in bat populations (Streicker et al., 2010). It remains to be determined whether more subtle features of the virus such as codon usage (He et al., 2017; Zhang et al., 2018) or regulation by host noncoding RNAs (Zhao et al., 2018) could impact virus-host interactions. Another more controversial theory is that recombination plays a role in lyssavirus evolution and host adaptation. The recombination detection program (RDP) (Martin, Murrell, Golden, Khoosal, & Muhire, 2015) has been used to identify some potential instances of recombination among lyssaviruses (Deviatkin & Lukashev, 2018; He et al., 2012), and it has been suggested that a recombination event between bat- and canid-associated rabies viruses of the Americas resulted in the emergence of the American indigenous variants associated with terrestrial wildlife, i.e., the SCSK and RRV variants of the USA (Ding, Xu, Sun, He, & He, 2017). Such an event, if proven, must be considered quite exceptional. The molecular process of lyssavirus genome replication and its site in the central nervous system dictate that such events would be extremely rare, consistent with minimal evidence for recombination from the vast majority of phylogenetic studies. Finally, through an ancestral state reconstruction of the genus based on a time-scaled phylogeny, it has been suggested that the lyssavirus genus emerged in the Palearctic approximately 27,000 years ago and subsequently spread to the rest of the world in three main dispersal events (Hayman, Fooks, Marston, & Garcia-R, 2016). This contradicts another theory postulating the emergence of all lyssaviruses from Africa based on the high diversity of the genus on this continent (Rupprecht, Kuzmin, & Meslin, 2017).

5.5.5 Impact of climate change

Another factor that will increasingly impact the spread of lyssaviruses is climate change. As global temperatures rise, the range of many of the hosts that harbor these viruses will inevitably be affected. In particular, concern has been expressed regarding the strong likelihood that vampire bats will expand their range northwards into southern regions of the USA (Hayes & Piaggio, 2018). Similarly, climatic changes that modify the distribution of insect species will likely have significant effects on the range of insectivorous bats in many parts of the world and potentially expand the range of their associated lyssaviruses. In Alaska, it is speculated that expansion of the red fox range and displacement of arctic foxes to more northern regions might reduce arctic fox populations in areas of human habitation and thereby reduce the incidence of reported rabies cases (Hueffer & Murphy, 2018). However, the red fox is well known as an efficient rabies reservoir in its own right and such predictions may not extend throughout its range. It should be recognized that such predictions about the impacts of climate change on disease distribution remain speculative at this time.

5.6 Concluding remarks

Molecular epidemiology of lyssaviruses provides important public health information on the source of human rabies infections, especially in countries where the disease is rare and may be acquired either from indigenous sources (Ma et al., 2018) or as a result of travel to countries with endemic rabies (Malerczyk, DeTora, & Gniel, 2011). Such cases have illustrated the long incubation periods sometimes observed in humans (Boland et al., 2014; Johnson, Fooks, & McColl, 2008). It has also facilitated the identification of new disease foci in wildlife (van Thiel et al., 2009), played a crucial role in better understanding disease emergence in countries previously considered rabies-free (De Benedictis et al., 2009; Lan et al., 2017), and improved understanding of the factors that contribute to endemic rabies (Cleaveland et al., 2014). Improved ability to fully characterize the genomes of these viruses, together with development of new tools to interrogate such sequence data, allows greater insight into their epidemiology. Such methods have also been applied to the vaccines employed in rabies control. Whole genome sequencing has enabled a re-evaluation of the developmental history of many of the vaccine strains used for ORV of wildlife (Geue et al., 2008). HTS methods have provided unparalleled opportunities to probe variability in vaccine stocks (Höper et al., 2015), an approach that may revolutionize quality control protocols for many vaccines, and enabled exploration of changes to the virus that have accompanied rare vaccine-induced rabies cases following ORV campaigns (Pfaff et al., 2018).

Despite success in the elimination of terrestrial rabies in Western Europe and its control in large parts of the Americas, the ability of the rabies virus to persist in a variety of wildlife carnivores dictates that ongoing vigilance is needed to protect human health from exposure and prevent re-introduction of the disease into dog populations. Illegal movement of unvaccinated animals (Hercules et al., 2018; Johnson, Freuling, Horton, Müller, & Fooks,

2011), unsustainable domestic animal vaccination programs, especially in the developing world, mass migration of humans and their domestic animals due to conflict and ongoing emergence of novel viral variants in wildlife (Favoretto et al., 2001) may all contribute to renewed disease challenges. Indeed if we accept recorded history as accurately describing the existence of rabies in ancient times, together with estimated dates of the emergence of current rabies virus lineages, it is apparent that rabies viruses have emerged more than once and that emergence of novel lyssavirus variants is a continuing threat. Ongoing surveillance and molecular phylogenetic investigation continue to extend the known diversity of the *Lyssavirus* genus around the world and with this increased knowledge a better understanding of the history of the genus and the evolutionary mechanisms in play that impact viral-host interactions should be achieved. Such insight will hopefully provide us with improved capabilities to further control these important pathogens.

Acknowledgments

I thank C. Fehlner-Gardiner and G. Mitchell for reviewing a first draft of this work.

References

Adelshin, R. V., Melnikova, O. V., Trushina, Y. N., Botvinkin, A. D., Borisova, T. I., Andaev, E. I., ... Balakhonov, S. V. (2015). A new outbreak of fox rabies at the Russian-Mongolian border. *Virologica Sinica*, *30*(4), 313–315. https://doi.org/10.1007/s12250-015-3609-0.

Ahmed, K., Phommachanh, P., Vorachith, P., Matsumoto, T., Lamaningao, P., Mori, D., ... Nishizono, A. (2015). Molecular epidemiology of rabies viruses circulating in two rabies endemic provinces of Laos, 2011–2012: Regional diversity in Southeast Asia. *PLoS Neglected Tropical Diseases*, *9*(3), e0003645.

Albas, A., Campos, A. C., Araujo, D. B., Rodrigues, C. S., Sodre, M. M., Durigon, E. L., & Favoretto, S. R. (2011). Molecular characterization of rabies virus isolated from non-haematophagous bats in Brazil. *Revista da Sociedade de Medicina Tropical*, *44*, 678–683.

Amarasinghe, G. K., Ayllón, M. A., Bào, Y., Basler, C. F., Bavari, S., Blasdell, K. R., ... Kuhn, J. H. (2019). Taxonomy of the order Mononegavirales: update 2019. *Archives of Virology*, *164*(7), 1967–1980. https://doi.org/10.1007/s00705-019-04247-4.

Amarilla, A. C. F., Pompei, J. C. A., Araujo, D. B., Vázquez, F. A., Galeano, R. R., Delgado, L. M., ... Favoretto, S. R. (2018). Re-emergence of rabies virus maintained by canid populations in Paraguay. *Zoonoses and Public Health*, *65*(1), 222–226. https://doi.org/10.1111/zph.12392.

Amouri, I. K., Kharmachi, H., Djebbi, A., Saadi, M., Hogga, N., Zakour, L. B., & Ghram, A. (2011). Molecular characterization of rabies virus isolated from dogs in Tunisia: Evidence of two phylogenetic variants. *Virus Research*, *158*(1), 246–250. https://doi.org/10.1016/j.virusres.2010.10.015.

Annand, E. J., & Reid, P. A. (2014). Clinical review of two fatal equine cases of infection with the insectivorous bat strain of Australian bat lyssavirus. *Australian Veterinary Journal*, *92*(9), 324–332.

Aréchiga Ceballos, N., Morón, S. V., Berciano, J. M., Nicolás, O., López, C. A., Juste, J., ... Echevarría, J. E. (2013). Novel Lyssavirus in Bat, Spain. *Emerging Infectious Disease Journal*, *19*(5), 793–795. https://doi.org/10.3201/eid1905.121071.

Aréchiga-Ceballos, N., Velasco-Villa, A., Shi, M., Flores-Chávez, S., Barrón, B., Cuevas-Domínguez, E., ... Aguilar-Setién, A. (2010). New rabies virus variant found during an epizootic in white-nosed coatis from the Yucatan Peninsula. *Epidemiology and Infection*, *138*(11), 1586–1589. https://doi.org/10.1017/S0950268810000762.

Arguin, P. M., Murray-Lillibridge, K., Miranda, M. E. G., Smith, J. S., Calaor, A. B., & Rupprecht, C. E. (2002). Serologic evidence of lyssavirus infections among bats, the Philippines. *Emerging Infectious Diseases*, *8*, 258–262.

Badrane, H., Bahloul, C., Perrin, P., & Tordo, N. (2001). Evidence of two *Lyssavirus* phylogroups with distinct pathogenicity and immunogenicity. *Journal of Virology*, 75(7), 3268–3276. https://doi.org/10.1128/jvi.75.7.3268-3276.2001.

Bannazadeh Baghi, H., Alinezhad, F., Kuzmin, I., & Rupprecht, C. E. (2018). A perspective on rabies in the Middle East—beyond neglect. *Veterinary Sciences*, 5(3), 67.

Banyard, A. C., Selden, D., Wu, G., Thorne, L., Jennings, D., Marston, D., … Fooks, A. R. (2018). Isolation, antigenicity and immunogenicity of Lleida bat lyssavirus. *Journal of General Virology*, 99(12), 1590–1599. https://doi.org/10.1099/jgv.0.001068.

Barbosa, T. F. S., Medeiros, D. B. d. A., Travassos da Rosa, E. S., Casseb, L. M. N., Medeiros, R., Pereira, A. d. S., & Nunes, M. R. (2008). Molecular epidemiology of rabies virus isolated from different sources during a bat-transmitted human outbreak occurring in Augusto Correa municipality, Brazilian Amazon. *Virology*, 370(2), 228–236. https://doi.org/10.1016/j.virol.2007.10.005.

Barton, H. D., Gregory, A. J., Davis, R., Hanlon, C. A., & Wisely, S. M. (2010). Contrasting landscape epidemiology of two sympatric rabies virus strains. *Molecular Ecology*, 19(13), 2725–2738. https://doi.org/10.1111/j.1365-294X.2010.04668.x.

Bauldauf, S. L. (2003). Phylogeny for the faint of heart: a tutorial. *Trends in Genetics*, 19, 345–351.

Bernardi, F., Nadin-Davis, S. A., Wandeler, A. I., Armstrong, J., Gomes, A. A. B., Lima, F. S., … Ito, F. H. (2005). Antigenic and genetic characterization of rabies viruses isolated from domestic and wild animals of Brazil identifies the hoary fox as a rabies reservoir. *Journal of General Virology*, 86, 3153–3162.

Biek, R., Henderson, C. I., Waller, L. A., Rupprecht, C. E., & Real, L. A. (2007). A high-resolution genetic signature of demographic and spatial expansion in epizootic rabies virus. *Proceedings of the National Academy of Sciences USA*, 104, 7993–7998.

Blehert, D. S. (2012). Fungal disease and the developing story of bat white-nose syndrome. *PLoS Pathogens*, 8(7), e1002779.

Boland, T. A., McGuone, D., Jindal, J., Rocha, M., Cumming, M., Rupprecht, C. E., … Rosenthal, E. S. (2014). Phylogenetic and epidemiologic evidence of multiyear incubation in human rabies. *Annals of Neurology*, 75(1), 155–160. https://doi.org/10.1002/ana.24016.

Boldbaatar, B., Inoue, S., Tuya, N., Dulam, P., Batchuluun, D., Sugiura, N., … Yamada, A. (2010). Molecular epidmeiology of rabies virus in Mongolia, 2005–2008. *Japanese Journal of Infectious Diseases*, 63, 358–363.

Borucki, M. K., Chen-Harris, H., Lao, V., Vanier, G., Wadford, D. A., Messenger, S., & Allen, J. E. (2013). Ultra-deep sequencing of intra-host rabies virus populations during cross-species transmission. *PLoS Neglected Tropical Diseases*, 7(11), e2555.

Bouckaert, R., Heled, J., Kühnert, D., Vaughan, T., Wu, C.-H., Xie, D., … Drummond, A. J. (2014). BEAST 2: A software platform for Bayesian evolutionary analysis. *PLoS Computational Biology*, 10(4), e1003537. https://doi.org/10.1371/journal.pcbi.1003537.

Bourhy, H., Kissi, B., Audry, L., Smreczak, M., Sadkowska-Todys, M., Kulonen, K., … Holmes, E. C. (1999). Ecology and evolution of rabies virus in Europe. *Journal of General Virology*, 80(10), 2545–2557. https://doi.org/10.1099/0022-1317-80-10-2545.

Bourhy, H., Reynes, J.-M., Dunham, E. J., Dacheux, L., Larrous, F., Thi Que Huong, V., … Holmes, E. C. (2008). The origin and phylogeography of dog rabies virus. *Journal of General Virology*, 89, 2673–2681.

Brunker, K., Lemey, P., Marston, D. A., Fooks, A. R., Lugelo, A., Ngeleja, C., … Biek, R. (2018). Landscape attributes governing local transmission of an endemic zoonosis: Rabies virus in domestic dogs. *Molecular Ecology*, 27(3), 773–788. https://doi.org/10.1111/mec.14470.

Brunker, K., Marston, D. A., Horton, D. L., Cleaveland, S., Fooks, A. R., Kazwala, R., … Hampson, K. (2015). Elucidating the phylodynamics of endemic rabies virus in eastern Africa using whole-genome sequencing. *Virus Evolution*, 1(1), 1–11.

Brunker, K., Nadin-Davis, S., & Biek, R. (2018). Genomic sequencing, evolution and molecular epidemiology of rabies virus. *In Vol. 37(2) Revue Scientifique Et Technique De L'Office International Des Epizooties: Rabies* (pp. 401–408). Paris, France: World Organisation for Animal Health.

Calderón, A., Guzmán, C., Mattar, S., Rodríguez, V., Acosta, A., & Martínez, C. (2019). Frugivorous bats in the Colombian Caribbean region are reservoirs of the rabies virus. *Annals of Clinical Microbiology and Antimicrobials*, 18(1), 11. https://doi.org/10.1186/s12941-019-0308-y.

Carnieli, P., Jr., Castilho, J. G., Oliveira, R. N., Brandão, P. E., & Batista, H. B. C. R. (2016). Identification of different species of mammalians involved in zoonoses as reservoirs or hosts by sequencing of the mitochondrial DNA cytochrome B gene. *Annual Research and Review in Biology, 10*(1), ARRB.25230.

Carnieli, P., Castilho, J. G., Fahl, W. d. O., Véras, N. M. C., Carrieri, M. L., & Kotait, I. (2009). Molecular characterization of Rabies Virus isolates from dogs and crab-eating foxes in Northeastern Brazil. *Virus Research, 141*(1), 81–89. https://doi.org/10.1016/j.virusres.2008.12.015.

Castilho, J. G., Canello, F. M., Scheffer, K. C., Achkar, S. M., Carrieri, M. L., & Kotait, I. (2008). Antigenic and genetic characterization of the first rabies virus isolated from the bat Eumops perotis in Brazil. *Revista do Instituto de Medicina Tropical de São Paulo, 50*, 95–99.

Castilho, J. G., Carnieli, P., Jr., Oliveira, R. N., Fahl, W. O., Cavalcante, R., Santana, A. A., ... Kotait, I. (2010). A comparative study of rabies virus isolates from hematophagous bats in Brazil. *Journal of Wildlife Diseases, 46*(4), 1335–1339. https://doi.org/10.7589/0090-3558-46.4.1335.

Castilho, J. G., Carnieli, P., Durymanova, E. A., Fahl, W. O., Oliveira, R. N., Macedo, C. I., ... Kotait, I. (2010). Human rabies transmitted by vampire bats: Antigenic and genetic characterization of rabies virus isolates from the Amazon region (Brazil and Ecuador). *Virus Research, 153*(1), 100–105. https://doi.org/10.1016/j.virusres.2010.07.012.

Chaves, L. B., Achkar, S. M., Rodrigues da Silva, A. C., Caporale, G. M. M., Cruz, P. S., Batista, A. M., ... de Gaspari, E. (2015). Monoclonal antibodies for characterization of rabies virus isolated from non-hematophagous bats in Brazil. *The Journal of Infection in Developing Countries, 9*(11), 1238–1249.

Chiou, H.-Y., Hsieh, C.-H., Jeng, C.-R., Chan, F.-T., EWang, H.-Y., & Pang, V. F. (2014). Molecular characterization of cryptically circulating rabies virus from ferret badgers, Taiwan. *Emerging Infectious Diseases, 20*(5), 790–798.

Choudhuri, S. (2014). Phylogenetic analysis. In *Bioinformatics for beginners: Genes, Genomes, Molecular Evolution, Databases and Analytical Tools* (1st ed., pp. 209–218): San Diego, USA: Academic Press.

Cleaveland, S., Beyer, H., Hampson, K., Haydon, D., Lankester, F., Lembo, T., ... Townsend, S. (2014). The changing landscape of rabies epidemiology and control. *Onderstepoort Journal of Veterinary Research, 81*(2). https://doi.org/10.4102/ojvr.v81i2.731.

Cliquet, F., & Aubert, M. (2004). Elimination of terrestrial rabies in western European countries. In *Paper presented at the Control of Infectious Animal Diseases by Vaccination, Buenos Aires*.

Coetzer, A., Anahory, I., Dias, P. T., Sabeta, C. T., Scott, T. P., Markotter, W., & Nel, L. H. (2017). Enhanced diagnosis of rabies and molecular evidence for the transboundary spread of the disease in Mozambique. *Journal of the South African Veterinary Association, 88*, a1397.

Coetzer, A., Gwenhure, L., Makaya, P., Markotter, W., & Nel, L. (2019). Epidemiological aspects of the persistent transmission of rabies during an outbreak (2010–2017) in Harare, Zimbabwe. *PLoS One, 14*(1), e0210018. https://doi.org/10.1371/journal.pone.0210018.

Colombi, D., Serra-Cobo, J., Métras, R., Apolloni, A., Poletto, C., López-Roig, M., ... Colizza, V. (2019). Mechanisms for lyssavirus persistence in non-synanthropic bats in Europe: insights from a modeling study. *Scientific Reports, 9*(1), 537. https://doi.org/10.1038/s41598-018-36485-y.

Condori-Condori, R. E., Streicker, D. G., Cabezas-Sanchez, C., & Velasco-Villa, A. (2013). Enzootic and epizootic rabies associated with vampire bats, Peru. *Emerging Infectious Diseases, 19*(9), 1463–1469.

Cordeiro, R. A., Duarte, N. F. H., Rolim, B. N., Soares Júnior, F. A., Franco, I. C. F., Ferrer, L. L., ... Sidrim, J. J. C. (2016). The importance of wild canids in the epidemiology of rabies in Northeast Brazil: A retrospective study. *Zoonoses and Public Health, 63*(6), 486–493. https://doi.org/10.1111/zph.12253.

Cullingham, C. I., Kyle, C. J., Pond, B. A., Rees, E. E., & White, B. N. (2009). Differential permeability of rivers to raccoon gene flow corresponds to rabies incidence in Ontario, Canada. *Molecular Ecology, 18*(1), 43–53. https://doi.org/10.1111/j.1365-294X.2008.03989.x.

Cullingham, C. I., Pond, B. A., Kyle, C. J., Rees, E. E., Rosatte, R. C., & White, B. N. (2008). Combining direct and indirect genetic methods to estimate dispersal for informing wildlife disease management decisions. *Molecular Ecology, 17*(22), 4874–4886. https://doi.org/10.1111/j.1365-294X.2008.03956.x.

David, D., Dveres, N., Yakobson, B. A., & Davidson, I. (2009). Emergence of dog rabies in the Northern region of Israel. *Epidemiology and Infection, 137*(4), 544–548. https://doi.org/10.1017/S0950268808001180.

David, D., Hughes, G. J., Yakobson, B. A., Davidson, I., Un, H., Aylan, O., ... Rupprecht, C. E. (2007). Identification of novel canine rabies virus clades in the Middle East and North Africa. *Journal of General Virology, 88*(3), 967–980. https://doi.org/10.1099/vir.0.82352-0.

Davis, P. L., Rambaut, A., Bourhy, H., & Holmes, E. C. (2007). The evolutionary dynamics of canid and mongoose rabies virus in southern Africa. *Archives of Virology, 152*(7), 1251–1258. https://doi.org/10.1007/s00705-007-0962-9.

Davis, R., Nadin-Davis, S. A., Moore, M., & Hanlon, C. (2013). Genetic characterization and phylogenetic analysis of skunk-associated rabies viruses in North America with special emphasis on the central plains. *Virus Research, 174*, 27–36.

De Benedictis, P., Capua, I., Mutinelli, F., Wernig, J. M., Arič, T., & Hostnik, P. (2009). Update on fox rabies in Italy and Slovenia. *WHO Rabies Bulletin Europe, 33*(1), 5–7.

De Benedictis, P., De Battisti, C., Dacheux, L., Marciano, S., Ormelli, S., Salomoni, A., ... Cattoli, G. (2011). Lyssavirus detection and typing using pyrosequencing. *Journal of Clinical Microbiology, 49*, 1932–1938.

De Benedictis, P., Sow, A., Fusaro, A., Veggiato, C., Talbi, C., Kaboré, A., ... Capua, I. (2010). Phylogenetic analysis of rabies viruses from Burkina Faso, 2007. *Zoonoses and Public Health, 57*(7–8), e42–e46. https://doi.org/10.1111/j.1863-2378.2009.01291.x.

Dellicour, S., Rose, R., Faria, N. R., Vieira, L. F. P., Bourhy, H., Gilbert, M., ... Pybus, O. G. (2017). Using viral gene sequences to compare and explain the heterogeneous spatial dynamics of virus epidemics. *Molecular Biology and Evolution, 34*(10), 2563–2571. https://doi.org/10.1093/molbev/msx176.

Dellicour, S., Rose, R., & Pybus, O. G. (2016). Explaining the geographic spread of emerging epidemics: a framework for comparing viral phylogenies and environmental landscape data. *BMC Bioinformatics, 17*(1), 82. https://doi.org/10.1186/s12859-016-0924-x.

Delmas, O., Holmes, E. C., Talbi, C., Larrous, F., Dacheux, L., Bouchier, C., & Bourhy, H. (2008). Genomic diversity and evolution of the Lyssaviruses. *PLoS One, 3*(4), e2057.

Deviatkin, A. A., & Lukashev, A. N. (2018). Recombination in the rabies virus and other lyssaviruses. *Infection, Genetics and Evolution, 60*, 97–102. https://doi.org/10.1016/j.meegid.2018.02.026.

Deviatkin, A. A., Lukashev, A. N., Poleshchuk, E. M., Dedkov, V. G., Tkachev, S. E., Sidorov, G. N., ... Shipulin, G. A. (2017). The phylodynamics of the rabies virus in the Russian Federation. *PLoS One, 12*(2), e0171855. https://doi.org/10.1371/journal.pone.0171855.

Dibia, I. N., Sumiarto, B., Susetya, H., Putra, A. A. G., Scott-Orr, H., & Mahardika, G. N. (2015). Phylogeography of the current rabies viruses in Indonesia. *Journal of Veterinary Science, 16*(4), 459–466.

Ding, N. -Z., Xu, D. -S., Sun, Y. -Y., He, H. -B., & He, C. -Q. (2017). A permanent host shift of rabies virus from Chiroptera to Carnivora associated with recombination. *Scientific Reports, 7*(1), 289. https://doi.org/10.1038/s41598-017-00395-2.

Drummond, A. J., Suchard, M. A., Xie, D., & Rambaut, A. (2012). Bayesian phylogenetics with BEAUti and the BEAST 1.7. *Molecular and Biological Evolution, 29*(8), 1969–1973. https://doi.org/10.1093/molbev/mss075.

Dzikwi, A. A., Kuzmin, I. V., Umoh, J. U., Kwaga, J. K. P., Ahmad, A. A., & Rupprecht, C. E. (2010). Evidence of Lagos bat virus circulation among Nigerian fruit bats. *Journal of Wildlife Diseases, 46*(1), 267–271. https://doi.org/10.7589/0090-3558-46.1.267.

Eggerbauer, E., Troupin, C., Passior, K., Pfaff, F., Höper, D., Neubauer-Juric, A., ... Freuling, C. M. (2017). Chapter Eight—The recently discovered Bokeloh Bat Lyssavirus: insights into its genetic heterogeneity and spatial distribution in Europe and the population genetics of its primary host. In M. Beer & D. Höper (Eds.), *Vol. 99. Advances in Virus Research* (pp. 199–232): Academic Press.

Ellison, J. A., Gilbert, A. T., Recuenco, S., Moran, D., Alvarez, D. A., Kuzmina, N. A., ... Rupprecht, C. E. (2014). Bat rabies in Guatemala. *PLoS Neglected Tropical Diseases, 8*(7), e3070. https://doi.org/10.1371/journal.pntd.0003070.

Escobar, L. E., Peterson, A. T., Favi, M., Yung, V., & Medina-Vogel, G. (2015). Bat-borne rabies in Latin America. *Revista do Instituto de Medicina Tropical de São Paulo, 57*, 63–72.

Escobar, L. E., Peterson, A. T., Favi, M., Yung, V., Pons, D. J., & Medina-Vogel, G. (2013). Ecology and geography of transmission of two bat-borne Rabies Lineages in Chile. *PLoS Neglected Tropical Diseases, 7*(12), e2577. https://doi.org/10.1371/journal.pntd.0002577.

Evans, J. S., Wu, G., Selden, D., Buczkowski, H., Thorne, L., Fooks, A. R., & Banyard, A. C. (2018). Utilisation of Chimeric Lyssaviruses to assess vaccine protection against highly divergent Lyssaviruses. *Viruses, 10*(3), 130.

Faria, N. R., Suchard, M. A., Rambaut, A., Streicker, D. G., & Lemey, P. (2013). Simultaneously reconstructing viral cross-species transmission history and identifying the underlying constraints. *Philosophical Transactions of the Royal Society B, 368*, 20120196.

Favoretto, S. R., de Mattos, C. C., de Mattos, C. A., Campos, A. C. A., Sacramento, D. R. V., & Durigon, E. L. (2013). The emergence of wildlife species as a source of human rabies infection in Brazil. *Epidemiology and Infection, 141*(7), 1552–1561. https://doi.org/10.1017/S0950268813000198.

Favoretto, S. R., de Mattos, C. C., Morais, N. B., Alves Araújo, F. A., & de Mattos, C. A. (2001). Rabies in marmosets (Callithrix jacchus), Ceará, Brazil. *Emerging Infectious Diseases, 7*, 1062–1065.

Fehlner-Gardiner, C., Nadin-Davis, S., Armstrong, J., Muldoon, F., Bachmann, P., & Wandeler, A. (2008). ERA vaccine-derived cases of rabies in wildlife and domestic animals in Ontario, Canada, 1989-2004. *Journal of Wildlife Diseases*, *44*(1), 71–85.

Feng, Y., Wang, W., Guo, J., Alatengheli, U., Li, Y., Yang, G., ... Tu, C. (2014). Disease outbreaks caused by steppe-type rabies viruses in China. *Epidemiology and Infection*, *143*(6), 1287–1291. https://doi.org/10.1017/S0950268814001952.

Field, H. E. (2018). Evidence of Australian bat lyssavirus infection in diverse Australian bat taxa. *Zoonoses and Public Health*, *65*, 742–748.

Fooks, A. (2004). The challenge of new and emerging lyssaviruses. *Expert Review of Vaccines*, *3*(4), 333–336. https://doi.org/10.1586/14760584.3.4.333.

Fooks, A. R., Cliquet, F., Finke, S., Freuling, C. M., Hemachudha, T., Mani, R. S., ... Banyard, A. C. (2017). Rabies. *Nature Reviews. Disease Primers*, *3*, 17091.

Fooks, A. R., McElhinney, L. M., Pounder, D. J., Finnegan, C. J., Mansfield, K. L., Johnson, N., ... Nathwani, D. (2003). Case report: Isolation of a European bat lyssavirus type 2a from a fatal human case of rabies encephalitis. *Journal of Medical Virology*, *71*(2), 281–289. https://doi.org/10.1002/jmv.10481.

Francis, J. R., Nourse, C., Vaska, V. L., Calvert, S., Northill, J. A., McCall, B., & Mattke, A. C. (2014). Australian bat lyssavirus in a child: the first reported case. *Pediatrics*, *133*(4), e1063–e1067.

Franka, R., Constantine, D. G., Kuzmin, I., Velasco-Villa, A., Reeder, S. A., Streicker, D., ... Rupprecht, C. E. (2006). A new phylogenetic lineage of Rabies virus associated with western pipistrelle bats (Pipistrellus hesperus). *Journal of General Virology*, *87*(8), 2309–2321. https://doi.org/10.1099/vir.0.81822-0.

Fraser, G. C., Hooper, P. T., Lunt, R. A., Gould, A. R., Gleeson, L. J., Hyatt, A. D., ... Kattenbelt, J. A. (1996). Encephalitis caused by a lyssavirus in fruit bats in Australia. *Emerging Infectious Diseases*, *2*, 327–331.

Freuling, C. M., Beer, M., Conraths, F. J., Finke, S., Hoffmann, B., Keller, B., ... Muller, T. (2011). Novel lyssavirus in Natterer's bat, Germany. *Emerging Infectious Diseases*, *17*(8), 1519–1522.

Garcés-Ayala, F., Aréchiga-Ceballos, N., Ortiz-Alcántara, J. M., González-Durán, E., Pérez-Agüeros, S. I., Méndez-Tenorio, A., ... Ramírez-González, J. E. (2017). Molecular characterization of atypical antigenic variants of canine rabies virus reveals its reintroduction by wildlife vectors in southeastern Mexico. *Archives of Virology*, *162*(12), 3629–3637. https://doi.org/10.1007/s00705-017-3529-4.

Geue, L., Schares, S., Schnick, C., Kliemt, J., Beckert, A., Freuling, C., ... Mueller, T. (2008). Genetic characterisation of attenuated SAD rabies virus strains used for oral vaccintion of wildlife. *Vaccine*, *26*, 3227–3235.

Glenn, T. C. (2011). Field guide to next-generation DNA sequencers. *Molecular Ecology Resources*, *11*, 759–769.

Goldsmith, E. W., Renshaw, B., Clement, C. J., Himschoot, E. A., Hundertmark, K. J., & Hueffer, K. (2016). Population structure of two rabies hosts relative to the known distribution of rabies virus variants in Alaska. *Molecular Ecology*, *25*(3), 675–688. https://doi.org/10.1111/mec.13509.

Gong, W., Jiang, Y., Za, Y., Zeng, Z., Shao, M., Fan, J., ... Tu, C. (2010). Temporal and spatial dynamics of rabies viruses in China and Southeast Asia. *Virus Research*, *150*(1), 111–118. https://doi.org/10.1016/j.virusres.2010.02.019.

Gould, A. R., Hyatt, A. D., Lunt, R., Kattenbelt, J. A., Hengstberger, S., & Blacksell, S. D. (1998). Characterisation of a novel lyssavirus isolated from Pteropid bats in Australia. *Virus Research*, *54*, 165–187.

Gould, A. R., Kattenbelt, J. A., Gumley, S. G., & Lunt, R. A. (2002). Characterisation of an Australian bat lyssavirus variant isolated from an insectivorous bat. *Virus Research*, *89*(1), 1–28. https://doi.org/10.1016/S0168-1702(02)00056-4.

Grubaugh, N. D., Ladner, J. T., Lemey, P., Pybus, O. G., Rambaut, A., Holmes, E. C., & Andersen, K. G. (2019). Tracking virus outbreaks in the twenty-first century. *Nature Microbiology*, *4*(1), 10–19. https://doi.org/10.1038/s41564-018-0296-2.

Guarino, H., Castilho, J. G., Souto, J., Oliveira, R. d. N., Carrieri, M. L., & Kotait, I. (2013). Antigenic and genetic characterization of rabies virus isolates from Uruguay. *Virus Research*, *173*(2), 415–420. https://doi.org/10.1016/j.virusres.2012.12.013.

Gunawardena, P. S., Marston, D. A., Ellis, R. J., Wise, E. L., Karawita, A. C., Breed, A. C., ... Fooks, A. R. (2016). Lyssavirus in Indian Flying Foxes, Sri Lanka. *Emerging Infectious Diseases*, *22*(8), 1456–1459. https://doi.org/10.3201/eid2208.151986.

Guyatt, K. J., Twin, J., Davis, P., Holmes, E. C., Smith, G. A., Smith, I. L., ... Young, P. L. (2003). A molecular epidemiological study of Australian bat lyssavirus. *Journal of General Virology*, *84*(2), 485–496. https://doi.org/10.1099/vir.0.18652-0.

Hanke, D., Freuling, C. M., Fischer, S., Hueffer, K., Hundertmark, K., Nadin-Davis, S., ... Hoper, D. (2016). Saptio-temporal analysis of the genetic diversity of Arctic rabies viruses and their reservoir hosts in Greenland. *PLoS Neglected Tropical Diseases, 10*(7), e0004779.

Hanlon, C. A. (2013). Rabies in terrestrial animals. In A. C. Jackson (Ed.), *Rabies: Scientific basis of the disease and its management* (3rd ed., pp. 179–213). San Diego: Academic Press.

Hanlon, C. A., Kuzmin, I. V., Blanton, J. D., Weldon, W. C., Manangan, J. S., & Rupprecht, C. E. (2005). Efficacy of rabies biologics against new lyssaviruses from Eurasia. *Virus Research, 111*, 44–54.

Harris, S. L., Johnson, N., Brookes, S. M., Hutson, A. M., Fooks, A. R., & Jones, G. (2008). The application of genetic markers for EBLV surveillance in European bat species. *Developments in Biologicals (Basel), 131*, 347–363.

Hassel, R., Vos, A., Clausen, P., Moore, S., van der Westhuizen, J., Khaiseb, S., ... Müller, T. (2018). Experimental screening studies on rabies virus transmission and oral rabies vaccination of the Greater Kudu (Tragelaphus strepsiceros). *Scientific Reports, 8*(1), 16599. https://doi.org/10.1038/s41598-018-34985-5.

Hatakeyama, K., Sadamasu, K., & Kai, A. (2011). Phylogenetic analysis of rabies viruses isolates from animals in Tokyo in the 1950s. *The Journal of the Japanese Association for Infectious Diseases, 85*, 238–243.

Hayes, M. A., & Piaggio, A. J. (2018). Assessing the potential impacts of a changing climate on the distribution of a rabies virus vector. *PLoS One, 13*(2), e0192887. https://doi.org/10.1371/journal.pone.0192887.

Hayman, D. T. S., Fooks, A. R., Horton, D., Suu-Ire, R., Breed, A. C., Cunningham, A. A., & Wood, J. L. N. (2008). Antibodies against Lagos Bat Virus in Megachiroptera from West Africa. *Emerging Infectious Disease Journal, 14*(6), 926–928. https://doi.org/10.3201/eid1406.071421.

Hayman, D. T. S., Fooks, A. R., Marston, D. A., & Garcia-R, J. C. (2016). The Global Phylogeography of Lyssaviruses—Challenging the 'Out of Africa' Hypothesis. *PLoS Neglected Tropical Diseases, 10*(12), e0005266. https://doi.org/10.1371/journal.pntd.0005266.

Hayman, D. T. S., Johnson, N., Horton, D. L., Hedge, J., Wakeley, P. R., Banyard, A. C., ... Fooks, A. (2011). Evolutionary history of rabies in Ghana. *PLoS Neglected Tropical Diseases, 5*(4), e1001. https://doi.org/10.1371/journal.pntd.0001001.

He, C. -Q., Meng, S. -L., Yan, H. -Y., Ding, N. -Z., He, H. -B., Yan, J. -X., & Xu, G. -L. (2012). Isolation and identification of a novel rabies virus lineage in China with natural recombinant nucleoprotein gene. *PLoS One, 7*(12), e49992. https://doi.org/10.1371/journal.pone.0049992.

He, W., Zhang, H. L., Zhang, Y. Z., Wang, R., Lu, S., Ji, Y., ... Su, S. (2017). Codon usage bias in the N gene of rabies virus. *Infection, Genetics and Evolution, 54*, 458–465. https://doi.org/10.1016/j.meegid.2017.08.012.

Hercules, Y., Bryant, N. J., Wallace, R. M., Nelson, R., Palumbo, G., Williams, J. N., ... Brown, C. (2018). Rabies in a dog imported from Egypt—Connecticut, 2017. *Morbidity and Mortality Weekly Review, 67*(50), 1388–1391.

Hirano, S., Itou, T., Shibuya, H., Kashiwazaki, Y., & Sakai, T. (2010). Molecular epidemiology of rabies virus isolates in Uganda. *Virus Research, 147*(1), 135–138. https://doi.org/10.1016/j.virusres.2009.10.003.

Holmes, E. C., Woelk, C. H., Kassis, R., & Bourhy, H. (2002). Genetic constraints and the adaptive evolution of rabies virus in nature. *Virology, 292*, 247–257.

Hooper, P. T., Lunt, R. A., Gould, A. R., Samaratunga, H., Hyatt, A. D., Gleeson, L. J., ... Murray, P. K. (1997). A new lyssavirus—the first endemic rabies-related virus recognized in Australia. *Bulletin du Institut Pasteur, 95*, 209–218.

Höper, D., Freuling, C. M., Müller, T., Hanke, D., von Messling, V., Duchow, K., ... Mettenleiter, T. C. (2015). High definition viral vaccine strain identity and stability testing using full-genome population data—The next generation of vaccine quality control. *Vaccine, 33*(43), 5829–5837. https://doi.org/10.1016/j.vaccine.2015.08.091.

Hornyák, Á., Juhász, T., Forró, B., Kecskeméti, S., & Bányai, K. (2018). Resurgence of rabies in Hungary during 2013–2014: An attempt to track the origin of identified strains. *Transboundary and Emerging Diseases, 65*(1), e14–e24. https://doi.org/10.1111/tbed.12658.

Horton, D. L., McElhinney, L., Freuling, C. M., Marston, D. A., Banyard, A. C., Goharriz, H., ... Fooks, A. R. (2015). Complex epidemiology of a zoonotic disease in a culturally diverse region: Phylogeography of rabies virus in the Middle East. *PLoS Neglected Tropical Diseases, 9*(3), e0003569.

Hu, S. -C., Hsu, C. -L., Lee, M. -S., Tu, Y. -C., Chang, J. -C., Wu, C. -H., ... Hsu, W. -C. (2018). Lyssavirus in Japanese Pipistrelle, Taiwan. *Emerging Infectious Diseases, 24*(4), 782–785. https://doi.org/10.3201/eid2404.171696.

Hueffer, K., & Murphy, M. (2018). Rabies in Alaska, from the past to an uncertain future. *International Journal of Circumpolar Health, 77*(1), 1475185. https://doi.org/10.1080/22423982.2018.1475185.

Jamil, K. M., Ahmed, K., Hossain, M., Matsumoto, T., Ali, M. A., Hossain, S., ... Nishizono, A. (2012). Arctic-like Rabies Virus, Bangladesh. *Emerging Infectious Diseases*, *18*(12), 2021–2024. https://doi.org/10.3201/eid1812.120061.

Johnson, N., Aréchiga-Ceballos, N., & Aguilar-Setien, A. (2014). Vampire Bat Rabies: Ecology, Epidemiology and Control. *Viruses*, *6*(5), 1911.

Johnson, N., Fooks, A., & McColl, K. (2008). Reexamination of human rabies case with long incubation, Australia. *Emerging Infectious Diseases*, *14*, 1950–1951.

Johnson, N., Freuling, C., Horton, D., Müller, T., & Fooks, A. R. (2011). Imported rabies, European Union and Switzerland, 2001–2010. *Emerging Infectious Diseases*, *17*, 753–754.

Johnson, N., Freuling, C., Marston, D. A., Tordo, N., Fooks, A. R., & Müller, T. (2007). Identification of European bat lyssavirus isolates with short genomic insertions. *Virus Research*, *128*, 140–143.

Johnson, N., Letshwenyo, M., Baipoledi, E. K., Thobokwe, G., & Fooks, A. R. (2004). Molecular epidemiology of rabies in Botswana: a comparison between antibody typing and nucleotide sequence phylogeny. *Veterinary Microbiology*, *101*(1), 31–38. https://doi.org/10.1016/j.vetmic.2004.03.007.

Johnson, N., Mansfield, K. L., Marston, D. A., Wilson, C., Goddard, T., Selden, D., ... Fooks, A. R. (2010). A new outbreak of rabies in rare Ethiopian wolves (Canis simensis). *Archives of Virology*, *155*(7), 1175–1177. https://doi.org/10.1007/s00705-010-0689-x.

Johnson, N., Un, H., Fooks, A. R., Freuling, C., Müller, T., Aylan, O., & Vos, A. (2009). Rabies epidemiology and control in Turkey: Past and present. *Epidemiology and Infection*, *138*(3), 305–312. https://doi.org/10.1017/S0950268809990963.

Kauhala, K., & Kowalczyk, R. (2011). Invasion of the raccoon dog *Nyctereutes procyonoides* in Europe: history of colonization, features behind its success, and threats to native fauna. *Current Zoology*, *57*, 584–598.

Kgaladi, J., Wright, N., Coertse, J., Markotter, W., Marston, D., Fooks, A. R., ... Nel, L. H. (2013). Diversity and Epidemiology of Mokola Virus. *PLoS Neglected Tropical Diseases*. *7*(10), e2511. https://doi.org/10.1371/journal.pntd.0002511.

Kilianski, A., Roth, P. A., Liem, A. T., Hill, J. M., Willis, K. L., Rossmaier, R. D., ... Rosenzweig, C. N. (2016). Use of unamplified RNA/cDNA-hybrid nanopore sequencing for rapid detection and characterization of RNA viruses. *Emerging Infectious Diseases*, *22*(8), 1448–1451. https://doi.org/10.3201/eid2208.160270.

Kobayashi, Y., Sato, G., Kato, M., Itou, T., Cunha, E. M. S., Silva, M. V., ... Sakai, T. (2007). Genetic diversity of bat rabies viruses in Brazil. *Archives of Virology*, *152*, 1995–2004.

Kobayashi, Y., Sato, G., Mochizuki, N., Hirano, S., Itou, T., Carvalho, A. A. B., ... Sakai, T. (2008). Molecular and geographic analyses of vampire bat-transmitted cattle rabies in central Brazil. *BMC Veterinary Research*, *4*(1), 44. https://doi.org/10.1186/1746-6148-4-44.

Kobayashi, Y., Suzuki, Y., Itou, T., Ito, F. H., Sakai, T., & Gojobori, T. (2011). Evolutionary history of dog rabies in Brazil. *Journal of General Virology*, *92*(1), 85–90. https://doi.org/10.1099/vir.0.026468-0.

Kotait, I., Oliveira, R. N., Carrieri, M. L., Castilho, J. G., Macedo, C. I., Pereira, P. M. C., ... Rupprecht, C. E. (2019). Non-human primates as a reservoir for rabies virus in Brazil. *Zoonoses and Public Health*, *66*(1), 47–59. https://doi.org/10.1111/zph.12527.

Kumar, S., Stecher, G., Li, M., Knyaz, C., & Tamura, K. (2018). MEGA X: Molecular Evolutionary Genetics Analysis across Computing Platforms. *Molecular Biology and Evolution*, *35*(6), 1547–1549. https://doi.org/10.1093/molbev/msy096.

Kuzmin, I. V., Botvinkin, A. D., McElhinney, L. M., Smith, J. S., Orciari, L. A., Hughes, G. J., ... Rupprecht, C. E. (2004). Molecular epidemiology of terrestrial rabies in the former Soviet Union. *Journal of Wildlife Diseases*, *40*(4), 617–631. https://doi.org/10.7589/0090-3558-40.4.617.

Kuzmin, I. V., Hughes, G. J., Botvinkin, A. D., Gribencha, S. G., & Rupprecht, C. E. (2008). Arctic and Arctic-like rabies viruses: distribution, phylogeny and evolutionary history. *Epidemiology and Infection*, *136*(4), 509–519. https://doi.org/10.1017/S095026880700903X.

Kuzmin, I. V., Hughes, G. J., Botvinkin, A. D., Orciari, L. A., & Rupprecht, C. E. (2005). Phylogenetic relationships of Irkut and West Caucasian bat viruses within the *Lyssavirus* genus and suggested quantitative criteria based on the N gene sequence for lyssavirus genotype definition. *Virus Research*, *111*, 28–43.

Kuzmin, I. V., Mayer, A. E., Niezgoda, M., Markotter, W., Agwanda, B., Breiman, R. F., & Rupprecht, C. E. (2010). Shimoni bat virus, a new representative of the Lyssavirus genus. *Virus Research*, *149*(2), 197–210. https://doi.org/10.1016/j.virusres.2010.01.018.

Kuzmin, I. V., Niezgoda, M., Carroll, D. S., Keeler, N., Hossain, M. J., Breiman, R. F., ... Rupprecht, C. E. (2006). Lyssavirus Surveillance in Bats, Bangladesh. *Emerging Infectious Disease journal, 12*(3), 486. https://doi.org/10.3201/eid1203.050333.

Kuzmin, I. V., Niezgoda, M., Franka, R., Agwanda, B., Markotter, W., Beagley, J. C., ... Rupprecht, C. E. (2008a). Lagos Bat Virus in Kenya. *Journal of Clinical Microbiology, 46*(4), 1451–1461. https://doi.org/10.1128/jcm.00016-08.

Kuzmin, I. V., Niezgoda, M., Franka, R., Agwanda, B., Markotter, W., Beagley, J. C., ... Rupprecht, C. E. (2008b). Possible emergence of West Caucasian bat virus in Africa. *Emerging Infectious Disease journal, 14*(12), 1887. https://doi.org/10.3201/eid1412.080750.

Kuzmin, I. V., Orciari, L. A., Arai, Y. T., Smith, J. S., Hanlon, C. A., Kameoka, Y., & Rupprecht, C. E. (2003). Bat lyssaviruses (Aravan and Khujand) from Central Asia: Phylogenetic relationships according to N, P and G gene sequences. *Virus Research, 97*(2), 65–79. https://doi.org/10.1016/S0168-1702(03)00217-X.

Kuzmin, I. V., Shi, M., Orciari, L. A., Yager, P. A., Velasco-Villa, A., Kuzmina, N. A., ... Rupprecht, C. E. (2012). Molecular inferences suggest multiple host shifts of rabies viruses from bats to mesocarniovores in Arizona during 2001–2009. *PLoS Pathogens, 8*(6), e1002786.

Kuzmin, I. V., Turmelle, A. S., Agwanda, B., Markotter, W., Niezgoda, M., Breiman, R. F., & Rupprecht, C. E. (2011). Commerson's Leaf-Nosed Bat (Hipposideros commersoni) is the Likely Reservoir of Shimoni Bat Virus. *Vector Borne and Zoonotic Diseases, 11*(11), 1465–1470. https://doi.org/10.1089/vbz.2011.0663.

Kuzmin, I. V., Wu, X., Tordo, N., & Rupprecht, C. E. (2008). Complete genomes of Aravan, Khujand, Irkut and West Caucasian bat viruses, with special attention to the polymerase gene and non-coding regions. *Virus Research, 136* (1-2), 81–90.

Kuzmina, N. A., Lemey, P., Kuzmin, I. V., Mayes, B. C., Ellison, J. A., Orciari, L. A., ... Rupprecht, C. E. (2013). The Phylogeography and Spatiotemporal Spread of South-Central Skunk Rabies Virus. *PLoS One, 8*(12), e82348. https://doi.org/10.1371/journal.pone.0082348.

Lam, T. T. -Y., Hon, C. -C., & Tang, J. W. (2010). Use of phylogenetics in the molecular epidemiology and evolutionary studies of viral infections. *Critical Reviews in Clinical Laboratory Sciences, 47*(1), 5–49. https://doi.org/10.3109/10408361003633318.

Lan, Y. -C., Wen, T. -H., Chang, C. -C., Liu, H. -F., Lee, P. -F., Huang, C. -Y., ... Chen, Y. -M. A. (2017). Indigenous Wildlife Rabies in Taiwan: Ferret Badgers, a Long Term Terrestrial Reservoir. *BioMed Research International, 2017*, 6. https://doi.org/10.1155/2017/5491640.

Lauring, A. S., & Andino, R. (2010). Quasispecies theory and the behavior of RNA viruses. *PLoS Pathogens, 6*(7), e1001005. https://doi.org/10.1371/journal.ppat.1001005.

Leonova, G. N., Belikov, S. I., Kondratov, I. G., Krylova, N. V., Pavlenko, E. V., Romanova, E. V., ... Petukhova, S. A. (2009). A fatal case of bat lyssavirus infection in Primorye Territory of the Russian Far East. *Rabies Bulletin Europe, 33*(4), 5–7.

Leopardi, S., Priori, P., Zecchin, B., Poglayen, G., Trevisiol, K., Lelli, D., ... De Benedictis, P. (2018). Active and passive surveillance for bat lyssaviruses in Italy revealed serological evidence for their circulation in three bat species. *Epidemiology and Infection. 147*, e63. https://doi.org/10.1017/S0950268818003072.

Leyrat, C., Ribeiro, E. A., Jr., Gerard, F. C. A., Ivanov, I., Rigrok, R. W. H., & Jamin, M. (2011). Structure, interactions with host cell and functions of rhabdovirus phosphoprotein. *Future Virology, 6*(4), 465–481.

Li, H., Guo, Z. Y., Zhang, J., Tao, X. Y., Zhu, W. Y., Tang, Q., & Liu, H. T. (2016). Whole genome sequencing and comparisons of different Chinese rabies virus lineages including the first complete genome of an arctic-like strain in China. *Biomedical and Environmental Sciences, 29*(5), 340–346. https://doi.org/10.3967/bes2016.044.

Lin, Y. -C., Chu, P. -Y., Chang, M. -Y., Hsiao, K. -L., Lin, J. -H., & Liu, H. -F. (2016). Spatial temporal dynamics and molecular evolution of re-emerging rabies virus in Taiwan. *International Journal of Molecular Sciences, 17*(3), 392.

Liu, Y., Zhang, S., Zhao, J., Zhang, F., & Hu, R. (2013). Isolation of Irkut virus from a *Murina leucogaster* bat in China. *PLoS Neglected Tropical Diseases, 7*(3), e2097.

Lojkić, I., Cac, Z., Bedeković, T., Lemo, N., Brstilo, M., Muller, T., & Freuling, C. M. (2012). Diversity of currently circulating rabies virus strains in Croatia. *Berliner und Münchener Tierärztliche Wochenschrift, 125*(5-6), 249–254.

Lumlertdacha, B., Boongird, K., Wanghongsa, S., Wacharapluesadee, S., Chanhome, L., Khawplod, P., ... Rupprecht, C. (2005). Survey for bat lyssaviruses, Thailand. *Emerging Infectious Diseases, 11*, 232–236.

Ma, X., Monroe, B. P., Cleaton, J. M., Orciari, L. A., Li, Y., Kirby, J. D., ... Blanton, J. D. (2018). Rabies surveillance in the United States during 2017. *Journal of the American Veterinary Medical Association, 253*(12), 1555–1568. https://doi.org/10.2460/javma.253.12.1555.

Macedo, C. I., Carnieli, P., Jr., Fahl, W. O., Lima, J. Y. O., Oliveira, R. N., Achkar, S. M., ... Kotait, I. (2010). Genetic characterization of rabies virus isolated from bovines and equines between 2007 and 2008, in the States of São Paulo and Minas Gerais. *Revista da Sociedade Brasileira de Medicina Tropical*, *43*, 116–120.

Madhusudana, S. N., Mani, R., Ashwin, Y. B., & Desai, A. (2013). Rabid Fox Bites and Human Rabies in a Village Community in Southern India: Epidemiological and Laboratory Investigations, Management and Follow-Up. *Vector Borne and Zoonotic Diseases*, *13*(5), 324–329. https://doi.org/10.1089/vbz.2012.1146.

Malerczyk, C., DeTora, L., & Gniel, D. (2011). Imported human rabies cases in Europe, the United States, and Japan, 1990 to 2010. *Journal of Travel Medicine*, *18*, 402–407.

Mansfield, K. L., McElhinney, L., Hübschle, O., Mettler, F., Sabeta, C., Nel, L. H., & Fooks, A. R. (2006). A molecular epidemiological study of rabies epizootics in kudu (Tragelaphus strepsiceros) in Namibia. *BMC Veterinary Research*, *2*(1), 2. https://doi.org/10.1186/1746-6148-2-2.

Mansfield, K. L., Racloz, V., McElhinney, L. M., Marston, D. A., Johnson, N., Rønsholt, L., ... Fooks, A. R. (2006). Molecular epidemiological study of Arctic rabies virus isolates from Greenland and comparison with isolates from throughout the Arctic and Baltic regions. *Virus Research*, *116*(1), 1–10. https://doi.org/10.1016/j.virusres.2005.08.007.

Mardis, E. R. (2008). Next-Generation DNA sequencing methods. *Annual Review of Genomics and Human Genetics*, *9*, 387–402.

Markotter, W., Kuzmin, I., Rupprecht, C. E., & Nel, L. H. (2008). Phylogeny of Lagos bat virus: Challenges for lyssavirus taxonomy. *Virus Research*, *135*(1), 10–21. https://doi.org/10.1016/j.virusres.2008.02.001.

Marston, D. A., Banyard, A. C., McElhinney, L. M., Freuling, C. M., Finke, S., de Lamballerie, X., ... Fooks, A. R. (2018). The lyssavirus host-specificity conundrum—rabies virus—the exception not the rule. *Current Opinion in Virology*, *28*, 68–73. https://doi.org/10.1016/j.coviro.2017.11.007.

Marston, D. A., Ellis, R. J., Horton, D. L., Kuzmin, I. V., Wise, E. L., McElhinney, L. M., ... Fooks, A. R. (2012). Complete Genome Sequence of Ikoma Lyssavirus. *Journal of Virology*, *86*(18), 10242–10243. https://doi.org/10.1128/jvi.01628-12.

Marston, D. A., Ellis, R. J., Wise, E. L., Aréchiga-Ceballos, N., Freuling, C. M., Banyard, A. C., ... Echevarría, J. E. (2017). Complete genome sequence of Lleida Bat Lyssavirus. *Genome Announcements*, *5*(2). https://doi.org/10.1128/genomeA.01427-16. e01427-01416.

Marston, D. A., Horton, D. L., Ngeleja, C., Hampson, K., McElhinney, L. M., Banyard, A. C., ... Lembo, T. (2012). Ikoma Lyssavirus, highly divergent novel Lyssavirus in an African Civet. *Emerging Infectious Diseases*, *18*(4), 664–667. https://doi.org/10.3201/eid1804.111553.

Marston, D. A., Horton, D. L., Nunez, J., Ellis, R. J., Orton, R. J., Johnson, N., ... Fooks, A. R. (2017). Genetic analysis of a rabies virus host shift event reveals within-host viral dynamics in a new host. *Virus Evolution*, *3*(2). https://doi.org/10.1093/ve/vex038.

Marston, D. A., McElhinney, L. M., Ali, Y. H., Intisar, K. S., Ho, S. M., Freuling, C., ... Fooks, A. R. (2009). Phylogenetic analysis of rabies viruses from Sudan provides evidence of a viral clade with a unique molecular signature. *Virus Research*, *145*(2), 244–250. https://doi.org/10.1016/j.virusres.2009.07.010.

Marston, D. A., McElhinney, L. M., Ellis, R. J., Horton, D. L., Wise, E. L., Leech, S. L., ... Fooks, A. R. (2013). Next generation sequencing of viral RNA genomes. *BMC Genomics*, *14*, 444.

Marston, D. A., McElhinney, L. M., Johnson, N., Műller, T., Conzelmann, K. K., Tordo, N., & Fooks, A. R. (2007). Comparative analysis of the full genome sequence of European bat lyssavirus type 1 and type 2 with other lyssaviruses and evidence for a conserved transcription termination and polyadenylation motif in the G-L 3′ non-translated region. *Journal of General Virology*, *88*, 1302–1314.

Martin, D. P., Murrell, B., Golden, M., Khoosal, A., & Muhire, B. (2015). RDP4: Detection and analysis of recombination patterns in virus genomes. *Virus Evolution*, *1*, vev003.

Matsumoto, T., Ahmed, K., Wimalaratne, O., Nanayakkara, S., Perera, D., Karunanayake, D., & Nishizono, A. (2011). Novel Sylvatic Rabies Virus Variant in Endangered Golden Palm Civet, Sri Lanka. *Emerging Infectious Diseases*, *17*(12), 2346–2349. https://doi.org/10.3201/eid1712.110811.

McElhinney, L. M., Marston, D. A., Freuling, C. M., Cragg, W., Stankov, S., Lalosević, D., ... Fooks, A. R. (2011). Molecular diversity and evolutionary history of rabies virus strains circulating in the Balkans. *Journal of General Virology*, *92*(9), 2171–2180. https://doi.org/10.1099/vir.0.032748-0.

McElhinney, L. M., Marston, D. A., Leech, S., Freuling, C. M., van der Poel, W. H. M., Echevarria, J., ... Fooks, A. R. (2013). Molecular Epidemiology of Bat Lyssaviruses in Europe. *Zoonoses and Public Health*, *60*(1), 35–45. https://doi.org/10.1111/zph.12003.

McElhinney, L. M., Marston, D. A., Wise, E. L., Freuling, C. M., Bourhy, H., Zanoni, R., ... Fooks, A. R. (2018). Molecular Epidemiology and Evolution of European Bat Lyssavirus 2. *International Journal of Molecular Sciences, 19*(1), 156.

Mélade, J., McCulloch, S., Ramasindrazana, B., Lagadec, E., Turpin, M., Pascalis, H., ... Dellagi, K. (2016). Serological evidence of Lyssaviruses among Bats on Southwestern Indian Ocean Islands. *PLoS One, 11*(8), e0160553. https://doi.org/10.1371/journal.pone.0160553.

Meng, S., Sun, Y., Wu, X., Tang, J., Xu, G., Lei, Y., ... Rupprecht, C. E. (2011). Evolutionary dynamics of rabies viruses highlights the importance of China rabies transmission in Asia. *Virology, 410*(2), 403–409. https://doi.org/10.1016/j.virol.2010.12.011.

Menozzi, B. D., de Novaes Oliveira, R., Paiz, L. M., Richini-Pereira, V. B., & Langoni, H. (2017). Antigenic and genotypic characterization of rabies virus isolated from bats (Mammalia: Chiroptera) from municipalities in São Paulo State, Southeastern Brazil. *Archives of Virology, 162*(5), 1201–1209. https://doi.org/10.1007/s00705-017-3220-9.

Messenger, S. L., Smith, J. S., & Rupprecht, C. E. (2002). Emerging epidemiology of bat-associated cryptic cases of rabies in humans in the United States. *Clinical Infectious Diseases, 35*, 738–747.

Metlin, A. E., Rybakov, S., Gruzdev, K., Neuvonen, E., & Huovilainen, A. (2007). Genetic heterogeneity of Russian, Estonian and Finnish field rabies viruses. *Archives of Virology, 152*(9), 1645–1654. https://doi.org/10.1007/s00705-007-1001-6.

Miao, F. M., Chen, T., Liu, Y., Zhang, S. F., Zhang, F., Li, N., & Hu, R. L. (2018). Emerging new phylogenetic groups of rabies virus in Chinese ferret badgers. *Biomedical and Environmental Sciences, 31*(6), 479–482.

Ming, P., Yan, J., Rayner, S., Meng, S., Xu, G., Tang, Q., ... Yang, X. (2010). A history estimate and evolutionary analysis of rabies virus variants in China. *Journal of General Virology, 91*(3), 759–764. https://doi.org/10.1099/vir.0.016436-0.

Mingo-Casas, P., Sandonís, V., Obón, E., Berciano, J. M., Vázquez-Morón, S., Juste, J., & Echevarría, J. E. (2018). First cases of European bat lyssavirus type 1 in Iberian serotine bats: Implications for the molecular epidemiology of bat rabies in Europe. *PLoS Neglected Tropical Diseases, 12*(4), e0006290. https://doi.org/10.1371/journal.pntd.0006290.

Mochizuki, N., Kobayashi, Y., Sato, G., Hirano, S., Itou, T., Ito, F. H., & Sakai, T. (2011). Determination and molecular analysis of the complete genome sequence of two wild-type rabies viruses isolated from a haematophagous bat and a frugivorous bat in Brazil. *Journal of Veterinary Medical Science, 73*(6), 759–766.

Moldal, T., Vikøren, T., Cliquet, F., Marston, D. A., van der Kooij, J., Madslien, K., & Ørpetveit, I. (2017). First detection of European bat lyssavirus type 2 (EBLV-2) in Norway. *BMC Veterinary Research, 13*(1), 216. https://doi.org/10.1186/s12917-017-1135-z.

Moore, P. R., Jansen, C. C., Graham, G. C., Smith, I. L., & Craig, S. B. (2010). Emerging tropical diseases in Australia. Part 3. Australian bat lyssavirus. *Annals of Tropical Medicine and Parasitology, 104*, 613–621.

Mørk, T., Bohlin, J., Fuglei, E., Åsbakk, K., & Tryland, M. (2011). Rabies in the arctic fox population, Svalbard, Norway. *Journal of Wildlife Diseases, 47*(4), 945–957. https://doi.org/10.7589/0090-3558-47.4.945.

Mørk, T., & Prestrud, P. (2004). Arctic Rabies—a review. *Acta Veterinaria Scandinavica, 45*, 1–9.

Muleya, W., Namangala, B., Mweene, A., Zulu, L., Fandamu, P., Banda, D., ... Ishii, A. (2012). Molecular epidemiology and a loop-mediated isothermal amplification method for diagnosis of infection with rabies virus in Zambia. *Virus Research, 163*(1), 160–168. https://doi.org/10.1016/j.virusres.2011.09.010.

Müller, T., Freuling, C. M., Wysocki, P., Roumiantzeff, M., Freney, J., Mettenleiter, T. C., & Vos, A. (2015). Terrestrial rabies control in the European Union: Historical achievements and challenges ahead. *The Veterinary Journal, 203*(1), 10–17.

Müller, T., Johnson, N., Freuling, C. M., Fooks, A. R., Selhorst, T., & Vos, A. (2007). Epidemiology of bat rabies in Germany. *Archives of Virology, 152*, 273–288.

Nadin-Davis, S., Alnabelseya, N., & Knowles, M. K. (2017). The phylogeography of *Myotis* bat-associated rabies viruses across Canada. *PLoS Neglected Tropical Diseases, 11*(5), e0005541.

Nadin-Davis, S., Buchanan, T., Nituch, L., & Fehlner-Gardiner, C. (2020). *A long-distance translocation initiated an outbreak of raccoon rabies in Hamilton, Ontario, Canada. PLoS Neglected Tropical Diseases.* in press.

Nadin-Davis, S. A. (2013). Molecular epidemiology. In A. C. Jackson (Ed.), *Rabies: Scientific Basis of the Disease and its Management* (3rd ed., pp. 123–177). Oxford, UK: Academic Press.

Nadin-Davis, S. A. (2019). Rapid identification of the Raccoon Rabies Virus Variant using a Real-time Reverse-Transcriptase Polymerase Chain Reaction. *Journal of Virological Methods. 273,* 113713.

Nadin-Davis, S. A., & Bingham, J. (2004). Europe as a source of rabies for the rest of the worls. In A. A. King, A. R. Fooks, M. Aubert, & A. I. Wandeler (Eds.), *Historical persepctive of rabies in Europe and the Mediterranean basin* (pp. 259–280). Paris: OIE.

Nadin-Davis, S. A., Colville, A., Trewby, H., Biek, R., & Real, L. (2017). Application of high-throughput sequencing to whole rabies viral genome characterisation and its use for re-evaluation of a raccoon strain incursion into the province of Ontario. *Virus Research, 232,* 123–133.

Nadin-Davis, S. A., Feng, Y., Mousse, D., Wandeler, A. I., & Aris-Brosou, S. (2010). Spatial and temporal dynamics of rabies virus variants in big brown bat populations across Canada: footprints of an emerging zoonosis. *Molecular Ecology, 19,* 2120–2136.

Nadin-Davis, S. A., Fu, Q., Trewby, H., Biek, R., Johnson, R. H., & Real, L. (2018). Geography but not alternative host species explain the spread of raccoon rabies in Vermont. *Epidemiology and Infection, 146,* 1977–1986.

Nadin-Davis, S. A., Guerrero, E., Knowles, M. K., & Feng, Y. (2012). DNA barcoding facilitates bat species identification for improved surveillance of bat-associated rabies across Canada. *Open Journal of Zoology, 5,* 27–37.

Nadin-Davis, S. A., & Loza-Rubio, E. (2006). The molecular epidemiology of rabies associated with chiropteran hosts in Mexico. *Virus Research, 117,* 215–226.

Nadin-Davis, S. A., & Real, L. A. (2011). Molecular phylogenetics of the Lyssaviruses-Insights from a coalescent approach. In *Vol. 79. Advances in Virus Research* (pp. 203–238). Burlington, USA: Academic Press.

Nadin-Davis, S. A., Sheen, M., & Wandeler, A. I. (2003). Use of discriminatory probes for strain typing of formalin-fixed, rabies virus-infected tissues by in situ hybridization. *Journal of Clinical Microbiology, 41*(9), 4343–4352. https://doi.org/10.1128/jcm.41.9.4343-4352.2003.

Nadin-Davis, S. A., Sheen, M., & Wandeler, A. I. (2012). Recent emergernce of the arctic rabies virus lineage. *Virus Research, 163,* 352–362.

Nadin-Davis, S. A., Simani, S., Armstrong, J., Fayaz, A., & Wandeler, A. I. (2003). Molecular and antigenic characterization of rabies viruses from Iran identifies variants with distinct epidemiological origins. *Epidemiology and Infection, 131,* 777–790.

Nadin-Davis, S. A., Torres, G., de Los Angeles Ribas, M., Guzman, M., de la Paz, R. C., Morales, M., & Wandeler, A. I. (2006). A molecular epidemiological study of rabies in Cuba. *Epidemiology and Infection, 134,* 1313–1324.

Nadin-Davis, S. A., Turner, G., Paul, J. P. V., Madhusudana, N., & Wandeler, A. I. (2007). Emergence of arctic-like rabies lineage in India. *Emerging Infectious Diseases, 13*(1), 111–116.

Nadin-Davis, S. A., Velez, J., Malaga, C., & Wandeler, A. I. (2008). A molecular epidemiological study of rabies in Puerto Rico. *Virus Research, 131,* 8–15.

Nanayakkara, S., Smith, J. S., & Rupprecht, C. E. (2003). Rabies in Sri Lanka: splendid isolation. *Emerging Infectious Diseases, 9,* 368–371.

Nel, L., Jacobs, J., Jaftha, J., von Teichman, B., & Bingham, J. (2000). New cases of Mokola virus infection in South Africa: a genotypic comparison of southern African virus isolates. *Virus Genes, 20,* 103–106.

Nel, L. H., Sabeta, C. T., von Teichman, B., Jaftha, J. B., Rupprecht, C. E., & Bingham, J. (2005). Mongoose rabies in southern Africa: a re-evaluation based on molecular epidemiology. *Virus Research, 109*(2), 165–173. https://doi.org/10.1016/j.virusres.2004.12.003.

Neubaum, M. A., Shankar, V., Douglas, M. R., Douglas, M. E., O'Shea, T. J., & Rupprecht, C. E. (2008). An analysis of correspondence between unique rabies virus variants and divergent big brown bat (Eptesicus fuscus) mitochondrial DNA lineages. *Archives of Virology, 153*(6), 1139. https://doi.org/10.1007/s00705-008-0081-2.

Nguyen, A. K. T., Nguyen, D. V., Ngo, G. C., Nguyen, T. T., Inoue, S., Yamada, A., … Nguyen, H. T. H. (2011). Molecular epidemiology of rabies virus in Vietnam (2006–2009). *Japanese Journal of Infectious Diseases, 64,* 391–396.

Nokireki, T., Sironen, T., Smura, T., Karkamo, V., Sihvonen, L., & Gadd, T. (2017). Second case of European bat lyssavirus type 2 detected in a Daubenton's bat in Finland. *Acta Veterinaria Scandinavica, 59*(1), 62. https://doi.org/10.1186/s13028-017-0331-y.

Nokireki, T., Tammiranta, N., Kokkonen, U. -M., Kantala, T., & Gadd, T. (2018). Tentative novel lyssavirus in a bat in Finland. *Transboundary and Emerging Diseases, 65*(3), 593–596. https://doi.org/10.1111/tbed.12833.

Oliveira, R. d. N., de Souza, S. P., Lobo, R. S. V., Castilho, J. G., Macedo, C. I., Carnieli, P. J., … Brandao, P. E. (2010). Rabies virus in insectivorous bats: implications of the diversity of the nucleoprotein and glycoprotein genes for molecular epidemiology. *Virology, 405,* 352–360.

Orłowska, A., & Żmudziński, J. F. (2014). Molecular epidemiology of rabies virus in Poland. *Archives of Virology, 159*(8), 2043–2050. https://doi.org/10.1007/s00705-014-2045-z.

Páez, A., Saad, C., Nůňez, C., & Bóshell, J. (2005). Molecular epidemiology of rabies in northern Colombia 1994–2003. Evidence for human and fox rabies associated with dogs. *Epidemiology and Infection, 133*, 529–536.

Pant, G. R., Lavenir, R., Wong, F. Y. K., Certoma, A., Larrous, F., Bhatta, D. R., ... Dacheux, L. (2013). Recent Emergence and Spread of an Arctic-Related Phylogenetic Lineage of Rabies Virus in Nepal. *PLoS Neglected Tropical Diseases, 7*(11), e2560. https://doi.org/10.1371/journal.pntd.0002560.

Paweska, J. T., Blumberg, L. H., Liebenberg, C., Hewlett, R. H., Grobbelaar, A. A., Leman, P. A., ... Swanepoel, R. (2006). Fatal human infection with rabies-related Duvenhage virus, South Africa. *Emerging Infectious Diseases, 12*, 1965–1967.

Pfaff, F., Müller, T., Freuling, C. M., Fehlner-Gardiner, C., Nadin-Davis, S., Robardet, E., ... Höper, D. (2018). In-depth genome analyses of viruses from vaccine-derived rabies cases and corresponding live-attenuated oral rabies vaccines. *Vaccine*. https://doi.org/10.1016/j.vaccine.2018.01.083.

Picard-Meyer, E., Barrat, J., Wasniewski, M., Wandeler, A., Nadin-Davis, S., Lowings, J. P., ... Cliquet, F. (2004). Epidemiology of rabid bats in France, 1989 to 2002. *Veterinary Record, 155*, 774–777.

Picard-Meyer, E., Beven, V., Hirchaud, E., Guillaume, C., Larcher, G., Robardet, E., ... Cliquet, F. (2019). Lleida Bat Lyssavirus isolation in Miniopterus schreibersii in France. *Zoonoses and Public Health, 66*(2), 254–258. https://doi.org/10.1111/zph.12535.

Picard-Meyer, E., Brookes, S. M., Barrat, J., Litaize, E., Patron, C., Biarnais, M., ... Cliquet, F. (2008). Experimental infection of foxes with European bat lyssaviruses type 1 and -2. *Developments in Biology (Basel), 131*, 339–345.

Picard-Meyer, E., Dubourg-Savage, M. -J., Arthur, L., Barataud, M., Bécu, D., Bracco, S., ... Cliquet, F. (2011). Active surveillance of bat rabies in France: A 5-year study (2004–2009). *Veterinary Microbiology, 151*(3), 390–395. https://doi.org/10.1016/j.vetmic.2011.03.034.

Picard-Meyer, E., Mrenoshki, S., Milicevic, V., Ilieva, D., Cvetkovikj, I., Cvetkovikj, A., ... Cliquet, F. (2013). Molecular characterisation of rabies virus strains in the Republic of Macedonia. *Archives of Virology, 158*(1), 237–240. https://doi.org/10.1007/s00705-012-1466-9.

Picard-Meyer, E., Robardet, E., Arthur, L., Larcher, G., Harbusch, C., Servat, A., & Cliquet, F. (2014). Bat Rabies in France: A 24-Year Retrospective Epidemiological Study. *PLoS One, 9*(6), e98622. https://doi.org/10.1371/journal.pone.0098622.

Picard-Meyer, E., Robardet, E., Moroz, D., Trotsenko, Z., Drozhzhe, Z., Biarnais, M., ... Cliquet, F. (2012). Molecular epidemiology of rabies in Ukraine. *Archives of Virology, 157*(9), 1689–1698. https://doi.org/10.1007/s00705-012-1351-6.

Picard-Meyer, E., Servat, A., Robardet, E., Moinet, M., Borel, C., & Cliquet, F. (2013). Isolation of Bokeloh bat lyssavirus in Myotis nattereri in France. *Archives of Virology, 158*(11), 2333–2340. https://doi.org/10.1007/s00705-013-1747-y.

Pinero, C., Dohmen, F. G., Beltran, F., Martinez, L., Novaro, L., Russo, S., ... Cisterna, D. M. (2012). High diversity of rabies viruses asociated with insectivorous bats in Argentina: presence of several independent enzootics. *PLoS Neglected Tropical Diseases, 6*(5), e1635.

Queiroz, L. H., Favoretto, S. R., Cunha, E. M. S., Campos, A. C. A., Lopes, M. C., de Carvalho, C., ... Durigon, E. L. (2012). Rabies in southeast Brazil: a change in the epidemiological pattern. *Archives of Virology, 157*(1), 93–105. https://doi.org/10.1007/s00705-011-1146-1.

Reddy, G. B. M., Krishnappa, S., Vinayagamurthy, B., Singh, R., Singh, K. P., Saminathan, M., ... Rahman, H. (2018). Molecular epidemiology of rabies virus circulating in domestic animals in India. *Virus Diseases, 29*(3), 362–368.

Reddy, R. V. C., Mohana Subramanian, B., Surendra, K. S. N. L., Babu, R. P. A., Rana, S. K., Manjari, K. S., & Srinivasan, V. A. (2014). Rabies virus isolates of India—Simultaneous existence of two distinct evolutionary lineages. *Infection, Genetics and Evolution, 27*, 163–172. https://doi.org/10.1016/j.meegid.2014.07.014.

Reynes, J. -M., Andriamandimby, S. F., Razafitrimo, G. M., Razainirina, J., Jeanmaire, E. M., Bourhy, H., ... Heraud, J. -M. (2011). Laboratory Surveillance of Rabies in Humans, Domestic Animals, and Bats in Madagascar from 2005 to 2010. *Advances in Preventive Medicine, 2011*, 6. https://doi.org/10.4061/2011/727821.

Reynes, J. -M., Molia, S., Audry, L., Hout, S., Ngin, S., Walston, J., & Bourhy, H. (2004). Serologic evidence of lyssavirus infection in bats, Cambodia. *Emerging Infectious Diseases, 10*, 2231–2234.

Ribadeau-Dumas, F., Cliquet, F., Gautret, P., Robardet, E., Le Pen, C., & Bourhy, H. (2016). Travel-associated rabies in pets and residual rabies risk, western Europe. *Emerging Infectious Diseases, 22*(7), 1268–1271.

Rihtarič, D., Hostnik, P., Grom, J., & Toplak, I. (2011). Molecular epidemiology of the rabies virus in Slovenia 1994–2010. *Veterinary Microbiology, 152*(1), 181–186. https://doi.org/10.1016/j.vetmic.2011.04.019.

Robardet, E., Ilieva, D., Iliev, E., Gagnev, E., Picard-Meyer, E., & Cliquet, F. (2013). Epidemiology and molecular diversity of rabies viruses in Bulgaria. *Epidemiology and Infection*, *142*(4), 871–877. https://doi.org/10.1017/S0950268813001556.

Robardet, E., Picard-Meyer, E., Dobroštana, M., Jaceviciene, I., Mähar, K., Muižniece, Z., ... Cliquet, F. (2016). Rabies in the Baltic States: Decoding a Process of Control and Elimination. *PLoS Neglected Tropical Diseases*, *10*(2), e0004432. https://doi.org/10.1371/journal.pntd.0004432.

Rocha, S. M., de Oliveira, S. V., Heinemann, M. B., & Gonçalves, V. S. P. (2017). Epidemiological Profile of Wild Rabies in Brazil (2002–2012). *Transboundary and Emerging Diseases*, *64*(2), 624–633. https://doi.org/10.1111/tbed.12428.

Root, J. J., Puskas, R. B., Fischer, J. W., Swope, C. B., Neubaum, M. A., Reeder, S. A., & Piaggio, A. J. (2009). Landscape genetics of raccoons (*Procyon lotor*) associated with ridges and valleys of Pennsylvania: Implications for oral rabies vaccination programs. *Vector Borne and Zoonotic Diseases*, *9*, 583–588.

Rosatte, R. C., Donovan, D., Allan, M., Bruce, L., Buchanan, T., Sobey, K., ... Wandeler, A. (2009). The control of raccoon rabies in Ontario, Canada: Proactive and reactive tactics, 1994–2007. *Journal of Wildlife Diseases*, *45*, 772–784.

Rupprecht, C., Kuzmin, I., & Meslin, F. (2017). Lyssaviruses and rabies: current conundrums, concerns, contradictions and controversies [version 1; referees: 2 approved]. *F1000Research*, *6*(184). https://doi.org/10.12688/f1000research.10416.1.

Sabeta, C., Blumberg, L. H., Miyen, J. M., Mohale, D., Shumba, W., & Wandeler, A. (2010). Mokola virus involved in a human contact (South Africa). *FEMS Immunology and Medical Microbiology*, *58*(1), 85–90. https://doi.org/10.1111/j.1574-695X.2009.00609.x.

Sabeta, C. T., Bingham, J., & Nel, L. H. (2003). Molecular epidemiology of canid rabies in Zimbabwe and South Africa. *Virus Research*, *91*(2), 203–211. https://doi.org/10.1016/S0168-1702(02)00272-1.

Sabeta, C. T., Janse van Rensburg, D. D., Phahladira, B., Mohale, D., Harrison-White, R. F., Esterhuyzen, C., & Williams, J. H. (2018). Rabies of canid biotype in wild dog (*Lycaon pictus*) and spotted hyaena (*Crocuta crocuta*) in Madikwe Game Reserve, South Africa in 2014–2015: Diagnosis, possible origins and implications for control. *Journal of the South African Veterinary Association*, *89*, a1517.

Sabeta, C. T., Mansfield, K. L., McElhinney, L. M., Fooks, A. R., & Nel, L. H. (2007). Molecular epidemiology of rabies in bat-eared foxes (Otocyon megalotis) in South Africa. *Virus Research*, *129*(1), 1–10. https://doi.org/10.1016/j.virusres.2007.04.024.

Sabeta, C. T., Markotter, W., Mohale, D. K., Shumba, W., Wandeler, A. I., & Nel, L. H. (2007). Mokola virus in domestic mammals, South Africa. *Emerging Infectious Diseases*, *13*, 1371–1373.

Sabeta, C. T., Shumba, W., Mohale, D. K., Miyen, J. M., Wandeler, A. I., & Nel, L. H. (2008). Mongoose rabies and the African civet in Zimbabwe. *The Veterinary Record*, *163*, 580.

Sadeuh-Mba, S. A., Momo, J. B., Besong, L., Loul, S., & Njouom, R. (2017). Molecular characterization and phylogenetic relatedness of dog-derived Rabies Viruses circulating in Cameroon between 2010 and 2016. *PLoS Neglected Tropical Diseases*, *11*(10), e0006041. https://doi.org/10.1371/journal.pntd.0006041.

Saito, M., Oshitani, H., Orbina, J. R. C., Tohma, K., de Guzman, A. S., Kamigaki, T., ... Quiambao, B. P. (2013). Genetic diversity and geographic distribution of genetically distinct rabies viruses in the Philippines. *PLoS Neglected Tropical Diseases*, *7*(4), e2144. https://doi.org/10.1371/journal.pntd.0002144.

Saitou, Y., Kobayashi, Y., Hirano, S., Mochizuki, N., Itou, T., Ito, F. H., & Sakai, T. (2010). A method for simultaneous detection and identification of Brazilian dog- and vampire bat-related rabies virus by reverse transcription loop-mediated isothermal amplification assay. *Journal of Virological Methods*, *168*, 13–17.

Scott, T., Fischer, M., Khaiseb, S., Freuling, C., Hoper, D., Hoffmann, B., ... Nel, L. (2013). Complete Genome and molecular epidemiological data infer the maintenance of rabies among kudu (*Tragelaphus strepsiceros*) in Namibia. *PLoS One*, *8*(3), e58739.

Seetahal, J. F. R., Velasco-Villa, A., Allicock, O. M., Adesiyun, A. A., Bissessar, J., Amour, K., ... Carrington, C. V. F. (2013). Evolutionary history and phylogeography of rabies viruses associated with outbreaks in Trinidad. *PLoS Neglected Tropical Diseases*, *7*(8), e2365. https://doi.org/10.1371/journal.pntd.0002365.

Seetahal, J. F. R., Vokaty, A., Vigilato, M. A. N., Carrington, C. V. F., Pradel, J., Louison, B., ... Rupprecht, C. E. (2018). Rabies in the Caribbean: a situational analysis and historic review. *Tropical Medicine and Infectious Disease*, *3*, e89.

Serra-Cobo, J., Amengual, B., Abellán, C., & Bourhy, H. (2002). European bat Lyssavirus infection in Spanish bat populations. *Emerging Infectious Diseases*, *8*, 413–420.

Serra-Cobo, J., López-Roig, M., Lavenir, R., Abdelatif, E., Boucekkine, W., Elharrak, M., ... Bourhy, H. (2018). Active sero-survey for European bat lyssavirus type-1 circulation in North African insectivorous bats. *Emerging Microbes & Infections*, *7*(1), 1–4. https://doi.org/10.1038/s41426-018-0214-y.

Serra-Cobo, J., López-Roig, M., Seguí, M., Sánchez, L. P., Nadal, J., Borrás, M., … Bourhy, H. (2013). Ecological factors associated with European Bat Lyssavirus Seroprevalence in Spanish Bats. *PLoS One, 8*(5), e64467. https://doi.org/10.1371/journal.pone.0064467.

Shankar, V., Orciari, L. A., de Mattos, C., Kuzmin, I. V., Pape, W. J., O'Shea, T. J., & Rupprecht, C. E. (2005). Genetic divergence of rabies viruses from bat species of Colorado, USA. *Vector-Boprne and Zoonotic Diseases, 5*(4), 330–341.

Shao, X. Q., Yan, X. J., Luo, G. L., Zhang, H. L., Chai, X. L., Wang, F. X., … Zhang, Y. Z. (2010). Genetic evidence for domestic raccoon dog rabies caused by Arctic-like rabies virus in Inner Mongolia, China. *Epidemiology and Infection, 139*(4), 629–635. https://doi.org/10.1017/S0950268810001263.

Šimić, I., Lojkić, I., Krešić, N., Cliquet, F., Picard-Meyer, E., Wasniewski, M., … Bedeković, T. (2018). Molecular and serological survey of lyssaviruses in Croatian bat populations. *BMC Veterinary Research, 14*(1), 274. https://doi.org/10.1186/s12917-018-1592-z.

Slate, D., Rupprecht, C. E., Rooney, J. A., Donovan, D., Lein, D. H., & Chipman, R. B. (2005). Status of oral rabies vaccination in wild carnivores in the United States. *Virus Research, 111*, 68–76.

Smith, G. C., Aegerter, J. N., Allnutt, T. R., MacNicoll, A. D., Learmount, J., Hutson, A. M., & Atterby, H. (2011). Bat population genetics and Lyssavirus presence in Great Britain. *Epidemiology and Infection, 139*(10), 1463–1469. https://doi.org/10.1017/S0950268810002876.

Smith, J. S., Orciari, L. A., Yager, P. A., Seidel, H. D., & Warner, C. K. (1992). Epidemiologic and historical relationships among 87 rabies virus isolates as determined by limited sequence analysis. *The Journal of Infectious Diseases, 166*, 296–307.

Smreczak, M., Orłowska, A., Marzec, A., Trębas, P., Müller, T., Freuling, C. M., & Żmudziński, J. F. (2018). Bokeloh bat lyssavirus isolation in a Natterer's bat, Poland. *Zoonoses and Public Health, 65*(8), 1015–1019. https://doi.org/10.1111/zph.12519.

Streicker, D. G., Altizer, S. M., Velasco-Villa, A., & Rupprecht, C. E. (2012). Variable evolutionary routes to host establishment across repeated rabies virus host shifts among bats. *Proceedings of the National Academy of Sciences USA, 109*(48), 19715–19720.

Streicker, D. G., Turmelle, A. S., Vonhof, M. J., Kuzmin, I. V., McCracken, G. F., & Rupprecht, C. E. (2010). Host phylogeny constrains cross-species emergence and establishment of rabies virus in bats. *Science, 329*, 676–679.

Streicker, D. G., Winternitz, J. C., Satterfield, D. A., Condori-Condori, R. E., Broos, A., Tello, C., … Valderrama, W. (2016). Host-pathogen evolutionary signatures reveal dynamics and future invasions of vampire bat rabies. *Proceedings of the National Academy of Sciences USA, 113*(39), 10926–10931.

Susetya, H., Sugiyama, M., Inagaki, A., Ito, N., Mudiarto, G., & Minamoto, N. (2008). Molecular epidemiology of rabies in Indonesia. *Virus Research, 135*(1), 144–149. https://doi.org/10.1016/j.virusres.2008.03.001.

Szanto, A. G., Nadin-Davis, S. A., Rosatte, R. C., & White, B. N. (2011). Genetic tracking of the raccoon variant of rabies virus in eastern North America. *Epidemics, 3*, 76–87.

Talbi, C., & Bourhy, H. (2011). La rage canine en Afrique au travers de l'analyse génétique, spatiale et temporelle des isolats. *Virologie, 15*, 307–318.

Talbi, C., Holmes, E. C., de Benedictis, P., Faye, O., Nakouné, E., Gamatié, D., … Bourhy, H. (2009). Evolutionary history and dynamics of dog rabies virus in western and central Africa. *Journal of General Virology, 90*(4), 783–791. https://doi.org/10.1099/vir.0.007765-0.

Talbi, C., Lemey, P., Suchard, M. A., Abdelatif, E., Elharrak, M., Jalal, N., … Bourhy, H. (2010). Phylodynamics and human-mediated dispersal of a zoonotic virus. *PLoS Pathogens, 6*(10), e1001166. https://doi.org/10.1371/journal.ppat.1001166.

Tao, X. -Y., Guo, Z. -Y., Li, H., Jiao, W. -T., Shen, X. -X., Zhu, W. -Y., … Tang, Q. (2015). Rabies cases in the west of China have two distinct origins. *PLoS Neglected Tropical Diseases, 9*(10), e0004140. https://doi.org/10.1371/journal.pntd.0004140.

Tao, X. -Y., Li, M. -L., Guo, Z. -Y., Yan, J. -H., & Zhu, W. -Y. (2019). Inner Mongolia: A potential portal for the spread of rabies to Western China. *Vector Borne and Zoonotic Diseases, 19*(1), 51–58. https://doi.org/10.1089/vbz.2017.2248.

Tao, X. -Y., Li, M. -L., Wang, Q., Baima, C., Hong, M., Li, W., … Zhu, W. -Y. (2019). The reemergence of human rabies and emergence of an Indian subcontinent lineage in Tibet, China. *PLoS Neglected Tropical Diseases, 13*(1), e0007036. https://doi.org/10.1371/journal.pntd.0007036.

Tao, X. -Y., Tang, Q., Rayner, S., Guo, Z. -Y., Li, H., Lang, S. -L., … Liang, G. -D. (2013). Molecular phylodynamic analysis indicates lineage displacement occurred in Chinese rabies epidemics between 1949 to 2010. *PLoS Neglected Tropical Diseases, 7*(7), e2294. https://doi.org/10.1371/journal.pntd.0002294.

Tasioudi, K. E., Iliadou, P., Agianniotaki, E. I., Robardet, E., Liandris, E., Doudounakis, S., ... Mangana-Vougiouka, O. (2014). Recurrence of animal rabies, Greece, 2012. *Emerging Infectious Disease Journal*, 20(2), 326–329. https://doi.org/10.3201/eid2002.130473.

Tenzin, Wacharapluesadee, S., Denduangboripant, J., Dhand, N. K., Dorji, R., Tshering, D., ... Ward, M. P. (2010). Rabies virus strains circulating in Bhutan: implications for control. *Epidemiology and Infection*, 139(10), 1457–1462. https://doi.org/10.1017/S0950268810002682.

Tian, H., Feng, Y., Vrancken, B., Cazelles, B., Tan, H., Gill, M. S., ... Dellicour, S. (2018). Transmission dynamics of re-emerging rabies in domestic dogs of rural China. *PLoS Pathogens*, 14(12), e1007392. https://doi.org/10.1371/journal.ppat.1007392.

Tohma, K., Saito, M., Demetria, C. S., Manalo, D. L., Quiambao, B. P., Kamigaki, T., & Oshitani, H. (2016). Molecular and mathematical modeling analyses of inter-island transmission of rabies into a previously rabies-free island in the Philippines. *Infection, Genetics and Evolution*, 38, 22–28. https://doi.org/10.1016/j.meegid.2015.12.001.

Tohma, K., Saito, M., Kamigaki, T., Tuason, L. T., Demetria, C. S., Orbina, J. R. C., ... Oshitani, H. (2014). Phylogeographic analysis of rabies viruses in the Philippines. *Infection, Genetics and Evolution*, 23, 86–94. https://doi.org/10.1016/j.meegid.2014.01.026.

Torres, C., Lema, C., Dohmen, F. G., Beltran, F., Novaro, L., Russo, S., ... Cisterna, D. M. (2014). Phylodynamics of vampire bat-transmitted rabies in Argentina. *Molecular Ecology*, 23, 2340–2352.

Traoré, A., Picard-Meyer, E., Mauti, S., Biarnais, M., Balmer, O., Samaké, K., ... Cliquet, F. (2016). Molecular characterization of Canine Rabies Virus, Mali, 2006–2013. *Emerging Infectious Disease Journal*, 22(5), 866–870. https://doi.org/10.3201/eid2205.150470.

Trewby, H., Nadin-Davis, S. A., Real, L. A., & Biek, R. (2017). Phylogeographic analysis of rabies virus incursions across US-Canada border. *Emerging Infectious Diseases*, 23(9), 1454–1461.

Troupin, C., Dacheux, L., Tanguy, M., Sabeta, C., Blanc, H., Bouchier, C., ... Bourhy, H. (2016). Large-scale phylogenomic analysis reveals the complex evolutionary history of rabies virus in multiple carnivore hosts. *PLoS Pathogens*, 12(12), e1006041.

Tsai, K. J., Hsu, W. C., Chuang, W. C., Chang, J. C., Tu, Y. C., Tsai, H. J., ... Lee, S. H. (2016). Emergence of a sylvatic enzootic formosan ferret badger-associated rabies in Taiwan and the geographical separation of two phylogenetic groups of rabies viruses. *Veterinary Microbiology*, 182, 28–34. https://doi.org/10.1016/j.vetmic.2015.10.030.

Turcitu, M. A., Barboi, G., Vuta, V., Mihai, I., Boncea, D., Dumitrescu, F., ... Freuling, C. M. (2010). Molecular epidemiology of rabies virus in Romania provides evidence for a high degree of heterogeneity and virus diversity. *Virus Research*, 150(1), 28–33. https://doi.org/10.1016/j.virusres.2010.02.008.

Van der Poel, W. H., Van der Heide, R., Verstraten, E. R., Takumi, K., Lina, P. H., & Kramps, J. A. (2005). European bat lyssaviruses, The Netherlands. *Emerging Infectious Diseases*, 11, 1854–1859.

Van Eeden, C., Markotter, W., & Nel, L. H. (2011). Molecular phylogeny of Duvenhage virus. *South African Journal of Science*, 107, 177.

van Thiel, P. -P. A. M., de Bie, R. M. A., Eftimov, F., Tepaske, R., Zaaijer, H. L., van Doornum, G. J. J., ... Kager, P. A. (2009). Fatal human rabies due to duvenhage virus from a Bat in Kenya: Failure of treatment with coma-induction, ketamine, and antiviral drugs. *PLoS Neglected Tropical Diseases*, 3(7), e428. https://doi.org/10.1371/journal.pntd.0000428.

Vázquez-Morón, S., Juste, J., Ibáñez, C., Berciano, J. M., & Echevarría, J. E. (2011). Phylogeny of European bat lyssavirus 1 in Eptesicus isabellinus bats, Spain. *Emerging Infectious Diseases*, 17, 520–523.

Velasco-Villa, A., Escobar, L. E., Sanchez, A., Shi, M., Streicker, D. G., Gallardo-Romero, N. F., ... Emerson, G. (2017). Successful strategies implemented towards the elimination of canine rabies in the Western Hemisphere. *Antiviral Research*, 143, 1–12. https://doi.org/10.1016/j.antiviral.2017.03.023.

Velasco-Villa, A., Mauldin, M. R., Shi, M., Escobar, L. E., Gallardo-Romero, N. F., Damon, I., ... Emerson, G. (2017). The history of rabies in the Western Hemisphere. *Antiviral Research*, 146, 221–232. https://doi.org/10.1016/j.antiviral.2017.03.013.

Velasco-Villa, A., Messenger, S. L., Orciari, L. A., Niezgoda, M., Blanton, J. D., Fukagawa, C., & Rupprecht, C. E. (2008). Identification of new rabies virus variant in Mexican immigrant. *Emerging Infectious Diseases*, 14(12), 1906–1908. https://doi.org/10.3201/eid1412.080671.

Velasco-Villa, A., Orciari, L. A., Juarez-Islas, V., Gomez-Sierra, M., Padilla-Medina, I., Flisser, A., ... Rupprecht, C. E. (2006). Molecular diversity of rabies viruses associated with bats in Mexico and other countries of the Americas. *Jounal of Clinical Microbiology*, 44(5), 1697–1710.

Velasco-Villa, A., Orciari, L. A., Souza, V., Juárez-Islas, V., Gomez-Sierre, M., Castillo, A., ... Rupprecht, C. E. (2005). Molecular epizootiology of rabies associated with terrestrial carnivores in Mexico. *Virus Research, 111*, 13–27.

Vos, A., Müller, T., Neubert, L., Zurbriggen, A., Botteron, C., Pöhle, D., ... Jackson, A. C. (2004). Rabies in red foxes (Vulpes vulpes) experimentally infected with European bat lyssavirus type 1. *Journal of Veterinary Medicine Series B, 51*, 327–332.

Winkler, W. G., & Jenkins, S. R. (1991). Raccoon rabies. In G. M. Baer (Ed.), *The natural history of rabies* (2nd ed., pp. 325–340). Boca Raton: CRC Press.

Wise, E. L., Marston, D. A., Banyard, A. C., Goharriz, H., Selden, D., Maclaren, N., ... Fooks, A. R. (2017). Passive surveillance of United Kingdom bats for lyssaviruses (2005–2015). *Epidemiology and Infection, 145*(12), 2445–2457. https://doi.org/10.1017/S0950268817001455.

Wright, E., Hayman, D. T. S., Vaughan, A., Temperton, N. J., Wood, J. L. N., Cunningham, A. A., ... Fooks, A. R. (2010). Virus neutralising activity of African fruit bat (Eidolon helvum) sera against emerging lyssaviruses. *Virology, 408*(2), 183–189. https://doi.org/10.1016/j.virol.2010.09.014.

Wu, X., Franka, R., Velasco-Villa, A., & Rupprecht, C. E. (2007). Are all lyssavirus genes equal for phylogenetic analyses? *Virus Research, 129*(2), 91–103. https://doi.org/10.1016/j.virusres.2007.06.022.

Yang, D. -K., Shin, E. K., Oh, Y. -I., Kang, H. -K., Lee, K. -W., Cho, S. -D., & Song, J. -Y. (2011). Molecular epidemiology of rabies virus circulating in South Korea, 1998–2010. *Journal of Veterinary Medical Science, 73*(8), 1077–1082.

Yao, H. -W., Yang, Y., Liu, K., Li, X. -L., Zuo, S. -Q., Sun, R. -X., ... Cao, W. -C. (2015). The spatiotemporal expansion of human rabies and its probable explanation in Mainland China, 2004–2013. *PLoS Neglected Tropical Diseases, 9*(2), e0003502. https://doi.org/10.1371/journal.pntd.0003502.

Zecchin, B., De Nardi, M., Nouvellet, P., Vernesi, C., Babbucci, M., Crestanello, B., ... Cattoli, G. (2019). Genetic and spatial characterization of the red fox (Vulpes vulpes) population in the area stretching between the Eastern and Dinaric Alps and its relationship with rabies and canine distemper dynamics. *PLoS One, 14*(3), e0213515. https://doi.org/10.1371/journal.pone.0213515.

Zhang, H. -L., Zhang, Y. -Z., Yang, W. -H., Tao, X. -Y., Li, H., Ding, J. -C., ... Tang, Q. (2014). Molecular epidemiology of reemergent rabies in Yunnan Province, Southwestern China. *Emerging Infectious Diseases, 20*(9), 1433–1442. https://doi.org/10.3201/eid2009.130440.

Zhang, S., Zhao, J., Liu, Y., Fooks, A. R., Zhang, F., & Hu, R. (2010). Characterization of a rabies virus isolate from a ferret badger (Melogale moschata) with unique molecular differences in glycoprotein antigenic site III. *Virus Research, 149*(2), 143–151. https://doi.org/10.1016/j.virusres.2010.01.010.

Zhang, X., Cai, Y., Zhai, X., Liu, J., Zhao, W., Ji, S., ... Zhou, J. (2018). Comprehensive analysis of codon usage on rabies virus and other lyssaviruses. *International Journal of Molecular Sciences, 19*(8), e2397.

Zhang, Y. -Z., Xiong, C. -L., Lin, X. -D., Zhou, D. -J., Jiang, R. -J., Xiao, Q. -Y., ... Fu, Z. F. (2009). Genetic diversity of Chinese rabies viruses: Evidence for the presence of two distinct clades in China. *Infection, Genetics and Evolution, 9*(1), 87–96. https://doi.org/10.1016/j.meegid.2008.10.014.

Zhao, J., Liu, Y., Zhang, S., Zhang, F., Wang, Y., Mi, L., ... Hu, R. (2014). Molecular characterization of three ferret badger (*Melogale moschata*) rabies virus isolates from Jiangxi province, China. *Archives of Virology, 159*(8), 2059–2067.

Zhao, J. H., Zhao, L. F., Liu, F., Jiang, H. Y., & Yang, J. L. (2019). Ferret badger rabies in Zhejiang, Jiangxi and Taiwan, China. *Archives of Virology, 164*(2), 579–584.

Zhao, P., Liu, S., Zhong, Z., Jiang, T., Weng, R., Xie, M., ... Xia, X. (2018). Analysis of expression profiles of long noncoding RNAs and mRNAs in brains of mice infected by rabies virus by RNA sequencing. *Scientific Reports, 8*(1), 11858. https://doi.org/10.1038/s41598-018-30359-z.

Zieger, U., Marston, D. A., Sharma, R., Chikweto, A., Tiwari, K., Sayyid, M., ... Horton, D. L. (2014). The phylogeography of rabies in Grenada, West Indies, and implications for control. *PLoS Neglected Tropical Diseases, 8*(10), e3251. https://doi.org/10.1371/journal.pntd.0003251.

Zulu, G. C., Sabeta, C. T., & Nel, L. H. (2009). Molecular epidemiology of rabies: Focus on domestic dogs (Canis familiaris) and black-backed jackals (Canis mesomelas) from northern South Africa. *Virus Research, 140*(1), 71–78. https://doi.org/10.1016/j.virusres.2008.11.004.

CHAPTER 6

Rabies in terrestrial animals

Thomas Müller, Conrad M. Freuling

Institute of Molecular Virology and Cell Biology, Friedrich-Loeffler-Institut, Federal Research Institute for Animal Health, Greifswald-Insel Riems, Germany

6.1 Introduction

There is probably no other word that triggers such a "spiral of fear" and different reactions among people than "rabies," independent of the diversity of life experiences, acquired knowledge, and occupation or avocation as well as geographic location and travel history. These reactions often result from the fact that rabies is a significant cause of human and animal mortality. Rabies is one of the oldest recognized zoonoses and defined as an acute progressive encephalomyelitis. Disease distribution encompasses all continents, with the exception of Antarctica. Recognized etiological agents consist of at least 18 recognized and putative lyssavirus species (Amarasinghe et al., 2018). While in principle all mammals are susceptible to rabies in varying degrees, the primary reservoirs reside in the orders Carnivora and Chiroptera. The lyssavirus reservoir in its entire complexity is cryptic and the plethora of lyssavirus species and viral variants maintained by a diversity of abundant reservoir hosts presents a formidable challenge to a strict concept of disease eradication (Rupprecht et al., 2008).

In this chapter, we focus on the prototype species for lyssaviruses, Rabies virus (RABV), in mammals other than chiroptera. RABV infections are responsible for more than 99% of all human rabies deaths, hence they pose a serious public health threat worldwide. RABVs are transmitted by a wide range of nonvolant mammalian hosts with a global distribution (World Health Organization, 2018); however, only a few species are able to maintain the virus independently. Therefore, this chapter especially elucidates the concept and the occurrence of RABV in reservoir and spillover hosts.

6.1.1 General pathogenesis

In rabies pathogenesis, the virus has to bypass the barrier of the skin, which usually occurs by the bite inflictions of diseased and infectious animals. Other routes of transmission such as by the oral route have been experimentally demonstrated (Charlton & Casey, 1979; Fischman & Ward, 1968), but do not have any relevance in rabies epidemiology. Notably, due to intermittent shedding and a species-specific resistance, not every bite of an infected animal leads to the development of rabies. Virus from infectious saliva may initially infect muscles or peripheral nerves directly (Fooks et al., 2017). After receptor-mediated entry into a peripheral neuron, the lyssavirus travels by retrograde transport in the neuron's axon within endosomal transport vesicles to the spinal cord (Gluska et al., 2014; Klingen, Conzelmann, & Finke, 2008) via either the dorsal root (sensory neuron) or the ventral root (motor neuron) ganglia (Begeman et al., 2017). Lyssaviruses effectively avoid and negatively regulate the activation of an immune response (Schnell, McGettigan, Wirblich, & Papaneri, 2010; Scott & Nel, 2016).

In the brain, further replication and trans-synaptic spread results in centrifugal dissemination within the CNS with multiple infected neurons. With increasing neuronal dysfunction, the onset of clinical signs begins (Fu & Jackson, 2005). The variable times from initial virus entry at the periphery, centripetal spread to the CNS, and subsequent massive replication leading to disease dictate the diversity in incubation periods reported, both in naturally and in experimentally infected animals. The duration of the clinical stage can also vary to up to 10 days, but usually ends with the death of the animal after coma and cardiac arrest. The specifics in rabies pathogenesis founded on both epidemiological and experimental data provide the basis for practical considerations in rabies control, for example, the practice of a 10-day observation period as recommended by the WHO (World Health Organization, 2018). If a dog, cat, or ferret potentially exposes a person or animal, and rabies needs to be ruled in or out, the biting animal may simply be observed for 10 days. If the animal remains alive and well during this 10-day period, the risk of rabies virus transmission from the potential exposure to this animal is negligible. If there were a risk of viral shedding and transmission, the animal would have manifested signs of clinical rabies (sometimes including sudden death) during the observation period. Any clinical sign or sudden death needs to be confirmed by laboratory investigations. Interestingly, confinement and observation of suspect animals had already been a veterinary hygienic measure before the pathogenesis of the disease was uncovered (Blancou, 2004).

6.1.2 Incubation period

The incubation period is the time elapsed between exposure to RABV, for example, through the bite of an infected animal, and the point when clinical symptoms and signs are first apparent. In general, the incubation period varies between a few days to several months (usually 2–3 months), depending on the type of RABV strain, the susceptibility and immune status of the host, the viral inoculation dose, the site of entry in the body, the density of motor endplates at the wound site, and the proximity of virus entry to the CNS (Hemachudha, Laothamatas, & Rupprecht, 2002; Hemachudha et al., 2013; Ugolini, 2011).

In fact, little is known about the incubation periods of animals due to naturally acquired rabies because it is not possible to trace back the time point of exposure under natural conditions, in particular for wildlife. Even for domestic animals and livestock, exposure events to RABV in endemic areas are vague and very often go unnoticed. Hence, most data for incubation periods in different animal species were obtained from experimental studies, but they

TABLE 6.1 Reported minimum and maximum incubation periods for rabies in selected experimentally infected mammals.

Species	Incubation period	Reference
Dogs	4–92	Bindrich and Olechnowski (1959), Hammami et al. (1999)
Cats	14–43	Richards (1962), Soulebot et al. (1981)
Cattle	10–15	Hudson, Weinstock, Jordan, and Bold-Fletcher (1996a)
Horses	6–27	Hudson, Weinstock, Jordan, and Bold-Fletcher (1996b)
Sheep	9–40	Hudson et al. (1996a), Soria Baltazar, Artois, and Blancou (1992)
Foxes	4–153	Black and Lawson (1970), Richards (1962)
Arctic foxes	7–37	Follmann, Ritter, and Baer (1988), Follmann, Ritter, and Hartbauer (2004)
Raccoon dogs	10–34	Freuling et al. (2017), Moore et al. (2017)
Raccoons	7–97	Niezgoda et al. (1991), Richards (1962)
Jackals	24–58	Yakobson et al. (1999)
Skunks	11–201	Charlton, Webster, and Casey (1991), Moore et al. (2017)
Mongooses	9–24	Chaparro and Esterhuysen (1993), Moore et al. (2017)
Coyotes	11–57	Richards (1962)

have the disadvantage that they reflect natural inoculation doses only to a limited extent (Table 6.1). Also, often relatively high inoculation doses were chosen to make sure control animals in such experimental studies died.

The high variability of incubation periods resulted in a commonly imposed quarantine period for exposed animals (mostly for dogs but also other animals) of six months. This period is based on epizootic and experimental evidence that if exposed animals are going to develop rabies from an exposure, there is a strong likelihood that it will occur within 6 months of the exposure.

6.1.3 Clinical signs

Once an animal is exposed to rabies, the probability of infection depends upon the individual, the species, the RABV variant, the amount of virus, and the severity and route of the exposure. Multiple deep bites to the head and face from a proven rabid animal would be more likely to cause infection than a superficial wound of a distal extremity (Turkmen et al., 2012).

The first clinical signs of rabies in infected animals may be nonspecific and include general lethargy, fever, poor appetite, vomiting, and anorexia. However, the clinical course is one way in that there is little waxing and waning of clinical signs, but a progressive deterioration in clinical condition. As rabies is strongly neurotropic, clinical signs are directly associated with effects from viral infection of the CNS. One of the first clinical signs may be changes in behavior and may consist of episodes of mild or dramatic abnormalities. For example, rabid animals may become more reclusive, somnolent, or attention-seeking than normal and only bite as soon as people get close to them in an attempt to assess their atypical behavior. Unpredicted and intermittent attacking of animate (humans or other animals), inanimate, or unseen or unapparent objects has been reported. Other animals tend to get aggressive biting at anything, especially when invoked by auditory, visual, or tactile stimuli. Two forms of rabies–the excitatory or "furious" form and the paralytic or "dumb" form—are believed to be characteristic for dogs (Lackay, Kuang, & Fu, 2008).

Most rabid animals show signs of CNS disturbance. The most reliable indicators are sudden and severe behavioral changes and unexplained paralysis that worsen over time. Behavioral changes can include sudden loss of appetite, signs of apprehension or nervousness, irritability, and hyperexcitability. The animal may seek solitude, or an otherwise unfriendly animal may become friendly. Uncharacteristic aggressiveness can develop, and wild animals may lose their fear of people. Animals that are normally nocturnal may be seen wandering around during the daytime.

The furious form of rabies is the classic "mad-dog" syndrome, although it is seen in all species. The animal becomes irritable and may viciously and aggressively use its teeth and claws with the slightest provocation. The posture is alert and anxious, with pupils dilated. Noise can provoke an attack. Such animals lose fear and caution of other animals. Young pups seek out human companionship and are overly playful, but will bite even when petted and become vicious within a few hours. As the disease progresses, seizures and lack of muscle coordination are common. Death is caused by progressive paralysis. Postmortem analyses often demonstrate foreign objects such as feces, straw, sticks, and stones in the stomach, particularly in dogs and ruminants. Also, animals frequently have self-inflicted wounds.

6.2 Lyssavirus infections in reservoir species

6.2.1 Basic concepts of reservoir host ecology for terrestrial rabies

In a broader sense, the term reservoir refers to an ecologic system in which an infectious agent survives independently. It encompasses all potential host populations, including that of any intermediate host or vector (if applicable), in the context of any environmental setting and ecological requisite required to maintain the agent indefinitely (Ashford, 2003). Vertebrate hosts that form an essential part of the system are reservoir hosts, which, once discovered, elucidate the complete life cycle of infectious diseases, providing effective prevention and control (Ashford, 2003). Usually, these reservoir hosts are perceived as primary hosts that maintain a pathogen but show no ill effects and serve as a source of infection.

In this sense, however, rabies and its causative agents clearly are the exception rather than the rule (Marston et al., 2018). Rabies reservoir species provide a source of susceptible individuals for RABV and are capable of sustained intraspecies virus maintenance within a geographic area; however, there is no "subclinical infection," carrier state, or chronic shedding of RABVs (Zhang et al., 2008). In fact, the various RABV variants are exquisitely adapted to a specific reservoir host population, within which they cause fatal infection but still achieve transmission to the next host before the infected host dies (Fooks et al., 2017). Distribution, abundance, population density, and contact rate are essential prerequisites for a reservoir species in order to support the virus-host relationship. In addition, virus maintenance may also take advantage of potentially competent coexisting species, which may serve as an efficient ancillary vector (Bell, 1980; Childs, Trimarchi, & Krebs, 1994).

Intriguingly, rabies is the most significant viral zoonosis associated with bats (Chiroptera). Chiroptera are reservoirs for at least 16 of the 18 species of lyssaviruses known today (World Health Organization, 2018) (see Chapters 5 and 7), and almost all bat lyssaviruses have caused fatal spillovers (cross-species transmission) into humans and terrestrial mammals (Johnson, Vos, et al., 2010; Messenger, Smith, Orciari, Yager, & Rupprecht, 2003; Van Thiel et al., 2008). However, RABV is the only lyssavirus species known to have multiple primary reservoir hosts in two different orders in the animal world, reflecting its astonishing ability for host switching and adaptation to gain advantage to given circumstances in the course of evolution (Badrane & Tordo, 2001; Marston et al., 2017).

Interestingly, bats are only primary hosts of RABV in the New World (World Health Organization, 2018), which may complicate disease elimination strategies in terrestrial animals in the Americas (Velasco-Villa, Mauldin, et al., 2017). Unlike all other lyssaviruses, however, a broad range of mammalian species primarily within the Carnivora order serve as reservoir hosts of RABV worldwide, causing huge endemics known as terrestrial rabies (Marston et al., 2017). Depending on their susceptibility, coevolution, and ranges of distribution, these mammalian reservoir hosts involve representatives of the families of *Canidae, Procyonidae, Herpestidae, Mephitidae, Viverridae*, and *Mustelidae* (Bourhy et al., 1999; Hanke et al., 2016; Kuzmin, Hughes, Botvinkin, Gribencha, & Rupprecht, 2008; McElhinney et al., 2006; Nel, Thomson, & von Teichman, 1993; Velasco-Villa, Mauldin, et al., 2017). Here, multiple lineages of RABV have coevolved in domestic and wild reservoir species and circulate within host conspecifics. So far, wildlife rabies reservoirs are primarily found in the Northern Hemisphere. The fact that the marmoset (*Callithrix jacchus*) has recently been identified as a new wild rabies reservoir host in Brazil (Favoretto, de Mattos, Morais, Araujo, & de Mattos, 2001) clearly indicates that improvement of rabies surveillance in other parts of the world will most likely expand the list of reservoir species in the near future.

In most regions of the world, unfortunately, multiple reservoirs are present that may or may not maintain species-adapted variants of RABV, as exemplified for North America and Europe. In countries where canine rabies is endemic, the high burden of dog-mediated rabies can mask the existence of additional rabies wildlife reservoirs on the one hand, but at the same time can favor sustained spillovers (host switch) into wildlife on the other hand. For the latter, Turkey is a good example (Marston et al., 2017; Vos et al., 2009).

6.2.2 Reservoir hosts for terrestrial rabies

6.2.2.1 *Canidae*

The Canidae represent a group of species in one family of the order Carnivora, suborder Caniformia (Sillero-Zubiri, 2018), that occur on almost all continents, where they have arrived either independently or were introduced by humans over extended periods. Members of this family are called canids and include 34 closely related species that diverged within the last 10 million years, including domestic dogs, wolves, coyotes, foxes, jackals, dingoes, and many other extant and extinct dog-like mammals (Sillero-Zubiri, 2018). Some but not all of these species play a crucial role in the maintenance and spread of RABV. Ten of the 34 member species of this family are known reservoir hosts for terrestrial RABV with different geographic distributions.

6.2.2.2 Domestic dogs

Domestic dogs (*Canis lupus familiaris*) in their broad genetic and morphologic diversity and worldwide distribution are the only recognized domestic reservoir host for terrestrial RABV and by far pose the greatest threat to public health (World Health Organization, 2018). Undoubtedly, the ancestor of modern dog breeds is the grey wolf; however, the geographic and temporal origins remain controversial (Frantz et al., 2016). The first appearances of dogs are dated to Europe ~15,000 years ago and in Far East Asia more than 12,500 years ago (Larson et al., 2012; Pionnier-Capitan et al., 2011). It is unknown at what point in time the altered social behavior of dogs as opposed to wolves created an ecological niche for the rabies virus to be transmissible among dogs.

Since its first written record in the Mesopotamian Codex of Eshnunna of Babylon circa 1930 BC (Dunlop & Williams, 1996), dog-mediated rabies has been a scourge both for its prevalence as well as its status as a dual public horror and biomedical travesty (Rupprecht et al., 2008).

Whether RABV maintenance and transmission to humans driven by domestic dogs was already widespread millennia ago remains elusive. However, the high occurrence of virus variants belonging to the cosmopolitan cluster of RABV strongly suggests that in recent history, human migration and exploration played a crucial role in the global spread of dog-mediated rabies (Troupin et al., 2016). Over a long period of time, rabies control in dogs seemed impossible. Only during the past 100 years have interventions targeted at this host species resulted in regional successes. Indeed, many historical and current precedents exist, demonstrating the viability of this concept both in developed and developing countries (see Lembo, Craig, Miles, Hampson, & Meslin (2013) for a review). Around the turn from the 19th to the 20th century, dog-mediated rabies was eliminated from a number of European countries (see Müller et al. (2012) for review), mainly through strict sanitary measures. This included notification of rabies, stray dog elimination, culling of suspect animals, dog muzzling requirements, tracing the movement of rabid dogs and their contacts, movement restrictions, and quarantine, all constituting "classical" measures to control dog rabies (Blancou, 2003; Théodoridès, 1986).

This changed with the advent of effective animal rabies vaccines at the beginning of the last century. Since then, mass immunization of dogs has become the mainstay of successful dog rabies control and eventual elimination (Coleman & Dye, 1996; Hampson et al., 2009). Intensive control efforts resulted in the virtual disappearance and elimination of dog-mediated rabies in certain parts of the world, as successfully demonstrated for Europe (Müller et al., 2012), North America (Rupprecht et al., 2008; Velasco-Villa et al., 2008), and Latin America (Vigilato et al., 2013). Despite these regional successes in dog rabies control, unfortunately, today domestic dogs still represent the most significant reservoir and vector for the disease. From a global perspective, more than 95% of human rabies deaths occur in underserved regions of middle- and low-income countries, particularly in Africa and Asia. In these regions, dog rabies is responsible annually for millions of suspect human exposures, the delivery of more than 14 million postexposure prophylaxis (PEP) regimens, and an estimated 70,000 human rabies deaths (Hampson et al., 2015; Taylor, Hampson, Fahrion, Abela-Ridder, & Nel, 2017). Even in developed countries such as India and China, human mortality through dog-transmitted rabies is high (Suraweera et al., 2012; Tang et al., 2005; Wu, Hu, Zhang, Dong, & Rupprecht, 2009). Dog-mediated human rabies is a neglected disease of poverty, affecting underprivileged communities and especially children younger than 15 years of age.

6.2.2.3 Foxes

Vulpes is a genus of the Canidae family, and the members of this genus are colloquially referred to as true foxes, meaning they form a proper clade (Macdonald & Sillero-Zubiri, 2004). Here, we refer to foxes when the word "fox" occurs on the common names of species. Foxes are omnivorous animals that are very social and live flexible lives. They are extremely adaptable and therefore are found all over the world, calling a wide range of terrains their home territory.

Red foxes

The red fox (*Vulpes vulpes*) can be regarded as the prototypic sylvatic reservoir host for RABV almost across its entire distribution range. Red foxes have one of the widest geographic distributions of all terrestrial mammals, with its range greatly expanded by human activity. Its current range extends from Western Europe and the Mediterranean basin including Northern Africa (Atlas Mountains) to the Middle East and the far eastern parts of the Eurasian continent, including the islands of Japan, through India and China. Multiple introductions of red foxes have occurred in North America and Australia, where it is an alien species.

The red fox is highly susceptible to infection (Blancou, 1988), although this may depend on the specific RABV variant, with fox-adapted variants being more virulent in foxes (Blancou, Aubert, Andral, & Artois, 1979). Together with the relatively short incubation period of only several weeks and the shedding of large amounts of infectious virus in saliva, these are factors that contribute to the maintenance of RABV in the fox population (Wachendörfer & Frost, 1980, 1992; Wandeler et al., 1974). Population turnover and behavioral traits such as mating and dispersal explain seasonal patterns of case frequencies, that is, peaks of cases in spring and autumn (Aubert, 1992; Wachendörfer & Frost, 1992).

Red fox-mediated rabies is endemic in many parts of the world. Since the middle of the 20th century, fox-mediated rabies has been a major public health problem in Europe (Wandeler, 2008) with different RABV lineages circulating in local fox populations (Bourhy et al., 1999; McElhinney et al., 2006). As a result, the majority ($\approx$70%) of rabies cases diagnosed in Europe were from foxes (EFSA, 2015). Due to the implementation of control measures using oral rabies vaccination (ORV) of foxes, the disease disappeared in large parts of Western and Central Europe (Freuling et al., 2013) while it is still endemic in Eastern Europe (EFSA, 2015; Müller & Freuling, 2018) (Fig. 6.1).

Historically, wildlife rabies has been reported only scarcely in great parts of Asia (Gruzdev, 2008). Data from enhanced rabies surveillance, however, indicate that as in Europe, the red fox seems to play an important role in the maintenance and epidemiology of the disease in Central Asia (Deviatkin et al., 2017; Kuzmin et al., 2004). Epidemiological studies confirmed the presence of fox rabies in Kazakhstan (Sultanov, Abdrakhmanov, Abdybekova, Karatayev, & Torgerson, 2016), Kirgizstan (Taichiev & Botvinkin, 2006), Siberia (Adelshin et al., 2015), and Mongolia (Boldbaatar et al., 2010; Botvinkin, Otgonbaatar, Tsoodol, & Kuzmin, 2008; Odontsetseg, Uuganbayar, Tserendorj, & Adiyasuren, 2009) (Fig. 6.2). Rabies

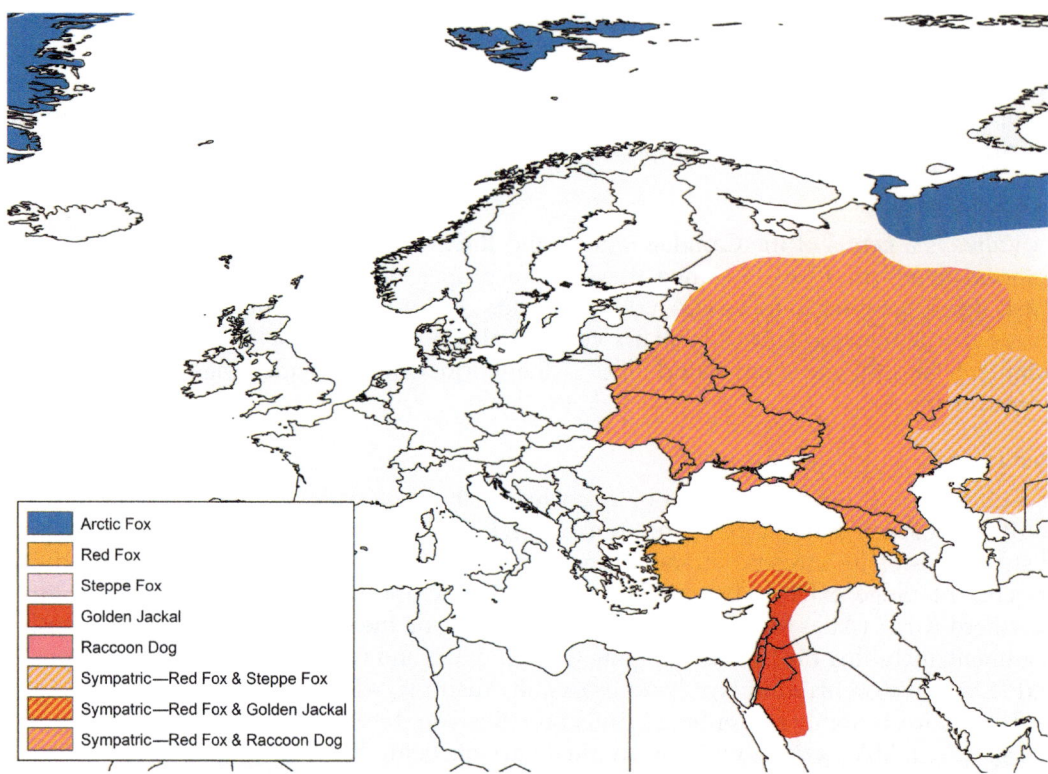

FIG. 6.1 Map of Europe and the Middle East showing the distribution of red fox-, raccoon dog-, and golden jackal-mediated rabies.

in foxes causing cases in livestock has also been reported in China (Liu et al., 2016; Tao et al., 2015). In countries in the Middle East, fox rabies has emerged in recent years (Horton et al., 2015; Seimenis, 2008). For example, since the beginning of the 1990s, fox-mediated rabies has been endemic in Oman (Hussain et al., 2013). How quickly epidemiological situations can change at the domestic animal—wildlife interface was demonstrated by the real-time observation of a sustained spillover of dog RABV variants into the red fox population, as experienced in Turkey (Marston et al., 2017). Initially, rabies in red foxes became endemic in the Aegean region of Turkey at the end of the 1990s (Johnson et al., 2003); however, within a few years, fox rabies spread across the western provinces of the country (Johnson, Fooks, Valtchovski, & Müller, 2006; Johnson, Un, Vos, Aylan, & Fooks, 2006).

Rabies in red foxes has also been reported in North America, although the situation there is a little different compared to other parts in the world. The only evidence of a rabies epidemic truly driven by red foxes was reported from Ontario, Canada, where, since the mid-1950s, an arctic variant of RABV was present in red fox populations that was eventually eliminated using ORV by the end of the 1990s (MacInnes et al., 2001; Rosatte et al., 2007). In other regions of North America where red foxes are sympatric, meaning they closely match the range of other reservoir species such as arctic foxes, skunks, and raccoons, they clearly represent a substantial proportion of rabid wildlife with an average proportion of animals tested positive in

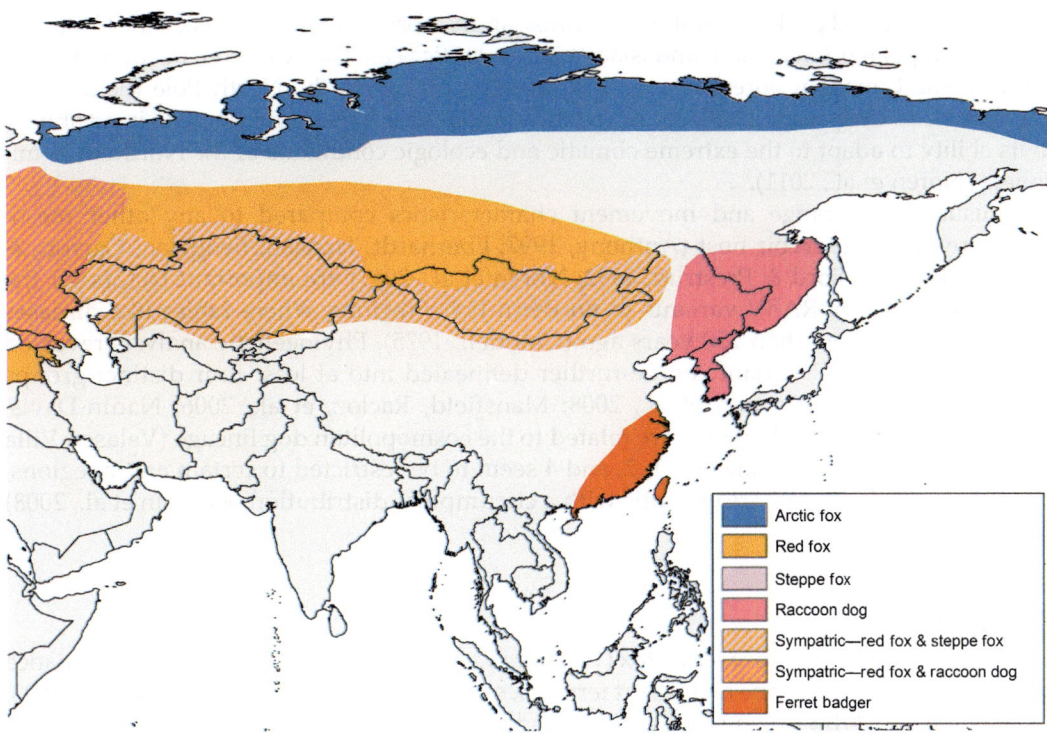

FIG. 6.2 Map of Asia depicting the presumed spread of wildlife-mediated rabies perpetuated by the red fox, the raccoon dog, the steppe fox, the arctic fox, and the ferret badger.

the United States of 6.2% (Ma, Monroe, Cleaton, Orciari, Li, et al., 2018a). In Alaska, where there is approximately an even divide between rabies-positive arctic foxes (52%) and red foxes (48%) (Kim et al., 2014), it is not clear if this species is a maintenance or spillover host for the arctic rabies virus in Alaska (Hueffer & Murphy, 2018). Elsewhere, the proportion of rabid red foxes ranged between 3% and 16% (Kelly & Sleeman, 2003; Wang et al., 2009), suggesting that they represent spillovers from other reservoir species. Interestingly, in 1993, a small cluster of rabies cases among foxes on Prince Edward Island, Canada, was shown to be of bat origin. Based on the close spatial and chronological detection of bat rabies in the three foxes from this island and also based on the presence, albeit in small numbers, of complete virions in the salivary glands of one fox, it was assumed that some degree of intraspecific transmission of the bat-associated RABV among the foxes had occurred (Daoust, Wandeler, & Casey, 1996).

These cross-species transmission events and host-switching events are critical and may have huge implications for the elimination of rabies in the Americas when the source of RABV is in reservoir bat species.

Arctic foxes

The Arctic fox (*Vulpes lagopus*), also known as the white fox, polar fox, or snow fox, is the only rabies reservoir species with a Holarctic distribution (Wozencraft, 2005). Arctic foxes live in Arctic tundra and pack ice habitats in Fennoscandia, northern Russia, Alaska, the Canadian Arctic Archipelago, Greenland, and islands in the Arctic Seas. They can even be found farther south in the Canadian boreal forests and on sea ice close to the North Pole (Feldhamer, Thompson, & Chapman, 2003). Its range throughout this harsh environment is attributed to its ability to adapt to the extreme climatic and ecologic conditions of the Northern hemisphere (Noren et al., 2011).

Unusual home range and movement characteristics compared to any other mesocarnivore rabies reservoir host (Anthony, 1997; Eberhardt, Hanson, Bengtson, Garrott, & Hanson, 1982; Frafjord & Prestrud, 1992; Noren et al., 2011) are important factors for the spread of arctic fox RABV variants in northern Polar regions, where rabies-like diseases were described more than 150 years ago (Crandell, 1975). Phylogenetic analysis revealed that the arctic RABV variant can be further delineated into at least four distinct groups (Hanke et al., 2016; Kuzmin et al., 2008; Mansfield, Racloz, et al., 2006; Nadin-Davis, Sheen, & Wandeler, 2012), which are related to the cosmopolitan dog lineage (Velasco-Villa et al., 2008). While arctic lineages 1, 2, and 4 seem to be restricted to certain arctic regions, viruses of arctic lineage 3 are enzootic with a circumpolar distribution (Kuzmin et al., 2008) (Figs. 6.2 and 6.4).

Other foxes

There is evidence that the Steppe fox (*Vulpes corsac*) is involved in rabies virus maintenance and circulation in the steppe and desert territories of Mongolia, Siberia (Botvinkin et al., 2008; Deviatkin et al., 2017; Kuzmin et al., 2004), and China (Feng et al., 2015) (Fig. 6.2).

In the Americas, besides the presence of RABV in red and arctic foxes, dog-related RABV variants have been shown to circulate in other species of foxes. Gray foxes (*Urocyon cinereoargenteus*) are considered a primary reservoir species responsible for maintaining gray fox-adapted RABV variants in Texas, Arizona, New Mexico, and Northwest Mexico (Ma, Monroe, Cleaton, Orciari, Li, et al., 2018a; Sidwa et al., 2005; Velasco-Villa, Escobar, et al., 2017). Also, the Crab-eating fox (*Cerdocyon thous*) and the Hoary fox (*Lycalopex vetulus*) are suspected to be reservoir hosts in Brazil (Antunes et al., 2018; Bernardi et al., 2005; Silva et al., 2009) (Fig. 6.4) while there is a small body of evidence that in Northwestern Peru, the Peruvian fox (*Lycalopex sechurae*) is maintaining an RABV variant adapted to this species (Velasco-Villa, Escobar, et al., 2017; Velasco-Villa, Mauldin, et al., 2017).

Rabies cases in bat-eared foxes (*Otocyon megalotis*) were first reported in South Africa, most likely via spillovers from the domestic dog. Sporadic cases were also documented in Namibia (Swanepoel et al., 1993). Based on surveillance data, it was suggested that this species is an independent carnivore reservoir of RABV in southern Africa (Bingham, 2005; Swanepoel et al., 1993). This was supported by molecular analyses of RABV from bat-eared foxes, of which the majority clustered genetically. Of note, some of the viruses isolated from bat-eared foxes were related to other wildlife or domestic rabies lineages, indicating limited spillovers (Sabeta, Mansfield, McElhinney, Fooks, & Nel, 2007).

As rabies surveillance in wildlife is inadequate in most parts of the world, for all these species of foxes other than red and arctic foxes, the question remains open as to whether they represent maintenance or spillover hosts.

6.2.2.4 Raccoon dog

The raccoon dog (*Nyctereutes procyonoides*), the only extant species in the genus *Nyctereutes* of the family Canidae, is indigenous to East Asia. Its original area of distribution is restricted to the eastern parts of China, Ussuria (Russia), Korea, and Japan. At the beginning of the 20th century, raccoon dogs of the *N. p. ussuriensis* subspecies were introduced into various territories and republics of the former Soviet Union in an attempt to make fur hunting more profitable. While most introductions failed because of unfavorable climatic conditions and food resources, successful introductions occurred in the European part of Russia and the Baltic states, from where this alien invasive species inexorably spread in all directions and conquered new habitats (Kauhala & Kowalczyk, 2011; Nowak, 1984). In Europe, the raccoon dog is now abundant throughout most of Central and Eastern Europe and Scandinavia, where it shows a sympatric occurrence with red foxes; an end of the spread is not in sight (Kauhala & Kowalczyk, 2011).

Their close genetic relation to true foxes might explain the high susceptibility of this species to infections from RABV. Experimental studies even revealed a higher susceptibility of raccoon dogs to endemic RABV lineages compared to red foxes. While the clinical course of infection differed only marginally, it appears that raccoon dogs shed RABV at higher frequencies than their vulpine counterparts (Botvinkin, Gribanova, & Nikifirova, 1983), which favors independent transmission cycles of RABV. There is strong evidence suggesting that the raccoon dog plays a role as reservoir in the epidemiology of rabies (Fig. 6.1). In Eastern Europe, since the end of the 1990s this species has become the second most commonly rabies-affected wild carnivore after the red fox (Niin, Laine, Guiot, Demerson, & Cliquet, 2008; Singer, Kauhala, Holmala, & Smith, 2009; Zienius, Pridotkas, Lelesius, & Sereika, 2011), and hence a contributor to the risk of rabies recurrence in countries within its range (EFSA, 2015). In Finland in the late 1980s, raccoon dogs were the cause of a rabies epizootic (Westerling, Anderons, Rimeicans, Lukauskas, & Dranseika, 2004) that was eliminated in 1991 (Sihvonen, 2001).

Maintenance of RABV by raccoon dogs has also been reported from the Far East, China, (Liu et al., 2016; Shao et al., 2011) and Korea (Kim et al., 2006; Oem, Kim, Kim, Lee, & Lee, 2014) (Fig. 6.2). In areas of Europe and Russia where there is sympatric occurrence with red foxes, RABVs maintained by the two species are phylogenetically very similar (Deviatkin et al., 2017). In terms of rabies control, fortunately, the available oral rabies vaccine baits developed for red foxes are also highly efficacious for this species (Bankovskiy, Safonov, & Kurilchuk, 2008; Cliquet et al., 2006, 2008; Schuster et al., 2001) as well as the bait distribution system and chosen vaccination strategies (Cliquet et al., 2012; Müller et al., 2015). Efforts to control rabies in raccoon dogs in other parts of the world have not yet been undertaken.

6.2.2.5 Jackals and coyotes

In the paraphyletic *Canis* genus, the coyote (*Canis latrans*) and the golden jackal (*Canis aureus*) are much more closely related to wolves and dogs than the African black-backed (*Canis mesomelas*) and side-striped (*Canis adustus*) jackals (Lindblad-Toh et al., 2005; Viranta, Atickem, Werdelin, & Stenseth, 2017). Whereas side-striped jackals have a sub-Saharan distribution, black-backed jackals are native to two geographically separated areas in Eastern and Southern Africa.

206 6. Rabies in terrestrial animals

Unlike the side-striped jackal, the black-backed jackal tends to shy clear of human settlements (Nattrass, Conradie, Drouilly, & O'Riain, 2017). Both species appear to be a reservoir for rabies in southern Africa (Bingham, 2005; Bingham & Foggin, 1993) (Fig. 6.3). Phylogenetic analysis of RABV isolates supports the hypothesis that black-backed jackals are capable of sustaining rabies cycles independent of domestic dogs (Cohen et al., 2007; Mansfield, McElhinney, et al., 2006; Zulu, Sabeta, & Nel, 2009). However, in some areas, the RABV virus isolates from jackals seem to be closely related to previously characterized viruses from dogs from the same geographic region, suggesting a common dog-jackal transmission cycle (Bellan et al., 2012). In Zimbabwe, the side-striped jackal is responsible for 80% of recorded cases of rabies, probably as a result of frequent contact with domestic dogs in human settlements, although there is no jackal-mediated rabies observed in the national parks of the country (Bingham, Foggin, Wandeler, & Hill, 1999). The spread of rabies in Namibia and northern South Africa in the 1950s and 1970s, respectively, is believed to have been prompted by independent transmission cycles in black-backed jackals, where it is endemic nowadays (Hübschle, 1988; Scott, Coetzer, & Nel, 2016; Swanepoel et al., 1993).

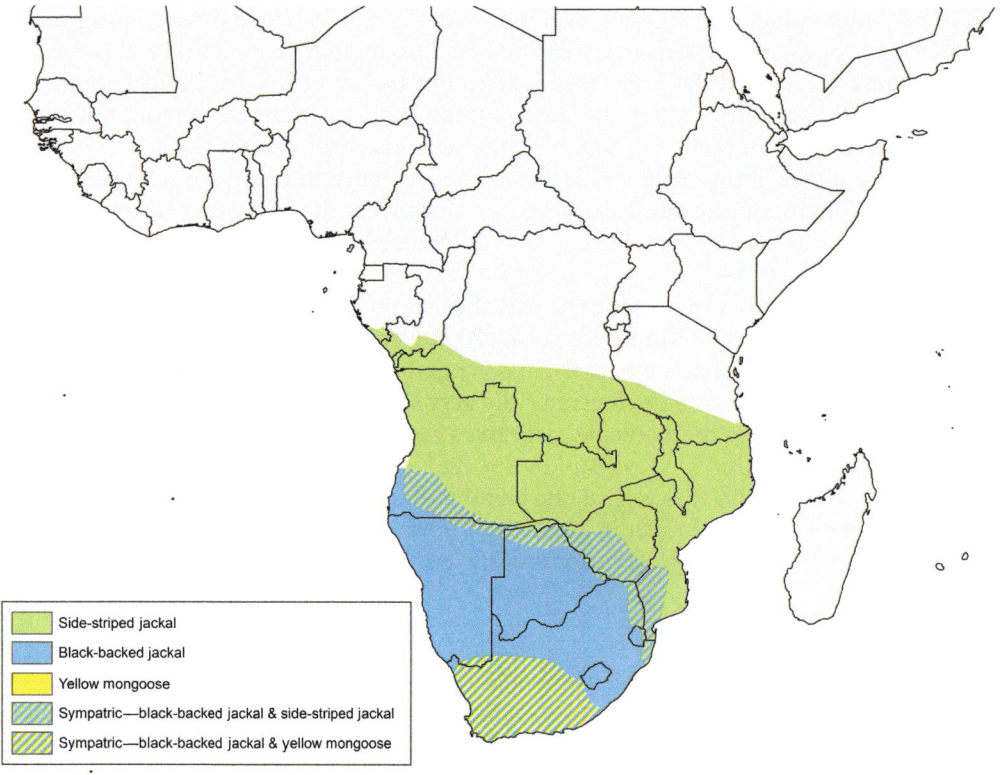

FIG. 6.3 Expected range of wildlife rabies in Africa driven by African black-backed and side-striped jackals as well as the Yellow mongoose.

Golden jackals (*C. aureus*) range from Africa to Europe, the Middle East, Central Asia, and Southeast Asia. Of note, genetic analyses support the delineation of *C. aureus* into *C. anthus* (African wolf) and *C. aureus* (Koepfli et al., 2015). Due to its tolerance of dry habitats and its omnivorous diet, *C. aureus* can live in a wide variety of habitats (Moehlman & Hayssen, 2018). In Europe, this species appears to be expanding its distribution area by colonizing new habitats from the Balkan Peninsula into Central Europe, with evidence of reproduction in Italy and Austria, and sightings in Germany, Slovakia, and the Czech Republic (Arnold et al., 2012). In Europe, only a few rabies cases were reported from this species (Johnson, Fooks, et al., 2006; Johnson, Un, et al., 2006). However, it was reported that golden jackals play a role in wildlife rabies in the Middle East (Seimenis, 2008) (Fig. 6.1). In Israel, surveillance data from past decades indicate that both foxes and golden jackals act as wildlife reservoirs, with a shift toward jackal rabies in recent years (Shimshony, 1997; Yakobson, David, & Aldomy, 2004). Starting in 2017, there has been an unprecedented increase of jackal-associated rabies in Israel (Yakobson, personal communication).

The coyote (*C. latrans*), the smaller relative of the grey wolf, is abundant throughout North America, including Mexico, and sporadically occurs in Central America. It fills much of the same ecological niche as the golden jackal in the Old World (Tigas, Van Vuren, & Sauvajot, 2002). This species is highly adaptable and readily tolerates living near humans. The first recorded epizootics of rabies in coyotes date from the beginning of the 20th century in California, Oregon, Nevada, and Utah (Velasco-Villa, Mauldin, et al., 2017). From 1988 through 1994, an epizootic of canine rabies transmitted by coyotes was reported in south Texas (Clark et al., 1994), which posed an imminent threat to the human and susceptible domestic pet populations (Sidwa et al., 2005). The implementation of an oral coyote rabies vaccination program in 1995 resulted in eventual elimination of the dog-coyote RABV variant (Meehan, 1995; Sidwa et al., 2005).

6.2.2.6 Raccoons

The common raccoon (*Procyon lotor*) is the largest member of the *Procyonidae*, a New World family of the order Carnivora. While the native range of the raccoon is Central and North America, it has been expanding its range in Central Europe and the Russian Federation after escapes by farmed animals or deliberate introductions (Beltran-Beck, Garcia, & Gortazar, 2012; Vos, Ortmann, Kretzschmar, Köhnemann, & Michler, 2012). Raccoons were also introduced and have established populations in the Caucasus region and Japan (Timm, Cuarón, Reid, Helgen, & González-Maya, 2016). Today, raccoons occur in four Asian (Japan, Georgia, Azerbaijan, Iran) and a further 20 European countries (Farashi, Kaboli, & Karami, 2013; Gehrt, 2019); however, its main distribution area outside the New World is in Germany (Fischer et al., 2015).

Despite their distribution in the northern hemisphere, raccoons are only recognized as rabies reservoirs and vectors of the disease in North America (Fig. 6.4). Raccoons were thought to be the only member of the *Procyonidae* that acts as a rabies reservoir. However, there are reports of rabies cases in the Kinkaju *(Potus flavus)*, another member of the Procyonidae, in Peru (Vargas-Linares, Romaní-Romaní, López-Ingunza, Arrasco-Alegre, & Yagui-Moscoso, 2014). This supports the assumption that this species, distributed over large areas in Latin America, may also play a role in the maintenance and transmission of the disease (Velasco-Villa, Mauldin, et al., 2017).

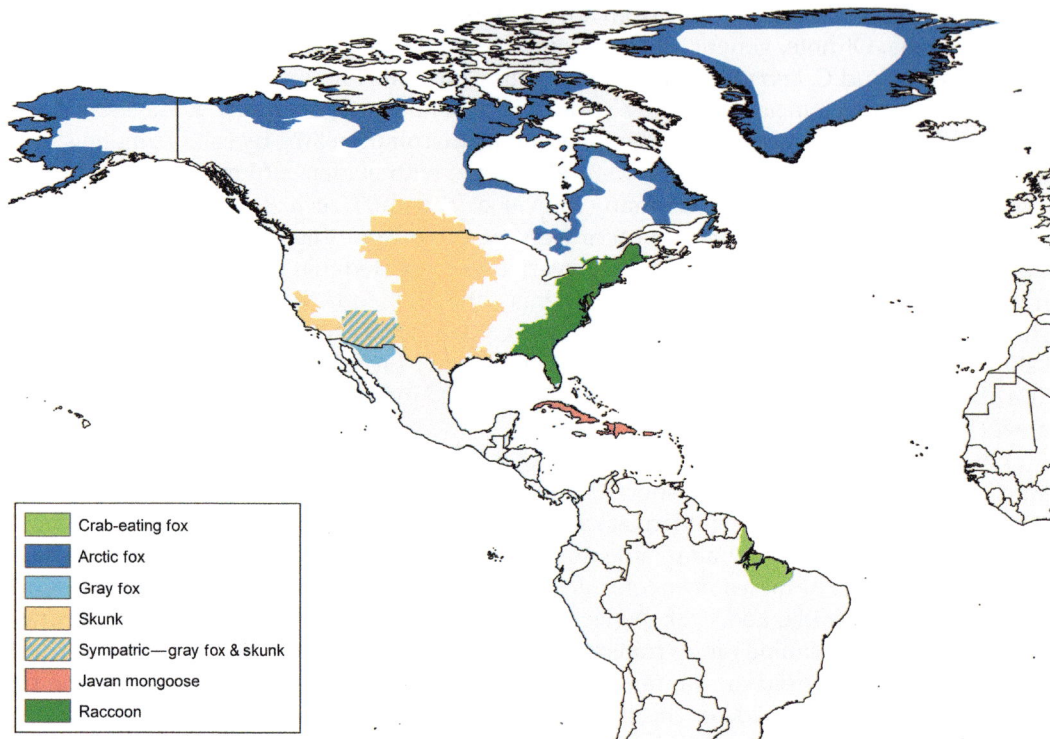

FIG. 6.4 Map of the Americas with the assumed distribution of wildlife rabies by major reservoir species such as the arctic fox, the North American raccoon, the gray fox, the crab-eating fox, the spotted skunk, the striped skunk, and the small Indian mongoose.

In the United States, raccoons are the most significant reservoir from a public and animal health perspective. Historically, raccoons as rabies vectors had a notable role only in the southeastern United States. The first record of a rabid raccoon in Florida was reported in 1947, and by the mid-1950s, it was apparent that a significant epizootic was in progress in the state (McLean, 1970, 1971). Translocation of incubating raccoons from Florida to West Virginia in the mid-1970s (Nettles, Shaddock, Sikes, & Reyes, 1979) initiated an epizootic event that progressively spread throughout the eastern United States (Raccoon Rabies Epizootic—United States, 1994) and into eastern Canada (Wandeler, 1999). It was shown that the responsible virus was a specific RABV variant (Badrane & Tordo, 2001; Nadin-Davis, 1998; Nadin-Davis, Colville, Trewby, Biek, & Real, 2017; Szanto, Nadin-Davis, Rosatte, & White, 2011) that seemed to be adapted to circulating in raccoons with specific pathological lesions (Hamir, Moser, & Rupprecht, 1996). Interestingly, phylogenetic analyses showed that the ancestral origin of this virus was a sustained spillover (host switch) from bats (Badrane & Tordo, 2001; Kuzmin et al., 2012).

While raccoons might become infected with other RABV variants through spillover, infected raccoons cause "spillovers" into a wide variety of other animals, including skunks, red and grey foxes, opossums, woodchucks, livestock, horses, and, most importantly, companion animals

(Wallace et al., 2014). Due to their behavioral flexibility (Daniels, Fanelli, Gilbert, & Benson-Amram, 2019), raccoons occur at especially high densities in suburban areas, offering potential interactions with companion animals. In fact, the raccoon variant is responsible for the highest number of rabid companion animals reported in the United States, and raccoons were four times more likely to transmit RABV to other species when compared to skunks (Wallace et al., 2014).

Experimental studies of rabies in raccoons showed a variable mortality, depending on inoculation dose and virus strain used (Table 6.1). While lower doses ($<10^{4.5}$ MICLD$_{50}$) led to an incubation period of 29–97 days postinfection (Niezgoda, Diehl, Hanlon, & Rupprecht, 1991), more than 80% of the naïve control animals challenged with $10^{6.9}$ MICLD$_{50}$ succumbed in a vaccine efficacy study (Gilbert, Johnson, Walker, et al., 2018). Here, incubation periods ranged between 10 and 14 days. In another study on the efficacy of a vaccine in free-ranging raccoons, 10 of 11 (91%) control raccoons succumbed within 30 days to an RABV challenge at a dose of $10^{4.9}$ MICLD$_{50}$ (Rupprecht et al., 1993).

Generally, survivors of experimental challenge infections were seronegative, thus questioning the field data of seroconversion rates of up to 30% of nonvaccinated raccoon populations. The origins of these observations have been intensively discussed (Hanlon, 2013). A challenge infection of these naturally acquired seropositive animals indicated a certain level of protection as opposed to serologically naïve animals from field areas (Blanton et al., 2018). One eminent challenge in the interpretation of seroprevalence is the variation of cut-off values used for serum neutralization tests, with lower cut-offs having a higher risk of false positive reactions. Unfortunately, many studies lack reported individual titers, and to this end, rabies is one example of the difficulties in interpreting serological testing for wildlife diseases (Gilbert et al., 2013).

The westward range expansion of the raccoon variant of the RABV is prevented by an oral rabies vaccination zone (Ma, Monroe, Cleaton, Orciari, Li, et al., 2018a). In 2017, both vaccinia-rabies glycoprotein recombinant vaccine baits (VRG) and adenovirus-rabies glycoprotein recombinant vaccine baits (ONRAB) were distributed. Post-ORV, the population immunity levels have averaged 30% across several years for VRG and have led to concern about the ability of ORV products to eliminate RABV circulation in raccoons (Elmore et al., 2017; Slate et al., 2009). Field trials using ONRAB suggest higher seroconversion rates as opposed to VRG (Fehlner-Gardiner et al., 2012; Gilbert, Johnson, Nelson, et al., 2018; Mainguy, Fehlner-Gardiner, Slate, & Rudd, 2013; Rosatte et al., 2009; Slate et al., 2014). However, analyses from ORV zones in New York only demonstrated a limited increase of seroprevalence after baiting with ONRAB (Pedersen et al., 2019).

6.2.2.7 Skunks

Skunks are mammals comprising genera (*Conepatus*, *Mephitis*, and *Spilogale*) of the family *Mephitidae*, which exclusively occur in North and South America (Wilson & Reeder, 2005). Of the 10 species of skunks, spotted skunks (*Spilogale putorius*) and striped skunks (*Mephitis mephitis*) seem to play an important role in the epidemiology of the disease. Both have a wide sympatric range extending through southern Canada, the United States, and northern Mexico.

Several epizootics of rabies in these two skunk species have been reported from southern and north-central regions of the United States as well as north-central Mexico (Dyer et al.,

2014; Smith, Orciari, Yager, Seidel, & Warner, 1992; Velasco-Villa et al., 2008) (Fig. 6.4). While recent outbreaks are mainly driven by dog-derived RABVs, early rabies outbreaks in skunks are believed to have been related to bat-derived skunk RABV variants (Velasco-Villa, Mauldin, et al., 2017). The latter has been corroborated by recent reports of repeated sustained cross-species transmission (host switch) of RABVs associated with big brown bats to skunks during the early 2000s in the Flagstaff area of Arizona (Blanton, Robertson, Palmer, & Rupprecht, 2009; Leslie et al., 2006). Phylogenetic analysis revealed that each of these outbreaks was caused by an independent introduction of bat RABV into populations of skunks and other carnivores (Kuzmin et al., 2012).

Nowadays, due to its strong ability to adapt to human-modified environments, the striped skunk is the primary reservoir species for RABV in the United States, maintaining at least three different RABV variants, that is, the south central, north central, and California skunk RABV variant (Ma, Monroe, Cleaton, Orciari, Li, et al., 2018a). Unfortunately, the control of rabies in skunk populations using oral rabies vaccination represents a challenge for various reasons (Vos et al., 2017; Wohlers, Lankau, Oertli, & Maki, 2018).

6.2.2.8 Mongooses

Mongooses (family Herpestidae) are small feliform carnivores native to southern Eurasia and mainland Africa. Of the 29 species with the common name mongoose, only the yellow mongoose (*Cynictis penicillata*) and the small Indian mongoose (*Herpestes auropunctatus*) are known to be reservoirs for RABV. The small Indian mongoose is native to Asia (e.g., India, Pakistan, Iran, Iraq) but was introduced in the late 19th and early 20th century to several other parts of the world, including the Carribean, mainly to protect sugar cane plantations from rats and snakes. In Grenada, Cuba, the Dominican Republic (and Haiti by extension), and Puerto Rico, the small Indian mongoose forms a significant, if not primary, reservoir host for rabies (Seetahal et al., 2018) (Fig. 6.4). The first major outbreak among mongoose in this region was reported in Puerto Rico in 1950 (Tierkel, Arbona, Rivera, & De Juan, 1952), but clinical observations in mongooses suggested rabies as early as the beginning of the 20th century (Everard & Everard, 1988). Molecular analyses indicate that mongoose rabies emerged from separate introductions of dog-maintained viruses on these Caribbean islands (Nadin-Davis, Velez, Malaga, & Wandeler, 2008; Velasco-Villa, Mauldin, et al., 2017; Zieger et al., 2014).

The reasons are unclear as to why there is sustained transmission among the Indian mongoose only on these four islands in the Caribbean Sea and not elsewhere. Presumably, a high population density on the affected islands and the historical absence of endemic RABV in the dog populations of other Caribbean islands are responsible for this situation. Also, high numbers of seropositive animals, as found in Grenada, appear to be indicative for a high infection rate combined with a high frequency of biting among mongooses (Everard & Baer, 1974). Rabies control on these four Caribbean islands using population reduction was practiced, although without sustained success (Everard & Everard, 1988). Oral vaccination of mongoose seems to offer the best promise of an alternative approach to the elimination of rabies in these island environments. Efficacious and safe vaccine candidates have been tested in the small Indian mongoose (Blanton et al., 2006; Ortmann et al., 2018; Vos et al., 2018), but mongoose-specific vaccine baits and optimal baiting strategies are still in the developmental stage.

While 12 species of mongoose occur in southern Africa (Gilchrist, Jennings, & Veron, 2018), only the yellow mongoose (*Cynictis penicillata*) (Fig. 6.3) is regularly involved in the endemic rabies cycle that predominates on the central plateau of South Africa (King, Meredith, & Thomson, 1993; Swanepoel et al., 1993). Rabies cases were also reported in the slender mongoose (*Galerella sanguinea*) in Zimbabwe in the 1970s, leading to the assumption that the slender mongoose was the reservoir host species for this RABV variant in that country (Foggin, 1988).

Experimental infections revealed differences between the African mongoose RABV variant and the canid RABV. When mongooses were experimentally infected with both canid RABV and African mongoose RABV, a significantly higher proportion of mongooses inoculated with the latter virus died (Chaparro & Esterhuysen, 1993). In contrast, studies in mice suggest that mongoose RABV strains are less pathogenic than dog RABV strains (Seo et al., 2017).

Several molecular epidemiological studies have confirmed that the African mongoose RABV is distinctly different from the canid variant from southern Africa (Nel et al., 2005; Nel et al., 1993; Troupin et al., 2016; Van Zyl, 2008; Van Zyl, Markotter, & Nel, 2010; von Teichman, Thomson, Meredith, & Nel, 1995). Also, the genetic differences provide evidence that the African mongoose RABV was present in southern Africa substantially before the arrival of the canid variant in the early 1900s (Van Zyl et al., 2010). Besides South Africa, the circulation of the African mongoose RABV variant was confirmed for Namibia (Mansfield, McElhinney, et al., 2006) and Zimbabwe (Sabeta et al., 2008). Interestingly, in Zimbabwe the mongoose variant was isolated several times from African civets (*Civettictis civetta*), supporting the historical implication that this species from the viviridae family might support the maintenance of mongoose rabies (Sabeta et al., 2008).

Elsewhere, there have been reports of rabies in mongooses along their distribution (Everard & Everard, 1988). In India, mongoose bites have led to several human rabies cases (Chhabra, Ichhpujani, Tewari, & Lal, 2004; Mani, Moorkoth, Balasubramanian, Devi, & Madhusudana, 2016; Ratho, Prasad, & Bindra, 1997; Singh et al., 2001). Also, in Sri Lanka, mongooses were the most affected wild animal species, suggesting a potential role as reservoir (Karunanayake et al., 2014), although sequence analyses cannot confirm a mongoose-specific genetic variant (Arai et al., 2001). Possibly, overwhelming dog-mediated rabies and the lack of surveillance in wild animals, including mongoose, camouflage independent mongoose-mediated rabies transmission.

6.2.2.9 Ferret badgers

Ferret badgers comprise five species of mustelids of the genus *Melogale*, with a geographical occurrence in Southeast Asia. Against the background of the dominating dog-mediated rabies cycle, independent rabies in wildlife reservoir species may be overlooked. In China, the cases of rabies in the Chinese ferret badgers (*Melogale moschata*) were only systematically recorded during the past decade (Wang, Tang, & Liang, 2014). According to molecular epidemiological studies, RABV in ferret badgers formed independent infection cycles that probably originated from trans-species infection of dog rabies during long-term epidemics (Liu et al., 2010; Miao et al., 2018; Zhang et al., 2013; Zhang et al., 2009). The detection of ferret badger-associated human rabies supports their importance for public health (Wang et al., 2014).

Although Taiwan was declared rabies-free in humans and domestic animals for five decades, in 2013 ferret badgers were diagnosed with rabies. Since then, a variety of wildlife species tested positive for rabies. Retrospective analyses confirmed that rabies was present in Taiwan prior to 2013 (Chang et al., 2015). This is corroborated by phylogenetic analyses, which demonstrate distinct genetic lineages on the island of Taiwan with the closest relation to lineages circulating in mainland China (Lan et al., 2017; Miao et al., 2018; Tsai et al., 2016) (Fig. 6.2). Since the discovery of ferret badger-associated rabies in Taiwan, studies were undertaken to evaluate the efficacy of oral rabies vaccines (Hsu et al., 2017; Zhao et al., 2014), and baits (Wallace et al., 2018).

It is unknown whether there is independent Chinese ferret badger-associated transmission of RABV in other countries or whether other members of the genus *Melogale* also support independent transmission.

6.2.2.10 Marmosets

In northeastern Brazil, RABV was isolated from the white-tufted marmoset (*C. jacchus*), an indigenous nonhuman primate (Kotait et al., 2018). In the absence of dog-mediated rabies, exposure to infected marmosets is a primary source of rabies infection in humans in coastal states (Rio Grande do Norte, Ceara, Piaui, and Pernambuco) of Brazil (Favoretto et al., 2001). In recent years, there has been a potential spread as well as an increasing importance of marmoset rabies (Antunes et al., 2018; Kotait et al., 2018).

It was shown that the virus isolated from the marmoset is genetically and antigenically distinct from other variants (Favoretto et al., 2001; Kotait et al., 2018), supporting this species being a reservoir for RABV. Phylogenetically, the marmoset RABV is more closely related to insectivorous bat RABV lineages, implying a historical spillover (host switch) infection from bats with subsequent adaptation to marmosets (Kotait et al., 2018).

6.3 Spillover (dead-end) hosts for terrestrial rabies

While only a few carnivorous animal species are primary reservoir hosts for rabies, all other nonvolant mammal species are considered to be potential spillover hosts, provided they are exposed to infectious virus. Spillover hosts are unable to maintain the virus independently, meaning that infection of these species with RABV is a dead end. Although onward transmission of the virus occurs in only a few exceptional cases, nevertheless they may pose a potential threat as vectors for RABV transmission to humans. Regardless of the reservoir hosts, rabies has been diagnosed in a plethora of other spillover hosts. In Europe, for example, where fox- and raccoon dog-mediated rabies were predominant between 1990 and 2018, 42% of all rabies cases were reported from spillover hosts, including both domestic animals and wildlife (Fig. 6.5). By comparison, in the United States, 24% of all rabies cases reported between 1990 and 2018 affected species other than the main reservoir hosts of terrestrial rabies, for example, raccoons and skunks (Fig. 6.6). In the following sections, the focus is on a few spillover species that are continuously the subject of discussion regarding their role in RABV transmission.

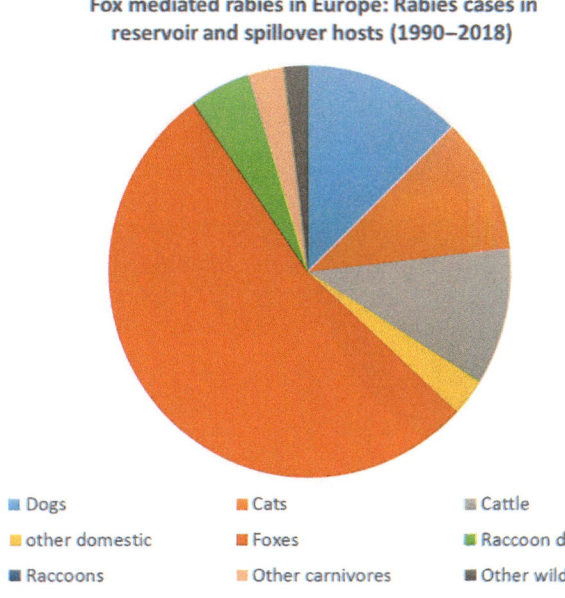

FIG. 6.5 Cases of rabies in reservoir and spillover hosts in Europe (1990–2018) resulting from fox- and raccoon dog-mediated rabies. *Data from WHO European rabies database (Friedrich-Loeffler-Institut, WHO Collaborating Centre for Rabies Surveillance and Research)*

6.3.1 Cats

With few exceptions, members of Felidae exploit a rather solitary existence. This may be one reason why no reservoirs have been described for any feline species, even when feral cats are present in high densities. This fact has to be taken into account when planning and implementing pet vaccination programs, which should focus on dogs.

While experimental studies suggests that cats (*Felis catus*) are more refractory to infection (Dean & Guevin, 1963; Vaughn, 1975), epidemiological data support that cat rabies is a result of spillover infections without a large degree of onward transmission (Vaughn, 1975). Data from the WHO Collaborating Centre for Rabies Surveillance and Research show that cat rabies on average accounted for 16% of all rabies cases in Poland, 25% in the Ukraine, and only 7% in Romania, with only limited temporal changes. In the United States, only about 1% of all submitted cats test positive for rabies, and cats account for between 4% and 6% of all reported rabies cases (Ma, Monroe, Cleaton, Orciari, Li, et al., 2018a). Wildlife feline species seem hardly affected by rabies, with few reported cases in cougars (*Felis concolor*), pumas (*Puma concolor*) (Krebs, Williams, Smith, Rupprecht, & Childs, 2003), lions (Bwangamoi, Rottcher, & Wekesa, 1990), and tigers (Pandit, 1950). Interestingly, the only exception to this is the bobcat (*Lynx rufus*), which is ranked third in other nonreservoir carnivores (Krebs et al., 2003).

6.3.2 Cattle

Cattle, like any other livestock, are dead-end hosts for rabies and as herbivorous animals are not well adapted to onward transmission of the virus (Vos et al., 2013). In many parts of the world, people rely on livestock for food and work, and therefore cattle play a significant

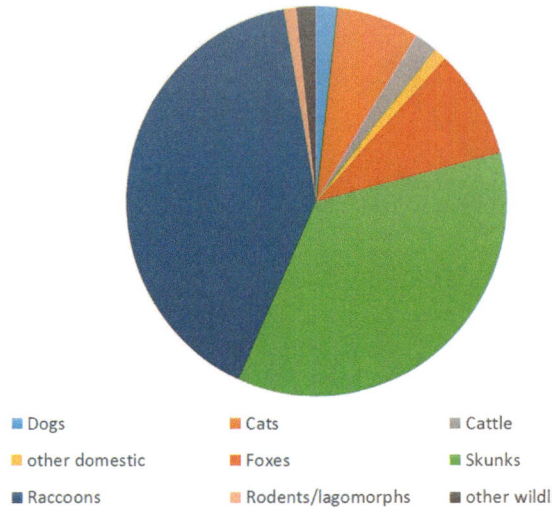

FIG. 6.6 Cases of rabies cases in reservoir and spillover hosts in the United States (2010–17) resulting from raccoon- and skunk-mediated rabies. *Data from Birhane, M. G., Cleaton, J. M., Monroe, B. P., Wadhwa, A., Orciari, L. A., Yager, P., ..., Wallace, R. M. (2017). Rabies surveillance in the United States during 2015.* Journal of the American Veterinary Medical Association, 250(10), 1117–1130. https://doi.org/10.2460/javma.250.10.1117; Blanton, J. D., Dyer, J., McBrayer, J., & Rupprecht, C. E. (2012). *Rabies surveillance in the United States during 2011.* Journal of the American Veterinary Medical Association, 241(6), 712–722. https://doi.org/10.2460/javma.241.6.712; Blanton, J. D., Palmer, D., Dyer, J., & Rupprecht, C. E. (2011). *Rabies surveillance in the United States during 2010.* Journal of the American Veterinary Medical Association, 239(6), 773–783. https://doi.org/10.2460/javma.239.6.773; Dyer, J. L., Wallace, R., Orciari, L., Hightower, D., Yager, P., & Blanton, J. D. (2013). *Rabies surveillance in the United States during 2012.* Journal of the American Veterinary Medical Association, 243(6), 805–815. https://doi.org/10.2460/javma.243.6.805; Dyer, J. L., Yager, P., Orciari, L., Greenberg, L., Wallace, R., Hanlon, C. A., & Blanton, J. D. (2014). *Rabies surveillance in the United States during 2013.* Journal of the American Veterinary Medical Association, 245(10), 1111–1123. https://doi.org/10.2460/javma.245.10.1111; Ma, X., Monroe, B. P., Cleaton, J. M., Orciari, L. A., Li, Y., Kirby, J. D., ..., Blanton, J. D. (2018a). *Rabies surveillance in the United States during 2017.* Journal of the American Veterinary Medical Association, 253(12), 1555–1568. https://doi.org/10.2460/javma.253.12.1555; Ma, X., Monroe, B. P., Cleaton, J. M., Orciari, L. A., Yager, P., Li, Y., ..., Wallace, R. M. (2018b). *Rabies surveillance in the United States during 2016.* Journal of the American Veterinary Medical Association, 252(8), 945–957. https://doi.org/10.2460/javma.252.8.945; Monroe, B. P., Yager, P., Blanton, J., Birhane, M. G., Wadhwa, A., Orciari, L., ..., Wallace, R. (2016). *Rabies surveillance in the United States during 2014.* Journal of the American Veterinary Medical Association, 248(7), 777–788. https://doi.org/10.2460/javma.248.7.777.

role in rural livelihoods and the economies of developing countries. They are providers of income and employment for producers and others working in sometimes complex value chains (Herrero et al., 2013). Cattle typically become infected by RABV after being bitten by reservoir hosts or, in Latin America, by vampire bats (*Desmodus rotundus*) causing paralytic rabies (Prieto & Baer, 1972). Patterns of rabies infection in livestock depend on the infection dynamics of disease reservoirs within a particular area and contact rates between cattle and reservoir populations (Haydon, Cleaveland, Taylor, & Laurenson, 2002). Cattle losses due to rabies are a major economic problem, but there is a particular shortage of published estimates of rabies incidence in livestock and the resulting economic cost, in particular cattle (Vos et al., 2013). However, published reports of bovine rabies in African and Asian countries, where the main source of disease transmission is the domestic dog, demonstrate that livestock losses from spillover can be substantial (Balako, Sisay, Hussein, & Asefa, 2018; Brookes et al., 2019; Feng et al., 2016; Lembo et al., 2008; Tenzin, Dhand, Dorjee, & Ward, 2011; Tenzin, Sharma, Dhand, Timsina, & Ward, 2010).

In any case, cattle are an important sentinel species for dog- and wildlife-mediated rabies, particularly in regions where surveillance in reservoir species is inadequate (Hikufe et al., 2019; Vos et al., 2013), whereas wildlife rabies appears to represent a greater risk to cattle than dog-mediated rabies.

6.3.3 Kudu

Since its first detection in the 1970s, rabies in the greater kudu (*Tragelaphus strepsiceros*) (Hassel, 2011; Hübschle, 1988; Swanepoel et al., 1993), a woodland antelope distributed throughout eastern and southern Africa (East, 1999), has been a mystery for its very limited geographical extent (mainly Namibia) and assumed mode of transmission (Scott et al., 2016).

It was speculated that horizontal transmission among kudu could be the reason for the observed epidemic waves in Namibia in the recent past (Barnard & Hassel, 1981; Barnard, Hassel, Geyer, & De Koker, 1982; Hübschle, 1988; Schneider, 1985; Scott, Hassel, & Nel, 2012). This would be remarkable in so far that it shakes a dogma regarding the mode of transmission of RABVs.

Although recent phylogenetic (Scott et al., 2013) and experimental studies (Hassel et al., 2018) support the possibility of onward horizontal transmission among kudu, it cannot plausibly explain the rapid spread of the disease in kudus over large territories. The fact that livestock, in particular cattle and other wildlife, is affected by rabies in the same areas where kudu rabies occurs (Hikufe et al., 2019) as well as earlier phylogenetic studies (Mansfield, McElhinney, et al., 2006; Mansfield, Racloz, et al., 2006) rather suggest that the jackal and kudu may form part of the same epidemiological cycle of rabies in Namibian wildlife. Thus, a combination of continuous spillover events from jackals to livestock and kudu and perhaps locally restricted horizontal transmission are more likely to explain the epidemiological pattern. Under these circumstances, the kudu cannot be regarded as an independent reservoir for rabies.

6.3.4 Wolves, African wild dogs, and other carnivores

Despite grey wolves (*Canis lupus lupus*) being genetically closest to dogs, it is widely accepted that they do not represent a reservoir for rabies and that cases in wolves represent incidental spillover infections (World Health Organization, 2018). There is no evidence for a species-specific RABV variant that is adapted for wolf-to-wolf transmission, and even the theoretically conceivable long-distance transmission of RABV by wolves has not been documented. Rather, because of their social structure and interactions, such infections often cause the extinction of an affected pack (Chapman, 1978). While some reports suggest that a rabies epidemic caused population decline in packs of grey wolves (Ballard & Krausman,

1997), others could not confirm such an effect on the population (Weiler, Garner, & Ritter, 1995).

Historically, starting from medieval times, severe epizootics in grey wolves resulting in numerous human victims were documented (Steele & Fernandez, 1991). Because of the reduction of the wolf population, the most prominent natural enemy of the fox, an increasing number of fox epizootics were observed in the 17th and 18th centuries (Blancou, 2004; Pastoret, Kappeler, & Aubert, 2004). It cannot be conclusively resolved whether the wolf was overrepresented due to its ferocity and humans' fear, or whether indeed there was wolf-mediated rabies at the time. Possibly, the population density of wolves was much higher and RABV variants may have been adapted to this transmission cycle, but would have disappeared with the extinction of the reservoir. Under the current epidemiological situation, wolf rabies cases have been reported from areas with a presence of rabies and wolves both in North America (Ballard & Krausman, 1997; Chapman, 1978; Johnson, 1995; Theberge, Forbes, Barker, & Bollinger, 1994) and Eurasia, for example, from the Russian Federation and states of the former Soviet Union (Cherkasskiy, 1988; Sidorov, Sidorova, & Poleshchuk, 2010; Sultanov et al., 2016; Taichiev & Botvinkin, 2006), Turkey (Johnson, Un, et al., 2010), Iran (Gholami et al., 2017), India (Isloor et al., 2014), Mongolia (Botvinkin et al., 2008), and China (Meng et al., 2010). In Europe between 1990 and 2018, rabies cases in wolves represented only a small fraction (0.3%) of the overall number of animals that tested positive.

Documented wolf attacks on humans are often associated with rabid animals (Linnell et al., 2002; McNay, 2002). Rabid wolves often travel over longer distances, and tend to bite larger numbers of people and domestic animals (Linnell et al., 2002). Victims of rabid wolf attacks, if not instantly killed, suffer from severe lesions, particularly in the head region (Baltazard & Ghodssi, 1954; Lojkic et al., 2009; Mishaeva, Votyakov, Velhin, Nekhai, & Titov, 2007; Shah & Jaswal, 1976; Turkmen et al., 2012). In Iran, the failure of vaccine-only PEP in wolf victims initiated the first field trials for PEP with serum (Bahmanyar, Fayaz, Noursalehi, Mohammadi, & Koprowski, 1976). Here, with 9%, the importance of the wolf as a vector for human rabies is exceptionally high (Gholami et al., 2017).

Historical figures from Russia from 1886 to 1927 show that wolf bites contributed to up to 4% of rabies treatments at Pasteur's stations (Botvinkin & Kosenko, 2004). Later data from several states of the former Soviet Union indicate a similar situation. Between 1976 and 1980, 3.5% of human rabies cases were caused by wolves (Sidorov et al., 2010), whereas in Kazakhstan, figures show that wolves only contribute 0.07% to the total number of bite injuries (Sultanov et al., 2016).

RABV infections, most likely as a result of spillovers by rabid domestic dogs, represent a threat to the existence of the Ethiopian wolf (*Canis simensis*) (Marino et al., 2017; Randall et al., 2004), one of the most threatened canid species in the world. Similarly, rabies caused massive die-offs in packs of the endangered African wild dog (*Lyacon pictus*) in Kenya (Kat, Alexander, Smith, & Munson, 1995), South Africa (Hofmeyr, Bingham, Lane, Ide, & Nel, 2000; Sabeta et al., 2018), and Botswana (Canning, Camphor, & Schroder, 2019). Vaccination strategies, also including the use of oral rabies vaccines, were already applied for the Ethiopian wolf (Knobel et al., 2008; Randall et al., 2006; Sillero-Zubiri et al., 2016), and have been considered for the African wild dog (Knobel, du Toit, & Bingham, 2002; Vial, Cleaveland, Rasmussen, & Haydon, 2006).

6.3.5 Rodents, opossums, and marsupials

Systematic surveillance in wild and synanthropic rodents in areas endemic for rabies across the world revealed only exceptional instances of dead-end spillover of RABV infection (World Health Organization, 2018). The same holds true for the opossum (Didelphis virginianus), which was shown to be rather resistant to experimental infection (Beamer, Mohr, & Barr, 1960).

Scientific data clearly indicate that these species are not primary hosts and do not play a role in the transmission or maintenance of RABV (World Health Organization, 2018). In fact, this argues in part for fundamental taxonomic differences in species susceptibility and viral-host response.

Acknowledgments

We would like to express our appreciation to all colleagues who worked on previous versions of this chapter. The technical support for creating maps by Patrick Wysocki (FLI) is gratefully acknowledged.

This chapter is dedicated to our former colleague and friend Dr. Hartmut Schlüter, who died in April 2019.

References

Adelshin, R., Melnikova, O., Trushina, Y. N., Botvinkin, A. D., Borisova, T. I., Andaev, E., … Balakhonov, S. (2015). A new outbreak of fox rabies at the Russian–Mongolian border. *Virologica Sinica, 30*. https://doi.org/10.1007/s12250-015-3609-0.

Amarasinghe, G. K., Arechiga Ceballos, N. G., Banyard, A. C., Basler, C. F., Bavari, S., Bennett, A. J., … Kuhn, J. H. (2018). Taxonomy of the order Mononegavirales: Update 2018. *Archives of Virology*. https://doi.org/10.1007/s00705-018-3814-x.

Anthony, R. M. (1997). Home ranges and movements of arctic fox (Alopex lagopus) in Western Alaska. *Arctic, 50*(2), 147–157.

Antunes, K. D., Matos, J. C. C., Mol, L. P., Oliveira, M. A., Arcebispo, T. L. M., Santos, V. G., … Silva, M. X. (2018). Descriptive analysis of rabies in wild animals in the state of Sergipe, Brazil. *Arquivo Brasileiro de Medicina Veterinaria e Zootecnia, 70*, 169–173.

Arai, Y. T., Takahashi, H., Kameoka, Y., Shiino, T., Wimalaratne, O., & Lodmell, D. L. (2001). Characterization of Sri Lanka rabies virus isolates using nucleotide sequence analysis of nucleoprotein gene. *Acta Virologica, 45*(5–6), 327–333.

Arnold, J., Humer, A., Heltai, M., Murariu, D., Spassov, N., & Hackländer, K. (2012). Current status and distribution of golden jackals *Canis aureus* in Europe. *Mammal Review, 42*(1), 1–11. https://doi.org/10.1111/j.1365-2907.2011.00185.x.

Ashford, R. W. (2003). When is a reservoir not a reservoir? *Emerging Infectious Diseases, 9*(11), 1495–1496. https://doi.org/10.3201/eid0911.030088.

Aubert, M. (1992). Epidemiology of fox rabies. In K. Bögel, F. X. Meslin, & M. Kaplan (Eds.), *Wildlife rabies control* (pp. 9–18): Kent Wells Medical Ltd.

Badrane, H., & Tordo, N. (2001). Host switching in Lyssavirus history from the Chiroptera to the Carnivora orders. *Journal of Virology, 75*(17), 8096–8104.

Bahmanyar, M., Fayaz, A., Noursalehi, S., Mohammadi, M., & Koprowski, H. (1976). Successful protection of humans exposed to rabies infection—Postexposure treatment with new human diploid cell rabies vaccine and antirabies serum. *Journal of the American Medical Association, 236*(24), 2751–2754.

Balako, G., Sisay, G., Hussein, M., & Asefa, D. (2018). Rabies outbreak among livestock in a pastoralist community, Southern Ethiopia. *Ethiopian Journal of Health Sciences, 28*(6), 805–808. https://doi.org/10.4314/ejhs.v28i6.16.

Ballard, W. B., & Krausman, P. R. (1997). Occurrence of rabies in wolves of Alaska. *Journal of Wildlife Diseases, 33*(2), 242–245.

Baltazard, M., & Ghodssi, M. (1954). Prevention of human rabies; treatment of persons bitten by rabid wolves in Iran. *Bulletin of the World Health Organization, 10*(5), 797–803.

Bankovskiy, D., Safonov, G., & Kurilchuk, Y. (2008). Immunogenicity of the ERA G 333 rabies virus strain in foxes and raccoon dogs. *Developments in Biologicals, 131,* 461–466.

Barnard, B. J. H., & Hassel, R. H. (1981). Rabies in kudus (Tragelaphus-strepsiceros) in South West Africa-Namibia. *Journal of the South African Veterinary Association, 52*(4), 309–314.

Barnard, B. J. H., Hassel, R. H., Geyer, H. J., & De Koker, W. C. (1982). Nonbite transmission of rabies in kudu (Tragelaphus strepsiceros). *Onderstepoort Journal of Veterinary Research, 49*(4), 191–192.

Beamer, P. D., Mohr, C. O., & Barr, T. R. B. (1960). Resistance of the opossum to rabies virus. *American Journal of Veterinary Research, 21*(82), 507–510.

Begeman, L., Geurtsvan Kessel, C., Finke, S., Freuling, C. M., Koopmans, M., Müller, T., ... Kuiken, T. (2017). Comparative pathogenesis of rabies in bats and carnivores, and implications for spillover to humans. *The Lancet Infectious Diseases, 18*(4), e147–e159. https://doi.org/10.1016/S1473-3099(17)30574-1.

Bell, G. P. (1980). A possible case of interspecific transmission of rabies in insectivorous bats. *Journal of Mammalogy, 61*(3), 528–530.

Bellan, S. E., Cizauskas, C. A., Miyen, J., Ebersohn, K., Kusters, M., Prager, K. C., ... Getz, W. M. (2012). Black-backed jackal exposure to rabies virus, canine distemper virus, and Bacillus anthracis in Etosha National Park, Namibia. *Journal of Wildlife Diseases, 48*(2), 371–381. https://doi.org/10.7589/0090-3558-48.2.371.

Beltran-Beck, B., Garcia, F. J., & Gortazar, C. (2012). Raccoons in Europe: Disease hazards due to the establishment of an invasive species. *European Journal of Wildlife Research, 58*(1), 5–15. https://doi.org/10.1007/s10344-011-0600-4.

Bernardi, F., Nadin-Davis, S. A., Wandeler, A. I., Armstrong, J., Gomes, A. A., Lima, F. S., ... Ito, F. H. (2005). Antigenic and genetic characterization of rabies viruses isolated from domestic and wild animals of Brazil identifies the hoary fox as a rabies reservoir. *Journal of General Virology, 86*(Pt 11), 3153–3162.

Bindrich, H., & Olechnowski, A. F. (1959). Untersuchungen über die Einwirkung von Hyarolonidase auf das Virus der Tollwut bei Hunden. *Archiv für Experimentelle Veterinärmedizin, 13,* 523–537.

Bingham, J. (2005). Canine rabies ecology in southern Africa. *Emerging Infectious Diseases, 11*(9), 1337–1342.

Bingham, J., & Foggin, C. M. (1993). Jackal rabies in Zimbabwe. *Onderstepoort Journal of Veterinary Research, 60*(4), 365–366.

Bingham, J., Foggin, C. M., Wandeler, A. I., & Hill, F. W. (1999). The epidemiology of rabies in Zimbabwe. 2. Rabies in jackals (*Canis adustus* and *Canis mesomelas*). *Onderstepoort Journal of Veterinary Research, 66*(1), 11–23.

Black, J. G., & Lawson, K. F. (1970). Sylvatic rabies studies in the silver fox. Susceptibility and immune response. *Canadian Journal of Comparative Medicine, 34,* 309–311.

Blancou, J. (1988). Ecology and epidemiology of fox rabies. *Reviews of Infectious Diseases, 10*(4), 606–609.

Blancou, J. (2003). *History of the surveillance and control of transmissible animal diseases.* Paris: World Organisation for Animal Health.

Blancou, J. (2004). Rabies in Europe and the Mediterranean Basin: From antiquity to the 19th century. In A. A. King, A. R. Fooks, M. Aubert, & A. I. Wandeler (Eds.), *Historical perspective of rabies in Europe and the Mediterranean Basin* (pp. 15–23). Paris: OIE.

Blancou, J., Aubert, M. F. A., Andral, L., & Artois, M. (1979). Rage expérimentale der renard roux (Vulpes vulpes). I. Sensibilié selon la voie d'infection et al dose infectante. *Revue De Medecine Veterinaire, 130*(7), 1001–1015.

Blanton, J. D., Meadows, A., Murphy, S., Manangan, J. S., Hanlon, C. A., Faber, M. L., ... Rupprecht, C. R. (2006). Vaccination of Small Asian Mongoose (Herpestes javanicus) Against Rabies. *Journal of Wildlife Diseases, 42*(3), 663–666.

Blanton, J. D., Niezgoda, M., Hanlon, C., Swope, C. B., Suckow, J. R., Saidy, B., ... Slate, D. (2018). Evaluation of oral rabie svaccination: Protection against rabie sin wild caught raccoons (procyo lotor). *Journal of Wildlife Diseases, 54* (520-527), 528.

Blanton, J. D., Robertson, K., Palmer, D., & Rupprecht, C. E. (2009). Rabies surveillance in the United States during 2008. *Journal of the American Veterinary Medical Association, 235*(6), 676–689. https://doi.org/10.2460/javma.235.6.676.

Boldbaatar, B., Inoue, S., Tuya, N., Dulam, P., Batchuluun, D., Sugiura, N., ... Yamada, A. (2010). Molecular epidemiology of rabies virus in Mongolia, 2005–2008. *Japanese Journal of Infectious Diseases, 63*(5), 358–363.

Botvinkin, A. D., Gribanova, L., & Nikifirova, T. A. (1983). Experimental rabies in the raccoon dog. *Zhurnal Mikrobiologii, Epidemiologii, i Immunobiologii, 12,* 37–40.

Botvinkin, A. D., & Kosenko, M. (2004). Rabies in the European parts of Russia, Belarus and the Ukraine. In A. A. King, A. R. Fooks, M. Aubert, & A. I. Wandeler (Eds.), *Historical perspective of rabies in Europe and the Mediterranean Basin* (pp. 47–77). Paris: OIE.

Botvinkin, A. D., Otgonbaatar, D., Tsoodol, S., & Kuzmin, I. V. (2008). Rabies in the Mongolian steppes. *Developments in Biologicals*, *131*, 199–205.

Bourhy, H., Kissi, B., Audry, L., Smreczak, M., Sadkowska-Todys, M., Kulonen, K., … Holmes, E. C. (1999). Ecology and evolution of rabies virus in Europe. *Journal of General Virology*, *80*(10), 2545–2557.

Brookes, V. J., Gill, G. S., Singh, B. B., Sandhu, B. S., Dhand, N. K., Aulakh, R. S., & Ward, M. P. (2019). Challenges to human rabies elimination highlighted following a rabies outbreak in bovines and a human in Punjab, India. *Zoonoses and Public Health*, *66*(3), 325–336. https://doi.org/10.1111/zph.12568.

Bwangamoi, O., Rottcher, D., & Wekesa, C. (1990). Rabies, microbesnoitiosis and sarcocystosis in a lion. *Veterinary Record*, *127*(16), 411.

Canning, G., Camphor, H., & Schroder, B. (2019). Rabies outbreak in African Wild Dogs (Lycaon pictus) in the Tuli region, Botswana: Interventions and management mitigation recommendations. *Journal for Nature Conservation*, *48*, 71–76.

Chang, J. C., Tsai, K. J., Hsu, W. C., Tu, Y. C., Chuang, W. C., Chang, C. Y., … Lee, S. H. (2015). Rabies Virus Infection in Ferret Badgers (Melogale moschata subaurantiaca) in Taiwan: A Retrospective Study. *Journal of Wildlife Diseases*, *51*(4), 923–928. https://doi.org/10.7589/2015-04-090.

Chaparro, F., & Esterhuysen, J. J. (1993). The role of the yellow mongoose (Cynictis penicillata) in the epidemiology of rabies in South Africa—preliminary results. *Onderstepoort Journal of Veterinary Research*, *60*(4), 373–377.

Chapman, R. C. (1978). Rabies—Decimation of a wolf pack in arctic Alaska. *Science*, *201*(4353), 365–367.

Charlton, K. M., & Casey, G. A. (1979). Experimental oral and nasal transmission of rabies virus in mice. *Canadian Journal of Comparative Medicine*, *43*(1), 10–15.

Charlton, K. M., Webster, W. A., & Casey, G. A. (1991). Skunk rabies. In In G. M. Baer (Ed.), *Vol. 2. The natural history of rabies* (pp. 307–324). New York, USA: Academic Press.

Cherkasskiy, B. L. (1988). Roles of the wolf and the raccoon dog in the ecology and epidemiology of rabies in the USSR. *Reviews of Infectious Diseases*, *10*(4), 634–636.

Chhabra, M., Ichhpujani, R. L., Tewari, K. N., & Lal, S. (2004). Human rabies in Delhi. *Indian Journal of Pediatrics*, *71*(3), 217–220.

Childs, J. E., Trimarchi, C. V., & Krebs, J. W. (1994). The epidemiology of bat rabies in New York State, 1988-92. *Epidemiology and Infection*, *113*(3), 501–511.

Clark, K. A., Neill, S. U., Smith, J. S., Wilson, P. J., Whadford, V. W., & McKirahan, G. W. (1994). Epizootic canine rabies transmitted by coyotes in south Texas. *Journal of the American Veterinary Medical Association*, *204*(4), 536–540.

Cliquet, F., Guiot, A. L., Munier, M., Bailly, J., Rupprecht, C. E., & Barrat, J. (2006). Safety and efficacy of the oral rabies vaccine SAG2 in raccoon dogs. *Vaccine*, *24*(20), 4386–4392.

Cliquet, F., Guiot, A. L., Schumacher, C., Maki, J., Cael, N., & Barrat, J. (2008). Efficacy of a square presentation of V-RG vaccine baits in red fox, domestic dog and raccoon dog. *Developments in Biologicals*, *131*, 257–264.

Cliquet, F., Robardet, E., Must, K., Laine, M., Peik, K., Picard-Meyer, E., … Niin, E. (2012). Eliminating rabies in Estonia. *PLoS Neglected Tropical Diseases*, *6*(2), e1535. https://doi.org/10.1371/journal.pntd.0001535.

Cohen, C., Sartorius, B., Sabeta, C., Zulu, G., Paweska, J., Mogoswane, M., … Blumberg, L. (2007). Epidemiology and molecular virus characterization of reemerging rabies, South Africa. *Emerging Infectious Diseases*, *13*(12), 1879–1886. https://doi.org/10.3201/eid1312.070836.

Coleman, P. G., & Dye, C. (1996). Immunization coverage required to prevent outbreaks of dog rabies. *Vaccine*, *14*(3), 185–186.

Crandell, R. A. (1975). Arctic fox rabies. In G. M. Baer (Ed.), *Vol. 2. The natural history of rabies* (1st ed., pp. 23–40).

Daniels, S. E., Fanelli, R. E., Gilbert, A., & Benson-Amram, S. (2019). Behavioral flexibility of a generalist carnivore. *Animal Cognition*. https://doi.org/10.1007/s10071-019-01252-7.

Daoust, P. Y., Wandeler, A. I., & Casey, G. A. (1996). Cluster of rabies cases of probable bat origin among red foxes in Prince Edward Island, Canada. *Journal of Wildlife Diseases*, *32*(2), 403–406.

Dean, D. J., & Guevin, V. H. (1963). Rabies vaccination of cats. *Journal of the American Veterinary Medical Association*, *142*(4), 367.

Deviatkin, A. A., Lukashev, A. N., Poleshchuk, E. M., Dedkov, V. G., Tkachev, S. E., Sidorov, G. N., … Shipulin, G. A. (2017). The phylodynamics of the rabies virus in the Russian Federation. *PLoS ONE*, *12*(2), e0171855. https://doi.org/10.1371/journal.pone.0171855.

Dunlop, R. H., & Williams, D. J. (1996). *Veterinary medicine: An illustrated history. Vol. 3*. Mosby.

Dyer, J. L., Yager, P., Orciari, L., Greenberg, L., Wallace, R., Hanlon, C. A., & Blanton, J. D. (2014). Rabies surveillance in the United States during 2013. *Journal of the American Veterinary Medical Association*, *245*(10), 1111–1123. https://doi.org/10.2460/javma.245.10.1111.

East, R. (1999). *African antelope database 1998*. Gland, Switzerland/Cambridge, UK: IUCN/SSC Antelope Specialist Group, IUCN.

Eberhardt, L. E., Hanson, W. C., Bengtson, J. L., Garrott, R. A., & Hanson, E. E. (1982). Arctic fox home range characteristics in an oil-development area. *Journal of Wildlife Management*, *46*(1), 183–190. https://doi.org/10.2307/3808421.

EFSA. (2015). Scientific opinion—Update on oral vaccination of foxes and raccoon dogs against rabies. *EFSA Journal*, *13*(7), 4164. https://doi.org/10.2903/j.efsa.2015.4164.

Elmore, S. A., Chipman, R. B., Slate, D., Huyvaert, K. P., Ver Cauteren, K. C., & Gilbert, A. T. (2017). Management and modeling approaches for controlling raccoon rabies: The road to elimination. *PLoS Neglected Tropical Diseases*, *11*(3), e0005249. https://doi.org/10.1371/journal.pntd.0005249.

Everard, C. O. R., & Baer, G. M. (1974). Epidemiology of mongoose rabies in Grenada. *Journal of Wildlife Diseases*, *10*(3), 190–196.

Everard, C. O. R., & Everard, J. D. (1988). Mongoose rabies. *Reviews of Infectious Diseases*, *10*(4), 610–614.

Farashi, A., Kaboli, M., & Karami, M. (2013). Predicting range expansion of invasive raccoons in northern Iran using ENFA model at two different scales. *Ecological Informatics*, *15*, 96–102. https://doi.org/10.1016/j.ecoinf.2013.01.001.

Favoretto, S. R., de Mattos, C. C., Morais, N. B., Araujo, F. A., & de Mattos, C. A. (2001). Rabies in Marmosets (*Callithrix jacchus*), Ceara, Brazil. *Emerging Infectious Diseases*, *7*(6), 1062–1065.

Fehlner-Gardiner, C., Rudd, R., Donovan, D., Slate, D., Kempf, L., & Badcock, J. (2012). Comparing ONRAB(R) and RABORAL V-RG(R) oral rabies vaccine field performance in raccoons and striped skunks, New Brunswick, Canada, and Maine, USA. *Journal of Wildlife Diseases*, *48*(1), 157–167. https://doi.org/10.7589/0090-3558-48.1.157.

Feldhamer, G. A., Thompson, B. C., & Chapman, J. A. (2003). *Wild mammals of North America: Biology, management, and conservation*. Chicago and London: The University of Chicago Press.

Feng, Y., Shi, Y., Yu, M., Xu, W., Gong, W., Tu, Z., … Tu, C. (2016). Livestock rabies outbreaks in Shanxi province, China. *Archives of Virology*, *161*(10), 2851–2854. https://doi.org/10.1007/s00705-016-2982-9.

Feng, Y., Wang, W., Guo, J., Alatengheli, Li, Y., Yang, G., … Tu, C. (2015). Disease outbreaks caused by steppe-type rabies viruses in China. *Epidemiology and Infection*, *143*(6), 1287–1291. https://doi.org/10.1017/S0950268814001952.

Fischer, M. L., Hochkirch, A., Heddergott, M., Schulze, C., Anheyer-Behmenburg, H. E., Lang, J., … Frantz, A. C. (2015). Historical invasion records can be misleading: Genetic evidence for multiple introductions of invasive raccoons (procyon lotor) in Germany. *PLoS ONE*, *10*(5), e0125441. https://doi.org/10.1371/journal.pone.0125441.

Fischman, H. R., & Ward, F. E. (1968). Oral transmission of rabies virus in experimental animals. *American Journal of Epidemiology*, *88*(1), 132–138.

Foggin, C. M. (1988). *Rabies and rabies-related viruses in Zimbabwe: Historical, virological and ecological aspects* [PhD thesis]. University of Zimbabwe.

Follmann, E. H., Ritter, D. G., & Baer, G. M. (1988). Immunization of arctic foxes (alopex lagopus) with oral rabies vaccine. *Journal of Wildlife Diseases*, *24*(3), 477–483.

Follmann, E. H., Ritter, D. G., & Hartbauer, D. W. (2004). Oral vaccination of captive arctic foxes with lyophilized SAG2 rabies vaccine. *Journal of Wildlife Diseases*, *40*(2), 328–334.

Fooks, A. R., Cliquet, F., Finke, S., Freuling, C., Hemachudha, T., Mani, R. S., … Banyard, A. C. (2017). Rabies. *Nature Reviews. Disease Primers*, *3*(17091), 1–19. https://doi.org/10.1038/nrdp.2017.91.

Frafjord, K., & Prestrud, P. (1992). Home range and movements of arctic foxes Alopex-lagopus in Svalbard. *Polar Biology*, *12*(5), 519–526.

Frantz, L. A., Mullin, V. E., Pionnier-Capitan, M., Lebrasseur, O., Ollivier, M., Perri, A., … Larson, G. (2016). Genomic and archaeological evidence suggest a dual origin of domestic dogs. *Science*, *352*(6290), 1228–1231. https://doi.org/10.1126/science.aaf3161.

Freuling, C. M., Eggerbauer, E., Finke, S., Kaiser, C., Kaiser, C., Kretzschmar, A., … Muller, T. (2017). Efficacy of the oral rabies virus vaccine strain SPBN GASGAS in foxes and raccoon dogs. *Vaccine*. https://doi.org/10.1016/j.vaccine.2017.09.093.

Freuling, C. M., Hampson, K., Selhorst, T., Schroder, R., Meslin, F. X., Mettenleiter, T. C., & Muller, T. (2013). The elimination of fox rabies from Europe: Determinants of success and lessons for the future. *Philosophical transactions of the Royal Society of London. Series B, Biological sciences, 368*(1623), 20120142. https://doi.org/10.1098/rstb.2012.0142.

Fu, Z. F., & Jackson, A. C. (2005). Neuronal dysfunction and death in rabies virus infection. *Journal of Neurovirology, 11*(1), 101–106. https://doi.org/10.1080/13550280590900445.

Gehrt, S. (2019). Procyon lotor (raccoon). In *Invasive species compendium*: CAB International. https://www.cabi.org/ISC/datasheetreport/67856.

Gholami, A., Massoudi, S., Kharazian Moghaddam, M., Ghazi Marashi, M., Marashi, M., Bashar, R., ... Shirzadi, M. R. (2017). The role of the gray wolf in rabies transmission in iran and preliminary assessment of an oral rabies vaccine in this animal. *Journal of Medical Microbiology and Infectious Diseases, 5*(3), 56–61. https://doi.org/10.29252/JoMMID.5.3.4.56.

Gilbert, A. T., Fooks, A. R., Hayman, D. T., Horton, D. L., Müller, T., Plowright, R., ... Rupprecht, C. E. (2013). Deciphering serology to understand the ecology of infectious diseases in wildlife. *Ecosystem Health, 6*. https://doi.org/10.1007/s10393-013-0856-0.

Gilbert, A. T., Johnson, S. R., Nelson, K. M., Chipman, R. B., Ver Cauteren, K. C., Algeo, T. P., ... Slate, D. (2018). Field trials of ontario rabies vaccine bait in the Northeastern USA, 2012-14. *Journal of Wildlife Diseases, 54*(4), 790–801. https://doi.org/10.7589/2017-09-242.

Gilbert, A. T., Johnson, S., Walker, N., Wickham, C., Beath, A., & Ver Cauteren, K. (2018). Efficacy of Ontario Rabies Vaccine Baits (ONRAB) against rabies infection in raccoons. *Vaccine, 36*(32 Pt B), 4919–4926. https://doi.org/10.1016/j.vaccine.2018.06.052.

Gilchrist, J. S., Jennings, A. P., & Veron, G. (2018). Family Herpestidae (Mongooses). In D. E. Wilson & R. A. Mittermeier (Eds.), *Handbook of the mammals of the world – Volume 1, Carnivores*: Lynx Edicions in association with Conservation International and IUCN.

Gluska, S., Zahavi, E. E., Chein, M., Gradus, T., Bauer, A., Finke, S., & Perlson, E. (2014). Rabies Virus Hijacks and accelerates the p75NTR retrograde axonal transport machinery. *PLoS Pathogens, 10*(8), e1004348. https://doi.org/10.1371/journal.ppat.1004348.

Gruzdev, K. N. (2008). The rabies situation in Central Asia. *Developments in Biologicals, 131*, 37–42.

Hamir, A. N., Moser, G., & Rupprecht, C. E. (1996). Clinicopathologic variation in raccoons infected with different street rabies virus isolates. *Journal of Veterinary Diagnostic Investigation, 8*(1), 31–37.

Hammami, S., Schumacher, C., Cliquet, F., Tlatli, A., Aubert, A., & Aubert, M. (1999). Vaccination of Tunisian dogs with the lyophilised SAG2 oral rabies vaccine incorporated into the DBL2 dog bait. *Veterinary Research, 30*(6), 607–613.

Hampson, K., Coudeville, L., Lembo, T., Sambo, M., Kieffer, A., Attlan, M., ... Global Alliance for Rabies Control Partners for Rabies Prevention. (2015). Estimating the global burden of endemic canine rabies. *PLoS Neglected Tropical Diseases, 9*(4), e0003709. https://doi.org/10.1371/journal.pntd.0003709.

Hampson, K., Dushoff, J., Cleaveland, S., Haydon, D. T., Kaare, M., Packer, C., & Dobson, A. (2009). Transmission dynamics and prospects for the elimination of canine rabies. *PLoS Biology, 7*(3), e53. https://doi.org/10.1371/journal.pbio.1000053.

Hanke, D., Freuling, C. M., Fischer, S., Hueffer, K., Hundertmark, K., Nadin-Davis, S., ... Höper, D. (2016). Spatiotemporal analysis of the genetic diversity of arctic rabies viruses and their reservoir hosts in Greenland. *PLoS Neglected Tropical Diseases, 10*(7), e0004779. https://doi.org/10.1371/journal.pntd.0004779.

Hanlon, C. (2013). Rabies in terrestrial animals. In A. C. Jackson (Ed.), *Vol. 3. Rabies: Scientific basis of the disease and its management* (3rd ed., pp. 179–213). New York: Academic Press.

Hassel, R. (2011). Rabies in kudu antelope (Tragelaphus strepsiceros) in Namibia: Past and Present. In *Paper presented at the 10th International Conference of the Southern and East African Rabies Group (SEARG) Maputo, Mosambique*.

Hassel, R., Vos, A., Clausen, P., Moore, S., van der Westhuizen, J., Khaiseb, S., ... Müller, T. (2018). Experimental screening studies on rabies virus transmission and oral rabies vaccination of the Greater Kudu (Tragelaphus strepsiceros). *Scientific Reports, 8*(1), 16599. https://doi.org/10.1038/s41598-018-34985-5.

Haydon, D. T., Cleaveland, S., Taylor, L. H., & Laurenson, M. K. (2002). Identifying reservoirs of infection: A conceptual and practical challenge. *Emerging Infectious Diseases, 8*(12), 1468–1473. https://doi.org/10.3201/eid0812.010317.

Hemachudha, T., Laothamatas, J., & Rupprecht, C. E. (2002). Human rabies: A disease of complex neuropathogenetic mechanisms and diagnostic challenges. *Lancet Neurology, 1*(2), 101–109.

Hemachudha, T., Ugolini, G., Wacharapluesadee, S., Sungkarat, W., Shuangshoti, S., & Laothamatas, J. (2013). Human rabies: Neuropathogenesis, diagnosis, and management. *Lancet Neurology, 12*(5), 498–513. https://doi.org/10.1016/S1474-4422(13)70038-3.

Herrero, M., Grace, D., Njuki, J., Johnson, N., Enahoro, D., Silvestri, S., & Rufino, M. (2013). The roles of livestock in developing countries. *Animal, 7*(Suppl 1), 3–18.

Hikufe, E. H., Freuling, C. M., Athingo, R., Shilongo, A., Ndevaetela, E. -E., Helao, M., … Maseke, A. (2019). Ecology and epidemiology of rabies in humans, domestic animals and wildlife in Namibia, 2011-2017. *PLoS Neglected Tropical Diseases, 13*(4), e0007355.

Hofmeyr, M., Bingham, J., Lane, E. P., Ide, A., & Nel, L. (2000). Rabies in African wild dogs (Lycaon pictus) in the Madikwe Game Reserve, South Africa. *Veterinary Record, 146*(2), 50–52.

Horton, D. L., McElhinney, L. M., Freuling, C. M., Marston, D. A., Banyard, A. C., Goharrriz, H., … Fooks, A. R. (2015). Complex epidemiology of a zoonotic disease in a culturally diverse region: Phylogeography of rabies virus in the middle east. *PLoS Neglected Tropical Diseases, 9*(3), e0003569. https://doi.org/10.1371/journal.pntd.0003569.

Hsu, A. P., Tseng, C. H., Barrat, J., Lee, S. H., Shih, Y. H., Wasniewski, M., … Tsai, H. J. (2017). Safety, efficacy and immunogenicity evaluation of the SAG2 oral rabies vaccine in Formosan ferret badgers. *PLoS One, 12*(10), e0184831. https://doi.org/10.1371/journal.pone.0184831.

Hübschle, O. J. (1988). Rabies in the kudu antelope (Tragelaphus strepsiceros). *Reviews of Infectious Diseases, 10*(Suppl 4), 629–633.

Hudson, L. C., Weinstock, D., Jordan, T., & Bold-Fletcher, N. O. (1996a). Clinical features of experimentally induced rabies in cattle and sheep. *Journal of veterinary medicine. B, Infectious diseases and veterinary public health, 43*(2), 85–95.

Hudson, L. C., Weinstock, D., Jordan, T., & Bold-Fletcher, N. O. (1996b). Clinical presentation of experimentally induced rabies in horses. *Zentralblatt für Veterinärmedizin Reihe B, 43*(5), 277–285.

Hueffer, K., & Murphy, M. (2018). Rabies in Alaska, from the past to an uncertain future. *International Journal of Circumpolar Health, 77*(1), 1475185. https://doi.org/10.1080/22423982.2018.1475185.

Hussain, M. H., Ward, M. P., Body, M., Al-Rawahi, A., Wadir, A. A., Al-Habsi, S., … Almaawali, M. G. (2013). Spatio-temporal pattern of sylvatic rabies in the Sultanate of Oman, 2006–2010. *Preventive Veterinary Medicine, 110*(3), 281–289. https://doi.org/10.1016/j.prevetmed.2013.01.001.

Isloor, S., Marissen, W. E., Veeresh, B. H., Nithin Prabhu, K., Kuzmin, I. V., Rupprecht, C. E., … Rahman, A. (2014). First case report of rabies in a wolf (Canis lupus pallipes) from India. *The Journal of Veterinary Medical Science, 1*(3), 1012.

Johnson, M. R. (1995). Rabies in wolves and its potential role in a Yellowstone wolf population. In *Ecology and conservation of wolves in a changing world* (pp. 431–439). Canadian Circumpolar Institute Occasional Publication No. 35.

Johnson, N., Black, C., Smith, J., Un, H., McElhinney, L. M., Aylan, O., & Fooks, A. R. (2003). Rabies emergence among foxes in Turkey. *Journal of Wildlife Diseases, 39*(2), 262–270. https://doi.org/10.7589/0090-3558-39.2.262.

Johnson, N., Fooks, A. R., Valtchovski, R., & Müller, T. (2006). Evidence for trans-border movement of rabies by wildlife reservoirs between countries in the Balkan Peninsular. *Veterinary Microbiology, 120*, 71–76.

Johnson, N., Un, H., Fooks, A. R., Freuling, C., Müller, T., Aylan, O., & Vos, A. (2010). Rabies epidemiology and control in Turkey: Past and present. *Epidemiology and Infection, 138*(3), 305–312. https://doi.org/10.1017/S0950268809990963.

Johnson, N., Un, H., Vos, A., Aylan, O., & Fooks, A. R. (2006). Wildlife rabies in Western Turkey: The spread of rabies through the western provinces of Turkey. *Epidemiology and Infection, 134*(2), 369–375. https://doi.org/10.1017/S0950268805005017.

Johnson, N., Vos, A., Freuling, C., Tordo, N., Fooks, A. R., & Muller, T. (2010). Human rabies due to lyssavirus infection of bat origin. *Veterinary Microbiology, 142*(3-4), 151–159. https://doi.org/10.1016/j.vetmic.2010.02.001.

Karunanayake, D., Matsumoto, T., Wimalaratne, O., Nanayakkara, S., Perera, D., Nishizono, A., & Ahmed, K. (2014). Twelve years of rabies surveillance in Sri Lanka, 1999-2010. *PLoS Neglected Tropical Diseases, 8*(10), e3205. https://doi.org/10.1371/journal.pntd.0003205.

Kat, P. W., Alexander, K. A., Smith, J. S., & Munson, L. (1995). Rabies and African wild dogs in Kenya. *Proceedings of the Royal Society of London Series B, 262*(1364), 229–233.

Kauhala, K., & Kowalczyk, R. (2011). Invasion of the raccoon dog Nyctereutes procyonoides in Europe: History of colonization, features behind its success, and threats to native fauna. *Current Zoology, 57*(5), 584–598.

Kelly, T. R., & Sleeman, J. M. (2003). Morbidity and mortality of red foxes (Vulpes vulpes) and gray foxes (Urocyon cinereoargenteus) admitted to the Wildlife Center of Virginia, 1993-2001. *Journal of Wildlife Diseases, 39*(2), 467–469. https://doi.org/10.7589/0090-3558-39.2.467.

Kim, B. I., Blanton, J. D., Gilbert, A., Castrodale, L., Hueffer, K., Slate, D., & Rupprecht, C. E. (2014). A conceptual model for the impact of climate change on fox rabies in Alaska, 1980-2010. *Zoonoses and Public Health, 61*(1), 72–80. https://doi.org/10.1111/zph.12044.

Kim, C. H., Lee, C. G., Yoon, H. C., Nam, H. M., Park, C. K., Lee, J. C., … Wee, S. H. (2006). Rabies, an emerging disease in Korea. *Journal of veterinary medicine. B, Infectious diseases and veterinary public health, 53*(3), 111–115. https://doi.org/10.1111/j.1439-0450.2006.00928.x.

King, A. A., Meredith, C. D., & Thomson, G. R. (1993). Canid and viverrid rabies viruses in South Africa. *Onderstepoort Journal of Veterinary Research, 60*(4), 295–299.

Klingen, Y., Conzelmann, K. K., & Finke, S. (2008). Double-labeled rabies virus: Live tracking of enveloped virus transport. *Journal of Virology, 82*(1), 237–245. https://doi.org/10.1128/JVI.01342-07.

Knobel, D. L., du Toit, J. T., & Bingham, J. (2002). Development of a bait and baiting system for delivery of oral rabies vaccine to free-ranging African wild dogs (Lycaon pictus). *Journal of Wildlife Diseases, 38*(2), 352–362.

Knobel, D. L., Fooks, A. R., Brookes, S. M., Randall, D. A., Williams, S. D., Argaw, K., … Laurenson, M. K. (2008). Trapping and vaccination of endangered Ethiopian wolves to control an outbreak of rabies. *Journal of Applied Ecology, 45*(1), 109–116.

Koepfli, K. -P., Pollinger, J., Godinho, R., Robinson, J., Lea, A., Hendricks, S., … Wayne, R. K. (2015). Genome-wide evidence reveals that African and Eurasian golden jackals are distinct species. *Current Biology, 25*(16), 2158–2165. https://doi.org/10.1016/j.cub.2015.06.060.

Kotait, I., Oliveira, R. N., Carrieri, M. L., Castilho, J. G., Macedo, C. I., Pereira, P. M. C., … Rupprecht, C. E. (2018). Nonhuman primates as a reservoir for rabies virus in Brazil. *Zoonoses and Public Health*. https://doi.org/10.1111/zph.12527.

Krebs, J. W., Williams, S. M., Smith, J. S., Rupprecht, C. E., & Childs, J. E. (2003). Rabies among infrequently reported mammalian carnivores in the United States, 1960-2000. *Journal of Wildlife Diseases, 39*(2), 253–261.

Kuzmin, I. V., Botvinkin, A. D., McElhinney, L. M., Smith, J. S., Orciari, L. A., Hughes, G. J., … Rupprecht, C. E. (2004). Molecular epidemiology of terrestrial rabies in the former Soviet Union. *Journal of Wildlife Diseases, 40*(4), 617–631.

Kuzmin, I. V., Hughes, G. J., Botvinkin, A. D., Gribencha, S. G., & Rupprecht, C. E. (2008). Arctic and Arctic-like rabies viruses: Distribution, phylogeny and evolutionary history. *Epidemiology and Infection, 136*(4), 509–519.

Kuzmin, I. V., Shi, M., Orciari, L. A., Yager, P. A., Velasco-Villa, A., Kuzmina, N. A., … Rupprecht, C. E. (2012). Molecular inferences suggest multiple host shifts of rabies viruses from bats to mesocarnivores in Arizona during 2001-2009. *PLoS Pathogens, 8*(6), e1002786. https://doi.org/10.1371/journal.ppat.1002786.

Lackay, S. N., Kuang, Y., & Fu, Z. F. (2008). Rabies in small animals. *Veterinary Clinics of North America: Small Animal Practice, 38*(4), 851–861. ix https://doi.org/10.1016/j.cvsm.2008.03.003.

Lan, Y. C., Wen, T. H., Chang, C. C., Liu, H. F., Lee, P. F., Huang, C. Y., … Chen, Y. A. (2017). Indigenous wildlife rabies in Taiwan: Ferret badgers, a long term terrestrial reservoir. *BioMed Research International, 2017*, 5491640. https://doi.org/10.1155/2017/5491640.

Larson, G., Karlsson, E. K., Perri, A., Webster, M. T., Ho, S. Y., Peters, J., … Lindblad-Toh, K. (2012). Rethinking dog domestication by integrating genetics, archeology, and biogeography. *Proceedings of the National Academy of Sciences of the United States of America, 109*(23), 8878–8883. https://doi.org/10.1073/pnas.1203005109.

Lembo, T., Craig, P. S., Miles, M. A., Hampson, K. R., & Meslin, F. -X. (2013). Zoonoses prevention, control and elimination in dogs. In C. Macpherson, F. -X. Meslin, & A. Wandeler (Eds.), *Dogs, zoonoses and public health* (2nd ed., pp. 205–258). Wallingford, UK: CAB International.

Lembo, T., Hampson, K., Haydon, D. T., Craft, M., Dobson, A., Dushoff, J., … Cleaveland, S. (2008). Exploring reservoir dynamics: A case study of rabies in the Serengeti ecosystem. *Journal of Applied Ecology, 45*(4), 1246–1257. https://doi.org/10.1111/j.1365-2664.2008.01468.x.

Leslie, M. J., Messenger, S., Rohde, R. E., Smith, J., Cheshier, R., Hanlon, C., & Rupprecht, C. E. (2006). Bat-associated rabies virus in skunks. *Emerging Infectious Diseases, 12*(8), 1274–1277.

Lindblad-Toh, K., Wade, C. M., Mikkelsen, T. S., Karlsson, E. K., Jaffe, D. B., Kamal, M., … Lander, E. S. (2005). Genome sequence, comparative analysis and haplotype structure of the domestic dog. *Nature, 438*, 803. https://doi.org/10.1038/nature04338. https://www.nature.com/articles/nature04338#supplementary-information.

Linnell, J. D. C., Andersen, R., Andersone, Z., Balciauskas, L., Blanco, J. C., Boitani, L., … Wabakken, P. (2002). The fear for wolves: A review of wolfs attacks on humans. *NINA Oppdragsmelding 731* (pp. 1–65). *Norsk institutt for naturforskning*.

Liu, Y., Zhang, S., Wu, X., Zhao, J., Hou, Y., Zhang, F., … Hu, R. (2010). Ferret badger rabies origin and its revisited importance as potential source of rabies transmission in Southeast China. *BMC Infectious Diseases, 10*. https://doi.org/10.1186/1471-2334-10-234.

Liu, Y., Zhang, H. P., Zhang, S. F., Wang, J. X., Zhou, H. N., Zhang, F., … Hu, R. L. (2016). Rabies outbreaks and vaccination in domestic camels and cattle in Northwest China. *PLoS Neglected Tropical Diseases, 10*(9), e0004890. https://doi.org/10.1371/journal.pntd.0004890.

Lojkic, I., Galic, M., Cac, Z., Jelic, I., Bedekovic, T., Lojkic, M., & Cvetnic, Z. (2009). Bites of a rabid wolf in 67-old man in north-eastern part of Croatia. *Rabies Bulletin Europe*, *33*(3), 5–7.

Ma, X., Monroe, B. P., Cleaton, J. M., Orciari, L. A., Li, Y., Kirby, J. D., … Blanton, J. D. (2018a). Rabies surveillance in the United States during 2017. *Journal of the American Veterinary Medical Association*, *253*(12), 1555–1568. https://doi.org/10.2460/javma.253.12.1555.

Macdonald, D. W., & Sillero-Zubiri, C. (2004). *The biology and conservation of wild canids*. Oxford: Oxford University Press.

MacInnes, C. D., Smith, S. M., Tinline, R. R., Ayers, N. R., Bachmann, P., Ball, D. G., … Voigt, D. R. (2001). Elimination of rabies from red foxes in eastern Ontario. *Journal of Wildlife Diseases*, *37*(1), 119–132. https://doi.org/10.7589/0090-3558-37.1.119.

Mainguy, J., Fehlner-Gardiner, C., Slate, D., & Rudd, R. J. (2013). Oral rabies vaccination in raccoons: Comparison of ONRAB(R) and RABORAL V-RG(R) vaccine-bait field performance in Quebec, Canada and Vermont, USA. *Journal of Wildlife Diseases*, *49*(1), 190–193. https://doi.org/10.7589/2011-11-342.

Mani, R. S., Moorkoth, A. P., Balasubramanian, P., Devi, K. L., & Madhusudana, S. N. (2016). Rabies following mongoose bite. *Indian Journal of Medical Microbiology*, *34*(2), 256–257. https://doi.org/10.4103/0255-0857.176848.

Mansfield, K. L., McElhinney, L., Hübschle, O., Mettler, F., Sabeta, C., Nel, L. H., & Fooks, A. R. (2006). A molecular epidemiological study of rabies epizooties in kudu (Tragelaphus strepsiceros) in Namibia. *BMC Veterinary Research*, *2*(2).

Mansfield, K. L., Racloz, V., McElhinney, L. M., Marston, D. A., Johnson, N., Ronsholt, L., … Fooks, A. R. (2006). Molecular epidemiological study of Arctic rabies virus isolates from Greenland and comparison with isolates from throughout the Arctic and Baltic regions. *Virus Research*, *116*(1-2), 1–10. https://doi.org/10.1016/j.virusres.2005.08.007.

Marino, J., Sillero-Zubiri, C., Deressa, A., Bedin, E., Bitewa, A., Lema, F., … Fooks, A. R. (2017). Rabies and distemper outbreaks in smallest ethiopian wolf population. *Emerging Infectious Diseases*, *23*(12), 2102–2104. https://doi.org/10.3201/eid2312.170893.

Marston, D. A., Banyard, A. C., McElhinney, L. M., Freuling, C. M., Finke, S., de Lamballerie, X., … Fooks, A. R. (2018). The lyssavirus host-specificity conundrum—Rabies virus—The exception not the rule. *Current Opinion in Virology*, *28*, 68–73. https://doi.org/10.1016/j.coviro.2017.11.007.

Marston, D. A., Horton, D. L., Nunez, J., Ellis, R. J., Orton, R. J., Johnson, N., … Fooks, A. R. (2017). Genetic analysis of a rabies virus host shift event reveals within-host viral dynamics in a new host. *Virus Evolution*, *3*(2), vex038. https://doi.org/10.1093/ve/vex038.

McElhinney, L. M., Marston, D., Johnson, N., Black, C., Matouch, O., Lalosevic, D., … Fooks, A. R. (2006). Molecular epidemiology of rabies viruses in Europe. *Developments in Biologicals*, *125*, 17–28.

McLean, R. G. (1970). Wildlife rabies in the United States: Recent history and current concepts. *Journal of Wildlife Diseases*, *6*(4), 229–235. discussion 247-228.

McLean, R. G. (1971). Rabies in raccoons in the Southeastern United States. *Journal of Infectious Diseases*, *123*(6), 680–681.

McNay, M. E. (2002). Wolf-human interactions in Alaska and Canada: A review of the case history. *Wildlife Society Bulletin*, *30*(3), 831–843.

Meehan, S. K. (1995). Rabies epizootic in coyotes combated with oral vaccination program. *Journal of the American Veterinary Medical Association*, *206*(8), 1097–1099.

Meng, S., Xu, G., Wu, X., Lei, Y., Yan, J., Nadin-Davis, S. A., … Rupprecht, C. E. (2010). Transmission dynamics of rabies in China over the last 40 years: 1969-2009. *Journal of Clinical Virology*, *49*(1), 47–52. https://doi.org/10.1016/j.jcv.2010.06.014.

Messenger, S. L., Smith, J. S., Orciari, L. A., Yager, P. A., & Rupprecht, C. E. (2003). Emerging pattern of rabies deaths and increased viral infectivity. *Emerging Infectious Diseases*, *9*(2), 151–154.

Miao, F. M., Chen, T., Liu, Y., Zhang, S. F., Zhang, F., Li, N., & Hu, R. L. (2018). Emerging new phylogenetic groups of rabies virus in Chinese ferret badgers. *Biomedical and environmental sciences: BES*, *31*(6), 479–482. https://doi.org/10.3967/bes2018.064.

Mishaeva, N., Votyakov, V., Velhin, S., Nekhai, M., & Titov, L. (2007). Complex rabies post-exposure prophylactic treatment after severe wolf bites in Belarus. *Rabies Bulletin Europe*, *31*(2), 6–10.

Moehlman, P. D., & Hayssen, V. (2018). *Canis aureus* (Carnivore: Canidae). *Mammalian Species*, *50*(957), 14–25. https://doi.org/10.1093/mspecies/sey002.

Moore, S., Gilbert, A., Vos, A., Freuling, C. M., Ellis, C., Kliemt, J., & Müller, T. (2017). Rabies virus antibodies from oral vaccination as a correlate of protection against lethal infection in wildlife. *Tropical Medicine and Infectious Disease, 2*(31). https://doi.org/10.3390/tropicalmed2030031.

Müller, T., Demetriou, P., Moynagh, J., Cliquet, F., Fooks, A. R., Conraths, F. J., ... Freuling, C. M. (2012). Rabies elimination in Europe—A success story. In A. R. Fooks & T. Müller (Eds.), *Rabies Control - Towards Sustainable Prevention at the Source, Compendium of the OIE Global Conference on Rabies Control, Incheon-Seoul, 7-9 September 2011, Republic of Korea* (pp. 31–44). Paris: OIE.

Müller, T., & Freuling, C. M. (2018). Rabies control in Europe: An overview of past, current and future strategies. *Revue scientifique et technique (International Office of Epizootics), 37*(2), 409–419. https://doi.org/10.20506/rst.37.2.2811.

Müller, T., Freuling, C. M., Wysocki, P., Roumiantzeff, M., Freney, J., Mettenleiter, T. C., & Vos, A. (2015). Terrestrial rabies control in the European Union: Historical achievements and challenges ahead. *Veterinary Journal, 203*(1), 10–17. https://doi.org/10.1016/j.tvjl.2014.10.026.

Nadin-Davis, S. A. (1998). Polymerase chain reaction protocols for rabies virus discrimination. *Journal of Virological Methods, 75*(1), 1–8.

Nadin-Davis, S. A., Colville, A., Trewby, H., Biek, R., & Real, L. (2017). Application of high-throughput sequencing to whole rabies viral genome characterisation and its use for phylogenetic re-evaluation of a raccoon strain incursion into the province of Ontario. *Virus Research, 232*, 123–133. https://doi.org/10.1016/j.virusres.2017.02.007.

Nadin-Davis, S. A., Sheen, M., & Wandeler, A. I. (2012). Recent emergence of the Arctic rabies virus lineage. *Virus Research, 163*(1), 352–362. https://doi.org/10.1016/j.virusres.2011.10.026.

Nadin-Davis, S. A., Velez, J., Malaga, C., & Wandeler, A. I. (2008). A molecular epidemiological study of rabies in Puerto Rico. *Virus Research, 131*(1), 8–15.

Nattrass, N., Conradie, B., Drouilly, M., & O'Riain, M. J. (2017). Understanding the black-backed jackal. In University of Cape Town Centre for Social Science Research (Ed.), *Instiute for communities and wildlife in Africa*. Cape Town, Republic of South Africa: Centre for Social Science Research.

Nel, L. H., Sabeta, C. T., Teichman, B. v., Jaftha, J. B., Rupprecht, C. E., & Bingham, J. (2005). Mongoose rabies in southern Africa: A re-evaluation based on molecular epidemiology. *Virus Research, 109*(2), 165–173.

Nel, L. H., Thomson, G. R., & von Teichman, B. F. (1993). Molecular epidemiology of rabies virus in South Africa. *Onderstepoort Journal of Veterinary Research, 60*(4), 301–306.

Nettles, V. F., Shaddock, J. H., Sikes, R. K., & Reyes, C. R. (1979). Rabies in translocated raccoons. *American Journal of Public Health, 69*(6), 601–602.

Niezgoda, M., Diehl, D., Hanlon, C. A., & Rupprecht, C. E. (1991). Pathogenesis of street rabies virus in raccoons. In *Paper presented at the 40th Annual Conference of the Wildlife Disease Association, Fort Collins, CO, USA*.

Niin, E., Laine, M., Guiot, A. L., Demerson, J. M., & Cliquet, F. (2008). Rabies in Estonia: Situation before and after the first campaigns of oral vaccination of wildlife with SAG2 vaccine bait. *Vaccine, 26*(29-30), 3556–3565.

Noren, K., Carmichael, L., Dalen, L., Hersteinsson, P., Samelius, G., Fuglei, E., ... Angerbjorn, A. (2011). Arctic fox Vulpes lagopus population structure: Circumpolar patterns and processes. *Oikos, 120*(6), 873–885. https://doi.org/10.1111/j.1600-0706.2010.18766.x.

Nowak, E. (1984). Verbreitungs- und Bestandsentwicklung des Marderhundes Nyctereutes procyonoides (Gray, 1834) in Europa. *Zeitschrift für Jagdwissenschaft, 30*, 137–154.

Odontsetseg, N., Uuganbayar, D., Tserendorj, S., & Adiyasuren, Z. (2009). Animal and human rabies in Mongolia. *Revue scientifique et technique (International Office of Epizootics), 28*(3), 995–1003.

Oem, J. K., Kim, S. H., Kim, Y. H., Lee, M. H., & Lee, K. K. (2014). Reemergence of rabies in the southern Han river region, Korea. *Journal of Wildlife Diseases, 50*(3), 681–688. https://doi.org/10.7589/2013-07-177.

Ortmann, S., Vos, A., Kretzschmar, A., Walther, N., Kaiser, C., Freuling, C., ... Müller, T. (2018). Safety studies with the oral rabies virus vaccine strain SPBN GASGAS in the small Indian mongoose (Herpestes auropunctatus). *BMC Veterinary Research, 14*(1), 90. https://doi.org/10.1186/s12917-018-1417-0.

Pandit, S. R. (1950). Two instances of proved rabies in the tiger. *Indian Medical Gazette, 85*(10), 441.

Pastoret, P. P., Kappeler, A., & Aubert, M. (2004). European rabies control and its history. In A. A. King, A. R. Fooks, M. Aubert, & A. I. Wandeler (Eds.), *Historical perspective of rabies in Europe and the Mediterranean Basin* (pp. 337–350). Paris: OIE.

Pedersen, K., Gilbert, A. T., Nelson, K. M., Morgan, D. P., Davis, A. J., Ver Cauteren, K. C., ... Chipman, R. B. (2019). Raccoon (Procyon Lotor) Response to Ontario Rabies Vaccine Baits (Onrab) in St. Lawrence County, New York, USA. *Journal of Wildlife Diseases*. https://doi.org/10.7589/2018-09-216.

Pionnier-Capitan, M., Bemilli, C., Bodu, P., Célérier, G., Ferrié, J.-G., Fosse, P., ... Vigne, J.-D. (2011). New evidence for Upper Palaeolithic small domestic dogs in South-Western Europe. *Journal of Archaeological Science, 38*(9), 2123–2140. https://doi.org/10.1016/j.jas.2011.02.028.

Prieto, J. F., & Baer, G. M. (1972). Outbreak of bovine paralytic rabies in Tuxtepec, Oaxaca, Mexico. *American Journal of Tropical Medicine and Hygiene, 21*(2), 219–225.

Raccoon Rabies Epizootic—United States. (1994). *MMWR. Morbidity and Mortality Weekly Report, 43*(15), 269–284.

Randall, D. A., Marino, J., Haydon, D. T., Sillero-Zubiri, C., Knobel, D. L., Tallents, L. A., ... Laurenson, M. K. (2006). An integrated disease management strategy for the control of rabies in Ethiopian wolves. *Biological Conservation, 131*(2), 151–162.

Randall, D. A., Williams, S. D., Kuzmin, I. V., Rupprecht, C. E., Tallents, L. A., Tefera, Z., ... Laurenson, M. K. (2004). Rabies in endangered Ethiopian wolves. *Emerging Infectious Diseases, 10*(12), 2214–2217.

Ratho, R. K., Prasad, S. R., & Bindra, M. S. (1997). Rabies after mongoose bite. *Journal of the Association of Physicians of India, 45*(4), 327.

Richards, S. H. (1962). *Rabies study data summary, 1953–60*. Retrieved from North Dakota.

Rosatte, R. C., Donovan, D., Davies, J. C., Allan, M., Bachmann, P., Stevenson, B., ... Lawson, K. (2009). Aerial distribution of ONRAB baits as a tactic to control rabies in raccoons and striped skunks in Ontario, Canada. *Journal of Wildlife Diseases, 45*(2), 363–374. https://doi.org/10.7589/0090-3558-45.2.363.

Rosatte, R. C., Power, M. J., Donovan, D., Davies, J. C., Allan, M., Bachmann, P., ... Muldoon, F. (2007). Elimination of arctic variant rabies in red foxes, metropolitan Toronto. *Emerging Infectious Diseases, 13*(1), 25–27. https://doi.org/10.3201/eid1301.060622.

Rupprecht, C. E., Barrett, J., Briggs, D., Cliquet, F., Fooks, A. R., Lumlertdacha, B., ... Wandeler, A. I. (2008). Can rabies be eradicated? *Developmental Biology, 131*, 95–121.

Rupprecht, C. E., Hanlon, C. A., Niezgoda, M., Buchanan, J. R., Diehl, D., & Koprowski, H. (1993). Recombinant rabies vaccines: Efficacy assessment in free-ranging animals. *Onderstepoort Journal of Veterinary Research, 60*(4), 463–468.

Sabeta, C. T., Janse van Rensburg, D., Phahladira, B., Mohale, D., Harrison-White, R. F., Esterhuyzen, C., & Williams, J. H. (2018). Rabies of canid biotype in wild dog (Lycaon pictus) and spotted hyaena (Crocuta crocuta) in Madikwe Game Reserve, South Africa in 2014-2015: Diagnosis, possible origins and implications for control. *Journal of the South African Veterinary Association, 89*(0), e1–e13. https://doi.org/10.4102/jsava.v89i0.1517.

Sabeta, C. T., Mansfield, K. L., McElhinney, L. M., Fooks, A. R., & Nel, L. H. (2007). Molecular epidemiology of rabies in bat-eared foxes (Otocyon megalotis) in South Africa. *Virus Research, 129*(1), 1–10.

Sabeta, C. T., Shumba, W., Mohale, D. K., Miyen, J. M., Wandeler, A. I., & Nel, L. H. (2008). Mongoose rabies and the African civet in Zimbabwe. *Veterinary Record, 163*(19), 580.

Schneider, H. P. (1985). Rabies in South Western Africa/Namibia. In E. Kuwert, C. Merieux, H. Koprowski, & K. Bögel (Eds.), *Rabies in the tropics* (pp. 520–535). Berlin: Springer.

Schnell, M. J., McGettigan, J., Wirblich, C., & Papaneri, A. (2010). The cell biology of rabies virus: Using stealth to reach the brain. *Nature Reviews Microbiology, 8*(1), 51–61.

Schuster, P., Müller, T., Vos, A., Selhorst, T., Neubert, L., & Pommerening, E. (2001). Comparative immunogenicity and efficacy studies with oral rabies virus vaccine SAD P5/88 in raccoon dogs and red foxes. *Acta Veterinaria Hungarica, 49*(3), 285–290.

Scott, T. P., Coetzer, A., & Nel, L. H. (2016). Rabies in Namibia, more than a horrendous disease: The social, environmental and economic challenges faced. In W. Sherman (Ed.), *Handbook of Africa* (pp. 183–209). Hauppauge: Nova Science Publ.

Scott, T. P., Fischer, M., Khaiseb, S., Freuling, C., Hoper, D., Hoffmann, B., ... Nel, L. H. (2013). Complete genome and molecular epidemiological data infer the maintenance of rabies among kudu (Tragelaphus strepsiceros) in Namibia. *PLoS ONE, 8*(3), e58739. https://doi.org/10.1371/journal.pone.0058739.

Scott, T. P., Hassel, R., & Nel, L. (2012). Rabies in kudu (Tragelaphus strepsiceros). *Berliner und Münchener Tierärztliche Wochenschrift, 125*(5-6), 236–241.

Scott, T. P., & Nel, L. H. (2016). Subversion of the immune response by rabies virus. *Viruses, 8*(8). https://doi.org/10.3390/v8080231.

Seetahal, J., Vokaty, A., Vigilato, M., Carrington, C., Pradel, J., Louison, B., ... Rupprecht, C. (2018). Rabies in the caribbean: A situational analysis and historic review. *Tropical Medicine and Infectious Disease, 3*(3), 89.

Seimenis, A. (2008). The rabies situation in the Middle East. *Developments in Biologicals, 131*, 43–53.

Seo, W., Servat, A., Cliquet, F., Akinbowale, J., Prehaud, C., Lafon, M., & Sabeta, C. (2017). Comparison of G protein sequences of South African street rabies viruses showing distinct progression of the disease in a mouse model of experimental rabies. *Microbes and Infection*. https://doi.org/10.1016/j.micinf.2017.05.005.

Shah, U., & Jaswal, G. S. (1976). Victims of a rabid wolf in india: Effect of severity and location of bites on development of rabies. *Journal of Infectious Diseases, 134*(1), 25–29.

Shao, X. Q., Yan, X. J., Luo, G. L., Zhang, H. L., Chai, X. L., Wang, F. X., … Zhang, Y. Z. (2011). Genetic evidence for domestic raccoon dog rabies caused by Arctic-like rabies virus in Inner Mongolia, China. *Epidemiology and Infection, 139*(4), 629–635. https://doi.org/10.1017/s0950268810001263.

Shimshony, A. (1997). Epidemiology of emerging zoonoses in Israel. *Emerging Infectious Diseases, 3*(2), 229–238. https://doi.org/10.3201/eid0302.970221.

Sidorov, G. N., Sidorova, D. G., & Poleshchuk, E. M. (2010). Rabies of wild mammals in Russia at the turn of the 20th and 21st centuries. *Biology Bulletin, 37*(7), 684–694. https://doi.org/10.1134/s1062359010070034.

Sidwa, T. J., Wilson, P. J., Moore, G. M., Oertli, E. H., Hicks, B. N., Rohde, R. E., & D.H., J. (2005). Evaluation of oral rabies vaccination programs for control of rabies epizootics in coyotes and gray foxes: 1995-2003. *Journal of the American Veterinary Medical Association, 227*(5), 785–792.

Sihvonen, L. (2001). Documenting freedom from disease and re-establishing a free status after a breakdown rabies. *Acta Veterinaria Scandinavica*, 89–91.

Sillero-Zubiri, C. (2018). Familiy Canidae (Dogs). In D. E. Wilson & R. A. Mittermeier (Eds.), *Vol. 1. Handbook of the mammals of the world – Volume 1, Carnivores* (pp. 352–448): Lynx Edicions in association with Conservation International and IUCN.

Sillero-Zubiri, C., Marino, J., Gordon, C. H., Bedin, E., Hussein, A., Regassa, F., … Fooks, A. R. (2016). Feasibility and efficacy of oral rabies vaccine SAG2 in endangered Ethiopian wolves. *Vaccine, 34*(40), 4792–4798. https://doi.org/10.1016/j.vaccine.2016.08.021.

Silva, M. L., Lima, Fda. S., de Barros Gomes, A. A., de Azevedo, S. S., Alves, C. J., Bernardi, F., & Ito, F. H. (2009). Isolation of rabies virus from the parotid salivary glands of foxes (*Pseudalopex vetulus*) from Paraíba State, Northeastern Brazil. *Brazilian Journal of Microbiology, 40*, 446–449.

Singer, A., Kauhala, K., Holmala, K., & Smith, G. C. (2009). Rabies in northeastern Europe—The threat from invasive raccoon dogs. *Journal of Wildlife Diseases, 45*(4), 1121–1137. https://doi.org/10.7589/0090-3558-45.4.1121.

Singh, J., Jain, D. C., Bhatia, R., Ichhpujani, R. L., Harit, A. K., Panda, R. C., … Sokhey, J. (2001). Epidemiological characteristics of rabies in Delhi and surrounding areas, 1998. *Indian Pediatrics, 38*(12), 1354–1360.

Slate, D., Algeo, T. P., Nelson, K. M., Chipman, R. B., Donovan, D., Blanton, J. D., … Rupprecht, C. E. (2009). Oral rabies vaccination in north america: Opportunities, complexities, and challenges. *PLoS Neglected Tropical Diseases, 3*(12), e549. https://doi.org/10.1371/journal.pntd.0000549.

Slate, D., Chipman, R. B., Algeo, T. P., Mills, S. A., Nelson, K. M., Croson, C. K., … Rupprecht, C. E. (2014). Safety and immunogenicity of Ontario Rabies Vaccine Bait (ONRAB) in the first us field trial in raccoons (Procyon lotor). *Journal of Wildlife Diseases, 50*(3), 582–595. https://doi.org/10.7589/2013-08-207.

Smith, J. S., Orciari, L. A., Yager, P. A., Seidel, H. D., & Warner, C. K. (1992). Epidemiologic and historial relationships among 87 rabies virus isolates as determined by limited sequence analysis. *Journal of Infectious Diseases, 166*, 296–307.

Soria Baltazar, R., Artois, M., & Blancou, J. (1992). Experimental infection of sheep with a rabies virus of canine origin: Study of the pathogenicity for that species. *Revue scientifique et technique (International Office of Epizootics), 11*(3), 829–836.

Soulebot, J. P., Brun, A., Chappuis, G., Guillemin, F., Petermann, H. G., Precausta, P., & Terre, J. (1981). Experimental rabies in cats: Immune response and persistence of immunity. *The Cornell Veterinarian, 71*(3), 311–325.

Steele, J. H., & Fernandez, P. J. (1991). History of rabies and global aspects. In G. M. Baer (Ed.), *The natural history of rabies* (pp. 1–24). New York: Academic Press.

Sultanov, A. A., Abdrakhmanov, S. K., Abdybekova, A. M., Karatayev, B. S., & Torgerson, P. R. (2016). Rabies in Kazakhstan. *PLoS Neglected Tropical Diseases, 10*(8), e0004889. https://doi.org/10.1371/journal.pntd.0004889.

Suraweera, W., Morris, S. K., Kumar, R., Warrell, D. A., Warrell, M. J., Jha, P., & Million Death Study, C. (2012). Deaths from symptomatically identifiable furious rabies in India: A nationally representative mortality survey. *PLoS Neglected Tropical Diseases, 6*(10). https://doi.org/10.1371/journal.pntd.0001847.

Swanepoel, R., Barnard, B. J., Meredith, C. D., Bishop, G. C., Bruckner, G. K., Foggin, C. M., & Hubschle, O. J. (1993). Rabies in southern Africa. *Onderstepoort Journal of Veterinary Research, 60*(4), 325–346.

Szanto, A. G., Nadin-Davis, S. A., Rosatte, R. C., & White, B. N. (2011). Genetic tracking of the raccoon variant of rabies virus in eastern North America. *Epidemics, 3*(2), 76–87. https://doi.org/10.1016/j.epidem.2011.02.002.

Taichiev, I. T., & Botvinkin, A. D. (2006). Epidemiology of fox rabies in the Kyrgyz Republic. *Rabies Bulletin Europe, 30*(3), 5–7.

Tang, X., Luo, M., Zhang, S., Fooks, A. R., Hu, R., & Tu, C. (2005). Pivotal role of dogs in rabies transmission, China. *Emerging Infectious Diseases, 11*(12), 1970–1972.

Tao, X. Y., Guo, Z. Y., Li, H., Jiao, W. T., Shen, X. X., Zhu, W. Y., ... Tang, Q. (2015). Rabies cases in the west of china have two distinct origins. *PLoS Neglected Tropical Diseases, 9*(10), e0004140. https://doi.org/10.1371/journal.pntd.0004140.

Taylor, L. H., Hampson, K., Fahrion, A., Abela-Ridder, B., & Nel, L. H. (2017). Difficulties in estimating the human burden of canine rabies. *Acta Tropica, 165*, 133–140. https://doi.org/10.1016/j.actatropica.2015.12.007.

Tenzin, Dhand, N. K., Dorjee, J., & Ward, M. P. (2011). Re-emergence of rabies in dogs and other domestic animals in eastern Bhutan, 2005-2007. *Epidemiology and Infection, 139*(2), 220–225. https://doi.org/10.1017/S0950268810001135.

Tenzin, Sharma, B., Dhand, N. K., Timsina, N., & Ward, M. P. (2010). Reemergence of rabies in Chhukha district, Bhutan, 2008. *Emerging Infectious Diseases, 16*(12), 1925–1930. https://doi.org/10.3201/eid1612.100958.

Theberge, J. B., Forbes, G. J., Barker, I. K., & Bollinger, T. (1994). Rabies in wolves of the great lakes region. *Journal of Wildlife Diseases, 30*(4), 563–566.

Théodoridès, J. (1986). *Histoire de la rage. Cave canem*. Paris: Fondation Singer-Polignac Masson.

Tierkel, E. S., Arbona, G., Rivera, A., & De Juan, A. (1952). Mongoose rabies in Puerto Rico. *Public Health Reports, 67*(3), 274–278.

Tigas, L. A., Van Vuren, D. H., & Sauvajot, R. M. (2002). Behavioral responses of bobcats and coyotes to habitat fragmentation and corridors in an urban environment. *Biological Conservation, 108*(3), 299–306.

Timm, R. M., Cuarón, A. D., Reid, F., Helgen, K., & González-Maya, J. F. (2016). Procyon lotor. In *The IUCN red list of threatened species*. https://doi.org/10.2305/IUCN.UK.2016-1.RLTS.T41686A45216638.en.

Troupin, C., Dacheux, L., Tanguy, M., Sabeta, C., Blanc, H., Bouchier, C., ... Bourhy, H. (2016). Large-scale phylogenomic analysis reveals the complex evolutionary history of rabies virus in multiple carnivore hosts. *PLoS Pathogens, 12*(12), e1006041. https://doi.org/10.1371/journal.ppat.1006041.

Tsai, K. J., Hsu, W. C., Chuang, W. C., Chang, J. C., Tu, Y. C., Tsai, H. J., ... Lee, S. H. (2016). Emergence of a sylvatic enzootic formosan ferret badger-associated rabies in Taiwan and the geographical separation of two phylogenetic groups of rabies viruses. *Veterinary Microbiology, 182*, 28–34. https://doi.org/10.1016/j.vetmic.2015.10.030.

Turkmen, S., Sahin, A., Gunaydin, M., Tatli, O., Karaca, Y., Turedi, S., & Gunduz, A. (2012). A wild wolf attack and its unfortunate outcome: Rabies and death. *Wilderness and Environmental Medicine, 23*(3), 248–250.

Ugolini, G. (2011). Rabies virus as a transneuronal tracer of neuronal connections. *Advances in Virus Research, 79*, 165–202. https://doi.org/10.1016/B978-0-12-387040-7.00010-X.

Van Thiel, P. P., van den Hoek, J. A., Eftiov, F., Tepaske, R., Zaarijer, H. J., Spanjaard, L., ... Kager, P. A. (2008). Fatal case of human rabies (Duvenhage virus) from a bat in Kenya: The Netherlands, December 2007. *Eurosurveillance, 13*(2). pii: 8007.

Van Zyl, N. (2008). *Molecular epidemiology of African mongoose rabies and Mokola virus*. Pretoria, Soouth Africa: University of Pretoria.

Van Zyl, N., Markotter, W., & Nel, L. H. (2010). Evolutionary history of African mongoose rabies. *Virus Research, 150* (1-2), 93–102. https://doi.org/10.1016/j.virusres.2010.02.018.

Vargas-Linares, E., Romaní-Romaní, F., López-Ingunza, R., Arrasco-Alegre, J., & Yagui-Moscoso, M. (2014). Rabia en Potos flavus identificados en el departamento de madre de dios, Perú. *Revista Peruana de Medicina Experimental y Salud Publica, 31*, 88–93.

Vaughn, J. B. (1975). Cat rabies. In G. M. Bear (Ed.), *Vol. ll. The Natural history of rabies* (pp. 139–155). New York, San Francisco, London: Academic Press.

Velasco-Villa, A., Escobar, L. E., Sanchez, A., Shi, M., Streicker, D. G., Gallardo-Romero, N. F., ... Emerson, G. (2017). Successful strategies implemented towards the elimination of canine rabies in the Western Hemisphere. *Antiviral Research, 143*, 1–12. https://doi.org/10.1016/j.antiviral.2017.03.023.

Velasco-Villa, A., Mauldin, M. R., Shi, M., Escobar, L. E., Gallardo-Romero, N. F., Damon, I., ... Emerson, G. (2017). The history of rabies in the Western Hemisphere. *Antiviral Research*. https://doi.org/10.1016/j.antiviral.2017.03.013.

Velasco-Villa, A., Reeder, S. A., Orciari, L. A., Yager, P. A., Franka, R., Blanton, J. D., ... Rupprecht, C. E. (2008). Enzootic rabies elimination from dogs and reemergence in wild terrestrial carnivores, United States. *Emerging Infectious Diseases, 14*(12), 1849–1854. https://doi.org/10.3201/eid1412.080876.

Vial, F., Cleaveland, S., Rasmussen, G., & Haydon, D. T. (2006). Development of vaccination strategies for the management of rabies in African wild dogs. *Biological Conservation, 131*(2), 180–192.

Vigilato, M. A., Clavijo, A., Knobl, T., Silva, H. M., Cosivi, O., Schneider, M. C., ... Espinal, M. A. (2013). Progress towards eliminating canine rabies: Policies and perspectives from Latin America and the Caribbean. *Philosophical transactions of the Royal Society of London. Series B, Biological sciences, 368*(1623), 20120143. https://doi.org/10.1098/rstb.2012.0143.

Viranta, S., Atickem, A., Werdelin, L., & Stenseth, N. C. (2017). Rediscovering a forgotten canid species. *BMC Zoology, 2*(1), 6. https://doi.org/10.1186/s40850-017-0015-0.

von Teichman, B. F., Thomson, G. R., Meredith, C. D., & Nel, L. H. (1995). Molecular epidemiology of rabies virus in South Africa: Evidence for two distinct virus groups. *Journal of General Virology, 76*(1), 73–82.

Vos, A., Freuling, C., Eskiizmirliler, S., Un, H., Aylan, O., Johnson, N., ... Askaroglu, H. (2009). Rabies in foxes, Aegean region, Turkey. *Emerging Infectious Diseases, 15*(10), 1620–1622.

Vos, A., Freuling, C. M., Hundt, B., Kaiser, C., Nemitz, S., Neubert, A., ... Müller, T. (2017). Oral vaccination of wildlife against rabies: Differences among host species in vaccine uptake efficiency. *Vaccine, 35*(32), 3938–3944. https://doi.org/10.1016/j.vaccine.2017.06.022.

Vos, A., Freuling, C., Ortmann, S., Kretzschmar, A., Mayer, D., Schliephake, A., & Muller, T. (2018). An assessment of shedding with the oral rabies virus vaccine strain SPBN GASGAS in target and nontarget species. *Vaccine, 36*(6), 811–817. https://doi.org/10.1016/j.vaccine.2017.12.076.

Vos, A., Ortmann, S., Kretzschmar, A. S., Köhnemann, B., & Michler, F. (2012). The raccoon (Procyon lotor) as potential rabies reservoir species in Germany: A risk assessment. *Berliner und Münchener Tierärztliche Wochenschrift, 125*(5-6), 228–235.

Vos, A., Un, H., Hampson, K., DeBalogh, K., Aylan, O., Freuling, C. M., ... Johnson, N. (2013). Bovine rabies in Turkey: Patterns of infection and implications for costs and control. *Epidemiology and Infection*, 1–9. https://doi.org/10.1017/S0950268813002811.

Wachendörfer, G., & Frost, J. W. (1980). Epizootiology and control of rabies in Central Europe. In E. Zimen (Ed.), *Biogeographica: Vol. 18. The red fox* (pp. 263-275): Dr. W. Junk.

Wachendörfer, G., & Frost, J. W. (1992). Epidemiology of red fox rabies: A review. In K. Bögel, F. X. Meslin, & M. Kaplan (Eds.), *Wildlife rabies control* (pp. 19–31). Kent: Wells Medical Ltd.

Wallace, R. M., Gilbert, A., Slate, D., Chipman, R., Singh, A., Cassie, W., & Blanton, J. D. (2014). Right place, wrong species: A 20-year review of rabies virus cross species transmission among terrestrial mammals in the United States. *PLoS One, 9*(10), e107539. https://doi.org/10.1371/journal.pone.0107539.

Wallace, R. M., Lai, Y., Doty, J. B., Chen, C. C., Vora, N. M., Blanton, J. D., ... Pei, K. J. C. (2018). Initial pen and field assessment of baits to use in oral rabies vaccination of Formosan ferret-badgers in response to the re-emergence of rabies in Taiwan. *PLoS One, 13*(1), e0189998. https://doi.org/10.1371/journal.pone.0189998.

Wandeler, A. I. (1999). Raccon rabies in eastern Ontario. *Canadian Veterinary Journal, 40*, 731.

Wandeler, A. I. (2008). The rabies situation in Western Europe. *Developments in Biologicals, 131*, 19–25.

Wandeler, A. I., Muller, J., Wachendorfer, G., Schale, W., Forster, U., & Steck, F. (1974). Rabies in wild carnivores in central Europe. III. Ecology and biology of the fox in relation to control operations. *Zentralblatt für Veterinärmedizin Reihe B. 21*(10), 765–773.

Wang, L., Tang, Q., & Liang, G. (2014). Rabies and rabies virus in wildlife in mainland China, 1990–2013. *International Journal of Infectious Diseases, 25*, 122–129. https://doi.org/10.1016/j.ijid.2014.04.016.

Wang, X., Werner, B. G., Konomi, R., Hennigan, D., Fadden, D., Caten, E., ... DeMaria, A. (2009). Animal rabies in Massachusetts, 1985-2006. *Journal of Wildlife Diseases, 45*(2), 375–387. https://doi.org/10.7589/0090-3558-45.2.375.

Weiler, G. J., Garner, G. W., & Ritter, D. G. (1995). Occurrence of rabies in a wolf population in Northeastern Alaska. *Journal of Wildlife Diseases, 31*(1), 79–82.

Westerling, B., Anderons, Z., Rimeicans, J., Lukauskas, K., & Dranseika, A. (2004). Rabies in the Baltics. In A. A. King, A. R. Fooks, M. Aubert, & A. I. Wandeler (Eds.), *Historical perspective of rabies in Europe and the Mediterranean Basin*. Paris: OIE.

Wilson, D. E., & Reeder, D. M. (2005). *Mammal species of the world: A taxonomic and geographic reference* (3rd ed.). Johns Hopkins University Press.

Wohlers, A., Lankau, E. W., Oertli, E. H., & Maki, J. (2018). Challenges to controlling rabies in skunk populations using oral rabies vaccination: A review. *Zoonoses and Public Health, 65*(4), 373–385. https://doi.org/10.1111/zph.12471.

World Health Organization. (2018). WHO expert consultation on rabies, third report. *World Health Organization Technical Report Series: 1012* (p. 195).

Wozencraft, W. C. (2005). Order Carnivora. In D. E. Wilson & D. M. Reeder (Eds.), *Mammal species of the world: A taxonomic and geographic reference* (3rd ed., pp. 532–628): Johns Hopkins University Press.

Wu, X., Hu, R., Zhang, Y., Dong, G., & Rupprecht, C. (2009). Reemerging rabies and lack of systemic surveillance in People's Republic of China. *Emerging Infectious Diseases, 15*(8), 1159–1164.

Yakobson, B. A., David, D., & Aldomy, F. (2004). Rabies in Israel and Jordan. In A. A. King, A. R. Fooks, M. Aubert, & A. I. Wandeler (Eds.), *Historical perspective of rabies in Europe and the Mediterranean Basin: A testament to rabies by Dr Arthur A. King* (pp. 171–183).

Yakobson, B. A., Kind, R. J., David, D., Sheichat, N., Rotenderg, D., Dveres, N., ... Orgad, U. (1999). Comparetive efficacy of two oral vaccines in captive jackals (*Canis aureus*). In *Paper presented at the 10th Annual Rabies in the Americas Meeting, Nov. 14-19, 1999, San Diego, USA*.

Zhang, Y. Z., Fu, Z. F., Wang, D. M., Zhou, J. Z., Wang, Z. X., Lv, T. F., ... Rupprecht, C. E. (2008). Investigation of the role of healthy dogs as potential carriers of rabies virus. *Vector Borne and Zoonotic Diseases, 8*(3), 313–319. https://doi.org/10.1089/vbz.2007.0209.

Zhang, S., Liu, Y., Hou, Y., Zhao, J., Zhang, F., Wang, Y., & Hu, R. (2013). Epidemic and maintenance of rabies in Chinese ferret badgers (Melogale moschata) indicated by epidemiology and the molecular signatures of rabies viruses. *Virologica Sinica, 28*(3), 146–151. https://doi.org/10.1007/s12250-013-3316-7.

Zhang, S., Tang, Q., Wu, X., Liu, Y., Zhang, F., Rupprecht, C. E., & Hu, R. (2009). Rabies in ferret badgers, southeastern China. *Emerging Infectious Diseases, 15*(6), 946–949. https://doi.org/10.3201/eid1506.081485.

Zhao, J., Liu, Y., Zhang, S., Fang, L., Zhang, F., & Hu, R. (2014). Experimental oral immunization of ferret badgers (Melogale moschata) with a recombinant canine adenovirus vaccine CAV-2-E3Delta-RGP and an attenuated rabies virus SRV9. *Journal of Wildlife Diseases, 50*(2), 374–377. https://doi.org/10.7589/2013-01-020.

Zieger, U., Marston, D. A., Sharma, R., Chikweto, A., Tiwari, K., Sayyid, M., ... Horton, D. L. (2014). The phylogeography of rabies in Grenada, West Indies, and implications for control. *PLoS Neglected Tropical Diseases, 8*(10), e3251. https://doi.org/10.1371/journal.pntd.0003251.

Zienius, D., Pridotkas, G., Lelesius, R., & Sereika, V. (2011). Raccoon dog rabies surveillance and post-vaccination monitoring in Lithuania 2006 to 2010. *Acta Veterinaria Scandinavica, 53*, 58. https://doi.org/10.1186/1751-0147-53-58.

Zulu, G. C., Sabeta, C. T., & Nel, L. H. (2009). Molecular epidemiology of rabies: Focus on domestic dogs (*Canis familiaris*) and black-backed jackals (*Canis mesomelas*) from northern South Africa. *Virus Research, 140*(1-2), 71–78. https://doi.org/10.1016/j.virusres.2008.11.004.

CHAPTER 7

Bat rabies

Ashley C. Banyard[a], April Davis[b], Amy T. Gilbert[c], Wanda Markotter[d]

[a]Animal and Plant Health Agency (Weybridge), Addlestone, Surrey, United Kingdom
[b]Rabies Laboratory, Wadsworth Center New York State Department of Health, Slingerlands, NY, United States [c]United States Department of Agriculture, Animal and Plant Health Inspection Service, Wildlife Services, National Wildlife Research Center, Fort Collins, CO, United States [d]Centre for Viral Zoonoses, Department of Medical Virology, Faculty of Health Sciences, University of Pretoria, Pretoria, South Africa

7.1 Introduction

The lyssaviruses are a group of viruses capable of causing the invariably fatal neurological disease known as rabies, and as such are of importance to both veterinary and public health. Rabies virus (RABV) is the archetypal lyssavirus and is the only viral pathogen for which infection, following the onset of clinical disease, is almost always fatal (Fooks et al., 2017). During clinical stages of disease, an infected animal vector often attempts to transmit the virus to others through the mechanistic action of a bite. In this way, the excretion of virus from the salivary glands of infected individuals renders bite contact as the most efficient route of transmission although other mechanisms have been described. Dogs (*Canis lupus familiaris*) are the species most commonly associated with RABV infection and although the disease often presents as a paralytic disorder the association of aggressive dogs with rabies remains at the forefront of the human psyche when considering the disease.

While dogs are the main reservoir of rabies globally, and from the perspective of human infection remain the primary concern, the role of bats (Order Chiroptera) as lyssavirus reservoirs is well established (Banyard & Fooks, 2017). The association of rabies with hematophagous bats (primarily *Desmodus rotundus*), linked to their unusual evolutionary adaptation to feed on blood, has embedded rabies in the human psyche as well as driven an irrational and unjustified fear of bats across many cultures. While the negative press around bats is unjustified, human infection following bat bite does occur but is rare in comparison to the incidence of human infections associated with circulation of RABV in dogs (Dato, Campagnolo, Long, & Rupprecht, 2016).

The increased global attention and surveillance programs targeting bats as reservoirs of viral zoonoses and the advent of rapid sequencing technologies has meant that viruses of bats, including novel lyssaviruses, are being described regularly. Alongside classical RABV, the lyssavirus genus currently includes 15 distinct viruses, including *Aravan lyssavirus (ARAV)*; *Australian bat lyssavirus (ABLV)*; *Bokeloh bat lyssavirus (BBLV)*; *Duvenhage lyssavirus (DUVV)*; *European bat-1 lyssavirus (EBLV-1)*; *European bat-2 lyssavirus (EBLV-2)*; *Gannoruwa bat lyssavirus (GBLV)*; *Ikoma lyssavirus (IKOV)*; *Irkut lyssavirus (IRKV)*; *Khujand lyssavirus (KHUV)*; *Lagos bat lyssavirus (LBV)*; *Lleida bat lyssavirus (LLEBV)*; *Mokola lyssavirus (MOKV)*; *Shimoni bat lyssavirus (SHIBV)*; and *West Caucasian bat lyssavirus (WCBV)*. For this group of viruses, the invariably fatal outcome of infection means that vaccine-mediated protection is of great significance and is directly relevant to at risk human and animal populations. Studies have demonstrated that genetic, and where defined antigenic, divergence is of importance when considering the level of crossprotection afforded by rabies vaccines. Existing rabies vaccines are all based on classical RABV strains. Antigenically, a number of lyssavirus species are sufficiently divergent from RABV to escape antibody mediated neutralization following vaccination with current rabies vaccines (Evans et al., 2018). Existing human and animal rabies vaccines are highly efficacious against all variants of classical RABV and have been commercially available for decades with little alteration to the seed strains that they are based upon. Following vaccination, demonstration of a neutralizing antibody titer over a defined threshold is understood to protect individuals from the development of disease when infection with classical RABV occurs (Warrell, 2012). However, antigenic divergence across the lyssavirus genus has demonstrated that the existing rabies vaccines are not sufficient to protect against all lyssavirus species.

Investigation into neutralization of lyssavirus species has revealed that there is antibody mediated protection against ARAV, ABLV, BBLV, DUVV, EBLV-1, EBLV-2, IRKV, KHUV, GBLV, although the level of neutralizing antibody required to protect against any particular lyssavirus species is undefined (Evans et al., 2018). As these viruses group phylogenetically and antigenically with classical RABV, this group of 10 lyssaviruses has been classified further into a phylogroup, based on genetic and antigenic-relatedness termed phylogroup I. Within the genus, two additional phylogroups have been defined where in vivo vaccination-challenge experiments have demonstrated that the antibody response generated from rabies vaccines is not sufficient to generate virus neutralizing antibodies that confer protection (Evans, Horton, Easton, Fooks, & Banyard, 2012). The phylogroup II viruses consist of the African lyssaviruses: LBV, MOK, and SHIBV, while phylogroup III is broadly defined as the remaining lyssaviruses, WCBV, IKOV, and LLEBV, which appear to have Eurasian and African distribution (Fig. 7.1).The lack of crossprotection between and within these phylogroups warrants a revision in the way these viruses are considered when describing vaccine escape or protection. This chapter details current knowledge regarding each of the lyssavirus species with respect to their infection of bats and highlights areas where knowledge gaps exist and further studies are warranted.

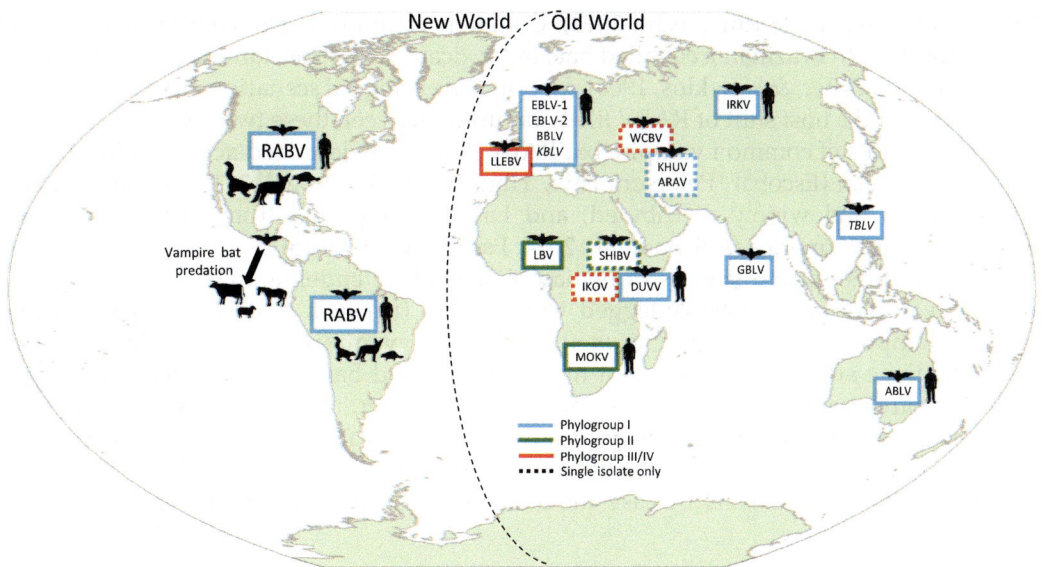

FIG. 7.1 Global distribution of lyssaviruses. Acronyms in boxes are placed according to detection of different lyssaviruses in different areas. The wide distribution of some lyssaviruses is generalized in this figure (e.g., LBV) and further details of distribution are present elsewhere. Viruses that have not been fully characterized or officially classified within the lyssavirus genus are italicized. Phylogroup definitions are shown as per the key and as within the text.

7.2 Bat rabies in the New World

7.2.1 Vampire bat rabies: Historical perspectives

Descriptions of human and cattle mortality associated with bat bites may have been described as early as the 16th century in Latin America (Baer, 1975), but laboratory diagnoses by Negri body detection were not reported from infected cattle in Brazil until the early 20th century (Carini, 1911). Carini linked the outbreaks to a wildlife source, given reports of cattle being bitten by bats and later dying and the rarity of cases in dogs at that time. Haupt and Rehaag reproduced RABV infection in a rabbit and guinea pig following inoculation of brain suspension from a nose-leaf bat that had been observed biting a cow yet had stomach contents containing blood (Haupt & Rehaag, 1921). During the 1920s, an outbreak in Trinidad led to reports of bat bites associated with cattle and human mortality, multiple isolations of RABV from common vampire bats (*D. rotundus*) and experimental transmission of common vampire bat RABV to rabbits, rhesus macaques (*Macacus mulatta*), cattle (*Bos* sp.), and a dog (Pawan, 1936b). Isolation of RABV during outbreaks in cattle was documented retrospectively in Mexico as early as 1932 and later from a vampire bat roosting within the vicinity of infected cattle in 1944, yet outbreaks of bovine paralytic rabies had been reported in Mexico as early as 1910 (Johnson, 1948).

Common vampire bat RABV shares a common ancestor with Brazilian free-tailed bat (*Tadarida brasiliensis*) RABV and is most closely related to the North American rather than the South American Brazilian free-tailed bat RABV lineage (Kuzmina et al., 2013; Streicker,

Altizer, Velasco-Villa, & Rupprecht, 2012). Common vampire bats have been documented sharing roosts with Brazilian free-tailed bats in Mexican caves (Constantine, 1967a; Constantine, Tierkel, Kleckner, & Hawkins, 1968), favoring an ecological scenario where cross species transmission and a host shift of RABV likely occurred between these two species. Numerous antigenic variants of common vampire bat RABV have been described across Latin America and the Caribbean (Escobar, Peterson, Favi, Yung, & Medina-Vogel, 2015): Variant 3 is the most common and widely distributed, and has been reported from thirteen countries throughout Latin America (Ellison et al., 2014; Escobar et al., 2015).

There are three related species of hematophagous bats endemic to the Neotropics: the common vampire bat, the hairy-legged vampire bat (*Diphylla ecaudata*), and the white-winged vampire bat (*Diaemus youngi*). All three vampire species belong to the Desmodontinae subfamily, which is a monophyletic and basal group among the highly diverse Neotropical family of phyllostomid bats. However, the common vampire bat has the most extensive geographic range, occurs in greatest abundance among these three species, and is the only species recognized as a reservoir of RABV (Gilbert, 2018) (Fig. 7.2). Spatiotemporal analysis of case incidence in livestock informs much of what we

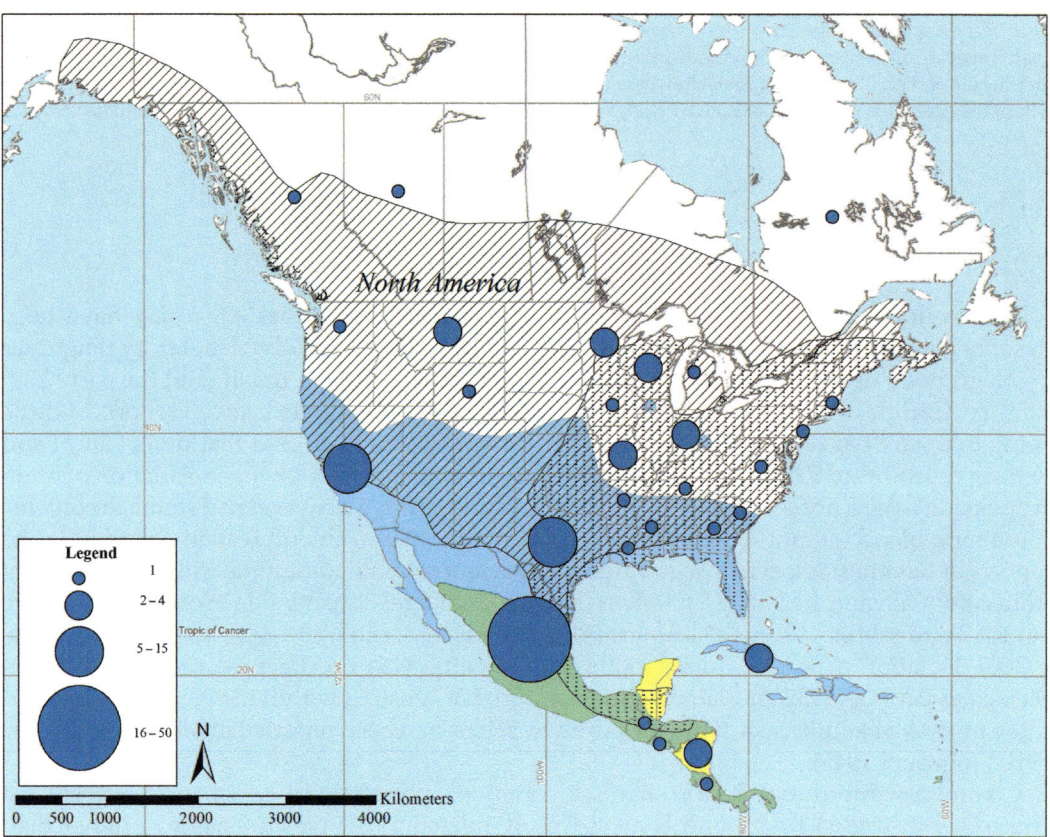

FIG. 7.2 See figure legend on opposite page.

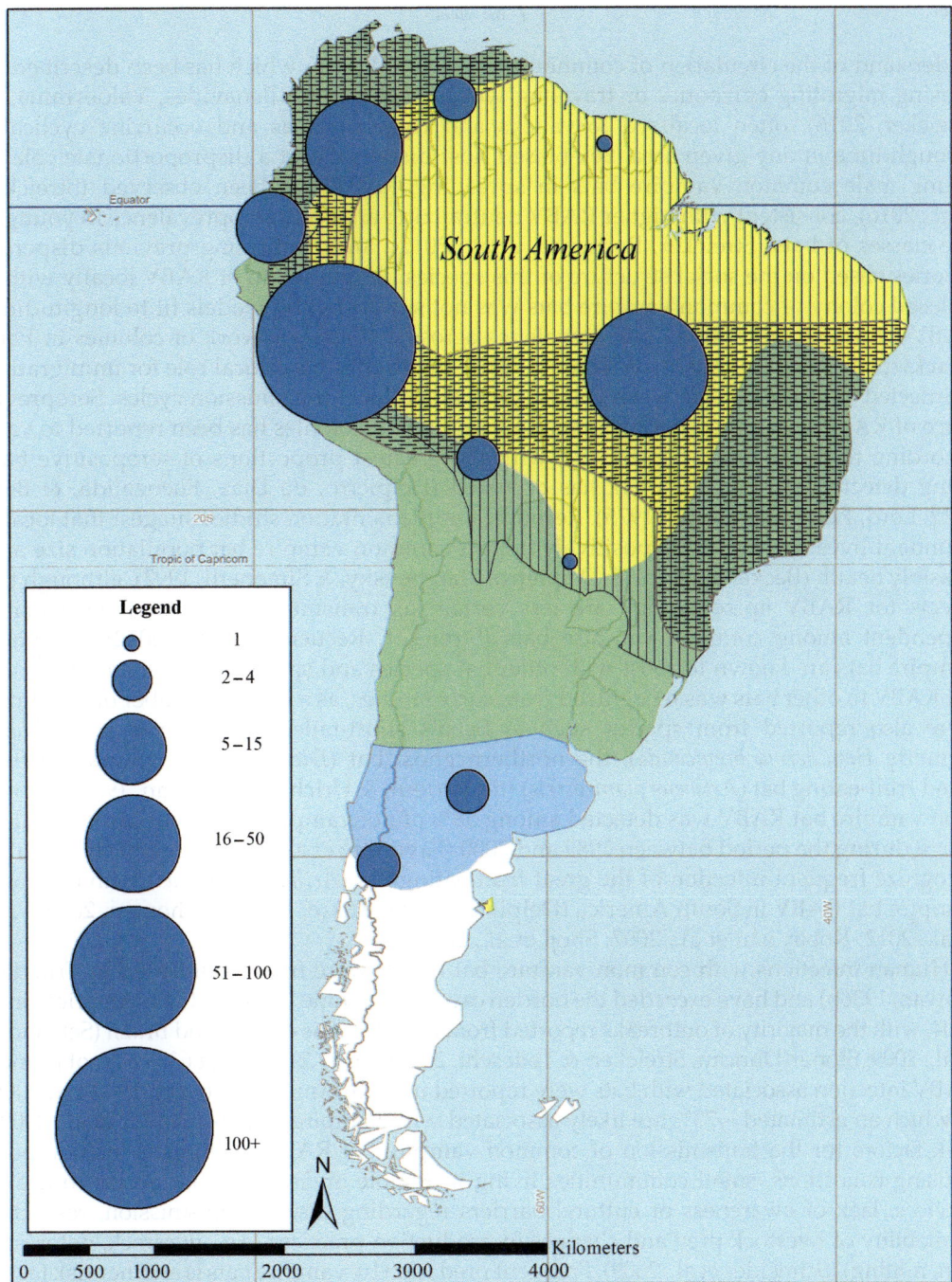

FIG. 7.2 Geographic ranges of major bat RABV reservoirs are shown along with state or province level bat RABV associated human mortality estimates during 1996–2016 for (A) North and Central America and (B) South America. Bat species ranges illustrated across both panels include the Brazilian free-tailed bat (*Tadarida brasiliensis*; blue shading), the common vampire bat (*Desmodus rotundus*; yellow shading) and the range overlap of these two species appears as green shading. Also shown only for North America are the ranges for the silver-haired bat (*Lasionycteris noctivagans*; diagonal hatching) and the tricolored bat (*Perimyotis subflavus*; dotted hatching). Bat species ranges illustrated only for South America (B) include the hairy-legged vampire bat (*Diphylla ecaudata*; horizontal hatching) and the white-winged vampire bat (*Diaemus youngi*; vertical hatching).

understand of the circulation of common vampire bat RABV, which has been described as causing migrating epizootics or traveling waves of infection (Benavides, Valderrama, & Streicker, 2016), often localized to valleys and riparian areas and occurring cyclically through time in any given area. Sex-biased host dispersal and a disproportionate role of young male common vampire bats in spreading RABV has been observed (Streicker et al., 2016), consistent with higher RABV neutralizing antibody seroprevalence in younger age classes of bats (Streicker, Recuenco, et al., 2012), and confirming previous dispersal theories based on the social structure of this species. Perpetuation of RABV locally within a closed colony of common vampire bats was not supported by models fit to longitudinal RABV-neutralizing antibody seroprevalence across a spatial network of colonies in Peru (Blackwood, Streicker, Altizer, & Rohani, 2013), highlighting a critical role for immigration of infected vampire bats between colonies to sustain local transmission cycles. Seroprevalence of RABV antibodies among common vampire bat colonies has been reported to vary according to the timing of epizootics, with the greatest proportions of seropositive bats being detected after an outbreak has occurred (Delpietro, de Diaz, Fuenzalida, & Bell, 1972; Lord, Fuenzalida, et al., 1975). Resource supplementation studies suggest that locally abundant livestock resources positively impact common vampire bat population size and possibly health (Becker et al., 2018; Delpietro, Marchevsky, & Simonetti, 1992), although the effects for RABV epizootiology are less certain as transmission may not be density-dependent among common vampire bats (Streicker, Recuenco, et al., 2012). Common vampire bats are known to roost with other bat species and spillover of common vampire bat RABV to other bats was recognized from early studies, as a limited number of isolations were also reported from species such as Seba's short-tailed bat (*Carollia perspicillata*; formerly *Hemiderma brevicauda*), the northern ghost bat (*Diclidurus albus*), and the flat-faced fruit-eating bat (*Artibeus planirostris*) (de Verteuil & Urich, 1936; Pawan, 1936b). Common vampire bat RABV was detected among 58% of nonvampire bat rabies cases typed in Brazil during the period between 1989 and 2000 (Favoretto et al., 2002). Several studies also recognize frequent infection of the great fruit-eating bat (*Artibeus lituratus*) with common vampire bat RABV in South America (Delpietro, Lord, Russo, & Gury-Dhomen, 2009; Fahl et al., 2012; Kobayashi et al., 2007; Shoji et al., 2004).

Human infections with common vampire bat RABV were first documented in Trinidad (Pawan, 1936b) and have exceeded the burden caused by canine RABV in Latin America since 2004, with the majority of outbreaks reported from remote areas of Peru and Brazil (Schneider et al., 2009; Stoner-Duncan, Streicker, & Tedeschi, 2014) (Fig. 7.2). At least 637 cases of human RABV infection associated with bats were reported in Latin America between 1990 and 2006, of which an estimated ~73% are likely associated with vampire bats (Schneider et al., 2009). Risk factors for the transmission of common vampire bat RABV to humans include poor housing conditions, small communities living in remote areas with poor access to health services, lack of awareness or cultural barriers regarding disease transmission, restricted availability of livestock prey and changes in production practices (e.g., livestock, deforestation, mining) (Schneider et al., 2009). Frequent predation by vampire bats is another risk factor and surveys have estimated that this number can vary from ~40% to 65% among human populations at risk in the Amazon (Gilbert et al., 2012; Schneider et al., 2009). Despite development of an efficacious vaccine for cattle, livestock vaccination rates remain low throughout Latin America and outbreaks of common vampire bat RABV continue to occur in cattle throughout the geographic range of the bats posing an infection risk to man and other animals (Johnson, Arechiga-Ceballos, & Aguilar-Setien, 2014). Estimates report an average of 100,000 head of cattle infected with RABV annually, representing an economic loss of ~$30 million USD, though these are likely an underreporting of the true burden and economic cost due to inadequate laboratory-based surveillance in most countries of Latin America (Acha & Malaga-Alba, 1988). Rabies-specific antibodies have been detected in unvaccinated at-risk humans and cattle in Latin America (Gilbert et al., 2012; Lord, Delpietro, Fuenzalida, De Diaz, & Lazaro, 1975), suggesting that not all human and livestock exposures to common vampire bat RABV are fatal.

Seasonal incidence of spillover RABV infections from common vampire bats to humans and cattle typically has been reported after the onset of the rainy season (Streicker et al., 2016). However, outbreaks during the dry season have also been reported for humans (Goncalves, Sa-Neto, & Brazil, 2002) and cattle (Prieto & Baer, 1972). Regardless of the season, several reports indicate stronger coincidence of depredation on humans during restricted availability or elimination of a livestock prey source (de Verteuil & Urich, 1936; Delpietro, Konolsaisen, Marchevsky, & Russo, 1994; Lopez, Miranda, Tejada, & Fishbein, 1992; McCarthy, 1989). Common vampire bats typically inhabit caves, abandoned mines and tunnels, wells, and tree hollow roosts located in close to proximity to water and it has been suggested that they remain loyal to local resources and use rivers and other waterways for foraging and short-distance local movements (Turner, 1975). In Argentina, three vampire bats (two females, one male) were recaptured from the same areas where they had been banded 12–15 years prior, demonstrating a capacity for exceptional longevity and local fidelity (Delpietro et al., 1992), despite an average lifespan of ~3 years (Linhart, 1973; Lord, Muradali, & Lazaro, 1976).

7.2.2 Rabies and insectivorous bats in the Americas

Despite the association of cattle outbreaks with vampire bat RABV during the early part of the 20th century, insectivorous bat RABV in the Americas was not well appreciated until decades later. In 1953, a 7-year-old male in Florida was bitten by a northern yellow bat (*Lasiurus intermedius*; formerly *Dasypterus floridanus*). The bat was captured and tested positive for RABV and PEP was administered to the boy who survived (Venters, Hoffert, Scatterday, & Hardy, 1954). Later that same year, a 39-year-old female was attacked by a hoary bat (*Lasiuruscinereus*, Lc) in Pennsylvania. Again laboratory testing confirmed infection and the bite victim received PEP and survived (Witte, 1954). The first human death attributed to insectivorous bat RABV was diagnosed retrospectively in a female who was bitten by an unidentified bat in Texas and died soon thereafter (Sulkin & Greve, 1954). A second case was documented a few years later in 1958, where a 53-year-old female in California died despite receiving PEP following the bite from a silver-haired bat (*Lasionycteris noctivagans*, Ln) (Humphrey, Kemp, & Wood, 1960). Following reports of these early cases, active surveillance studies were initiated by several research teams, which documented a diverse number of bat species naturally infected with RABV and a wide geographic distribution of infected bats in the United States and Canada (Avery & Tailyour, 1960; Beauregard & Stewart, 1964; Burns, Farinacci, & Murnane, 1956a; Constantine, 1967b; Girard,

Hitchcock, Edsall, & MacCready, 1965; Schneider et al., 1957; Sullivan, Grimes, Eads, Menzies, & Irons, 1954; Venters et al., 1954). While bite contact remains the principal mode of bat RABV transmission, potential for aerosol transmission was also documented by isolation of virus from air samples and elaborate experiments establishing transmission to captive carnivores in caves harboring large colonies of Brazilian free-tailed bats (Constantine, 1962; Winkler, 1968), although findings were never adequately scientifically corroborated.

Bat RABV was circulating prior to the arrival of Europeans in the late 1400s (Hayman, Fooks, Marston, & Garcia, 2016; Kuzmina et al., 2013), and Brazilian free-tailed bat and vampire bat RABVs appear to have a basal phylogenetic relationship within the monophyletic group of bat RABV. Besides the common vampire bat (Family Phyllostomidae), most bat RABV reservoirs are from family Vespertilionidae, with fewer known from the Molossidae (Constantine, 2009; Escobar et al., 2015; Gilbert, 2018). Among widely distributed reservoir hosts, Brazilian free-tailed bat RABV circulates in distinct North and South American cycles, whereas RABV circulation in hoary bats from Canada through to Argentina appears to comprise a single cycle (Pinero et al., 2012). Chile is the only country in South America where Brazilian free-tailed bats and hoary bats are reported to be the principal bat RABV reservoirs, likely a result of the limited geographic range of common vampire bats in this country (Escobar et al., 2013; Yung, Favi, & Fernandez, 2012) (Fig. 7.2B).

Bat RABV reservoirs exhibit a continuum of ecological and behavioral traits. Highly gregarious Brazilian free-tailed bats form the densest aggregations of any mammal and are highly specialized for long-distance migrations (Constantine, 1967a; McCracken, 2003). In contrast, several *Eptesicus* and *Myotis* reservoirs of RABV form much smaller colonies and display fission-fusion colony behaviors and more localized dispersal (Kerth & Konig, 1999; Willis & Brigham, 2004). There are also several RABV reservoirs among the solitary and migratory lasiurine bats (*Lasiurus* spp.) (Cryan, 2003), consistent with the idea that roosting density must not be a critical determinant of RABV reservoir competence (Fisher, Streicker, & Schnell, 2018; Streicker, Recuenco, et al., 2012). There is reduced seasonality among tropical bat species with more continuous parturition in comparison to heterothermic temperate bats with highly synchronized life histories, and the rate of RABV evolution in tropical and subtropical bats is four times faster than in temperate bats (Streicker, Lemey, Velasco-Villa, & Rupprecht, 2012). Perpetuation of RABV in temperate bat reservoirs appears to be driven both by annual synchronized pulses of susceptible young during the summer months and overwintering of infections and reduced host mortality during the winter months, which then seed infection cycles upon arousal in the spring (Davis et al., 2016; George et al., 2011). A longitudinal study of RABV neutralizing antibody dynamics across multiple colonies of big brown bats (*Eptesicus fuscus*) from Colorado demonstrated a nonlinear increase in seroconversion rate during a calendar year, consistent with the modeling studies (O'Shea, Bowen, Stanley, Shankar, & Rupprecht, 2014). The RABV perpetuation dynamics in tropical or subtropical reservoir hosts besides common vampire bats have been understudied with the exception of the Brazilian free-tailed bat (Constantine, Tierkel, et al., 1968; Steece & Altenbach, 1989). As reported for common vampire bats (Blackwood et al., 2013), spatial population structure and connectivity are likely important features supporting RABV perpetuation and epizootic spread among tropical reservoir hosts, especially among species that do not aggregate in large colonies.

Spillover of bat RABVs within the bat community is relatively common, but asymmetrical across species (Pinero et al., 2012; Streicker et al., 2010; Streicker, Altizer, et al., 2012). Spillover of bat RABV among North American reservoirs is favored by genetic relatedness between hosts and geographic range overlap (Streicker et al., 2010). Cross-species transmission of insectivorous bat RABV occurs infrequently to wild and domestic carnivores (Condori-Condori, Streicker, Cabezas-Sanchez, & Velasco-Villa, 2013; Constantine, 2009; de Mattos, Favi, Yung, Pavletic, & de Mattos, 2000; Delpietro et al., 2009; Ellison et al., 2013; McQuiston, Yager, Smith, & Rupprecht, 2001), yet hotspots can have significant impact as observed among striped skunk (*Mephitis mephitis*) and gray fox (*Urocyon cinereoargenteus*) populations in Arizona (Kuzmin et al., 2012; Leslie et al., 2006) and the common marmoset (*Callithrix jacchus*) in Brazil (Favoretto et al., 2006).

The frequency of human post-exposure prophylaxis in the US appears lower in areas where only bat RABV circulates compared to areas where carnivore RABVs circulate (Christian, Blanton, Auslander, & Rupprecht, 2009), yet the majority of indigenously acquired human rabies cases in the US in recent time has been linked to insectivorous bat RABV. Between 1958 and 2018 in the US and Canada, 70% (50 of 71) of human rabies cases with an epidemiological link to bats were associated with silver-haired bat, tricolored bat (*Perimyotis subflavus*, Ps), or Brazilian free-tailed bat (Tb) RABV (De Serres, Dallaire, Cote, & Skowronski, 2008; Ma et al., 2018).

Transmission of silver-haired bat and/or tricolored bat RABV to humans has been more intensively studied because neither species is known to roost peridomestically in contrast to Brazilian free-tailed bats, and case histories of patients suggest a pattern of cryptic exposures associated with these two species (Messenger, Smith, & Rupprecht, 2002). One study established a unique ability of Ln/Ps bat RABV to replicate in non-neuronal cells and at lower temperatures compared to dog RABV, suggesting an adaptive strategy given relatively superficial bat bites, although a gross difference in pathogenicity among multiple Ln/Ps bat and dog RABV isolates by intramuscular inoculation in mice was not reported (Dietzschold et al., 2000; Morimoto et al., 1996). A recent paper suggested lower pathogenicity of *Eptesicus furinalis* and *Myotis nigricans* bat RABVs for mice, compared to common vampire bat and canine variants, following intradermal (footpad) inoculation (Fuoco et al., 2018). However, our understanding of the comparative pathogenicity of bat RABV relative to carnivore RABV is still in its infancy (Begeman et al., 2018). Cases of human survival following central nervous system infection with bat RABV are rare (Hattwick, Weis, Stechschulte, Baer, & Gregg, 1972; Willoughby Jr. et al., 2005). A total of four cases were associated with organ transplantation from a single donor infected with *T. brasiliensis* RABV (Srinivasan et al., 2005).

7.3 Bat rabies in the Old World

7.3.1 Eurasian lyssaviruses

The detection of lyssaviruses across Eurasia continues with numerous novel lyssavirus species being detected in the last decade (Shipley et al., 2019). Here, the salient features in the discovery and assessment of lyssaviruses within Europe and Asia in a chronological order of discovery are overviewed.

7.3.1.1 European bat-1 lyssavirus

The original association of European bats with a virus capable of causing rabies was made following the discovery of a virus that caused rabies disease in unidentified bats in the former Republic of Yugoslavia and Germany in the 1950s (Mohr, 1957; Nikolic & Jelesic, 1956; Schatz et al., 2013). Following this report, it was many years before, through development of serological methods (Schneider & Cox, 1994) and later genetic methods (Tordo, Poch, Ermine, Keith, & Rougeon, 1988) of virus differentiation, that distinct lyssavirus species would be defined. Both partial and full genome sequencing has suggested the molecular divergence of EBLV-1 from the other lyssavirus species at between 500 and 750 years (Davis et al., 2005; Davis, Bourhy, & Holmes, 2006). EBLV-1 has only been reported in Europe, although it is likely present across the entire known geographical range of the host species from which it is most commonly associated (i.e., Eptesicus spp.). Of the thirty-nine species of *Eptesicus* currently defined, *E. serotinus isabellinus* and *E. serotinus* are most commonly found infected with EBLV-1. From a viral perspective, two lineages of EBLV-1 have been described (Amengual, Whitby, King, Cobo, & Bourhy, 1997) with EBLV-1a being reported in France, northern Germany, the Netherlands, and Poland. A second lineage, EBLV-1b has been reported in France, southern Germany, Poland, and Spain (Smreczak, Trebas, Orlowska, & Mudzinski, 2008). EBLV-1 continues to be discovered in new countries, including a recent detection in England among Serotine bats in 2018 (APHA, unpublished data). Indeed, the recognized geographic range of *E. serotinus* includes much of Western Europe, as far north as Denmark and Sweden, as far south of Central Europe as North Africa, and into the Himalayas and north to Korea in the east and is thought to be expanding its habitat across Europe (Schatz et al., 2014). Bats in the genus *Eptesicus* are not commonly migratory, although movements of up to 330 km have been recorded for some species from Eastern Europe.

Interestingly, although EBLV-1 is predominantly associated with *Eptesicus* sp., it has also been reported from a variety of bat species in Spain, including the greater horseshoe bat (*Rhinolophus ferrumequinum*), the greater mouse-eared bat (*Myotis myotis*), Natterer's bat (*Myotis nattereri*), and the common bent-winged bat (formerly *Miniopterus schreibersii* now *M. natalensis*) (Amengual, Bourhy, Lopez-Roig, & Serra-Cobo, 2007; SerraCobo, Amengual, Abellan, & Bourhy, 2002). Seropositivity in individuals that have been captured repeatedly over successive active and passive surveillance campaigns suggests that many bats survive exposure (Brookes et al., 2005; Echevarria, Avellon, Juste, Vera, & Ibanez, 2001; Harris et al., 2009; SerraCobo et al., 2002). However, in contrast, capture-recapture studies have demonstrated fluctuating antibody titers with the potential for serological detection limits of the test involved preventing accurate assessment of EBLV-1 exposure status (Amengual et al., 2007; Amengual, Bourhy, Lopez-Roig, & Serra-Cobo, 2008). Reported, clinical disease of EBLV-1 infection in bats manifests as abnormal behavior, including weakness, erratic behavior, spasms, and sometimes paralysis. Where clinically-ill bats are grounded, there is opportunity for transmission to terrestrial species and incidents of such transmission have been documented through EBLV-1 cases involving a Beech marten (*Martes foina*) (Muller et al., 2004), sheep (*Ovis aries*) (Tjornehoj, Fooks, Agerholm, & Ronsholt, 2006), and domestic cats (*Felis catus*) (Dacheux et al., 2009).

Spillover into human populations has also occurred and there have been three reported human rabies cases associated with EBLV-1 infection, although only one has been scientifically corroborated. The two human cases that lack characterization of either the virus or the bat species involved both occurred in the Lugansk province in the Ukraine. The first involved a 15-year-old female that developed and died from rabies in 1977 following a bat bite (Anonymous, 1986). The second involved a 34-year-old male who developed and died from clinical rabies approximately two months after being bitten by a bat in the same area in 2002 (Botvinkin, Selnikova, Anotonova, Moiseeva, & Nesterenko, 2006). Unfortunately, neither the bat involved, not human samples were retained in either case and as such the nature of the infecting virus and the bat species that caused the infection remain undefined and it is possible that these human deaths were caused by any one of a number of lyssaviruses discovered since these early reports.

One human case that was sufficiently characterized to determine the infecting agent involved an 11-year-old female from Belgorod, Russia, who was bitten by a bat on the mouth and developed atypical hydrophobia (Selimov et al., 1989, 1991). The bat was never identified, but the virus that caused rabies in this case was characterized and initially named Yuli virus. The virus was isolated following intracerebral (IC) inoculation of suckling mice with a 10% brain suspension from the patient. Assessment of the virus using a panel of monoclonal antibodies (mAbs) to lyssavirus nucleocapsid protein confirmed association with lyssaviruses and at the time was termed an antigenic variant of the European Duvenhage virus (Selimov et al., 1989). This virus was later determined to be EBLV-1 (Bourhy, Kissi, Lafon, Sacramento, & Tordo, 1992).

7.3.1.2 European bat-2 lyssavirus

EBLV-2 is predominantly associated with Daubenton's bats (*Myotis daubentonii*) and is considered enzootic within Daubenton's bat populations where detected. EBLV-2 infection has been detected across six European countries, including Finland (3 rabid bats) (Jakava-Viljanen, Lilley, Kyheroinen, & Huovilainen, 2010), the Netherlands (4 rabid bats) (Nieuwenhuis, 1988), Germany (6 rabid bats) (Freuling et al., 2008), Switzerland (4 rabid bats) (Amengual et al., 1997), and the UK. Within the UK, cases are split across Scotland (4 rabid bats), Wales (1 rabid bats), and England (22 rabid bats) with the multiple bat fatalities reported in 2018 contributing to the disproportionate detection of EBLV-2 in the UK (Wise et al., 2017; APHA, unpublished data). EBLV-2 was originally isolated from a Pond bat (*Myotis dasycneme*) in the Netherlands. Two cases of human infection with EBLV-2 have been reported (Fooks, McElhinney, et al., 2003; Lumio et al., 1986) and, as with human infection with EBLV-1, the clinical presentation was similar to that observed for human infection with classical RABV (Johnson et al., 2010). EBLV-2 infection in bats also results in disease indistinguishable from rabies. Typically, the infected animal is grounded, agitated, and aggressive (McElhinney et al., 2013). In diseased bats, EBLV-2 is generally first detected in the brain and to a lesser extent in other organs, including the tongue and salivary glands (Johnson et al., 2006). Little is understood about the perpetuation of EBLV-2 in its presumed reservoir host and where and how transmission occurs in the natural environment remains unclear (Banyard & Fooks, 2011; Johnson et al., 2008).

7.3.1.3 Aravan bat lyssavirus

Only a single isolate of ARAV has been described to date, being isolated from an apparently healthy lesser mouse-eared bat (*Myotis blythi*) in 1991. The bat was captured as part of a wider study involving the examination of 269 bats collected in the Osh region of Kyrgyzstan between 1988 and 1992 (Arai, Kuzmin, Kameoka, & Botvinkin, 2003). As with other lyssaviruses, where only a single isolate exists, little is understood regarding the epidemiology and host range of the virus.

7.3.1.4 Khujand bat lyssavirus

KHUV was discovered in 2001 in a grounded whiskered bat (*Myotis mystacinus*) found near the town of Khujand, Tajikistan (Botvinkin et al., 2003). Subsequent characterization of the isolated virus revealed that it represented a new lyssavirus species. Little is known about the epidemiology of this lyssavirus in bats as only a single isolation has been made. Further, as the taxonomy of whiskered bats has been modified, it is likely that the bat species from which KHUV was isolated may actually be the steppe whiskered bat, *Myotis aurascens* (Benda, Gazaryan, & Vallo, 2016).

7.3.1.5 West Caucasian bat lyssavirus

WCBV was isolated from a common bent-wing bat (*M. natalensis*) that was captured, in an apparently healthy state, during departure for nocturnal foraging at a cave entrance, approximately 100 km southeast of the town of Krasnodar on the border between Russia and Georgia. The WCBV isolate was obtained by inoculation of bat brain homogenate into mice that succumbed displaying disease consistent with lyssavirus infection before death at 9–13 days post intracerebral infection (Botvinkin et al., 2003). The virus is highly divergent, with no serological crossreactivity to other lyssaviruses (Horton et al., 2010; Kuzmin et al., 2008b; Kuzmin, Hughes, Botvinkin, Orciari, & Rupprecht, 2005). Neutralizing antibodies against WCBV in *Miniopterus* bats collected in Kenya (Kuzmin, Niezgoda, et al., 2008b) indicate that WCBV or some other antigenically similar virus also circulates in Africa and in the rest of the Old World where these bats are abundant.

7.3.1.6 Irkut bat lyssavirus

IRKV was originally detected in what was thought to be a greater tube-nosed bat (*Murina leucogaster*), although the identification of the bat was never confirmed. The bat involved was captured in an apartment in Irkutsk, Russia, in 2002 and died after 10 days in captivity with disease signs, including exhaustion, inappetance, and malaise. Diagnostic testing revealed lyssavirus specific antigen and nucleic acid and the virus was named IRKV (Botvinkin et al., 2003; Kuzmin et al., 2005). Following this detection, a human fatality from IRKV in Russia was reported although again the bat species involved was not confirmed (Leonova et al., 2009). The virus deemed the causative agent of this human infection was initially termed Ozernoe virus, but repeated detections of Irkut in China (Liu et al., 2013) and genetic assessment of these viruses have demonstrated that the virus was Irkut (Liu et al., 2013). Furthermore, in 2018, a rabid dog that had been involved in a human bite case was determined to be infected with IRKV (Chen et al., 2018). Follow-up studies demonstrated that IRKV could infect domestic dogs and cats and cause disease indistinguishable from rabies. This event demonstrated the risk of spillover infection and the potential for cross-species transmission events that could pose a high threat to human and animal health if IRKV were to become established in a terrestrial carnivore host (Shipley et al., 2019).

7.3.1.7 Bokeloh bat lyssavirus

The discovery of BBLV in a Natterer's bat (*Myotis nattereri*) from Germany in 2009 was unusual as the active and passive surveillance programs across the European Union had studied European bat populations extensively for several years without detection of this virus. Interestingly, while novel lyssaviruses are often only represented by single detections, BBLV has been reported on a total of eight occasions since its initial discovery in 2009 (Eggerbauer et al., 2017; Freuling et al., 2013;Freuling et al., 2011; Picard-Meyer, Borel, et al., 2013; Picard-Meyer, Servat, et al., 2013; Smreczak et al., 2018), extending the geographic range of this virus across Germany and France and into Poland. Genetic analysis of isolates described to date has demonstrated two different viral lineages of BBLV (Eggerbauer et al., 2017) with no apparent host basis for this differentiation. Detections have predominantly occurred from Natterer's bats although one case in Germany was isolated from a common pipistrelle bat (*Pipistrellus pipistrellus*) demonstrating infection of other bat species. It remains unclear whether this observation represents a spillover event or cross-species transmission of virus between bat species (Eggerbauer et al., 2017). The detection of BBLV in a Natterer's bat in Poland further extended the known geographical range for this virus (Smreczak et al., 2018) although the evolution and factors determining host range of BBLV remain puzzling (Eggerbauer et al., 2017; Shipley et al., 2019).

7.3.1.8 Lleida bat lyssavirus

The initial report of LLEBV was made following detection of nucleic acid alone (Ceballos et al., 2013). Viral RNA was detected in brain material from a common bent-wing bat (*Miniopterus* sp.) in Spain, with virus isolated from a cervical spinal cord homogenate from the original carcass several years later by intracerebral inoculation into mice (Banyard et al., 2018). Genetic and antigenic assessment of this lyssavirus suggested that LLEBV is the most genetically divergent lyssavirus discovered, with protection from existing rabies vaccines being severely limited (Banyard et al., 2018). Following this initial detection, a second isolate has been reported from the same species of bat in France (Picard-Meyer et al., 2019).

7.3.1.9 Unclassified viruses

Evidence for another distinct lyssavirus, termed Kotalahti bat lyssavirus (KBLV), in European bats was reported following the detection of virus antigen and RNA in a Brandt's bat (*Myotis brandtii*) in Finland in 2017. The bat was severely decomposed and assessment of virus antigen in the brain was inconclusive. Molecular tools detected lyssavirus RNA and the nucleocapsid gene was amplified. Sequence analysis determined that the genetic segment was most closely related to Khujand (80.1%) across the area assessed (Nokireki, Tammiranta, Kokkonen, Kantala, & Gadd, 2018). No virus isolate has been recovered and as such no virus isolate exists for KBLV. For this reason, KBLV remains unclassified.

7.4 African lyssaviruses

Of the 16 officially classified lyssavirus species, five have only been detected on the African continent with representatives of all three proposed phylogroups. The association of LBV, DUVV, and SHIBV with bats is clear, but the reservoir host species for MOKV and IKOV remains uncertain. In general, rabies cases are grossly underdiagnosed in Africa (Mallewa et al., 2007). Detection is sporadic and, due to limited surveillance, the epidemiological situation regarding lyssavirus infection across Africa is very limited. Diagnostic techniques are based on the conserved nucleoprotein and characterization to identify the specific virus causing the infection seldom occurs. Most infections are assumed to be due to RABV (Coertse et al., 2017), which remains prolific in dog populations across the continent (Hampson et al., 2015).

7.4.1 Lagos bat virus

LBV was first detected in pooled brain samples from the African straw-colored fruit bats (*Eidolon helvum*) in Lagos, Nigeria, in 1956 (Boulger & Porterfield, 1958). Experimentally, an absence of Negri bodies in the brains of infected mice and a lack of virus neutralization by rabies immune serum meant that the relationship of LBV to classical RABV was initially disregarded (Boulger & Porterfield, 1958). Later studies demonstrated reactivity in complement fixation and virus neutralization tests and demonstrated a link between the LBV isolate and RABV (Shope et al., 1970). From this point, the concept of rabies-related lyssaviruses originated. Since 1956, there have been several detections of LBV with 70% of cases reported from South Africa and reflecting active surveillance efforts in specific areas. The virus circulates among bats in sub-Saharan Africa with reports from South Africa, Nigeria, Central African Republic, Guinea, Senegal, Zimbabwe, Ethiopia, Kenya, and Ghana. LBV infection is most commonly associated with fruit bats, including Wahlberg's epauletted fruit bat (*Epomophorus wahlbergi*), the African straw-colored fruit bat, and the Egyptian fruit bat (*Rousettus aegyptiacus*) (Dzikwi et al., 2010; Hayman et al., 2008, 2010, 2012; Markotter, Randles, et al., 2006). A single isolate was reported from a Peter's dwarf epauletted fruit bat (*Micropterus pusillus*) and an insectivorous bat, the Gambian slit-faced bat (*Nycteris gambiensis*) (Swanepoel et al., 1993). Spillover events of LBV from bats into terrestrial mammals have been reported, albeit infrequently (Markotter, Kuzmin, et al., 2006; Markotter, Randles, et al., 2006) with no documented infection of humans (Markotter et al., 2008). The detection of LBV in a rabid African water mongoose (*Atilux paludinosus*) was the first isolation of LBV from a terrestrial wildlife species (Markotter, Kuzmin, et al., 2006). Previous isolations in terrestrial mammals were all from domestic animals, including cats (Foggin, 1982; King & Crick, 1988) and dogs (Markotter et al., 2008; Mebatsion, Cox, & Frost, 1992).

Studies have revealed a high seroprevalence of antibodies reactive against LBV in *E. helvum* and *R. aegyptiacus* in several African countries (Dzikwi et al., 2010; Freuling et al., 2015; Hayman et al., 2008; Kuzmin et al., 2008a; Wright et al., 2010). Seroprevalence estimates range from 14% to 67% in *E. helvum* and 29% to 46% in *R. aegyptiacus*, with adult *R. aegyptiacus* having a higher seroprevalence (60%) than subadults (31%). Several studies have reported serological evidence of lyssavirus exposure to LBV in diverse bat species and additional geographical locations outside the African continent, including Comoros, Cambodia, La Reunion, Madagascar, Mauritius, Mayotte, and Seychelles (Melade et al., 2016; Reynes et al., 2004, 2011).

Despite relatively high seroprevalence, LBV is very seldom isolated. High seroprevalence may mean exposure with seroconversion and recovery or seroconversion and latent infection rather than active infection. Studies have reported an absence of detectable LBV from 931 oral swabs and 1182 brains from healthy bats by nested RT-PCR, but generated sequences from brain material of one dead bat from which LBV was isolated (Kuzmin, Niezgoda, et al., 2008b). Others tested 796 oral swabs from *E. helvum* with no detection of RNA (Hayman et al., 2012) although seropositivity in both *E. helvum* and the Gambian epauletted fruit bat (*Epomophorus gambianus*) in Nigeria (Dzikwi et al., 2010; Hayman et al., 2018) and the Democratic Republic of Congo (Kalemba et al., 2017) has been reported. As seen with insectivorous European bats, in a clinically sick bat, high levels of RNA can be detected in the brain but also in salivary glands, tongue, and in salivary excreta, consistent with transmission routes described for other lyssaviruses (Kuzmin, Niezgoda, et al., 2008a).

The discovery of additional LBV isolates and their genome analysis have revealed that the LBV phylogeny is more complex than originally thought (Kuzmin et al., 2010; Kuzmin, Niezgoda, et al., 2008b; Nadin-Davis, Abdel-Malik, Armstrong, & Wandeler, 2002) with a high intrinsic species diversity comprising four main lineages. A Senegalese (1985), a Kenyan (2007), and a French isolate (of either Togolese or Egyptian origin, 1999) are highly similar (>99% nucleotide identity across the N gene) and constitute lineage A. The original isolate (Nigeria, 1956) is genetically distant, potentially following tissue culture adaptation, and constitutes lineage B whereas a third lineage (C) is made up of isolates from the Central African Republic, Zimbabwe, and South Africa (Markotter, Randles, et al., 2006). A fourth lineage (D) has been characterized from *R. aegyptiacus* in Kenya with only 79.5%–80.9% similarity to lineages A, B, and C (Kuzmin et al., 2010). These four distinct lineages are geographically clustered with the isolates from South Africa showing very little sequence variation, despite isolations occurring over a 25-year period (Markotter et al., 2008). Divergence between lineages is high with lineage A sharing <80% identity with the other lineages across a fragment of the N gene, a percentage cut-off value previously suggested as suitable for lyssavirus species differentiation (Bourhy, Kissi, & Tordo, 1993; Kissi, Tordo, & Bourhy, 1995). However, phylogenetic assessment of alternative regions of the genome demonstrates less divergence between lineages in comparison to the N gene.

7.4.2 Duvenhage virus

In 1970, a virus was isolated in South Africa from a man who died of rabies following a bat bite on the lip 35 days previously (Meredith, Prossouw, & Koch, 1971). The virus was serologically related to RABV and named DUVV after the individual who died. The bat was never positively identified, but from the description and the abundance of bat species in the area the common bent winged bat (*M. natalensis*) was implicated (Meredith et al., 1971). Further isolations from insectivorous bats occurred in South Africa in 1981 and 2012 (King, Meredith, & Thomson, 1994; Markotter, unpublished) and in Zimbabwe in 1986 (Foggin, 1982). The Egyptian slit-faced bat (*Nycteris thebaica*) was positively linked to the 1986 and 2012 isolates, with no confirmation of *Miniopterus* spp. involvement in DUVV epidemiology as of yet. Two human infections with DUVV have been reported. In 2006, a 77-year-old male was scratched on the face by an unidentified bat in the North West Province in South Africa and he developed symptoms 27 days later and died within 14 days (Paweska et al., 2006). The second case was a 34-year-old Dutch female who died after a bat flew against her face in Tsavo West National Park, Kenya (van Thiel et al., 2008). She noticed two small superficial wounds on her face and washed and cleaned them. She developed symptoms 23 days after the incident when she returned to the Netherlands and died 20 days later. In both these cases, no post-exposure prophylaxis was initiated. The bat species were never positively identified but described as small potentially referring to an insectivorous species. No spillover infections of DUVV to other animals have been reported.

Neutralizing antibodies (30%) against DUVV have been reported in *N. thebaica* in Swaziland (Markotter, Monadjem, & Nel, 2013) and a *Miniopterus* bat in Kenya (Kuzmin, Niezgoda, et al., 2008b). Serological evidence of DUVV exposure from outside the African continent has also been reported in bats from Comoros, La Reunion, Madagascar, Mauritius, Mayotte, and Seychelles (Melade et al., 2016). This survey included *Miniopterus, Chaerephon, Mormopterus, Otomops, Pteropus, Rousettus,* and *Hipposideros* spp., which includes frugivorous and insectivorous bats. Phylogenetic analyses have indicated that DUVV isolates form two distinct lineages: one comprising sequences from the southern African isolates and a second the Kenya isolate (Van Eeden, Markotter, & Nel, 2011).

7.4.3 Shimoni bat lyssavirus

A single isolate of SHIBV was reported in 2009, from the brain of a dead striped leaf-nosed bat (*Hipposideros vittatus* originally thought to be *H. commersoni*) found in a cave in Kenya. The virus aligns with phylogroup II, being classified phylogenetically between MOKV and LBV. Comparative serology in *R. aegyptiacus* and *H. vittatus* in sympatric roosts in Kenya indicated that the seroprevalence in *H. vittatus* to SHIBV is the same in the presence or absence of *R. aegyptiacus*. The data suggest that *H. vittatus* is the reservoir host for SHIBV (Kuzmin et al., 2011). Serological evidence of SHIBV exposure has also been reported from *E. helvum* in the Democratic Republic of Congo, though only from bat sera, which also were also reactive against LBV (Kalemba et al., 2017).

7.4.4 African lyssaviruses not associated with bats

Two additional lyssavirus species have been isolated on the African continent from mammals other than bats. IKOV has only been isolated once from an African civet (*Civettictis civetta*) in 2009 in the Serengeti in Tanzania (Marston et al., 2012). The virus aligns with phylogroup III along with WCBV and LLEBV. The reservoir host is unknown and there is no active surveillance at this time for this lyssavirus species. Both in vitro and in vivo experimentation have demonstrated no protection offered by existing rabies vaccines against IKOV (Horton et al., 2014).

MOKV was first isolated from pooled organs from shrews (*Crocidura flavescens manni*) in the Mokola forest, Ibadan, Nigeria in 1968 (Kemp, Causey, Moore, Odelola, & Fabiyi, 1972; Shope et al., 1970) and since then has also been isolated from domestic cats in South Africa (Coertse et al., 2017; Nel et al., 2000; Schneider, Barnard, & Schneider, 1985): shrews (*Crocidura* sp.) in Cameroon in 1974 (Le-Gonidec, Rickenbach, Robin, & Heme, 1978; Swanepoel, 1994); domestic cats and a dog in Zimbabwe in 1981 and 1982 (Foggin, 1982); the rusty-bellied brush-furred rat (*Lophuromys sikapusi*) from the Central Africa Republic in 1983 (Swanepoel, 1994); and between 1989 and 1990 in domestic cats in Ethiopia (Mebatsion et al., 1992). Two human infections have also been reported in Nigeria in 1969 and 1971, although only one of these was fatal (Familusi, Osunkoya, Moore, Kemp, & Fabiyi, 1972), whereas the other isolation was likely a laboratory contaminant (Familusi & Moore, 1972). Most MOKV isolations have been from cats (Coertse et al., 2017), which represent a dead end infection (Meredith, Nel, & von Teichman, 1996; Sabeta et al., 2007). The reservoir of MOKV might be a species that interacts with cats, but in the absence of surveillance, there is no evidence to support or refute this hypothesis (Kgaladi et al., 2013). MOKV has been isolated from shrews on three occasions and once from a rodent. These isolations in conjunction with pathogenicity studies that have demonstrated virus shedding in the saliva of shrews and rodents invite speculation that these animals could be possible reservoir hosts (Kemp et al., 1972). MOKV is closely related to LBV with some serological crossreactivity previously being reported. Crossneutralization shown by LBV seropositive bat sera (Dzikwi et al., 2010; Kuzmin, Niezgoda, et al., 2008a) suggests that bats cannot yet be ruled out as reservoirs; however, the lack of surveillance targeting rodents, shrews, and other potential sylvatic reservoirs in Africa is notable and therefore the reservoir host of MOKV remains speculative.

7.5 Asian and Australian lyssaviruses

Within Australia, a lyssavirus has been associated with both infection of both humans and animals. Further, the discovery of novel lyssaviruses in Sri Lanka and Thailand has also been reported. These viruses are described now.

7.5.1 Australian bat lyssavirus

With the exception of an 1867 rabies outbreak in Tasmanian dogs and rare cases of imported human rabies, Australia was considered rabies free until the 1996 discovery of a rabies-like virus in a juvenile female black flying fox (*Pteropus alecto*) in Northern New South Wales, Australia (Fraser et al., 1996; McColl et al., 2002). In a retrospective study, a second juvenile female black flying fox that was euthanized in 1995 also tested positive for ABLV (Fraser et al., 1996). Since these initial cases, field studies have documented ABLV infections in all four species of Megachiroptera found in Australia and one species of insectivorous bats, the yellow-bellied sheath-tailed bat (*Saccolaimus flaviventris*) (Prada, Boyd, Baker, Jackson, & O'Dea, 2019). Two variants of ABLV have been described among the Megachiroptera, yet only one variant has been documented from the Microchiroptera (Prada et al., 2019; Weir, Annand, Reid, & Broder, 2014). Although black flying foxes can be found throughout Asia, ABLV has only been isolated from Australian bats, three humans, and two horses (Arguin et al., 2002; Prada et al., 2019; Shinwari et al., 2014; Si et al., 2016). A comparison with all known lyssaviruses shows that genetically and serologically, ABLV is most closely related to RABV and Gannoruwa bat lyssavirus (GBLV). These similarities are advantageous as current rabies biologics are crossprotective for pre- and post-exposure treatment.

Using ABLV and cell lines from the brain and kidney of black flying foxes, researchers have investigated various mechanisms that may be involved in the immune response to infection, including autophagy (Peng et al., 2016). These studies demonstrate that during lyssavirus infection of bats, the induction of autophagy may be a protective mechanism of the host to decrease viral replication thereby improving the prospect of survival.

7.5.2 Gannoruwa bat lyssavirus

Sixty-two grounded bats were collected during 2014–2015 in Gannoruwa, Sri Lanka, and tested for rabies. Of the 62 bats, 4 (6.5% *Pteropus medius*) tested positive for lyssavirus infection by direct fluorescent antibody test. Virus was isolated in cell culture from three of the four samples. All samples tested negative on a real-time RT-PCR rabies assay, but positive using a pan-lyssavirus RT-PCR test. Whole-genome sequencing was performed and the novel lyssavirus was named Gannoruwa bat lyssavirus (GBLV) and appears most closely related to RABV and ABLV (Gunawardena et al., 2016).

7.5.3 Unclassified Australasian viruses

Taiwan bat lyssavirus (TBLV) was first reported in 2016 as part of an active surveillance program. Between 2014 and 2017, 332 bats from 13 species were submitted for rabies testing. Two of the 332 (0.6%) bats tested positive by direct fluorescent antibody test and by RT-PCR. Both bats were identified as Japanese house bats (*Pipistrellus abramus*). Virus was isolated from the brain of the bats and further characterized by electron microscopy. Virus was also isolated from the salivary glands of both bats. The ICTV has not determined the status of TBLV as a lyssavirus; however, TBLV shares closest phylogenetic relationship to EBLV-1 and DUVV (Hu et al., 2018).

7.6 Experimental studies with lyssaviruses in bats

In response to the 1950s discovery of insectivorous bat rabies in the United States, researchers have attempted to understand rabies pathogenesis in bats and the mechanisms of transmission both between bats and potential mechanisms of transmission from bats to terrestrial carnivores (Grimes, Eads, & Irons, 1955; Schneider et al., 1957). Here we detail experimental efforts to understand the pathogenesis of lyssavirus infection in bats.

7.6.1 New World lyssavirus studies in bats

7.6.1.1 *Insectivorous bat RABV*

Early in vivo studies with bats targeted the species most commonly reported with RABV infection in the United States: little brown bats, big brown bats, and Brazilian free-tailed bats. While all three species are commensal bats, they differ significantly in their behavioral biology and ecology. For example, a large colony of Brazilian free-tailed bats may consist of more than 10 million bats, while a large colony of big or little big brown bats will be significantly smaller, usually in the order of <10,000 bats (Constantine, 1967b; Dietz, Nill, & Von Helversen, 2009). Additionally, Brazilian free-tailed bats do not hibernate during winter months and are capable of migrating thousands of miles. In contrast, both *M. lucifugus* and *E. fuscus* enter periods of extensive torpor and generally exhibit more localized movements (Dietz et al., 2009). An early pathogenesis study inoculated *M. lucifugus*, *E. fuscus*, and *P. subflavus* bats with either a bat or canine RABV, varying both the dose and inoculation technique (Stamm, Kisling, & Eidson, 1956). As with earlier studies, the IC route resulted all experimental animals developing rabies. Additionally, in 20% of the bats inoculated IC with a bat-derived RABV, virus was detected in the saliva and was shed intermittently for 18 days. The fatality rate and incubation time following intramuscular (IM) inoculation was correlated with dose. By documenting the susceptibility of bats to bat RABV and confirming the presence of virus in the saliva, this study was one the first groups to demonstrate the transmission potential for bat to bat infection (Stamm et al., 1956).

A few years later, Sulkin, Krutzsch, Wallis, and Allen (1957) inoculated a canine RABV isolate via the IM route into Brazilian free-tailed bats and both IM or into the brown adipose tissue of little brown bats. The study demonstrated viral dissemination to the brain, brown fat, and salivary glands (Sulkin et al., 1957). In *M. lucifugus*, the mortality rate post inoculation into brown fat was greater than seen following IM inoculation. Additionally, viral dissemination to the salivary glands increased following brown fat inoculation compared to IM inoculation. Viral dissemination to the brown fat suggested a possible mechanism for virus maintenance and was linked to the potential for virus overwintering during periods of torpor.

Numerous experimental studies have been undertaken with insectivorous bats since Stamm and Sulkin's early pathogenesis work. Intracranial inoculation (IC) continues to be the most reliable route of inoculation with close to a 100% fatality rate and decreased incubation period (Constantine & Woodall, 1966; Dietz et al., 2009; Moreno & Baer, 1980; Reagan & Brueckner, 1951). Viral shedding in the saliva following IC inoculation of bats with homologous RABV was assessed and detected independently in Brazilian free-tailed bats, red bats, big brown bats, hoary bats, silver-haired bats and leaf-nosed bats (*Macrotus waterhousii*) in the context of transmission studies to carnivores and rodents (Constantine, 1966a; Constantine, Solomon, & Woodall, 1968; Constantine & Woodall, 1966).

The inclusion of big brown bats in rabies studies is common as they are a widely distributed reservoir of RABV in North America and relatively easy to capture and acclimate to captivity. The IM inoculation of adult big brown bats with a homologous RABV variant resulted in 80% mortality with 10% ($n=2$) testing positive for viral nucleic acid in saliva using PCR although isolation of live virus was not successful. Furthermore, the majority (85%) of bats seroconverted within 2 weeks of inoculation (Jackson et al., 2008). In another study, big brown bats were inoculated IM on three separate occasions with a varied dose of a homologous bat RABV over a period of 489–620 days (depending on treatment groups) (Turmelle, Jackson, Green, McCracken, & Rupprecht, 2010). In this experiment, fatality rates were similar between groups inoculated with titers ranging from $10^{1.9}$ to $10^{4.9}$ MICLD$_{50}$ but decreased substantially at the lower doses of 10^{09} and $10^{-0.1}$ MICLD$_{50}$. Mortality following a secondary inoculation was decreased despite using higher dose of inoculum. However, mortality was seen in the groups previously inoculated with the lower doses. Only one bat developed rabies following a tertiary inoculation with $10^{4.9}$ MICLD$_{50}$. Similar to previous studies, the majority (70%) of bats seroconverted at some point during the study; however, viral shedding data were not reported. This study demonstrated survival and seroconversion following exposure to homologous RABV, but also revealed that repeated exposures may facilitate an anamnestic response and contribute to long-term survival as is likely expected and as observed with other viral infections (Turmelle, Jackson, et al., 2010).

A further study compared the pathogenicity of two big brown bat RABV isolates, EfV1 and EfV2 in their natural reservoir host (Davis, Gordy, & Bowen, 2013). Although all the bats inoculated with EfV1 survived, 72% of the bats inoculated with EfV2 developed rabies. Of the bats that developed rabies, four (50%; $n=4/8$) shed virus in their saliva as detected by virus isolation. The majority (90%) of bats had demonstrable virus neutralizing antibodies (VNA), including all the surviving bats inoculated with EfV2. However, 10% ($n=3$) of the bats inoculated with EfV1 remained seronegative for the entire study. The difference in pathogenicity between two big brown bat variants was unexpected, yet it demonstrates the diversity between bat RABV subvariants and the challenges with reproducibility between studies. In comparison, mice were inoculated IM with either EfV1 or EfV2 and only one mouse (10%) inoculated with EfV2 developed rabies, revealing the limitations that may occur in nonhost models. A second element of the study was to evaluate the response to RABV in bats that had naturally occurring VNA. Five bats brought into captivity were seropositive and inoculated IM with EfV2; all remained healthy and developed a rapid amnestic response following inoculation.

As described previously, cross-species transmission of RABV can occur among mammals but adaptation to a novel species is infrequent, highlighting the influence of the species barrier from a pathobiological perspective. The silver-haired bat RABV variant (SHBV) is commonly implicated in human cases acquired within North America.

Previous studies have demonstrated the pathogenicity of SHBV both in cell culture and mouse models (Dietzschold et al., 2000; Morimoto et al., 1996; Yan et al., 2001). However, increased pathogenicity of this variant was not observed in bat infection models. Following IM or SC inoculation with either a little brown bat RABV, big brown bat RABV, or silver-haired bat RABV, the greatest mortality occurred in little brown bats and big brown bats that were exposed to a homologous RABV. The reduced mortality following exposure to the silver haired RABV was unexpected. Similarly, silver-haired bats inoculated IM or SC with a homologous RABV were more likely to develop rabies as opposed to those inoculated with a RABV from a big or little brown bat (Davis et al., 2016; Davis, Gordy, et al., 2013; Davis, Jarvis, Pouliott, Morgan, & Rudd, 2013). Additionally, viral shedding and dissemination were greater in bats that developed rabies following SC inoculation. Although mortality was similar following IM and SC inoculation, the resultant increased incubation time following SC inoculation may allow greater dissemination and thus increased potential or transmission. A positive relationship between incubation period and amount of virus shed in salivary glands had previously been suggested during a Brazilian free-tailed bat infection with homologous RABV (Baer & Bales, 1967). A slower disease course may enable greater spread to peripheral shedding sites and facilitate the opportunity for and efficiency of transmission events.

7.6.1.2 Hematophagous bat RABV

Unlike the potentially cryptic infection associated with insectivorous bat RABV, the transmission route of RABV from vampire bats to other mammals is often clearly visible through the generation of a wound and seepage of blood from the site due to the anticoagulant properties of common vampire bat saliva. In 1980, Moreno and Baer evaluated the pathogenesis of vampire bat RABV in common vampire bats comparing route and dose (Moreno & Baer, 1980). It was not surprising that 100% of the bats inoculated IC with the highest dose of virus developed rabies. In the IM and SC groups, mortality correlated positively with dose. However, bats that were inoculated IM were more likely to develop rabies than the group inoculated with the same dose via the SC route. Bats inoculated SC demonstrated a longer incubation than those inoculated IM or IC. Depending on the inoculation route, virus was present in the saliva of bats up to 8 days prior to the onset of clinical signs. An early report proposed that healthy vampire bats could shed and transmit rabies without developing clinical signs although reliance on diagnostic antigen detection methods that are now contraindicated may have skewed outputs in this regard (Pawan, 1936a). However, the Moreno and Baer study did not support the carrier state theory in vampire bats as all bats that shed virus in their saliva went on to develop rabies (Moreno & Baer, 1980). Further studies have reported excretion of rabies from healthy vampire bats although this observation has never been scientifically corroborated (Aguilar-Setien et al., 2005; Setien et al., 1998).

Finally, frugivorous bats (*Artibeus intermedius*) have been assessed for susceptibility to infection with a vampire bat RABV strain and were inoculated via the IC, IM, or SC routes. The induction of VNAs and viral shedding in the saliva were assessed. In contrast to previous studies, none of the bats shed virus in their saliva but all seroconverted with VNA. Additionally, mortality in this study was considerably lower in nonhomologous species with only 33% in bats inoculated IC and none of the bats inoculated IM or SC developing rabies (Obregon-Morales et al., 2017).

7.6.2 Old World lyssavirus studies in bats

7.6.2.1 EBLV-1

In 1997, a Dutch Zoo reported the discovery of an EBLV-1 infected Egyptian flying fox, prompting further investigation into susceptibility and transmissibility among Egyptian flying foxes. Following IC inoculation, 63% of Egyptian flying foxes developed EBLV-1 infection following experimental inoculation with an EBLV-1 variant isolated from an Egyptian flying fox (Van der Poel et al., 2000). Interestingly, observed mortality was significantly lower (13%; $n=2$), when using an EBLV-1 inoculum isolated from an *E. serotinus*, perhaps indicating host adaptation.

When the big brown bat was inoculated IC with EBLV-1, the fatality rate was 67% (Franka et al., 2008) while a 100% fatality rate occurred in serotine bats by the same route (Freuling et al., 2009). Although the different mortality rates following IC inoculation could be attributed a species barrier, in this study, the mortality rate was conversely greater in big brown bats (44%) as opposed to serotine bats (14%) following an IM inoculation (Franka et al., 2008; Freuling et al., 2009). Interestingly, none of the serotine bats seroconverted following exposure to EBLV-1, yet several of Egyptian fruit bats and big brown bats became seropositive following IC, IM, and SC inoculation. Additionally, virus was not present in the saliva of serotine bats following inoculation but could be demonstrated in big brown bats up to 2 days prior to clinical signs. Seroconversion and salivary excretion suggest EBLV-1 may be more pathogenic in heterologous hosts; however, it could also demonstrate viral clearance or tolerance in the homologous host.

7.6.2.2 EBLV-2

In contrast to the studies with EBLV-1, IC infection resulted in clinical signs in all Daubenton's bats inoculated with EBLV-2 (Johnson et al., 2008). Although none of the bats inoculated IM or intranasally (IN) developed rabies, 14% ($n=1$) of the bats died following SC inoculation. Interestingly, and similar to studies with EBLV-1 in *E. serotinus*, none of the bats seroconverted during the experiment. However, virus was shed in the saliva of the bat inoculated SC up to 2 days prior to clinical signs (Johnson et al., 2008).

Pathogenesis studies of EBLV-2 with Daubenton's bats have demonstrated that IC inoculation leads to a rapid development of disease. Although inoculation via peripheral routes, such as IM and IN did not lead to mortality or seroconversion of the animals exposed, one of seven bats inoculated by the subdermal route developed disease. The detection of viral RNA in the oral cavity of the single animal that succumbed to infection suggests that biting or allogrooming are likely the most effective route of virus transmission, although further evidence is required to corroborate this assumption (Johnson et al., 2008). Scratches or bites might explain the exposure and fatal infection of two bat biologists with histories of encounters with Daubenton's bats (Fooks, Brookes, Johnson, McElhinney, & Hutson, 2003). However, in a number of in vivo studies, there have been no reports of an infected bat biting another bat and that bat developing disease, though it is very likely that the housing conditions described in many captive animal containment facilities do not adequately mimic or facilitate the natural social behaviors of the bat species under study (Abbott et al., 2020; Gilbert et al., 2020). The potential for infection via low transmissibility rates may confer an advantage to these viruses. Indeed, if transmissibility rates are high, disease may occur and individuals succumb, reducing the potential for further spread. The dissemination of EBLV-2 within experimentally infected bats is identical to that reported in bats infected with RABV and neuroinvasion activates the same innate immune responses (Johnson et al., 2006), both suggesting a similar pathology. However, the apparent host restriction and existing epidemiology of EBLV-2, as with other Old World lyssaviruses, suggests that these viruses are fundamentally different from RABV, although whether this constraint is virological or ecological is yet to be defined (Vos et al., 2007).

7.6.2.3 ABLV

Since the discovery of ABLV in 1996, several field studies have been performed with bats and terrestrial animals to determine susceptibility (McCall, Field, Smith, Storie, & Harrower, 2005; McColl, Chamberlain, Lunt, Newberry, & Westbury, 2007). Based on surveillance data, it is clear that a native fruit bat species, the gray-headed flying fox (*Pteropus poliocephalus*), is a definitive host (Warrilow et al., 2003). When experimentally exposed, IM inoculation resulted in three of seven (43%) animals developing clinical signs of disease, including muscle weakness, trembling, and limb paralysis (McCall et al., 2005). ABLV was detected in the brain of the three animals, yet none developed VNA. Conversely, all remaining animals survived infection and developed neutralizing antibodies against ABLV. Additionally, ABLV inoculation experiments have been conducted to assess the susceptibility of companion animals to infection (McColl et al., 2002, 2007). Although a number of subjects, both dogs and cats, showed occasional neurologic signs, clinical signs did not progress and all survived during the post-inoculation observation period. All animals seroconverted and no ABLV antigen or viral genome was detectable in oral swabs or tissue samples tested. This suggests a limited scope for cross-species transmission although further studies in other species are warranted.

7.6.2.4 ARAV, KHUV, IRKV, WCBV

The detection of several bat lyssaviruses in Asia and Africa has encouraged further virus discovery projects in bats (Mani et al., 2017; Simic et al., 2018; Smreczak et al., 2018). To date, only a single isolate of each WCBV, KHUV, ARAV has been reported, whereas a small number of cases of IRKV have been described (Chen et al., 2018; Liu et al., 2013).Where in vivo experimentation has been attempted, pathogenesis studies with *E. fuscus* assessed infection with ARAV, KHUV, IRKV, and WCBV (Hughes et al., 2006; Kuzmin, Franka, & Rupprecht, 2008). Observed mortality rates were 75%, 60%, 55%, and 29%, respectively, and the incubation time was significantly shorter in the group inoculated with IRKV. Following oral instillation of virus, all bats remained healthy with none developing disease (Kuzmin, Franka, & Rupprecht, 2008). The presence of VNA was assessed in the bats inoculated with ARAV, KHUV, or IRKV. VNAs were absent in bats that succumbed but those that survived to the end of the experiment were seropositive upon termination. One caveat in the WCBV experiment is the reuse of bats that had previously been inoculated, but survived infection, with IRKV or RABV during an earlier study. The impact of this approach remains unclear although it is widely accepted that neutralizing antibodies developed in response to infection with a phylogroup I lyssavirus are unable to neutralize more divergent viruses such as WCBV and as such previous exposure may not have impacted on the results for challenge with WCBV. However, a more thorough approach to investigating these viruses in their presumptive reservoir hosts is warranted.

7.6.2.5 LBV

Studies with LBV in bats have been limited with only a single IC challenge study. Here, several different LBV isolates were inoculated into Straw-colored fruit bats (*E. helvum*) in an assessment of comparative pathogenicity and excretion. The study was limited to only a few bats per LBV lineage but appeared to demonstrate variation in ability of different isolates to be excreted in saliva. Disease presentation was as expected with clinical signs of rabies developing in all bats although the time to development of disease differed slightly between isolates (Suu-Ire et al., 2018). Further studies are warranted both with LBV and the other African lyssaviruses in bats to try and understand mechanisms of perpetuation in presumptive reservoir hosts as well as to assess whether viruses that haven't yet been associated with bats (e.g., MOKV and IKOV) are able cause disease in African bats.

7.7 Important knowledge gaps and challenges to lyssavirus research

7.7.1 Aerosol transmission

Natural mechanisms of virus transmission between conspecifics have been a conundrum for decades and continue to be a concern through the potential for cross-species transmission that may lead to host shifts of bat lyssaviruses to terrestrial carnivores. The low-level seroprevalence described in numerous studies, especially those involving serosurveillance for lyssavirus-neutralizing antibodies in Old World bat populations where RABV is thought not to circulate, further questions mechanisms of exposure versus infection that leads to productive disease. Aerosol transmission of viruses is an interesting transmission route for which evidence remains, although scientific corroboration is often lacking. In an attempt to elucidate potential transmission mechanisms, early studies were undertaken in caves in which terrestrial animals were cohabitated below large bat colonies. To evaluate aerosol transmission, caged coyotes (*Canis latrans*) and gray foxes (*Urocyon cinereoargenteus*) were placed in caves inhabited by a colony of over 20 million bats. The experimental animals were monitored extensively, and the caged animals developed clinical rabies (Constantine, 1962) highlighting the potential for aerosol transmission of RABV (Constantine, 1966b). However, it is not clear that biting can be ruled out as a potential transmission mechanism. Additionally, the cave environment was considered inhospitable based on oppressive temperatures, humidity, and ammonia levels due to the number of bats within the colony. The poor ventilation, heat and humidity (80–96 F and between 50%–77% humidity), ammonia (up to 100 ppm), and carbon dioxide (up to 7600 ppm) levels could be caustic irritants to the eyes and mucous membranes. It is unknown if the observations reported in this study are truly of relevance to the picture of virus transmission and such studies require further corroboration. However, RABV was detected from the nasal mucosa of Brazilian free-tailed bats and aerosol transmission may somehow relate to the relatively high (i.e., >50%) natural seroprevalence against RABV observed among Brazilian free-tailed bats (Constantine, Emmons, & Woodie, 1972; Constantine, Solomon, & Woodall, 1968; Steece & Altenbach, 1989; Turmelle, Allen, et al., 2010).

To evaluate aerosol transmission under controlled laboratory conditions, researchers exposed bats to an intranasal inoculation or aerosol exposure (Davis, Rudd, & Bowen, 2007; Franka et al., 2008; Freuling et al., 2009; Johnson et al., 2008). In studies using insectivorous bats, scientists instilled either EBLV-1 or EBLV-2 into nasal passages. None of the bats in these experiments developed rabies or seroconverted. To evaluate the susceptibility of aerosol transmission in Brazilian free-tailed bats, the species which make up the large colonies described in Constantine's earlier cave studies, Davis et al. (2007) exposed this species and big brown bats to aerosolized RABV using a Glass-Col inhalation system. Although none of the animal developed rabies, all seroconverted following aerosol exposure. However, when bats were challenged using an IM inoculation 6 months later, the fatality rate was 42% highlighting the complexities in correlating VNA and protection. Conversely, four mice exposed to aerosolized RABV developed clinical signs between 8 and 17 days post exposure, again illustrating complexities of rabies transmission and host susceptibility. Moreover, the observation that bats in the three studies using IN or aerosol exposure failed to develop rabies suggests that even where potentially possible, this route of infection is likely an inefficient route of transmission. However, further studies where exposure of mucous membranes to low levels of virus inocula are warranted.

7.7.2 The role of torpor in potential maintenance of bat lyssavirus infections

Bats enter torpor where temperatures dictate and can remain in a state of temporary hibernation for considerable periods (Dietz et al., 2009). The association of viral pathogens with bats and the infection status of those infected prior to entering torpor is of interest. Early studies with flaviviruses have demonstrated a potentially complex overwintering mechanism of viral infection in hibernating bats. Studies observed decreased viral growth and extended incubation times in torpid animals suggesting the presence of mechanisms that allow the virus and host to survive throughout hibernation (Herbold, Heuschele, Berry, & Parsons, 1983; La Motte Jr, 1958). For RABV, a series of experiments in the late 1950s highlighted potential circumstances for viral persistence by virtue of body temperature as well as suggesting a possible role for the adipose tissue of hibernating bats in the maintenance of

RABV (Sulkin, 1962; Sulkin et al., 1957; Sulkin, Allen, Sims, Krutzsch, & Kim, 1960). However, a more recent study was unsuccessful in corroborating an involvement of brown fat in virus maintenance through periods of torpor (Kuzmin, Botvinkin, & Shaimardanov, 1994). More recently, hibernation studies with experimentally inoculated bats demonstrate that torpor results in an extended incubation period, likely the result of decreased cellular and viral metabolism (Davis et al., 2016). Studies addressing these potential virus:host interactions are incredibly difficult to address through both the protected nature of bats across much of the globe alongside knowledge gaps in bat immunology that likely contribute significantly to outcomes of infection.

7.7.3 Inoculation route and dose

There is no standard route of inoculation, dose, or bat RABV routinely used in bat rabies studies, likely due to the large number of bat lyssaviruses that circulate throughout the world (Reid & Jackson, 2001). Additionally, it is difficult to estimate the dose a bat could potentially transmit during the mechanistic activity of a bite given their size. However, the use of multiple inoculation routes and homologous and heterologous bat RABV in experimental infection studies have provided a basic insight into the mechanism of rabies transmission between bats. As described earlier, aerosol and oral exposure to rabies may not be the most optimal route for rabies transmission between bats. The various enzymes and antibodies present in the oropharyngeal cavity and the gastrointestinal tract result in an inhospitable environment for many pathogens. However, aerosol exposure may result in the development of antibodies, but the level of protection is unknown and studies have suggested that VNA do not always correlate with protection. IC inoculation has a high fatality rate but is not consistent with a realistic situation. With few exceptions, the fatality rate of bats inoculated intramuscularly consistently shows a mortality rate between 0% and 50%. However, this appears to be dependent on both dose and bat RABV. Indeed, two variants isolated from big brown bats have significantly different outcomes in homologous species. Based on experimental results, subcutaneous exposure is also likely to be a common route of transmission, resulting in a decreased mortality rate and increased incubation period when compared to IM exposure. Regardless of the route of dose inoculated, bats appear to be most susceptible to a homologous RABV. Alongside experimental route and dose, the impact of any viral passage between isolation, through characterization and experimental inoculation remains unclear. Additionally, addressing genetic diversity using nucleic acid sequencing technologies has been challenging due to the inability to differentiate between genetic mutation on messenger RNA versus genome and antigenome strand templates.

7.7.4 Lyssavirus excretion

Not all reported bat studies have assessed viral shedding in the saliva. However, those that did demonstrated that viral shedding does occur but is intermittent and animals can start shedding virus anywhere within 24h of clinical presentation up to 3 weeks prior to developing clinical signs. Outputs from studies must be treated with caution where molecular testing alone has been used to assess virus shedding as alongside the potential for laboratory contamination, the detection of nucleic acid does not necessarily correlate with the presence of live virus. The amount of virus shed in bat saliva has been difficult to measure as studies have used different techniques to determine the titer in saliva. It is difficult to extrapolate RT-PCR Ct values to an infectious viral load as RT-PCR measures all viral RNA, not just infectious virus. Other studies have used clinical signs of the mouse inoculation test or solely the presence or absence of virus in cell culture to identify virus in saliva. One study quantified the amount of virus excreted in the saliva of bats and reported titers of 0.9×10^3 to 1.8×10^3 in cell culture (Aguilar-Setien et al., 2005). Based on experimental studies, the infectious dose required for transmission is unknown and is likely to involve several factors, including variant and location of bite. One small murine study attempted to address the minimal infectious dose for lyssaviruses and concluded that, although difficult to demonstrate it is likely that if one viable infectious particle manages to avoid clearance in non-neuronal cells and infect a peripheral nerve, it may be sufficient to cause productive infection (Banyard et al., 2014). It remains unknown whether naturally occurring compounds found in saliva influence viral survival.

7.7.5 The caveats of using wild caught bats in laboratory studies

Most bats in experimental studies are wild caught with the potential for having been previously exposure to a lyssavirus. Although most investigators perform serology prior to inoculation, the lack of VNA does not always correlate with exposure history and the sensitivity of virus neutralization tests, alongside the often small volumes of sera available to test make outputs hard to interpret from the perspective of previous exposure. In a multiple low-dose repeated exposure experiment, a decreased mortality rate following a second or third inoculation several months after the primary exposure was noted (Turmelle, Jackson, et al., 2010). However, survival following a viral challenge was also dependent on the route and dose of the inoculum. An earlier study monitoring rabies in large bat colonies found a low prevalence of active infection but the presence of IgM antibodies in juveniles and high prevalence of IgG antibodies in adults (Steece & Altenbach, 1989). Taken together, these studies suggest early and repeated low-level exposures to RABV may result in the development some level of long-term immunity. Because seroreversion does occur in bats, it could be expected that a significant portion of the bats caught for lyssavirus studies have had prior exposure.

One of the most significant problems with performing rabies studies in bats is the lack of immunological naïve animals. Unlike other animal models, breeding insectivorous bats in captivity is difficult, given their biological and ecological/environmental requirements. Prior to entering a captive colony, bats are usually bled to determine previous rabies exposure as demonstrated by the presence of VNA. However, the absence of VNA does not always signify a lack of previous exposure. All the studies described earlier used wild-caught bats. To date, there has only been one study using insectivorous bats born in captivity with no known exposure (Davis, Jarvis, Pouliott, & Rudd, 2013). These bats were born to wild caught bats that were pregnant at the time of capture. In this study, RABV naïve adult big brown bats were inoculated with one of two homologous or a heterologous RABV. As in previous studies, the mortality rate with a homologous RABV was higher whereas the mortality following exposure to a heterologous RABV was considerably lower. However, none of the bats seroconverted following the first inoculation and there was a 75% mortality when the surviving bats were challenged suggesting a single exposure may not provide adequate protection from a subsequent challenge. Alternatively, the amount of virus in the experimental inocula may have been greater than likely be transferred during a natural bite exposure. In this study, maternal antibodies against RABV would have cleared long before the adult naïve adult bats were exposed to rabies. If inoculation had been performed in the presence of maternal antibodies, the mortality may have been less while priming the immune system for a future exposure to RABV. The potential role of memory B cells and/or memory T cells in both single and repeated exposures requires investigation.

7.7.6 Complexities of bat surveillance initiatives

Surveillance has consistently demonstrated that healthy bats can be seropositive for lyssavirus exposure although the approach taken to surveillance can impact on outputs (Bowen et al., 2013; Constantine, Solomon, & Woodall, 1968; Klug, Turmelle, Ellison, Baerwald, & Barclay, 2011; Kuzmin, Niezgoda, et al., 2008b; O'Shea et al., 2014; Schneider et al., 1957; Steece & Altenbach, 1989; Zieger et al., 2017). In the years since the first positive bat was identified in the US, the seroprevalence has remained consistent in healthy bats caught for surveillance. Kuzmin, Niezgoda, et al. (2008a) tested over 1100 bats and identified one (<0.1%) bat positive for Lagos Bat Virus (LBV) (Kuzmin, Niezgoda, et al., 2008a). Rabies surveillance at a windfarm reported that less than 1% of the bats were rabies positive (Klug et al., 2011). Surveillance of RABV among 672 bats collected in Guatemala demonstrated that approximately 0.3% of bats were infected (Ellison et al., 2014). However, three studies, two of which were done to determine the distribution of ABLV in Australia, found between 0% and 9% of the bats tested were positive for a lyssavirus (Field, 2018; Warrilow et al., 2003). The third study performed in Colorado reported 6% ($n=2/35$ bats) of the tested bats were positive for RABV (Shankar et al., 2005). The most recent survey from the CDC reports a rabies positivity rate of 5.9% based on passive surveillance (Ma et al., 2018). However, passive surveillance estimates represent a compilation of reports from rabies laboratories located within each state and is skewed toward the sampling of sick or injured bats. Thus, while not representative of prevalence in natural populations of bats, the passive-surveillance based estimates are an accurate indicator of transmission risk since the sampling occurs following human or animal contact.

Serosurveillance of wild caught bats generally support the findings of captive bat studies and provide further insight of lyssavirus maintenance in bat colonies. One of the earliest studies found much higher levels of IgM in juvenile bats when compared to adults, yet both had high levels of anti-rabies IgG antibodies (Steece & Altenbach, 1989). Seropositivity for RABV and EBLV-1 increased over the summer, which has been documented by subsequent studies (Constantine, Tierkel, et al., 1968; O'Shea et al., 2014; Robardet et al., 2017; Steece & Altenbach, 1989; Turmelle, Allen, et al., 2010). However, seropositivity was not always related to seasonal variation (Bowen et al., 2013; Suu-Ire et al., 2017). Seropositivity rates differ among bat species, which is not surprising given variable ecology among species (Hayman et al., 2008). Some studies have reported a greater rate of seropositivity in female bats over males, which could be expected given the large maternity colonies in which females raise offspring (Kuzmin, Niezgoda, et al., 2008b; O'Shea et al., 2014).

Concurrent infection with two lyssaviruses has not been reported; however, exposure to multiple lyssaviruses has been documented in serosurveys of wild-caught bats in Africa. Indeed, some bats were seropositive for LBV and MOKV, which is unexpected as MOKV has only been diagnosed in terrestrial mammals although the likelihood that serological crossreactivity and exposure to as yet undefined lyssaviruses is unclear (Kuzmin, Niezgoda, et al., 2008a; Wright et al., 2010). At least two studies have reported bats with antibodies to both SHIBV and LBV although again with the caveat that cross-species neutralization likely occurs (Kalemba et al., 2017; Kuzmin et al., 2011).

Serological surveillance can also be valuable in areas in which terrestrial rabies is endemic, yet bat lyssaviruses have not been reported. In Bangladesh, 288 bats were tested for the presence of VNA; 212 of the 288 bats were also tested for lyssaviruses using the dFAT. All bats were negative for lyssaviruses infection but the sera from three bats demonstrated neutralization against Khujand (Kuzmin et al., 2006). Although canine rabies is endemic in India, the presence of bat lyssaviruses has not been identified. To assess the risk of bat lyssaviruses in India, 164 bats were tested for rabies and the serum of 78 of those bats for VNA. None of the bats were rabies antigen positive and four (5%) were positive for antirabies VNA (Mani et al., 2017). The detection of VNA against WCBV in Kenyan Miniopterus bats 4 years following the discovery of WCBV near the Russian-Georgia border may signify a wider geographical distribution (Botvinkin et al., 2003). Miniopterus maintain a broad geographical range and can migrate more than 100 km, characteristics that may allow incremental lyssavirus movement over long distances (Amengual et al., 2007; Serra-Cobo et al., 2013). A similar situation may have resulted in the presence of anti DUVV antibodies in bats on islands off the Mozambique coast. Although the distances are too far traditional migration, heavy winds could affect movements (Melade et al., 2016). Although terrestrial rabies was eradicated from Croatia in 2014, the presence of bat lyssaviruses was unknown. Simic et al. (2018) tested 350 bats for the presence of antibodies against EBLV-1, 20 (5.7%) of which were positive (Simic et al., 2018). They also tested oral swabs for the presence of lyssavirus excretion and all samples were negative. Zieger et al., 2017 reported 7.2% of bats tested on Grenada, an island nation where mongoose rabies is endemic but without a history of bat rabies, were seropositive (Zieger et al., 2017). The route of exposure is not known, but the authors suggest the movement of bats from locations where bat lyssaviruses are endemic.

7.7.7 Abortive infection and carrier state in bats

There has been considerable discussion around the significance of rabies VNA in bats and other mammals and what the results from serological assessments of both wild-caught and experimentally infected bats actually represent. The term abortive infection has been used to describe an unproductive infection due to lack of viral replication (Kaplan, 1969), the production of virus but inability to produce adequate amounts to establish an infection (Bell, 1964), or clinical infection but one from which the animal can recover, with or without sequelae

(Smith, 1981). The mechanisms responsible for abortive infections are unclear but are likely the result of unrecognized viral/host dynamics.

Abortive infections have been described for multiple viral infections, including herpes, poliomyelitis, and influenza (Kaplan, 1976; Wyatt, 2007). Abortive rabies infection was used to describe the recovery of mice and rabbits inoculated IC with rabies (Bell, 1964). More recently, abortive infection has been used to describe the finding of antirabies VNA in bats although detection of VNA does not necessarily correlate with detection of RABV antigen in the brain or saliva. Although abortive infection is often discussed in relation to seropositivity in bats, multiple studies have also demonstrated VNA in clinically healthy humans, cattle, hyenas, rabbits, rats, and carnivores, including dogs, hyenas, jaguars, and pumas (Bell, 1964; Black & Wiktor, 1986; Gilbert et al., 2012, 2015; Gomme, Wirblich, Addya, Rall, & Schnell, 2012; Jorge et al., 2010; Smith, 1981).

The suggestion of bats as carriers of rabies goes back to a manuscript by Pawan (1936a) in which saliva from healthy vampire bats was IC inoculated into mice resulting in clinical signs compatible with rabies (Pawan, 1936a). A subsequent study suggested the clinical manifestations seen in these mice were the result of Rio Bravo virus as opposed to RABV infection (Constantine & Woodall, 1964). Using modern, more sensitive technologies, including real-time RT-PCR, there have been reports of viral shedding from bats and canids that remained healthy or recovered from clinical RABV infection (Aguilar-Setien et al., 2005; Robardet et al., 2017). However, it should be noted that virus was only able to be detected using hnRT-PCR in the Robardet studies and virus was unable to be isolated in cell culture. Aguilar-Setien et al. (2005) isolated virus in cell culture from healthy bats but only following IC inoculation using a high titered RABV. Translating these results to real-world applications is difficult as other studies were unable to repeat their findings.

7.7.8 Future experimental approaches

The understanding gained through inoculation and assessment of lyssavirus infection in bats has been significantly limited by a lack of knowledge regarding the biology and immunology of the host. Greater knowledge of bat immunobiology is required before stronger conclusions between the interaction of different viral pathogens and bats can be recognized. With current information, hypotheses can be developed based on the experimental and field studies performed over the past 70 years. Bats may receive their initial exposure to a lyssavirus as juveniles, during a time at which they may have some protection afforded by maternal antibodies. As bats are long-lived mammals, they are likely to undergo repeated exposure during their life span, potentially being exposed to multiple antigenically divergent viruses, as demonstrated by reported seroconversion and seroreversion. Although bats may have differing levels of susceptibility, clearly they are susceptible to lyssavirus infection despite being the presumptive natural reservoir associated with most lyssaviruses. With this understanding, it is difficult to theorize which bats will go on to develop clinical infection. Is it the result of undetectable immune suppression or advancing age and immune involution? Do transmitted dose and site of exposure impact on clinical outcome? Is exposure to novel variants a driver for induction of clinical disease? All of these questions and more remain.

Using a mouse model to study bat lyssaviruses is inherently limiting given the lack of genetic relatedness, differences in ecology, life history, and immune function. However, maintaining bats in captivity is resource and time intensive, which limits the ability of many scientists. Additionally, one needs to consider the potential for previous exposure when commencing lyssavirus studies in bats. The advantage of studying bat lyssaviruses in mice is the plethora of well-established assays, reagents, biologics, and even genetically similar animals that can be used to identify mechanisms that are integral in lyssavirus infection. Although the ability to study lyssaviruses in bats has been hampered by the lack of technologies currently available for traditional lab animal models, laboratories are developing immortalized cell lines, assays to identify upregulation of immune response genes, and have sequenced the entire genome of at least one bat species (Crameri et al., 2009; Virtue, Marsh, Baker, & Wang, 2011; Zhang et al., 2017). These innovations, and the ones in development, will enable scientists to further investigate the dynamics of the viral/host relationship.

7.8 Future prospects for controlling lyssaviruses in bats

The status of bats globally as being protected legally or species of conservation interest means that efforts to target bats as part of any form of preventative treatment are severely limited. For bat disease control, efforts are targeted toward the common vampire bat due to the impact on human and livestock health and the associated economic burden (Anderson et al., 2012; Schneider et al., 2009). The diversity of bat RABV reservoirs in the Americas is recognised as a constraint for the prospect of eliminating RABV from bats. Further challenges, including both the diverse ecologies of reservoir hosts and logistical issues in tailoring delivery systems to access and vaccinate bats, also exist. While progress has been made to establish the immunogenicity and efficacy of commonly used oral rabies vaccines, vaccine delivery systems rely on either direct application to individual bats or indirect grooming and contact behaviors of colonial species.

Historically, control methods for vampire bats have consisted of both nonlethal and lethal methods, frequently applied as reactive strategies during or following an outbreak to reduce bite incidence and the opportunity for RABV transmission to livestock and humans. Nonlethal methods may involve using repellents or physical barriers (e.g., mosquito nets, window screening), while specific lethal methods involve the use of anticoagulants either applied directly to captured vampire bats or to livestock prey (Constantine, 1970; Linhart, 1975). Nonspecific lethal methods, including roost destruction by gassing or the use of explosives, have long been recognized as ineffective with the incidental cumulative mortality of millions of beneficial nontarget bat species being highly undesirable (Linhart, 1975; Lord, 1988; O'Shea, Cryan, Hayman, Plowright, & Streicker, 2016; WHO, 2005). The application of strychnine to cattle wounds was one of the earliest specific lethal methods for local population reduction of vampire bats (de Verteuil & Urich, 1936; Greenhall, 1963). The anticoagulants diphenadione and chlorophacinone were subsequently developed and subjected to field evaluation during the early 1970s, and localized vampire bat population reduction associated with the use of these substances was effective in reducing bite incidence to cattle (Linhart, Flores-Crespo, & Mitchell, 1972; Thompson, Mitchell, & Burns, 1972). Later that decade, the anticoagulant warfarin was also evaluated in a field study and recommended as another anticoagulant tool (Flores-Crespo, Fernandez, De Anda-Lopez, Velarde, & Anaya, 1979). Proactive local population reduction strategies were also developed to prevent the invasion of migrating epizootics into localized areas (Fornes et al., 1974; Lord, 1988), although the evidence for the effectiveness of this technique has been infrequently reported in the literature. The World Health Organization (WHO) Expert Consultation on Rabies has consistently advocated vaccination of at-risk cattle, but revised their position on population reduction of vampire bats by anticoagulants. An early report recognized that specific methods involving application of anticoagulants to vampire bats was the only available method at that time (WHO, 2005), yet the use of such methods was characterized as obsolete in a later report (WHO, 2013). The most recent report acknowledges that anticoagulant methods for local vampire bat population reduction may prevent RABV infections in cattle and should be reviewed and updated to reflect the best available science and technology (WHO, 2018). In an ecological study of several vampire bat colonies in Peru, seroprevalence of RABV-neutralizing antibodies was greater among colonies subject to periodic lethal control compared to unmanaged colonies, suggesting potential exacerbation of RABV spread resulting from localized control efforts, although this study did not discriminate the effects of specific versus nonspecific methods (Streicker, Recuenco, et al., 2012). In the absence of other specific and effective methods, lethal control using anticoagulants remains the only advocated method for local control of vampire bat populations.

Oral vaccination of bats, akin to the approach taken with wild carnivores, has been evaluated in several experimental trials with captive common vampire bats, but field evaluation of the method has not occurred to date and there are no oral rabies vaccines licensed for use with bats. It should be noted that control of RABV in vampire bats may not control the problem of continued depredation on livestock and other associated infections (Constantine, 1988). There have been a suite of vaccine efficacy studies evaluating doses and delivery methods of vaccinia-rabies recombinant virus (VRG) to vampire bats (Almeida, Martorelli, Aires, Barros, & Massad, 2008; Almeida, Martorelli, Aires, Sallum, & Massad, 2005). In one study, intramuscular, scarification, and direct oral instillation of VRG were more efficacious than aerosol vaccination route (Aguilar-Setien et al., 2002). Indirect routes of vaccination, such as applying vaccine to an individual bat within a roosting group, demonstrated variable experimental efficacy (Almeida et al., 2005, 2008). The immunogenicity of a raccoon pox-rabies recombinant virus was recently evaluated in Brazilian free-tailed bats (Stading et al., 2016), and the efficacy of oronasal and indirect routes of vaccination was evaluated in big brown bats (Stading et al., 2017) and demonstrated immunogenicity when vaccination via the oronasal route was practiced (Gilbert et al., 2020).

Novel methods of vaccinating bats and other wildlife hosts have been proposed, and can be classified as mechanical or biological (Bakker et al., 2019). Novel mechanical methods include the concept of aerosolizing vaccines or other pharmaceuticals for application to roosting populations of bats to combat spread and impact of the fungus (*Pseudogynmnoascus destructans*) associated with White Nose Syndrome and mass mortality in certain New World temperate bat species (Garner, 2018). Sublethal aerosol exposures is one explanation that may be associated with the relatively high natural population immunity observed in Brazilian free-tailed bats that form massive roosting assemblages (Constantine et al., 1972; Constantine, Tierkel, et al., 1968; Steece & Altenbach, 1989; Turmelle, Allen, et al., 2010). Experimental work demonstrated the immunogenicity, but not efficacy, of aerosol RABV delivery to bats (Davis et al., 2007). Enhancement of indirect transmission building upon social network theory could be another mechanical advancement in vaccine delivery to bat populations, but focusing delivery to highly connected individuals (Rushmore et al., 2014). Biological control agents are also in development to combat emerging infectious diseases, relying on a conceptual foundation of weakly transmissible host-specific viruses to carry foreign genetic material from pathogen proteins of interest (Basinski, Nuismer, & Remien, 2019; Murphy, Redwood, & Jarvis, 2016). Although regulatory hurdles for transmissible vaccines may be as or more stringent compared to inactivated or nontransmissible live vaccine counterparts, more basic science is urgently needed to develop and evaluate appropriate host-specific constructs for bats, both to combat RABV and other high-profile zoonoses associated with bats. The apparent host specificity, efficient transmission, and ability to infect seropositive individuals may make bat herpesviruses worth consideration as potential vaccine vectors (Murphy et al., 2016; Shabman et al., 2016; Subudhi et al., 2018).

The concept of mass pre-exposure prophylaxis for at risk human populations in Latin America has been discussed at least since 1991 (Schneider et al., 2009), yet to date has only been attempted in Peru (Kessels et al., 2017). This rabies prevention strategy for at-risk populations is also advocated by the recent 2018 WHO report (WHO, 2018).

7.9 Conclusions

The targeted elimination of dog-mediated human rabies by 2030 will significantly reduce the number of human rabies cases globally. However, the existing undefined impact of other lyssaviruses in human rabies infection may mean that elimination of classical RABV will highlight pockets of lyssavirus infection in existing RABV endemic areas. If other lyssavirus infections do come to light, then questions around vaccine protection will rapidly become relevant to control. The divergence across the genus and the resulting lack of vaccine drive neutralization may require further investigation. Existing studies have highlighted the degree of crossneutralization afforded by existing rabies vaccines. However, gaps in the activity of neutralizing antibodies against divergent viruses have demonstrated where, should a cross-species transmission event occur, requirements for new vaccines exist. Importantly, it is clear that for protection against multiple divergent lyssaviruses, presentation of a number of different glycoproteins is required. While the likelihood of a divergent bat lyssavirus crossing the species barrier to establish ongoing transmission within terrestrial carnivores is negligible, the possibility exists and as such the need to continue to investigate the presence of these viruses in volant and nonvolant populations remains.

Acknowledgments

We would like to thank PAHO for assistance in collating data to generate the maps in Fig. 7.2 and Justin Fischer for map creation. ACB was financially supported by the UK Department for Environment, Food and Rural Affairs (Defra), the Scottish Government, and the Welsh Government (grant numbers SE0431 and SV3500). ACB was also part funded by the European Union's Horizon 2020 research and innovation program under RABYD-VAX grant agreement no. 733176.

References

Abbott, R. C., Saindon, L., Falendysz, E. A., Greenberg, L., Orciari, L., Subbian Satheshkumar, P., & Rocke, T. E. (2020). Rabies outbreak in captive big brown bats (*Eptesicus fuscus*) used in a white-nose syndrome vaccine trial. *Journal of Wildlife Diseases, 56*(1), 197–202. https://doi.org/10.7589/2018-10-258.

Acha, P. N., & Malaga-Alba, A. (1988). Economic losses due to *Desmodus rotundus*. In A. M. Greenhall & U. Schmidt (Eds.), *Natural history of vampire bats* (pp. 207–214). Boca Raton, Florida: CRC Press.

Aguilar-Setien, A., Leon, Y. C., Tesoro, E. C., Kretschmer, R., Brochier, B., & Pastoret, P. P. (2002). Vaccination of vampire bats using recombinant vaccinia-rabies virus. *Journal of Wildlife Diseases, 38*(3), 539–544. https://doi.org/10.7589/0090-3558-38.3.539.

Aguilar-Setien, A., Loza-Rubio, E., Salas-Rojas, M., Brisseau, N., Cliquet, F., Pastoret, P. P., ... Kretschmer, R. (2005). Salivary excretion of rabies virus by healthy vampire bats. *Epidemiology and Infection, 133*(3), 517–522.

Almeida, M. F., Martorelli, L. F., Aires, C. C., Barros, R. F., & Massad, E. (2008). Vaccinating the vampire bat Desmodus rotundus against rabies. *Virus Research, 137*(2), 275–277. https://doi.org/10.1016/j.virusres.2008.07.024.

Almeida, M. F., Martorelli, L. F., Aires, C. C., Sallum, P. C., & Massad, E. (2005). Indirect oral immunization of captive vampires, *Desmodus rotundus*. *Virus Research, 111*(1), 77–82.

Amengual, B., Bourhy, H., Lopez-Roig, M., & Serra-Cobo, J. (2007). Temporal dynamics of European bat Lyssavirus type 1 and survival of Myotis myotis bats in natural colonies. *PLoS One, 2*(6), e566. https://doi.org/10.1371/journal.pone.0000566.

Amengual, B., Bourhy, H., Lopez-Roig, M., & Serra-Cobo, J. (2008). Active monitoring of EBLV infection in natural colonies of the mouse-eared Bat (Myotis myotis). *Developmental Biology (Basel), 131*, 547–553.

Amengual, B., Whitby, J. E., King, A., Cobo, J. S., & Bourhy, H. (1997). Evolution of European bat lyssaviruses. *Journal of General Virology, 78*(Pt 9), 2319–2328.

Anderson, A., Shwiff, S., Gebhardt, K., Ramirez, A., Shwiff, S., Kohler, D., & Lecuona, L. (2012). Economic evaluation of vampire bat (*Desmodus rotundus*) rabies prevention in Mexico. *Transboundary and Emerging Diseases*, 1–7. https://doi.org/10.1111/tbed.12007.

Anonymous. (1986). Bat rabies in the Union of Socialist Soviet Republics. *Rabies Bulletin Europe, 10*, 12–14.

Arai, Y. T., Kuzmin, I. V., Kameoka, Y., & Botvinkin, A. D. (2003). New lyssavirus genotype from the Lesser Mouse-eared Bat (Myotis blythi), Kyrghyzstan. *Emerging Infectious Diseases, 9*(3), 333–337. https://doi.org/10.3201/eid0903.020252.

Arguin, P. M., Murray-Lillibridge, K., Miranda, M. E., Smith, J. S., Calaor, A. B., & Rupprecht, C. E. (2002). Serologic evidence of Lyssavirus infections among bats, the Philippines. *Emerging Infectious Diseases, 8*(3), 258–262.

Avery, R. J., & Tailyour, J. M. (1960). The isolation of the rabies virus from insectivorous bats in British Columbia. *Canadian Journal of Comparative Medicine and Veterinary Science, 24*(5), 143–146.

Baer, G. M. (1975). Bovine paralytic rabies and rabies in the vampire bat. In G. M. Baer (Ed.), *Vol. II. The natural history of rabies* (pp. 155–175). New York: Academic Press.

Baer, G. M., & Bales, G. L. (1967). Experimental rabies infection in the Mexican freetail bat. *Journal of Infectious Diseases, 117*(1), 82–90.

Bakker, K. M., Rocke, T. E., Osorio, J. E., Abbott, R. C., Tello, C., Carrera, J. E., ... & Streicker, D. G. (2019). Fluorescent biomarkers demonstrate prospects for spreadable vaccines to control disease transmission in wild bats. *Nature Ecology & Evolution, 3*(12), 1697–1704.

Banyard, A. C., & Fooks, A. R. (2011). Rabies and rabies-related lyssaviruses. In *Oxford textbook of Zoonoses: Biology, clinical practice and public health control* (pp. 398–422). Oxford: Oxford University Press.

Banyard, A. C., & Fooks, A. R. (2017). The impact of novel lyssavirus discovery. *Microbiology Australia, 38*(1), 18–21.

Banyard, A. C., Healy, D. M., Brookes, S. M., Voller, K., Hicks, D. J., Nunez, A., & Fooks, A. R. (2014). Lyssavirus infection: 'Low dose, multiple exposure' in the mouse model. *Virus Research, 181*, 35–42. https://doi.org/10.1016/j.virusres.2013.12.029.

Banyard, A. C., Selden, D., Wu, G., Thorne, L., Jennings, D., Marston, D., ... Fooks, A. R. (2018). Isolation, antigenicity and immunogenicity of Lleida bat lyssavirus. *Journal of General Virology*. https://doi.org/10.1099/jgv.0.001068.

Basinski, A. J., Nuismer, S. L., & Remien, C. H. (2019). A little goes a long way: Weak vaccine transmission facilitates oral vaccination campaigns against zoonotic pathogens. *PLoS Neglected Tropical Diseases, 13*(3), e0007251. https://doi.org/10.1371/journal.pntd.0007251.

Beauregard, M., & Stewart, R. C. (1964). Bat rabies in Ontario. *Canadian Journal of Comparative Medicine and Veterinary Science, 28*(2), 43–45.

Becker, D. J., Czirjak, G. A., Volokhov, D. V., Bentz, A. B., Carrera, J. E., Camus, M. S., ... Streicker, D. G. (2018). Livestock abundance predicts vampire bat demography, immune profiles and bacterial infection risk. *Philosophical Transactions of the Royal Society of London. Series B, Biological Sciences, 373*(1745). https://doi.org/10.1098/rstb.2017.0089.

Begeman, L., GeurtsvanKessel, C., Finke, S., Freuling, C. M., Koopmans, M., Muller, T., ... Kuiken, T. (2018). Comparative pathogenesis of rabies in bats and carnivores, and implications for spillover to humans. *The Lancet Infectious Diseases, 18*(4), e147–e159. https://doi.org/10.1016/S1473-3099(17)30574-1.

Bell, J. F. (1964). Abortive rabies infection. I. Experimental production in white mice and general discussion. *The Journal of Infectious Diseases, 114*, 249–257. https://doi.org/10.1093/infdis/114.3.249.

Benavides, J. A., Valderrama, W., & Streicker, D. G. (2016). Spatial expansions and travelling waves of rabies in vampire bats. *Proceedings of the Royal Society of London, Series B: Biological Sciences, 283*(1832), 20160328.

Benda, P., Gazaryan, S., & Vallo, P. (2016). On the distribution and taxonomy of bats of the Myotis mystacinus morphogroup from the Caucasus region (Chiroptera: Vespertilionidae). *Turkish Journal of Zoology, 40*, 842–863. https://doi.org/10.3906/zoo-1505-47.

Black, D., & Wiktor, T. J. (1986). Survey of raccoon hunters for rabies antibody titers: Pilot study. *The Journal of the Florida Medical Association, 73*(7), 517–520.

Blackwood, J. C., Streicker, D. G., Altizer, S., & Rohani, P. (2013). Resolving the roles of immunity, pathogenesis, and immigration for rabies persistence in vampire bats. *Proceedings of the National Academy of Sciences of the United States of America, 110*(51), 20837–20842. https://doi.org/10.1073/pnas.1308817110.

Botvinkin, A., Selnikova, O. P., Anotonova, L. A., Moiseeva, A. B., & Nesterenko, E. Y. (2006). New human rabies case caused from a bat bite in the Ukraine. *Rabies Bulletin Europe, 3*, 5–7.

Botvinkin, A. D., Poleschuk, E. M., Kuzmin, I. V., Borisova, T. I., Gazaryan, S. V., Yager, P., & Rupprecht, C. E. (2003). Novel lyssaviruses isolated from bats in Russia. *Emerging Infectious Diseases, 9*(12), 1623–1625.

Boulger, L. R., & Porterfield, J. S. (1958). Isolation of a virus from Nigerian fruit bats. *Transactions of the Royal Society of Tropical Medicine and Hygiene, 52*(5), 421–424.

Bourhy, H., Kissi, B., Lafon, M., Sacramento, D., & Tordo, N. (1992). Antigenic and molecular characterization of bat rabies virus in Europe. *Journal of Clinical Microbiology, 30*(9), 2419–2426.

Bourhy, H., Kissi, B., & Tordo, N. (1993). Molecular diversity of the Lyssavirus genus. *Virology, 194*(1), 70–81. https://doi.org/10.1006/viro.1993.1236.

Bowen, R. A., O'Shea, T. J., Shankar, V., Neubaum, M. A., Neubaum, D. J., & Rupprecht, C. (2013). Prevalence of neutralizing antibodies to rabies virus in serum of seven species of insectivorous bats from Colorado and New Mexico. *Journal of Wildlife Diseases, 49*(2), 367–374.

Brookes, S. M., Aegerter, J. N., Smith, G. C., Healy, D. M., Jolliffe, T., Swift, S. M., ... Fooks, A. R. (2005). Prevalence of antibodies to European Bat Lyssavirus type-2 in Scottish bats. *Emerging Infectious Diseases, 11*(4), 572–578.

Burns, K. F., Farinacci, C. F., & Murnane, T. G. (1956a). Insectivorous bats naturally infected with rabies in the southwestern United States. *American Journal of Public Health, 46*, 1089–1097.

Carini, A. (1911). Sur une grande epizootie de rage. *Annales de l'Institut Pasteur, 25*, 843–846.

Ceballos, N. A., Moron, S. V., Berciano, J. M., Nicolas, O., Lopez, C. A., Juste, J., ... Echevarria, J. E. (2013). Novel lyssavirus in bat, Spain. *Emerging Infectious Diseases, 19*(5), 793–795. https://doi.org/10.3201/eid1905.121071.

Chen, T., Miao, F. M., Liu, Y., Zhang, S. F., Zhang, F., Li, N., & Hu, R. L. (2018). Possible transmission of Irkut virus from dogs to humans. *Biomedical and Environmental Sciences, 31*(2), 146–148. https://doi.org/10.3967/bes2018.017.

Christian, K. A., Blanton, J. D., Auslander, M., & Rupprecht, C. E. (2009). Epidemiology of rabies post-exposure prophylaxis—United States of America, 2006-2008. *Vaccine, 27*(51), 7156–7161. https://doi.org/10.1016/j.vaccine.2009.09.028.

Coertse, J., Markotter, W., le Roux, K., Stewart, D., Sabeta, C. T., & Nel, L. H. (2017). New isolations of the rabies-related Mokola virus from South Africa. *BMC Veterinary Research, 13*(1), 37. https://doi.org/10.1186/s12917-017-0948-0.

Condori-Condori, R. E., Streicker, D. G., Cabezas-Sanchez, C., & Velasco-Villa, A. (2013). Enzootic and epizootic rabies associated with vampire bats, Peru. *Emerging Infectious Diseases, 19*(9). https://doi.org/10.3201/eid1809.130083.

Constantine, D. G. (1962). Rabies transmission by non-bite route. *Public Health Reports, 77*, 287–289.

Constantine, D. G. (1966a). Transmission experiments with bat rabies isolates: Bite transmission of rabies to foxes and coyote by free-tailed bats. *American Journal of Veterinary Research, 27*(116), 20–23.

Constantine, D. G. (1966b). Transmission experiments with bat rabies isolates: Responses of certain Carnivora to rabies virus isolated from animals infected by nonbite route. *American Journal of Veterinary Research, 27*(116), 13–15.

Constantine, D. G. (1967a). Activity patterns of the Mexican free-tailed bat. *University of New Mexico Publications in Biology, 7*, 1–79.

Constantine, D. G. (1967b). Bat rabies in the southwestern United States. *Public Health Reports, 82*(10), 867–888.

Constantine, D. G. (1970). Bats in relation to the health, welfare, and economy of man. In W. A. Wimsatt (Ed.), *Vol. 2. Biology of bats* (pp. 319–449). New York: Academic Press.

Constantine, D. G. (1988). Transmission of pathogenic microorganisms by vampire bats. In A. M. Greehall & U. Schmidt (Eds.), *Natural history of vampire bats* (pp. 167–189). Boca Raton: CRC Press.

Constantine, D. G. (2009). *Bat rabies and other lyssavirus infections*. Retrieved from Fort Collins: https://pubs.usgs.gov/circ/circ1329/pdf/circ1329.pdf.

Constantine, D. G., Emmons, R. W., & Woodie, J. D. (1972). Rabies virus in nasal mucosa of naturally infected bats. *Science, 175*(27), 1255–1256.

Constantine, D. G., Solomon, G. C., & Woodall, D. F. (1968). Transmission experiments with bat rabies isolates: Responses of certain carnivores and rodents to rabies viruses from four species of bats. *American Journal of Veterinary Research, 29*(1), 181–190.

Constantine, D. G., Tierkel, E. S., Kleckner, M. D., & Hawkins, D. M. (1968). Rabies in New Mexico cavern bats. *Public Health Reports, 83*(4), 303–316.

Constantine, D. G., & Woodall, D. F. (1964). Latent infection of Rio Bravo virus in salivary glands of bats. *Public Health Reports, 79*, 1033–1039.

Constantine, D. G., & Woodall, D. F. (1966). Transmission experiments with bat rabies isolates: Reactions of certain Carnivora, opossum, rodents, and bats to rabies virus of red bat origin when exposed by bat bite or by intrasmuscular inoculation. *American Journal of Veterinary Research, 27*(116), 24–32.

Crameri, G., Todd, S., Grimley, S., McEachern, J. A., Marsh, G. A., Smith, C., … Wang, L. F. (2009). Establishment, immortalisation and characterisation of pteropid bat cell lines. *PLoS One, 4*(12), e8266. https://doi.org/10.1371/journal.pone.0008266.

Cryan, P. M. (2003). Seasonal distribution of migratory tree bats (Lasiurus and Lasionycteris) in North America. *Journal of Mammalogy, 84*(2), 579–593.

Dacheux, L., Larrous, F., Mailles, A., Boisseleau, D., Delmas, O., Biron, C., … Bourhy, H. (2009). European bat Lyssavirus transmission among cats, Europe. *Emerging Infectious Diseases, 15*(2), 280–284.

Dato, V. M., Campagnolo, E. R., Long, J., & Rupprecht, C. E. (2016). A systematic review of human bat rabies virus variant cases: Evaluating unprotected physical contact with claws and teeth in support of accurate risk assessments. *PLoS One, 11*(7), e0159443. https://doi.org/10.1371/journal.pone.0159443.

Davis, A. D., Gordy, P. A., & Bowen, R. A. (2013). Unique characteristics of bat rabies viruses in big brown bats (Eptesicus fuscus). *Archives of Virology, 158*(4), 809–820. https://doi.org/10.1007/s00705-012-1551-0.

Davis, A. D., Jarvis, J. A., Pouliott, C., & Rudd, R. J. (2013). Rabies virus infection in Eptesicus fuscus bats born in captivity (naive bats). *PLoS One, 8*(5), e64808. https://doi.org/10.1371/journal.pone.0064808.

Davis, A. D., Jarvis, J. A., Pouliott, C. E., Morgan, S. M., & Rudd, R. J. (2013). Susceptibility and pathogenesis of little brown bats (Myotis lucifugus) to heterologous and homologous rabies viruses. *Journal of Virology, 87*(16), 9008–9015. https://doi.org/10.1128/JVI.03554-12.

Davis, A. D., Morgan, S. M., Dupuis, M., Poulliott, C. E., Jarvis, J. A., Franchini, R., … Rudd, R. J. (2016). Overwintering of rabies virus in silver haired bats (Lasionycteris noctivagans). *PLoS One, 11*(5), e0155542. https://doi.org/10.1371/journal.pone.0155542.

Davis, A. D., Rudd, R. J., & Bowen, R. A. (2007). Effects of aerosolized rabies virus exposure on bats and mice. *Journal of Infectious Diseases, 195*(8), 1144–1150.

Davis, P. L., Bourhy, H., & Holmes, E. C. (2006). The evolutionary history and dynamics of bat rabies virus. *Infection, Genetics and Evolution, 6*(6), 464–473.

Davis, P. L., Holmes, E. C., Larrous, F., Van der Poel, W. H., Tjornehoj, K., Alonso, W. J., & Bourhy, H. (2005). Phylogeography, population dynamics, and molecular evolution of European bat lyssaviruses. *Journal of Virology, 79*(16), 10487–10497.

de Mattos, C. A., Favi, M., Yung, V., Pavletic, C., & de Mattos, C. C. (2000). Bat rabies in urban centers in Chile. *Journal of Wildlife Diseases, 36*(2), 231–240.

De Serres, G., Dallaire, F., Cote, M., & Skowronski, D. M. (2008). Bat rabies in the United States and Canada from 1950 through 2007: Human cases with and without bat contact. *Clinical Infectious Diseases, 46*(9), 1329–1337. https://doi.org/10.1086/586745.

de Verteuil, E., & Urich, F. W. (1936). The study and control of paralytic rabies transmitted by vampire bats in Trinidad, B.W.I. *Transactions of the Royal Society of Tropical Medicine and Hygiene, 29*, 317–347.

Delpietro, H., de Diaz, A. M., Fuenzalida, E., & Bell, J. F. (1972). Determinacion de la tasa de ataque de rabia en murciélagos. *Boletín de la Oficina Sanitaria Panamericana, 73*, 222.

Delpietro, H., Konolsaisen, F., Marchevsky, N., & Russo, G. (1994). Domestic cat predation on vampire bats (*Desmodus rotundus*) while foraging on goats, pigs, cows and human beings. *Applied Animal Behaviour Science, 39*, 141–150.

Delpietro, H. A., Lord, R. D., Russo, R. G., & Gury-Dhomen, F. (2009). Observations of sylvatic rabies in Northern Argentina during outbreaks of paralytic cattle rabies transmitted by vampire bats (Desmodus rotundus). *Journal of Wildlife Diseases, 45*(4), 1169–1173. https://doi.org/10.7589/0090-3558-45.4.1169.

Delpietro, H. A., Marchevsky, N., & Simonetti, E. (1992). Relative population densities and predation of the common vampire bat (*Desmodus rotundus*) in natural and cattle-raising areas in northeast Argentina. *Preventive Veterinary Medicine, 14*, 13–20.

Dietz, C., Nill, D., & Von Helversen, O. (2009). *Bats of Britain. Europe and Northwest Africa.*: A & C Black Publishers Ltd.

Dietzschold, B., Morimoto, K., Hooper, D. C., Smith, J. S., Rupprecht, C. E., & Koprowski, H. (2000). Genotypic and phenotypic diversity of rabies virus variants involved in human rabies: Implications for postexposure prophylaxis. *Journal of Human Virology, 3*(1), 50–57.

Dzikwi, A. A., Kuzmin, I. I., Umoh, J. U., Kwaga, J. K., Ahmad, A. A., & Rupprecht, C. E. (2010). Evidence of Lagos bat virus circulation among Nigerian fruit bats. *Journal of Wildlife Diseases, 46*(1), 267–271. https://doi.org/10.7589/0090-3558-46.1.267.

Echevarria, J. E., Avellon, A., Juste, J., Vera, M., & Ibanez, C. (2001). Screening of active lyssavirus infection in wild bat populations by viral RNA detection on oropharyngeal swabs. *Journal of Clinical Microbiology, 39*(10), 3678–3683. https://doi.org/10.1128/JCM.39.10.3678-3683.2001.

Eggerbauer, E., Troupin, C., Passior, K., Pfaff, F., Hoper, D., Neubauer-Juric, A., … Freuling, C. M. (2017). The recently discovered Bokeloh bat lyssavirus: Insights into its genetic heterogeneity and spatial distribution in Europe and the population genetics of its primary host. *Advances in Virus Research, 99*, 199–232. https://doi.org/10.1016/bs.aivir.2017.07.004.

Ellison, J. A., Gilbert, A. T., Recuenco, S., Moran, D., Alvarez, D. A., Kuzmina, N., … Rupprecht, C. E. (2014). Bat rabies in Guatemala. *PLoS Neglected Tropical Diseases, 8*(7), e3070. https://doi.org/10.1371/journal.pntd.0003070.

Ellison, J. A., Johnson, S. R., Kuzmina, N., Gilbert, A., Carson, W. C., VerCauteren, K. C., & Rupprecht, C. E. (2013). Multidisciplinary approach to epizootiology and pathogenesis of bat rabies viruses in the United States. *Zoonoses and Public Health, 60*(1), 46–57. https://doi.org/10.1111/zph.12019.

Escobar, L. E., Peterson, A. T., Favi, M., Yung, V., & Medina-Vogel, G. (2015). Bat-borne rabies in Latin America. *Revista do Instituto de Medicina Tropical de São Paulo, 57*(1), 63–72. https://doi.org/10.1590/S0036-46652015000100009.

Escobar, L. E., Peterson, A. T., Favi, M., Yung, V., Pons, D. J., & Medina-Vogel, G. (2013). Ecology and geography of transmission of two bat-borne rabies lineages in Chile. *PLoS Neglected Tropical Diseases, 7*(12), e2577. https://doi.org/10.1371/journal.pntd.0002577.

Evans, J. S., Horton, D. L., Easton, A. J., Fooks, A. R., & Banyard, A. C. (2012). Rabies virus vaccines: Is there a need for a pan-lyssavirus vaccine? *Vaccine, 30*(52), 7447–7454. https://doi.org/10.1016/j.vaccine.2012.10.015.

Evans, J. S., Wu, G., Selden, D., Buczkowski, H., Thorne, L., Fooks, A., & Banyard, A. (2018). Utilisation of Chimeric Lyssaviruses to assess vaccine protection against highly divergent Lyssaviruses. *Viruses, 10*(3), 130.

Fahl, W. O., Carnieli, P., Jr., Castilho, J. G., Carrieri, M. L., Kotait, I., Iamamoto, K., … Brandao, P. E. (2012). Desmodus rotundus and Artibeus spp. bats might present distinct rabies virus lineages. *The Brazilian Journal of Infectious Diseases, 16*(6), 545–551. https://doi.org/10.1016/j.bjid.2012.07.002.

Familusi, J. B., & Moore, D. L. (1972). Isolation of a rabies related virus from the cerebrospinal fluid of a child with 'aseptic meningitis'. *The African Journal of Medical Sciences, 3*(1), 93–96.

Familusi, J. B., Osunkoya, B. O., Moore, D. L., Kemp, G. E., & Fabiyi, A. (1972). A fatal human infection with Mokola virus. *American Journal of Tropical Medicine and Hygiene, 21*(6), 959–963.

Favoretto, S. R., Carrieri, M. L., Cunha, E. M., Aguiar, E. A., Silva, L. H., Sodre, M. M., … Kotait, I. (2002). Antigenic typing of Brazilian rabies virus samples isolated from animals and humans, 1989-2000. *Revista do Instituto de Medicina Tropical de São Paulo, 44*(2), 91–95. https://doi.org/10.1590/s0036-46652002000200007.

Favoretto, S. R., de Mattos, C. C., de Morais, N. B., Carrieri, M. L., Rolim, B. N., Silva, L. M., ... de Mattos, C. A. (2006). Rabies virus maintained by dogs in humans and terrestrial wildlife, Ceara State, Brazil. *Emerging Infectious Diseases*, *12*(12), 1978–1981. https://doi.org/10.3201/eid1212.060429.

Field, H. E. (2018). Evidence of Australian bat lyssavirus infection in diverse Australian bat taxa. *Zoonoses and Public Health*. https://doi.org/10.1111/zph.12480.

Fisher, C. R., Streicker, D. G., & Schnell, M. J. (2018). The spread and evolution of rabies virus: Conquering new frontiers. *Nature Reviews. Microbiology*, *16*(4), 241–255, https://doi.org/10.1038/nrmicro.2018.11.

Flores-Crespo, R., Fernandez, S. S., De Anda-Lopez, D., Velarde, F. I., & Anaya, R. M. (1979). Intramuscular inoculation of cattle with warfarin: A new technique for the control of vampire bats. *Bulletin of the Panamerican Health Organization*, *13*(2), 147–161.

Foggin, C. M. (1982). Atypical rabies virus in cats and a dog in Zimbabwe. *Veterinary Record*, *110*(14), 338.

Fooks, A. R., Brookes, S. M., Johnson, N., McElhinney, L. M., & Hutson, A. M. (2003). European bat lyssaviruses: An emerging zoonosis. *Epidemiology and Infection*, *131*(3), 1029–1039.

Fooks, A. R., Cliquet, F., Finke, S., Freuling, C., Hemachudha, T., Mani, R. S., ... Banyard, A. C. (2017). Rabies. *Nature Reviews. Disease Primers*, *3*(17091).

Fooks, A. R., McElhinney, L. M., Pounder, D. J., Finnegan, C. J., Mansfield, K., Johnson, N., ... Nathwani, D. (2003). Case report: Isolation of a European bat lyssavirus type 2a from a fatal human case of rabies encephalitis. *Journal of Medical Virology*, *71*(2), 281–289. https://doi.org/10.1002/jmv.10481.

Fornes, A., Lord, R. D., Kuns, M. L., Larghi, O. P., Fuenzalida, E., & Lazara, L. (1974). Control of bovine rabies through vampire bat control. *Journal of Wildlife Diseases*, *10*(4), 310–316.

Franka, R., Johnson, N., Muller, T., Vos, A., Neubert, L., Freuling, C., ... Fooks, A. R. (2008). Susceptibility of North American big brown bats (Eptesicus fuscus) to infection with European bat lyssavirus type 1. *Journal of General Virology*, *89*(Pt 8), 1998–2010.

Fraser, G. C., Hooper, P. T., Lunt, R. A., Gould, A. R., Gleeson, L. J., Hyatt, A. D., ... Kattenbelt, J. A. (1996). Encephalitis caused by a Lyssavirus in fruit bats in Australia. *Emerging Infectious Diseases*, *2*(4), 327–331.

Freuling, C., Grossmann, E., Conraths, F. J., Schameitat, A., Kliemt, J., Auer, E., ... Muller, T. (2008). First isolation of EBLV-2 in Germany. *Veterinary Microbiology*, *131*(1-2), 26–34.

Freuling, C., Vos, A., Johnson, N., Kaipf, I., Denzinger, A., Neubert, L., ... Muller, T. (2009). Experimental infection of serotine bats (Eptesicus serotinus) with European bat lyssavirus type 1a. *Journal of General Virology*, *90*(Pt 10), 2493–2502.

Freuling, C. M., Abendroth, B., Beer, M., Fischer, M., Hanke, D., Hoffmann, B., ... Muller, T. (2013). Molecular diagnostics for the detection of Bokeloh bat lyssavirus in a bat from Bavaria, Germany. *Virus Research*, *177*(2), 201–204. https://doi.org/10.1016/j.virusres.2013.07.021.

Freuling, C. M., Beer, M., Conraths, F. J., Finke, S., Hoffmann, B., Keller, B., ... Muller, T. (2011). Novel lyssavirus in Natterer's bat, Germany. *Emerging Infectious Diseases*, *17*(8), 1519–1522. https://doi.org/10.3201/eid1708.110201.

Freuling, C. M., Binger, T., Beer, M., Adu-Sarkodie, Y., Schatz, J., Fischer, M., ... Muller, T. (2015). Lagos bat virus transmission in an Eidolon helvum bat colony, Ghana. *Virus Research*, *210*, 42–45. https://doi.org/10.1016/j.virusres.2015.07.009.

Fuoco, N. L., Fernandes, E. R., Dos Ramos Silva, S., Luiz, F. G., Ribeiro, O. G., & Santos Katz, I. S. (2018). Street rabies virus strains associated with insectivorous bats are less pathogenic than strains isolated from other reservoirs. *Antiviral Research*, *160*, 94–100. https://doi.org/10.1016/j.antiviral.2018.10.023.

Garner, S. (November 22, 2018). How to vaccinate a wild bat. *Scientific American*. Retrieved from: https://blogs.scientificamerican.com/observations/how-to-vaccinate-a-wild-bat/.

George, D. B., Webb, C. T., Farnsworth, M. L., O'Shea, T. J., Bowen, R. A., Smith, D. L., ... Rupprecht, C. E. (2011). Host and viral ecology determine bat rabies seasonality and maintenance. *Proceedings of the National Academy of Sciences of the United States of America*, *108*(25), 10208–10213. https://doi.org/10.1073/pnas.1010875108.

Gilbert, A. (2018). Rabies virus vectors and reservoir species. *Revue scientifique et technique (International Office of Epizootics)*, *37*(2), 371–384.

Gilbert, A. T., McCracken, G. F., Sheeler, L. L., Muller, L. I., O'Rourke, D., Kelch, W. J., & New, J. C., Jr. (2015). Rabies surveillance among bats in Tennessee, USA, 1996-2010. *Journal of Wildlife Diseases*, *51*(4), 821–832. https://doi.org/10.7589/2014-12-277.

Gilbert, A. T., Petersen, B. W., Recuenco, S., Niezgoda, M., Gomez, J., Laguna-Torres, V. A., & Rupprecht, C. (2012). Evidence of rabies virus exposure among humans in the Peruvian Amazon. *The American Journal of Tropical Medicine and Hygiene*, *87*(2), 206–215. https://doi.org/10.4269/ajtmh.2012.11-0689.

Gilbert, A. T., Wu, X., Jackson, F. R., Franka, R., McCracken, G. F., & Rupprecht, C. E. (2020). Safety, immunogenicity, and efficacy of intramuscular and oral delivery of ERA-G333 recombinant rabies virus vaccine to big brown bats (*Eptesicus fuscus*). *Journal of Wildlife Diseases*, in press. https://doi.org/10.7589/2019-04-108.

Girard, K. F., Hitchcock, H. B., Edsall, G., & MacCready, R. A. (1965). Rabies in bats in southern New England. *The New England Journal of Medicine*, 272(2), 75–80.

Gomme, E. A., Wirblich, C., Addya, S., Rall, G. F., & Schnell, M. J. (2012). Immune clearance of attenuated rabies virus results in neuronal survival with altered gene expression. *PLoS Pathogens*, 8(10), e1002971. https://doi.org/10.1371/journal.ppat.1002971.

Goncalves, M. A., Sa-Neto, R. J., & Brazil, T. K. (2002). Outbreak of aggressions and transmission of rabies in human beings by vampire bats in northeastern Brazil. *Revista da Sociedade Brasileira de Medicina Tropical*, 35(5), 461–464.

Greenhall, A. M. (1963). Use of mist-nets and strychnine for vampire control in Trinidad. *Journal of Mammalogy*, 44, 396–399.

Grimes, J. E., Eads, R. B., & Irons, J. V. (1955). An additional species of insectivorous bat naturally infected with rabies. *American Journal of Tropical Medicine and Hygiene*, 4(3), 554–556.

Gunawardena, P. S., Marston, D. A., Ellis, R. J., Wise, E. L., Karawita, A. C., Breed, A. C., ... Fooks, A. R. (2016). Lyssavirus in Indian flying foxes, Sri Lanka. *Emerging Infectious Diseases*, 22(8), 1456–1459. https://doi.org/10.3201/eid2208.151986.

Hampson, K., Coudeville, L., Lembo, T., Sambo, M., Kieffer, A., Attlan, M., ... Global Alliance for Rabies Control Partners for Rabies Prevention. (2015). Estimating the global burden of endemic canine rabies. *PLoS Neglected Tropical Diseases*, 9(4), e0003709. https://doi.org/10.1371/journal.pntd.0003709.

Harris, S. L., Aegerter, J. N., Brookes, S. M., McElhinney, L. M., Jones, G., Smith, G. C., & Fooks, A. R. (2009). Targeted surveillance for European bat lyssaviruses in English bats (2003-06). *Journal of Wildlife Diseases*, 45(4), 1030–1041. https://doi.org/10.7589/0090-3558-45.4.1030.

Hattwick, M. A., Weis, T. T., Stechschulte, C. J., Baer, G. M., & Gregg, M. B. (1972). Recovery from rabies. A case report. *Annals of Internal Medicine*, 76(6), 931–942.

Haupt, H., & Rehaag, H. (1921). Durch Fledermäuse verbreitete seuchenhafte Tollwut unter Viehbeständen in Santa Catharina (Süd-Brasilien). *Infektionskrankheiten Parasitare Krankheiten und Hygiene der Haustiere*, 22, 104–127.

Hayman, D. T. S., Emmerich, P., Yu, M., Wang, L. F., Suu-Ire, R., Fooks, A. R., ... Wood, J. L. (2010). Long-term survival of an urban fruit bat seropositive for Ebola and Lagos bat viruses. *PLoS One*, 5(8). https://doi.org/10.1371/journal.pone.0011978.

Hayman, D. T. S., Fooks, A. R., Horton, D., Suu-Ire, R., Breed, A. C., Cunningham, A. A., & Wood, J. L. (2008). Antibodies against Lagos bat virus in megachiroptera from West Africa. *Emerging Infectious Diseases*, 14(6), 926–928.

Hayman, D. T. S., Fooks, A. R., Marston, D. A., & Garcia, R. J. (2016). The global phylogeography of lyssaviruses—Challenging the 'Out of Africa' hypothesis. *PLoS Neglected Tropical Diseases*, 10(12), e0005266. https://doi.org/10.1371/journal.pntd.0005266.

Hayman, D. T. S., Fooks, A. R., Rowcliffe, J. M., McCrea, R., Restif, O., Baker, K. S., ... Wood, J. L. (2012). Endemic Lagos bat virus infection in Eidolon helvum. *Epidemiology and Infection*, 140(12), 2163–2171. https://doi.org/10.1017/S0950268812000167.

Hayman, D. T. S., Luis, A. D., Restif, O., Baker, K. S., Fooks, A. R., Leach, C., ... Webb, C. T. (2018). Maternal antibody and the maintenance of a lyssavirus in populations of seasonally breeding African bats. *PLoS One*, 13(6), e0198563. https://doi.org/10.1371/journal.pone.0198563.

Herbold, J. R., Heuschele, W. P., Berry, R. L., & Parsons, M. A. (1983). Reservoir of St. Louis encephalitis virus in Ohio bats. *American Journal of Veterinary Research*, 44(10), 1889–1893.

Horton, D. L., Banyard, A. C., Marston, D. A., Wise, E., Selden, D., Nunez, A., ... Fooks, A. R. (2014). Antigenic and genetic characterization of a divergent African virus, Ikoma lyssavirus. *The Journal of General Virology*, 95(Pt 5), 1025–1032. https://doi.org/10.1099/vir.0.061952-0.

Horton, D. L., McElhinney, L. M., Marston, D. A., Wood, J. L., Russell, C. A., Lewis, N., ... Smith, D. J. (2010). Quantifying antigenic relationships among the Lyssaviruses. *Journal of Virology*, 84, 11841–11848.

Hu, S. C., Hsu, C. L., Lee, M. S., Tu, Y. C., Chang, J. C., Wu, C. H., ... Hsu, W. C. (2018). Lyssavirus in Japanese Pipistrelle, Taiwan. *Emerging Infectious Diseases*, 24(4), 782–785. https://doi.org/10.3201/eid2404.171696.

Hughes, G. J., Kuzmin, I. V., Schmitz, A., Blanton, J., Manangan, J., Murphy, S., & Rupprecht, C. E. (2006). Experimental infection of big brown bats (Eptesicus fuscus) with Eurasian bat lyssaviruses Aravan, Khujand, and Irkut virus. *Archives of Virology*, 151(10), 2021–2035. https://doi.org/10.1007/s00705-005-0785-0.

Humphrey, G. L., Kemp, G. E., & Wood, E. G. (1960). A fatal case of rabies in a woman bitten by an insectivorous bat. *Public Health Reports, 75*, 317–326.

Jackson, F. R., Turmelle, A. S., Farino, D. M., Franka, R., McCracken, G. F., & Rupprecht, C. E. (2008). Experimental rabies virus infection of big brown bats (Eptesicus fuscus). *Journal of Wildlife Diseases, 44*(3), 612–621. https://doi.org/10.7589/0090-3558-44.3.612.

Jakava-Viljanen, M., Lilley, T., Kyheroinen, E. M., & Huovilainen, A. (2010). First encounter of European bat lyssavirus type 2 (EBLV-2) in a bat in Finland. *Epidemiology and Infection, 138*(11), 1581–1585.

Johnson, H. N. (1948). Derriengue; vampire bat rabies in Mexico. *American Journal of Hygiene, 47*(2), 189–204.

Johnson, N., Arechiga-Ceballos, N., & Aguilar-Setien, A. (2014). Vampire bat rabies: Ecology, epidemiology and control. *Viruses, 6*(5), 1911–1928. https://doi.org/10.3390/v6051911.

Johnson, N., McKimmie, C. S., Mansfield, K. L., Wakeley, P. R., Brookes, S. M., Fazakerley, J. K., & Fooks, A. R. (2006). Lyssavirus infection activates interferon gene expression in the brain. *Journal of General Virology, 87*(Pt 9), 2663–2667.

Johnson, N., Vos, A., Freuling, C., Tordo, N., Fooks, A. R., & Muller, T. (2010). Human rabies due to lyssavirus infection of bat origin. *Veterinary Microbiology, 142*(3-4), 151–159.

Johnson, N., Vos, A., Neubert, L., Freuling, C., Mansfield, K. L., Kaipf, I., … Fooks, A. R. (2008). Experimental study of European bat lyssavirus type-2 infection in Daubenton's bats (Myotis daubentonii). *Journal of General Virology, 89*(Pt 11), 2662–2672.

Jorge, R. S., Pereira, M. S., Morato, R. G., Scheffer, K. C., Carnieli, P., Jr., Ferreira, F., … May-Junior, J. A. (2010). Detection of rabies virus antibodies in Brazilian free-ranging wild carnivores. *Journal of Wildlife Diseases, 46*(4), 1310–1315. https://doi.org/10.7589/0090-3558-46.4.1310.

Kalemba, L. N., Niezgoda, M., Gilbert, A. T., Doty, J. B., Wallace, R. M., Malekani, J. M., & Carroll, D. S. (2017). Exposure to lyssaviruses in bats of the Democratic Republic of the Congo. *Journal of Wildlife Diseases, 53*(2), 408–410. https://doi.org/10.7589/2016-06-122.

Kaplan, C. (1976). Rabies. *The British Journal of Clinical Practice, 30*(11-12), 208–211.

Kaplan, M. M. (1969). Epidemiology of rabies. *Nature, 221*(5179), 421–425.

Kemp, G. E., Causey, O. R., Moore, D. L., Odelola, A., & Fabiyi, A. (1972). Mokola virus. Further studies on IbAn 27377, a new rabies-related etiologic agent of zoonosis in Nigeria. *American Journal of Tropical Medicine and Hygiene, 21*(3), 356–359.

Kerth, G., & Konig, B. (1999). Fission, fusion and nonrandom associations in female Bechstein's bats (*Myotis bechsteinii*). *Behaviour, 136*(9), 1187–1202.

Kessels, J. A., Recuenco, S., Navarro-Vela, A. M., Deray, R., Vigilato, M. A., Ertl, H., … Briggs, D. (2017). Pre-exposure rabies prophylaxis: A systematic review. *Bulletin of the World Health Organization, 95*(3), 210–219.

Kgaladi, J., Wright, N., Coertse, J., Markotter, W., Marston, D., Fooks, A. R., … Nel, L. H. (2013). Diversity and epidemiology of Mokola virus. *PLoS Neglected Tropical Diseases, 7*(10), e2511. https://doi.org/10.1371/journal.pntd.0002511.

King, A., & Crick, J. (1988). Rabies-related viruses. In J. B. Campbell & K. M. Charlton (Eds.), *Rabies* (pp. 177–200). Boston: Kluwer Academic Publishers.

King, A. A., Meredith, C. D., & Thomson, G. R. (1994). The biology of southern African lyssavirus variants. *Current Topics in Microbiology and Immunology, 187*, 267–295.

Kissi, B., Tordo, N., & Bourhy, H. (1995). Genetic polymorphism in the rabies virus nucleoprotein gene. *Virology, 209*(2), 526–537. https://doi.org/10.1006/viro.1995.1285.

Klug, B. J., Turmelle, A. S., Ellison, J. A., Baerwald, E. F., & Barclay, R. M. (2011). Rabies prevalence in migratory tree-bats in Alberta and the influence of roosting ecology and sampling method on reported prevalence of rabies in bats. *Journal of Wildlife Diseases, 47*(1), 64–77. https://doi.org/10.7589/0090-3558-47.1.64.

Kobayashi, Y., Sato, G., Kato, M., Itou, T., Cunha, E. M., Silva, M. V., … Sakai, T. (2007). Genetic diversity of bat rabies viruses in Brazil. *Archives of Virology, 152*(11), 1995–2004. https://doi.org/10.1007/s00705-007-1033-y.

Kuzmin, I. V., Botvinkin, A. D., & Shaimardanov, R. T. (1994). Experimental lyssavirus infection in chiropters. *Voprosy Virusologii, 39*(1), 17–21.

Kuzmin, I. V., Bozick, B., Guagliardo, S. A., Kunkel, R., Shak, J. R., Tong, S., & Rupprecht, C. E. (2011). Bats, emerging infectious diseases, and the rabies paradigm revisited. *Emerging Health Threats Journal, 4*, 7159. https://doi.org/10.3402/ehtj.v4i0.7159.

Kuzmin, I. V., Franka, R., & Rupprecht, C. E. (2008). Experimental infection of big brown bats (Eptesicus fuscus) with West Caucasian bat virus (WCBV). *Developmental Biology (Basel)*, *131*, 327–337.

Kuzmin, I. V., Hughes, G. J., Botvinkin, A. D., Orciari, L. A., & Rupprecht, C. E. (2005). Phylogenetic relationships of Irkut and West Caucasian bat viruses within the Lyssavirus genus and suggested quantitative criteria based on the N gene sequence for lyssavirus genotype definition. *Virus Research*, *111*(1), 28–43.

Kuzmin, I. V., Mayer, A. E., Niezgoda, M., Markotter, W., Agwanda, B., Breiman, R. F., & Rupprecht, C. E. (2010). Shimoni bat virus, a new representative of the Lyssavirus genus. *Virus Research*, *149*(2), 197–210.

Kuzmin, I. V., Niezgoda, M., Carroll, D. S., Keeler, N., Hossain, M. J., Breiman, R. F., … Rupprecht, C. E. (2006). Lyssavirus surveillance in bats, Bangladesh. *Emerging Infectious Diseases*, *12*(3), 486–488.

Kuzmin, I. V., Niezgoda, M., Franka, R., Agwanda, B., Markotter, W., Beagley, J. C., … Rupprecht, C. E. (2008a). Lagos bat virus in Kenya. *Journal of Clinical Microbiology*, *46*(4), 1451–1461.

Kuzmin, I. V., Niezgoda, M., Franka, R., Agwanda, B., Markotter, W., Beagley, J. C., … Rupprecht, C. E. (2008b). Possible emergence of West Caucasian bat virus in Africa. *Emerging Infectious Diseases*, *14*(12), 1887–1889.

Kuzmin, I. V., Shi, M., Orciari, L. A., Yager, P. A., Velasco-Villa, A., Kuzmina, N. A., … Rupprecht, C. E. (2012). Molecular inferences suggest multiple host shifts of rabies viruses from bats to mesocarnivores in Arizona during 2001-2009. *PLoS Pathogens*, *8*(6), e1002786. https://doi.org/10.1371/journal.ppat.1002786.

Kuzmina, N. A., Kuzmin, I. V., Ellison, J. A., Taylor, S. T., Bergman, D. L., Dew, B., & Rupprecht, C. E. (2013). A reassessment of the evolutionary timescale of bat rabies viruses based upon glycoprotein gene sequences. *Virus Genes*, *47*(2), 305–310. https://doi.org/10.1007/s11262-013-0952-9.

La Motte, L. C., Jr. (1958). Japanese B Encephalitis in bats during simulated hibernation. *American Journal of Epidemiology*, *67*(1), 101–108.

Le-Gonidec, G., Rickenbach, A., Robin, Y., & Heme, G. (1978). Isolation of a strain of Mokola virus in Cameroon. *Annals Microbiologie (Paris)*, *129*(2), 245–249.

Leonova, G. N., Belikov, S. I., Kondratov, I. G., Krylova, N. V., Pavlenko, E. V., Tiunov, M. P., & Tkachev, S. E. (2009). A fatal case of bat lyssavirus infection in Primorye Territory of the Russian Far East. *WHO Rabies Bulletin Europe*, *33*, 5–8.

Leslie, M. J., Messenger, S., Rohde, R. E., Smith, J., Cheshier, R., Hanlon, C., & Rupprecht, C. E. (2006). Bat-associated rabies virus in Skunks. *Emerging Infectious Diseases*, *12*(8), 1274–1277.

Linhart, S. B. (1973). Age determination and occurrence of incremental growth lines in the dental cementum of the common vampire bat (*Desmodus rotundus*). *Journal of Mammalogy*, *54*(2), 493–496.

Linhart, S. B. (1975). The biology and control of vampire bats. In G. M. Baer (Ed.), *Vol. II. The natural history of rabies* (pp. 221–241). New York: Academic Press.

Linhart, S. B., Flores-Crespo, R., & Mitchell, G. C. (1972). Control of vampire bats by topical application of an anticoagulant, chlorophacinone. *Boletín de la Oficina Sanitaria Panamericana*, *72*(2), 31–38.

Liu, Y., Li, N., Zhang, S., Zhang, F., Lian, H., Wang, Y., … Hu, R. (2013). Analysis of the complete genome of the first Irkut virus isolate from China: Comparison across the Lyssavirus genus. *Molecular Phylogenetics and Evolution*, *69*(3), 687–693. https://doi.org/10.1016/j.ympev.2013.07.008.

Lopez, A., Miranda, P., Tejada, E., & Fishbein, D. B. (1992). Outbreak of human rabies in the Peruvian jungle. *Lancet*, *339*(8790), 408–411.

Lord, R. D. (1988). Control of vampire bats. In A. M. Greenhall & U. Schmidt (Eds.), *Natural history of vampire bats* (pp. 215–226). Boca Raton, Florida: CRC Press.

Lord, R. D., Delpietro, H., Fuenzalida, E., De Diaz, A. M., & Lazaro, L. (1975). Presence of rabies neutralizing antibodies in wild carnivores following an outbreak of bovine rabies. *Journal of Wildlife Diseases*, *11*(2), 210–213.

Lord, R. D., Fuenzalida, E., Delpietro, H., Larghi, O. P., de Diaz, A. M., & Lazaro, L. (1975). Observations on the epizootiology of vampire bat rabies. *Bulletin of the Pan American Health Organization*, *9*(3), 189–195.

Lord, R. D., Muradali, F., & Lazaro, L. (1976). Age composition of vampire (*Desmodus rotundus*) bats in northern Argentina and southern Brazil. *Journal of Mammalogy*, *57*, 573–575.

Lumio, J., Hillbom, M., Roine, R., Ketonen, L., Haltia, M., Valle, M., … Lahdevirta, J. (1986). Human rabies of bat origin in Europe. *Lancet*, *1*(8477), 378. https://doi.org/10.1016/s0140-6736(86)92336-6.

Ma, X., Monroe, B. P., Cleaton, J. M., Orciari, L. A., Li, Y., Kirby, J. D., … Blanton, J. D. (2018). Rabies surveillance in the United States during 2017. *Journal of the American Veterinary Medical Association*, *253*(12), 1555–1568. https://doi.org/10.2460/javma.253.12.1555.

Mallewa, M., Fooks, A. R., Banda, D., Chikungwa, P., Mankhambo, L., Molyneux, E., … Solomon, T. (2007). Rabies encephalitis in malaria-endemic area, Malawi, Africa. *Emerging Infectious Diseases*, *13*(1), 136–139.

Mani, R. S., Dovih, D. P., Ashwini, M. A., Chattopadhyay, B., Harsha, P. K., Garg, K. M., … Madhusudana, S. N. (2017). Serological evidence of lyssavirus infection among bats in Nagaland, a North-Eastern State in India. *Epidemiology and Infection, 145*(8), 1635–1641. https://doi.org/10.1017/S0950268817000310.

Markotter, W., Kuzmin, I., Rupprecht, C. E., Randles, J., Sabeta, C. T., Wandeler, A. I., & Nel, L. H. (2006). Isolation of Lagos bat virus from water mongoose. *Emerging Infectious Diseases, 12*(12), 1913–1918.

Markotter, W., Monadjem, A., & Nel, L. H. (2013). Antibodies against Duvenhage virus in insectivorous bats in Swaziland. *Journal of Wildlife Diseases, 49*(4), 1000–1003. https://doi.org/10.7589/2012-10-257.

Markotter, W., Randles, J., Rupprecht, C. E., Sabeta, C. T., Taylor, P. J., Wandeler, A. I., & Nel, L. H. (2006). Lagos bat virus, South Africa. *Emerging Infectious Diseases, 12*(3), 504–506.

Markotter, W., Van Eeden, C., Kuzmin, I. V., Rupprecht, C. E., Paweska, J. T., Swanepoel, R., … Nel, L. H. (2008). Epidemiology and pathogenicity of African bat lyssaviruses. *Developmental Biology (Basel), 131*, 317–325.

Marston, D. A., Ellis, R. J., Horton, D. L., Kuzmin, I. V., Wise, E. L., McElhinney, L. M., … Fooks, A. R. (2012). Complete genome sequence of Ikoma lyssavirus. *Journal of Virology, 86*(18), 10242–10243. https://doi.org/10.1128/JVI.01628-12.

McCall, B. J., Field, H. E., Smith, G. A., Storie, G. J., & Harrower, B. J. (2005). Defining the risk of human exposure to Australian bat lyssavirus through potential non-bat animal infection. *Communicable Diseases Intelligence, 29*(2), 202–205.

McCarthy, T. J. (1989). Human depredation by vampire bats (*Desmodus rotundus*) following a hog cholera campaign. *The American Journal of Tropical Medicine and Hygiene, 40*(3), 320–322.

McColl, K. A., Chamberlain, T., Lunt, R. A., Newberry, K. M., Middleton, D., & Westbury, H. A. (2002). Pathogenesis studies with Australian bat lyssavirus in grey-headed flying foxes (Pteropus poliocephalus). *Australian Veterinary Journal, 80*(10), 636–641.

McColl, K. A., Chamberlain, T., Lunt, R. A., Newberry, K. M., & Westbury, H. A. (2007). Susceptibility of domestic dogs and cats to Australian bat lyssavirus (ABLV). *Veterinary Microbiology, 123*(1-3), 15–25.

McCracken, G. F. (2003). *Estimates of population sizes in summer colonies of Brazilian free-tailed bats (Tadarida brasiliensis)*. (USGS/BRD/ITR-2003-003). Retrieved from (Chapter 3): https://pubs.usgs.gov/itr/2003/0003/report.pdf.

McElhinney, L. M., Marston, D. A., Leech, S., Freuling, C. M., van der Poel, W. H., Echevarria, J., … Fooks, A. R. (2013). Molecular epidemiology of bat lyssaviruses in Europe. *Zoonoses and Public Health, 60*(1), 35–45. https://doi.org/10.1111/zph.12003.

McQuiston, J. H., Yager, P. A., Smith, J. S., & Rupprecht, C. E. (2001). Epidemiologic characteristics of rabies virus variants in dogs and cats in the United States, 1999. *Journal of the American Veterinary Medical Association, 218*(12), 1939–1942.

Mebatsion, T., Cox, J. H., & Frost, J. W. (1992). Isolation and characterization of 115 street rabies virus isolates from Ethiopia by using monoclonal antibodies: Identification of 2 isolates as Mokola and Lagos bat viruses. *Journal of Infectious Diseases, 166*(5), 972–977.

Melade, J., McCulloch, S., Ramasindrazana, B., Lagadec, E., Turpin, M., Pascalis, H., … Dellagi, K. (2016). Serological evidence of lyssaviruses among bats on Southwestern Indian Ocean Islands. *PLoS One, 11*(8), e0160553. https://doi.org/10.1371/journal.pone.0160553.

Meredith, C. D., Nel, L. H., & von Teichman, B. F. (1996). Further isolation of Mokola virus in South Africa. *Veterinary Record, 138*(5), 119–120.

Meredith, C. D., Prossouw, A. P., & Koch, H. P. (1971). An unusual case of human rabies thought to be of chiropteran origin. *South African Medical Journal, 45*(28), 767–769.

Messenger, S. L., Smith, J. S., & Rupprecht, C. E. (2002). Emerging epidemiology of bat-associated cryptic cases of rabies in humans in the United States. *Clinical Infectious Diseases, 35*(6), 738–747. https://doi.org/10.1086/342387.

Mohr, W. (1957). Die Tollwut. *Medizinische Klinik, 52*, 1057–1060.

Moreno, J. A., & Baer, G. M. (1980). Experimental rabies in the vampire bat. *American Journal of Tropical Medicine and Hygiene, 29*(2), 254–259.

Morimoto, K., Patel, M., Corisdeo, S., Hooper, D. C., Fu, Z. F., Rupprecht, C. E., … Dietzschold, B. (1996). Characterization of a unique variant of bat rabies virus responsible for newly emerging human cases in North America. *Proceedings of the National Academy of Sciences of the United States of America, 93*(11), 5653–5658. https://doi.org/10.1073/pnas.93.11.5653.

Muller, T., Cox, J., Peter, W., Schafer, R., Johnson, N., McElhinney, L. M., … Fooks, A. R. (2004). Spill-over of European bat lyssavirus type 1 into a stone marten (Martes foina) in Germany. *Journal of Veterinary Medicine. B, Infectious Diseases and Veterinary Public Health, 51*(2), 49–54. https://doi.org/10.1111/j.1439-0450.2003.00725.x.

Murphy, A. A., Redwood, A. J., & Jarvis, M. A. (2016). Self-disseminating vaccines for emerging infectious diseases. *Expert Review of Vaccines, 15*(1), 31–39.

Nadin-Davis, S. A., Abdel-Malik, M., Armstrong, J., & Wandeler, A. I. (2002). Lyssavirus P gene characterisation provides insights into the phylogeny of the genus and identifies structural similarities and diversity within the encoded phosphoprotein. *Virology, 298*(2), 286–305.

Nel, L., Jacobs, J., Jaftha, J., von Teichman, B., Bingham, J., & Olivier, M. (2000). New cases of Mokola virus infection in South Africa: A genotypic comparison of Southern African virus isolates. *Virus Genes, 20*(2), 103–106.

Nieuwenhuis, H. U. (1988). The rabies situation in The Netherlands. *Parassitologia, 30*(1), 123–128.

Nikolic, M., & Jelesic, Z. (1956). Isolation of rabies virus from insectivorous bats in Yugoslavia. *Bulletin of the World Health Organization, 14*, 801–804.

Nokireki, T., Tammiranta, N., Kokkonen, U. M., Kantala, T., & Gadd, T. (2018). Tentative novel lyssavirus in a bat in Finland. *Transboundary and Emerging Diseases, 65*(3), 593–596. https://doi.org/10.1111/tbed.12833.

Obregon-Morales, C., Aguilar-Setien, A., Perea Martinez, L., Galvez-Romero, G., Martinez-Martinez, F. O., & Arechiga-Ceballos, N. (2017). Experimental infection of Artibeus intermedius with a vampire bat rabies virus. *Comparative Immunology, Microbiology and Infectious Diseases, 52*, 43–47. https://doi.org/10.1016/j.cimid.2017.05.008.

O'Shea, T. J., Bowen, R. A., Stanley, T. R., Shankar, V., & Rupprecht, C. E. (2014). Variability in seroprevalence of rabies virus neutralizing antibodies and associated factors in a Colorado population of big brown bats (Eptesicus fuscus). *PLoS One, 9*(1), e86261. https://doi.org/10.1371/journal.pone.0086261.

O'Shea, T. J., Cryan, P. M., Hayman, D., Plowright, R., & Streicker, D. G. (2016). Multiple mortality events in bats: A global review. *Mammal Review, 46*, 175–190.

Pawan, J. L. (1936a). Rabies in the vampire bat of Trinidad, with special reference to the clinical course and the latency of infection. *Annals of Tropical Medicine and Parasitology, 30*, 410–422.

Pawan, J. L. (1936b). The transmission of paralytic rabies in Trinidad by the vampire bat (*Desmodus rotundus murinus* Wagner, 1840). *Annals of Tropical Medicine and Parasitology, 30*, 137–156.

Paweska, J. T., Blumberg, L. H., Liebenberg, C., Hewlett, R. H., Grobbelaar, A. A., Leman, P. A., … Swanepoel, R. (2006). Fatal human infection with rabies-related Duvenhage virus, South Africa. *Emerging Infectious Diseases, 12*(12), 1965–1967. https://doi.org/10.3201/eid1212.060764.

Peng, J., Zhu, S., Hu, L., Ye, P., Wang, Y., Tian, Q., … Guo, X. (2016). Wild-type rabies virus induces autophagy in human and mouse neuroblastoma cell lines. *Autophagy, 12*(10), 1704–1720.

Picard-Meyer, E., Beven, V., Hirchaud, E., Guillaume, C., Larcher, G., Robardet, E., … Cliquet, F. (2019). Lleida Bat Lyssavirus isolation in Miniopterus schreibersii in France. *Zoonoses and Public Health, 66*(2), 254–258. https://doi.org/10.1111/zph.12535.

Picard-Meyer, E., Borel, C., Moinet, M., Servat, A., Rasquin, P., & Cliquet, F. (2013). Isolation of the novel BBLV Lyssavirus in Natterer's bat in France. *Bulletin Epidémiologique—Santé animale, Alimentation, 53*.

Picard-Meyer, E., Servat, A., Robardet, E., Moinet, M., Borel, C., & Cliquet, F. (2013). Isolation of Bokeloh bat lyssavirus in Myotis nattereri in France. *Archives of Virology*. https://doi.org/10.1007/s00705-013-1747-y.

Pinero, C., Gury Dohmen, F., Beltran, F., Martinez, L., Novaro, L., Russo, S., … Cisterna, D. M. (2012). High diversity of rabies viruses associated with insectivorous bats in Argentina: Presence of several independent enzootics. *PLoS Neglected Tropical Diseases, 6*(5), e1635. https://doi.org/10.1371/journal.pntd.0001635.

Prada, D., Boyd, V., Baker, M., Jackson, B., & O'Dea, M. (2019). Insights into Australian bat lyssavirus in insectivorous bats of Western Australia. *Tropical Medicine and Infectious Disease, 4*(1). https://doi.org/10.3390/tropicalmed4010046.

Prieto, J. F., & Baer, G. M. (1972). An outbreak of bovine paralytic rabies in Tuxtepec, Oaxaca, Mexico. *The American Journal of Tropical Medicine and Hygiene, 21*(2), 219–225. https://doi.org/10.4269/ajtmh.1972.21.219.

Reagan, R. L., & Brueckner, A. L. (1951). Transmission of a strain of rabies virus to the large brown bat (Eptesicus fuscus) and to the cave bat (Myotis lucifugus). *The Cornell Veterinarian, 41*(3), 295–298.

Reid, J. E., & Jackson, A. C. (2001). Experimental rabies virus infection in Artibeus jamaicensis bats with CVS-24 variants. *Journal of Neurovirology, 7*(6), 511–517.

Reynes, J. M., Andriamandimby, S. F., Razafitrimo, G. M., Razainirina, J., Jeanmaire, E. M., Bourhy, H., & Heraud, J. M. (2011). Laboratory surveillance of rabies in humans, domestic animals, and bats in madagascar from 2005 to 2010. *Advances in Preventive Medicine, 2011*, 727821. https://doi.org/10.4061/2011/727821.

Reynes, J. M., Molia, S., Audry, L., Hout, S., Ngin, S., Walston, J., & Bourhy, H. (2004). Serologic evidence of lyssavirus infection in bats, Cambodia. *Emerging Infectious Diseases, 10*(12), 2231–2234. https://doi.org/10.3201/eid1012.040459.

Robardet, E., Borel, C., Moinet, M., Jouan, D., Wasniewski, M., Barrat, J., … Picard-Meyer, E. (2017). Longitudinal survey of two serotine bat (Eptesicus serotinus) maternity colonies exposed to EBLV-1 (European Bat Lyssavirus type 1): Assessment of survival and serological status variations using capture-recapture models. *PLoS Neglected Tropical Diseases, 11*(11), e0006048. https://doi.org/10.1371/journal.pntd.0006048.

Rushmore, J., Caillaud, D., Hall, R. J., Stumpf, R. M., Meyers, L. A., & Altizer, S. (2014). Network-based vaccination improves prospects for disease control in wild chimpanzees. *Journal of the Royal Society, Interface*, *11*(97), 20140349. https://doi.org/10.1098/rsif.2014.0349.

Sabeta, C. T., Markotter, W., Mohale, D. K., Shumba, W., Wandeler, A. I., & Nel, L. H. (2007). Mokola virus in domestic mammals, South Africa. *Emerging Infectious Diseases*, *13*(9), 1371–1373.

Schatz, J., Fooks, A., McElhinney, L., Horton, D., Echevarria, J., Vazquez-Moron, S., … Freuling, C. (2013). Bat rabies surveillance in Europe. *Zoonoses and Public Health*, *60*(1), 22–34.

Schatz, J., Ohlendorf, B., Busse, P., Pelz, G., Dolch, D., Teubner, J., … Freuling, C. M. (2014). Twenty years of active bat rabies surveillance in Germany: A detailed analysis and future perspectives. *Epidemiology and Infection*, *142*(6), 1155–1166.

Schneider, L. G., Barnard, B. J. H., & Schneider, H. P. (1985). Application of monoclonal antibodies for epidemiological investigations and oral vaccination studies: I—African viruses. In E. Kuwert, C. Merieux, H. Koprowski, & K. Bogel (Eds.), *Rabies in the tropics* (pp. 49–53). Berlin: Springer-Verlag.

Schneider, L. G., & Cox, J. H. (1994). Bat lyssaviruses in Europe. *Current Topics in Microbiology and Immunology*, *187*, 207–218.

Schneider, M. C., Romijn, P. C., Uieda, W., Tamayo, H., da Silva, D. F., Belotto, A., … Leanes, L. F. (2009). Rabies transmitted by vampire bats to humans: An emerging zoonotic disease in Latin America? *Revista Panamericana de Salud Pública*, *25*(3), 260–269.

Schneider, N. J., Scatterday, J. E., Lewis, A. L., Jennings, W. L., Venters, H. D., & Hardy, A. V. (1957). Rabies in bats in Florida. *American Journal of Public Health*, *47*, 983–989.

Selimov, M. A., Smekhov, A. M., Antonova, L. A., Shablovskaya, E. A., King, A. A., & Kulikova, L. G. (1991). New strains of rabies-related viruses isolated from bats in the Ukraine. *Acta Virologica*, *35*(3), 226–231.

Selimov, M. A., Tatarov, A. G., Botvinkin, A. D., Klueva, E. V., Kulikova, L. G., & Khismatullina, N. A. (1989). Rabies-related Yuli virus; identification with a panel of monoclonal antibodies. *Acta Virologica*, *33*(6), 542–546.

SerraCobo, J., Amengual, B., Abellan, C., & Bourhy, H. (2002). European bat Lyssavirus infection in Spanish bat populations. *Emerging Infectious Diseases*, *8*(4), 413–420.

Serra-Cobo, J., Lopez-Roig, M., Segui, M., Sanchez, L. P., Nadal, J., Borras, M., … Bourhy, H. (2013). Ecological factors associated with European bat lyssavirus seroprevalence in spanish bats. *PLoS One*, *8*(5), e64467. https://doi.org/10.1371/journal.pone.0064467.

Setien, A. A., Brochier, B., Tordo, N., De Paz, O., Desmettre, P., Peharpre, D., & Pastoret, P. P. (1998). Experimental rabies infection and oral vaccination in vampire bats (Desmodus rotundus). *Vaccine*, *16*(11-12), 1122–1126. S0264-410X(98)80108-4 [pii].

Shabman, R. S., Shrivastava, S., Tsibane, T., Attie, O., Jayaprakash, A., Mire, C. E., … Basler, C. F. (2016). Isolation and characterization of a novel gammaherpesvirus from a microbat cell line. *mSphere*, *1*(1). https://doi.org/10.1128/mSphere.00070-15.

Shankar, V., Orciari, L. A., De Mattos, C., Kuzmin, I. V., Pape, W. J., O'Shea, T. J., & Rupprecht, C. E. (2005). Genetic divergence of rabies viruses from bat species of Colorado, USA. *Vector Borne and Zoonotic Diseases*, *5*(4), 330–341. https://doi.org/10.1089/vbz.2005.5.330.

Shinwari, M. W., Annand, E. J., Driver, L., Warrilow, D., Harrower, B., Allcock, R. J., … Diallo, I. S. (2014). Australian bat lyssavirus infection in two horses. *Veterinary Microbiology*, *173*(3-4), 224–231. https://doi.org/10.1016/j.vetmic.2014.07.029.

Shipley, R., Wright, E., Selden, D., Wu, G., Aegerter, J., Fooks, A. R., & Banyard, A. C. (2019). Bats and viruses: Emergence of novel lyssaviruses and association of bats with viral zoonoses in the EU. *Tropical Medicine and Infectious Disease*, *7*(4), 1. https://doi.org/10.3390/tropicalmed4010031.

Shoji, Y., Kobayashi, Y., Sato, G., Itou, T., Miura, Y., Mikami, T., … Sakai, T. (2004). Genetic characterization of rabies viruses isolated from frugivorous bat (Artibeus spp.) in Brazil. *The Journal of Veterinary Medical Science*, *66*(10), 1271–1273. https://doi.org/10.1292/jvms.66.1271.

Shope, R. E., Murphy, F. A., Harrison, A. K., Causey, O. R., Kemp, G. E., Simpson, D. I., & Moore, D. L. (1970). Two African viruses serologically and morphologically related to rabies virus. *Journal of Virology*, *6*(5), 690–692.

Si, D., Marquess, J., Donnan, E., Harrower, B., McCall, B., Bennett, S., & Lambert, S. (2016). Potential exposures to Australian bat lyssavirus notified in Queensland, Australia, 2009-2014. *PLoS Neglected Tropical Diseases*, *10*(12), e0005227. https://doi.org/10.1371/journal.pntd.0005227.

Simic, I., Lojkic, I., Kresic, N., Cliquet, F., Picard-Meyer, E., Wasniewski, M., … Bedekovic, T. (2018). Molecular and serological survey of lyssaviruses in Croatian bat populations. *BMC Veterinary Research*, *14*(1), 274. https://doi.org/10.1186/s12917-018-1592-z.

Smith, J. S. (1981). Mouse model for abortive rabies infection of the central nervous system. *Infection and Immunity*, *31*(1), 297–308.

Smreczak, M., Orlowska, A., Marzec, A., Trebas, P., Muller, T., Freuling, C. M., & Zmudzinski, J. F. (2018). Bokeloh bat lyssavirus isolation in a Natterer's bat, Poland. *Zoonoses and Public Health*. https://doi.org/10.1111/zph.12519.

Smreczak, M., Trebas, P., Orlowska, A., & Mudzinski, J. F. (2008). Rabies surveillance in Poland (1992-2006). *Developmental Biology (Basel)*, *131*, 249–256.

Srinivasan, A., Burton, E. C., Kuehnert, M. J., Rupprecht, C., Sutker, W. L., Ksiazek, T. G., … Rabies in Transplant Recipients Investigation Team. (2005). Transmission of rabies virus from an organ donor to four transplant recipients. *New England Journal of Medicine*, *352*(11), 1103–1111. https://doi.org/10.1056/NEJMoa043018.

Stading, B., Ellison, J. A., Carson, W. C., Satheshkumar, P. S., Rocke, T. E., & Osorio, J. E. (2017). Protection of bats (Eptesicus fuscus) against rabies following topical or oronasal exposure to a recombinant raccoon poxvirus vaccine. *PLoS Neglected Tropical Diseases*, *11*(10), e0005958. https://doi.org/10.1371/journal.pntd.0005958.

Stading, B. R., Osorio, J. E., Velasco-Villa, A., Smotherman, M., Kingstad-Bakke, B., & Rocke, T. E. (2016). Infectivity of attenuated poxvirus vaccine vectors and immunogenicity of a raccoonpox vectored rabies vaccine in the Brazilian Free-tailed bat (Tadarida brasiliensis). *Vaccine*, *34*(44), 5352–5358. https://doi.org/10.1016/j.vaccine.2016.08.088.

Stamm, D. D., Kisling, R. E., & Eidson, M. E. (1956). Experimental rabies infection in insectivorous bats. *The Journal of Infectious Diseases*, *98*, 10–14.

Steece, R., & Altenbach, J. S. (1989). Prevalence of rabies specific antibodies in the Mexican free-tailed bat (Tadarida brasiliensis mexicana) at Lava Cave, New Mexico. *Journal of Wildlife Diseases*, *25*(4), 490–496.

Stoner-Duncan, B., Streicker, D. G., & Tedeschi, C. M. (2014). Vampire bats and rabies: Toward an ecological solution to a public health problem. *PLoS Neglected Tropical Diseases*, *8*(6), e2867. https://doi.org/10.1371/journal.pntd.0002867.

Streicker, D. G., Altizer, S. M., Velasco-Villa, A., & Rupprecht, C. E. (2012). Variable evolutionary routes to host establishment across repeated rabies virus host shifts among bats. *Proceedings of the National Academy of Sciences of the United States of America*, *109*(48), 19715–19720. https://doi.org/10.1073/pnas.1203456109.

Streicker, D. G., Lemey, P., Velasco-Villa, A., & Rupprecht, C. E. (2012). Rates of viral evolution are linked to host geography in bat rabies. *PLoS Pathogens*, *8*(5), e1002720. https://doi.org/10.1371/journal.ppat.1002720.

Streicker, D. G., Recuenco, S., Valderrama, W., Gomez Benavides, J., Vargas, I., Pacheco, V., … Altizer, S. (2012). Ecological and anthropogenic drivers of rabies exposure in vampire bats: Implications for transmission and control. *Proceedings of the Biological Sciences*, *279*(1742), 3384–3392. https://doi.org/10.1098/rspb.2012.0538.

Streicker, D. G., Turmelle, A. S., Vonhof, M. J., Kuzmin, I. V., McCracken, G. F., & Rupprecht, C. E. (2010). Host phylogeny constrains cross-species emergence and establishment of rabies virus in bats. *Science*, *329*(5992), 676–679. https://doi.org/10.1126/science.1188836 329/5992/676 [pii].

Streicker, D. G., Winternitz, J. C., Satterfield, D. A., Condori-Condori, R. E., Broos, A., Tello, C., … Valderrama, W. (2016). Host-pathogen evolutionary signatures reveal dynamics and future invasions of vampire bat rabies. *Proceedings of the National Academy of Sciences of the United States of America*, *113*(39), 10926–10931. https://doi.org/10.1073/pnas.1606587113.

Subudhi, S., Rapin, N., Dorville, N., Hill, J. E., Town, J., Willis, C. K. R., … Misra, V. (2018). Isolation, characterization and prevalence of a novel Gammaherpesvirus in Eptesicus fuscus, the North American big brown bat. *Virology*, *516*, 227–238. https://doi.org/10.1016/j.virol.2018.01.024.

Sulkin, S. E. (1962). Bat rabies: Experimental demonstration of the "reservoiring mechanism" *American Journal of Public Health*, *52*, 489–498.

Sulkin, S. E., Allen, R., Sims, R., Krutzsch, P. H., & Kim, C. H. (1960). Studies on the pathogenesis of rabies in insectivorous bats. II. Influence of environmental temperature. *Journal of Experimental Medicine*, *112*, 595–617.

Sulkin, S. E., & Greve, M. J. (1954). Human rabies caused by bat bite. *Texas State Journal of Medicine*, *50*(8), 620–621.

Sulkin, S. E., Krutzsch, P. H., Wallis, C., & Allen, R. (1957). Role of brown fat in pathogenesis of rabiues in insectivorous bats (Tadaria b. mexicana). *Proceedings of the Society for Experimental Biology and Medicine*, *96*, 461–464.

Sullivan, T. D., Grimes, J. E., Eads, R. B., Menzies, G. C., & Irons, J. V. (1954). Recovery of rabies virus from colonial bats in Texas. *Public Health Reports*, *69*(8), 766–768.

Suu-Ire, R., Begeman, L., Banyard, A. C., Breed, A. C., Drosten, C., Eggerbauer, E., … Horton, D. L. (2018). Pathogenesis of bat rabies in a natural reservoir: Comparative susceptibility of the straw-colored fruit bat (Eidolon helvum) to three strains of Lagos bat virus. *PLoS Neglected Tropical Diseases*, *12*(3), e0006311.

Suu-Ire, R., Fooks, A., Banyard, A., Selden, D., Amponsah-Mensah, K., Riesle, S., ... Cunningham, A. (2017). Lagos bat virus infection dynamics in free-ranging straw-colored fruit bats (Eidolon helvum). *Tropical Medicine and Infectious Disease, 2*(3), 25.

Swanepoel, R. (1994). Rabies. In J. A. W. Coetzer, G. R. Thompson, & R. C. Tustin (Eds.), *Infectious diseases of livestock with special reference to Southern Africa* (pp. 493–553). Cape Town: Oxford University Press/NECC.

Swanepoel, R., Barnard, B. J., Meredith, C. D., Bishop, G. C., Bruckner, G. K., Foggin, C. M., & Hubschle, O. J. (1993). Rabies in southern Africa. *Onderstepoort Journal of Veterinary Research, 60*(4), 325–346.

Thompson, R. D., Mitchell, G. C., & Burns, R. J. (1972). Vampire bat control by systemic treatment of livestock with an anticoagulant. *Science, 177*(51), 806–808.

Tjornehoj, K., Fooks, A. R., Agerholm, J. S., & Ronsholt, L. (2006). Natural and experimental infection of sheep with European bat lyssavirus type-1 of Danish bat origin. *Journal of Comparative Pathology, 134*(2-3), 190–201.

Tordo, N., Poch, O., Ermine, A., Keith, G., & Rougeon, F. (1988). Completion of the rabies virus genome sequence determination: Highly conserved domains among the L (polymerase) proteins of unsegmented negative-strand RNA viruses. *Virology, 165,* 565–576.

Turmelle, A. S., Allen, L. C., Jackson, F. R., Kunz, T. H., Rupprecht, C., & McCracken, G. F. (2010). Ecology of rabies virus exposure in colonies of Brazilian free-tailed bats (*Tadarida brasiliensis*) at natural and man-made roosts in Texas. *Vector Borne and Zoonotic Diseases, 10,* 165–175. https://doi.org/10.1089/vbz.2008.0163.

Turmelle, A. S., Jackson, F. R., Green, D., McCracken, G. F., & Rupprecht, C. E. (2010). Host immunity to repeated rabies virus infection in big brown bats. *Journal of General Virology, 91*(Pt 9), 2360–2366.

Turner, D. C. (1975). *The vampire bat: A field study in behavior and ecology.* Baltimore, Maryland: The Johns Hopkins University Press.

Van der Poel, W. H., Van der Heide, R., Van Amerongen, G., Van Keulen, L. J., Wellenberg, G. J., Bourhy, H., ... Osterhaus, A. D. (2000). Characterisation of a recently isolated lyssavirus in frugivorous zoo bats. *Archives of Virology, 145*(9), 1919–1931.

Van Eeden, C., Markotter, W., & Nel, L. (2011). Molecular phylogeny of duvenhage virus. *South African Journal of Science, 107,* 11–12.

van Thiel, P. P., van den Hoek, J. A., Eftimov, F., Tepaske, R., Zaaijer, H. J., Spanjaard, L., ... Kager, P. A. (2008). Fatal case of human rabies (Duvenhage virus) from a bat in Kenya: The Netherlands, December 2007. *Euro Surveillance, 13*(2).

Venters, H. D., Hoffert, W. R., Scatterday, J. E., & Hardy, A. V. (1954). Rabies in bats in Florida. *American Journal of Public Health, 44,* 182–185.

Virtue, E. R., Marsh, G. A., Baker, M. L., & Wang, L. F. (2011). Interferon production and signaling pathways are antagonized during henipavirus infection of fruit bat cell lines. *PLoS One, 6*(7), e22488. https://doi.org/10.1371/journal.pone.0022488.

Vos, A., Kaipf, I., Denzinger, A., Fooks, A. R., Johnson, N., & Muller, T. (2007). European bat lyssaviruses: An ecological enigma. *Acta Chiropterologica, 9,* 283–296.

Warrell, M. J. (2012). Current rabies vaccines and prophylaxis schedules: Preventing rabies before and after exposure. *Travel Medicine and Infectious Disease, 10*(1), 1–15. https://doi.org/10.1016/j.tmaid.2011.12.005.

Warrilow, D., Harrower, B., Smith, I. L., Field, H., Taylor, R., Walker, C., & Smith, G. A. (2003). Public health surveillance for Australian bat lyssavirus in Queensland, Australia, 2000-2001. *Emerging Infectious Diseases, 9*(2), 262–264.

Weir, D. L., Annand, E. J., Reid, P. A., & Broder, C. C. (2014). Recent observations on Australian bat lyssavirus tropism and viral entry. *Viruses, 6*(2), 909–926. https://doi.org/10.3390/v6020909.

WHO. (2005). *WHO Expert Consultation on Rabies, first report.* Geneva, Switzerland: WHO. Retrieved from: https://apps.who.int/iris/handle/10665/43262.

WHO. (2013). *WHO Expert Consultation on Rabies, second report.* Geneva, Switzerland: WHO. Retrieved from: http://www.who.int/iris/handle/10665/85346.

WHO. (2018). *WHO Expert Consultation on Rabies, third report.* Geneva, Switzerland: WHO Retrieved from: https://apps.who.int/iris/handle/10665/272364.

Willis, C. K. R., & Brigham, R. M. (2004). Roost switching, roost sharing and social cohesion: Forest-dwelling big brown bats, *Eptesicus fuscus,* conform to the fission-fusion model. *Animal Behaviour, 68,* 495–505.

Willoughby, R. E., Jr., Tieves, K. S., Hoffman, G. M., Ghanayem, N. S., Amlie-Lefond, C. M., Schwabe, M. J., ... Rupprecht, C. E. (2005). Survival after treatment of rabies with induction of coma. *The New England Journal of Medicine, 352*(24), 2508–2514. https://doi.org/10.1056/NEJMoa050382.

Winkler, W. G. (1968). Airborne rabies virus isolation. *Journal of Wildlife Diseases, 4*(2), 37–40.

Wise, E. L., Marston, D. A., Banyard, A. C., Goharriz, H., Selden, D., Maclaren, N., ... Fooks, A. R. (2017). Passive surveillance of United Kingdom bats for lyssaviruses (2005-2015). *Epidemiology and Infection, 145*(12), 2445–2457. https://doi.org/10.1017/S0950268817001455.

Witte, E. J. (1954). Bat rabies in Pennsylvania. *American Journal of Public Health and the Nation's Health, 44*(2), 186–187. https://doi.org/10.2105/ajph.44.2.186.

Wright, E., Hayman, D. T., Vaughan, A., Temperton, N. J., Wood, J. L., Cunningham, A. A., ... Fooks, A. R. (2010). Virus neutralising activity of African fruit bat (Eidolon helvum) sera against emerging lyssaviruses. *Virology, 408*(2), 183–189.

Wyatt, J. (2007). Rabies-update on a global disease. *The Pediatric Infectious Disease Journal, 26*(4), 351–352. https://doi.org/10.1097/01.inf.0000258776.47697.97.

Yan, X., Prosniak, M., Curtis, M. T., Weiss, M. L., Faber, M., Dietzschold, B., & Fu, Z. F. (2001). Silver-haired bat rabies virus variant does not induce apoptosis in the brain of experimentally infected mice. *Journal of Neurovirology, 7*(6), 518–527.

Yung, V., Favi, M., & Fernandez, J. (2012). Typing of the rabies virus in Chile, 2002-2008. *Epidemiology and Infection, 140*(12), 2157–2162. https://doi.org/10.1017/S0950268812000520.

Zhang, Q., Zeng, L. P., Zhou, P., Irving, A. T., Li, S., Shi, Z. L., & Wang, L. F. (2017). IFNAR2-dependent gene expression profile induced by IFN-alpha in Pteropus alecto bat cells and impact of IFNAR2 knockout on virus infection. *PLoS One, 12*(8), e0182866. https://doi.org/10.1371/journal.pone.0182866.

Zieger, U., Cheetham, S., Santana, S. E., Leiser-Miller, L., Matthew-Belmar, V., Goharriz, H., & Fooks, A. R. (2017). Natural exposure of bats in Grenada to rabies virus. *Infection Ecology and Epidemiology, 7*(1), 1332935. https://doi.org/10.1080/20008686.2017.1332935.

CHAPTER 8

Human disease

Alan C. Jackson

Professor of Medicine (Neurology), University of Manitoba, Winnipeg, MB, Canada

8.1 Introduction

Since antiquity, rabies has been one of the most feared diseases. Human rabies remains an important public health problem in many developing countries where dog rabies is endemic (Fooks et al., 2014). Worldwide there are at least 59,000 human deaths each year due to rabies and the vast majority are related to endemic canine rabies (Hampson et al., 2015). Beginning in the 1990s, up to six human cases of rabies were diagnosed per year in the United States, and many of these infections were acquired indigenously from unrecognized exposures to insectivorous bats (Noah et al., 1998). A significant number of additional rabies cases probably go unrecognized in resource-rich countries such as in the United States and Canada because (1) human rabies is uncommon in these locations and physicians are not familiar with the disease, (2) undiagnosed acute and fatal neurologic illnesses are common, and (3) there may be no history of an animal exposure.

8.2 Exposures, incubation period, and prodromal symptoms

The infectious cycle of rabies virus (RABV) is perpetuated mainly through animal bites and the deposition of RABV-laden saliva into subcutaneous tissues and muscles. With respect to human rabies, worldwide, dogs are by far the most common and important rabies vector; bats are most important in the Americas, although there are reservoirs in various terrestrial animals. Other types of nonbite exposures, including contamination of an open wound, scratch, abrasion, or mucous membrane by saliva or central nervous system (CNS) tissue from an infected animal, are quite common, although they are rarely responsible for transmission of RABV. Handling and skinning of infected carcasses and consumption of raw infected meat have resulted in transmission of RABV (Kureishi, Xu, Wu, & Stiver, 1992; Tariq, Shafi,

Jamal, & Ahmad, 1991; Wallerstein, 1999). Rarely, but notably, inhalation of aerosolized RABV in caves containing millions of bats (Constantine, 1962) or in laboratories (Tillotson, Axelrod, & Lyman, 1977; Winkler, Fashinell, Leffingwell, Howard, & Conomy, 1973) has resulted in human rabies. At least eight cases of rabies have resulted from transplantation (human-to-human) of RABV-infected corneas (Table 8.1). Rabies also developed in a patient from India 16 days after corneal transplantation, but the source of the infection was unknown in this case (Masthi et al., 2012) and it is unlikely that it was due to the transplanted cornea. In other reports, transmission did not occur after corneal transplantation from a donor with rabies in France (Sureau, Portnoi, Rollin, Lapresle, & Chaouni-Berbich, 1981) and from another donor in Germany, in which there were two cornea transplant recipients (Johnson, Brookes, Fooks, & Ross, 2005). In 2004, transplantations of organs and a vascular artery segment in Texas were associated with transmission of RABV and the development of fatal rabies in four recipients (Srinivasan et al., 2005; Table 8.2). The donor for these cases presented with gastrointestinal symptoms, throat pain, intermittent periods of confusion and agitation, and he had mild fever and ballistic trunk movements (Burton et al., 2005). The initial CT head scan showed a small subarachnoid hemorrhage. In retrospect, it is highly doubtful that this clinical presentation could be explained by a small subarachnoid hemorrhage. Subsequently, there was neurologic deterioration and a repeat CT head scan showed a large subarachnoid hemorrhage with evidence of herniation. He progressed to brain death, and his organs (lungs, kidneys, and liver) and iliac vessels were harvested (Burton et al., 2005). The four recipients of the liver, kidneys, and an iliac artery segment (for a liver transplant) developed clinical rabies within a month and died. The donor had anti-RABV antibodies in serum at the time of death and three of the four recipients had antibodies on postoperative days 35 and 36 (Srinivasan et al., 2005). Immunosuppression of the recipients in order to prevent organ rejection results in a favorable environment for viral replication and spread. Only later, it was determined that the donor had been bitten by a bat, and antigenic typing indicated that the

TABLE 8.1 Human rabies cases transmitted by corneal transplantation.

Location	Year	Age of patient (recipient)	Time to death (days)	References
United States	1978	37	50	Houff et al. (1979)
France	1979	36	41	Galian et al. (1980)
Thailand	1981	41	22	Thongcharoen et al. (1981)
Thailand	1981	25	33	Thongcharoen et al. (1981)
India	1987	62	15	Gode and Bhide (1988)
India	1988	48	264[a]	Gode and Bhide (1988)
Iran	1994	40	27	Javadi, Fayaz, Mirdehghan, and Ainollahi (1996)
Iran	1994	35	41	Javadi et al. (1996)

[a] Patient received two doses of rabies vaccine about 1 month after the transplant.

TABLE 8.2 Cases of human rabies associated with organ transplantation.

	Sex/age	Organ transplanted	Onset of clinical rabies posttransplantation (days)	References
Donor in United States	Male/20	–	–	Burton et al. (2005) and Srinivasan et al. (2005)
Recipient 1	Male/53	Liver	21	Burton et al. (2005) and Srinivasan et al. (2005)
Recipient 2	Female/50	Kidney	27	Burton et al. (2005) and Srinivasan et al. (2005)
Recipient 3	Male/18	Kidney	27	Burton et al. (2005) and Srinivasan et al. (2005)
Recipient 4	Female/55	Iliac artery segment	27	Burton et al. (2005) and Srinivasan et al. (2005)
Donor in Germany	Female/26	–	–	Maier et al. (2010)
Recipient 1	Female/46	Lung	6 weeks	Maier et al. (2010)
Recipient 2	Male/72	Kidney	5 weeks	Maier et al. (2010)
Recipient 3	Male/47	Kidney/pancreas	5 weeks	Maier et al. (2010)
Donor in United States	Male/20	–	–	Vora et al. (2013)
Recipient	Male/49	Kidney	18 months	Vora et al. (2013)
Donor in Kuwait	Male/28	–	–	Al Mousawi (2015), Elsiesy et al. (2015), and Saeed and Al-Mousawi (2017)
Recipient 1 in Kuwait	Female/5	Kidney	15 weeks	Al Mousawi (2015), Elsiesy et al. (2015), and Saeed and Al-Mousawi (2017)
Recipient 2 in Kuwait	Unknown	Kidney	8–10 weeks	Al Mousawi (2015) and Elsiesy et al. (2015)
Recipient 3 in Saudi Arabia	Unknown	Heart	Unknown	Al Mousawi (2015) and Elsiesy et al. (2015)
Recipient 4 in Saudi Arabia	Male	Liver	Unknown	Al Mousawi (2015) and Elsiesy et al. (2015)
Donor in China	Male/6	–	–	Gong et al. (2017) and Zhou et al. (2016)
Recipient 1	Male/55	Kidney	42 days	Gong et al. (2017) and Zhou et al. (2016)

Continued

TABLE 8.2 Cases of human rabies associated with organ transplantation—cont'd

	Sex/age	Organ transplanted	Onset of clinical rabies posttransplantation (days)	References
Recipient 2	Male/43	Kidney	48 days	Gong et al. (2017) and Zhou et al. (2016)
Donor in China	Male/2	–	–	Chen et al. (2017, 2018)
Recipient 1	Female/29	Kidney	40 days	Chen et al. (2017, 2018)
Recipient 2	Female/47	Kidney	43 days	Chen et al. (2017, 2018)
Donor in China	Unknown	–	–	Lu, Zhu, and Wu (2018)
Recipient 1	Unknown	Heart	Unknown	Lu et al. (2018)
Recipient 2	Unknown	Kidney	41 days	Lu et al. (2018)
Recipient 3	Unknown	Kidney	45 days	Lu et al. (2018)

RABV variant was associated with Brazilian (Mexican) free-tail bats (Krebs, Mandel, Swerdlow, & Rupprecht, 2005). All four transplantation recipients had histopathologic features of encephalitis with cytoplasmic inclusions characteristic of Negri bodies and RABV antigen was detected in neurons with immunohistochemical staining from multiple areas of the CNS. RABV antigen was also observed in peripheral nerves of the transplanted kidneys, liver, and arterial graft. Transmission occurred again from a donor to organ transplant recipients in Germany that resulted in three fatal cases in 2005 (Maier et al., 2010; Table 8.1). More recently, there have been seven rabies cases transmitted from three different organ donors in China (Table 8.1). It is now very clear and it should be emphasized that tissues or organs should not be transplanted from a donor who dies from an undiagnosed neurologic disease, because the risk of transmitting unsuspected infectious agents, including RABV, is unacceptably high.

Most other reported cases of human-to-human transmission have not been well documented. Two patients with rabies from Ethiopia were described and their only known exposure was contact with family members who died of rabies (Fekadu et al., 1996). In this report, a 41-year-old female died of rabies 33 days after her 5-year-old son died of rabies; he had bitten his mother on her little finger. A 5-year-old boy presented with rabies 36 days after his mother died of rabies; he had repeatedly received kisses from his mother on his mouth during her illness. Sexual transmission of RABV has not been documented. Although natural human-to-human transmission of rabies likely occurs very rarely, anyone in direct contact with rabies patients, including family members and health care workers, should employ barrier nursing techniques in order to minimize the risk of transmission of the virus via saliva or other secretions (Remington, Shope, & Andrews, 1985). There is evidence of transplacental transmission of RABV in a single report from Turkey (Sipahioglu & Alpaut, 1985).

The incubation period for human rabies is usually 20–90 days after exposure, although occasionally disease develops after only a few days (Anderson, Nicholson, Tauxe, & Winkler, 1984)

and rare cases have occurred a year or more following exposure. Three immigrants from Laos, the Philippines, and Mexico developed rabies in the United States due to RABV strains from their countries of origin with incubation periods of at least 11 months and 4 and 6 years, which were based on the time of their immigration (Smith, Fishbein, Rupprecht, & Clark, 1991). A case of rabies in a 10-year-old Vietnamese girl in Australia in 1990 was also likely acquired at least 5 years earlier (Bek, Smith, Levy, Sullivan, & Rubin, 1992; McColl et al., 1993). The incubation period (from exposure to onset of disease) in rabies is longer and more variable than for most other infectious diseases, which may cause considerable emotional stress to the patient. Very long incubation periods raise the possibility of another unrecognized or forgotten exposure in rabies endemic areas. Severe multiple bites and facial bites are associated with shorter incubation periods (Warrell & Warrell, 1991), although there is a lack of a correlation between the site of the bite and the incubation period (Dupont & Earle, 1965). There may be no history of a bite exposure because it was unrecognized, particularly with insectivorous bat bites because they may be very small (Jackson & Fenton, 2001; Fig. 8.1), or because a bite was either forgotten or no inquiry was made while the patient was still lucid. In the United States and Canada, 20% of patients who acquire rabies from bats have no history of bat contact, and only 38%

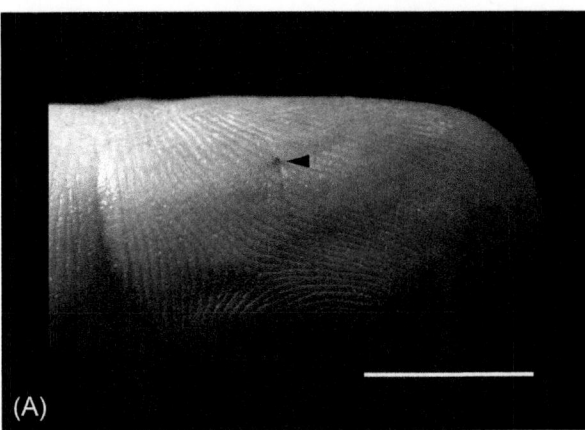

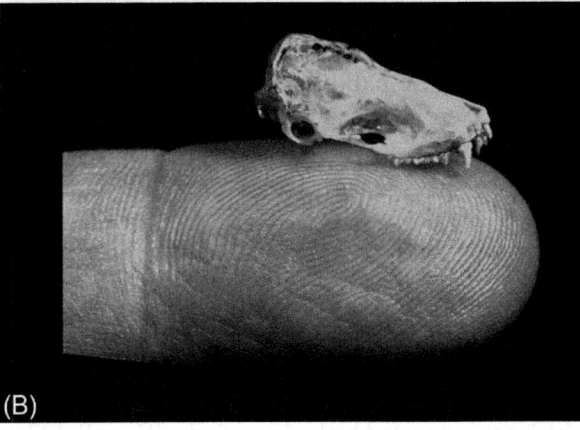

FIG. 8.1 Small puncture wound *(arrowhead)* involving the right ring finger of a bat biologist (A) caused by a defensive bite from a canine tooth of a silver-haired bat (*Lasionycteris noctivagans*) (Bar = 10mm). Skull of a silver-haired bat (B) (length of 17.1mm) is resting on a distal phalanx, which demonstrates the small size of the bat and its teeth. *Reproduced from Jackson, A. C., & Fenton, M. B. (2001). Human rabies and bat bites (letter). Lancet, 357(9269), 1714. Copyright © 2001, Elsevier.*

have a history of a bite or scratch (Jackson, 2011). With known bite exposures from rabid animals, the following has been observed in untreated persons who develop rabies: 50%–80% occurrence after head bites, 15%–40% after hand or arm bites, and 3%–10% after leg bites. The risk is about 0.1% for contamination of minor wounds with saliva, including scratches (Hattwick, 1974). The biologic bases for these observations are unclear, but a number of factors may be responsible, including the density of RABV receptors in affected tissues, the degree of innervation in tissues in different anatomical locations, the quantity of virus inoculated, and the properties of the RABV variant. Some individuals with RABV exposures may have inapparent RABV infection and develop naturally acquired immunity (Doege & Northrop, 1974). Low titers of RABV neutralizing antibodies (VNA) have been detected in Canadian Inuit hunters (7 of 20) and their wives (2 of 11) (Orr, Rubin, & Aoki, 1988). A survey in fox trappers in northern Alaska identified a 68-year-old aboriginal male who had trapped for about 47 years and had a VNA concentration of 2.30IU/mL; he did not recall ever being bitten by a fox. Black and Wiktor (1986) also observed low titers of rabies VNA in 17% (5 of 30) of Florida raccoon hunters, but not in a control group of hunters. VNA were detected in five humans in Amazonian communities in Peru (at high risk of vampire bat depredation), who had not previously received rabies vaccine; four had a history of a bat bite (Gilbert et al., 2012). In four, the VNA titers ranged from 0.4 to 0.6IU/mL, whereas one had a high titer of 2.8IU/mL. Surprisingly, VNA were also detected in 6.6% (15 of 226) of unimmunized students and faculty members of a veterinary medical school at the inception of a rabies vaccine trial (Ruegsegger, Black, & Sharpless, 1961). These studies suggest that exposure of humans to RABV under natural conditions can rarely result in result in immunization (natural vaccination) without the development of clinical disease.

Nonspecific prodromal symptoms of rabies, including fever, chills, malaise, fatigue, insomnia, anorexia, headache, anxiety, and irritability, may last for up to 10 days prior to the onset of neurologic symptoms (Warrell, 1976). About 30%–70% of patients develop pain, paresthesias, and/or pruritus at or close to the site of the bite, and the bite wound has often healed by the time these symptoms develop (Dupont & Earle, 1965; Hattwick, 1974). The pruritus may result in severe excoriations from scratching. Retro-orbital pain also occurred as an early symptom in some patients with transmission by corneal transplantation. Local neurologic symptoms may reflect infection and associated inflammation involving local peripheral sensory ganglia (dorsal root or trigeminal ganglia) (Mitrabhakdi et al., 2005). The initial neurologic symptoms may occasionally occur at a site distant from the bite, although the pathogenetic basis for this phenomenon is not clear. Two patients bitten on their toes developed rabies with early severe itching of their ears (Hemachudha, 1994). Tremor has also been described involving the bitten extremity (Warrell, 1976).

8.3 Clinical forms of disease

8.3.1 Encephalitic rabies

About 80% of patients develop an encephalitic or classical (also called *furious*) form of rabies and about 20% have a paralytic form of disease. In encephalitic rabies, patients have episodes of generalized arousal or hyperexcitability, which are separated by lucid periods (Warrell, 1976), and these features reflect brain involvement with the infection. Intermittent episodes may occur with confusion, hallucinations, agitation, and aggressive behavior, which typically last for periods of 1–5 min (Hattwick, 1974; Hemachudha, 1997; Warrell & Warrell, 1991). The episodes may occur spontaneously or be precipitated by a variety of sensory stimuli (tactile, auditory, visual, or olfactory). Biting behavior of patients with rabies has been described (Dupont & Earle, 1965; Emmons et al., 1973; Warrell, 1976), but it is unusual. Fever is common and may be quite high (over 42°C/107°F) and there may be signs of autonomic dysfunction, including hypersalivation, lacrimation, sweating, piloerection (gooseflesh), and dilated pupils. The autonomic dysfunction may result from the infection directly involving the autonomic nervous system centers or pathways in the hypothalamus, spinal cord and/or autonomic ganglia. Parasympathetic stimulation may increase the production of saliva above the normal volume of about 1 L/24 h. Often patients appear frightened with wide palpebral fissures, dilated pupils, and an open mouth (Nicholson, 1994). Movement disorders have been noted (Warrell, 1976). Seizures, including convulsions, may occur, but they are not common or prominent and they usually occur late in the illness. A case of rabies was recently reported from the United States with status epilepticus (Villamar, Smith, Wilson, & Smith, 2017), but this case was very poorly documented with many serious flaws in the report and it is very unlikely this case was actually rabies (Jackson & Del Bigio, 2018). Cranial nerve signs may be present, including ophthalmoplegia, facial weakness, impaired swallowing, and tongue weakness. There may also be nuchal rigidity, reflecting leptomeningeal inflammation.

About 50%–80% of patients develop hydrophobia, which is a characteristic and the most specific manifestation of rabies. Hydrophobia is not a feature of any other diseases. The term *hydrophobia* is derived from the Greek word meaning "fear of water." Patients may initially experience pain in the throat or difficulty swallowing. On attempts to swallow, they experience contractions of the diaphragm, sternocleidomastoids, scalenes and other accessory muscles of inspiration, which last for about 5–15 s and may be associated with epigastric pain (Fig. 8.2). These symptoms may be followed by contraction of neck muscles, resulting in flexion or extension of the neck and rarely with opisthotonic posturing. There may be associated retching, vomiting, coughing, aspiration into the trachea, grimacing, convulsions, and hypoxia (Editorial, 1975). Patients may die during severe spasms with the development of cardiorespiratory arrest if supportive care measures are not initiated (Warrell & Warrell, 1991). During the spasms, there is an associated feeling of terror, often without associated pain. Patients avoid drinking for long periods of time, even despite intense thirst, resulting in dehydration. Subsequently, the sight, sound, or even mention of water (or liquids) may trigger these spasms, indicating that hydrophobia is reinforced by conditioning (Warrell et al., 1976). Hydrophobic spasms may also occur spontaneously, particularly later in the course of the illness. A draft of air on the skin or the breath of an examiner may have the same effect, which has been termed *aerophobia*, and a variety of other stimuli, including water splashed on the skin, attempts by the patient to speak and stimulation from bright lights or loud sounds, also may precipitate spasms (Warrell, 1976). Patients may wear heavy clothing in order to avoid drafts. The fan test, elicited by fanning a current of air across the face and observing the patient for spasms of the pharyngeal and neck muscles, has been used as a bedside diagnostic test for the presence of aerophobia (Wilson, Hettiarachchi, & Wijesuriya, 1975). Sobbing respiration (like a child who has been crying) with a two-stage (sniff-sniff) inspiration followed by a slow, full expiration has been described (Pearson, 1976). Later these spasms

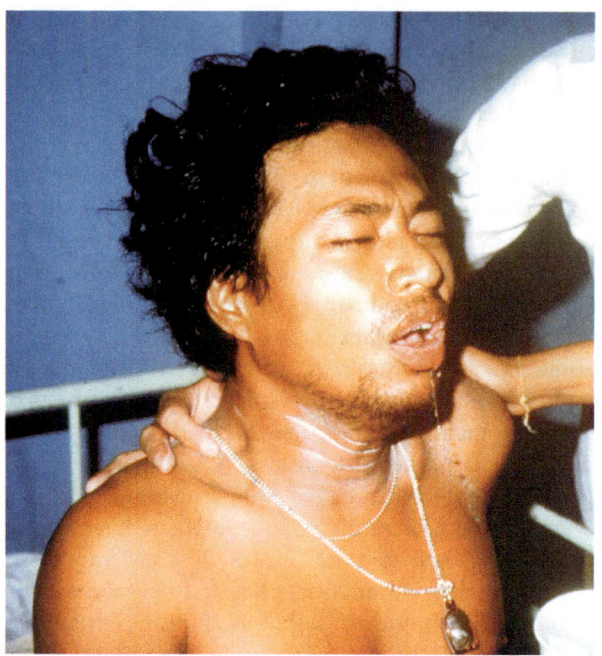

FIG. 8.2 Hydrophobic spasm of inspiratory muscles associated with terror in a patient with furious rabies encephalitis attempting to swallow water. *Copyright D. A. Warrell, Oxford, United Kingdom.*

merge with the development of periodic, apneustic or ataxic breathing as the patient's level of consciousness deteriorates (Warrell et al., 1976). The hydrophobia of rabies is likely due to selective infection of neurons that inhibit the inspiratory motor neurons in the region of the nucleus ambiguus in the brainstem (Warrell, 1976; Warrell et al., 1976). This results in exaggeration of defensive reflexes that protect the respiratory tract. Vocal cord weakness may result in a change in the voice, and patients may make bark-like sounds. Increased libido and hypersexuality may be early manifestations of rabies in females (Dutta, 1996; Gardner, 1970; Senthilkumaran, Balamurgan, Sweni, Menezes, & Thirumalaikolundusubramanian, 2011; Tian, Chen, & Yan, 2019), likely due to the limbic system involvement. Priapism (painful spontaneous erections) and spontaneous ejaculations occasionally occur in males in rabies, and they may occur early in the disease and be the presenting clinical feature (Bhandari & Kumar, 1986; Talaulicar, 1977; Tian et al., 2019; Udwadia, Udwadia, Rao, & Kapadia, 1988). Some clinical features in rabies have a greater association with particular RABV variants. For example, bat-acquired rabies more frequently has tremor and myoclonus, whereas dog-acquired rabies more frequently has hydrophobia and aerophobia (Udow, Marrie, & Jackson, 2013). In encephalitic rabies, there is often progression to severe flaccid paralysis, coma, and multiple organ failure. The paralysis that develops either in association with or after the development of coma should not be confused with paralytic rabies in which the muscle weakness develops, in contrast, early in the course of the illness (see Section 8.3.2). Rabies is almost always fatal, and death often occurs within 14 days of the onset of clinical manifestations, although the time of death may be influenced by critical-care measures. Experimental studies indicate that there is disintegration of the sleep-wake cycle that progresses to brain death, which occurs prior to cardiac arrest (Gourmelon, Briet, Clarencon, Court, & Tsiang, 1991; Tesoriero, Del, & Bentivoglio, 2019).

A wide variety of medical complications can develop in patients with rabies. Many of these complications may also occur in critically ill patients with other acute neurological disorders, but some are likely related to the widespread infection in the CNS with systemic (extraneural) organ involvement due to infection of autonomic or sensory neurons (Jackson et al., 1999). Cardiopulmonary complications are the most common and important. Respiratory complications include hyperventilation, hypoxemia, respiratory depression with apnea, atelectasis, and aspiration with secondary pneumonia (Hattwick, 1974). Sinus tachycardia is a common cardiac feature, and the degree of the tachycardia is often greater than that expected for the degree of fever (Warrell et al., 1976). Cardiac arrhythmias (including wandering atrial/nodal pacemaker, sinus bradycardia, and supraventricular or ventricular ectopic beats), hypotension, heart failure and cardiac arrest may occur (Hattwick, 1974; Warrell et al., 1976). Cardiac arrhythmias may account for the sudden death of patients who are alert and do not have advanced neurologic signs of rabies. Cardiac manifestations may reflect infection involving the autonomic nervous system or the myocardium with a myocarditis (Cheetham, Hart, Coghill, & Fox, 1970; Jackson et al., 1999; Metze & Feiden, 1991; Park, Crane, Pal, & Cagnina, 2019; Raman, Prosser, Spreadbury, Cockcroft, & Okubadejo, 1988; Ross & Armentrout, 1962). Either hyperthermia or hypothermia may be present, which may reflect hypothalamic involvement of the infection. Gastrointestinal hemorrhage, especially hematemesis, is a common complication (Kureishi et al., 1992). Endocrine complications include both inappropriate secretion of antidiuretic hormone and diabetes insipidus (Bhatt, Hattwick, Gerdsen, Emmons, & Johnson, 1974; Hattwick, 1974).

8.3.2 Paralytic rabies

In paralytic rabies, flaccid muscle weakness develops early in the course of the disease and the weakness is prominent. Patients are frequently misdiagnosed with this clinical form of the disease, especially if a history of an animal bite is not obtained. The earliest description of paralytic rabies was recorded in 1887 (Gamaleia, 1887). Paralytic rabies has also been called *dumb rabies*. Patients may be literally dumb or mute due to laryngeal muscle weakness, but the term *dumb rabies* usually refers to the quieter clinical features and prominent weakness rather than specifically to the presence of anarthria (Editorial, 1978; Mills, Swanepoel, Hayes, & Gelfand, 1978). The development of paralytic rabies is not related to the anatomical site of the bite (Tirawatnpong et al., 1989), and the incubation period is similar to that in encephalitic rabies. Patients are usually alert with a normal mental status, at the onset of this clinical form of rabies. The weakness often begins in the bitten extremity and spreads to involve the other extremities, sometimes in an ascending pattern. Muscle fasciculations may be present (Phuapradit, Manatsathit, Warrell, & Warrell, 1985). The facial muscles are frequently weak bilaterally. Associated bilateral deafness has been reported (Phuapradit et al., 1985). Although patients may have local pain, paresthesias or pruritus at the site of the bite, the sensory examination is usually normal in patients with paralytic rabies. The clinical picture may be confused with the Guillain-Barré syndrome, including both the acute inflammatory demyelinating polyradiculopathy and the more severe motor-sensory neuropathy of acute onset with predominant axonal involvement (called the *axonal* Guillain-Barré syndrome) (Feasby et al., 1986; Griffin et al., 1996; Sheikh et al., 2005). Sphincter involvement, especially with urinary incontinence, is common in paralytic rabies, but this is not a feature of the Guillain-Barré syndrome (Asbury & Cornblath, 1990). In addition, pain and sensory disturbances may occur in paralytic rabies. Myoedema has been reported as a sign observed in paralytic rabies, but not in encephalitic rabies (Hemachudha, Phanthumchinda, Phanuphak, & Manutsathit, 1987). However, myoedema has not been confirmed as an important sign of paralytic rabies in other reports. In myoedema, percussion of a muscle (e.g., deltoid or thigh muscle) with a tendon hammer results in local mounding of the muscle without propagated contractions and with electrical silence; the mounding disappears over a few seconds. Myoedema is thought to be a normal physiological phenomenon and its presence does not indicate neuromuscular pathology (Hornung & Nix, 1992). Hence, the importance of this sign in rabies requires future clarification. Bulbar and respiratory muscles eventually become weak in paralytic rabies, resulting in death. Hydrophobia is more unusual in the paralytic form of the disease, although mild inspiratory spasms are commonly observed (Hemachudha et al., 1988). Survival in paralytic rabies is usually longer than in encephalitic rabies (Hemachudha, Wacharapluesadee, Mitrabhakdi, Morimoto, & Lewis, 2005). It is unclear if the hydrophobic spasms *per se* lead to death in the first few days of illness in encephalitic disease (Editorial, 1978) or if they reflect a more life-threatening distribution of the brain infection.

An unusual human outbreak of rabies affecting over 70 people occurred in Trinidad between 1929 and 1937 with transmission of the virus from vampire bats (Hurst & Pawan, 1931; Pawan, 1939; Waterman, 1959). All patients in this outbreak had the paralytic form of the disease. This led to diagnostic uncertainty and initially poliomyelitis and botulism were suspected. Nine miners died of paralytic rabies transmitted by vampire bats in British Guiana (presently Guyana) in 1953 (Nehaul, 1955). Similarly, seven children died of paralytic rabies in Surinam in 1973–74 and vampire bats were probably also the responsible vector (Verlinde, Li-Fo-Sjoe, Versteeg, & Dekker, 1975). However, RABV transmitted by vampire bats does not always produce paralytic rabies. A 1990 outbreak of human rabies in Peru with transmission from vampire bats exclusively produced cases of encephalitic rabies (Lopez, Miranda, Tejada, & Fishbein, 1992). Recent outbreaks of human rabies with transmission from vampire bats have occurred in the Amazon region of northern Brazil in 2004. Apparently, all of the 21 cases had the paralytic form of rabies, but only limited clinical and pathological information has been reported (da Rosa et al., 2006; Fernandes et al., 2011). Furthermore, it has been observed that a dog may bite two individuals and one develops encephalitic rabies and the other develops paralytic rabies (Hemachudha et al., 1988; Wilde & Chutivongse, 1988).

The pathogenetic basis for the two different clinical forms of rabies has not been determined (Hemachudha et al., 2005). In a small series, there were no marked differences in the regional distribution of RABV antigen or in the inflammatory changes (Tirawatnpong et al., 1989). However, at the time of death, the distribution of the viral infection may be much more widespread and not closely reflect the distribution at the time of the patient's presentation with paralytic rabies. Electrophysiologic studies have indicated that peripheral nerve involvement, including demyelination, likely contributes to the weakness in paralytic rabies (Mitrabhakdi et al., 2005). It is curious that an earlier serum-neutralizing antibody response was observed in patients with encephalitic rabies than in those with paralytic rabies (Hemachudha, 1994). There is evidence that patients with paralytic rabies have defects in immune responsiveness, including lack of lymphocyte proliferative responses to RABV antigen (Hemachudha et al., 1988) and lower levels of serum cytokines, including interleukin-6 and the soluble interleukin-2 receptor, than patients with encephalitic rabies (Hemachudha, Panpanich, Phanuphak, Manatsathit, & Wilde, 1993). In contrast, a Chinese case that was misdiagnosed as axonal Guillain-Barré syndrome (Griffin et al., 1996) had pathologic changes that were most marked in the ventral spinal nerve roots without prominent inflammation or motor neuron degeneration (Sheikh et al., 2005). This may have occurred either because axonal degeneration can be an early morphologic consequence of RABV-infected motor neurons or because axonal degeneration may be caused by immune injury. In support of the latter hypothesis, a case of encephalitic rabies was treated with high-dose intravenous rabies immune globulin and developed severe paralysis (Hemachudha et al., 2003). However, the lack of anti-RABV antibodies in some paralytic rabies cases argues that antibody-mediated injury to nerves is not the only mechanism resulting in paralysis in rabies.

8.4 Investigations

8.4.1 Imaging studies

Computed tomographic (CT) studies of the brain usually are normal in rabies (Faoagali, De Buse, Strutton, & Samaratunga, 1988; Mrak & Young, 1993; White et al., 1994), although hypodense cortical lesions (Sow et al., 1996) and nonenhancing basal ganglia hypodensities (Awasthi, Parmar, Patankar, & Castillo, 2001) have been described. There have been reports of magnetic resonance imaging (MRI) studies of the brain with normal findings (Mrak & Young, 1993; Sing & Soo, 1996), but imaging may show lesions that are usually located in gray matter areas of the brain parenchyma, including the brainstem (Awasthi et al., 2001; Hantson et al., 1993; Laothamatas, Sungkarat, & Hemachudha, 2011; Pleasure & Fischbein, 2000). For example, increased signals were observed on T_2-weighted images in the medulla and pons with only minimal gadolinium enhancement in these areas in a patient from California infected by a RABV strain associated with Brazilian (Mexican) free-tailed bats (Pleasure & Fischbein, 2000). Gadolinium enhancement of cervical nerve roots was described in a patient with paralytic rabies (Laothamatas, Hemachudha, Tulyadechanont, & Mitrabhakdi, 1997). Gadolinium enhancement involving the medulla and hypothalamus was also described in the same report in another patient with paralytic rabies. This indicates imaging evidence of brain infection, which has been shown in histopathologic studies at the time of death (Chopra, Banerjee, Murthy, & Pal, 1980). MRI findings in both the brain and spinal cord were found to be similar in a small number of patients with encephalitic and paralytic rabies (Laothamatas, Hemachudha, Mitrabhakdi, Wannakrairot, & Tulayadaechanont, 2003).

8.4.2 Laboratory studies

The electroencephalogram may be normal or show nonspecific abnormalities in human rabies. Slow-wave activity has been observed as well as periodic (Komsuoglu, Dora, & Kalabay, 1981) and epileptiform activity. Electrophysiological studies showed evidence of peripheral nerve and/or anterior horn cell involvement in a small series, and features of peripheral nerve demyelination were observed in the paralytic cases (Mitrabhakdi et al., 2005). Electrophysiologic evidence of a primary axonal neuropathy was found in two patients with paralytic rabies in another report (Prier, Gibert, Bodros, Vachon, et al., 1979; Prier, Gibert, Bodros, & Krymolieres, 1979). Hematologic and biochemical tests are usually normal, although hyponatremia may occur secondary to inappropriate secretion of antidiuretic hormone. Cerebrospinal fluid (CSF) analysis often becomes abnormal in human rabies. A CSF pleocytosis (elevated number of white cells) was found in 59% of cases in the first week of illness and in 87% after the first week (Anderson et al., 1984). The white cell count is usually less than 100 cells/μL and the leukocytes are predominantly mononuclear cells. The CSF protein concentration may be mildly elevated and glucose is usually in the normal range, although low CSF glucose levels have occasionally been reported (Chotmongkol, Vuttivirojana, & Cheepblangchai, 1991; Roine et al., 1988). Serum-neutralizing antibodies against RABV are not usually present in unimmunized patients until the second week of the illness, and patients may die of rabies without developing a detectable serum antibody level (Anderson et al., 1984; Hattwick, 1974; Kasempimolporn, Hemachudha, Khawplod, & Manatsathit, 1991). Antibody had not developed in serum by 10 days after the onset of clinical symptoms in five of 18 (28%) patients with rabies in the United States (Noah et al., 1998). One patient, who had received interferon therapy, had not developed antibodies by the time of death 24 days after the onset of symptoms (Sibley et al., 1981). RABV antibodies develop in the CSF later than in the serum and the CSF titer is lower. Very high titers of RABV antibodies in the CSF have been interpreted as evidence of rabies encephalitis in vaccinated patients (Alvarez et al., 1994; Hattwick, 1974; Madhusudana, Nagaraj, Uday, Ratnavalli, & Kumar, 2002; Porras et al., 1976; Tillotson et al., 1977). RABV may occasionally be isolated from saliva and rarely from the CSF or urine sediment (Anderson et al., 1984). Virus isolation is more likely during early disease before neutralizing antibodies appear, because they produce "autosterilization" of tissues. RABV antigen may be demonstrated antemortem by using the fluorescent antibody technique in frozen sections from skin biopsies (Fig. 8.3). A skin biopsy should be obtained containing hair follicles (minimum of 10) by using a full-thickness punch biopsy (5–6 mm in diameter), which is typically taken from the posterior region of the neck at the hairline. Many sections of the biopsy specimen, which should include several hair follicles, are examined with fluorescent antibody staining for RABV antigen that is found in adjacent small sensory

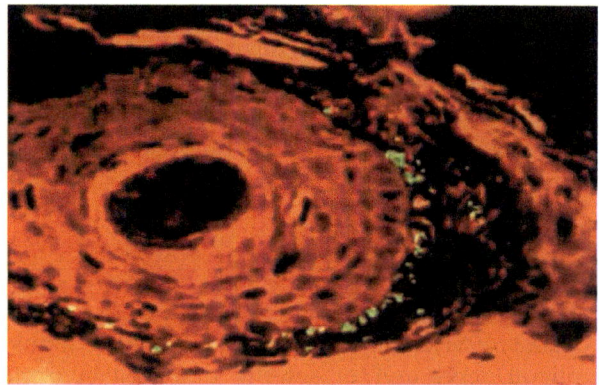

FIG. 8.3 Human hair follicle from nuchal skin biopsy. Nerve fibers surrounding the follicle are stained by specific fluorescence indicating the presence of RABV antigen (direct fluorescence antibody method on a frozen section; 250× magnification). *From Trimarchi, C. V., & Nadin-Davis, S. A. (2007). Diagnostic evaluation. In A. C. Jackson & W. H. Wunner (Eds.), Rabies (2nd ed.) (pp. 411–469). London: Elsevier. Copyright © 2007, Elsevier.*

nerves (Bryceson et al., 1975; Warrell et al., 1988). Antigen detection has also been performed on corneal impression smears, but the sensitivity of the method is low and false-positive results may occur (Anderson et al., 1984; Koch, Sagartz, Davidson, & Lawhaswasdi, 1975; Mathuranayagam & Rao, 1984; Noah et al., 1998; Warrell et al., 1988).

8.4.2.1 Detection of RABV RNA

Small amounts of RABV RNA from saliva, brain tissue, or CSF can be amplified using the reverse transcriptase polymerase chain reaction (RT-PCR), and this technique has proven to be a valuable diagnostic tool for rabies. RT-PCR was initially used on CSF specimens and subsequently on saliva to confirm a diagnosis of rabies (Crepin et al., 1998; Kamolvarin et al., 1993; McColl et al., 1993). Saliva for RT-PCR analysis can be collected with a sterile eyedropper pipette. In a study on both saliva and CSF samples from nine patients with confirmed rabies, the premortem diagnosis of rabies was confirmed by positive RT-PCR in five of nine patients (56%) in saliva and in only two of nine patients (22%) in CSF (Crepin et al., 1998). In comparison, skin biopsies were positive for RABV antigen using the fluorescent antibody technique in six of seven patients (86%). These findings led to a recommendation that both skin biopsy and saliva specimens be obtained for testing with immunofluorescence and RT-PCR, respectively. RABV RNA can also be detected in skin biopsies using RT-PCR (Dacheux et al., 2008). Of 20 human rabies cases diagnosed before death in the United States between 1980 and 1996, RABV RNA was detected in saliva from all 10 patients who had the test performed, including three who had negative viral isolation from saliva (Noah et al., 1998). It should be emphasized that negative tests for the detection of antigens or RNA never exclude a diagnosis of rabies, and, if clinical suspicion is high, then these tests should be repeated.

8.4.2.2 Brain tissues

The presence of Negri bodies in neurons is a pathologic hallmark of rabies observed on routine histologic staining, but these characteristic inclusion bodies in the cytoplasm of infected neurons (see Chapter 10) may be absent. The diagnosis of rabies in humans using brain biopsies has not been assessed adequately, but RABV antigen was detected in brain tissues obtained by biopsy from three of three cases in the United States from 1980 to 1996 (Noah et al., 1998). Postmortem brain tissue may be obtained by a needle (e.g., Vim-Silverman or trucut needle) aspiration technique through either the orbit or foramen magnum (Sow et al., 1996; Tong, Leung, & Lam, 1999; Warrell, 1996) and assessed for viral isolation, RABV antigen or RABV RNA, although false-negative results may occur. Hence, a full autopsy may not be required to confirm a diagnosis when rabies is clinically suspected but unconfirmed antemortem. Rabies may not be diagnosed until postmortem neuropathologic examination of the brain is performed because this diagnosis was not considered by the patient's physicians (Geyer et al., 1997; King et al., 1978; Munoz et al., 1996; Parker et al., 2003; Silverstein et al., 2003). A range of diagnostic investigations may be performed on postmortem human tissues, including virus isolation, the fluorescent antibody test [on fresh or formalin-fixed, paraffin-embedded (Whitfield et al., 2001) specimens], immunoperoxidase staining for RABV antigen or in situ hybridization for RABV RNA (Jackson & Wunner, 1991), or detection of RABV RNA by using reverse transcriptase polymerase chain reaction amplification (see Section 8.4.2.1).

8.5 Differential diagnosis

For patients and their relatives who are unable to recall an animal exposure, even when questioned directly, it may prove more difficult to make a diagnosis of rabies. Most cases without a clear history of rabies exposure are due to bat RABVs. There may be a history of recent travel in a rabies endemic area; dog bites are usually recognized, but may not receive appropriate medical attention. Rabies is most commonly misdiagnosed as either a psychiatric or laryngopharyngeal disorder. The disease may also present with bizarre neuropsychiatric symptoms mimicking conditions such as schizophrenic psychosis or acute mania (Goswami, Shankar, Channabasavanna, & Chattopadhyay, 1984).

Patients often become quite fearful about the possibility of developing rabies after an animal bite or exposure. Rabies hysteria is a conversion disorder (classified as a somatoform disorder) in which patients exhibit clinical features similar to rabies with unconscious motivation that involves poorly understood neural networks (Ron, 2001; Wilson et al., 1975), which should not be confused with malingering (feigning) in which there is deception by the patient. Rabies hysteria is probably the most difficult differential diagnosis. In general, it is characterized by a shorter incubation period (often a few hours or a day or two) than rabies, an early onset of inability of the patient to communicate, bizarre spasms, spitting out of water taken in the mouth with no actual attempt at swallowing, barking, biting, aggressive behavior directed toward health care workers, lack of fever and neurologic signs, and a long clinical course with recovery. Village practitioners in endemic areas may establish a reputation that they can cure rabies due to recovery of patients with rabies hysteria (Wilson et al., 1975). However, it should be emphasized that the clinical picture may be so bizarre in patients with rabies that they may be misdiagnosed as having hysteria (Bisseru, 1972).

Other viral encephalitides may show behavioral disturbances with fluctuations in the level of consciousness. However, hydrophobic spasms are not observed in these conditions and it is unusual for a conscious patient to have prominent brainstem signs in other encephalitides. Herpes *simiae* (B virus) encephalomyelitis, which is transmitted by monkey bites, is often associated with a shorter incubation period than rabies (e.g., 3–5 days); vesicles may be present at the site of the bite (also in the monkey's oral cavity), and recovery may occur (Whitley, 2014). Anti-N-methyl-D-aspartate receptor (anti-NMDA) encephalitis occurs in young patients (especially females) and is characterized by behavioral changes, autonomic instability, hypoventilation, and seizures, and it has recently been recognized that this autoimmune disease rivals viral etiologies as a cause of encephalitis (Gable, Sheriff, Dalmau, Tilley, & Glaser, 2012). This disease may clinically resemble rabies and immunotherapy can be very effective. Hence, recognition and appropriate diagnostic investigations are important. Two cases of rabies in the United States were misdiagnosed as Creutzfeldt-Jakob disease (Geyer et al., 1997) and both of these patients had a rapidly progressive neurological illness with prominent myoclonus.

Tetanus, a disease caused by the neurotoxin from the bacteria *Clostridium tetani*, may develop in association with a dirty wound caused by an animal bite. Tetanus has a shorter incubation period (usually 3–21 days) than rabies and, unlike rabies, it is characterized by sustained muscle rigidity involving axial muscles, including paraspinal, abdominal, masseter (trismus), laryngeal and respiratory muscles, with superimposed brief recurrent muscle spasms (Brook, 2014). In tetanus, the mental state is not affected, there is no CSF pleocytosis, and the prognosis is much better than in rabies.

Postvaccinal encephalomyelitis is another important differential diagnosis, particularly in patients who have been immunized with a vaccine derived from neural tissues (e.g., Semple vaccine). Postvaccinal encephalomyelitis usually develops within 2 weeks of initiation of vaccination, which is helpful in the differential diagnosis. Local sensory symptoms (paresthesias, pain, and pruritus), alternating intervals of agitation and lucidity and hydrophobia are clinical features that strongly suggest a diagnosis of rabies rather than postvaccinal encephalomyelits.

Paralytic rabies resembles the Guillain-Barré syndrome, including both acute inflammatory demyelinating polyradiculopathy and acute motor-sensory axonal neuropathy. In a recent pathologic series of the latter (Griffin et al., 1996), one case (case number 1 in the report) was subsequently demonstrated to have paralytic rabies (Sheikh et al., 2005). Local symptoms at the site of the bite, piloerection, early or persistent bladder dysfunction and fever are more suggestive of paralytic rabies. The Guillain-Barré syndrome may occasionally occur as a postvaccinal complication from rabies vaccines derived from neural tissues, particularly with the suckling mouse brain vaccine (Toro, Vergara, & Roman, 1977).

8.6 Rabies due to other Lyssavirus species

In addition to RABV, which is *Lyssavirus* species 1, there are six other Lyssavirus species and five have been associated with cases of human rabies: Mokola virus (species 3), Duvenhage virus (species 4), European bat lyssavirus 1 (species 5), European bat lyssavirus 2 (species 6), and Australian bat lyssavirus (species 7) (Table 8.3). In addition, one human case was reported due to Irkut virus infection (Leonova et al., 2009), which has not yet been designated a species. They are commonly called *rabies-like* or *rabies-related viruses*. Lagos bat virus (species 2), which was first isolated from fruit-eating bats in Nigeria, is the only species that has not been associated with human disease. Although not yet designated species, Aravan virus and Khujand virus are lyssaviruses isolated from bats in Central Asia, whereas Irkut virus and West Caucasian bat virus are lyssaviruses isolated from bats in Russia (Botvinkin et al., 2003; Kuzmin et al., 2003; Kuzmin, Hughes, Botvinkin, Orciari, & Rupprecht, 2005) and of these four lyssaviruses only Irkut virus has been reported to be associated with human disease (see Chapters 5 and 7).

8.6.1 Duvenhage virus

In 1970, a 31-year-old man from rural South Africa developed an illness with fever, excessive sweating, hydrophobia, and spasms of his face, arms, and torso that were precipitated by being touched (Meredith et al., 1971). He also exhibited confusion, irritability, and marked aggressiveness. He died after an illness lasting about 5 days. He lived outside the recognized enzootic and epizootic areas for rabies. He had been bitten on the lip by a bat while sleeping about 4 weeks earlier. The virus isolated from his brain was a new virus and was

TABLE 8.3 Reported human rabies cases due to other *Lyssavirus* species.

Virus (species)	Year	Location	Age of patient	References
Mokola (3)[a]	1968	Nigeria	3.5	Familusi and Moore (1972)
Mokola (3)	1971	Nigeria	6	Familusi, Osunkoya, Moore, Kemp, and Fabiyi (1972)
Duvenhage (4)	1970	South Africa	31	Meredith, Rossouw, and Van Praag Koch (1971)
Duvenhage (4)	2006	South Africa	77	Paweska et al. (2006)
Duvenhage (4)	2007	Kenya	34	van Thiel et al. (2009)
European bat lyssavirus 1 (5)	1985	Russia	11	Selimov et al. (1989)
European bat lyssavirus 2 (6)	1985	Finland	30	Roine et al. (1988)
European bat lyssavirus 2 (6)	2002	Scotland	55	Johnson et al. (2012) and Nathwani et al. (2003)
Australian bat lyssavirus (7)	1996	Australia	39	Samaratunga, Searle, and Hudson (1998)
Australian bat lyssavirus (7)	1998	Australia	37	Hanna et al. (2000)
Australian bat lyssavirus (7)	2013	Australia	8	Francis et al. (2014)
Irkut (pending)	2007	Russia	20	Leonova et al. (2009)

[a] *It is doubtful that this patient's clinical picture was actually caused by Mokola virus infection.*

characterized and named *Duvenhage virus* (species 4). This patient's clinical illness was indistinguishable from that caused by RABV (species 1).

In February, 2006 a 77-year-old male was scratched on the cheek by an insectivorous bat in North West Province, South Africa (Paweska et al., 2006). He did not receive postexposure treatment and he became ill 27 days later with an influenza-like illness and hallucinations. On his third day of illness he was admitted to hospital with fever, generalized rigidity, and involuntary grimacing, and the next day he had generalized seizures. He died on day 14 of his illness. Duvenhage virus was identified by RT-PCR on antemortem saliva and postmortem brain tissue and confirmed with sequencing of the nucleoprotein amplicons.

In 2007 a bat flew in the face of a 34-year-old Dutch female in Kenya; she sustained two small superficial wounds on the right side of her nose and she did not receive postexposure rabies prophylaxis (van Thiel et al., 2009). Twenty-three days later she experienced malaise, dizziness, muscle aches, and headache, and two days later she had difficulty with speech and swallowing. She later had fever, hypersalivation, limb weakness, and a seizure. She died 45 days after the bat exposure (on day 20 of her hospital admission), after receiving aggressive medical interventions. Duvenhage virus was confirmed antemortem by PCR studies on a nuchal skin biopsy.

8.6.2 Mokola virus

Mokola virus was first isolated from shrews in Nigeria (Shope et al., 1970). In 1968, a 3½-year-old girl from Nigeria presented with a sudden onset of fever and convulsions (Familusi & Moore, 1972). She rapidly made a complete recovery. There were no cells in her CSF, and the CSF protein and glucose were normal. Mokola virus was isolated from her CSF, although the shrew isolate of Mokola virus was handled in the same laboratory during the same time period. Cross-contamination of specimens in the laboratory remains a possible explanation for this viral isolation. The patient's neutralizing antibody titers were very low and disappeared within several months. The febrile convulsion was unlikely related to Mokola virus infection.

A 6-year-old girl died in Nigeria in 1971 after a 6-day illness (Familusi et al., 1972). She presented with drowsiness, confusion, and weakness involving her extremities and trunk and progressed to coma. Her CSF was normal without a pleocytosis. At autopsy, there were large eosinophilic inclusion bodies in the cytoplasm of neurons and Mokola virus was isolated from her brain. Shrews were known to be plentiful around the house where she lived, although there was no documented evidence that she had actually been bitten. Mokola virus infection was associated with meningoencephalitis in this case without the typical features of brainstem involvement seen in encephalitic rabies.

8.6.3 European bat lyssavirus 1

In 1985, an 11-year-old girl from Belgorod, Russia, was bitten on the lower lip by an unidentified bat and died with signs of rabies (Selimov et al., 1989). The viral isolate was called *Yuli virus* and classified as European bat lyssavirus type 1 (species 5) (Bourhy, Kissi, Lafon, Sacramento, & Tordo, 1992). There was an earlier fatal case in a 15-year-old female in Voroshilovgrad (now Lugansk), Ukraine, in 1977 that developed after a bat bite (Anonymous, 1986). However, no viral isolate is available for molecular characterization, and it is uncertain if the infection was actually caused by European bat lyssavirus type 1 or another lyssavirus. A similar situation applies in the case of a 34-year-old male who died with clinical rabies (with hypersalivation and hydrophobia) 45 days after a bit bite in the Lugansk province, Ukraine, in 2002, which is only 50 km from the site of the 1977 case (Botvinkin et al., 2005).

8.6.4 European bat lyssavirus 2

In 1985, a 30-year-old zoologist from Finland developed numbness in his right arm and neck with leg weakness (Roine et al., 1988). His CSF was normal without a pleocytosis. Subsequently, he developed myoclonus of his legs, agitation, hyperexcitability, inspiratory spasms, dysarthria, dysphagia, and hypersalivation. He had a delirium that progressed to coma. Diabetes insipidus occurred and he died 23 days after the onset of the illness.

He had never been vaccinated against rabies and had been bitten by bats in several countries over the prior 5-year period, including an exposure in southern Finland 51 days prior to the onset of his symptoms. A virus was isolated that resembled the enzootic European bat RABV isolates and it was classified as European bat lyssavirus type 2 (species 6) (Bourhy et al., 1992). This patient also had a clinical illness that was indistinguishable from rabies associated with species 1.

In 2002 a 55-year-old bat conservationist presented with hematemesis and a 5-day history of arm paresthesias, left arm and shoulder pain, and difficulty swallowing (Fooks et al., 2003; Nathwani et al., 2003). About 19 weeks prior, he had been bitten on his left finger by a Daubenton's bat in Angus, Scotland, although he also had a remote history of other bat bites. On admission to hospital in Dundee, Scotland, he was febrile and had gaze-evoked nystagmus, dysarthria, truncal, limb, and gait ataxia, areflexia in the arms and hyperreflexia in the legs, and his behavior was inappropriately familiar. CT and MRI scans did not show significant abnormalities. CSF showed a normal cell count and a mildly elevated CSF protein (58 mg/dL). He was treated with intravenous immunoglobulin and also subsequently with high-dose methylprednisolone and cyclophosphamide (Fooks et al., 2003). On day 5 of hospitalization he became acutely confused, agitated, and aggressive. He was sedated and a repeat CSF examination showed a pleocytosis and CSF protein elevated at 1.09 mg/dL. His mental state subsequently deteriorated, and his limbs became flaccid; he died on day 14 of hospitalization. Heminested RT-PCR was positive for lyssavirus RNA in saliva obtained on day 9 of hospitalization, with high homology with previous EBLV type 2a isolates obtained from bats in the United Kingdom. No RABV antibodies were detected in sera or CSF, and virus could not be isolated from saliva, skin biopsies, or CSF. The lyssavirus was cultured from postmortem brain tissue. This was the first case of indigenous human rabies in the United Kingdom in 100 years and this case has had important public health implications for bat exposures in the region (Shipley et al., 2019).

8.6.5 Australian bat lyssavirus

In 1996, a 39-year-old female from Australia died after a 20-day illness (Samaratunga et al., 1998). She cared for fruit bats and had sustained numerous scratches to her left arm over 4 weeks prior to the onset of her illness, and she was likely bitten by a yellow-bellied sheathtail bat, *Saccolaimus flaviventris* (an insectivorous bat), in her care (Hanna et al., 2000). She developed progressive left arm weakness. Her CSF showed 100 white cells/μL (80% mononuclear cells and 20% polymorphonuclear leukocytes). She deteriorated with diplopia, dysarthria, dysphagia, and ataxia. She later developed progressive limb and facial weakness with reduced deep tendon reflexes and fluctuations in her level of consciousness prior to her death. Small eosinophilic cytoplasmic inclusions were observed in neurons in gray matter areas. RT-PCR amplification of RNA extracted from brain tissue and CSF indicated that she was infected with a virus identical to Australian bat lyssavirus (species 7) that had been identified previously in flying foxes, which are fruit-eating bats. This patient had typical brainstem involvement of rabies, which quickly progressed to diffuse brain involvement.

In 1998, a 37-year-old woman from Mackay, Queensland, was admitted to hospital with a 5-day history of fever, paresthesias around the dorsum of her left hand, pain about the left shoulder girdle, and sore throat with difficulty swallowing (Hanna et al., 2000). There were pharyngeal spasms, evidence of autonomic instability, and progressive neurologic deterioration. She died 19 days after the onset of the illness. Twenty-seven months prior to the onset of her illness (during 1996) she was bitten at the base of her left little finger by a flying fox (fruit bat) in the course of removing the bat from the back of a young child. She did not receive rabies postexposure prophylaxis. Heminested PCR analyses on multiple tissues and saliva were positive for the flying-fox (*Pteropus* spp.) variant of Australian bat lyssavirus (Hanna et al., 2000). Although *Lyssavirus* infections of flying foxes have only recently been recognized in Australia, RABV infection was recognized in a gray-head flying fox (*Pteropus poliocephalus*) that died in India in 1978 (Pal et al., 1980) and also in two dog-faced fruit bats (*Cyanopterus brachyotis*) from Thailand (Smith, Lawhaswasdi, Vick, & Stanton, 1967). It is unclear exactly when and how Australian bat lyssavirus obtained its foothold in Australian frugivorous and insectivorous bats, but it is clear that the virus poses a threat to human health in this region.

In 2013, an 8-year-boy from Australia was scratched by a bat on his left forearm (Francis et al., 2014). About 8 weeks later he presented with a 2-day history of fever, anorexia, and worsening abdominal pain. He subsequently developed prominent behavioral outbursts with aggression. He developed spasms of his legs, abdomen, and face associated with pupillary dilatation, tachycardia, and hypertension. He required endotracheal intubation. CSF analysis was normal without pleocytosis. MR brain scan was normal other than some small foci within the periventricular white matter. He developed severe dysautonomia, including sinoatrial node dysfunction resulting in sinus arrest. He had no spontaneous movements or reflexes and his pupils became fixed and dilated. His electroencephalogram became isoelectric in the absence of sedatives and he died on day 26 of his hospital admission. Lyssavirus RNA was detected in saliva and CSF and serum lyssavirus antibodies were also detected.

8.6.6 Irkut virus

In 2007, a 20-year-old female died of rabies in the Primorye Territory, which is in the Russian Far East (Leonova et al., 2009). On August 10, 2007, she was bitten on the lip by a bat of unknown species, and did not receive any postexposure therapy. A month later on September 10th she developed fever, headache, vomiting, diplopia, and head and hand tremor. She was admitted to hospital on the next day with bulbar symptoms. She progressed into a deep stupor with flaccid paresis and died after 11 days. Molecular characterization of a virus isolated from her brain identified it as a bat lyssavirus called Irkut virus, which had previously been isolated from a dead greater tubenosed bat (*Murina leucogaster*) in Irkutsk (Botvinkin et al., 2003).

8.7 Conclusions

The clinical features of rabies develop after a long incubation period, typically lasting 20–90 days, and a few days of prodromal symptoms. Encephalitic rabies occurs in 80% of cases;

paralytic rabies in 20%. Hydrophobia is the distinctive feature of encephalitic rabies. Diagnostic laboratory tests may confirm a clinical suspicion of rabies. Other non-RABV lyssaviruses may cause a clinical illness indistinguishable from rabies.

References

Al Mousawi, M. S. A. (2015). Transmission of rabies virus from a deceased organ donor to six recipients in Kuwait and Saudi Arabia. In *Presented at the 2015 Organ Donation Congress in Seoul, South Korea on October 18, 2015*.

Alvarez, L., Fajardo, R., Lopez, E., Pedroza, R., Hemachudha, T., Kamolvarin, N., … Baer, G. M. (1994). Partial recovery from rabies in a nine-year-old boy. *The Pediatric Infectious Disease Journal, 13*, 1154–1155.

Anderson, L. J., Nicholson, K. G., Tauxe, R. V., & Winkler, W. G. (1984). Human rabies in the United States, 1960 to 1979: Epidemiology, diagnosis, and prevention. *Annals of Internal Medicine, 100*, 728–735.

Anonymous. (1986). Bat rabies in the Union of Soviet Socialist Republics. *Rabies Bulletin Europe, 10*, 12–14.

Asbury, A. K., & Cornblath, D. R. (1990). Assessment of current diagnostic criteria for Guillain-Barre syndrome. *Annals of Neurology, 27*(Suppl), S21–S24.

Awasthi, M., Parmar, H., Patankar, T., & Castillo, M. (2001). Imaging findings in rabies encephalitis. *American Journal of Neuroradiology, 22*(4), 677–680.

Bek, M. D., Smith, W. T., Levy, M. H., Sullivan, E., & Rubin, G. L. (1992). Rabies case in New South Wales, 1990: Public health aspects. *Medical Journal of Australia, 156*, 596–600.

Bhandari, M., & Kumar, S. (1986). Penile hyperexcitability as the presenting symptom of rabies. *British Journal of Urology, 58*, 224–233.

Bhatt, D. R., Hattwick, M. A. W., Gerdsen, R., Emmons, R. W., & Johnson, H. N. (1974). Human rabies: Diagnosis, complications, and management. *American Journal of Diseases of Children, 127*, 862–869.

Bisseru, B. (1972). Human rabies. In B. Bisseru (Ed.), *Rabies* (pp. 385–453). London: William Heinemann Medical Books.

Black, D., & Wiktor, T. J. (1986). Survey of raccoon hunters for rabies antibody titers: Pilot study. *Journal of the Florida Medical Association, 73*, 517–520.

Botvinkin, A. D., Poleschuk, E. M., Kuzmin, I. V., Borisova, T. I., Gazaryan, S. V., Yager, P., & Rupprecht, C. E. (2003). Novel lyssaviruses isolated from bats in Russia. *Emerging Infectious Diseases, 9*(12), 1623–1625.

Botvinkin, A. D., Selnikova, O. P., Antonova, L. A., Moiseeva, A. B., Nesterenko, E. Y., & Gromashevsky, L. V. (2005). Human rabies case caused from a bat bite in Ukraine. *Rabies Bulletin Europe, 29*(3), 5–7.

Bourhy, H., Kissi, B., Lafon, M., Sacramento, D., & Tordo, N. (1992). Antigenic and molecular characterization of bat rabies virus in Europe. *Journal of Clinical Microbiology, 30*(9), 2419–2426.

Brook, I. (2014). Tetanus. In W. M. Scheld, R. J. Whitley, & C. Marra (Eds.), *Infections of the central nervous system* (4th ed., pp. 634–658). Philadelphia: Lippincott Williams & Wilkins.

Bryceson, A. D. M., Greenwood, B. M., Warrell, D. A., Davidson, N. M., Pope, H. M., Lawrie, J. H., … Wilcox, G. E. (1975). Demonstration during life of rabies antigen in humans. *Journal of Infectious Diseases, 131*(1), 71–74.

Burton, E. C., Burns, D. K., Opatowsky, M. J., El-Feky, W. H., Fischbach, B., Melton, L., … Klintmalm, G. (2005). Rabies encephalomyelitis: Clinical, neuroradiological, and pathological findings in 4 transplant recipients. *Archives of Neurology, 62*(6), 873–882. Retrieved from: http://www.medscape.com/.

Cheetham, H. D., Hart, J., Coghill, N. F., & Fox, B. (1970). Rabies with myocarditis: Two cases in England. *Lancet, 1*, 921–922.

Chen, J., Liu, G., Jin, T., Zhang, R., Ou, X., Zhang, H., … Zhang, R. (2018). Epidemiological and genetic characteristics of rabies virus transmitted through organ transplantation. *Frontiers in Cellular and Infection Microbiology, 8*, 86.

Chen, S., Zhang, H., Luo, M., Chen, J., Yao, D., Chen, F., … Chen, T. (2017). Rabies virus transmission in solid organ transplantation, China, 2015-2016. *Emerging Infectious Diseases, 23*(9), 1600–1602.

Chopra, J. S., Banerjee, A. K., Murthy, J. M. K., & Pal, S. R. (1980). Paralytic rabies: A clinico-pathological study. *Brain, 103*, 789–802.

Chotmongkol, V., Vuttivirojana, A., & Cheepblangchai, M. (1991). Unusual manifestation in paralytic rabies. *Southeast Asian Journal of Tropical Medicine and Public Health, 22*, 279–280.

Constantine, D. G. (1962). Rabies transmission by nonbite route. *Public Health Reports, 77*, 287–289.

Crepin, P., Audry, L., Rotivel, Y., Gacoin, A., Caroff, C., & Bourhy, H. (1998). Intravitam diagnosis of human rabies by PCR using saliva and cerebrospinal fluid. *Journal of Clinical Microbiology, 36*(4), 1117–1121.

da Rosa, E. S. T., Kotait, I., Barbosa, T. F. S., Carrier, M. L., Brandão, P. E., Pinheiro, A. S., ... Vasconcelos, P. F. C. (2006). Bat-transmitted human rabies outbreaks, Brazilian Amazon. *Emerging Infectious Diseases, 12*(8), 1197–1202.

Dacheux, L., Reynes, J. M., Buchy, P., Sivuth, O., Diop, B. M., Rousset, D., ... Bourhy, H. (2008). A reliable diagnosis of human rabies based on analysis of skin biopsy specimens. *Clinical Infectious Diseases, 47*(11), 1410–1417.

Doege, T. C., & Northrop, R. L. (1974). Evidence for inapparent rabies infection. *Lancet, 2*, 826–829.

Dupont, J. R., & Earle, K. M. (1965). Human rabies encephalitis. A study of forty-nine fatal cases with a review of the literature. *Neurology, 15*, 1023–1034.

Dutta, J. K. (1996). Excessive libido in a woman with rabies. *Postgraduate Medical Journal, 72*, 554.

Editorial. (1975). Diagnosis and management of human rabies. *British Medical Journal, 3*(5986), 721–722.

Editorial. (1978). Dumb rabies. *Lancet, 2*(8098), 1031–1032.

Elsiesy, H., Hussain, I., Abaalkhail, F. A., Al-Hamoudi, W. K., Al Sebayel, M. I., Broering, D. C., et al. (2015). Rabies outbreak involving four transplant recipients in Kuwait and Saudi Arabia. In *Platform presentation at the XXVIth international meeting on research advances and rabies control in the Americas in Fort Collins, Colorado, USA on October 5, 2015*.

Emmons, R. W., Leonard, L. L., DeGenaro, F., Jr., Protas, E. S., Bazeley, P. L., Giammona, S. T., & Sturckow, K. (1973). A case of human rabies with prolonged survival. *Intervirology, 1*(1), 60–72.

Familusi, J. B., & Moore, D. L. (1972). Isolation of a rabies related virus from the cerebrospinal fluid of a child with "aseptic meningitis". *The African Journal of Medical Sciences, 3*, 93–96.

Familusi, J. B., Osunkoya, B. O., Moore, D. L., Kemp, G. E., & Fabiyi, A. (1972). A fatal human infection with Mokola virus. *American Journal of Tropical Medicine and Hygiene, 21*, 959–963.

Faoagali, J. L., De Buse, P., Strutton, G. M., & Samaratunga, H. (1988). A case of rabies. *Medical Journal of Australia, 149*, 702–707.

Feasby, T. E., Gilbert, J. J., Brown, W. F., Bolton, C. F., Hahn, A. F., Koopman, W. F., & Zochodne, D. W. (1986). An acute axonal form of Guillain-Barre polyneuropathy. *Brain, 109*, 1115–1126.

Fekadu, M., Endeshaw, T., Alemu, W., Bogale, Y., Teshager, T., & Olson, J. G. (1996). Possible human-to-human transmission of rabies in Ethiopia. *Ethiopian Medical Journal, 34*(2), 123–127.

Fernandes, E. R., de Andrade, H. F. J., Lancellotti, C. L., Quaresma, J. A., Demachki, S., da Costa Vasconcelos, P. F., & Duarte, M. I. (2011). In situ apoptosis of adaptive immune cells and the cellular escape of rabies virus in CNS from patients with human rabies transmitted by *Desmodus rotundus*. *Virus Research, 156*(1–2), 121–126.

Fooks, A. R., Banyard, A. C., Horton, D. L., Johnson, N., McElhinney, L., & Jackson, A. C. (2014). Current status of rabies and prospects for elimination. *Lancet, 384*(9951), 1389–1399.

Fooks, A. R., McElhinney, L. M., Pounder, D. J., Finnegan, C. J., Mansfield, K., Johnson, N., ... Nathwani, D. (2003). Case report: Isolation of a European bat lyssavirus type 2a from a fatal human case of rabies encephalitis. *Journal of Medical Virology, 71*(2), 281–289.

Francis, J. R., Nourse, C., Vaska, V. L., Calvert, S., Northill, J. A., McCall, B., & Mattke, A. C. (2014). Australian Bat Lyssavirus in a child: The first reported case. *Pediatrics, 133*(4), e1063–e1067.

Gable, M. S., Sheriff, H., Dalmau, J., Tilley, D. H., & Glaser, C. A. (2012). The frequency of autoimmune N-methyl-D-aspartate receptor encephalitis surpasses that of individual viral etiologies in young individuals enrolled in the California Encephalitis Project. *Clinical Infectious Diseases, 54*(7), 899–904.

Galian, A., Guerin, J. M., Lamotte, M., Le Charpentier, Y., Mikol, J., Dureaux, J. B., ... Sureau, P. (1980). Human-to-human transmission of rabies via a corneal transplant—France. *Morbidity and Mortality Weekly Report, 29*, 25–26.

Gamaleia, N. (1887). Etude sur la rage paralytique chez l'homme. *Annales de L'Institut Pasteur (Paris), 1*, 63–83.

Gardner, A. M. N. (1970). An unusual case of rabies (letter). *Lancet, 2*, 523.

Geyer, R., Van Leuven, M., Murphy, J., Damrow, T., Sastry, L., Miller, S., ... Stehr-Green, P. A. (1997). Human rabies—Montana and Washington, 1997. *Morbidity and Mortality Weekly Report, 46*(33), 770–774.

Gilbert, A. T., Peterson, B. W., Recuenco, S., Niezgoda, M., Gómez, J., Laguna-Torres, V. A., & Rupprecht, C. (2012). Evidence of rabies virus exposure among humans in the Peruvian Amazon. *American Journal of Tropical Medicine and Hygiene, 87*(2), 206–215.

Gode, G. R., & Bhide, N. K. (1988). Two rabies deaths after corneal grafts from one donor (Letter). *Lancet, 2*, 791.

Gong, C., Li, X., Luo, M., Zhang, Z., Wang, Q., Wang, Q., ... Wu, J. (2017). Laboratory investigation of the rabies transmission through organ transplantation in China (letter). *Journal of Infection, 74*(417), 427–431.

Goswami, U., Shankar, S. K., Channabasavanna, S. M., & Chattopadhyay, A. (1984). Psychiatric presentations in rabies: A clinico-pathologic report from south India with a review of literature. *Tropical and Geographical Medicine, 36*(1), 77–81.

Gourmelon, P., Briet, D., Clarencon, D., Court, L., & Tsiang, H. (1991). Sleep alterations in experimental street rabies virus infection occur in the absence of major EEG abnormalities. *Brain Research, 554,* 159–165.

Griffin, J. W., Li, C. Y., Ho, T. W., Tian, M., Gao, C. Y., Xue, P., … Asbury, A. K. (1996). Pathology of the motor-sensory axonal Guillain-Barre syndrome. *Annals of Neurology, 39,* 17–28.

Hampson, K., Coudeville, L., Lembo, T., Sambo, M., Kieffer, A., Attlan, M., … Dushoff, J. (2015). Estimating the global burden of endemic canine rabies. *PLoS Neglected Tropical Diseases, 9*(4), e0003709.

Hanna, J. N., Carney, I. K., Smith, G. A., Tannenberg, A. E. G., Deverill, J. E., Botha, J. A., … Searle, J. W. (2000). Australian bat lyssavius infection: A second human case, with a long incubation period. *Medical Journal of Australia, 172*(12), 597–599.

Hantson, P., Guerit, J. M., de Tourtchaninoff, M., Deconinck, B., Mahieu, P., Dooms, G., … Brucher, J. M. (1993). Rabies encephalitis mimicking the electrophysiological pattern of brain death. A case report. *European Neurology, 33,* 212–217.

Hattwick, M. A. W. (1974). Human rabies. *Public Health Reviews, 3,* 229–274.

Hemachudha, T. (1994). Human rabies: Clinical aspects, pathogenesis, and potential therapy. In C. E. Rupprecht, B. Dietzschold, & H. Koprowski (Eds.), *Current topics in microbiology and immunology. Lyssaviruses* (pp. 121–143). Berlin: Springer-Verlag, pp. 121–143

Hemachudha, T. (1997). Rabies. In K. L. Roos (Ed.), *Central nervous system infectious diseases and therapy* (pp. 573–600). New York: Marcel Dekker.

Hemachudha, T., Panpanich, T., Phanuphak, P., Manatsathit, S., & Wilde, H. (1993). Immune activation in human rabies. *Transactions of the Royal Society of Tropical Medicine and Hygiene, 87,* 106–108.

Hemachudha, T., Phanthumchinda, K., Phanuphak, P., & Manutsathit, S. (1987). Myoedema as a clinical sign in paralytic rabies (Letter). *Lancet, 1,* 1210.

Hemachudha, T., Phanuphak, P., Sriwanthana, B., Manutsathit, S., Phanthumchinda, K., Siriprasomsup, W., … Kaoroptham, S. (1988). Immunologic study of human encephalitic and paralytic rabies: Preliminary report of 16 patients. *American Journal of Medicine, 84,* 673–677.

Hemachudha, T., Sunsaneewitayakul, B., Mitrabhakdi, E., Suankratay, C., Laothamathas, J., Wacharapluesadee, S., … Wilde, H. (2003). Paralytic complications following intravenous rabies immune globulin treatment in a patient with furious rabies (letter). *International Journal of Infectious Diseases, 7*(1), 76–77.

Hemachudha, T., Wacharapluesadee, S., Mitrabhakdi, E., Morimoto, K., & Lewis, R. A. (2005). Pathophysiology of human paralytic rabies. *Journal of Neurovirology, 11*(1), 93–100.

Hornung, K., & Nix, W. A. (1992). Myoedema. A clinical and electrophysiological evaluation. *European Neurology, 32*(3), 130–133.

Houff, S. A., Burton, R. C., Wilson, R. W., Henson, T. E., London, W. T., Baer, G. M., … Sever, J. L. (1979). Human-to-human transmission of rabies virus by corneal transplant. *New England Journal of Medicine, 300,* 603–604.

Hurst, E. W., & Pawan, J. L. (1931). An outbreak of rabies in Trinidad without history of bites, and with the symptoms of acute ascending myelitis. *Lancet, 2,* 622–628.

Jackson, A. C. (2011). Update on rabies. *Research and Reports in Tropical Medicine, 2,* 31–43.

Jackson, A. C., & Del Bigio, M. R. (2018). Reader response: Rabies encephalitis presenting with new-onset refractory status epilepticus (letter). *Neurology Clinical Practice, 8*(5), 370–371.

Jackson, A. C., & Fenton, M. B. (2001). Human rabies and bat bites (letter). *Lancet, 357*(9269), 1714.

Jackson, A. C., & Wunner, W. H. (1991). Detection of rabies virus genomic RNA and mRNA in mouse and human brains by using in situ hybridization. *Journal of Virology, 65*(6), 2839–2844.

Jackson, A. C., Ye, H., Phelan, C. C., Ridaura-Sanz, C., Zheng, Q., Li, Z., … Lopez-Corella, E. (1999). Extraneural organ involvement in human rabies. *Laboratory Investigation, 79*(8), 945–951.

Javadi, M. A., Fayaz, A., Mirdehghan, S. A., & Ainollahi, B. (1996). Transmission of rabies by corneal graft. *Cornea, 15*(4), 431–433.

Johnson, N., Brookes, S. M., Fooks, A. R., & Ross, R. S. (2005). Review of human rabies cases in the UK and in Germany. *Veterinary Record, 157*(22), 715.

Johnson, N., Brookes, S. M., Healy, D. M., Spencer, Y., Hicks, D., Nunez, A., … Fooks, A. R. (2012). Pathology associated with a human case of rabies in the United Kingdom caused by European bat lyssavirus type-2. *Intervirology, 55*(5), 391–394.

Kamolvarin, N., Tirawatnpong, T., Rattanasiwamoke, R., Tirawatnpong, S., Panpanich, T., & Hemachudha, T. (1993). Diagnosis of rabies by polymerase chain reaction with nested primers. *Journal of Infectious Diseases, 167,* 207–210.

Kasempimolporn, S., Hemachudha, T., Khawplod, P., & Manatsathit, S. (1991). Human immune response to rabies nucleocapsid and glycoprotein antigens. *Clinical and Experimental Immunology*, *84*(2), 195–199.

King, D. B., Sangalang, V. E., Manuel, R., Marrie, T., Pointer, A. E., & Thomson, A. D. (1978). A suspected case of human rabies - Nova Scotia. *Canadian Diseases Weekly Report*, *4*, 49–51.

Koch, F. J., Sagartz, J. W., Davidson, D. E., & Lawhaswasdi, K. (1975). Diagnosis of human rabies by the cornea test. *American Journal of Clinical Pathology*, *63*, 509–515.

Komsuoglu, S. S., Dora, F., & Kalabay, O. (1981). Periodic EEG activity in human rabies encephalitis (letter). *Journal of Neurology, Neurosurgery and Psychiatry*, *44*, 264–265.

Krebs, J. W., Mandel, E. J., Swerdlow, D. L., & Rupprecht, C. E. (2005). Rabies surveillance in the United States during 2004. *Journal of the American Veterinary Medical Association*, *227*(12), 1912–1925.

Kureishi, A., Xu, L. Z., Wu, H., & Stiver, H. G. (1992). Rabies in China: Recommendations for control. *Bulletin of the World Health Organization*, *70*, 443–450.

Kuzmin, I. V., Hughes, G. J., Botvinkin, A. D., Orciari, L. A., & Rupprecht, C. E. (2005). Phylogenetic relationships of Irkut and West Caucasian bat viruses within the Lyssavirus genus and suggested quantitative criteria based on the N gene sequence for lyssavirus genotype definition. *Virus Research*, *111*(1), 28–43.

Kuzmin, I. V., Orciari, L. A., Arai, Y. T., Smith, J. S., Hanlon, C. A., Kameoka, Y., & Rupprecht, C. E. (2003). Bat lyssaviruses (Aravan and Khujand) from Central Asia: Phylogenetic relationships according to N, P and G gene sequences. *Virus Research*, *97*(2), 65–79.

Laothamatas, J., Hemachudha, T., Mitrabhakdi, E., Wannakrairot, P., & Tulayadaechanont, S. (2003). MR imaging in human rabies. *American Journal of Neuroradiology*, *24*(6), 1102–1109.

Laothamatas, J., Hemachudha, T., Tulyadechanont, S., & Mitrabhakdi, E. (1997). Neuroimaging in paralytic rabies. *Ramathibodi Medical Journal*, *20*(3), 149–156.

Laothamatas, J., Sungkarat, W., & Hemachudha, T. (2011). Neuroimaging in rabies. *Advances in Virus Research*, *79*, 309–327.

Leonova, G. N., Belikov, S. I., Kondratov, I. G., Krylova, N. V., Pavlenko, E. V., Romanova, E. V., ... Petukhova, S. A. (2009). A fatal case of bat lyssavirus infection in Primorye Territory of the Russian Far East. *Rabies Bulletin Europe*, *33*(4), 5–8.

Lopez, R. A., Miranda, P. P., Tejada, V. E., & Fishbein, D. B. (1992). Outbreak of human rabies in the Peruvian jungle. *Lancet*, *339*(8790), 408–411.

Lu, X. X., Zhu, W. Y., & Wu, G. Z. (2018). Rabies virus transmission via solid organs or tissue allotransplantation. *Infectious Diseases of Poverty*, *7*(1), 82–0467.

Madhusudana, S. N., Nagaraj, D., Uday, M., Ratnavalli, E., & Kumar, M. V. (2002). Partial recovery from rabies in a six-year-old girl (letter). *International Journal of Infectious Diseases*, *6*(1), 85–86.

Maier, T., Schwarting, A., Mauer, D., Ross, R. S., Martens, A., Kliem, V., ... Drosten, C. (2010). Management and outcomes after multiple corneal and solid organ transplantations from a donor infected with rabies virus. *Clinical Infectious Diseases*, *50*(8), 1112–1119.

Masthi, N. R., Raviprakash, D., Gangasagara, S. B., Sriprakash, K. S., Ashwin, B. Y., Ullas, P. T., & Madhusudhana, S. N. (2012). Rabies in a blind patient: Confusion after corneal transplantation. *National Medical Journal of India*, *25*(2), 83–84.

Mathuranayagam, D., & Rao, P. V. (1984). Antemortem diagnosis of human rabies by corneal impression smears using immunofluorescent technique. *Indian Journal of Medical Research*, *79*, 463–467.

McColl, K. A., Gould, A. R., Selleck, P. W., Hooper, P. T., Westbury, H. A., & Smith, J. S. (1993). Polymerase chain reaction and other laboratory techniques in the diagnosis of long incubation rabies in Australia. *Australian Veterinary Journal*, *70*(3), 84–89.

Meredith, C. D., Rossouw, A. P., & Van Praag Koch, H. (1971). An unusual case of human rabies thought to be of chiropteran origin. *South African Medical Journal*, *45*(28), 767–769.

Metze, K., & Feiden, W. (1991). Rabies virus ribonucleoprotein in the heart (letter). *New England Journal of Medicine*, *324*, 1814–1815.

Mills, R. P., Swanepoel, R., Hayes, M. M., & Gelfand, M. (1978). Dumb rabies: Its development following vaccination in a subject with rabies. *Central African Journal of Medicine*, *24*(6), 115–117.

Mitrabhakdi, E., Shuangshoti, S., Wannakrairot, P., Lewis, R. A., Susuki, K., Laothamatas, J., & Hemachudha, T. (2005). Difference in neuropathogenetic mechanisms in human furious and paralytic rabies. *Journal of the Neurological Sciences*, *238*(1–2), 3–10.

Mrak, R. E., & Young, L. (1993). Rabies encephalitis in a patient with no history of exposure. *Human Pathology, 24*, 109–110.

Munoz, J. L., Wolff, R., Jain, A., Sabino, J., Jacquette, G., Rapoport, M., ... Morse, D. L. (1996). Human rabies—Connecticut, 1995. *Morbidity and Mortality Weekly Report, 45*(10), 207–209.

Nathwani, D., McIntyre, P. G., White, K., Shearer, A. J., Reynolds, N., Walker, D., ... Fooks, A. R. (2003). Fatal human rabies caused by European bat lyssavirus type 2a infection in Scotland. *Clinical Infectious Diseases, 37*(4), 598–601.

Nehaul, B. B. G. (1955). Rabies transmitted by bats in British Guiana. *American Journal of Tropical Medicine and Hygiene, 4*, 550–553.

Nicholson, K. G. (1994). Human rabies. In R. R. McKendall & W. G. Stroop (Eds.), *Neurological disease and therapy Handbook of neurovirology* (pp. 463–480). New York: Marcel Dekker.

Noah, D. L., Drenzek, C. L., Smith, J. S., Krebs, J. W., Orciari, L., Shaddock, J., ... Childs, J. E. (1998). Epidemiology of human rabies in the United States, 1980 to 1996. *Annals of Internal Medicine, 128*(11), 922–930.

Orr, P. H., Rubin, M. R., & Aoki, F. Y. (1988). Naturally acquired serum rabies neutralizing antibody in a Canadian Inuit population. *Arctic Medical Research, 47*(Suppl. 1), 699–700.

Pal, S. R., Arora, B., Chhuttani, P. N., Broor, S., Choudhury, S., Joshi, R. M., & Ray, S. D. (1980). Rabies virus infection of a flying fox bat, *Pteropus poliocephalus* in Chandigarh, Northern India. *Tropical and Geographical Medicine, 32*(3), 265–267.

Park, S. C., Crane, I. M., Pal, K., & Cagnina, R. E. (2019). Rabies encephalitis with myocarditis mimicking ST-elevation myocardial infarction. *Open Forum Infectious Diseases, 6*(6). ofz260.

Parker, R., McKay, D., Hawes, C., Daly, P., Bryce, E., Doyle, P., ... Naus, M. (2003). Human rabies, British Columbia—January 2003. *Canada Communicable Disease Report, 29*(16), 137–138.

Pawan, J. L. (1939). Paralysis as a clinical manifestation in human rabies. *Annals of Tropical Medicine and Parasitology, 33*, 21–29.

Paweska, J. T., Blumberg, L. H., Liebenberg, C., Hewlett, R. H., Grobbelaar, A. A., Leman, P. A., ... Swanepoel, R. (2006). Fatal human infection with rabies-related Duvenhage virus, South Africa. *Emerging Infectious Diseases, 12*(12), 1965–1967.

Pearson, C. A. (1976). Rabies (letter). *Lancet, 1*, 206.

Phuapradit, P., Manatsathit, S., Warrell, M. J., & Warrell, D. A. (1985). Paralytic rabies: Some unusual clinical presentations. *Journal of the Medical Association of Thailand, 68*, 106–110.

Pleasure, S. J., & Fischbein, N. J. (2000). Correlation of clinical and neuroimaging findings in a case of rabies encephalitis. *Archives of Neurology, 57*(12), 1765–1769.

Porras, C., Barboza, J. J., Fuenzalida, E., Adaros, H. L., Oviedo, A. M., & Furst, J. (1976). Recovery from rabies in man. *Annals of Internal Medicine, 85*, 44–48.

Prier, S., Gibert, C., Bodros, A., & Krymolieres, F. (1979). Neurophysiological changes in non-vaccinated rabies patients (letter). *Lancet, 1*, 620.

Prier, S., Gibert, C., Bodros, A., Vachon, F., Atanasiu, P., & Masson, M. (1979). Les neuropathies de la rage humaine: etude clinique et electrophysiologique de deux cas [Human rabies neuropathies: Clinical and electrophysiological study in two cases]. *Revue Neurologique, 135*, 161–168.

Raman, G. V., Prosser, A., Spreadbury, P. L., Cockcroft, P. M., & Okubadejo, O. A. (1988). Rabies presenting with myocarditis and encephalitis. *Journal of Infection, 17*, 155–158.

Remington, P. L., Shope, T., & Andrews, J. (1985). A recommended approach to the evaluation of human rabies exposure in an acute-care hospital. *Journal of the American Medical Association, 254*, 67–69.

Roine, R. O., Hillbom, M., Valle, M., Haltia, M., Ketonen, L., Neuvonen, E., ... Lahdevirta, J. (1988). Fatal encephalitis caused by a bat-borne rabies-related virus: Clinical findings. *Brain, 111*, 1505–1516.

Ron, M. (2001). Explaining the unexplained: Understanding hysteria (editorial). *Brain, 124*(Pt 6), 1065–1066.

Ross, E., & Armentrout, S. A. (1962). Myocarditis associated with rabies: Report of a case. *New England Journal of Medicine, 266*, 1087–1089.

Ruegsegger, J. M., Black, J., & Sharpless, G. R. (1961). Primary antirabies immunization of man with HEP flury virus vaccine. *American Journal of Public Health, 51*, 706–716.

Saeed, B., & Al-Mousawi, M. (2017). Rabies acquired through kidney transplantation in a child: A case report. *Experimental and Clinical Transplantation, 3*, 355–357.

Samaratunga, H., Searle, J. W., & Hudson, N. (1998). Non-rabies lyssavirus human encephalitis from fruit bats: Australian bat lyssavirus (pteropid lyssavirus) infection. *Neuropathology and Applied Neurobiology, 24*(4), 331–335.

Selimov, M. A., Tatarov, A. G., Botvinkin, A. D., Klueva, E. V., Kulikova, L. G., & Khismatullina, N. A. (1989). Rabies-related Yuli virus; identification with a panel of monoclonal antibodies. *Acta Virologica, 33*, 542–546.

Senthilkumaran, S., Balamurgan, N., Sweni, S., Menezes, R. G., & Thirumalaikolundusubramanian, P. (2011). Hypersexuality in a 28-year-old woman with rabies. *Archives of Sexual Behavior, 40*, 1327–1328.

Sheikh, K. A., Ramos-Alvarez, M., Jackson, A. C., Li, C. Y., Asbury, A. K., & Griffin, J. W. (2005). Overlap of pathology in paralytic rabies and axonal Guillain-Barré syndrome. *Annals of Neurology, 57*(5), 768–772.

Shipley, R., Wright, E., Selden, D., Wu, G., Aegerter, J., Fooks, A. R., & Banyard, A. C. (2019). Bats and viruses: Emergence of novel lyssaviruses and association of bats with viral zoonoses in the EU. *Tropical Medicine and Infectious Disease, 4*(1), E31.

Shope, R. E., Murphy, F. A., Harrison, A. K., Causey, O. R., Kemp, G. E., Simpson, D. I. H., & Moore, D. L. (1970). Two African viruses serologically and morphologically related to rabies virus. *Journal of Virology, 6*, 690–692.

Sibley, W. A., Ray, C. G., Petersen, E., Ryan, K., Graham, A. R., Gibbs, M. A., … Sacks, J. J. (1981). Human rabies acquired outside the United States from a dog bite. *Morbidity and Mortality Weekly Report, 43*, 537–540.

Silverstein, M. A., Salgado, C. D., Bassin, S., Bleck, T. P., Lopes, M. B., Farr, B. M., … Miller, G. B. (2003). First human death associated with raccoon rabies—Virginia, 2003. *Morbidity and Mortality Weekly Report, 52*(45), 1102–1103.

Sing, T. M., & Soo, M. Y. (1996). Imaging findings in rabies. *Australasian Radiology, 40*(3), 338–341.

Sipahioglu, U., & Alpaut, S. (1985). Transplacental rabies in a human. *Mikrobiyoloji Bulteni, 19*, 95–99 [Turkish].

Smith, J. S., Fishbein, D. B., Rupprecht, C. E., & Clark, K. (1991). Unexplained rabies in three immigrants in the United States: A virologic investigation. *New England Journal of Medicine, 324*(4), 205–211.

Smith, P. C., Lawhaswasdi, K., Vick, W. E., & Stanton, J. S. (1967). Isolation of rabies virus from fruit bats in Thailand. *Nature, 216*(113), 384.

Sow, P. S., Diop, B. M., Ndour, C. T. Y., Soumare, M., Ndoye, B., Faye, M. A., … Collseck, A. M. (1996). Occipital cerebral aspiration ponction: Technical procedure to take a brain specimen for postmortem virological diagnosis of human rabies in Dakar. *Medecine Et Maladies Infectieuses, 26*(5), 534–536.

Srinivasan, A., Burton, E. C., Kuehnert, M. J., Rupprecht, C., Sutker, W. L., Ksiazek, T. G., … Zaki, S. R. (2005). Transmission of rabies virus from an organ donor to four transplant recipients. *New England Journal of Medicine, 352*(11), 1103–1111.

Sureau, P., Portnoi, D., Rollin, P., Lapresle, C., & Chaouni-Berbich, A. (1981). Prevention of inter-human rabies transmission after corneal graft. *Comptes Rendus de l'Académie des Sciences - Series III, Sciences de la Vie, 293*(13), 689–692 [French].

Talaulicar, P. M. S. (1977). Persistent priapism in rabies. *British Journal of Urology, 49*, 462.

Tariq, W. U. Z., Shafi, M. S., Jamal, S., & Ahmad, M. (1991). Rabies in man handling infected calf (letter). *Lancet, 337*, 1224.

Tesoriero, C., Del, G. F., & Bentivoglio, M. (2019). Sleep and brain infections. *Brain Research Bulletin, 145*, 59–74.

Thongcharoen, P., Wasi, C., Sirikavin, S., Boonthai, P., Bedavanij, A., Dumavibhat, P., … Tantawachakit, S. (1981). Human-to-human transmission of rabies via corneal transplant—Thailand. *Morbidity and Mortality Weekly Report, 30*, 473–474.

Tian, Z., Chen, Y., & Yan, W. (2019). Clinical features of rabies patients with abnormal sexual behaviors as the presenting manifestations: A case report and literature review. *BMC Infectious Diseases, 19*(1), 679–4252.

Tillotson, J. R., Axelrod, D., & Lyman, D. O. (1977). Rabies in a laboratory worker—New York. *Morbidity and Mortality Weekly Report, 26*, 183–184.

Tirawatnpong, S., Hemachudha, T., Manutsathit, S., Shuangshoti, S., Phanthumchinda, K., & Phanuphak, P. (1989). Regional distribution of rabies viral antigen in central nervous system of human encephalitic and paralytic rabies. *Journal of the Neurological Sciences, 92*, 91–99.

Tong, T. R., Leung, K. M., & Lam, A. W. S. (1999). Trucut needle biopsy through superior orbital fissure for diagnosis of rabies. *Lancet, 354*(9196), 2137–2138.

Toro, G., Vergara, I., & Roman, G. (1977). Neuroparalytic accidents of antirabies vaccination with suckling mouse brain vaccine: Clinical and pathologic study of 21 cases. *Archives of Neurology, 34*, 694–700.

Udow, S. J., Marrie, R. A., & Jackson, A. C. (2013). Clinical features of dog- and bat-acquired rabies in humans. *Clinical Infectious Diseases, 57*(5), 689–696.

Udwadia, Z. F., Udwadia, F. E., Rao, P. P., & Kapadia, F. (1988). Penile hyperexcitability with recurrent ejaculations as the presenting manifestation of a case of rabies. *Postgraduate Medical Journal, 64*, 85–86.

van Thiel, P. P., de Bie, R. M., Eftimov, F., Tepaske, R., Zaaijer, H. L., van Doornum, G. J., … Kager, P. A. (2009). Fatal human rabies due to Duvenhage virus from a bat in Kenya: Failure of treatment with coma-induction, ketamine, and antiviral drugs. *PLoS Neglected Tropical Diseases, 3*(7), e428.

Verlinde, J. D., Li-Fo-Sjoe, E., Versteeg, J., & Dekker, S. M. (1975). A local outbreak of paralytic rabies in Surinam children. *Tropical and Geographical Medicine, 27*(2), 137–142.

Villamar, M. F., Smith, J. H., Wilson, D., & Smith, V. D. (2017). Rabies encephalitis presenting with new-onset refractory status epilepticus (NORSE). *Neurology Clinical Practice, 7*(5), 421–424.

Vora, N. M., Basavaraju, S. V., Feldman, K. A., Paddock, C. D., Orciari, L., Gitterman, S., ... Kuehnert, M. J. (2013). Raccoon rabies virus variant transmission through solid organ transplantation. *Journal of the American Medical Association, 310*(4), 398–407.

Wallerstein, C. (1999). Rabies cases increase in the Philippines. *BMJ, 318*(7194), 1306.

Warrell, D. A. (1976). The clinical picture of rabies in man. *Transactions of the Royal Society of Tropical Medicine and Hygiene, 70*, 188–195.

Warrell, M. J. (1996). Rabies. In G. C. Cook (Ed.), *Manson's tropical diseases* (20th ed., pp. 700–720). London: W.B. Saunders.

Warrell, D. A., Davidson, N. M., Pope, H. M., Bailie, W. E., Lawrie, J. H., Ormerod, L. D., ... Lewis, P. (1976). Pathophysiologic studies in human rabies. *American Journal of Medicine, 60*(2), 180–190.

Warrell, M. J., Looareesuwan, S., Manatsathit, S., White, N. J., Phuapradit, P., Vejjajiva, A., ... Warrell, D. A. (1988). Rapid diagnosis of rabies and post-vaccinal encephalitides. *Clinical and Experimental Immunology, 71*, 229–234.

Warrell, D. A., & Warrell, M. J. (1991). Rabies. In H. P. Lambert (Ed.), *Infections of the central nervous system* (pp. 317–328). Philadelphia: B.C. Decker.

Waterman, J. A. (1959). Acute ascending rabic myelitis. Rabies—Transmitted by bats to human beings and animals. *Caribbean Medical Journal, 21*, 46–74.

White, M., Davis, A., Rawlings, J., Neill, S., Hendricks, K., Simpson, D., ... Echeverri, G. B. (1994). Human rabies—Texas and California, 1993. *Morbidity and Mortality Weekly Report, 43*(6), 93–96.

Whitfield, S. G., Fekadu, M., Shaddock, J. H., Niezgoda, M., Warner, C. K., & Messenger, S. L. (2001). A comparative study of the fluorescent antibody test for rabies diagnosis in fresh and formalin-fixed brain tissue specimens. *Journal of Virological Methods, 95*(1–2), 145–151.

Whitley, R. J. (2014). B virus. In W. M. Scheld, R. J. Whitley, & C. Marra (Eds.), *Infections of the central nervous system* (4th ed., pp. 204–209). Philadelphia: Lippincott Williams & Wilkins.

Wilde, H., & Chutivongse, S. (1988). Rabies: Current management in Southeast Asia. *Medical Progress, 15*, 14–23.

Wilson, J. M., Hettiarachchi, J., & Wijesuriya, L. M. (1975). Presenting features and diagnosis of rabies. *Lancet, 2*, 1139–1140.

Winkler, W. G., Fashinell, T. R., Leffingwell, L., Howard, P., & Conomy, J. P. (1973). Airborne rabies transmission in a laboratory worker. *Journal of the American Medical Association, 226*, 1219–1221.

Zhou, H., Zhu, W., Zeng, J., He, J., Liu, K., Li, Y., ... Yu, H. (2016). Probable rabies virus transmission through organ transplantation, China, 2015. *Emerging Infectious Diseases, 22*(8), 1348–1352.

CHAPTER 9

Pathogenesis

Alan C. Jackson

Professor of Medicine (Neurology), University of Manitoba, Winnipeg, MB, Canada

9.1 Introduction

Rabies virus (RABV) is a highly neurotropic virus that spreads along neural pathways and invades the central nervous system (CNS), where it causes an acute infection. Most of what we know about the events that take place during rabies infection has been learned from experimental models using animals. Fixed laboratory strains of RABV and rodent models have commonly been used, because they are easier to handle and less expensive, although the events in these models may not closely mimic the disease under natural conditions either in humans or in rabies vectors. There are a number of sequential steps that occur after peripheral inoculation of RABV from an animal bite, which is the most common mechanism of transmission (Fig. 9.1). The steps include replication in peripheral tissues, spread along peripheral nerves and the spinal cord to the brain, dissemination within the CNS, and centrifugal spread from the CNS along nerves to various organs, including the salivary glands. Each of the pathogenetic steps will be discussed in this chapter. In addition, mechanisms of immune-mediated pathology and neuronal dysfunction in rabies will be addressed.

9.2 Virus entry into the nervous system

9.2.1 Earliest events

Early studies in rabies pathogenesis, which were performed in order to establish the pathways and rate of viral spread, involved amputation of the tail or leg of an animal proximal to the site of inoculation with a "fixed" or "street" (wild-type) strain of RABV. The development of clinical rabies could be prevented with amputation and the timing of the procedure was shown to be critical. In later studies, neurectomy of the sciatic nerve was performed instead of amputation and similar results were observed (Baer, Shantha, & Bourne, 1968;

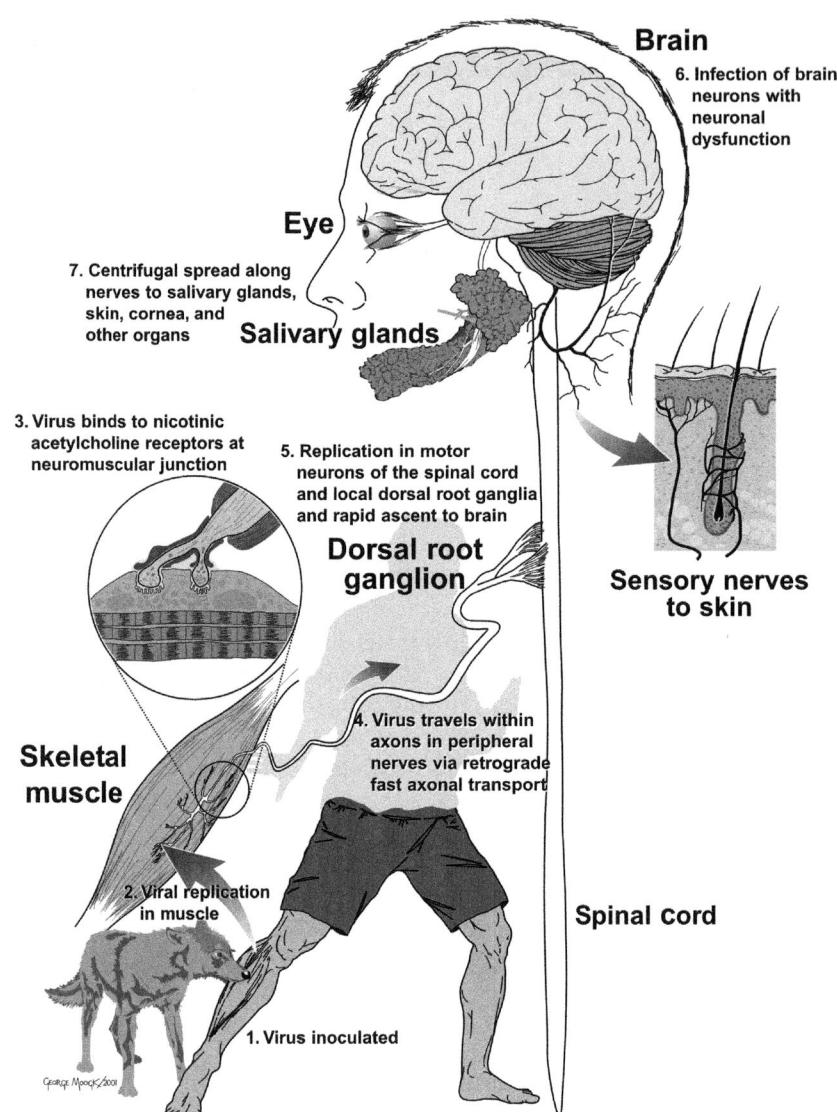

FIG. 9.1 Schematic diagram showing the sequential steps in the pathogenesis of rabies after an animal bite.

Baer, Shanthaveerappa, & Bourne, 1965). These experiments clearly demonstrated that there was an incubation period in rabies during which there was time-dependent movement of virus along peripheral nerves from the site of inoculation to the CNS and models using street RABV supported the idea that the virus remains at or near the site of entry for most of the long incubation period (Baer & Cleary, 1972). However, the time periods in which the procedures were life saving in rodents infected with fixed RABV strains were relatively short (Baer et al., 1965; Dean, Evans, & McClure, 1963), suggesting a different mechanism of viral entry for fixed viruses than in natural rabies (due to street viruses).

Under natural conditions, humans and animals may experience long and variable incubation periods following a bite exposure. This may play a role in maintaining enzootic rabies, especially in high-density, high-contact populations where there is a tendency for the disease to "burn" itself out by rapidly reducing the number of susceptible animals. In humans, the incubation period is usually between 20 and 90 days, although incubation periods rarely may be as short as a few days or longer than a year (Smith, Fishbein, Rupprecht, & Clark, 1991). There is uncertainty about the events that occur during this incubation period. There has been speculation that macrophages may sequester RABV in vivo because persistent in vitro infections of human and murine monocytic cell lines and of primary murine bone marrow macrophages have been demonstrated with different RABV strains (Ray, Ewalt, & Lodmell, 1995). However, this has not yet been demonstrated in animal models. The most useful experimental animal studies examining the events that take place during the incubation period were performed in striped skunks using a Canadian isolate of street RABV obtained from skunk salivary glands (Charlton, Nadin-Davis, Casey, & Wandeler, 1997). These studies, which used reverse transcriptase—polymerase chain reaction (RT-PCR) amplification, showed that viral genomic RNA was frequently present in the inoculated muscle (detected in four of nine skunks), but not in either spinal ganglia or the spinal cord when skunks were sacrificed 62–64 days postinoculation. Immunohistochemical studies performed prior to the development of clinical disease showed evidence of infection of extrafusal muscle fibers and occasional fibrocytes at the site of inoculation. Although it is unclear, the infection of muscle fibers may be a critical pathogenetic step for the virus to gain access to the peripheral nervous system. In a highly susceptible host after intramuscular inoculation, RABV-infected suckling hamsters showed early infection of striated muscle cells near the site of inoculation and, shortly afterward, neuromuscular and neurotendinal spindles became infected near the site of inoculation, which was followed by evidence of infection of small nerves within muscles, tendons, and adjoining connective tissues (Murphy, Bauer, Harrison, & Winn, 1973). However, these events occurred within a few days of inoculation and do not mimic the situation with the longer incubation periods seen in natural infections.

In mouse models, early infection of muscle or other extraneural tissues was not observed following inoculation of fixed RABV strains (Coulon et al., 1989; Johnson, 1965). Virus-specific RNA was not detected with RT-PCR amplification in the masseter muscle of adult mice between 6 and 30 h after inoculation of the challenge virus standard (CVS) strain of fixed RABV in the muscle, although viral RNA was identified in trigeminal ganglia at 18 h and in the brainstem at 24 h after inoculation (Shankar, Dietzschold, & Koprowski, 1991). These studies strongly suggest that RABV is capable of direct entry into peripheral nerves without a replicative cycle in extraneural cells during the short incubation period. This is likely the mechanism of viral entry in rodent models using fixed strains of RABV, accounting for the short period of time (a few days) during which amputation or neurectomy is protective after peripheral inoculation of fixed RABV (Baer et al., 1965; Dean et al., 1963). Unfortunately, these models do not provide information about events that take place during the long incubation period of natural rabies.

9.2.2 Superficial and nonbite exposures

The vast majority of human rabies cases that occur without a history of an exposure are thought to be due to unrecognized or forgotten bites. Molecular characterization of the RABV strains has indicated that they are most frequently from the variant found in silver-haired bats and tricolored bats (formerly called eastern pipistrelle bats) in the USA (Noah et al., 1998), which are small bats. Experimental studies on the silver-haired bat rabies virus (SHBRV) indicate that the virus replicates well at lower than normal body temperatures (34°C) and is associated with higher infectivity in cell types present in the dermis, including fibroblasts and epithelial cells, than with coyote street virus (Morimoto et al., 1996). Hence, the SHBRV was likely selected for efficient local replication in the dermis, which could explain the success of this variant. Experimental studies have not yet been reported in bats, which are associated with more superficial exposures involving skin and subcutaneous tissues. The resulting pathways of viral spread may be different than after deeper exposures involving muscle.

Humans have rarely been infected by bat viruses via the airborne route either in caves where millions of bats roost (Constantine, 1962) or in laboratory accidents by aerosolized RABV (Tillotson, Axelrod, & Lyman, 1977; Winkler, Fashinell, Leffingwell, Howard, & Conomy, 1973). Viral entry by the olfactory and oral routes is much less common than by bites. Relatively little experimental work has been undertaken with routes of viral entry other than one simulating a bite exposure using inoculation techniques. The nasal mucosa has been shown to act as a site of viral entry by suckling guinea pigs that have inhaled street RABV (Hronovsky & Benda, 1969). RABV antigen was initially detected in nasal mucosa cells 6 days later. Early brain infection was prominent in the olfactory bulbs, suggesting that RABV spread into the brain by an olfactory pathway. Similar results were obtained using a variety of RABV strains in mice and hamsters (Fischman & Schaeffer, 1971). RABV antigen has been observed in olfactory receptor cells of naturally infected Brazilian free-tailed bats obtained from a cave, suggesting that the nasal mucosa is a portal of entry in natural infection of bats by airborne RABV in caves (Constantine, Emmons, & Woodie, 1972). Experimental studies showing transmission of rabies to a variety of species of carnivorous animals caged in a cave containing millions of bats supported infection by the airborne route (Constantine, 1962). However, it is thought that the presence of a very large number (millions) of bats in an unventilated area is necessary for airborne transmission of RABV.

Oral transmission of RABV might occur naturally by consumption of carcasses of rabid animals by wildlife and may also be important when humans eat raw dog meat (Wallerstein, 1999). Low susceptibility was observed when mice (Charlton & Casey, 1979a) and skunks (Charlton & Casey, 1979c) were given CVS or street RABV either by the oral route or intestinal instillation. Mice, hamsters, guinea pigs, and rabbits of different ages were infected with CVS either orally or by gastric tube administration (Fischman & Ward, 1968). In CVS-infected weanling mice and hamsters that were infected by this route, RABV antigen was not observed in intestinal mucosal cells, but was found in neurons in Auerbach's and Meissner's plexuses of the stomach and intestine (Fischman & Schaeffer, 1971). These findings suggest that viral entry by the oral route likely occurs via breaks in the integrity of the gastrointestinal mucosa. However, the importance of oral transmission in natural rabies of animals remains uncertain.

9.3 RABV receptors

One striking feature of RABV is its almost exclusive neurotropism. The RABV glycoprotein is thought to be of prime importance in this process by binding to neurospecific receptors (see Chapter 2). These receptors have a role in normal cell function and are hijacked by viruses to gain entry into cells. At least four RABV receptors have been proposed and it is very likely that additional ones will be identified in the future.

9.3.1 Nicotinic acetylcholine receptor

The nicotinic acetylcholine receptor (nAChR) was the first identified receptor for RABV (Lentz, Burrage, Smith, Crick, & Tignor, 1982). RABV antigen was detected at sites coincident with the nAChR in infected cultured chick myotubes from chicken embryos and also shortly after immersion of mouse diaphragms in a suspension of RABV. It was evident from these studies that the distribution of viral antigen detected by fluorescent antibody staining at sites in neuromuscular junctions corresponded to the distribution of nAChRs. The receptors were stained with the rhodamine-conjugated antagonist α-bungarotoxin. Pretreatment of myotubes with either the irreversible binding nicotinic cholinergic antagonist α-bungarotoxin or the reversible binding *d*-tubocurarine reduced the number of myotubes that became infected with RABV. Studies in other laboratories showed that pretreatment of cultured rat myotubes with α-bungarotoxin had an inhibitory effect on infection (Tsiang, de la Porte, Ambroise, Derer, & Koenig, 1986). Binding of radiolabeled RABV to purified *Torpedo* acetylcholine receptor was also inhibited by nicotinic antagonists, but not by atropine (a muscarinic antagonist) (Lentz, Benson, Klimowicz, Wilson, & Hawrot, 1986). Monoclonal antibodies raised against a peptide containing residues 190–203 of the RABV glycoprotein also inhibited binding of the RABV glycoprotein and α-bungarotoxin to the AChR (Bracci et al., 1988). Both RABV and neurotoxins bind to residues 173–204 of the α_1-subunit of the AchR and the highest-affinity virus-binding determinants are located within residues 179–192 (Lentz, 1990). These studies have provided strong evidence that RABV binds to nicotinic acetylcholine receptors in neuromuscular junctions.

Snake venom neurotoxins are polypeptides that bind with high affinity to nAChRs and competitively block the depolarizing action of acetylcholine. When the amino acid sequence of the RABV glycoprotein was compared with that of snake venom neurotoxins, a significant sequence similarity was found between a segment (residues 151–238) of the RABV glycoprotein and the entire long neurotoxin sequence (71–74 residues) (Lentz, Wilson, Hawrot, & Speicher, 1984). The glycoprotein showed identity with residues at the end of loop 2 of the long neurotoxin (the "toxic loop"), which is a long central loop projecting from the molecule that is highly conserved among all of the neurotoxins. This suggests that this region of the RABV glycoprotein is likely a recognition site for the acetylcholine receptor (Lentz, 1985).

Lentz and coworkers indicated that binding of RABV to AChRs would localize and concentrate the virus on postsynaptic cells, which would facilitate subsequent uptake and transfer of virus to peripheral motor nerves (Lentz et al., 1982). Studies performed in chick spinal cord muscle cocultures showed that the CVS strain of RABV and AChR tracers colocalized at neuromuscular junctions and nerve terminals, which provided evidence that the neuromuscular junction is the major site of entry into neurons (Lewis, Fu, & Lentz, 2000). Concentration of RABV at neuromuscular junctions may increase the likelihood of neuronal uptake. Subsequently, colocalization with endosome tracers indicated that the virus resides in an early endosome compartment. There is also supporting ultrastructural evidence that RABV particles enter nerve terminals by endocytosis (Charlton & Casey, 1979b; Iwasaki & Clark, 1975). The acidic interior of the endosome triggers fusion of the viral membrane with the endosome membrane, which allows the viral nucleocapsid to escape into the cytoplasm. However, it has not yet been resolved whether the viral uncoating actually takes place in nerve terminals or in the cell body (perikaryon) after transport in the axon.

Although RABV infection with fixed strains is restricted to a small number of cell types in vivo, fixed viruses can infect a much larger variety of cell types in vitro (Reagan & Wunner, 1985). There is evidence that carbohydrate moieties, phospholipids, highly sialylated gangliosides, and other membrane-associated proteins might contribute to the cellular membrane receptor structure for RABV (Broughan & Wunner, 1995; Conti, Superti, & Tsiang, 1986; Superti et al., 1986; Superti, Seganti, Tsiang, & Orsi, 1984).

Variations in animal susceptibility to RABV infection have been recognized for many years. When infected by intramuscular inoculation, foxes are highly sensitive to RABV infection, dogs are less sensitive, and opossums are highly resistant (Baer, Bellini, & Fishbein, 1990). The difference in susceptibility between the red fox and the opossum could reflect the quantity of acetylcholine receptors in muscle (Baer, Shaddock, Quirion, Dam, & Lentz, 1990). A striking difference in the muscle content, B_{max}, of nicotinic acetylcholine receptors was found with 180.5 fmol/mg protein present in red foxes and only 11.4 fmol/mg protein present in opossums, which was a highly significant difference ($P < .001$). No difference was observed in the binding affinity, K_d. In addition, radiolabeled RABV bound much better to fox muscles than opossum muscles. Hence, the susceptibility of different animal species to RABV may, at least in part, be related to the quantity of nicotinic acetylcholine receptors in their muscles.

Although important in the neuromuscular junction, nAChR has not yet been determined whether it is also an important RABV receptor in the CNS. Binding of RABV to nicotinic acetylcholine receptors in the brain could cause neuronal dysfunction. An antiRABV glycoprotein monoclonal antibody was used to generate (by immunization) an antiidiotypic antibody, B9, that selectively binds to nAChRs (Hanham, Zhao, & Tignor, 1993). Immunostaining of neuronal elements in the brains of RABV-infected mice with the B9 antibody was greatly reduced. This suggests that RABV binds to nAChRs in the brain. Hueffer and coworkers observed that the neurotoxin-like region of the RABV glycoprotein inhibited acetylcholine responses of α4β2 nicotinic receptors in vitro, as did full-length ectodomain of the RABV glycoprotein (Hueffer et al., 2017). The same peptides significantly altered a nicotinic receptor-induced behavior in C. elegans and increased locomotor activity levels when injected into the lateral ventricle of mice. The authors speculated that these results provide a potential mechanism for the behavioral changes in RABV infection.

9.3.2 Neural cell adhesion molecule (NCAM) receptor

The NCAM receptor, which is a cell adhesion glycoprotein of the immunoglobulin superfamily on their cell surface, has been identified as RABV receptor since it was not reported on the surface of resistant cell lines (Thoulouze et al., 1998). Incubation of susceptible cells with RABV decreased surface expression of NCAM and had no effect on other integral proteins of the cell membrane, whereas another virus, vaccinia virus, did not affect surface NCAM expression. This is consistent with internalization of RABV-NCAM receptor complexes during viral entry by adsorptive endocytosis. RABV infection was also inhibited when NCAM receptor was blocked with heparan sulfate, which is a natural ligand physiologically, and by either polyclonal or monoclonal antibodies directed against NCAM receptor (Thoulouze et al., 1998). Furthermore, soluble NCAM neutralized RABV infection, indicating that occupation of the receptor site on virus particles prevented binding to the RABV receptors on target cells. When resistant L cells were transfected with NCAM cDNA, the cells became susceptible to RABV infection. Hence, there is robust in vitro evidence that NCAM is a RABV receptor.

When primary cortical cultures were prepared from NCAM receptor-deficient ("knock-out") and their wild-type littermate mice (Thoulouze et al., 1998) and infected with CVS, a significantly lower mean number of cells became infected in NCAM receptor-deficient cultures ($7.8 \pm 3.9\%$) than in wild-type cultures ($18.6 \pm 8.9\%$) ($P < .005$). In vivo, after inoculation of CVS into the masseter muscle of NCAM receptor-deficient and wild-type mice, significantly less RABV antigen was found in the brainstem/cerebellum, diencephalon, and cerebral cortex in NCAM receptor-deficient than in wild-type mice, indicating that viral spread was less efficient without NCAM receptor. After inoculation of CVS into hindlimb muscles, the mean survival of NCAM receptor-deficient mice was 13.6 days compared to 10.2 days in wild-type mice ($P = .002$), indicating that the disease progressed slower without NCAM receptor. The absence of NCAM receptor in vivo only mildly delayed the death of mice. Interestingly, this suggests that there must be other functionally important RABV receptors in the CNS in addition to NCAM receptor. The NCAM receptor is localized in presynaptic membranes and, hence, it is well positioned for internalization of RABV by receptor-mediated endocytosis into vesicles (Lafon, 2005). Subsequently, there is retrograde transport of either these vesicles carrying dissociated rabies virions or of viral nucleocapsids, which are released after uncoating of the virus with fusion of the viral envelope.

9.3.3 Low-affinity p75 neurotrophin receptor

A report that the low-affinity p75 neurotrophin receptor ($p75^{NTR}$) is a receptor for street RABV further suggests multiple candidates for the RABV receptor (Tuffereau, Benejean, Blondel, Kieffer, & Flamand, 1998). When a random-primed cDNA library from the mRNA of neuroblastoma cells (NG108) was used to transfect COS7 cells, a single plasmid was identified after subcloning, which, when transfected into BSR cells, bound soluble RABV glycoprotein. The 1.3-kb insert of this plasmid showed high amino acid sequence homology with both rat and human $p75^{NTR}$. Most cell lines of nonneuronal cell origin, including BSR cells, are not permissive for street RABV infection. However, the BSR cells with stable expression of $p75^{NTR}$ were able to bind soluble RABV glycoprotein. A fox street RABV isolate was also able to infect $p75^{NTR}$-expressing BSR cells, but relatively few untransfected control BSR cells. BSR cells expressing $p75^{NTR}$ were only slightly more susceptible to infection with CVS and $p75^{NTR}$-expressing BSR cells were 3 to 10 times more susceptible to CVS infection than control BSR cells in the presence of 10% serum. Subsequently, Tuffereau et al. (2007) performed further studies in cultured adult mouse dorsal root ganglion neurons and reported that, although $p75^{NTR}$ is a receptor for soluble RABV glycoprotein in transfected cells of heterologous systems, a RABV glycoprotein-interaction is not necessary for RABV infection of primary neurons.

Since CVS, like street RABV strains, is highly neuronotropic in vivo, one would expect that CVS would also use the same receptors as street RABV in vivo. In addition, evaluation of nonadapted street RABV infection of mice would be difficult because, for example, a high incidence of spontaneous recovery with neurologic sequelae has been observed after peripheral inoculation of mice with a fox isolate of street virus (Jackson, Reimer, & Ludwin, 1989). When $p75^{NTR}$-deficient mice were infected intracerebrally with CVS, similar clinical features of disease and pathologic changes were observed in the brain as in mice expressing $p75^{NTR}$ (Jackson & Park, 1999). $p75^{NTR}$ is not present at the neuromuscular junction, and it is mainly present in the dorsal horn of the spinal cord, suggesting that it could be involved in trafficking of RABV by a sensory pathway (Lafon, 2005). Ligand-$p75^{NTR}$ complexes are normally internalized by clathrin-coated pits into endosomes (Butowt & Von Bartheld, 2003). Lafon (2005) has speculated that $p75^{NTR}$ may play an important role in the retrograde transport of RABV by forming a RABV-$p75^{NTR}$ complex that is transported into the cell in endocytic compartments, possibly following caveolae transcytosis.

9.3.4 Metabotropic glutamate receptor subtype 2

Metabotropic glutamate receptor subtype 2 (mGluR2) is a member of the G protein-coupled receptor family that is abundant in the CNS. Wang and colleagues have recently demonstrated that mGluR2 is a functional cellular entry receptor that interacts directly with the RABV glycoprotein to mediate virus entry into cells (Wang et al., 2018). RABV infection was markedly decreased after mGluR2 siRNA knockdown in cells and antibodies to mGluR2 blocked RABV infection in cells in vitro (Wang et al., 2018). Also, the mGluR2 ectodomain soluble protein neutralized the infectivity of RABV both in vitro and in vivo in mice. Hence, mGluR2 should be considered another RABV receptor.

9.4 Spread to the CNS

Centripetal spread of RABV to the CNS occurs within motor and perhaps also sensory axons of peripheral nerves. Colchicine, a microtubule-disrupting agent active for tubulin-containing cytoskeletal structures, is an effective inhibitor of fast axonal transport in the sciatic nerve of rats (Tsiang, 1979). When colchicine was applied locally to the sciatic nerve using elastomer cuffs to obtain high local concentrations of the drug, adverse systemic effects were avoided. Propagation of RABV was prevented, providing strong evidence that RABV spreads from sites of peripheral inoculation to the CNS by retrograde fast axonal transport. Human dorsal root ganglia neurons in a compartmentalized cell culture system were used to show that viral retrograde transport occurs at a rate of between 50 and 100 mm/day (Tsiang, Ceccaldi, & Lycke, 1991). There is evidence that the RABV phosphoprotein, which is a member of the ribonucleocapsid complex (see Chapter 2), interacts with dynein light chain 8 (LC8). Dynein LC8 is a component of both cytoplasmic myosin V and dynein that are involved in actin-based transport (important in early steps of viral entry) and microtubule-based transport (for fast axonal transport) in neurons, respectively (Jacob, Badrane, Ceccaldi, & Tordo, 2000; Raux, Flamand, & Blondel, 2000). This led to speculation that RABV phosphoprotein-dynein interaction may be of fundamental importance in axonal transport of RABV. However, studies performed in young mice have shown that deletions of the dynein light-chain-binding region of recombinant SAD-L16, which contained the genetic sequence of the Street Alabama Dufferin (SAD)-B19 strain, resulted in mutant viruses that demonstrated only minor effects on viral spread after peripheral inoculation and they remained neuroinvasive and neurovirulent (Mebatsion, 2001; Rasalingam, Rossiter, Mebatsion, & Jackson, 2005). Mazarakis et al. (2001) have demonstrated that RABV glycoprotein-pseudotyped lentivirus (equine infectious anemia virus)-based vectors enhance gene transfer to neurons by facilitating retrograde axonal transport. Hence, the RABV glycoprotein may play a more important role than the phosphoprotein.

Other investigators used mouse and hamster models to demonstrate early and at least near-simultaneous involvement of motor neurons in the spinal cord and primary sensory neurons in dorsal root ganglia (Johnson, 1965; Murphy, Bauer, et al., 1973; Jackson & Reimer, 1989; Coulon et al., 1989). After inoculation of mice in the masseter muscle with CVS, early infection was observed in trigeminal ganglia (Jackson, 1991b; Shankar et al., 1991). Studies using RT-PCR amplification showed that infection was detectable in trigeminal ganglia (18 h post-inoculation) before the brainstem (24 h post-inoculation) (Shankar et al., 1991). However, elegant transneuronal tracer methods using CVS in rats (Tang, Rampin, Giuliano, & Ugolini, 1999) and studies in rhesus monkeys (Kelly & Strick, 2000) have not shown early infection of primary sensory neurons. Two days after inoculation of CVS into the bulbospongiosus muscle of rats, the distribution of RABV antigen was limited to ipsilateral bulbospongiosus motor neurons in the spinal cord (Tang et al., 1999). One day later (3 days post-inoculation), there was evidence of transfer of antigen to interneurons in the dorsal gray commissure, intermediate zone, and sacral parasympathetic nucleus and also to external urethral sphincter motor neurons; at this time there was no labeling of primary sensory neurons in local dorsal root ganglia. This study indicates that a motor pathway rather than a sensory pathway is important in the spread of RABV to the CNS. It is unclear if the different results obtained in earlier studies are due to differences in the animal models, including the species of the host and the route of inoculation.

There are differences in the clinical manifestations of rabies acquired from dogs and bats (Udow, Marrie, & Jackson, 2013). For example, bat-acquired rabies more frequently has tremor and myoclonus, whereas dog-acquired rabies more frequently has hydrophobia and aerophobia. Although no experimental studies have evaluated the pathways that bat viruses spread after superficial inoculation (e.g., into the skin), it is expected that there would be important differences in the pathways of virus spread in the host versus a deep bite by a dog involving skeletal muscles (Begeman et al., 2018; Udow et al., 2013), which may account, at least in part, for the different clinical manifestations in infection with these RABV variants.

9.5 Spread within the CNS

Once CNS neurons in the spinal cord or brainstem become infected in rodent models, there is rapid dissemination of RABV infection along neuroanatomical pathways. RABV also spreads within the CNS, as in the peripheral nervous system, by fast axonal transport. Evidence was provided for axonal transport using stereotaxic brain inoculation in rats (Gillet, Derer, & Tsiang, 1986) and by the administration of colchicine, which inhibited virus transport within the CNS (Ceccaldi, Ermine, & Tsiang, 1990; Ceccaldi, Gillet, & Tsiang, 1989). Studies on cultured rat dorsal root ganglia neurons showed that anterograde fast axonal transport of RABV is in the range of 100–400 mm/day (Tsiang, Lycke, Ceccaldi, Ermine, & Hirardot, 1989). However, the importance of this is unclear because transneuronal tracing studies with CVS in rhesus monkeys have indicated that the spread of RABV occurs exclusively by retrograde axonal transport with trans-synaptic transport of RABV also occurring exclusively in the retrograde direction (Kelly & Strick, 2000). Studies performed with RABV glycoprotein gene-deficient recombinant RABV showed limited spread in the brains of mice after intracerebral inoculation (Etessami et al., 2000). After stereotaxic inoculation of the recombinant virus into the rat striatum, infection remained restricted to initially infected neurons and there was no evidence of trans-synaptic spread to secondary neurons. Hence, the RABV glycoprotein is necessary for trans-synaptic spread of RABV from one neuron to another.

Ultrastructural studies in a skunk model indicated that most viral budding occurs on synaptic or adjacent plasma membranes of dendrites, with less prominent budding from the plasma membrane of the perikaryon (Charlton & Casey, 1979b). Most virions were found partially engulfed by an invaginated membrane of an adjacent axon terminal, indicating transneuronal dendroaxonal transfer of virus. Virions were also occasionally observed budding freely into the intercellular space.

After footpad inoculation of mice with CVS, there was early involvement of neurons in the brainstem tegmentum and deep cerebellar nuclei (Jackson & Reimer, 1989). Subsequently, the infection spread to involve cerebellar Purkinje cells and neurons in the diencephalon, basal ganglia, and cerebral cortex. RABV, like Borna disease virus (Carbone, Duchala, Griffin, Kincaid, & Narayan, 1987), spread to the hippocampus relatively late after peripheral inoculation. RABV predominantly infected pyramidal neurons of the hippocampus, with relative sparing of neurons in the dentate gyrus in adult mice (Jackson & Reimer, 1989). The basis for cell selectivity is uncertain, although Gosztonyi and Ludwig (2001) speculated that if N-methyl-D-aspartate (NMDA) NR1 receptors are involved as RABV receptors, then cell selectivity can be explained by the fact that RABV spreads only by retrograde (not by anterograde) fast axonal transport. Therefore, the virus cannot infect dentate granule cells by the perforant path and mossy fibers from CA3 that predominantly have α-amino-3-hydroxy-5-methyl-4-isoxazole propionate (AMPA) and kainate receptors rather than NMDA receptors. Although RABV is highly neuronotropic, skunk RABV has been observed to infect Bergmann glia in the cerebellum more prominently than Purkinje cells in experimentally infected skunks (Jackson, Phelan, & Rossiter, 2000). In street virus–infected skunks, initial infection was present in the lumbar spinal cord and transit to the brain occurred via a variety of long ascending and descending fiber tracts, including rubrospinal, corticospinal, spinothalamic, spino-olivary, vestibulospinal/spinovestibular, reticulospinal/spinoreticular, cerebellospinal/spinocerebellar, and dorsal column pathways (Charlton, Casey, Wandeler, & Nadin-Davis, 1996).

9.6 Spread from the CNS

Centrifugal spread or viral spread from the CNS to peripheral sites along neuronal routes is essential for transmission of RABV to its natural hosts. Salivary gland infection is necessary for the transfer of infectious oral fluids by rabid vectors. The salivary glands receive parasympathetic innervation by the facial (via the submandibular ganglion or Langley's ganglion in some animals) and glossopharyngeal (via the otic ganglion) nerves, sympathetic innervation via the superior (or cranial) cervical ganglion and afferent (sensory) innervation (Emmelin, 1967). Unilateral excision of a portion of the lingual nerve and the cranial cervical ganglion of dogs and foxes resulted in very low viral titers in denervated salivary glands compared with contralateral salivary glands after street RABV infection (Dean et al., 1963). Evidence of widespread infection of salivary gland epithelial cells is a result of viral spread along multiple terminal axons rather than spread between epithelial cells (Charlton, Casey, & Campbell, 1983). RABV antigen was found concentrated in the apical region of mucous acinar cells and ultrastructural studies showed that viral matrices were present in the basal region and there was viral budding on the apical plasma membrane into the acinar lumen and into the intercellular canaliculi and, occasionally, onto membranes of secretory granules (Balachandran & Charlton, 1994). Viral titers in salivary glands may be higher than in CNS tissues (Dierks, 1975).

In addition to salivary gland infection, evidence was found in a suckling hamster model of centrifugal spread involving the central, peripheral, and autonomic nervous systems in many peripheral sites (Murphy, Harrison, Winn, & Bauer, 1973). Infection was observed in the ganglion cell layer of the retina and in corneal epithelial cells, which are innervated by sensory afferents via the trigeminal nerve. Epithelial cells in both superficial and deep layers of the cornea were found to be infected (Balachandran & Charlton, 1994). Detection of RABV antigen in corneal impression smears has been used as a diagnostic test for human rabies (Koch, Sagartz, Davidson, & Lawhaswasdi, 1975), and RABV has been transmitted by corneal transplantation in humans (see Chapters 8 and 12). Infection may be found in free sensory nerve endings of tactile hair in a skin biopsy, which is one of the best diagnostic methods of confirming an antemortem diagnosis of rabies in humans (see Chapters 8 and 12). Antigen may be demonstrated in small nerves around hair follicles or in epithelial cells of hair follicles in the skin, which is taken from the nape of the neck because it is rich in hair follicles. Widespread infection may be observed in sensory nerve end organs in the oral and nasal cavities, including the olfactory epithelium and taste buds in the tongue. Because preganglionic spinal neurons are not prominently involved in rabies, it is unclear the actual role autonomic pathways that play in centrifugal spread of RABV and whether this pathway is only taken very late after intraspinal viral spread (Hemachudha et al., 2013; Ugolini, 2011).

Studies in both natural and experimental rabies have demonstrated infection involving neurons in a variety of extraneural organs, including the adrenal medulla, cardiac ganglia and plexuses in the luminal gastrointestinal tract, major salivary glands, liver and exocrine pancreas (Balachandran & Charlton, 1994; Debbie & Trimarchi, 1970; Jackson et al., 1999). In addition, there is infection involving a variety of non-neuronal cells, including acini in major salivary glands in rabies vectors, epithelium of the tongue, cardiac and skeletal muscle, hair follicles, and even pancreatic islets (Debbie & Trimarchi, 1970; Murphy, Harrison, et al., 1973; Balachandran & Charlton, 1994; Jackson et al., 1999). There are a few reports of myocarditis in human cases of rabies (Araujo, de Brito, & Machado, 1971; Cheetham, Hart, Coghill, & Fox, 1970; Ross & Armentrout, 1962).

9.7 Animal models of RABV neurovirulence

Viral neurovirulence can be defined as the capacity of a virus to cause disease of the nervous system, especially the CNS. Analysis of neurovirulence has frequently been approached in experimental models by comparing infections in a host with closely related viruses (e.g., different RABV strains or a parent RABV and a variant) (Jackson, 1991a). The ability of a virus to spread to the CNS from a peripheral site, or *neuroinvasiveness*, is an important component of neurovirulence after natural routes of viral entry. The route of inoculation is often very important in evaluating neurovirulence experimentally. Intracerebral inoculation is commonly used for convenience and a number of peripheral sites also have been used in different models, including footpad, intramuscular, intraperitoneal, and intraocular inoculation. Species, age, and the immune status of the host have also proved to be important factors in neurovirulence (Flamand et al., 1984). Monoclonal antibody-resistant (MAR) variant viruses were selected in vitro from CVS and ERA laboratory strains of RABV with neutralizing antiglycoprotein antibodies (Dietzschold et al., 1983; Seif, Coulon, Rollin, & Flamand, 1985). Mutations involving antigenic site III are located between amino acid residues 330 and 338 of the CVS and ERA glycoprotein. Variants with a single amino acid change at position 333, with loss of either arginine (Dietzschold et al., 1983; Seif et al., 1985) or lysine (Tuffereau et al., 1989), have been shown to have diminished virulence in mice after intracerebral inoculation, whereas variants with amino acid changes at other positions remain neurovirulent. Comparisons of avirulent variants with their parent viruses in mouse and rat models using different routes of inoculation have been a useful approach in understanding the biological bases of RABV neurovirulence. Both MAR variants RV194-2 (Dietzschold et al., 1983) and Av01 (Coulon, Rollin, Aubert, & Flamand, 1982) have substitution of a glutamine for the arginine of CVS at position 333 of the glycoprotein.

Avirulent RABV variants, but not the parent CVS strain, have been shown to cause infection in extraneural sites close to the site of inoculation in different models. For example, Av01 infected the anterior epithelium of the lens after inoculation into the anterior chamber of the eye in rats (Kucera, Dolivo, Coulon, & Flamand, 1985). Similarly, RV194-2 inoculated into the tongue of mice and rats produced local infection involving epithelial tissues, glandular cells and muscles (Torres-Anjel, Montano-Hirose, Cazabon, Oakman, & Wiktor, 1984). In these models, the variant viruses demonstrated less restricted cellular tropism than the more highly neuronotropic parental CVS strain.

Two independent studies in mice showed no impairment in neuroinvasiveness after peripheral inoculation of AvO1 or RV194-2 (Coulon et al., 1989; Jackson, 1991b). An excellent model was developed for studying the pathways of viral spread to the brain by inoculating RABV into the anterior chamber of the eye in rats (Kucera et al., 1985). There are six potential neural pathways for viral spread to occur between the eye and brain. RABV was localized in tissues using immunofluorescent staining. After inoculation of CVS, viral antigen was initially detected at 24h in the ipsilateral ciliary ganglion and later in the Edinger-Westphal nucleus of the oculomotor nerve (parasympathetic pathway). At 48h, virus also spread to the ipsilateral ganglion of the trigeminal nerve (an afferent sensory pathway) and to neurons of the contralateral area praetectalis medialis, which projects to the retina via preopticoretinal fibers. In contrast, Av01 propagated in the trigeminal pathway but not in either parasympathetic or preopticoretinal fibers. Neurons in the trigeminal ganglion also were infected at 48 h, indicating a similar rate of spread. Thus, avirulent AvO1 spreads to the brain in this model using more limited pathways than its virulent parent virus.

Intracerebral inoculation is a crude technique in which the inoculum spreads throughout the cerebrospinal fluid (CSF) spaces, including the ventricular system and subarachnoid space (Mims, 1960). A stereotaxic apparatus can deliver an inoculum into a precise location in the brain. Av01 was surprisingly found to be neurovirulent after stereotaxic inoculation into the neostriatum or cerebellum of adult mice (Yang & Jackson, 1992), although Av01 infected fewer neurons and deaths occurred later than after stereotaxic inoculation with CVS (Jackson, 1994). After inoculation of Av01 into the striatum, the infection was widespread in the brain and there were morphologic changes of apoptosis in neurons (A.C. Jackson, unpublished observations) and also infiltration with inflammatory cells. Serum-neutralizing antibodies against RABV were produced later and at lower levels than after intracerebral inoculation. Av01 is likely neurovirulent after stereotaxic brain inoculation because this route produces both a direct site of viral entry into the CNS and a low level of immune stimulation.

Since centrifugal spread of CVS is limited, comparisons of the spread of CVS and variants from the CNS have not been as useful as for comparisons of spread to the CNS and within the CNS. In the model of Kucera et al. (1985) using intraocular inoculation of rats, CVS spread from the nuclei of the accessory optic system to ganglionic cells of the retina in both eyes, whereas Av01 did not show evidence of centrifugal spread in this model.

9.8 Structural damage caused by RABV infection in the CNS

Despite the dramatic and severe clinical neurological signs in rabies, the neuropathological findings are usually quite mild, especially under natural conditions (see Chapter 10). Yet, experimental infection with fixed viruses, on the other hand, induces the expression of innate immune molecules, extensive inflammatory cells into the CNS, and apoptosis in laboratory animals (Sarmento, Li, Howerth, Jackson, & Fu, 2005). The infiltration of inflammatory cells and induction of apoptosis correlates with the enhancement of blood-brain-barrier (BBB) permeability and the attenuation of the virus (Kuang, Lackay, Zhao, & Fu, 2009). It is thus hypothesized that induction of the innate immune responses is one of the important mechanisms of RABV attenuation (Kuang et al., 2009; Wang et al., 2005). These studies are summarized here.

9.8.1 Innate immune responses

The innate immune system, also known as nonspecific immune system, provided immediate defense against infections. Although the brain was thought traditionally as an immune-privileged site, resident CNS cells, including microglia, astrocytes, and neurons, are known now capable of initiating innate immune responses (Carson, 2002; Reiss, Chesler, Hodges, Ireland, & Chen, 2002). Within the brain, these innate responses are critical in establishing protective immunity, and the defenses mounted by these cell types are the first to engage and counter viruses or other infectious agents. Innate immune responses also recruit leukocytes into the CNS and establish a microenvironment that can potentially direct the activity of infiltrating cells. The induction of innate immune gene expression has been reported in mice infected with fixed viruses and it was found by using RT-PCR that IL-6, IFN-γ, and TNF-α were upregulated in the CNS of mice at 4–6 days after infection with CVS-F3 (Phares, Kean, Mikheeva, & Hooper, 2006). Using Affymetrix microarrays, Prehaud, Megret, Lafage, and Lafon (2005) found that infection with fixed RABV CVS strain induced the expression of innate immune response genes in a human postmitotic neuron-derivative cell line, NT2-N (Prehaud et al., 2005). These genes include beta interferon (IFN-β), chemokines (CCL-5, CXCL-10), and inflammatory cytokines (IL-6, TNF-α, IL-1α). The same virus (CVS) induced the expression of Toll-like receptors (TLR) (McKimmie, Johnson, Fooks, & Fazakerley, 2005), IFN-β, IL-6, and Mx1 (Johnson et al., 2006) in the CNS of mice after either intracerebral or intramuscular inoculation. However, many of the innate immune genes (TLRs, chemokines, and cytokines) were upregulated only in the CNS of mice infected with fixed virus (CVS-B2c), but not in mice infected with street RABV (SHBRV, a virus derived from silver haired bats) when the gene expression was analyzed with Affymetrix microarrays (Wang et al., 2005). RT-PCR confirmed that indeed many of these genes are upregulated in mice infected with fixed virus, but not in mice infected with street RABV. Furthermore, the protein levels for some of the chemokines and cytokines were also found to be increased in the CNS of mice infected with fixed virus (B2c), not in mice infected with street virus (DRV, a virus derived from a Mexican dog) (Kuang et al., 2009). These studies indicate that fixed RABV induces while street RABV evades the innate immune responses (Wang et al., 2005). Induction of innate immune responses leads to the extensive infiltration of inflammatory cells into the CNS of mice infected with fixed RABV, while infiltration of inflammatory cells is scarce in the CNS of mice infected with street RABV (Kuang et al., 2009; Wang et al., 2005).

9.8.2 Apoptosis

Apoptosis is a process by which cells undergo physiologic cell death in response to diverse stimuli. It is a normal process in embryonic development, maturation of the immune system, and in normal tissue turnover (Buja, Eigenbrodt, & Eigenbrodt, 1993; Thompson, 1995). Morphologically, apoptosis is characterized by nuclear and cytoplasmic condensation of single parenchymal cells followed by fragmentation of the nuclear chromatin and the subsequent formation of multiple fragments of condensed nuclear material and cytoplasm (Buja et al., 1993). Phagocytosis of this material occurs, although an inflammatory reaction is normally absent. In contrast, cellular death due to necrosis is characterized by preservation of cell outlines and there is variable swelling of the cell and its organelles. Cellular fragmentation occurs as a late event in necrosis. There are derangements in energy and substrate metabolism in necrosis that result in breaks in the plasma membrane and organellar membranes. Apoptosis, on the other hand, is associated with endonuclease-mediated cleavage of the DNA of nuclear chromatin, resulting in DNA fragments with sizes in multiples of a single nucleosome length (180 base pairs). The internucleosomal cleavage of the DNA in apoptosis results in a "ladder" appearance of the DNA upon agarose gel electrophoresis, whereas in necrosis there is less specific degradation of DNA into a "smear" containing fragments of various sizes following electrophoresis.

Apoptotic cell death likely plays an important pathogenetic role in a wide variety of viral infections, including those produced by a large number of RNA and DNA viruses and apoptosis occurs in the CNS of humans and experimental animals in many of these infections (Allsopp & Fazakerley, 2000; Hardwick, 1997; Roulston, Marcellus, & Branton, 1999). Strong evidence of apoptotic cell death was found in both cultured cells and neurons in experimental mouse rabies models infected by intracerebral inoculation of fixed RABV strains (Jackson, 1999; Jackson & Park, 1998; Jackson & Rossiter, 1997). In vitro studies using CVS-infected cultured rat prostatic adenocarcinoma (AT3) cells showed striking morphologic changes, revealing apoptosis, at the levels of both light and electron microscopy, whereas AT3 cells transfected with the *bcl*-2 gene (an antiapoptosis gene) did not demonstrate apoptotic changes (Jackson & Rossiter, 1997). Terminal deoxynucleotidyltransferase-mediated dUTP-digoxigenin nick end labeling (TUNEL) staining was also demonstrated in infected AT3 cells, indicating evidence of oligonucleosomal DNA fragmentation typical of apoptosis. In addition, in vitro infection of mouse neuroblastoma (N18) cells with CVS was associated with apoptosis (Theerasurakarn & Ubol, 1998). In vitro studies have also shown that the ERA strain of fixed RABV replicates and induces apoptosis in mouse spleen lymphocytes and the human T-lymphocyte cell line Jurkat (Thoulouze, Lafage, Montano-Hirose, & Lafon, 1997) and that cell death was concomitant with expression of the viral glycoprotein. Whereas CVS induces apoptosis in mouse embryonic hippocampal neurons, the extent of apoptosis and pathogenicity was studied in primary neuron cultures infected with two stable variants of CVS-24, CVS-B2c and CVS-N2c (Morimoto, Hooper, Spitsin, Koprowski, & Dietzschold, 1999). It was found that the extent of apoptosis in adult mice was actually lower in primary neuron cultures infected with the more pathogenic variant CVS-N2c than with the less pathogenic variant CVS-B2c, indicating an inverse relationship between and pathogenicity and apoptosis. Guigoni and Coulon (2002) observed that primary cultures of CVS-infected purified rat spinal motoneurons did not show major evidence of apoptosis over a period of 7 days, while infected purified hippocampal neurons showed apoptosis in over 90% of neurons within 3 days, indicating that different neuronal cell types respond differently to RABV infection. CVS and Pasteur virus (PV) strains induce only limited apoptosis whereas two vaccine strains, Evelyn-Rokitnicki-Abelseth (ERA) and SN-10, induce strong apoptosis in the human neuroblastoma SK-N-SH cell line and in lymphoblastoid Jurkat cells (Baloul & Lafon, 2003; Lay, Prehaud, Dietzschold, & Lafon, 2003; Prehaud, Lay, Dietzschold, & Lafon, 2003; Thoulouze et al., 1997; Thoulouze et al., 2003). Hence, there are both virus-dependent and cell-dependent mechanisms for induction of apoptosis. Furthermore, RABV-induced apoptosis is activated by caspase-dependent and caspase-independent pathways (Sarmento, Tseggai, Dhingra, & Fu, 2006; Thoulouze et al., 2003). There is activation of caspase 8 and caspase 3, but not caspase 9, and poly ADP-ribose polymerase (PARP) is cleaved, confirming activation of downstream caspases and involvement of the extrinsic apoptotic pathway (Kassis, Larrous, Estaquier, & Bourhy, 2004; Sarmento et al., 2006; Ubol, Sukwattanapan, & Utaisincharoen, 1998). Apoptosis-inducing factor is a proapoptotic signal transducing molecule that was shown in infection to be upregulated and translocated from the cytoplasm to the nucleus, where it binds to DNA and provokes chromatin condensation, indicating activation of a caspase-independent pathway (Sarmento et al., 2006).

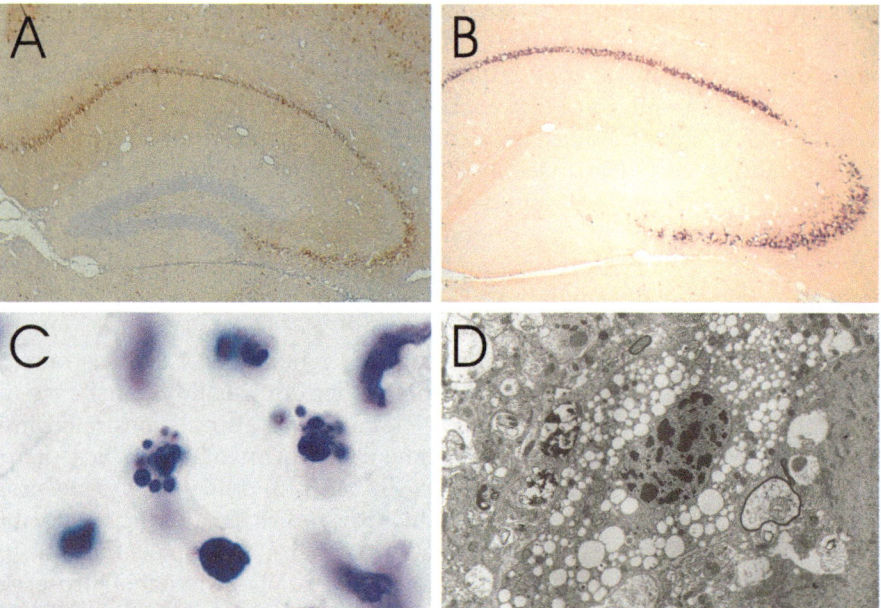

FIG. 9.2 (A) Immunostaining for RABV antigen in the hippocampus of an adult mouse 7 days after intracerebral inoculation with CVS showing antigen in pyramidal neurons and in cortical neurons; neurons in the dentate gyrus do not demonstrate staining. (B) TUNEL staining in the hippocampus of a mouse 7 days after intracerebral inoculation with CVS showing marked staining is present in pyramidal neurons but not in neurons of the dentate gyrus (B). Note the similarity in the distribution of TUNEL staining in (B) and RABV antigen in (A). (C) Neurons in the cerebral cortex 8 days after inoculation with CVS showing multiple condensations of nuclear chromatin in two cells. (D) Hippocampal pyramidal neuron showing a pattern of irregular chromatin condensation and marked cytoplasmic vacuolation. (A: immunoperoxidase-hematoxylin; B: TUNEL staining; C: cresyl violet staining; D: transmission electron microscopy; magnifications: A, B, ×27; C, ×1220; D, ×4870. *Adapted with permission from Jackson, A. C., & Rossiter, J. P. (1997). Apoptosis plays an important role in experimental rabies virus infection.* Journal of Virology, 71(7), 5603–5607 Copyright © 1997, American Society for Microbiology.

In adult mice infected intracerebrally with CVS, specific morphologic changes associated with apoptosis were observed in neurons, particularly in pyramidal neurons of the hippocampus and cortical neurons, and there was positive TUNEL staining in the same regions (Jackson & Rossiter, 1997) (Fig. 9.2). Double-labeling studies indicated that infected neurons actually underwent apoptosis. However, not all infected neurons (e.g., Purkinje cells) demonstrated these morphologic features of apoptosis or positive TUNEL staining. Increased expression of the proapoptotic Bax protein was observed in neurons in areas where apoptosis was prominent (Jackson & Rossiter, 1997). Studies in *bax*-deficient mice showed that neuronal apoptosis was less marked with similar clinical disease as in wild-type littermates, indicating that the Bax protein plays an important role in modulating RABV-induced apoptosis under specific experimental conditions (Jackson, 1999).

Both CVS- and SAD-L16 (a vaccine strain based on SAD-B19)-infected suckling mice show widespread and severe morphologic changes of apoptosis with positive TUNEL staining and activation of caspase 3, a downstream caspase, after intracerebral and peripheral routes of inoculation (Jackson & Park, 1998; Rasalingam, Rossiter, & Jackson, 2005; Rasalingam, Rossiter, Mebatsion, & Jackson, 2005) (Fig. 9.3). In suckling mice infected with CVS via

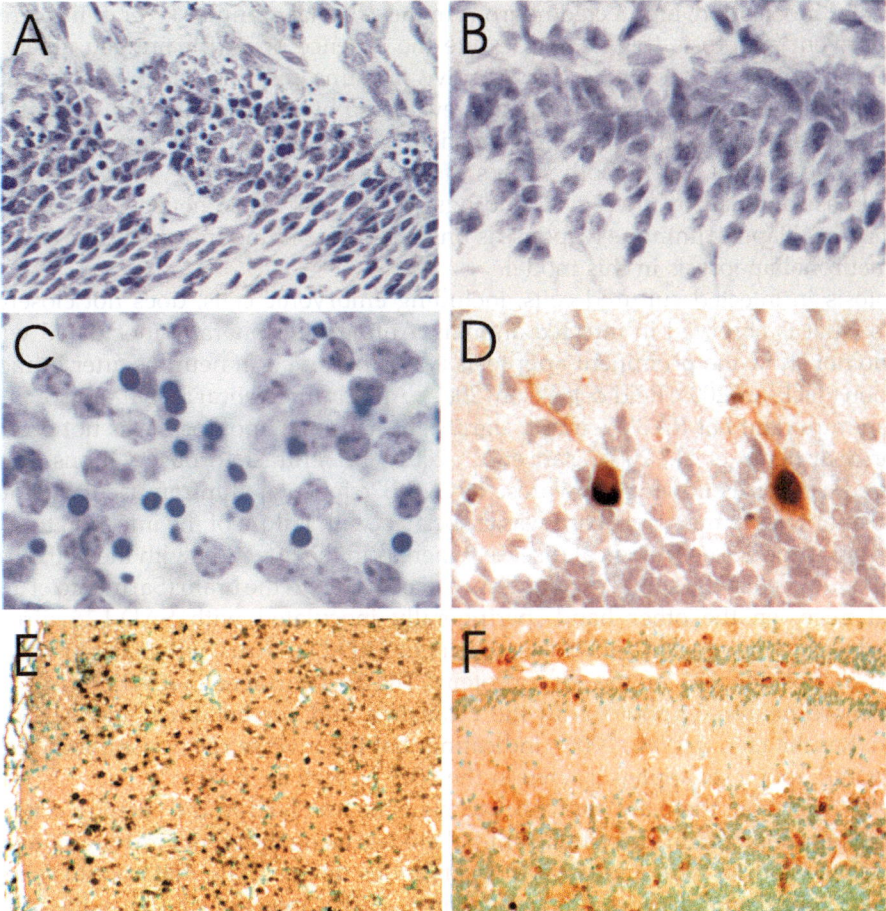

FIG. 9.3 Brain sections after intracerebral inoculation with CVS-11 of 6-day-old mice (A–C) and of 7-day-old mice with L16 (D–F). (A) Nuclear chromatin condensations in multiple cells in the external granular layer of the cerebellum in a CVS-infected suckling mouse. (B) External granular layer of the cerebellum of an uninfected mouse of the same age showing the absence of typical apoptotic morphology. (C) Multiple neurons in the dentate gyrus of the hippocampus of a CVS-infected suckling mouse showing chromatin condensations involving entire nuclei. (D) Immunoperoxidase staining for activated caspase-3 and (E, F) TUNEL staining in L16–infected mouse brains. (D) Activated caspase 3 staining is present in Purkinje cells of the cerebellum 4 days p.i. and (E) TUNEL staining is present in in many neurons in the cerebral cortex and (F) in neurons in the cerebellar external and internal granular layers 4 and 6 days p.i., respectively. A–C: cresyl violet staining; D: caspase 3 immunostaining; E, F: TUNEL staining-methyl green; magnifications: A, ×400; B, D, ×480; C, ×880; E, ×115; F, ×180. A–C, Adapted from Jackson, A. C., & Park, H. (1998). Apoptotic cell death in experimental rabies in suckling mice. Acta Neuropathologica, 95(2), 159–164 with kind permission of Springer Science and Business Media, D–F, Adapted with permission from Rasalingam, P., Rossiter, J. P., & Jackson, A. C. (2005). Recombinant rabies virus vaccine strain SAD-L16 inoculated intracerebrally in young mice produces a severe encephalitis with extensive neuronal apoptosis. Canadian Journal of Veterinary Research, 69(2), 100–105 Copyright © 2005, Canadian Veterinary Medical Association.

intracerebral inoculation, uninfected neurons in the external granular layer of the cerebellum also underwent apoptosis (Fig. 9.3A) despite the absence of RABV antigen, likely due to indirect mechanisms. The role of the adaptive immune response in producing neuronal apoptosis with intracerebral inoculation was evaluated by comparing the infections in adult C57BL/6J mice with nude mice (T-cell deficient) and *Rag1* mice (T- and B-cell deficient) (Rutherford & Jackson, 2004). Both strains of immunodeficient mice showed very similar clinical disease and neuropathological findings, including marked neuronal apoptosis, indicating that the adaptive immune response is unlikely to be of fundamental importance in producing neuronal apoptosis in this model.

Apoptosis in infected cultured cells, including embryonic cells, does not closely correspond to what is observed in infected animals. Animals peripherally inoculated with CVS strains do not show the prominent apoptosis that is observed in neurons after intracerebral inoculation (Jackson, 2003; Reid & Jackson, 2001). After intracerebral inoculation of mice with SHBV, an important bat RABV variant, significant neuronal apoptosis was not observed in the brain (Sarmento et al., 2005; Yan et al., 2001), in contrast to observations with fixed (attenuated) strains. Following a low dose of CVS-B2c inoculated intramuscularly into mice, neuronal apoptosis in the spinal cord was associated with failure of the infection to spread to the brain and produce neurological disease, whereas in the infection with the SHBV, apoptosis was not induced in the spinal cord and spread occurred to the brain (Sarmento et al., 2005). Neonatal mice, on the other hand, peripherally inoculated with SAD-L16 virus, were compared with mice infected with the less virulent SAD-D29 virus, which has an attenuating mutation at position 333 of the glycoprotein. The less virulent SAD-D29 virus actually induced more neuronal apoptosis in the brainstem and cerebellum than SAD-L16 virus (Jackson, Rasalingam, & Weli, 2006), indicating that the inverse relationship between pathogenicity and apoptosis applies in vivo in the CNS as well as in vitro.

In RABV infection, there are complex mechanisms involved in the ensuing cell death or survival of neurons, both in vitro and in animal models using different viral strains and routes of inoculation. Both in vitro and in vivo observations demonstrate that apoptosis may be a protective rather than a pathogenic mechanism in RABV infections because the less pathogenic viruses induce more apoptosis than the more pathogenic viruses in vitro and also in vivo using peripheral routes of inoculation (Jackson et al., 2006; Morimoto et al., 1999; Prehaud et al., 2003; Sarmento et al., 2005; Yan et al., 2001).

There is a report demonstrating apoptosis in a single human rabies case (Adle-Biassette et al., 1996), but morphologic evidence of neuronal apoptosis has generally not been prominent in natural rabies in humans or animals. Juntrakul, Ruangvejvorachai, Shuangshoti, Wacharapluesadee, and Hemachudha (2005) reported that TUNEL positive cells were observed throughout the neuroaxis in seven cases of human rabies, but this may have been due to nonspecific staining. Also, morphologic evidence of neuronal apoptosis was not assessed or illustrated in this report. Jackson and coworkers did not find evidence of neuronal apoptosis in 12 human rabies cases using histological analysis, TUNEL staining, and staining for cleaved caspase-3 (Jackson, Randle, Lawrance, & Rossiter, 2008).

9.8.3 Degeneration of neuronal processes

Scott and coworkers have comprehensively evaluated CVS infection in adult transgenic mice expressing the yellow fluorescent protein (YFP; H clone) using hindlimb footpad inoculation of CVS (Scott, Rossiter, Andrew, & Jackson, 2008). In these mice, YFP expression is driven in a subpopulation of neurons using the *thy1* vector, and there are strong fluorescent signals in dendrites, axons, and presynaptic nerve terminals (Feng et al., 2000). Conventional histopathology showed mild inflammatory changes without significant degenerative neuronal changes, but at late clinical time points with the development of severe clinical neurological disease, fluorescence microscopy showed marked abnormalities, especially beading and/or swelling, in dendrites and axons of layer V cortical pyramidal neurons, severe involvement of axons in the brainstem and the inferior cerebellar peduncle, and severe abnormalities affecting axons of cerebellar mossy fibers (Fig. 9.4). The structural changes take a few days to develop, likely because they are mediated by abnormal axoplasmic structural protein function. Toluidine blue-stained resin sections and electron microscopy showed vacuolation in cortical neurons that corresponded to swollen mitochondria, and vacuolation in the neuropil of the cerebral cortex. Axonal swellings, key markers of axonal degeneration, were observed. Vacuolation was also observed in axons and in presynaptic nerve endings. These morphological changes are sufficient to explain the severe clinical disease with a fatal outcome.

The morphologic changes in axons have a striking similarity to the neurodegenerative changes that occur in diabetic sensory and autonomic neuropathy, in which a key feature is the presence of axonal swellings that are composed of accumulations of mitochondria and cytoskeletal proteins (e.g., neurofilaments) (Lauria et al., 2003; Schmidt et al., 1997). Diabetes-induced oxidative stress in sensory neurons and peripheral nerves is demonstrated by increased production of reactive oxygen species (ROS) (Nishikawa et al., 2000; Russell et al., 2002), lipid peroxidation (Obrosova et al., 2002), and protein nitrosylation (Obrosova et al., 2005)

9.8.4 Blood-brain-barrier (BBB)

The BBB is a separation of circulating blood from the brain extracellular fluid in the CNS. It occurs along all capillaries and consists of tight junctions around the capillaries (Hamilton, Foss, & Leach, 2007). Endothelial cells restrict the diffusion of microscopic objects and large or hydrophilic molecules into the cerebrospinal fluid (CSF), while allowing the diffusion of small hydrophobic molecules (O_2, CO_2, hormones) (Gloor et al., 2001). Cells of the barrier actively transport metabolic products such as glucose across the barrier with specific proteins. The loss of BBB integrity during CNS infection and autoimmunity has generally been associated with the development of neurological signs. The BBB, the lack of lymphatic drainage in the CNS, low MHC expression, and elevated levels of immunosuppressive molecules collectively contribute to a state of immunological isolation relative to the rest of the body (Davis, Rall, & Schnell, 2015; O'Donnell & Rall, 2010).

It was found that infection of attenuated RABV CVS-F3 increased BBB permeability and CNS inflammation in the absence of neurological sequelae, leading to virus clearance from the CNS (Phares et al., 2006). The loss of BBB integrity is associated with the expression of several chemokines/cytokines and the accumulation of CD4- and CD19-positive cells in

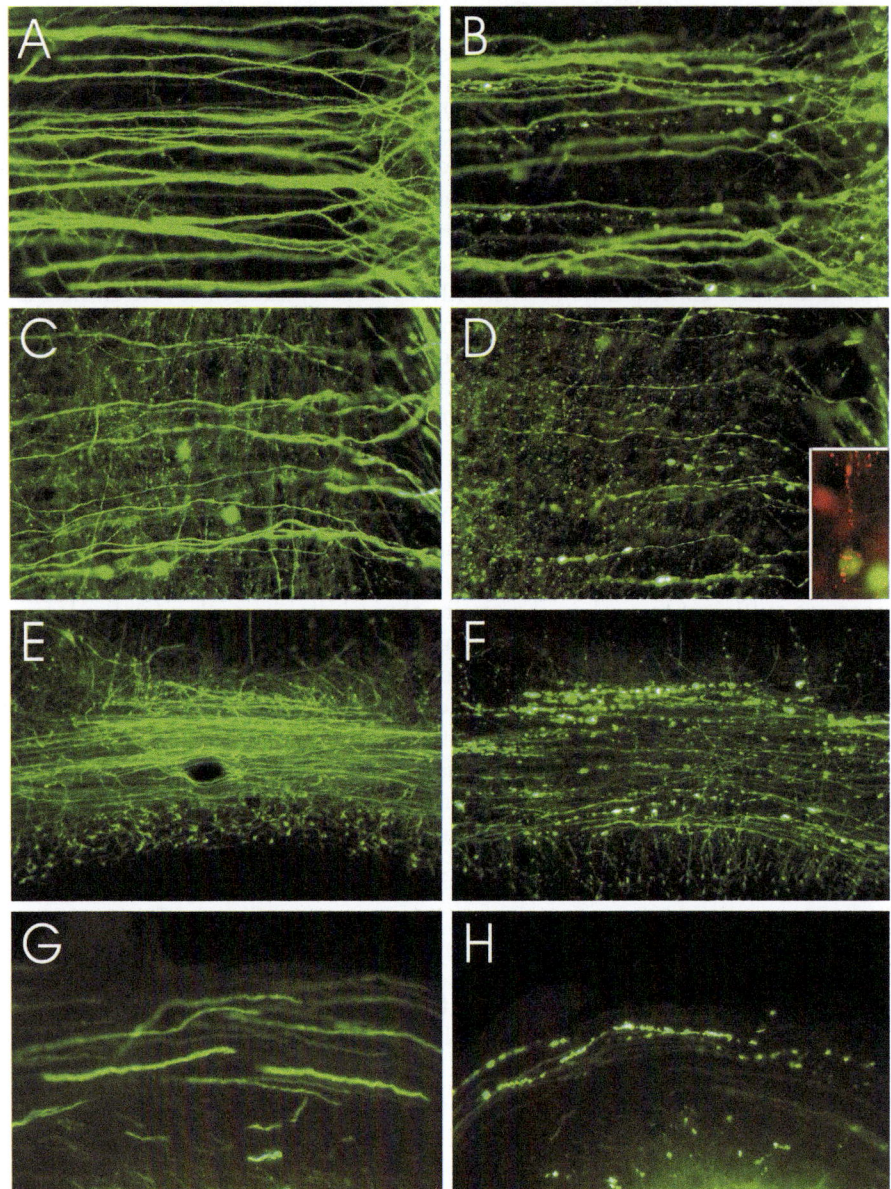

FIG. 9.4 Fluorescence microscopy showing dendrites (A and B) and axons (C and D) of layer V pyramidal neurons in the cerebral cortex of mock-infected (A and C) and moribund CVS-infected (B, D, and D inset) YFP mice. In infected mice, beading is observed in a minority of dendrites (B), while more axons are involved (D). There are no abnormalities in the dendrites (A) or axons (C) of mock-infected mice. Axons in mock-infected mice are slightly varicose (C), which is characteristic of these fibers. Fluorescence microscopy shows RABV antigen (red) in the perikaryon and dendrite of a YFP-expressing neuron (D inset). Morphology of the cerebellar mossy fibers of mock-infected (E) and moribund CVS-infected YFP mice (F). Mossy fiber axons in the cerebellar commissure of moribund mice show severe beading (F), whereas no abnormalities were observed in mock-infected mice (E). Axons in the inferior cerebellar peduncles are normal in mock-infected mice (G) and show marked beading in CVS-infected moribund mice (H). A–D, ×420; D inset, ×400; E, F ×140; G, H, ×380. *Adapted with permission from Scott, C. A., Rossiter, J. P., Andrew, R. D., & Jackson, A. C. (2008). Structural abnormalities in neurons are sufficient to explain the clinical disease and fatal outcome in experimental rabies in yellow fluorescent protein-expressing transgenic mice.* Journal of Virology, *82(1), 513–521; doi: https://doi.org/10.1128/JVI.01677-07 Copyright © 2008, American Society for Microbiology.*

the CNS, particularly in the cerebellum. It was further demonstrated that street RABV (for example, SHBRV) can induces a strong virus-specific immune response in the periphery, but unable to enhance the BBB permeability. As a consequence, immune effectors cannot be delivered into the CNS, leading to the death of the infected animals (Roy & Hooper, 2007). In this study, the authors used PLSJL mice that are less susceptible to SHBRV infection than the 129/SvEv mice, largely due to the elevated capacity of PLSJL mice to mediate BBB permeability changes in response to the infection. Treatment of the SHBRV-infected mice with the steroid hormone dehydroepiandrosterone (DHEA) reduced the BBB permeability, resulting in increased mortality. On the other hand, immunization of SHBRV-infected mice with myelin basic protein (MBP) that induces extensive BBB permeability and CNS inflammation results in greater virus clearance and improved survival (Roy & Hooper, 2007; Spitsin et al., 2008). Subsequent studies have demonstrated that only infection with fixed viruses (CVS-F3, CVS-B2c, HEP) leads to the enhancement of BBB permeability while infection with street RABV (SHBRV, DRV) does not (Kuang et al., 2009; Roy & Hooper, 2008; Zhao, Toriumi, Kuang, Chen, & Fu, 2009). Fixed RABV not only can lead to clearance of fixed viruses in the CNS, but also can clear street RABV from the CNS (Faber et al., 2009; Li et al., 2012; Wang et al., 2011). It has been demonstrated that these fixed viruses enhance the BBB permeability, allowing primed B cells enter into the CNS and producing virus-neutralizing antibody in situ to clear RABV from the CNS (Hooper, Phares, Fabis, & Roy, 2009; Roy & Hooper, 2007).

Overall, these studies demonstrate that induction of innate immunity is one of the important mechanisms for RABV attenuation. Infection with small doses of fixed RABV can induce innate immune responses, including the expression of innate immune genes, infiltration of inflammatory cells, induction of apoptosis, and enhancement of BBB permeability. It also indicates that fixed RABV can induce neurological diseases via immune-mediated pathogenesis particularly when infected with large doses (Sarmento et al., 2005). On the other hand, evasion of innate immunity by street RABV is one of the pathogenetic mechanisms for rabies. However, what leads to the death of the infected individuals is not entirely clear.

9.9 Brain dysfunction in rabies

The dramatic and severe clinical neurological signs in rabies with only mild neuropathological findings under natural conditions (see Chapter 10) led to the hypothesis that rabies results from neuronal dysfunction rather than structural damage (Tsiang, 1982). Many experimental studies have been performed to gain an understanding of the bases of this neuronal dysfunction. Although no fundamental underlying defect has been identified to explain this dysfunction, major areas of research in this area will be summarized.

9.9.1 Neuropeptide synthesis during rabies

Although studies performed in vitro have shown that RABV has little or no inhibitory effect on cellular RNA and protein synthesis (Ermine & Flamand, 1977; Madore & England, 1977; Tuffereau & Martinet-Edelist, 1985), in vivo studies using CVS-24–infected rats showed that there was progressive reduction in the expression of the noninducible housekeeping gene that encodes glyceraldehyde-3-phosphate dehydrogenase and the late response gene that encodes proenkephalin, possibly due to the global suppression of cellular protein synthesis related to extensive synthesis of RABV mRNA (Fu et al., 1993). This occurred in association with induction of immediate-early-response genes (*erg-1*, *junB*, and *c-fos*) in the hippocampus and cerebral cortex, where there was colocalization of expression of these genes with viral mRNA expression. In another study, infection of mice with CVS-N2c resulted in downregulation of about 90% of genes in the normal brain at more than four-fold lower levels by using subtraction hybridization (Prosniak, Hooper, Dietzschold, & Koprowski, 2001). Only about 1.4% of genes became upregulated, including genes involved in regulation of cell metabolism, protein synthesis, and growth and differentiation. However, Weihe et al. (2008) reported that RABV infection of mice caused a strong induction of calcitonin gene-related peptide (CGRP), vasoactive intestinal peptide (VIP), and somatostatin. Surprisingly, the induction of these peptides even occurred in neurons that are not infected with RABV, for example, neurons in the dentate gyrus. It has been proposed that the strong RABV-induced upregulation of CGRP in the brain may be associated with immune evasion since CGRP can have an inhibitory effect on antigen presentation. It is unknown whether the expression of neuropeptides is affected in natural rabies.

9.9.2 Defective neurotransmission

9.9.2.1 *Acetylcholine*

A hypothesis that defective cholinergic neurotransmission might be the basis for neuronal dysfunction in rabies led to the investigation of specific binding to muscarinic acetylcholine receptors in CVS peripherally infected rat brains. ^{3}H-labeled antagonist, quinuclidinyl benzylate (QNB), was used as an indication of defective neurotransmission (Tsiang, 1982). Binding of ^{3}H-labeled QNB to AChRs in infected brain homogenates was decreased by 96 h after infection compared with controls and the binding was markedly decreased at 120 h, 10–20 h before death was expected to occur. The greatest reduction in binding was found in the hippocampus and smaller reductions were observed in the cerebral cortex and in the caudate nucleus.

When cholinergic neurotransmission was examined in mice infected intracerebrally with CVS and compared with mock-infected control mice, the enzymatic activities of choline acetyltransferase and acetylcholinesterase, which are required for the synthesis and degradation of acetylcholine, respectively, were similar in the cerebral cortex and hippocampus of moribund CVS-infected and control mice (Jackson, 1993). In contrast to the findings in infected rats, QNB binding to muscarinic acetylcholine receptors, which was assessed with ^{3}H-labeled QNB using Scatchard plots, was not significantly different in the cerebral cortex or hippocampus of CVS-infected and uninfected control mice. These findings cast doubt on the importance of RABV binding to muscarinic acetylcholine receptors in the brain. However, it is possible that differences in the species (mouse versus rat) or in the route of inoculation (peripheral versus intracerebral) account for the differences in the results of the two studies.

In naturally infected rabid dogs, specific binding of ^{3}H-labeled QNB was reduced in the hippocampus (35%) and in the brainstem (27%), but not in other brain regions, compared with uninfected control dogs (Dumrongphol, Srikiatkhachorn, Hemachudha,

Kotchabhakdi, & Govitrapong, 1996). The results were similar whether the clinical disease was of the furious or dumb form. K_d values were increased, indicating a decrease in receptor affinity, and B_{max} values, reflecting receptor content, were unchanged in rabid dogs. Curiously, increased K_d values were found to be similar in the hippocampus whether or not RABV antigen was detectable at that site. These findings argue against alteration of muscarinic receptor binding as a specific consequence of RABV infection of neurons. They suggest an unknown indirect mechanism for altered receptor affinity that is not related to clinical manifestations of disease or the local viral load.

9.9.2.2 Serotonin

Insofar as defective neurotransmission involving other neurotransmitters could be important in the pathogenesis of rabies, the role of serotonin has been examined with the great interest. Serotonin has a wide distribution in the brain and it is important in the control of sleep and wakefulness, pain perception, memory and a variety of behaviors (Julius, 1991). Alterations of sleep stages have been recognized in experimental rabies in mice (Gourmelon, Briet, Clarencon, Court, & Tsiang, 1991; Gourmelon, Briet, Court, & Tsiang, 1986) (see Section 9.3). Again, ligand binding to serotonin (5-HT) receptor subtypes was studied in the brains of CVS-infected rats (Ceccaldi, Fillion, Ermine, Tsiang, & Fillion, 1993). In this case, binding to 5-HT_1 receptor sites using [^{3}H] 5-HT was not affected in the hippocampus, but there was a marked decrease in B_{max} in the cerebral cortex 5 days after inoculation of CVS into the masseter muscles. In the presence of drugs that mask 5-HT_{1A}, 5-HT_{1B}, and 5-HT_{1C} receptors, [^{3}H] 5-HT binding was reduced by 50% in the cerebral cortex 3 days after inoculation, whereas binding of ligands specific for 5-HT_{1A} and 5-HT_{1B} receptor sites was not affected. These results indicate that RABV infection must affect other 5-HT receptors in the cerebral cortex. Furthermore, the reduced binding was demonstrated before RABV antigen was detected in the cerebral cortex. Hence, the effect of RABV on receptor binding is unlikely due to either direct or indirect effects of viral replication in cortical neurons. There are important serotonergic projections from the dorsal raphe nuclei in the brainstem to the cerebral cortex and early infection of the midbrain raphe nuclei in experimental rabies in skunks has been documented (Smart & Charlton, 1992). Is it possible that the reduced binding of serotonin to the 5-HT receptors is an indirect effect of the infection at noncortical sites by unknown mechanisms? Or is it part of a physiological response to the stress produced by the infection? In support of impaired serotonergic neurotransmission in rabies, potassium-evoked release of [^{3}H] 5-HT-labeled synaptosomes from the cerebral cortex of CVS-infected rats was decreased 31% compared with controls (Bouzamondo, Ladogana, & Tsiang, 1993). Hence, there is evidence of both impaired release and impaired binding of serotonin, possibly playing an important role in producing the neuronal dysfunction in rabies.

9.9.2.3 γ-Amino-n-butyric acid

Impairments of both release and uptake of γ-amino-n-butyric acid (GABA) have been found in CVS-infected primary rat cortical neuronal cultures (Ladogana, Bouzamondo, Pocchiari, & Tsiang, 1994). A 45% reduction of [^{3}H] GABA uptake was found 3 days after infection, which coincided with the time of peak viral growth in the cultures. Kinetic analysis revealed major reductions in V_{max}, indicating a decrease in the number of fully active GABA transport sites. There were no significant changes in K_m in infected cultures in comparison to controls, reflecting the affinity of the GABA transport system for its substrate. Potassium- and veratridine-induced [^{3}H] GABA release was increased in infected cultures by 98% and 35%, respectively, compared with controls. The importance of these abnormalities in both the uptake and release of GABA on rabies pathogenesis in vivo has yet to be determined.

9.9.3 Electrophysiological alterations

In addition to effects on neurotransmission, viruses may have important effects on the electrophysiological properties of neurons. Electroencephalographic (EEG) recordings of mice infected with CVS showed that the initial changes were alterations of sleep stages, including the disappearance of rapid-eye-movement (REM) sleep and the development of pseudoperiodic facial myoclonus (Gourmelon et al., 1986). Later, there was a generalized slowing of the EEG recordings (at 2–4 cycles per second). Terminally, there was an extinction of hippocampal slow activity with flattening of cortical activity. Brain electrical activity terminated about 30 minutes before cardiac arrest, indicating that cerebral death in experimental rabies occurs prior to failure of vegetative functions. Street virus–infected mice showed progressive disappearance of all sleep stages with a concomitant increase in the duration of waking stages (indicating insomnia) and these changes occurred before the development of clinical signs of rabies (Gourmelon et al., 1991). There was an absence of EEG abnormalities in street virus–infected mice that lasted through the preagonal phase of the disease. Since pathologic changes are more marked in neurons infected with fixed RABV than street RABV strains, these observations are consistent with the idea that functional impairment of brain neurons is much more important in street RABV infection than in infection with fixed RABV strains.

9.9.4 Ion channels

Defective neurotransmission is not the only potential explanation for functional impairment of neurons in rabies. Viral infections might also have important effects on ion channels of neurons. Studies were performed in vitro using RABV (RC-HL strain) infection of mouse neuroblastoma NA cells and the whole-cell patch clamp technique (Iwata, Komori, Unno, Minamoto, & Ohashi, 1999). The infection reduced the functional expression of voltage-dependent sodium channels and inward rectifier potassium channels, and there was a decreased resting membrane potential reflecting membrane depolarization. There was no change in the expression of delayed rectifier potassium channels, indicating that nonselective dysfunction of ion channels had not occurred. The reduction in the number of sodium channels and inward rectifier potassium channels could prevent infected neurons from firing action potentials and generating synaptic potentials, resulting in functional impairment.

RABV (RC-HL strain) infection of NG108-15 cells in vitro was not shown to alter the functional expression of voltage-dependent calcium ion channels (Iwata, Unno, Minamoto, Ohashi, & Komori, 2000). NG108-15 cells express both α_2-adrenoreceptors and muscarinic receptors. Induced voltage-dependent calcium ion channel current inhibition with noradrenaline (for α_2-adrenoreceptors) was decreased significantly in RABV infection, whereas carbachol (for muscarinic receptors) inhibition remained unchanged. Since α_2-adrenoreceptor-mediated inhibition of voltage-dependent calcium ion current serves as a brake mechanism to keep neurons from releasing their neurotransmitters beyond physiological requirements, the impaired modulation by α_2-adrenoreceptors could possibly contribute to clinical features of rabies, including hyperexcitability and aggressive behavior (Iwata et al., 2000).

9.9.5 Nitric oxide

Nitric oxide (NO) is a short-lived gaseous radical that acts as a biologic mediator for diverse cell types. It is produced by many different cells and mediates a variety of functions, including vasodilation, neurotransmission, immune cytotoxicity, production of synaptic plasticity in the brain and neurotoxicity (Lowenstein, Dinerman, & Snyder, 1994; Nathan, 1992). NO is released by the enzyme nitric oxide synthase (NOS), which also produces other reactive oxides of nitrogen (Nathan, 1992). There are three isoforms of NOS: neuronal NOS (nNOS, also NOS-1), inducible NOS (iNOS, also NOS-2), and endothelial NOS (eNOS, also NOS-3). nNOS is constitutively expressed and inducible by cytokines, including IFN-χ, TNF-α, and IL-12, whereas iNOS is inducible with lipopolysaccharides, IFN-χ and TNF-α.

NO plays a variety of roles in different viral infections (Reiss & Komatsu, 1998). In some viral infections (e.g., with Sindbis virus), inhibition of NOS results in increased mortality of infected mice, suggesting that NO plays a protective role in the pathogenesis of the viral infection (Tucker, Griffin, Choi, Bui, & Wesselingh, 1996). During infection with vesicular stomatitis virus (a rhabdovirus), NO has been shown to inhibit viral replication and promote viral clearance and recovery of infected mice (Komatsu, Bi, & Reiss, 1996).

Induction of iNOS mRNA occurred in mice infected experimentally with street RABV (Koprowski et al., 1993). iNOS mRNA was detected using RT-PCR amplification in the brains of three of six paralyzed mice, 9–14 days after inoculation of RABV in the masseter muscle. iNOS mRNA expression was induced rapidly in the brains of the rabid mice. It was speculated that NO and/or other endogenous neurotoxins may mediate the neuronal dysfunction in rabies and other infectious diseases (Koprowski et al., 1993; Zheng et al., 1993). The onset of clinical signs in RABV-infected rats and the clinical progression of the disease correlated with increasing quantities of NO in the brain to levels up to 30-fold more than in controls, which was determined using spin trapping of NO and electron paramagnetic resonance spectroscopy (Hooper et al., 1995). iNOS was detected by immunostaining in CVS-infected rats in many cells throughout the brain near blood vessels, which were identified as microglia and macrophages (Van Dam et al., 1995). CVS-24-infected rats developed a reduction in nNOS activity with reductions in nNOS mRNA and nNOS immunoreactivity and an increase in iNOS activity in the brain in a time-dependent manner (Akaike et al., 1995). Choline acetyltransferase activity in the brain remained unchanged, indicating that the decrease in nNOS activity did not reflect generalized neuronal loss. The NO produced by macrophages may be neurotoxic because its reaction with the superoxide anion O_2^- leads to the formation of peroxynitrate, which is a reactive oxidizing agent capable of causing tissue damage (Akaike et al., 1995). Ubol, Sukwattanapan, and Maneerat (2001) found that mice treated with the iNOS inhibitor, aminoguanidine (AG), delayed the death of CVS-11-infected mice by 1.0 to 1.6 days (depending on the dose). A delay in RABV replication was observed in the AG-treated mice. The role of NO in rabies pathogenesis clearly needs further study since it exerts both beneficial and detrimental effects and complex mechanisms are likely involved.

9.9.6 Excitotoxicity

Excitatory amino acids (e.g., glutamate) have been recognized to play a role in neuronal injury in a variety of neurological diseases, including stroke, epilepsy and neurodegenerative disorders. Recently, there is evidence that neurotropic viruses, including human immunodeficiency virus (Kaul & Lipton, 2004; Nath et al., 2000) and Sindbis virus (Darman et al., 2004; Nargi-Aizenman et al., 2004; Nargi-Aizenman & Griffin, 2001), induce neuronal injury through excitotoxic mechanisms. There has been recent speculation that the N-methyl-D-aspartate (NMDA) receptor may be one of the RABV receptors (Gosztonyi & Ludwig, 2001). Tsiang and coworkers reported that the noncompetitive NMDA antagonists ketamine and/or MK-801 inhibited RABV infection in primary neuron cultures, inhibited RABV genome transcription and restricted viral spread in an experimental model of rabies in rats (Lockhart, Tordo, & Tsiang, 1992; Lockhart, Tsiang, Ceccaldi, & Guillemer, 1991; Tsiang, Ceccaldi, Ermine, Lockhart, & Guillemer, 1991). The in vitro doses of ketamine and MK-801 used were much higher than required for stimulation of glutamate receptors, indicating that other mechanisms of actions, including antiviral effects, were likely of primary importance. In more recent studies, Weli, Scott, Ward, and Jackson (2006) observed that CVS-infected cortical and hippocampal mouse embryonic neurons showed loss of trypan blue exclusion, morphologic apoptotic features, and activated caspase 3 expression, indicating apoptosis and that the NMDA antagonists, ketamine (125 μM) and MK-801 (60 μM), had no significant neuroprotective effect. Glutamate-stimulated increases of intracellular calcium were reduced in CVS-infected hippocampal neurons compared with mock-infected neurons. Ketamine (120 mg/kg/d intraperitoneally) given to adult ICR mice infected with CVS via the hindlimb footpad produced no beneficial effects. Hence, there was no supportive evidence that excitotoxicity plays an important role in RABV infection or that ketamine is a useful therapeutic agent in this experimental model of rabies.

9.9.7 Oxidative stress

Oxidative stress plays a role in neurodegeneration in a variety of diseases, including Parkinson's disease and Alzheimer's disease, and amyotrophic lateral sclerosis (Andersen, 2004; Dexter et al., 1989; Giasson et al., 2000; Pedersen et al., 1998; Sayre et al., 1997). Oxidative stress in viral infections has also been recognized (Schwarz, 1996). Reactive oxygen species (ROS), which can be generated by mitochondria or the family of NADPH oxidases, modulate the permissiveness of cells to viral replication, regulate host inflammatory and immune responses, and cause oxidative damage to both host tissues and progeny virus (Valyi-Nagy & Dermody, 2005). Oxidative injury is observed in experimental acute encephalitis caused by herpes simplex virus 1 in mice (Milatovic et al., 2002; Schachtele, Hu, Little, & Lokensgard, 2010; Valyi-Nagy, Olson, Valyi-Nagy, Montine, & Dermody, 2000). RABV infection of DRG neurons can persist for more than 20 days without any cytopathic effects observed in the cultures (Tsiang, Ceccaldi, & Lycke, 1991), so they are a good cell type to evaluate the effects of RABV infection on neuronal processes (axons). RABV infection of many other primary neurons results in apoptotic cell death (Morimoto et al., 1999; Weli et al., 2006).

Jackson, Kammouni, Zherebitskaya, and Fernyhough (2010) evaluated immunostaining in CVS- and mock-infected cultures of DRG neurons derived from adult mice for neuron-specific β-tubulin, RABV antigen, and for amino acid adducts of 4-hydroxy-2-nonenal

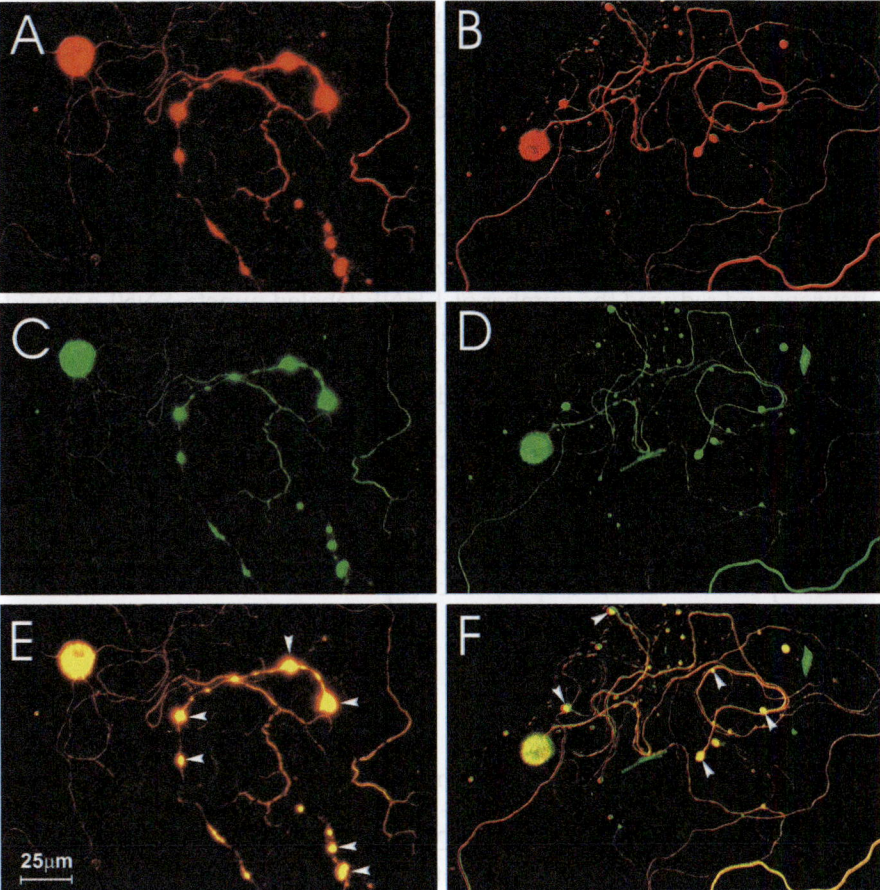

FIG. 9.5 CVS infection causes formation of axonal swellings in DRG cultures. Fluorescence microscopy showing CVS-infected mouse DRG neurons at 72 h p.i. β-tubulin is a marker of DRG neuronal cell bodies and axons (red) and expression of β-tubulin in CVS-infected neurons showed multiple axonal swellings (A, B). Staining for β-tubulin III (A) shows one (large spherical body) at 72 h p.i. (A). Axonal swellings are well established at 72 h p.i. (A-F; indicated by arrowheads in E, F). RABV antigen is strongly expressed in the neuronal cell bodies, axons, and axonal swellings at 72 h p.i. (C, E). Staining for 4-hydroxy-2-nonenal (4-HNE) (green) showed expression in the axons of CVS-infected neurons and showed accumulation in regions with axonal swellings (D, F). In CVS-infected neurons, merging of signals (yellow) for β-tubulin (E) and 4-HNE (F) showed there was strong expression of these elements in axons and in axonal swellings (arrowheads). *Adapted with permission from Jackson, A. C., Kammouni, W., Zherebitskaya, E., & Fernyhough, P. (2010). Role of oxidative stress in rabies virus infection of adult mouse dorsal root ganglion neurons.* Journal of Virology, 84(9), 4697–4705; doi: https://doi.org/10.1128/JVI.02654-09 Copyright ©2010, American Society for Microbiology.

(4-HNE), which is a marker of lipid peroxidation and, hence, oxidative stress (Fig. 9.5). Neuronal viability (by trypan blue exclusion), TUNEL staining, and axonal growth were also assessed in the cultures. CVS infected 33%–54% of cultured DRG neurons, similar to the findings of other investigators. Neuronal viability and TUNEL staining were similar in CVS- and mock-infected DRG neurons. There were significantly more 4-HNE-labeled puncta at 2 and 3 days postinfection (p.i.) in CVS-infected cultures than in mock infection.

Axonal outgrowth was reduced at these time points in CVS infection vs. mock-infected cultures. Axonal swellings with 4-HNE-labeled puncta were also associated with aggregations of actively respiring mitochondria, and recently it has been shown that 4-HNE directly impairs mitochondrial function in cultured DRG neurons (Akude, Zherebitskaya, Chowdhury, Girling, & Fernyhough, 2010).

Kammouni et al. (2012) evaluated whether the inducible transcription factor nuclear factor (NF)-κB acts as a critical bridge linking CVS infection and oxidative stress. CVS infection induced expression of NF-κB p50 subunit versus mock infection on Western immunoblotting. Ciliary neurotrophic factor, a potent activator of NF-κB, had no effect on mock-infected rat DRG neurons and reduced the number of 4-HNE-labeled puncta. SN50, a peptide inhibitor of NF-κB, and CVS infection had an additive effect in producing axonal swellings, indicating that NF-κB is neuroprotective. The fluorescent signal for subunit p50 was quantitatively evaluated in the nucleus and cytoplasm of mock- and CVS-infected rat DRG neurons. At 24h post-infection (p.i.), there was a significant increase in the nucleus:cytoplasm ratio, indicating increased transcriptional activity of NF-κB, perhaps as a response to stress. However, at both 48 and 72h p.i., there was significantly reduced nuclear localization of NF-κB. CVS infection may induce oxidative stress by inhibiting nuclear activation of NF-κB. A RABV protein may directly inhibit NF-κB activity, which occurs in infections with hepatitis C virus, cowpox, raccoonpox, some strains of vaccinia virus, and African swine fever virus.

9.9.8 Mitochondrial dysfunction

Because mitochondrial dysfunction is the major cause of oxidative stress, mitochondrial function was evaluated in CVS- vs. mock-infected cells (Alandijany, Kammouni, Roy Chowdhury, Fernyhough, & Jackson, 2013). CVS infection substantially increased maximal uncoupled respiration and Complex IV respiration and Complex I and Complex IV activities, but did not affect Complexes II-III or citrate synthase activities. Increases in Complex I activity, but not Complex IV activity, correlated with susceptibility of the cells to CVS infection. CVS infection maintained coupled respiration and rate of proton leak, indicating a tight mitochondrial coupling. Possibly as a result of enhanced Complex activity and efficient coupling, a high mitochondrial membrane potential was generated. CVS infection reduced the intracellular ATP level and altered the cellular redox state as indicated by a high NADH/NAD+ ratio. A higher rate of ROS generation occurred in CVS-infected neurons in the presence of mitochondrial substrates and inhibitors. Hence, CVS infection induces mitochondrial dysfunction leading to ROS overgeneration and oxidative stress.

In order to evaluate whether a RABV protein targets mitochondria and triggers dysfunction, mitochondrial extracts of mouse neuroblastoma cells were analyzed with a proteomics approach (Kammouni et al., 2015). Peptides were identified belonging to the RABV nucleocapsid protein (N), phosphoprotein (P), and glycoprotein (G), and the extract was most highly enriched with P. P was also detected by immunoblotting in CVS-infected purified mitochondrial extracts and also in Complex I immunoprecipitates from the extracts, but not in mock-infected extracts (Kammouni et al., 2015). A plasmid expressing P in cells increased Complex I activity and increased ROS generation, whereas expression of other CVS proteins did not.

Recombinant plasmids encoding various P gene segments were analyzed. Expression of a peptide from amino acid 139–172 increased Complex I activity and ROS generation similar to expression of the entire P protein, whereas peptides that did not contain this region did not increase Complex I activity or induce ROS generation (Kammouni et al., 2015). This indicated that a region of the CVS P interacts with Complex I in mitochondria causing mitochondrial dysfunction, increased generation of ROS, and oxidative stress.

Putative critical amino acid positions in the CVS P between 139 and 172 have been identified using a saturation alanine mutagenesis screen and by creating point mutations involving serine residues that decrease Complex I activity and decrease ROS production (Kammouni, Wood, & Jackson, 2017b). Two CVS recombinant viruses with serine to alanine mutations at positions 162 (A162r) and 166 (A166r) did not increase Complex I activity or ROS generation and also did not induce axonal swellings or inhibit axonal growth in DRG neurons (Kammouni et al., 2017b) (Fig. 9.6). The P of two street RABV variants and Mokola (MOK) virus was also evaluated (Kammouni, Wood, & Jackson, 2017a). The P of these viruses, like CVS, induces an increase in Complex I activities and ROS levels in transfected cells. Although the sequence homology of P is only 45% with MOK (higher for street viruses) and CVS, serine

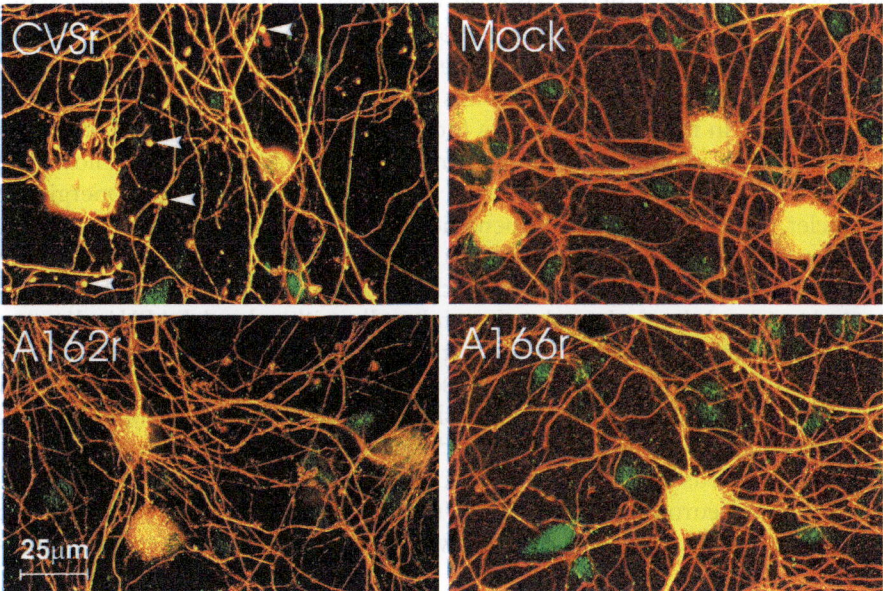

FIG. 9.6 Seventy-two hours after Infection of rat dorsal root ganglion neurons (MOI of 10) showing colocalization of immunofluorescent staining for tubulin (red) and 4-HNE (green) in axonal swellings (arrowheads) on the merged images (yellow), indicating oxidative stress, in recombinant CVS (CVSr) infection, but not with infection with A162r, A166r, or mock infection. Hence, the A162r and A166r mutants (both with serine to alanine changes) do not induce oxidative stress. *Reproduced with permission from Kammouni, W., Wood, H., & Jackson, A. C. (2017). Serine residues at positions 162 and 166 of the rabies virus phosphoprotein are critical for the induction of oxidative stress in rabies virus infection. Journal of Neurovirology, 23(3), 358–368; doi: https://doi.org/10.1007/s13365-016-0506-8 Copyright ©2017, Springer Nature.*

residues are conserved at positions 162 and 166, suggesting their potential importance in oxidative stress. Hence, RABV infection is a mitochondrial disorder initiated by interaction of the RABV P and Complex I. S162 and S166 are critical sites in the P for this interaction and the resulting mitochondrial dysfunction produces oxidative stress in neurons causing acute degenerative changes affecting neuronal processes resulting in a severe and fatal clinical disease.

9.9.9 Bases for behavioral changes

The neuroanatomical bases for the behavioral changes in animals with rabies have not yet been well characterized. Limbic system infection and dysfunction are suspected to play an important role in the behavioral changes, including alertness, loss of natural timidity, aberrant sexual behavior, and aggressiveness (Johnson, 1971). However, experimental rabies studies in these models have not been particularly helpful in giving insights into the neuroanatomical substrate for behavioral changes because these changes are not normally observed in rodent models and hippocampal infection actually occurs relatively late after peripheral routes of inoculation (Jackson & Reimer, 1989). The neural mechanisms of aggressive behavior are not well understood. Aggressive behavior is associated with lesions in a variety of locations in the brain, including the posterior olfactory bulbs, the ventromedial nucleus of the hypothalamus and the septal area (Isaacson, 1989). Offensive aggression, which is often impulsive and seemingly unprovoked, has been associated with low CNS serotonergic activity and also increased testosterone in humans and animal studies (Kalin, 1999). Aggressive behavior is essential in most rabies vectors for horizontal transmission of the virus to other hosts by biting. Early and selective brainstem infection in rabies would allow centrifugal spread of the virus to salivary glands as well as involvement of the serotonergic system in the raphe nuclei, resulting in aggressive behavior of animals with adequate cognitive and motor function in order to execute successful viral transmission by biting. Few studies have been performed in natural models of rabies in which aggressive behavior is exhibited. In the best available study, striped skunks inoculated peripherally with a skunk RABV isolate were compared with skunks infected with CVS (Smart & Charlton, 1992). The street virus–infected skunks exhibited aggressive responses to presentation of a stick in their cages, whereas this behavior was not observed in CVS-infected skunks. Heavy accumulations of viral antigen were found in the midbrain raphe nuclei, red nucleus, dorsal motor nucleus of the vagus and hypoglossal nucleus in street virus–infected skunks, but not in CVS-infected skunks. Impaired serotonin neurotransmission from the raphe nuclei in the brainstem, however, may account for the development of aggressive behavior in natural vectors of rabies.

9.10 Recovery from rabies and chronic RABV infection

Although rabies is usually considered a uniformly fatal disease, it has been recognized that animals sometimes may recover from rabies. Recovery from rabies has also been called *abortive rabies*, which can occur either with or without neurologic sequelae (Bell, 1975). There have been a large number of reports of survival after the development of neurologic illness, particularly in experimental animals (Jackson, 1997). Because of limitations on laboratory diagnostic tests performed during life, a conclusive diagnosis of rabies is only rarely made in natural cases that recover. Animals clinically suspected of having rabies are usually killed and they do not have an opportunity to recover. In a series of five reports from the Pasteur Institute of Southern India, the unusual case of a chronically infected dog has been described. A 14-year-old boy died with hydrophobia 48 days after he stepped on a dog and was bitten in November 1965 (Veeraraghavan et al., 1967; Veeraraghavan et al., 1968; Veeraraghavan et al., 1969; Veeraraghavan et al., 1970; Veeraraghavan, Gajanana, & Rangasami, 1967). The dog was observed at the Pasteur Institute until it died in February 1969 (Veeraraghavan et al., 1970). During that period, RABV was isolated from daily saliva samples taken from the dog on 13 occasions between January and May 1966 (Veeraraghavan, Gajanana, Rangasami, Kumari, et al., 1967) and once in January 1967 after the dog was given a course of prednisolone (Veeraraghavan et al., 1968). RABV was not isolated postmortem from the dog's brain, spinal cord or salivary glands, although fluorescent antibody staining showed RABV antigen in its brain and spinal cord (Veeraraghavan et al., 1970). No antiRABV antibodies were detected in the dog's blood at any time (Veeraraghavan et al., 1968; Veeraraghavan et al., 1969; Veeraraghavan et al., 1970; Veeraraghavan, Gajanana, Rangasami, Kumari, et al., 1967). Although this is an extremely interesting and unusual series of reports, it is unlikely that this seronegative dog excreted a virulent RABV that was responsible for the boy's death. The boy may have become infected from an undocumented rabies exposure months or even years earlier (Smith et al., 1991). There was a poor correlation of this laboratory's results with viral isolation and antigen detection in saliva samples and in CNS tissues from the dog. This might be explained by the presence of neutralizing antibodies in tissues, but this dog was seronegative. The viral isolations from saliva samples could be explained by crosscontamination of specimens in the laboratory. Because of a number of inconsistencies in these reports, the validity of this series of reports remains uncertain.

In another report, five dogs in Ethiopia are described that remained healthy for up to 72 months after the first isolation of RABV from their saliva (Fekadu, 1972; Fekadu, 1975). However, exposures from these dogs did not result in any human cases of rabies. In a follow-up study, secretion of RABV was documented in the saliva of a dog experimentally infected with an Ethiopian strain of dog RABV for up to 6 months after its recovery from rabies (Fekadu, Shaddock, & Baer, 1981).

Finally, remarkable cases of experimental rabies in two cats have been reported (Murphy et al., 1980). Cat 1 developed paralysis, most marked in its hindlimbs, 17 days after inoculation of a RABV strain isolated from a big brown bat. The cat showed slow progressive recovery until 100 weeks after inoculation, when it developed progressive neurologic deterioration with aggressive behavior and weakness and atrophy; it was killed for further study 136 weeks after inoculation. Cat 2 remained well for 120 weeks after inoculation, before it developed progressive neurologic deterioration; it was killed at 136 weeks after inoculation. There were high titers of neutralizing antibody in the serum and CSF of both cats. RABV was not isolated from the saliva or tissue suspensions from these cats, but was isolated from the brain of cat 2 by explant culture techniques. RABV antigen was detected at multiple sites in the CNS and viral inclusions were found in neurons at four sites in the brain of cat 2. Degenerative neuronal changes were noted and there were extensive inflammatory changes in both cats (Fig. 9.7). Perl, Bell, Moore, and Stewart (1977) also reported a similar recrudescent form

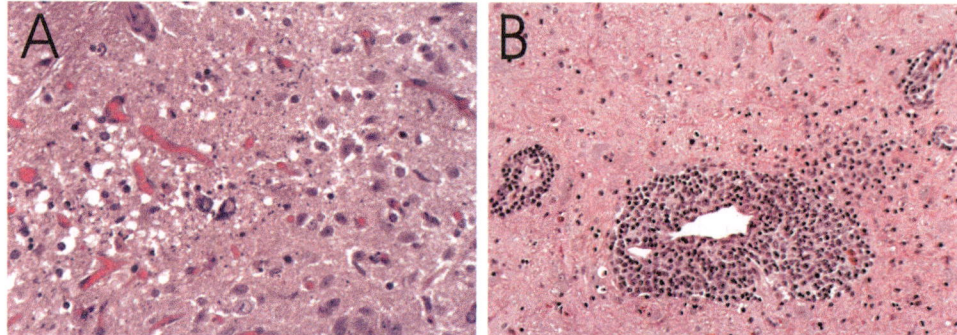

FIG. 9.7 Medial geniculate body of the thalamus of cat 1, which was killed at 136 weeks post-infection, showing degenerative neuronal changes with vacuolation (A) and massive perivascular lymphocytic and plasmacytic infiltration (B), which were seen throughout the brain. Hematoxylin and eosin; magnifications: A, ×115; B, ×60. *Courtesy of Dr. Frederick A. Murphy, University of Texas Medical Branch, Galveston, TX.*

of rabies in a cat experimentally infected with a bat RABV isolate. At necropsy, there were features of chronic encephalitis. RABV could not be isolated from CNS tissues, probably because of the presence of neutralizing antibodies. These well-documented extraordinary cases indicate that chronic RABV infection may occur rarely, at least under experimental conditions. However, it is doubtful whether chronic rabies infections have any significance in the natural history of rabies, including a role in perpetuation of rabies in natural reservoirs. If animals with chronic rabies are unable to transmit the virus and are incompetent vectors, then this chronic state may not have any biological importance in nature.

Early studies on rabies pathogenesis in vampire bats, which were performed in Trinidad, suggested that bats might be chronically infected with RABV and secrete infectious RABV over periods lasting up to several months (Pawan, 1936). These early studies were performed before modern virological methods became available and suffered from inadequate diagnostic evaluations, which was largely limited to examination of tissues for Negri bodies. Infections with a variety of other bat viruses, including Rio Bravo virus, may have been misdiagnosed as RABV (Constantine, 1988; Moreno & Baer, 1980). More recent experimental studies have shown that vampire bats have variable incubation periods lasting up to 4 weeks and then develop an acute disease with excretion of virus in the saliva that is not prolonged (Moreno & Baer, 1980). A study of Brazilian free-tailed bats from a dense cave population in New Mexico revealed that 69% of the bats had neutralizing RABV antibodies, but only 0.5% had active infection as assessed by direct fluorescent antibody testing of the brain (Steece & Altenbach, 1989). Hence, seroconversion likely occurs in many naturally infected bats, although it is unknown whether any central nervous system involvement normally occurs in this setting, and fatal infections may be relatively infrequent. In Spain, serotine bats (*Eptesicus serotinus*) in which EBLV-1 infection has been recognized were recently studied with RT-PCR amplification of oropharyngeal swabs and simultaneous brain samples. Of 33 bats, a positive RT-PCR result was found in 13 (39%) oropharyngeal swabs and 5 (15%) brains and the positive brains were usually associated with clinical disease (Echevarria, Avellon, Juste, Vera, & Ibanez, 2001). Unless the infection was associated with a previously unrecognized pattern of viral spread in the host, viral RNA was cleared from the brain but not from extraneural tissues in many of these bats. Of course, a positive RT-PCR result does not indicate the presence of infectious virus at a site, rather, it may be a marker of remote infection in some cases.

9.11 Conclusions

Rabies is a normally fatal viral infection of the nervous system in humans and animals with characteristic clinical manifestations. Considerable progress has been made in understanding the pathogenesis of rabies. RABV is highly neurotropic. It binds to the nAChR at the neuromuscular junction and it spreads by axonal transport via peripheral nerves to the CNS, where it causes widespread infection in neurons within the CNS. The combination of virus-induced behavioral changes in rabies vectors and centrifugal spread of the virus to salivary glands allows efficient transmission of the infection. An understanding of RABV neurovirulence is emerging from basic studies of virus variants in a variety of animal models. A single amino acid change in the RABV glycoprotein at position 333 has dramatic effects on the outcome of infection and it affects both the efficiency of viral spread involving different afferent and efferent pathways and cellular tropisms. The precise events at the site of viral entry during the long incubation period of rabies remain poorly understood. The fundamental basis for neuronal dysfunction in rabies has not yet been determined, although there are several hypotheses under active study at the present time. Recent evidence indicates the importance of oxidative stress due to mitochondrial dysfunction in RABV infection. A better understanding of rabies pathogenesis will, hopefully, lead to advances in the treatment of rabies and other viral diseases.

References

Adle-Biassette, H., Bourhy, H., Gisselbrecht, M., Chretien, F., Wingertsmann, L., Baudrimont, M., … Gray, F. (1996). Rabies encephalitis in a patient with AIDS: A clinicopathological study. *Acta Neuropathologica, 92*(4), 415–420.

Akaike, T., Weihe, E., Schaefer, M., Fu, Z. F., Zheng, Y. M., Vogel, W., … Dietzschold, B. (1995). Effect of neurotropic virus infection on neuronal and inducible nitric oxide synthase activity in rat brain. *Journal of Neurovirology, 1*(1), 118–125.

Akude, E., Zherebitskaya, E., Chowdhury, S. K. R., Girling, K., & Fernyhough, P. (2010). 4-Hydroxy-2-nonenal induces mitochondrial dysfunction and aberrant axonal outgrowth in adult sensory neurons that mimics features of diabetic neuropathy. *Neurotoxicology Research, 17*(1), 28–38.

Alandijany, T., Kammouni, W., Roy Chowdhury, S. K., Fernyhough, P., & Jackson, A. C. (2013). Mitochondrial dysfunction in rabies virus infection of neurons. *Journal of Neurovirology, 19*(6), 537–549.

Allsopp, T. E., & Fazakerley, J. K. (2000). Altruistic cell suicide and the specialized case of the virus-infected nervous system. *Trends in Neurological Sciences, 23*(7), 284–290.

Andersen, J. K. (2004). Oxidative stress in neurodegeneration: Cause or consequence? *Nature Medicine, 10*(Suppl), S18–S25.

Araujo, M. D. F., de Brito, T., & Machado, C. G. (1971). Myocarditis in human rabies. *Revista do Instituto de Medicina Tropical de São Paulo, 13*, 99–102.

Baer, G. M., Bellini, W. J., & Fishbein, D. B. (1990). Rhabdoviruses. In B. N. Fields, D. M. Knipe, R. M. Chanock, M. S. Hirsch, J. L. Melnick, T. P. Monath, & B. Roizman (Eds.), *Virology: Volume 1* (2nd ed., pp. 883–930). New York: Raven Press.

Baer, G. M., & Cleary, W. F. (1972). A model in mice for the pathogenesis and treatment of rabies. *Journal of Infectious Diseases, 125*, 520–527.

Baer, G. M., Shaddock, J. H., Quirion, R., Dam, T. V., & Lentz, T. L. (1990). Rabies susceptibility and acetylcholine receptor (Letter). *Lancet, 335*, 664–665.

Baer, G. M., Shantha, T. R., & Bourne, G. H. (1968). The pathogenesis of street rabies virus in rats. *Bulletin of the World Health Organization, 38*, 119–125.

Baer, G. M., Shanthaveerappa, T. R., & Bourne, G. H. (1965). Studies on the pathogenesis of fixed rabies virus in rats. *Bulletin of the World Health Organization, 33*, 783–794.

Balachandran, A., & Charlton, K. (1994). Experimental rabies infection of non-nervous tissues in skunks (*Mephitis mephitis*) and foxes (*Vulpes vulpes*). *Veterinary Pathology, 31*, 93–102.

Baloul, L., & Lafon, M. (2003). Apoptosis and rabies virus neuroinvasion. *Biochimie, 85*(8), 777–788.

Begeman, L., Geurtsvan Kessel, C., Finke, S., Freuling, C. M., Koopmans, M., Muller, T., … Kuiken, T. (2018). Comparative pathogenesis of rabies in bats and carnivores, and implications for spillover to humans. *The Lancet Infectious Diseases, 18*(4), e147–e159. https://doi.org/10.1016/S1473-3099 (17)30574–1.

Bell, J. F. (1975). Latency and abortive rabies. In G. M. Baer (Ed.), *The natural history of rabies* (pp. 331–354). New York: Academic Press.

Bouzamondo, E., Ladogana, A., & Tsiang, H. (1993). Alteration of potassium-evoked 5-HT release from virus-infected rat cortical synaptosomes. *NeuroReport, 4*(5), 555–558.

Bracci, L., Antoni, G., Cusi, M. G., Lozzi, L., Niccolai, N., Petreni, S., … Neri, P. (1988). Antipeptide monoclonal antibodies inhibit the binding of rabies virus glycoprotein and alpha-bungarotoxin to the nicotinic acetylcholine receptor. *Molecular Immunology, 25*, 881–888.

Broughan, J. H., & Wunner, W. H. (1995). Characterization of protein involvement in rabies virus binding to BHK-21 cells. *Archives of Virology, 140*, 75–93.

Buja, L. M., Eigenbrodt, M. L., & Eigenbrodt, E. H. (1993). Apoptosis and necrosis: Basic types and mechanisms of cell death. *Archives of Pathology and Laboratory Medicine, 117*, 1208–1214.

Butowt, R., & Von Bartheld, C. S. (2003). Connecting the dots: Trafficking of neurotrophins, lectins and diverse pathogens by binding to the neurotrophin receptor p75NTR. *European Journal of Neuroscience, 17*(4), 673–680.

Carbone, K. M., Duchala, C. S., Griffin, J. W., Kincaid, A. L., & Narayan, O. (1987). Pathogenesis of Borna disease in rats: Evidence that intra-axonal spread is the major route for virus dissemination and the determinant for disease incubation. *Journal of Virology, 61*, 3431–3440.

Carson, M. J. (2002). Microglia as liaisons between the immune and central nervous systems: Functional implications for multiple sclerosis. *Glia, 40*(2), 218–231.

Ceccaldi, P.-E., Ermine, A., & Tsiang, H. (1990). Continuous delivery of colchicine in the rat brain with osmotic pumps for inhibition of rabies virus transport. *Journal of Virological Methods, 28*, 79–84.

Ceccaldi, P.-E., Fillion, M.-P., Ermine, A., Tsiang, H., & Fillion, G. (1993). Rabies virus selectively alters 5-HT_1 receptor subtypes in rat brain. *European Journal of Pharmacology, 245*(2), 129–138.

Ceccaldi, P. E., Gillet, J. P., & Tsiang, H. (1989). Inhibition of the transport of rabies virus in the central nervous system. *Journal of Neuropathology and Experimental Neurology, 48*, 620–630.

Charlton, K. M., & Casey, G. A. (1979a). Experimental oral and nasal transmission of rabies virus in mice. *Canadian Journal of Comparative Medicine, 43*, 10–15.

Charlton, K. M., & Casey, G. A. (1979b). Experimental rabies in skunks: Immunofluorescence light and electron microscopic studies. *Laboratory Investigation, 41*, 36–44.

Charlton, K. M., & Casey, G. A. (1979c). Experimental rabies in skunks: Oral, nasal, tracheal and intestinal exposure. *Canadian Journal of Comparative Medicine, 43*, 168–172.

Charlton, K. M., Casey, G. A., & Campbell, J. B. (1983). Experimental rabies in skunks: Mechanisms of infection of the salivary glands. *Canadian Journal of Comparative Medicine, 47*, 363–369.

Charlton, K. M., Casey, G. A., Wandeler, A. I., & Nadin-Davis, S. (1996). Early events in rabies virus infection of the central nervous system in skunks (*Mephitis mephitis*). *Acta Neuropathologica, 91*, 89–98.

Charlton, K. M., Nadin-Davis, S., Casey, G. A., & Wandeler, A. I. (1997). The long incubation period in rabies: Delayed progression of infection in muscle at the site of exposure. *Acta Neuropathologica, 94*(1), 73–77.

Cheetham, H. D., Hart, J., Coghill, N. F., & Fox, B. (1970). Rabies with myocarditis: Two cases in England. *Lancet, 1*, 921–922.

Constantine, D. G. (1962). Rabies transmission by nonbite route. *Public Health Reports, 77*, 287–289.

Constantine, D. G. (1988). Transmission of pathogenic organisms by vampire bats. In A. M. Greenhall & U. Schmidt (Eds.), *Natural history of vampire bats* (pp. 167–189). Boca Raton, Florida: CRC Press.

Constantine, D. G., Emmons, R. W., & Woodie, J. D. (1972). Rabies virus in nasal mucosa of naturally infected bats. *Science, 175*, 1255–1256.

Conti, C., Superti, F., & Tsiang, H. (1986). Membrane carbohydrate requirement for rabies virus binding to chicken embryo related cells. *Intervirology, 26*(3), 164–168.

Coulon, P., Derbin, C., Kucera, P., Lafay, F., Prehaud, C., & Flamand, A. (1989). Invasion of the peripheral nervous systems of adult mice by the CVS strain of rabies virus and its avirulent derivative AvO1. *Journal of Virology, 63*, 3550–3554.

Coulon, P., Rollin, P., Aubert, M., & Flamand, A. (1982). Molecular basis of rabies virus virulence. I. Selection of avirulent mutants of the CVS strain with anti-G monoclonal antibodies. *Journal of General Virology, 61*, 97–100.

Darman, J., Backovic, S., Dike, S., Maragakis, N. J., Krishnan, C., Rothstein, J. D., … Kerr, D. A. (2004). Viral-induced spinal motor neuron death is non-cell-autonomous and involves glutamate excitotoxicity. *Journal of Neuroscience, 24*(34), 7566–7575.

Davis, B. M., Rall, G. F., & Schnell, M. J. (2015). Everything you always wanted to know about rabies virus (but were afraid to ask). *Annual Review of Virology, 2*(1), 451–471.

Dean, D. J., Evans, W. M., & McClure, R. C. (1963). Pathogenesis of rabies. *Bulletin of the World Health Organization, 29*, 803–811.

Debbie, J. G., & Trimarchi, C. V. (1970). Pantropism of rabies virus in free-ranging rabid red fox *Vulpes fulva. Journal of Wildlife Diseases, 6*, 500–506.

Dexter, D. T., Carter, C. J., Wells, F. R., Javoy-Agid, F., Agid, Y., Lees, A., … Marsden, C. D. (1989). Basal lipid peroxidation in substantia nigra is increased in Parkinson's disease. *Journal of Neurochemistry, 52*(2), 381–389.

Dierks, R. E. (1975). Electron microscopy of extraneural rabies infection. In G. M. Baer (Ed.), *The natural history of rabies* (pp. 303–318). New York: Academic Press.

Dietzschold, B., Wunner, W. H., Wiktor, T. J., Lopes, A. D., Lafon, M., Smith, C. L., & Koprowski, H. (1983). Characterization of an antigenic determinant of the glycoprotein that correlates with pathogenicity of rabies virus. *Proceedings of the National Academy of Sciences of the United States of America, 80*, 70–74.

Dumrongphol, H., Srikiatkhachorn, A., Hemachudha, T., Kotchabhakdi, N., & Govitrapong, P. (1996). Alteration of muscarinic acetylcholine receptors in rabies viral-infected dog brains. *Journal of the Neurological Sciences, 137*(1), 1–6.

Echevarria, J. E., Avellon, A., Juste, J., Vera, M., & Ibanez, C. (2001). Screening of active lyssavirus infection in wild bat populations by viral RNA detection on oropharyngeal swabs. *Journal of Clinical Microbiology, 39*(10), 3678–3683.

Emmelin, N. (1967). Nervous control of salivary glands. In C.F. Code (Ed.), Handbook of physiology, Section 6, Volume II (pp. 595-632). Washington, DC: American Physiological Society.

Ermine, A., & Flamand, A. (1977). RNA syntheses in BHK_{21} cells infected by rabies virus. *Annals of Microbiology, 128*, 477–488.

Etessami, R., Conzelmann, K. K., Fadai-Ghotbi, B., Natelson, B., Tsiang, H., & Ceccaldi, P. E. (2000). Spread and pathogenic characteristics of a G-deficient rabies virus recombinant: An in vitro and in vivo study. *Journal of General Virology, 81*, 2147–2153.

Faber, M., Li, J., Kean, R. B., Hooper, D. C., Alugupalli, K. R., & Dietzschold, B. (2009). Effective preexposure and postexposure prophylaxis of rabies with a highly attenuated recombinant rabies virus. *Proceedings of the National Academy of Sciences of the United States of America, 106*(27), 11300–11305.

Fekadu, M. (1972). Atypical rabies in dogs in Ethiopia. *Ethiopian Medical Journal, 10*, 79–86.

Fekadu, M. (1975). Asymptomatic non-fatal canine rabies (Letter). *Lancet, 1*, 569.

Fekadu, M., Shaddock, J. H., & Baer, G. M. (1981). Intermittent excretion of rabies virus in the saliva of a dog two and six months after it had recovered from experimental rabies. *American Journal of Tropical Medicine and Hygiene, 30*, 1113–1115.

Feng, G., Mellor, R. H., Bernstein, M., Keller-Peck, C., Nguyen, Q. T., Wallace, M., … Sanes, J. R. (2000). Imaging neuronal subsets in transgenic mice expressing multiple spectral variants of GFP. *Neuron, 28*(1), 41–51.

Fischman, H. R., & Schaeffer, M. (1971). Pathogenesis of experimental rabies as revealed by immunofluorescence. *Annals of the New York Academy of Sciences, 177*, 78–97.

Fischman, H. R., & Ward, F. E. (1968). Oral transmission of rabies virus in experimental animals. *American Journal of Epidemiology, 88*(1), 132–138.

Flamand, A., Coulon, P., Pepin, M., Blancou, J., Rollin, P., & Portnoi, D. (1984). Immunogenic and protective power of avirulent mutants of rabies virus selected with neutralizing monoclonal antibodies. In R. M. Chanock & R. A. Lerner (Eds.), *Modern approaches to vaccines: Molecular and chemical basis of virus virulence and immunogenicity* (pp. 289–294). Cold Spring Harbor, New York: Cold Spring Harbor Laboratory.

Fu, Z. F., Weihe, E., Zheng, Y. M., Schafer, M. -H., Sheng, H., Corisdeo, S., … Dietzschold, B. (1993). Differential effects of rabies and Borna disease viruses on immediate-early- and late-response gene expression in brain tissues. *Journal of Virology, 67*, 6674–6681.

Giasson, B. I., Duda, J. E., Murray, I. V., Chen, Q., Souza, J. M., Hurtig, H. I., … Lee, V. M. (2000). Oxidative damage linked to neurodegeneration by selective alpha-synuclein nitration in synucleinopathy lesions. *Science, 290*(5493), 985–989.

Gillet, J. P., Derer, P., & Tsiang, H. (1986). Axonal transport of rabies virus in the central nervous system of the rat. *Journal of Neuropathology and Experimental Neurology, 45*, 619–634.

Gloor, S. M., Wachtel, M., Bolliger, M. F., Ishihara, H., Landmann, R., & Frei, K. (2001). Molecular and cellular permeability control at the blood-brain barrier. *Brain Research Reviews, 36*(2-3), 258–264.

Gosztonyi, G., & Ludwig, H. (2001). Interactions of viral proteins with neurotransmitter receptors may protect or destroy neurons. *Current Topics in Microbiology and Immunology, 253*, 121–144.

Gourmelon, P., Briet, D., Clarencon, D., Court, L., & Tsiang, H. (1991). Sleep alterations in experimental street rabies virus infection occur in the absence of major EEG abnormalities. *Brain Research, 554*, 159–165.

Gourmelon, P., Briet, D., Court, L., & Tsiang, H. (1986). Electrophysiological and sleep alterations in experimental mouse rabies. *Brain Research, 398*, 128–140.

Guigoni, C., & Coulon, P. (2002). Rabies virus is not cytolytic for rat spinal motoneurons in vitro. *Journal of Neurovirology, 8*(4), 306–317.

Hamilton, R. D., Foss, A. J., & Leach, L. (2007). Establishment of a human in vitro model of the outer blood-retinal barrier. *Journal of Anatomy, 211*(6), 707–716.

Hanham, C. A., Zhao, F., & Tignor, G. H. (1993). Evidence from the anti-idiotypic network that the acetylcholine receptor is a rabies virus receptor. *Journal of Virology, 67*, 530–542.

Hardwick, J. M. (1997). Virus-induced apoptosis. *Advances in Pharmacology, 41*, 295–336.

Hemachudha, T., Ugolini, G., Wacharapluesadee, S., Sungkarat, W., Shuangshoti, S., & Laothamatas, J. (2013). Human rabies: Neuropathogenesis, diagnosis, and management. *Lancet Neurology, 12*(5), 498–513.

Hooper, D. C., Ohnishi, S. T., Kean, R., Numagami, Y., Dietzschold, B., & Koprowski, H. (1995). Local nitric oxide production in viral and autoimmune diseases of the central nervous system. *Proceedings of the National Academy of Sciences of the United States of America, 92*(12), 5312–5316.

Hooper, D. C., Phares, T. W., Fabis, M. J., & Roy, A. (2009). The production of antibody by invading B cells is required for the clearance of rabies virus from the central nervous system. *PLoS Neglected Tropical Diseases, 3*(10), e535.

Hronovsky, V., & Benda, R. (1969). Development of inhalation rabies infection in suckling guinea pigs. *Acta Virologica, 13*, 198–202.

Hueffer, K., Khatri, S., Rideout, S., Harris, M. B., Papke, R. L., Stokes, C., & Schulte, M. K. (2017). Rabies virus modifies host behaviour through a snake-toxin like region of its glycoprotein that inhibits neurotransmitter receptors in the CNS. *Scientific Reports, 7*(1), 12818.

Isaacson, R. L. (1989). The neural and behavioural mechanisms of aggression and their alteration by rabies and other viral infections. In O. Thraenhart & H. Koprowski (Eds.), *rabies control: Proceedings of the Second International IMVI ESSEN/WHO Symposium on "New Developments in Rabies Control", Essen, 5–7 July 1988; and, Report of the WHO Consul+tation on Rabies, Essen, 8 July 1988 WHO Consultation on Rabies* (pp. 17–23). Royal Tunbridge Wells, Kent: Wells Medical.

Iwasaki, Y., & Clark, H. F. (1975). Cell to cell transmission of virus in the central nervous system. II. Experimental rabies in mouse. *Laboratory Investigation, 33*, 391–399.

Iwata, M., Komori, S., Unno, T., Minamoto, N., & Ohashi, H. (1999). Modification of membrane currents in mouse neuroblastoma cells following infection with rabies virus. *British Journal of Pharmacology, 126*(8), 1691–1698.

Iwata, M., Unno, T., Minamoto, N., Ohashi, H., & Komori, S. (2000). Rabies virus infection prevents the modulation by α_2-adrenoceptors, but not muscarinic receptors, of Ca^{2+} channels in NG108-15 cells. *European Journal of Pharmacology, 404*(1–2), 79–88.

Jackson, A. C. (1991a). Analysis of viral neurovirulence. In J. Brosius & R. T. Fremeau (Eds.), *Molecular genetic approaches to neuropsychiatric diseases* (pp. 259–277). San Diego: Academic Press.

Jackson, A. C. (1991b). Biological basis of rabies virus neurovirulence in mice: Comparative pathogenesis study using the immunoperoxidase technique. *Journal of Virology, 65*(1), 537–540.

Jackson, A. C. (1993). Cholinergic system in experimental rabies in mice. *Acta Virologica, 37*(6), 502–508.

Jackson, A. C. (1994). Animal models of rabies virus neurovirulence. In C. E. Rupprecht, B. Dietzschold, & H. Koprowski (Eds.), *Lyssaviruses: Vol. 187, Current topics in microbiology and immunology* (pp. 85–93). (pp. 85–93). Berlin: Springer-Verlag.

Jackson, A. C. (1997). Rabies. In N. Nathanson, R. Ahmed, F. Gonzalez-Scarano, D. E. Griffin, K. Holmes, F. A. Murphy, & H. L. Robinson (Eds.), *Viral pathogenesis* (pp. 575–591). Philadelphia: Lippincott-Raven.

Jackson, A. C. (1999). Apoptosis in experimental rabies in *bax*-deficient mice. *Acta Neuropathologica, 98*(3), 288–294.

Jackson, A. C. (2003). Neuronal apoptosis in experimental rabies: Role of the route of viral entry. *Neurology, 60*(Suppl. 1), A102.

Jackson, A. C., Kammouni, W., Zherebitskaya, E., & Fernyhough, P. (2010). Role of oxidative stress in rabies virus infection of adult mouse dorsal root ganglion neurons. *Journal of Virology, 84*(9), 4697–4705.

Jackson, A. C., & Park, H. (1998). Apoptotic cell death in experimental rabies in suckling mice. *Acta Neuropathologica, 95*(2), 159–164.

Jackson, A. C., & Park, H. (1999). Experimental rabies virus infection of p75 neurotrophin receptor-deficient mice. *Acta Neuropathologica, 98*(6), 641–644.

Jackson, A. C., Phelan, C. C., & Rossiter, J. P. (2000). Infection of Bergmann glia in the cerebellum of a skunk experimentally infected with street rabies virus. *Canadian Journal of Veterinary Research, 64*(4), 226–228.

Jackson, A. C., Randle, E., Lawrance, G., & Rossiter, J. P. (2008). Neuronal apoptosis does not play an important role in human rabies encephalitis. *Journal of Neurovirology, 14*(5), 368–375.

Jackson, A. C., Rasalingam, P., & Weli, S. C. (2006). Comparative pathogenesis of recombinant rabies vaccine strain SAD-L16 and SAD-D29 with replacement of Arg333 in the glycoprotein after peripheral inoculation of neonatal mice: Less neurovirulent strain is a stronger inducer of neuronal apoptosis. *Acta Neuropathologica, 111*(4), 372–378.

Jackson, A. C., & Reimer, D. L. (1989). Pathogenesis of experimental rabies in mice: An immunohistochemical study. *Acta Neuropathologica, 78*(2), 159–165.

Jackson, A. C., Reimer, D. L., & Ludwin, S. K. (1989). Spontaneous recovery from the encephalomyelitis in mice caused by street rabies virus. *Neuropathology and Applied Neurobiology, 15*(5), 459–475.

Jackson, A. C., & Rossiter, J. P. (1997). Apoptosis plays an important role in experimental rabies virus infection. *Journal of Virology, 71*(7), 5603–5607.

Jackson, A. C., Ye, H., Phelan, C. C., Ridaura-Sanz, C., Zheng, Q., Li, Z., … Lopez-Corella, E. (1999). Extraneural organ involvement in human rabies. *Laboratory Investigation, 79*(8), 945–951.

Jacob, Y., Badrane, H., Ceccaldi, P. E., & Tordo, N. (2000). Cytoplasmic dynein LC8 interacts with lyssavirus phosphoprotein. *Journal of Virology, 74*(21), 10217–10222.

Johnson, R. T. (1965). Experimental rabies: Studies of cellular vulnerability and pathogenesis using fluorescent antibody staining. *Journal of Neuropathology and Experimental Neurology, 24*, 662–674.

Johnson, R. T. (1971). The pathogenesis of experimental rabies. In Y. Nagano & F. M. Davenport (Eds.), *Rabies* (pp. 59–75). Baltimore: University Park Press.

Johnson, N., McKimmie, C. S., Mansfield, K. L., Wakeley, P. R., Brookes, S. M., Fazakerley, J. K., & Fooks, A. R. (2006). Lyssavirus infection activates interferon gene expression in the brain. *Journal of General Virology, 87*(Pt 9), 2663–2667.

Julius, D. (1991). Molecular biology of serotonin receptors. *Annual Review of Neuroscience, 14*, 335–360.

Juntrakul, S., Ruangvejvorachai, P., Shuangshoti, S., Wacharapluesadee, S., & Hemachudha, T. (2005). Mechanisms of escape phenomenon of spinal cord and brainstem in human rabies. *BMC Infectious Diseases, 5*(1), 104.

Kalin, N. H. (1999). Primate models to understand human aggression. *Journal of Clinical Psychiatry, 60*(Suppl. 15), 29–32.

Kammouni, W., Hasan, L., Saleh, A., Wood, H., Fernyhough, P., & Jackson, A. C. (2012). Role of nuclear factor-κB in oxidative stress associated with rabies virus infection of adult rat dorsal root ganglion neurons. *Journal of Virology, 86*(15), 8139–8146.

Kammouni, W., Wood, H., & Jackson, A. C. (2017a). Lyssavirus phosphoproteins increase mitochondrial complex I activity and levels of reactive oxygen species. *Journal of Neurovirology, 23*(5), 756–762.

Kammouni, W., Wood, H., & Jackson, A. C. (2017b). Serine residues at positions 162 and 166 of the rabies virus phosphoprotein are critical for the induction of oxidative stress in rabies virus infection. *Journal of Neurovirology, 23*(3), 358–368.

Kammouni, W., Wood, H., Saleh, A., Appolinario, C. M., Fernyhough, P., & Jackson, A. C. (2015). Rabies virus phosphoprotein interacts with mitochondrial Complex I and induces mitochondrial dysfunction and oxidative stress. *Journal of Neurovirology, 21*(4), 370–382.

Kassis, R., Larrous, F., Estaquier, J., & Bourhy, H. (2004). Lyssavirus matrix protein induces apoptosis by a TRAIL-dependent mechanism involving caspase-8 activation. *Journal of Virology, 78*(12), 6543–6555.

Kaul, M., & Lipton, S. A. (2004). Signaling pathways to neuronal damage and apoptosis in human immunodeficiency virus type 1-associated dementia: Chemokine receptors, excitotoxicity, and beyond. *Journal of Neurovirology, 10*(Suppl. 1), 97–101.

Kelly, R. M., & Strick, P. L. (2000). Rabies as a transneuronal tracer of circuits in the central nervous system. *Journal of Neuroscience Methods, 103*(1), 63–71.

Koch, F. J., Sagartz, J. W., Davidson, D. E., & Lawhaswasdi, K. (1975). Diagnosis of human rabies by the cornea test. *American Journal of Clinical Pathology, 63*, 509–515.

Komatsu, T., Bi, Z., & Reiss, C. S. (1996). Interferon-γ induced type I nitric oxide synthase activity inhibits viral replication in neurons. *Journal of Neuroimmunology, 68*(1-2), 101–108.

Koprowski, H., Zheng, Y. M., Heber-Katz, E., Fraser, N., Rorke, L., Fu, Z. F., ... Dietzschold, B. (1993). In vivo expression of inducible nitric oxide synthase in experimentally induced neurologic disease. *Proceedings of the National Academy of Sciences of the United States of America, 90*(7), 3024–3027.

Kuang, Y., Lackay, S. N., Zhao, L., & Fu, Z. F. (2009). Role of chemokines in the enhancement of BBB permeability and inflammatory infiltration after rabies virus infection. *Virus Research, 144*(1–2), 18–26.

Kucera, P., Dolivo, M., Coulon, P., & Flamand, A. (1985). Pathways of the early propagation of virulent and avirulent rabies strains from the eye to the brain. *Journal of Virology, 55*(1), 158–162.

Ladogana, A., Bouzamondo, E., Pocchiari, M., & Tsiang, H. (1994). Modification of tritiated γ-amino-n-butyric acid transport in rabies virus-infected primary cortical cultures. *Journal of General Virology, 75*(3), 623–627.

Lafon, M. (2005). Rabies virus receptors. *Journal of Neurovirology, 11*(1), 82–87.

Lauria, G., Morbin, M., Lombardi, R., Borgna, M., Mazzoleni, G., Sghirlanzoni, A., & Pareyson, D. (2003). Axonal swellings predict the degeneration of epidermal nerve fibers in painful neuropathies. *Neurology, 61*(5), 631–636.

Lay, S., Prehaud, C., Dietzschold, B., & Lafon, M. (2003). Glycoprotein of nonpathogenic rabies viruses is a major inducer of apoptosis in human Jurkat T cells. *Annals of the New York Academy of Sciences, 1010*, 577–581.

Lentz, T. L. (1985). Rabies virus receptors. *Trends in Neurological Sciences, 8*, 360–364.

Lentz, T. L. (1990). Rabies virus binding to an acetylcholine receptor α-subunit peptide. *Journal of Molecular Recognition, 3*, 82–88.

Lentz, T. L., Benson, R. J. J., Klimowicz, D., Wilson, P. T., & Hawrot, E. (1986). Binding of rabies virus to purified Torpedo acetylcholine receptor. *Molecular Brain Research, 387*, 211–219.

Lentz, T. L., Burrage, T. G., Smith, A. L., Crick, J., & Tignor, G. H. (1982). Is the acetylcholine receptor a rabies virus receptor? *Science, 215*(4529), 182–184.

Lentz, T. L., Wilson, P. T., Hawrot, E., & Speicher, D. W. (1984). Amino acid sequence similarity between rabies virus glycoprotein and snake venom curaremimetic neurotoxins. *Science, 226*, 847–848.

Lewis, P., Fu, Y., & Lentz, T. L. (2000). Rabies virus entry at the neuromuscular junction in nerve-muscle cocultures. *Muscle and Nerve, 23*(5), 720–730.

Li, J., Ertel, A., Portocarrero, C., Barkhouse, D. A., Dietzschold, B., Hooper, D. C., & Faber, M. (2012). Postexposure treatment with the live-attenuated rabies virus (RV) vaccine TriGAS triggers the clearance of wild-type RV from the Central Nervous System (CNS) through the rapid induction of genes relevant to adaptive immunity in CNS tissues. *Journal of Virology, 86*(6), 3200–3210.

Lockhart, B. P., Tordo, N., & Tsiang, H. (1992). Inhibition of rabies virus transcription in rat cortical neurons with the dissociative anesthetic ketamine. *Antimicrobial Agents and Chemotherapy, 36*, 1750–1755.

Lockhart, B. P., Tsiang, H., Ceccaldi, P. E., & Guillemer, S. (1991). Ketamine-mediated inhibition of rabies virus infection in vitro and in rat brain. *Antiviral Chemistry and Chemotherapy, 2*, 9–15.

Lowenstein, C. J., Dinerman, J. L., & Snyder, S. H. (1994). Nitric oxide: A physiologic messenger. *Annals of Internal Medicine, 120*, 227–237.

Madore, H. P., & England, J. M. (1977). Rabies virus protein synthesis in infected BHK-21 cells. *Journal of Virology, 22*, 102–112.

Mazarakis, N. D., Azzouz, M., Rohll, J. B., Ellard, F. M., Wilkes, F. J., Olsen, A. L., ... Mitrophanous, K. A. (2001). Rabies virus glycoprotein pseudotyping of lentiviral vectors enables retrograde axonal transport and access to the nervous system after peripheral delivery. *Human Molecular Genetics, 10*(19), 2109–2121.

McKimmie, C. S., Johnson, N., Fooks, A. R., & Fazakerley, J. K. (2005). Viruses selectively upregulate Toll-like receptors in the central nervous system. *Biochemical and Biophysical Research Communications, 336*(3), 925–933.

Mebatsion, T. (2001). Extensive attenuation of rabies virus by simultaneously modifying the dynein light chain binding site in the P protein and replacing Arg333 in the G protein. *Journal of Virology, 75*(23), 11496–11502.

Milatovic, D., Zhang, Y., Olson, S. J., Montine, K. S., Roberts, L. J., Morrow, J. D., ... Valyi-Nagy, T. (2002). Herpes simplex virus type 1 encephalitis is associated with elevated levels of F2-isoprostanes and F4-neuroprostanes. *Journal of Neurovirology, 8*(4), 295–305.

Mims, C. A. (1960). Intracerebral injections and the growth of viruses in the mouse brain. *British Journal of Experimental Pathology, 41*, 52–59.

Moreno, J. A., & Baer, G. M. (1980). Experimental rabies in the vampire bat. *American Journal of Tropical Medicine and Hygiene, 29*(2), 254–259.

Morimoto, K., Hooper, D. C., Spitsin, S., Koprowski, H., & Dietzschold, B. (1999). Pathogenicity of different rabies virus variants inversely correlates with apoptosis and rabies virus glycoprotein expression in infected primary neuron cultures. *Journal of Virology, 73*(1), 510–518.

Morimoto, K., Patel, M., Corisdeo, S., Hooper, D. C., Fu, Z. F., Rupprecht, C. E., ... Dietzschold, B. (1996). Characterization of a unique variant of bat rabies virus responsible for newly emerging human cases in North America. *Proceedings of the National Academy of Sciences of the United States of America, 93*(11), 5653–5658.

Murphy, F. A., Bauer, S. P., Harrison, A. K., & Winn, W. C. (1973). Comparative pathogenesis of rabies and rabies-like viruses: Viral infection and transit from inoculation site to the central nervous system. *Laboratory Investigation, 28*, 361–376.

Murphy, F. A., Bell, J. F., Bauer, S. P., Gardner, J. J., Moore, G. J., Harrison, A. K., & Coe, J. E. (1980). Experimental chronic rabies in the cat. *Laboratory Investigation, 43*, 231–241.

Murphy, F. A., Harrison, A. K., Winn, W. C., & Bauer, S. P. (1973). Comparative pathogenesis of rabies and rabies-like viruses: Infection of the central nervous system and centrifugal spread of virus to peripheral tissues. *Laboratory Investigation, 29*, 1–16.

Nargi-Aizenman, J. L., & Griffin, D. E. (2001). Sindbis virus-induced neuronal death is both necrotic and apoptotic and is ameliorated by N-methyl-D-aspartate receptor antagonists. *Journal of Virology, 75*(15), 7114–7121.

Nargi-Aizenman, J. L., Havert, M. B., Zhang, M., Irani, D. N., Rothstein, J. D., & Griffin, D. E. (2004). Glutamate receptor antagonists protect from virus-induced neural degeneration. *Annals of Neurology, 55*(4), 541–549.

Nath, A., Haughey, N. J., Jones, M., Anderson, C., Bell, J. E., & Geiger, J. D. (2000). Synergistic neurotoxicity by human immunodeficiency virus proteins Tat and gp120: Protection by memantine. *Annals of Neurology, 47*(2), 186–194.

Nathan, C. (1992). Nitric oxide as a secretory product of mammalian cells. *Federation of American Societies for Experimental Biology Journal, 6*, 3051–3064.

Nishikawa, T., Edelstein, D., Du, X. L., Yamagishi, S., Matsumura, T., Kaneda, Y., ... Brownlee, M. (2000). Normalizing mitochondrial superoxide production blocks three pathways of hyperglycaemic damage. *Nature, 404*(6779), 787–790.

Noah, D. L., Drenzek, C. L., Smith, J. S., Krebs, J. W., Orciari, L., Shaddock, J., ... Childs, J. E. (1998). Epidemiology of human rabies in the United States, 1980 to 1996. *Annals of Internal Medicine, 128*(11), 922–930.

Obrosova, I. G., Pacher, P., Szabo, C., Zsengeller, Z., Hirooka, H., Stevens, M. J., & Yorek, M. A. (2005). Aldose reductase inhibition counteracts oxidative-nitrosative stress and poly(ADP-ribose) polymerase activation in tissue sites for diabetes complications. *Diabetes, 54*(1), 234–242.

Obrosova, I. G., Van, H. C., Fathallah, L., Cao, X. C., Greene, D. A., & Stevens, M. J. (2002). An aldose reductase inhibitor reverses early diabetes-induced changes in peripheral nerve function, metabolism, and antioxidative defense. *FASEB Journal, 16*(1), 123–125.

O'Donnell, L. A., & Rall, G. F. (2010). Blue moon neurovirology: The merits of studying rare CNS diseases of viral origin. *Journal of Neuroimmune Pharmacology, 5*(3), 443–455.

Pawan, J. L. (1936). Rabies in the vampire bat of Trinidad, with special reference to the clinical course and the latency of infection. *Annals of Tropical Medicine and Parasitology, 30*, 401–422.

Pedersen, W. A., Fu, W., Keller, J. N., Markesbery, W. R., Appel, S., Smith, R. G., ... Mattson, M. P. (1998). Protein modification by the lipid peroxidation product 4-hydroxynonenal in the spinal cords of amyotrophic lateral sclerosis patients. *Annals of Neurology, 44*(5), 819–824.

Perl, D. P., Bell, J. F., Moore, G. J., & Stewart, S. J. (1977). Chronic recrudescent rabies in a cat. *Proceedings of the Society for Experimental Biology and Medicine, 155*, 540–548.

Phares, T. W., Kean, R. B., Mikheeva, T., & Hooper, D. C. (2006). Regional differences in blood-brain barrier permeability changes and inflammation in the apathogenic clearance of virus from the central nervous system. *Journal of Immunology, 176*(12), 7666–7675.

Prehaud, C., Lay, S., Dietzschold, B., & Lafon, M. (2003). Glycoprotein of nonpathogenic rabies viruses is a key determinant of human cell apoptosis. *Journal of Virology, 77*(19), 10537–10547.

Prehaud, C., Megret, F., Lafage, M., & Lafon, M. (2005). Viral infection switches TLR-3-positive human neurons to become strong producers of beta interferon. *Journal of Virology, 79*(20), 12893–12904.

Prosniak, M., Hooper, D. C., Dietzschold, B., & Koprowski, H. (2001). Effect of rabies virus infection on gene expression in mouse brain. *Proceedings of the National Academy of Sciences of the United States of America, 98*(5), 2758–2763.

Rasalingam, P., Rossiter, J. P., & Jackson, A. C. (2005). Recombinant rabies virus vaccine strain SAD-L16 inoculated intracerebrally in young mice produces a severe encephalitis with extensive neuronal apoptosis. *Canadian Journal of Veterinary Research, 69*(2), 100–105.

Rasalingam, P., Rossiter, J. P., Mebatsion, T., & Jackson, A. C. (2005). Comparative pathogenesis of the SAD-L16 strain of rabies virus and a mutant modifying the dynein light chain binding site of the rabies virus phosphoprotein in young mice. *Virus Research, 111*(1), 55–60.

Raux, H., Flamand, A., & Blondel, D. (2000). Interaction of the rabies virus P protein with the LC8 dynein light chain. *Journal of Virology, 74*(21), 10212–10216.

Ray, N. B., Ewalt, L. C., & Lodmell, D. L. (1995). Rabies virus replication in primary murine bone marrow macrophages and in human and murine macrophage-like cell lines: Implications for viral persistence. *Journal of Virology, 69*, 764–772.

Reagan, K. J., & Wunner, W. H. (1985). Rabies virus interaction with various cell lines is independent of the acetylcholine receptor: Brief report. *Archives of Virology, 84*, 277–282.

Reid, J. E., & Jackson, A. C. (2001). Experimental rabies virus infection in *Artibeus jamaicensis* bats with CVS-24 variants. *Journal of Neurovirology, 7*(6), 511–517.

Reiss, C. S., Chesler, D. A., Hodges, J., Ireland, D. D., & Chen, N. (2002). Innate immune responses in viral encephalitis. *Current Topics in Microbiology and Immunology, 265*, 63–94.

Reiss, C. S., & Komatsu, T. (1998). Does nitric oxide play a critical role in viral infections? *Journal of Virology, 72*(6), 4547–4551.

Ross, E., & Armentrout, S. A. (1962). Myocarditis associated with rabies: Report of a case. *New England Journal of Medicine, 266*, 1087–1089.

Roulston, A., Marcellus, R. C., & Branton, P. E. (1999). Viruses and apoptosis. *Annual Review of Microbiology, 53*, 577–628.

Roy, A., & Hooper, D. C. (2007). Lethal silver-haired bat rabies virus infection can be prevented by opening the blood-brain barrier. *Journal of Virology, 81*(15), 7993–7998.

Roy, A., & Hooper, D. C. (2008). Immune evasion by rabies viruses through the maintenance of blood-brain barrier integrity. *Journal of Neurovirology, 14*(5), 401–411.

Russell, J. W., Golovoy, D., Vincent, A. M., Mahendru, P., Olzmann, J. A., Mentzer, A., & Feldman, E. L. (2002). High glucose-induced oxidative stress and mitochondrial dysfunction in neurons. *FASEB Journal, 16*(13), 1738–1748.

Rutherford, M., & Jackson, A. C. (2004). Neuronal apoptosis in immunodeficient mice infected with the challenge virus standard strain of rabies virus by intracerebral inoculation. *Journal of Neurovirology, 10*(6), 409–413.

Sarmento, L., Li, X., Howerth, E., Jackson, A. C., & Fu, Z. F. (2005). Glycoprotein-mediated induction of apoptosis limits the spread of attenuated rabies viruses in the central nervous system of mice. *Journal of Neurovirology, 11*(6), 571–581.

Sarmento, L., Tseggai, T., Dhingra, V., & Fu, Z. F. (2006). Rabies virus-induced apoptosis involves caspase-dependent and caspase-independent pathways. *Virus Research, 121*, 144–151.

Sayre, L. M., Zelasko, D. A., Harris, P. L., Perry, G., Salomon, R. G., & Smith, M. A. (1997). 4-Hydroxynonenal-derived advanced lipid peroxidation end products are increased in Alzheimer's disease. *Journal of Neurochemistry, 68*(5), 2092–2097.

Schachtele, S. J., Hu, S., Little, M. R., & Lokensgard, J. R. (2010). Herpes simplex virus induces neural oxidative damage via microglial cell Toll-like receptor-2. *Journal of Neuroinflammation, 7*(1), 35.

Schmidt, R. E., Dorsey, D., Parvin, C. A., Beaudet, L. N., Plurad, S. B., & Roth, K. A. (1997). Dystrophic axonal swellings develop as a function of age and diabetes in human dorsal root ganglia. *Journal of Neuropathology and Experimental Neurology, 56*(9), 1028–1043.

Schwarz, K. B. (1996). Oxidative stress during viral infection: A review. *Free Radical Biology and Medicine, 21*(5), 641–649.

Scott, C. A., Rossiter, J. P., Andrew, R. D., & Jackson, A. C. (2008). Structural abnormalities in neurons are sufficient to explain the clinical disease and fatal outcome in experimental rabies in yellow fluorescent protein-expressing transgenic mice. *Journal of Virology, 82*(1), 513–521.

Seif, I., Coulon, P., Rollin, P. E., & Flamand, A. (1985). Rabies virulence: Effect on pathogenicity and sequence characterization of rabies virus mutations affecting antigenic site III of the glycoprotein. *Journal of Virology, 53*, 926–935.

Shankar, V., Dietzschold, B., & Koprowski, H. (1991). Direct entry of rabies virus into the central nervous system without prior local replication. *Journal of Virology, 65*, 2736–2738.

Smart, N. L., & Charlton, K. M. (1992). The distribution of challenge virus standard rabies virus versus skunk street rabies virus in the brains of experimentally infected rabid skunks. *Acta Neuropathologica, 84*, 501–508.

Smith, J. S., Fishbein, D. B., Rupprecht, C. E., & Clark, K. (1991). Unexplained rabies in three immigrants in the United States: A virologic investigation. *New England Journal of Medicine, 324*(4), 205–211.

Spitsin, S., Portocarrero, C., Phares, T. W., Kean, R. B., Brimer, C. M., Koprowski, H., & Hooper, D. C. (2008). Early blood-brain barrier permeability in cerebella of PLSJL mice immunized with myelin basic protein. *Journal of Neuroimmunology, 196*(1–2), 8–15.

Steece, R., & Altenbach, J. S. (1989). Prevalence of rabies specific antibodies in the Mexican free-tailed bat (*Tadarida brasiliensis mexicana*) at Lava Cave, New Mexico. *Journal of Wildlife Diseases, 25*(4), 490–496.

Superti, F., Hauttecoeur, B., Morelec, M. J., Goldoni, P., Bizzini, B., & Tsiang, H. (1986). Involvement of gangliosides in rabies virus infection. *Journal of General Virology, 67*, 47–56.

Superti, F., Seganti, L., Tsiang, H., & Orsi, N. (1984). Role of phospholipids in rhabdovirus attachment to CER cells. Brief report. *Archives of Virology, 81*(3–4), 321–328.

Tang, Y., Rampin, O., Giuliano, F., & Ugolini, G. (1999). Spinal and brain circuits to motoneurons of the bulbospongiosus muscle: Retrograde transneuronal tracing with rabies virus. *Journal of Comparative Neurology, 414*(2), 167–192.

Theerasurakarn, S., & Ubol, S. (1998). Apoptosis induction in brain during the fixed strain of rabies virus infection correlates with onset and severity of illness. *Journal of Neurovirology, 4*(4), 407–414.

Thompson, C. B. (1995). Apoptosis in the pathogenesis and treatment of disease. *Science, 267*(5203), 1456–1462.

Thoulouze, M. I., Lafage, M., Montano-Hirose, J. A., & Lafon, M. (1997). Rabies virus infects mouse and human lymphocytes and induces apoptosis. *Journal of Virology, 71*(10), 7372–7380.

Thoulouze, M. I., Lafage, M., Schachner, M., Hartmann, U., Cremer, H., & Lafon, M. (1998). The neural cell adhesion molecule is a receptor for rabies virus. *Journal of Virology, 72*(9), 7181–7190.

Thoulouze, M. I., Lafage, M., Yuste, V. J., Baloul, L., Edelman, L., Kroemer, G., ... Lafon, M. (2003). High level of Bcl-2 counteracts apoptosis mediated by a live rabies virus vaccine strain and induces long-term infection. *Virology, 314*(2), 549–561.

Tillotson, J. R., Axelrod, D., & Lyman, D. O. (1977). Rabies in a laboratory worker—New York. *Morbidity and Mortality Weekly Report, 26*, 183–184.

Torres-Anjel, M. J., Montano-Hirose, J., Cazabon, E. P. I., Oakman, J. K., & Wiktor, T. J. (1984). A new approach to the pathobiology of rabies virus as aided by immunoperoxidase staining. In *27th Annual Proceedings of the American Association of Veterinary Laboratory Diagnosticians.* pp. 1–26.

Tsiang, H. (1979). Evidence for an intraaxonal transport of fixed and street rabies virus. *Journal of Neuropathology and Experimental Neurology, 38*, 286–296.

Tsiang, H. (1982). Neuronal function impairment in rabies-infected rat brain. *Journal of General Virology, 61*(2), 277–281.

Tsiang, H., Ceccaldi, P.-E., Ermine, A., Lockhart, B., & Guillemer, S. (1991). Inhibition of rabies virus infection in cultured rat cortical neurons by an N-methyl-D-aspartate noncompetitive antagonist, MK-801. *Antimicrobial Agents and Chemotherapy, 35*(3), 572–574.

Tsiang, H., Ceccaldi, P. E., & Lycke, E. (1991). Rabies virus infection and transport in human sensory dorsal root ganglia neurons. *Journal of General Virology, 72*(5), 1191–1194.

Tsiang, H., de la Porte, S., Ambroise, D. J., Derer, M., & Koenig, J. (1986). Infection of cultured rat myotubes and neurons from the spinal cord by rabies virus. *Journal of Neuropathology and Experimental Neurology, 45*, 28–42.

Tsiang, H., Lycke, E., Ceccaldi, P.-E., Ermine, A., & Hirardot, X. (1989). The anterograde transport of rabies virus in rat sensory dorsal root ganglia neurons. *Journal of General Virology, 70*, 2075–2085.

Tucker, P. C., Griffin, D. E., Choi, S., Bui, N., & Wesselingh, S. (1996). Inhibition of nitric oxide synthesis increases mortality in Sindbis virus encephalitis. *Journal of Virology, 70*, 3972–3977.

Tuffereau, C., Benejean, J., Blondel, D., Kieffer, B., & Flamand, A. (1998). Low-affinity nerve-growth factor receptor (P75NTR) can serve as a receptor for rabies virus. *European Molecular Biology Organization Journal, 17*(24), 7250–7259.

Tuffereau, C., Leblois, H., Benejean, J., Coulon, P., Lafay, F., & Flamand, A. (1989). Arginine or lysine in position 333 of ERA and CVS glycoprotein is necessary for rabies virulence in adult mice. *Virology, 172*, 206–212.

Tuffereau, C., & Martinet-Edelist, C. (1985). Shut-off of cellular RNA after infection with rabies virus. *Comptes Rendus de l'Academie des Sciences—Series III: Sciences de la Vie, 300*, 597–600.

Tuffereau, C., Schmidt, K., Langevin, C., Lafay, F., Dechant, G., & Koltzenburg, M. (2007). The rabies virus glycoprotein receptor p75NTR is not essential for rabies virus infection. *Journal of Virology, 81*(24), 13622–13630.

Ubol, S., Sukwattanapan, C., & Maneerat, Y. (2001). Inducible nitric oxide synthase inhibition delays death of rabies virus-infected mice. *Journal of Medical Microbiology, 50*(3), 238–242.

Ubol, S., Sukwattanapan, C., & Utaisincharoen, P. (1998). Rabies virus replication induces Bax-related, caspase dependent apoptosis in mouse neuroblastoma cells. *Virus Research, 56*(2), 207–215.

Udow, S. J., Marrie, R. A., & Jackson, A. C. (2013). Clinical features of dog- and bat-acquired rabies in humans. *Clinical Infectious Diseases, 57*(5), 689–696.

Ugolini, G. (2011). Rabies virus as a transneuronal tracer of neuronal connections. *Advances in Virus Research, 79*, 165–202.

Valyi-Nagy, T., & Dermody, T. S. (2005). Role of oxidative damage in the pathogenesis of viral infections of the nervous system. *Histology and Histopathology, 20*(3), 957–967.

Valyi-Nagy, T., Olson, S. J., Valyi-Nagy, K., Montine, T. J., & Dermody, T. S. (2000). Herpes simplex virus type 1 latency in the murine nervous system is associated with oxidative damage to neurons. *Virology, 278*(2), 309–321.

Van Dam, A. M., Bauer, J., Manahing, W. K. H., Marquette, C., Tilders, F. J. H., & Berkenbosch, F. (1995). Appearance of inducible nitric oxide synthase in the rat central nervous system after rabies virus infection and during experimental allergic encephalomyelitis but not after peripheral administration of endotoxin. *Journal of Neuroscience Research, 40*, 251–260.

Veeraraghavan, N., Gajanana, A., & Rangasami, R. (1967). Hydrophobia among persons bitten by apparently healthy animals. In *The Pasteur Institute of Southern India, Coonoor: Annual Report of the Director 1965 and Scientific Report 1966* (pp. 90–91). Madras: Diocesan Press.

Veeraraghavan, N., Gajanana, A., Rangasami, R., Kumari, C., Saraswathi, K. C., Devaraj, R., et al. (1967). Studies on the salivary excretion of rabies virus by the dog from Surandai. In *The Pasteur Institute of Southern India, Coonoor: Annual Report of the Director 1965 and Scientific Report 1966* (pp. 91–97). Madras: Diocesan Press.

Veeraraghavan, N., Gajanana, A., Rangasami, R., Oonnunni, P. T., Saraswathi, K. C., Devaraj, R., et al. (1969). Studies on the salivary excretion of rabies virus by the dog from Surandai. In *The Pasteur Institute of Southern India, Coonoor: Annual Report of the Director 1967 and Scientific Report 1968* (pp. 68–70). Madras: Diocesan Press.

Veeraraghavan, N., Gajanana, A., Rangasami, R., Oonnunni, P. T., Saraswathi, K. C., Devaraj, R., et al. (1970). Studies on the salivary excretion of rabies virus by the dog from Surandai. In *The Pasteur Institute of Southern India, Coonoor: Annual Report of the Director 1968 and Scientific Report 1969* (p. 66). Madras: Diocesan Press.

Veeraraghavan, N., Gajanana, A., Rangasami, R., Saraswathi, K. C., Devaraj, R., & Hallan, K. M. (1968). Studies on the salivary excretion of rabies virus by the dog from Surandai. In *The Pasteur Institute of Southern Indian, Coonoor: Annual Report of the Director 1966 and Scientific Report 1967* (pp. 71–78). Madras: Diocesan Press.

Wallerstein, C. (1999). Rabies cases increase in the Philippines. *BMJ, 318*(7194), 1306.

Wang, Z. W., Sarmento, L., Wang, Y., Li, X. Q., Dhingra, V., Tseggai, T., … Fu, Z. F. (2005). Attenuated rabies virus activates, while pathogenic rabies virus evades, the host innate immune responses in the central nervous system. *Journal of Virology, 79*(19), 12554–12565.

Wang, J., Wang, Z., Liu, R., Shuai, L., Wang, X., Luo, J., … Bu, Z. (2018). Metabotropic glutamate receptor subtype 2 is a cellular receptor for rabies virus. *PLoS Pathogens, 14*(7). e1007189.

Wang, H., Zhang, G., Wen, Y., Yang, S., Xia, X., & Fu, Z. F. (2011). Intracerebral administration of recombinant rabies virus expressing GM-CSF prevents the development of rabies after infection with street virus. *PLoS One, 6*(9), e25414.

Weihe, E., Bette, M., Preuss, M. A., Faber, M., Schafer, M. K., Rehnelt, J., et al. (2008). Role of virus-induced neuropeptides in the brain in the pathogenesis of rabies. In B. Dodet, A. R. Fooks, T. Müller, & N. Tordo (Eds.), *Developments in Biologicals: Vol. 131. Towards the elimination of rabies in Eurasia* (131st ed., pp. 73–81). Kager: Basel.

Weli, S. C., Scott, C. A., Ward, C. A., & Jackson, A. C. (2006). Rabies virus infection of primary neuronal cultures and adult mice: Failure to demonstrate evidence of excitotoxicity. *Journal of Virology, 80*(20), 10270–10273.

Winkler, W. G., Fashinell, T. R., Leffingwell, L., Howard, P., & Conomy, J. P. (1973). Airborne rabies transmission in a laboratory worker. *Journal of the American Medical Association, 226*, 1219–1221.

Yan, X., Prosniak, M., Curtis, M. T., Weiss, M. L., Faber, M., Dietzschold, B., & Fu, Z. F. (2001). Silver-haired bat rabies virus variant does not induce apoptosis in the brain of experimentally infected mice. *Journal of Neurovirology, 7*(6), 518–527.

Yang, C., & Jackson, A. C. (1992). Basis of neurovirulence of avirulent rabies virus variant Av01 with stereotaxic brain inoculation in mice. *Journal of General Virology, 73*(4), 895–900.

Zhao, L., Toriumi, H., Kuang, Y., Chen, H., & Fu, Z. F. (2009). The roles of chemokines in rabies virus infection: Overexpression may not always be beneficial. *Journal of Virology, 83*(22), 11808–11818.

Zheng, Y. M., Schafer, M. K.-H., Weihe, E., Sheng, H., Corisdeo, S., Fu, Z. F., … Dietzschold, B. (1993). Severity of neurological signs and degree of inflammatory lesions in the brains of rats with Borna disease correlate with the induction of nitric oxide synthase. *Journal of Virology, 67*, 5786–5791.

CHAPTER 10

Pathology

John P. Rossiter[a], Alan C. Jackson[b]

[a]Department of Pathology and Molecular Medicine, Queen's University and Kingston Health Sciences Centre, Kingston, ON, Canada [b]Professor of Medicine (Neurology), University of Manitoba, Winnipeg, MB, Canada

10.1 Introduction

Investigation of the pathological changes in the central and peripheral nervous systems and extraneural organs of human rabies cases as well as naturally and experimentally infected animals has provided an important foundation for ongoing study of the pathogenesis of rabies (see Chapter 9). Following an exposure with deposition of rabies virus at or near the site of a bite from a rabid animal, the virus spreads centripetally toward the central nervous system (CNS) by retrograde fast axonal transport through the peripheral nervous system, typically to the spinal cord. Rapid neuron-to-neuron trans-synaptic viral dissemination within the spinal cord and brain results in a polioencephalomyelitis (i.e., an inflammatory disease predominantly involving the gray matter of the brain and spinal cord (Love, Wiley, & Lucas, 2015)). This is followed by centrifugal spread away from the CNS along peripheral nerve pathways, with resulting infection of salivary glands, skin, heart, and other viscera.

Many of the cardinal pathological features of rabies virus infection were first described over about a 40-year period extending from the early 1870s to the early 1900s (Abba & Bormans, 1905; Babes, 1892; Benedikt, 1878; Gowers, 1877; Negri, 1903a, 1909; Nepveu, 1872; Pasteur, Chamberland, Roux, & Thuillier, 1881; Ramón y Cajal & Garcia, 1904; Schaffer, 1888; Van Gehuchten & Nelis, 1900). However, throughout the subsequent century, especially following the introduction of electron microscopy and immunohistochemistry, pathological studies have continued to provide key insights into our understanding of this dreaded disease.

10.2 Macroscopic findings

Macroscopic examination of the brain in rabies victims is frequently unremarkable, or shows a spectrum of relatively mild and nonspecific changes (Love et al., 2015; Nieberg & Blumberg, 1972; Perl & Good, 1991; Sukru-Aksel, 1958). There is often mild cerebral edema, but severe cerebral swelling and associated brain herniation are not features of rabies. There may be congestion of leptomeningeal and parenchymal blood vessels, sometimes associated with multiple petechiae (Lowenberg, 1928; Tangchai, Yenbutr, & Vejjajiva, 1970). Frank subarachnoid or parenchymal hemorrhage is not a recognized feature of rabies. Thickening of the basal leptomeninges due to prominent inflammatory cell infiltration has been reported in a few cases of rabies in children (Tangchai et al., 1970). Although the brain parenchyma is typically grossly unremarkable, focal changes are seen in some cases, perhaps related to prolongation of survival with critical care measures (Rubin, Sullivan, Summers, Gregg, & Sikes, 1970). Multifocal gray and white matter tissue softening and discoloration were found in an immunosuppressed male with prolonged survival (Walker, Thiessen, Graeb, Moore, & Mackenzie, 2016). A variety of macroscopic abnormalities, including discoloration of the cortical mantle, generalized softening of deep gray matter and infarction of the insular cortex, were described in four transplant recipients infected with rabies virus from a common donor (Burton et al., 2005). Softening and congestion of the amygdalae and extensive laminar necrosis of the cerebral cortex were described in a case of bat-transmitted human rabies, where the patient received human rabies immune globulin and antiviral therapy and died 33 days after the onset of symptoms. Death was attributed to a direct viral effect rather than anoxic brain injury or autolysis of "respirator brain" (Dolman & Charlton, 1987). Leptomeningeal and vascular congestion may also be seen in the spinal cord, and may be intense (Gowers, 1877; Lowenberg, 1928; Tangchai et al., 1970). In their classic study of the 1929–30 Trinidad outbreak of paralytic rabies, Hurst and Pawan (1932) described the victims' spinal cords as having the "consistency of butter," likely reflecting extensive tissue injury.

10.3 Pathology in the central nervous system

10.3.1 Overview

Despite the catastrophic clinical outcome of rabies virus encephalomyelitis, the histopathological changes observed in the CNS are typically relatively mild, with varying degrees of mononuclear inflammatory cell infiltration of the leptomeninges, perivascular cuffing, microglial activation with formation of "Babes nodules," and neuronophagia. Moreover, this combination of features is not unique to rabies and can be seen in a variety of other viral encephalitides (Love et al., 2015). However, Negri bodies, eosinophilic cytoplasmic viral inclusions that are unique to rabies, are identified in infected neurons in many cases. The extent of the infection of the CNS by rabies virus is best highlighted by immunostaining for rabies virus antigen.

10.3.2 Inflammation

Some degree of inflammatory cell infiltration of the leptomeninges is usually seen in human rabies cases, although the extent and intensity can vary greatly (Dupont & Earle, 1965; Perl & Good, 1991; Tangchai et al., 1970). The infiltrates are typically composed predominantly of lymphocytes and monocytes, with smaller numbers of plasma cells, but neutrophils can predominate when inflammation is intense, especially in fulminant childhood cases (Perl & Good, 1991; Tangchai et al., 1970). In the classic study by Dupont and Earle (1965) of 49 cases of human rabies encephalitis, frank leptomeningitis was found in three cases, all of them children. Tangchai et al. (1970) observed meningitis in 4 of 24 cases, again all children, in whom the clinical course was more fulminant by comparison with their other cases. The meninges of the brainstem are frequently involved, especially in cases where leptomeningeal inflammation is sparse overall (Perl & Good, 1991). Paradoxically, in paralytic rabies cases inflammatory cell infiltration of the spinal meninges is typically relatively sparse, despite intense inflammation within the adjacent spinal cord in many of the reported cases (Chopra, Banerjee, Murthy, & Pal, 1980; Hurst & Pawan, 1932).

Perivascular mononuclear inflammatory cell infiltrates (Fig. 10.1), consisting predominantly of lymphocytes and monocytes, are seen in the great majority of human rabies cases,

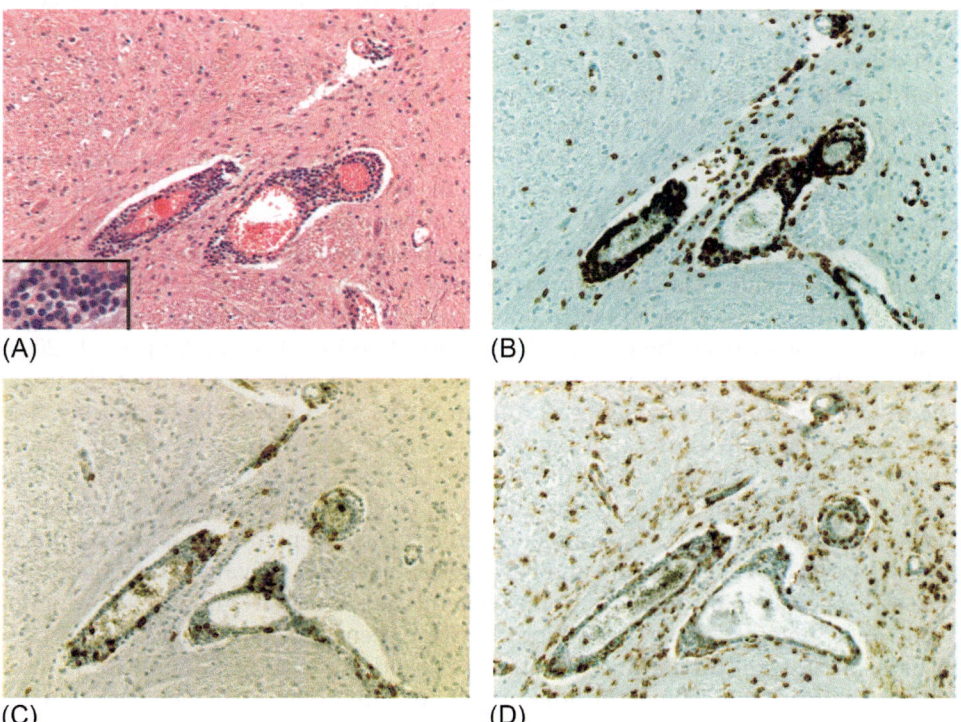

FIG. 10.1 Perivascular mononuclear inflammatory cell infiltrates in the brainstem (medulla oblongata) of a human rabies case. (A) Hematoxylin-Phloxine-Saffron (HPS) stained section (magnification 80×). Inset shows a higher power view of an infiltrate that is composed predominantly of lymphocytes (magnification 240×). (B–D) Adjacent immunoperoxidase stained sections showing: (B) CD3-positive T lymphocytes, (C) CD20-positive B lymphocytes, and (D) CD68-positive monocyte/macrophages in a perivascular distribution and microglia/macrophages in the neuropil of the brainstem.

but their density and distribution can vary greatly between cases. This perivascular "cuffing" is seen predominantly in gray matter, especially in the brainstem and spinal cord, with relative sparing of white matter. Dupont and Earle (1965) observed perivascular cuffing in 48 of their 49 cases, in the following locations and approximate proportions of cases: medulla and pons (38%), spinal cord (35%), cerebral cortex (26%), hippocampus (14%), thalamus (29%), basal ganglia (26%), and cerebellum (14%). In the paralytic rabies series of Hurst and Pawan (1932) (3 cases) and Chopra et al. (1980) (11 cases), dense perivascular infiltrates of lymphocytes and some neutrophils were seen in the anterior and posterior horn gray matter of most cases. This was especially prominent in the lumbar and lower thoracic segments in the cases of Chopra et al. (1980), and was associated with extension of inflammation along perivascular spaces into the adjacent white matter. By contrast, inflammation was considerably less severe in the brain and predominantly involved the medulla (both series) and dorsal half of the pons (Hurst & Pawan, 1932).

In the case of a patient who died 17 days into the clinical course of encephalitic rabies, Iwasaki and coworkers found that 50% to 70% of perivascular mononuclear cells were CD3 immunopositive T-lymphocytes (Iwasaki, Sako, Tsunoda, & Ohara, 1993; Iwasaki & Tobita, 2002). Approximately one-third of these were CD4-positive helper T cells. Only occasional CD20-positive B cells were observed and the remaining perivascular cells were CD68-positive monocyte/macrophage lineage cells. More than half of the T lymphocytes were found in the CNS parenchyma surrounding the perivascular spaces. However, in a rabies patient who only survived for 9 days, there was virtually no inflammation or tissue injury, despite the finding of numerous neurons containing Negri bodies (see later), which emphasizes an important point that fatal encephalitic rabies may not necessarily be accompanied by significant inflammation (Iwasaki et al., 1993). The degree of inflammation may at least be partially influenced by the strain of rabies virus. In dogs experimentally inoculated with an Ethiopian dog rabies virus strain, widespread inflammation, neuronal degeneration, and neuronophagia (see later) were seen, whereas such lesions were generally much less severe in animals infected with a Mexican dog virus strain (Fekadu, Chandler, & Harrison, 1982). An especially florid and widespread encephalitic pattern was reported in a patient presenting with paralytic rabies caused by canine virus transmitted by a fox bite (Suja et al., 2004).

In 1892, Babes described microscopic accumulations of cells surrounding chromatolytic and degenerating neurons in a series of human rabies cases. He called these foci "les ilots inflammatoires pericellulaires de la rage" (pericellular inflammatory islets of rabies) and "nodules rabiques" (rabidic nodules) (Babes, 1892). Subsequently referred to in the literature as "Babes nodules," these microglial nodules (Fig. 10.2) are composed predominantly of activated microglia/monocytes and are seen in other viral encephalitides and other infectious disorders (Love et al., 2015). Dupont and Earle (1965) observed activated microglia in many of their cases, especially in the medulla, and in several cases the microglia had a predominantly "rod-cell" morphology. Classic Babes nodules were seen in 42% of their cases. In paralytic rabies cases, microglial proliferation, both diffuse and nodular, was found throughout the spinal gray matter and focally in the adjacent white matter in most cases. Marked microglial activation was seen in the medulla and dorsal half of the pons in several of these cases (Chopra et al., 1980; Hurst & Pawan, 1932).

In a study of rabies-infected dogs sacrificed soon after onset of paralytic or furious illness, Shuangshoti et al. (2013) observed a striking inflammatory response (microglial nodules and/

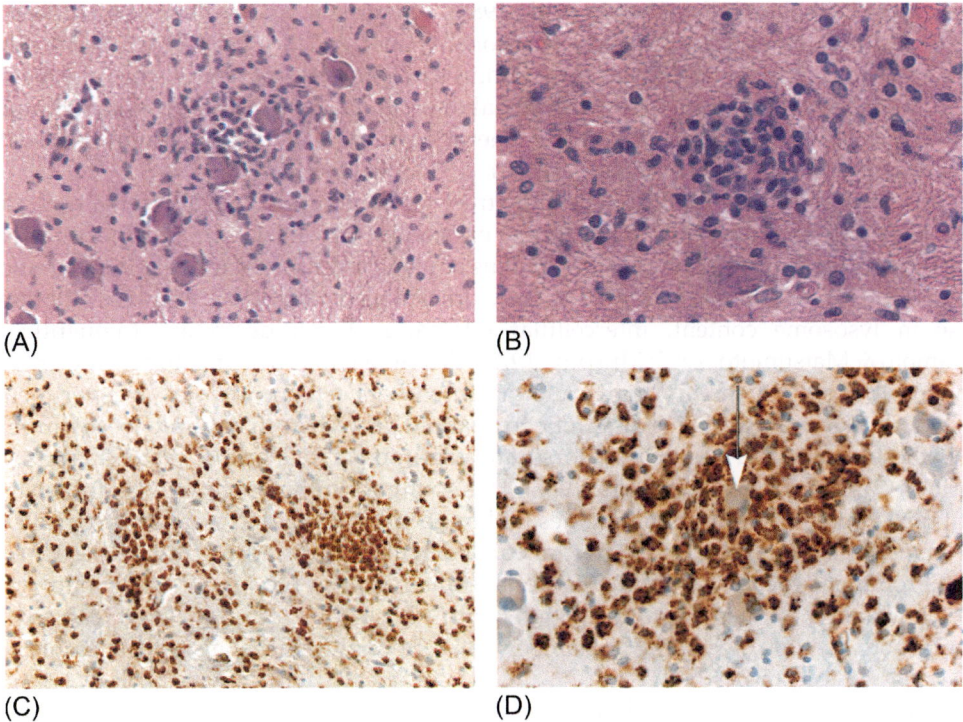

FIG. 10.2 Microglial nodules (Babes nodules) in the medulla oblongata of a human rabies case. (A, B) HPS stained sections (magnifications A 160×, B 260×). (C, D) CD68-immunoperoxidase stained sections showing (C) diffuse and nodular microglial proliferation (magnification 85×) and (D) a microglial nodule containing a neuron (arrow) undergoing neuronophagia (magnification 170×).

or marked T-lymphocyte predominant perivascular cuffing) throughout the brainstem of animals with the paralytic form. This contrasted with lesser degrees of inflammation in other brain areas and in the spinal cords of the paralytic animals, and throughout the CNS of furious cases.

10.3.3 Cell injury and cell death

Neuronophagia (Fig. 10.2D), a microscopic pattern characterized by accumulations of activated microglia/macrophages in the process of phagocytosing degenerating and/or dying neurons, is seen in many rabies cases. Once again, however, the severity and anatomical extent of neuronophagia and resulting neuronal loss can vary greatly between cases. Dupont and Earle (1965) observed neuronophagia in 57% of their cases. The neurons within these foci often have a shrunken appearance, with condensed cytoplasm and pyknotic nuclei (Perl & Good, 1991). Central chromatolysis is a cytological pattern of swelling of the neuronal cell body, disruption and dispersal of Nissl granules from the central part of the perikaryon, and peripheral displacement of the nucleus, that is classically seen in response to axonal injury and may also be seen in rabies. In some paralytic rabies cases, there was extensive neuronal degeneration and loss in the anterior and posterior horns of the spinal cord and to a lesser extent in the medulla (Chopra et al., 1980), whereas in others there was marked neuronal central chromatolysis, but only occasional (Hurst & Pawan, 1932) or absent (Sheikh et al., 2005) neuronophagia. Central chromatolysis of spinal motor neurons has also been described in encephalitic rabies cases (Mitrabhakdi et al., 2005). In addition to chromatolysis, vacuolation of neuronal cytoplasm and degenerative changes in nuclear chromatin have been reported (Lowenberg, 1928; Reisman, Alpers, & Cooper, 1933).

In rodents experimentally infected with street virus, neurons typically remain relatively intact, with little alteration in the structure of organelles. However, with fixed virus infection using intracerebral inoculation in adult animals or using any route of inoculation in immature animals, widespread neuronal injury with cytoplasmic condensation, multivesiculation, increase in lysosome content, intercellular edema, and cell death are frequently seen (Miyamoto & Matsumoto, 1967; Murphy, 1977). Prominent microvacuolation of the gray matter neuropil, especially cerebral cortex and thalamus, has been documented in experimentally infected skunks and foxes with the Arctic fox variant, as well as in naturally occurring infection in these species and also in cow, horse, and cat (Bundza & Charlton, 1988; Charlton, 1984; Charlton, Casey, Webster, & Bundza, 1987). This spongiform change closely resembled that of traditional spongiform encephalopathies, although the vacuolation was less extensive than that found in skunks experimentally inoculated with scrapie agent (Bundza & Charlton, 1988). Prominent vacuolation of the neuronal soma has been reported in mice following intracerebral or intramuscular inoculation with the CVS strain of fixed rabies virus (Greenwood, Newton, Pearson, & Schamber, 1997). Spongiform change is not a feature of human rabies cases.

The role of apoptotic cell death in the pathogenesis of experimentally induced and naturally occurring rabies infection has been investigated in considerable detail. The data strongly indicate that neuronal apoptosis does not play an important pathogenetic role in human rabies encephalitis or in canine infection with "street" rabies virus (Jackson, Randle, Lawrance, & Rossiter, 2008; Suja, Mahadevan, Madhusudana, & Shankar, 2011).

Juntrakul, Ruangvejvorachai, Shuangshoti, Wacharapluesadee, and Hemachudha (2005) observed cytoplasmic cytochrome-c immunoreactivity in neurons and numerous TUNEL-labeled (terminal deoxynucleotidyltransferase-mediated DNA nick-end labeling) cells in many regions of the CNS in a series of 10 human rabies cases. Foci of TUNEL-positive neurons were also observed in the brainstem and hippocampus of a patient who developed rabies encephalitis on a background of AIDS (Adle-Biassette et al., 1996). However, neither of these two reports showed any cytological evidence of apoptosis, such as cell shrinkage, nuclear karyorrhexis, and formation of apoptotic bodies, and apoptotic cell death cannot be reliably diagnosed in the absence of such morphological features. Jackson et al. (2008) evaluated brain tissue sections from the cerebral cortex, hippocampus, and brainstem in 12 human rabies cases for evidence of neuronal apoptosis, using morphological assessment, TUNEL staining, and immunohistochemical staining for activated (cleaved) caspase-3, a downstream effector of apoptosis. There was a complete lack of morphological evidence of apoptosis in neurons, including neurons with strong immunohistochemical staining for rabies virus antigen. There was TUNEL staining of scattered non-neuronal cells within the neuropil of these cases and focally of apoptotic perivascular inflammatory cells, but no evidence of TUNEL staining of neurons. Likewise, in all of these cases activated caspase-3 was not seen in neurons, although there was multifocal immunostaining of the processes of activated microglia in 9 of 12 cases

(Jackson et al., 2008). Suja and coworkers in their studies of canine (Suja et al., 2011; Suja, Mahadevan, Madhusudhana, Vijayasarathi, & Shankar, 2009) and human (Suja et al., 2011) street virus infected brains found no morphological, TUNEL, or DNA laddering evidence of neuronal apoptosis in multiple neuroanatomical areas. In canine brains, a few TUNEL-labeled perivascular microglial cells, vascular endothelial cells, and white matter glial cells were seen, whereas in human brains occasional TUNEL-positive inflammatory cells were seen in the hippocampus and medulla oblongata (Suja et al., 2011).

In animal models, prominent neuronal apoptosis has been identified in the brains of adult and immature mice following intracerebral inoculation with the CVS strain of fixed rabies virus (Fu & Jackson, 2005; Jackson & Park, 1998; Jackson & Rossiter, 1997; Theerasurakarn & Ubol, 1998; Yan et al., 2001). However, following peripheral inoculation of a fruit-eating bat species with rabies challenge virus standard (CVS) strain, apoptosis was not observed (Reid & Jackson, 2001). In 6-week old mice inoculated in the hindlimb footpad with CVS strain, TUNEL labeling showed no definite neuronal staining and only scattered positive non-neuronal cells (predominantly inflammatory) in the cerebral cortex, while immunostaining for activated caspase-3 showed only occasional non-neuronal cells in the cerebral cortex of moribund animals (Scott, Rossiter, Andrew, & Jackson, 2008). Extensive TUNEL-positivity was detected in the brains of mice intracerebrally inoculated with street (wild-type) rabies virus strains in one study (Ubol & Kasisith, 2000), but other investigators found little or none with bat rabies virus infections (Sarmento, Li, Howerth, Jackson, & Fu, 2005; Yan et al., 2001). A compelling experimentally based model for a "subversive neuroinvasive strategy of rabies virus" in naturally occurring infection has been advanced. Preservation of the integrity of the neuronal network by avoidance of neuronal apoptosis, together with induction of apoptosis in potentially protective T lymphocytes, permits dissemination of the virus, its excretion in saliva and transmission by bite to another host (Baloul & Lafon, 2003; Lafon, 2011).

10.3.4 Negri and lyssa bodies

In the early 1900s, Adelchi Negri undertook a series of detailed studies of rabies virus-infected animal brains and described the characteristic neuronal intracytoplasmic inclusions that now bear his name (Negri, 1903a, 1903b). Although he mistakenly interpreted these intracytoplasmic bodies as a protozoan species, he established their detection as being specifically diagnostic for rabies virus infection. Following his premature death from tuberculosis at age 37, his work was summarized by his wife (Negri-Luzzani, 1913). For many decades before the introduction of electron microscopy and immunofluorescence staining of viral material, light microscopic identification of Negri bodies remained the predominant pathological method for diagnosing rabies encephalomyelitis. Furthermore, Negri's work stimulated numerous studies, extending well into the second half of the 20th century, on the nature of Negri bodies and their significance in the pathogenesis of rabies (see Kristensson, Dastur, Manghani, Tsiang, and Bentivoglio (1996) and Perl and Good (1991) for detailed reviews). In recent years, there is increasing experimental evidence that Negri bodies may be sites of viral transcription and replication, with exploitation of cellular biology for this purpose (Lahaye et al., 2009; Menager et al., 2009). However, it is noteworthy that while Negri bodies are a pathognomic finding in most cases of "street" rabies viral infection, they are almost never detected with "fixed" virus strains.

On hematoxylin and eosin stained sections, Negri bodies appear as dense, well-defined, oval or round, eosinophilic cytoplasmic inclusions (Fig. 10.3). They are typically 2–10 μm in diameter, but may range from 0.5 to 27 μm in size (Negri, 1903a; Nieberg & Blumberg, 1972; Perl & Good, 1991). There is considerable interspecies variation in the average size of Negri bodies, ranging from small in rabbits and raccoons, to large in dogs, guinea pigs and skunks, and very large in cows (Perl & Good, 1991). Within an individual neuron Negri bodies may be single or multiple and they are typically located in the perikaryon, but may occasionally be found in dendrites and axons.

Using the Mann methylene blue and eosin staining method (Lepine & Atanasiu, 1996), Negri observed a small, basophilic, and granular "innerkörperchen" (inner body) within the Negri body inclusions (Negri, 1903a, 1909). Negri and subsequent investigators emphasized the presence of this inner body/basophilic granule/basophilic stippling, to the extent that in 1925 Goodpasture (1925) coined the term "lyssa bodies" to distinguish other inclusions

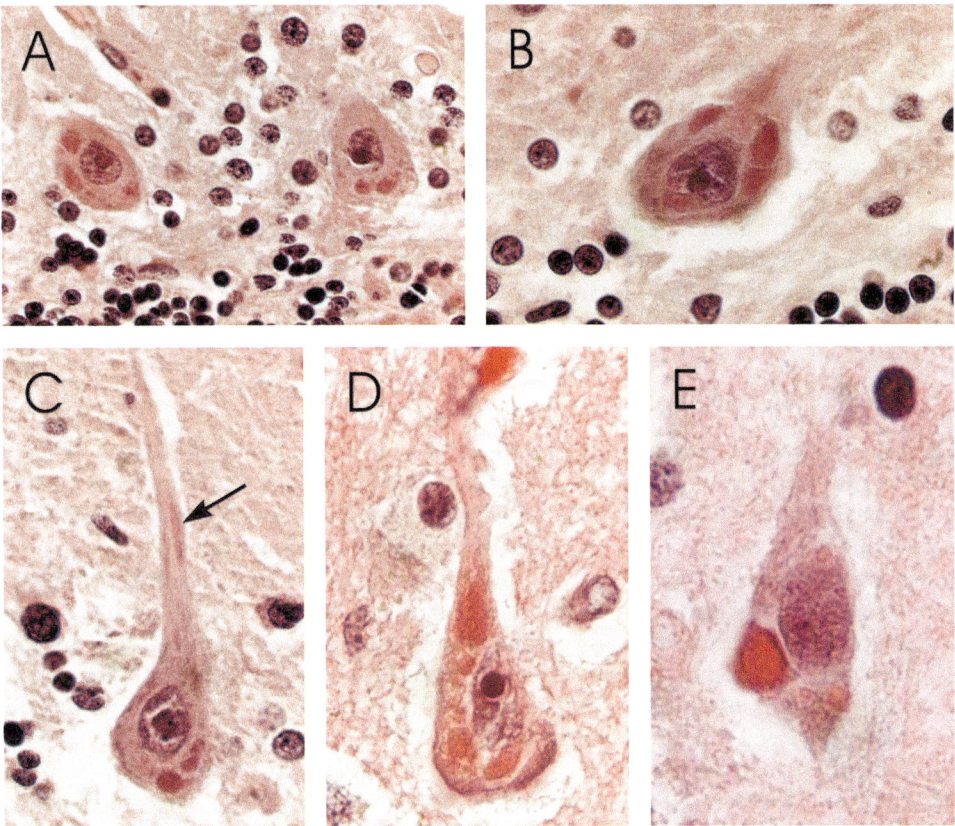

FIG. 10.3 Hematoxylin and eosin (HE) stained sections showing Negri bodies in the perikarya of (A–C) cerebellar Purkinje cells and (D, E) pyramidal neurons in the cerebral cortex of human rabies cases. The arrow in (C) indicates a Negri body in an apical dendrite (magnifications, A 315×, B 460×, C 550×, D 730×, E 865×).

that lacked these staining features from classic Negri bodies. The impetus to make this distinction was to recognize the presence of the other small eosinophilic cytoplasmic inclusions that lacked an internal structure, found in healthy neurons in a variety of animal species. The concern was that the other inclusion bodies could be misinterpreted as rabies virus inclusions (Goodpasture, 1925; Szlachta & Habel, 1953; Tierkel, 1973b). However, this is less of an issue with human tissue as lyssa bodies are typically more numerous than Negri bodies (Mrak & Young, 1994; Sung, Hayano, Mastri, & Okagaki, 1976). They are frequently more irregularly shaped and less clearly demarcated from the surrounding cell cytoplasm than Negri bodies, although they share ultrastructural features with Negri bodies (Iwasaki & Tobita, 2002; Mrak & Young, 1994; Perl & Good, 1991) and are immunoreactive for rabies virus antigen (see below).

Negri bodies have been shown in 50% to 90% of street rabies infections in different series of human rabies cases (Dupont & Earle, 1965; Herzog, 1945; Jogai, Radotra, & Banerjee, 2000; Negri-Luzzani, 1913), influenced in part by the species, the extent of tissue sampling, and possibly by the duration of clinical disease. Dupont and Earle (1965) observed Negri bodies in 71% of their cases, and Negri bodies unassociated with inflammation in 15% of cases. Although Negri bodies may be found in virtually any neuronal population in the CNS or peripheral nerve ganglia, they tend to be most numerous and largest in larger neurons (Tangchai et al., 1970), especially in hippocampal pyramidal neurons and cerebellar Purkinje cells. Dupont and Earle (1965) observed Negri bodies in the following locations and approximate proportions of their series of 59 cases: cerebellum (60%), hippocampus (43%), medulla (14%), pontine nuclei (12%), spinal cord (10%), cerebral cortex (7%), midbrain (7%), basal ganglia (5%), thalamus (2%) and peripheral nerve ganglia (5%), and (Tangchai et al., 1970) in their series of 24 human rabies cases: cerebellum (54%), hippocampus (50%), brainstem (50%), hypothalamus (42%), thalamus and hypothalamus (42%), cerebral cortex (21%), and spinal cord (21%). In their monumental study of over one thousand biologically positive rabies cases from a diverse range of species, Tustin and Smit (1962) found that Negri bodies were present in 71% of cases when the hippocampus alone was examined histologically. Remarkably, numerous and diffusely distributed Negri bodies have been reported in two immunocompromised patients, one of whom had AIDS (Adle-Biassette et al., 1996) and the other a renal transplant recipient with prolonged survival (Walker et al., 2016). In a series of experimentally infected dogs, an increasing proportion of neurons contained Negri bodies as the disease progressed (Marinesco & Storesco, 1931). More numerous and widely distributed Negri bodies have also been observed with increasing duration of survival in some human case series (Sandhyamani, Roy, Gode, & Kalla, 1981). Negri bodies are more likely to be found in areas of the CNS where there is little inflammation and are less frequently seen in degenerating neurons and/or in association with inflammatory foci (Dupont & Earle, 1965; Iwasaki & Tobita, 2002; Marinesco & Storesco, 1931; Sukru-Aksel, 1958).

The ultrastructure of Negri bodies and rabies virions has been investigated in human rabies cases (Adle-Biassette et al., 1996; de Brito, Araujo, & Tiriba, 1973; Gonzalez-Angulo, Marquez-Monter, Feria-Velasco, & Zavala, 1970; Iwasaki, Liu, Yamamoto, & Konno, 1985; Leech, 1971; Manghani, Dastur, Nanavaty, & Patel, 1986; Morecki & Zimmerman, 1969; Mrak & Young, 1993; Sandhyamani et al., 1981) and in experimentally inoculated animals (Charlton & Casey, 1979; Fekadu et al., 1982; Hottle, Morgan, Peers, & Wyckoff, 1951; Iwasaki & Clark,

1975; Iwasaki, Ohtani, & Clark, 1975; Miyamoto & Matsumoto, 1965; Murphy, Bauer, Harrison, & Winn, 1973; Perl, Callaway, & Hicklin, 1972). In addition, many aspects of rabies virus replication and virus/host cell interaction have been studied ultrastructurally in vitro (Davies, Englert, Sharpless, & Cabasso, 1963; Hummeler, Koprowski, & Wiktor, 1967; Iwasaki & Clark, 1977; Iwasaki & Minamoto, 1982; Iwasaki, Wiktor, & Koprowski, 1973; Lewis & Lentz, 1998; Matsumoto & Kawai, 1969; Matsumoto, Schneider, Kawai, & Yonezawa, 1974). Negri bodies show a similar spectrum of ultrastructural features in both human and animal material, being composed of large aggregates of granulo-filamentous matrix material (matrix) and varying numbers of viral particles (Fig. 10.4). The matrix consists of randomly oriented viral nucleocapsids (Hummeler, Tomassini, Sokol, Kuwert, & Koprowski, 1968; Schneider et al., 1973). In "immature" inclusions, it has a filamentous appearance with visible substructural coiling of

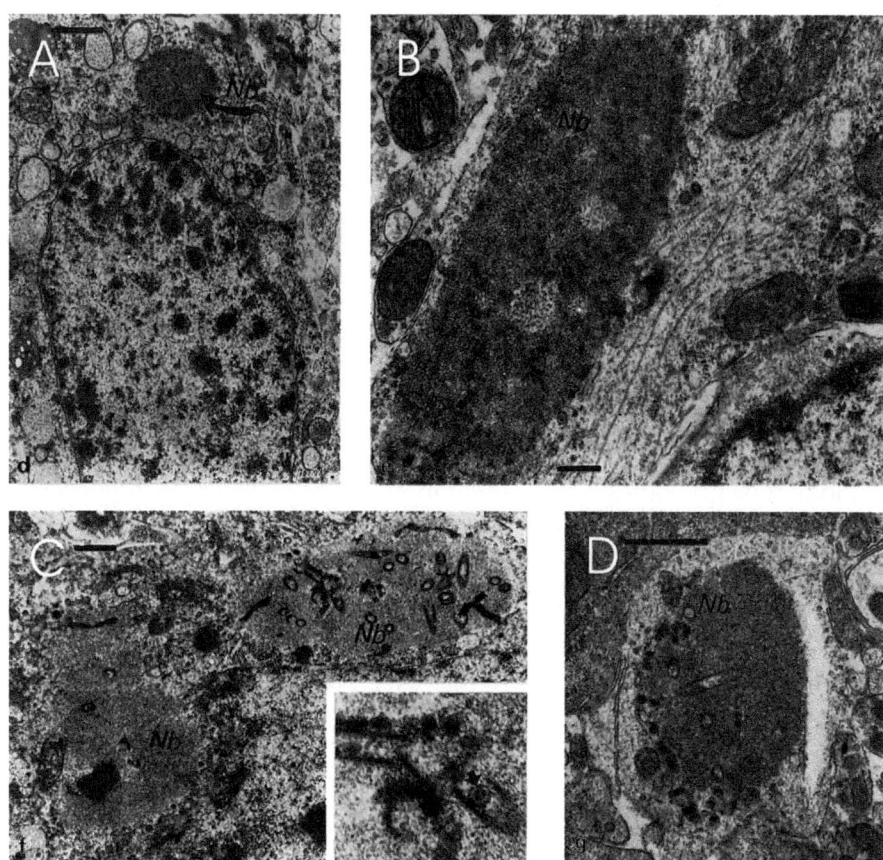

FIG. 10.4 Ultrastructural features of Negri bodies. (A) A circular Negri body (Nb) in the perikaryon of a hippocampal neuron from a human patient. (B) A large elongated Negri body in a dendrite of a cortical neuron from a mouse. (C, D) Negri bodies in infected mouse brains containing bullet-shaped virions. The inset in C shows a virion with its core. Scale bars correspond in A, C, D to 1 µm, in B to 0.5 µm. *Reproduced from Kristensson, K., Dastur, D. K., Manghani, D. K., Tsiang, H., & Bentivoglio, M. (1996). Rabies: Interactions between neurons and viruses. A review of the history of Negri inclusion bodies. Neuropathology and Applied Neurobiology, 22, 179–187 with permission.*

the nucleocapsid strands, but with maturation and increasing density, the individual strands become increasingly difficult to resolve, such that the matrix has a more granular and electron-dense appearance (Matsumoto et al., 1974; Miyamoto & Matsumoto, 1967; Murphy, Bauer, et al., 1973). Bullet-shaped or tubular virions are typically associated with the matrix accumulations (for more detailed descriptions, see Perl and Good (1991), Iwasaki and Tobita (2002), and Chapter 2), but may be absent in some Negri bodies (Morecki & Zimmerman, 1969; Perl & Good, 1991; Sandhyamani et al., 1981). In a study of experimentally infected rhesus monkey tissue, Perl and colleagues (Perl et al., 1972; Perl & Good, 1991) described three basic ultrastructural configurations of Negri bodies. In the first type, seen most often in the thalamus and caudate nuclei, many bullet-shaped rabies virions were found around the periphery of the matrix, where they were often attached to dilations in the endoplasmic reticulum. Bullet-shaped virions were also frequently located within deep invaginations of the matrix contour, accounting for the inner bodies (innerkörperchen) seen by light microscopy. However, entrapment of ribosomes and endoplasmic reticulum has been proposed as an alternative explanation (Iwasaki & Tobita, 2002). In the second type of inclusion, found predominantly in brainstem, cerebellar, and spinal neurons, tubular virions were dispersed throughout the matrix, whereas in the third type (in hippocampal neurons) virions were typically not seen within aggregates of matrix.

Lyssa bodies, without the inner body, exhibit essentially the same ultrastructural features as Negri bodies (Matsumoto, 1963; Miyamoto & Matsumoto, 1965). Furthermore, a much more extensive distribution of lyssa bodies is observed in viral infection by electron microscopy than can be resolved by routine light microscopy. Lyssa bodies, which form small aggregates of virions and matrix, are undetectable by light microscopy in neuronal perikarya and also in dendrites and axons (Charlton & Casey, 1979; Fekadu et al., 1982; Iwasaki et al., 1985; Jenson, Rabin, Wende, & Melnick, 1967; Murphy, Bauer, et al., 1973; Murphy, Harrison, Winn, & Bauer, 1973; Perl & Good, 1991). Small aggregates of matrix and virions have also been observed in astrocytes (Fekadu et al., 1982; Iwasaki & Clark, 1975; Matsumoto, 1963; Perl & Good, 1991) and oligodendrocytes (Perl & Good, 1991).

In the later stages of rabies virus assembly, the virions acquire a lipid bilayer envelope by budding through host cell plasma membranes into the extracellular space (see Chapter 2). Virions also may bud intracytoplasmically through membranes of the endoplasmic reticulum or, less frequently, Golgi apparatus or outer lamella of the nuclear envelope (Charlton & Casey, 1979; Gosztonyi, 1994; Iwasaki et al., 1975, 1985; Iwasaki & Clark, 1975; Murphy, Bauer, et al., 1973; Murphy, Harrison, et al., 1973). Importantly, cell surface viral budding may not be associated with adjacent Negri bodies or nucleocapsid matrix (Iwasaki & Tobita, 2002). Also, neuron-to-neuron transmission of rabies virus, especially at synaptic junctions, has been clearly established ultrastructurally (Burrage, Tignor, & Smith, 1983; Charlton & Casey, 1979; Iwasaki et al., 1975, 1985).

Experimental investigations employing in vitro infection of neuronal and non-neuronal cell lines with CVS strain of fixed rabies virus indicate that Negri body-like structures that develop in the infected cells are likely sites of viral transcription and replication (Lahaye et al., 2009) with viral exploitation of cellular compartmentalization and cellular proteins, especially the innate immune response receptor Toll-like receptor 3 (TLR3), to promote viral replication and potentially evade apoptosis (Menager et al., 2009). However, it is noteworthy that although Negri bodies are a characteristic cytological feature in many cases of infection with "street" rabies virus strains, they are almost never observed with "fixed" virus strains (Kristensson et al., 1996). This raises the question of the exact correlation between the Negri body-like structures characterized in the foregoing in vitro studies using fixed virus strains and the classic Negri bodies seen in wild-type infection. The fact that Negri bodies are almost never found following in vivo infection with fixed strains indicates that, whereas their presence is not essential for a fatal outcome, they may be a reflection of a more efficient "subversive neuroinvasive strategy" (Lafon, 2011) in naturally occurring infection, but not in the context of experimental infection with attenuated fixed virus strains.

10.3.5 Degeneration of neuronal processes

The fact that the pathological abnormalities described here, i.e., perivascular inflammation, microglial activation, neuronophagia and Negri bodies, may be minimal or even absent in some fatal cases (Jogai et al., 2000), indicates that these are not essential neuropathological accompaniments of neurological dysfunction in rabies virus infection. This suggests that morphological correlates of neuronal dysfunction in rabies, should they exist, are likely to be most apparent at a fine structural scale. The neuropil microvacuolation observed by Charlton and colleagues in experimentally infected skunks and foxes, consisted of membrane-bound vacuoles in neuronal processes, predominantly dendrites, and has some similarities to excitotoxic amino acid-induced dendritic swelling (Charlton, 1984; Charlton et al., 1987). Li, Sarmento, and Fu (2005) found severe disorganization and destruction of axons and dendrites, with relative preservation of the neuronal cell bodies, in mice infected with a pathogenic rabies virus strain (N2C virus derived from CVS-24), but not with an attenuated strain (SN-10 derived from the SAD B19 vaccine strain). There was complete loss of neurofilament and MAP-2 (microtubule-associated protein 2) immunoreactivity in neurons infected with the pathogenic strain. These investigators consequently proposed that pathogenic rabies virus infection may induce degeneration of neuronal processes by disrupting cytoskeletal integrity and that this may form the basis for neuronal dysfunction in rabies (Li et al., 2005). In a Golgi technique study of cortical pyramidal neurons of rabies infected mice, Torres-Fernandez, Yepes, and Gomez (2007) observed decreased soma size and morphological changes in dendrites, including loss of spines, in approximately 13%, 8%, and 32% of neurons following intracerebral street-virus, intramuscular street-virus, and intramuscular CVS fixed-virus inoculations, respectively. The potential pathogenetic role of structural changes in neuronal processes was assessed by Scott et al. (2008) in transgenic mice expressing yellow fluorescent protein (YFP) in subpopulations of CNS neurons following peripheral inoculation with the challenge virus standard (CVS-11) fixed rabies virus strain. Although histopathological changes were minimal in paraffin-embedded brain sections from moribund CVS-infected animals in this study, fluorescence microscopy showed marked beading and fragmentation of the dendrites and axons of pyramidal neurons in layer V of the cerebral cortex, cerebellar mossy fibers, and of axons in brainstem tracts. On resin-embedded sections, numerous vacuoles were seen in the perikarya and proximal dendrites of pyramidal neurons in the cerebral cortex and CA1 hippocampal sector and also throughout the neuropil of the cerebral cortex. Ultrastructurally, there were swollen mitochondria within the perkarya, dendrites, and axons of many cortical pyramidal neurons and also frequently at sites of dendritic and axonal beading (Fig. 10.5). However, affected cortical neurons, which also typically contained nucleocapsid material, did not otherwise show overt degenerative features, such as perikaryal swelling, loss of plasma, or nuclear membrane integrity or abnormal chromatin condensation. The neuropil vacuolation seen by light microscopy on resin sections corresponded ultrastructurally with swollen neuronal processes and very distended presynaptic nerve endings, with many of the neuropil vacuoles containing swollen mitochondria and membranous-type debris (Fig. 10.5). It was concluded that this spectrum of structural changes was sufficient to explain the severe clinical disease and fatal outcome in this experimental rabies model (Scott et al., 2008). There is evidence that this neuronal process degeneration is a result of

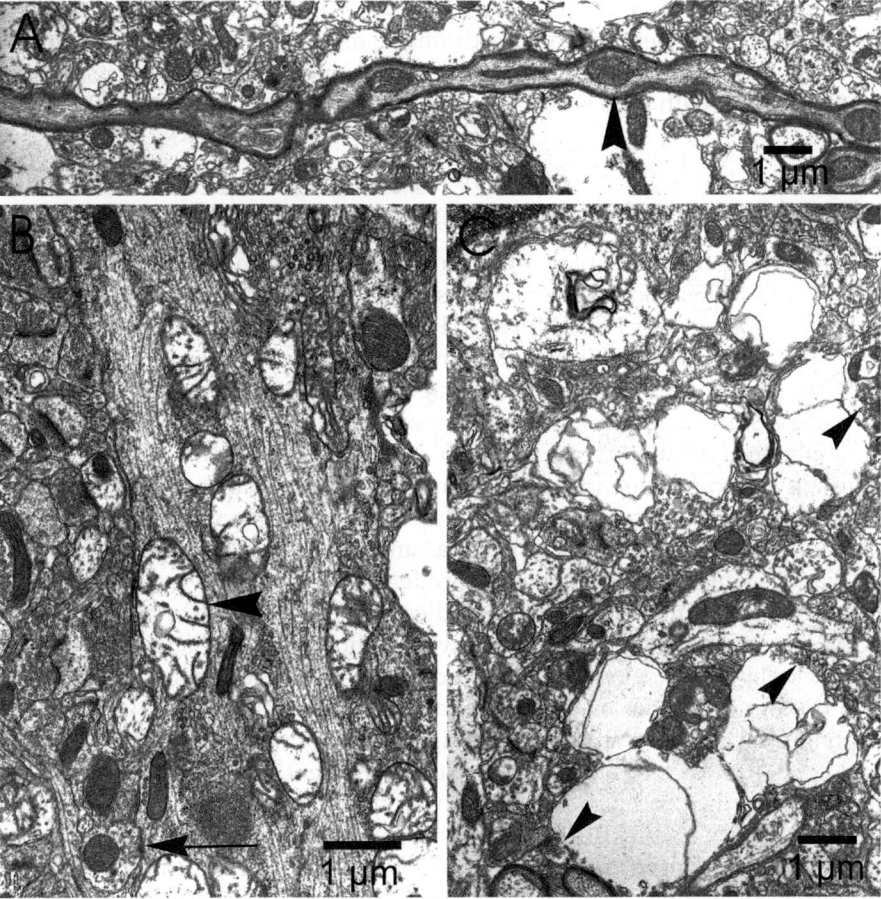

FIG. 10.5 Electron micrographs of the cerebral cortex of a moribund CVS infected mouse showing (A) an axon of a pyramidal neurons containing swollen mitochondria (arrowhead) corresponding with areas of beading; (B) a dendrite containing swollen mitochondria (arrowhead); and (C) vacuolated presynaptic nerve endings containing synaptic vesicles (arrowheads). *Reproduced from Scott, C. A., Rossiter, J. P., Andrew, R. D., & Jackson, A. C. (2008). Structural abnormalities in neurons are sufficient to explain the clinical disease and fatal outcome in experimental rabies in yellow fluorescent protein-expressing transgenic mice. Journal of Virology, 82(1), 513–521 with permission.*

oxidative stress (Jackson, 2016; Jackson, Kammouni, Zherebitskaya, & Fernyhough, 2010; Kammouni et al., 2012) (see Chapter 9).

In an investigation of mice intracerebrally infected with an MRV strain of street rabies virus, Song et al. (2013) saw no obvious cytopathological changes in the cell bodies and dendrites of infected hippocampal CA1 pyramidal neurons by conventional histology and immunolabeling for MAP-2. However, using fluorescence staining with Alexa Fluor 488 phalloidin for filamentous (F)-actin, they observed a significant decrease in the number of dendritic spines and the presence of many "rope-like" structures along the dendrites, reflecting F-actin reorganization. Western blot analysis of infected hippocampal tissue indicated depolymerizaton of F-actin. Furthermore, in MRV-infected cultures of primary hippocampal neurons, these investigators observed a dramatic decrease in the number and size of dendritic spines. Using a modified Golgi-Cox method, Monroy-Gomez, Santamaria, and Torres-Fernandez (2018) demonstrated a dramatic reduction in the number and length of dendrites of neurons in the ventral horn of the spinal cord of mice following hindlimb intramuscular inoculation with CVS strain.

10.3.6 Distribution of rabies virus antigen

The development and use of immunofluorescence (Bingham & van der Merwe, 2002; Charlton & Casey, 1979; Goldwasser & Kissling, 1958; Johnson, Swoveland, & Emmons, 1980; Murphy, Bauer, et al., 1973; Murphy, Harrison, et al., 1973) and immunoperoxidase techniques (Fekadu, Greer, Chandler, & Sanderlin, 1988; Iwasaki et al., 1985; Jackson & Reimer, 1989; Last, Jardine, Smit, & van der Lugt, 1994; Tirawatnpong et al., 1989) for the detection of rabies virus antigen represented important methodological advances for both the diagnosis and investigation of rabies virus infection. In 1958, Goldwasser and Kissling (1958) used immunofluorescence microscopy to demonstrate rabies virus antigen in infected brain tissue and they established that Negri bodies contain viral antigen. The fluorescent antibody test (FAT) subsequently became established as an important routine laboratory method for the rapid diagnosis of rabies (Bingham & van der Merwe, 2002).

Comprehensive immunohistochemical studies of the distribution of rabies virus antigen in human CNS material have been relatively few in number, but have shown a fairly consistent pattern. Polyclonal or monoclonal antiribonucleoprotein/nucleocapsid antibodies have typically been used in these studies (Iwasaki et al., 1985; Jogai et al., 2000; Johnson et al., 1980; Tirawatnpong et al., 1989). Rabies virus antigen (RVAg) was found throughout the brain and spinal cord in most cases. It was consistently present in far more neurons than those containing Negri bodies and also in cases in which no Negri bodies were detected, despite a meticulous search. In a series of 20 cases with a clinical diagnosis of rabies, a histopathological diagnosis could be made in only 17 cases, whereas all of the cases exhibited positive RVAg-immunohistochemical staining (Jogai et al., 2000). RVAg was typically seen in the cytoplasm of neuronal perikarya and in dendrites and axons and appeared as blob-like masses (10–20 μm) and granules (1–3 μm) (Fig. 10.6), with the larger masses corresponding with Negri bodies seen on hematoxylin and eosin-stained sections (Jogai et al., 2000). The intensity of staining may vary from cell to cell and some neurons showed diffuse staining of their cytoplasm and processes (Feiden, Feiden, Gerhard, Reinhardt, & Wandeler, 1985; Iwasaki et al.,

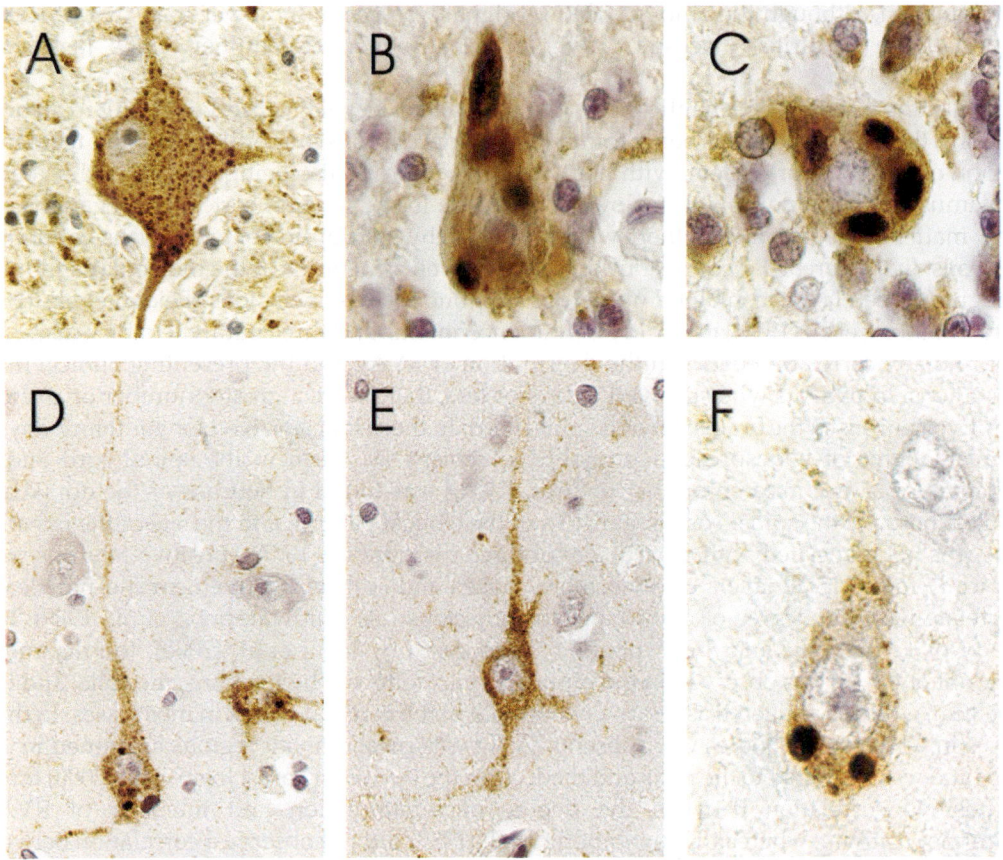

FIG. 10.6 Immunoperoxidase staining for rabies virus antigen (mouse monoclonal antirabies virus nucleocapsid protein IgG) in human rabies cases. (A) Motor neuron in anterior horn of spinal cord; (B, C) cerebellar Purkinje cells; (D–F) pyramidal neurons in cerebral cortex. The larger immunolabeled masses correspond with Negri bodies (magnifications, A 256×, B 535×, C 567×, D 300×, E 290×, F 516×).

1985; Johnson et al., 1980). RVAg was also observed in processes in the neuropil remote from cell bodies as oval or spindle-shaped masses (Iwasaki et al., 1985) and was inconsistently present in some astrocytes and oligodendrocytes (Feiden et al., 1985; Jogai et al., 2000; Tirawatnpong et al., 1989).

In a quantitative study of neuronal infection in a human case, neurons with Negri bodies contained larger mean amounts of RVAg than those that did not. Of a variety of neuronal populations, cerebellar Purkinje cells and periaqueductal gray matter neurons showed the largest percentage area for both Negri bodies and RVAg signal, whereas neurons in the trochlear nucleus had a much smaller area of Negri bodies, despite a similar RVAg signal (Jackson, Ye, Ridaura-Sanz, & Lopez-Corella, 2001). Rabies virus genomic RNA and mRNA has been detected by in situ hybridization in brain tissue from human rabies cases and in both CVS- and street virus-infected mouse brains. The distribution of virus RNA was similar to that of viral antigen, although the amount of RNA signal was generally lower than that of RVAg, especially in dendrites (Jackson, 1992; Jackson, Reimer, & Wunner, 1989; Jackson & Wunner, 1991).

In a study of three encephalitic rabies cases, Feiden et al. (1985) observed RVAg in all brain regions, the highest amounts being found in the hippocampus, hypothalamus, and tegmental region of the lower brainstem, with many positive neurons also being present in the ventral thalamus and basal portion of the lower brainstem. In the cerebral cortex, basal ganglia and gray matter of the spinal cord, there was a more patchy distribution of fewer virus-containing neurons. In the cerebellar cortex, Purkinje cells as well as neurons in the molecular and internal granule cell layers contained numerous immunopositive inclusions (Feiden et al., 1985). Tirawatnpong et al. (1989), in a study of four encephalitic and three paralytic rabies cases, did not find any correlation between the distribution of RVAg and the presenting clinical manifestations. In patients who survived 7 days or less, there were a greater number of antigen-positive neurons in the brainstem and spinal cord. In those that survived longer than 7 days, a similar degree of widespread neuronal involvement was seen in the spinal cord and in supratentorial and infratentorial structures. RVAg was found in neurons of the dorsal and ventral horns of the spinal cord, regardless of the clinical pattern, and the site of the infecting bite was not associated with a particular antigen distribution. RVAg-positive neurons were typically found in all layers of the cerebral cortex and cortical involvement did not clearly correlate with the degree of disturbance of consciousness (Tirawatnpong et al., 1989). In a study of 12 paralytic and 8 encephalitic rabies cases by Jogai et al. (2000), the maximum amount of RVAg was observed in the hippocampus, followed by the pons, medulla, and cerebellum, whereas antigen was relatively minimal in the cerebral cortex in most cases. In three cases in which Negri bodies were absent, RVAg was present in all regions examined in one case and was restricted to the pons and medulla in the two other cases. Jogai et al. (2000) found a positive correlation between the degree of inflammation and intensity of RVAg-immunopositivity, whereas Tirawatnpong et al. (1989) did not observe a correlation between either the anatomical distribution or amount of RVAg and inflammation in their cases. In a series of nine human rabies cases (and also in six naturally infected "furious" canines), Suja et al. (2011) observed RVAg in almost all brain areas, but with striking involvement of lateral and ventral group of thalamic nuclei, basal ganglia, and limbic structures. There was labeling of multiple Negri bodies and antigen was also observed in oligodendrocytes and the processes of fibrous astrocytes (Suja et al., 2011).

The immunohistochemical distribution of RVAg in major brain regions of 13 naturally infected domestic and wild mammalian species has been reported by Stein, Rech, Harrison, and Brown (2010). Animal models have been especially valuable for investigating the centripetal and centrifugal spread of rabies virus to and from the CNS and for establishing the anatomic sequence of spread within the CNS (Baer, Shanthaveerappa, & Bourne, 1965; Charlton & Casey, 1979; Charlton, Casey, Wandeler, & Nadin-Davis, 1996; Coulon et al., 1989; Dean, Evans, & McClure, 1963; Huygelen, 1960; Huygelen & Mortelmans, 1959; Jackson & Reimer, 1989; Kliger & Bernkopf, 1943; Murphy, Bauer, et al., 1973; Murphy, Harrison, et al., 1973; Schneider, 1969a, 1969b; Schneider & Hamann, 1969; Shankar, Dietzschold, & Koprowski, 1991; Smart & Charlton, 1992). Schneider (1969a) established an ascending pattern of infection in the mouse by sequential immunofluorescence and infectivity titration studies of parts of the CNS. In experimentally infected mice, rats and hamsters, spread of virus from the lumbar spinal cord to the brainstem happens within hours (Baer et al., 1965; Baer, Shantha, & Bourne, 1968; Murphy, Bauer, et al., 1973; Murphy, Harrison, et al., 1973). Murphy (1977) described an "ascending wave of rabies infection in the brain," as demonstrated by a decreasing immunofluorescent gradient from brainstem to forebrain at earlier stages of infection, and massive accumulations of RVAg throughout the brain and spinal cord at terminal stages (Murphy, Bauer, et al., 1973; Murphy, Harrison, et al., 1973). In skunks inoculated into hindlimb foot muscle with street rabies virus, Charlton and Casey (1979) observed granular RVAg immunoflourescence in scattered neurons in the spinal cord, medulla oblongata, pons, and cerebellum in one animal killed at 10 days and another at 14 days post-inoculation. In all other skunks killed on day 14 or later, there was intense immunofluorescence in gray matter throughout the brain and spinal cord. Analysis of the early events in this skunk model indicated viral entrance and replication at L2 and L3 levels of the spinal cord, local spread by propriospinal neurons and early and rapid spread to the brain via long ascending and descending fiber tracts (Charlton et al., 1996). RVAg-positive fine particles, aggregates, and filaments were found in most neurons and fine granules aligned within axons were seen in the white matter (Charlton et al., 1996; Charlton & Casey, 1979; Smart & Charlton, 1992).

Following hindlimb footpad inoculation of mice with fixed rabies virus, RVAg was initially detected by immunoperoxidase staining on day 4 in lumbar dorsal root ganglia and gray matter of the lumbar spinal cord, with much greater involvement on the side of inoculation (Jackson & Reimer, 1989). A few positive neurons were also seen in the brainstem tegmentum at this time. On day 5, there were more infected neurons in the lumbar and sacral cord segments, including ventral horn neurons contralateral to the inoculation side. There was heavy involvement of the brainstem tegmentum, prominent infection of the deep cerebellar nuclei, but only a few positive Purkinje cells and no RVAg in the cerebral cortex or hippocampus. On day 6, there was again more signal in the spinal cord and brainstem tegmentum. Numerous Purkinje cells and a few cerebellar internal granule cells were immunopositive. Many neurons in the cerebral cortex were also now infected, but there was still no evidence of hippocampal infection. Hippocampal RVAg was initially seen in the CA3 region on day 8, with spread to CA1 and CA4 and sparse involvement of the dentate gyrus by day 10. The quantity of RVAg in the CNS decreased progressively in surviving mice after day 10, with death occurring between days 10 and 18 (Jackson & Reimer, 1989). One conclusion from this study was that the hippocampus is not a good location for the detection of early CNS infection following peripheral inoculation. A similar conclusion was reached on the basis of fluorescent antibody testing (FAT) of 252 rabies-positive brains from a diverse range of naturally infected species by (Bingham & van der Merwe, 2002). Whereas Negri bodies, when present, are typically most readily found in the hippocampus, cerebellar Purkinje cells and pyramidal neurons of the cerebral cortex (Tierkel, 1973a), the hippocampus, cerebellum, and different parts of the cerebrum were negative by FAT in 4.9%, 4.5%, and 3.9%–11.1%, respectively, of Bingham and van der Merwes' series. By contrast, the only structures that were FAT-positive in all 252 animal brains were the thalamus, pons, and medulla oblongata, with a consistent abundance of RVAg in the thalamus (Bingham & van der Merwe, 2002).

Suja et al. (2011) reported significant differences in the patterns of RVAg immunohistochemical staining between mice experimentally inoculated with street virus versus CVS strain. Following intramuscular inoculation with street virus, there was widespread infection, with a caudocranial gradient and intraneuronal aggregation of viral antigen into multiple

Negri body-like globular masses and extensive dendritic spread. By contrast, in CVS-infected animals, there was diffuse cytoplasmic staining of neurons and minimal dendritic spread.

In their study of rabies-infected dogs sacrificed soon after onset of paralytic or furious illness, Shuangshoti et al. (2013) observed a caudal-to-rostral gradient (high-to-low) in RVAg burden as follows: spinal cord, brainstem, cerebellum, caudate and thalamus, hippocampus, cerebrum. There was a greater quantity of RVAg in most regions of the CNS, including the spinal cord, in furious versus paralytic animals. In one dog with paralytic rabies, RVAg was only found in the spinal cord. In a subsequent study of this canine material, Shuangshoti et al. (2016) investigated the quantity of RVAg virus antigen within neuronal cell bodies and processes. They found that following the appearance of RVAg in neuronal cell bodies there was a delay in cell process involvement, consistent with viral replication in the cell body and subsequent transport into the neuronal processes. There were greater quantities of RVAg in the processes of furious versus paralytic animals, controlling for equivalent degrees of neuronal cell body involvement, suggesting that intracellular transport of rabies virus may be slower in the paralytic form.

10.4 Pathology in the peripheral nervous system

Spread of rabies virus through the peripheral nervous system (PNS) plays an essential role in the centripetal and centrifugal phases of rabies infection. There are associated inflammatory, reactive, and degenerative changes in many of the structural components of the PNS, including neuronal cell bodies in sensory and autonomic ganglia and their capsular/satellite cells, as well as in sensory and motor axons and their enveloping Schwann cells. This results in varying degrees of degeneration and loss of sensory and autonomic neurons (neuronopathy), reactive proliferation of their satellite cells, and demyelination and Wallerian degeneration of nerve fibers in spinal nerve roots and peripheral nerves. Historically, there has been particular interest in the diagnostic utility of histological changes in the peripheral nerve ganglia.

10.4.1 Changes in sensory and autonomic ganglia

The proliferation of capsule (satellite) cells was briefly described in 1872 by Nepveau in the Gasserian ganglion (sensory ganglion of the trigeminal nerve) of a human rabies case, but it was Van Gehuchten and Nelis in 1900 who first recognized the importance of ganglionic lesions as a diagnostic marker of rabies. They described marked proliferation of capsular cells surrounding chromatolytic neurons in spinal and cranial nerve ganglia of animal and human rabies cases (Fig. 10.7). This capsular cell reaction, together with varying degrees of interstitial lymphocytic infiltration, resulted in grossly apparent enlargement and increased firmness of involved spinal ganglia (Iwasaki & Tobita, 2002; Perl & Good, 1991; Van Gehuchten & Nelis, 1900). These changes, often referred to in the literature as "Van Gehuchten and Nelis lesions" or "Van Gehuchten nodules" were reconfirmed in a number of later studies (Hardenbergh, 1916; Marinesco & Storesco, 1931; Mitrabhakdi et al., 2005; Sung et al., 1976; Tangchai & Vejjajiva, 1971).

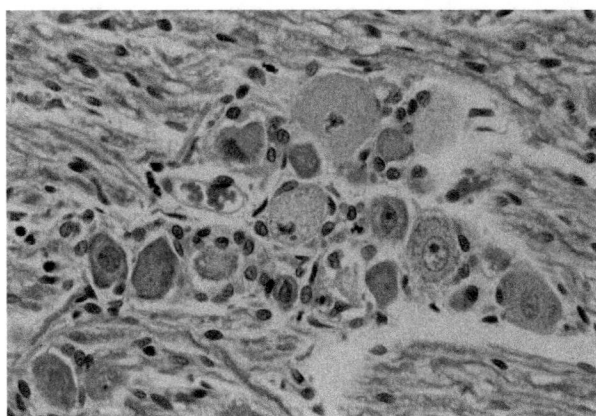

FIG. 10.7 Van Gehuchten nodule (Van Gehuchten and Nelis lesion) in rat trigeminal ganglion following inoculation of street virus into the ipsilateral mental nerve. There is a group of chromatolytic neurons accompanied by proliferation of satellite cells and a sparse lymphocytic infiltrate (HE stain, magnification 285×). *Reproduced from Iwasaki, Y., & Tobita, M. (2002). Pathology. In A.C. Jackson & W. H. Wunner (Eds.), Rabies (pp. 283–306). San Diego: Academic Press with permission.*

In a series of 52 human rabies cases, Herzog (1945) could not find Negri bodies in the hippocampus in almost 50% of the cases, whereas he observed capsular cell proliferation and neuronal degenerative changes in the ganglion nodosum of the vagus nerve in all 52 cases, with or without accompanying focal or diffuse inflammatory cell infiltration. In a study of 9 human rabies cases, Tangchai and Vejjajiva (1971) reported leukocyte infiltration in 3, hypertrophy and proliferation of capsular cells in 5, neuronal degeneration in 5, and extreme vascular congestion in 2 cases. There was also increased stromal and epineurial collagen in the ganglia. Many of the neurons were moderately swollen, with pale finely vacuolated cytoplasm, and some fragmented neurons undergoing neuronophagia by histiocytes were seen. Some neurons also contained round acidophilic cytoplasmic inclusions (5–10 µm), although these lacked inner bodies (Tangchai & Vejjajiva, 1971). Numerous Negri bodies have been found in the Gasserian (Garcia-Tamayo, Avila-Mayor, & Anzola-Perez, 1972) and dorsal root ganglia (Sung et al., 1976) in some cases. Severe lymphocytic inflammation and necrosis has been observed in the inferior cervical sympathetic ganglion of an encephalitic rabies case with prolonged survival (Sandhyamani et al., 1981). Mitrabhakdi et al. (2005) reported electrophysiological abnormalities consistent with dorsal root ganglionopathy in one encephalitic rabies patient and two paralytic patients who had severe prodromal paresthesiae. Postmortem analysis showed severe ganglionitis, with infiltration predominantly by CD3-positive T lymphocytes.

In experimentally infected immature hamsters, Murphy, Bauer, et al. (1973) and Murphy, Harrison, et al. (1973) first observed RVAg immunofluorescence in small numbers of ipsilateral lumbar dorsal root ganglion neurons 60–70 h following hindlimb inoculation and in lumbar autonomic ganglia by 72 h. The RVAg initially had a "dustlike" distribution, with rapid subsequent progression to involvement of nearly all dorsal root ganglia by brilliant aggregate fluorescence. This was most dense at the peripheral margins of individual ganglion cells, with myelinated axon hillocks being especially heavily infected. Satellite cells did not contain RVAg. Ultrastructurally, large masses of nucleocapsid material were seen within ganglion cells, particularly at their margins, with comparatively small numbers of virus particles budding from intracytoplasmic membranes deeper within the cytoplasm (Murphy, Bauer, et al., 1973; Murphy, Harrison, et al., 1973). In skunks, Charlton and Casey (1979) found RVAg immunofluorescence in scattered lumbar dorsal root ganglia 10 days following hindlimb inoculation, at which time no fluorescence was seen in peripheral nerve fibers. By 14 days post-inoculation, many dorsal root neurons contained antigen and linear arrays of granular fluorescent RVAg were seen in axons of hindlimb and forelimb peripheral nerves. After 14 days, variable, but increasing degrees of neuronal chromatolysis, neuronophagia and inflammatory cell infiltration were seen in the trigeminal and dorsal root ganglia (Charlton & Casey, 1979).

In adult mice inoculated in the right hindlimb footpad with the CVS strain, Rossiter, Hsu, and Jackson (2009) first observed RVAg in a right lumbo-sacral dorsal root ganglion of one of three infected animals at 3 days post inoculation (p.i.) and by 3.5 p.i. antigen could be traced within the dorsal roots and spinal cord. At day 4.5 p.i. antigen was first seen in left sided dorsal root ganglia and at later time points there was bilateral infection of ganglia, Multifocal mononuclear inflammatory cells infiltrates of dorsal ganglia were first observed at 4 days p.i., and these infiltrates subsequently became more prominent. Degenerating gangliocytes became increasingly frequent after 4 days p.i. There was a spectrum of light microscopic changes, with varying degrees of nuclear eccentricity and irregularity, cytoplasmic chromatolysis and vacuolation, to more advanced degeneration with neuronophagia. The neuronal nuclei tended to be paler and their nuclei smaller and less intensely stained those in mock-infected mice. However, there was no evidence of karyorrhectic chromatin condensation characteristic of apoptosis and TUNEL staining or immunolabeling for activated caspase-3 was not observed in gangliocytes of CVS- or mock-infected mice. These degenerative changes preferentially involved gangliocytes that had a medium-to-large soma size and that were intermixed with morphologically normal neurons. The satellite (capsule) cells surrounding degenerating gangliocytes tended to be larger and more numerous than those around morphologically intact neurons. Ultrastructurally there was a spectrum of changes in gangliocytes that included features characteristic of "the axotomy response" (nuclear eccentricity, central accumulation of mitochondria and Golgi apparatus and preferential localization of rough endoplasmic reticulum toward the periphery of perikarya), the appearance of numerous autophagic compartments and aggregation of intermediate filaments, while the neurons retained relatively intact mitochondria and plasma membranes. At later stages the neuronal cytoplasm was highly vacuolated, but some cells still contained relatively intact mitochondria (Rossiter et al., 2009). The findings suggest that autophagy played an important role in gangliocyte infection.

Following right hindlimb intramuscular inoculation of adult mice with street rabies virus 1088 strain, Kimitsuki et al. (2017) observed chromatolysis, vacuolation, and neuronophagia of ipsilateral lumbosacral dorsal root gangliocytes at day 5 p.i. There were similar changes on the left at day 8 p.i., by which time there was mild-to-moderate mononuclear inflammatory cell infiltration in the right-sided ganglion. By electron microscopy, many virus particles were seen in relation to the rough endoplasmic reticulum of right-sided gangliocytes at day 8 p.i.

10.4.2 Changes in spinal nerve roots and peripheral nerves

Pathological studies of spinal nerve roots and peripheral nerves in human rabies cases have been fewer and have tended to be less comprehensive than those of other parts of the nervous system. Knutti (1929) observed from slight focal to extensive necrosis in dorsal nerve roots in a case of paralytic rabies, with only minor changes in the ventral roots. Hurst and Pawan (1932) noted lymphocytic infiltrates in the connective tissue sheath of the sciatic nerve in one of their paralytic cases. In peripheral nerves from 9 encephalitic rabies cases, including material from the facial region, upper and lower limbs, Tangchai and Vejjajiva (1971) observed diffuse perivenous, subepineurial, and subperineurial mononuclear inflammatory cell infiltration (3 cases), degeneration of nerve fibers (7 cases), proliferation and hypertrophy of Schwann cells (5 cases), and subepineurial and -perineurial edema (3 cases). Similar changes were seen in dorsal spinal nerve roots, but were milder. Chopra et al. (1980) studied a total of 4 spinal nerves and 17 peripheral nerves in their series of 11 paralytic rabies cases. In all of the spinal nerves, there was both Wallerian degeneration and segmental demyelination, but inflammatory cell infiltration was seen in only one of the spinal nerve specimens. The peripheral nerves showed variable degrees of segmental demyelination and remyelination, loss of myelinated fibers, and Wallerian degeneration. In 9 of 17 nerves, segmental demyelination was the primary lesion. There was no apparent relationship between the degree of spinal or peripheral nerve pathology and the duration of incubation or clinical illness (Chopra et al., 1980). Mitrabhakdi et al. (2005) observed heavy lymphocytic infiltration of dorsal and ventral roots in two paralytic cases, but by contrast only a mild degree of inflammation in dorsal and ventral roots in encephalitic rabies cases.

A predominant pattern of acute motor axonal neuropathy involving ventral spinal roots and peripheral nerves, in the absence of prominent inflammation or motor neuron degeneration, has been found in a paralytic rabies case, which was initially diagnosed and reported as an axonal Guillain-Barré syndrome (Sheikh et al., 2005). RVAg was observed in multiple lumbar anterior horn cells and in their dendrites in this case, as well as in a large proportion of ventral root myelinated axons. Ultrastructurally, mature viral particles were seen in some axons in the ventral root exit zone. Double-label immunostaining showed colocalization of human IgG and C3d complement activation marker with RVAg on ventral root axons, supporting the possibility that the pathogenesis of paralytic rabies may include immune-mediated axonal degeneration (Sheikh et al., 2005).

In experimentally infected hamsters, Murphy (1977) first observed RVAg in peripheral nerves proximal to the inoculation site only concomitant with or later than RVAg detection in ipsilateral spinal ganglia or spinal cord, with individual axons showing a fine linear dust-like pattern of immunofluorescence. With disease progression, the majority of axons in peripheral nerves of the inoculated hindlimb, and subsequently throughout the body, showed this pattern. Ultrastructurally, there was concentration of virus particles at nodes of Ranvier, reflecting budding from the high density of membranous organelles at these sites. By contrast, in internodal regions, virus particles only budded individually or in small groups from plasma membranes, resulting in the presence of virions between the axonal and adjacent Schwann cell plasma membranes (Jenson, Rabin, Bentinck, & Melnick, 1969; Murphy, Bauer, et al., 1973; Murphy, Harrison, et al., 1973). Murphy, Bauer, et al. (1973) never found viral particles in Schwann cells, whereas Atanasiu and Sisman (1967) did observe Schwann cell infection. In sciatic nerves of mice experimentally inoculated with canine rabies virus, degeneration of approximately 40% of myelinated axons was found, while only occasional degenerating unmyelinated axons were observed (Teixeira et al., 1986). Axonal degeneration and severe demyelination, possibly immunologically mediated, was found in the trigeminal and facial nerves of experimentally infected rats with street rabies virus (Minguetti, Hofmeister, Hayashi, & Montano, 1997). Kimitsuki et al. (2017) observed axonal degeneration and mononuclear inflammatory cell infiltrates in lumbosacral dorsal roots and ipsilateral fasciculus gracilis of the spinal cord of mice, 8 days after hindlimb intramuscular inoculation with street rabies virus 1088 strain.

10.5 Pathology involving the inoculation site, eye, and extraneural organs

Localized replication of rabies virus in extraneural tissue at the inoculation site, including within skeletal muscle, may be an important feature of rabies virus pathogenesis preceding centripetal spread through peripheral nerves to the CNS. Later in the course of infection, centrifugal spread through both somatic sensory and autonomic divisions of the peripheral nervous system results in involvement of a broad range of extraneural tissues and organs, including lacrimal and salivary glands, cornea, skin, heart, gastrointestinal tract, and adrenal glands.

10.5.1 Changes at inoculation site

Rabies infection usually results from inoculation of saliva into deep soft tissues through a bite from a rabid animal. Direct uptake of virus by sensory nerve endings has been documented in several studies (Baer et al., 1968; Coulon et al., 1989; Dean et al., 1963; Kucera, Dolivo, Coulon, & Flamand, 1985; Shankar et al., 1991). However, in many cases, initial local replication within skeletal muscle at the inoculation site likely precedes entry into peripheral nerves at motor end plates and neuromuscular and neurotendinal spindles (Charlton & Casey, 1979; Murphy, Bauer, et al., 1973; Murphy, Harrison, et al., 1973). Furthermore, viral sequestration and delayed progression of infection in skeletal muscle fibers at the site of exposure appears to be a key factor in rabies cases with a long incubation period (Baer & Cleary, 1972; Charlton, Nadin-Davis, Casey, & Wandeler, 1997). Following experimental intramuscular inoculation of immature hamsters, the first evidence of viral replication was seen as RVAg in individual nearby striated muscle fibers, rapidly followed by involvement of groups of fibers (Murphy, Bauer, et al., 1973; Murphy, Harrison, et al., 1973). Ultrastructurally, moderate numbers of virus particles were found budding from the membranes of the sarcolemma and sarcoplasmic reticulum, in the absence of any associated inflammatory response or significant cytopathological changes. This was rapidly followed by detection of RVAg in neuromuscular and neurotendinal spindles (the sensory stretch receptors found deep within skeletal muscles and tendons respectively), with the unmyelinated nerve endings wrapping these proprioreceptors also frequently containing RVAg. Later in the course of infection, following centrifugal spread, infected neuromuscular spindles were found in other parts of the body and were especially numerous in the subcutaneous musculature of the nose (Murphy, Bauer, et al., 1973; Murphy, Harrison, et al., 1973).

In the case of bat-transmitted rabies, it appears that the particular ability of bat rabies strains to infect and replicate within epithelial cells and fibroblasts accounts for the observation that even superficial bat-inflicted wounds have a high probability of causing clinical disease (Love et al., 2015; Morimoto et al., 1996).

10.5.2 Ocular pathology

Given that the retina is a direct extension of the CNS, it is perhaps surprising how few detailed accounts of ocular pathology exist in the literature. Haltia, Tarkkanen, and Kivela (1989) described the ocular pathology in the case of a 30-year-old man who developed rabies after receiving several bites by a bat. These authors observed lymphocytic and plasma cell infiltrates in the ciliary body and focally in the choroid, focal loss of retinal pigment epithelium, perivascular inflammation in the retinal nerve fiber layer, focal endothelial destruction and occlusion of retinal veins, destruction of many retinal ganglion cells, and partial loss of bipolar cells. RVAg was seen in the cytoplasm of many of the surviving ganglion cells. In experimentally infected rabbits, Dejean (1937) observed corneal sensory loss and clouding, retinal venous congestion, choroidal hemorrhages, vitreous clouding, and the presence of Negri-like bodies in retinal ganglion cells. Murphy, Bauer, et al. (1973) and Murphy, Harrison, et al. (1973) observed large aggregates of RVAg in the retinal ganglion cell layer of nearly every terminally infected hamster and focal dustlike antigen in the inner and outer retinal nuclear layers and corneal epithelium in some animals.

Centrifugal spread of virus via sensory fibers to corneal epithelium underlies the use of immunofluorescent staining of corneal impressions for antemortem diagnosis of rabies (Schneider, 1969c) and also explains the rare instances of rabies virus transmission through infected corneal transplants (Gode & Bhide, 1988; Houff et al., 1979).

10.5.3 Changes in extraneural organs

Centrifugal viral spread to cutaneous nerve endings surrounding hair follicles (especially in the head region) forms the basis for antemortem diagnosis by means of immunostained nuchal skin biopsies, in a large proportion of animal (Blenden, Bell, Tsao, & Umoh, 1983) and human rabies cases (Blenden, Creech, & Torres-Anjel, 1986; Bryceson et al., 1975). In some cases, RVAg is also found in epidermal cells (Bago, Revilla-Fernandez, Allerberger, & Krause, 2005; Balachandran & Charlton, 1994; Jackson et al., 1999).

Centrifugal spread to the major salivary glands, resulting in production of saliva containing high titers of rabies virus, is a central feature of bite transmission of rabies by natural vectors, such as dog (Goldwasser, Kissling, Carski, & Hosty, 1959), fox (Balachandran & Charlton, 1994; Dierks, Murphy, & Harrison, 1969), and skunk (Balachandran & Charlton, 1994). Ulastructurally, budding of numerous virions from the apical membranes of mucogenic cells and their release into intercellular canaliculi has been documented (Balachandran & Charlton, 1994; Dierks et al., 1969). In human rabies cases, RVAg was found in acini of minor salivary glands of the tongue, as well as in skeletal muscle fibers of the tongue, but there was no significant involvement of acini in the major salivary glands (Jackson et al., 1999; Li, Feng, & Ye, 1995).

Widespread distribution of RVAg has been observed in autonomic nerve plexuses related to multiple organs, including cardiac ganglia and the submucosal plexus of Meissner and

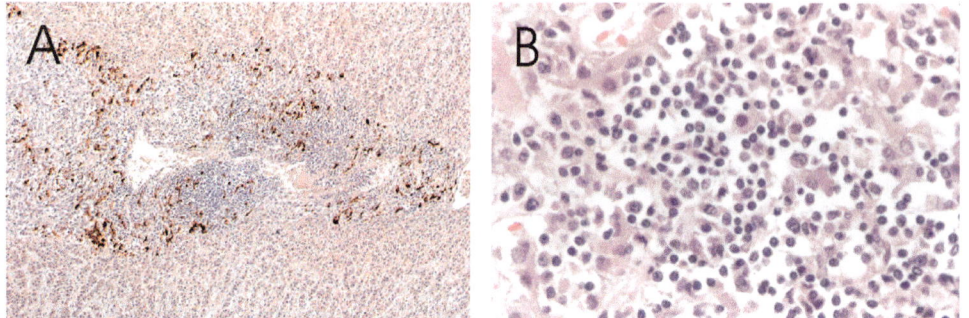

FIG. 10.8 Adrenal gland showing abundant rabies virus antigen in the adrenal medulla with sparing of the adrenal cortex (A). Mononuclear inflammatory infiltrate in the adrenal medulla (B). A, immunoperoxidase-hematoxylin; B, HE stain. magnifications, A 80×; B 140×. *Reproduced from Jackson, A. C., Ye, H., Phelan, C. C., Ridaura-Sanz, C., Zheng, Q., Li, Z., ..., Lopez-Corella, E. (1999). Extraneural organ involvement in human rabies. Laboratory Investigation, 79(8), 945–951 with permission.*

myenteric plexus of Auerbach in the gastrointestinal tract, in both animal (Debbie & Trimarchi, 1970; Fischman & Schaeffer, 1971; Murphy, Bauer, et al., 1973; Murphy, Harrison, et al., 1973) and human material (Jackson et al., 1999; Jogai, Radotra, & Banerjee, 2002). There is typically an associated mild mononuclear cell inflammatory response (Jackson et al., 1999), although more prominent inflammation and degeneration of enteric ganglion cells has been observed in some cases (Love, 1944). The adrenal medulla, as an extension of the sympathetic nervous system, has been found to be frequently involved, with many cells containing RVAg in both animals (Debbie & Trimarchi, 1970; Fischman & Schaeffer, 1971; Murphy, Bauer, et al., 1973; Murphy, Harrison, et al., 1973) and humans (Jackson et al., 1999; Jogai et al., 2002), and may be accompanied by moderate-to-severe inflammation (medullitis) (Fig. 10.8) (Almeida, Teixeira, de Oliveira, Brandao, & Gobbi, 1986; Jackson et al., 1999; Lopez-Corella, Ridaura-Sanz, & Samayoa-Palma, 1997; Love, 1944).

Clinical and/or pathological evidence of cardiac involvement is a recognized feature of some human rabies cases (Burton et al., 2005; Cheetham, Hart, Coghill, & Fox, 1970; Raman, Prosser, Spreadbury, Cockcroft, & Okubadejo, 1988; Ross & Armentrout, 1962; Roux et al., 1976). Myocarditis, characterized by multifocal muscle fiber degeneration/necrosis, has been observed in some cases (Burton et al., 2005; Cheetham et al., 1970; Ross & Armentrout, 1962). RVAg within cardiac myocytes has also been reported in cases with mild or no associated inflammation (Jackson et al., 1999; Metze & Feiden, 1991).

10.6 Summary and conclusions

As an almost invariably fatal infection of the CNS, rabies shares a number of histopathological features with other viral encephalitides, such as leptomeningeal and perivascular mononuclear inflammatory cell infiltration, microglial activation, and neuronophagia. However, in some cases, inflammatory changes and neuronal cell death are minimal or even absent, indicating that these are not essential contributors to a fatal outcome in rabies. The presence of Negri body viral inclusions in the cytoplasm of neurons is a unique and diagnostic feature in many cases of infection with street rabies virus strains. Recent data indicate that Negri bodies may be sites of viral transcription and replication, with viral "hijacking" of cellular compartmentalization and proteins, especially the innate immune response receptor TLR3, to promote replication and potentially evade the innate immune response and apoptosis.

There is strong evidence that neuronal apoptosis does not play an important pathogenetic role in human rabies encephalitis or in most instances of animal infection with "street" rabies virus. In paralytic rabies, by comparison with encephalitic cases, there tends to be, at least in some cases, more severe involvement of the spinal cord and brainstem and a greater degree of injury in spinal nerve roots and peripheral nerves. Ultrastructural and immunohistochemical studies of human and animal rabies material have contributed greatly to an understanding of rabies pathogenesis, including centripetal and centrifugal phases of viral spread through the peripheral nervous system, with resulting involvement of autonomic nerve plexuses in multiple organs and viral spread to extraneural organs, with the major salivary glands being of particular importance in the transmission of rabies virus by natural vectors. Degenerative structural changes in neuronal processes, including swelling of mitochondria in axons and dendrites, loss of dendritic spines, and vacuolation of presynaptic nerve endings, may be sufficient to explain severe clinical disease and fatal outcome in experimental rabies.

References

Abba, F., & Bormans, A. (1905). Sur le diagnostic histologique de la rage. *Annales de L'Institut Pasteur (Paris)*, 19, 49.
Adle-Biassette, H., Bourhy, H., Gisselbrecht, M., Chretien, F., Wingertsmann, L., Baudrimont, M., … Gray, F. (1996). Rabies encephalitis in a patient with AIDS: A clinicopathological study. *Acta Neuropathologica*, 92(4), 415–420.
Almeida, H. d. O., Teixeira, V. d. P., de Oliveira, G., Brandao, M. d. C., & Gobbi, H. (1986). Medulite supra-renalica em casos de raiva humana [adrenal medullitis in cases of human rabies] [portuguese]. *Memórias do Instituto Oswaldo Cruz*, 81(4), 439–442.
Atanasiu, P., & Sisman, J. (1967). Morphological aspects of rabies virus (French). *Bulletin de l Office International des Epizooties*, 67(3), 521–533.
Babes, V. M. (1892). Sur certains caracteres des lesions histologiques de la rage. *Annales de L'Institut Pasteur*, 6, 209–223.
Baer, G. M., & Cleary, W. F. (1972). A model in mice for the pathogenesis and treatment of rabies. *Journal of Infectious Diseases*, 125, 520–527.
Baer, G. M., Shantha, T. R., & Bourne, G. H. (1968). The pathogenesis of street rabies virus in rats. *Bulletin of the World Health Organization*, 38, 119–125.
Baer, G. M., Shanthaveerappa, T. R., & Bourne, G. H. (1965). Studies on the pathogenesis of fixed rabies virus in rats. *Bulletin of the World Health Organization*, 33, 783–794.
Bago, Z., Revilla-Fernandez, S., Allerberger, F., & Krause, R. (2005). Value of immunohistochemistry for rapid ante mortem rabies diagnosis. *International Journal of Infectious Diseases*, 9(6), 351–352.
Balachandran, A., & Charlton, K. (1994). Experimental rabies infection of non-nervous tissues in skunks (*Mephitis mephitis*) and foxes (*Vulpes vulpes*). *Veterinary Pathology*, 31, 93–102.
Baloul, L., & Lafon, M. (2003). Apoptosis and rabies virus neuroinvasion. *Biochimie*, 85(8), 777–788.
Benedikt, M. (1878). Zur pathologischen Anatomie der Lyssa. *Virchows Archiv für Pathologische Anatomie und Physiologie und für Klinische Medizin*, 72, 425–431.
Bingham, J., & van der Merwe, M. (2002). Distribution of rabies antigen in infected brain material: Determining the reliability of different regions of the brain for the rabies fluorescent antibody test. *Journal of Virological Methods*, 101(1-2), 85–94.

Blenden, D. C., Bell, J. F., Tsao, A. T., & Umoh, J. U. (1983). Immunofluorescent examination of the skin of rabies-infected animals as a means of early detection of rabies virus antigen. *Journal of Clinical Microbiology, 18*(3), 631–636.

Blenden, D. C., Creech, W., & Torres-Anjel, M. J. (1986). Use of immunofluorescence examination to detect rabies virus antigen in the skin of humans with clinical encephalitis. *Journal of Infectious Diseases, 154*, 698–701.

Bryceson, A. D. M., Greenwood, B. M., Warrell, D. A., Davidson, N. M., Pope, H. M., Lawrie, J. H., ... Wilcox, G. E. (1975). Demonstration during life of rabies antigen in humans. *Journal of Infectious Diseases, 131*(1), 71–74.

Bundza, A., & Charlton, K. M. (1988). Comparison of spongiform lesions in experimental scrapie and rabies in skunks. *Acta Neuropathologica, 76*, 275–280.

Burrage, T. G., Tignor, G. H., & Smith, A. L. (1983). Immunoelectron microscopic localization of rabies virus antigen in central nervous system and peripheral tissue using low- temperature embedding and protein A-gold. *Journal of Virological Methods, 7*(5-6), 337–350.

Burton, E. C., Burns, D. K., Opatowsky, M. J., El-Feky, W. H., Fischbach, B., Melton, L., ... Klintmalm, G. (2005). Rabies encephalomyelitis: Clinical, neuroradiological, and pathological findings in 4 transplant recipients. *Archives of Neurology, 62*(6), 873–882. Retrieved from: http://www.medscape.com/.

Charlton, K. M. (1984). Rabies: Spongiform lesions in the brain. *Acta Neuropathologica, 63*, 198–202.

Charlton, K. M., & Casey, G. A. (1979). Experimental rabies in skunks: Immunofluorescence light and electron microscopic studies. *Laboratory Investigation, 41*, 36–44.

Charlton, K. M., Casey, G. A., Wandeler, A. I., & Nadin-Davis, S. (1996). Early events in rabies virus infection of the central nervous system in skunks (*Mephitis mephitis*). *Acta Neuropathologica, 91*, 89–98.

Charlton, K. M., Casey, G. A., Webster, W. A., & Bundza, A. (1987). Experimental rabies in skunks and foxes: Pathogenesis of the spongiform lesions. *Laboratory Investigation, 57*, 634–645.

Charlton, K. M., Nadin-Davis, S., Casey, G. A., & Wandeler, A. I. (1997). The long incubation period in rabies: Delayed progression of infection in muscle at the site of exposure. *Acta Neuropathologica, 94*(1), 73–77.

Cheetham, H. D., Hart, J., Coghill, N. F., & Fox, B. (1970). Rabies with myocarditis: Two cases in England. *Lancet, 1*, 921–922.

Chopra, J. S., Banerjee, A. K., Murthy, J. M. K., & Pal, S. R. (1980). Paralytic rabies: A clinico-pathological study. *Brain, 103*, 789–802.

Coulon, P., Derbin, C., Kucera, P., Lafay, F., Prehaud, C., & Flamand, A. (1989). Invasion of the peripheral nervous systems of adult mice by the CVS strain of rabies virus and its avirulent derivative AvO1. *Journal of Virology, 63*, 3550–3554.

Davies, M. C., Englert, M. E., Sharpless, G. R., & Cabasso, V. J. (1963). The electron microscopy of rabies virus in cultures of chicken embryo tissues. *Virology, 21*, 642–651.

de Brito, T., Araujo, M. D. F., & Tiriba, A. (1973). Ultrastructure of the Negri body in human rabies. *Journal of the Neurological Sciences, 20*, 363–372.

Dean, D. J., Evans, W. M., & McClure, R. C. (1963). Pathogenesis of rabies. *Bulletin of the World Health Organization, 29*, 803–811.

Debbie, J. G., & Trimarchi, C. V. (1970). Pantropism of rabies virus in free-ranging rabid red fox *Vulpes fulva*. *Journal of Wildlife Diseases, 6*, 500–506.

Dejean, C. (1937). Les modifications du fond d'oeil dans la rage chez le lapin. *Bulletin des Sociétés d'Ophtalmologie de France, 50*, 247–254.

Dierks, R. E., Murphy, F. A., & Harrison, A. K. (1969). Extraneural rabies virus infection. Virus development in fox salivary gland. *American Journal of Pathology, 54*(2), 251–273.

Dolman, C. L., & Charlton, K. M. (1987). Massive necrosis of the brain in rabies. *Canadian Journal of Neurological Sciences, 14*(2), 162–165.

Dupont, J. R., & Earle, K. M. (1965). Human rabies encephalitis. A study of forty-nine fatal cases with a review of the literature. *Neurology, 15*, 1023–1034.

Feiden, W., Feiden, U., Gerhard, L., Reinhardt, V., & Wandeler, A. (1985). Rabies encephalitis: Immunohistochemical investigations. *Clinical Neuropathology, 4*, 156–164.

Fekadu, M., Chandler, F. W., & Harrison, A. K. (1982). Pathogenesis of rabies in dogs inoculated with an Ethiopian rabies virus strain. Immunofluorescence, histologic and ultrastructural studies of the central nervous system. *Archives of Virology, 71*, 109–126.

Fekadu, M., Greer, P. W., Chandler, F. W., & Sanderlin, D. W. (1988). Use of the avidin-biotin peroxidase system to detect rabies antigen in formalin-fixed paraffin-embedded tissues. *Journal of Virological Methods, 19*, 91–96.

Fischman, H. R., & Schaeffer, M. (1971). Pathogenesis of experimental rabies as revealed by immunofluorescence. *Annals of the New York Academy of Sciences, 177*, 78–97.

Fu, Z. F., & Jackson, A. C. (2005). Neuronal dysfunction and death in rabies virus infection. *Journal of Neurovirology, 11*(1), 101–106.

Garcia-Tamayo, J., Avila-Mayor, A., & Anzola-Perez, E. (1972). Rabies virus neuronitis in humans. *Archives of Pathology, 94*, 11–15.

Gode, G. R., & Bhide, N. K. (1988). Two rabies deaths after corneal grafts from one donor (Letter). *Lancet, 2*, 791.

Goldwasser, R. A., & Kissling, R. E. (1958). Fluorescent antibody staining of street and fixed rabies virus antigens. *Proceedings of the Society for Experimental Biology and Medicine, 98*, 219–223.

Goldwasser, R. A., Kissling, R. E., Carski, T. R., & Hosty, T. S. (1959). Fluorescent antibody staining of rabies virus antigens in the salivary glands of rabid animals. *Bulletin of the World Health Organization, 20*, 579–588.

Gonzalez-Angulo, A., Marquez-Monter, H., Feria-Velasco, A., & Zavala, B. J. (1970). The ultrastructure of Negri bodies in Purkinje neurons in human rabies. *Neurology, 20*, 323–328.

Goodpasture, E. W. (1925). A study of rabies, with reference to a neural transmission of the virus in rabbits, and the structure and significance of Negri bodies. *American Journal of Pathology, 1*, 547–584.

Gosztonyi, G. (1994). Reproduction of lyssaviruses: ultrastructural composition of lyssavirus and functional aspects of pathogenesis. In C. E. Rupprecht, B. Dietzschold, & H. Koprowski (Eds.), *Current topics in microbiology and immunology, Volume 187: Lyssaviruses* (pp. 43–68). Berlin: Springer-Verlag.

Gowers, W. R. (1877). The pathological anatomy of hydrophobia. *Transactions of the Pathological Society of London, 28*, 10–23.

Greenwood, R. J., Newton, W. E., Pearson, G. L., & Schamber, G. J. (1997). Population and movement characteristics of radio-collared striped skunks in North Dakota during an epizootic of rabies. *Journal of Wildlife Diseases, 33*(2), 226–241.

Haltia, M., Tarkkanen, A., & Kivela, T. (1989). Rabies: Ocular pathology. *British Journal of Ophthalmology, 73*, 61–67.

Hardenbergh, J. B. (1916). The reliability of cell proliferation changes in the diagnosis of rabies. *Journal of the American Veterinary Medical Association, 49*, 663.

Herzog, E. (1945). Histologic diagnosis of rabies. *Archives of Pathology, 39*, 279–280.

Hottle, G. A., Morgan, G., Peers, J. H., & Wyckoff, R. W. G. (1951). The electron microscopy of rabies inclusion (Negri) bodies. *Proceedings of the Society for Experimental Biology and Medicine, 77*, 721–723.

Houff, S. A., Burton, R. C., Wilson, R. W., Henson, T. E., London, W. T., Baer, G. M., ... Sever, J. L. (1979). Human-to-human transmission of rabies virus by corneal transplant. *New England Journal of Medicine, 300*, 603–604.

Hummeler, K., Koprowski, H., & Wiktor, T. J. (1967). Structure and development of rabies virus in tissue culture. *Journal of Virology, 1*(1), 152–170.

Hummeler, K., Tomassini, N., Sokol, F., Kuwert, E., & Koprowski, H. (1968). Morphology of the nucleoprotein component of rabies virus. *Journal of Virology, 2*(10), 1191–1199.

Hurst, E. W., & Pawan, J. L. (1932). A further account of the Trinidad outbreak of acute rabic myelitis: Histology of the experimental disease. *Journal of Pathology and Bacteriology, 35*, 301–321.

Huygelen, C. (1960). Further observations on the pathogenesis of rabies in guinea-pigs after experimental infection with the Flury strain. *Antonie Van Leeuwenhoek, 26*, 66–72.

Huygelen, C., & Mortelmans, J. (1959). Quantitative determination of the dissemination of Flury rabies virus in the central nervous system of the guinea-pig after intramuscular inoculation in the hind leg. *Antonie Van Leeuwenhoek, 25*, 265–271.

Iwasaki, Y., & Clark, H. F. (1975). Cell to cell transmission of virus in the central nervous system. II. Experimental rabies in mouse. *Laboratory Investigation, 33*, 391–399.

Iwasaki, Y., & Clark, H. F. (1977). Rabies virus infection in mouse neuroblastoma cells. *Laboratory Investigation, 36*, 578–584.

Iwasaki, Y., Liu, D. S., Yamamoto, T., & Konno, H. (1985). On the replication and spread of rabies virus in the human central nervous system. *Journal of Neuropathology and Experimental Neurology, 44*, 185–195.

Iwasaki, Y., & Minamoto, N. (1982). Scanning and freeze-fracture electron microscopy of rabies virus infection in murine neuroblastoma cells. *Comparative Immunology, Microbiology and Infectious Diseases, 5*(1-3), 1–8.

Iwasaki, Y., Ohtani, S., & Clark, H. F. (1975). Maturation of rabies virus by budding from neuronal cell membrane in suckling mouse brain. *Journal of Virology, 15*(4), 1020–1023.

Iwasaki, Y., Sako, K., Tsunoda, I., & Ohara, Y. (1993). Phenotypes of mononuclear cell infiltrates in human central nervous system. *Acta Neuropathologica, 85*(6), 653–657.

Iwasaki, Y., & Tobita, M. (2002). Pathology. In A. C. Jackson & W. H. Wunner (Eds.), *Rabies* (pp. 283–306). San Diego: Academic Press.

Iwasaki, Y., Wiktor, T. J., & Koprowski, H. (1973). Early events of rabies virus replication in tissue cultures: An electron microscopic study. *Laboratory Investigation, 28*, 142–148.

Jackson, A. C. (1992). Detection of rabies virus mRNA in mouse brain by using in situ hybridization with digoxigenin-labelled RNA probes. *Molecular and Cellular Probes, 6*(2), 131–136.

Jackson, A. C. (2016). Diabolical effects of rabies encephalitis. *Journal of Neurovirology, 22*(1), 8–13.

Jackson, A. C., Kammouni, W., Zherebitskaya, E., & Fernyhough, P. (2010). Role of oxidative stress in rabies virus infection of adult mouse dorsal root ganglion neurons. *Journal of Virology, 84*(9), 4697–4705.

Jackson, A. C., & Park, H. (1998). Apoptotic cell death in experimental rabies in suckling mice. *Acta Neuropathologica, 95*(2), 159–164.

Jackson, A. C., Randle, E., Lawrance, G., & Rossiter, J. P. (2008). Neuronal apoptosis does not play an important role in human rabies encephalitis. *Journal of Neurovirology, 14*(5), 368–375.

Jackson, A. C., & Reimer, D. L. (1989). Pathogenesis of experimental rabies in mice: An immunohistochemical study. *Acta Neuropathologica, 78*(2), 159–165.

Jackson, A. C., Reimer, D. L., & Wunner, W. H. (1989). Detection of rabies virus RNA in the central nervous system of experimentally infected mice using in situ hybridization with RNA probes. *Journal of Virological Methods, 25*(1), 1–11.

Jackson, A. C., & Rossiter, J. P. (1997). Apoptosis plays an important role in experimental rabies virus infection. *Journal of Virology, 71*(7), 5603–5607.

Jackson, A. C., & Wunner, W. H. (1991). Detection of rabies virus genomic RNA and mRNA in mouse and human brains by using in situ hybridization. *Journal of Virology, 65*(6), 2839–2844.

Jackson, A. C., Ye, H., Phelan, C. C., Ridaura-Sanz, C., Zheng, Q., Li, Z., ... Lopez-Corella, E. (1999). Extraneural organ involvement in human rabies. *Laboratory Investigation, 79*(8), 945–951.

Jackson, A. C., Ye, H., Ridaura-Sanz, C., & Lopez-Corella, E. (2001). Quantitative study of the infection in brain neurons in human rabies. *Journal of Medical Virology, 65*(3), 614–618.

Jenson, A. B., Rabin, E. R., Bentinck, D. C., & Melnick, J. L. (1969). Rabiesvirus neuronitis. *Journal of Virology, 3*, 265–269.

Jenson, A. B., Rabin, E. R., Wende, R. D., & Melnick, J. L. (1967). A comparative light and electron microscopic study of rabies and Hart Park virus encephalitis. *Experimental and Molecular Pathology, 7*(1), 1–10.

Jogai, S., Radotra, B. D., & Banerjee, A. K. (2000). Immunohistochemical study of human rabies. *Neuropathology, 20*(3), 197–203.

Jogai, S., Radotra, B. D., & Banerjee, A. K. (2002). Rabies viral antigen in extracranial organs: A post-mortem study. *Neuropathology and Applied Neurobiology, 28*(4), 334–338.

Johnson, K. P., Swoveland, P. T., & Emmons, R. W. (1980). Diagnosis of rabies by immunofluorescence in trypsin-treated histologic sections. *The Journal of the American Medical Association, 244*, 41–43.

Juntrakul, S., Ruangvejvorachai, P., Shuangshoti, S., Wacharapluesadee, S., & Hemachudha, T. (2005). Mechanisms of escape phenomenon of spinal cord and brainstem in human rabies. *BMC Infectious Diseases, 5*(1), 104.

Kammouni, W., Hasan, L., Saleh, A., Wood, H., Fernyhough, P., & Jackson, A. C. (2012). Role of nuclear factor-κB in oxidative stress associated with rabies virus infection of adult rat dorsal root ganglion neurons. *Journal of Virology, 86*(15), 8139–8146.

Kimitsuki, K., Yamada, K., Shiwa, N., Inoue, S., Nishizono, A., & Park, C. H. (2017). Pathological lesions in the central nervous system and peripheral tissues of ddY mice with street rabies virus (1088 strain). *The Journal of Veterinary Medical Science, 79*(6), 970–978.

Kliger, I. J., & Bernkopf, H. (1943). The path of dissemination of rabies virus in the body of normal and immunized mice. *British Journal of Experimental Pathology, 24*, 15–21.

Knutti, R. E. (1929). Acute ascending paralysis and myelitis due to the virus of rabies. *The Journal of the American Medical Association, 93*, 754–758.

Kristensson, K., Dastur, D. K., Manghani, D. K., Tsiang, H., & Bentivoglio, M. (1996). Rabies: Interactions between neurons and viruses. A review of the history of Negri inclusion bodies. *Neuropathology and Applied Neurobiology, 22*, 179–187.

Kucera, P., Dolivo, M., Coulon, P., & Flamand, A. (1985). Pathways of the early propagation of virulent and avirulent rabies strains from the eye to the brain. *Journal of Virology, 55*(1), 158–162.

Lafon, M. (2011). Evasive strategies in rabies virus infection. *Advances in Virus Research, 79*, 33–53.

Lahaye, X., Vidy, A., Pomier, C., Obiang, L., Harper, F., Gaudin, Y., & Blondel, D. (2009). Functional characterization of Negri bodies (NBs) in rabies virus infected cells: Evidence that NBs are sites of viral transcription and replication. *Journal of Virology, 83*(16), 7948–7958.

Last, R. D., Jardine, J. E., Smit, M. M., & van der Lugt, J. J. (1994). Application of immunoperoxidase techniques to formalin-fixed brain tissue for the diagnosis of rabies in southern Africa. *Onderstepoort Journal of Veterinary Research, 61*(2), 183–187.

Leech, R. W. (1971). Electron-microscopic study of the inlusion body in human rabies. *Neurology, 21*(1), 91–94.

Lepine, P., & Atanasiu, P. (1996). Histopathological diagnosis. In World Health Organization, F. -X. Meslin, M. M. Kaplan, & H. Koprowski (Eds.), *Laboratory techniques in rabies* (4th ed., pp. 66–79). Geneva: World Health Organization.

Lewis, P., & Lentz, T. L. (1998). Rabies virus entry into cultured rat hippocampal neurons. *Journal of Neurocytology, 27*(8), 559–573.

Li, X. Q., Sarmento, L., & Fu, Z. F. (2005). Degeneration of neuronal processes after infection with pathogenic, but not attenuated, rabies viruses. *Journal of Virology, 79*(15), 10063–10068.

Li, Z., Feng, Z., & Ye, H. (1995). Rabies viral antigen in human tongues and salivary glands. *The Journal of Tropical Medicine and Hygiene, 98*(5), 330–332.

Lopez-Corella, E., Ridaura-Sanz, C., & Samayoa-Palma, J. E. (1997). Human rabies. Systemic pathology in 33 autopsies. *Laboratory Investigation, 76*, 140A.

Love, S., Wiley, C. A., & Lucas, S. (2015). Viral diseases. In S. Love, A. Perry, J. Ironside, & H. Budka (Eds.), *Greenfield's neuropathology* (9th ed., pp. 1087–1191). Boca Raton, FL: CRC Press.

Love, S. V. (1944). Paralytic rabies: Review of the literature and report of a case. *Journal of Pediatrics, 24*, 312–325.

Lowenberg, K. (1928). Rabies in man. Microscopic observations. *Archives of Neurology and Psychiatry, 19*, 638–646.

Manghani, D. K., Dastur, D. K., Nanavaty, A. N., & Patel, R. (1986). Pleomorphism of fine structure of rabies virus in human and experimental brain. *Journal of the Neurological Sciences, 75*, 181–193.

Marinesco, G., & Storesco, G. (1931). Etudes sur la pathologie de la rage. *Archives Roumaines de Pathologie Expérimentales et de Microbiologie, 4*, 243–288.

Matsumoto, S. (1963). Electron microscope studies of rabies virus in mouse brain. *Journal of Cell Biology, 19*, 565–591.

Matsumoto, S., & Kawai, A. (1969). Comparative studies on development of rabies virus in different host cells. *Virology, 39*(3), 449–459.

Matsumoto, S., Schneider, L. G., Kawai, A., & Yonezawa, T. (1974). Further studies on the replication of rabies and rabies-like viruses in organized cultures of mammalian neural tissues. *Journal of Virology, 14*(4), 981–996.

Menager, P., Roux, P., Megret, F., Bourgeois, J. P., Le Sourd, A. M., Danckaert, A., ... Lafon, M. (2009). Toll-like receptor 3 (TLR3) plays a major role in the formation of rabies virus Negri bodies. *PLoS Pathogens, 5*(2), e1000315.

Metze, K., & Feiden, W. (1991). Rabies virus ribonucleoprotein in the heart (Letter). *New England Journal of Medicine, 324*, 1814–1815.

Minguetti, G., Hofmeister, R. M., Hayashi, Y., & Montano, J. A. (1997). Ultrastructure of cranial nerves of rats inoculated with rabies virus. *Arquivos de Neuro-Psiquiatria, 55*(4), 680–686.

Mitrabhakdi, E., Shuangshoti, S., Wannakrairot, P., Lewis, R. A., Susuki, K., Laothamatas, J., & Hemachudha, T. (2005). Difference in neuropathogenetic mechanisms in human furious and paralytic rabies. *Journal of the Neurological Sciences, 238*(1-2), 3–10.

Miyamoto, K., & Matsumoto, S. (1965). The nature of the Negri body. *Journal of Cell Biology, 27*, 677–682.

Miyamoto, K., & Matsumoto, S. (1967). Comparative studies between pathogenesis of street and fixed rabies infection. *Journal of Experimental Medicine, 125*, 447–474.

Monroy-Gomez, J., Santamaria, G., & Torres-Fernandez, O. (2018). Overexpression of MAP2 and NF-H associated with dendritic pathology in the spinal cord of mice infected with rabies virus. *Viruses, 10*(3), v10030112.

Morecki, R., & Zimmerman, H. M. (1969). Human rabies encephalitis. Fine structure study of cytoplasmic inclusions. *Archives of Neurology, 20*(6), 599–604.

Morimoto, K., Patel, M., Corisdeo, S., Hooper, D. C., Fu, Z. F., Rupprecht, C. E., ... Dietzschold, B. (1996). Characterization of a unique variant of bat rabies virus responsible for newly emerging human cases in North America. *Proceedings of the National Academy of Sciences of the United States of America, 93*(11), 5653–5658.

Mrak, R. E., & Young, L. (1993). Rabies encephalitis in a patient with no history of exposure. *Human Pathology, 24,* 109–110.

Mrak, R. E., & Young, L. (1994). Rabies encephalitis in humans: Pathology, pathogenesis and pathophysiology. *Journal of Neuropathology and Experimental Neurology, 53,* 1–10.

Murphy, F. A. (1977). Rabies pathogenesis: Brief review. *Archives of Virology, 54,* 279–297.

Murphy, F. A., Bauer, S. P., Harrison, A. K., & Winn, W. C. (1973). Comparative pathogenesis of rabies and rabies-like viruses: Viral infection and transit from inoculation site to the central nervous system. *Laboratory Investigation, 28,* 361–376.

Murphy, F. A., Harrison, A. K., Winn, W. C., & Bauer, S. P. (1973). Comparative pathogenesis of rabies and rabies-like viruses: Infection of the central nervous system and centrifugal spread of virus to peripheral tissues. *Laboratory Investigation, 29,* 1–16.

Negri, A. (1903a). Beitrag zum Studium der Aetiologie der Tollwuth. *Zeitschrift für Hygiene und Infektionskrankheiten, 43,* 507–528.

Negri, A. (1903b). Zur Aetiologie der Tollwuth. Die diagnose der Tollwuth auf Grund der Neuen Befunde. *Zeitschrift für Hygiene und Infektionskrankheiten, 44,* 519.

Negri, A. (1909). Uber die Morphologie und der Entwicklungszyklus des Parasiten der Tollwut (Neurocytes hydrophobiae Calkins). *Zeitschrift für Hygiene und Infektionskrankheiten, 63,* 421–440.

Negri-Luzzani, L. (1913). Le diagnostic de la rage par la demonstration du parasite specifique. Resultats de dix ans d'experiences. *Annales de L'Institut Pasteur (Paris), 27,* 1039–1064.

Nepveu, M. (1872). Un cas de rage. *Comptes Rendus des Séances et Mémoires de la Société de Biologie, 4,* 133.

Nieberg, K. C., & Blumberg, J. M. (1972). Viral encephalitides. In J. Minckler (Ed.), *Pathology of the nervous system* (pp. 2266–2323). New York: McGraw-Hill Book Company.

Pasteur, L., Chamberland, C. E., Roux, E., & Thuillier, L. (1881). Sur la rage. *Comptes Rendus de l'Académie des Sciences, 92,* 1259–1260.

Perl, D. P., Callaway, C. S., & Hicklin, M. (1972). An ultrastructural study of Negri bodies in experimental rabies following prolonged incubation. *Journal of Neuropathology and Experimental Neurology, 31,* 172.

Perl, D. P., & Good, P. F. (1991). The pathology of rabies in the central nervous system. In G. M. Baer (Ed.), *The natural history of rabies* (2nd ed., pp. 163–190). Boca Raton, Florida: CRC Press.

Raman, G. V., Prosser, A., Spreadbury, P. L., Cockcroft, P. M., & Okubadejo, O. A. (1988). Rabies presenting with myocarditis and encephalitis. *Journal of Infection, 17,* 155–158.

Ramón y Cajal, S., & Garcia, D. (1904). Las lesiones del retículo de las células nerviosas en la rabia. *Trabajos del Laboratorio de Investigaciones biológicas de la Universidad de Madrid, 3,* 213.

Reid, J. E., & Jackson, A. C. (2001). Experimental rabies virus infection in *Artibeus jamaicensis* bats with CVS-24 variants. *Journal of Neurovirology, 7*(6), 511–517.

Reisman, D., Alpers, B. J., & Cooper, D. A. (1933). Hydrophobia. Report of two fatal cases with pathologic studies in one. *Archives of Internal Medicine, 51,* 643–655.

Ross, E., & Armentrout, S. A. (1962). Myocarditis associated with rabies: Report of a case. *New England Journal of Medicine, 266,* 1087–1089.

Rossiter, J. P., Hsu, L., & Jackson, A. C. (2009). Selective vulnerability of dorsal root ganglia neurons in experimental rabies after peripheral inoculation of CVS-11 in adult mice. *Acta Neuropathologica, 118*(2), 249–259.

Roux, F., Bourgeade, A., Salaun, J. J., Bondurand, A., Ette, M., & Bertrand, E. (1976). L'atteinte cardiaque dans la rage humaine [Cardiac involvement in human rabies]. *Coeur et Médecine Interne, 15*(1), 37–44.

Rubin, R. H., Sullivan, L., Summers, R., Gregg, M. B., & Sikes, R. K. (1970). A case of human rabies in Kansas: Epidemiologic, clinical, and laboratory considerations. *Journal of Infectious Diseases, 122,* 318–322.

Sandhyamani, S., Roy, S., Gode, G. R., & Kalla, G. N. (1981). Pathology of rabies: A light- and electron-microscopical study with particular reference to the changes in cases with prolonged survival. *Acta Neuropathologica, 54,* 247–251.

Sarmento, L., Li, X., Howerth, E., Jackson, A. C., & Fu, Z. F. (2005). Glycoprotein-mediated induction of apoptosis limits the spread of attenuated rabies viruses in the central nervous system of mice. *Journal of Neurovirology, 11*(6), 571–581.

Schaffer, K. (1888). Histologische Untersuchung eines Falles von Lyssa. *Archiv für Psychiatrie und Nervenkrankheiten, 19,* 45–63.

Schneider, L. G. (1969a). Die Pathogenese der Tollwut bei der Maus. I. Die Virusausbreitung vom Infektionsort zum Zentralnervensystem. *Zentralblatt fur Bakteriologie, 211,* 281–308.

Schneider, L. G. (1969b). Die Pathogenese der Tollwut bei der Maus. II. Die Virusausbreitung innerhalb des ZNS. *Zentralblatt fur Bakteriologie, 212,* 1–13.

Schneider, L. G. (1969c). The cornea test; a new method for the intra-vitam diagnosis of rabies. *Zentralblatt Fur Veterinarmedizin: Reihe B, 16*(1), 24–31.

Schneider, L. G., Dietzschold, B., Dierks, R. E., Matthaeus, W., Enzmann, P. J., & Strohmaier, K. (1973). Rabies group-specific ribonucleoprotein antigen and a test system for grouping and typing of rhabdoviruses. *Journal of Virology, 11*(5), 748–755.

Schneider, L. G., & Hamann, I. (1969). Die Pathogenese der Tollwut bei der Maus. III. Die zentrifugale Virusausbreitung und die Virusgeneralisierung im Organismus. *Zentralblatt fur Bakteriologie, 212,* 13–41.

Scott, C. A., Rossiter, J. P., Andrew, R. D., & Jackson, A. C. (2008). Structural abnormalities in neurons are sufficient to explain the clinical disease and fatal outcome in experimental rabies in yellow fluorescent protein-expressing transgenic mice. *Journal of Virology, 82*(1), 513–521.

Shankar, V., Dietzschold, B., & Koprowski, H. (1991). Direct entry of rabies virus into the central nervous system without prior local replication. *Journal of Virology, 65,* 2736–2738.

Sheikh, K. A., Ramos-Alvarez, M., Jackson, A. C., Li, C. Y., Asbury, A. K., & Griffin, J. W. (2005). Overlap of pathology in paralytic rabies and axonal Guillain-Barré syndrome. *Annals of Neurology, 57*(5), 768–772.

Shuangshoti, S., Thepa, N., Phukpattaranont, P., Jittmittraphap, A., Intarut, N., Tepsumethanon, V., … Hemachudha, T. (2013). Reduced viral burden in paralytic compared to furious canine rabies is associated with prominent inflammation at the brainstem level (In press). *BMC Veterinary Research, 9*(1), 31.

Shuangshoti, S., Thorner, P. S., Teerapakpinyo, C., Thepa, N., Phukpattaranont, P., Intarut, N., … Hemachudha, T. (2016). Intracellular spread of rabies virus Is reduced in the paralytic form of canine rabies compared to the furious form. *PLoS Neglected Tropical Diseases, 10*(6), e0004748.

Smart, N. L., & Charlton, K. M. (1992). The distribution of challenge virus standard rabies virus versus skunk street rabies virus in the brains of experimentally infected rabid skunks. *Acta Neuropathologica, 84,* 501–508.

Song, Y., Hou, J., Qiao, B., Li, Y., Xu, Y., Duan, M., … Sun, L. (2013). Street rabies virus causes dendritic injury and F-actin depolymerization in the hippocampus. *Journal of General Virology, 94,* 276–283.

Stein, L. T., Rech, R. R., Harrison, L., & Brown, C. C. (2010). Immunohistochemical study of rabies virus within the central nervous system of domestic and wildlife species. *Veterinary Pathology, 47*(4), 630–636.

Suja, M. S., Mahadevan, A., Madhusudana, S. N., & Shankar, S. K. (2011). Role of apoptosis in rabies viral encephalitis: A comparative study in mice, canine, and human brain with a review of literature. *Pathology Research International, 2011,* 374286.

Suja, M. S., Mahadevan, A., Madhusudhana, S. N., Vijayasarathi, S. K., & Shankar, S. K. (2009). Neuroanatomical mapping of rabies nucleocapsid viral antigen distribution and apoptosis in pathogenesis in street dog rabies—An immunohistochemical study. *Clinical Neuropathology, 28*(2), 113–124.

Suja, M. S., Mahadevan, A., Sundaram, C., Mani, J., Sagar, B. C., Hemachudha, T., … Shankar, S. K. (2004). Rabies encephalitis following fox bite—Histological and immunohistochemical evaluation of lesions caused by virus. *Clinical Neuropathology, 23*(6), 271–276.

Sukru-Aksel, I. (1958). Pathologische Anatomie der Lyssa. In O. Lubarsch, F. Henke, & R. Rossle (Eds.), *Handbuch der Speziellen pathologischen Anatomie und Histologie* (pp. 417–435). Berlin: Springer-Verlag.

Sung, J. H., Hayano, M., Mastri, A. R., & Okagaki, T. (1976). A case of human rabies and ultrastructure of the Negri body. *Journal of Neuropathology and Experimental Neurology, 35,* 541–559.

Szlachta, H. L., & Habel, R. E. (1953). Inclusions resembling Negri bodies in the brains of nonrabid cats. *The Cornell Veterinarian, 43*(2), 207–212.

Tangchai, P., & Vejjajiva, A. (1971). Pathology of the peripheral nervous system in human rabies: A study of nine autopsy cases. *Brain, 94,* 299–306.

Tangchai, P., Yenbutr, D., & Vejjajiva, A. (1970). Central nervous system lesions in human rabies: A study of twenty-four cases. *Journal of the Medical Association of Thailand, 53,* 471–488.

Teixeira, F., Aranda, F. J., Castillo, S., Perez, M., Del Peon, L., & Hernandez, O. (1986). Experimental rabies: Ultrastructural quantitative analysis of the changes in the sciatic nerve. *Experimental and Molecular Pathology, 45*(3), 287–293.

Theerasurakarn, S., & Ubol, S. (1998). Apoptosis induction in brain during the fixed strain of rabies virus infection correlates with onset and severity of illness. *Journal of Neurovirology, 4*(4), 407–414.

Tierkel, E. S. (1973a). Laboratory techniques in rabies: Rapid microscopic examination for negri bodies and preparation of specimens for biological test. *World Health Organization Monograph Series, 23,* 41–55.

Tierkel, E. S. (1973b). Rapid microscopic examination for Negri bodies and preparation of specimens for biological test. In M. M. Kaplan & H. Koprowski (Eds.), *Laboratory techniques in rabies* (3rd ed., pp.1–55). Geneva: World Health Organization.

Tirawatnpong, S., Hemachudha, T., Manutsathit, S., Shuangshoti, S., Phanthumchinda, K., & Phanuphak, P. (1989). Regional distribution of rabies viral antigen in central nervous system of human encephalitic and paralytic rabies. *Journal of the Neurological Sciences, 92*, 91–99.

Torres-Fernandez, O., Yepes, G. E., & Gomez, J. E. (2007). Neuronal dendritic morphology alterations in the cerebral cortex of rabies-infected mice: A Golgi study (Spanish). *Biomédica, 27*(4), 605–613.

Tustin, R. C., & Smit, J. D. (1962). Rabies in South Africa. An analysis of histological examination. *Journal of the South African Veterinary Medical Association, 33*, 295–310.

Ubol, S., & Kasisith, J. (2000). Reactivation of Nedd-2, a developmentally down-regulated apoptotic gene, in apoptosis induced by a street strain of rabies virus. *Journal of Medical Microbiology, 49*(11), 1043–1046.

Van Gehuchten, A., & Nelis, C. (1900). Les lesions histologiques de la rage chez les animaux et chez l'homme. *Bulletin de l'Académie royale de médecine de Belgique, 14*, 31–66.

Walker, G., Thiessen, B., Graeb, D., Moore, G. R., & Mackenzie, I. R. A. (2016). An unusual case of rabies encephalitis. *Canadian Journal of Neurological Sciences, 43*(6), 8521–8554.

Yan, X., Prosniak, M., Curtis, M. T., Weiss, M. L., Faber, M., Dietzschold, B., & Fu, Z. F. (2001). Silver-haired bat rabies virus variant does not induce apoptosis in the brain of experimentally infected mice. *Journal of Neurovirology, 7*(6), 518–527.

CHAPTER 11

Immunology

Monique Lafon
Institut Pasteur, Paris, France

11.1 Introduction

Viruses are obligate parasites. Successful completion of the virus cycle and subsequent transmission to a new host relies upon the evolution of strategies that allow the virus to hijack the host's cellular machinery, modulate host cell survival and escape host lines of defense. Rabies virus (RABV), a neurotropic virus causing fatal encephalitis, is transmitted in the saliva of an infected animal (mainly dogs but also bats and other animal vectors) after bites or scratches. After entry at the neuromuscular junction or passage through the synaptic cleft, RABV particles propagate toward the cell body by retrograde transport using axonal vesicles (Klingen, Conzelmann, & Finke, 2008). Virus replication occurs in the cell bodies and dendrites (Ugolini, 1995, 2010) from which newly formed viral particles are released (Bauer et al., 2014). RABV infects neurons (both motor and sensory) almost exclusively and travels from one neuron to the next from the spinal cord to the brainstem, from where it reaches the salivary glands via cranial nerves. Once in the salivary glands, RABV is excreted in saliva and can be then transmitted to a new host.

During its journey, RABV faces host defenses at different steps: at first, RABV particles delivered in the skin or muscle by the bite are rapidly detected by the early line of defense, the innate immune response, which contributes to both eliminate microbes locally and to set up a specific immune response (B and T cells) in the periphery (extraneurally). After the infection of muscles cells and its entry into nerves, the virus has to cope with the innate immune response launched by the infected muscle and infected neurons that have the capacity to counter the infection. Once infection is settled in the neurons, the infected neurons are protected from the destruction by infiltrating T cells and by mechanisms limiting the inflammation of neuronal tissue. In addition, the central control of immunological homeostasis operated by the NS resulting in an inappropriate downregulation of the immune responsiveness in the periphery that might also facilitate the propagation of the virus in the NS (Fig. 11.1A).

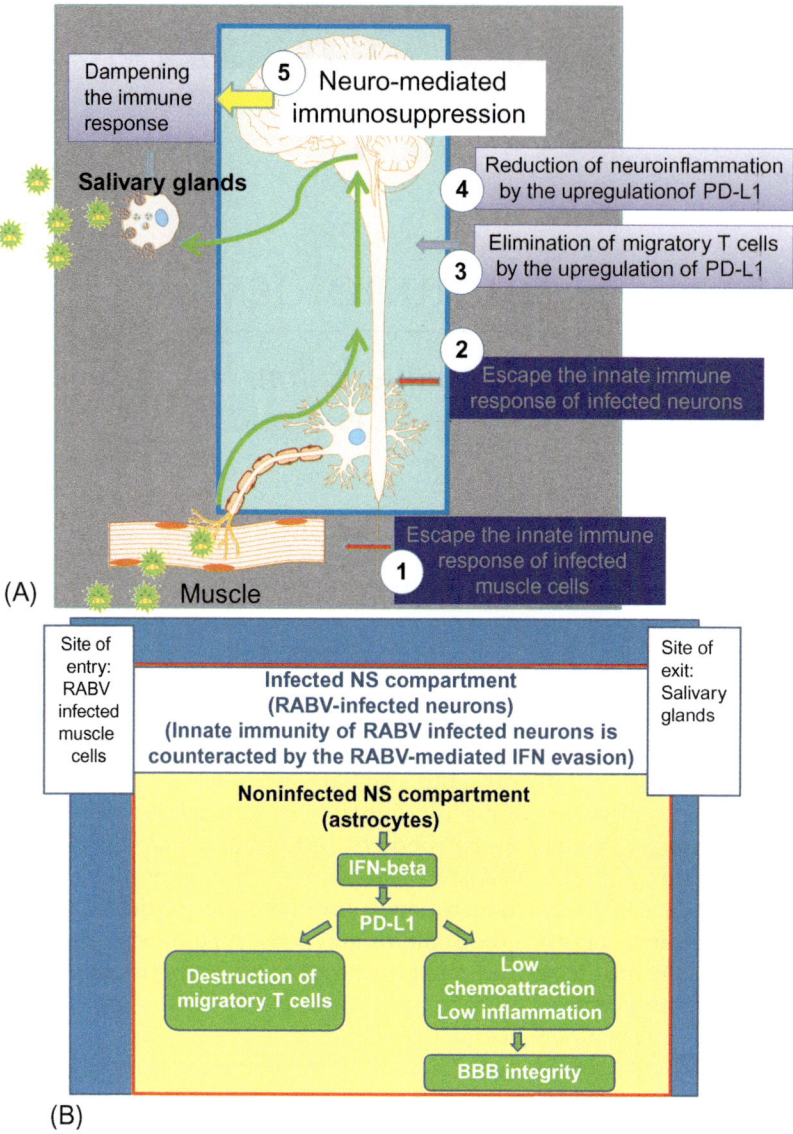

FIG. 11.1 (A) Successful achievement of the RABV cycle relies on the preservation of the neuronal network. Once RABV has entered the NS, its progression is not interrupted by the host defense mechanisms that are inefficient. This might result from the intrinsic capacity of this virus to: (1) evade the innate immune response launched by the infected muscle cells, (2) evade the innate immune response launched by the infected neurons, (3) eliminate the protective T cells migrating into the NS by the upregulation of immunoevasive molecule such as PD-L1, (4) limit the inflammation in NS tissues, and (5) trigger a neuro-mediated immunosuppression, which dampens the immune response in periphery. (B) Dual role of type I IFN in RABV infection may result from the compartmentalization of the NS. One can distinguish between the network of the infected neurons, in which RABV evades the IFN response by viral protein-mediated mechanisms (white rectangular box), from the rest of the NS (yellow box) made of glial cells in which type I IFN can be produced (heterocellular IFN). Production of heterocellular type I IFN (IFN-beta) is not controlled because the glial cells are not productively infected. RABV infection propagates rapidly in the neuronal network in which IFN production is reduced, whereas the heterocellular IFN-beta might promote RABV infection by upregulating the expression of type I IFN-dependent PD-L1 (and HLAG) proteins functioning both as anti-inflammatory molecules and as immunosubversive molecules allowing the elimination of migratory T cells.

Preservation of the integrity of the neuronal network up to the brainstem provides the opportunity for the virus to reach the salivary glands and be transmitted to a new host.

The knowledge gained of the interactions of RABV with the immune responses have been acquired mainly in models of experimental rabies in mice using laboratory-adapted RABV strains and sometimes street RABV strains injected by intramuscular or intraplantar (footpad) route to mimic natural transmission by a bite. Fatal rabies encephalitis can be reproduced in this model using challenge virus standard (CVS), in particular. This virus invades the spinal cord and brain regions and causes fatal encephalitis (Camelo, Lafage, & Lafon, 2000; Park, Kondo, et al., 2006; Xiang, Knowles, McCarrick, & Ertl, 1995). Some mutant strains of RABV with attenuated pathogenicity cause only transient infection of the NS (Galelli, Baloul, & Lafon, 2000; Hooper et al., 1998; Irwin, Wunner, Ertl, & Jackson, 1999; Weiland, Cox, Meyer, Dahme, & Reddehase, 1992; Xiang et al., 1995). This is the case for Pasteur virus (PV), which results in a nonfatal abortive disease characterized by a transient and restricted infection of the NS followed by irreversible limb paralysis (Galelli et al., 2000). This is the case also of recombinant laboratory RABV strains designed in the hope of developing new live rabies vaccines (Barkhouse, Faber, & Hooper, 2015; Faber et al., 2009; Gnanadurai et al., 2015; Schutsky et al., 2013). These strains have been so attenuated that they have lost the features of virulent RABV strain: they stimulate strong immune responses, have lost both neuroinvasiveness and the capacity to evade the defense responses of the host. Their study reveals the properties evolved by a virulent RABV to invade the NS and to escape the defensive mechanisms of the host.

11.2 RABV innate immune response

11.2.1 RABV triggering of host innate immune response and virus evasive mechanisms

The innate immune response is the first line of defense against infectious agents. It involves the release of type 1 interferons (IFN-α/β), inflammatory cytokines and chemokines, the activation of complement and the attraction of macrophages, neutrophils and NK cells into infected tissues. This innate immune response is triggered in the first hours following the entry of pathogens and is not pathogen specific. It contrasts with the adaptive immune response that is tailored to a specific pathogen and requires several days to develop. The innate immune system can sense the presence of microorganisms through "pattern recognition receptors" (PRRs) that recognize danger signals and "pathogen-associated molecular patterns" (PAMPs) expressed by microbes. Toll-like receptors (TLRs) or retinoic-acid-inducible gene (RIG) like receptors (RLRs) are important PRRs for the recognition of viral dsRNAs and ssRNAs. The TLRs are a family of 13 members. The RLRs family, mainly involved in virus detection, consists of three proteins: RIG protein 1, (RIG-I), melanoma differentiation-associated gene-5 (MDA-5), and laboratory of genetics and physiology 2 (LGP2) proteins. Some of these receptors are at the surface of the cells, detecting the presence of danger signals present in the extracellular milieu. This is the case for TLR2 and TLR4. Other receptors are expressed in the cytoplasm (RLRs) or in endosomal vesicles (TLR3, 7–9, and 13) allowing the detection of danger signals produced in the early steps of the entry or replication of intracellular pathogens. Recruitment of particular receptors depends upon the motifs that they bind to and the localization of the receptors. For example, TLR3 senses only dsRNA with a length higher than 40–50 base pairs, a constraint allowing the formation of complex homodimers gathering two molecules of TLR3 with the dsRNA (Liu et al., 2008). RLRs sense viral RNAs, but only those present in the cytoplasm and encoding a triphosphate at the 5′-end (Hornung et al., 2006). Resulting signal transduction cascades involve TRIF, Myd88, or IPS-1 as adaptors of TLR3, of TLRs other than TLR3 and of RLRs, respectively. They trigger production of chemokines, inflammatory cytokines, and antiviral molecules such as IFNs. In particular, following TLR7 or TLR9 stimulation, a subpopulation of blood cells, the plasmacytoid dendritic cells (pDCs) are an important source of type I IFN. The pDCs participate in both an early antiviral response and later, after their maturation and acquisition of the antigen-presentation capacity, in the activation of T cells (Colonna, Trinchieri, & Liu, 2004).

RABV is known to trigger a RIG-I mediated innate immune response in the cells it infects (Hornung et al., 2006) by detecting the 5′-triphosphate base pairing of the viral genome (Pichlmair et al., 2006). Viruses have evolved sophisticated strategies to escape the innate immune response (Randall & Goodbourn, 2008; Versteeg & Garcia-Sastre, 2010). This is the case for RABV (Rieder & Conzelmann, 2009). The N and the P proteins of RABV are multifunctional proteins involved in RNA synthesis and in counteracting the host innate immune response. The N protein limits RIG-I-signaling (Masatani et al., 2010a, 2010b), whereas the P protein inhibits IRF3 and IRF7 phosphorylation (Brzozka, Finke, & Conzelmann, 2005; Rieder et al., 2011), suppresses STAT1 nuclear translocation (Brzozka, Finke, & Conzelmann, 2006; Vidy, El Bougrini, Chelbi-Alix, & Blondel, 2007) and sequesters an antiviral protein, the promyelocytic leukemia (PML) protein, in the cytoplasm (Blondel, Kheddache, Lahaye, Dianoux, & Chelbi-Alix, 2010). As a result, downregulation of the IFN response can be observed in vitro. For example, in RABV-infected human postmitotic neurons (NT2-N), transcription of *IFN-beta* gene is seen as early as 6h post infection, and IFN-beta protein is produced during the first 24h post infection, whereas transcription and production decline thereafter (Prehaud, Megret, Lafage, & Lafon, 2005). However, the evasive mechanism triggered by RABV applies in RABV-infected cells only, since it depends upon the presence of N and/or P proteins in the cytoplasm of the target cell. The evasive mechanism of the IFN response therefore cannot apply in noninfected cells such as sentinel cells (macrophages, pDCs) in the periphery, nor in glial cells such as astrocytes in the NS, allowing these uninfected cells to still mount innate immune responses and produce heterocellular type 1 IFN (Lafon, 2011; Pfefferkorn et al., 2016).

11.2.2 Innate immune response in the periphery

RABV is inoculated in the skin, subcutaneous tissues, or in muscle by bites or scratches. The entry of the virus by the host is rapidly detected by host defense mechanisms in the periphery. The IFN response triggered at the site of entry has an antiviral effect. Evidence has been indirectly obtained by comparing the viral load in the thigh muscle of two groups of mice, parental mice and mice lacking the type I IFN receptor (IFNAR) after injection of the hindlimb with CVS. The viral load measured by the accumulation of viral RNA was increased in mice lacking IFNAR compared to parental strain of mice (Chopy, Detje, Lafage, Kalinke, & Lafon, 2011). This observation suggests that some viral particles might be readily eliminated at this early step of infection.

The cells-producing type I IFN at the site of the injection are likely sentinel cells (pDCs or macrophages). Infection of these cells is dispensable to mount an innate immune response, as shown in vitro when macrophages could be activated by addition of inactivated RABV to the culture (Nakamichi, Inoue, Takasaki, Morimoto, & Kurane, 2004). Nevertheless, there is in vitro experimental evidence that RABV can infect bone-marrow-derived conventional DCs (cDCs) and macrophages in vitro. Despite nonproductive infection, RABV triggers the production of IFNs, cytokines, and chemokines in these cells (Faul et al., 2010; Nakamichi et al., 2004). In cell culture, maturation of cDCs in the presence of RABV is controlled by IFN, which production might rely on the recognition of intracytoplasmic RABV RNAs through RIG-I and mda-5 receptors and not TLR7 (Faul et al., 2010), a characteristic of cDCs (Eisenacher, Steinberg, Reindl, & Krug, 2007).

However, it seems that virulent RABV strains trigger weaker DCs activation versus attenuated strains. Experiments were performed with highly attenuated recombinant RABV (rRABV) genetically modified to allow the expression of chemokines or multiple copies of G protein in a search for more effective rabies vaccines. These RABVs trigger a stronger activation of cDCs in the periphery than the parental rRABV strains (Li, McGettigan, Faber, Schnell, & Dietzschold, 2008; Pulmanausahakul et al., 2001; Wen et al., 2011). Experiments comparing the capacity of dog RABV strains (DRV-NG11 from a dog in Nigeria, or DRV from a dog in Mexico RABV) and of highly attenuated recombinant RABV strains (CVS-B2C or TrisGAS strain) to activate DCs, showed that in contrast to the attenuated RABV strain, the dog RABV strains poorly stimulate the DCs in vitro and as a consequence trigger a lower antibody immune response than the attenuated RABV strains (Gnanadurai et al., 2015; Yang et al., 2015). The poor activation of DCs by the dog RABV strain from Mexico results from the low binding of this virus to DCs (Yang et al., 2015). These data suggest that DC activation inversely correlates with the pathogenicity of the RABV strains.

At the site of injection, muscle cells can be infected as observed in experimental rabies in skunks with a wild-type RABV strain (Charlton & Casey, 1979, 1981). In vitro, muscle infection could be reproduced in the mouse muscle myoblast G-8 cells using the Japanese Nishigahara RABV strain and its derivative, the Ni-CE strain causing lethal and subclinical infections, respectively, after intramuscular injection in mice. Compared to NI-CE, only the Nishigahara RABV strain triggers stable viral replication in muscle cells (Yamaoka et al., 2013, 2017). It has been demonstrated that the muscle cell infection by the Nishigahara RABV strain and entry into the peripheral nerve are promoted through the INF antagonist of the phosphoprotein of the Nishigahara RABV strain, mainly by the N-terminally truncated isoforms of the phosphoprotein (Okada et al., 2016; Yamaoka et al., 2013, 2017). These observations suggest that the capacity of a RABV strain to infect muscle cells correlates with the pathogenicity of the strain.

Therefore, it seems that virulent RABV strains use selected mechanisms allowing them to escape, at least partially, the local host innate immune response early after they have entered the body. These mechanisms could contribute to limit early elimination of viral particles and promote virus entry into the NS.

11.2.3 Innate immune response in the NS

Like most tissues in the organism, the NS expresses different types of receptors capable to sense danger and pathogen signals (Boivin, Coulombe, & Rivest, 2002; Bottcher et al., 2003; Koedel et al., 2004; McKimmie, Johnson, Fooks, & Fazakerley, 2005; Nguyen, Julien, & Rivest, 2002). Central neurons express TLR1–4 as well as TLR 7 and 8 (Barajon et al., 2009; Kim et al., 2007; Ma et al., 2006; Ma, Haynes, Sidman, & Vartanian, 2007; Prehaud et al., 2005; Tang et al., 2007). They express the RLRs (RIG-I and Mda-5) (Chopy, Pothlichet, et al., 2011; Lafon, Megret, Lafage, & Prehaud, 2006; Menager et al., 2009; Peltier, Simms, Farmer, & Miller, 2010), but not the RLR LGP2, which seems to be actively degraded in neurons (Chopy, Pothlichet, et al., 2011). Peripheral nerve plexuses and peripheral nerves (dorsal root ganglion sensory neurons and fibers of sciatic nerves) express TLRs (TLR3, 4, and 7) with prominent expression of TLR3 (Barajon et al., 2009; Cameron et al., 2007; Goethals, Ydens, Timmerman, & Janssens, 2010). Besides astrocytes, a main producer of IFN–beta during viral infections, including RABV infection, neurons take an active part in the innate immune response being both responders to IFN and IFN producers, secreting type I IFN [predominantly IFN-beta in the brain, but no IFN-alpha or Type III IFN lambda] (Delhaye et al., 2006; Kallfass et al., 2012; Pfefferkorn et al., 2016; Prehaud et al., 2005; Sommereyns, Paul, Staeheli, & Michiels, 2008).

11.2.3.1 *RABV evasion of the IFN response in infected neurons*

After infection, human neurons can mount a classical primary IFN response with activation of IRF3 and NF-kappa B, as well as a secondary IFN response, (activation of STATs and IRF7), leading to the production of cytokines (IL-6, TNF-alpha) and chemokines (CXCL10 and CCL5) (Chopy, Detje, et al., 2011; Chopy, Pothlichet, et al., 2011; Prehaud et al., 2005).

Dampening the IFN response favors RABV infection as demonstrated with the death of mice intracerebrally infected with P protein RABV mutants lacking the capacity to decrease the host IFN response (Ito et al., 2010) and with the earlier death of mice lacking IFNAR specifically in the NS, compared to parental mice after CVS intramuscular injection (Chopy, Detje, et al., 2011). These two experiments indicate that RABV infection is sensible to IFN signaling and that P protein-mediated IFN evasion is efficient in the NS. Moreover, it has been shown that virulence, at least for a Japanese vaccine strain (Nishigahara RABV strain), depends upon the capacity of this strain to evade the innate immune response and this process is controlled by the ability of the P and N protein to evade the innate immune response (Masatani et al., 2010b; Shimizu, Ito, Sugiyama, & Minamoto, 2006). Thus, evasion of the IFN response in infected neurons may be critical for RABV progression in the NS through the neuronal network allowing the virus to reach the brainstem and the salivary glands.

11.2.3.2 *RABV limits the inflammatory response in the NS*

Inflammation is a key component of host responses to cell damage or microbial entry leading to the production of inflammatory mediators, including complement, adhesion molecules, cyclo-oxygenase enzymes and their products, as well as cytokines and chemokines. Release of these toxic factors has dramatic consequences when the site of inflammation is in the NS, where severe involvement can lead to significant NS pathology with neuronal death (Brown & Neher, 2010). In the brain, both neurons and glial cells can mount antiviral, inflammatory, and chemokine responses. Astrocytes can respond to the presence of innate immune stimulus in the brain by producing proinflammatory cytokines and chemokines (Park, Lee, et al., 2006). Nevertheless, an important role is taken by microglia in the induction of neuroinflammation, a feature that may reflect the density or the subcellular localization of the innate immune receptors (Bsibsi et al., 2006).

Transcriptome and proteomic analysis of the inflammatory response triggered in the NS of mice by various virulent strains of RABV showed that RABV infection stimulates the expression of chemokines (CCL5, CCL2, CCL9, and CXCL9) and inflammatory cytokines (IL-6, IL-12) (Baloul, Camelo, & Lafon, 2004; Camelo et al., 2000; Chopy, Pothlichet, et al., 2011; Sugiura et al., 2011; Wang et al., 2005). However, the inflammatory reaction in the RABV-infected NS is transient with the expression of a majority of markers being rapidly downregulated in the spinal cord and with a slight delay in the brain (Chopy, Pothlichet, et al., 2011).

Cells expressing inflammatory markers in the RABV-infected NS such as TNF-alpha or IL-1 are in infected neurons, but in neighboring glial or endothelial cells (Marquette et al., 1996; Nuovo, DeFaria, Chanona-Vilchi, & Zhang, 2004; Van Dam et al., 1995).

When compared with other encephalitic virus infections such as caused by Borna virus, RABV triggers only limited inflammation (Fu et al., 1993; Shankar et al., 1992). Moreover, comparison of the inflammatory reaction triggered by RABV strains of various degree of pathogenicity indicates that the more pathogenic strains trigger weaker inflammatory responses (Baloul & Lafon, 2003; Hicks et al., 2009; Laothamatas et al., 2008; Wang et al., 2005). For example, transcriptome analysis performed in the NS of mice infected with wild-type (street) RABV strains, such as a dog RABV strain isolated in China or a bat RABV isolate from North America (silver haired bat isolate) showed that innate immune response is stimulated but to a more limited extent compared with those triggered by laboratory strains (Sugiura et al., 2011; Wang et al., 2005; Zhao et al., 2011).

With the sole exception of experiments in which mice inoculated with rRABV encoding chemokines died because of the excessive influx of monocytes and T cells into the brain (Zhao, Toriumi, Kuang, Chen, & Fu, 2009), most experimental evidence shows that inflammation does not promote RABV infection, but instead limits the propagation of the virus through the NS. Immunization of mice with proinflammatory myelin basic protein (MBP) prior to RABV infection improved the survival to a challenge with a virulent bat RABV strain and, conversely, treatment with a steroid hormone decreasing brain inflammation and with minocycline, a tetracycline derivate with anti-inflammatory properties, increased the mortality rate (Jackson, Scott, Owen, Weli, & Rossiter, 2007; Roy & Hooper, 2007). Also, over expression of TNF-alpha by a recombinant RABV attenuates RABV replication by inducing a strong T-cell infiltration and microglial activation (Faber et al., 2005). It is likely that this low inflammatory reaction in the infected NS contributes to keeping intact the BBB, a condition that correlates with RABV pathogenicity, with nonpathogenic RABV strains triggering an early and transient opening of the BBB, but not pathogenic strains (Chai, He, Zhou, Lu, & Fu, 2014; Miao et al., 2017; Phares, Kean, Mikheeva, & Hooper, 2006; Roy, Phares, Koprowski, & Hooper, 2007).

Altogether, these data indicate that virulent RABV strains trigger a moderate inflammatory response in the NS, and suggest that regulatory mechanisms are set up in the course of the infection to reduce the RABV-induced inflammation of the NS.

Limitation of neuroinflammation occurs by several mechanisms. RABV infection avoids neuronal apoptosis (Lafon, 2011) and rarely infects glial cells or only transiently (Pfefferkorn et al., 2016); two intrinsic features of the infection contributing to limit neuroinflammation. In addition, in the course of RABV infection the expression of anti-inflammatory molecules is upregulated in the NS. This is the case for the anti-inflammatory soluble proteins TNFR1 and 2, which can interfere with the binding of TNF to its receptors (Chopy, Pothlichet, et al., 2011). This is also the case for the suppressors of cytokine signaling (SOCS), a family of proteins that negatively control cytokine signal transduction, with SOCS-1 being upregulated in the brain of RABV infected dogs in noninfected cells in close vicinity of infected neurons (Nuovo et al., 2004). More importantly, RABV upregulates the expression of HLA-G, a nonclassical MHC molecule and B7-H1, (also named PD-L1) the ligand of PD-1, (programmed death protein-1), in neurons, and also in the infected NS, in the case of PD-L1 (Lafon, 2011; Lafon et al., 2008, 2005). Besides, their immune-tolerant properties, which are exploited by RABV, (see later), HLAG and PD-L1 molecules are now also considered as providing negative feedback that limits tissue inflammation (Carosella, Moreau, Aractingi, & Rouas-Freiss, 2001; Francisco et al., 2009; Phares, Stohlman, Hinton, Atkinson, & Bergmann, 2010). This is the case in particular for PD-L1, which dampens the expression of proinflammatory molecules (such as iNos and TNF-alpha) during viral encephalitis (Phares et al., 2010) and activates T regulatory cells (Tregs), which limits inflammation in neurodegenerative diseases (Francisco et al., 2009; Sheean et al., 2018). Whereas HLA-G influences the cytokine balance toward a Th2 pattern by promoting the secretion of IL-4, IL-3, and IL-10 and by downregulating the production of IFN-gamma and TNF-alpha (Carosella et al., 2001).

Limitation of inflammation by RABV infection might be permitted by (1) reducing the entry in the NS of mononuclear leukocytes, monocytes, and macrophages, (2) maintaining the impermeability of the BBB, and (3) minimizing the release of neurotoxic molecules that can compromise NS function and host survival. These conditions should preserve not only the integrity of the infected neuronal network, but also the life of the host, allowing the virus to reach the brainstem and the salivary glands before the premature death of the infected host.

11.3 RABV adaptive immune response

Building an adaptive immune response against a microbe, even a neurotropic virus that rapidly enters the NS after its inoculation in muscle, always occurs in the periphery and never in the NS, which is devoid of lymphoid organs (Galea, Bechmann, & Perry, 2007). The triggering of the adaptive immune response takes place in the lymphoid organs such as the lymph nodes or spleen relies on the activation of pDCs and of type 1 IFN that they produce in a TLR7- and 9-dependent manner after encountering the microbe (Colonna et al., 2004; Diebold et al., 2003; Steinman, 1991).

The $CD4^+$ T lymphocytes recognize foreign antigens that have been processed through the MHC class II exogenous presentation pathway by activated DCs. Once presented by the MHC, the peptides of the digested foreign antigen are recognized by T cells bearing the appropriate T-cell receptor (TCR) and CD4 molecule. Signaling via the TCR and CD4 molecule triggers activation and differentiation of T cells into two functional subsets, the T helper 1

(Th1) and T helper 2 (Th2) cells. The distinction of the two subsets, which is clearer in the mouse than in the human immune system, is based on the cytokines they secrete: interferon-gamma is the signature cytokine for Th1 cells, whereas interleukin-4 (IL-4) is the signature cytokine for the Th2 cells. Generation of Th1 cells is under the control of IL-12 produced by macrophages and DCs. Th1 cells limit the proliferation of pathogens via IFN- gamma production and provide help for antibody production by B lymphocytes. $CD8^+$ T cells, in contrast to $CD4^+$ T cells, recognize foreign antigens that have been processed by the endogenous pathway of cells expressing MHC class I molecules. Infected cells export pathogen peptides embedded in the groove of MHC class I molecules to the cell surface. The peptide-charged infected cells activate T cells expressing the CD8 accessory surface molecules and the appropriate TCR. Activated $CD8^+$ T lymphocytes produce IFN-γ and kill the infected cells via cytotoxicity by means of perforin and granzyme release and/or Fas-mediated lysis.

11.3.1 RABV specific immune response in the periphery

After the injection of the encephalitic RABV strain, CVS, in the hind limb of mice, the size of the draining popliteal lymph nodes and those of spleen increase. Draining lymph nodes are populated with activated T cells expressing the marker of activation CD69. Activation can also be observed among peripheral blood lymphocytes (Vuaillat et al., 2008).

A strong B-cell response is mounted in the spleen. When mice were injected with a less pathogenic virus (the PV strain), similar activation of T cells was observed in lymph nodes and blood suggesting that adaptive immune response is independent of the virulence of the RABV strain. Indeed, when mice were injected with an encephalitic RABV bat strain (silver-haired bat rabies virus, SHBRV) or with a less pathogenic virus (CVS-F3, mutant of CVS encoding a mutation in the G protein), the resulting adaptive immune responses (neutralizing antibodies, $CD4^+$, $CD8^+$ T cells response) were not different (Roy & Hooper, 2007). However, in the case of an experimental infection of dogs with a dog RABV strain which does not activate DCs, the immune response was impaired (Gnanadurai et al., 2015). This suggests that RABV dog strain and RABV bat strain do not stimulate equally the immune response in the periphery. The hypothesis that the activation if the immune response depends upon the nature of the RABV strain may explain why some patients died without having develop an antibody response in the periphery, whereas others die of rabies despite having mounted an immune response in the periphery attested by the presence of neutralizing antibodies in their blood (Hemachudha, 1994; Hunter et al., 2010). The adaptive immune response triggered by RABV strains in the periphery, which is likely an event that occurs late after the virus has already entered the NS, might be unrelated to RABV pathogenicity. However, it cannot been excluded that the adaptive immune response can play a role at the late final stages of infection, when virus particles are produced in the salivary glands. This was suggested by the observation that in an experimental model of rabies in skunks, cyclophosphamide-induced immunosuppression was found to increase the infection of the salivary glands (Charlton, Casey, & Campbell, 1984).

11.3.2 RABV provokes the killing of migratory T cells

Most infections of the NS are controlled by infiltrating T cells. This is, for example, observed during the course of West Nile virus brain infection, where CD8$^+$ T cells attracted by the chemokines produced by inflammatory cells in the infected NS are a critical factor for controlling the infection (Klein et al., 2005; Zhang, Chan, Lu, Diamond, & Klein, 2008). In rabies, sterilization of the infection by T cells is inefficient, and is specifically inactivated by the virus (Lafon, 2008). Immunohistochemical studies performed on rabies autopsy cases revealed that the cells undergoing death were leukocytes and not neurons (Hemachudha et al., 2005; Tobiume et al., 2009). This observation was reproduced in mice infected with the encephalitic RABV strain CVS. Immunocytochemistry of brain and spinal cord slices revealed that despite a heavy load of viral antigens, infected neurons do not undergo death. In contrast, the migrating T cells (CD3+) were apoptotic (Baloul et al., 2004; Baloul & Lafon, 2003; Kojima et al., 2009; Lafon, 2005; Rossiter, Hsu, & Jackson, 2009). Moreover, pathogenicity of the CVS strain was similar in immunocompetent mice Balb/c mice and in Nu/Nu Balb/c mice, indicating that T cells do not control the outcome of encephalitic rabies (Lafon, 2005). In striking contrast, deprivation of T cells transformed an abortive infection into an encephalitic rabies similar to that caused by the encephalitic strain CVS infection, showing that T cells is a critical factor in the restriction of the NS infection caused by an abortive RABV strain. Indeed, when apoptosis was analyzed in the spinal cord of immunocompetent mice infected with the abortive RABV strain PV, killing of T cells was not observed; instead, infected neurons died (Galelli et al., 2000). Altogether, these observations indicate that T cells have a protective potential to control RABV infection in the NS, nevertheless their capacity to control RABV infection is impeded with the encephalitic RABV strain. The mechanisms by which the encephalitic RABV strain evades the host T-cell response was further studied as described now.

11.3.2.1 *Entry of lymphocytes in the RABV-infected NS*

Mononuclear leukocytes, monocytes, and macrophages are recruited to the NS in pathological conditions, including infections by neurotropic viruses (Davoust, Vuaillat, Androdias, & Nataf, 2008). Once activated, the T and B cells and macrophages from the periphery expressing surface adhesion molecules have the capacity to enter the NS in postcapillary venules (Engelhardt, 2008). This entry is independent of BBB permeabilization that modulates the entry of solutes and not cells (Bechmann, Galea, & Perry, 2007). The absence of T-cell protection against an infection by the encephalitic RABV strain might be related to a blockage of T cells access to the NS. This is likely not the case because after infection with an encephalitic RABV strain, blood T cells expressed markers of activation (CD69) and were highly positive for collapsing response mediator protein 2 (CRMP2), a marker of T-cell polarization and migration. The brain was enriched with this type of cell, indicating that RABV-activated T cells have migratory properties (Vuaillat et al., 2008). Thus, activation and entry into the NS are not limiting factors for T-cell protective function. When infiltration of T cells in the NS was compared in mice infected either with an abortive or an encephalitic RABV strain (Baloul et al., 2004), the parenchyma became invaded by infiltrating T cells similarly in the two groups of mice. However, this phenomenon was interrupted after a few days of infection by an encephalitic strain, whereas CD3$^+$ T cells accumulation in PV infected NS was continuous. Disappearance of T cells in the CVS-infected brain and an increase in number of apoptotic cells in the NS were concomitant events. These observations strongly suggest that encephalitic RABV strains, but not abortive strains, trigger unfavorable conditions for T-cell survival in the infected NS.

Neutralizing antibodies have been described as a critical factor for protection against RABV (Hooper et al., 1998; Montano-Hirose et al., 1993; Wiktor et al., 1984; Wunner, Dietzschold, Curtis, & Wiktor, 1983). The entry of B cells into the RABV-infected NS and the local secretion of antibody contribute to the clearance of attenuated RABV from NS (Hooper, Phares, Fabis, & Roy, 2009). It is striking to note that during the course of encephalitic RABV infection B cells are almost undetectable in brain (Camelo et al., 2000; Kojima et al., 2010), suggesting that restricted entry or specific destruction of migratory B cells could also contribute to RABV virulence.

11.3.2.2 Destruction of T cells in the RABV-infected NS

Tumors evade immune surveillance by multiple mechanisms, including the inhibition of tumor-specific T-cell immunity. In order to escape attack from protective T cells, tumor cells upregulate expression of certain surface molecules such as PD-L1, Fas-L, and HLA-G, which triggers death signaling in activated T cells expressing the corresponding ligands PD-1 for PD-L1, Fas for FasL and CD8—among others—for HLA-G (Dong et al., 2002; Gratas et al., 1998; Rouas-Freiss, Moreau, Menier, & Carosella, 2003). Studies evaluating whether RABV-infected neurons upregulate immunosubversive molecules to kill activated T cells following an evasive strategy similar to that selected by tumors cells have been undertaken both in vivo and in vitro. In vitro, RABV infection was found to upregulate the expression of HLA-G at the surface of human neurons (Lafon et al., 2005; Megret et al., 2007). In vivo, comparison of experimental rabies in mice caused by CVS, which kills T cells, or by PV, which does not kill T cells, leads to the finding that the CVS-infected NS, but not the PV-infected NS, upregulates the expression of FasL. In mice lacking a functional FasL, there was less T-cell apoptosis in the NS than in control mice. Remarkably, RABV morbidity and mortality were reduced in these mice. Destruction of T cells through the Fas/FasL pathway can be enhanced by indoleamine 2, 3 dioxygenase (IDO), which RABV upregulates expression in the infected neurons and brain (Prehaud et al., 2005; Zhao et al., 2011). The enzyme IDO converts extracellular tryptophan into kynurenine, thereby reducing its concentration in the microenvironment that in turn markedly enhances the sensitivity of any nearby T cell for Fas-ligand induced apoptosis (Kwidzinski et al., 2003).

In addition, RABV-infected brain upregulates the expression of another immunosubversive molecule, PD-L1 (Chopy, Pothlichet, et al., 2011; Lafon et al., 2008). Whereas noninfected NS was almost devoid of PD-L1 expression, RABV infection triggers neural PD-L1 expression that increases as the infection progresses. Infected neurons and also noninfected neural cells, including astrocyte-like cells, were shown to be positive for PD-L1. RABV infection of PD-L1-deficient mice (PD-L1$^{-/-}$ mice) resulted in a notable reduction in clinical signs and mortality. Reduction of RABV virulence in PD-L1$^{-/-}$ mice was concomitant of a reduction of CD8$^+$ T-cell apoptosis among the migratory T cells.

Altogether these experiments indicate that despite the triggering of a classical adaptive immune response in the periphery and the infiltration of the lymphocytes into the infected NS, the protection, which could have been conferred in the NS by this immune response, is substantially impeded by RABV infection (Lafon, 2011).

11.4 RABV infection triggers a CNS-mediated immune unresponsiveness

The dampening of immune protection already triggered by RABV is completed by a central immunosuppression caused by the neuronal reflex control of immunity triggered by the NS facing an excess of inflammation in an attempt to restore general homeostasis.

RABV infection by a pathogenic strain induces an immune unresponsiveness (Camelo, Lafage, Galelli, & Lafon, 2001; Hirai et al., 1992; Kasempimolporn, Saengseesom, Mitmoonpitak, Akesowan, & Sitprija, 1997; Kasempimolporn, Tirawatnapong, Saengseesom, Nookhai, & Sitprija, 2001; Perry, Hotchkiss, & Lodmell, 1990; Torres-Anjel, Volz, Torres, Turk, & Tshikuka, 1988; Tshikuka, Torres-Anjel, Blenden, & Elliott, 1992; Wiktor, Doherty, & Koprowski, 1977a, 1977b) characterized by the impairment of T-cells functions with an alteration of cytokine pattern, an inhibition of T-cells proliferation, and the destruction of immune cells without modify immune cells proportion (CD4/CD8 ratio constant) in the lymphoid organs (Perry et al., 1990). This leads to the atrophy of the spleen and the thymus of RABV-infected mammals. TNF-alpha receptor has been found to play a role in RABV immune unresponsiveness, since immune cells lacking the TNF alpha p55 receptor were less immunosuppressed compared to the wild type (Camelo et al., 2000). Most importantly, infection of the brain is required since immune unresponsiveness does not occur after the infection of the NS with an abortive RABV strain, which infects the spinal cord only (Camelo et al., 2001). This suggests that the property of the NS that centrally controls the immune response in the periphery might be triggered (Tracey, 2009). NS modulates the immune functions through two main immune-neuroendocrine pathways: the hypothalamo-pituitary (HPA) axis and the autonomous NS (ANS) composed of sympathetic and parasympathetic nerves fibers (Johnston & Webster, 2009). The homeostatic reflex is activated after the brain senses the presence of an excess of inflammatory cytokines such as TNF-alpha, IL-1β, or IL-6 in the periphery, by neuronal (mainly through local afferent fibers of the vagus nerve) and by humoral pathway (Johnston & Webster, 2009). The NS in centers localized in frontal, hypothalamic, and brainstem process this input.

This general immune unresponsiveness controlled by the NS may be advantageous for RABV propagation since a mouse strain having a less efficient HPA axis is less susceptible to rabies (Roy & Hooper, 2007). This central immunosuppression may limit peripheral control of infection in the muscle or the salivary glands (see Fig. 11.1).

Thus, RABV infection not only actively inhibits the T-cell response and inflammation in the NS by upregulating PD-L1 and FasL molecules, but also benefits from the intrinsic capacity of NS to trigger central immunosuppression in order to maintain whole body homeostasis.

11.5 Paradoxical role of IFN in RABV virulence

A series of experiments clearly indicate that RABV infection is sensitized to the activation of IFN signaling and that P protein-mediated IFN evasion is efficient. Nevertheless, in the course of infection, the IFN induction in the whole RABV-infected NS is far from being abrogated (Chopy, Detje, et al., 2011; Chopy, Pothlichet, et al., 2011; Lafon et al., 2008; Li et al., 2012; Miao et al., 2017; Sugiura et al., 2011; Wang et al., 2005). Indeed, after injection of RABV

(CVS) into the hindlimbs, a progressive infection within the spinal cord and the brain is accompanied by a robust innate immune response characterized by a type 1 IFN response. This is not a peculiar property of the laboratory strains, since similar observation was made after infection with a highly virulent RABV strains, the DOG-4 strain (Li et al., 2012) or the Chinese dog BD06 RABV strain (Miao et al., 2017).

It may not be surprising that IFN can be produced in the NS during infection because the mechanisms evolved by RABV to escape the IFN response are restricted to infected neurons, the only cell type expressing the P and N proteins. These mechanisms cannot operate in glial cells because they do not express any viral proteins, glial cells being rarely infected in vivo (Iwasaki & Clark, 1975). Glia are efficient innate responders (Park, Lee, et al., 2006) and, in particular, they are IFN responders (Lafon et al., 2005). They do not need to be infected to mount an innate immune response as shown by treating microglial cultures with inactivated RABV (Nakamichi et al., 2005). Indeed, in the brain of RABV naturally infected dogs or experimentally infected mice, the cells expressing cytokines were not the infected neurons, but noninfected nearby cells with glial or macrophage morphology (Marquette et al., 1996; Nuovo et al., 2004; Van Dam et al., 1995), suggesting that noninfected glial cells such as astrocytes may be IFN responders and producers of heterocellular IFN, which is produced by neighboring cells and not by infected cells (Pfefferkorn et al., 2016). In this case, we can distinguish the infected part of the NS consisting of the network of infected neurons, in which the IFN response is limited by the evasive mechanisms of the virus, from the noninfected part made of the glial cells that are resistant to the RABV evasive mechanisms and in which a heterocellular IFN response can settle. One can wonder what the function of the heterocellular IFN is. It can be speculated the heterocellular IFN makes non-neuronal cells refractory to infection. RABV neuronotropism (tropism to neurons) and virus progression could be facilitated by this means as demonstrated for poliovirus, which is another neuronotropic virus (Ida-Hosonuma et al., 2005; Kuss, Etheredge, & Pfeiffer, 2008; Pfeiffer, 2010).

Lessons from the use of recombinant forms of IFN-beta as treatment in relapsing forms of multiple sclerosis highlight that IFN promotes the production of anti-inflammatory molecules and reduces the trafficking of inflammatory cells across the endothelium of brain capillaries (review in Kieseier (2011)). If similar functions were demonstrated in RABV-infected brain, it could be proposed that IFN reduces the trafficking of inflammatory cells, and contributes to the low inflammatory environment triggered by RABV by upregulating the production of anti-inflammatory molecules.

Beside intrinsic antiviral or anti-inflammatory properties, type I IFN also controls the expression of a large number of genes (ISG, interferon stimulated genes) (Takeuchi & Akira, 2010). Among those genes are the PD-L1 and the nonclassical MHC Class I molecule HLA-G, two genes in which expression is upregulated in RABV infection in an IFN-dependent manner (Chopy, Pothlichet, et al., 2011; Lafon et al., 2008, 2005). PD-L1 has been demonstrated to contribute to the killing of migratory T cells in RABV infection (Lafon et al., 2008). In addition, PD-L1 and HLA-G have anti-inflammatory functions (Carosella et al., 2001; Phares et al., 2010).

To explain the dual role of IFN in RABV infection, it can be proposed that heterocellular IFN produced in the NS in the course of RABV infection contributes to the low inflammatory environment set up by RABV infection and to the RABV-mediated killing of migratory T cells, whereas IFN antiviral effect takes place in infected neurons (Fig. 11.1B).

11.6 Conclusions

Thus, RABV has selected a series of mechanisms to escape the host immune surveillance possibly explaining why, in the absence of postexposure treatment (PET), rabies is one of the very few human infections with a near 100% mortality rate. Despite these well-adapted viral strategies to escape the immune response, RABV infection can be limited if vaccine is injected promptly after exposure suggesting that the viral-mediated paralysis of the host immune response requires some time, which can be exploited for postexposure treatment. However, the efficacy of rabies PET requires public education; prompt wound cleansing, supplies of efficient vaccines, and availability of rabies immunoglobulins. Half of the victims being children, preexposure vaccination of young individuals should be considered in an attempt to improve the global health of humans. In addition, improved knowledge of the immune evasive mechanisms evolved by RABV to infect the NS may help identify new therapeutical targets such as the central neural immune reflex or neuroinflammation.

References

Baloul, L., Camelo, S., & Lafon, M. (2004). Up-regulation of Fas ligand (FasL) in the central NS: A mechanism of immune evasion by rabies virus. *Journal of Neurovirology, 10*(6), 372–382.

Baloul, L., & Lafon, M. (2003). Apoptosis and rabies virus neuroinvasion. *Biochimie, 85*(8), 777–788.

Barajon, I., Serrao, G., Arnaboldi, F., Opizzi, E., Ripamonti, G., Balsari, A., & Rumio, C. (2009). Toll-like receptors 3, 4, and 7 are expressed in the enteric NS and dorsal root ganglia. *The Journal of Histochemistry and Cytochemistry, 57* (11), 1013–1023. doi: jhc.2009.953539 [pii] https://doi.org/10.1369/jhc.2009.953539.

Barkhouse, D. A., Faber, M., & Hooper, D. C. (2015). Pre- and post-exposure safety and efficacy of attenuated rabies virus vaccines are enhanced by their expression of IFNgamma. *Virology, 474*, 174–180. https://doi.org/10.1016/j.virol.2014.10.025.

Bauer, A., Nolden, T., Schroter, J., Romer-Oberdorfer, A., Gluska, S., Perlson, E., & Finke, S. (2014). Anterograde glycoprotein-dependent transport of newly generated rabies virus in dorsal root ganglion neurons. *Journal of Virology, 88*(24), 14172–14183. https://doi.org/10.1128/JVI.02254-14.

Bechmann, I., Galea, I., & Perry, V. H. (2007). What is the blood-brain barrier (not)? *Trends in Immunology, 28*(1), 5–11. doi: S1471-4906(06)00329-2 [pii] https://doi.org/10.1016/j.it.2006.11.007.

Blondel, D., Kheddache, S., Lahaye, X., Dianoux, L., & Chelbi-Alix, M. K. (2010). Resistance to rabies virus infection conferred by the PMLIV isoform. *Journal of Virology, 84*(20), 10719–10726. doi: JVI.01286-10 [pii] https://doi.org/10.1128/JVI.01286-10.

Boivin, G., Coulombe, Z., & Rivest, S. (2002). Intranasal herpes simplex virus type 2 inoculation causes a profound thymidine kinase dependent cerebral inflammatory response in the mouse hindbrain. *The European Journal of Neuroscience, 16*(1), 29–43.

Bottcher, T., von Mering, M., Ebert, S., Meyding-Lamade, U., Kuhnt, U., Gerber, J., & Nau, R. (2003). Differential regulation of toll-like receptor mRNAs in experimental murine central NS infections. *Neuroscience Letters, 344*(1), 17–20.

Brown, G. C., & Neher, J. J. (2010). Inflammatory neurodegeneration and mechanisms of microglial killing of neurons. *Molecular Neurobiology, 41*(2–3), 242–247. https://doi.org/10.1007/s12035-010-8105-9.

Brzozka, K., Finke, S., & Conzelmann, K. K. (2005). Identification of the rabies virus alpha/beta interferon antagonist: Phosphoprotein P interferes with phosphorylation of interferon regulatory factor 3. *Journal of Virology, 79*(12), 7673–7681.

Brzozka, K., Finke, S., & Conzelmann, K. K. (2006). Inhibition of interferon signaling by rabies virus phosphoprotein P: Activation-dependent binding of STAT1 and STAT2. *Journal of Virology, 80*(6), 2675–2683.

Bsibsi, M., Persoon-Deen, C., Verwer, R. W., Meeuwsen, S., Ravid, R., & Van Noort, J. M. (2006). Toll-like receptor 3 on adult human astrocytes triggers production of neuroprotective mediators. *Glia, 53*(7), 688–695. https://doi.org/10.1002/glia.20328.

Camelo, S., Lafage, M., Galelli, A., & Lafon, M. (2001). Selective role for the p55 Kd TNF-alpha receptor in immune unresponsiveness induced by an acute viral encephalitis. *Journal of Neuroimmunology, 113*(1), 95–108.

Camelo, S., Lafage, M., & Lafon, M. (2000). Absence of the p55 Kd TNF-alpha receptor promotes survival in rabies virus acute encephalitis. *Journal of Neurovirology, 6*(6), 507–518.

Cameron, J. S., Alexopoulou, L., Sloane, J. A., DiBernardo, A. B., Ma, Y., Kosaras, B., ... Vartanian, T. (2007). Toll-like receptor 3 is a potent negative regulator of axonal growth in mammals. *The Journal of Neuroscience, 27*(47), 13033–13041.

Carosella, E. D., Moreau, P., Aractingi, S., & Rouas-Freiss, N. (2001). HLA-G: A shield against inflammatory aggression. *Trends in Immunology, 22*(10), 553–555.

Chai, Q., He, W. Q., Zhou, M., Lu, H., & Fu, Z. F. (2014). Enhancement of blood-brain barrier permeability and reduction of tight junction protein expression are modulated by chemokines/cytokines induced by rabies virus infection. *Journal of Virology, 88*(9), 4698–4710. https://doi.org/10.1128/JVI.03149-13.

Charlton, K. M., & Casey, G. A. (1979). Experimental rabies in skunks: Immunofluorescence light and electron microscopic studies. *Laboratory Investigation, 41*(1), 36–44.

Charlton, K. M., & Casey, G. A. (1981). Experimental rabies in skunks: Persistence of virus in denervated muscle at the inoculation site. *Canadian Journal of Comparative Medicine, 45*(4), 357–362.

Charlton, K. M., Casey, G. A., & Campbell, J. B. (1984). Experimental rabies in skunks: Effects of immunosuppression induced by cyclophosphamide. *Canadian Journal of Comparative Medicine, 48*(1), 72–77.

Chopy, D., Detje, C. N., Lafage, M., Kalinke, U., & Lafon, M. (2011). The type I interferon response bridles rabies virus infection and reduces pathogenicity. *Journal of Neurovirology, 17*(4), 353–367. https://doi.org/10.1007/s13365-011-0041-6.

Chopy, D., Pothlichet, J., Lafage, M., Megret, F., Fiette, L., Si-Tahar, M., & Lafon, M. (2011). Ambivalent role of the innate immune response in rabies virus pathogenesis. *Journal of Virology, 85*(13), 6657–6668. https://doi.org/10.1128/JVI.00302-11.

Colonna, M., Trinchieri, G., & Liu, Y. J. (2004). Plasmacytoid dendritic cells in immunity. *Nature Immunology, 5*(12), 1219–1226. https://doi.org/10.1038/ni1141.

Davoust, N., Vuaillat, C., Androdias, G., & Nataf, S. (2008). From bone marrow to microglia: Barriers and avenues. *Trends in Immunology, 29*(5), 227–234. doi: S1471-4906(08)00088-4 [pii] https://doi.org/10.1016/j.it.2008.01.010.

Delhaye, S., Paul, S., Blakqori, G., Minet, M., Weber, F., Staeheli, P., & Michiels, T. (2006). Neurons produce type I interferon during viral encephalitis. *Proceedings of the National Academy of Sciences of the United States of America, 103*(20), 7835–7840.

Diebold, S. S., Montoya, M., Unger, H., Alexopoulou, L., Roy, P., Haswell, L. E., ... Reis e Sousa, C. (2003). Viral infection switches non-plasmacytoid dendritic cells into high interferon producers. *Nature, 424*(6946), 324–328. https://doi.org/10.1038/nature01783.

Dong, H., Strome, S. E., Salomao, D. R., Tamura, H., Hirano, F., Flies, D. B., ... Chen, L. (2002). Tumor-associated PD-L1 promotes T-cell apoptosis: A potential mechanism of immune evasion. *Nature Medicine, 8*(8), 793–800.

Eisenacher, K., Steinberg, C., Reindl, W., & Krug, A. (2007). The role of viral nucleic acid recognition in dendritic cells for innate and adaptive antiviral immunity. *Immunobiology, 212*(9–10), 701–714. doi: S0171-2985(07)00112-X [pii] https://doi.org/10.1016/j.imbio.2007.09.007.

Engelhardt, B. (2008). The blood-central NS barriers actively control immune cell entry into the central NS. *Current Pharmaceutical Design, 14*(16), 1555–1565.

Faber, M., Bette, M., Preuss, M. A., Pulmanausahakul, R., Rehnelt, J., Schnell, M. J., ... Weihe, E. (2005). Overexpression of tumor necrosis factor alpha by a recombinant rabies virus attenuates replication in neurons and prevents lethal infection in mice. *Journal of Virology, 79*(24), 15405–15416. 79/24/15405 [pii] https://doi.org/10.1128/JVI.79.24.15405-15416.2005.

Faber, M., Li, J., Kean, R. B., Hooper, D. C., Alugupalli, K. R., & Dietzschold, B. (2009). Effective preexposure and postexposure prophylaxis of rabies with a highly attenuated recombinant rabies virus. *Proceedings of the National Academy of Sciences of the United States of America, 106*(27), 11300–11305. https://doi.org/10.1073/pnas.0905640106.

Faul, E. J., Wanjalla, C. N., Suthar, M. S., Gale, M., Wirblich, C., & Schnell, M. J. (2010). Rabies virus infection induces type I interferon production in an IPS-1 dependent manner while dendritic cell activation relies on IFNAR signaling. *PLoS Pathogens, 6*(7). e1001016 https://doi.org/10.1371/journal.ppat.1001016.

Francisco, L. M., Salinas, V. H., Brown, K. E., Vanguri, V. K., Freeman, G. J., Kuchroo, V. K., & Sharpe, A. H. (2009). PD-L1 regulates the development, maintenance, and function of induced regulatory T cells. *The Journal of Experimental Medicine, 206*(13), 3015–3029. https://doi.org/10.1084/jem.20090847.

Fu, Z. F., Weihe, E., Zheng, Y. M., Schafer, M. K., Sheng, H., Corisdeo, S., … Dietzschold, B. (1993). Differential effects of rabies and Borna disease viruses on immediate-early- and late-response gene expression in brain tissues. *Journal of Virology, 67*(11), 6674–6681.

Galea, I., Bechmann, I., & Perry, V. H. (2007). What is immune privilege (not)? *Trends in Immunology, 28*(1), 12–18. doi: S1471-4906(06)00326-7 [pii] https://doi.org/10.1016/j.it.2006.11.004.

Galelli, A., Baloul, L., & Lafon, M. (2000). Abortive rabies virus central nervous infection is controlled by T lymphocyte local recruitment and induction of apoptosis. *Journal of Neurovirology, 6*(5), 359–372.

Gnanadurai, C. W., Yang, Y., Huang, Y., Li, Z., Leyson, C. M., Cooper, T. L., … Fu, Z. F. (2015). Differential host immune responses after infection with wild-type or lab-attenuated rabies viruses in dogs. *PLoS Neglected Tropical Diseases, 9*(8). e0004023 https://doi.org/10.1371/journal.pntd.0004023.

Goethals, S., Ydens, E., Timmerman, V., & Janssens, S. (2010). Toll-like receptor expression in the peripheral nerve. *Glia, 58*(14), 1701–1709. https://doi.org/10.1002/glia.21041.

Gratas, C., Tohma, Y., Barnas, C., Taniere, P., Hainaut, P., & Ohgaki, H. (1998). Up-regulation of Fas (APO-1/CD95) ligand and down-regulation of Fas expression in human esophageal cancer. *Cancer Research, 58*(10), 2057–2062.

Hemachudha, T. (1994). Human rabies: Clinical aspects, pathogenesis, and potential therapy. *Current Topics in Microbiology and Immunology, 187*, 121–143.

Hemachudha, T., Wacharapluesadee, S., Mitrabhakdi, E., Wilde, H., Morimoto, K., & Lewis, R. A. (2005). Pathophysiology of human paralytic rabies. *Journal of Neurovirology, 11*(1), 93–100.

Hicks, D. J., Nunez, A., Healy, D. M., Brookes, S. M., Johnson, N., & Fooks, A. R. (2009). Comparative pathological study of the murine brain after experimental infection with classical rabies virus and European bat lyssaviruses. *Journal of Comparative Pathology, 140*(2-3), 113–126.

Hirai, K., Kawano, H., Mifune, K., Fujii, H., Nishizono, A., Shichijo, A., & Mannen, K. (1992). Suppression of cell-mediated immunity by street rabies virus infection. *Microbiology and Immunology, 36*(12), 1277–1290.

Hooper, D. C., Morimoto, K., Bette, M., Weihe, E., Koprowski, H., & Dietzschold, B. (1998). Collaboration of antibody and inflammation in clearance of rabies virus from the central NS. *Journal of Virology, 72*(5), 3711–3719.

Hooper, D. C., Phares, T. W., Fabis, M. J., & Roy, A. (2009). The production of antibody by invading B cells is required for the clearance of rabies virus from the central NS. *PLoS Neglected Tropical Diseases, 3*(10), e535.

Hornung, V., Ellegast, J., Kim, S., Brzozka, K., Jung, A., Kato, H., … Hartmann, G. (2006). 5′-Triphosphate RNA is the ligand for RIG-I. *Science, 314*(5801), 994–997.

Hunter, M., Johnson, N., Hedderwick, S., McCaughey, C., Lowry, K., McConville, J., … Fooks, A. R. (2010). Immunovirological correlates in human rabies treated with therapeutic coma. *Journal of Medical Virology, 82*(7), 1255–1265. https://doi.org/10.1002/jmv.21785.

Ida-Hosonuma, M., Iwasaki, T., Yoshikawa, T., Nagata, N., Sato, Y., Sata, T., … Koike, S. (2005). The alpha/beta interferon response controls tissue tropism and pathogenicity of poliovirus. *Journal of Virology, 79*(7), 4460–4469. 79/7/4460 [pii] https://doi.org/10.1128/JVI.79.7.4460-4469.2005.

Irwin, D. J., Wunner, W. H., Ertl, H. C., & Jackson, A. C. (1999). Basis of rabies virus neurovirulence in mice: Expression of major histocompatibility complex class I and class II mRNAs. *Journal of Neurovirology, 5*(5), 485–494.

Ito, N., Moseley, G. W., Blondel, D., Shimizu, K., Rowe, C. L., Ito, Y., … Sugiyama, M. (2010). Role of interferon antagonist activity of rabies virus phosphoprotein in viral pathogenicity. *Journal of Virology, 84*(13), 6699–6710. doi: JVI.00011-10 [pii] https://doi.org/10.1128/JVI.00011-10.

Iwasaki, Y., & Clark, H. F. (1975). Cell to cell transmission of virus in the central NS. II. Experimental rabies in mouse. *Laboratory Investigation, 33*(4), 391–399.

Jackson, A. C., Scott, C. A., Owen, J., Weli, S. C., & Rossiter, J. P. (2007). Therapy with minocycline aggravates experimental rabies in mice. *Journal of Virology, 81*, 6248–6253.

Johnston, G. R., & Webster, N. R. (2009). Cytokines and the immunomodulatory function of the vagus nerve. *British Journal of Anaesthesia, 102*(4), 453–462. doi: aep037 [pii] https://doi.org/10.1093/bja/aep037.

Kallfass, C., Ackerman, A., Lienenklaus, S., Weiss, S., Heimrich, B., & Staeheli, P. (2012). Visualizing production of beta interferon by astrocytes and microglia in brain of La Crosse virus-infected mice. *Journal of Virology, 86*(20), 11223–11230. https://doi.org/10.1128/JVI.01093-12.

Kasempimolporn, S., Saengseesom, W., Mitmoonpitak, C., Akesowan, S., & Sitprija, V. (1997). Cell-mediated immunosuppression in mice by street rabies virus not restored by calcium ionophore or PMA. *Asian Pacific Journal of Allergy and Immunology, 15*(3), 127–132.

Kasempimolporn, S., Tirawatnapong, T., Saengseesom, W., Nookhai, S., & Sitprija, V. (2001). Immunosuppression in rabies virus infection mediated by lymphocyte apoptosis. *Japanese Journal of Infectious Diseases, 54*(4), 144–147.

Kieseier, B. C. (2011). The mechanism of action of interferon-beta in relapsing multiple sclerosis. *CNS Drugs, 25*(6), 491–502. https://doi.org/10.2165/11591110-000000000-000004.

Kim, D., Kim, M. A., Cho, I. H., Kim, M. S., Lee, S., Jo, E. K., ... Lee, S. J. (2007). A critical role of toll-like receptor 2 in nerve injury-induced spinal cord glial cell activation and pain hypersensitivity. *The Journal of Biological Chemistry, 282*(20), 14975–14983. doi: M607277200 [pii] https://doi.org/10.1074/jbc.M607277200.

Klein, R. S., Lin, E., Zhang, B., Luster, A. D., Tollett, J., Samuel, M. A., ... Diamond, M. S. (2005). Neuronal CXCL10 directs CD8+ T-cell recruitment and control of West Nile virus encephalitis. *Journal of Virology, 79*(17), 11457–11466. 79/17/11457 [pii] https://doi.org/10.1128/JVI.79.17.11457-11466.2005.

Klingen, Y., Conzelmann, K. K., & Finke, S. (2008). Double-labeled rabies virus: Live tracking of enveloped virus transport. *Journal of Virology, 82*(1), 237–245. doi: JVI.01342-07 [pii] https://doi.org/10.1128/JVI.01342-07.

Koedel, U., Rupprecht, T., Angele, B., Heesemann, J., Wagner, H., Pfister, H. W., & Kirschning, C. J. (2004). MyD88 is required for mounting a robust host immune response to *Streptococcus pneumoniae* in the CNS. *Brain, 127*(Pt 6), 1437–1445.

Kojima, D., Park, C. H., Satoh, Y., Inoue, S., Noguchi, A., & Oyamada, T. (2009). Pathology of the spinal cord of C57BL/6J mice infected with rabies virus (CVS-11 strain). *The Journal of Veterinary Medical Science, 71*(3), 319–324.

Kojima, D., Park, C. H., Tsujikawa, S., Kohara, K., Hatai, H., Oyamada, T., ... Inoue, S. (2010). Lesions of the central NS induced by intracerebral inoculation of BALB/c mice with rabies virus (CVS-11). *The Journal of Veterinary Medical Science, 72*(8), 1011–1016 doi: JST.JSTAGE/jvms/09-0550 [pii].

Kuss, S. K., Etheredge, C. A., & Pfeiffer, J. K. (2008). Multiple host barriers restrict poliovirus trafficking in mice. *PLoS Pathogens, 4*(6). e1000082 https://doi.org/10.1371/journal.ppat.1000082.

Kwidzinski, E., Bunse, J., Kovac, A. D., Ullrich, O., Zipp, F., Nitsch, R., & Bechmann, I. (2003). IDO (indolamine 2,3-dioxygenase) expression and function in the CNS. *Advances in Experimental Medicine and Biology, 527*, 113–118.

Lafon, M. (2005). Modulation of the immune response in the NS by rabies virus. *Current Topics in Microbiology and Immunology, 289*, 239–258.

Lafon, M. (2008). Immune evasion, a critical strategy for rabies virus. *Developmental Biology (Basel), 131*, 413–419.

Lafon, M. (2011). Evasive strategies in rabies virus infection. *Advances in Virus Research, 79*, 33–53. https://doi.org/10.1016/B978-0-12-387040-7.00003-2.

Lafon, M., Megret, F., Lafage, M., & Prehaud, C. (2006). The innate immune facet of brain: Human neurons express TLR-3 and sense viral dsRNA. *Journal of Molecular Neuroscience, 29*(3), 185–194. https://doi.org/10.1385/JMN:29:3:185.

Lafon, M., Megret, F., Meuth, S. G., Simon, O., Velandia Romero, M. L., Lafage, M., ... Wiendl, H. (2008). Detrimental contribution of the immuno-inhibitor PD-L1 to rabies virus encephalitis. *Journal of Immunology, 180*(11), 7506–7515.

Lafon, M., Prehaud, C., Megret, F., Lafage, M., Mouillot, G., Roa, M., ... Carosella, E. D. (2005). Modulation of HLA-G expression in human neural cells after neurotropic viral infections. *Journal of Virology, 79*(24), 15226–15237. https://doi.org/10.1128/JVI.79.24.15226-15237.2005.

Laothamatas, J., Wacharapluesadee, S., Lumlertdacha, B., Ampawong, S., Tepsumethanon, V., Shuangshoti, S., ... Hemachudha, T. (2008). Furious and paralytic rabies of canine origin: Neuroimaging with virological and cytokine studies. *Journal of Neurovirology, 14*(2), 119–129.

Li, J., Ertel, A., Portocarrero, C., Barkhouse, D. A., Dietzschold, B., Hooper, D. C., & Faber, M. (2012). Postexposure treatment with the live-attenuated rabies virus (RV) vaccine TriGAS triggers the clearance of wild-type RV from the central NS (CNS) through the rapid induction of genes relevant to adaptive immunity in CNS tissues. *Journal of Virology, 86*(6), 3200–3210. https://doi.org/10.1128/JVI.06699-11.

Li, J., McGettigan, J. P., Faber, M., Schnell, M. J., & Dietzschold, B. (2008). Infection of monocytes or immature dendritic cells (DCs) with an attenuated rabies virus results in DC maturation and a strong activation of the NFkappaB signaling pathway. *Vaccine, 26*(3), 419–426. doi: S0264-410X(07)01262-5 [pii] https://doi.org/10.1016/j.vaccine.2007.10.072.

Liu, L., Botos, I., Wang, Y., Leonard, J. N., Shiloach, J., Segal, D. M., & Davies, D. R. (2008). Structural basis of toll-like receptor 3 signaling with double-stranded RNA. *Science, 320*(5874), 379–381. doi: 320/5874/379 [pii] https://doi.org/10.1126/science.1155406.

Ma, Y., Haynes, R. L., Sidman, R. L., & Vartanian, T. (2007). TLR8: An innate immune receptor in brain, neurons and axons. *Cell Cycle, 6*(23), 2859–2868.

Ma, Y., Li, J., Chiu, I., Wang, Y., Sloane, J. A., Lu, J., ... Vartanian, T. (2006). Toll-like receptor 8 functions as a negative regulator of neurite outgrowth and inducer of neuronal apoptosis. *The Journal of Cell Biology, 175*(2), 209–215. doi: jcb.200606016 [pii] https://doi.org/10.1083/jcb.200606016.

Marquette, C., Van Dam, A. M., Ceccaldi, P. E., Weber, P., Haour, F., & Tsiang, H. (1996). Induction of immunoreactive interleukin-1 beta and tumor necrosis factor-alpha in the brains of rabies virus infected rats. *Journal of Neuroimmunology, 68*(1–2), 45–51.

Masatani, T., Ito, N., Shimizu, K., Ito, Y., Nakagawa, K., Abe, M.,... Sugiyama, M. (2010a). Amino acids at positions 273 and 394 in rabies virus nucleoprotein are important for both evasion of host RIG-I-mediated antiviral response and pathogenicity. *Virus Research*, doi: S0168-1702(10)00352-7 [pii] https://doi.org/10.1016/j.virusres.2010.09.016.

Masatani, T., Ito, N., Shimizu, K., Ito, Y., Nakagawa, K., Sawaki, Y.,... Sugiyama, M. (2010b). Rabies virus nucleoprotein functions to evade activation of the RIG-I-mediated antiviral response. *Journal of Virology, 84*(8), 4002–4012. doi: JVI.02220-09 [pii] https://doi.org/10.1128/JVI.02220-09.

McKimmie, C. S., Johnson, N., Fooks, A. R., & Fazakerley, J. K. (2005). Viruses selectively upregulate toll-like receptors in the central NS. *Biochemical and Biophysical Research Communications, 336*(3), 925–933.

Megret, F., Prehaud, C., Lafage, M., Moreau, P., Rouas-Freiss, N., Carosella, E. D., & Lafon, M. (2007). Modulation of HLA-G and HLA-E expression in human neuronal cells after rabies virus or herpes virus simplex type 1 infections. *Human Immunology, 68*(4), 294–302. https://doi.org/10.1016/j.humimm.2006.12.003.

Menager, P., Roux, P., Megret, F., Bourgeois, J. P., Le Sourd, A. M., Danckaert, A.,... Lafon, M. (2009). Toll-like receptor 3 (TLR3) plays a major role in the formation of rabies virus Negri bodies. *PLoS Pathogens, 5*(2), e1000315. https://doi.org/10.1371/journal.ppat.1000315.

Miao, F. M., Zhang, S. F., Wang, S. C., Liu, Y., Zhang, F., & Hu, R. L. (2017). Comparison of immune responses to attenuated rabies virus and street virus in mouse brain. *Archives of Virology, 162*(1), 247–257. https://doi.org/10.1007/s00705-016-3081-7.

Montano-Hirose, J. A., Lafage, M., Weber, P., Badrane, H., Tordo, N., & Lafon, M. (1993). Protective activity of a murine monoclonal antibody against European bat lyssavirus 1 (EBL1) infection in mice. *Vaccine, 11*(12), 1259–1266.

Nakamichi, K., Inoue, S., Takasaki, T., Morimoto, K., & Kurane, I. (2004). Rabies virus stimulates nitric oxide production and CXC chemokine ligand 10 expression in macrophages through activation of extracellular signal-regulated kinases 1 and 2. *Journal of Virology, 78*(17), 9376–9388.

Nakamichi, K., Saiki, M., Sawada, M., Takayama-Ito, M., Yamamuro, Y., Morimoto, K., & Kurane, I. (2005). Rabies virus-induced activation of mitogen-activated protein kinase and NF-kappaB signaling pathways regulates expression of CXC and CC chemokine ligands in microglia. *Journal of Virology, 79*(18), 11801–11812.

Nguyen, M. D., Julien, J. P., & Rivest, S. (2002). Innate immunity: The missing link in neuroprotection and neurodegeneration? *Nature Reviews. Neuroscience, 3*(3), 216–227.

Nuovo, G. J., DeFaria, D. L., Chanona-Vilchi, J. G., & Zhang, Y. (2004). Molecular detection of rabies encephalitis and correlation with cytokine expression. *Modern Pathology*.

Okada, K., Ito, N., Yamaoka, S., Masatani, T., Ebihara, H., Goto, H.,... Sugiyama, M. (2016). Roles of the rabies virus phosphoprotein isoforms in pathogenesis. *Journal of Virology, 90*(18), 8226–8237. https://doi.org/10.1128/JVI.00809-16.

Park, C. H., Kondo, M., Inoue, S., Noguchi, A., Oyamada, T., Yoshikawa, H., & Yamada, A. (2006). The histopathogenesis of paralytic rabies in six-week-old C57BL/6J mice following inoculation of the CVS-11 strain into the right triceps surae muscle. *The Journal of Veterinary Medical Science, 68*(6), 589–595.

Park, C., Lee, S., Cho, I. H., Lee, H. K., Kim, D., Choi, S. Y.,... Lee, S. J. (2006). TLR3-mediated signal induces proinflammatory cytokine and chemokine gene expression in astrocytes: Differential signaling mechanisms of TLR3-induced IP-10 and IL-8 gene expression. *Glia, 53*(3), 248–256. https://doi.org/10.1002/glia.20278.

Peltier, D. C., Simms, A., Farmer, J. R., & Miller, D. J. (2010). Human neuronal cells possess functional cytoplasmic and TLR-mediated innate immune pathways influenced by phosphatidylinositol-3 kinase signaling. *Journal of Immunology, 184*(12), 7010–7021. doi: jimmunol.0904133 [pii] https://doi.org/10.4049/jimmunol.0904133.

Perry, L. L., Hotchkiss, J. D., & Lodmell, D. L. (1990). Murine susceptibility to street rabies virus is unrelated to induction of host lymphoid depletion. *Journal of Immunology, 144*(9), 3552–3557.

Pfefferkorn, C., Kallfass, C., Lienenklaus, S., Spanier, J., Kalinke, U., Rieder, M.,... Staeheli, P. (2016). Abortively infected astrocytes appear to represent the Main source of interferon Beta in the virus-infected brain. *Journal of Virology, 90*(4), 2031–2038. https://doi.org/10.1128/JVI.02979-15.

Pfeiffer, J. K. (2010). Innate host barriers to viral trafficking and population diversity: Lessons learned from poliovirus. *Advances in Virus Research, 77*, 85–118. doi: B978-0-12-385034-8.00004-1 [pii] https://doi.org/10.1016/B978-0-12-385034-8.00004-1.

Phares, T. W., Kean, R. B., Mikheeva, T., & Hooper, D. C. (2006). Regional differences in blood-brain barrier permeability changes and inflammation in the apathogenic clearance of virus from the central NS. *Journal of Immunology, 176*(12), 7666–7675.

Phares, T. W., Stohlman, S. A., Hinton, D. R., Atkinson, R., & Bergmann, C. C. (2010). Enhanced antiviral T cell function in the absence of PD-L1 is insufficient to prevent persistence but exacerbates axonal bystander damage during viral encephalomyelitis. *Journal of Immunology, 185*(9), 5607–5618. doi: jimmunol.1001984 [pii] https://doi.org/10.4049/jimmunol.1001984.

Pichlmair, A., Schulz, O., Tan, C. P., Naslund, T. I., Liljestrom, P., Weber, F., & Reis e Sousa, C. (2006). RIG-I-mediated antiviral responses to single-stranded RNA bearing 5′-phosphates. *Science, 314*(5801), 997–1001.

Prehaud, C., Megret, F., Lafage, M., & Lafon, M. (2005). Virus infection switches TLR-3-positive human neurons to become strong producers of beta interferon. *Journal of Virology, 79*(20), 12893–12904. https://doi.org/10.1128/JVI.79.20.12893-12904.2005.

Pulmanausahakul, R., Faber, M., Morimoto, K., Spitsin, S., Weihe, E., Hooper, D. C., ... Dietzschold, B. (2001). Overexpression of cytochrome C by a recombinant rabies virus attenuates pathogenicity and enhances antiviral immunity. *Journal of Virology, 75*(22), 10800–10807. https://doi.org/10.1128/JVI.75.22.10800-10807.2001.

Randall, R. E., & Goodbourn, S. (2008). Interferons and viruses: an interplay between induction, signalling, antiviral responses and virus countermeasures. *The Journal of General Virology, 89*(Pt 1), 1–47. doi: 89/1/1 [pii] https://doi.org/10.1099/vir.0.83391-0.

Rieder, M., Brzozka, K., Pfaller, C. K., Cox, J. H., Stitz, L., & Conzelmann, K. K. (2011). Genetic dissection of interferon-antagonistic functions of rabies virus phosphoprotein: Inhibition of interferon regulatory factor 3 activation is important for pathogenicity. *Journal of Virology, 85*(2), 842–852. doi: JVI.01427-10 [pii] https://doi.org/10.1128/JVI.01427-10.

Rieder, M., & Conzelmann, K. K. (2009). Rhabdovirus evasion of the interferon system. *Journal of Interferon & Cytokine Research, 29*(9), 499–509.

Rossiter, J. P., Hsu, L., & Jackson, A. C. (2009). Selective vulnerability of dorsal root ganglia neurons in experimental rabies after peripheral inoculation of CVS-11 in adult mice. *Acta Neuropathologica, 118*, 249–259.

Rouas-Freiss, N., Moreau, P., Menier, C., & Carosella, E. D. (2003). HLA-G in cancer: A way to turn off the immune system. *Seminars in Cancer Biology, 13*(5), 325–336.

Roy, A., & Hooper, D. C. (2007). Lethal silver-haired bat rabies virus infection can be prevented by opening the blood-brain barrier. *Journal of Virology, 81*(15), 7993–7998.

Roy, A., Phares, T. W., Koprowski, H., & Hooper, D. C. (2007). Failure to open the blood-brain barrier and deliver immune effectors to central NS tissues leads to the lethal outcome of silver-haired bat rabies virus infection. *Journal of Virology, 81*(3), 1110–1118.

Schutsky, K., Curtis, D., Bongiorno, E. K., Barkhouse, D. A., Kean, R. B., Dietzschold, B., ... Faber, M. (2013). Intramuscular inoculation of mice with the live-attenuated recombinant rabies virus TriGAS results in a transient infection of the draining lymph nodes and a robust, long-lasting protective immune response against rabies. *Journal of Virology, 87*(3), 1834–1841. https://doi.org/10.1128/JVI.02589-12.

Shankar, V., Kao, M., Hamir, A. N., Sheng, H., Koprowski, H., & Dietzschold, B. (1992). Kinetics of virus spread and changes in levels of several cytokine mRNAs in the brain after intranasal infection of rats with Borna disease virus. *Journal of Virology, 66*(2), 992–998.

Sheean, R. K., McKay, F. C., Cretney, E., Bye, C. R., Perera, N. D., Tomas, D., ... Turner, B. J. (2018). Association of Regulatory T-cell expansion with progression of amyotrophic lateral sclerosis: A study of humans and a transgenic mouse model. *JAMA Neurology, 75*(6), 681–689. https://doi.org/10.1001/jamaneurol.2018.0035.

Shimizu, K., Ito, N., Sugiyama, M., & Minamoto, N. (2006). Sensitivity of rabies virus to type I interferon is determined by the phosphoprotein gene. *Microbiology and Immunology, 50*(12), 975–978.

Sommereyns, C., Paul, S., Staeheli, P., & Michiels, T. (2008). IFN-lambda (IFN-lambda) is expressed in a tissue-dependent fashion and primarily acts on epithelial cells in vivo. *PLoS Pathogens, 4*(3). e1000017 https://doi.org/10.1371/journal.ppat.1000017.

Steinman, R. M. (1991). The dendritic cell system and its role in immunogenicity. *Annual Review of Immunology, 9*, 271–296. https://doi.org/10.1146/annurev.iy.09.040191.001415.

Sugiura, N., Uda, A., Inoue, S., Kojima, D., Hamamoto, N., Kaku, Y., ... Yamada, A. (2011). Gene expression analysis of host innate immune responses in the central NS following lethal CVS-11 infection in mice. *Japanese Journal of Infectious Diseases, 64*(6), 463–472.

Takeuchi, O., & Akira, S. (2010). Pattern recognition receptors and inflammation. *Cell, 140*(6), 805–820. doi: S0092-8674 (10)00023-1 [pii] https://doi.org/10.1016/j.cell.2010.01.022.

Tang, S. C., Arumugam, T. V., Xu, X., Cheng, A., Mughal, M. R., Jo, D. G., ... Mattson, M. P. (2007). Pivotal role for neuronal toll-like receptors in ischemic brain injury and functional deficits. *Proceedings of the National Academy of Sciences of the United States of America, 104*(34), 13798–13803. doi: 0702553104 [pii] https://doi.org/10.1073/pnas.0702553104.

Tobiume, M., Sato, Y., Katano, H., Nakajima, N., Tanaka, K., Noguchi, A., ... Sata, T. (2009). Rabies virus dissemination in neural tissues of autopsy cases due to rabies imported into Japan from the Philippines: Immunohistochemistry. *Pathology International, 59*(8), 555–566.

Torres-Anjel, M. J., Volz, D., Torres, M. J., Turk, M., & Tshikuka, J. G. (1988). Failure to thrive, wasting syndrome, and immunodeficiency in rabies: A hypophyseal/hypothalamic/thymic axis effect of rabies virus. *Reviews of Infectious Diseases, 10*(Suppl 4), S710–S725.

Tracey, K. J. (2009). Reflex control of immunity. *Nature Reviews. Immunology, 9*(6), 418–428. https://doi.org/10.1038/nri2566.

Tshikuka, J. G., Torres-Anjel, M. J., Blenden, D. C., & Elliott, S. C. (1992). The microepidemiology of wasting syndrome, a common link to diarrheal disease, cancer, rabies, animal models of AIDS, and HIV-AIDS YHAIDS. The feline leukemia virus and rabies virus models. *Annals of the New York Academy of Sciences, 653*, 274–296.

Ugolini, G. (1995). Specificity of rabies virus as a transneuronal tracer of motor networks: Transfer from hypoglossal motoneurons to connected second-order and higher order central NS cell groups. *The Journal of Comparative Neurology, 356*(3), 457–480.

Ugolini, G. (2010). Advances in viral transneuronal tracing. *Journal of Neuroscience Methods*. Epub ahead of print. doi: S0165-0270(09)00623-2 [pii] https://doi.org/10.1016/j.jneumeth.2009.12.001.

Van Dam, A. M., Bauer, J., Man, A., Hing, W. K., Marquette, C., Tilders, F. J., & Berkenbosch, F. (1995). Appearance of inducible nitric oxide synthase in the rat central NS after rabies virus infection and during experimental allergic encephalomyelitis but not after peripheral administration of endotoxin. *Journal of Neuroscience Research, 40*(2), 251–260.

Versteeg, G. A., & Garcia-Sastre, A. (2010). Viral tricks to grid-lock the type I interferon system. *Current Opinion in Microbiology, 13*(4), 508–516. doi: S1369-5274(10)00070-6 [pii] https://doi.org/10.1016/j.mib.2010.05.009.

Vidy, A., El Bougrini, J., Chelbi-Alix, M. K., & Blondel, D. (2007). The nucleocytoplasmic rabies virus P protein counteracts interferon signaling by inhibiting both nuclear accumulation and DNA binding of STAT1. *Journal of Virology, 81*(8), 4255–4263.

Vuaillat, C., Varrin-Doyer, M., Bernard, A., Sagardoy, I., Cavagna, S., Chounlamountri, I., ... Giraudon, P. (2008). High CRMP2 expression in peripheral T lymphocytes is associated with recruitment to the brain during virus-induced neuroinflammation. *Journal of Neuroimmunology, 193*(1–2), 38–51.

Wang, Z. W., Sarmento, L., Wang, Y., Li, X. Q., Dhingra, V., Tseggai, T., ... Fu, Z. F. (2005). Attenuated rabies virus activates, while pathogenic rabies virus evades, the host innate immune responses in the central NS. *Journal of Virology, 79*(19), 12554–12565.

Weiland, F., Cox, J. H., Meyer, S., Dahme, E., & Reddehase, M. J. (1992). Rabies virus neuritic paralysis: Immunopathogenesis of nonfatal paralytic rabies. *Journal of Virology, 66*(8), 5096–5099.

Wen, Y., Wang, H., Wu, H., Yang, F., Tripp, R. A., Hogan, R. J., & Fu, Z. F. (2011). Rabies virus expressing dendritic cell-activating molecules enhances the innate and adaptive immune response to vaccination. *Journal of Virology, 85*(4), 1634–1644. doi: JVI.01552-10 [pii] https://doi.org/10.1128/JVI.01552-10.

Wiktor, T. J., Doherty, P. C., & Koprowski, H. (1977a). In vitro evidence of cell-mediated immunity after exposure of mice to both live and inactivated rabies virus. *Proceedings of the National Academy of Sciences of the United States of America, 74*(1), 334–338.

Wiktor, T. J., Doherty, P. C., & Koprowski, H. (1977b). Suppression of cell-mediated immunity by street rabies virus. *The Journal of Experimental Medicine, 145*(6), 1617–1622.

Wiktor, T. J., Macfarlan, R. I., Reagan, K. J., Dietzschold, B., Curtis, P. J., Wunner, W. H., ... Mackett, M. (1984). Protection from rabies by a vaccinia virus recombinant containing the rabies virus glycoprotein gene. *Proceedings of the National Academy of Sciences of the United States of America, 81*(22), 7194–7198.

Wunner, W. H., Dietzschold, B., Curtis, P. J., & Wiktor, T. J. (1983). Rabies subunit vaccines. *The Journal of General Virology, 64*(Pt 8), 1649–1656.

Xiang, Z. Q., Knowles, B. B., McCarrick, J. W., & Ertl, H. C. (1995). Immune effector mechanisms required for protection to rabies virus. *Virology, 214*(2), 398–404.

Yamaoka, S., Ito, N., Ohka, S., Kaneda, S., Nakamura, H., Agari, T., ... Sugiyama, M. (2013). Involvement of the rabies virus phosphoprotein gene in neuroinvasiveness. *Journal of Virology*, *87*(22), 12327–12338. https://doi.org/10.1128/JVI.02132-13.

Yamaoka, S., Okada, K., Ito, N., Okadera, K., Mitake, H., Nakagawa, K., & Sugiyama, M. (2017). Defect of rabies virus phosphoprotein in its interferon-antagonist activity negatively affects viral replication in muscle cells. *The Journal of Veterinary Medical Science*, *79*(8), 1394–1397. https://doi.org/10.1292/jvms.17-0054.

Yang, Y., Huang, Y., Gnanadurai, C. W., Cao, S., Liu, X., Cui, M., & Fu, Z. F. (2015). The inability of wild-type rabies virus to activate dendritic cells is dependent on the glycoprotein and correlates with its low level of the de novo-synthesized leader RNA. *Journal of Virology*, *89*(4), 2157–2169. https://doi.org/10.1128/JVI.02092-14.

Zhang, B., Chan, Y. K., Lu, B., Diamond, M. S., & Klein, R. S. (2008). CXCR3 mediates region-specific antiviral T cell trafficking within the central NS during West Nile virus encephalitis. *Journal of Immunology*, *180*(4), 2641–2649 doi: 180/4/2641 [pii].

Zhao, L., Toriumi, H., Kuang, Y., Chen, H., & Fu, Z. F. (2009). The roles of chemokines in rabies virus infection: Overexpression may not always be beneficial. *Journal of Virology*.

Zhao, P., Zhao, L., Zhang, T., Qi, Y., Wang, T., Liu, K., ... Xia, X. (2011). Innate immune response gene expression profiles in central NS of mice infected with rabies virus. *Comparative Immunology, Microbiology and Infectious Diseases*. doi: S0147-9571(11)00077-4 [pii] https://doi.org/10.1016/j.cimid.2011.09.003.

CHAPTER 12

Laboratory diagnosis of rabies

Lorraine M. McElhinney[a], Denise A. Marston[a], Megan Golding[a], Susan A. Nadin-Davis[b]

[a]Animal & Plant Health Agency (Weybridge), Addlestone, Surrey, United Kingdom [b]Canadian Food Inspection Agency, Ottawa Laboratory-Fallowfield, Ottawa, ON, Canada

12.1 Laboratory-based rabies diagnostic testing

A robust diagnostic and surveillance program combined with a national disease notification system is pivotal to achieve the "Zero human Rabies deaths by 2030" target set by the United Against Rabies Collaboration in a tripartite agreement between the World Health Organization (WHO), the World Organization for Animal Health (OIE), and the Food and Agriculture Organization (FAO) in collaboration with Global Alliance for Rabies Control (GARC) (World Health Organization, 2015). Historically, a rabies laboratory may have been simply asked to perform rabies testing on clinical submissions. However, a modern rabies laboratory now performs a range of functions, particularly if serving as regional or national reference laboratories (NRLs). Activities can include both passive and targeted surveillance of animals or humans (postmortem and antemortem), confirmatory or reference functions, production of control material, regulatory testing, training, research, test validation, typing, proficiency testing, and maintenance of a virus repository. To provide confidence in such activities, an increasing number of NRLs are seeking accreditation and certification to international Quality Assurance standards. In addition, laboratories are increasingly asked to comply with stringent biosafety and biosecurity requirements, which vary depending on regional regulations.

Rabies is a notifiable disease in many, but not all countries. The case history may raise the index of suspicion in both animals and humans, e.g., bite, illegal pet travel, contact with infected wildlife, including bats. Under a notifiable system, when presented with a suspect case, clinicians or veterinarians contact the competent authority in their country to initiate an investigation.

Early detection of rabies is based on the observation of clinical signs, which are rarely pathognomonic, and usually appear after the variable incubation period and differ depending upon the species affected. Differential diagnosis can be difficult due to common sequelae resulting from various diseases and conditions, including transmissible spongiform encephalopathies, tetanus, listeriosis, acute disability resulting from trauma, poisoning and other viral nonsuppurative encephalitides (Dimaano, Scholand, Alera, & Belandres, 2011). Paralytic rabies in humans is often mistaken for Guillain-Barré Syndrome (Hemachudha & Mitrabhakdi, 2000; Solomon et al., 2005). Other infections can also mask or mimic rabies leading to clinical misdiagnosis (Mallewa et al., 2007). Thus, reliable and quality-assured laboratory-based diagnostic assays are essential for confirming the presence of rabies virus or other lyssaviruses. There are currently no diagnostic tests available to detect lyssaviruses before the onset of clinical signs. However, there are a number of highly reliable, sensitive, and specific assays that have now been established globally in quality-assured diagnostic laboratories. Such tests are pivotal for ensuring rapid case management with respect to human and animal rabies exposures, including the implementation of lifesaving public and animal health control, including effective pre- and postexposure prophylaxis (Ma et al., 2018). Rapid *positive* diagnosis supports the case for an urgent public health investigation and medical care for all clinical and potentially exposed persons, as well as examination and rabies booster vaccination of currently vaccinated pets and livestock. It also allows clinicians to cease the investigations or interventions initiated to cover other disorders in the differential diagnosis. A *negative* diagnosis can be as important as a positive diagnosis by ensuring other conditions are immediately considered in clinical cases and ruling out the need for rabies control prophylaxis for both human and animal exposures. In the absence of diagnostic testing, public and animal health control, involving significant resources, may need to be applied in every suspect case to minimize the risk of rabies deaths.

Currently, rabies diagnosis can only be confidently confirmed by laboratory tests conducted postmortem on central nervous system (CNS) tissue removed from the cranium. Postmortem diagnostic assays for rabies in animals have been standardized as far as is possible with detailed protocols published in the OIE's manual of diagnostic tests and vaccines for terrestrial animals and WHO's text book Laboratory Techniques in Rabies (OIE, 2018b; World Health Organization, 2018a). Twinning and outreach programs, offered via the networks of OIE Rabies Reference Laboratories and WHO Rabies Collaborating Centers, have also assisted in globally standardizing rabies diagnosis.

12.1.1 Indications for rabies testing

The single most important role of rabies diagnostic testing is the protection of human and domestic animal health. In this regard, the indication for rabies testing is directly related to an exposed human or domestic animal. This is often referred to as passive surveillance in that the submitted animals are selected on the basis of observed clinical signs and their encounters with humans and animals.

In 2017, 4077 confirmed rabies cases were reported in Europe, representing approximately 8% of the 51,421 animals tested for rabies in 30/32 participating European countries (Source: WHO Rabies Bulletin Europe). In the United States in 2017, 93,651 samples were screened for rabies of which 4454 (4.8%) were confirmed as positive for rabies (Ma et al., 2018). The majority of the reported cases in Europe were detected in domestic animals (dogs, cats, and cattle accounted for ~60%) and foxes (29%). Bats (42 cases) accounted for only 1% of the reported rabies cases in Europe in 2017. In contrast, wildlife species accounted for the majority (4055,

91.0%) of the confirmed rabies cases in the United States in 2017 (including 1275 (28.6%) in raccoons, 939 (21.1%) in skunks, and 314 (7.0%) in foxes). Bat rabies (1433) accounted for nearly a third of the cases (32.2%) in the United States and domestic rabies was reported in only 8.4% of cases (276 (6.2%) in cats, 62 (1.4%) in dogs, and 36 (0.8%) in cattle) (Ma et al., 2018). In Russia, of the 12,514 tests undertaken, rabies was confirmed in 2151 (17.2%) submissions within 61 of 85 Federal regions, with wild canid rabies (fox and raccoon dog) accounting for the biggest proportion of cases (Shulpin et al., 2018).

The distribution and epidemiology of rabies cases will thus vary by area and to a certain degree will be reliant upon and reflect sampling bias. Some countries have never initiated wildlife or bat surveillance programs, while in other regions, e.g., those that are free of canine rabies, wildlife/bats, may account for the biggest proportion of submissions (e.g., United States, Canada, Western Europe, and Australia) (Fehlner-Gardiner, 2018; Ma et al., 2018; Prada, Boyd, Baker, Jackson, & O'Dea, 2019; Schatz et al., 2013). Test results on the submitted animals in passive surveillance schemes are thus not representative of population-based testing in the various species. Whereas active surveillance is necessary to gain good epizootiological knowledge on rabies, limited information can be enhanced through careful specimen acceptance policies. An optimal understanding of local and regional transmission patterns will help guide risk assessment for potentially exposed humans and animals, especially in situations where the exposing animal is not available for diagnostic testing or observation (Anonymous, 2011).

In addition to the testing of animals that commonly contribute to disease transmission, rabies surveillance can be enhanced by the acceptance of uncommonly tested animals with neurological signs compatible with rabies, such as deer, bear, beaver, badger, or other potential spill over species. For heightened surveillance, vigilance and typing of viral strains (or lyssavirus species) may identify new and emerging viruses and hosts. This is particularly important in areas where rabies control efforts are ongoing through methods, such as oral vaccination or trap-vaccinate-and-release.

In passive or active bat surveillance programs, the speciation of the individuals is extremely important because the risk of rabies may vary according to bat species, and the viral variants that are found in these animals are diverse. When rabies is confirmed, the diagnosis allows initiation of human postexposure prophylaxis for potentially exposed persons and the management of potentially exposed domestic animals. The difficulty with human and domestic animal encounters with these animals is that bats are comparatively small bodied. A potential exposure, such as a bite, from these animals may be ignored by humans because the actual trauma from a bite may be painless or unremarkable. In some countries, such as the United Kingdom, clinicians offer precautionary post exposure prophylaxis following any bite from a bat. However, in other countries, some humans may not have awoken from slumber when they were bitten by a bat, or may have been unable to report a bite if noticed (e.g., small child) leading to untreated "cryptic" rabies cases detected only by postmortem virus typing.

It is important to understand the risk of rabies following bites from pet animals in different countries and whether euthanasia and testing is demanded for reasons other than a genuine suspicion of disease. Observation or quarantine may be considered more appropriate than euthanasia and testing (Anonymous, 2011).

However, in some developed countries where wildlife rabies is present, if a concern about a potential exposure is voiced to health authorities and the pet is euthanized immediately after the biting incident, it is often tested for rabies in the absence of clinical signs. Thus, some diagnostic laboratories may test numerous healthy but unwanted domestic dogs and cats. In countries free of terrestrial rabies, the travel movements and compliance of the pet are also important when determining risk and need for rabies testing.

12.1.2 Biosafety and shipment of diagnostic specimens

All activities related to the handling of animals and samples for rabies diagnosis should be performed using appropriate biosafety practices to avoid exposure (e.g., bites, licks, sharps injury), as well as potential exposures to infected tissues or fluids. During sample preparation and testing, the most likely risk of exposure is through accidental penetrating injuries with contaminated laboratory equipment or exposure of mucous membranes or broken skin to infectious tissue or fluids. The highest viral concentrations are found in central nervous system tissue, salivary glands, and saliva, but any innervated tissue may be a source of virus exposure. People preparing specimens for submission and laboratory personnel involved in diagnostic activities can protect themselves against exposure to rabies through appropriate Personal Protection Equipment (PPE). The level of protection required by law will differ globally and likely reflects both the biological properties of the virus and the disease status in the country. Local health and safety guidelines must be followed by each laboratory depending on the risk levels assigned in their country, but general guidelines on biosafety and biosecurity risk assessments can be found in the latest edition of the OIE Terrestrial Manual (OIE, 2018b).

In developed rabies-endemic countries, rabies virus is more likely to be handled as a Biosafety Level 2 (BSL2) pathogen. However, in other countries, and in particular rabies-free countries, the high mortality rate and fear of release from the laboratory into rabies susceptible species results in BSL3 or higher standards being placed on rabies laboratories. Some countries may also control the handling of rabies virus and infected material under antiterrorism and national security regulations. In the United Kingdom, the highest biosafety level for animal pathogens is assigned to rabies virus and specific licenses are required to handle live virus within highly expensive and regulated high containment laboratories.

The following section outlines some minimal and preferable measures to reduce or mitigate exposure to virus, where circumstances and resources allow. The initial capture or confinement of a rabies-suspect animal can be quite problematic. Rabid animals can be unpredictable, sometimes alternating from an apparently weakened neurologic state to one of dangerous aggression. Almost all jurisdiction of animal control is local; this can present substantial difficulties for the potentially exposed person or owner of an exposed domestic animal, particularly if the rabies-suspect animal is a wildlife species. Some local animal control officials may assist with situations involving only stray or rabies-suspect dogs, while some may assist where the problematic animal is any domestic species. It is rare to find personnel with experience of any animal control situation whether it involves a domestic species, or bat, skunk, raccoon, fox, or other wildlife species. In the United States, individual members of the public often resort to independently confining or capturing a potentially rabid animal and then making arrangements for its decapitation and submission to a laboratory.

Depending on the size of the suspect animal and capacity of the testing laboratory, the whole carcass, the head of the animal, or sampled brain material may be submitted. Local regulations and disease status will dictate the procedures that officials are likely to undertake. In many rabies laboratories, handling and submitting small animal carcasses is manageable and preferable to ensure optimal brain sampling. However, manipulating large animal carcasses in containment laboratories with regulated waste disposal can be logistically problematic and expensive. Hence for high numbers of submissions and large carcasses, particularly in rabies endemic regions, removing brain material from suspect animals under field conditions, may be the only practicable option. To ensure the safety of the procedure and the quality of the material, this must be undertaken by highly trained staff. Where possible, the PPE procedures outlined for laboratory necropsies should be followed.

The diagnostic specimen must be packaged appropriately to preclude leakage of infectious material or deterioration of the sample due to extreme heat or freeze-thaw cycles. The specimen should be placed within an inner unbreakable leak-proof container and then a secondary leak-proof container (e.g., sealed plastic bag, screw cap plastic container). The double-enclosed specimen is then placed in a leak-proof outer shipping packaging with suitable cushioning material, which is often a sturdy insulated fiber board or polystyrene box (Styrofoam cooler) with an outer cardboard pack. The OIE Terrestrial Manual 2018 describes the necessary guidelines to be followed, e.g., International Air Transport Association (IATA) conditions for transporting infectious substances by air (OIE, 2018b). Adequate cooling material, such as frozen cold packs, should be placed around the double-enclosed specimen. Dry ice (solid carbon dioxide) must never be placed within the primary or secondary containers to avoid the risk of explosion. Absorbent material is mandatory to absorb condensation from cold packs and to contain any possible leakage from the sample. The outside of the container is often labeled with an International Biohazard sticker and "Exempt Animal Specimen" designation or with a UN3373 Biological Substance, Category B sticker (Category A sticker is required if cultured rabies virus is transported). Each laboratory that performs rabies diagnosis typically offers a rabies examination submission form either in hard copy or online. If multiple samples are being submitted at the same time, it is often necessary to use a separate form for each specimen. Submission forms should be packed separately from the samples and are typically placed in a sealed plastic bag to ensure that they will not become contaminated during transit or opening of the specimen. Express shipping is often the recommended method for delivery of a sample to testing laboratories.

During necropsy, the use of heavy rubber gloves, laboratory gown and waterproof apron, boots, surgical masks, protective sleeves, and a face shield is highly recommended. Fume hoods or microbiological safety cabinets (MSCs) are useful in providing protection from odor, other pathogens and bone fragments, but they are not mandatory in all countries. Skin, respiratory, and eye protection is necessary for safe removal of brain tissue from animals submitted for testing. An oscillating saw rather than hammer and chisel or hand saw is recommended to avoid operator injury and the dispersal of bone fragments and aerosolized brain tissue. While MSCs are not applicable for field settings, operators should secure the area within which necropsies take place to protect staff from interruption, e.g., while handling sharps and to minimize the risk to others from exposure to virus.

During laboratory activities, PPE often consists of disposable examination gloves, laboratory coats, sharps precautions, eye protection, where possible the use of MSCs. During the diagnostic process, all sample materials, positive control material and slides should be manipulated with attention to minimize aerosols or airborne droplet spray. While aerosol transmission of rabies is extremely rare, it may be possible via exposed mucosal membranes (Johnson, Phillpotts, & Fooks, 2006). The manipulation and staining of slides and cleanup of the microscope and associated areas should be done with care to preclude potential exposures to rabies virus through glass chips and shards from slides. The use of glass, where possible, should be avoided. Disinfectants should be locally validated before they are employed to decontaminate surface areas and equipment. All equipment should be decontaminated before removal from laboratories for servicing or maintenance.

The World Health Organization (World Health Organization, 2018b) highly recommends preexposure immunization against rabies for personnel routinely involved in these activities. Preexposure vaccination provides priming immunization against rabies so that if an exposure occurs, postexposure management is relatively simple: just day 0 and day 3 vaccine administrations, without the need for rabies immunoglobulin. Moreover, preexposure vaccination may also provide protection against an unrecognized or "cryptic" exposure to rabies. Access to rabies diagnostic areas should be restricted to immunized personnel with the area secured and labeled accordingly. Some rabies laboratories, depending on local health and safety guidelines, may demand minimum levels of antibody titers for individual staff monitored by occupational health screens (serological testing) (Manning et al., 2008), e.g., the WHO 0.5 IU/mL level suggesting successful seroconversion following vaccination.

All sample materials processed in the field or laboratory settings are considered medical waste and must be disposed of in accordance with local guidelines.

12.1.3 Quality assurance

Valid test results underpin all rabies diagnosis, surveillance, and trade. To offer assurance and confidence in reported test results, an increasing number of rabies laboratories are accredited to the international standard for testing laboratories (ISO/IEC 17025:2005 or 2017) or working toward that standard. The OIE Manual 2018 has dedicated a full chapter to quality assurance in a veterinary testing laboratory (OIE, 2018a). Ultimately, to ensure mutual recognition of test validity, competent staff must be seen to adhere to validated procedures using fit-for-purpose reagents, facilities, and equipment within a good sample management system. The test method applied must be validated and appropriate for the sample submitted and the customer requirements.

Quality-assured laboratories must invest in a quality management system. This incorporates a robust Laboratory Information Management System (LIMS) for recording sample submission data, test data, and reporting information and covers the whole system from receipt to report. In addition to LIMS, the laboratory should have a register for where samples are stored, calibration and service history of all equipment used, training and competence records of staff, source and validity of reagents, written protocols, and a plan for auditing all aspects of the system. In addition to the quality management system, all tests must be validated locally to ensure that any deviations in equipment or reagents do not unduly affect the sensitivity and specificity of the test. Specificity checks are especially important, as most commercially available reagents are predominantly tested against only a limited number of rabies virus (RABV) strains. Laboratories must ensure that their reagents can detect locally circulating RABV or other lyssaviruses. When various clinical specimens are tested, e.g., in antemortem human diagnosis, the laboratory must validate the processes and the tests employed with the different sample matrices. Reagents and control material must be tested as fit for purpose and the use of control cards is encouraged. Control cards for reference material can be set up easily and reviewed regularly (e.g., quarterly) to ensure that the controls and tests are performing with high levels of consistency. Batch-to-batch testing is essential to confirm the quality of any new batches of reagents (commercial or in house) using local methods and local strains. Rabies serology tests used for trade (pet movements) require the use of control cards (OIE reference serum, virus titration) and most quality-assured molecular tests will regularly require the review of test uncertainty data to ensure that controls have not deteriorated during storage (see real-time RT-PCR).

The establishment of method performance specifications should also provide evidence that the accuracy, precision, analytical sensitivity, and analytical specificity of the procedure and the associated reagents are adequate to meet the regulatory requirements in some countries for human testing and patient clinical decision processes (e.g., animal testing that dictates patient care).

The tests described in detail by the OIE are examples of assays that have proven to be robust and highly reproducible in many laboratories. Prior to establishing them in a rabies laboratory for the first time, sensitivity and specificity must be established using local staff, equipment, viruses, and reagents. The rabies tests included in the OIE Manual (OIE, 2018b) and are not prescribed, however, and validated variations (e.g., different primers, alternative detection systems) can be employed. Laboratories that are accredited to ISO17025 are required to provide evidence of test validation and fitness for purpose. Any modifications to test procedures must be justified by robust validation data and parallel testing to retain accreditation. However, while all EU NRLs hold ISO17025 accreditation for rabies testing, not all NRLs do and modifications (intentional or accidental) may go unnoticed. Introducing new reagents without comprehensive validation may also unduly affect the sensitivity or specificity of the tests.

An annual rabies diagnostic proficiency program has been available for voluntary participation in the United States since 1994. Among the diagnostic laboratories enrolled, performance has been acceptable on strongly positive and negative slides; discrepancies occur with very weakly positive specimens (Powell, 1997). However, there is no mandatory remedial training or evaluation of laboratories that may perform poorly on these challenging samples. If the procedure is "modified by the laboratory," meaning any change to the assay that could affect its performance specifications for sensitivity, specificity, accuracy, or precision, and so on, these may be done but at present are not regulated nor subject to oversight in nonaccredited laboratories. Laboratory modification that could affect performance specifications include but are not limited to: (1) changes in specimen handling instructions; (2) incubation times or temperatures; (3) changes in specimen or reagent dilution (each laboratory determines this internally); (4) using different positive and negative control material (each laboratory determines this internally); (5) using a different antibody reagent (i.e., monoclonal versus polyclonal, in house versus commercial) (each laboratory determines this internally); (6) changes in or elimination of a procedural step (e.g., single reader vs recommended two readers—each laboratory determines this internally); (7) changes in equipment (lack of calibration or servicing); (8) changes in personnel (without appropriate training and maintenance of competence checks). As described earlier, introducing control cards to monitor reagents will help identify issues relating to test and operator performance.

Since 2009, the European Union Reference Laboratory has organized an international interlaboratory trial program for laboratories wishing to evaluate their proficiency in rabies testing (Robardet, Picard-Meyer, Andrieu, Servat, & Cliquet, 2011). These trials now occur on a biennial basis and consist of panels that evaluate a laboratory's proficiency in DFA test, RTCIT, RT-PCR, and real-time RT-PCR. It should be noted however that in addition to classical rabies virus the panels include lyssaviruses such as the Australian bat lyssavirus (ABLV), and European lyssaviruses such as EBLV-1, EBLV-2, and BBLV. The proficiency results demonstrate a growing level of consistency and standardization in the participating laboratories with relatively few labs reporting discrepancies. Only 2 of the 45 participating laboratories recorded an overall discordant result in the 2019 interlaboratory trial (Source: EURL ANSES Inter-Laboratory Test for Rabies Diagnosis—report 10-2019). No discordant results have been observed for conventional RT-PCR for four consecutive trials and discordant results for real-time RT-PCR tend to relate to restricted specificity (not able to detect all lyssaviruses). Although perhaps not optimized for a worldwide clientele, the laboratories participating can state the lyssaviruses that can and cannot be detected by the methods applied, so the results are interpreted according to this information.

An interlaboratory proficiency test for Latin America and the Caribbean rabies laboratories demonstrated a significant lack of standardization in the tests employed between the laboratories with 91% of labs reporting at least one discordant result (Clavijo et al., 2017). Answers from the associated questionnaire pointed to widespread technical differences and poor quality assurance systems. Such proficiency schemes are powerful tools to generate action plans aimed at improving confidence levels in regional and national rabies diagnostic reporting and resilience to new and emerging threats. As new lyssaviruses continue to be discovered and with significant increases in human and pet travel, it is important for reference laboratories to validate and establish pan-lyssavirus tools.

12.1.4 Optimal sampling for rabies diagnosis

For all approaches to rabies diagnosis, optimal sampling is key to a successful detection of the pathogen. The brainstem (e.g., medulla) and the cerebellum are recommended due to the localized accumulation of virus during disease. While in most countries the carcass (small mammals) or head of the suspect animal is submitted to the laboratory for diagnosis, this is not always practical, e.g., for large species, high volumes or where cold chain/limited resources prevent it. In such cases, removal of brain material in the field is sometimes necessary (Jarvis, Brown, Appler, Fitzgerald, & Davis, 2019). However, removing specific regions of the brain safely and accurately under field conditions can often be challenging. Although viral antigen is widespread throughout the brain in most rabid animals, in some specimens the distribution is limited to the brainstem region alone (Bingham & van der Merwe, 2002). These observations have led to the recommendation that a complete cross section of the brainstem at the level of the medulla, pons, or midbrain must be examined for a definitive diagnosis.

TABLE 12.1 Limits of detection for rabies virus, antigen, and RNA in decomposed carcasses (McElhinney et al., 2014).

Temperature	Rabies virus (RTCIT)	Rabies antigen (DFA test)	Rabies RNA (hnRT-PCR)
4°C	18 days	36 days	70 days
25°C	3 days	12 days	48 days
35°C	3 days	3 days	48 days

Ideally, carcasses or brain material are submitted within 24 h of death and kept refrigerated during transportation and prior to testing. However, in some countries, cold chain transportation or storage is unreliable or unavailable. Animal carcasses may also be found and submitted in varying states of decomposition. While positive results from decomposed submissions may be reliable, any negative results would be invalid as test sensitivity may be compromised. A study to determine the limits of detection of rabies virus, antigen, and RNA was conducted using infected mouse carcasses left for varying times at three temperatures (McElhinney, Marston, Brookes, & Fooks, 2014). Infectious virus could be detected for up to 3 days at 35°C and up to 18 days at cold temperatures. As expected, viral antigen was more stable than virus while viral RNA could be detected even in putrefied samples (Table 12.1). A faster decomposition was suggested for tissues removed from carcasses and stored/transported suboptimally compared to brain material removed from the decomposed carcass just prior to testing. Cold chain transportation and storage is thus essential as is the choice of test employed on decomposed submissions.

In the absence of cold chains, other means of preserving diagnostic samples have been explored. Preservation of brain tissue homogenates dried on filter paper (Wacharapluesadee, Phumesin, Lumlertdaecha, & Hemachudha, 2003), or Whatman Flinders Technology Associates (FTA) cards (Nadin-Davis, Sheen, & Wandeler, 2012; Picard-Meyer, Barrat, & Cliquet, 2007), has been shown to be of diagnostic value using molecular screening for rabies diagnosis and full rabies genome sequence data have even been recovered directly from an FTA card and used for epidemiological studies (Goharriz et al., 2017). It is essential that material is fully dried on the FTA cards and that they are not overloaded to ensure complete virus inactivation and allow for safe transport at room temperature in routine postal service.

Nonneural specimens from dogs, including oral swabs and whisker and hair follicles, which are less prone to decomposition than brain material, were screened for rabies viral RNA with some success (Wacharapluesadee et al., 2012). Active bat surveillance campaigns also often include screening of oral swabs by virus isolation or molecular tests (Schatz et al., 2013). However, the use of such sample types, where virus excretion is transient, reduces the test sensitivity compared to that obtained with brain tissue, and thus it appears that substitution of brain with these other sample types is not currently recommended for provision of accurate diagnosis.

12.1.5 Laboratory reporting practices

In countries with developed systems of reporting, the suspicion of rabies is notifiable and rabies diagnostic laboratories will report both submitted and confirmed suspect cases of rabies (humans and animals) to regional or national animal and human health authorities. Some reports are issued automatically via LIMS to the local authorities and some will be reported by phone or email. Some laboratories will also report results to the submitter. However, in some countries, such as the United Kingdom, "suspect" rabies diagnostic testing can only be undertaken by the national reference laboratory (APHA) at the specific request of the animal or health authorities and direct submissions from members of the public, clinicians, or veterinarians are not permitted. Results from UK suspect cases are thus only reported directly to the relevant authority. This restriction reflects the notifiable nature of both the suspicion and confirmation of rabies in the United Kingdom. Other countries have different local regulations regarding submissions, which may reflect regional networks under the authorization of a central or national reference laboratory. Rabies laboratories usually collate monthly or quarterly test figures for National Animal and Human Health departments who in turn can submit national summary reports to international bodies such as OIE and WHO. Suspect and surveillance test data from European laboratories are also collated by the EURL (ANSES, France), the European Food Safety Authority (EFSA), and European Centre for Disease Prevention and Control (ECDC).

Although the denominator number of each species tested is critically important (there cannot be any cases if no animals have been tested), these numbers are often more difficult to find and may not be readily available to the public. Some data are only available following publication of a review in a journal and while welcome to see some data in the public domain, it is often difficult to know if the review represents official locally reported figures or just those the authors are aware of. In some circumstances, only the confirmed cases are reported as individual positives and the lack of data on the negatives means that prevalence/incidence data are lost. This is particularly true, when the testing is performed as part of a surveillance program and not in response to an individual with clinical signs of the disease. In Europe, all test data are requested (negatives and positives) for all types of submissions (passive, active, suspect, death in quarantine, etc.) and all species (although unfortunately bats are commonly grouped as if one species). Additionally, spillover into dead-end hosts such as horses and cattle are often reported with other domestic animals but may reflect rabies transmission from wildlife rather than dogs. Rabies prevention efforts would be much enhanced by better global compilation of testing data, both negative and positive cases, preferably by a case occurrence location as precise as longitude and latitude. With this information, one would be able to assess the intensity and appropriateness of surveillance among the local rabies reservoir species, and, in comparison, the spillover into nonreservoir species. Identification of the strain or lyssavirus species involved is also of importance, particularly in areas that were previously free of disease, to understand the source and nature of the confirmed case. Such typing can detect vaccine induced cases, spillover cases or emerging viruses. It is hoped that the use of LIMS in a growing number of quality-assured labs will ease the future burden of holistic reporting to and from regulatory authorities.

12.2 History of rabies diagnostic tests

Clinical descriptions of human and animal rabies date back four millennia, even to a 23rd-century BC document from Babylon (King, Fooks, Aubert, & Wandeler, 2004). The first described laboratory-based diagnostic method came about through the microscopic identification of Negri bodies in brain tissue in 1903 by an Italian pathologist and microbiologist after which they are named (Negri, 1903). Negri's observation led to a histological staining approach (Seller's stain), performed on wet slide impressions of fresh unfixed samples of hippocampus and sometimes other parts of the brain (Sellers, 1923; Young & Sellers, 1927). However, Negri bodies are not always present in an infected animal, but, at the time, their presence provided relatively good specificity in comparison to a presumptive diagnosis based solely on clinical signs. Staining for Negri bodies is rarely undertaken routinely in modern rabies laboratories.

Another relatively historical diagnostic method used for several decades was the practice of using mice to attempt virus isolation in submitted material (Koprowski, 1966, 1973). While still in use in some rabies laboratories, for ethical and practical reasons, the cheaper and faster in vitro isolation of rabies virus and other lyssaviruses in mammalian cell lines largely replaced the in vivo tests in rabies diagnostic reference laboratories. However, it was the establishment of fluorescence microscopy and availability of fluorescein-labeled antibodies that ushered in the direct Fluorescent Antibody Test (FAT, DFA test, or dFA), which is still considered the "gold standard" rabies diagnostic technique and is used globally as a primary rabies diagnostic assay due to its fast processing time, high specificity, and relatively low cost (Goldwasser & Kissling, 1958). Nucleic-acid-based detection of rabies virus was first introduced to rabies diagnostic laboratories in 1991 (Sacramento, Bourhy, & Tordo, 1991; Smith, Fishbein, Rupprecht, & Clark, 1991), and this is now routinely employed in quality-assured laboratory settings. Validated assays, including conventional (gel-based) Reverse Transcriptase Polymerase Chain Reaction (RT-PCR) and Real-Time RT-PCR, are now accepted by the OIE for primary rabies diagnostics and are fast replacing conventional rabies techniques such as the virus isolation assays, particularly for high-throughput screening and surveillance (Marston et al., 2019; OIE, 2018b). More recently, techniques such as the direct rapid immunohistochemical test (DRIT) and rabies lateral flow assays (LFA) have been evaluated to enable rapid rabies diagnosis in remote or field settings where access to fluorescent microscopes may not be feasible (Patrick et al., 2019; Servat, Robardet, & Cliquet, 2019).

In addition to detection, genetic sequencing or typing of the viral strains or lyssavirus species involved in confirmed cases of rabies is becoming part of the routine diagnostic service offered by many regional or national reference laboratories.

12.3 Detection of viral antigen

12.3.1 The direct fluorescent antibody test

The direct fluorescent antibody test (DFA test, dFAT or FAT) is the most widely used postmortem test for rabies diagnosis and is recommended by both WHO and OIE (Goldwasser & Kissling, 1958; World Health Organization, 2018a). It is also used to confirm the presence of rabies virus antigen in cell culture (RTCIT) or in brain tissue of mice that have been inoculated for diagnosis (MIT). Touch impression smears are prepared from the brainstem region and cerebellum and are fixed in high-grade cold acetone and then stained with a specific antibody or antibodies conjugated to fluorescein isothiocyanate (FITC). In the DFA test, the aggregates of lyssavirus nucleocapsid protein are identified by their specific apple-green fluorescence using fluorescent microscopy. The DFA test is highly sensitive and specific and under ideal conditions, a definitive result can be available within a few hours of sample receipt. However, the sensitivity of the DFA test is highly dependent on the quality of the brain material submitted, the specificity of the conjugated antibodies, the reliability of the microscope, and the skills of the diagnostic staff. The reliability of DFA test can be significantly affected by autolysis, lyssavirus species and may be lowered in samples from vaccinated animals.

Imprecise necropsy can adversely affect the DFA test. Reduced sensitivity is probable if regions other than the brainstem (midbrain, pons, and medulla) and cerebellum are tested. Autolysis due to decomposition may result in a disruption in the antibody-binding sites preventing detection (false negatives) or the presence of autofluorescence making it difficult to differentiate specific from nonspecific fluorescence (false positives). A lack of standardization with respect to processing brain material and the quality of the specimen tested can lead to false-negative results (Hanlon, Smith, & Anderson, 1999; McElhinney et al., 2014; Rudd, Smith, Yager, Orciari, & Trimarchi, 2005).

Indeterminate results may occur when employing the DFA test and confirmatory testing using alternative methods, such as nucleic acid detection, is highly recommended (Appler, Brunt, Jarvis, & Davis, 2019).

12.3.1.1 *Fluorescent conjugate selection, preparation, and evaluation*

The FITC-conjugated antibodies employed in a laboratory for the DFA test may be developed in house or commercially supplied. Some laboratories use a cocktail of monoclonal antibodies raised in mice against prominent antigens (e.g., nucleoprotein or glycoprotein), while others use a polyclonal antirabies antiserum raised in suitable species such as rabbits or goats against whole inactivated virus. Most commercially available reagents are directed against rabies virus and have varying success against other lyssaviruses. Hence, any FITC reagent must be fully validated using locally circulating lyssaviruses and for the relevant diagnostic specimens to ensure optimum sensitivity and specificity.

All new batches of rabies reagents must be tested as fit for purpose prior to use. For rabies FITC-labeled conjugates used in the DFA test, the reagent must be titrated to determine a working dilution and consistency with previous batches. Although the inclusion of a counterstain is optional, the use of Evans Blue, specifically formulated for immunofluorescent assays, provides the advantage of contrast coloring of tissue on a slide (contrasts with the apple green of the positive immunofluorescent result), quenching of background fluorescence, and also visual confirmation that the test reagent was actually added to a slide. The amount of counterstain added to a test reagent is determined by titration when the working dilution of the conjugate is determined.

In general, the working dilution for each conjugate is determined through initial evaluation of serial twofold dilutions, for example, at 1:10, 1:20, and 1:40 or more. These should be prepared exactly as the reagent will be used for diagnostic samples, for example, with the same diluent, tubes, syringe filters, and counterstain concentration. Two or more diagnosticians should read and record results independently. The results should then be compared to arrive at a consensus for the optimal dilution providing crisp +4 staining, meaning a glaring, apple-green brilliance, with minimal background fluorescence. The working dilution can be more precisely determined by additional examination of the performance of limited dilutions around the end point. Because antigen presentation and antibody avidity and affinity vary with different virus samples, and viral inclusions appear quite differently with different reagents, the fitness for purpose check must include testing the new batch against the range of locally circulating viruses (including lyssaviruses if relevant). Control material should be stored in aliquots to avoid repeated freeze-thaw cycles and maintained at −40°C or below.

Stock solutions of testing reagents may be stored as frozen aliquots at a minimum of −20°C or preferably lower. Reagents at the working dilution can be filtered prior to use and may be stored at +4°C but discarded if not used within 7 days. With disposable filter units that attach directly to a syringe, the working dilution of the conjugate can be filtered as it is added to test slides. The diluted test reagents ready for dispensing onto test slides can be stored at +4°C in the syringe with the attached filter unit, as long as reagent is prevented from drying on the filter or tip of the dispensing syringe through sealing with a syringe tip or plastic wrap.

The two sets of slides prepared from each specimen should be evaluated with FITC-conjugated antirabies virus antibodies (e.g., two different monoclonal antibody pools or hyperimmune serum conjugate). While hyperimmune serum conjugates consist of the greatest diversity of antirabies virus antibodies and hence the broadest potential for reaction with diverse rabies virus and related viruses, these preparations have a higher innate risk of nonspecific reactivity. This risk can be managed through the use of a specificity control reagent. The specificity control for a hyperimmune rabies reagent is an FITC-labeled serum reagent produced in the same animal host as the rabies reagent but directed to an agent other than rabies virus. In contrast, monoclonal antibody reagents rarely manifest nonspecific reactivity. However, they present the risk of not reacting with variants that lack the specific epitope(s) to which the monoclonal antibodies are directed. Thus, it is recommended to use reagents prepared from two different pools of monoclonal antibodies to minimize the risk of nonrecognition of any one variant. Specificity controls for monoclonal reagents consist of FITC-labeled mouse monoclonal antibodies that are of the same isotype and protein concentration as the rabies reagents but directed to an agent other than rabies virus.

12.3.1.2 *Immunofluorescence test protocol*

After acetone fixation, the test slides and control slides are air dried at room temperature. As recommended, two different antirabies virus conjugates may be added, one to each set of slides, by dispensing through a syringe fitted with a 0.45-µm low protein binding filter. The slides are incubated for 30 min at 37°C in a high humidity chamber. After staining, the slides are briefly rinsed under a stream of PBS, and then soaked in PBS for 3–5 min with attention that control slides and slides from each test animal are in separate rinse containers to avoid crosscontamination. The slides are then soaked again in new PBS for a second 3–5 min interval. Slides are carefully blotted to remove excess liquid (no rinsing is necessary), then briefly air dried. A small amount of 20% glycerol-Tris buffered saline pH 9.0 is placed onto coverslips arranged on absorbent paper. The air-dried slides are inverted and placed on the coverslips with mounting media, thus allowing excess mounting media to be wicked into the absorbent paper when light pressure is applied to the back of the slides. Slides should be read by fluorescence microscopy within 2 h of cover slipping.

The minimum rabies diagnostic procedure recommends the examination of 40 fields at a magnification of 200× or more for each conjugate and brain area tested by 2 laboratory diagnosticians. Thus, for basic diagnosis, 40 fields × 2 conjugates × 2 brain areas sampled × 2

FIG. 12.1   Depicting specific apple green staining of EBLV-1 infected brain cells.

observers equals 320 observations for definitive diagnosis. If fluorescence is observed, fluorescing inclusions should be examined at 400 × magnification for resolution of very fine dust-like inclusions and recognition of some types of nonspecific staining (Fig. 12.1).

12.3.2 Direct rapid immunohistochemical test

The Direct Rapid Immunohistochemical Test (DRIT) is an OIE-approved primary diagnostic assay (OIE, 2018b) with a similar sensitivity and specificity as the DFA test on suspect animal brains from a wide variety of animal species in Africa, North America, Europe, and Asia (Dürr et al., 2008; Lembo et al., 2006; Madhusudana, Subha, Thankappan, & Ashwin, 2012; Middel, Fehlner-Gardiner, Pulham, & Buchanan, 2017). The DRIT can be used in a laboratory or field-based setting and since 2005, the United States Department of Agriculture's Wildlife Services (USDA WS) program has used the DRIT to test >94,000 samples collected from wildlife in strategic rabies management areas. The DRIT can be completed in under 1 h and is fast becoming a cost-effective alternative to DFA testing, particularly in field-based surveillance (Patrick et al., 2019; Rupprecht et al., 2018). Touch impressions of brain tissue are collected from suspect animals and fixed in 10% buffered formalin. Rather than FITC-labeled antibodies, the DRIT impressions are stained using biotinylated rabies virus-specific monoclonal or polyclonal antibodies, then incubated with a streptavidin-peroxidase enzyme and a chromogen reporter (such as acetyl 3-amino-9-ethylcarbazole) to detect viral nucleoprotein inclusions within infected tissue.

A significant advantage over the DFA test is the use of a light microscope, rather than an expensive fluorescence microscope (Dürr et al., 2008; Ehimiyein et al., 2014). Results are easier to interpret than the apple green fluorescence of the DFA test, although the use of two operators where possible is highly recommended. While reagents must be refrigerated prior to use, the incubation during the assay is undertaken at room temperature enabling field-based use. The biotinylated antibodies are available from the South African OIE rabies reference laboratory (Onderstepoort).

12.3.3 Lateral flow assays

Promising results were initially obtained for the use of a lateral flow assay (LFA) to detect rabies virus in brain, saliva, and cell culture with a sensitivity of 91.7% and specificity of 100% (95.8% CI) compared to the DFA test (Kang et al., 2007). A number of LFAs were subsequently developed and six commercially available LFAs were assessed for analytical sensitivity and specificity (Eggerbauer et al., 2016). Using field samples, the sensitivities ranged from 0% up to 100% and none of the tests investigated proved to be satisfactory. In a more recent interlaboratory trial, two commercially available LFAs were assessed and while 100% sensitivity was reached for the RABV strains, the devices failed to adequately detect bat lyssaviruses (Servat et al., 2019). Validation is thus warranted using locally circulating viruses. The assays are thus yet to be approved by the OIE and caution should be exercised when employing the devices for rabies diagnosis.

There is a strong desire for immediately available diagnostic (disease detection) and serologic (vaccine response verification) testing, typically at a veterinary practice, humane association, animal control, or wildlife agency, but also for use in humans (Jayakumar & Padmanaban, 1994; Madhusudana, Paul, Abhilash, & Suja, 2004; Perrin, Gontier, Lecocq, & Bourhy, 1992; Vasanth, Madhusudana, Abhilash, Suga, & Muhamuda, 2004; Zanluca et al., 2011), and there are prototype tests being developed toward this end (Kasempimolporn, Saengseesom, Huadsakul, Boonchang, & Sitprija, 2011; Servat et al., 2012). The ultimate problem is that rabies demands exquisite sensitivity so as not to miss a public health threat to humans or their domestic animals. At a minimum, these types of tests would be useful as an initial screening tool. Several strategies could be considered. One approach would be to develop the test method so that it is as specific as possible, meaning that the target yields strongly positive results that would need no further confirmation. All other results would need definitive testing by traditional diagnostic methods. The other approach would be to screen with the highest possible sensitivity, knowing that some true negatives would be identified as potentially positive. Samples with positive results by screening could be accepted as a presumptive positive or be sent for confirmatory or definitive testing. The technology exists for these tests, but the process of refinement, regulatory approval, and then acceptance by the end users will be more complex than with most animal diseases, due to public health implications, as well as the emotion and concern that rabies elicits from many individuals.

12.3.4 Immunohistochemistry on formalin-fixed paraffin embedded tissues

If brain material from a potentially rabid animal is fixed in formalin, significant diagnostic delay will occur as the tissue requires time to be completely chemically fixed. Samples of the tissue need to be trimmed and embedded in paraffin blocks so that sections may be cut with a microtome and placed on microscope slides. These slides may then be subjected to routine histopathology staining, but a specific immunohistochemical procedure will be needed to detect lyssavirus antigen. A research technique based on this approach requires significant expertise (Fekadu, Greer, Chandler, & Sanderlin, 1988; Hamir, Moser, Fu, Dietzschold, & Rupprecht, 1995) and can be undertaken in only a few capable laboratories. One method based upon detection of an avidin biotin complex (Balachandran & Charlton, 1994) has been used on several occasions to diagnose animal submissions to the National Reference

Laboratory for Rabies in Ottawa, Canada. The process may take up to a week from the time that tissue in formalin is submitted. Typically, the situations in which this method for diagnosis becomes necessary are complex and due to the time required to release results histopathological testing is often more useful for research or archival studies than for routine diagnosis.

12.3.5 Enzyme-linked immunosorbent assay

A monoclonal antibody (MAb)-based capture enzyme-linked immunosorbent assay (ELISA) is also available for the diagnosis of rabies in suspect specimens (Xu et al., 2007). The ELISA (named WELYSSA) uses a cocktail of four mouse monoclonal antibodies directed against the rabies virus nucleocapsid and was validated using representatives of the lyssavirus genus and from various geographic origins and phylogenetic lineages. The assay has been shown to be highly sensitive and specific and may prove useful for large epidemiological surveys, but local validation is essential to ensure optimal sensitivity and specificity. The assay is only approved by the OIE in limited circumstances.

12.4 Molecular methods of viral RNA detection

12.4.1 Advantages and disadvantages of molecular methods

In most cases, molecular methods detect viral RNA using an amplification strategy referred to as the reverse transcriptase polymerase chain reaction (RT-PCR), in contrast to the traditional approach of detecting viral protein in the DFA test. Due to the difference in the nature of the target molecule in the two types of assays, differences in their performance need to be considered.

Variation at the level of the nucleic acid genome is significantly greater than at the protein level due to genetic code redundancy (Bourhy, Kissi, & Tordo, 1993; Kuzmin, Hughes, Botvinkin, Orciari, & Rupprecht, 2005). Consequently, failure to detect a virus present in a sample (false-negative result) can be potentially more frequent using molecular methods, which usually rely on the hybridization of relatively short segments of nucleic acid (oligonucleotide) to a minimum of two separate locations in the genomic target, compared to antibody-antigen binding strategies that form the basis of the DFA detection method. Since the epitope(s) detected by an antibody are more likely to be conserved than a particular nucleotide sequence, serological methods have an advantage when one requires a broadly reactive test capable of detecting a wide range of lyssaviruses. Moreover, molecular-based methods are generally more time consuming and costly than the DFA test. However, there are conserved regions within the genomic sequence of lyssaviruses; therefore, pan-lyssavirus molecular assays are not only possible, but multiple examples exist. In their favor, molecular-based methods based on an amplification process are highly sensitive. However, this can exacerbate the potential for false-positive results due to sample crosscontamination either during tissue processing or the amplification process itself if rigorous attention to the operational requirements for such assays is not considered (Kwok & Higuchi, 1989). Consequently, routine rabies diagnosis on fresh brain tissue, collected in the terminal stages of disease when levels of viral antigen in the brain are high, still continues to be performed by the DFA test, which remains the gold standard postmortem diagnostic test for rabies due to its rapidity and cost effectiveness (OIE, 2018b; World Health Organization, 2018a). However, there are situations where the DFA test performance is less than optimal and where molecular methods can, if applied carefully and correctly, provide either a confirmatory or an alternate diagnostic capability. Moreover, when this procedure generates a positive amplification result, sequence characterization of the product and comparison with other reference viruses can elucidate the variant type, in what species it is most commonly transmitted, and from which geographic area closely related viruses circulate.

The utility and application of various nucleic acid amplification tests have been reviewed (Wacharapluesadee & Hemachudha, 2010), and this information is updated here.

The DFA test procedure rapidly loses sensitivity when applied to brain tissue that is substantially decomposed. Controlled observations have shown that in such a situation a molecular method of detection such as RT-PCR can be greatly more sensitive (David et al., 2002; McElhinney et al., 2014) and capable of diagnosing specimens that would otherwise be scored as unfit for testing using the DFA test (Prabhu et al., 2018). RT-PCR is superior to the mouse inoculation test for detection of rabies virus in samples maintained under various conditions of storage for long periods (Lopes, Venditti, & Queiroz, 2010) and is now a preferred method due to animal welfare considerations. Moreover, in some jurisdictions, when an animal has had human contact but is scored as rabies DFA test negative, this result must be confirmed with an alternate test; since molecular methods can be completed more quickly than either virus culture or the MIT, and with superior sensitivity (Picard-Meyer et al., 2004) they are extremely useful for the routine confirmation of DFA test results. Genetic tests can also be used as an adjunct to the DFA test when unexpected or unusual fluorescent staining patterns are observed and confirmation of virus presence is required; sometimes, a combination of tests is required to reach a consensus on the disposition of a particular case (McColl et al., 1993). The issue of discordant results between many of these tests can be problematic, however, when a genetic test such as a RT-PCR is the only method that suggests presence of rabies virus. Distinguishing between a false-positive RT-PCR result and a false-negative result for the other assays in such a situation may be a highly complex process with no clear-cut resolution. The best recourse is to avoid such situations wherever possible by adherence to the guidelines recommended for the performance of PCRs (Cooper & Poinar, 2000). These include careful processing of tissues using clean, sterile instruments and supplies in each case; use of physically separate areas for performing tissue extraction, PCR and post-PCR analysis; and use of dedicated pipettes in each of these areas to avoid sample crosscontamination.

12.4.2 RNA extraction

Before applying molecular methods to a sample, total RNA or total nucleic acids must first be recovered from the tissue to be tested. One commonly used method utilizes a commercial reagent known as TRIzol, a phenol/guanidine isothiocyanate solution based on earlier acidic phenol methods of RNA extraction. This reagent rapidly inactivates any nuclease present and quickly dissolves soft tissues such as brain, making it especially suitable for application in rabies diagnosis. After addition of chloroform to the mixture to facilitate a liquid-phase separation, RNA recovered in the aqueous phase is readily precipitated by addition of isopropanol. This method is reasonably simple and amenable to moderate throughput in terms of sample numbers.

Other methods that are sometimes used rely on commercially available kits that avoid the use of noxious chemicals and the requirement for RNA precipitation; for example, kits that provide a silica-membrane spin column approach for recovery of total RNA from a wide variety of tissue types. More recently, the application of magnetic bead technology to nucleic acid purification has resulted in the development of platforms using a 96-well plate format to enable high-throughput sample processing. Several companies market instruments employing this technology. It relies on the binding of nucleic acids to magnetic bead particles followed by multiple washing steps to remove other contaminants before elution of highly purified nucleic acid from the particles.

12.4.3 Reverse transcription-polymerase chain reaction methods

Since the emergence of PCR technology (Saiki & Erlich, 1989), many studies have described its application to the development of highly sensitive diagnostic methods for many pathogens. Assays that detect rabies viruses and other lyssaviruses involve an RT-PCR in which conversion of the viral RNA target to a cDNA copy using the enzyme reverse transcriptase is followed by amplification of a target sequence defined by a pair of synthetic oligonucleotides or primers (World Health Organization, 2018a). It is usually applied postmortem to fresh brain material from a suspect animal or human. The technique may also be applied to saliva, skin samples (often collected antemortem from suspect humans), salivary glands, and, indeed, to virtually any other tissue or sample, e.g., urine collected intra vitam, though with due consideration of the likelihood of virus being present in such samples incorporated into test result interpretation. Compared to the DFA test, RT-PCR processing time may be lengthy and relatively expensive, depending on the details of the assay employed, and the exquisite sensitivity of the technique can have a downside due to the increased risk of sample crosscontamination and hence false-positive findings. Despite these considerations, this technique is finding increased application for lyssavirus diagnosis.

The principle of the PCR depends on the use of two synthetic oligonucleotides that can hybridize to opposite strands of a dsDNA target and are oriented in such a way that when they prime new DNA synthesis, the newly created DNA strands overlap in sequence. The reaction, catalyzed by a thermostable DNA polymerase, requires repeated thermocycling thus: first, high heat (95°C) to denature the DNA template into single strands, then a lower temperature (usually 45–60°C) to allow annealing of the oligonucleotides to their target sequences, and finally an incubation (usually 72°C) to allow the annealed oligonucleotides to prime DNA synthesis using a mix of all four deoxynucleotide triphosphates (dNTPs) as substrates. This cycling is repeated for 25–40 cycles and a successful PCR produces a double-stranded DNA product (amplicon) of specific length, defined at its two ends by the primers used in the reaction. The extent of amplification of the target sequence, usually by more than 100,000 fold, is the basis for the assay's exquisite sensitivity.

Critical to the design of an assay is careful consideration of the viral group to be targeted and this will depend on the nature of the viruses that are likely to be encountered in a certain geographical area. For instance, in the Americas, primers that successfully amplify all known rabies viruses may be sufficiently broad in scope for most situations because any indigenously acquired lyssavirus infection will be due to classical rabies virus. In other areas, particularly Africa and Europe, the presence of more divergent lyssaviruses dictates the need for assays that detect many lyssavirus species. In all these situations, comprehensive information regarding the nucleotide sequence diversity of the region of the viral genome to be targeted is required to assist in the design of primers that will support the development of assays with the required specificity and inclusivity.

Since the first description of PCR as a tool for lyssavirus detection, the technology has evolved such that the methods employed to detect the amplification product have improved significantly. While the traditional methods use agarose gel electrophoresis to ascertain the presence/absence of a product of the expected size, this further extended the reporting time of the test and increased the possibility of sample crosscontamination due to the additional handling of products often present in high concentrations. This becomes particularly problematic with the application of nested PCRs in which a second round of amplification is performed on the product of a first round PCR to maximize assay sensitivity. The emergence of methods that detect amplicon production in real time, through the use of various fluorogenic approaches, removed the need for product analysis by gel electrophoresis and thereby both streamlined the method and reduced the likelihood of false-positive results since post-PCR handling of the reactions was no longer necessary. These assays also enhanced sensitivity to levels at or higher than those attainable by nested PCR methods.

Table 12.2 lists some of the more commonly used RT-PCR assays described in the literature over several years and the salient features of many of these assays are further detailed in the following sections.

12.4.3.1 Primer design

Since the N gene is one of the more conserved regions of the lyssavirus genome (Le Mercier, Jacob, & Tordo, 1997), this gene has been the target of virtually all efforts to develop broadly crossreactive PCRs for rabies virus detection while a small number of assays have targeted conserved regions of the L gene for this purpose. Many early studies in this area aimed to amplify relatively long segments of this gene, up to the complete open reading frame, to allow direct comparison of the gene sequence with the antigenic properties of its encoded nucleoprotein. While these studies highlighted the value of PCR for viral detection, they were not focused primarily on use of this technology for routine diagnostic application, which is optimized when relatively short sequence is targeted. Constraints on the design of PCR primers used for diagnostic methods include the need to accommodate the genetic diversity of the viruses to be targeted as well as limiting the length of the PCR product (amplicon) to a few hundred base pairs to optimize the amplification process. Fooks et al. (2009) provide a comprehensive listing of a wide variety of primers used for RT-PCR detection of lyssaviruses and several of these are described later.

Many investigators have identified sequences at or close to the N gene start codon as very useful primer binding sites (see Table 12.2), which, when paired with primers targeting neighboring sequences, have been used for robust assay development. A small number of

TABLE 12.2 Conventional RT-PCR assay details designed to be applicable to a broad range of lyssaviruses.

Assay gene target and format	Primer	Sequence (5'–3')	Location in genome[a]/ orientation	Reference
N, RT-PCR	N1	TTTGAGACTGCTCCTTTT	589–605/+	Sacramento et al. (1991)
	N2	CCCATATAGCATCCTAC	1013–1029/−	
N, RT-PCR	N7	ATGTAACACCTCTACAATG	55–73/+	Bourhy et al. (1993)
	N8	AGTCTCTTCAGCCATCTC	1568–1585/−	
N, hnRT-PCR	JW12	ATGTAACACCYCTACAATG	55–73/+	Heaton et al. (1997), OIE (2018a)
	JW6 UNI	CARTTVGCRCACATYTTRTG	641–660/−	
	JW10 UNI	GTCATYARWGTRTGRTGYTC	641–660/−	
N, nRT-PCR	10g	CTACAATGGATGCCGAC	66–82/+	Orciari et al. (2001), Smith, Orciari, Yager, Seidel, and Warner (1992)
	504	TATACTCGAATCATGATGAATGGAGGTCGACT	1290–1317/+	
	105	TTCTTATGAGTCACTCGAATATGTCTTGTTTAG	1394–1424/−	
	304	TTGACGAAGATCTTGCTCAT	1514–1533/−	
N, RT-PCR	RabNfor	TTGTRGAYCAATATGAGTACAA	135–156/+	Nadin-Davis (1998)
	RabNrev	CCGGCTCAAACATTCTTCTTA	876–896/−	
N, nRT-PCR	GRAB 1F	AARATNGTRGARCAYCACAC	538–557/+	Vázquez-Morón, Avellon, and Echevarría (2006)
	GRAB 2F	AARATGTGYGCIAAYTGGAG	574–593/+	
	GRAB 1R	GCRTTSGANGARTAAGGAGA	911–892/−	
	GRAB 2R	TCYTGHCCIGGCTCRAACAT	833–814/−	
L, RT-PCR Pan rhabdovirus	PVO3	CCADMCBTTTTGYCKYARRCCTTC	7526–7503/+	Bourhy, Cowley, Larrous, Holmes, and Walker (2005)
	PV04	RAAGGYAGRTTTTTYKCDYTRATG	7068–7088/−	

[a]Based on the sequence of the Pasteur virus reference strain (GenBank Accession M13215). A comprehensive list of all assays, including species specific methods, can be found in Fooks et al. (2009).

base mismatches between the primer and its target sequence may not necessarily prevent their annealing. The stringency of primer annealing is dependent on the annealing temperature; the lower the annealing temperature, the less stringent is the annealing process, and the more mismatches can be accommodated. The caveat to this, however, is that as the annealing temperature is lowered, the chance of primer annealing to poorly related sequences rises and substantial nonspecific primer binding occurs. Ultimately, this leads to a highly nonspecific reaction of little value as a diagnostic test. The position of mismatches also impacts the extent to which they may hinder proper primer annealing; in particular, a mismatch close to the 3' terminus of the primer is often highly detrimental to the PCR since the 3' end of the primer must anneal strongly in order to prime new DNA synthesis. One strategy often employed to overcome variability within the targeted sequence is to use a combination of primers of different sequences that can anneal to the same position within different target sequences. This was employed in the design of the JW6/JW10 primer set to target multiple lyssaviruses (Heaton et al., 1997). Subsequent modification of this assay involved design of the degenerate primers, JW6UNI and JW10 UNI, in which two or more bases inserted into certain positions of each oligonucleotide during synthesis facilitated detection of lyssaviruses of all three phylogroups (OIE, 2018b). Many lyssavirus assays now incorporate degenerate primers to ensure broad coverage.

This approach can be taken one step further by employing a combination of primers with distinct specificities so that both lyssavirus detection and species or variant identification can be achieved in the same assay. In the United Kingdom and Europe, such a strategy has been employed for detection and differentiation of rabies viruses and European bat lyssaviruses (Black et al., 2000; Picard-Meyer et al., 2004). Several other investigators have also described multiplex PCRs that identify multiple rabies virus variants based upon the size of the amplicons produced (Nadin-Davis, Huang, & Wandeler, 1996; Nel, Bingham, Jacobs, & Jaftha, 1998; Rohde, Neill, Clark, & Smith, 1997). While few studies have compared results using different RT-PCRs, a study exploring detection of Indian rabies viruses using three sets of primers described by other investigators found complete concordance in their performance (Babu, Manoharan, Ramadass, & Chandran, 2012), thereby supporting the robustness in design of these reagents.

Despite more limited sequence information available to support development of assays targeting highly conserved sections of the L gene encoding the polymerase, primers that successfully amplified a section of this gene for a number of Rhabdoviruses representing several genera have been designed (Bourhy et al., 2005), and a PCR targeting this same conserved L gene region has been used for human rabies diagnosis in Africa and Asia (Dacheux et al., 2008). Most laboratories performing RT-PCR on a routine basis will maintain several primer combinations and may apply two or more primer pairs to evaluate a particular specimen, giving due consideration to the source of the sample and the viruses to which the specimen could have been exposed. As additional nucleotide sequence information becomes available for other portions of the lyssavirus genome, other targets for a broadly crossreactive PCR assay may be identified.

12.4.3.2 Reverse transcription

The first crucial step of a RT-PCR is the copying of the rabies virus RNA as a complementary DNA (cDNA) strand, achieved using dNTP substrates and a reverse transcriptase enzyme. Many different commercial preparations of this enzyme, ranging from preparations of natural enzymes prepared from retrovirus cultures to highly engineered versions of the enzyme with improved copying fidelity and processivity, have been employed for this step. For a robust diagnostic method, the latter enzyme preparations are preferred due to their improved performance despite their higher cost.

The initial annealing step involves incubation of a mixture of primers and RNA, first at a brief denaturation temperature (65–70°C) followed by a short period (5–10 min) at an appropriate temperature for primer binding to the target. Since the rabies virus life cycle includes production of full-length negative and positive sense copies of its genome, as well as significant amounts of mRNA, sequence-specific primers targeting either positive (messenger) or negative (genomic) sense sequences or both can be used to prime cDNA synthesis. Alternatively, random primers, comprising a mixture of 6–8 base oligonucleotides of random sequence, are sometimes used in place of sequence-specific primers for the annealing step.

Their advantage is that they can bind to several places along the length of the target sequence and can thus generate a wider range of copies than sequence-specific primers. This can improve sensitivity especially for samples of poor integrity. Since random primers will also prime cDNA synthesis of host RNA more dNTP substrates will be consumed and extended extension times may be required to ensure maximal cDNA copying of viral template. The annealing temperature employed must account for the nature of the primers employed at this step as random primers will have a much lower melting temperature then sequence-specific primers. In one-step RT-PCRs, in which all reagents for the entire process of cDNA synthesis and PCR amplification are included at the start of the reaction, both forward and reverse primers are necessarily included at this stage and thus available for cDNA priming.

The extension phase during which cDNA is generated proceeds at a fixed temperature according to the optimal requirements of the enzyme employed (ranging between 35 and 50°C) for 30 min or longer. The reaction is normally terminated by heating at 70–95°C to inactivate the enzyme prior to amplification.

12.4.3.3 Polymerase chain reaction

A critical component beyond the primers and dNTP substrates is the enzyme used to drive the synthesis. Early PCR methods employed the thermostable enzyme *Taq* DNA polymerase but a variety of other thermostable enzymes, many with engineered modifications to improve copying fidelity and processivity, are now available commercially. Precise thermocycling conditions to be used in an assay will be dictated by the melting temperature of the primers as well as the characteristics of the DNA polymerase employed. Several other factors can impact the sensitivity of the RT-PCR. While it is possible for an RT-PCR to generate products of several kilobases, the efficiency of the amplification diminishes with increased length due to limitations at both the reverse transcription and PCR stages. Accordingly, PCRs targeting small sections of the lyssavirus genome (usually between 100 and 200bp) are the most sensitive for viral detection, particularly when assaying significantly degraded RNA. The number of amplification cycles performed can also significantly impact test sensitivity. In an ideal reaction, the amount of PCR product doubles after each cycle, so increasing the number of cycles increases the product yield exponentially. Generally, 25–40 thermocycles of PCR are employed. Use of cycle numbers higher than this generally becomes unproductive due to the gradual loss of enzymatic activity of the DNA polymerase, despite use of thermostable enzymes relatively resistant to high temperature, and the depletion of the reaction components.

Nested PCR, in which the product of a PCR is subjected to a second round of amplification using primers internal to those employed for the first round (Kamolvarin et al., 1993), is another approach that has been used to improve assay sensitivity. A heminested PCR (hn PCR) (Heaton et al., 1997; Picard-Meyer et al., 2004) employs one of the first-round primers in combination with an internal primer in the second PCR. Nested strategies increase the sensitivity of the assay enormously (Elmgren, Nadin-Davis, Muldoon, & Wandeler, 2002) and a number of investigators have reported the use of this approach for the detection of multiple lyssavirus species (Coertse, Weyer, Nel, & Markotter, 2010; Vázquez-Morón et al., 2006). However, they come at the cost of greatly increasing the chance of a false-positive result unless stringent precautions are taken to prevent carryover contamination of the sample.

12.4.3.4 PCR product detection

Traditionally PCR products were detected after electrophoresis through agarose by staining with ethidium bromide, a dye that intercalates between DNA bases and that is readily visualized under UV light. This allows the approximate size of the amplicon to be determined with reference to standard DNA markers. Alternative, less toxic dyes now commercially available can replace ethidium bromide and yield comparable sensitivity. Other methods of detection, which have been employed in the past, include hybridization of the amplicon to rabies-specific oligonucleotide/DNA probes either after Southern transfer from the gel to a membrane (Heaton et al., 1997; Heaton, McElhinney, & Lowings, 1999) or by ELISA-based methods (AravindhBabu, Manoharan, & Ramadass, 2014; Black et al., 2000; Whitby, Heaton, Whitby, O'sullivan, & Johnstone, 1997). Probe-based methods confirm the specific nature of the amplicon (Heaton et al., 1999) and, in rare instances, have identified nonspecific bands of a size similar to that of the expected product as false positives (Trimarchi & Smith, 2002). A method that employed rapid pyrosequencing technology for both detection and characterization of amplicons generated by a pan-lyssavirus RT-PCR has also been described (De Benedictis et al., 2011). Most of these approaches have now been superseded by real-time RT-PCR technologies.

12.4.4 Real-time RT-PCRs for lyssavirus detection

By combining the amplification and detection of nucleic acid within a closed tube system, real-time PCR platforms offer a more rapid and reliable indication of the presence of lyssavirus RNA in suspect samples than conventional gel-based RT-PCR. While two-step real-time RT-PCRs have been described (Hayman, Banyard, et al., 2011), the current trend is to a closed tube strategy, in which the entire RT-PCR occurs as a one-step reaction; this reduces both the risk of sample crosscontamination leading to false-positive results and assay turn-around time (Marston et al., 2019; Wadhwa et al., 2017; Wakeley et al., 2005). In these systems, which require dedicated thermal cycler instrumentation, production of the amplicon is directly linked to increased fluorescence of a fluorophore included in the reaction.

The two most popular chemistries that have evolved to support this technology have both been applied to the development of lyssavirus assays and some of the more commonly applied tests are listed in Table 12.3 and described in the following sections.

12.4.4.1 The 5' nuclease assay

The chemistry of this assay, originally referred to as TaqMan, employs a dual-labeled probe (DLP) as an integral part of the reaction. This probe comprises a synthetic oligonucleotide, which binds to internal sequence on one strand of the amplicon and is labeled at both its 5'-and 3'-ends. The 5' terminal of the DLP is covalently attached to a reporter dye that fluoresces when irradiated at a certain wavelength. A quencher moiety is covalently attached to the DLP's 3'-end such that emissions by the reporter dye are effectively quenched. When target sequence is present in the PCR, specific product is generated and the DLP binds to its cognate sequence during the annealing step just prior to strand extension from one of the PCR primers. During DNA synthesis, the 5' nuclease activity of the DNA polymerase degrades the probe, reporter and quencher become dissociated from each other, and the presence of PCR

TABLE 12.3 Real-time RT-PCR assays for lyssavirus detection.

Assay gene target and format	Primer/probe	Sequence (5'-3')	Location in genome[a]/orientation	Reference
N, SYBR	JW12 N165-146	ATG TAA CAC CYC TAC AAT G GCA GGG TAY TTR TAC TCA TA	55–73/+ 165–146/−	Hayman, Banyard, et al. (2011), Marston et al. (2019)
L, SYBR	Pan-Lyssa-7531F Pan-Lyssa-7749R	TTC TTC GCT YTR ATG TCW TGG AA ATG RTT GTT CCA CTT YTC ATA RTC	7074–7096/+ 7292–7269/−	Fischer et al. (2014)
N, TaqMan LN34	Primer F1 Primer F2 Primer R Probe LN34 Probe LN34 lago	ACGCTTAACAACCAGATCAAAGAA ACGCTTAACAACAAAATCADAGAAG CMGGGTAYTTRTAYTCATAYTGRTC (FAM) AA+C+ACCY+C+T+ACA+A+TGGA (BHQ1) (FAM) AA+C+ACTA+C+T+ACA+A+TGGA (BHQ1)	1–24/+ 1–25/+ 140–164/− 60–76 59–75	Wadhwa et al. (2017)
N, TaqMan L, TaqMan	RABVRPN1 RABVFPN2 RABVPN RABVRPL1 RABVFPL2 RABVPL	GCTCTGGGCTGGTGTCGTTC ACGGGGACTTCCCGCTCAG 6FAM-CGAGCCARGGCAGGAGACTGCGG–BBQ GGTTTCCGGDGCYGTDCCTC CCTAGGGGAGACYTTGCCRT 6FAM-CCCGTCAYATAGGGTCRGCTCARGGGC–BBQ	707–726/+ 880–899/− 819–841 9472–9491/+ 9660–9679/− 9561–9587	Faye et al. (2017)
N, TaqMan (all rabies viruses)	RABVD1-F RABVD1-R1 RABVD1-R2 RABVD1 probe	ATGTAACACCYCTACAATG GCMGGRTAYTTRTAYTCATA GGCMGGRTAYTTRTAYTCAT 6FAM-CGAYAAGA/ZEN/TTGTATTYAARGTCAAKAATCA GGT-3IABkFQ	55–73/+ 146–165/− 147–166/− 78–111	Nadin-Davis (2019), Nadin-Davis, Sheen, and Wandeler (2009)

[a] Based on the sequence of the Pasteur virus reference strain (GenBank Accession M13215). Additional assays, including more species-specific methods, can be found in Fooks et al. (2009).
+ signs in the LN34 probes precede LNA (locked nucleotide) bases.

product is detected as fluorescence emitted by the reporter dye at a defined wavelength. Levels of this signal rise during the course of the reaction until the reagents in the reaction are exhausted and a plateau is reached. If no amplicon is produced during the PCR, the unbound DLP remains intact, no increase in fluorescent signal is recorded, and the sample is scored as negative.

Whereas this approach provides exquisite specificity to the assay, the design of primers and probes that are broadly crossreactive can be challenging due to nucleotide sequence variation observed between members of the genus. Depending on their location and frequency base mismatches between the target and primer/probe sequences can substantially reduce assay sensitivity or even preclude detection completely. The advantage of probe-based assays is that they can accommodate multiple targets (multiplexing) within a single reaction by use of probes coupled to different fluorophores with distinct emission wavelengths that are readily differentiated.

In a relatively early study, Black, Lowings, Smith, Heaton, and McElhinney (2002) reported an ambitious effort to both detect and discriminate between lyssavirus genotypes 1–6 using TaqMan technology. The authors reported the successful detection and assignment of 106 lyssaviruses using these reagents, with no crossreactivity with 18 nonlyssavirus isolates. Another assay based on primers used in a hnRT-PCR method to detect all African lyssaviruses was converted into a real-time format with addition of a probe and shown to detect all members of a limited cohort representing the diversity of this group (Coertse et al., 2010). The real-time format of this assay was proven to be even more sensitive than the standard RT-PCR for the diagnosis of severely decomposed tissues (Markotter et al., 2015).

A simplified protocol derived from the study by Black et al. (2002) that detects and discriminates lyssaviruses of genotypes 1, 5, and 6 (RABV, EBLV-1, and EBLV-2) employs primers JW12 and N165-146 to produce a 111-bp amplicon that is differentiated using three distinct genotyping probes, each labeled with a different reporter dye (Wakeley et al., 2005). Of 62 lyssaviruses evaluated by this method, all were readily detected and typed except for 1 American bat rabies virus, which was only weakly detected, possibly due to 3 mismatches between the rabies virus probe and the target sequence for this isolate.

In the Americas, efforts to apply TaqMan technology to detection of rabies viruses circulating in North American reservoir hosts initially highlighted the challenges in designing such assays due to viral genetic diversity (Hughes, Smith, Hanlon, & Rupprecht, 2004). Subsequently, researchers explored the utility of three different assays, targeting separate regions of the N gene, to detect all rabies viruses of the Americas (Nadin-Davis et al., 2009). One of these assays (RABVD1), which employed a modified version of the assay described by Wakeley et al. (2005), detected 203 samples representative of all major rabies virus lineages circulating globally; high C_T values yielded by a few samples were due to poor RNA sample integrity as measured by a β-actin internal control assay. A variation on this assay has been extensively evaluated for routine rabies diagnosis in New York state and found to compare favorably with the DFA test (Dupuis, Brunt, Appler, Davis, & Rudd, 2015) as well as identifying rabies in a small percentage of animal submissions originally deemed unfit for testing by the DFA test (Appler et al., 2019). An additional TaqMan assay that specifically identifies the raccoon variant of rabies virus can be multiplexed with the RABVD1 real-time RT-PCR for use in areas of North America where this viral type circulates (Nadin-Davis, 2019) (Fig. 12.2).

Another study to develop a TaqMan assay that detects all rabies viruses, employed the method described by Wakeley et al. (2005) together with a second assay that targets sequence slightly downstream within the N gene ORF (Hoffmann et al., 2010). While neither assay individually detected all 93 viruses in the study panel, together the 2 assays were 100% successful. Faye et al. (2017) described another TaqMan assay targeting both N and L sequences, which consistently detected all African rabies viruses but did not crossreact with any other lyssavirus species tested; however, further evaluation of this method did indicate that the N gene assay failed to detect certain rabies virus lineages circulating in other parts of the globe.

Other TaqMan assays that target specific groups of lyssaviruses have been described. One such assay uses two biotype specific forward primers, a common reverse primer and two

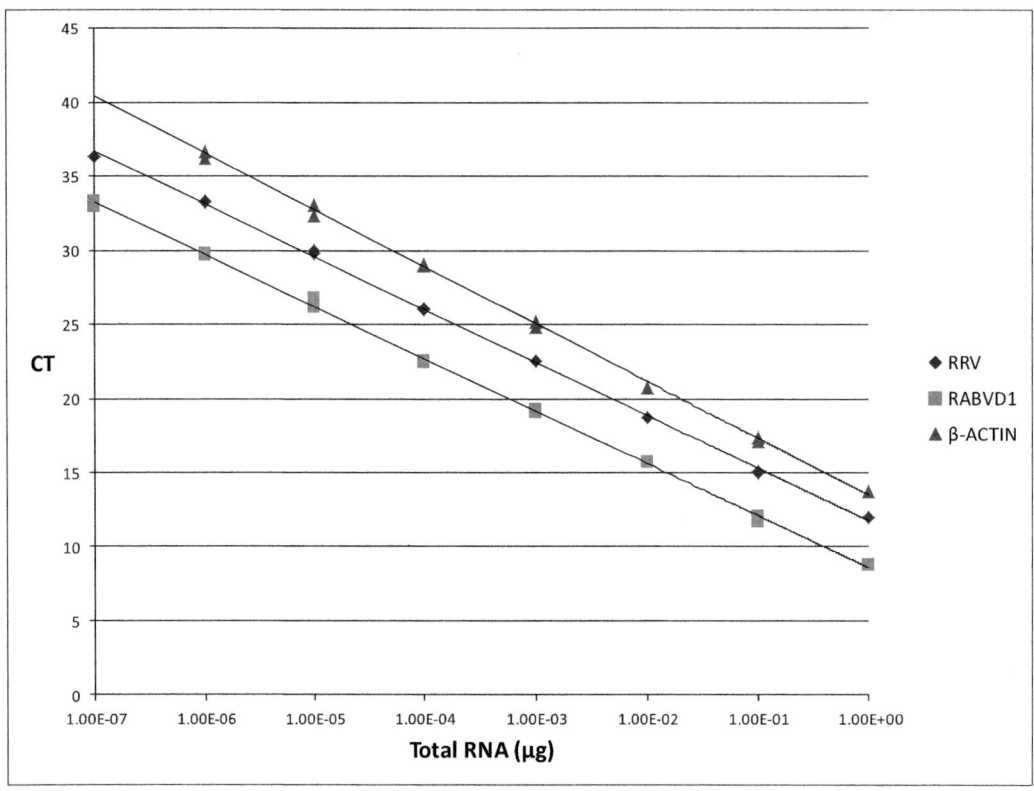

FIG. 12.2 Example of a TaqMan multiplex real-time RT-PCR targeting three distinct sequences: RABVD1, a generic rabies virus sequence detected using a FAM-labeled probe; RRV, a sequence specific to the raccoon variant of rabies virus detected using a Cy3-labeled probe; β-ACTIN, a sequence of the housekeeping beta-actin mRNA used as an internal control detected using a Cy5-labeled probe. Tenfold serial dilutions of total RNA extracted from a sample infected with the raccoon variant of rabies virus was used to generate these standard curves.

distinct DLPs to detect and discriminate the flying fox and insectivorous biotypes of ABLV (Smith, Northill, Harrower, & Smith, 2002). A later optimized version of this method was specific to ABLVs with no crossreactivity with other lyssaviruses (Foord et al., 2006). Assays to detect and discriminate rabies virus and EBLV-1 in samples from across Poland were reported (Orlowska, Smreczak, Trebas, & Zmudzinski, 2008), while assays that detect and discriminate Eurasian bat lyssaviruses were applied on an experimental basis only (Hughes et al., 2006). A TaqMan assay, targeting sequences close to the 3′-end of the N gene open reading frame (ORF), was developed based on rabies virus N gene sequence data acquired over several years during a nationwide survey in Thailand. While this method detected a wide range of field isolates from Thailand and several other Asian countries (Wacharapluesadee et al., 2008), it failed to detect the Challenge Virus Standard (CVS) strain due to significant sequence mismatch of the probe with the target. This study explored the effects of sequence mismatches between the reagents and target and showed that the TaqMan assay was permissive for multiple mismatches, though with some impact on reaction efficiency and sensitivity. Recently a pan-lyssavirus TaqMan real-time RT-PCR named LN34 (Wadhwa et al., 2017), which employs a mix of degenerate primers and probes, has undergone extensive validation in 14 laboratories and reported to be highly effective in detection of all lyssaviruses tested (Gigante et al., 2018).

Several studies have reported that TaqMan assays are more sensitive, by up to 100 fold, than even the hnRT-PCR methods previously employed (Foord et al., 2006; Markotter et al., 2015; Nadin-Davis et al., 2009; Orlowska et al., 2008; Smith et al., 2002). Of the studies that have directly compared the sensitivity of real-time RT-PCR methods and the DFA test, the molecular method is comparable or superior especially for autolysed samples (Dupuis et al., 2015).

12.4.4.2 Assays using intercalating dyes

This chemistry employs a DNA-intercalating dye, usually SYBR Green, to detect the production of all dsDNA within the reaction. As amplicon is produced by the action of the two PCR primers, increasing amounts of the dye intercalate with the product resulting in increasing levels of fluorescence. Whereas this is a relatively inexpensive option for method development, the caveat is that unless the reaction is carefully designed and performed the production of nonspecific dsDNA products can potentially yield false-positive results. In order to overcome this potential issue, the melting temperature (T_M) of the amplicon is also calculated by performance of a melting curve analysis, a process that can readily be automated in most commercial instruments employed for real-time PCR assays.

Assays initially developed using this technology often used reagents derived from previous RT-PCR methods, and ranged from those detecting a subgroup of lyssaviruses, e.g., rabies viruses circulating in India (Nagaraj et al., 2006) or EBLV-1 in Europe (Picard-Meyer et al., 2015) to those that attempted to detect all known lyssaviruses (Hayman, Banyard, et al., 2011; Suin et al., 2014). Some of these assays were two-step protocols (Nagaraj et al., 2006); indeed, that of Suin et al. (2014) involved a first-round standard RT-PCR followed by real-time PCR with two sets of internal primers and detection with SYBR Green. The pan-lyssavirus assay developed by Hayman, Banyard, et al. (2011) adapted two primers (JW12 and N165-146) used previously for TaqMan assays to a SYBR Green chemistry using both melting curve analysis and gel analysis of the amplicons to confirm the specificity of the results. This real-time RT-PCR method was more sensitive than the hnRT-PCR previously described for detection of a broad range of lyssaviruses (Hayman, Banyard, et al., 2011). This assay was further improved by utilizing the primer pair in a one-step SYBR assay, which affords both sensitivity and specificity across the lyssavirus genus and has been validated for diagnostic use (Marston et al., 2019; OIE, 2018b). Typical results are shown in Fig. 12.3.

Variations on published methods that seek to ensure better coverage of specific rabies virus variants have been described (Wang et al., 2014). Indeed alternative modification of the method of Wakeley et al. (2005) added a second assay targeting a conserved region of the L gene to ensure diagnostic consistency; this method also described secondary analyses involving two multiplex PCRs to characterize any lyssaviruses detected (Fischer et al., 2014). In a rather different strategy employing both platform types, a method for detection of all lyssaviruses that uses a TaqMan approach for rabies virus detection and a SYBR Green assay for detection of all other lyssaviruses has also been described (Dacheux et al., 2016).

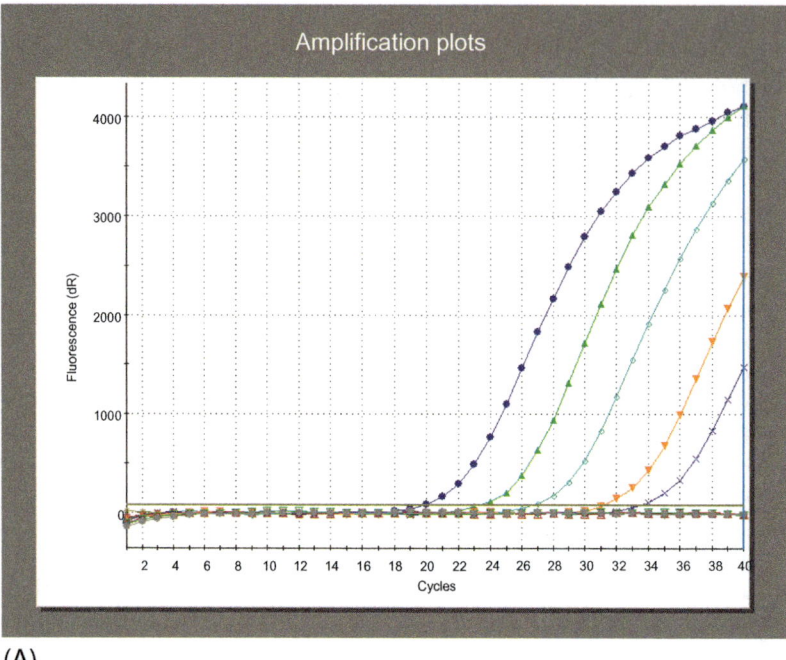

(A)

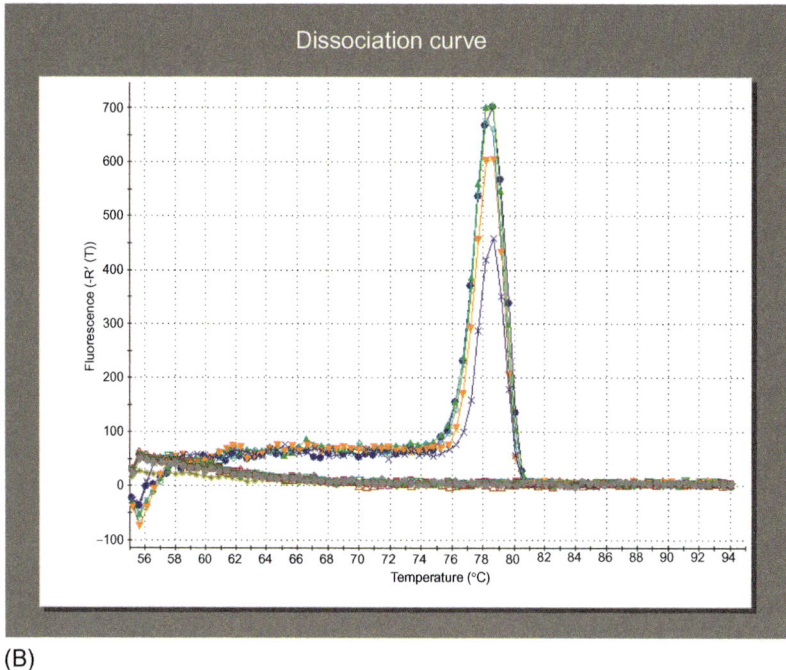

(B)

FIG. 12.3 Example amplification plot (A) and dissociation curve (B) from a 10-fold serial dilution series of RABV RNA extracted from a positive brain sample using a SYBR Green real-time RT-PCR.

12.4.4.3 Choice of assay

It is apparent that many real-time RT-PCR assays have been reported for lyssavirus detection, some of which were selectively developed for individual virus strains or certain virus lineages while others were developed to detect a wider range of rabies virus strains and/or lyssavirus species (Table 12.3). Reagents specifically formulated for the two distinct types of assay are detailed in the relevant publications and when used according to their specifications can yield equivalent results regardless of the instrumentation or mastermix employed (Picard-Meyer et al., 2015; Picard-Meyer, Peytavin de Garam, Schereffer, Robardet, & Cliquet, 2019). Beyond providing the substrates and enzymes required for the RT-PCR, many include additional features such as the inclusion of well standardization dyes (e.g., ROX) and strategies to eliminate contaminating amplicons from the reaction (e.g., UNG glycosylase). Most current protocols employ one-step kits, which reduce the opportunity for contamination making the system "closed tube" and increase sensitivity compared to the two-step alternatives.

Despite the many advantages of real-time PCR as a diagnostic tool, some technical aspects of the assay are important to consider when applying this technology. It can be difficult to define the appropriate end point of the assay (i.e., the cycle number beyond which samples cannot be considered positive or must be deemed suspicious) given the potential for nonspecific degradation of TaqMan probes at late cycles and the difficulty in correlating results with those of other gold standard tests, which may not be as sensitive as real-time PCR methods. To minimize problems related to these and other issues, strict adherence to the "minimal information for publication of quantitative real-time PCR experiments" (MIQE) guidelines for real-time PCR assay development and reporting (Bustin et al., 2009) are encouraged. In some situations, repeat testing with a fresh sample may be required to reach a conclusive result.

All published assays will have limitations as regards their diagnostic range as minor strain variations or novel lyssaviruses will not have formed part of the primer/probe design. It is therefore essential to consider the lyssavirus lineages and strains in each region when choosing and establishing this methodology for routine diagnostic use. For a laboratory conducting surveillance on brain material and expecting high numbers of negative samples, the use of the cheaper SYBR green real-time PCR would be recommended. The SYBR green approach would also be optimal when conducting scanning surveillance where novel or divergent lyssaviruses may be present. Indeed, emerging or novel virus strains may render the highly specific probe-based assays ineffective (Fooks et al., 2009). The more specific probe-based assays may be preferable in a laboratory conducting lyssavirus diagnosis on suspect samples, which are expected to have a restricted lyssavirus species range, contain low viral loads and/or require rapid typing.

12.4.4.4 Required controls

As in any diagnostic assay, use of appropriate controls is essential to proper interpretation of the results. For molecular assays, the following controls are strongly recommended:

Mock extraction control

A known rabies-negative sample should be processed for RNA extraction in parallel with the sample under investigation so as to control for any inadvertent contamination of the specimens through aerosol generation or reagent contamination. Ideally the negative sample should comprise rabies-negative brain tissue although a water sample can also be used if such tissue is not readily available.

Positive and negative PCR controls

Upon PCR set-up, each assay run should include at least one water sample as a negative control and RNA from one or more rabies-positive samples for use as positive controls. Any reaction run in which either the negative or positive controls fail must be considered invalid and the PCR repeated. When many samples are being examined use of a number of negative controls interspersed with the samples is recommended.

Control for template integrity

Evaluation of a sample for its suitability for PCR can be a useful control if sample integrity is in question. Smith, McElhinney, Heaton, Black, and Lowings (2000) described a ribosomal RNA (rRNA) internal control that is suitable for 14 different mammalian species and which can be incorporated into a lyssavirus RT-PCR to assess template quality. A slightly modified version of this method, performed separately from the lyssavirus RT-PCR, has proven useful for evaluation of samples processed at the National Reference Centre for Rabies in Ottawa, Canada (Nadin-Davis et al., 2009, 2012). Any sample, which fails to amplify such internal controls, should be declared unfit for testing. Such controls could also be invaluable to applied research studies; for example, to investigate conflicting results of studies that explored the utility of RT-PCR for the detection of rabies RNA in brain material stored in 50% glycerol saline (Biswal, Ratho, & Mishra, 2007; Muleya et al., 2012).

The use of controls is an important aspect of any real-time RT-PCR assay since these assays tend to be more susceptible to inhibitory factors than standard RT-PCRs. Assays that detect host 18S rRNA (Coertse et al., 2010) or β-actin mRNA are most often incorporated as internal controls for lyssavirus RNA detection by 5′ nuclease assays (Hughes et al., 2004; Nadin-Davis et al., 2009; Wakeley et al., 2005). Hoffmann and colleagues have explored the use of several other controls to evaluate RNA extraction and PCR inhibitory effects (Hoffmann, Depner, Schirrmeier, & Beer, 2006; Hoffmann et al., 2010).

12.4.5 Application of molecular methods to fixed tissues

Fresh brain material is the ideal tissue for rabies diagnosis, but in some situations brain tissue is fixed prior to suspicion of rabies in the differential diagnosis, and such specimens are occasionally submitted for testing. While immunohistochemical methods have been applied for diagnosis of such samples, molecular techniques that use labeled DNA or RNA probes to detect virus in tissue sections in situ have also been developed drawing extensively from those prior methods. An in situ hybridization method that uses digoxigenin-labeled probes to detect and type rabies virus transcripts in formalin-fixed tissues received by the National Reference Centre for Rabies in Canada has been described (Nadin-Davis, Sheen, & Wandeler, 2003). Other laboratories have also reported on the utility of in situ hybridization as a confirmatory test for DFA analysis of fixed tissues (Warner, Whitfield, Fekadu, & Ho, 1997). Although commercial kits for labeling of probes are available, there is no commercial source for rabies-specific probes or sequences. So, probes must be developed in-house, with appropriate attention given to the impact that lyssavirus diversity will have on the ability of a probe to anneal to its target sequence. Moreover, in situ hybridization methods are very labor intensive due to the many incubation periods and blocking steps needed, as well as the initial time required for preparation of tissue sections, and so these procedures are performed in exceptional circumstances only.

Other approaches have explored the application of RT-PCR protocols for virus RNA detection in this sample type. In a comparison of the DFA test and RT-PCR, Kulonen, Fekadu, Whitfield, and Warner (1999) reported that both could accurately detect rabies virus in 12 of 24 paraffin-embedded Finnish samples that had been Carnoy-fixed for 1–24 h provided that very short RT-PCR products (139 bp) were targeted while an RT-PCR that generated a 304-bp amplicon detected only 67% of the positive samples. In contrast, rabies virus sequences were detected in brain tissues fixed in formalin for 1 day to 1 week and then stored as paraffin-embedded tissue for 1 month to 16 years at 30°C (Wacharapluesadee, Ruangvejvorachai, & Hemachudha, 2006). That study also noted that the most efficient RT-PCR assay was one that detected the smallest product (150 bp), and products larger than 400 bp could not be generated in any sample. In a study using pathological specimens from Sri Lanka, rabies virus detection by IHC and molecular methods using fresh brain samples was directly comparable. However, using molecular methods on the formalin-fixed paraffin-embedded (FFPE) samples used for IHC directly yielded poor results with only 2 of the 13 samples yielding a positive amplification using TaqMan RT-PCR and non for hnRT-PCR (Beck et al., 2017). These studies clearly showed that as the length of the formalin fixation increased, generation of RT-PCR products became increasingly difficult, and only short products could be made with any frequency. The fixation and paraffin-embedding processes cause degradation and modification of the RNA in the tissue and result in RNA crosslinking to other cellular components, thereby rendering RNA extraction inefficient and adversely affecting reverse transcription; this results in an inability to produce amplicons longer than a few hundred base pairs.

Notwithstanding these challenges, this approach opens up the possibility of retroactive rabies diagnosis on FFPE samples stored for significant periods of time. It does require careful extraction of RNA from tissue sections following tissue deparaffinization and treatment with proteinase K. Moreover, the need to amplify short amplicons for optimal target detection makes this tissue type an ideal candidate for evaluation using real-time RT-PCR techniques. Currently, studies are in progress to evaluate the application of such methods to the detection of rabies viruses in FFPE samples (Condori et al., 2020).

12.4.6 Application of RT-PCR in research

Molecular methods do, of course, have enormous value in studies on rabies pathogenesis, and this application is described in more detail elsewhere (see Chapter 8). Standard RT-PCR methods have been used to examine organ distribution of virus in experimental (Charlton, Nadin-Davis, Casey, & Wandeler, 1997; Vos et al., 2004) or naturally infected animals (Serra-Cobo, Amengual, Abellán, & Bourhy, 2002), but real-time RT-PCR is particularly advantageous in such studies because, by its quantitative nature, it can provide measurements of actual viral load in such samples. Accordingly, real-time RT-PCR has been employed to determine postmortem levels of virus present in various tissues of infected humans

(Panning et al., 2010) or animals (Brookes et al., 2007; Freuling et al., 2009; Johnson et al., 2008), and to determine titer of virus present in saliva samples from patients undergoing experimental treatment (Nadin-Davis et al., 2009). For epidemiological studies of rabies in wildlife, especially in bat populations of Europe where many species are protected, RT-PCR testing for the presence of lyssaviruses in RNA recovered from oral swabs has been useful (Echevarría, Avellón, Juste, Vera, & Ibáñez, 2001; Picard-Meyer et al., 2011).

12.4.7 Other molecular methods for rabies virus detection

Amplification methods distinct from PCR have been developed for detection of specific nucleic acid sequences, but just a few studies have reported studies to explore the application of such methods to rabies diagnosis (Fooks et al., 2009). Three isothermal methods have been promoted as inexpensive assays for rabies virus diagnosis because the technology is not technically demanding, the reagents cost just a few dollars per test, and they can potentially be used in the field. These considerations make such tests of interest to developing countries lacking health system infrastructure, particularly in more remote areas where rabies continues to pose a significant threat to human health.

The first of these methods, known as nucleic-acid sequence-based amplification (NASBA), employs three enzymes, avian myeloblastosis virus reverse transcriptase, *Escherichia coli* RNase H, and T7 RNA polymerase, together with a set of primers, to support isothermal amplification of an RNA template. Wacharapluesadee and Hemachudha (2001) explored the application of this technology to detect rabies virus RNA by targeting a 180-base viral sequence within the central region of the rabies virus N gene, which was subsequently recognized by a reporter probe binding to internal sequence. With respect to sensitivity, this assay appeared to compare favorably in comparison with standard RT-PCR.

Another novel amplification strategy is the reverse-transcription loop-mediated isothermal amplification (RT-LAMP) method. The LAMP method, first described by Notomi et al. (2000), employs a DNA polymerase with high strand displacement activity (e.g., Bst DNA polymerase large fragment) with two inner and two outer primers. Each of the two inner primers binds to two distinct sequences, of opposite sense, on the target molecule. During a 1 h 65°C incubation, a complex series of reactions involving all primers generate several stem-loop structures so that the final reaction yields several products of various sizes when examined by gel electrophoresis. The specificity of these products can be confirmed by restriction endonuclease analysis, which should generate a small number of digestion products of predicted sizes. Optimal efficient amplification is achieved by targeting DNA sequences <300 bp in length and including additional loop primers in the reaction (Nagamine, Hase, & Notomi, 2002). The use of reverse transcriptase together with the DNA polymerase allows the assay to be applied to RNA targets.

The use of RT-LAMP assays for rabies virus detection has been explored in the Philippines (Boldbaatar et al., 2009) and in Brazil where it was able to discriminate between vampire bat- and dog-associated rabies viruses but failed to detect certain other field isolates (Saitou et al., 2010). Successful application of this technology to the detection of several African rabies viruses was also reported (Hayman, Johnson, et al., 2011; Muleya et al., 2012). Biotinylated versions of some of the primers of the former study were also incorporated into a lateral flow device to facilitate product detection (Hayman, Johnson, et al., 2011). The sensitivity of this RT-LAMP assay has been reported to compare very favorably with that of a RT-PCR while another study comparing an RT-LAMP assay with both the DFA test and a SYBR Green real-time RT-PCR suggested this methodology has promise for rabies virus detection despite exhibiting slightly lower sensitivity and specificity than the PCR method (Naji, Fadajan, Afshar, & Fazeli, 2020). However, some of the challenges inherent to these assays must be borne in mind. The design of broadly reactive RT-LAMP assays capable of detecting a wide range of lyssaviruses would be very challenging, given that multiple primers within a short stretch of target sequence are needed and especially given the observation that a single mismatch at the end of a primer can critically interfere with the assay (Boldbaatar et al., 2009).

While these alternative molecular methods may be realistic at a local level if they can be validated for regionally circulating rabies viruses of limited genetic variability, one must keep in mind that in Africa, where many of these assays are being evaluated, many rabies virus variants as well as other lyssavirus species circulate widely. To date none of these assays have been validated against significant numbers of virus samples and much work and refinement is needed before they can be considered an appropriate alternative to the DFA test or RT-PCR methods. Another technique, which does show broad-scale promise, is based on recombinase polymerase amplification (RPA) in which a recombinase facilitates isothermal primer binding to a DNA target through a strand displacement process followed by strand extension by DNA polymerase. Multiple rounds of this process result in amplification of a dsDNA target while inclusion of a reverse transcription step allows detection of RNA templates. Similar to a TaqMan RT-PCR, this assay employs forward and reverse sense primers and a distinct type of DLP for product detection through increasing fluorescence. An initial study demonstrated the potential of RT-RPA for rabies virus detection (Coertse, Weyer, Nel, & Markotter, 2019; Schlottau, Freuling, Müller, Beer, & Hoffmann, 2017). This method also detected small numbers of rabies viruses from other continents as well as many other lyssavirus species and the potential development of pan-lyssavirus assays that employ this technique may be considered in the future.

Another experimental approach to lyssavirus detection involves the use of microarray technology, which, when linked to sequence-independent PCR amplification of template, shows promise for both detection and genotyping applications, and which could be a most useful tool for the identification of novel lyssaviruses (Gurrala et al., 2009). However, while this approach could be a useful research tool, extensive refinement to the method would be required before it could be considered as a viable diagnostic alternative.

12.5 Detection of live virus

Detection of the presence of live virus (virus isolation) is possible through inoculation of a brain homogenate into young mice (mouse inoculation test, MIT), or into live cell culture (Rabies Tissue Culture Isolation Test—RTCIT). The MIT involves the intracerebral inoculation of a brain homogenate from the suspected animal into suckling or juvenile mice followed by a period of observation and confirmatory testing of any mice that succumb to the disease by dFA (Koprowski, 1966, 1973). The observation period is usually 21 days with clinical signs often developing within 14 days and much sooner for the mouse adapted laboratory strains used as a positive control. However, depending on viral load or the lyssavirus species/strain involved, the clinical signs may not appear for many days/weeks. The observation period may thus be extended to 28 days or beyond when longer incubation periods are suspected (e.g., for non-RABV lyssaviruses in clinical material). As a result, clinicians and veterinarians rarely rely upon this test for clinical decisions and case management. Due to animal welfare reasons and the establishment of sensitive and specific in vitro methods, MIT is currently not recommended for routine primary diagnosis. However, it is still a valuable tool for isolating or amplifying viral strains when in vitro approaches fail, to generate standardized control material and for research purposes.

Techniques involving the in vitro isolation of rabies and other lyssaviruses in mammalian cell lines (RTCIT) have largely replaced the MIT in diagnostic reference laboratories. Mouse neuroblastoma cells are inoculated with diagnostic material either in 8 chamber slides or 96-well microtiter plates and left to incubate for 72 h. For low viral load material, the material may need to under a number of passages. Hence, as for MIT, the delay in obtaining results from RTCIT precludes its use as a primary diagnostic tool. Moreover, if the homogenate contains an additional pathogen that causes mice to succumb or coinfects cell cultures, additional methods to confirm the nature of the pathogen are required.

12.6 Antemortem diagnosis of rabies

Antemortem diagnosis of rabies in humans is possible during the clinical illness, the major constraint being the brevity of the clinical period and the transient excretion of virus. Rapid and accurate diagnosis is therefore important with respect to both patient management and identification of the need for postexposure prophylaxis of patient contacts. Typically, each day results in a significant decline in neurological status, ending invariably with death. For human antemortem diagnosis of rabies, the best sample appropriate for dFA testing is skin biopsy material from the nape of the neck that includes several innervated hair follicles (Crepin et al., 1998). Initial evaluation consists of cryostat sectioning of the frozen skin biopsy, fixation in acetone, staining with rabies virus specific fluorescent conjugate, and examination under fluorescence microscopy. Positive observations typically occur where the skin follicles are innervated. This evaluation is complete within several hours of receipt of the sample but is dependent upon the availability of a cryostat. It is a highly specific test, but a negative finding on this sample with this method cannot definitively rule out rabies; subsequent samples from the patient may ultimately be positive.

Other useful samples are fresh saliva, serum, and cerebrospinal fluid (CSF). Serum and CSF are evaluated for antibodies by two methods. The indirect fluorescent antibody test can be completed quickly (i.e., within a few hours), but it is a binding test, not a functional test for antibody and it can yield false-positive results, especially in patients with encephalitis due to unrelated etiologies. The rapid fluorescent focus inhibition test (RFFIT) takes one full day (20+ hours) and the fluorescent antibody virus neutralization test (FAVN) takes 48 h to complete, but they are functional assays in which rabies virus neutralizing antibody reactivity is assessed by preventing rabies virus infection of cells (see Chapter 13).

However, in most patients, antibodies develop late in the clinical course, so the patient may remain seronegative throughout most of the investigation. Virus neutralization assays are only of benefit diagnostically if they provide evidence of antibodies in an unvaccinated patient. In most cases rabies immunoglobulin and vaccine are frequently administered upon first suspecting rabies.

In parallel with these tests, the subcutaneous tissues from the skin biopsy and the saliva are usually subjected to RT-PCR assays. Several studies have reported that a testing combination involving DFA test applied to skin biopsy tissue and standard or hnRT-PCR applied to saliva and skin biopsy could accurately identify most human cases (Brito et al., 2011; Crepin et al., 1998; Dacheux et al., 2008; Elmgren et al., 2002; Macedo et al., 2006; Smith et al., 2003). Some studies that explored the use of saliva only for antemortem rabies testing by molecular methods have shown that, whereas the detection rate can be quite high, it does not reach 100% (Hemachudha & Wacharapluesadee, 2004; Nagaraj et al., 2006; Wacharapluesadee & Hemachudha, 2001), either due to sporadic secretion of virus into this fluid and/or presence of very low levels of virus. Serial sampling of suspect patients is optimal for definitive diagnosis and must be considered before rendering a negative diagnosis using this sample type alone. Dacheux et al. (2008) suggested that a 100% sensitivity rate can be achieved if samples collected from a patient on at least three different days were tested. If the patient remains alive but ill for more than 2–3 weeks or recovers quickly, the differential diagnosis is less likely to include rabies.

The rate of detection of rabies virus in other sample types, including urine or CSF samples, appears to be consistently lower (Dacheux et al., 2008), although by testing a combination of such fluid samples accurate diagnosis has frequently been achieved. Interestingly, however, it has been noted that patients with paralytic rabies can give false-negative results by such tests, perhaps due to the timing of sample collection and/or the very limited titers of virus present in these sample types in patients exhibiting this form of the disease (Hemachudha & Wacharapluesadee, 2004; Wacharapluesadee & Hemachudha, 2002). Moreover, at the National Reference Centre for Rabies in Canada, and at the Animal and Plant Health Agency in the United Kingdom, CSF has consistently been found to be a poor sample type for human rabies diagnosis using molecular techniques (Elmgren et al., 2002).

In recent years, real-time RT-PCR methods have gained increasing acceptance for human antemortem diagnosis (Nadin-Davis et al., 2009) due to their reduced turnaround times and improved sensitivity, especially when testing saliva samples (Nagaraj et al., 2006). The type of such tests to be applied to human diagnosis should consider any recorded potential exposures as well as the patient's travel history so as to ensure that the test will detect any lyssavirus to which the patient could have been exposed. Such considerations reinforce the need for pan-lyssavirus diagnostic tests, particularly for those patients who present in the developed world and for whom global travel is common. Real-time RT-PCR has been employed to detect rabies virus intra vitam in saliva, corneal swab, and sputum samples taken from a patient infected after organ transplantation (Panning et al., 2010).

Although performance of antemortem diagnosis in domestic animals has not often been considered, animal rights considerations seeking to avoid unnecessary euthanasia of animals for rabies testing may exert pressure to develop robust animal antemortem testing regimens in the future. Saengseesom, Mitmoonpitak, Kasempimolporn, and Sitprija (2007) report a small study to explore the utility of saliva and CSF samples for antemortem diagnosis in dogs. Again, saliva was the better sample type for virus detection but did not yield a 100% detection level. More recent studies have further explored the use of RT-PCR methods for intra vitam detection of rabies in bodily fluids such as urine, saliva, and milk from suspect animals (Dandale et al., 2012, 2013). Compared to postmortem testing of brain tissue by DFA test all these results yielded values significantly below 100% sensitivity with real-time methods achieving higher success rates than nested RT-PCRs. Antemortem skin biopsy testing using a hnRT-PCR has also been explored and suggested to be a more useful alternative to saliva testing though again with less than 100% sensitivity compared to a postmortem test on brain tissue (Singh & Ahmad, 2018).

Thus, while antemortem diagnosis of rabies in animals can identify positive cases, negative results from such tests cannot be considered definitive; as a result such procedures are not practical under the majority of circumstances at this time. Without advanced supportive medical care, a rabid animal will succumb to the disease quickly. By the time diagnostic results would be available on a battery of samples similar to those described for humans, the animal would most likely be moribund or dead. Moreover, negative findings early in the clinical course of a rabid animal may mislead public health management of potentially exposed persons.

12.7 Conclusions

Even with the proper application of many of the methods described here, rabies control and its eventual elimination from an area remains challenging. Each laboratory-confirmed rabies positive case is the tip of the iceberg as it will often have had contact with either humans and/or numerous other animals, many of which are never submitted and diagnosed, particularly in areas where rabies is harbored by wildlife. Thus, in most systems that rely on passive submissions for diagnostic testing, the true incidence of the disease is unknown and ultimately demonstrating its elimination through control programs is difficult. Moreover, as canine rabies continues to circulate in many countries the potential for its reintroduction, via translocation of diseased "rescue dogs," into developed countries in which the disease has been eliminated, has been described in several reports (Hercules et al., 2018; Ribadeau-Dumas et al., 2016). Such a threat may be further exacerbated by an inadequate vaccination rate ($<70\%$) of pet dogs in the introduced country and the often significant time delay to identify that the responsible viral variant is not known to occur in that country. Of course, the impact of rabies on human health remains the principal driver toward elimination of this disease due to the huge costs that it can confer onto public health systems as well as the emotional distress that the disease imparts. A single rabies case in the wrong place, such as a county fair or pet shop, can result in the triage of large numbers of people for potential exposure to this animal and the need for postexposure prophylaxis (Noah et al., 1996). Despite the clear public health threat, fiscal constraints often jeopardize implementation of the infrastructure critical to rabies prevention, a problem especially severe for governments of developing countries. Accordingly, continued development of reliable, accurate, cost-effective, and timely primary diagnostic tools is needed in the fight to combat this disease.

Acknowledgments

LMM, DAM, and MG were financially supported by the UK Department for Environment, Food and Rural Affairs (Defra), Scottish Government, and Welsh Government by grant SV3500 and by European Virus Archive global (EVAg) project that has received funding from the European Union's Horizon 2020 research and innovation program under grant agreement No 653316.

References

Anonymous. (2011). Compendium of animal rabies prevention and control. *Morbidity and Mortality Weekly Report. Recommendations and Reports*, 60, 1–17.

Appler, K., Brunt, S., Jarvis, J. A., & Davis, A. D. (2019). Clarifying indeterminate results on the rabies direct fluorescent antibody test using real-time reverse transcriptase polymerase chain reaction. *Public Health Reports*, 134(1), 57–62.

AravindhBabu, R., Manoharan, S., & Ramadass, P. (2014). Diagnostic evaluation of RT-PCR–ELISA for the detection of rabies virus. *VirusDisease*, 25(1), 120–124.

Babu, R. A., Manoharan, S., Ramadass, P., & Chandran, N. (2012). Evaluation of RT-PCR assay for routine laboratory diagnosis of rabies in post mortem brain samples from different species of animals. *Indian Journal of Virology*, 23(3), 392–396.

Balachandran, A., & Charlton, K. (1994). Experimental rabies infection of non-nervous tissues in skunks (*Mephitis mephitis*) and foxes (*Vulpes vulpes*). *Veterinary Pathology*, 31(1), 93–102. https://doi.org/10.1177/030098589403100112.

Beck, S., Gunawardena, P., Horton, D., Hicks, D., Marston, D., Ortiz-Pelaez, A., … Núñez, A. (2017). Pathobiological investigation of naturally infected canine rabies cases from Sri Lanka. *BMC Veterinary Research*, 13(1), 99.

Bingham, J., & van der Merwe, M. (2002). Distribution of rabies antigen in infected brain material: Determining the reliability of different regions of the brain for the rabies fluorescent antibody test. *Journal of Virological Methods*, 101(1–2), 85–94. https://doi.org/10.1016/s0166-0934(01)00423-2.

Biswal, M., Ratho, R., & Mishra, B. (2007). Usefulness of reverse transcriptase-polymerase chain reaction for detection of rabies RNA in archival samples. *Japanese Journal of Infectious Diseases*, 60(5), 298.

Black, E. M., Lowings, J. P., Smith, J., Heaton, P. R., & McElhinney, L. M. (2002). A rapid RT-PCR method to differentiate six established genotypes of rabies and rabies-related viruses using TaqMan™ technology. *Journal of Virological Methods*, 105(1), 25–35.

Black, E. M., McElhinney, L. M., Lowings, J. P., Smith, J., Johnstone, P., & Heaton, P. R. (2000). Molecular methods to distinguish between classical rabies and the rabies-related European bat lyssaviruses. *Journal of Virological Methods*, 87(1–2), 123–131.

Boldbaatar, B., Inoue, S., Sugiura, N., Noguchi, A., Orbina, J., Demetria, C., … Yamada, A. (2009). Rapid detection of rabies virus by reverse transcription loop-mediated isothermal amplification. *Japanese Journal of Infectious Diseases*, 62(3), 187–191.

Bourhy, H., Cowley, J., Larrous, F., Holmes, E., & Walker, P. (2005). Phylogenetic relationships among rhabdoviruses inferred using the L polymerase gene. *Journal of General Virology*, 86(10), 2849–2858.

Bourhy, H., Kissi, B., & Tordo, N. (1993). Molecular diversity of the Lyssavirus genus. *Virology (New York, NY)*, 194(1), 70–81.

Brito, M. G. d., Chamone, T. L., Silva, F. J. d., Wada, M. Y., Miranda, A. B. d., Castilho, J. G., … Lemos, F. L. (2011). Antemortem diagnosis of human rabies in a veterinarian infected when handling a herbivore in Minas Gerais, Brazil. *Revista do Instituto de Medicina Tropical de São Paulo*, 53(1), 39–44.

Brookes, S. M., Klopfleisch, R., Müller, T., Healy, D. M., Teifke, J. P., Lange, E., … Kaden, V. (2007). Susceptibility of sheep to European bat lyssavirus type-1 and-2 infection: A clinical pathogenesis study. *Veterinary Microbiology*, 125(3–4), 210–223.

Bustin, S. A., Benes, V., Garson, J. A., Hellemans, J., Huggett, J., Kubista, M., … Shipley, G. L. (2009). The MIQE guidelines: Minimum information for publication of quantitative real-time PCR experiments. *Clinical Chemistry*, 55(4), 611–622.

Charlton, K., Nadin-Davis, S., Casey, G., & Wandeler, A. (1997). The long incubation period in rabies: Delayed progression of infection in muscle at the site of exposure. *Acta Neuropathologica*, 94(1), 73–77.

Clavijo, A., Freire de Carvalho, M. H., Orciari, L. A., Velasco-Villa, A., Ellison, J. A., Greenberg, L., ... Del Rio-Vilas, V. J. (2017). An inter-laboratory proficiency testing exercise for rabies diagnosis in Latin America and the Caribbean. *PLoS Neglected Tropical Diseases*, *11*(4), e0005427. https://doi.org/10.1371/journal.pntd.0005427.

Coertse, J., Weyer, J., Nel, L. H., & Markotter, W. (2010). Improved PCR methods for detection of African rabies and rabies-related lyssaviruses. *Journal of Clinical Microbiology*, *48*(11), 3949–3955.

Coertse, J., Weyer, J., Nel, L. H., & Markotter, W. (2019). Reverse transcription recombinase polymerase amplification assay for rapid detection of canine associated rabies virus in Africa. *PLoS One*, *14*(7), e0219292.

Condori, R. E., Niezgoda, M., Lopez, G., Matos, C. A., Mateo, E. D., Gigante, C., ... Li, Y. (2020). Using the LN34 pan-Lyssavirus real-time RT-PCR assay for rabies diagnosis and rapid genetic typing from formalin-fixed human brain tissue. *Viruses*, *12*(1), v12010120.

Cooper, A., & Poinar, H. N. (2000). Ancient DNA: Do it right or not at all. *Science*, *289*(5482), 1139.

Crepin, P., Audry, L., Rotivel, Y., Gacoin, A., Caroff, C., & Bourhy, H. (1998). Intravitam diagnosis of human rabies by PCR using saliva and cerebrospinal fluid. *Journal of Clinical Microbiology*, *36*(4), 1117–1121.

Dacheux, L., Larrous, F., Lavenir, R., Lepelletier, A., Faouzi, A., Troupin, C., ... Bourhy, H. (2016). Dual combined real-time reverse transcription polymerase chain reaction assay for the diagnosis of lyssavirus infection. *PLoS Neglected Tropical Diseases*, *10*(7), e0004812.

Dacheux, L., Reynes, J. -M., Buchy, P., Sivuth, O., Diop, B. M., Rousset, D., ... Nareth, C. (2008). A reliable diagnosis of human rabies based on analysis of skin biopsy specimens. *Clinical Infectious Diseases*, *47*(11), 1410–1417.

Dandale, M., Singh, C., Ramneek, V., Deka, D., Bansal, K., & Sood, N. (2013). Sensitivity comparison of nested RT-PCR and TaqMan real time PCR for intravitam diagnosis of rabies in animals from urine samples. *Vet World*, *6*(4), 189–192.

Dandale, M., Singh, C., Ramneek, V., Deka, D., Sandhu, B., Bansal, K., & Sood, N. (2012). Nested RT-PCR for ante mortem diagnosis of rabies from body secretion/excretion of animals suspected for rabies. *WORLD*, *5*(11), 690–693.

David, D., Yakobson, B., Rotenberg, D., Dveres, N., Davidson, I., & Stram, Y. (2002). Rabies virus detection by RT-PCR in decomposed naturally infected brains. *Veterinary Microbiology*, *87*(2), 111–118.

De Benedictis, P., De Battisti, C., Dacheux, L., Marciano, S., Ormelli, S., Salomoni, A., ... Cattoli, G. (2011). Lyssavirus detection and typing using pyrosequencing. *Journal of Clinical Microbiology*, *49*(5), 1932–1938. https://doi.org/10.1128/jcm.02015-10.

Dimaano, E. M., Scholand, S. J., Alera, M. T. P., & Belandres, D. B. (2011). Clinical and epidemiological features of human rabies cases in the Philippines: A review from 1987 to 2006. *International Journal of Infectious Diseases*, *15*(7), e495–e499.

Dupuis, M., Brunt, S., Appler, K., Davis, A., & Rudd, R. (2015). Comparison of automated quantitative reverse transcription-PCR and direct fluorescent-antibody detection for routine rabies diagnosis in the United States. *Journal of Clinical Microbiology*, *53*(9), 2983–2989.

Dürr, S., Naïssengar, S., Mindekem, R., Diguimbye, C., Niezgoda, M., Kuzmin, I., ... Zinsstag, J. (2008). Rabies diagnosis for developing countries. *PLoS Neglected Tropical Diseases*, *2*(3), e206. https://doi.org/10.1371/journal.pntd.0000206.

Echevarría, J. E., Avellón, A., Juste, J., Vera, M., & Ibáñez, C. (2001). Screening of active lyssavirus infection in wild bat populations by viral RNA detection on oropharyngeal swabs. *Journal of Clinical Microbiology*, *39*(10), 3678–3683.

Eggerbauer, E., de Benedictis, P., Hoffmann, B., Mettenleiter, T. C., Schlottau, K., Ngoepe, E. C., ... Muller, T. (2016). Evaluation of six commercially available rapid immunochromatographic tests for the diagnosis of rabies in brain material. *PLoS Neglected Tropical Diseases*, *10*(6), e0004776. https://doi.org/10.1371/journal.pntd.0004776.

Ehimiyein, A., Niezgoda, M., Orciari, L., Osinubi, M., Ehimiyein, I., Adawa, D., ... Rupprecht, C. (2014). Efficacy of a direct rapid immunohistochemical test (DRIT) for rabies detection in Nigeria. *African Journal of Biomedical Research*, *17*(2), 101–107.

Elmgren, L. D., Nadin-Davis, S. A., Muldoon, F. T., & Wandeler, A. I. (2002). Diagnosis and analysis of a recent case of human rabies in Canada. *Canadian Journal of Infectious Diseases and Medical Microbiology*, *13*(2), 129–133.

Faye, M., Dacheux, L., Weidmann, M., Diop, S. A., Loucoubar, C., Bourhy, H., ... Faye, O. (2017). Development and validation of sensitive real-time RT-PCR assay for broad detection of rabies virus. *Journal of Virological Methods*, *243*, 120–130.

Fehlner-Gardiner, C. (2018). Rabies control in North America-past, present and future. *Revue scientifique et technique (International Office of Epizootics)*, *37*(2), 421–437.

Fekadu, M., Greer, P., Chandler, F., & Sanderlin, D. (1988). Use of the avidin-biotin peroxidase system to detect rabies antigen in formalin-fixed paraffin-embedded tissues. *Journal of Virological Methods*, *19*(2), 91–96.

Fischer, M., Freuling, C. M., Müller, T., Wegelt, A., Kooi, E. A., Rasmussen, T. B., ... Beer, M. (2014). Molecular double-check strategy for the identification and characterization of European lyssaviruses. *Journal of Virological Methods*, *203*, 23–32.

Fooks, A. R., Johnson, N., Freuling, C. M., Wakeley, P. R., Banyard, A. C., McElhinney, L. M., ... Weiss, R. A. (2009). Emerging technologies for the detection of rabies virus: Challenges and hopes in the 21st century. *PLoS Neglected Tropical Diseases*, *3*(9), e530.

Foord, A., Heine, H., Pritchard, L., Lunt, R., Newberry, K., Rootes, C., & Boyle, D. (2006). Molecular diagnosis of lyssaviruses and sequence comparison of Australian bat lyssavirus samples. *Australian Veterinary Journal*, *84*(7), 225–230.

Freuling, C., Vos, A., Johnson, N., Kaipf, I., Denzinger, A., Neubert, L., ... Tordo, N. (2009). Experimental infection of serotine bats (*Eptesicus serotinus*) with European bat lyssavirus type 1a. *Journal of General Virology*, *90*(10), 2493–2502.

Gigante, C. M., Dettinger, L., Powell, J. W., Seiders, M., Condori, R. E. C., Griesser, R., ... Breckenridge, M. (2018). Multi-site evaluation of the LN34 pan-lyssavirus real-time RT-PCR assay for post-mortem rabies diagnostics. *PLoS One*, *13*(5), e0197074.

Goharriz, H., Marston, D., Sharifzoda, F., Ellis, R., Horton, D., Khakimov, T., ... Bazarov, M. (2017). First complete genomic sequence of a rabies virus from the Republic of Tajikistan obtained directly from a Flinders Technology Associates card. *Genome Announcements*, *5*(27), e00515–e00517.

Goldwasser, R., & Kissling, R. (1958). Fluorescent antibody staining of street and fixed rabies virus antigens. *Proceedings of the Society for Experimental Biology and Medicine*, *98*(2), 219–223.

Gurrala, R., Dastjerdi, A., Johnson, N., Nunez-Garcia, J., Grierson, S., Steinbach, F., & Banks, M. (2009). Development of a DNA microarray for simultaneous detection and genotyping of lyssaviruses. *Virus Research*, *144*(1–2), 202–208.

Hamir, A., Moser, G., Fu, Z., Dietzschold, B., & Rupprecht, C. (1995). Immunohistochemical test for rabies: Identification of a diagnostically superior monoclonal antibody. *The Veterinary Record*, *136*(12), 295–296.

Hanlon, C. A., Smith, J. S., & Anderson, G. R. (1999). Recommendations of a national working group on prevention and control of rabies in the United States. Article II: Laboratory diagnosis of rabies. The National Working Group on Rabies Prevention and Control. *Journal of the American Veterinary Medical Association*, *215*, 1444–1446.

Hayman, D. T., Banyard, A. C., Wakeley, P. R., Harkess, G., Marston, D., Wood, J. L., ... Fooks, A. R. (2011). A universal real-time assay for the detection of Lyssaviruses. *Journal of Virological Methods*, *177*(1), 87–93.

Hayman, D. T., Johnson, N., Horton, D. L., Hedge, J., Wakeley, P. R., Banyard, A. C., ... Fooks, A. R. (2011). Evolutionary history of rabies in Ghana. *PLoS Neglected Tropical Diseases*, *5*(4), e1001.

Heaton, P. R., JohnstonE, P., MCElhinney, L. M., Cowley, R., O'Sullivan, E., & Whitby, J. E. (1997). Heminested PCR assay for detection of six genotypes of rabies and rabies-related viruses. *Journal of Clinical Microbiology*, *35*(11), 2762–2766.

Heaton, P. R., McElhinney, L. M., & Lowings, J. P. (1999). Detection and identification of rabies and rabies-related viruses using rapid-cycle PCR. *Journal of Virological Methods*, *81*(1–2), 63–69.

Hemachudha, T., & Mitrabhakdi, E. (2000). Rabies. In L. E. Davis & P. G. E. Kennedy (Eds.), *Infectious diseases of the central nervous system* (pp. 401–444): Butterworth Heinemann Oxford.

Hemachudha, T., & Wacharapluesadee, S. (2004). Antemortem diagnosis of human rabies. *Clinical Infectious Diseases: An Official Publication of the Infectious Diseases Society of America*, *39*(7), 1085.

Hercules, Y., Bryant, N. J., Wallace, R. M., Nelson, R., Palumbo, G., Williams, J. N., ... Brown, C. (2018). Rabies in a dog imported from Egypt—Connecticut, 2017. *MMWR. Morbidity and Mortality Weekly Report*, *67*(50), 1388–1391. https://doi.org/10.15585/mmwr.mm6750a3.

Hoffmann, B., Depner, K., Schirrmeier, H., & Beer, M. (2006). A universal heterologous internal control system for duplex real-time RT-PCR assays used in a detection system for pestiviruses. *Journal of Virological Methods*, *136*(1–2), 200–209.

Hoffmann, B., Freuling, C., Wakeley, P., Rasmussen, T. B., Leech, S., Fooks, A., ... Müller, T. (2010). Improved safety for molecular diagnosis of classical rabies viruses by use of a TaqMan real-time reverse transcription-PCR "double check" strategy. *Journal of Clinical Microbiology*, *48*(11), 3970–3978.

Hughes, G., Kuzmin, I., Schmitz, A., Blanton, J., Manangan, J., Murphy, S., & Rupprecht, C. (2006). Experimental infection of big brown bats (*Eptesicus fuscus*) with Eurasian bat lyssaviruses Aravan, Khujand, and Irkut virus. *Archives of Virology*, *151*(10), 2021–2035.

Hughes, G., Smith, J., Hanlon, C., & Rupprecht, C. (2004). Evaluation of a TaqMan PCR assay to detect rabies virus RNA: Influence of sequence variation and application to quantification of viral loads. *Journal of Clinical Microbiology, 42*(1), 299–306.

Jarvis, J. A., Brown, K. T., Appler, K. A., Fitzgerald, D. P., & Davis, A. D. (2019). Rabies necropsy techniques in large and small animals. *Journal of Visualized Experiments*, (149). https://doi.org/10.3791/59574.

Jayakumar, R., & Padmanaban, V. D. (1994). A dipstick dot enzyme immunoassay for detection of rabies antigen. *Zentralblatt für Bakteriologie, 280*(3), 382–385.

Johnson, N., Phillpotts, R., & Fooks, A. (2006). Airborne transmission of lyssaviruses. *Journal of Medical Microbiology, 55*(6), 785–790.

Johnson, N., Vos, A., Neubert, L., Freuling, C., Mansfield, K. L., Kaipf, I., ... Franka, R. (2008). Experimental study of European bat lyssavirus type-2 infection in Daubenton's bats (*Myotis daubentonii*). *Journal of General Virology, 89*(11), 2662–2672.

Kamolvarin, N., Tirawatnpong, T., Rattanasiwamoke, R., Tirawatnpong, S., Panpanich, T., & Hemachudha, T. (1993). Diagnosis of rabies by polymerase chain reaction with nested primers. *Journal of Infectious Diseases, 167*(1), 207–210.

Kang, B., Oh, J., Lee, C., Park, B. K., Park, Y., Hong, K., ... Song, D. (2007). Evaluation of a rapid immunodiagnostic test kit for rabies virus. *Journal of Virological Methods, 145*(1), 30–36. https://doi.org/10.1016/j.jviromet.2007.05.005.

Kasempimolporn, S., Saengseesom, W., Huadsakul, S., Boonchang, S., & Sitprija, V. (2011). Evaluation of a rapid immunochromatographic test strip for detection of Rabies virus in dog saliva samples. *Journal of Veterinary Diagnostic Investigation, 23*(6), 1197–1201.

King, A. A., Fooks, A. R., Aubert, M., & Wandeler, A. (2004). *Historical perspective of rabies in Europe and the Mediterranean Basin*. Paris: OIE.

Koprowski, H. (1966). *Laboratory techniques in rabies. Mouse inoculation test* (1st ed.). World Health Organization.

Koprowski, H. (1973). *Laboratory techniques in rabies. The mouse inoculation test* (3rd ed.). World Health Organization.

Kulonen, K., Fekadu, M., Whitfield, S., & Warner, C. (1999). An evaluation of immunofluorescence and PCR. Methods for detection of rabies in archival Carnoy-fixed, paraffin-embedded brain tissue. *Journal of Veterinary Medicine Series B, 46*(3), 151–156.

Kuzmin, I. V., Hughes, G. J., Botvinkin, A. D., Orciari, L. A., & Rupprecht, C. E. (2005). Phylogenetic relationships of Irkut and West Caucasian bat viruses within the Lyssavirus genus and suggested quantitative criteria based on the N gene sequence for lyssavirus genotype definition. *Virus Research, 111*(1), 28–43.

Kwok, S., & Higuchi, R. (1989). Avoiding false positives with PCR. *Nature, 339*, 237.

Le Mercier, P., Jacob, Y., & Tordo, N. (1997). The complete Mokola virus genome sequence: Structure of the RNA-dependent RNA polymerase. *Journal of General Virology, 78*(7), 1571–1576.

Lembo, T., Niezgoda, M., Velasco-Villa, A., Cleaveland, S., Ernest, E., & Rupprecht, C. E. (2006). Evaluation of a direct, rapid immunohistochemical test for rabies diagnosis. *Emerging Infectious Diseases, 12*(2), 310–313.

Lopes, M. C., Venditti, L. L. R., & Queiroz, L. H. (2010). Comparison between RT-PCR and the mouse inoculation test for detection of rabies virus in samples kept for long periods under different conditions. *Journal of Virological Methods, 164*(1–2), 19–23.

Ma, X., Monroe, B. P., Cleaton, J. M., Orciari, L. A., Li, Y., Kirby, J. D., ... Blanton, J. D. (2018). Rabies surveillance in the United States during 2017. *Journal of the American Veterinary Medical Association, 253*(12), 1555–1568.

Macedo, C. I., Carnieli, P., Jr., Brandão, P. E., Rosa, E. S., Oliveira, R. d. N., Castilho, J. G., ... Carrieri, M. L. (2006). Diagnosis of human rabies cases by polymerase chain reaction of neck-skin samples. *Brazilian Journal of Infectious Diseases, 10*(5), 341–345.

Madhusudana, S. N., Paul, J. P. V., Abhilash, V. K., & Suja, M. S. (2004). Rapid diagnosis of rabies in humans and animals by a dot blot enzyme immunoassay. *International Journal of Infectious Diseases, 8*(6), 339–345.

Madhusudana, S. N., Subha, S., Thankappan, U., & Ashwin, Y. B. (2012). Evaluation of a direct rapid immunohistochemical test (dRIT) for rapid diagnosis of rabies in animals and humans. *Virologica Sinica, 27*(5), 299–302. https://doi.org/10.1007/s12250-012-3265-6.

Mallewa, M., Fooks, A. R., Banda, D., Chikungwa, P., Mankhambo, L., Molyneux, E., ... Solomon, T. (2007). Rabies encephalitis in malaria-endemic area, Malawi, Africa. *Emerging Infectious Diseases, 13*(1), 136–139.

Manning, S. E., Rupprecht, C. E., Fishbein, D., Hanlon, C. A., Lumlertdacha, B., Guerra, M., ... Jenkins, S. R. (2008). Human rabies prevention—United States, 2008: Recommendations of the advisory committee on immunization practices. *MMWR—Recommendations and Reports, 57*(RR-3), 1–28.

Markotter, W., Coertse, J., Le Roux, K., Peens, J., Weyer, J., Blumberg, L., & Nel, L. H. (2015). Utility of forensic detection of rabies virus in decomposed exhumed dog carcasses. *Journal of the South African Veterinary Association, 86*(1), 01–05.

Marston, D. A., Jennings, D. L., MacLaren, N. C., Dorey-Robinson, D., Fooks, A. R., Banyard, A. C., & McElhinney, L. M. (2019). Pan-lyssavirus real time RT-PCR for rabies diagnosis. *Journal of Visualized Experiments*. 149. https://doi.org/10.3791/59709.

McColl, K., Gould, A., Selleck, P., Hooper, P., Westbury, H., & Smith, J. (1993). Polymerase chain reaction and other laboratory techniques in the diagnosis of long incubation rabies in Australia. *Australian Veterinary Journal*, 70(3), 84–89.

McElhinney, L. M., Marston, D. A., Brookes, S. M., & Fooks, A. R. (2014). Effects of carcase decomposition on rabies virus infectivity and detection. *Journal of Virological Methods*, 207, 110–113.

Middel, K., Fehlner-Gardiner, C., Pulham, N., & Buchanan, T. (2017). Incorporating direct rapid immunohistochemical testing into large-scale wildlife rabies surveillance. *Tropical Medicine and Infectious Disease*, 2(3), 21.

Muleya, W., Namangala, B., Mweene, A., Zulu, L., Fandamu, P., Banda, D., … Ishii, A. (2012). Molecular epidemiology and a loop-mediated isothermal amplification method for diagnosis of infection with rabies virus in Zambia. *Virus Research*, 163(1), 160–168.

Nadin-Davis, S. A. (1998). Polymerase chain reaction protocols for rabies virus discrimination. *Journal of Virological Methods*, 75(1), 1–8.

Nadin-Davis, S. (2019). Rapid identification of the raccoon rabies virus variant using a real-time reverse-transcriptase polymerase chain reaction. *Journal of Virological Methods*, 273, 113713.

Nadin-Davis, S. A., Huang, W., & Wandeler, A. I. (1996). The design of strain-specific polymerase chain reactions for discrimination of the raccoon rabies virus strain from indigenous rabies viruses of Ontario. *Journal of Virological Methods*, 57(2), 141–156.

Nadin-Davis, S. A., Sheen, M., & Wandeler, A. I. (2003). Use of discriminatory probes for strain typing of formalin-fixed, rabies virus-infected tissues by in situ hybridization. *Journal of Clinical Microbiology*, 41(9), 4343–4352.

Nadin-Davis, S. A., Sheen, M., & Wandeler, A. I. (2012). Recent emergence of the Arctic rabies virus lineage. *Virus Research*, 163(1), 352–362.

Nadin-Davis, S. A., Sheen, M., & Wandeler, A. I. (2009). Development of real-time reverse transcriptase polymerase chain reaction methods for human rabies diagnosis. *Journal of Medical Virology*, 81(8), 1484–1497.

Nagamine, K., Hase, T., & Notomi, T. (2002). Accelerated reaction by loop-mediated isothermal amplification using loop primers. *Molecular and Cellular Probes*, 16(3), 223–229.

Nagaraj, T., Vasanth, J. P., Desai, A., Kamat, A., Madhusudana, S., & Ravi, V. (2006). Ante mortem diagnosis of human rabies using saliva samples: Comparison of real time and conventional RT-PCR techniques. *Journal of Clinical Virology*, 36(1), 17–23.

Naji, E., Fadajan, Z., Afshar, D., & Fazeli, M. (2020). Comparison of reverse transcription loop-mediated isothermal amplification method with SYBR green real-time RT-PCR and direct fluorescent antibody test for diagnosis of rabies. *Japanese Journal of Infectious Diseases*, 73(1), 19–25.

Negri, A. (1903). Beitrag zum studium der aetiologie der tollwuth. *Medical Microbiology and Immunology*, 43(1), 507–528.

Nel, L. H., Bingham, J., Jacobs, J. A., & Jaftha, J. B. (1998). A nucleotide-specific polymerase chain reaction assay to differentiate rabies virus biotypes in South Africa. *Onderstepoort Journal of Veterinary Research*, 65, 297–303.

Noah, D. L., Smith, M. G., Gotthardt, J. C., Krebs, J. W., Green, D., & Childs, J. E. (1996). Mass human exposure to rabies in New Hampshire: Exposures, treatment, and cost. *American Journal of Public Health*, 86(8_Pt_1), 1149–1151.

Notomi, T., Okayama, H., Masubuchi, H., Yonekawa, T., Watanabe, K., Amino, N., & Hase, T. (2000). Loop-mediated isothermal amplification of DNA. *Nucleic Acids Research*, 28(12), e63.

OIE. (2018a). *Manual of diagnostic tests and vaccines for terrestrial animals* (8th ed., pp. 64–71). Paris: Office international des epizooties.

OIE. (2018b). *Manual of diagnostic tests and vaccines for terrestrial animals* (8th ed., pp. 578–612). Paris: Office international des epizooties.

Orciari, L. A., Niezgoda, M., Hanlon, C. A., Shaddock, J. H., Sanderlin, D. W., Yager, P. A., & Rupprecht, C. E. (2001). Rapid clearance of SAG-2 rabies virus from dogs after oral vaccination. *Vaccine*, 19(31), 4511–4518. https://doi.org/10.1016/S0264-410X(01)00186-4.

Orlowska, A., Smreczak, M., Trebas, P., & Zmudzinski, J. F. (2008). Comparison of real-time PCR and heminested RT-PCR methods in the detection of rabies virus infection in bats and terrestrial animals. *Bulletin of the Veterinary Institute in Pulawy*, 52(3), 313–318.

Panning, M., Baumgarte, S., Pfefferle, S., Maier, T., Martens, A., & Drosten, C. (2010). Comparative analysis of rabies virus reverse transcription-PCR and virus isolation using samples from a patient infected with rabies virus. *Journal of Clinical Microbiology*, 48(8), 2960–2962.

Patrick, E. M., Bjorklund, B. M., Kirby, J. D., Nelson, K. M., Chipman, R. B., & Rupprecht, C. E. (2019). Enhanced rabies surveillance using a direct rapid immunohistochemical test. *Journal of Visualized Experiments, 146*, e59416.

Perrin, P., Gontier, C., Lecocq, E., & Bourhy, H. (1992). A modified rapid enzyme immunoassay for the detection of rabies and rabies-related viruses: RREID-lyssa. *Biologicals, 20*(1), 51–58.

Picard-Meyer, E., Barrat, J., & Cliquet, F. (2007). Use of filter paper (FTA®) technology for sampling, recovery and molecular characterisation of rabies viruses. *Journal of Virological Methods, 140*(1–2), 174–182.

Picard-Meyer, E., Bruyere, V., Barrat, J., Tissot, E., Barrat, M., & Cliquet, F. (2004). Development of a hemi-nested RT-PCR method for the specific determination of European Bat Lyssavirus 1: Comparison with other rabies diagnostic methods. *Vaccine, 22*(15–16), 1921–1929.

Picard-Meyer, E., Dubourg-Savage, M. -J., Arthur, L., Barataud, M., Bécu, D., Bracco, S., ... Moinet, M. (2011). Active surveillance of bat rabies in France: A 5-year study (2004–2009). *Veterinary Microbiology, 151*(3–4), 390–395.

Picard-Meyer, E., Peytavin de Garam, C., Schereffer, J. L., Marchal, C., Robardet, E., & Cliquet, F. (2015). Cross-platform evaluation of commercial real-time SYBR green RT-PCR kits for sensitive and rapid detection of European bat lyssavirus type 1. *BioMed Research International, 2015*.

Picard-Meyer, E., Peytavin de Garam, C., Schereffer, J. L., Robardet, E., & Cliquet, F. (2019). Evaluation of six TaqMan RT-rtPCR kits on two thermocyclers for the reliable detection of rabies virus RNA. *Journal of Veterinary Diagnostic Investigation, 31*(1), 47–57. https://doi.org/10.1177/1040638718818223.

Powell, J. (1997). *Proficiency testing in the rabies diagnostic laboratory.* Paper presented at the Abstracts of the eighth annual rabies in the Americas conference, Kingston, Ontario.

Prabhu, K., Isloor, S., Veeresh, B., Rathnamma, D., Sharada, R., Das, L., ... Rahman, S. (2018). Application and comparative evaluation of fluorescent antibody, immunohistochemistry and reverse transcription polymerase chain reaction tests for the detection of rabies virus antigen or nucleic acid in brain samples of animals suspected of rabies in India. *Veterinary Sciences, 5*(1), 24.

Prada, D., Boyd, V., Baker, M., Jackson, B., & O'Dea, M. (2019). Insights into Australian bat lyssavirus in insectivorous bats of Western Australia. *Tropical Medicine and Infectious Disease, 4*(1), 46.

Ribadeau-Dumas, F., Cliquet, F., Gautret, P., Robardet, E., Le Pen, C., & Bourhy, H. (2016). Travel-associated rabies in pets and residual rabies risk, Western Europe. *Emerging Infectious Diseases, 22*(7), 1268–1271. https://doi.org/10.3201/eid2207.151733.

Robardet, E., Picard-Meyer, E., Andrieu, S., Servat, A., & Cliquet, F. (2011). International interlaboratory trials on rabies diagnosis: An overview of results and variation in reference diagnosis techniques (fluorescent antibody test, rabies tissue culture infection test, mouse inoculation test) and molecular biology techniques. *Journal of Virological Methods, 177*(1), 15–25.

Rohde, R. E., Neill, S. U., Clark, K. A., & Smith, J. S. (1997). Molecular epidemiology of rabies epizootics in Texas. *Clinical and Diagnostic Virology, 8*(3), 209–217.

Rudd, R. J., Smith, J. S., Yager, P. A., Orciari, L. A., & Trimarchi, C. V. (2005). A need for standardized rabies-virus diagnostic procedures: Effect of cover-glass mountant on the reliability of antigen detection by the fluorescent antibody test. *Virus Research, 111*(1), 83–88.

Rupprecht, C., Xiang, Z., Servat, A., Franka, R., Kirby, J., & Ertl, H. (2018). Additional progress in the development and application of a direct, rapid immunohistochemical test for rabies diagnosis. *Veterinary Sciences, 5*(2), 59.

Sacramento, D., Bourhy, H., & Tordo, N. (1991). PCR technique as an alternative method for diagnosis and molecular epidemiology of rabies virus. *Molecular and Cellular Probes, 5*(3), 229–240.

Saengseesom, W., Mitmoonpitak, C., Kasempimolporn, S., & Sitprija, V. (2007). Real-time PCR analysis of dog cerebrospinal fluid and saliva samples for ante-mortem diagnosis of rabies. *Southeast Asian Journal of Tropical Medicine and Public Health, 38*(1), 53–57.

Saiki, R. K., & Erlich, H. A. (1989). *PCR technology: Principles and applications for DNA amplification.* New York: Stockton.

Saitou, Y., Kobayashi, Y., Hirano, S., Mochizuki, N., Itou, T., Ito, F. H., & Sakai, T. (2010). A method for simultaneous detection and identification of Brazilian dog-and vampire bat-related rabies virus by reverse transcription loop-mediated isothermal amplification assay. *Journal of Virological Methods, 168*(1–2), 13–17.

Schatz, J., Fooks, A., McElhinney, L., Horton, D., Echevarria, J., Vázquez-Moron, S., ... Freuling, C. (2013). Bat rabies surveillance in Europe. *Zoonoses and Public Health, 60*(1), 22–34.

Schlottau, K., Freuling, C. M., Müller, T., Beer, M., & Hoffmann, B. (2017). Development of molecular confirmation tools for swift and easy rabies diagnostics. *Virology Journal, 14*(1), 184.

Sellers, T. (1923). Status of rabies in the United States in 1921. *American Journal of Public Health, 13*(9), 742–747.

Serra-Cobo, J., Amengual, B., Abellán, C., & Bourhy, H. (2002). European bat lyssavirus infection in Spanish bat populations. *Emerging Infectious Diseases, 8*(4), 413–420.

Servat, A., Picard-Meyer, E., Robardet, E., Muzniece, Z., Must, K., & Cliquet, F. (2012). Evaluation of a Rapid Immunochromatographic Diagnostic Test for the detection of rabies from brain material of European mammals. *Biologicals, 40*(1), 61–66.

Servat, A., Robardet, E., & Cliquet, F. (2019). An inter-laboratory comparison to evaluate the technical performance of rabies diagnosis lateral flow assays. *Journal of Virological Methods, 272*, 113702.

Shulpin, M., Nazarov, N., Chupin, S., Korennoy, F., Metlin, A. Y., & Mischenko, A. (2018). Rabies surveillance in the Russian Federation. *Revue scientifique et technique (International Office of Epizootics), 37*(2), 483–495.

Singh, C., & Ahmad, A. (2018). Molecular approach for ante-mortem diagnosis of rabies in dogs. *The Indian Journal of Medical Research, 147*(5), 513.

Smith, J. S., Fishbein, D. B., Rupprecht, C. E., & Clark, K. (1991). Unexplained rabies in three immigrants in the United States. A virologic investigation. *New England Journal of Medicine, 324*(4), 205–211.

Smith, J., McElhinney, L. M., Heaton, P. R., Black, E. M., & Lowings, J. P. (2000). Assessment of template quality by the incorporation of an internal control into a RT-PCR for the detection of rabies and rabies-related viruses. *Journal of Virological Methods, 84*(2), 107–115.

Smith, J., McElhinney, L., Parsons, G., Brink, N., Doherty, T., Agranoff, D., … Fooks, A. R. (2003). Case report: Rapid ante-mortem diagnosis of a human case of rabies imported into the UK from the Philippines. *Journal of Medical Virology, 69*(1), 150–155.

Smith, I. L., Northill, J. A., Harrower, B. J., & Smith, G. A. (2002). Detection of Australian bat lyssavirus using a fluorogenic probe. *Journal of Clinical Virology, 25*(3), 285–291.

Smith, J. S., Orciari, L. A., Yager, P. A., Seidel, H. D., & Warner, C. K. (1992). Epidemiologic and historical relationships among 87 rabies virus isolates as determined by limited sequence analysis. *Journal of Infectious Diseases, 166*(2), 296–307.

Solomon, T., Marston, D., Mallewa, M., Felton, T., Shaw, S., McElhinney, L. M., … Kwong, G. N. M. (2005). Paralytic rabies after a two week holiday in India. *BMJ, 331*(7515), 501–503.

Suin, V., Nazé, F., Francart, A., Lamoral, S., De Craeye, S., Kalai, M., & Van Gucht, S. (2014). A two-step lyssavirus real-time polymerase chain reaction using degenerate primers with superior sensitivity to the fluorescent antigen test, *BioMed Research International, 2014*, 256175.

Trimarchi, C. V., & Smith, J. S. (2002). Diagnostic evaluation. In A. C. Jackson & W. H. Wunner (Eds.), *Rabies* (pp. 307–349). San Diego: Academic Press.

Vasanth, J. P., Madhusudana, S. N., Abhilash, V. K., Suga, M. S., & Muhamuda, K. (2004). Development and evaluation of an enzyme immunoassay for rapid diagnosis of rabies in humans and animals. *Indian Journal of Pathology & Microbiology, 47*(4), 574–578.

Vázquez-Morón, S., Avellon, A., & Echevarría, J. (2006). RT-PCR for detection of all seven genotypes of Lyssavirus genus. *Journal of Virological Methods, 135*(2), 281–287.

Vos, A., Müller, T., Neubert, L., Zurbriggen, A., Botteron, C., Pöhle, D., … Jackson, A. (2004). Rabies in red foxes (*Vulpes vulpes*) experimentally infected with European bat lyssavirus type 1. *Journal of Veterinary Medicine Series B, 51*(7), 327–332.

Wacharapluesadee, S., & Hemachudha, T. (2001). Nucleic-acid sequence based amplification in the rapid diagnosis of rabies. *The Lancet, 358*(9285), 892–893.

Wacharapluesadee, S., & Hemachudha, T. (2002). Urine samples for rabies RNA detection in the diagnosis of rabies in humans. *Clinical Infectious Diseases, 34*(6), 874–875.

Wacharapluesadee, S., & Hemachudha, T. (2010). Ante- and post-mortem diagnosis of rabies using nucleic acid-amplification tests. *Expert Review of Molecular Diagnostics, 10*(2), 207–218.

Wacharapluesadee, S., Phumesin, P., Lumlertdaecha, B., & Hemachudha, T. (2003). Diagnosis of rabies by use of brain tissue dried on filter paper. *Clinical Infectious Diseases, 36*(5), 674–675.

Wacharapluesadee, S., Ruangvejvorachai, P., & Hemachudha, T. (2006). A simple method for detection of rabies viral sequences in 16-year old archival brain specimens with one-week fixation in formalin. *Journal of Virological Methods, 134*(1–2), 267–271.

Wacharapluesadee, S., Sutipanya, J., Damrongwatanapokin, S., Phumesin, P., Chamnanpood, P., Leowijuk, C., & Hemachudha, T. (2008). Development of a TaqMan real-time RT-PCR assay for the detection of rabies virus. *Journal of Virological Methods, 151*(2), 317–320.

Wacharapluesadee, S., Tepsumethanon, V., Supavonwong, P., Kaewpom, T., Intarut, N., & Hemachudha, T. (2012). Detection of rabies viral RNA by TaqMan real-time RT-PCR using non-neural specimens from dogs infected with rabies virus. *Journal of Virological Methods, 184*(1–2), 109–112.

Wadhwa, A., Wilkins, K., Gao, J., Condori, R. E. C., Gigante, C. M., Zhao, H., … Velasco-Villa, A. (2017). A pan-lyssavirus Taqman real-time RT-PCR assay for the detection of highly variable rabies virus and other lyssaviruses. *PLoS Neglected Tropical Diseases, 11*(1), e0005258.

Wakeley, P., Johnson, N., McElhinney, L., Marston, D., Sawyer, J., & Fooks, A. (2005). Development of a real-time, TaqMan reverse transcription-PCR assay for detection and differentiation of lyssavirus genotypes 1, 5, and 6. *Journal of Clinical Microbiology, 43*(6), 2786–2792.

Wang, L., Liu, Y., Zhang, S., Wang, Y., Zhao, J., Miao, F., & Hu, R. (2014). A SYBR-green I quantitative real-time reverse transcription-PCR assay for rabies viruses with different virulence. *Virologica Sinica, 29*(2), 131–132.

Warner, C. K., Whitfield, S. G., Fekadu, M., & Ho, H. (1997). Procedures for reproducible detection of rabies virus antigen mRNA and genome in situ in formalin-fixed tissues. *Journal of Virological Methods, 67*(1), 5–12.

Whitby, J., Heaton, P., Whitby, H., O'sullivan, E., & Johnstone, P. (1997). Rapid detection of rabies and rabies-related viruses by RT-PCR and enzyme-linked immunosorbent assay. *Journal of Virological Methods, 69*(1–2), 63–72.

World Health Organization. (2015). *Global elimination of dog-mediated human rabies. The time is now*. Available at: http://apps.who.int/iris/bitstream/10665/204621/1/WHO_HTM_NTD_NZD_2016.02_eng.pdf?ua=1. Paper presented at the Rabies Global Conference, Geneva.

World Health Organization. (2018a). *Laboratory techniques in rabies* (5th ed.). Geneva.

World Health Organization. (2018b). Rabies vaccines: WHO position paper, April 2018–recommendations. *Vaccine, 36*(37), 5500–5503.

Xu, G., Weber, P., Hu, Q., Xue, H., Audry, L., Li, C., … Bourhy, H. (2007). A simple sandwich ELISA (WELYSSA) for the detection of lyssavirus nucleocapsid in rabies suspected specimens using mouse monoclonal antibodies. *Biologicals, 35*(4), 297–302. https://doi.org/10.1016/j.biologicals.2006.10.002.

Young, C., & Sellers, T. (1927). Laboratory: A new method for staining Negri bodies of rabies. *American Journal of Public Health, 17*(10), 1080–1081.

Zanluca, C., dos Passos Aires, L. R., Mueller, P. P., dos Santos, V. V., Carrieri, M. L., Pinto, A. R., & Zanetti, C. R. (2011). Novel monoclonal antibodies that bind to wild and fixed rabies virus strains. *Journal of Virological Methods, 175*(1), 66–73.

CHAPTER 13

Measures of rabies immunity

Susan M. Moore, Chandra R. Gordon

Rabies Laboratory/Veterinary Diagnostic Laboratory, College of Veterinary Medicine, Kansas State University, Manhattan, KS, United States

13.1 Introduction

The measurement of rabies immunity plays a significant role in the history of rabies control and prevention. The efficacy of early rabies vaccines was partly evaluated through rabies serology using the mouse neutralization test (MNT). The MNT, an in vivo method, was the gold standard for many years. Research concerning rabies pathogenesis, immunity, and vaccine development as well as routine immunity monitoring of rabies-vaccinated people has relied upon the current gold standard test, the rapid fluorescent focus inhibition test (RFFIT), or modifications of it. In recent decades, new vaccines (e.g., adjuvanted, DNA/RNA), new biologics (monoclonal cocktails and purified equine rabies immunoglobulin to replace human rabies immunoglobulin), and new/shortened/reduced vaccine regimens have been developed. These have driven adaptions of the RFFIT and the development of new assays to specifically and precisely measure rabies immunity in the context of these changes in rabies prophylaxis. In addition, the initiative announced in 2015 to eliminate human rabies deaths caused by canine rabies by 2030 requires rabies serology surveillance in endemic developing countries. To accomplish this efficiently, the assays should be assessable and able to produce high-quality results. Along with these new developments, there has been an increase in regulatory oversight to assure appropriate considerations for rabies biologics and regimen approval. These changes make it all the more imperative that rabies immunity assays are well understood and properly performed while the results are accurately interpreted.

There are a number of variables to be considered to assess the immune status of an individual host or among a population. For the rabies virus (RABV), important variables include the source, dose, and route of potential natural exposure. For vaccination, important variables include the vaccine type, potency, and virus strain; the vaccination route and schedule; and individual host factors (e.g., major histocompatibility genes). Although perhaps often overlooked, it is essential to have a basic understanding of the laboratory methods used to measure and assess the host's immune status. The precision, accuracy, sensitivity, and specificity of a method must be well defined and understood. For clinical samples, results must be reported in the context of a consequential diagnostic value for each method so that a particular test result for a patient can be interpreted in relation to the patient's history and clinical management. If the laboratory method parameters are not clearly and objectively defined, conclusions based on test results from various methods may be inherently misleading. Moreover, the expertise and qualification/certification of the laboratory performing the methods are important considerations. Independent evaluation of the proficiency of the laboratory provides confidence in the results. Lastly, the purpose of the testing must be matched with the performance characteristics of the chosen method to assure meaningful interpretation. If RFFIT (Smith, Yager, & Baer, 1973), developed for measuring vaccine response in serum samples, is applied for the analysis of biologic products such as human or equine rabies immune globulin (RIG) or RABV neutralizing monoclonal antibodies, the method will most likely need modifications and thus also subsequent method validation. Also, for this purpose, the laboratory performing the testing must meet good manufacturing practice (GMP) quality standards and operate a functioning quality management system (QMS). A laboratory performing rabies immunity assays for routine monitoring of the human response to the rabies vaccination would have to adhere to local quality standards for human clinical laboratories.

Rabies immunity is comprised of both the cellular and humoral arms of the immune system (see Chapter 11). Cellular immunity assays such as flow cytometric techniques for CD (cluster of differentiation) markers and cytokines tend to be reserved for research studies due to the restrictions of sample management and expense. Humoral immunity is commonly assessed by serology. Serology is the study of the immunological properties of blood serum or other bodily fluids. For the most part, serology is the investigation of antibodies in serum, although the immunity assessment may be conducted on cerebrospinal fluid and other sources of fluid. Antibodies are produced by plasma cells, which may be specifically activated in response to antigens such as those from viruses and bacteria to protect the host. The primary action of an antibody is to bind to antigen. The secondary or effector actions of antibodies include neutralization and opsonization of infectious agents, and activation of other immune mediators (see Fig. 13.1). Complement activation and antibody-dependent cellular cytotoxity (ADCC) are other effector functions that rely on the binding action of antibodies. Not all antibodies have effector actions. Some antibodies that bind to an antigen may not result in a biological effect because they are not effective in eliciting a secondary effect. Effector actions occur in accordance with the individual characteristics of a specific antibody structure and depend upon the class, subclass, or variable region of an antibody. In a competent host, exposure to an antigen will activate multiple immune cell clones and result in the production of a polyclonal antibody response. As the response develops, affinity maturation occurs, resulting in the higher specificity of the antibodies produced.

RABV-specific antibodies are produced by the immune system in response to infection or vaccination in vivo, or by immune cells or molecular methods in vitro. The reasons for performing rabies serology can range from infection diagnosis to an investigation of epitope specificity of an anti-RABV glycoprotein monoclonal antibody. The characterization of an antibody's affinity, specificity, quantity, and neutralizing function; complement binding function; and class/subclass is achieved by various methods. Many serological techniques developed over the decades differ not only in their ability to detect the function, affinity, and

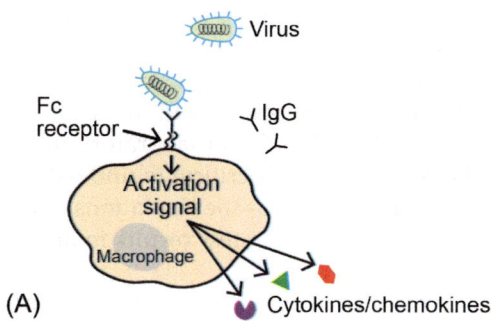
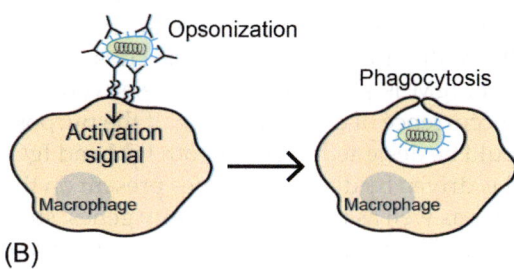
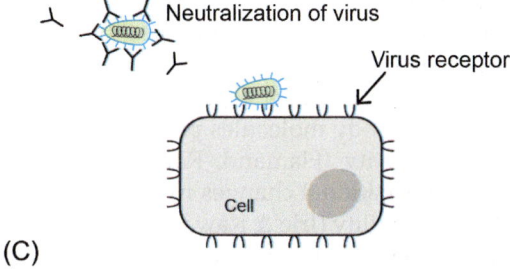

FIG. 13.1 The effector functions of antibodies include: (A) activation of immune cells such as macrophages to produce cytokines and chemokines through Fc receptor binding; (B) Opsonization of infectious organisms induces phagocytosis of the organisms through Fc receptor binding; and (C) Neutralization of virus though binding of proteins used for attachment and entry of the virus, thereby blocking infection of the cell.

specificity of RABV antibodies, but also in the ease and practicality with which they are performed. The ability to select an appropriate method and appropriately interpret test results is dependent on the knowledge of the specific strengths, weaknesses, and limitations of available methods. Numerous reports indicate that protection against rabies is largely dependent upon the presence of RABV-neutralizing antibodies (RVNA) (Dietzschold, 1993; Finke & Conzelmann, 2005; Hooper, 2005; Hooper et al., 1998). Thus, assays to detect and quantify RVNA such as RFFIT (Smith et al., 1973) and the fluorescent antibody virus neutralization test (FAVN) (Cliquet, Aubert, & Sagne, 1998) are the methods recommended for quantitation purposes in rabies serology. While these remain the gold standard methods for rabies serology, there is a need for methods that can be used in resource-poor areas where rabies is most prevalent. Evaluating and defining through fit-for-purpose analysis methods that are simple and cost-effective can be reasonably applied. Antigen-binding assays have proven to be useful for the detection of specific isotypes of RABV antibodies, either using whole virions or specific viral proteins as antigen(s). The decision to use a specific assay should start with the purpose of testing and the intended application of results. Other factors to consider are the assay complexity, the degree of precision and/or accuracy, the specificity, and the range of detection. In addition, the availability of laboratory materials, instruments, and safety equipment also must be considered. It is critical to understand exactly what aspect of RABV-specific antibodies is measured as well as the limitations of the assay in order to select a suitable test as well as interpret and use the test results in an appropriate manner.

Investigative serology focuses on the detection and measurement of immune components in blood (usually serum), including immunoglobulins of several subclasses directed against specific epitopes. The presence of rabies-specific IgM and IgG antibodies is dependent upon the time point in the course of the humoral immune response after exposure to an antigen. Thus, the timing of blood sampling will affect the ability to detect the specific class of antibody. In the initial or primary antibody response, IgM is produced first in relatively low levels, followed by higher levels of IgG after the occurrence of class switching. If the purpose of the assay is to detect the initial response, it should be designed to detect both IgM and IgG. The specificity of the immunoglobulin produced is driven by distinct epitopes present on the rabies viral proteins used to generate the antibodies as well as the host immunity genes, such as the MHC genes. Consequently, the epitopes of the virus used in the assay play a major role in the assay specificity. Exposure to RABV, whether through vaccination with inactivated virus or through exposure, induces the formation of antibodies potentially against all viral proteins, but predominantly against the RABV glycoprotein (G) and nucleoprotein (N). Studies of monoclonal antibodies (MAbs) capable of neutralizing RABV indicate that these MAbs are directed against a number of epitopes on the G of RABV (Buthelezi et al., 2016; Tordo, 1996). RABV neutralization requires a minimum number of antibody molecules per G spike to induce steric hindrance of the virus-receptor binding activity (Flamand, Raux, Gaudin, & Ruigrok, 1993). Another mechanism may involve conformational changes in the G protein, ultimately resulting in the loss of virion receptor binding ability (Irie & Kawai, 2002). The humoral immune response elicited by RABV vaccination consists of a mixture of polyclonal antibodies that influences a variety of complex neutralization mechanisms.

The detection of antibodies specific for rabies viral antigens can be achieved by methods such as precipitation, agglutination, immunoelectrophoresis, radioimmunoassay, enzyme-linked immunosorbent assays (ELISA), western blots, indirect immunofluorescence, immunoelectron microscopy, and immunochromatographic and serum neutralization assays. All these assays depend on an antibody-antigen interaction to detect the presence of an antibody. Two basic types of assays are used: (1) assays involving primary binding activity between antibodies and antigens, and (2) functional assays to measure the neutralization actions of antibodies. Although other components and products of the immune system are involved, protection from clinical rabies after infection relies heavily on the presence of RVNA. Therefore, methods to detect and quantify antibodies that can functionally neutralize RABV are recommended to quantify the level of immunity after rabies vaccination. Evaluation of the results of any laboratory test must include consideration of the purpose of testing such as surveillance, diagnosis, manufacturing, immune protection, etc. The validity of the results includes fit for purpose.

13.2 Rabies serology methods

The detection and measurement of immune components in blood are the basic goals of investigative and diagnostic serology. The underlying principle in serologic assays is the ability to detect the antibody-antigen reaction. The methods to accomplish this can be categorized in several ways: functional antibody versus binding antibody, simple versus complex, expensive versus inexpensive, rapid versus time consuming, and screening versus confirmatory. Historically, the functional serum neutralization methods are the gold standard for RABV antibody measurement, for the important reason that it is the RABV-neutralizing antibodies raised in response to vaccination that provide protection from infection. However, with advances in both rabies biologics and laboratory testing, the need for the modification of methods and the development of accessible methods has increased. Thus, it is more and more important to understand the technical performance and the essential differences between methods. To that end, in this section various available methods are described and the next section will address fitness-for-purpose and assay selection.

13.2.1 Serum neutralization assays

Rabies serum neutralization (SN) assays are distinguished by the ability to detect the neutralization activity of specific antibodies in vitro and therefore attempts to measure the potential protective action of these antibodies in vivo. The technical performance of RABV neutralization assays requires the use of infectious virus, requiring performance in a high-level biosafety facility. In general, they are labor intensive and time consuming. Depending on the number of samples to be tested and modifications of the basic assay, SN methods can be costly or cost efficient as well as low throughput or high throughput, and can serve the purposes of screening or confirmatory. There are two rabies SN assays recognized by the World Health Organization (WHO) and the World Organisation for Animal Health (OIE) to measure RVNA: the RFFIT, described in 1973 by Smith et al. (1973), and the FAVN, developed in 1997 by Cliquet et al. RVNA measurement by serum neutralization assays is based on the same principle as the MNT, extensively employed in early rabies serology work. The MNT involves the injection of test serum dilutions in mice followed by a challenge with a standard dose of RABV, with the read-out being mortality among the mice (Atanasiu, 1973). Although this is truly a "real" measurement of the protective function of RVNA in the serum, the biological variation of individual mouse immunity as well as the possible interference of other immune effectors inherently make it exceedingly variable and difficult to standardize. The in vitro methods such as RFFIT are commonly used with resulting higher precision and improved sensitivity.

Both the RFFIT and the FAVN tests consist of the incubation of serial dilutions of heat-inactivated serum with a fixed amount of live RABV for 60–90 min at 37°C. The measurement of residual virus infectivity is accomplished by the detection of virus in a cell culture via a labeled anti-RABV virus antibody and subsequent calculation of the quantitative titer by the number of microscopic fields containing virus-infected cells. The classical RFFIT method is conducted in multichamber slides (see Fig. 13.2). Serum is serially diluted five-fold and tested in each well. Variations of RFFIT include the use of microtiter plates in place of slides and the use of two-fold or three-fold dilutions. The challenge RABV dose used in the RFFIT

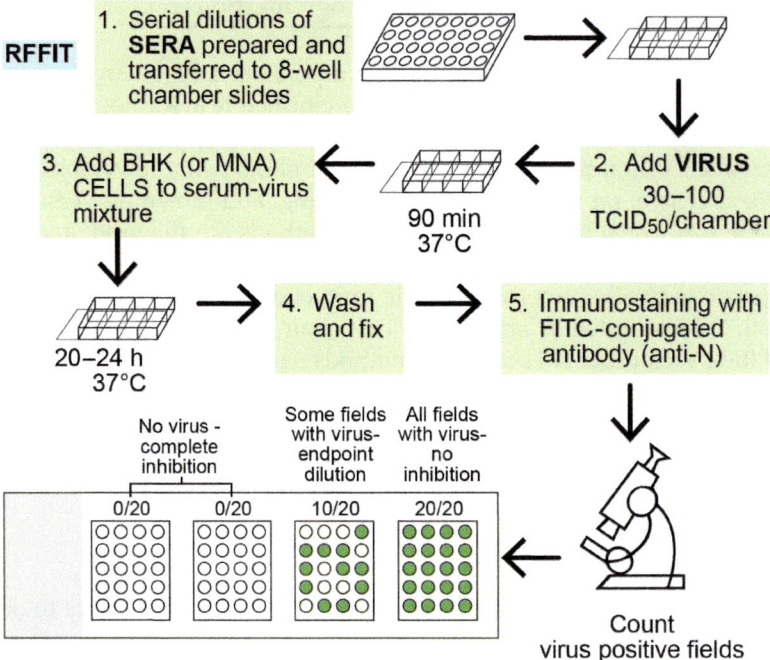

FIG. 13.2 RFFIT procedure. Serum is serially diluted in a 96-well plate and transferred into eight-well chamber slides. The rabies challenge virus is added to the diluted serum and the slides are incubated at 37°C for 90 min, after which baby hamster kidney (BHK) or mouse neuroblastoma (MNA) cells are added to each of the wells. The slides are incubated at 37°C in a 2%–5% CO_2 incubator for 20–24 h. The wells containing an adherent monolayer of cells are washed and the cells are fixed with 80% cold acetone. An FITC-conjugated antirabies antibody directed against the RABV N is added in order to detect virus-infected cells. In eight-well chamber slides, 20 fields of each well are examined, using a fluorescent microscope, for the presence of fluorescence in the cells that indicates the presence of nonneutralized RABV. The titer of RVNA in the serum sample being analyzed is defined as the dilution at which 50% of the observed microscopic fields contain one or more infected cells.

should be between 30 and 100 TCID50, 50% tissue culture infective dose. After the virus is added to the diluted serum, the slides are incubated at 37°C for 90 min, after which baby hamster kidney (BHK) or mouse neuroblastoma (MNA) cells are added to each of the wells. Diethylaminoethyl-Dextran (DEAE-Dextran) has been used, typically at a 0.01 μg/mL concentration, in some variations of RFFIT to enhance the susceptibility of the cells to RABV infection (Kaplan, Wiktor, Maes, Campbell, & Koprowski, 1967). The slides are generally incubated at 37°C in a 2%–5% CO_2 incubator for 20–24 h, although the incubation period is extended to 48 h in some variations of the method conducted in microtiter plates. The wells containing an adherent monolayer of cells are washed, and the cells are fixed with 80% cold acetone. An FITC-conjugated anti-RABV antibody directed against the RABV nucleoprotein (N) is added in order to detect virus-infected cells under fluorescence microscopy. In eight-well chamber slides, 20 fields of each well are examined for the presence of fluorescence in the cells, an absence of which indicates that antibodies in the sample neutralized the virus and the presence of which indicates a lack of antibodies. The titer of RVNA in the serum sample being analyzed is defined as the dilution at which 50% of the observed microscopic fields contain one or more infected cells. Mathematical calculation using the Reed and Muench formula, the Spearman-Karber formula, or Probit analysis will determine the exact quantitative titer of RVNA in the serum sample. Alternatively, the quantitative titer of RVNA can be more simply defined, but with less precision, as the highest serum dilution where 100% viral inhibition occurred, thus indicating that there were no infected cells at that dilution and all subsequent higher dilutions exhibit infected cells (Aubert, 1996; Habel, 1996).

Transcribing a serum dilution value into a standardized and more globally recognized measure of IU/mL is achieved by a simple calculation, wherein the value from a serum sample being tested is compared to the serum dilution value of a reference serum standard containing a specific amount (potency) of RVNA (Velleca & Forrester, 1981). Converting the titer result into IU/mL is especially important to standardize the value if the results are to be compared between laboratories and time because titer values are dependent on the virus dose in that batch of testing and that particular laboratory. The IU/mL value controls for the variability between batches of testing and between laboratories. This is particularly applicable when laboratories modify the standard assay steps (e.g., virus strain/dose, reference serum, cell type, platform, microscopic reading method). See Table 13.1, which

TABLE 13.1 The results of an RFFIT exchange (nine international RFFIT laboratory comparisons of a six-sample panel) demonstrate that the measurements in IU/mL value were more precise (average 31.3% CV%) than the titer value (average 47.4% CV%).

	Sample #	Laboratory identification code									Ave	CV %	Range	
		A	B	C	D	E	F	G	H	I			Low	High
IU/mL values reported	1	2.30	3.22	3.04	3.33	2.00	3.10	2.61	2.27	3.17	*2.78*	*17.7*	2.00	3.33
	2	2.07	2.70	2.87	2.67	1.90	3.03	3.46	1.60	3.23	*2.62*	*24.2*	1.60	3.46
	3	0.40	0.55	0.48	0.67	0.40	0.60	1.10	0.25	0.60	*0.56*	*43.0*	0.25	1.10
	4	0.40	0.40	0.47	0.50	0.40	0.63	0.66	0.28	0.63	*0.49*	*26.8*	0.28	0.66
	5	0.40	0.39	0.40	0.42	0.10	0.63	0.60	0.20	0.67	*0.42*	*44.9*	0.20	0.67
	6	0.00	0.00	0.09	0.10	0.10	0.10	0.05	0.03	0.20	*0.08*	*84.2*	0.03	0.20
	AVE											*31.3*		
Titer values reported	1	287	161	203	128	280	280	349	287	229	*245.00*	*28.7*	128	349
	2	279	135	192	102	270	273	413	203	236	*233.66*	*39.2*	102	413
	3	51	28	32	25	50	54	133	32	41	*49.60*	*66.6*	25	133
	4	56	20	30	13	56	55	78	36	46	*43.27*	*47.7*	13	78
	5	55	19	27	16	19	54	74	26	50	*37.63*	*55.1*	16	74
	6	5	5	6	4	5	9	7	3	11	*6.12*	*40.9*	3	11
	AVE											*47.4*		

The variable range per sample measurement across laboratories is wider when titer values are compared. The results across laboratories also show good comparison in IU/mL, in consideration that some RFFIT laboratories in the group modify the standard RFFIT procedure (modified RFFIT). Average and CV values are given in bold and italic to separate from laboratory values.

demonstrates that while different laboratories produce highly variable titer values (high CV %) for the same sample, when converted into IU/mL values, the precision (CV%) is greatly improved. The quality of the test components as well as the skill and expertise of the technician conducting the test, including the analysis of the microscopic readout, can substantially affect the precision of RFFIT test results. To simplify and reduce the subjectivity of the microscopic counting step, the FAVN method uses four replicates of serum using three-fold dilutions in microtiter wells. It also scores each well as either positive or negative for the presence of RABV-infected cells after a 48-h incubation.

Just as a direct comparison of the RFFIT and FAVN demonstrated no statistically significant differences in results when conducted in a laboratory adhering to good quality assurance standards (Briggs et al., 1998), interlaboratory exchange sample panels demonstrate that different laboratories performing a modification of the original RFFIT can produce comparable results (see Table 13.1). Even within a laboratory, it is advisable to perform routine method comparison. A laboratory method comparison at the Kansas State University Rabies Laboratory in 2018 demonstrated correlation between RFFIT and FAVN, with an average percent recovery (against known IU/mL values) of 111% and 87%, respectively. The precision and repeatability of virus neutralization test results can be controlled by strict adherence to the dose and strain of the challenge virus used and the source of the standard reference serum. Early published reports that compared different laboratory RFFIT results reported that the use of a high infective dose of challenge virus resulted in reduced sensitivity for testing low-titered sera, whereas a low viral dose of challenge virus could result in lower precision when testing high-titered sera such as RABV immunoglobulin (RIG) preparations (Fitzgerald, Baer, Cabasso, & Vallancourt, 1975). In addition, the use of an equine RIG as the reference standard to determine IU/mL values resulted in significantly different titer results than when a human RIG reference standard was used (Lyng, Bentzon, & Fitzgerald, 1989). Measuring RVNA from patients vaccinated with a vaccine prepared with a parent virus strain heterologous to the challenge virus strain in the RFFIT (usually CVS-11) can result in lower titers than if a homologous challenge strain is used (Moore, Ricke, Davis, & Briggs, 2005). RABV neutralization tests identify the presence of all classes of immunoglobulin in a sample (both IgM and IgG). Therefore, they will be able to detect the early production of RABV antibodies after exposure or vaccination, but they may not be as sensitive as the IFA (Smith, 1991). Because the virus neutralization testing method depends on the measurement of residual or "nonneutralized" RABV infecting the cells, the presence of interference factors in the sera or culture media that adversely affect cell (and ultimately viral) growth will mimic virus neutralization by nonspecifically inhibiting viral growth (Rudd, Appler, & Wong, 2013). Any inhibition of viral growth not directly due to a rabies-specific neutralizing antibody will give a false positive result. Because of this and the potential for cross-reactive antibodies or other immune-modulating proteins to interfere in SN methods in the lower serum dilutions, it is of utmost importance to challenge the assay and define the lower limit of detection (LOD) and the lower limit of quantification (LLOQ) to be able to interpret the results accurately. All too often, low positive results that have not been confirmed as specifically caused by RVNA are interpreted as true positive, allowing for misleading conclusions. The automation of many of the steps in SN assays can improve accuracy and precision, including the addition of media to the plates, serial dilution of the serum samples, the addition of virus and anti-RABV conjugate, and some more tedious steps such as plate washing.

Automated reading of FAVN and RFFIT reduces the work time required for the microscopic analysis readout and aids in the minimization of errors (Peharpre et al., 1999). Advances in automated readers that combine imaging sensitivity and adaptability with robotics and software analysis allow for both high throughput and standardization for rabid, high-quality SN results (Burgado et al., 2018). The expense of the equipment required to undertake automated reading of the RFFIT or FAVN, the requirement for a consistent cell monolayer, and the need for a good quality FITC conjugate limit the practicality of this enhancement, especially for laboratories that do not conduct large numbers of tests. As an alternative to microscopic fluorescence measurement, a microneutralization test (RAMIN), the indirect immunoperoxidase virus neutralization (IPVN) technique, and the modified FAVN employ a mouse anti-RABV antibody and a peroxidase antimouse conjugate, enabling automated reading by a spectrophotometer (Cardoso, Silva, Albas, Ferreira, & Perri, 2004; Hostnik, 2000; Mannen et al., 1987). In each of these studies, a good correlation was confirmed between traditional RABV neutralization methods and the modifications that were made to each test. Other modifications take advantage of molecular techniques to prepare recombinant viruses to use in place of the standard challenge virus, CVS-11, for standardization of fluorescence or for adaptability to detect different specificities of antibodies. Modified CVS-11 expressing green fluorescent protein (GFP) eliminates the need for FITC-conjugated anti-RABV antibodies (Burgado et al., 2018; Khawplod et al., 2005). Using this modified CVS-11 in combination with flow cytometry to detect any residual virus present after incubation with serum reportedly increases the sensitivity. This is because each cell is individually assessed for viral infectivity, creating a more precise percentage of viral inhibition; however, it can be of limited accessibility due to the cost of instrumentation (Bordignon et al., 2002). Other adaptions of the RFFIT include the use of conjugates developed to meet the need for a consistent supply in a specific country or region. An example is the development of an antiphosphoprotein (P)-FITC conjugate. Because the established anti-N conjugates are not available in South Korea, this group prepared a viable replacement by the development and validation of an anti-P conjugate (Um et al., 2017). Another adaptation that allows the use of a reagent already available for another assay to be used in the RFFIT is an immunohistochemistry assay (Madhusudana et al., 2014). This assay was developed to address the expense of antirabies fluorochrome antibody conjugates and use of a fluorescence microscope. It uses the same biotinylated anti-N monoclonal antibodies prepared for the DRIT (Lembo et al., 2006). There is a growing need to be able to determine the specificities of polyclonal and monoclonal rabies antibodies. This can be accomplished by modifying the RFFIT by use of different challenge virus rabies strains, lyssaviruses, or escape mutants (Fallahi, Wandeler, & Nadin-Davis, 2016; Serra-Cobo et al., 2018). The pseudotype method developed by Wright et al. (2008) utilizes pseudotype viruses—lentivirus vectors expressing the RABV glycoprotein and a reporter (e.g., lacZ, GFP) can be used to determine the specificities of polyclonal and monoclonal rabies antibodies (Wright et al., 2008). By expressing the glycoprotein from different RABV strains, a panel of pseudotypes can be used in cross-species comparison studies. Because the pseudoviruses are replication-incompetent particles, this method is applicable in areas where high-level biocontainment facilities are not available. Multiple applications of this assay demonstrate its accessibility and adaptability (Moeschler, Locher, Conzelmann, Kramer, & Zimmer, 2016; Nie et al., 2017). With monoclonal rabies immunoglobulin in use and in development, the need to determine the specificity of the neutralization ability of the product is important. Accomplishing this through modification of the SN assay with the use of pseudotypes as the challenge virus is easier and has greater ability for standardization than using wild-type strains, which can be difficult to grow in tissue culture cells (Fallahi et al., 2016). A challenge for some wanting to use serum neutralization assays can be sample volume. Adapting the RFFIT or FAVN for this purpose is possible, but this affects the linear range because the tiny volume is more affected by sample evaporation and thus concentration during the neutralization incubation period, but careful performance of a validation procedure can prove useful for this purpose (Smith & Gilbert, 2017).

In recent years, the replacement of the National Institute of Health (NIH) test for vaccine potency (an in vivo challenge potency assay performed in mice) has been under consideration. Reasons for this include poor reproducibility and precision as well as the initiative to reduce the use of live animals in laboratory assays (Combes & Balls, 2014). Several assays have been proposed. The most successful and promising include:

- An in vitro immunogenicity assay rather than challenge in mice, which is a modification of the RFFIT/FAVN following the European Pharmacopeia method for RIG potency for higher precision (Kramer, Schildger, Behrensdorf-Nicol, Hanschmann, & Duchow, 2009).
- Two in vitro antigen capture ELISA assays:
  - Immuno-capture ELISA using one monoclonal to antigenic site III for both capture and detection (Gibert et al., 2013).
  - Immuno-capture ELISA using two monoclonals, one to antigenic site III for capture and the other to antigenic site II for detection (Chabaud-Riou et al., 2017; Morgeaux et al., 2017).

The proposed methods have shown good precision and reproducibility. Discussion and proposals for the use of in vitro serologic methods for rabies vaccine potency testing were covered in detail in a National Toxicology Program Interagency Center for the Evaluation of Alternative Toxicological Methods (NICEATM) and International Alliance for Biological Standardization (IABS) meeting (Poston et al., 2019). It was reported that the immuno-capture method described in the Chabaud-Riou paper, based on its superior performance in comparison with the NIH, is under consideration for implementation by the European Directorate for the Quality of Medicines and Health Care (EDQM). In the conclusion, it was noted that without the cooperation and commitment of international regulatory agencies as well as manufacturers regarding the approval of in vitro serological methods for this use, global implementation is difficult, to say the least.

13.2.2 ELISA, IFA, and immunochromatographic assays

Binding assays are methods that detect or measure immunoglobulin molecules by their ability to bind specifically to their target antigen. This binding can be detected by agglutination, precipitation, the use of secondary detection systems usually bound to a color development system for visualization, quantitation by optical density (OD), or fluorescent measurement. ELISA assays are the most commonly used binding assay. ELISA assays may be based on indirect, competitive, and blocking approaches. Other binding assays are immunochromatographic (e.g., lateral flow) and indirect fluorescent antibody (IFA). Western blots are used to identify the fine specificities of antibodies. Antigen binding assays such as ELISAs and IFAs are rapid, simple, and often do not require the manipulation of infectious RABV during the assay, although antigen preparation may involve live virus. These assays rely on the interaction of the antibody and antigen, regardless of the ability of the antibody to neutralize RABV, and they are useful for the detection of RABV binding antibodies. An assay with whole virus as the target antigen may be useful to identify the presence of RABV antibodies specific for the different antigens on the RABV such as to detect exposure to rabies whether through exposure or vaccination. Conversely, purified viral proteins can be used to distinguish the specific composition of antibodies that may be present. For example, an assay using purified RABV glycoprotein as the antigen can better estimate the neutralizing antibody present than an assay using purified phosphoprotein because the neutralizing antibody is directed against the glycoprotein epitopes (Grassi, Wandeler, & Peterhans, 1989). Binding assays are able to identify the subclass of RABV antibodies, for example, IgM and IgG, by using a conjugated antisubclass Ig antibody as the secondary antibody. This type of assay can provide additional information on the kinetics of the immune response (see Fig. 13.3).

The IFA technique involves adding test serum to slides fixed with RABV-infected cells. Rabies virus antibodies in the serum bind to antigens on RABV proteins present in the infected cells and are subsequently detected by FITC-labeled anti-IgG or anti-IgM. A fluorescence microscope is required to evaluate the slides for the presence of labeled antibodies. Quantification of the antibodies can be accomplished by serial dilution of the serum to determine the antibody titer. Because infected cells are used as the source of RABV antigens, the potential exists for both antibodies with specificities to RABV antigens and to cellular antigens to be detected. The possibility of detecting antibody binding that is not specific to the

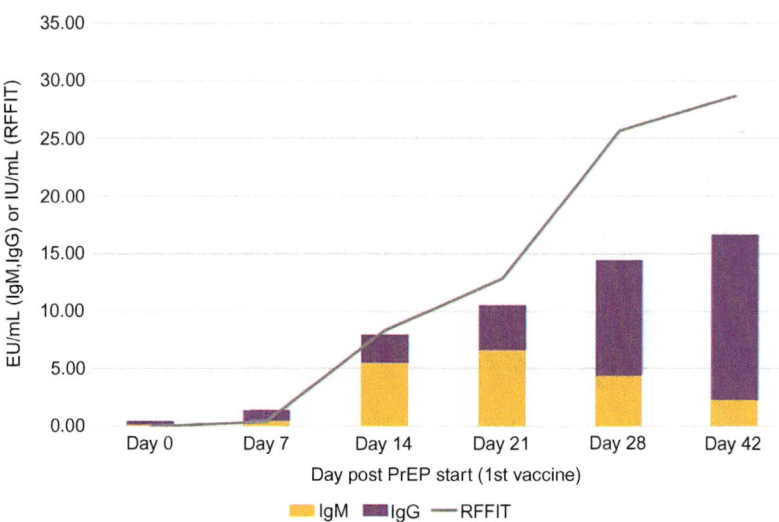

FIG. 13.3 Average IgM and IgG rabies antibody levels measured by indirect ELISA and RVNA measured by RFFIT in 10 human subjects on days 0, 7, 14, 21, 28, and 42 after rabies preexposure vaccination series initiation show the relative contribution of the neutralizing activity by IgM and IgG antibody classes as the humoral immune response develops.

RABV antigen (i.e., autoantibodies, antibodies to cellular antigens, etc.) is important to consider when evaluating the results.

Early ELISA methods such as the one described by Nicholson and Prestage (1982) used inactivated whole virus as the antigen and anti-IgG as the secondary antibody. This technique offered greater specificity over the IFA because the only source for binding is the whole virus coated to the surface of the well, not in a cell where there are other antigens present as possible targets for interfering antibodies. The secondary antibody employed by ELISA methods may be species-specific, but if staphylococcus Protein A or G is employed, the method can be applied to samples from a number of species because Protein A/G binds to the Fc portion of the IgG of many species (Goding, 1978; Page & Thorpe, 2002).

Other types of ELISAs include competitive (cELISA) and blocking ELISA. Both methods involve the use of a labeled RABV antibody that is used to either compete with (cELISA) or to detect antigen not blocked by (blocking ELISA) the RABV antibodies in the test sample. In blocking ELISA, the incubation of the serum sample with the inactivated RABV antigen on the well surface is followed by the addition of an enzyme-labeled anti-RABV antibody. Any unbound (unblocked) inactivated RABV is bound by the labeled antibody. The amount of enzyme-labeled antibody is detected by adding a conjugate for color development; the amount of RABV antibody in the serum sample is inversely related to intensity of color development. The level of antibody in the test serum can be quantitated by the use of a standard curve and an optical density (OD) reader. Competition for RABV binding between the anti-RABV antibodies in a serum sample with a labeled anti-RABV monoclonal antibody is the basis of the cELISA (see Fig. 13.4). Similar to blocking ELISA, the labeled antibody is

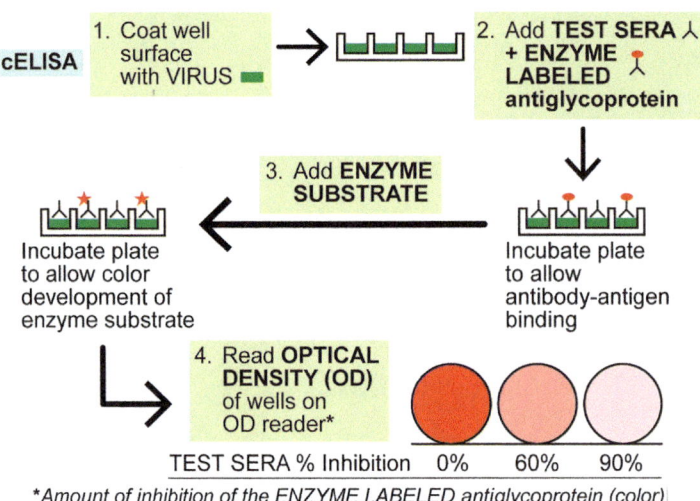

FIG. 13.4 Competitive ELISA. A labeled rabies antibody competes with the rabies antibodies in the test sample. The test serum sample and an enzyme-labeled antirabies antibody is incubated with the inactivated rabies antigen on the well surface. The amount of enzyme-labeled antibody is detected by adding a conjugate for color development; the amount of rabies antibody in the serum sample is inversely related to the intensity of color development. The level of antibody in the test serum can be quantitated by the use of a standard curve and an OD reader.

measured and used to determine the level of anti-RABV antibody in the sample. Both these methods can reduce the effect of nonspecific binding because the antibody that is measured is a purified reagent antibody. The use of different reagent monoclonal antibodies in these assays allows the detection of various specificities of RABV antibodies that may be present in the sample (i.e., the use of a labeled anti-RABV N will detect anti-N in the sample and the use of a labeled anti-RABV G will detect anti-G in the sample). Such an assay was utilized to measure the potency of equine rabies immunoglobulin product in a study by Korimbocus, Dehay, Tordo, Cano, and Morgeaux (2016). The F(ab')2 molecules in the product were mixed with a monoclonal targeting antigenic site II in the trimeric form (natural) of the rabies glycoprotein. The assay results matched the MNT with a correlation of $r=0.751$; however, the cELISA results were systematically higher and more variable when compared to the RFFIT. Blocking and competitive ELISAs have attained application in monitoring oral rabies vaccine programs due to their sensitivity, specificity, and standardization (Moore et al., 2017; Wasniewski et al., 2013; Wasniewski et al., 2016). The ability to compare results across programs and regions leads to better understanding of the success of the programs over time and place.

An electrochemiluminescent (ECL) adaption of the blocking ELISA method employs microtiter plates fitted with a series of electrodes at the bottom of the wells. Applying an electrical current across the electrodes causes the generation of a luminescent signal by the chemical energy ligand-binding reactions. Quantitation of the signal converts the measurement to antibody concentration. This method has been applied to the measurement of proteins and has the potential for greater sensitivity and faster results compared to the traditional ELISA method (Guglielmo-Viret, Attree, Blanco-Gros, & Thullier, 2005; Ma, Niezgoda, Blanton, Recuenco, & Rupprecht, 2012).

Immunochromatographic assays to detect RABV antibodies are useful for fieldwork and in areas where a low-tech screening method is required. By adapting the concepts of ELISA to an absorbent test strip, the testing process is simplified to progress across a straight line by having the sample interacting with the assay reagents sequentially. A version of this method for detecting the RABV antibody requires an initial step of mixing the serum sample with inactivated virus before adding the mixture to the absorbent material of the test strip. The mixture then flows across to encounter a labeled anti-RABV antibody that will bind to any unbound inactivated RABV. The mixture continues to flow toward two areas (lines) of the strip bound with detection antibodies; the first detection antibody is specific for the RABV, and the second detection antibody is specific for the labeled anti-RABV antibody. By using this design, if the mixture contains inactivated RABV bound with labeled anti-RABV antibody, there will be a color development at the first strip, indicating that the sample did not contain enough anti-RABV antibody to bind the inactivated RABV in the mixture. If the sample contains an anti-RABV antibody, then the labeled antibody will pass the first strip and be bound by the labeled anti-RABV antibody, causing a color development at the second strip. The results will be either positive (only second strip visible) or negative (first and second strip visible) for the presence of an anti-RABV antibody in the test serum sample. The level of antibody to define positive or negative is set and defined by the design of the assay and can only be altered by the concentration or dilution of the serum sample. Its simplicity and portability allows use by operators with a minimal amount of education and training; however, the need to pay attention to the proper performance of the assay is still applicable as with a test performed in a laboratory setting to assure accurate results. The lateral flow assay is useful for point-of-care situations where an initial rapid screening result would determine whether further action, such as additional testing or vaccination, was necessary (Nishizono et al., 2012). The application of an immunochromatographic point-of-care device is described in a comparison study of the seroprevalance of rabies antibodies in dogs in the Philippines, Thailand, and Japan (Manalo et al., 2016). This paper also describes the validation of the device against RFFIT, demonstrating similar NPV, PPV, concordance, and reproducibility among different laboratories.

ELISA and other binding assays have several advantages, including the fact that they are rapid, require little expertise, do not need high-level biohazard facilities to be performed, and several steps of the procedure can be automated (i.e., serial dilution of the sera, addition of reagents, and optical density reading). Additionally, software packages are available to calculate antibody concentration, thus allowing for objective reading and interpretation. The disadvantages of ELISA methods include the restrictive nature of the conjugated antibody or protein A/G that limits the isotype of immunoglobulin detected. The use of species-specific anti-IgG (or IgM) confines the utility of the assay to a certain species. Additionally, although the use of protein A increases test application to several species, it does not react with all forms of IgG3 and therefore will lead to an underestimation of the level of RABV-specific antibodies in serum containing higher proportions of IgG3 antibodies (Carpenter, 1997). The degree of nonspecific binding detected by an ELISA will depend on the purity of the antigen preparation and the efficiency of the coating step because immunoglobulins will nonspecifically adhere to glass, plastic, contaminating material (i.e., mycoplasma), or cell culture components. Quantification of IgG antibodies that bind to RABV will not precisely demonstrate the level of protective virus-neutralizing antibodies present in the sera. Therefore, the application of ELISA for attempting to measure the amount of RVNA requires utilizing a cutoff value that can reasonably be assured to represent correlation to RVNA in the diversity of humoral immune responses in the target population (Moore, Pralle, Engelman, Hartschuh, & Smith, 2016). Reporting of ELISA results using IU/mL values is not reflective of this unit of measurement as defined by the WHO, where 1 International Unit of neutralizing activity is present per mg of protein. Therefore, the use of IU/mL to describe an antigen-binding assay for an RABV titer result is misleading. Because not all binding antibodies neutralize virus, whether for RABV or other pathogens, titers obtained from antigen-binding assays are not biologically identical to RVNA titers; therefore, caution must be taken when evaluating test results. The early antibody response consists primarily of IgM. Thus, the use of indirect ELISA methods using anti-IgG as the detection (secondary) mechanism will not measure the early IgM response (see Table 13.2). The use of ELISA in place of a serum neutralization assay (as has been traditionally used) for efficacy in clinical trial studies may lead to assumptions of inferiority when the day 7 and day 14 results are compared with results from previous studies (Moore et al., 2016). Fig. 13.3 illustrates the difference in antibody kinetics evident when the results of this type of ELISA are compared to serum neutralization results. These average results of a group of individuals show the relative contribution of IgM and IgG to serum neutralization measurement; the relationship will vary between individuals while also depending on each individual's ability to produce highly neutralizing antibodies. ELISA tests are recognized by WHO as acceptable for antibody detection or measurement when RFFIT or FAVN testing is not feasible (World Health Organization, 2018). The OIE also recognizes an

TABLE 13.2 Average rabies antibody levels as defined by assay type.

		Day 0	Day 14	Day 28	Day 90	All days
ELISA	EU/mL	0.157	6.704	7.248	5.210	6.388
	Std error	0.6059	23.7799	23.0968	16.0411	21.2420
RFFIT	IU/mL	0.12	36.96	12.63	5.71	18.43
	Std error	0.6437	144.1476	50.2041	27.1077	90.3475

The average with standard error of all the subjects' measurements by day demonstrate the kinetics of rabies vaccine response as measured by RFFIT (RVNA IU/mL) and by ELISA (antirabies glycoprotein EU/mL). Significant differences ($P < .05$) between method results were noted at days 14 and 30 as well as overall.

indirect ELISA for some purposes (Fooks, 2018); other ELISA methods are not prescribed tests for rabies, but they may be useful for monitoring vaccination campaigns in wildlife populations. In such cases, the ELISA kit should be validated for the purpose of testing in the species under study. The inherent differences should be taken into account upon interpretation by informed health care providers, whether veterinary or human, or researchers for the optimal prevention of rabies.

13.3 Assay selection

Knowing the purpose of the testing as well as the performance characteristics of the available methods is basic for selecting the one that will give valid and meaningful results. In reality, this can mean the available methods may not be the best method of the purpose, but knowing the limitations can aid in proper interpretation and application of the results. RABV antibodies are measured for diverse reasons. The reasons for these measurements will influence the requirements for method sensitivity, specificity, precision, accuracy, linear range, limit of detection, and the robustness of the method as well as the need for laboratory certification to specific quality. The requirements for specific reagents, instrumentation, or facilities as well as expertise may vary according to the particular methods. The consequence of selecting an improper method can be as simple as getting a result that does not answer an academic question, thus leading the research down a wrong path, or as complex as providing incomplete or misleading information that will be used to make essential health care decisions, whether veterinary, medical, or public health, for the prevention of clinical rabies. For example, if there is an encephalitis suggestive of clinical rabies, evaluating a sample from a human or animal with an assay that can only detect IgG antibodies could be insensitive or misleading because IgM antibodies, which are produced before IgG, may remain undetected. Thus, a negative test result would be misleading. Besides the consequence of using an unsuitable method for individual diagnosis, ambiguous results add potentially incorrect information to the body of data compiled for typical antibody responses in rabies patients and in rabies-vaccinated individuals (humans and animals).

Laboratory tests for RABV antibodies are used for research, human vaccination decisions, pet travel permits, wildlife vaccination program evaluations, and pharmaceutical product development and licensure. Also, rabies serology tests can be used to measure vaccine potency.

No one method will be the ideal fit for all purposes. The method that will "fit" must be defined by the characteristics of RABV antibodies that are most important or by the parameter of interest. For example, to research the difference between monoclonal antibodies produced against the glycoprotein of the ERA RABV strain, a serum neutralization assay is essential if the ultimate purpose of the monoclonal is therapeutic. The challenge virus used in the serum neutralization assay should be a consideration. If the purpose of the monoclonal antibody is use as a therapeutic agent, then the challenge virus used should be one that is most closely related to the RABV variants that are enzootic in the regions where the biologic is intended for use. Moreover, if the monoclonal antibody is intended for eventual licensure, the laboratory method (and the performing laboratory) selected must be approved, validated, and recognized by a licensing authority. Conversely, if the purpose of the monoclonal antibody for use in diagnostic testing is to differentiate ERA-infected brains from brains infected with other strains, then the method best used to illustrate the difference in monoclonal antibodies would be an IFA using ERA-infected cells. There is a growing need to detect and measure rabies antibody specificity as well as potency against different RABV variants and other lyssaviruses, such as for monoclonal products and RNA/DNA vaccine responses. Thus, the challenge virus or viral antigen used in the selected assay is a critical factor.

Below are specific assay requirements that apply to some common reasons for measuring RABV antibodies:

- Sensitivity: detection of low levels in an initial response to infection or vaccination/ability to measure low levels of IgM and IgG (clinical diagnosis, evaluation of postexposure treatment, some research purposes).
- Standardized: for comparable results between laboratories and over time (clinical trial testing, human testing for vaccine response either for postexposure or preexposure, oral-bait program evaluation, pet travel).
- Cost effective: to obtain screening results from large numbers of samples.
- Adaptable for detecting different immunoglobulin subclasses: research and clinical applications.
- Adaptable for detecting specificities or antibodies from different species: research and surveillance schemes.
- Approved by regulatory authorities: biologic product testing, pet travel.
- Low technology or low-level biocontainment facilities: field research or in developing countries.

Consideration of the sample is also a factor in choosing the proper assay. Attempting to measure a F(ab')2 product with an ELISA with a secondary detection system that relies on binding to the Fc portion of the immunoglobulin would be futile. A blocking or competitive ELISA or a serum neutralization assay would be a better "fit" for this purpose because these assays do not rely on the complete structure of the immunoglobulin, only the antigen-binding portion, F(ab')2, for detection. The condition of the sample plays a large role in the resulting quality of the result produced. In SN assays, components in the sample that are toxic to the tissue culture cells can provide unique challenges, especially in field samples for surveillance in wildlife. Methods to alleviate this problem have been addressed in various ways, such as utilizing a cell monolayer and removing the toxic serum after the serum-virus incubation (Bedekovic et al., 2013). ELISA methods have proved to be useful in reducing the interference of "dirty" field samples (Bedekovic et al., 2016). The use of filter paper to obtain and transport field samples has also been initiated (Wasniewski, Barrat, Combes, Guiot, & Cliquet, 2014). Nonspecific inhibition is also a problem in human samples tested using SN methods. Factors that cause these nonspecific reactions are elusive, but one study identified cross-reactive antibodies in acutely ill patients with other infectious diseases (Rudd et al., 2013). Interestingly, bioactive compounds that provide chemical protection (innate immunity molecules) against microorganisms have been shown to inhibit RABV (Cunha Neto Rdos et al., 2015). Possibly such molecules may be present in serum and cause nonspecific inhibition in SN assays.

In human laboratory medicine, it is not uncommon to screen samples using a sensitive assay to identify positive samples from negative samples and then follow the sensitive screening tool with confirmatory testing with a more specific assay to identify the true positive samples and exclude the false positives. Several methods can be used effectively for screening purposes. Depending on the screening goal, assays such as ELISA using whole virus antigen, lateral flow with a positive or negative readout, and IFA can identify samples that potentially contain RABV-specific antibodies. Testing with a western blot technique can confirm the specificity of the antibodies detected in the screening assay or testing with a serum neutralization (SN) method can confirm the neutralizing function of the antibodies. A screening method with lower accuracy (result may not be the true value), but higher precision (repeat measurements are clustered closely although they may not be near the true value) may be more useful for oral baiting surveillance—if it is quick, standardized, and simple—than a more accurate method that is cumbersome, time consuming, and more variable. For the purpose of evaluating the oral vaccine baiting campaigns, determination of individual "protection" is less important than herd immunity levels and the ability to confidently compare results between laboratories and over time.

The most effective rabies prevention programs include rabies serology in various steps, such as registering and licensing new rabies vaccines, monitoring the immune status of humans and animals, implementing animal vaccination campaigns, undertaking rabies surveillance and epidemiology, and, finally, maintaining rabies-free zones. It therefore follows that selecting the best "fit-for-purpose" serological assay as well as ensuring that the test is performed properly and that adequate quality assurance procedures are in place is vital for the success of the program. Valid rabies serology results affect international trade agreements, the establishment and maintenance of new and existing rabies-free zones, the ever-needed development of new vaccines, and the evaluation of research or surveillance results published in scientific journals. Knowledge of the method's performance characteristics and proper performance of the assay using established quality assurance procedures are the most effective way to ensure valid results that are repeatable and applicable to the purpose.

13.4 Assuring quality results

Quality results depend on the assay performance characteristics ("fit for purpose"), the expertise of the technician, the laboratory facility, the equipment and materials, and notably the sample itself. Performance characteristics are defined by method validation, which consists of a thorough plan of experiments to examine the accuracy, precision, specificity, sensitivity, linear range, and robustness of the method for a particular purpose (ICH, FDA, WHO, OIE). The expertise of the technician depends on training and experience, and is proven by proficiency testing to demonstrate the ability to perform the assay to defined standards. The laboratory facility, including space, environment, and water quality, among other factors, must meet minimum criteria to assure that the testing is performed in a quality controlled state. Audits to establish standards and continual monitoring for the ability to maintain standards is essential for confidence in the test results produced.

Essential reagents, materials, and equipment must have consistent quality. This is obtained by using only approved reagents and materials as well as only approved equipment as defined by the method validation. Just as there are consequences for the selection of an unsuitable assay for a given purpose, inadequate training or expertise of personnel, improperly maintained facilities or equipment, low-quality samples, and erroneous conclusions can be made if the capabilities and limitations of the assay, laboratory, or sample are not considered when interpreting the results. The sample quality as well as sample type are critical to the selection of the test method as well as the interpretation of the results. The laboratory needs to take steps to ensure these factors are addressed and all risks are mitigated through a quality system implemented by the laboratory.

The only way to know if a method has the performance characteristics that "fit the purpose" for which it will be used is to define the test method through validation. A method with acceptable accuracy and precision levels for the measurement of antibodies in a potency range of 0.1 to 10.0 IU/mL in a serum matrix cannot claim the same accuracy and precision levels for higher-potency samples or samples in a different matrix or body fluid without validation experiments to evaluate these adaptions of the method. The method parameters important for a qualitative assay are sensitivity, specificity, and predictive value. In addition to sensitivity and specificity, a quantitative assay requires definition of accuracy (closeness to the true value), precision (repeatability of the measure), linearity, and reportable range. Validations performed are for specific purposes such as the evaluation of clinical trial samples for a human monoclonal antibody combination for the postexposure treatment of rabies and vaccine potency evaluation (Kostense et al., 2012; Kramer, Bruckner, Daas, & Milne, 2010). Robustness evaluation describes the ability of the method to perform to set criteria during normal variations in laboratory conditions, including normal variations of equipment performance, reagent lots, or between different personnel. Biologic variation (in sample and in biologic components of the assay) must be considered separately from analytical variation. For example, two test results from the same sample may vary solely on the basis of the receptivity of cells to virus infection; cells used last week may have different virus infectivity characteristics than the cells used in subsequent testing. Susceptibility of MNA cells to RABV infection can vary depending on cell passage number (Pouliott et al., 2017). This variance in susceptibility is one factor affecting RABV antibody measurement by serum neutralization assays and highlights the need to control and manage these factors. The variation from these types of factors is separate from other sources of variation. For repeat measures of the same sample, there are statistical tools to set the expected variance and for determining what variance is evidence of a significant difference, such as minimum significance ratio (MSR) (Khan & Findlay, 2009). Measurement or detection of RABV antibodies can be influenced by interference. Interference can be caused by cross-reacting antibodies, nonspecific binding, and the matrix effect (hemolysis, lipemia, or "dirty" samples, etc.).

Interference can occur not only with the antibody of interest in the sample but also in the interaction of the detected or competing antibodies in the assay. Naturally occurring proteins in samples, such as albumins, fibrinogen, and complement factors, can result in assay interference (Selby, 1999). Results from samples with interfering factors can be misleading if the effects of these interfering factors are not considered. In most cases, interference will occur at low levels and will not cause measurement problems at higher dilutions or in samples with high potency because specific binding is stronger than the weaker interference reaction. When interference is suspected or needs to be ruled out, samples may be evaluated by an alternative method in which the effect of interfering factors is minimized so that specific activity may be detected and measured.

The lower limit of detection (LOD) is affected by interference and the assay parameters. If the purpose of testing is to determine the presence or absence of RABV antibodies, it is critical to define the lowest level of antibodies that an assay can reliably detect. But if the ability to accurately and precisely measure low levels of RABV antibodies is important, as in the evaluation of passive RABV immunoglobulin levels in postexposure treatment, then defining an assay's lower limit of quantitation (LLOQ) is required. Cut-off values assigned to an assay depend both on the LOD or LLOQ and the purpose of testing (e.g., detection of antibody or measurement of the level associated with protection). If the application of the rabies serology testing is to identify low levels of RABV antibodies and exclude false negative test results, the cut-off level should be low, but this may yield some false positive test results. Conversely, a higher cut-off value (i.e., above a level that might allow some false positive test results) would identify only true positive test results that would be acceptable if the purpose of the method is to reliably identify only those individuals. The trade-off is that a high cut-off level would increase the number of false negatives (i.e., exclude some true positives that are low). The probability of false positive and false negatives is related to the precision of the assay. Assays with a high variability particularly at the cut-off level would exclude some true positive samples with potency values close to the cut-off level and conversely identify some true negatives as positive. Upon repeat testing, these samples could generate either positive or negative test results.

The matrix of the sample can affect the LOD and LLOQ for a specific method. Therefore, whenever the sample matrix is altered, reevaluation of this parameter is required. Any change that impacts the sensitivity of an assay will also change the LOD (see Table 13.3). Indeed, any change in the procedure or sample may require revalidation to determine the effect on the established performance characteristics. Method variations listed in Table 13.4 (for either binding assays or neutralization assays) can be used to customize a method for certain purposes, such as measurement of antibodies from a particular species or within a range of potency values, but the changes implemented to customize an assay may also result in changes in the performance characteristics of a method.

Immunity can be measured by different methods. It is natural to compare the results from different methods, and it is important to consider how the comparison is made. Although it is very common to evaluate agreement between methods by a correlation coefficient, conclusions based on this value are improper. The optimum way to conduct a method comparison is to calculate the "mean and standard deviation of the between method differences" (Altman, 1991). It is not enough to just generate and examine the data; it is essential to apply appropriate statistical tools. A functional understanding of statistics or collaboration with a

TABLE 13.3 A panel of serum samples consisting of dilution of an international rabies reference serum (SRIG), internal RVNA positive standards, and an RVNA negative serum were tested in the same laboratory in three independent runs in duplicate, each with varying challenge virus doses ($TCID_{50}$).

Sample ID	IU/mL values					Reciprocal titer values				
	High $TCID_{50}$	Med $TCID_{50}$	Low $TCID_{50}$	Average	CV%	High $TCID_{50}$	Med $TCID_{50}$	Low $TCID_{50}$	Average	CV%
SRIG 2.0 IU/mL	2.06	1.91	1.80	**1.92**	**7.2**	55	66	133	**85**	**49.1**
SRIG 0.5 IU/mL	0.42	0.39	0.43	**0.41**	**5.0**	11	13	32	**19**	**60.6**
Internal std-1	12.20	13.10	15.37	**13.56**	**12.0**	328	456	1042	**609**	**62.6**
Internal std-2	2.18	1.92	1.96	**2.02**	**8.2**	57	66	158	**94**	**58.9**
Internal std-3	0.44	0.42	0.52	**0.46**	**11.7**	12	14	39	**22**	**68.8**
Internal std-4	NN	0.06	0.06	**0.07**	**15.4**	NN	2	4	**3**	**44.2**
SRIG 0.2 IU/mL	0.27	0.28	0.16	**0.24**	**30.2**	7	10	12	**10**	**24.3**
SRIG 0.1 IU/mL	0.09	0.10	0.08	**0.09**	**10.5**	2	4	6	**4**	**50.8**
neg serum	NN	NN	NN	**0.06**	**45.7**	NN	NN	NN	**2**	**0.0**
TCID50	75	30	10			75	30	10		

Average IU/mL and reciprocal titer values are displayed. NN indicate no neutralization of virus was observed and no value calculated. Loss of sensitivity at the 0.1 IU/mL level was observed when the high virus dose was used. IU/mL values are comparable across virus doses ($TCID_{50}$) used while the titer values were not. Average and CV values are given in bold to separate from laboratory values.

TABLE 13.4 "Fit for Purpose" method variations that can be applied to neutralization or antigen binding assays.

Neutralization assays	Antigen binding assays
• Strain of challenge virus/pseudotype virus • Dose of challenge virus • Cell type • Serial dilution scheme • Calculation formula • Detection system 　○ Fluorescent-labeled antibody 　○ Enzyme-labeled antibody 　○ Modified challenge virus (e.g., green fluorescent protein)	• Source of antigen—selection of virus strain • Type of antigen—virus protein(s) 　○ Whole virus 　○ Purified protein • Platform—slides, plates, or beads • Detection system variables 　○ Species specific or nonspecies specific 　○ Immunoglobulin class

statistician is often essential for these exercises. The application of statistics to evaluations of immunoassay performance is a specialized area of competence, and it is of particular importance when the assay will be used to determine acceptance of biologics (Findlay et al., 2000).

As previously mentioned, there are some critical components that are essential to consider, identify, and control to ensure precise and accurate measurements for serum neutralization assays, such as strain and dose of challenge virus, cell type, and reference standards. For results to be comparable over time from the same laboratory and possibly between laboratories, these components must be standardized. Whenever any critical steps or components are changed, as may be necessary for a specific purpose, the modified method will require revalidation. A standard reference rabies immune globulin serum (SRIG) provides a defined potency standard in international units per mL (IU/mL). By comparison of the SRIG result to the test sample result, the value of the test sample is standardized and comparable. But if the SRIG used is not the same or not calibrated against a known standard, discrepancies can occur (Yu et al., 2012). The value of the test sample is standardized through comparison with the SRIG result in that assay at that time. It is essential for the SRIG to be precisely described for each batch of test results. If the standards are not identical or not calibrated against a known standard, results cannot be directly compared between assay runs and between laboratories. A standard reference serum of equine source may perform differently than human SRIG such that batches of test results will yield different values depending upon the control serum (Haase, Seinsche, & Schneider, 1985). The potency assigned to an SRIG by one method may not be the same in a different method and cannot be automatically assumed. For example, a control serum at 0.5 IU/mL in the RFFIT method may perform at 0.7 equivalent units per mL (EU/mL) in an ELISA-based assay. If this standard with a known performance by RFFIT was directly applied as the standard for an ELISA and assumed to perform at 0.5 EU/mL, the ELISA results would bias toward the exclusion of some samples that might meet an RFFIT 0.5 IU/mL value. Control serum, such as an SRIG, needs to be fully characterized by a new method and its potency needs to be assigned in units applicable to that particular method (Moore & Hanlon, 2010). A comparison of two international SRIG products, WHO first international rabies immune globulin and WHO second international rabies immune globulin, over several years shows that reference serum can lose potency over time, with the first RIG lower in potency by 2.5% in 1997 to 19% lower in 2012 by RFFIT, yet higher in potency by ELISA. Interestingly, the difference in potency by RFFIT was affected by the type of tissue culture cells used; a minor difference was noted using MNA cells in comparison with BHK cells. This illustrates the importance of calibration and monitoring of the RIG in use in a particular laboratory and for a particular assay as well as how a minor difference in assay components may affect performance. If the challenge virus of an assay is substantially different than the virus source for a vaccine, the serologic results from clinical trials may underestimate responses to the vaccine (Brookes, Healy, & Fooks, 2006; Moore et al., 2005). The same is true for antigen binding assays where the virus strain and type (whole or protein) used in the detection system should ideally be the same in order to obtain the most informative results.

Despite the potential negative effect of a change in how a method is performed, there are good reasons to introduce variations to a procedure. These may include the need to measure RABV antibodies from a specific species, which may require a change in the detection system or the need to measure the potency of samples that are beyond the normal linear range of testing and hence may require a predilution to achieve a different range of sample dilutions to be tested. Method validation reveals the robustness and limitations of the assay and its performance characteristics. In addition to method validation, conducting continual monitoring of method performance increases the chances that potential problems will be quickly identified. Regular participation in proficiency programs is one way to monitor the performance of the method and also assists in the identification of drifts and trends. In addition, the performance/robustness of the assay should be evaluated in individual laboratories. These evaluations are necessary to determine acceptable ranges (Yager & Moore, 2015). Some variables that should be examined are:

- Variations in time and temperature of heat inactivation.
- Storage conditions of samples and of essential materials.
- Variations of temperature and time of assay steps.
- Variations in reading between technologists or types of automated readers.
- Variations in type of calculations for conversion of raw readings into standardized values.

If good quality control practices are in place, results may be comparable between laboratories, even when there are differences in procedure. Conversely, even if laboratories are following the same protocol and using the same components, the agreement in results for the same sample can vary based on method variables related to environment, personnel training, and equipment performance (See Table 13.1). Likewise, methods can be standardized, but unless the laboratories are adhering to the same quality assurance standards, the results may still demonstrate greater variability than is ideal. For example, different laboratories performing RFFIT with the use of different virus doses can produce similar results but have different levels of sensitivity (see Table 13.2). Acceptance criteria for precision and accuracy are different depending upon the type of assay. Cell-based assays such as serum neutralization are inherently more variable and thus are allowed greater variability than binding assays. The precision of binding assays is generally expected to be in the range of 5%–20% while cell-based assays may be allowed a precision variability of 30% and up to 50% (Bioanalytical Method Validation Guidance for Industry, 2018; Chaloner-Larsson, Anderson, & Egan, 1997). In general, for serological titration assays, a two-fold difference in replicate measurements is commonly recognized as the upper level of reproducibility (Wood & Durham, 1980). The precision of an assay should be taken into account when reviewing rabies serology results in relation to the survival of experimental challenge, interlaboratory comparisons, and proficiency testing as well as when establishing acceptable levels for proof of sero-conversion or an adequate response to rabies vaccination.

13.5 Defining "adequate" or "minimum" response to rabies vaccination

It is well known that vaccination resulting in the production of RABV-neutralizing antibodies (RVNA) serves to prevent rabies in persons who have been in contact with a rabid animal. People who have an increased risk of rabies exposure are vaccinated preexposure to provide protection for unnoticed exposures and to reduce the vaccination schedule upon known exposure. This population should have periodic RVNA titer checks to evaluate the need for booster vaccinations. There are two major sources of guidelines in regard to an adequate response to rabies vaccination: the WHO Expert Committee on Rabies and the Advisory Committee on Immunization Practices (ACIP). Because the acceptable levels given by these two guidelines are currently different and there is lack of understanding of how these levels were obtained and what they mean, there is confusion in the medical and veterinary fields about how to interpret rabies serology results in regard to booster vaccination decisions. Guidelines for the prevention of rabies, a fatal disease, should be clear and unambiguous. The optimal level for protection in a patient who recently completed a postexposure series may be different than the level that shows continuing immunity in an individual who has been preexposure vaccinated more than 2 years ago due to the timing of the blood sampling alone (Van Nieuwenhove, Damanet, & Soentjens, 2019). Health professionals need clear guidelines to reference in making life-saving decisions about preventative rabies treatment, for both pre- and postexposure situations. Increasingly, some pet owners and veterinarians are opposing booster vaccination scheduled for pets and turning to rabies serology as proof of sero-conversion as a correlate for protection in a regulatory sense. As this situation advances, regulatory officials need to have a comprehensive understanding of these data on neutralizing antibody protection and the measurement methods that can be used. Lastly, vaccine and rabies immune globulin licensing requirements include rabies serology data, with studies setting supportable acceptance levels for RVNA. Guideline instructions, recommendations, laws, and regulations should clearly define the acceptable RVNA level in terms applicable to the recommended laboratory methods and clarify the situations where a different level may apply.

Because rabies is preventable by vaccination that produces RABV RVNA (other immune mediators may be at play but are not readily measured), the use of methods to quantitate RVNA is preferred when testing for vaccine response in humans. Currently, the most utilized method for this purpose is the RFFIT (and modified RFFIT methods). Not only should the method for this purpose be confirmed to measure RVNA, it should also be standardized to allow comparison to other laboratories and to established guidelines for human vaccination and vaccine manufacturer's instructions. The method needs to provide results that can be related to the guidelines and/or regulations in regard to units of measure and the ability to detect the level stated. No specific RVNA level has been identified as representing absolute protection under all circumstances and in all hosts against all RABV variant infections.

Data regarding proof of rabies protection afforded by vaccination as well as the predictive ability of rabies serology results are primarily obtained from published rabies challenge studies in animals. A review of dog and cat challenge studies which included measurement of the vaccine response by rabies serology by Aubert (1992) concluded the assurance of protection as defined by rabies serology is tied to the extent to which the assigned cut-off exceeds the protective levels demonstrated in dogs and cats, 0.2 and 0.1 IU/mL, respectively. Aubert's conclusion was supported by a previous report by Bunn and Ridpath (1984), which reported the probability of survival at statistical analysis of the survival rate against the serology results of animals in a challenge study. The probability of survival increased as the RVNA level increased to approximately 1.0 IU/mL, with a survival probability of approximately 99% at the level of 0.5 IU/mL. There is a paucity of published data regarding the longevity of the immune response and survival prediction based on rabies serology. A study by Lawson and Crawley (1972) provides additional information. The authors challenged vaccinated dogs and cats at 5 and 4 years, respectively, postvaccination and reported that 92% of dogs and 100% of cats survived; 54% of dogs and 87% of cats had detectable RVNA before challenge. Further, in a report investigating published records on the ability of rabies serology to predict survival in wildlife, few conclusions could be drawn due to the highly variable study designs and test methods/cut-off values utilized (Moore et al., 2017). Also, in this article a study into the relationship between survival rates with standardized rabies serology results, both RFFIT and blocking ELISA, determined a reasonably close association; however, species differences were noted.

RVNA levels attained by the majority of subjects in human vaccine clinical trials formed the basis for the levels recognized as the minimal adequate response in vaccinated humans. The levels recommended by the WHO and the current ACIP are different. This difference is not new, but a misunderstanding of the different levels is common and indeed present, even in the published literature. The ACIP recommends the use of RFFIT because other methods (i.e., ELISA, IFA, immunochromatographic) do not measure RVNA specifically and therefore cannot be correlated to RFFIT. Because applying the WHO level will result in more vaccinated individuals falling within the "need for booster" group and ELISA methods may produce results not correlated with the guideline levels, these important components relied on for vaccination decisions require clarification. Immunization against rabies with the vaccine produced by Pasteur in 1885, and which remained largely unchanged until the advent of tissue culture vaccines, involved multiple vaccinations and was not without serious consequences (Steele, 1975). In contrast, modern tissue culture vaccines are safe and effective (Wunner & Briggs, 2010). The safety of the vaccine combined with the knowledge of the protective effects of RVNA upon rabies exposure led to the practice of preexposure vaccination for people at frequent or continuous risk of exposure (Centers for Disease Control and Prevention, 2008). Initially, it was suggested that this population receive a booster vaccination every 2 years to ensure ongoing "protection" by RVNA (Centers for Disease Control, 1976). RVNA was measured in the early days of rabies vaccine development by the MNT, a cumbersome in vivo test. Concerns about the adverse effects of too-frequent vaccination and the implementation of a rapid test (RFFIT) for RVNA influenced a change in the recommendation from periodic booster vaccination to boosting only when the RVNA level falls below a level representing an adequate response (Centers for Disease Control, 1980).

The level of 0.5 IU/mL was recommended at the Joint WHO/IABS Symposium on the Standardization of Rabies Vaccines for Human Use Produced in Tissue Culture held in Marburg, West Germany, in November 1977 (Bogel, 1978). After the results of several international human rabies vaccine trials were presented, recommendations were given by specific working groups. The working group for vaccine potency requirements of reduced immunization schedules and preexposure vaccination stated: "The group suggests that the serum be tested 4 weeks after the last inoculation and at that time a minimum value of 0.5 IU/mL be attained to demonstrate seroconversion" (Bogel, 1978). Based on the subsequent reports of RABV antibody levels attained after pre- or postexposure vaccination series, the level of 0.5 IU/mL was accepted as proof of seroconversion. The designated level in the ACIP was also based on RFFIT results from human vaccine trials and the observation that no nonspecific inhibition (false positive) reactions were found in serum dilutions above 1:5 (personal communication, Jean Smith). This led to the conclusion that if a specific RVNA titer result was detected at this level, then seroconversion had been achieved. The level described in the ACIP is approximately 0.1 IU/mL in the RFFIT as originally described (Moore & Hanlon, 2010).

Therefore, both levels, 0.5 and 0.1 IU/mL, although different by five-fold, are based on the same rationale, which is the specific detection of RVNA. The difference is the degree of confidence that the designated level can be assured to be a true measurement and not a false positive. The WHO does not define the assay used and recognizes the differences in testing methods and laboratory capabilities. The variance in rabies serology results was described in early publications on methods; wide differences in RVNA levels were obtained by MNT and RFFIT and by different laboratories (Bogel, 1978; Fitzgerald, Gallagher, Hunter, Spivey, & Seligmann Jr., 1978). The ACIP states that rabies serology should be performed by the RFFIT, thereby designating a single approved method. Although the ACIP defines a method, it does not define the adequate level of RVNA in standardized terms. "Complete neutralization at a serum dilution of 1:5 in the RFFIT" can represent different titers and IU/mL values in laboratories that perform "modified" RFFIT assays (see Tables 13.1 and 13.2). Besides standardizing the value reported for rabies serology, defining the specific parameters and standard reagents that comprise acceptable methods for RVNA measurement would aid in interpreting results for the determination of the need for booster vaccination.

Understanding the method and then interpretation of the result is essential for the optimal management of humans and animals (Moore et al., 2017). Indirect ELISA methods detect and measure the presence of RAB-specific antibodies based on their binding ability; they do not measure the neutralizing ability of the antibodies (Irie & Kawai, 2002). The immune response to rabies vaccination involves antibody production that is polyclonal. In each individual, a polyclonal response may include clonal antibodies that vary in affinity, avidity, and ability to neutralize the RABV. Therefore, the relationship between the level of binding antibodies and neutralizing antibodies cannot be predicted and is not linear. In one individual, an ELISA result may be higher than the RFFIT result and in another individual, the opposite can be true (Moore et al., 2016). Applying an "adequate" level of virus-neutralizing antibodies, as described by the WHO and ACIP, to ELISA results will not be accurate in every individual. The validity of a method is unique to each laboratory and the parameters of a validation should be carefully considered. Method validation documents performance standards and includes identification and verification of the lower limit of quantitation (LOQ). This is the lowest level that will produce accurate and precise results. The LOQ is the level that, by implication based on the history stated above, both the WHO and ACIP recognize as an adequate response to rabies vaccination. Performance of a method within each laboratory that generates a rabies serology result and consideration of the rationale behind the two different definitions of adequate rabies vaccine response will need to be considered toward the development of clear language concerning the clinical management of persons or animals at risk for exposure or following an exposure. In a review of three human rabies vaccine clinical trial studies, the percent of nonspecific/cross-reactive results reported in subjects not rabies vaccinated ranged from 1.3% to 4.3% when the cut-off for negative results was set at 0.1 IU/mL, but this percentage was reduced to 0% to 0.4% when the cut-off was set at 0.4 IU/mL. These finding provide confidence to the use of 0.5 IU/mL as a robust level correlated with protection in animal studies and with low probability of nonspecificity.

The proportion of individuals in various populations who are either below the WHO or ACIP guidelines of a minimum response to vaccination has not changed from initially published studies. A review of rabies serology results from sequential student classes at three veterinary colleges between 2005 and 2011, which were tested at the Kansas State University

Rabies Laboratory demonstrated the consistency of rabies vaccine response as measured by rabies serology over 7 years (Moore, 2015). The percentage of populations failing to meet the ACIP level ranged from 0% to 8%, and the percentage of populations failing to meet the WHO level ranged from 15% to 25%. These proportions are similar to what is reported in the peer-reviewed literature from clinical vaccine studies of preexposure vaccinated individuals, where 2%–7% of vaccinated people failed to show complete neutralization at a 1:5 serum dilution after 2 years, and approximately 75% of persons remain sero-converted (>0.5 IU/mL) for 5–10 years after preexposure vaccination (Mansfield et al., 2016; Rodrigues et al., 1987; Strady et al., 1998).

13.6 Regulatory compliance

The regulatory landscape can be complicated for a testing laboratory asked to complete testing for vaccine development, clinical trials, surveillance testing, or even product testing. The regulations are not entirely clear at times regarding when compliance is required and which standards are needed. Thus subject to interpretation, this leaves testing facilities to decide how to comply with the various regulations and standards. Laboratories must have a robust quality system to meet the diverse expectations of those using and interpreting the results. If the testing is for drug development or evaluation, there are complexities in bringing a product to market in a highly regulated environment to which quality requirements change based on the product's stage of development (Grappin, 2017). Not only do the requirements change based upon the stage of development or the reason for testing, but they can also change based on national or regional requirements. As an example, FDA GLP and OECD GLP have recognizable differences within the verbiage, content, and interpretation of the regulations. Some may be more stringent or more prescriptive toward one aspect of the regulation (Huntsinger, 2008). The end result for all standards followed is to ensure the experimental information generated is of sufficient quality and integrity to support the reports or results submitted for testing. These differences, while not affecting the quality of the results or the data, can result in issues when submitting studies globally. Fewer regulatory requirements may be needed to comply with general single patient testing or vaccine surveillance testing. However, a robust quality system is needed to ensure that defendable results are given to the client or regulatory agencies, no matter the purpose of the testing.

Testing laboratories take on the responsibility to ensure that the most appropriate analytical method is used to provide the needed information to the client. These results must meet the regulatory scrutiny of those interpreting the results in order to ensure that correct decisions are being made based upon the data. When discussing a robust quality system, more than just passing controls must be in place. Not only are the standards different, as in GLP, GCP, and GMP, but each country has its own version of GLP, GCP, and GMP standards. Comparing the US FDA GMP to the EU GMP (Table 13.5) illustrates the differences as well as the challenges faced by the testing laboratories and researchers. The EU is made up of 27 individual member states, all with their own national legislation derived from EU directives. Each EU member state has its own regulatory authority with GMP inspectors at different

TABLE 13.5 EU and US—Good manufacturing practices—Summary comparison of the regulations as related to a testing laboratory.

EU—GMP directives	US—Code of Fed. Regs
Art. 6 Quality management	*No comparable regulation*
Art. 7 Personnel	211.25 Personnel qualifications
	211.28 Personnel responsibilities (Sub Part B)
Art. 8 Premises and equipment	211.42–211.58 Buildings and equipment (Sub Part C)
	211.63–211.72 Equipment (Sub Part D)
Art. 9 Documentation	211.180–211.198 Records and reports (Sub Part J)
Art. 10 Production	211.100–211.115 Production and process control (Sub Part F)
Art. 11 Quality control	211.22 Responsibilities of quality control unit (Sub Part B)
	211.160–211.176 Laboratory controls (Sub Part 1)
Art. 12 Work contracted out	*No comparable regulation*
Art. 13 Complaints and product recall	211.98 Complaint Files
	211.18 General requirements (ref. Recalls) (Sub Part J)
Art. 14 Self-inspections	*No comparable regulation*
No directly comparable regulation	Part II Electronic records and signatures

levels of experience, training, and exposure to the testing and pharmaceutical industry. The United States is not without issues because most of the Food and Drug Administration (FDA) investigators have in the past had very little or no experience in the pharmaceutical industry before joining the FDA (QP.Solutions, n.d.). These factors, together with more minor cultural differences, have resulted in a different EU versus FDA approach to the way in which pharmaceutical companies are inspected and the way GMP is enforced. The FDA has traditionally been most concerned with compliance and enforcement. Inspection approaches have been based on reviews of documentary evidence rather than observation of actual practices, which has been the European style. More and more research is being performed across the globe. Studies are not only being undertaken for the countries of origin, but are also submitted to other countries for approval.

With the importance of producing quality products and processes as well as the increase in regulatory scrutiny of all aspects of product development and manufacturing, researchers and testing laboratories have to understand what regulatory environment is applicable for each phase of development of a product (Fig. 13.5). Testing laboratories will need a robust quality management system that can adapt to be compliant with all phases of GXP (Fig. 13.5) as well as be compliant with different country's regulatory requirements. Drug development from the beginning of research until the product is on the market generally will range from 12 to 25 years, but could exceed this timeline. The responsibility of the researcher or the testing laboratory is critical to the success of the product within the regulatory review and

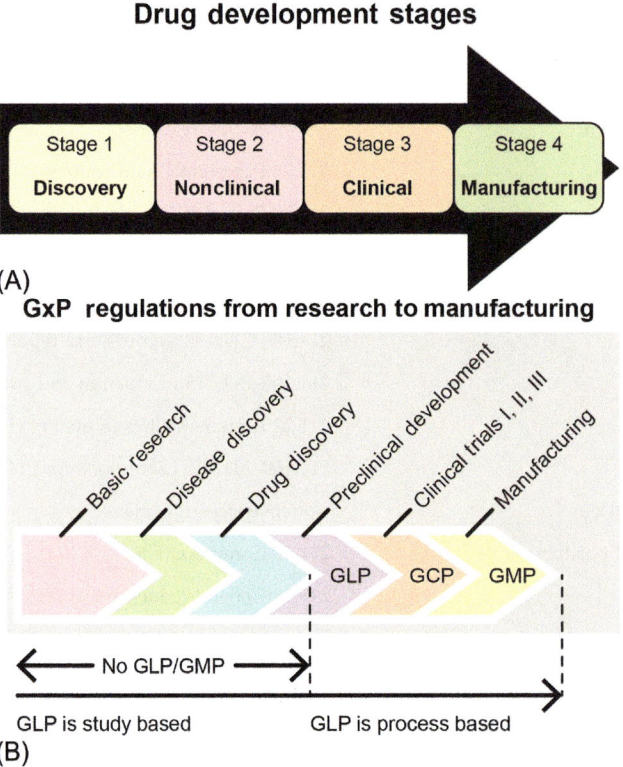

FIG. 13.5 (A) Drug development is delineated into specific stages, each with its own requirement of regulatory oversight. (B) The intended use of the data to be generated will define the recommended regulatory level of testing for a product or drug.

approval of the product. The responsibility is for not only providing quality results but also meeting all aspects of the standards applicable to the development of the research. Within the different GXP standards (GLP, GCLP, GCP, GMP), all have sections that define compliance for the organization, personnel, training, facilities, validation, consumables, documentation, procedures, reporting, contracts, subcontracted work, proficiency testing, environmental monitoring, record retention that includes long-term archival procedures, and quality control/quality assurance oversight via a quality unit or quality team. The difference can be as minor as terminology, such as the name of the position with overall leadership to the study or production. Other differences between standards, which are more critical to the compliance of the study, such as the omission of a formalized corrective action/preventive action (CAPA System) in one standard, but required in another standard can pose obstacles for the testing laboratories or researchers endeavoring to ensure compliance as the study progresses through the development stages (DeRoo, 2014). The different requirements and content create many different challenges for researchers and testing laboratories throughout the process. A robust quality system created to meet all the requirements of each phase of product development is necessary to allow for a seamless transition from one stage to another during drug development.

Research work completed during the initial stages of development poses unique challenges as to knowing the extent of quality needed during research and discovery. There is essentially no standard requirement needed for these stages. However, fundamental needs for quality are still necessary. These challenges generally stem from researchers not being trained in the fundamental aspects of laboratory quality. Many have been trained in experimental design but lack the basic knowledge and understanding of the need for documentation and traceability. Time constraints and the belief that implementing a GXP quality system creates an unnecessary burden of paperwork prevent most researchers from following quality requirements during the pre-GLP stages (DeRoo, 2014). Experimental design is often thought to meet GXP-ish requirements or provide the study to be completed in the "spirit of GXP." Without following the regulatory guidelines, compliance to any given standard is impossible to meet. Research work during these stages may be based on unsubstantiated assumptions about the process or test methods used, the nature of the reagents and materials used, and the instrument or equipment used. There is the possibility that the primary documentation generated may not have been verified or reviewed to ensure that all the information being captured by each analyst provides adequate detail and traceability for use in product development reports.

As the development phases move from discovery to preclinical development, the requirements will move from pre-GLP to GLP to GCP to GMP (Fig. 13.5). Not only does this change the standards, but it also changes the sample matrix from product to animal to human and back to product testing. The transitioning through the different phases does not always move forward, as we may need to go back to a previous phase to access any issues found or testing discrepancies. In simplest terms, GLP is to ensure a compliant study dealing with characterizations of the product and animal model testing, GCP is to ensure a compliant clinical trial that protects the human participates, and GMP is to ensure compliant manufacturing of the product. Implementing the GLP basic framework during the pre-GLP stages will assist with the transition to the GLP phase and beyond. We must ensure that the data is accurate and verifiable as well as document traceability. Instruments must be calibrated, the test methods must be validated, personnel must be trained, and controls must be traceable or verified against a known national or international control standard. The additions of the GLP framework will assist with providing the robust data needed for future patents, applications for NIH grants, and/or the transition to GLP product testing. Study reports submitted for approval will be reviewed against historical data, known disease cycles, currently approved drug performance, and the purpose of the new drug. The comparison of the historical data and the current study data needs to be easily reviewed and accessed for effectiveness. Implementing a GXP quality system will allow for comparisons of past studies and disease research against the new data seamlessly. This is accomplished by establishing "new" processes against the "old" processes such as new test method validation against a gold standard, in house laboratory controls validated against national and/or international controls, participation in external proficiency testing programs if available, and validation of all new equipment. If an external proficiency testing program is not available, the laboratory should perform internal blind testing for each analyst to assess ability as well as to ensure that consistent, accurate, and reliable results are reported or split samples with another laboratory or researcher performing the same or a comparable test method. Proficiency testing data also allow for the researcher or laboratory leadership to routinely review for subtle shifts and trends that, over time, would affect patient and study results. Tracking and trending control data will allow for the tracking of normal variation, and quick observation of any abnormal variations. A robust quality system will allow for the control of these differences, such as changes from one lot number to the next for reagents, cells, controls, and materials as well as different analysts performing the test. Variations will be slight within the trending and tracking, as the differences in the process will not shift the data outside the acceptance criteria. Some variation may be the result of causes that are not normally present in the method, such as incorrect reagents, incorrect sample type, deviations from the procedure, and inaccurate measurements. These types of variations need to be investigated to determine the cause and correct the issue to prevent it from happening again. Any trends or shifts outside the acceptance criteria for control charts must be investigated to ensure that results reported to clients are actual and correct. These investigations must follow a set process to ensure all aspects of testing are reviewed and accessed against established perimeters for equipment, personnel, methods, reagents, controls, temperature, and materials. Investigations are a fundamental component of all quality standards within the GXP regulations.

13.7 Conclusions

Rabies serology has provided necessary data for vaccine and biologics evaluation/licensing, sero-surveillance and epidemiology studies, research into humoral immunity, and establishment of regulations for pet import/export, among other uses. Rabies prevention and control efforts thus far could not have attained the state of progress without these data. The future holds the promise of improved vaccines in combination with rabies monoclonal antibody cocktails to provide endemic developing countries with rabies prophylaxis biologics to reduce the burden of human rabies deaths. New vaccines for control of wildlife are also in development. These new biologics require the rabies serology field to advance at pace with the development of high-throughput and accessible methods, all while not losing sight of the high level of quality currently in place. The value of accurate and meaningful rabies serology results cannot be overstated. Accurate and meaningful results are only obtained by choosing the method "fit for purpose," which requires a thorough understanding of the method's performance characteristics and limitations. Strict adherence to the key components of the method chosen will assure quality results. Rabies antibody acceptance levels should ideally be defined in a standard format (IU/mL) and in consideration of methods that are validated for this purpose, giving due consideration to the method's performance characteristics, especially concerning variability at the lower limit of detection. Different regulatory standards and guidelines exist across the world for vaccines, biologics, and diagnostics kit licensing and use. Each standard has a common goal: to ensure the integrity of the laboratory data, protect human welfare, and provide safe and effective products. These goals can only be met if all aspects of the testing and phase developments are held to the appropriate regulatory standards and monitored throughout the development process. From the beginning of the work at the bench level to the end of the process, implementation of the standards apply. This often continues during clinical use of products intended for the improvement or protection of human and animal health, including direct assessment of, and assessment of host responses to, rabies vaccines and sources of polyclonal and monoclonal products for the prevention of rabies.

References

Altman, D. G. (1991). Statistical analysis of comparison between laboratory methods. *Journal of Clinical Pathology, 44* (8), 700–701.

Atanasiu, P. (1973). Quantitative assay and potency test of antirabies serum and immunoglobulin. *Monograph Series World Health Organization, 23*, 314–318.

Aubert, M. F. (1992). Practical significance of rabies antibodies in cats and dogs. *Revue Scientifique et Technique, 11*(3), 735–760.

Aubert, M. F. (1996). Methods for the calculation of titres. In F. X. Meslin, M. M. Kaplan, & H. Koprowski (Eds.), *Laboratory Techniques in Rabies* (4th ed., pp. 445–459). Geneva: World Health Organization (Reprinted from: Not in File).

Bedekovic, T., Lemo, N., Lojkic, I., Mihaljevic, Z., Jungin, A., Cvetnic, Z., … Hostnik, P. (2013). Modification of the fluorescent antibody virus neutralisation test—Elimination of the cytotoxic effect for the detection of RABV neutralising antibodies. *Journal of Virological Methods, 189*(1), 204–208. https://doi.org/10.1016/j.jviromet.2013.01.022.

Bedekovic, T., Simic, I., Kresic, N., Lojkic, I., Mihaljevic, Z., Sucec, I., … Hostnik, P. (2016). Evaluation of ELISA for the detection of RABV antibodies from the thoracic liquid and muscle extract samples in the monitoring of fox oral vaccination campaigns. *BMC Vet. Res, 12*, 76. https://doi.org/10.1186/s12917-016-0701-0.

Bioanalytical Method Validation Guidance for Industry. (2018). Retrieved from Silver Springs: https://www.fda.gov/regulatory-information/search-fda-guidance-documents/bioanalytical-method-validation-guidance-industry.

Bogel, K. (1978). Proposed International Reference Rabies Vaccine (HDC-Origin) and the potency tests used to test these products. In *Joint WHO/IABS symposium on the standardization of rabies vaccines for human use produced in tissue culture [rabies III]* (40 ed., pp. 267–271). Basel: S. Karger (Reprinted from: Not in File).

Bordignon, J., Comin, F., Ferreira, S. C., Caporale, G. M., Lima Filho, J. H., & Zanetti, C. R. (2002). Calculating RABV neutralizing antibodies titres by flow cytometry. *Revista do Instituto de Medicina Tropical de São Paulo, 44*(3), 151–154. pii: S0036-46652002000300007.

Briggs, D. J., Smith, J. S., Mueller, F. L., Schwenke, J., Davis, R. D., Gordon, C. R., … Rupprecht, C. E. (1998). A comparison of two serological methods for detecting the immune response after rabies vaccination in dogs and cats being exported to rabies-free areas. *Biologicals, 26*(4), 347–355. pii: S1045-1056(98)90162-2 https://doi.org/10.1006/biol.1998.0162.

Brookes, S. M., Healy, D. M., & Fooks, A. R. (2006). Ability of rabies vaccine strains to elicit cross-neutralising antibodies. *Developmental Biology (Basel), 125*, 185–193.

Bunn, T. O., & Ridpath, H. D. (1984). *The relationship between rabies antibody titers in dogs and cats and protection from challenge* (pp. 43–45). No. 11, U. S. Department of Health, Education and Welfare, Public Health.

Burgado, J., Greenberg, L., Niezgoda, M., Kumar, A., Olson, V., Wu, X., & Satheshkumar, P. S. (2018). A high throughput neutralization test based on GFP expression by recombinant RABV. *PLoS Neglected Tropical Diseases, 12*(12) e0007011. https://doi.org/10.1371/journal.pntd.0007011.

Buthelezi, S. G., Dirr, H. W., Chakauya, E., Chikwamba, R., Martens, L., Tsekoa, T. L., … Vandermarliere, E. (2016). The Lyssavirus glycoprotein: A key to cross-immunity. *Virology, 498*, 250–256. https://doi.org/10.1016/j.virol.2016.08.034.

Cardoso, T. C., Silva, L. H., Albas, A., Ferreira, H. L., & Perri, S. H. (2004). Rabies neutralizing antibody detection by indirect immunperoxidase serum neutralization assay performed on chicken embryo related cell line. *Memórias do Instituto Oswaldo Cruz, 99*(5), 531–534. pii: S0074-02762004000500013.

Carpenter, A. B. (1997). Enzyme linked immuno-sorbent assay. In N. R. Rose, E. C. DeMacario, J. Fahey, H. Fiedman, & G. M. Pen (Eds.), *Manual of Clinical Laboratory Immunology* (5th ed., pp. 2–9). Boston: Little Brown (Reprinted from: Not in File).

Centers for Disease Control. (1976). *Rabies Recommendations of the Public Health Service Advisory committee on Immunization Practices.* (Rep. No. Vol. 25/No. 51). Atlanta, GA.

Centers for Disease Control. (1980). *Rabies Prevention Recommendation of the Immunization Practices Advisory Committee (ACIP).* (Rep. No. Vol. 29/No. 23). Atlanta, GA.

Centers for Disease Control and Prevention. (2008). *Human Rabies Prevention—United States, 2008 Recommendatoins of the Advisory Committee On Immunization Practices.* (Rep. No. Vol. 57/RR-3). Atlanta, GA.

Chabaud-Riou, M., Moreno, N., Guinchard, F., Nicolai, M. C., Niogret-Siohan, E., Seve, N., … Riou, P. (2017). G-protein based ELISA as a potency test for rabies vaccines. *Biologicals, 46*, 124–129. https://doi.org/10.1016/j.biologicals.2017.02.002.

Chaloner-Larsson, G., Anderson, R., & Egan, A. (1997). A WHO guide to goo manufacturing practice (GMP) requirements. In *Part: Validation*. Geneva: World Health Organization (Reprinted from: Not in File).

Cliquet, F., Aubert, M., & Sagne, L. (1998). Development of a fluorescent antibody virus neutralisation test (FAVN test) for the quantitation of rabies-neutralising antibody. *Journal of Immunological Methods, 212*(1), 79–87. pii: S0022175997002123.

Combes, R. D., & Balls, M. (2014). The three Rs—Opportunities for improving animal welfare and the quality of scientific research. *Alternatives to Laboratory Animals, 42*(4), 245–259. https://doi.org/10.1177/026119291404200406.

Cunha Neto Rdos, S., Vigerelli, H., Jared, C., Antoniazzi, M. M., Chaves, L. B., da Silva Ade, C., ... Pimenta, D. C. (2015). Synergic effects between ocellatin-F1 and bufotenine on the inhibition of BHK-21 cellular infection by the RABV. *Journal of Venomous Animals and Toxins including Tropical Diseases, 21*, 50. https://doi.org/10.1186/s40409-015-0048-1.

DeRoo, D. (2014). *GLPs and GMPs: When are they necessary?* Retrieved from https://www.namsa.com/wp-content/uploads/2015/10/WP_Requirements-for-GLP-and-GMP-Testing.pdf.

Dietzschold, B. (1993). Antibody-mediated clearance of viruses from the mammalian central nervous system. *Trends in Microbiology, 1*(2), 63–66.

Fallahi, F., Wandeler, A. I., & Nadin-Davis, S. A. (2016). Characterization of epitopes on the RABV glycoprotein by selection and analysis of escape mutants. *Virus Research, 220*, 161–171. https://doi.org/10.1016/j.virusres.2016.04.019.

Findlay, J. W., Smith, W. C., Lee, J. W., Nordblom, G. D., Das, I., DeSilva, B. S., ... Bowsher, R. R. (2000). Validation of immunoassays for bioanalysis: A pharmaceutical industry perspective. *Journal of Pharmaceutical and Biomedical Analysis, 21*(6), 1249–1273. pii: S0731708599002447.

Finke, S., & Conzelmann, K. K. (2005). Replication strategies of RABV. *Virus Research, 111*(2), 120–131. https://doi.org/10.1016/j.virusres.2005.04.004.

Fitzgerald, E. A., Baer, G. M., Cabasso, V. F., & Vallancourt, R. F. (1975). A collaborative study on the potency testing of antirabies globulin. *Journal of Biological Standardization, 3*(3), 273–278.

Fitzgerald, E. A., Gallagher, M., Hunter, W. S., Spivey, R. F., & Seligmann, E. B., Jr. (1978). Laboratory evaluation of the immune response to rabies vaccine. *Journal of Biological Standardization, 6*(2), 101–109.

Flamand, A., Raux, H., Gaudin, Y., & Ruigrok, R. W. (1993). Mechanisms of RABV neutralization. *Virology, 194*(1), 302–313. pii: S0042-6822(83)71261-4 https://doi.org/10.1006/viro.1993.1261.

Fooks, A. (2018). *Rabies (infection with RABV and other Lyssaviruses). In OIE Terrestrial Manual*. Available at: https://www.oie.int/standard-setting/terrestrial-manual/access-online/.

Gibert, R., Alberti, M., Poirier, B., Jallet, C., Tordo, N., & Morgeaux, S. (2013). A relevant in vitro ELISA test in alternative to the in vivo NIH test for human rabies vaccine batch release. *Vaccine, 31*(50), 6022–6029. https://doi.org/10.1016/j.vaccine.2013.10.019.

Grappin, J. (Producer). (2017, 5/1/2019). Retrieved from https://www.eag.com/resources/whitepapers/medical-device-regulations/.

Goding, J. W. (1978). Use of staphylococcal protein a as an immunological reagent. *Journal of Immunological Methods, 20*, 241–253. https://doi.org/10.1016/0022-1759(78)90259-4.

Grassi, M., Wandeler, A. I., & Peterhans, E. (1989). Enzyme-linked immunosorbent assay for determination of antibodies to the envelope glycoprotein of RABV. *Journal of Clinical Microbiology, 27*(5), 899–902.

Guglielmo-Viret, V., Attree, O., Blanco-Gros, V., & Thullier, P. (2005). Comparison of electrochemiluminescence assay and ELISA for the detection of *Clostridium botulinum* type B neurotoxin. *Journal of Immunological Methods, 301*(1–2), 164–172. pii: S0022-1759(05)00110-9 https://doi.org/10.1016/j.jim.2005.04.003.

Haase, M., Seinsche, D., & Schneider, W. (1985). The mouse neutralization test in comparison with the rapid fluorescent focus inhibition test: Differences in the results in rabies antibody determinations. *Journal of Biological Standardization, 13*(2), 123–128.

Habel, K. (1996). Habel test for potency. In F. X. Meslin, M. M. Kaplan, & H. Koprowski (Eds.), *Laboratory techniques in rabies* (4th ed., pp. 369–373). Geneva: World Health Organization (Reprinted from: Not in File).

Hooper, D. C. (2005). The role of immune responses in the pathogenesis of rabies. *Journal of Neurovirology, 11*(1), 88–92. https://doi.org/10.1080/13550280590900418.

Hooper, D. C., Morimoto, K., Bette, M., Weihe, E., Koprowski, H., & Dietzschold, B. (1998). Collaboration of antibody and inflammation in clearance of RABV from the central nervous system. *Journal of Virology, 72*(5), 3711–3719.

Hostnik, P. (2000). The modification of fluorescent antibody virus neutralization (FAVN) test for the detection of antibodies to RABV. *Journal of Veterinary Medicine. B, Infectious Diseases and Veterinary Public Health, 47*(6), 423–427.

Huntsinger, D. W. (2008). OECD and USA GLP applications. *Annali dell'Istituto Superiore di Sanità, 44*(4), 403–406.

Irie, T., & Kawai, A. (2002). Studies on the different conditions for RABV neutralization by monoclonal antibodies #1-46-12 and #7-1-9. *Journal of General Virology*, 83(Pt 12), 3045–3053.

Kaplan, M. M., Wiktor, T. J., Maes, R. F., Campbell, J. B., & Koprowski, H. (1967). Effect of polyions on the infectivity of RABV in tissue culture: Construction of a single-cycle growth curve. *Journal of Virology*, 1(1), 145–151.

Khan, N. K., & Findlay, J. W. (2009). Assay design. In N. K. Khan & J. W. Findlay (Eds.), *Ligand-binding assays: Development, validation, and implementation in the drug development arena* (pp. 196–219). Hoboken: John Wiley and Sons [Reprinted from: Not in File].

Khawplod, P., Inoue, K., Shoji, Y., Wilde, H., Ubol, S., Nishizono, A., ... Morimoto, K. (2005). A novel rapid fluorescent focus inhibition test for RABV using a recombinant RABV visualizing a green fluorescent protein. *Journal of Virological Methods*, 125(1), 35–40. pii: S0166-0934(04)00388-X https://doi.org/10.1016/j.jviromet.2004.12.003.

Korimbocus, J., Dehay, N., Tordo, N., Cano, F., & Morgeaux, S. (2016). Development and validation of a quantitative competitive ELISA for potency testing of equine anti rabies sera with other potential use. *Vaccine*, 34(28), 3310–3316. https://doi.org/10.1016/j.vaccine.2016.04.086.

Kostense, S., Moore, S., Companjen, A., Bakker, A. B., Marissen, W. E., von, E. R., ... Goudsmit, J. (2012). Validation of the rapid fluorescent focus inhibition test (RFFIT) for RABV neutralizing antibodies in clinical samples. *Antimicrobial Agents and Chemotherapy*. https://doi.org/10.1128/AAC.06179-11 pii: AAC.06179-11.

Kramer, B., Bruckner, L., Daas, A., & Milne, C. (2010). Collaborative study for validation of a serological potency assay for rabies vaccine (inactivated) for veterinary use. *Pharmeuropa Bio & Scientific Notes*, 2010(2), 37–55.

Kramer, B., Schildger, H., Behrensdorf-Nicol, H. A., Hanschmann, K. M., & Duchow, K. (2009). The rapid fluorescent focus inhibition test is a suitable method for batch potency testing of inactivated rabies vaccines. *Biologicals*, 37(2), 119–126. pii: S1045-1056(09)00002-5 https://doi.org/10.1016/j.biologicals.2009.01.001.

Lawson, K. F., & Crawley, J. F. (1972). The ERA strain of rabies vaccine. *Canadian Journal of Comparative Medicine*, 36(4), 339–344.

Lembo, T., Niezgoda, M., Velasco-Villa, A., Cleaveland, S., Ernest, E., & Rupprecht, C. E. (2006). Evaluation of a direct, rapid immunohistochemical test for rabies diagnosis. *Emerging Infectious Diseases*, 12(2), 310–313. https://doi.org/10.3201/eid1202.050812.

Lyng, J., Bentzon, M. W., & Fitzgerald, E. A. (1989). Potency assay of antibodies against rabies. A report on a collaborative study. *Journal of Biological Standardization*, 17(3), 267–280.

Ma, X., Niezgoda, M., Blanton, J. D., Recuenco, S., & Rupprecht, C. E. (2012). Evaluation of a new serological technique for detecting RABV antibodies following vaccination. *Vaccine*. pii: S0264-410X(12)00892-4 https://doi.org/10.1016/j.vaccine.2012.06.037.

Madhusudana, S. N., Malavalli, B. V., Thankappan, U. P., Sundramoorthy, S., Belludi, A. Y., Pulagumbaly, S. B., & Sanyal, S. (2014). Development and evaluation of a new immunohistochemistry-based test for the detection of RABV neutralizing antibodies. *Human Vaccines & Immunotherapeutics*, 10(5), 1359–1365. https://doi.org/10.4161/hv.28042.

Manalo, D. L., Yamada, K., Watanabe, I., Miranda, M. E., Lapiz, S. M., Tapdasan, E., ... Nishizono, A. (2016). A comparative study of the RAPINA and the virus-neutralizing test (RFFIT) for the estimation of antirabies-neutralizing antibody levels in dog samples. *Zoonoses and Public Health*. https://doi.org/10.1111/zph.12313.

Mannen, K., Mifune, K., Reid-Sanden, F. L., Smith, J. S., Yager, P. A., Sumner, J. W., ... Baer, G. M. (1987). Microneutralization test for RABV based on an enzyme immunoassay. *Journal of Clinical Microbiology*, 25(12), 2440–2442.

Mansfield, K. L., Andrews, N., Goharriz, H., Goddard, T., McElhinney, L. M., Brown, K. E., & Fooks, A. R. (2016). Rabies pre-exposure prophylaxis elicits long-lasting immunity in humans. *Vaccine*, 34(48), 5959–5967. https://doi.org/10.1016/j.vaccine.2016.09.058.

Moeschler, S., Locher, S., Conzelmann, K. K., Kramer, B., & Zimmer, G. (2016). Quantification of Lyssavirus-neutralizing antibodies using vesicular stomatitis virus Pseudotype particles. *Viruses*, 8(9). https://doi.org/10.3390/v8090254.

Moore, S. M., & Hanlon, C. A. (2010). Rabies-specific antibodies: Measuring surrogates of protection against a fatal disease. *PLoS Neglected Tropical Diseases*, 4(3), e595. https://doi.org/10.1371/journal.pntd.0000595.

Moore, S. M., Pralle, S., Engelman, L., Hartschuh, H., & Smith, M. (2016). Rabies vaccine response measurement is assay dependent. *Biologicals*, 44(6), 481–486. pii: S1045-1056(16)30106-3 https://doi.org/10.1016/j.biologicals.2016.09.007.

Moore, S. M., Ricke, T. A., Davis, R. D., & Briggs, D. J. (2005). The influence of homologous vs. heterologous challenge virus strains on the serological test results of RABV neutralizing assays. *Biologicals*, 33(4), 269–276. pii: S1045-1056(05)00057-6 https://doi.org/10.1016/j.biologicals.2005.06.005.

Moore, S. M., Gilbert, A., Vos, A., Freuling, C. M., Ellis, C., Kliemt, J., & Muller, T. (2017). RABV antibodies from oral vaccination as a correlate of protection against lethal infection in wildlife. *Tropical Medicine and Infectious Disease, 2*(31). https://doi.org/10.3390/tropicalmed2030031.

Moore, S. M. (2015). *Rabies Serology: Relationship between assay type, interpretation, and application of results* [Doctor of Philosophy Dissertation]. Manhattan, Kansas: Kansas State University. Retrieved from http://hdl.handle.net/2097/18931 (2015-04-15T18:50:55Z).

Morgeaux, S., Poirier, B., Ragan, C. I., Wilkinson, D., Arabin, U., Guinet-Morlot, F., ... Chapsal, J. M. (2017). Replacement of in vivo human rabies vaccine potency testing by in vitro glycoprotein quantification using ELISA—Results of an international collaborative study. *Vaccine, 35*(6), 966–971. https://doi.org/10.1016/j.vaccine.2016.12.039.

Nicholson, K. G., & Prestage, H. (1982). Enzyme-linked immunosorbent assay: A rapid reproducible test for the measurement of rabies antibody. *Journal of Medical Virology, 9*(1), 43–49.

Nie, J., Wu, X., Ma, J., Cao, S., Huang, W., Liu, Q., ... Wang, Y. (2017). Development of in vitro and in vivo RABV neutralization assays based on a high-titer pseudovirus system. *Scientific Reports, 7*, 42769. https://doi.org/10.1038/srep42769.

Nishizono, A., Yamada, K., Khawplod, P., Shiota, S., Perera, D., Matsumoto, T., ... Ahmed, K. (2012). Evaluation of an improved rapid neutralizing antibody detection test (RAPINA) for qualitative and semiquantitative detection of rabies neutralizing antibody in humans and dogs. *Vaccine, 30*(26), 3891–3896. pii: S0264-410X(12)00511-7 https://doi.org/10.1016/j.vaccine.2012.04.003.

Page, M., & Thorpe, R. (2002). Purification of IgG using protein a or protein G. In J. M. Walker (Ed.), *The protein protocols handbook* (pp. 993–994): Humana Press.

Peharpre, D., Cliquet, F., Sagne, E., Renders, C., Costy, F., & Aubert, M. (1999). Comparison of visual microscopic and computer-automated fluorescence detection of RABV neutralizing antibodies. *Journal of Veterinary Diagnostic Investigation, 11*(4), 330–333.

Poston, R., Hill, R., Allen, C., Casey, W., Gatewood, D., Levis, R., ... Allen, D. (2019). Achieving scientific and regulatory success in implementing nonanimal approaches to human and veterinary rabies vaccine testing: A NICEATM and IABS workshop report. *Biologicals, 60*, 8–14. https://doi.org/10.1016/j.biologicals.2019.06.005.

Pouliott, C., Dupuis, M., Appler, K., Brunt, S., Rudd, R., & Davis, A. (2017). Susceptibility of neuroblastoma cells to RABV may be affected by passage number. *Journal of Virological Methods, 247*, 28–31. https://doi.org/10.1016/j.jviromet.2017.05.005.

QP.Solutions. *GMPs in the EU and USA—an overview and comparison*. Retrieved from http://qp.solutions/pharmaceutical-gmp/gmps-eu-usa-overview-comparison/.

Rodrigues, F. M., Mandke, V. B., Roumiantzeff, M., Rao, C. V., Mehta, J. M., Pavri, K. M., & Poonawalla, C. (1987). Persistence of rabies antibody 5 years after pre-exposure prophylaxis with human diploid cell antirabies vaccine and antibody response to a single booster dose. *Epidemiology and Infection, 99*(1), 91–95.

Rudd, R. J., Appler, K. A., & Wong, S. J. (2013). Presence of cross-reactions with other viral encephalitides in the indirect fluorescent-antibody test for diagnosis of rabies. *Journal of Clinical Microbiology, 51*(12), 4079–4082. https://doi.org/10.1128/JCM.01818-13.

Selby, C. (1999). Interference in immunoassay. *Annals of Clinical Biochemistry, 36*(Pt 6), 704–721.

Serra-Cobo, J., Lopez-Roig, M., Lavenir, R., Abdelatif, E., Boucekkine, W., Elharrak, M., ... Bourhy, H. (2018). Active sero-survey for European bat lyssavirus type-1 circulation in north African insectivorous bats. *Emerging Microbes and Infections, 7*(1), 213. https://doi.org/10.1038/s41426-018-0214-y.

Smith, J. (1991). Rabies serology. In G. M. Baer (Ed.), *The Natural History of Rabies* (2nd ed., pp. 235–252). Boca Raton: CRC Press [Reprinted from: Not in File].

Smith, J. S., Yager, P. A., & Baer, G. M. (1973). A rapid reproducible test for determining rabies neutralizing antibody. *Bulletin of the World Health Organization, 48*(5), 535–541.

Smith, T. G., & Gilbert, A. T. (2017). Comparison of a micro-neutralization test with the rapid fluorescent focus inhibition test for measuring RABV neutralizing antibodies. *Tropical Medicine and Infectious Disease, 2*(3). https://doi.org/10.3390/tropicalmed2030024.

Steele, J. (1975). History of rabies. In G. M. Baer (Ed.), *Vol. I. The natural history of rabies* (1st ed., pp. 1–28). New York: Academic Press [Reprinted from: Not in File].

Strady, A., Lang, J., Lienard, M., Blondeau, C., Jaussaud, R., & Plotkin, S. A. (1998). Antibody persistence following preexposure regimens of cell-culture rabies vaccines: 10-year follow-up and proposal for a new booster policy. *The Journal of Infectious Diseases, 177*(5), 1290–1295.

Tordo, N. (1996). Characteristics and molecular biology of the RABV. In F. X. Meslin, M. M. Kaplan, & H. Koprowski (Eds.), *Laboratory techniques in rabies* (4th ed., pp. 28–51). Geneva: World Health Organization [Reprinted from: Not in File].

Um, J., Chun, B. C., Lee, Y. S., Hwang, K. J., Yang, D. K., Park, J. S., & Kim, S. Y. (2017). Development and evaluation of an anti-RABV phosphoprotein-specific monoclonal antibody for detection of rabies neutralizing antibodies using RFFIT. *PLoS Neglected Tropical Diseases, 11*(12), e0006084. https://doi.org/10.1371/journal.pntd.0006084.

Van Nieuwenhove, M. D. M., Damanet, B., & Soentjens, P. (2019). Timing of intradermal rabies pre-exposure prophylaxis injections: Immunological effect on vaccination response. *Military Medicine.* https://doi.org/10.1093/milmed/usz048.

Velleca, W. M., & Forrester, F. T. (1981). *Laboratory methods for detecting rabies.* Atlanta: U.S. Department of Health and Human Services Public Health Service Centers for Disease Control.

Wasniewski, M., Almeida, I., Baur, A., Bedekovic, T., Boncea, D., Chaves, L. B., … Cliquet, F. (2016). First international collaborative study to evaluate rabies antibody detection method for use in monitoring the effectiveness of oral vaccination programmes in fox and raccoon dog in Europe. *Journal of Virological Methods, 238,* 77–85. pii: S0166-0934(16)30340-8 https://doi.org/10.1016/j.jviromet.2016.10.006.

Wasniewski, M., Barrat, J., Combes, B., Guiot, A. L., & Cliquet, F. (2014). Use of filter paper blood samples for rabies antibody detection in foxes and raccoon dogs. *Journal of Virological Methods, 204,* 11–16. https://doi.org/10.1016/j.jviromet.2014.04.005.

Wasniewski, M., Guiot, A. L., Schereffer, J. L., Tribout, L., Mahar, K., & Cliquet, F. (2013). Evaluation of an ELISA to detect rabies antibodies in orally vaccinated foxes and raccoon dogs sampled in the field. *Journal of Virological Methods, 187*(2), 264–270. pii: S0166-0934(12)00414-4 https://doi.org/10.1016/j.jviromet.2012.11.022.

World Health Organization. (2018). *WHO expert consultation on rabies.* third report Geneva: World Health Organization.

Wood, R. J., & Durham, T. M. (1980). Reproducibility of serological titers. *Journal of Clinical Microbiology, 11*(6), 541–545.

Wright, E., Temperton, N. J., Marston, D. A., McElhinney, L. M., Fooks, A. R., & Weiss, R. A. (2008). Investigating antibody neutralization of lyssaviruses using lentiviral pseudotypes: A cross-species comparison. *Journal of General Virology, 89*(Pt 9), 2204–2213. 89/9/2204 [pii] https://doi.org/10.1099/vir.0.2008/000349-0.

Wunner, W. H., & Briggs, D. J. (2010). Rabies in the 21 century. *PLoS Neglected Tropical Diseases, 4*(3), e591. https://doi.org/10.1371/journal.pntd.0000591.

Yager, M. L., & Moore, S. (2015). The rapid fluorescent focus inhibition test. C. E. Rupprecht & T. Nagarajan (Eds.), *Rabies diagnosis, research, and prevention* (pp. 199–215). (1st ed.). Vol. 2(pp. 199–215). San Diego: Academic Press.

Yu, P. C., Noguchi, A., Inoue, S., Tang, Q., Rayner, S., & Liang, G. D. (2012). Comparison of RFFIT tests with different standard sera and testing procedures. *Virologica Sinica, 27,* 187–193.

CHAPTER 14

Human and animal vaccines

Thirumeni Nagarajan[a], Hildegund C.J. Ertl[b]

[a]R&D center, Vaccines Division, Biological E Limited, Hyderabad, India [b]The Wistar Institute, Philadelphia, PA, United States

14.1 Introduction

Globally, rabies occurs in more than 150 countries and territories. As a zoonosis, more than 2 billion people live in areas in which rabies is enzootic (Hampson et al., 2015). Worldwide, millions of potential viral exposures are registered, resulting in tens of thousands of human deaths, with most occurring in the Old World. Rabies has one of the highest case fatality rates among infectious diseases, with over 98% of global human deaths attributed to exposure to infected dogs (Hampson et al., 2009). Human and animal rabies vaccines play a critical role in protection against rabies. Rabies control requires effective risk management assessment systems that include surveillance, reliable decentralized diagnostic laboratories, humane animal management based upon the prevalence of disease in a locality, continuing education of public health professionals trained on the use of appropriate pre-exposure prophylaxis (PrEP) and post-exposure prophylaxis (PEP) and animal vaccination, enhanced intersectorial communication between animal and human health professionals, with appropriate data reporting, sharing, and action. Without an adequate risk management system in place, there is a risk that PEP is used inappropriately. History has demonstrated that for a singular strategy of governments purchasing an ever-needed supply of human rabies vaccine is akin to an "incurable wound" and increases dramatically the financial burden of disease prevention in regions that can least afford to pay (Zinsstag et al., 2009). Clearly, establishing and enforcing national risk assessment systems to evaluate the use of PEP is cost beneficial and provides a more effective strategy for the overall prevention and control of rabies.

14.2 History of rabies vaccines

Rabies, a disease whose name originates from the Sanskrit word of *rabhas*, which translates into "to do violence," has been known as a fatal disease since at least 2300 BC. There was and still is no cure for rabies and avoidance of rabid animals was the only means of prevention. This was until a vaccine was developed and tested in 1885 by Louis Pasteur in a young boy named Joseph Meisner, who had been bitten by a rabid dog. The vaccine consisted of a crude solution of dried spinal cord tissue from rabbits infected with a rabies virus stock, which originated from a rabid cow and then was serially passaged 80 times in rabbits. The boy, who received 13 injection of this vaccine, survived without reported sequelae and the vaccine with some modifications was subsequently used until vaccines, in which rabies virus was inactivated by chemical methods, replaced it. The most commonly used, inactivated rabies vaccine was developed in 1911 by Lieutenant-Colonel Sir David Semple in India. He used virus from the brain of rabies virus-infected sheep that was inactivated with phenol. Semple-type vaccines, based on rabies virus derived from nerve tissue of infected animals, were used for decades. Although the treatment in general prevented rabies, it did so at a cost. Immune responses to the nerve tissue present in the vaccine preparations caused debilitating and sometimes fatal neurological disease in up to one out of 80–200 vaccine recipients. To reduce this side effect, rabies vaccines were subsequently generated from brains of suckling mice or newborn rats with not yet fully matured nervous systems. The World Health Organization (WHO) no longer recommends the use of nerve tissue-derived rabies vaccines and nearly all countries have discontinued their use.

Duck embryo vaccine derived from 7-day-old embryonated duck eggs was developed in the 1960s. Advances in mammalian tissue culture allowed Hilary Koprowski, Stanley Plotkin, and Tadeusz Wiktor to develop a human diploid cell line WI-38-derived, inactivated rabies vaccine (Wiktor, Plotkin, & Koprowski, 1978), called human diploid cell vaccine (HDCV), which was licensed in 1976 in Europe and in 1980 in the US. This and similar vaccines, as well as duck or chicken egg-derived vaccines, are currently used worldwide for rabies PEP and PrEP.

Passive immunization with rabies immunoglobulin (RIG) in combination with active immunization was first tested in 1953 in a village in Iran that was attacked by a rabid wolf (Habel & Koprowski, 1955). The clinical trial confirmed that treatment with RIG is essential to prevent vaccine failures in cases of severe exposures to rabies virus.

The initial rabies vaccines were developed without means to assess their likely efficacy. In the 1960s, potency tests were firmly established for preclinical testing of rabies vaccines (Habel, 1966). These potency tests are very cumbersome and provide highly variable results. They require large number of mice and efforts are underway to replace them with in vitro cell culture-based alternatives for both batch and final product potency testing and subsequent prequalification.

14.3 Pre-exposure prophylaxis

Rabies is a vaccine-preventable zoonotic disease. PrEP is recommended for anyone who is at continual, frequent, or increased risk for exposure to lyssaviruses, as a result of their occupation, activity or residence, such as laboratory workers dealing with isolating or propagating lyssaviruses, veterinarians, and animal handlers. Travelers to high-risk areas should also consider PrEP (Rupprecht et al., 2010; Warrell & Warrell, 2015; WHO, 2013). The principal modality of protection against disease is mediated by virus-neutralizing antibodies (VNAs), which clear rabies virus (RABV) before it reaches the central nervous system (CNS). Vaccines given before rabies exposure are in part effective through residual circulating VNAs and in part by allowing for rapid recall of antibody responses following a booster immunization that is recommended in case of an exposure to a rabid animal. PrEP simplifies treatment after exposure as it circumvents the need for passive immunizations and reduces the number of vaccine doses needed for protection. It provides other benefits; at least in theory it may protect against unrecognized exposures and may minimize risks caused by potential time delays between exposure and treatment (Rupprecht et al., 2010; Warrell & Warrell, 2015; WHO TRS, 2013). PrEP is usually given through the intramuscular (IM) route on days 0, 7, and 21 or 28, although shorter schedules have been proposed (Wieten et al., 2013). The intradermal (ID) route is seen as an economical and acceptable alternative to IM vaccination (Madhusudana & Mani, 2014), but requires appropriate medical training. Its cost effectiveness may also be limited by the use of only a portion of the vaccine within a vial; the remaining vaccine needs to be refrigerated and used the same day or it must be discarded.

14.4 Post-exposure prophylaxis

The WHO classifies rabies exposure broadly into major categories of contact (WHO TRS, 2013), which determine the type of treatment that is needed. Category I is touching an animal or receiving licks on intact skin—treatment is not recommended for this type of exposure. Category II exposure, which is caused by minor scratches or nibbles that do not result in bleeding, requires wound cleaning and vaccination. All transdermal or mucosal exposures, including bites and licks of broken skin or mucosal membranes, are considered category III exposures, which require thorough washing and disinfection of the wound, infiltration of the wound with RIG, and vaccination. Individuals that had received the complete regimen of a WHO-approved rabies vaccine previously need to receive a booster immunization; they should not receive treatment with RIG. Treatment should be initiated as soon as possible to prevent the virus from reaching the CNS.

Annually, more than 15 million people worldwide are estimated to receive PEP (Hampson et al., 2015). Although timely treatment with rabies vaccine and RIG is nearly 100% effective in prevention of rabies, PEP after the onset of symptomatic rabies is not effective (Sadeghi, Moallem, Yousefi-Abdolmaleki, & Montazeri, 2015; WHO TRS, 2013). IM and ID vaccination regimens for rabies PEP of human subjects are presented in Table 14.1.

14.5 Rabies virus strains for vaccine production

RABV-specific VNAs are directed to the RABV glycoprotein (G protein), a trimeric type I membrane protein that forms a spike extending 8.3nm from the viral membrane. G protein has three domains: an ectodomain, a transmembrane domain, and a cytoplasmic domain (Gaudin, Ruigrok, Tuffereau, Knossow, & Flamand, 1992). The ectodomain is

TABLE 14.1 IM and ID vaccination regimens for rabies postexposure prophylaxis of human subjects.

PEP	Route	Regimens	WHO; recommended/ alternative	Vaccination schedule (day)						Visits	Vaccine doses
				0	3	7	14	21	28		
Primary; unvaccinated subjects	IM	4 dose; Adapted Essen	Recommended	1	1	1	←	1	→	4	4
		2-1-1; Zagreb		2				1	1	3	4
	ID	2-site 1 week IPC		2	2	2				3	1–3
		2-site Thai Red Cross	Alternative	2	2	2			2	4	<2–4
		4-site 1 week		4	4	4				3	1.5–3
		4-site 1 month		4		2			3	3	<2
		4-site 1 week (optional day 28)		4		2			(1)	2	1.5
Booster; vaccinated subjects	IM	2 visits	Recommended	1	1					2	2
	ID	2 visits, 0.1 mL		1	1					2	<1
		4-site, 1 day		4						1	0.5–1

Adapted from Warrell, M. J. (2019). Rabies post-exposure vaccination in 2 visits within a week: A 4-site intradermal regimen. Vaccine 37(9), 1131–1136. https://doi.org/10.1016/j.vaccine.2019.01.019 with permission.

responsible for infectivity, neuroinvasiveness, and induction of VNAs (Langevin, Jaaro, Bressanelli, Fainzilber, & Tuffereau, 2002).

The RABV strains recommended for vaccine production are "fixed" viruses (as opposed to wild-type or "street" viruses) grown in the neural tissue of rabbits, sheep, goats, mice or rats, or in cell cultures, including continuous cell lines. Fixed RABV strains, such as Challenge Virus Standard (CVS), Flury low egg passage (LEP), Flury high egg passage (HEP), Kelev, Evelyn Rokitniki Abelseth (ERA), Vnukovo-32, Street Alabama Dufferin (SAD), Pasteur Virus (PV), and Pitmann Moore (PM), have or are being used for production of inactivated animal rabies vaccines (Reculard, 1996). However, the WHO does not recommend or endorse a specific RABV strain for vaccine production (WHO, 2005). Besides the use of the few proven historical seeds, derivation of new strains and isolates is encouraged, to maximize safety and minimize costs (Jackson, 2013; World Health Organization, 2013). These strains resemble each other in their ability to confer uniform protection against members of lyssavirus species 1. However, they may differ from each other in growth parameters in cell culture. For instance, PV gave the highest titers when bovine hamster kidney (BHK)-21 cells grown on microcarriers were infected with a multiplicity of infection (MOI) of 0.3 in bioreactors (Kallel, Rourou, Majoul, & Loukil, 2003). The highest titers obtained for Flury LEP occurred when BHK-21 cells were infected with a MOI of 0.01 in a static culture system (Tao et al., 2011). The Flury LEP strain is widely used for making rabies vaccines for humans and animals, because it can achieve high titers when grown in cell culture and it is highly immunogenic (Koprowski & Cox, 1948). The G proteins of CVS and ERA RABV strains appear as doublets when resolved by sodium dodecyl sulfate (SDS)-polyacrylamide gel electrophoresis (PAGE) under reducing conditions (Langevin & Tuffereau, 2002; Prehaud, Coulon, LaFay, Thiers, & Flamand, 1988). The G proteins are singlets for all other fixed RABV strains. Increasing G protein expression in viral seed strains may improve the immunogenicity of inactivated rabies vaccines. For example, a Flury LEP engineered to encode an additional G gene (rLEP-G) produced strikingly higher levels of G protein in cell culture and showed similar in vitro growth and biosafety characteristics. An inactivated vaccine based on the rLEP-G virus induced significantly higher antibody titers in vaccinated subjects than were induced by Flury LEP, suggesting that rLEP-G is an improved seed virus candidate for manufacture of inactivated rabies vaccine (Tao et al., 2011). Similarly, an inactivated rabies vaccine based on a highly attenuated triple RABV G protein variant of SAD (SPBAANGAS-GAS-GAS) might permit an antigen sparing strategy with reduced production costs (Faber, Dietzschold, & Li, 2009).

Comprehensive genetic characterization by full genome sequencing such as next-generation sequencing (Höper et al., 2015) of rabies vaccine virus is recommended (World Health Organization, 2013) to ensure strain identity. Vaccine manufacturers should sequence their seed strains and provide individual accession numbers through GenBank (Finke, Karger, Freuling, & Müller, 2012; Shi et al., 2010). It must be borne in mind that although nucleotide sequence analysis provides information on the specificity of the antibody response that would be induced by this sequences it does not guarantee efficacy, which depends on a number of other parameters such as vaccine dose and the vaccine's ability to trigger an innate and adaptive immune responses (Rupprecht et al., 2008).

14.6 Production of rabies vaccines

Whole inactivated RABV particles are highly immunogenic and form the basis for most human and animal rabies vaccines. Some rabies vaccines for animals are based on recombinant viral vectors derived from pox or adenoviruses and some countries use attenuated live RABV for wildlife immunization. Quality and magnitude of vaccine-induced antibody responses depend largely on the integrity of the glycoprotein (G), which in order to be optimally immunogenic should be present as a membrane-anchored protein in a repetitive rigid form. This can be achieved by using whole virus particles for immunization (Dietzschold, Faber, & Schnell, 2003) or viral recombinant vaccines. Only vaccines that are pure, potent, safe, affordable, and capable of providing stable and long-lasting immunity are recommended for mass vaccination of animals, which constitutes the most effective method of controlling and eliminating rabies. The manufacture of inactivated animal rabies vaccine that are affordable for resource-poor countries where rabies is most common necessitates the choice of inexpensive upstream and downstream processes and improved formulation strategies involving adjuvants that do not significantly add to the cost of production. The various aspects of the manufacturing process, such as the choice of cell line, culture system, downstream processes, formulation strategies, and potency testing, vary based on the product (Fig 14.1). The WHO

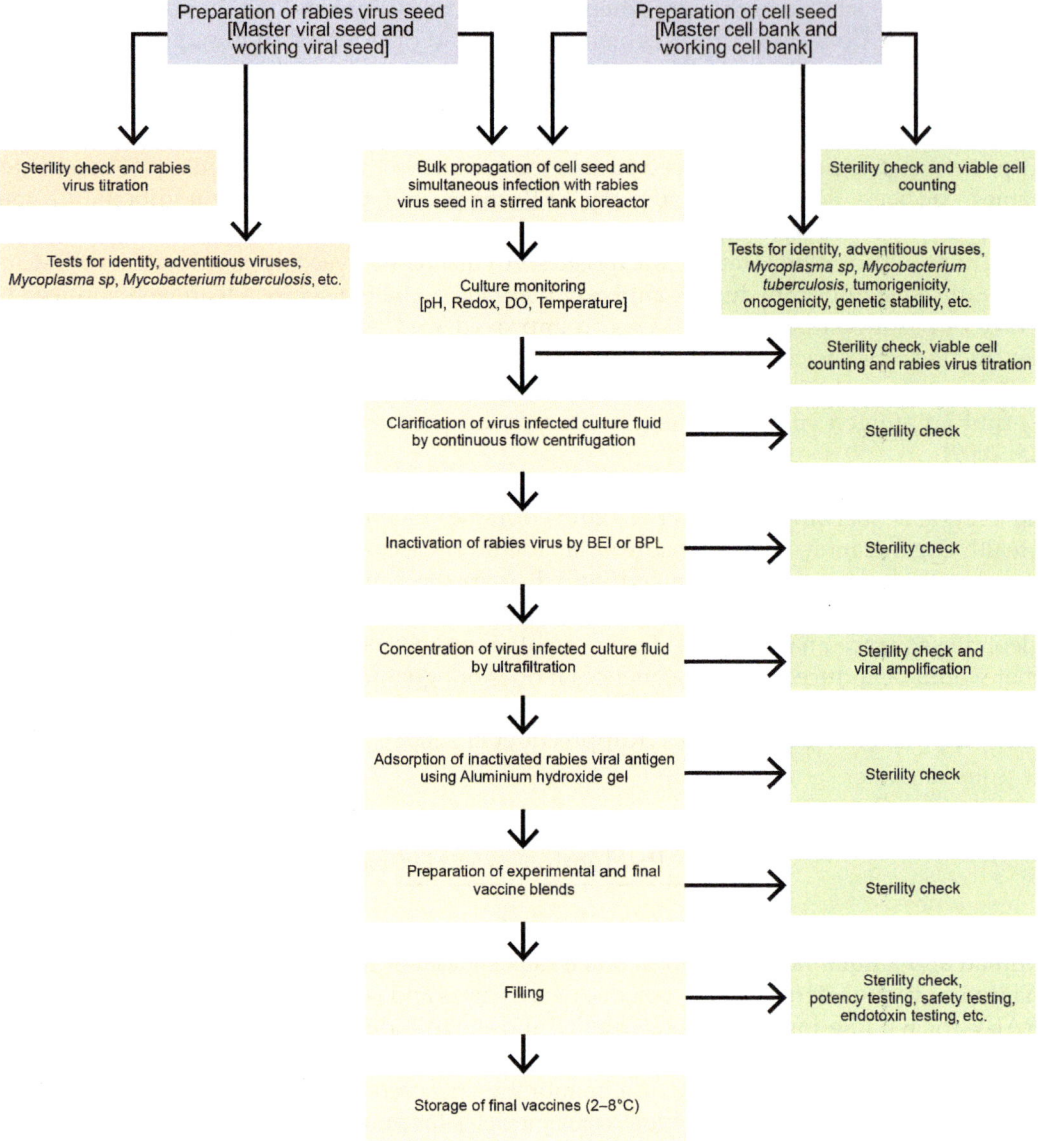

FIG. 14.1 Generic flowchart of upstream and downstream processes in production of commercial cell culture rabies vaccine for animals.

has published standard recommendations for rabies vaccine production (WHO, 2005). Ultimately, it is the responsibility of national governments to approve, license, and monitor human and animal rabies vaccines and production facilities in their own countries. Modern rabies vaccines for human and animal use are produced in vitro bypassing the need to infect live animals (Fig. 14.2).

Parenteral rabies vaccine	Cell substrates		
	Primary	Diploid	Continuous
Human	√ (Chicken, CEF) 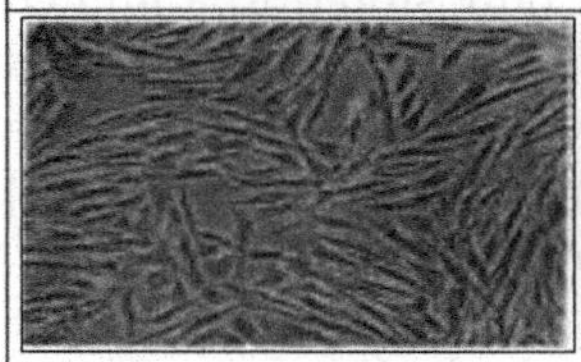	√ (Human, MRC-5) 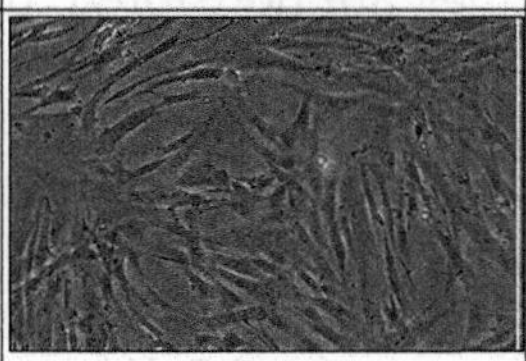	√ (Monkey, VERO) 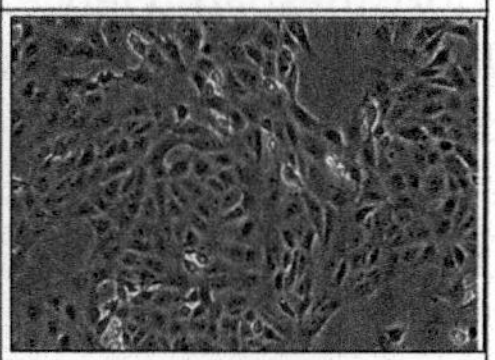
Veterinary	X	X	√ (Syrian Hamster, BHK-21) 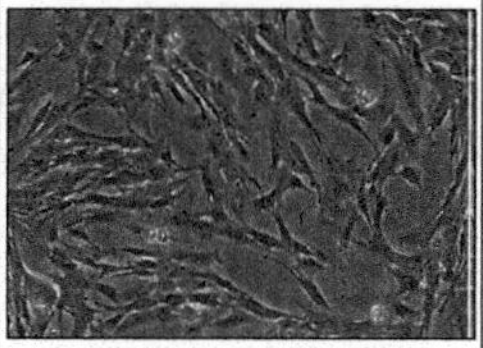

FIG. 14.2 Cell substrates used in production of commercial cell culture rabies vaccine for humans and animals.

14.7 Cell substrates for rabies virus propagation

Substrate refers to the cultured cells that are used to produce the desired biological product. Historically, cells derived from animal tissues acted as primary substrates for production of viral vaccines. Cell properties and their growth characteristics have a bearing on the quality of biologics produced in them (WHO, 1998). Within animal cell substrates, there are a number of cell types that are currently used by manufacturers for the production of rabies vaccines, including primary cells, diploid cells, and continuous cell lines. The challenge in standardizing cell substrates is to strike a balance between the desire for an efficient production system and the requirement to minimize risks. Obviously, manufacturing a safe product necessitates thorough characterization of a cell substrate, validation of the manufacturing process for removal or inactivation of potential adventitious agents and testing of the bulk and final product both for vaccine antigen and impurities. The WHO prequalified human rabies vaccines originate from primary cells (chicken embryo fibroblasts [CEF]; Rabipur, GSK), diploid cells (normal human fetal fibroblasts [MRC-5]; Imovax, Sanofi), and continuous cell lines (African Green Monkey Kidney cells [Vero]; Verorab, Sanofi). Most animal rabies vaccines produced from continuous cell lines are prepared in baby hamster kidney cells (BHK-21) some of which can grow in suspension culture, and Vero cells and immortalized hamster fibroblasts (Nil-2) cells that can grow as a monolayer culture (Reculard, 1996). The BHK-21 cell line is widely used for production of viral vaccines because of the significant advantages, including a well-documented origin and history (Hay, Caputo, & Chen, 1994; Rudd & Trimarchi, 1987). The animal rabies vaccines produced in BHK-21 cell line have been used for decades and have been proven to be safe and efficacious.

14.8 Primary cells

Primary cells are derived directly from an animal source. They retain the characteristics of the tissue from which they originate. They do not have tumorigenic properties and have a short lifespan. Hence, they are not stored, or only stored to a limited extent as cell banks. The most important source of primary cells for the production of human rabies vaccines is the avian embryo. CEFs derived from a 10-day-old embryo of specific pathogen-free (SPF) eggs are used for virus production (Jordan & Sandig, 2014). The purified CEF vaccine (PCECV) is produced using the Flury LEP RABV strain in compliance with applicable pharmacopoeia and WHO requirements (Barth & Franke, 1996). It reaches or exceeds the minimum potency requirement of ≥ 2.5 IU per single IM dose. The vaccine strain has no deviations from published genetic sequences (Finke et al., 2012). The PCECV is marketed globally as Rabipur or under different brand names in some countries (Giesen, Gniel, & Malerczyk, 2015). Nevertheless, CEFs do have certain limitations such as need for continuous harvesting, inconsistent starting material, risk of variation in permissiveness for the target virus, and risk of contamination with potential adventitious agents (Barrett, Mundt, Kistner, & Howard, 2009).

14.9 Diploid cells

Diploid cells have a finite in vitro lifespan and contain the full complement of genetic material. Like primary cells, they often retain many characteristics of the cell types from which they were derived (Barrett et al., 2009). Those that are in use for vaccine production are essentially free of all known adventitious agents (Ma et al., 2015). However, they can become infected during in vitro propagation. Diploid cells used for vaccine manufacture have to be maintained in a Good Manufacturing Practice (GMP) facility. The use of human diploid cells for virus propagation helps to mitigate the difficulties associated with the use of primary tissue culture. The essential argument in favor of the use of diploid cells is the fact that they undergo senescence and are nontumorigenic. They allow for multiple expansion passages of material obtained from well-characterized master and working cell banks in a closed system (Jordan & Sandig, 2014). However, they suffer from several disadvantages such as difficulty to upscale in bioreactors especially using microcarriers and a need for demanding growth media and difficulties to propagate without serum (Barrett et al., 2009). The two well-known human diploid cell strains, WI-38 and MRC-5, serve as the international reference strains (Hayflick, 1989; Jacobs, Jones, & Baille, 1970). The WHO recommends human diploid cells as one of the safest cell culture substrates for the production of viral vaccines. The HDCV manufacture involves PM 1503 3M strain of fixed RABV derived from a strain originally isolated by Pasteur and maintained by the National Institute of Health (NIH), USA. The safety and ability of the HDCV to efficiently induce high titers of RABV-specific VNAs were proven in several clinical trials (Cox, Klietmann, & Schneider, 1978; Cox & Schneider, 1976; Klietmann, Domres, & Cox, 1978; Kuwert, Marcus, Werner, Iwand, & Thraenhart, 1978;

Nicholson & Turner, 1978). The HDCV represents a milestone in human vaccine production and it permits, with a few injections, prophylaxis for persons at risk and efficient PEP (Hicks, Fooks, & Johnson, 2012). Rabies VNA persist upon HDCV immunization and a boost elicits anamnestic responses. HDCV induces cross-neutralizing antibodies that protect vaccinated subjects against phylogroup I lyssaviruses (Brookes, Parsons, Johnson, McElhinney, & Fooks, 2005; Fayaz et al., 2011). HDCVs have been licensed all over the world. The paramount advantage of primary and finite cells for vaccine production is that an enormous amount of regulatory experience has accumulated over several decades since the 1930s. Unfortunately, worldwide use of HDCV is limited by their high cost of production. Continuous cell lines can solve some of the limitations associated with diploid cells but may introduce other challenges (Jordan & Sandig, 2014).

14.10 Continuous cell lines

Continuous cell lines that originate from animal tissues serve as important cell substrates for the production of various types of biological pharmaceuticals. Unlike primary and diploid cells, they have tumorigenic potential and an infinite life span. Nevertheless, a number of studies suggest that cells below certain passage numbers are not tumorigenic (Furesz, Fanok, Contreras, & Becker, 1989; Horaud, 1992; Levenbook, Petricciani, & Elisberg, 1984; Vincent-Falquet et al., 1989). They are amenable for bulk propagation of viruses at lower production costs (Grachev, 1990). They allow for preparation of well-defined seed lot systems consisting of master, working, and extended working cell banks (Barone, Fracchia, Pascuali, & Proglio, 1997). One of the most frequently utilized mammalian cell lines for the vaccine production is the Vero cell line, which was established from the kidney tissue of an African Green Monkey (AGM) (Osada et al., 2014). It is a continuous cell line that is widely accepted by regulatory authorities for viral vaccine manufacture (Barrett et al., 2009). It is chosen mainly because it achieves high virus yields. Batches that lack adventitious agents are available. Its anchorage-dependent nature demands propagation on large culture surfaces such as roller bottles, microcarriers, Cell Factories, CellSTACK, CellCubes, fixed-bed bioreactors, etc. The purified Vero cell-derived rabies vaccine (PVRV) was introduced into clinical practice several decades ago and was an important step forward in the prevention of rabies (Rupprecht, Hanlon, & Slate, 2004). PVRV is more suitable for industrial scale-up than HDCV. The PVRV (Verorab) is licensed for use in over 100 countries and over 40 million doses have been administered (Toovey, 2007). Verorab has been assessed in a large number of clinical studies by considering 0.5 IU/mL antibody titer as the threshold of immunogenicity. This threshold correlates well with protection from clinical rabies. Verorab meets the WHO criteria for IM PEP for both the Essen (5 doses; 1-1-1-1-1) and Zagreb (4 doses; 2-1-1) regimens (WHO, 2005), and its immunogenicity remains unaffected by simultaneous administration of RIG (Lang et al., 1998). An improved serum-free, PVRV-Next Generation (PVRV-NG) vaccine was prepared from the inactivated PM strain of RABV common to Verorab and Imovax vaccines. It is compliant with the European Pharmacopoeia and the specifications defined by the WHO and the United States Food and Drug Administration (USFDA). In a Phase III clinical trial, it was confirmed that the

PVRV-NG vaccine, an evolution over the reference vaccine Verorab, is both safe and immunogenic, offering a new alternative for the prophylaxis of rabies (Li et al., 2013).

14.11 Production systems

The need to produce affordable vaccines largely decides the choice of production system, which should be amenable for large-scale manufacture and allow for ramping up process parameters with ease. Reusable or disposable roller bottles (RBs) are commonly used for cultivation of adherent cells. Nevertheless, handling large numbers of RBs is labor intensive and carries a relatively high risk of contamination. In addition, vaccine production using the traditional RBs leads to low vaccine yields (Yu, Huang, Zhang, Tang, & Liang, 2012). Despite that, the vaccine industry has taken advantage of the positive features of large-scale production in RB for viral vaccine production.

Multilayer cultivation systems possess increased surface areas and allow for easier large-scale production compared to RBs. Stacked devices, such as CellCube (Corning), CellSTACK (Corning), and Cell Factory (Thermo Scientific), reduce requirement of incubator space and manual handling and offer better process control. However, automation is required when it comes to upscaling. Lower operator skill level and lower investment costs make them affordable option for manufacturers with limited facility complexity. Modified solid-phase devices, in the form of microcarriers (e.g., Cytodex 3) and macrocarriers (e.g., Fibra-Cel disks), have been developed to overcome limitations associated with static and roller culture systems. Microcarrier culture using a perfusion-based stirred tank bioreactor (Kallel et al., 2006) and macrocarrier culture using a packed-bed bioreactor (Hassanzadeh, Zavareh, Shokrgozar, Ramezani, & Fayaz, 2011) have been developed. Both have been used for RABV antigen production. Microcarriers offer high surface area to volume ratio and efficient monitoring and control of key process parameters. They achieve high cell densities and allow for scale-up (Birch, 1999). Cell cultivation using microcarriers allows easy separation of media from cells during the process enabling improved virus yields (Gallo-Ramirez, Nikolay, Genzel, & Reichl, 2015). A general drawback of microcarrier-based processes is the need for inoculation with high cell numbers, and the expensive and disposable nature of microcarriers. Moreover, the use of media free from serum often results in poor cell attachment (Frazzati-Gallina, Paoli, Mourão-Fuches, Jorge, & Pereira, 2001), necessitating exogenous supplementation of recombinant adhesion factors.

Most of the cultivation systems applied to vaccine production offer restricted inoculation and harvesting options. They allow for limited monitoring of pH and oxygen and limited control of cultivations parameters, such as temperature and sometimes feeding rates. However, due to the simplicity and robustness of such systems, some vaccine manufacturers still rely on these cultivation systems. Such systems have also found their application in the generation of cell seeds for large-scale production in microcarrier systems or the generation of virus seeds to infect bioreactors (Jackson, 2013). All of the production systems described here are less preferred for animal rabies vaccine manufacture because of high cost. Therefore, it would be ideal to secure cells adapted for growth in suspension culture so that they can be bulk propagated with ease and at a reduced cost using a stirred tank bioreactor (Atanasiu, Ribeiro, & Tsiang, 1972; Chapman, Ramshaw, & Crick, 1973). A BHK CZ cell line, a derivative of the

BHK-21 cell line, has desirable features for industrial applications, including the ability to grow in suspension and sediment completely in approximately 14h, a prerequisite for infecting cells grown in suspension, and their ability to grow in medium with low levels of serum. Inactivated rabies vaccines for domestic animals that are produced in BHK cells grown in a stirred tank bioreactor are marketed worldwide.

14.12 Viral inactivation

Rabies viral particles can be inactivated by gamma or UV radiation and by chemical agents such as acetylethyleneimine (AEI), binary ethyleneimine (BEI), and β-propiolactone (BPL). Chemical agents that act on the viral nucleic acid with little or no effect on the viral protein coat are preferred to physical agents by vaccine manufacturers. Classical chemical inactivants such as phenol and formaldehyde are not recommended because they may reduce the immunogenicity of the vaccine and produce toxic or irritant substances (Reculard, 1996). Currently, BEI and BPL are the inactivants of choice for animal rabies vaccines. Preference for BEI as a viral inactivant relates to the fact that it is highly stable, easy to prepare, inexpensive, very effective, less hazardous to handle than many other inactivants, and results in a potent and stable vaccine (Bahnemann, 1990; Mondal, Neelima, Seetha Rama Reddy, Ananda Rao, & Srinivasan, 2005). By comparison, BPL is more expensive and less stable at 37°C than BEI. However, it is still widely used for viral inactivation because it is highly effective. In principle, RABV can be inactivated either before or after concentration, but care must be exercised to avoid formation of aggregates if virus is inactivated before concentration by ultrafiltration. Viral inactivation is a critical step in the manufacture of rabies vaccine because the presence of infectious virus in the final vaccine can cause rabies resulting in death. There have been incidents when rabies vaccines have been recalled from the field because they were suspected to contain infectious virus (CDC, 2004). This emphasizes the importance of optimizing inactivation conditions and monitoring inactivation kinetics. The inactivation process for veterinary rabies vaccine has been validated by studying the inactivation kinetics for BEI (Mondal et al., 2005).

14.13 Downstream processing

Regulatory requirements to ensure vaccine purity and safety depend on the particular use of a product (Wolff & Reichl, 2011). The development process for animal vaccines generally has less stringent regulatory and preclinical trial requirements. Adhering to these requirements constitutes the largest expense in human vaccine development. Development of animal vaccines usually has a shorter time from initial phase to market launch, thus increasing the return on investment required for research and development. The manufacture of inactivated animal rabies vaccines does not involve elaborate downstream processing, as is the case for human vaccine production, again reducing the time and cost involved. However, all animal rabies vaccines should meet the minimal requirements of national regulatory authorities in terms of purity, safety, potency, duration of immunity, efficacy, stability, antigen content, identity, effective inactivation, and sterility (OIE, 2019). Due to less elaborate downstream processing, animal rabies vaccines may contain several undesirable impurities, such as BSA, host cell DNA, and viral aggregates. For example, BSA has the risk of increasing immediate type allergic reactions, and sometimes death (Ohmori et al., 2005). Host cell DNA has the potential for transfer of either activated cellular and/or viral oncogenes (particularly if the cell substrate is tumorigenic) and production of infectious viruses from the viral nucleic acid and aberrant gene expression by insertion of sequences into sensitive gene control regions (Vitrology, 2008). It was shown that effective clearance of host cell DNA could be achieved using nuclease without compromising, stability, immunogenicity, and efficacy of a rabies vaccine (Li et al., 2014).

14.14 RABV purification

Sucrose density gradient ultracentrifugation (rate zonal ultracentrifugation) and column chromatography are suited for RABV purification. Density gradient ultracentrifugation is a well-known and established classical purification technique, generally used to purify bulk viruses. Briefly, the principle of the method is to separate particles according to differences in density and thereby, when applied to viruses, to separate them from lighter and heavier cellular material. The main advantage of this technique is its ability to combine concentration and purification steps in a single unit operation. Also, the technique offers a good resolution for separating full virions from empty capsids, which is difficult to achieve by chromatography. Although it is theoretically possible to band $\sim 10^{15}$ virus particles in a single procedure with large-scale continuous ultracentrifugation, the technique ends up being laborious and expensive to scale up (Altaras et al., 2005). Despite these limitations, several manufacturers still use rate zonal ultracentrifugation for RABV purification.

Scalable chromatographic separation techniques are considered potential alternative for large-scale downstream processing. Chromatography has been widely used for capture, concentration, and purification of viruses using three different arrangements of the stationary phase: packed beds, membrane adsorbers, and monoliths. For vaccine purification, packed porous beds suffer from two main disadvantages: limited flow rate imposed by the compromise between pressure drop and mass transfer resistances and, in most cases, low dynamic binding capacity, because the surface available for adsorption under normal contact times is limited to the external surface of the adsorbent particle (Nestola et al., 2015). Adsorption of viral particles to a solid phase, in fact, is a convenient and practical choice for fractionating and recovering viruses from impurities originating from cells and culture media. The chromatographic separation is driven by the selective physicochemical interactions between the viruses and closest impurities and the solid phase. The adsorption methods offer several important advantages: high flow rates can be used, thus limiting processing time, preserving biological activity of labile viruses and allowing for scale-up at a relatively low cost (Andreadis, Roth, Le Doux, Morgan, & Yarmush, 1999). However, it is important to remember that the design of suitable selective chromatographic protocols for virus purification must take into account the structure, physical, and chemical surface properties of the viruses (Braas, Searle, Slater, & Lyddiatt, 1996). As one example, affinity column chromatography using Cellufine sulfate is widely used since it allows high recovery of RABV antigens (Kulkarni, Sahai, Gunale, & Dhere, 2017).

14.15 Formulation

The goal of vaccination is to generate a strong immune response that provides long-term protection against infection. In general, inactivated vaccines are formulated with relatively higher dosage of antigens to ensure high potency of the final product (Habel, 1973). An adjuvant is commonly added because inactivated vaccines are unable to amplify the antigen (Montomoli et al., 2011). Most rabies vaccines used in humans do not contain an adjuvant. Adjuvanted animal rabies vaccines are usually presented in liquid form, whereas the vaccines meant for human use are presented both in freeze-dried and liquid forms. The key features that distinguish human rabies vaccines from animal rabies vaccines are represented in Table 14.2. Ideally, a rabies vaccine should be able to prime a T helper cell (Th) 1/Th2 balanced and

TABLE 14.2 Comparison of tissue culture derived human and veterinary rabies vaccines.

Particulars	Rabies vaccine	
	Human	Veterinary
Vaccine type	Freeze dried/liquid	Liquid
Adjuvant	Only liquid vaccine has aluminum hydroxide gel	Aluminum hydroxide gel
Preservative	Thiomersal (+/−)	Thiomersal (+)
Dose form	Single dose	Single dose/multidose
Dose volume	0.5 or 1.0 mL/dose	1.0 mL/dose
Route of administration	IM/SC/ID (only freeze dried vaccine)	IM/SC
Vaccine virus strain	PV/PM/Flury LEP	CVS/Flury LEP/SAD/PM
Cell substrate	CEF/VERO/MRC-5	BHK-21
Cell type	Anchorage dependent	Anchorage independent/suspension
Production system	Roller culture/microcarrier culture—bioreactor	Suspension culture—bioreactor
Inactivant	BPL	BEI
Purification technique	Rate zonal ultracentrifugation/column chromatography	No elaborate downstream processing
Potency	≥ 2.5 IU/dose	≥ 1.0 IU/dose
Relative potency by GP estimation	Applicable	Not applicable
Freedom from BSA	Applicable	Not applicable
Freedom from host cell DNA	Applicable only for VERO cell vaccine	Not applicable
Duration of immunity	1 year	3 years for nonendemic countries; 1 year for endemic countries
Shelf life	36 months/48 months	36 months
Cost	Expensive	Inexpensive

efficient immune response to achieve viral clearance (Ren et al., 2010). Aluminum salt is an adjuvant commonly used in the manufacture of inactivated animal rabies vaccine to improve its efficacy. Although safe and well tolerated, aluminum salt induces a strongly biased Th2 response and is a relatively weak adjuvant. In addition, aluminum salt-based adjuvants are limited in their use in that they preclude lyophilization or freezing. Novel vaccine formulations are extremely desirable, since currently used rabies vaccines need to be formulated with several microgram quantities of antigen and require multiple inoculations that induce a delayed VNA response coupled with weak cellular immunity. An effective adjuvant may render rabies vaccines more efficacious and allow for dose sparing with fewer inoculations achieving faster antibody responses at a reduced cost. In this regard, novel adjuvants such as Toll-like receptor (TLR) ligands and nanoparticles hold promise for formulating rabies vaccines (Asgary et al., 2016; Montaner et al., 2012; Wijaya et al., 2017; Zhang et al., 2016).

A cytosine phosphate-linked guanine (CpG) oligonucleotide (ODN) preferentially activated canine lymphocytes expressing TLR9 when used in combination with aluminum adjuvant. This combination acted as a potent adjuvant in dogs by inducing enhanced antibody titers and Th1 cellular immune response (Ren et al., 2010). A similar combination allowed in mice for a reduction of the five-dose to three-dose PEP. It also protected mice with lower doses of the vaccine (Yu et al., 2018). A non-CpG ODN (IMT504) allowed for fewer inoculations, significant dose sparing of vaccine, rapid antibody production, and protection (Montaner et al., 2012). An ISCOM based, Matrix-M Vet, adjuvant was shown to elicit both humoral and cellular immune responses using lower vaccine doses. Rabies vaccines (Rabix and Rabifel) for dogs and cats formulated with Matrix-M Vet adjuvant (Isconova, Sweden) were launched in Russia (VetBioChem, Russia). Green synthesized silver nanoparticles have been tested in combination with a rabies vaccine in animals where they were shown to be safe (Asgary et al., 2016). A TLR3 adjuvant, PIKA, a stabilized chemical analogue of double-stranded RNA, enhanced both humoral and cellular immunity in response to a rabies vaccine. It was well tolerated and more immunogenic than a commercial rabies vaccines when an accelerated regimen was tested in a phase 1 study in healthy human adults (Wijaya et al., 2017). Nevertheless, only extensive clinical studies as appropriate can confirm whether such adjuvants will permit greater access to current highly effective life-saving rabies vaccines.

14.16 Potency testing

Many variables might influence quality of a rabies vaccine, which makes extensive quality control (QC) of each lot essential and mandatory (Hendriksen, 2009). Current QC of inactivated rabies vaccines focuses mainly on the potency of the final product for batch release and assessment of stability to monitor process consistency but also to identify subpotent lots (Verch, Trausch, & Shank-Retzlaff, 2018). The potency of a rabies vaccine refers to its ability to prevent infection of a host or host cells by eliciting VNA and memory B-cell responses (McVey, Galvin, & Olson, 2003; Taffs, 2001). The potency of inactivated rabies vaccines for animal and human use is conventionally tested by a mouse protection test, which was originally developed at the NIH. The principle of the NIH test is the immunization of mice with different vaccine concentrations, followed by intracerebral challenge with 5 to 50 LD_{50}/dose of the CVS strain. The protection rates (ED_{50}) of standard and test vaccines are calculated and the potency of the test vaccine is given in International Units (IU).

The classical NIH method for potency testing of inactivated rabies vaccines is slow, expensive, requires large numbers of animals and involves significant pain, distress, inhumane end points, and safety issues for laboratory workers. Due to the fact that different seed virus strains are used for the production of rabies vaccines, difficulties arise from the different strain relationship between the test vaccine virus strain as well as the challenge virus strain and the reference vaccine virus strain (Blancou et al., 1989). The underlying animal variability leads to high assay variability (Wunderli, Dreesen, Miller, & Baer, 2003).

Animal welfare concerns, as well as scientific considerations, have led to a "3Rs concept" that comprises the refinement of animal procedures, the reduction of animal numbers, and the replacement of animal models (Romberg et al., 2012). Several methods for rabies vaccine potency testing have been reported as possible replacements for the NIH test, including serologic assays, in vitro antigen quantification methods such as ELISA, single radial immunodiffusion (SRID), luciferase immunoprecipitation system (LIPS), electrochemiluminescent (ECL) assay, time-resolved fluoroimmunoassay (TRFIA) and antigen-binding tests (ABT), an in vivo potency challenge test with refinement (use of analgesics, anesthetics and humane end points to minimize pain and distress, noninvasive challenge route), reduction of numbers of mice per dilution, numbers of dilutions tested, elimination of duplicate testing and testing of multiple batches at one time (de Moura, de Araujo, Cabello, Romijn, & Leite, 2009; Kramer, Schildger, Behrensdorf-Nicol, Hanschmann, & Duchow, 2009). Other methods include serological potency tests using ELISAs (Wang et al., 2018) or modified RFFIT (Moreira, Freitas, Machado, Almeida, & Moura, 2019).

14.17 Safety issues and mitigation strategies

Vaccination can cause adverse reactions, some of which are common to all vaccines while others are vaccine specific. Substances most often implicated in vaccine allergies include egg, gelatin, and yeast (Chung, 2014; Wood et al., 2008). In general, all vaccines, including human rabies vaccines, contain both active and inactive ingredients both of which can cause adverse reactions, including anaphylaxis and allergy. Vaccine ingredients include foreign proteins (BSA, HSA) used during manufacture, viral inactivants (BPL), preservatives (Thiomersal), stabilizers (HSA), adjuvants (aluminum phosphate gel), antibiotics (neomycin), and antigen itself (Wiedermann-Schmidt & Maurer, 2005). Four types of hypersensitivity reactions such as types I, II, III, and IV are known, of which types I to III play a role in connection with vaccination (Abbas, Lichtman, & Pober, 2000). Addition of HSA during RABV antigen production and to the finished product essentially as stabilizer is a common practice (Plotkin, Orenstein, & Offit, 2008; WHO, 2007). Extreme care must be taken such that the HSA in the final product does not undergo much change and therefore will not be able to cause adverse reactions in the recipients (European Medicines Agency, 2011; WHO, 1998; WHO, 2007). In addition, there are concerns about HSA with regard to its possibility of transmitting adventitious agents, including prions (Anderson, Baer, Frazier, & Quinnan, 1987; Finke et al., 2012; Mark, 2006; Wiedermann-Schmidt & Maurer, 2005).

HDCV may cause type I and III hypersensitivity reactions (Anderson et al., 1987; Dreesen, Bernard, Parker, Deutsch, & Brown, 1986; Fishbein et al., 1993; Swanson et al., 1987) attributed mainly to the presence of BPL-modified HSA, which can induce IgE (type 1) and form immune complexes (type III). There is also experimental evidence that HSA can suppress vaccine-induced production of tumor necrosis factor (TNF) (Marth & Kleinhappl, 2001). The risk of type I and III hypersensitivity reactions can be mitigated by inactivating RABV using BPL after removal of HSA (Arora, Moeller, & Froeschle, 2004; Briggs et al., 1996; Fishbein et al., 1989). The PCECV originating from primary CEFs contains micrograms quantity of egg proteins and is therefore contraindicated in patients who are hypersensitive to eggs (Dobardzic et al., 2007); they should be treated with vaccines that that contain no egg components, such as HDCV and PVRV. Alternatively, such patients may be tested for egg hypersensitivity. If the test is negative, PCECV may be administered in a clinical setting in which anaphylaxis can be readily recognized and managed. Gelatin has been implicated in both IgE (Chung, 2014) and non-IgE mediated allergic reactions to vaccines (Moylett & Hanson, 2004). The PCECV has <12 mg of gelatin and sensitized children tend to react (Moylett & Hanson, 2004; Sakaguchi et al., 1999). Patients with a history of gelatin hypersensitivity should be evaluated by an allergist before administration of gelatin-containing vaccines (Chernin, Swender, & Hostoffer Jr., 2011) and may require a rabies vaccine devoid of gelatin (Huang et al., 2018). Nevertheless, appropriate use of desensitization procedures can minimize the risk of anaphylaxis and many gelatin allergic patients are able to receive vaccines containing gelatin when medically necessary. Awareness of the potential for IgE-mediated sensitivity to rabies vaccines and their components is important to allow for their appropriate use (Gupta, Sindher, Saltzman, & Heimall, 2014). Sporadic incidences of adverse reactions following rabies vaccination indicate the importance of having a systemic approach and full understanding about the patients with known history of allergic conditions (Fang et al., 2016; Kang et al., 2015).

Other adverse events to rabies vaccines include local reactions and mild systemic reactions such as fever or headaches. Neurological adverse events such as Guillain-Barré syndrome or seizures have been described after rabies vaccination (Chakravarty, 2001; Mortiere & Falcone, 1997), but they are exceedingly rare with modern vaccines.

14.18 Oral rabies vaccination

Immunization of wildlife animals, which serve as the RABV's reservoir, can reduce rabies in humans, pets, and farm animals. Poisoning or trapping to control the spread of rabies was attempted early on but failed. As widespread parenteral rabies immunization of free-ranging animals is logistically close to impossible, oral rabies vaccine were developed in the 1960s (Baer, Abelseth, & Debbie, 1971). Oral rabies vaccination (ORV) is a socially acceptable disease control method for wildlife reservoirs. The initial requirements of such vaccines were safety, efficacy, and low cost, allowing animals to get immunized upon oral uptake of rabies vaccine-laden baits (Baer, 1988). Vaccination coverage of approximately 70% of the vector population is estimated to be sufficient to block viral transmission (Nyberg et al., 1992). The first field trials with a live RABV vaccine of the ERA strain placed into packets within chicken head baits took place in Switzerland, focusing on environmental safety, especially in nontarget species such as rodents and efficacy for red foxes (Steck, Wandeler, Bichsel, Capt, & Schneider, 1982). Thereafter, Germany used the attenuated SAD Bern RABV strain, called SAD B19, to develop their own seed virus and replaced the use of hand-delivered chicken head baits with bait components and attractants that could be easily automated and allowed for aerial distribution (Muller, Stohr, Teuffert, & Stohr, 1993). Soon, many European countries were using ORV for prevention and control of wildlife rabies (Aubert, Masson, Artois, & Barrat, 1994). A new vaccine produced from SAD Bern, by selection of escape mutants from neutralizing monoclonal antibody pressure, resulted in a highly attenuated rabies virus vaccine, called SAG2 (Cliquet & Aubert, 2004). SAG-2 is an attenuated RABV derived from the SAD-Bern strain (B19) with 2 nucleotide mutations in its glycoprotein sequence that change the amino acid in position 333. SAG2 has been widely used in Europe and led to wildlife rabies elimination in several European countries. It should be noted that no vaccine-induced rabies cases were reported in Europe after distribution of more than 20 million doses of SAG2-containing baits. Scientists in the US developed a vaccinia virus recombinant expressing the RABV glycoprotein, called VR-G, which upon completion of safety and efficacy studies is being used for immunization of raccoons, skunks, foxes, and jackals (Maki et al., 2017; Rupprecht et al., 1986). Another recombinant vaccine based on a replication-competent adenovirus of human serotype 5 also expressing the rabies virus glycoprotein was developed in Canada. It was shown to be safe in nontarget species (Knowles et al., 2009) and to induce VNA responses in raccoons, foxes, and skunks (Brown, Rosatte, Fehlner-Gardiner, Ellison, et al., 2014; Brown et al., 2012; Fehlner-Gardiner et al., 2012; McCoy et al., 2007). By the end of the 20th century, large portions of Western Europe and southern Ontario were becoming free of wildlife rabies virus transmission by the strategic use of ORV (MacInnes et al., 2001; Stohr & Meslin, 1996). However, limitations in effectiveness in certain species, and concerns of residual pathogenicity from first-generation ORVs remain as hurdles in the field and require new biotechnological approaches (Fehlner-Gardiner et al., 2008; Muller et al., 2009).

14.19 Wildlife

The safety of rabies vaccines for wildlife immunization should be evaluated in rodents, wildlife, and domestic animals (Faber et al., 2009). ORV using modified-live RABV or recombinant viruses has been highly successful in different reservoir species. The first animal targeted was the European red fox (*Vulpes vulpes*) followed by the raccoon dog (*Nyctereutes procyonoides*) (Steck et al., 1982). Subsequently, the concept of oral rabies baiting was investigated for raccoons (*Procyon lotor*) (Rupprecht et al., 1986; Rupprecht, Dietzschold, Cox, & Schneider, 1989), coyotes (*Canis latrans*) (Fearneyhough et al., 1998; Meehan, 1995), gray foxes (*Urocyon cinereoargenteus*), striped skunks (*Mephitis mephitis*) (Rosatte, Power, MacInnes, & Campbell, 1992), small Indian mongooses (*Herpestes auropunctatus*) (Blanton et al., 2006; Vos et al., 2013), and domestic dogs (*Canis lupus domesticus*) (Aylan & Vos, 1998; Haddad et al., 1994; Rupprecht et al., 2005). ORV has eliminated rabies in 12 European countries (Cliquet & Aubert, 2004; Cliquet, Picard-Meyer, & Robardet, 2014; Freuling et al., 2013) and is currently being used in the majority of rabies-affected European countries (Borg, 2013). Not all animal species respond equally well to vaccination by the oral route;

some species like the striped skunk seem to be extremely refractory to ORV, irrespective of the construct or the amount of virus present in the bait. The seroconversion rates in raccoons upon ORV are lower than in gray foxes and coyotes. These could be due to two reasons; vaccine is not as immunogenic in raccoons suggesting the need for an adjuvant and/or vaccine spillage suggesting the need for a more viscous vaccine mixture. N,N,N-Trimethylated chitosan (TMC) increases the viscosity of the vaccine and potentially acts as adjuvant to improve the immune response in raccoons (*Procyon lotor*) (Fry, Van Dalen, Hurley, & Nash, 2012). Wolves may not easily consume commercial baits. For this species, goat meat baits seem to have the highest uptake compared to rodent and intestine baits (Sillero-Zubiri et al., 2016). Further studies concerning the effectiveness of ORV in accordance with the National Standard Assay for Veterinary Biologic Products in dogs and raccoon dogs are needed (Choi et al., 2015). A recent study increased our understanding on the existence of diversity of variants in oral rabies vaccines widely used in Europe as well as the presence of a mix of at least two different variants in all tested batches. Routine sequencing analyses should be conducted to reveal the potential reversion of attenuated rabies vaccines to a virulent form and shifts in virus populations during the vaccine manufacturing process (Cliquet et al., 2015).

14.20 Domestic animals

The purpose of rabies vaccination of domestic animals is to protect individual animals when they are exposed to RABV and to reduce transmission to humans. To eliminate rabies from dogs in an endemic area, at least 70% of the population needs to be vaccinated (Coleman & Dye, 1996). Parenteral vaccination is the method of choice for owned dogs (i.e., dogs with a person that claims responsibility, according to the OIE definition) as a cost-effective measure in preventing human rabies (Zinsstag et al., 2009). Catching free roaming dogs is easier if they are owned. Therefore, dog ownership is an important factor in determining the percentage of dogs vaccinated during a campaign (Davlin & Vonville, 2012). Parenteral vaccination of stray or owned but uncontrolled dogs is more difficult, laborious, and expensive. Such dogs could receive ORV following international guidelines for rabies control in dogs. The cost of ORV is higher than that of the parenteral vaccination. This could be reduced by the use of aerial distribution in certain well-defined and restricted areas, but "hand-out" models to dogs may be the most practical form of delivery due to continued concerns about safety in nontarget species (Darkaoui et al., 2014). A combination of parenteral and oral vaccination may help to increase the vaccination coverage in the canine populations, which could lead to a strong reduction in human rabies.

Currently two oral rabies vaccines, i.e., SAG2 and VR-G, are recommended by the OIE for oral dog vaccination (OIE, 2011). SAG2 is a double deletion mutant derived from the SAD Berne strain of rabies. It is highly attenuated and induces solid protection in dogs. SAG-2 is registered for the control of canine rabies in India and has been mainly evaluated in Tunisia, Mexico, South Africa, and Indonesia, demonstrating its efficacy for dog vaccination in the field. The level of VNA induced by SAG-2 is generally low in dogs after ORV and not all dogs develop detectable VNA. VR-G, a recombinant Vaccinia virus expressing RABV glycoprotein (G), has been successfully used for control of fox rabies in Europe and coyote and raccoon rabies control in the USA. The safety and immunogenicity of a newly constructed rabies vaccine strain (ERAGS strain) was assessed upon oral or IM administering of growing pigs and sows. The vaccine was safe and none of the inoculated pigs developed clinical signs of rabies over 28 days, and RABV was not detected in tissue samples by FAT or RT-RCR. Additionally, the vaccinated pigs developed significant VNA titers against RABV, indicating that the ERAGS strain may be immunogenic in pigs. Thus, the ERAGS strain is a new, prospective candidate for a pig rabies vaccine (Yang, Kim, Choi, Lee, & Cho, 2016). In the future, reverse genetics applications may offer even safer and efficacious RABV vaccines for a broader range of species (Schnell, Mebatsion, & Conzelmann, 1994).

Traditional rabies vaccines were linked to the development of sarcomas in cats (Saba, 2017). An alternative vaccine based on a recombinant canary pox vector expressing the RABV glycoprotein was developed. This vaccine, called Purevax Feline Rabies, if given to young cats at 12 weeks followed by an initial boost 1 year later and subsequent boosts in 3-year intervals, provides protective immunity (Jas, Coupier, Toulemonde, Guigal, & Poulet, 2012).

14.21 Monitoring oral rabies vaccination

The effectiveness of ORV is assessed through direct observation and testing of trapped or hunted animals (Cliquet et al., 2010) for biomarkers (e.g., tetracycline, iophenoxic acid) present in the vaccine bait and/or RABV-specific VNAs (Darkaoui et al., 2014). The two current OIE-prescribed serological reference methods, the FAVNT and the RFFIT (OIE, 2012), are based on cell culture and are therefore sensitive to any cytotoxic products and contaminating agents present in field samples (Cliquet et al., 2010). By comparison, ELISAs are less time-consuming, easier techniques and preferred for assessing the serological response in countries undertaking mass vaccination of dogs and oral vaccination of wildlife (Yakobson et al., 2014; Zienius, Pridotkas, Lelesius, & Sereika, 2011) or for samples obtained from animal cadavers, including fluids from thoracic cavity or extracts from muscle, which are often of poor quality (Bedeković et al., 2016).

14.22 Future directions

Human deaths due to rabies are reported mainly from endemic areas of Asia and Africa despite the availability of vaccines that are highly effective if given in case of a severe exposure with RIG. The following tools are being developed to reduce rabies in humans and animals.

Improved vaccine strains: Improved vaccine strains that are more immunogenic due to increases in copy numbers of the G protein may reduce the need for multiple immunizations of humans and thereby reduce the cost of vaccination.

User-friendly delivery systems: Parenteral vaccination through IM route is widely practiced. Prefilled syringes (PFS) for liquid rabies vaccine can aid in the delivery of more precise volumes and minimize handling of the vaccine thereby reducing waste and overall cost.
Improved vaccine formulations: Vaccine formulations with novel adjuvants may allow for antigen and dose sparing.
Abbreviated vaccination regimens: Regimens for vaccinating human subjects, e.g., dog bite victims, using ID immunizations or more shots per visit may minimize the numbers of clinic visits.
Versatile combination vaccine for dogs: Induction of herd immunity to RABV and immunocontraception, a humane way of population reduction in dogs, through introduction of a bivalent vaccine targeting rabies virus and a hormone of reproduction may reduce numbers of stray dogs.
Universal baits for ORV: Design of universal baits suitable for a plethora of wild and domestic animal species can bring about reduction in cost of production and better field applicability.
A One Health focus on canine rabies elimination: Considering the eradication of smallpox and rinderpest, and considerations for similar strategies targeting polio, measles, etc., a World Health Assembly resolution would enable the technical ability and economic resources necessary to achieve the above proposals, with enhanced collaboration among FAO, OIE, WHO, and other principal stakeholders, to remove the single largest factor in the global burden of rabies by creation of herd immunity using modern vaccination strategies.

References

Abbas, A. K., Lichtman, A. H., & Pober, J. (2000). Disease caused by immune response: Hypersensitivity and autoimmunity. In *Cellular and molecular immunology* (pp. 404–415). Philadelphia: Saunders.

Altaras, N. E., Aunins, J. G., Evans, R. K., Kamen, A., Konz, J. O., & Wolf, J. J. (2005). Production and formulation of adenovirus vectors. *Advances in Biochemical Engineering/Biotechnology, 99*, 193–260.

Anderson, M. C., Baer, H., Frazier, D. J., & Quinnan, G. V. (1987). The role of specific IgE and beta-propiolactone in reactions resulting from booster doses of human diploid cell rabies vaccine. *The Journal of Allergy and Clinical Immunology, 80*(6), 861–868.

Andreadis, S. T., Roth, C. M., Le Doux, J. M., Morgan, J. R., & Yarmush, M. L. (1999). Large-scale processing of recombinant retroviruses for gene therapy. *Biotechnology Progress, 15*(1), 1–11.

Arora, A., Moeller, L., & Froeschle, J. (2004). Safety and immunogenicity of a new chromatographically purified rabies vaccine in comparison to the human diploid cell vaccine. *Journal of Travel Medicine, 11*(4), 195–199.

Asgary, V., Shoari, A., Baghbani-Arani, F., Sadat Shandiz, S. A., Khosravy, M. S., Janani, A., … Cohan, R. A. (2016). Green synthesis and evaluation of silver nanoparticles as adjuvant in rabies veterinary vaccine. *International Journal of Nanomedicine, 11*, 3597–3605. https://doi.org/10.2147/IJN.S109098. eCollection 2016.

Atanasiu, P., Ribeiro, M., & Tsiang, H. (1972). Antirabies vaccines from tissue culture obtained with the Pasteur strain. Results of vaccination. *Annales de l'Institut Pasteur (Paris), 123*(3), 427–441.

Aubert, M. F., Masson, E., Artois, M., & Barrat, J. (1994). Oral wildlife rabies vaccination field trials in Europe, with recent emphasis on France. *Current Topics in Microbiology and Immunology, 187*, 219–243.

Aylan, O., & Vos, A. (1998). Efficacy studies with SAD B19 in Turkish dogs. *Journal of Etlik Veterinary Microbiology, 9*, 93–102.

Baer, G. M. (1988). Oral rabies vaccination: An overview. *Reviews of Infectious Diseases, 10*(Suppl 4), S644–S648.

Baer, G. M., Abelseth, M. K., & Debbie, J. G. (1971). Oral vaccination of foxes against rabies. *American Journal of Epidemiology, 93*(6), 487–490.

Bahnemann, H. G. (1990). Inactivation of viral antigens for vaccine preparation with particular reference to the application of binary ethylenimine. *Vaccine, 8*(4), 299–303.

Barone, D., Fracchia, S., Pascuali, E., & Proglio, F. (1997). ICH-4 guidelines for the quality and safety of cell substrate used in the production of pharmaceuticals. In *Paper presented at the Italian-German Biotech Forum: Pharmaceutical Biotechnology*.

Barrett, P. N., Mundt, W., Kistner, O., & Howard, M. K. (2009). Vero cell platform in vaccine production: Moving towards cell culture-based viral vaccines. *Expert Review of Vaccines, 8*, 607–618.

Barth, R., & Franke, V. (1996). Purified chick-embryo cell vaccine for humans. In F. X. Meslin, M. M. Kaplan, & H. Koprowski (Eds.), *Laboratory techniques in rabies*. (4th ed., pp. 290–29) Geneva: WHO..

Bedeković, T., Šimić, I., Krešić, N., Lojkić, I., Mihaljević, Ž., Sučec, I., ... Hostnik, P. (2016). Evaluation of ELISA for the detection of rabies virus antibodies from the thoracic liquid and muscle extract samples in the monitoring of fox oral vaccination campaigns. *BMC Veterinary Research, 12*, 76. https://doi.org/10.1186/s12917-016-0701-0.

Birch, J. R. (1999). Suspension culture, animal cells. In M. C. Flickinger & S. W. Drew (Eds.), *Bioprocess technology: Fermentation, biocatalysis and bioseparation* (pp. 2509–2516): Wiley.

Blancou, J., Aubert, M. F., Cain, E., Selve, M., Thraenhart, O., & Bruckner, L. (1989). Effect of strain differences on the potency testing of rabies vaccines in mice. *Journal of Biological Standardization, 17*(3), 259–266.

Blanton, J. D., Meadows, A., Murphy, S. M., Manangan, J., Hanlon, C. A., Faber, M. L., ... Rupprecht, C. E. (2006). Vaccination of small Asian mongoose (Herpestes javanicus) against rabies. *Journal of Wildlife Diseases, 42*(3), 663–666.

Borg, T. (2013). Commission implementing decision of 29 November 2013 approving annual and mutiannual programmes and the financial contribution from the Union for the eradication, control and monitoring of certain animal diseases and zoonoses presented by the member states for 2014 and the following years. *Official Journal of the European Union, 2013*, 101–110.

Braas, G., Searle, P. F., Slater, N. K., & Lyddiatt, A. (1996). Strategies for the isolation and purification of retroviral vectors for gene therapy. *Bioseparation, 6*, 211–228.

Briggs, D. J., Dreesen, D. W., Morgan, P., Chin, J. E., Seedle, C. D., Cryz, L., ... Cryz, S. J. (1996). Safety and immunogenicity of Lyssavac Berna human diploid cell rabies vaccine in healthy adults. *Vaccine, 14*(14), 1361–1365.

Brookes, S. M., Parsons, G., Johnson, N., McElhinney, L. M., & Fooks, A. R. (2005). Rabies human diploid cell vaccine elicits cross-neutralising and cross-protecting immune responses against European and Australian bat lyssaviruses. *Vaccine, 23*(32), 4101–4109.

Brown, L. J., Rosatte, R. C., Fehlner-Gardiner, C., Bachmann, P., Ellison, J. A., Jackson, F. R., ... Donovan, D. (2014). Oral vaccination and protection of red foxes (Vulpes vulpes) against rabies using ONRAB, an adenovirus-rabies recombinant vaccine. *Vaccine, 32*(8), 984–989. https://doi.org/10.1016/j.vaccine.2013.12.015.

Brown, L. J., Rosatte, R. C., Fehlner-Gardiner, C., Taylor, J. S., Davies, J. C., & Donovan, D. (2012). Immune response and protection in raccoons (Procyon lotor) following consumption of baits containing ONRAB®, a human adenovirus rabies glycoprotein recombinant vaccine. *Journal of Wildlife Diseases, 48*(4), 1010–1020. https://doi.org/10.7589/2012-01-023.

CDC. (2004). Manufacturer's recall of human rabies vaccine. *MMWR Dispatch, 53*, 287–289.

Chakravarty, A. (2001). Neurologic illness following post-exposure prophylaxis with purifiled chick embryo cell antirabies vaccine. *The Journal of the Association of Physicians of India, 49*, 927–928.

Chapman, W. G., Ramshaw, I. A., & Crick, J. (1973). Inactivated rabies vaccine produced from the Flury LEP strain of virus grown in BHK-21 suspension cells. *Applied Microbiology, 26*(6), 858–862.

Chernin, L. R., Swender, D., & Hostoffer, R. W., Jr. (2011). Cracking the shell on egg-hypersensitive patients and egg-containing vaccines. *The Journal of the American Osteopathic Association, 111*(10 Suppl 6), S5–S6.

Choi, J., Yang, D. K., Kim, H. H., Jo, H. Y., Choi, S. S., Kim, J. T., ... Kim, H. W. (2015). Application of recombinant adenoviruses expressing glycoprotein or nucleoprotein of rabies virus to Korean raccoon dogs. *Clinical and Experimental Vaccine Research, 4*, 189–194.

Chung, E. H. (2014). Vaccine allergies. *Clinical and Experimental Vaccine Research, 3*, 50–57.

Cliquet, F., & Aubert, M. (2004). Elimination of terrestrial rabies in Western European countries. *Developmental Biology (Basel), 119*, 185–204.

Cliquet, F., Freuling, C., Smreczak, M., van der Poel, W. H. M., Horton, D., Fooks, A. R., ... Müller, T. (2010). Development of harmonised schemes for monitoring and reporting of rabies in animals in the European Union. EFSA Scientific Report (p. 60).

Cliquet, F., Picard-Meyer, E., Mojzis, M., Dirbakova, Z., Muizniece, Z., Jaceviciene, I., … Celer, V. (2015). In-depth characterization of live vaccines used in Europe for oral rabies vaccination of wildlife. *PLoS One, 10*, e01041537.

Cliquet, F., Picard-Meyer, E., & Robardet, E. (2014). Rabies in Europe: What are the risks? *Expert Review of Anti-Infective Therapy, 12*, 905–908.

Coleman, P. G., & Dye, C. (1996). Immunization coverage required to prevent outbreaks of dog rabies. *Vaccine, 14*, 185–186.

Cox, J. H., Klietmann, W., & Schneider, L. G. (1978). Human rabies immunoprophylaxis using HDC (MRC-5) vaccine. *Developments in Biological Standardization, 40*, 105–108.

Cox, J. H., & Schneider, L. G. (1976). Prophylactic immunization of humans against rabies by intradermal inoculation of human diploid cell culture vaccine. *Journal of Clinical Microbiology, 3*, 96–101.

Darkaoui, S., Boué, F., Demerson, J. M., Fassi Fihri, O., Yahia, K. I., & Cliquet, F. (2014). First trials of oral vaccination with rabies SAG2 dog baits in Morocco. *Clinical and Experimental Vaccine Research, 3*(2), 220–226. https://doi.org/10.7774/cevr.2014.3.2.220.

Davlin, S. L., & Vonville, H. M. (2012). Canine rabies vaccination and domestic dog population characteristics in the developing world: A systematic review. *Vaccine, 30*(24), 3492–3502. https://doi.org/10.1016/j.vaccine.2012.03.069.

de Moura, W. C., de Araujo, H. P., Cabello, P. H., Romijn, P. C., & Leite, J. P. (2009). Potency evaluation of rabies vaccine for human use: The impact of the reduction in the number of animals per dilution. *Journal of Virological Methods, 158*(1–2), 84–92. https://doi.org/10.1016/j.jviromet.2009.01.017.

Dietzschold, B., Faber, M., & Schnell, M. J. (2003). New approaches to the prevention and eradication of rabies. *Expert Review of Vaccines, 2*(3), 399–406. https://doi.org/10.1586/14760584.2.3.399.

Dobardzic, A., Izurieta, H., Woo, E. J., Iskander, J., Shadomy, S., Rupprecht, C., … Braun, M. M. (2007). Safety review of the purified chick embryo cell rabies vaccine: Data from the Vaccine Adverse Event Reporting System (VAERS), 1997-2005. *Vaccine, 25*(21), 4244–4251.

Dreesen, D. W., Bernard, K. W., Parker, R. A., Deutsch, A. J., & Brown, J. (1986). Immune complex-like disease in 23 persons following a booster dose of rabies human diploid cell vaccine. *Vaccine, 4*(1), 45–49.

European Medicines Agency. (2011). *Guidelines on the warning on transmissible agents in summary of product characteristics (SmPCs) and package leaflets for plasma-derived medicinal products*. London: European Medicine Agency.

Faber, M., Dietzschold, B., & Li, J. (2009). Immunogenicity and safety of recombinant rabies viruses used for oral vaccination of stray dogs and wildlife. *Zoonoses and Public Health, 56*(6-7), 262–269. https://doi.org/10.1111/j.1863-2378.2008.01215.x.

Fang, Y., Liu, M. Q., Chen, L., Zhu, Z. G., Zhu, Z. R., & Hu, Q. (2016). Rabies post-exposure prophylaxis for a child with severe allergic reaction to rabies vaccine. *Human Vaccines & Immunotherapeutics, 12*(7), 1802–1804. https://doi.org/10.1080/21645515.2016.1143158.

Fayaz, A., Simani, S., Janani, A., Farahtaj, F., Biglari, P., Howeizi, N., … Eslami, N. (2011). Antibody persistence, 32 years after post-exposure prophylaxis with human diploid cell rabies vaccine (HDCV). *Vaccine, 29*, 3742–3745.

Fearneyhough, M. G., Wilson, P. J., Clark, K. A., Smith, D. R., Johnston, D. H., Hicks, B., … Moore, G. M. (1998). Results of an oral rabies vaccination program for coyotes. *Journal of the American Veterinary Medical Association, 212*(4), 498–502.

Fehlner-Gardiner, C., Nadin-Davis, S., Armstrong, J., Muldoon, F., Bachmann, P., & Wandeler, A. (2008). Era vaccine-derived cases of rabies in wildlife and domestic animals in Ontario, Canada, 1989-2004. *Journal of Wildlife Diseases, 44*(1), 71–85.

Fehlner-Gardiner, C., Rudd, R., Donovan, D., Slate, D., Kempf, L., & Badcock, J. (2012). Comparing ONRAB® and RABORAL V-RG® oral rabies vaccine field performance in raccoons and striped skunks, New Brunswick, Canada, and Maine, USA. *Journal of Wildlife Diseases, 48*(1), 157–167.

Finke, S., Karger, A., Freuling, C., & Müller, T. (2012). Assessment of inactivated human rabies vaccines: Biochemical characterization and genetic identification of virus strains. *Vaccine, 30*, 3603–3609.

Fishbein, D. B., Dreesen, D. W., Holmes, D. F., Pacer, R. E., Ley, A. B., Yager, P. A., … Kemp, D. T. (1989). Human diploid cell rabies vaccine purified by zonal centrifugation: A controlled study of antibody response and side effects following primary and booster pre-exposure immunizations. *Vaccine, 7*, 437–442.

Fishbein, D. B., Yenne, K. M., Dreesen, D. W., Teplis, C. F., Mehta, N., & Briggs, D. J. (1993). Risk factors for systemic hypersensitivity reactions after booster vaccinations with human diploid cell rabies vaccine: A nationwide prospective study. *Vaccine, 11*, 1390–1394.

Frazzati-Gallina, N. M., Paoli, R. L., Mourão-Fuches, R. M., Jorge, S. A., & Pereira, C. A. (2001). Higher production of rabies virus in serum-free medium cell cultures on microcarriers. *Journal of Biotechnology, 92*, 67–72.

Freuling, C. M., Hampson, K., Selhorst, T., Schröder, R., Meslin, F. X., Mettenleiter, T. C., & Müller, T. (2013). The elimination of fox rabies from Europe: Determinants of success and lessons for the future. *Philosophical Transactions of the Royal Society of London. Series B, Biological Sciences, 368*(1623). 20120142 https://doi.org/10.1098/rstb.2012.0142.

Fry, T., Van Dalen, K., Hurley, J., & Nash, P. (2012). Mucosal adjuvants to improve wildlife rabies vaccination. *Journal of Wildlife Diseases, 48*, 1042–1046.

Furesz, J., Fanok, A., Contreras, G., & Becker, B. (1989). Tumorigenicity testing of various cell substrates for production of biologicals. *Developments in Biological Standardization, 70*, 233–243.

Gallo-Ramirez, L. E., Nikolay, A., Genzel, Y., & Reichl, U. (2015). Bioreactor concepts for cell culture-based viral vaccine production. *Expert Review of Vaccines, 14*, 1185–1195.

Gaudin, Y., Ruigrok, R. W., Tuffereau, C., Knossow, M., & Flamand, A. (1992). Rabies virus glycoprotein is a trimer. *Virology, 187*(2), 627–632.

Giesen, A., Gniel, D., & Malerczyk, C. (2015). 30 Years of rabies vaccination with Rabipur: A summary of clinical data and global experience. *Expert Review of Vaccines, 14*(3), 351–367. https://doi.org/10.1586/14760584.2015.1011134.

Grachev, V. P. (1990). World Health Organization attitude concerning the use of continuous cell lines as substrates for production of human virus vaccines. In A. Mizrahi (Ed.), *Advances in biotechnological processes: Vol. 14. Viral vaccines* (pp. 37–67): Wiley-Liss.

Gupta, M., Sindher, S., Saltzman, R., & Heimall, J. (2014). Evaluation and safe administration of rabies vaccine to a child presumably allergic to the gelatin content of the PCECV RabAvert. *The Journal of Allergy and Clinical Immunology. In Practice*, 1–2.

Habel, K. (1973). *Laboratory techniques in rabies: General considerations in vaccine production. Vol. 23* (pp. 189–191) World Health Organization.

Habel, K. (1966). Laboratory techniques in rabies. Modified Habel test for potency. *Monograph Series. World Health Organization, 23*, 144.

Habel, K., & Koprowski, H. (1955). Laboratory data supporting the clinical trial of anti-rabies serum in persons bitten by a rabid wolf. *Bulletin of the World Health Organization, 13*(5), 773–779.

Haddad, N., Ben Khelifa, R., Matter, H., Kharmachi, H., Aubert, M. F., Wandeler, A., & Blancou, J. (1994). Assay of oral vaccination of dogs against rabies in Tunisia with the vaccinal strain SADBern. *Vaccine*, (4), 307–309.

Hampson, K., Coudeville, L., Lembo, T., Sambo, M., Kieffer, A., Attlan, M., … Global Alliance for Rabies Control Partners for Rabies Prevention (2015). Estimating the global burden of endemic canine rabies. *PLoS Neglected Tropical Diseases, 9*(4), e0003709. https://doi.org/10.1371/journal.pntd.0003709.

Hampson, K., Dushoff, J., Cleaveland, S., Haydon, D. T., Kaare, M., Packer, C., & Dobson, A. (2009). Transmission dynamics and prospects for the elimination of canine rabies. *PLoS Biology, 7*(3), e53. https://doi.org/10.1371/journal.pbio.1000053.

Hassanzadeh, S. M., Zavareh, A., Shokrgozar, M. A., Ramezani, A., & Fayaz, A. (2011). High vero cell density and rabies virus proliferation on fibracel disks versus cytodex-1 in spinner flask. *Pakistan Journal of Biological Sciences, 14*(7), 441–448.

Hay, R. J., Caputo, J., & Chen, T. R. (1994). *ATCC cell lines and hybridomas* (8th ed, p. 640).

Hayflick, L. (1989). History of cell substrates used for human biologicals. *Developments in Biological Standardization, 70*, 11–26.

Hendriksen, C. F. (2009). Replacement, reduction and refinement alternatives to animal use in vaccine potency measurement. *Expert Review of Vaccines, 8*(3), 313–322. https://doi.org/10.1586/14760584.8.3.313.

Hicks, D. J., Fooks, A. R., & Johnson, N. (2012). Developments in rabies vaccines. *Clinical and Experimental Immunology, 169*(3), 199–204. https://doi.org/10.1111/j.1365-2249.2012.04592.x.

Höper, D., Freuling, C. M., Müller, T., Hanke, D., von Messling, V., Duchow, K., … Mettenleiter, T. C. (2015). High definition viral vaccine strain identity and stability testing using full-genome population data—The next generation of vaccine quality control. *Vaccine, 33*(43), 5829–5837. https://doi.org/10.1016/j.vaccine.2015.08.091.

Horaud, F. (1992). Absence of viral sequences in the WHO-Vero Cell Bank. A collaborative study. *Developments in Biological Standardization, 76*, 43–46.

Huang, S., Zhu, Z., Cai, L., Zhu, Z., Zhang, M., Hu, Q., & Fang, Y. (2018). Analysis on the risks of severe adverse events in rabies post-exposure prophylaxis and appropriate decision-making procedure. *Human Vaccines & Immunotherapeutics*, 1–5. https://doi.org/10.1080/21645515.2018.1533779.

Jackson, A. C. (Ed.), (2013). *Rabies: Scientific basis of the disease and its management* (3rd ed.). Oxford, UK: Elsevier Academic Press.

Jacobs, J. P., Jones, C. M., & Baille, J. P. (1970). Characteristics of a human diploid cell designated MRC-5. *Nature, 227,* 168–170.

Jas, D., Coupier, C., Toulemonde, C. E., Guigal, P. M., & Poulet, H. (2012). Three-year duration of immunity in cats vaccinated with a canarypox-vectored recombinant rabiesvirus vaccine. *Vaccine, 30*(49), 6991–6996. https://doi.org/10.1016/j.vaccine.2012.09.068.

Jordan, I., & Sandig, V. (2014). Matrix and backstage: Cellular substrates for viral vaccines. *Viruses, 6*(4), 1672–1700. https://doi.org/10.3390/v6041672.

Kallel, H., Diouani, M. F., Loukil, H., Trabelsi, K., Snoussi, M. A., Majoul, S., … Dellagi, K. (2006). Immunogenicity and efficacy of an in-house developed cell-culture derived veterinarian rabies vaccine. *Vaccine, 24*(22), 4856–4862. https://doi.org/10.1016/j.vaccine.2006.03.012.

Kallel, H., Rourou, S., Majoul, S., & Loukil, H. (2003). A novel process for the production of a veterinary rabies vaccine in BHK-21 cells grown on microcarriers in a 20-l bioreactor. *Applied Microbiology and Biotechnology, 61*(5-6), 441–446. https://doi.org/10.1007/s00253-003-1245-3.

Kang, H., Qi, Y., Wang, H., Zheng, X., Gao, Y., … Xia, X. (2015). Chimeric rabies virus-like particles containing membrane-anchored GM-CSF enhances the immune response against rabies virus. *Viruses, 7,* 1134–1152.

Klietmann, W., Domres, B., & Cox, J. H. (1978). Rabies post-exposure treatment and side-effects in man using HDC (MRC 5) vaccine. *Developments in Biological Standardization, 40,* 109–113.

Knowles, M. K., Nadin-Davis, S. A., Sheen, M., Rosatte, R., Mueller, R., & Beresford, A. (2009). Safety studies on an adenovirus recombinant vaccine for rabies (AdRG1.3-ONRAB) in target and non-target species. *Vaccine, 27*(47), 6619–6626. https://doi.org/10.1016/j.vaccine.2009.08.005.

Koprowski, H., & Cox, H. R. (1948). Studies on chick embryo adapted rabies virus; culture characteristics and pathogenicity. *Journal of Immunology, 60*(4), 533–554.

Kramer, B., Schildger, H., Behrensdorf-Nicol, H. A., Hanschmann, K. M., & Duchow, K. (2009). The rapid fluorescent focus inhibition test is a suitable method for batch potency testing of inactivated rabies vaccines. *Biologicals, 37*(2), 119–126.

Kulkarni, P. S., Sahai, A., Gunale, B., & Dhere, R. M. (2017). Development of a new purified vero cell rabies vaccine (Rabivax-S) at the serum institute of India Pvt Ltd. *Expert Review of Vaccines, 16*(4), 303–311. https://doi.org/10.1080/14760584.2017.1294068.

Kuwert, E. K., Marcus, I., Werner, J., Iwand, A., & Thraenhart, O. (1978). Some experiences with human diploid cell strain-(HDCS) rabies vaccine in pre- and post-exposure vaccinated humans. *Developments in Biological Standardization, 40,* 79–88.

Lang, J., Attanath, P., Quiambao, B., Singhasivanon, V., Chanthavanich, P., Montalban, C., … Sabcharoen, A. (1998). Evaluation of the safety, immunogenicity, and pharmacokinetic profile of a new, highly purified, heat-treated equine rabies immunoglobulin, administered either alone or in association with a purified, Vero-cell rabies vaccine. *Acta Tropica, 70*(3), 317–333.

Langevin, C., Jaaro, H., Bressanelli, S., Fainzilber, M., & Tuffereau, C. (2002). Rabies virus glycoprotein (RVG) is a trimeric ligand for the N-terminal cysteine-rich domain of the mammalian p75 neurotrophin receptor. *The Journal of Biological Chemistry, 277*(40), 37655–37662. https://doi.org/10.1074/jbc.M201374200.

Langevin, C., & Tuffereau, C. (2002). Mutations conferring resistance to neutralization by a soluble form of the neurotrophin receptor (p75NTR) map outside of the known antigenic sites of the rabies virus glycoprotein. *Journal of Virology, 76*(21), 10756–10765.

Levenbook, I. S., Petricciani, J. C., & Elisberg, B. L. (1984). Tumorigenicity of Vero cells. *Journal of Biological Standardization, 12*(4), 391–398.

Li, R., Huang, L., Li, J., Mo, Z., He, B., Wang, Y., … Pichon, S. (2013). A next-generation, serum-free, highly purified Vero cell rabies vaccine is safe and as immunogenic as the reference vaccine Verorab® when administered according to a post-exposure regimen in healthy children and adults in China. *Vaccine, 31*(50), 5940–5947. https://doi.org/10.1016/j.vaccine.2013.10.043.

Li, S. M., Bai, F. L., Xu, W. J., Yang, Y. B., An, Y., Li, T. H., … Wang, W. F. (2014). Removing residual DNA from Vero-cell culture-derived human rabies vaccine by using nuclease. *Biologicals, 42*(5), 271–276. https://doi.org/10.1016/j.biologicals.2014.06.005.

Ma, B., He, L. F., Zhang, Y. L., Chen, M., Wang, L. L., Yang, H. W., … Zheng, C. Y. (2015). Characteristics and viral propagation properties of a new human diploid cell line, Walvax-2, and its suitability as a candidate cell substrate for vaccine production. *Human Vaccines & Immunotherapeutics, 11*(4), 998–1009. https://doi.org/10.1080/21645515.2015.1009811.

MacInnes, C. D., Smith, S. M., Tinline, R. R., Ayers, N. R., Bachmann, P., Ball, D. G., ... Voigt, D. R. (2001). Elimination of rabies from red foxes in eastern Ontario. *Journal of Wildlife Diseases, 37*(1), 119–132.

Madhusudana, S. N., & Mani, R. S. (2014). Intradermal vaccination for rabies prophylaxis: Conceptualization, evolution, present status and future. *Expert Review of Vaccines, 13*(5), 641–655.

Maki, J., Guiot, A. L., Aubert, M., Brochier, B., Cliquet, F., Hanlon, C. A., ... Lankau, E. W. (2017). Oral vaccination of wildlife using a vaccinia-rabies-glycoprotein recombinant virus vaccine (RABORAL V-RG®): A global review. *Veterinary Research, 48*(1), 57.

Mark, C. (2006). Large variations in the ovalbumin content in six European influenza vaccines. *Pharmaceutical Sciences Notes, 2006*, 27–29.

Marth, E., & Kleinhappl, B. (2001). Albumin is a necessary stabilizer of TBE-vaccine to avoid fever in children after vaccination. *Vaccine, 20*(3-4), 532–537.

McCoy, K., Tatsis, N., Korioth-Schmitz, B., Lasaro, M. O., Hensley, S. E., Lin, S. W., ... Ertl, H. C. (2007). Effect of preexisting immunity to adenovirus human serotype 5 antigens on the immune responses of nonhuman primates to vaccine regimens based on human- or chimpanzee-derived adenovirus vectors. *Journal of Virology, 81*, 6594–6604.

McVey, D. S., Galvin, J. E., & Olson, S. C. (2003). A review of the effectiveness of vaccine potency control testing. *International Journal for Parasitology, 33*(5–6), 507–516.

Meehan, S. K. (1995). Rabies epizootic in coyotes combated with oral vaccination program. *Journal of the American Veterinary Medical Association, 206*(8), 1097–1099.

Mondal, S. K., Neelima, M., Seetha Rama Reddy, K., Ananda Rao, K., & Srinivasan, V. A. (2005). Validation of the inactivant binary ethylenimine for inactivating rabies virus for veterinary rabies vaccine production. *Biologicals, 33*(3), 185–189. https://doi.org/10.1016/j.biologicals.2005.05.003.

Montaner, A. D., De Nichilo, A., Rodriguez, J. M., Hernando-Insua, A., Flo, J., Lopez, R. A., ... Elias, F. (2012). IMT504: A new and potent adjuvant for rabies vaccines permitting significant dose sparing. *World Journal of Vaccines, 2*, 182–188.

Montomoli, E., Piccirella, S., Khadang, B., Mennitto, E., Camerini, R., & De Rosa, A. (2011). Current adjuvants and new perspectives in vaccine formulation. *Expert Review of Vaccines, 10*(7), 1053–1061. https://doi.org/10.1586/erv.11.48.

Moreira, W. C., Freitas, J. F. S., Machado, N. S., Almeida, A. E. C. C., & Moura, W. C. (2019). Development and prevalidation of a quantitative multi-dose serological assay for potency testing of inactivated rabies vaccines for human use. *Journal of Virological Methods, 263*, 54–59. https://doi.org/10.1016/j.jviromet.2018.10.003.

Mortiere, M. D., & Falcone, A. L. (1997). An acute neurologic syndrome temporally associated with postexposure treatment of rabies. *Pediatrics, 100*(4), 720–721.

Moylett, E. H., & Hanson, I. C. (2004). Mechanistic actions of the risks and adverse events associated with vaccine administration. *The Journal of Allergy and Clinical Immunology, 114*(5), 1010–1020, quiz 1021.

Muller, T., Batza, H. J., Beckert, A., Bunzenthal, C., Cox, J. H., Freuling, C. M., ... Mettenleiter, T. C. (2009). Analysis of vaccine-virus-associated rabies cases in red foxes (Vulpes vulpes) after oral rabies vaccination campaigns in Germany and Austria. *Archives of Virology, 154*(7), 1081–1091. https://doi.org/10.1007/s00705-009-0408-7.

Muller, T., Stohr, K., Teuffert, J., & Stohr, P. (1993). Experiences with the aerial distribution of baits for the oral immunization of foxes against rabies in eastern Germany. *Deutsche Tierärztliche Wochenschrift, 100*(5), 203–207.

Nestola, P., Peixoto, C., Silva, R. R., Alves, P. M., Mota, J. P., & Carrondo, M. J. (2015). Improved virus purification processes for vaccines and gene therapy. *Biotechnology and Bioengineering, 112*, 843–857.

Nicholson, K. G., & Turner, G. S. (1978). Studies with human diploid cell strain rabies vaccine and human antirabies immunoglobulin in man. *Developments in Biological Standardization, 40*, 115–120.

Nyberg, M., Kulonen, K., Neuvonen, E., Ek-Kommonen, C., Nuorgam, M., & Westerling, B. (1992). An epidemic of sylvatic rabies in Finland—Descriptive epidemiology and results of oral vaccination. *Acta Veterinaria Scandinavica, 33*, 43–57.

Ohmori, K., Masuda, K., Maeda, S., Kaburagi, Y., Kurata, K., Ohno, K., ... Sakaguchi, M. (2005). IgE reactivity to vaccine components in dogs that developed immediate-type allergic reactions after vaccination. *Veterinary Immunology and Immunopathology, 104*(3-4), 249–256. https://doi.org/10.1016/j.vetimm.2004.12.003.

OIE. (2011). Criteria for the use of parenteral and oral immunization of dogs (C. Schumacher). In *Global conference on rabies control: Seoul, 2011.*

OIE. (2012). *Manual of diagnostic tests and vaccines for terrestrial animals (mammals, birds and bees)* (7th ed.). Paris: World Health Organization for Animal Health.

OIE. (2019). *Manual of diagnostic tests and vaccines for terrestial animals*. Chapter 1.1.8 Paris: OIE.

Osada, N., Kohara, A., Yamaji, T., Hirayama, N., Kasai, F., Sekizuka, T., ... Hanada, K. (2014). The genome landscape of the African green monkey kidney-derived vero cell line. *DNA Research, 21*, 673–683.

Plotkin, S. A., Orenstein, W. A., & Offit, P. A. (2008). Rabies vaccines. In *Vaccines* (5th ed., pp. 687–714). Philadelphia: Elsevier, Saunders.

Prehaud, C., Coulon, P., LaFay, F., Thiers, C., & Flamand, A. (1988). Antigenic site II of the rabies virus glycoprotein: Structure and role in viral virulence. *Journal of Virology, 62*(1), 1–7.

Reculard, P. (1996). Cell-culture vaccines for veterinary use. In F. M. Meslin, M. M. Kaplan, & H. Koprowski (Eds.), *Laboratory techniques in rabies* (pp. 314–323). Geneva: WHO.

Ren, J., Sun, L., Yang, L., Wang, H., Wan, M., Zhang, P., ... Wang, L. (2010). A novel canine favored CpG oligodeoxynucleotide capable of enhancing the efficacy of an inactivated aluminum-adjuvanted rabies vaccine of dog use. *Vaccine, 28*(12), 2458–2464. https://doi.org/10.1016/j.vaccine.2009.12.077.

Romberg, J., Lang, S., Balks, E., Kamphuis, E., Duchow, K., Loos, D., ... Jungback, C. (2012). Potency testing of veterinary vaccines: The way from in vivo to in vitro. *Biologicals, 40*(1), 100–106. https://doi.org/10.1016/j.biologicals.2011.10.004.

Rosatte, R. C., Power, M. J., MacInnes, C. D., & Campbell, J. B. (1992). Trap-vaccinate-release and oral vaccination for rabies control in urban skunks, raccoons and foxes. *Journal of Wildlife Diseases, 28*, 562–571.

Rudd, R. J., & Trimarchi, C. V. (1987). Comparison of sensitivity of BHK-21 and murine neuroblastoma cells in the isolation of a street strain rabies virus. *Journal of Clinical Microbiology, 25*(8), 1456–1458.

Rupprecht, C. E., Barrett, J., Briggs, D., Cliquet, F., Fooks, A. R., Lumlertdacha, B., ... Wandeler, A. I. (2008). Can rabies be eradicated? *Developmental Biology (Basel), 131*, 95–121.

Rupprecht, C. E., Briggs, D., Brown, C. M., Franka, R., Katz, S. L., Kerr, H. D., ... CDC (2010). Use of a reduced (4-dose) vaccine schedule for postexposure prophylaxis to prevent human rabies: Recommendations of the advisory committee on immunization practices. *MMWR Recommendations and Reports, 59*, 1–9.

Rupprecht, C. E., Dietzschold, B., Cox, J. H., & Schneider, L. G. (1989). Oral vaccination of raccoons (procyon-lotor) with an attenuated (Sad-B19) rabies virus-vaccine. *Journal of Wildlife Diseases, 25*, 548–554.

Rupprecht, C. E., Hanlon, C. A., Blanton, J., Manangan, J., Morrill, P., Murphy, S., ... Dietzschold, B. (2005). Oral vaccination of dogs with recombinant rabies virus vaccines. *Virus Research, 111*(1), 101–105. https://doi.org/10.1016/j.viruses.2005.03.017.

Rupprecht, C. E., Hanlon, C. A., & Slate, D. (2004). Oral vaccination of wildlife against rabies: Opportunities and challenges in prevention and control. *Developments in Biologicals, 119*, 173–184.

Rupprecht, C. E., Wiktor, T. J., Johnston, D. H., Hamir, A. N., Dietzschold, B., Wunner, W. H., ... Koprowski, H. (1986). Oral immunization and protection of raccoons (Procyon lotor) with a vaccinia-rabies glycoprotein recombinant virus vaccine. *Proceedings of the National Academy of Sciences of the United States of America, 83*(20), 7947–7950.

Saba, C. F. (2017). Vaccine-associated feline sarcoma: Current perspectives. *Veterinary Medicine (Auckland), 8*, 13–20. https://doi.org/10.2147/VMRR.S116556. eCollection 2017.

Sadeghi, M., Moallem, S. A., Yousefi-Abdolmaleki, E., & Montazeri, M. (2015). The rabies early death phenomenon: A report of ineffective administration of rabies vaccine during symptomatic disease. *Indian Journal of Critical Care Medicine, 19*(7), 422–424. https://doi.org/10.4103/0972-5229.160292.

Sakaguchi, M., Hori, H., Hattori, S., Irie, S., Imai, A., Yanagida, M., ... Inouye, S. (1999). IgE reactivity to alpha1 and alpha2 chains of bovine type 1 collagen in children with bovine gelatin allergy. *The Journal of Allergy and Clinical Immunology, 104*(3 Pt 1), 695–699.

Schnell, M. J., Mebatsion, T., & Conzelmann, K. K. (1994). Infectious rabies viruses from cloned cDNA. *The EMBO Journal, 13*(18), 4195–4203.

Shi, L. T., Yu, Y. X., Liu, J. H., Tang, J. R., Wu, X. H., Cao, S. C., ... Dong, G. M. (2010). Analysis of full-length gene sequence of a rabies vaccine strain CTN-1 for human use in China. *Bing Du Xue Bao, 26*(3), 195–201.

Sillero-Zubiri, C., Marino, J., Gordon, C. H., Bedin, E., Hussein, A., Regassa, F., ... Fooks, A. R. (2016). Feasibility and efficacy of oral rabies vaccine SAG2 in endangered Ethiopian wolves. *Vaccine, 34*(40), 4792–4798. https://doi.org/10.1016/j.vaccine.2016.08.021.

Steck, F., Wandeler, A., Bichsel, P., Capt, S., & Schneider, L. (1982). Oral immunisation of foxes against rabies. A field study. *Zentralblatt für Veterinärmedizin. Reihe B, 29*(5), 372–396.

Stohr, K., & Meslin, F. M. (1996). Progress and setbacks in the oral immunisation of foxes against rabies in Europe. *The Veterinary Record, 139*(2), 32–35.

Swanson, M. C., Rosanoff, E., Gurwith, M., Deitch, M., Schnurrenberger, P., & Reed, C. E. (1987). IgE and IgG antibodies to beta-propiolactone and human serum albumin associated with urticarial reactions to rabies vaccine. *The Journal of Infectious Diseases, 155*(5), 909–913.

Taffs, R. E. (2001). Potency tests of combination vaccines. *Clinical Infectious Diseases, 33*(Suppl 4), S362–S366.

Tao, L., Ge, J., Wang, X., Wen, Z., Zhai, H., Hua, T., … Bu, Z. (2011). Generation of a recombinant rabies Flury LEP virus carrying an additional G gene creates an improved seed virus for inactivated vaccine production. *Virology Journal, 8*, 454. https://doi.org/10.1186/1743-422X-8-454.

Toovey, S. (2007). Preventing rabies with the Verorab vaccine: 1985-2005 Twenty years of clinical experience. *Travel Medicine and Infectious Disease, 5*(6), 327–348.

Verch, T., Trausch, J. J., & Shank-Retzlaff, M. (2018). Principles of vaccine potency assays. *Bioanalysis, 10*(3), 163–180. https://doi.org/10.4155/bio-2017-0176.

Vincent-Falquet, J. C., Peyron, L., Souvras, M., Moulin, J. C., Tektoff, J., & Patet, J. (1989). Qualification of working cell banks for the Vero cell line to produce licensed human vaccines. *Developments in Biological Standardization, 70*, 153–156.

Vitrology. (2008). Vitrology Biotech Retrieved, February 15 2012, from www.vitrologybiotech.com.

Vos, A., Kretzschmar, A., Ortmann, S., Lojkic, I., Habla, C., Müller, T., … Schuster, P. (2013). Oral vaccination of captive small Indian mongoose (Herpestes auropunctatus) against rabies. *Journal of Wildlife Diseases, 49*(4), 1033–1036. https://doi.org/10.7589/2013-02-035.

Wang, Z., Sun, Y., Wu, X., Carroll, D. S., Lv, W., You, L., … Meng, S. (2018). Development of a relative potency test using ELISA for human rabies vaccines. *Biologicals, 55*, 59–62. https://doi.org/10.1016/j.biologicals.2018.06.003.

Warrell, M. J., & Warrell, D. A. (2015). Rabies: The clinical features, management and prevention of the classic zoonosis. *Clinical Medicine (London, England), 15*(1), 78–81. https://doi.org/10.7861/clinmedicine.14-6-78.

WHO. (1998). *Requirements for the use of animal cells as in vitro substrates for the production of biologicals WHO technical report series 878* (pp. 19–52). Geneva: WHO.

WHO. (2005). *WHO Expert committee on biological standardization* (p. 154). Geneva: WHO.

Wiedermann-Schmidt, U., & Maurer, W. (2005). Hilfs- und Zusatzstoffe von ImpfstoffenMedizinischeRelevanz. *Wiener KlinischeWochenschrift, 117*, 510.

Wieten, R. W., Leenstra, T., van Thiel, P. P., van Vugt, M., Stijnis, C., Goorhuis, A., & Grobusch, M. P. (2013). Rabies vaccinations: Are abbreviated intradermal schedules the future? *Clinical Infectious Diseases, 56*(3), 414–419. https://doi.org/10.1093/cid/cis853.

Wijaya, L., Tham, C. Y. L., Chan, Y. F. Z., Wong, A. W. L., Li, L. T., Wang, L. F., … Low, J. G. (2017). An accelerated rabies vaccine schedule based on toll-like receptor 3 (TLR3) agonist PIKA adjuvant augments rabies virus specific antibody and T cell response in healthy adult volunteers. *Vaccine, 35*(8), 1175–1183. https://doi.org/10.1016/j.vaccine.2016.12.031.

Wiktor, T. J., Plotkin, S. A., & Koprowski, H. (1978). Development and clinical trials of the new human rabies vaccine of tissue culture (human diploid cell) origin. *Developments in Biological Standardization, 40*, 3–9.

Wolff, M., & Reichl, U. (2011). Downstream processing of cell culture-derived virus particles. *Expert Review of Vaccines, 10*(10), 1451–1475.

Wood, R. A., Berger, M., Dreskin, S. C., Setse, R., Engler, R. J., Dekker, C. L., … Hypersensitivity Working Group (2008). An algorithm for treatment of patients with hypersensitivity reactions after vaccines. *Pediatrics, 122*, e771–e777.

World Health Organization. (2007). Recommendations for inactivated rabies vaccine for human use produced in cell substrates and embryonated eggs. In *World Health Organization technical report series. WHO expert consultation on rabies* (pp. 83–132). Geneva: World Health Organization.

World Health Organization. (2013). WHO expert consultation on rabies. Second report. *World Health Organization technical report series* (982, pp. 1–139).

Wunderli, P. S., Dreesen, D. W., Miller, T. J., & Baer, G. M. (2003). Effect of heterogeneity of rabies virus strain and challenge route on efficacy of inactivated rabies vaccines in mice. *American Journal of Veterinary Research, 64*(4), 499–505.

Yakobson, B., Goga, I., Freuling, C. M., Fooks, A. R., Gjinovci, V., Hulaj, B., … Müller, T. (2014). Implementation and monitoring of oral rabies vaccination of foxes in Kosovo between 2010 and 2013—An international and intersectorial effort. *International Journal of Medical Microbiology, 304*(7), 902–910. https://doi.org/10.1016/j.ijmm.2014.07.009.

Yang, D. K., Kim, H. H., Choi, S. S., Lee, S. H., & Cho, I. S. (2016). A recombinant rabies virus (ERAGS) for use in a bait vaccine for swine. *Clinical and Experimental Vaccine Research, 5*(2), 169–174. https://doi.org/10.7774/cevr.2016.5.2.169.

Yu, P., Huang, Y., Zhang, Y., Tang, Q., & Liang, G. (2012). Production and evaluation of a chromatographically purified Vero cell rabies vaccine (PVRV) in China using microcarrier technology. *Human Vaccines & Immunotherapeutics, 8*, 1230–1235.

Yu, P., Yan, J., Wu, W., Tao, X., Lu, X., Liu, S., & Zhu, W. (2018). A CpG oligodeoxynucleotide enhances the immune response to rabies vaccination in mice. *Virology Journal, 15*(1), 174. https://doi.org/10.1186/s12985-018-1089-1.

Zhang, Y., Zhang, S., Li, W., Hu, Y., Zhao, J., Liu, F., … Li, L. (2016). A novel rabies vaccine based-on toll-like receptor 3 (TLR3) agonist PIKA adjuvant exhibiting excellent safety and efficacy in animal studies. *Virology, 489*, 165–172.

Zienius, D., Pridotkas, G., Lelesius, R., & Sereika, V. (2011). Raccoon dog rabies surveillance and post-vaccination monitoring in Lithuania 2006 to 2010. *Acta Veterinaria Scandinavica, 53*, 58. https://doi.org/10.1186/1751-0147-53-58.

Zinsstag, J., Durr, S., Penny, M. A., Mindekem, R., Roth, F., Menendez Gonzalez, S., … Hattendorf, J. (2009). Transmission dynamics and economics of rabies control in dogs and humans in an African city. *Proceedings of the National Academy of Sciences of the United States of America, 106*(35), 14996–15001. https://doi.org/10.1073/pnas.0904740106.

CHAPTER 15

Next generation of rabies vaccines

Hildegund C.J. Ertl
The Wistar Institute, Philadelphia, PA, United States

15.1 Introduction

Rabies is caused by lyssaviruses, which are simple negative-stranded RNA viruses that encode five structural proteins, i.e., the nucleoprotein (NP), the glycoprotein (G), the phosphoprotein (P), the matrix protein (M), and the polymerase (L). Rabies virus, a member of phylogroup I of lyssaviruses, is the most common cause of the disease, which can also, albeit very rarely, be caused by other lyssaviruses, including those of phylogroups II and III, which are antigenically distinct from phylogroup I. Correlates of protection are well defined and virus-neutralizing antibodies (VNAs) present in serum at titers equal to or above 0.5 international units (IU)/mL are assumed to provide protection (Wunderli, Shaddock, Schmid, Miller, & Baer, 1991), although this only applies to infections with rabies virus, which forms the basis of commercially available vaccines and rabies immunoglobulin (RIG) and not necessarily to infections with other lyssaviruses. VNAs are solely directed against the G protein. For postexposure prophylaxis (PEP) of severe exposures (World Health Organization [WHO] classification III), active immunization has to be combined with passive immunization with RIG. Rabies has the highest fatality rate of all known pathogens and, with a small number of exceptions, humans that develop symptomatic rabies will inevitably die. Although efficacious vaccines are available, mortality due to rabies remains high and the disease claims the lives of an estimated 55,000–60,000 humans each year. Worldwide rabies virus is mainly transmitted by dogs and fatal infections are disproportionately high in developing countries and in children (https://www.who.int/ith/diseases/rabies/en/) (Esposito, Picciolli, Semino, & Principi, 2013).

Although ignorance of appropriate treatment such as wound cleaning and disinfection or use of alternative ineffective treatments such as traditional herbal medicines may contribute to the high death rate, economic factors play a major role. Current vaccines require several injections given sequentially. Administration of PEP is thus time consuming and may result in loss of wages combined with travel costs and expenses for medical treatment.

RIG, which is required for protection upon severe exposure, is especially expensive and in very short supply, and thus, habitually underutilized. The development of alternative, less expensive vaccines, which would ideally reduce the need for RIG, is thus warranted. Such vaccines could replace the vaccines currently used for PEP. Alternatively, provided they are inexpensive and induce sustained immunity after a single dose, they could be used for childhood preexposure vaccination (Pre-P) in highly endemic areas. This possible application is being explored in the Amazon Basin in Peru to prevent the raising human death toll due to vampire bat-transmitted rabies virus (Estrategia nacional de zoonosis-MINSA, 2015).

15.2 Current rabies vaccine regimens

Rabies vaccinations follow guidelines formulated by expert panels of the WHO and finetuned by local regulatory agencies such as the Centers for Disease Control and Prevention in the US (World Health Organization, 2018).

Pre-P is only given to individuals at high risk for exposure such as veterinarians, animal handlers, rabies laboratory workers, and international travelers who are likely to come in contact with animals in countries where rabies is prevalent, and vaccine or RIG may not be readily available. The initial immunization used to consist of three doses given on days 0, 7, and 21 or 28. This has recently been changed (Kessels et al., 2017) to two doses given either intramuscularly (IM) or at a lower amount intradermally (ID) into 2 sites on days 0 and 7. A three-dose regimen is recommended for individuals with immunodeficiencies. For humans at continued high risk, periodic testing for antibody titers at intervals of 6–24 months is recommended and a booster immunization with a single dose of the vaccine is indicated once titers fall below 0.5 international units (IU)/mL, which is considered to be the minimal titer of rabies virus-specific antibodies that reliably represents seroconversion in humans and animals, which is indicative of protective immunity.

In most cases, humans are vaccinated after exposure. Initial thorough cleaning and disinfecting of the bite wound are essential to remove remaining virus-containing saliva. Immunization should be initiated as soon as possible in previously unvaccinated individuals. In the US, vaccine is given IM into one site on days 0, 3, 7 followed by a fourth dose anytime between days 14–28. As an alternative IM regimen WHO recommends vaccinations to two sites on days 0 followed by one site injections on days 7 and 21. The less costly ID immunization should be given into two sites on days 0, 3, and 7. In case of a severe exposure, the first dose is combined with RIG given into and around the bite site. The recommended dose of human RIG is 20 mg/kg; equine RIG should be used at 40 mg/kg. Higher doses should not be used as they may interfere with active immunization. Some guidelines have been revised concerning whether to give leftover RIG, which could not be injected into the wound, into an intramuscular site (Hampson et al., 2018) (see Chapter 16). Previously vaccinated individuals require a booster immunization. WHO recommends several types of booster immunizations for previously vaccinated individuals that are exposed to rabies virus: A 1-site ID vaccine on days 0 and 3, a 4-site ID vaccine on day 0, or a 1-site IM vaccine on days 0 and 3. RIG should not be given for severe exposures of previously vaccinated humans.

15.3 Incidence and risk for rabies and vaccine failures

Due to mandatory dog rabies vaccination combined, in part, with wildlife immunization programs, rabies has become rare in the Americas and Europe but remains common in Asia and Africa with an estimated annual mortality rate of 55,000–60,000 humans per year. Most cases (99%) are caused by rabid dogs and approximately 40% of the deaths affect children below the age of 15. Rabies occurs in more than 150 countries worldwide and each year more than 15 million humans receive rabies PEP. It has been estimated that in some countries up to 40% of humans require PEP during childhood, which suggests that in such areas preventative childhood vaccination should be considered to reduce the incidence of rabies and the cost of PEP.

Appropriate wound cleaning and the full course of PEP in nearly all cases prevent the development of symptomatic rabies. Bites by bats, which often cause minor lesions, can be overlooked and thus left untreated; they are the major cause of rabies in the US (Blanton, Palmer, Dyer, & Rupprecht, 2011). In developing countries, the cost of PEP, which in Africa and Asia exceeds the monthly income of an average family, discourages its use. Vaccine failures after full PEP given in a timely fashion correctly according to WHO recommendations are rare to nonexistent in healthy individuals but have been reported in humans with immunodeficiencies (Tantawichien, Jaijaroensup, Khawplod, & Sitprija, 2001).

15.4 Correlates of protection

Most of the rabies virus proteins such as the G, N, and P proteins contain epitopes for recognition by $CD4^+$ or $CD8^+$ T cells (Desmézières et al., 1999; Ertl et al., 1989; Larson, Wunner, Otvos, & Ertl, 1991). VNAs are exclusively directed to the G protein. The 65–67 kd G protein, which contains ~500 amino acids, has 2–6 potential sites for N-glycosylation, 12–16 conserved cysteine residues, 2 to 3 hydrophobic heptad repeats, a transmembrane domain, and a short cytoplasmic domain (Schneider & Diringer, 1976). Its structure has not yet been resolved by X-ray crystallography, although it is known that the protein is displayed in form of noncovalently linked trimers on the virion's surface. The G protein is a major determinant for pathogenicity and single amino acid substitution can strongly attenuate the virulence of rabies virus. The G protein also attaches to cellular receptors and as such determines the virus' tropism for neurons. Most importantly, the G protein expresses, depending on the strain, 3–5 conformation-dependent antibody-binding sites each with multiple partially overlapping epitopes for VNAs (Luo et al., 1998). Additional linear epitopes have been defined (Mansfield et al., 2008). Protection against rabies virus strongly correlates with titers of circulating VNAs and titers of 0.5 IU/mL determined by a validated and standardized assay such as a rapid immunofluorescent focus inhibition test (RFFIT) (Gelosa & Borroni, 1990) suffice to prevent an infection. $CD8^+$ T cells do not contribute to protection, whereas $CD4^+$ T cells are indirectly involved by their essential role in providing help for induction of long-lived plasma cells producing affinity-matured antibodies (Xiang, Knowles, McCarrick, & Ertl, 1995).

Molecular analyses suggest that circulating rabies viruses have evolved within the last 1500 years (Nadin-Davis & Real, 2011), although ancestral rabies viruses may have evolved in animals other than mammals millions of years ago. Rabies virus, like all negative-sense RNA viruses, shows high mutation rates and genetically distinct isolates characterize outbreaks in distinct regions or species. Accordingly, most monoclonal neutralizing antibodies fail to inhibit a wide range of rabies virus isolates and cross-reactive monoclonal antibodies appear to be rare (Müller et al., 2009). Notwithstanding, rabies vaccines based on a limited number of strains such as the challenge virus standard (CVS) or Pasteur virus (PV) strain and RIG, which is induced by the same vaccine strains, are efficacious worldwide against rabies virus and other phylogroup I lyssaviruses demonstrating that vaccines to rabies virus do not have to be updated periodically and do not require adjustments to regionally prevalent strains. Current vaccines against rabies virus are not efficacious against phylogroup II and III lyssaviruses, which are all transmitted by bats. Human cases caused by these viruses are very rare with a total of 12 reported human cases between 1970 and 2007 (Evans, Horton, Easton, Fooks, & Banyard, 2012). It is thus unlikely that vaccine manufacturers will be motivated to undertake the costly endeavor to change current rabies vaccines into pan-lyssavirus vaccines.

15.5 Novel vaccines to rabies

The continued high incidence of rabies in developing countries, the economic burden of Pre-P and PEP, and the globally limited availability of RIG warrant development of novel cost-effective rabies vaccines. Just about every vaccine prototype ranging from peptides to plant-derived vaccines has been evaluated and many showed efficacy in experimental animals (Table 15.1), but only a few progressed into clinical trials thus far (Table 15.2). In this chapter, we will briefly discuss some of the more promising prototypes. Monoclonal antibodies either derived from human B cells or from mouse hybridomas, the latter genetically modified to humanize the immunoglobulin constant region, are being developed to eventually replace RIG (Smith, Wu, Franka, & Rupprecht, 2011) and are not discussed. Vaccines are typically divided into live-attenuated and inactivated vaccines. This division does not suffice for most of the modern experimental vaccines, which are either based on genetic attenuation or modifications that enhance immunogenicity or on individual viral proteins, so-called subunit vaccines, which in case of rabies vaccines are typically based on the viral G protein or parts thereof. Subunit vaccines can consist of synthetic peptides or protein fragments carrying crucial B cell epitopes. Full-length proteins can be isolated from a variety of expression systems, including mammalian cells, insect cells, yeast cells, or plants. Subunit vaccines can also reflect genetic vaccines, which encode the rabies virus G protein. In addition, with advances in our knowledge of the interplay between innate and adaptive immune responses and the identification of pathogen recognition receptors as crucial initiators of inflammatory reactions (Coban, Ishii, & Akira, 2009), which are a prerequisite for activation of T and B cells, novel adjuvants have been developed and tested in conjunction with traditional rabies vaccines. Such adjuvants can be added to a vaccine as it is being assessed in clinical trials or their sequences can be directly incorporated into a genetic vaccine. Traditional rabies vaccines are delivered IM, SC, and ID to humans. Alternative routes of immunization such as intranasal or oral routes are being considered.

TABLE 15.1 Preclinical results with novel rabies vaccines.

Vaccines	Animal model	Number of doses	Immunogenicity	Efficacy	References
Genetically engineered attenuated rabies virus					
P protein-deleted	Mice, nonhuman primates	1 or 2	Accelerated and enhanced compared to traditional vaccines	Protection (Pre-P)	Cenna et al. (2009), Morimoto, Shoji, and Inoue (2005), Shoji et al. (2004)
M protein-deleted	Mice, nonhuman primates	1	Accelerated and enhanced compared to traditional vaccines	Protection (Pre-P)	Cenna et al. (2009), Ito et al. (2005)
Genetically engineered inactivated rabies virus					
Inactivated rabies virus with 2G protein copies	Mice, dogs	1	Enhanced compared to traditional vaccines	Protection (Pre-P)	Liu et al. (2014)
Traditional rabies vaccines with adjuvant					
Alum	Mice	1	Controversial results	n.t.	Lin and Perrin (1999), Shi et al. (2018)
CpGODN	Mice	3	Similar compared to traditional vaccines with fewer doses	n.t.	Wang et al. (2008)
ISCOMATRIX	Nonhuman primates	3–8	Increased VNA titers compared to traditional vaccines	n.t.	DiStefano et al. (2013)
TLR-9 agonist	Nonhuman primates	3–8	Increased VNA titers compared to traditional vaccines	n.t.	DiStefano et al. (2013)
Peptide vaccines					
Linear epitope vaccine in Freunds' Adjuvant	Goats	3	Low VNA titers	n.t.	Niederhäuser et al. (2008)
Mimotope vaccine	Mice	4	Noninferiority to the commercial vaccine	Parial protection (Pre-P)	Houimel and Dellagi (2009)
Multiepitope vaccine linked to gp69	Mice, dogs	3	Inferior to commercial vaccine	n.t.	Niu et al. (2016)

Continued

TABLE 15.1 Preclinical results with novel rabies vaccines—cont'd

Vaccines	Animal model	Number of doses	Immunogenicity	Efficacy	References
Protein vaccines					
Produced in mammalian cells	Mice	2	Induction of antibodies	Protection (Pre-P)	Fontana, Kratje, Etcheverrigaray, and Prieto (2015)
Produced in insect cells	Mice	2	Induction of VNAs	Protection (Pre-P)	Ramya et al. (2011)
Produced in tobacco plants	Mice	1–4	Induction of antibodies	Protection (Pre-P)	Ashraf et al. (2005)
Produced in maize	Sheep (oral)	1	Induction of VNAs	Protection (Pre-P)	Loza-Rubio et al. (2012)
Pseudotyped baculovirus	Mice	2	Induction of VNAs	Parial protection (Pre-P)	Wu et al. (2014)
Pseudotyped Newcastle disease virus	Dogs, Cats	3	Induction of VNAs	Protection (Pre-P)	Ge et al. (2011)
Genetic vaccines					
DNA vaccines with or without adjuvants or a viral vector boost	Mice, Nonhuman Primates, Rabbits, Horses, Dogs, Cats	1–5	Induction of VNAs	Partial or Complete Protection (Pre-P, PEP)	Fischer, Minke, Dufay, Baudu, and Audonnet (2003), Kaur, Saxena, Rai, and Bhatnagar (2010), Lodmell et al. (2002), Osorio et al. (1999), Xiang and Ertl (1995), Xiang et al. (1994)
RNA vaccine	Mice, Pigs	2–3	Induction of VNAs	Protection (Pre-P)	Schnee et al. (2016)
Recombinant vaccinia virus	Multiple	1	Induction of VNAs	Protection (Pre-P)	Weyer, Rupprecht, and Nel (2009)
Recombinant modified vaccinia ankara	Mice	1	Induction of VNAs	Partial protection (Pre-P)	Weyer et al. (2007)
Replication-defective adenovirus vectors	Mice, Nonhuman primates	1	Induction of VNAs	Protection (Pre-P)	Xiang et al. (2006, 2014)

n.t. = not tested.

TABLE 15.2 Clinical results with novel rabies vaccines.

Vaccine prototype	Amount of vaccine per dose	Timing of vaccination	Clinical trial stage	Immune response	Adverse events	References
Inactivated rabies vaccine + TLR-3 agonist (Pika vaccine)	2 IU	days 0, 3, 7 (2-2-1)	Phase I and II completed, Phase III ongoing	Noninferior to a 4 dose regimen (1-1-1-1) of a higher amount of Rabipur	Mild local and systemic reactions	Wijaya et al. (2017)
Insect cell-derived G protein-based VLPs	(?)	days 0, 3, 7	Phase I and II completed, Phase III ongoing	Unknown	Unknown	Unpublished
Modified spinach expressing epitopes of the G and N protein	20–150 g of spinach, orally	days 0, 14, 28	Phase I completed	Antibodies but no VNA titers till after a boost with a commercial vaccine	None	Yusibov et al. (2002)
RNA vaccine	80–640 μg	days 0, 7, 28 or days 0, 7, 28, 56	Phase I completed	low VNA titers in some individuals, poor responses after boost	Moderate-to-severe local and systemic reactions	Alberer et al. (2017)

15.5.1 Live attenuated rabies virus vaccines

In general, live vaccines are more immunogenic than killed vaccines, which relates to higher antigenic loads for stimulation of adaptive immune responses. Reverse genetics allows for attenuation of rabies virus through manipulation of its genome and thus for the development of potentially safe, attenuated live rabies vaccines. Rabies virus can be attenuated by several means. Deletion of the P protein, which serves as a cofactor for the viral polymerase, prevents viral replication (Cenna et al., 2008; Shoji et al., 2004). Furthermore, the P protein subverts immune responses by inhibiting IFN type I signaling thus attenuating rabies virus-specific immune responses. P protein-deleted rabies viruses, although they express markedly less G protein on the surface of infected cells compared to wild-type virus, induce more potent immune responses compared to inactivated virus (Cenna et al., 2009). Responses can be further improved by incorporating a second G protein gene in between the genes encoding the M protein and L protein (Cenna et al., 2008). The P protein-deleted virus lacks pathogenicity even after direct intracerebral injection into immunodeficient mice. Growth kinetics of recombinant virus show a strong reduction in titers, which may provide problems for cost-effective manufacturing.

M protein-deleted rabies virus, which shows growth kinetics similar to those of P protein-deleted rabies virus, expresses higher levels of G protein on in vitro infected cells and accordingly induces more potent VNA responses in mice (Cenna et al., 2009; Ito et al., 2005). In nonhuman primates, the M protein-deleted rabies virus given at two doses induces higher antibody, including VNA responses compared to two doses of HDCV (Cenna et al., 2009). M protein-deleted virus fails to cause disease in immunodeficient mice.

Although vaccines based on attenuated rabies virus may be suitable for Pre-P or PEP, several problems remain to be addressed. Foremost, reduced growth may cause manufacturing problems and thus render the vaccine too costly. Also, it is likely that even highly attenuated live rabies vaccines that show no pathogenicity even upon intracerebral inoculation into immunodeficient mice may not appeal to the public or fail to gain regulatory approval.

15.5.2 Inactivated genetically modified rabies vaccine

A rabies virus encoding two copies of the G protein was developed. This virus had very favorable growth characteristics but retained virulence in mice. Nevertheless, upon inactivation, the double G-protein virus outperformed a traditional rabies vaccine by inducing higher antibody titers and achieving superior protection at low vaccine doses (Liu et al., 2014). This vaccine may thus be suited to replace current vaccines for Pre-P, and once studies show protection in animals that are vaccinated after challenge for PEP.

15.5.3 Inactivated adjuvanted traditional rabies vaccines

Inactivated rabies vaccines are not potently immunogenic and, therefore, require multiple doses until protective VNA titers are achieved. Rabies vaccines used in the US, i.e., Imovax Rabies, an HDCV produced by Sanofi Pasteur and RabAvert, a purified chick embryo cell vaccine from Novartis, do not contain adjuvant such as alum. In fact, studies in experimental animals indicate that addition of alum does not improve the vaccine's immunogenicity (Lin & Perrin, 1999). Nevertheless, these results were contradicted by a subsequent study (Shi et al., 2018). Formulations containing rabies vaccine mixed with CpG-oligodeoxynucleotides, which trigger innate immune responses through Toll-like receptor (TLR) 9, showed enhanced and accelerated VNA responses in animals (Wang et al., 2008). Two adjuvants, ISCOMATRIX and IMO-2170, were tested with a rabies vaccine in nonhuman primates (DiStefano et al., 2013), where they were shown to allow for dose sparing. Another adjuvant, a TLR-3 agonist, was tested in human volunteers with no prior immunity to rabies virus (Wijaya et al., 2017). In a phase III trial following promising phase I and II trial results, the adjuvanted rabies vaccine, called PIKA vaccine, used at 2 international units (IU) per dose of the inactivated rabies vaccine was compared to Novartis' Rabipur at 7.4 IU per dose. The PIKA vaccine was given into 2 sites on days 0 and 3 and into 1 site on day 7, while Rabipur was given into 1 site on days 0, 3, 7, and 14. The PIKA vaccine was well tolerated. The trial showed noninferiority of the PIKA vaccine. Additional trials will be needed to assess if the PIKA vaccine performs better than the control vaccine if both are given at similar doses using comparable regimens.

15.5.4 Peptide vaccines

Linear B-cell epitopes have been identified in the rabies virus G proteins that can be expressed by synthetic peptide vaccines. Peptide vaccines, although exceptionally safe, are commonly poorly immunogenic and induce very narrow B-cell responses. Indeed, one study in goats with a peptide expressing a linear epitope of the rabies virus G protein only induced low titers of VNAs that failed to neutralize an escape mutant (Niederhäuser et al. (2008)). Mimotopes of the rabies virus G protein isolated from a random constrained hexapeptide phage display library and corresponding to antigenic site III induced a modest VNA response in mice. This study did not address the potential problem of viral escape or the overall breadth of the antibody response (Houimel & Dellagi, 2009). A multiepitope vaccine linked to a canine heatshock protein was immunogenic in mice and dogs but only protected ~75% of mice against challenge with rabies virus (Niu et al., 2016).

G and N protein peptides expressed by the coat protein of alfalfa mosaic virus grown in tobacco plants or spinach induced an immune response in mice (Yusibov et al., 2002). Spinach-derived virus when tested in rabies vaccine-immune humans elicited a weak recall response. Immunization of mice with spinach expressing peptides of the G protein was also poorly immunogenic and provided limited protection against subsequent challenge of experimental animals (Yusibov et al., 2002). Overall, the high variability of the viral G protein contraindicates vaccines that induce very narrow antibody responses to only one or a few epitopes.

15.5.5 Protein vaccines

The rabies virus G protein has been purified from a number of expression systems including those based on mammalian cells, insect cells, yeast, and plants. Yeast-derived protein failed to induce protective immune responses in mice (Klepfer et al., 1993), most likely reflecting poor folding of the final product. Mammalian cell- and baculovirus-infected insect cell-derived G proteins were shown to be immunogenic in mice (Fu et al., 1993). One baculovirus-derived G protein that spontaneously forms micelles (nanoparticles) has undergone phase I and II testing in humans. A phase III trial (CTRI/2016/08/007137) (Desai, 2016) is being conducted that compares 3 doses of the protein vaccine given at 50 µg/dose on days 0, 3 and 7 to 5 doses of Rabipur (2.5 IU/dose) given on days 0, 3, 7, 14, and 28. One would think that data from the initial trials were promising enough to warrant a phase III trial, but this remains an assumption as results for either of the trials have not yet been published.

G protein has been expressed in plants such as maize, spinach, tobacco leaves, and carrots, which upon ingestion of raw material or injection of purified protein induced a detectable VNA response and complete or partial protection against challenge in mice (Ashraf et al., 2005; Loza-Rubio, Rojas, Gómez, Olivera, & Gómez-Lim, 2008; Modelska et al., 1998; Rojas-Anaya, Loza-Rubio, Olivera-Flores, & Gomez-Lim, 2009). Again, the structural complexity of the rabies virus G protein, the requirement for its correct folding to elicit broad VNA responses, and the need for extensive purifications prior to injection into humans make it unlikely that protein-based rabies vaccines will replace current vaccines. The use of edible vaccines such as spinach or maize genetically engineered to express G protein remains an attractive option that continues to face challenges.

A small clinical trial was undertaken with a rabies vaccine based on spinach leaves that had been infected with alfalfa mosaic virus modified to express antigenic parts of the G and N protein.

Humans who had or had not been previously been vaccinated with a traditional rabies vaccine were fed raw spinach leaves 3 times in 14-day intervals. None of the individuals developed rabies virus-specific VNAs (Yusibov et al., 2002).

As a variation of traditional protein vaccines, viral vectors have been modified to express the rabies G protein on their surface. This type of pseudotyping was explored with baculovirus (Wu et al., 2014) and Newcastle disease virus (Ge et al., 2011), and both were shown to be immunogenic in experimental animals. The approach has advantages such as relative ease of production and purification of the pseudotyped viruses, the presence of pathogen-associated molecular patterns within the virus that obliviate the need for adjuvants and depending on the type of pseudotyped virus a potential for replication of the vaccine within the host. Although the latter is appealing for it would allow for dose sparing, it also poses risk factors.

15.5.6 Genetic vaccines

Genetic vaccines to rabies virus that encode the G protein express the immunogen once the vaccine carrier has transduced or infected a cell. The advantage of genetic vaccines is that they are highly versatile. Furthermore, the mammalian expression system allows for faithful expression of the G protein and its correct folding and, depending on the system, genetic vaccines can be highly immunogenic and cost effective. The disadvantage of genetic vaccines is that production of immunogenic levels of the transgene product takes time. The incubation period of rabies varies tremendously ranging from a few days to several years. Upon severe exposures, which typically result in shorter incubation times, rapid induction of protective VNA titers is of the essence. This may not be achieved by genetic vaccines, which nevertheless provide attractive alternatives for Pre-P.

15.5.7 DNA vaccines

DNA vaccines are very easy to construct and are well tolerated in humans. Numerous studies with DNA vaccine to rabies virus reported induction of protective VNA titers using various route of immunizations in experimental animals ranging from mice, horses, rabbits, cats, and dogs to monkeys (Bahloul, Jacob, Tordo, & Perrin, 1998; Bahloul et al., 2006; Lodmell, Parnell, Bailey, Ewalt, & Hanlon, 2001; Lodmell, Parnell, Weyhrich, & Ewalt, 2003; Tesoro Cruz et al., 2008; Tesoro Cruz, Hernández González, Alonso Morales, & Aguilar-Setién, 2006; Xiang et al., 1995, 1994). Some studies reported protection with DNA vaccines given to already-infected animals (Bahloul et al., 2003; Lodmell, Parnell, Bailey, Ewalt, & Hanlon, 2002; Tesoro Cruz et al., 2008). Unfortunately, in humans, DNA vaccines were found to be poorly immunogenic. Addition of genetic or traditional adjuvants (Garg, Kaur, Saxena, Prasad, & Bhatnagar, 2017; Xiang & Ertl, 1995), prime boost regimens in which DNA vaccines are typically used for priming followed by a booster immunization with a recombinant viral vector (Lodmell & Ewalt, 2001) or novel delivery methods such an injection of DNA followed by electroporation (Sardesai & Weiner, 2011), increase the immunogenicity of DNA in animals and for some antigens also in humans. To what degree such modifications affect the vaccine's safety remains to be investigated in more depth. Prime-boost regimens, although highly effective, may not be suited for PEP and their cost effectiveness for preventative vaccination is unlikely. The suitability of electroporation, which increases immune responses by enhancing transduction rates, for immunization in developing countries remains to be explored.

Efficacy was also reported for Sindbis virus-based DNA vaccines (Saxena et al., 2008), which unlike conventional DNA vaccine generate self-replicate RNA transcripts, and thus achieve superior protein expression levels.

15.5.8 RNA vaccines

RNA vaccines are another form of "naked" genetic vaccines. An mRNA vaccine encoding the rabies virus G protein was tested in human volunteers. The vaccine was given at escalating doses ID or IM by needle-syringe or with a special needle-free device. Some individuals were boosted 1 year later. The majority of patients reported systemic adverse events after vaccination with 12% reporting serious adverse events. Only 1 of the 42 individuals who received the vaccine by needle-syringe achieved titers of or above 0.5 IU, while most but not all achieved such titers when given the highest dose of the mRNA vaccine by the needle-free injection devise. By 1 year after vaccination, none of the individuals who had been primed with the needle-free devise and had been recalled for a boost had titers ≥ 0.5 IU and most had no detectable titers. After the boost using the same vaccine with the same injection procedure, only 57% of individuals developed titers ≥ 0.5 IU.

In summary, the mRNA rabies vaccine performed poorly in humans and it is not clear if it can be improved to achieve protective and sustained antibody titers to this highly immunogenic virus.

15.5.9 Viral vector vaccines

Viral vector vaccines carry an expression cassette encoding the vaccine antigen within their genome. Viral vector vaccines are by definition infectious vaccines as production of the vaccine antigen is achieved in situ upon infection of cells. Some viral vectors are based on attenuated viruses such as vaccinia virus or Modified Vaccinia Ankara (MVA) while others, such as adenoviral vectors, are genetically altered to render them replication defective. The most commonly explored viral vectors for rabies G protein are based on attenuated poxviruses or different strains of E1-deleted and, hence, replication-defective adenoviruses.

15.5.10 Recombinant poxviruses

A vaccinia virus expressing the full-length G protein of the Evelyn Rokitnicki Abelseth (ERA) strain was one of the first viral vectors that was produced and tested. The vaccine termed VR-G showed adequate efficacy in animals after oral administration and is used for vaccination of wildlife animals (Brochier et al., 1996). Due to its high reactogenicity, it is not suited for use in humans. Vectors based on the more attenuated MVA are less immunogenic and fail to elicit an immune response after oral application (Weyer, Rupprecht, Mans, Viljoen, & Nel, 2007).

15.5.11 Recombinant adenoviruses

In most adenoviral vaccine vectors derived from human or simian serotypes, the E1 domain, which encodes polypeptides essential for viral replication, are deleted, thus rendering the virus replication defective. An expression cassette is then cloned into the deleted E1 domain. Alternatively, a foreign gene can be cloned into the E3 domain, which encodes polypeptides that are not essential for viral replication. E3-deleted vectors remain replication competent. Adenoviruses are ubiquitous DNA viruses, which can be found in many species, although individual serotypes are species specific. Fifty-one different serotypes that can infect humans have been identified serologically; additional types 52–68 were characterized by genomic sequencing. Twenty-five serotypes have been isolated from simians, nine from bovines, six from sheep, and two from dogs. Theoretically, all of these adenoviruses can be vectored, although thus far efforts have focused on human serotypes HAdV-5, -26, and -35 (commonly referred to as AdHu5, AdHu26, etc.), simian serotypes SAdV-3, -23, -24, -25, and -63, canine adenovirus serotype 2 (CAV-2), and bovine adenovirus serotype 3 (BAdV-3). Human serotype vectors such as HAdV-5 are unsuited for clinical use as most humans become naturally infected with multiple serotypes of HAdV viruses and, consequently, develop VNAs, which interfere with active immunization (Chen et al., 2010). VNAs to adenoviruses derived from other species are rare in humans thus favoring their development as vaccine vectors (Chen et al., 2010; Xiang et al., 2006). A number of different E1-deleted adenovirus vectors have been tested preclinically, including those based on HAdV-5 (Xiang, Yang, Wilson, & Ertl, 1996), -26 (Chen et al., 2010), SAdV-23, -24, and -25 (Lasaro & Ertl, 2009). Vectors are highly immunogenic and provide protection to subsequent challenges. The SAdV-25 vector expressing the rabies G protein of the ERA strain (termed AdC68rab.gp) was explored in nonhuman primates where it was found to induce high and sustained rabies virus-specific VNA responses and protection against challenge. The vaccine was efficacious even when given at a modest dose. Upon challenge, vaccinated nonhuman primates mounted a vigorous recall response (Xiang, Greenberg, Ertl, & Rupprecht, 2014). AdC68rab.gp was not effective if given after exposure to rabies virus. The adenovirus vector backbone was further modified by replacing parts of the endogenous E4 domain with matching sequences from AdHV-5, a strategy which tends to increase production yields upon growth in cells that transcomplement the deleted E1 with that of HAdV-5 virus (Wang et al., 2018). Methods for thermostabilization have been developed (Dulal et al., 2016) and the vaccine is currently being tested for a phase I clinical trial. It is estimated that the vaccine once produced for mass vaccination would cost < \$4 (Wang et al., 2018), which, if efficacious and safe, would make a cost-efficient alternative to current rabies vaccines for Pre-P.

15.6 The "ideal" rabies vaccine

Let us dream and envisage an ideal rabies vaccine. Such a vaccine would have the following characteristics:

- Immunogenicity/Efficacy: The vaccine would after a single-dose induce protective titers of VNAs. Antibodies would increase rapidly that RIG no longer needs to be given to most individuals with severe exposure unless they experienced bites to the head, are small children or are immunocompromised. Antibody titers after vaccination would be sustained and increase after a booster immunization. Comparing the efficacy of novel vaccines to that of currently licensed vaccines will be a challenge. For Pre-P, one can rely on induction of VNAs as a surrogate measure for protection, but for PEP one would eventually have to test a novel vaccine in humans with a severe exposure, replacing a vaccine that is known to prevent death with one that may or may not protect will pose an ethical dilemma.

- Safety: A novel vaccine must be as safe as current rabies vaccines. Adjuvants may increase the vaccines reactogenicity and attenuated live rabies vaccines as well as replication-competent viral vector vaccine may not meet safety requirements. Surprisingly, even the rabies RNA vaccine was shown to be fairly reactogenic in human clinical trials.
- Cost: An ideal rabies vaccine would be cheap—a full Pre-P regimen would cost no more than $4 to allow for its inclusion into childhood immunization programs in rabies endemic resource poor countries. The cost of a PEP regimen could be reduced by a vaccine that induces a very rapid antibody response by allowing for a reduction in RIG. But again, exploring this in clinical trials will be an ethical challenge.
- Thermostability: The vaccine, which will mainly be used in resource poor countries, would be stable at ambient temperature for a few days and it would be stable in a refrigerator for the duration of its shelf live.
- Routes of Immunization: There has long been a push to develop vaccines that can be given orally. This is an obviously crucial factor for rabies immunization of wild-live and stray dogs. It could facilitate large-scale Pre-P immunization programs, but it may not be a major advantage for PEP where intensive wound cleaning and in case of a severe exposure administration of RIG require a visit to a healthcare provider.

Table 15.3 summarizes which one of the novel vaccine prototypes are likely to meet the requirements for an improved rabies vaccine for Pre-P or PEP. It should be stressed that one does not need to use the same type of vaccine for both. For PEP a protective VNA response has to come up very rapidly; this would not be a requirement for a Pre-P vaccine where instead induction of sustained antibody and long-term memory responses are essential. Lack of adverse events is crucial for a Pre-P vaccine designed for mass childhood immunization—it is advantageous for a PEP vaccine for which nevertheless moderate adverse events could be tolerated. Cost is always an issue but to realistically implement widespread Pre-P in children living in highly endemic areas the vaccine would have to be cost effective.

15.7 Summary

Rabies, which is well controlled in developed countries, has become a neglected disease. Dog rabies remains enzootic through most of the developing world where it causes an estimated 55,000–60,000 human deaths each year, a number that is likely an underestimate due to misdiagnosis and underreporting. Efficacious vaccines that could markedly reduce human fatalities are available, but as they are costly and of poor immunogenicity requiring multiple doses they will continue to be underutilized. Alternative vaccines are required. A number of

TABLE 15.3 The ideal rabies vaccine.

Vaccine prototype	Rapid onset of VNA titers	Sustained VNA titers	Immunological memory	Safety	Low cost	Thermostability
	Crucial for PEP	Crucial for Pre-P		Crucial for Pre-P Important for PEP		Important for PEP and Pre-P
Genetically engineered attenuated rabies virus	++	+++	++++	+/−	+/−	+++
Genetically engineered inactivated rabies virus	+++	++++	++++	++++	+/−	+++
Traditional rabies vaccines with adjuvant	+++	++++	++++	++	+/−	+++
Peptide vaccines	−	−	−	++++	?	+++
Protein vaccines	+++	++	++	depends on adjuvant	?	+++
DNA vaccines	−	+	+	++++	++++	++++
RNA vaccine	−	−	−	++	?	++++
Recombinant poxviruses	−	++++	++++	−	++++	++++
Replication-defective adenoviruses	−	++++	++++	++++	++++	++++

vaccines have undergone preclinical testing but of those adjuvanted rabies vaccines or inactivated rabies virus genetically modified to express two copies of the glycoprotein are likely to be suitable for PEP where rapid induction of protective immune responses is essential. A number of other vaccine platforms, especially genetic vaccines based on highly immunogenic carriers such as E1-deleted adenoviruses, may be suitable for preventative childhood vaccination and should be explored further.

References

Alberer, M., Gnad-Vogt, U., Hong, H. S., Mehr, K. T., Backert, L., Finak, G., … von Sonnenburg, F. (2017). Safety and immunogenicity of a mRNA rabies vaccine in healthy adults: An open-label, non-randomised, prospective, first-in-human phase 1 clinical trial. *Lancet*, *390*(10101), 1511–1520. https://doi.org/10.1016/S0140-6736(17)31665-3.

Ashraf, S., Singh, P. K., Yadav, D. K., Shahnawaz, M., Mishra, S., Sawant, S. V., & Tuli, R. (2005). High level expression of surface glycoprotein of rabies virus in tobacco leaves and its immunoprotective activity in mice. *Journal of Biotechnology*, *119*(1), 1–14. https://doi.org/10.1016/j.jbiotec.2005.06.009.

Bahloul, C., Ahmed, S. B. H., B'chir, B. I., Kharmachi, H., Hayouni, E. A., & Dellagi, K. (2003). Post-exposure therapy in mice against experimental rabies: A single injection of DNA vaccine is as effective as five injections of cell culture-derived vaccine. *Vaccine, 22*(2), 177–184.

Bahloul, C., Jacob, Y., Tordo, N., & Perrin, P. (1998). DNA-based immunization for exploring the enlargement of immunological cross-reactivity against the lyssaviruses. *Vaccine, 16*(4), 417–425.

Bahloul, C., Taieb, D., Diouani, M. F., Ahmed, S. B. H., Chtourou, Y., B'chir, B. I., ... Dellagi, K. (2006). Field trials of a very potent rabies DNA vaccine which induced long lasting virus neutralizing antibodies and protection in dogs in experimental conditions. *Vaccine, 24*(8), 1063–1072. https://doi.org/10.1016/j.vaccine.2005.09.016.

Blanton, J. D., Palmer, D., Dyer, J., & Rupprecht, C. E. (2011). Rabies surveillance in the United States during 2010. *Journal of the American Veterinary Medical Association, 239*(6), 773–783. https://doi.org/10.2460/javma.239.6.773.

Brochier, B., Aubert, M. F., Pastoret, P. P., Masson, E., Schon, J., Lombard, M., ... Desmettre, P. (1996). Field use of a vaccinia-rabies recombinant vaccine for the control of sylvatic rabies in Europe and North America. *Revue Scientifique et Technique, 15*(3), 947–970.

Cenna, J., Hunter, M., Tan, G. S., Papaneri, A. B., Ribka, E. P., Schnell, M. J., ... McGettigan, J. P. (2009). Replication-deficient rabies virus-based vaccines are safe and immunogenic in mice and nonhuman primates. *The Journal of Infectious Diseases, 200*(8), 1251–1260. https://doi.org/10.1086/605949.

Cenna, J., Tan, G. S., Papaneri, A. B., Dietzschold, B., Schnell, M. J., & McGettigan, J. P. (2008). Immune modulating effect by a phosphoprotein-deleted rabies virus vaccine vector expressing two copies of the rabies virus glycoprotein gene. *Vaccine, 26*(50), 6405–6414. https://doi.org/10.1016/j.vaccine.2008.08.069.

Chen, H., Xiang, Z. Q., Li, Y., Kurupati, R. K., Jia, B., Bian, A., ... Ertl, H. C. J. (2010). Adenovirus-based vaccines: Comparison of vectors from three species of adenoviridae. *Journal of Virology, 84*(20), 10522–10532. https://doi.org/10.1128/JVI.00450-10.

Coban, C., Ishii, K. J., & Akira, S. (2009). Immune interventions of human diseases through toll-like receptors. *Advances in Experimental Medicine and Biology, 655*, 63–80. https://doi.org/10.1007/978-1-4419-1132-2_7.

Desai, M. (2016). *CTRI/2016/08/007137 Immunogenicity and safety study of Rabies G protein Vaccine administered as a simulated post-exposure immunization in healthy volunteers*. Retrieved from http://www.ctri.nic.in/Clinicaltrials/pdf_generate.php?trialid=13580&EncHid=&modid=&compid=%27,%2713580det%27.

Desmézières, E., Jacob, Y., Saron, M. F., Delpeyroux, F., Tordo, N., & Perrin, P. (1999). Lyssavirus glycoproteins expressing immunologically potent foreign B cell and cytotoxic T lymphocyte epitopes as prototypes for multivalent vaccines. *The Journal of General Virology, 80*(Pt 9), 2343–2351. https://doi.org/10.1099/0022-1317-80-9-2343.

DiStefano, D., Antonello, J. M., Bett, A. J., Medi, M. B., Casimiro, D. R., & ter Meulen, J. (2013). Immunogenicity of a reduced-dose whole killed rabies vaccine is significantly enhanced by ISCOMATRIX™ adjuvant, Merck amorphous aluminum hydroxylphosphate sulfate (MAA) or a synthetic TLR9 agonist in rhesus macaques. *Vaccine, 31*(42), 4888–4893. https://doi.org/10.1016/j.vaccine.2013.07.034.

Dulal, P., Wright, D., Ashfield, R., Hill, A. V. S., Charleston, B., & Warimwe, G. M. (2016). Potency of a thermostabilised chimpanzee adenovirus Rift Valley fever vaccine in cattle. *Vaccine, 34*(20), 2296–2298. https://doi.org/10.1016/j.vaccine.2016.03.061.

Ertl, H. C., Dietzschold, B., Gore, M., Otvos, L., Larson, J. K., Wunner, W. H., & Koprowski, H. (1989). Induction of rabies virus-specific T-helper cells by synthetic peptides that carry dominant T-helper cell epitopes of the viral ribonucleoprotein. *Journal of Virology, 63*(7), 2885–2892.

Esposito, S., Picciolli, I., Semino, M., & Principi, N. (2013). Dog and cat bite-associated infections in children. *European Journal of Clinical Microbiology & Infectious Diseases, 32*(8), 971–976. https://doi.org/10.1007/s10096-013-1840-x.

Estrategia nacional de zoonosis-MINSA (2015). *Resultados plan de vacunación antirrábica de pre-exposición en comunidades en riesgo de rabia de la Region Amazonas: Perú 2011–2014*. Lima: Ministerio de Salud del Perú. 2014.

Evans, J. S., Horton, D. L., Easton, A. J., Fooks, A. R., & Banyard, A. C. (2012). Rabies virus vaccines: Is there a need for a pan-lyssavirus vaccine? *Vaccine, 30*(52), 7447–7454. https://doi.org/10.1016/j.vaccine.2012.10.015.

Fischer, L., Minke, J., Dufay, N., Baudu, P., & Audonnet, J. C. (2003). Rabies DNA vaccine in the horse: Strategies to improve serological responses. *Vaccine, 21*(31), 4593–4596.

Fontana, D., Kratje, R., Etcheverrigaray, M., & Prieto, C. (2015). Immunogenic virus-like particles continuously expressed in mammalian cells as a veterinary rabies vaccine candidate. *Vaccine, 33*(35), 4238–4246. https://doi.org/10.1016/j.vaccine.2015.03.088.

Fu, Z. F., Rupprecht, C. E., Dietzschold, B., Saikumar, P., Niu, H. S., Babka, I., ... Koprowski, H. (1993). Oral vaccination of racoons (Procyon lotor) with baculovirus-expressed rabies virus glycoprotein. *Vaccine, 11*(9), 925–928.

Garg, R., Kaur, M., Saxena, A., Prasad, R., & Bhatnagar, R. (2017). Alum adjuvanted rabies DNA vaccine confers 80% protection against lethal 50 LD50 rabies challenge virus standard strain. *Molecular Immunology, 85,* 166–173. https://doi.org/10.1016/j.molimm.2017.02.011.

Ge, J., Wang, X., Tao, L., Wen, Z., Feng, N., Yang, S., ... Bu, Z. (2011). Newcastle disease virus-vectored rabies vaccine is safe, highly immunogenic, and provides long-lasting protection in dogs and cats. *Journal of Virology, 85*(16), 8241–8252. https://doi.org/10.1128/JVI.00519-11.

Gelosa, L., & Borroni, G. (1990). Serological determination of rabies antibodies in vaccinated subjects. *Microbiologica, 13*(3), 257–262.

Hampson, K., Abela-Ridder, B., Bharti, O., Knopf, L., Léchenne, M., Mindekem, R., ... Trotter, C. (2018). Modelling to inform prophylaxis regimens to prevent human rabies. *Vaccine.* https://doi.org/10.1016/j.vaccine.2018.11.010.

Houimel, M., & Dellagi, K. (2009). Peptide mimotopes of rabies virus glycoprotein with immunogenic activity. *Vaccine, 27*(34), 4648–4655. https://doi.org/10.1016/j.vaccine.2009.05.055.

Ito, N., Sugiyama, M., Yamada, K., Shimizu, K., Takayama-Ito, M., Hosokawa, J., & Minamoto, N. (2005). Characterization of M gene-deficient rabies virus with advantages of effective immunization and safety as a vaccine strain. *Microbiology and Immunology, 49*(11), 971–979.

Kaur, M., Saxena, A., Rai, A., & Bhatnagar, R. (2010). Rabies DNA vaccine encoding lysosome-targeted glycoprotein supplemented with Emulsigen-D confers complete protection in preexposure and postexposure studies in BALB/c mice. *FASEB Journal: Official Publication of the Federation of American Societies for Experimental Biology, 24*(1), 173–183. https://doi.org/10.1096/fj.09-138644.

Kessels, J. A., Recuenco, S., Navarro-Vela, A. M., Deray, R., Vigilato, M., Ertl, H., ... Briggs, D. (2017). Pre-exposure rabies prophylaxis: A systematic review. *Bulletin of the World Health Organization, 95*(3), 210–219C. https://doi.org/10.2471/BLT.16.173039.

Klepfer, S. R., Debouck, C., Uffelman, J., Jacobs, P., Bollen, A., & Jones, E. V. (1993). Characterization of rabies glycoprotein expressed in yeast. *Archives of Virology, 128*(3–4), 269–286.

Larson, J. K., Wunner, W. H., Otvos, L., & Ertl, H. C. (1991). Identification of an immunodominant epitope within the phosphoprotein of rabies virus that is recognized by both class I- and class II-restricted T cells. *Journal of Virology, 65*(11), 5673–5679.

Lasaro, M. O., & Ertl, H. C. J. (2009). New insights on adenovirus as vaccine vectors. *Molecular Therapy, 17*(8), 1333–1339. https://doi.org/10.1038/mt.2009.130.

Lin, H., & Perrin, P. (1999). Influence of aluminum adjuvant to experimental rabies vaccine. *Chinese Journal of Experimental and Clinical Virology, 13*(2), 133–135.

Liu, X., Yang, Y., Sun, Z., Chen, J., Ai, J., Dun, C., ... Guo, X. (2014). A recombinant rabies virus encoding two copies of the glycoprotein gene confers protection in dogs against a virulent challenge. *PLoS One, 9*(2), e87105. https://doi.org/10.1371/journal.pone.0087105.

Lodmell, D. L., & Ewalt, L. C. (2001). Post-exposure DNA vaccination protects mice against rabies virus. *Vaccine, 19*(17–19), 2468–2473.

Lodmell, D. L., Parnell, M. J., Bailey, J. R., Ewalt, L. C., & Hanlon, C. A. (2001). One-time gene gun or intramuscular rabies DNA vaccination of non-human primates: Comparison of neutralizing antibody responses and protection against rabies virus 1 year after vaccination. *Vaccine, 20*(5–6), 838–844.

Lodmell, D. L., Parnell, M. J., Bailey, J. R., Ewalt, L. C., & Hanlon, C. A. (2002). Rabies DNA vaccination of non-human primates: Post-exposure studies using gene gun methodology that accelerates induction of neutralizing antibody and enhances neutralizing antibody titers. *Vaccine, 20*(17–18), 2221–2228.

Lodmell, D. L., Parnell, M. J., Weyhrich, J. T., & Ewalt, L. C. (2003). Canine rabies DNA vaccination: A single-dose intradermal injection into ear pinnae elicits elevated and persistent levels of neutralizing antibody. *Vaccine, 21*(25–26), 3998–4002.

Loza-Rubio, E., Rojas, E., Gómez, L., Olivera, M. T. J., & Gómez-Lim, M. A. (2008). Development of an edible rabies vaccine in maize using the Vnukovo strain. *Developments in Biologicals, 131,* 477–482.

Loza-Rubio, E., Rojas-Anaya, E., López, J., Olivera-Flores, M. T., Gómez-Lim, M., & Tapia-Pérez, G. (2012). Induction of a protective immune response to rabies virus in sheep after oral immunization with transgenic maize, expressing the rabies virus glycoprotein. *Vaccine, 30*(37), 5551–5556. https://doi.org/10.1016/j.vaccine.2012.06.039.

Luo, T. R., Minamoto, N., Hishida, M., Yamamoto, K., Fujise, T., Hiraga, S., ... Kinjo, T. (1998). Antigenic and functional analyses of glycoprotein of rabies virus using monoclonal antibodies. *Microbiology and Immunology, 42*(3), 187–193.

Mansfield, K. L., Johnson, N., Nunez, A., Hicks, D., Jackson, A. C., & Fooks, A. R. (2008). Up-regulation of chemokine gene transcripts and T-cell infiltration into the central nervous system and dorsal root ganglia are characteristics of experimental European bat lyssavirus type 2 infection of mice. *Journal of Neurovirology, 14*(3), 218–228. https://doi.org/10.1080/13550280802008297.

Modelska, A., Dietzschold, B., Sleysh, N., Fu, Z. F., Steplewski, K., Hooper, D. C., ... Yusibov, V. (1998). Immunization against rabies with plant-derived antigen. *Proceedings of the National Academy of Sciences of the United States of America, 95*(5), 2481–2485.

Morimoto, K., Shoji, Y., & Inoue, S. (2005). Characterization of P gene-deficient rabies virus: Propagation, pathogenicity and antigenicity. *Virus Research, 111*(1), 61–67. https://doi.org/10.1016/j.virusres.2005.03.011.

Müller, T., Dietzschold, B., Ertl, H., Fooks, A. R., Freuling, C., Fehlner-Gardiner, C., ... Kieny, M. P. (2009). Development of a mouse monoclonal antibody cocktail for post-exposure rabies prophylaxis in humans. *PLoS Neglected Tropical Diseases. 3*(11). https://doi.org/10.1371/journal.pntd.0000542.

Nadin-Davis, S. A., & Real, L. A. (2011). Molecular phylogenetics of the lyssaviruses–insights from a coalescent approach. *Advances in Virus Research, 79*, 203–238. https://doi.org/10.1016/B978-0-12-387040-7.00011-1.

Niederhäuser, S., Bruegger, D., Zahno, M. -L., Vogt, H. -R., Peterhans, E., Zanoni, R., & Bertoni, G. (2008). A synthetic peptide encompassing the G5 antigenic region of the rabies virus induces high avidity but poorly neutralizing antibody in immunized animals. *Vaccine, 26*(52), 6749–6753. https://doi.org/10.1016/j.vaccine.2008.10.020.

Niu, Y., Liu, Y., Yang, L., Qu, H., Zhao, J., Hu, R., ... Liu, W. (2016). Immunogenicity of multi-epitope-based vaccine candidates administered with the adjuvant Gp96 against rabies. *Virologica Sinica, 31*(2), 168–175. https://doi.org/10.1007/s12250-016-3734-4.

Osorio, J. E., Tomlinson, C. C., Frank, R. S., Haanes, E. J., Rushlow, K., Haynes, J. R., & Stinchcomb, D. T. (1999). Immunization of dogs and cats with a DNA vaccine against rabies virus. *Vaccine, 17*(9–10), 1109–1116.

Ramya, R., Mohana Subramanian, B., Sivakumar, V., Senthilkumar, R. L., Sambasiva Rao, K. R. S., & Srinivasan, V. A. (2011). Expression and solubilization of insect cell-based rabies virus glycoprotein and assessment of its immunogenicity and protective efficacy in mice. *Clinical and Vaccine Immunology: CVI, 18*(10), 1673–1679. https://doi.org/10.1128/CVI.05258-11.

Rojas-Anaya, E., Loza-Rubio, E., Olivera-Flores, M. T., & Gomez-Lim, M. (2009). Expression of rabies virus G protein in carrots (Daucus carota). *Transgenic Research, 18*(6), 911–919. https://doi.org/10.1007/s11248-009-9278-8.

Sardesai, N. Y., & Weiner, D. B. (2011). Electroporation delivery of DNA vaccines: Prospects for success. *Current Opinion in Immunology, 23*(3), 421–429. https://doi.org/10.1016/j.coi.2011.03.008.

Saxena, S., Dahiya, S. S., Sonwane, A. A., Patel, C. L., Saini, M., Rai, A., & Gupta, P. K. (2008). A sindbis virus replicon-based DNA vaccine encoding the rabies virus glycoprotein elicits immune responses and complete protection in mice from lethal challenge. *Vaccine, 26*(51), 6592–6601. https://doi.org/10.1016/j.vaccine.2008.09.055.

Schnee, M., Vogel, A. B., Voss, D., Petsch, B., Baumhof, P., Kramps, T., & Stitz, L. (2016). An mRNA vaccine encoding rabies virus glycoprotein induces protection against lethal infection in mice and correlates of protection in adult and newborn pigs. *PLoS Neglected Tropical Diseases, 10*(6), e0004746. https://doi.org/10.1371/journal.pntd.0004746.

Schneider, L. G., & Diringer, H. (1976). Structure and molecular biology of rabies virus. *Current Topics in Microbiology and Immunology, 75*, 153–180.

Shi, W., Kou, Y., Xiao, J., Zhang, L., Gao, F., Kong, W., ... Zhang, Y. (2018). Comparison of immunogenicity, efficacy and transcriptome changes of inactivated rabies virus vaccine with different adjuvants. *Vaccine, 36*(33), 5020–5029. https://doi.org/10.1016/j.vaccine.2018.07.006.

Shoji, Y., Inoue, S., Nakamichi, K., Kurane, I., Sakai, T., & Morimoto, K. (2004). Generation and characterization of P gene-deficient rabies virus. *Virology, 318*(1), 295–305. https://doi.org/10.1016/j.virol.2003.10.001.

Smith, T. G., Wu, X., Franka, R., & Rupprecht, C. E. (2011). Design of future rabies biologics and antiviral drugs. *Advances in Virus Research, 79*, 345–363. https://doi.org/10.1016/B978-0-12-387040-7.00016-0.

Tantawichien, T., Jaijaroensup, W., Khawplod, P., & Sitprija, V. (2001). Failure of multiple-site intradermal postexposure rabies vaccination in patients with human immunodeficiency virus with low CD4+ T lymphocyte counts. *Clinical Infectious Diseases, 33*(10), E122–E124. https://doi.org/10.1086/324087.

Tesoro Cruz, E., Feria Romero, I. A., López Mendoza, J. G., Orozco Suárez, S., Hernández González, R., Favela, F. B., … Aguilar-Setién, A. (2008). Efficient post-exposure prophylaxis against rabies by applying a four-dose DNA vaccine intranasally. *Vaccine, 26*(52), 6936–6944. https://doi.org/10.1016/j.vaccine.2008.09.083.

Tesoro Cruz, E., Hernández González, R., Alonso Morales, R., & Aguilar-Setién, J. A. (2006). Rabies DNA vaccination by the intranasal route in dogs. *Developments in Biologicals, 125*, 221–231.

Wang, X., Bao, M., Wan, M., Wei, H., Wang, L., Yu, H., … Wang, L. (2008). A CpG oligodeoxynucleotide acts as a potent adjuvant for inactivated rabies virus vaccine. *Vaccine, 26*(15), 1893–1901. https://doi.org/10.1016/j.vaccine.2008.01.043.

Wang, C., Dulal, P., Zhou, X., Xiang, Z., Goharriz, H., Banyard, A., … Douglas, A. D. (2018). A simian-adenovirus-vectored rabies vaccine suitable for thermostabilisation and clinical development for low-cost single-dose pre-exposure prophylaxis. *PLoS Neglected Tropical Diseases, 12*(10), e0006870. https://doi.org/10.1371/journal.pntd.0006870.

Weyer, J., Rupprecht, C. E., Mans, J., Viljoen, G. J., & Nel, L. H. (2007). Generation and evaluation of a recombinant modified vaccinia virus Ankara vaccine for rabies. *Vaccine, 25*(21), 4213–4222. https://doi.org/10.1016/j.vaccine.2007.02.084.

Weyer, J., Rupprecht, C. E., & Nel, L. H. (2009). Poxvirus-vectored vaccines for rabies–a review. *Vaccine, 27*(51), 7198–7201. https://doi.org/10.1016/j.vaccine.2009.09.033.

Wijaya, L., Tham, C. Y. L., Chan, Y. F. Z., Wong, A. W. L., Li, L. T., Wang, L. -F., … Low, J. G. (2017). An accelerated rabies vaccine schedule based on toll-like receptor 3 (TLR3) agonist PIKA adjuvant augments rabies virus specific antibody and T cell response in healthy adult volunteers. *Vaccine, 35*(8), 1175–1183. https://doi.org/10.1016/j.vaccine.2016.12.031.

World Health Organization. (2018). WHO Expert Consultation on Rabies (Third Report). World Health Organization Technical Report Series, (982), 1–139, back cover.

Wu, T. -L., Li, H., Faust, S. M., Chi, E., Zhou, S., Wright, F., … Ertl, H. C. J. (2014). CD8+ T cell recognition of epitopes within the capsid of adeno-associated virus 8-based gene transfer vectors depends on vectors' genome. *Molecular Therapy, 22*(1), 42–51. https://doi.org/10.1038/mt.2013.218.

Wunderli, P. S., Shaddock, J. H., Schmid, D. S., Miller, T. J., & Baer, G. M. (1991). The protective role of humoral neutralizing antibody in the NIH potency test for rabies vaccines. *Vaccine, 9*(9), 638–642.

Xiang, Z., & Ertl, H. C. (1995). Manipulation of the immune response to a plasmid-encoded viral antigen by coinoculation with plasmids expressing cytokines. *Immunity, 2*(2), 129–135.

Xiang, Z. Q., Greenberg, L., Ertl, H. C., & Rupprecht, C. E. (2014). Protection of non-human primates against rabies with an adenovirus recombinant vaccine. *Virology, 450–451*, 243–249. https://doi.org/10.1016/j.virol.2013.12.029.

Xiang, Z. Q., Knowles, B. B., McCarrick, J. W., & Ertl, H. C. (1995). Immune effector mechanisms required for protection to rabies virus. *Virology, 214*(2), 398–404. https://doi.org/10.1006/viro.1995.0049.

Xiang, Z. Q., Li, Y., Cun, A., Yang, W., Ellenberg, S., Switzer, W. M., … Ertl, H. C. (2006). Chimpanzee adenovirus antibodies in humans, sub-Saharan Africa. *Emerging Infectious Diseases, 12*(10), 1596–1599. https://doi.org/10.3201/eid1210.060078.

Xiang, Z. Q., Spitalnik, S., Tran, M., Wunner, W. H., Cheng, J., & Ertl, H. C. (1994). Vaccination with a plasmid vector carrying the rabies virus glycoprotein gene induces protective immunity against rabies virus. *Virology, 199*(1), 132–140. https://doi.org/10.1006/viro.1994.1105.

Xiang, Z. Q., Yang, Y., Wilson, J. M., & Ertl, H. C. (1996). A replication-defective human adenovirus recombinant serves as a highly efficacious vaccine carrier. *Virology, 219*(1), 220–227. https://doi.org/10.1006/viro.1996.0239.

Yusibov, V., Hooper, D. C., Spitsin, S. V., Fleysh, N., Kean, R. B., Mikheeva, T., … Koprowski, H. (2002). Expression in plants and immunogenicity of plant virus-based experimental rabies vaccine. *Vaccine, 20*(25–26), 3155–3164.

CHAPTER 16

Public health management of humans at risk

Deborah J. Briggs, Susan M. Moore

Department of Diagnostic Medicine, College of Veterinary Medicine, Kansas State University, Manhattan, KS, United States

16.1 Introduction

Human rabies vaccines are among the most efficacious vaccines being produced to prevent infectious diseases (WHO, 2017). In spite of this fact, human rabies deaths continue to occur at an alarming rate of one death every 10–15 min (Fooks, 2018). Approximately half of global rabies deaths occur in children under 15 years of age and over 99% of all human deaths are attributable to exposure to an infected dog (WHO, 2018c). Although symptomatic rabies is almost invariably fatal, because of the nature of the pathogenesis of the disease, rabies can be prevented even after an exposure has occurred (Pieracci et al., 2019). This characteristic of rabies, along with the fact that all of the tools are available to prevent human fatalities, raises the question as to why people still die of this horrific disease.

Over the past decade, several international initiatives have been launched aimed at reducing the global burden and improving public health management for the prevention of human rabies (GARC, 2018; Minghui, Stone, Semedo, & Nel, 2018; OIE, 2019; Vigilato, Molina-Flores, Del Rio Vilas, Pompei, & Cosivi, 2018; WHO, 2018a). It is clear that the most efficient strategy to prevent rabies in humans would be to eliminate the major source of infection by eliminating the transmission of rabies virus in the canine population (Elser, Hatch, Taylor, Nel, & Shwiff, 2018; Rupprecht, Kuzmin, Yale, Nagarajan, & Meslin, 2019; WHO, 2018f). Historically, public health programs aimed at eliminating canine rabies have been successful in Europe, North America, and Latin America and the same types of multidiscipline strategies used in these regions are beginning to make notable progress in Africa and Asia where most canine-mediated human rabies cases occur (Cleaveland et al., 2018; LeRoux et al., 2018; Rupprecht et al., 2019). Programs aimed at preventing canine mediated human rabies must be multidimensional in order to be successful (OIE, 2019; Rupprecht et al., 2019; WHO, 2018f).

Targeting mass canine vaccination only, in lieu of providing PEP for humans, is unethical. However, providing greater access to PEP without establishing and building sustainable rabies prevention programs that aim to eliminate virus transmission in dogs will lead to continually escalating budgetary requirements need to purchase rabies biologicals as awareness about rabies prevention increases throughout the population (Mindekem et al., 2017).

Recently, the World Health Organization (WHO) reviewed scientific evidence from several clinical trials evaluating the effectiveness of rabies vaccine regimens aimed at reducing the cost of both preexposure vaccination (PreP) and postexposure prophylaxis (PEP) (WHO, 2018c, 2018f). The result of the review of these data by global rabies experts, health economists, and WHO personnel confirmed the fact that the number of doses required for PreP and PEP can be safely reduced with no loss of efficacy (WHO, 2018c). Newer more cost-effective regimens may provide incentives for national governments to invest in human rabies prevention activities in the future (Tarantola, Tejiokem, & Briggs, 2018).

Improving access to human rabies biologicals, especially in rural areas, is a major obstacle to overcome in many low-income countries. Every five years, the Global Alliance Vaccine Initiative (GAVI) re-evaluates its five-year vaccine investment strategy (VIS) for the world's most impoverished countries. In 2017, after a lengthy and detailed examination, GAVI made a decision to include human rabies vaccines for PEP in their endemic disease prevention portfolio (GAVI, 2019). As a result, beginning in 2021 for the first time in history, GAVI-eligible countries will have an opportunity to apply for life-saving human rabies vaccines to improve access to PEP as part of their overall public health program.

Being aware of how to prevent rabies, even after an exposure has occurred, is key to saving human lives. Published data indicate that the age group generally considered to be most at risk of contracting rabies from exposure to infected dogs are children below the age of 15 (WHO, 2018c, 2018d). The one exception to children being the age group most at risk is in China where most rabies deaths occur in farmers in rural areas (Wei et al., 2019). Nevertheless, whether the most at-risk age group is children or adults, saving lives depends on the victim of a bite from a rabid dog or someone close to the bite victim knowing how to treat wounds and where to seek administration of PEP. These issues involve improving the educational awareness in communities at risk of exposure. Transferring knowledge to at risk communities involves understanding their culture and being sensitive as to the best methods to improve access to life saving information in each setting.

The cost of the ever-present risk of canine rabies to human health is enormous and most of the financial outlay comes from administration of human PEP. In fact, $1.7 billion is estimated to be spent on direct cost of PEP and $1.3 billion is estimated to be attributed to indirect costs of PEP (Elser et al., 2018; Hampson et al., 2015) Translated to a more human scale of understanding as to the suffering that rabies causes, it is important to know that the cost of one PEP treatment amounts to 31 days of wages in Asia and 51 days of wages in Africa (Abela-Ridder, de Balogh, Kessels, Dieuzy-Labaye, & Torres, 2018). Global elimination of the major source of human rabies, that is the continued circulation of rabies virus in the dog population of canine rabies endemic countries, would eliminate over 99% of all human rabies cases. Thus, the major aim of public health officials should be global elimination of canine rabies by implementing an integrated systematic approach that includes many components including improving access to human rabies biologicals in remote regions (Rupprecht et al., 2018; Rupprecht et al., 2019; Tarantola, Blanchi, et al., 2018; Tarantola, Ly, et al., 2018).

16.2 Obstacles to preventing rabies in humans

Considering the fact that rabies is one of the most preventable infectious diseases known to mankind, deaths still occur due to many factors including but not limited to: Inadequate or a lack of national rabies prevention programs in place; a lack of understanding the pathogenesis of the disease; a delay in seeking treatment after an exposure has occurred; inadequate treatment after an exposure has occurred; and limited or no access to rabies biologicals for many reasons (Hemachudha et al., 1999; Shantavasinkul et al., 2010; Wilde, 2007).

One of the first steps for a country to put in place in their efforts to prevent human rabies is to establish a National Rabies Prevention Program (OIE, 2019; PRP, 2018). The design and establishment of a National Rabies Prevention Program should include officials from both the human and animal side of public health as well as experts in the field of education, law enforcement, sanitation, advocacy, and community affairs at a minimum. In some cases, including the expertise of a medical anthropologist in the development of rabies prevention programs designed for specific populations that have diverse and different cultural backgrounds and belief systems can be of great benefit.

National Rabies Prevention Programs are a necessary first step in providing the political support required for local public health officials to implement procedural changes that will lead to a reduction in the risk of exposure to rabies in communities. Without political support at a national and local level, it is virtually impossible to encourage local communities to institute lasting changes that will reduce their risk of rabies.

As mentioned earlier, children living in most canine rabies-endemic regions constitute the largest group at risk of exposure to rabies through contact with dogs. When attacked, children often suffer severe injuries in the face and hand areas (Taylor, Costa, & Briggs, 2013). Both anatomical regions are highly innervated and therefore prone to shorter incubation times for development of clinical disease. When children are cautioned by their parents or guardians not to play with dogs and consequently disobey this advice and suffer from minor wounds that break the skin but are not immediately noticeable by adults, they may choose not to tell about the exposure to avoid being punished. PreP for children may be considered for specific populations at very high risk and located in remote regions where access to rabies biologicals is extremely limited to nonexistent (WHO, 2018c). Vaccinating all children living in canine rabies-endemic regions could lead to a shortage of rabies vaccine, and is not cost effective for national medical budgets aiming at the best utilization of their limited resources and therefore is not recommended. However, targeted use of PreP may be useful in some remote regions of the world where rabies biologicals are not available and rabies is endemic and difficult to prevent, for example in remote regions of Latin America where vampire bat rabies continues to pose a threat to humans (Kessels et al., 2017).

16.3 Exposure

Exposure to rabies requires immediate medical intervention as the risk of developing disease without treatment is high and once clinical signs are evident, death is inevitable (Smith, Wu, Fooks, Ma, & Banyard, 2019). Unknown and/or unreported exposures and delays in seeking medical treatment after exposure have resulted in infection and death and are among the most frequent causes of human rabies treatment failures (Hemachudha et al., 1999; Wilde, 2007; Wilde et al., 1996).

Recognizing that an exposure has occurred is the first step to understanding how to prevent disease and death (Table 16.1). Almost all human exposures to rabies occur when virus-infected saliva infiltrates open wounds caused by bites or scratches inflicted by the infected animal. As mentioned, more than 99% of all human rabies cases result from exposure to a rabid dog. Aerosol spread of rabies to humans and other mammals has been reported, but this type of exposure is rare (Gibbons, 2002; Winkler, Fashinell, Leffingwell, Howard, & Conomy, 1973). Exposure has also occurred through transplantation of infected tissues or organs from human donors that were not recognized as being infected with rabies (Lu, Zhu, & Wu, 2018; Saeed & Al-Mousawi, 2017; Zhang et al., 2018). Anecdotal human-to-human transmission of rabies virus, outside of direct transmission through infected tissue and organ material, has been reported but not scientifically confirmed (Fekadu et al., 1996).

WHO has stated that rabies virus has been recovered in the saliva, tears, urine, and nervous tissue of human rabies cases and exposure to any of these tissues or body fluids should be considered to be an exposure (WHO, 2018c).

Laboratory confirmation of an exposure to rabies can be a tremendous aid in reducing the medical costs associated with prevention of human rabies (Franka & Wallace, 2018). This is because if a "true exposure" to a rabid mammal can be ruled out based on solid scientific data obtained from a reliable diagnostic laboratory, it will not be necessary to administer PEP. Thus, every rabies endemic country should have a designated National Laboratory where rabies diagnostics can be undertaken. Countries that are free of rabies should have contingency plans in place where specimens from suspect rabid animal can also be submitted for examination in a timely manner. National laboratory diagnostic systems are pillars of human

TABLE 16.1 Categories of exposures and recommended treatment according to the World Health Organization and the Advisory Committee on Immunization Practices (ACIP) (Manning et al., 2008; WHO, 2018e).

Category	Type of exposure	Treatment
I	Touching or feeding animals; animal links on intact skin NO EXPOSURE	Wash intact skin with soap and water. NO PEP required.
II	Nibbling of uncovered skin; minor scratches or abrasions without bleeding EXPOSURE	Wash wounds with soap and water for approximately 15 min. Treat cleansed wound with virudical agent. *WHO recommends* post-exposure vaccination without rabies immune globulin. *ACIP recommends* post-exposure vaccination, including administration of rabies immune globulin.
III	Single or multiple transdermal bites or scratches; contamination of mucous membrane or broken skin with saliva from animal licks; exposures due to direct contact with bats	Wash wounds with soap and water for approximately 15 min. Treat cleansed wound with virudical agent. Administer post-exposure vaccination, including rabies immune globulin.

rabies prevention programs but are too often overlooked as being important tools in the effort to prevent human rabies deaths. National rabies laboratories serve several functions, including most importantly monitoring the epidemiology of rabies throughout the country and confirming or ruling out the necessity to administer PEP, thus saving precious medical budgets from unnecessary expense. It is therefore a worthwhile investment to establish at least one reliable rabies diagnostic laboratory within a country to aid in the prevention of human rabies. There are strategies in place through OIE and WHO to assist in the development of collaboration between laboratory systems in countries just initiating the process of instituting rabies prevention programs and countries with well-established diagnostic programs (Fooks et al., 2009; Fooks, Drew, & Tu, 2016).

The incubation time after an exposure occurs until patients develop clinical rabies is dependent upon many factors including: the severity of the bite wound; the anatomical location of the wound(s); the amount and variant of the rabies virus that was inoculated into open wounds; and the timeliness of PEP (WHO, 2018c). WHO has stated that in the absence of PEP, the probability of developing clinical rabies after an exposure to the head region is 55%, to the upper extremity region is 22%, to the trunk region is 9%, and to a lower limb is 12% (Shim, Hampson, Cleaveland, & Galvani, 2009).

16.4 Vaccination protocols to prevent rabies

Rabies vaccines are among the most effective human vaccines that have ever been developed in that they can be administered either before or after an exposure has occurred and can also be administered either intramuscularly (IM) or intradermally (ID) (Kessels, Tarantola, Salahuddin, Blumberg, & Knopf, 2019; WHO, 2018c). The intradermal route of administering PEP can reduce the volume of vaccine required and vaccine cost from 60%–85% (Salahuddin, Gohar, & Baig-Ansari, 2016; WHO, 2018c). All rabies vaccines administered IM require the full vial to be injected and rabies vaccines that are recommended by WHO to be administered ID require 0.1 mL to be administered regardless as to whether the vial is a 1 mL or 0.5 mL vial (WHO, 2017, 2018c, 2018d). Rabies vaccine should be administered into the deltoid muscle in adults and in the anterolateral thigh in infants. Information on correct administration of ID rabies vaccines is available on the WHO website (WHO, 2014). A recent review of published data on the effectiveness of administering rabies vaccines IM and ID, reported that both routes of administration are immunogenic and effective (Denis et al., 2018). Lower titers have been reported after ID administration compared to IM administration in some clinical studies; however, this was not clinically significant. The conclusions included the fact that development of rabies after PEP is extremely rare and almost always involves a deviation from WHO recommended protocol.

16.4.1 Preexposure vaccination

PreP is recommended for all personnel at constant risk of exposure to rabies (Manning et al., 2008). There are several reasons to utilize PreP as a strategy to protect persons at constant or high risk of exposure to rabies virus. First and foremost, PreP can help to protect an individual against an unrecognized exposure to rabies virus that may occur as a result of one's vocation, hobby, travel itinerary, or as a result of unrecognized exposure due to the prevalence of rabies circulating in animals inhabiting the region in which they live. Secondly, persons that have been previously vaccinated, either PreP or PEP, will have a shortened vaccine regimen for PEP after an exposure to rabies has occurred. Thirdly, no RIG is required for previously vaccinated individuals that subsequently require PEP. The recommended vaccination series for previously vaccinated individuals consists of a simplified booster series of vaccine and requires only one or two visits to the vaccination center thus saving indirect and well as direct costs (WHO, 2017, 2018c).

As mentioned previously, PreP coverage of entire populations of individuals is not economically prudent, nor is it necessary for prevention of human rabies. Recognizing that an exposure to rabies has occurred and seeking PEP immediately afterward will save lives in almost all cases of human exposure. Widespread use of PreP should be carefully evaluated prior to initiation. For example, PreP may be useful in regions where rabies endemicity is high, availability of rabies biologicals is limited or nonexistent, and where unrecognized exposures occur, for example, PreP would reduce the risk of vampire bat rabies in remote regions of Latin America where human deaths have been reported in clusters (Gilbert et al., 2012; Medeiros et al., 2016).

The recommended doses for PreP has recently been reduced from a series of three to a series of two medical visits (WHO, 2018c, 2018d) (Table 16.2). PreP can be administered as a series of two doses of vaccine given either IM or ID. If PreP is administered by the IM route, one IM dose us administered on each of days 0 and 7. These recent reductions in the number of doses recommended for PreP were thoroughly and carefully evaluated by WHO and global experts in the field of rabies (Kessels et al., 2019; Tarantola, Blanchi, et al., 2018; Tarantola, Ly, et al., 2018).

Occasionally, travelers requiring PreP may not have time to complete the entire regimen prior to leaving. In that case, WHO acknowledges that there may be some benefit in travelers receiving 1 dose of rabies vaccine prior to departure (WHO, 2018c). However, WHO recommends that, in these patients, a second dose should be administered as soon as possible and within one year of the initial dose of vaccine. In the event of an exposure in patients that have received only one dose of PreP, the full PEP regimen, including RIG, should be administered (WHO, 2018c).

TABLE 16.2 Pre-exposure vaccination regimens recommended by the World Health Organization (WHO, 2018b).

Route	Number of doses of vaccine	Number of clinic visits	Duration of course	Schedule	Notes
IM[a] or ID[b]	3	3	21 or 28 days	One dose on each of days 0, 3, and either day 21 or day 28	
IM or ID	2	2	7 days	One dose on each of days 0 and 7	

[a] *IM—Intramuscular administration. Vaccine should be administered into the upper deltoid region of the upper arm in patients ages ≥ 2 years of age and into the anterolateral thigh of children < 2 years of age.*

[b] *ID—Intradermal administration. Vaccine should be administered into the skin layer over the deltoid in patients ages ≥ 2 years of age and into the anterolateral thigh of children < 2 years of age and should produce an "orange peel" appearance if injected correctly.*

16.4.2 Postexposure vaccination

16.4.2.1 Before vaccination begins

Almost all human rabies exposures occur when virus-infected saliva enters open wounds incurred during the trauma of bite and scratches inflicted by a rabid animal. After an exposure occurs and before administering rabies vaccine, it is critically important to cleanse all wounds and abrasions that have broken the skin and to administer rabies immune globulin (RIG) appropriately. Conducting primary wound care as soon as possible after an exposure has occurred can help to reduce the risk of development of clinical rabies (Shaughnessy, 1954). The WHO recommends that all wounds and scratches incurred during an exposure be thoroughly washed with soap or detergent and copious amounts of water for approximately 15 min (WHO, 2017, 2018c). If possible, an iodine-containing (or other viricidal) preparation should be applied to the wound once it has been cleansed. If soap or detergent is not immediately available, flushing the wound with water should be instituted. Experimental evidence indicates that washing a wound helps to reduce rabies virus infection by eliminating or inactivating viral particles that may have been inoculated into the tissue at the time of exposure and is therefore considered a critical component of rabies PEP and should not be overlooked. In addition, tetanus prophylaxis and antibacterial treatment (although antibiotics are not administered routinely, they are almost always indicated for high-risk wounds) should also be initiated for all animal bites that cause tissue damage (WHO, 2018c, 2018d).

16.4.2.2 Rabies immune globulin

RIG is administered as a passive immune treatment in order to provide immediate access to rabies-virus neutralizing antibodies until the patient's immune system can begin to produce its own neutralizing antibody after vaccination. A patient undergoing primary PEP will generally produce measurable antibodies within 7 to 14 days after the initial dose of rabies vaccine has been administered (Habel, 1954).

After all wounds have been cleansed, WHO recommends that RIG should be administered to all patients that have incurred wounds that have broken the skin surface (WHO, 2018c, 2018d). WHO has defined three categories of exposures (Table 16.1). Wounds that have caused a break in the skin surface with bleeding are considered to be "Category III" exposures and all patients incurring Category III wounds should receive RIG. The Advisory Committee on Immunization Practices (ACIP) recommends that RIG be administered to patients that have incurred both Category II and Category III exposures (Manning et al., 2008).

RIG is produced from three sources: Human rabies immune globulin (HRIG) is produced from human plasma donors; equine rabies immune globulin (ERIG) is produced from vaccinated horses; and monoclonal production of antibodies (Mabs) is produced through modern cell culture technology (De Benedictis et al., 2016; Sparrow et al., 2018). Development of effective Mabs has been ongoing for several years and may someday replace the use of HRIG and ERIG but at the time of this publication, licensed Mabs for human use in prevention of rabies are limited to one product (Sparrow et al., 2018; UMMS, 2019). This product is a single Mab that was licensed in India in 2017 and has proven to be safe and effective in clinical trials. This Mab has also been shown to have neutralization ability to a broad panel of globally prevalent rabies virus isolates. Clearly, Mab have major advantages over HRIG and ERIG in that they can be produced using large-scale molecular technology and eliminate the use of animals for production purposes as well as can reduce the inadvertent risk of transferring viruses and other infections in the plasma from human donors.

The dose for administration of HRIG is 20 International Units (IU) per Kg of body weight of the patient and the dose for ERIG is 40 IU per KG of body weight. Regardless of which RIG is used, the RIG should be administered into and around wound sites in order to neutralize any rabies virus that was not eliminated by wound cleansing. Since animal bite wounds, especially in small children, can be severe and inflicted in multiple sites, the recommended dose of RIG may not be of sufficient volume to inject all wounds. In such a case, the dose should be diluted in a sufficient volume of sterile saline solution (2–3 fold) to inject all of the wounds (MSF, 2019; WHO, 2018e). No more than the recommended dose of RIG per Kg of body weight should be administered and persons previously vaccinated with an efficacious CCV should not receive RIG if they are subsequently re-exposed and require PEP.

The WHO recommends that the entire dose of RIG, or as much as anatomically feasible (avoiding compartment syndrome) be infiltrated into or as close as possible to the wound (s) or exposure sites. WHO has recently updated their recommendations to state that any additional RIG that is not able to be infiltrated into the wound area should not be injected at a separate anatomical site (WHO 2018b, 2018c, 2018e). However, the WHO recommends that if there is a high likelihood that there are additional small wounds (e.g. if a child does not report all wounds), exposure was to bats or exposure was other than through a bite, injection of the remaining RIG volume intramuscularly as close as possible to the presumed exposure site, to the degree that is anatomically feasible, is indicated. The same applies for mucosal exposure with no wound, and rinsing with RIG can be considered. In the case of suspected exposure to RABV in aerosols, an intramuscular injection of RIG is nevertheless recommended. Currently, the ACIP recommendations for prevention of human rabies are under review. At the time of the publication of this article, ACIP recommends that as much RIG as is anatomically possible, (being careful to avoid causing compartment syndrome), be infiltrated into wound(s) and exposure sites. ACIP recommends that additional RIG that cannot be infiltrated into wound sites should be injected into a site distant from the vaccination site. In the case of suspected exposure to rabies virus in aerosols, an intramuscular injection of RIG is also recommended by ACIP (Manning et al., 2008).

16.4.2.3 PEP vaccination regimens

Rabies vaccines can be administered as PEP by either the IM or by the ID route. One ID dose of vaccine contains a volume of 0.1 mL and one IM dose of vaccine contains either 0.5 or 1.0 mL, depending on the product used (Kessels et al., 2019). Both routes of administration produce effective protection against rabies and a recent review of literature on IM and ID PEP has concluded that there is no clinical significance between PEP administered by either route (Denis et al., 2018). The PEP regimens recommended by WHO are listed in Table 16.3.

16.4.3 Previously vaccinated individuals and booster vaccinations

As mentioned earlier, previously vaccinated persons, whether they received PreP or PEP, should not receive RIG if they are re-exposed to rabies. The protocol for administering PEP to previously vaccinated persons is limited to a short series of boosters (Table 16.3). Published data from clinical trials and from actual exposures over the 30 years since CCV were developed indicate that rabies vaccines are long lasting and even when titers fall below 0.5 IU/mL

TABLE 16.3 Post-exposure vaccination regimens recommended by the World Health Organization for patients not previously vaccinated (WHO, 2018e).

Route	Number of doses of vaccine	Number of clinic visits	Duration of course	Schedule	Notes
Persons NOT previously vaccinated					
IM[a]	5	5	28 days	1 dose on each of days 0, 3, 7, 14, 28	Wound cleansing prior to administration of vaccine
IM	4	4	14 – 28 days	1 dose on each of days 0, 3, 7, and between days 14-28	Wound cleansing prior to administration of vaccine
IM	4	3	21 days	2 doses on day 0, 1 dose on days 7 and 21	Wound cleansing prior to administration of vaccine
ID[b]	8	4		2 doses on each of days 0, 3, 7, 14, and 28	Wound cleansing prior to administration of vaccine
ID	6		7 days	2 doses on each of days 0, 3, and 7	Wound cleansing prior to administration and vaccine
Persons previously vaccinated with either post-exposure vaccination or pre-exposure vaccination					
IM	2	2	3 days	1 dose on each of days 0, and 3	Wound cleansing prior to administration of vaccine. NO RIG is required
ID	4	2	3 days	2 doses on each of days 0, and 3	Wound cleansing prior to administration of vaccine. NO RIG is required
ID[c]	4	1	1 day	4 sites on day 0	Wound cleansing prior to administration of vaccine. NO RIG is required

[a] IM—Intramuscular administration. Vaccine should be administered into the upper deltoid region of the upper arm in patients ages ≥2 years of age and into the anterolateral thigh of children < 2 years of age.
[b] ID—Intradermal administration. Vaccine should be administered into the skin layer over the deltoid in patients ages ≥2years of age and into the anterolateral thigh of children < 2 years of age and should produce an "orange peel" appearance if injected correctly.
[c] ID—Four intradermal doses of vaccine should be administered into the skin on four sites. In ages ≥2 years, the sites include left and right deltoid and suprascapular areas. In children < 2 years of age, the sites include right and left anterolateral thighs and right and left suprascapular areas.

in previously vaccinated individuals, a series of one or more booster doses of vaccine will induce an rapid anamnestic response (WHO, 2017). This occurs no matter if the initial vaccine was administered IM or ID and regardless of whether the booster was administered IM or ID and independently as to whether the previously vaccinated person had a detectable titer or not (Brown et al., 2008; Khawplod et al., 2002; Suwansrinon et al., 2006).

Routine annual boosters of rabies vaccine for previously vaccinated individuals at little or no risk of exposure are not recommended. There are select groups of individuals that should be monitored as to their level of circulating neutralizing antibody including but not limited to: Technicians working in rabies vaccine production facilities; rabies diagnostic and research laboratories; veterinarians working in rabies-endemic regions; and others whose occupation, hobby, or location puts them at increased risk of exposure. For these individuals, annual or biannual monitoring of titers will provide guidance as to when boosters should be administered according to the public health recommendations and regulations where the individual works, lives, or where the individual is in constant risk. Most regulatory agencies involved in rabies vaccine production or diagnostic laboratories working with live rabies virus or potentially infected tissues require their staff to maintain a titer above a level of 0.5 IU/mL. When titers fall below this level, these workers are usually given a routine booster of one dose of vaccine. This is generally sufficient to bring the titer up to the recommended level of 0.5 IU/mL.

16.5 Special populations

16.5.1 Pregnancy

Pregnancy is not a contraindication to administration of PEP if it is confirmed that an exposure to a known or suspect rabid animal has occurred (Manning et al., 2008; WHO, 2017, 2018c). In three retrospective studies conducted in Thailand and India, there were no reported fetal abnormalities associated with rabies vaccination (Abazeed & Cinti, 2007; Arya, 1990; Sudarshan et al., 1999). Data from these studies and from pharmacovigilance reporting collected over the decades by international rabies vaccine manufactures indicate that PEP in women is safe and should never be withheld in the event of an exposure.

16.5.2 Immunocompromised patients

All patients that may be immunocompromised and require PEP should consult an infectious disease specialist and serological testing between 2 and 4 weeks after PEP should be conducted to assess whether the patient has responded to vaccination is highly recommended. Data from clinical trials involving special populations that may be immunosuppressed have shown that rabies vaccination has elicited neutralizing antibodies above 0.5 IU/mL. This level of neutralizing antibodies is considered to be proof that the patient has responded to rabies PEP.

In a study in Iran, potentially immunosuppressed patients with Category II and III exposures to known or suspect rabid animals received PEP and did not succumb to rabies (Rahimi et al., 2015). These patients included persons with the following conditions: pregnancy; diabetes type 1; diabetes type 2; chronic infection with hepatitis B virus; various types of cancer and immunocompromised due to receiving corticosteroids for rheumatoid arthritis and for lupus erythematosus. Lower titers were reported in patients with cancer and diabetes II, but all patients developed an immune response above 0.5 IU by day 14 after vaccination was initiated. (Rahimi et al., 2015).

Clinical data on the immunological response to PEP in HIV-positive patients are dependent upon several factors, including their level of CD4+ cells (Pancharoen et al., 2001; Tantawichien, Jaijaroensup, Khawplod, & Sitprija, 2001; Thisyakorn, Pancharoen, & Wilde, 2001). Published studies have reported that HIV-infected children < 5 years of age with at least 25% normal CD4+ cells and children > 5 years of age through adults with CD4+ cell counts above 200 cells/mm^3 will have an adequate immune response after PEP. HIV-infected patients that have experienced Category II or III exposures, that are not on Anti-retroviral therapy (ART), and do not have minimum CD4+ cell counts should receive full PEP, including RIG, even if they have been immunized previously. Thorough wound cleansing is essential in all patients that have been exposed to rabies, it is an especially critical step in the prevention of rabies in immunocompromised patients (WHO, 2017).

16.5.3 Travelers

Consultation with a reliable travel medicine clinic can provide information as to whether PreP should be considered as part of the pretravel preparations. Some considerations as to whether PreP should be included in pretravel preparations include: epidemiology of rabies in the area and risk of exposure; remoteness of the region of travel; access to rabies biologicals; and intended length of stay. For example, backpackers visiting canine rabies-endemic regions with difficult access to rabies vaccines, cavers visiting remote regions where bats are roosting, and other travelers that may have direct encounters with potentially rabid animals are candidates for receiving PreP (WHO, 2017).

16.6 Adverse reactions to cell culture vaccines

Rabies vaccine is a life-saving vaccine and should never be withheld in the event that an exposure has occurred. Adverse reactions have been reported after vaccination and therefore, it is advisable for patients to remain under medical observation for a period of time after vaccine has been administered (generally 30 min). Patients that have had a history of severe hypersensitivity to any of the components or excipients that are included in one rabies vaccine should receive a different rabies vaccine if possible. However, even if an adverse event has been reported during the PEP course, the vaccine series should not be halted in the event that a patient has been exposed to rabies. Patients that have experienced adverse reactions during PEP should immediately seek medical advice.

Both local and systemic reactions have been recorded after the administration of CCVs in clinical trials (Ambrozaitis, Laiskonis, Balciuniene, Banzhoff, & Malerczyk, 2006; Pengsaa et al., 2009; Sudarshan et al., 2012; Tarantola, Blanchi, et al., 2018; Tarantola, Ly, et al., 2018). Generally, these studies reported local reactions including: pain; itchiness; redness and/or swelling at the injections site in 35%–45% of the enrolled subjects. Common systemic reactions were reported in 10%–15% of subjects including: fever; myalgia; malaise; headaches; dizziness; hives; and rash.

There is no contraindication for administration of PEP in persons that are receiving treatment with chloroquine or hydroxyl-chloroquine and PEP can be administered by either IM or ID route (WHO, 2018c). For travelers, it is recommended that PreP be completed prior to administration of antimalarials if at all possible.

16.7 Interchangeability and coadministration of vaccines

It may be necessary to change the brand of rabies vaccine during the PEP series due to the unavailability of one vaccine or to avoid further adverse reactions, should they occur. In the

US, one study reported that an anamnestic response occurred when persons vaccinated previously with Human Diploid Cell Vaccine (HDCV) were subsequently boosted with Purified Chick Embryo Cell Vaccine (PCECV), thus indicating that the vaccines were interchangeable (Briggs et al., 2000). When possible, it is generally recommended that the same brand of vaccine should be used throughout PreP and PEP. When it is impossible to do so, PEP should be continued with another CCV that has been produced according to WHO recommendations.

Clinical trials evaluating the immune response of rabies vaccines when coadministrated with other vaccines, including DPT and Japanese encephalitis vaccine, have shown that rabies vaccines are effective when coadministered (Lang, Feroldi, & Vien, 2009; Pengsaa et al., 2009). Therefore, WHO has indicated that rabies vaccines can be coadministered with other inactivated and live vaccines but separate syringes and different injections sites should be used (WHO, 2018c).

16.8 Educational initiatives

Human rabies prevention is all about educational awareness. Published reports highlight the fact that people living at risk of rabies in many different settings could benefit from improved educational awareness programs (Amparo et al., 2019; Glasgow, Worme, Keku, & Forde, 2019; San Jose, Magsino, & Bundalian Jr., 2019; Tiwari, Vanak, O'Dea, & Robertson, 2018). Understanding what constitutes an exposure to rabies, how to implement proper wound cleansing after a bite or scratch wound has occurred, and the urgent need to seek professional medical advice after an exposure occurs is of critical importance in the prevention of human rabies. Effective educational programs detailing how to prevent human rabies for all sectors of society would help save lives including: Government officials; medical professionals; educators; law enforcement agencies; community leaders; and citizens especially children.

New educational initiatives in Asia and African countries are incorporating novel approaches to improving awareness in children by integrating rabies prevention information in the school curriculum. This has proven to be a valuable method by which to improve awareness at an early age in canine rabies endemic countries (Amparo et al., 2019; LeRoux et al., 2018; Pai et al., 2018). However, these types of programs only work where organized educational systems are in place. There is still a need to improve rabies awareness in regions where children are not routinely educated in classroom settings. Developing successful educational programs must include an understanding of local cultures and mores and be respectful to the hierarchy of community leadership (Aenishaenslin et al., 2019). Providing educational material that is understandable at a community level in one country will not necessarily be understandable at a community level in another country. Therefore, it is prudent to involve local leaders and educators in order to develop and distribute rabies prevention programs and materials (Barbosa, Costa et al., 2018; Fenelon et al., 2017; Wallace et al., 2017). In the past several years, the global initiative "World Rabies Day," held annually on September 28, has been successful in involving many nations on all levels of society to bring attention to the understanding of rabies prevention at local, national, and international level (GARC, 2019).

16.9 Reducing vaccination regimens

The continuing global burden of rabies is a public health tragedy having an annual estimate of over 60,000 deaths per year and a cost of $1.7 billion in direct costs associated with PEP plus another $1.3 billion in indirect costs associated with seeking and delivering systems for administration of PEP (Hampson et al., 2015). The most efficient means by which to reduce this annual burden would be to eliminate canine rabies, the cause of over 99% of all human deaths. However, as mentioned previously, it is unethical to focus solely on investing in canine rabies elimination programs at the expense of saving human lives through improving access to PEP in regions where canine rabies continues to be a public health threat. WHO Rabies Modelling Consortium (2019) evaluated several rabies PEP regimens to determine the most economical PEP regimens that could help reduce the cost of PEP while improving access to human rabies vaccines in low-income countries that are eligible for funding from the GAVI, the vaccine alliance helping to provide vaccines to low-income countries. The results of this study indicated that more than one million lives will be lost in the 67 rabies-endemic countries eligible for GAVI funding between 2020 and 2035. Expanded availability of rabies vaccines for humans, through the use of a reduced one-week ID PEP regimen would prevent an estimated 489,000 deaths during the 15-year period. The WHO Rabies Modelling Consortium recommended use of the 1-week ID PEP regimen, currently recommended by WHO, as the most economical regimen to reduce the burden of human rabies (Table 16.2) (Tarantola, Blanchi, et al., 2018; Tarantola, Ly, et al., 2018; WHO, 2018c).

When CCVs were first developed and recommended for PEP by WHO, the vaccination protocol included only IM administration. The regimen for PEP extended over a 30-day period with the option of having a sixth booster dose of vaccine to be administered on day 90 (WHO, 1984). As clinical data were collected over a 40-year period, the ID route of administration was determined to be equally effective as the IM route of administration, saving an estimated 60%–80% of the vaccine per patient and reducing the cost of PEP (Kamoltham et al., 2003). Additionally, since CCVs were developed and licensed, the time required to administer PEP regimen has been reduced from the initial period of 90 days to 21 days and recently to a one-week PEP regimen (WHO, 1984, 2018c). After four decades of evaluating new human rabies vaccines, protocols, and administration routes, the question is whether PEP regimens using the currently licensed CCVs can be further reduced without affecting the efficacy of vaccination. A recent review highlighted several factors that should be considered when designing clinical studies evaluating the potential of changing PEP regimens (Tarantola, Blanchi, et al., 2018; Tarantola, Ly, et al., 2018). One of the most important issues when evaluating new vaccination regimes is to determine whether the new regimen under consideration will be more beneficial in some aspect than the currently recommended PEP regimens, particularly regarding cost effectiveness as all of the regimens for PEP as recommended by WHO are efficacious but some are considerably more cost effective than others (WHO Rabies Modelling Consortium, 2019; WHO, 2018f, 2018g).

16.10 Conclusions and future

A group of international health experts, government officials, NGOs, and other partners have set a goal of zero canine-mediated human rabies deaths by the year 2030 (Abela-Ridder et al., 2018). These organizations have developed and endorsed a global strategy that includes three objectives: Increase awareness of rabies; improve access to rabies biologicals; and eliminate the spread of canine rabies through mass vaccination programs (WHO, 2018g).

Clearly, administration of the recommended PEP protocol must be the first consideration when a human life is at risk after an exposure has occurred. However, focusing only on providing PEP without focusing on eliminating rabies in canines, the source of over 99% of human deaths, will never resolve the public health threat of rabies to humans and will only increase the financial burden of rabies globally. A "One Health" approach to the prevention of human health is essential and if this approach is undertaken, it will result in the most effective rabies prevention program that can be employed to reduce the number of human deaths. Historically, national canine rabies elimination programs have proven to be effective and when utilized in conjunction with well-designed risk assessment programs, the number of PEPs will decrease rather than increase as awareness improves (Freire de Carvalho et al., 2018; Seetahal et al., 2018; Vigilato et al., 2018). Risk assessment programs, evaluating whether a patient is at risk after a potential exposure has occurred must be established in order to avoid unnecessary use of PEP. Additionally, the national laws governing rabies prevention should align with improved epidemiological surveillance, educational awareness, canine vaccination programs, etc. For example, when surveillance is in place, it is possible to accurately monitor the epidemiology of canine rabies in a region. As mass canine vaccination programs reduce and finally eliminate the circulation of rabies in the dog population of specific regions, medical personnel should be confident enough to delay PEP while the offending dog has been observed for 10 days or for the length of observation period required in each country. National laws can be adapted to meet the changing epidemiology of canine rabies and can help to provide medical experts with the assurance that risk assessments prior to administration of PEP are worthwhile. When national laws requiring physicians to administer PEP to all bitten patients regardless of risk assessment and regardless of the changing epidemiology and surveillance information in regions where mass canine vaccination programs are successful, it is difficult to convince national health officers that the financial burden of administering PEP can be reduced as an incentive to use the One Health comprehensive approach to preventing human rabies deaths.

In the future, more efforts need to be focused on how to get rabies biologicals into remote regions where they are needed to save lives. New technology involving mobile phones, drones and the establishment of remote health stations may be of some use to improve access. This area of public health, i.e., improving access to medical assistance and supplies, is of growing concern for prevention of many diseases, including rabies. Combining delivery methods of several types of life-saving biological products may be one strategy to consider. With ever-improving modern digital methods of communication and delivery systems, novel strategies must be developed to reach the goal of 0 deaths caused by canine transmitted human rabies by 2030 (see Chapter 21).

References

Abazeed, M. E., & Cinti, S. (2007). Rabies prophylaxis for pregnant women. *Emerging Infectious Diseases*, *13*(12), 1966–1967. https://doi.org/10.3201/eid1312.070157.

Abela-Ridder, B., de Balogh, K., Kessels, J. A., Dieuzy-Labaye, I., & Torres, G. (2018). Global rabies control: The role of international organisations and the Global Strategic Plan to eliminate dog-mediated human rabies. *Revue Scientifique et Technique*, *37*(2), 741–749. https://doi.org/10.20506/rst.37.2.2837.

References

Aenishaenslin, C., Brunet, P., Levesque, F., Gouin, G. G., Simon, A., Saint-Charles, J., ... Ravel, A. (2019). Understanding the connections between dogs, health and inuit through a mixed-methods study. *EcoHealth, 16*(1), 151–160. https://doi.org/10.1007/s10393-018-1386-6.

Ambrozaitis, A., Laiskonis, A., Balciuniene, L., Banzhoff, A., & Malerczyk, C. (2006). Rabies post-exposure prophylaxis vaccination with purified chick embryo cell vaccine (PCECV) and purified vero cell rabies vaccine (PVRV) in a four-site intradermal schedule (4-0-2-0-1-1): An immunogenic, cost-effective and practical regimen. *Vaccine, 24*, 4116–4121.

Amparo, A. C. B., Mendoza, E. C. B., Licuan, D. A., Valenzuela, L. M., Madalipay, J. D., Jayme, S. I., & Taylor, L. H. (2019). Impact of integrating rabies education into the curriculum of public elementary schools in Ilocos Norte, Philippines on rabies knowledge, and animal bite incidence. *Frontiers in Public Health, 7*, 119. https://doi.org/10.3389/fpubh.2019.00119.

Arya, S. C. (1990). Rabies antibody profile among pregnant females administered therapeutic postexposure vaccine during pregnancy, their neonates and infants. *Vaccine, 8*(4), 410–411.

Barbosa Costa, G., Gilbert, A., Monroe, B., Blanton, J., Ngam Ngam, S., Recuenco, S., & Wallace, R. (2018). The influence of poverty and rabies knowledge on healthcare seeking behaviors and dog ownership, Cameroon. *PLoS One. 13*(6), e0197330. https://doi.org/10.1371/journal.pone.0197330.

Briggs, D. J., Dreesen, D. W., Nicolay, U., Chin, J. E., Davis, R., Gordon, C., & Banzhoff, A. (2000). Purified chick embryo cell culture rabies vaccine: Interchangeability with human diploid cell culture rabies vaccine and comparison of one versus two-dose post-exposure booster regimen for previously immunized persons. *Vaccine, 19*(9-10), 1055–1060.

Brown, D., Featherstone, J. J., Fooks, A. R., Gettner, S., Lloyd, E., & Schweiger, M. (2008). Intradermal pre-exposure rabies vaccine elicits long lasting immunity. *Vaccine, 26*(31), 3909–3912. https://doi.org/10.1016/j.vaccine.2008.04.081.

Cleaveland, S., Thumbi, S. M., Sambo, M., Lugelo, A., Lushasi, K., Hampson, K., & Lankester, F. (2018). Proof of concept of mass dog vaccination for thecontrol and elimination of canine rabies. *Revue Scientifique et Technique, 37*(2), 559–568. https://doi.org/10.20506/rst.37.2.2824.

De Benedictis, P., Minola, A., Rota Nodari, E., Aiello, R., Zecchin, B., Salomoni, A., ... Corti, D. (2016). Development of broad-spectrum human monoclonal antibodies for rabies post-exposure prophylaxis. *EMBO Molecular Medicine, 8* (4), 407–421. https://doi.org/10.15252/emmm.201505986.

Denis, M., Knezevic, I., Wilde, H., Hemachudha, T., Briggs, D., & Knopf, L. (2018). An overview of the immunogenicity and effectiveness of current human rabies vaccines administered by intradermal route. *Vaccine*. https://doi.org/10.1016/j.vaccine.2018.11.072.

Elser, J. L., Hatch, B. G., Taylor, L. H., Nel, L. H., & Shwiff, S. A. (2018). Towards canine rabies elimination: Economic comparisons of three project sites. *Transboundary and Emerging Diseases, 65*(1), 135–145. https://doi.org/10.1111/tbed.12637.

Fekadu, M., Endeshaw, T., Alemu, W., Bogale, Y., Teshager, T., & Olson, J. G. (1996). Possible human-to-human transmission of rabies in Ethiopia. *Ethiopian Medical Journal, 34*(2), 123–127.

Fenelon, N., Dely, P., Katz, M. A., Schaad, N. D., Dismer, A., Moran, D., ... Wallace, R. M. (2017). Knowledge, attitudes and practices regarding rabies risk in community members and healthcare professionals: Petionville, Haiti, 2013. *Epidemiology and Infection, 145*(8), 1624–1634. https://doi.org/10.1017/S0950268816003125.

Fooks, A. R. (2018). Conclusions Rabies. *Revue Scientifique et Technique, 37*(2), 761–769. https://doi.org/10.20506/rst.37.2.2839.

Fooks, A. R., Drew, T. W., & Tu, C. (2016). Laboratory twinning to build capacity for rabies diagnosis. *The Veterinary Record, 178*(10), 231–232. https://doi.org/10.1136/vr.i456.

Fooks, A. R., Johnson, N., Freuling, C. M., Wakeley, P. R., Banyard, A. C., McElhinney, L. M., ... Muller, T. (2009). Emerging technologies for the detection of rabies virus: challenges and hopes in the 21st century. *PLoS Neglected Tropical Diseases, 3*(9), e530. https://doi.org/10.1371/journal.pntd.0000530.

Franka, R., & Wallace, R. (2018). Rabies diagnosis and surveillance in animals in the era of rabies elimination. *Revue Scientifique et Technique, 37*(2), 359–370. https://doi.org/10.20506/rst.37.2.2807.

Freire de Carvalho, M., Vigilato, M. A. N., Pompei, J. A., Rocha, F., Vokaty, A., Molina-Flores, B., ... Del Rio Vilas, V. J. (2018). Rabies in the Americas: 1998-2014. *PLoS Neglected Tropical Diseases. 12*(3), e0006271, https://doi.org/10.1371/journal.pntd.0006271.

GARC. (2018). 0 by 30 our catalytic response. Retrieved from https://rabiesalliance.org/policy/united_against_rabies.

GARC. (2019). Get Ready for World Rabies Day. Retrieved from https://rabiesalliance.org/world-rabies-day.

GAVI. (2019). Retrieved from, https://www.gavi.org/about/strategy/vaccine-investment-strategy/. Accessed 1 June 2019.

Gibbons, R. V. (2002). Cryptogenic rabies, bats, and the question of aerosol transmission. *Annals of Emergency Medicine, 39*(5), 528–536.

Gilbert, A. T., Petersen, B. W., Recuenco, S., Niezgoda, M., Gomez, J., Laguna-Torres, V. A., & Rupprecht, C. (2012). Evidence of rabies virus exposure among humans in the Peruvian Amazon. *The American Journal of Tropical Medicine and Hygiene, 87*(2), 206–215. https://doi.org/10.4269/ajtmh.2012.11-0689.

Glasgow, L., Worme, A., Keku, E., & Forde, M. (2019). Knowledge, attitudes, and practices regarding rabies in Grenada. *PLoS Neglected Tropical Diseases. 13*(1)e0007079, https://doi.org/10.1371/journal.pntd.0007079.

Habel, K. (1954). Antiserum in the prophylaxis of rabies. *Bulletin of the World Health Organization, 10*, 781–788.

Hampson, K., Coudeville, L., Lembo, T., Sambo, M., Kieffer, A., Attlan, M., Barrat, ... Global Alliance for Rabies Control Partners for Rabies, Prevention (2015). Estimating the global burden of endemic canine rabies. *PLoS Neglected Tropical Diseases. 9*(4), e0003709, https://doi.org/10.1371/journal.pntd.0003709.

Hemachudha, T., Mitrabhakdi, E., Wilde, H., Vejabhuti, A., Siripataravanit, S., & Kingnate, D. (1999). Additional reports of failure to respond to treatment after rabies exposure in Thailand. *Clinical Infectious Diseases, 28*(1), 143–144. https://doi.org/10.1086/517179.

Kamoltham, T., Singhsa, J., Promsaranee, U., Sonthon, P., Mathean, P., & Thinyounyong, W. (2003). Elimination of human rabies in a canine endemic province in Thailand: five-year programme. *Bulletin of the World Health Organization, 81*(5), 375–381.

Kessels, J., Tarantola, A., Salahuddin, N., Blumberg, L., & Knopf, L. (2019). Rabies post-exposure prophylaxis: A systematic review on abridged vaccination schedules and the effect of changing administration routes during a single course. *Vaccine.* https://doi.org/10.1016/j.vaccine.2019.01.041.

Kessels, J. A., Recuenco, S., Navarro-Vela, A. M., Deray, R., Vigilato, M., Ertl, H., ... Briggs, D. (2017). Pre-exposure rabies prophylaxis: A systematic review. *Bulletin of the World Health Organization, 95*(3), 210–219c. https://doi.org/10.2471/blt.16.173039.

Khawplod, P., Benjavongkulchai, M., Limusanno, S., Chareonwai, S., Kaewchompoo, W., Tantawichien, T., & Wilde, H. (2002). Four-site intradermal postexposure boosters in previously rabies vaccinated subjects. *Journal of Travel Medicine, 9*(3), 153–155.

Lang, J., Feroldi, E., & Vien, N. C. (2009). Pre-exposure purified vero cell rabies vaccine and concomitant routine childhood vaccinations: 5-year post-vaccination follow-up study of an infant cohort in Vietnam. *Journal of Tropical Pediatrics, 55*(1), 26–31. https://doi.org/10.1093/tropej/fmm100.

LeRoux, K., Stewart, D., Perrett, K. D., Nel, L. H., Kessels, J. A., & Abela-Ridder, B. (2018). Rabies control in KwaZulu-Natal, South Africa. *Bulletin of the World Health Organization, 96*(5), 360–365. https://doi.org/10.2471/BLT.17.194886.

Lu, X. X., Zhu, W. Y., & Wu, G. Z. (2018). Rabies virus transmission via solid organs or tissue allotransplantation. *Infectious Diseases of Poverty. 7*(1)82. https://doi.org/10.1186/s40249-018-0467-7.

Manning, S. E., Rupprecht, C. E., Fishbein, D., Hanlon, C. A., Lumlertdacha, B., Guerra, M., ... Hull, H. F., & Advisory Committee on the Immunization Practices, Centers for Disease Control and Prevention (2008). Human rabies prevention—United States, 2008: recommendations of the Advisory Committee on Immunization Practices. *MMWR - Recommendations and Reports, 57*(RR-3), 1–28.

Medeiros, R., Jusot, V., Houillon, G., Rasuli, A., Martorelli, L., Kataoka, A. P., ... Tordo, N. (2016). Persistence of rabies virus-neutralizing antibodies after vaccination of rural population following vampire bat rabies outbreak in Brazil. *PLoS Neglected Tropical Diseases. 10*(9), e0004920. https://doi.org/10.1371/journal.pntd.0004920.

Mindekem, R., Lechenne, M. S., Naissengar, K. S., Oussiguere, A., Kebkiba, B., Moto, D. D., ... Zinsstag, J. (2017). Cost description and comparative cost efficiency of post-exposure prophylaxis and canine mass vaccination against rabies in N'Djamena, Chad. *Frontiers in Veterinary Science, 4*, 38. https://doi.org/10.3389/fvets.2017.00038.

Minghui, R., Stone, M., Semedo, M. H., & Nel, L. (2018). New global strategic plan to eliminate dog-mediated rabies by 2030. *The Lancet Global Health, 6*(8), e828–e829. https://doi.org/10.1016/S2214-109X(18)30302-4.

MSF. (2019). Essential Drugs/Vaccines, immunoglobulins and sera. Retrieved from https://medicalguidelines.msf.org/viewport/EssDr/english/human-rabies-immunoglobulin-hrig-16688380.html

OIE. (2019). FAO, OIE and WHO launch a guide for countries on taking a One Health approach to addressing zoonotic diseases. Retrieved from http://www.oie.int/en/for-the-media/press-releases/detail/article/fao-oie-and-who-launch-a-guide-for-countries-on-taking-a-one-health-approach-to-addressing-zoonoti/?utm_source=Press+. Access date: June 2019.

Pai, D., Kamath, A. T., Panduranga, K. P., Kamath, R., Chakravarthy, K. P., Nayak, R., ... Kumar, S. (2018). Survey of knowledge of school children towards the prevalence, severity, management of maxillofacial injuries, and rescue skills in the event of a dog bite. *Journal of the Indian Society of Pedodontics and Preventive Dentistry*, 36(4), 334–338. https://doi.org/10.4103/JISPPD.JISPPD_1110_17.

Pancharoen, C., Thisyakorn, U., Tantawichien, T., Jaijaroensup, W., Khawplod, P., & Wilde, H. (2001). Failure of pre- and postexposure rabies vaccinations in a child infected with HIV. *Scandinavian Journal of Infectious Diseases*, 33(5), 390–391.

Pengsaa, K., Limkittikul, K., Sabchareon, A., Ariyasriwatana, C., Chanthavanich, P., Attanath, P., & Malerczyk, C. (2009). A three-year clinical study on immunogenicity, safety, and booster response of purified chick embryo cell rabies vaccine administered intramuscularly or intradermally to 12- to 18-month-old Thai children, concomitantly with Japanese encephalitis vaccine. *The Pediatric Infectious Disease Journal*, 28(4), 335–337. https://doi.org/10.1097/INF.0b013e3181906351.

Pieracci, E. G., Pearson, C. M., Wallace, R. M., Blanton, J. D., Whitehouse, E. R., Ma, X., ... Olson, V. (2019). Vital Signs: Trends in Human Rabies Deaths and Exposures - United States, 1938-2018. *MMWR Morbidity and Mortality Weekly Report*, 68(23), 524–528. https://doi.org/10.15585/mmwr.mm6823e1.

PRP. (2018). Canine rabies blueprint: A blueprint for control of rabies in dog populations. Retrieved from https://caninerabiesblueprint.org/.

Rahimi, P., Vahabpour, R., Aghasadeghi, M. R., Sadat, S. M., Howaizi, N., Mostafavi, E., ... Fallahian, V. (2015). Neutralizing antibody response after intramuscular purified vero cell rabies vaccination (PVRV) in Iranian patients with specific medical conditions. *PLoS One*. 10(10), e0139171. https://doi.org/10.1371/journal.pone.0139171.

Rupprecht, C. E., Bannazadeh Baghi, H., Del Rio Vilas, V. J., Gibson, A. D., Lohr, F., Meslin, F. X., ... Gamble, L. (2018). Historical, current and expected future occurrence of rabies in enzootic regions. *Revue Scientifique et Technique*, 37(2), 729–739. https://doi.org/10.20506/rst.37.2.2836.

Rupprecht, C. E., Kuzmin, I. V., Yale, G., Nagarajan, T., & Meslin, F. X. (2019). Priorities in applied research to ensure programmatic success in the global elimination of canine rabies. *Vaccine*. https://doi.org/10.1016/j.vaccine.2019.01.015.

Saeed, B., & Al-Mousawi, M. (2017). Rabies acquired through kidney transplantation in a child: A case report. *Experimental and Clinical Transplantation*, 15(3), 355–357. https://doi.org/10.6002/ect.2017.0046.

Salahuddin, N., Gohar, M. A., & Baig-Ansari, N. (2016). Reducing cost of rabies post exposure prophylaxis: Experience of a tertiary care hospital in Pakistan. *PLoS Neglected Tropical Diseases*. 10(2), e0004448. https://doi.org/10.1371/journal.pntd.0004448.

San Jose, R., Magsino, P. J., & Bundalian, R., Jr. (2019). Pet owners' awareness on RA 9482 (Anti-Rabies Act of 2007) in Magalang, Pampanga Philippines. *Heliyon*. 5(5)e01759https://doi.org/10.1016/j.heliyon.2019.e01759.

Seetahal, J. F. R., Vokaty, A., Vigilato, M. A. N., Carrington, C. V. F., Pradel, J., Louison, B., ... Rupprecht, C. E. (2018). Rabies in the Caribbean: A situational analysis and historic review. *Tropical Medicine and Infectious Disease*. 3(3) https://doi.org/10.3390/tropicalmed3030089.

Shantavasinkul, P., Tantawichien, T., Wacharapluesadee, S., Jeamanukoolkit, A., Udomchaisakul, P., Chattranukulchai, P., ... Hemachudha, T. (2010). Failure of rabies postexposure prophylaxis in patients presenting with unusual manifestations. *Clinical Infectious Diseases*, 50(1), 77–79. https://doi.org/10.1086/649873.

Shaughnessy (1954). Treatment of wounds inflicted by rabid animals. *Bulletin of the World Health Organization*, 10, 805–813.

Shim, E., Hampson, K., Cleaveland, S., & Galvani, A. P. (2009). Evaluating the cost-effectiveness of rabies post- exposure prophylaxis: A case study in Tanzania. *Vaccine*, 27(51), 7167–7172. https://doi.org/10.1016/j.vaccine.2009.09.027.

Smith, S. P., Wu, G., Fooks, A. R., Ma, J., & Banyard, A. C. (2019). Trying to treat the untreatable: experimental approaches to clear rabies virus infection from the CNS. *The Journal of General Virology*. https://doi.org/10.1099/jgv.0.001269.

Sparrow, E., Torvaldsen, S., Newall, A. T., Wood, J. G., Sheikh, M., Kieny, M. P., & Abela-Ridder, B. (2018). Recent advances in the development of monoclonal antibodies for rabies post exposure prophylaxis: A review of the current status of the clinical development pipeline. *Vaccine*. https://doi.org/10.1016/j.vaccine.2018.11.004.

Sudarshan, M. K., Madhusudana, S. N., Mahendra, B. J., Ashwathnarayana, D. H., Jayakumary, M., & Gangaboriah (1999). Post exposure rabies prophylaxis with Purified Verocell Rabies Vaccine: a study of immunoresponse in pregnant women and their matched controls. *Indian Journal of Public Health*, 43(2), 76–78.

Sudarshan, M. K., Narayana, D. H., Madhusudana, S. N., Holla, R., Ashwin, B. Y., Gangaboraiah, B., & Ravish, H. S. (2012). Evaluation of a one week intradermal regimen for rabies post-exposure prophylaxis: results of a randomized, open label, active-controlled trial in healthy adult volunteers in India. *Human Vaccines & Immunotherapeutics*, 8(8), 1077–1081. https://doi.org/10.4161/hv.20471.

Suwansrinon, K., Wilde, H., Benjavongkulchai, M., Banjongkasaena, U., Lertjarutorn, S., Boonchang, S., ... Sitprija, V. (2006). Survival of neutralizing antibody in previously rabies vaccinated subjects: a prospective study showing long lasting immunity. *Vaccine, 24*(18), 3878–3880. https://doi.org/10.1016/j.vaccine.2006.02.027.

Tantawichien, T., Jaijaroensup, W., Khawplod, P., & Sitprija, V. (2001). Failure of multiple-site intradermal postexposure rabies vaccination in patients with human immunodeficiency virus with low CD4+ T lymphocyte counts. *Clinical Infectious Diseases, 33*(10), E122–E124. https://doi.org/10.1086/324087.

Tarantola, A., Blanchi, S., Cappelle, J., Ly, S., Chan, M., In, S., ... Y, J. (2018). Rabies postexposure prophylaxis noncompletion after dog bites: Estimating the unseen to meet the needs of the underserved. *American Journal of Epidemiology, 187*(2), 306–315. https://doi.org/10.1093/aje/kwx234.

Tarantola, A., Ly, S., Chan, M., In, S., Peng, Y., Hing, C., ... Mary, J. Y. (2018). Intradermal rabies post-exposure prophylaxis can be abridged with no measurable impact on clinical outcome in Cambodia, 2003–2014. *Vaccine.* https://doi.org/10.1016/j.vaccine.2018.10.054.

Tarantola, A., Tejiokem, M. C., & Briggs, D. J. (2018). Evaluating new rabies post-exposure prophylaxis (PEP) regimens or vaccines. *Vaccine.* https://doi.org/10.1016/j.vaccine.2018.10.103.

Taylor, L. H., Costa, P., & Briggs, D. J. (2013). Public health management of humans at risk. In A. C. Jackson (Ed.), *Rabies scientific basis of the disease and its management* (pp. 543–573). (3rd ed.). San Diego: Academic Press.

Thisyakorn, U., Pancharoen, C., & Wilde, H. (2001). Immunologic and virologic evaluation of HIV-1-infected children after rabies vaccination. *Vaccine, 19*(11–12), 1534–1537.

Tiwari, H. K., Vanak, A. T., O'Dea, M., & Robertson, I. D. (2018). Knowledge, attitudes and practices towards dog-bite related rabies in para-medical staff at rural primary health centres in Baramati, western India. *PLoS One. 13*(11), e0207025. https://doi.org/10.1371/journal.pone.0207025.

UMMS. (2019). Serum Institute of India launches Rabishield, developed in partnership with UMMS. Retrieved from https://umassmed.edu/news/news-archives/2017/10/serum-institute-of-india-launches-rabishield-developed-in-partnership-with-umms/

Vigilato, M. A. N., Molina-Flores, B., Del Rio Vilas, V. J., Pompei, J. C., & Cosivi, O. (2018). Canine rabies elimination: Governance principles. *Revue Scientifique et Technique, 37*(2), 703–709. https://doi.org/10.20506/rst.37.2.2859.

Wallace, R. M., Mehal, J., Nakazawa, Y., Recuenco, S., Bakamutumaho, B., Osinubi, M., ... Wamala, J. (2017). The impact of poverty on dog ownership and access to canine rabies vaccination: results from a knowledge, attitudes and practices survey, Uganda 2013. *Infectious Diseases of Poverty. 6*(1)97. https://doi.org/10.1186/s40249-017-0306-2.

Wei, Y., Liu, X., Li, D., Chen, S., Xu, J., Chen, K., & Yang, Z. (2019). *Bulletin of the World Health Organization, 97*, 51–58. https://doi.org/10.2471/BLT.18.217372.

WHO. (1984). WHO Expert Committee on Rabies. Seventh Report. Retrieved from Geneva: https://apps.who.int/iris/bitstream/handle/10665/38724/WHO_TRS_709.pdf?sequence=1&isAllowed=y. Accessed June 2019.

WHO. (2014). WHO Guideline for Pre and Post Exposure Prophylaxis in Humans. Retrieved from https://www.who.int/rabies/PEP_Prophylaxis_guideline_15_12_2014.pdf?ua=1. Accessed 2019.

WHO. (2017). WHO Immunological Basis for Immunization Series Module 17: Rabies Update 2017. Retrieved from Geneva: http://www.who.int/immunization/documents/policies/WHO_IVB_ISBN9789241513371/en/. Accessed June 2019.

WHO. (2018a). Driving progress toward rabies elimination. In: *Results of Gavi's learning agenda and 2nd International meeting of the Pan-African Rabies Control Network (PARACON). Meeting Report 12-14 September 2018*. Retrieved from https://www.who.int/rabies/resources/WHO-CDS-NTD-NZD-2019.02/en/ Accessed June 2019.

WHO. (2018b). Rabies vaccines and immunglobulins: WHO Position. Summary of 2017 updates. Retrieved from: https://apps.who.int/iris/bitstream/handle/10665/259855/WHO-CDS-NTD-NZD-2018.04-eng.pdf;jsessionid=207CF7B8C3C4FB9FB10F18CABD31A38A?sequence=1. Accessed October 2019.

WHO. (2018c). Rabies Vaccines: WHO Position Paper—April 2018. *Weekly Epidemiological Record, 16*(93), 201–220.

WHO. (2018d). Rabies vaccines: WHO position paper, April 2018—Recommendations. *Vaccine, 36*(37), 5500–5503. https://doi.org/10.1016/j.vaccine.2018.06.061.

WHO. (2018e). *WHO Expert Consulaton on Rabies: Third Report.* Retrieved from Geneva. https://apps.who.int/iris/bitstream/handle/10665/272364/9789241210218-eng.pdf?ua=1. Accessed 1 June 2019.

WHO. (2018f). WHO Expert Consultation on Rabies. Retrieved from Geneva, Switzerland: Retrieved from: https://www.who.int/rabies/resources/who_trs_1012/en/. Accessed June 2019.

WHO. (2018g). Zero by 30: the global strategic plan to end human deaths from dog-mediated rabies by 2030. Retrieved from https://www.who.int/rabies/resources/9789241513838/en/. Accessed June 2019.

WHO Rabies Modelling Consortium. (2019). The potential effect of improved provision of rabies post-exposure prophylaxis in Gavi-eligible countries: A modelling study. *The Lancet Infectious Diseases, 19*(1), 102–111. https://doi.org/10.1016/S1473-3099(18)30512-7.

Wilde, H. (2007). Failures of post-exposure rabies prophylaxis. *Vaccine, 25*(44), 7605–7609. https://doi.org/10.1016/j.vaccine.2007.08.054.

Wilde, H., Sirikawin, S., Sabcharoen, A., Kingnate, D., Tantawichien, T., Harischandra, P. A., ... Sitprija, V. (1996). Failure of postexposure treatment of rabies in children. *Clinical Infectious Diseases, 22*(2), 228–232.

Winkler, W. G., Fashinell, T. R., Leffingwell, L., Howard, P., & Conomy, P. (1973). Airborne rabies transmission in a laboratory worker. *JAMA, 226*(10), 1219–1221.

Zhang, J., Lin, J., Tian, Y., Ma, L., Sun, W., Zhang, L., ... Zhang, L. (2018). Transmission of rabies through solid organ transplantation: a notable problem in China. *BMC Infectious Diseases. 18*(1)273. https://doi.org/10.1186/s12879-018-3112-y.

Further reading

Ravish, H. S., Sudarshan, M. K., Madhusudana, S. N., Annadani, R. R., Narayana, D. H., Belludi, A. Y., ... Vijayashankar, V. (2014). Assessing safety and immunogenicity of post-exposure prophylaxis following interchangeability of rabies vaccines in humans. *Human Vaccines & Immunotherapeutics, 10*(5), 1354–1358. https://doi.org/10.4161/hv.28064.

CHAPTER 17

Therapy of human rabies

Alan C. Jackson

Professor of Medicine (Neurology), University of Manitoba, Winnipeg, MB, Canada

Preventative therapy for rabies after exposures is highly effective if current recommendations are followed (Manning et al., 2008; World Health Organization, 2018). However, treatment of patients with rabies remains disappointing. There has been an increased number of survivors reported in recent years, particularly from India (Table 17.1). Unfortunately, many of the survivors have severe neurological sequelae with poor quality of life and some subsequently die of complicating medical illnesses over a period of months to years after the acute illness. An arbitrary period to deem a patient a survivor from rabies is six months after the acute illness. Claiming survival of a patient immediately upon discharge from a critical care unit is premature (Caicedo et al., 2015; Jackson & Garland, 2015) because many medical complications may still occur that are directly related to the critical illness.

In 2001, a conference was held that included physicians who have experience in the management of human rabies and researchers with expertise in rabies pathogenesis. The opinions of the participants were published in a viewpoint article in *Clinical Infectious Diseases*, including therapeutic options when aggressive therapy is considered desirable (Jackson et al., 2003). It was felt that a combination of specific therapies should be considered, including rabies vaccine, rabies immune globulin, monoclonal antibodies (in the future), ribavirin, interferon-α, and ketamine, which is a dissociative anesthetic agent that is a noncompetitive antagonist of the *N*-methyl-D-aspartate (NMDA) receptor; the pros and cons of using each agent are discussed in this review. Previous studies performed in vitro and in an experimental animal model suggested ketamine may be a useful therapeutic agent (Lockhart, Tordo, & Tsiang, 1992; Lockhart, Tsiang, Ceccaldi, & Guillemer, 1991). However, subsequent experimental work did not support the conclusions of this earlier study (Weli, Scott, Ward, & Jackson, 2006). Many factors should be carefully considered before making a final decision whether to embark or not on an aggressive approach to management of a patient with rabies, which are summarized in Table 17.2. A palliative approach may be more appropriate for other patients.

TABLE 17.1 Cases of human rabies with recovery.[a]

	Location	Year of onset	Age of patient/sex	Transmission	Immunization prior to onset	Outcome	Reference
1	United States	1970	6/M	Bat bite	Duck embryo vaccine	Complete recovery	Hattwick et al. (1972)
2	Argentina	1972	45/F	Dog bites	Suckling mouse brain vaccine	Moderate sequelae	Porras et al. (1976)
3	United States	1977	32/M	Laboratory (vaccine strain)	Pre-exposure vaccination	Severe sequelae	Tillotson et al. (1977b), Tillotson et al. (1977a)
4	Mexico	1992	9/M	Dog bites	Postexposure vaccination (combination)	Severe sequelae[b]	Alvarez et al. (1994)
5	India	2000	6/F	Dog bites	Postexposure vaccination (combination)	Severe sequelae[c]	Madhusudana et al. (2002)
6	United States	2004	15/F	Bat bite	No postexposure therapy	Mild sequelae	Willoughby Jr. et al. (2005), Hu et al. (2007)
7	Brazil	2008	15/M	Vampire bat bite	Postexposure vaccination	Severe sequelae	Ministerio da Saude in Brazil (2008)
8	Turkey	2008	17/M	Dog bites	Postexposure vaccination (one dose)	Complete recovery	Karahocagil et al. (2013)
9	India	2010	8/M	Dog bite	Postexposure vaccination and rabies immunoglobulin	Severe sequelae (remained alive 5 years post acute illness)	Netravathi et al. (2015)
10	India	2011	17/M	Dog bite	Postexposure vaccination	Severe sequelae	de Souza and Madhusudana (2014)
11	South Africa	2012	4/M	Dog bites	Postexposure vaccination	Severe sequelae	Weyer et al. (2016)
12	Chile	2013	25/M	Dog bite(s)	Postexposure vaccination (one dose)	Moderate sequelae	Galvez et al. (2013)
13	India	2014	16/M	Dog bite(s)	Postexposure vaccination	Severe sequelae	Thakur (2014), Kumar, Ahmad, and Dutta (2015)

TABLE 17.1 Cases of human rabies with recovery[a]—cont'd

	Location	Year of onset	Age of patient/ sex	Transmission	Immunization prior to onset	Outcome	Reference
14	India	2014	6/M	Dog bites	Postexposure vaccination and equine rabies immune globulin	Severe sequelae	Karande et al. (2015)
15	India	2014	13/M	Dog bite	Postexposure vaccination	Severe sequelae	Manoj, Mukherjee, Johri, and Kumar (2016), Rao, Pimpalwar, Mukherjee, and Yadu (2017)
16	India	2015	10/M	Dog bite	Postexposure vaccination	Unknown	Damodar, Mani, and Prathyusha (2019)[d]
17	India	2015	5/M	Dog bites	Postexposure vaccination	Unknown	Damodar et al. (2019)[d]
18	India	2015	18/F	Dog bite	Postexposure vaccination and equine rabies immune globulin	Mild sequelae	Damodar et al. (2019)[d]
19	India	2015	10/M	Dog bites	Postexposure vaccination	Severe sequelae[e]	Mani et al. (2019)
20	India	2016	5/F	Dog bite	Postexposure vaccination	Severe sequelae[f]	Mani et al. (2019)
21	India	2017	26/M	Dog bite	Postexposure vaccination	Moderate sequelae	Mani et al. (2019)
22	India	2017	9/M	Dog bite	Postexposure vaccination and equine rabies immune globulin	Mild sequelae	Mani et al. (2019)
23	India	2017	4/M	Dog bite	Postexposure vaccination and equine rabies immune globulin	Severe sequelae	Mani et al. (2019)
24	India	2017	3/F	Dog bite	Postexposure vaccination	Moderate sequelae	Mani et al. (2019)

Continued

TABLE 17.1 Cases of human rabies with recovery[a]—cont'd

	Location	Year of onset	Age of patient/ sex	Transmission	Immunization prior to onset	Outcome	Reference
25	India	2017	5/F	Dog bites	Postexposure vaccination and human rabies immune globulin	Severe sequelae	Mani et al. (2019)
26	Brazil	2017	14/M	Vampire bat bites	Unknown	Severe sequelae	Herriman (2018), Tanner (2018)
27	India	2018	8/M	Dog bite	Postexposure vaccination	Severe sequelae	Bokade, Gajimwar, Meshram, and Wathore (2019)

[a] *Recovery of cases with atypical features of rabies without the development of rabies virus neutralizing antibodies has not been not included because they are likely not cases of rabies (Holzmann-Pazgal et al., 2010; Wiedeman et al., 2012). Cases reported by Rawat and Rao (2011) and by Apanga, Awoonor-Williams, Acheampong, and Adam (2016) were not sufficiently well documented for inclusion. Patients known to survive less than 6 months have been excluded (albeit an arbitrary period).*
[b] *Patient died less than four years after developing rabies with marked neurological sequelae (L. Alvarez, personal communication).*
[c] *Patient died about two years after developing rabies with marked neurological sequelae (S. Mahusudana, personal communication).*
[d] *Additional information provided by Dr. R.S. Mani, National Institute of Mental Health & Neurosciences, Bangalore, India.*
[e] *Patient died 6 months after developing rabies with marked neurological sequelae.*
[f] *Patient died 8 months after developing rabies with marked neurological sequelae.*

TABLE 17.2 Factors for and against initiation of an aggressive approach for the treatment of rabies.

Factors favoring initiation of an aggressive approach	Factors against initiation of an aggressive approach
Young age	Older age
Previously healthy	Medical comorbidities/immune compromise
One or more doses of rabies vaccine administered prior to the onset of clinical symptoms	No previous history of administration of rabies vaccine
Early clinical rabies (e.g., local sensory symptoms)	Late clinical rabies (e.g., quadriparesis or coma)
Presence of neutralizing antirabies virus antibodies in serum and/or cerebrospinal fluid	Absence of neutralizing antirabies virus antibodies in serum and/or cerebrospinal fluid
Diagnostic tests negative for rabies virus antigen and rabies virus RNA	Diagnostic tests positive for rabies virus antigen and rabies virus RNA
Access to critical care facilities	Lack of access to critical care facilities
Acceptance that the outcome could result in severe neurological deficits	Lack of acceptance that the outcome could result in severe neurological deficits

Reproduced with permission from Jackson, A. C. (2019). Treatment of rabies. In T. W. Post (Ed.), UpToDate. Waltham, MA: UpToDate (Accessed on January 1, 2020). Copyright © 2019 UpToDate, Inc. For more information visit www.uptodate.com.

17.1 Human cases with recovery from rabies

Survival from rabies has now been well documented in 27 patients (Table 17.1), and all but one of these patients received one or more doses of rabies vaccine prior to the onset of clinical disease. The first six survivors from rabies will be discussed here in detail. The first recovery from rabies, which has been one of the few cases with survival without significant neurological sequelae, occurred in 1970 (Hattwick, Weis, Stechschulte, Baer, & Gregg, 1972). Matthew Winkler, a 6-year-old boy from Ohio, was bitten on his left thumb by a big brown bat (*Eptesicus fuscus*), which was later shown to be rabid. Vaccination was initiated with duck embryo rabies vaccine beginning 4 days after the bite. Shortly after completing the multidose therapy (20 days after the bite), he became ill with fever and meningeal signs. His CSF showed 125 white cells/µL (75% mononuclear cells and 25% polymorphonuclear leukocytes) and the CSF protein was elevated. He developed abnormal behavior and later lapsed into a coma. He had focal neurological signs and seizures and developed cardiac and respiratory complications. He subsequently showed progressive improvement and had a good neurologic recovery. A brain biopsy was consistent with encephalitis. His serum neutralization titer against rabies virus peaked at 1:63,000 at 3 months. This titer was much higher than observed secondary to vaccination. He also had very high titers of neutralizing antibodies in the CSF, which are not observed with vaccination. Rabies virus was not isolated from brain tissue, CSF, or saliva, probably as a result of viral neutralization related to the high antibody levels.

The second case with recovery was a 45-year-old woman who sustained multiple deep bites to her left arm from a dog in Argentina in 1972 (Porras et al., 1976). The dog developed neurologic signs and died 4 days later. The patient received 14 daily doses of suckling mouse brain rabies vaccine beginning 10 days after the bites, which were followed by two booster doses. Twenty-one days after the bites (at the time of her twelfth vaccine dose), she developed left arm paresthesias, which subsequently spread and became accompanied by pain; vaccination was continued. She was admitted to hospital with quadriparesis and hyperreflexia 31 days after the bites. She had limb weakness, tremor in her upper extremities (greater on the left), cerebellar signs (asynergia, ataxia, dysmetria and dysdiadochokinesia), generalized myoclonus, and hyperreflexia in her lower extremities. Her CSF showed 5 cells/µL and CSF protein was mildly elevated at 0.65 g/L. Her serum neutralization titer against rabies virus peaked at 1:640,000 at about 3 months and she also had very high titers of neutralizing antibodies in CSF. Rabies virus was not isolated from her saliva or CSF, and corneal impression smears were negative for rabies virus antigen. Neurologic deterioration occurred shortly after she received each of the two booster doses of rabies vaccine and included altered mental status, generalized seizures, dysphagia, and quadriparesis. She showed neurological improvement over the next few months. Thirteen months after the onset of her symptoms, her recovery was reported as "nearly complete." However, there was no description of her residual neurological deficits (Porras et al., 1976). The unusual neurologic features of this patient, as well as the clinical worsening after booster doses of the suckling mouse brain rabies vaccine were administered, raise the question of whether encephalomyelitis due to the rabies vaccine played a significant role in this patient's clinical picture.

The third case occurred in a 32-year-old laboratory technician in New York in 1977 who was preimmunized with duck embryo rabies vaccine (Tillotson, Axelrod, & Lyman, 1977a;

Tillotson, Axelrod, & Lyman, 1977b). About 5 months prior to his illness, he had a rabies virus neutralizing antibody titer of 1:32. He worked with live rabies virus vaccine strains and he was likely exposed to an aerosol of rabies virus about two weeks prior to the onset of his illness. He experienced initial malaise, headache, fever, chills, and nausea and then lethargy with intermittent delirium. He was admitted to hospital in Albany, New York, 6 days after the onset of his symptoms with expressive aphasia, hyperreflexia, and primitive reflexes. CSF showed 230 white cells/μL (95% mononuclear cells), and CSF protein was elevated at 1.17 g/L. The day after admission to hospital, he deteriorated and went into a deep coma. His serum neutralizing antibody titer increased from 1:32 to 1:64,000 and subsequently increased to 1:175,000 over a 10-day period during his illness (Tillotson et al., 1977a). He also developed a high titer of CSF antibodies. Rabies virus antigen was not detectable in a skin biopsy or in corneal impression smears. Four months after the onset of his illness, he was ambulatory, but he had residual aphasia and spasticity (Tillotson et al., 1977a). This was the first report of a case of rabies in a preimmunized individual and only the fourth well-documented case with transmission due to airborne exposure to the virus.

The fourth case, from 1992, occurred in a 9-year-old boy in Mexico (Alvarez et al., 1994). This boy sustained severe facial bites from a dog and received local wound treatment. On the day after the bites, vaccination was initiated with VERO rabies vaccine, but passive immunization with rabies immune globulin was not given. Nineteen days after the bites, he developed fever and dysphagia. He subsequently had a variety of abnormal neurological signs and convulsions. He never developed hydrophobia or inspiratory spasms. He was admitted to hospital and subsequently became comatose. CSF showed 184 cells/μL (65% mononuclear cells). He required mechanical ventilation for several days. Rabies virus was not isolated from saliva and rabies virus antigen was not found in either a skin biopsy or corneal impression smears. His peak serum neutralizing antibody titer was 1:34,800 (39 days after the bite) and he had a very high CSF antibody titer. He had severe neurologic sequelae, including quadriparesis and visual impairment. Although he recovered for a period, he died almost 4 years later (L. Alvarez, personal communication).

The fifth case was a 6-year-old girl who was bitten on the face and hands by a dog in India and the dog died 4 days later (Madhusudana, Nagaraj, Uday, Ratnavalli, & Kumar, 2002). She received three doses of rabies purified chick embryo cell vaccine (PCECV) on days 0, 3, and 7, but no local wound treatment was given and rabies immune globulin was not administered. She developed clinical features of rabies 14 days after the bites, which included fever, dysphagia to liquids, and visual hallucinations. A rare neurologic complication to the PCECV was considered, and she was given methylprednisolone and one dose of rabies human diploid cell vaccine. She subsequently developed hypersalivation and focal motor seizures, and she became comatose. A MR scan showed T_2-weighted hyperintense signals in the cerebral cortex, basal ganglia, and brainstem. She had a CSF pleocytosis. Her peak serum neutralizing antibody titer was 1:312,000 (7800 IU/mL) after 110 days of illness, and she had a CSF antibody titer of 1:182,000 (4550 IU/mL) at this time. Rabies virus was not isolated, and both skin biopsies and corneal tests were negative for rabies virus antigen. She had severe neurologic sequelae, including rigidity and involuntary movements of her limbs and she had frequent opisthotonic postures. She died about 2 years later (S. Madhusudana, personal communication).

The sixth case occurred in Wisconsin in 2004 (Willoughby Jr. et al., 2005). A previously healthy 15-year-old female was bitten by a bat on her left index finger while attending a church service and she subsequently released the bat. The wound was washed with peroxide, but she did not seek medical attention at that time. About one month after the bite she developed numbness and tingling of her left hand, and over the next 3 days, she developed diplopia related to bilateral partial sixth-nerve palsies, unsteadiness, and nausea and vomiting. MRI brain was normal. On her fourth day of illness CSF showed 23 white cells/μL (93% lymphocytes), and CSF protein was mildly elevated at 50 mg/dL. She subsequently developed fever (38.8 °C), nystagmus, left arm tremor, and hypersalivation, and at about that time the history of the bat bite was obtained. The patient was transferred to a tertiary care hospital in Milwaukee 5 days after the onset of neurologic symptoms. A repeat MRI scan was normal. Neutralizing antirabies virus antibodies were detected in serum and CSF on the first hospital day (initially 1:102 and 1:47, respectively) and subsequently increased (to 1:1183 and 1:1300, respectively). These antibody levels were very high, especially during the early clinical course of her disease. Nuchal skin biopsies were negative for rabies virus antigen, rabies virus RNA was not detected in the skin biopsies or in saliva by RT-PCR, and viral isolation on saliva was negative. The patient was intubated and put into a drug-induced coma, which included the noncompetitive NMDA antagonist ketamine at 48 mg/kg/day as a continuous infusion and intravenous midazolam for 7 days. There was a deliberate attempt to maintain a burst-suppression pattern on her electroencephalogram, and supplemental phenobarbital was given. She also received intravenous ribavirin and amantadine 200 mg per day administered enterally. She improved and was discharged from hospital with neurologic deficits, and she has subsequently shown further progressive neurologic improvement (Hu, Willoughby Jr., Dhonau, & Mack, 2007). The therapeutic approach has been dubbed the "Milwaukee protocol" and strongly promoted for therapy of rabies patients.

Since 2004, there have been 22 reported cases with survival, including 16 (80%) from India (all since 2010 and 11 since 2015) and most of them had severe neurological sequelae, but there are a minority of survivors with mild-to-moderate sequelae (Table 17.1). Improvements in critical care are undoubtedly responsible, at least in part, for the increased number of patients surviving rabies in recent years.

There are also two reported cases with rabies virus antibodies, but without neutralizing antirabies virus antibodies in serum and CSF; they should not be considered survivors of rabies. The first was a female who lacked the typical clinical features of rabies and did not require intensive care (Holzmann-Pazgal et al., 2010). She had fever, headache, nuchal rigidity, disorientation, and limb weakness, and had a CSF pleocytosis and enlarged lateral ventricles on MR imaging. She developed only a low titer of rabies virus neutralizing antibodies in sera (up to 1:14) after receiving human rabies immune globulin and one dose of rabies vaccine and no detectable neutralizing antibodies in CSF (Holzmann-Pazgal et al., 2010). Other diagnostic tests for rabies (detection of rabies virus antigen and RNA) were negative. The second case was an 8-year-old female from California (Wiedeman et al., 2012). She experienced sore throat and vomiting. Later, over a few days, she developed swallowing difficulties. A few days later, she developed abdominal pain and neck and back pain, and then on the next day she had sore throat and abdominal pain and was noted to be confused. She deteriorated rapidly and required endotracheal intubation. CSF showed 6 leukocytes/μL with a protein of 62 mg/dL. Over the next few days, she developed ascending flaccid paralysis, decreased level of consciousness, and fever. MR imaging of brain showed multiple T2 and FLAIR signal abnormalities in cortical and subcortical regions and in the periventricular white matter. Electrophysiological studies were consistent with a demyelinating and predominantly motor polyneuropathy. She had rabies virus-specific IgG and IgM in her serum and CSF, but she did not develop rabies virus neutralizing antibodies. All other diagnostic tests for rabies were also negative. She showed progressive improvement after just over two weeks in hospital and she was discharged home after another 5 weeks in rehabilitation. Neither of these cases had typical clinical features of rabies. The atypical clinical features plus the lack or minimal development of rabies virus neutralizing antibodies suggest that it is unlikely that these two patients recovered from rabies. The exact etiology and pathogenetic mechanisms involved in the illnesses of these two patients remain elusive, but they should not be considered rabies survivors.

17.1.1 Milwaukee protocol

The therapy given to a rabies patient in Milwaukee in 2004 (Willoughby Jr. et al., 2005) has been called the Milwaukee protocol, but this protocol has changed over time despite the lack of therapeutic success. The main component of the protocol is therapeutic coma, which was correctly predicted to lack efficacy in the accompanying editorial with the case report (Jackson, 2005). An evaluation of the components of the protocol has recently been critically analyzed in detail with the conclusion that major reconsideration is needed for most of the recommendations. The protocol lacks a firm scientific rationale (Zeiler & Jackson, 2016) and at least 53 failures have now been documented (Table 17.3) plus an additional six cases from three countries who died after receiving therapy with favipiravir (Willoughby & Epstein, 2019). Claimed successes of the Milwaukee protocol include patients who have died and

TABLE 17.3 Cases of human rabies with treatment failures that used the main components of the "Milwaukee Protocol."

Case no.	Year of death	Age/sex	Virus source	Country	References
1	2005	47 Male	Kidney and pancreas transplant (dog variant)	Germany	Maier et al. (2010)
2	2005	46 Female	Lung transplant (dog variant)	Germany	Maier et al. (2010)
3	2005	72 Male	Kidney transplant (dog variant)	Germany	Maier et al. (2010)
4	2005	Unknown	Dog	India	Bagchi (2005)
5	2005	7 Male	Vampire bat	Brazil	a
6	2005	20–30 Female	Vampire bat	Brazil	a
7	2006	33 Male	Dog	Thailand	Hemachudha et al. (2006)
8	2006	16 Male	Bat	USA (Texas)	Houston Chronicle (2006)

TABLE 17.3 Cases of human rabies with treatment failures that used the main components of the "Milwaukee Protocol"—cont'd

Case no.	Year of death	Age/sex	Virus source	Country	References
9	2006	10 Female	Bat	USA (Indiana)	Christenson et al. (2007)
10	2006	11 Male	Dog (Philippines)	USA (California)	Christenson et al. (2007), Aramburo et al. (2011)
11	2007	73 Male	Bat	Canada (Alberta)	McDermid et al. (2008)
12	2007	55 Male	Dog (Morocco)	Germany	Drosten (2007)
13	2007	34 Female	Bat (Kenya)	The Netherlands	van Thiel et al. (2009)
14	2008	5 Male	Dog	Equatorial Guinea	Rubin et al. (2009)
15	2008	55 Male	Bat	USA (Missouri)	Pue et al. (2009), Turabelidze, Pue, Grim, and Patrick (2009)
16	2008	9 Female	Cat (vampire bat variant)	Colombia	Caicedo et al. (2015), Jackson and Garland (2015)
17	2008	15 Male	Vampire bat	Colombia	Badillo, Mantilla, and Pradilla (2009)
18	2008	Unknown	Bat	Brazil	Vargas, Romano, and Merchan-Hamann (2019)[b]
19	2009	37 Female	Dog (South Africa)	Northern Ireland	Hunter et al. (2010)
20	2009	42 Male	Dog (India)	USA (Virginia)	Blanton, Palmer, and Rupprecht (2010)
21	2010	11 Female	Cat	Romania	Luminos et al. (2011)
22	2010	45 Male	Dog	India	Mohan et al. (2014)
23	2011	41 Female	Dog (Guinea-Bissau)	Portugal	Santos et al. (2012)
24	2011	25 Male	Dog (Afghanistan)	USA (New York)	Javaid et al. (2012), Nat, Nat, Sharma, Pothineni, and Amzuta (2014)
25	2011	32 Male	Dog	India	Manesh et al. (2018)
26	2012	63 Male	Brown bat	USA (Massachusetts)	Greer, Robbins, Lijewski, Gonzalez, and McGuone (2013)
27	2012	9 Male	Marmoset	Brazil	NE 10 (2012)
28	2012	Unknown	Bat	Brazil	Vargas et al. (2019)[b]
29	2012	Unknown	Deer	Brazil	Vargas et al. (2019)[b]
30	2012	41 Male	Dog (Dominican Republic)	Canada (Ontario)	Branswell (2012)

Continued

TABLE 17.3 Cases of human rabies with treatment failures that used the main components of the "Milwaukee Protocol"—cont'd

Case no.	Year of death	Age/sex	Virus source	Country	References
31	2012	29 Male	Dog (Mozambique)	South Africa	Times Live (2012), IAfrica.com (2012)
32	2012	58 Female	Dog (India)	United Kingdom	Pathak et al. (2014)
33	2013	28 Male	Dog variant with no known exposure (Guatemala)	USA (Texas)	Wallace et al. (2014)
34	2013	30 Male[c]	Dog (China)	Taiwan	Chen et al. (2015)
35	2013	8 Male	Bat	Australia	Francis et al. (2014)
36	2013	Unknown	Marmoset	Brazil	Vargas et al. (2019)[b]
37	2014	24 Male	Dog	India	Ramya (2014)
38	2014	Male	Liver transplant (dog variant)	Saudi Arabia	Elsiesy et al. (2015)
39	2014	32 Female	Dog (India)	Israel	Cohen et al. (2016)
40	2014	23 Male	Dog	India	Manesh et al. (2018)
41	2014	Unknown	Dog	Brazil	Vargas et al. (2019)[b]
42	2015	55 Male	Kidney transplant (dog variant)	China	Gong et al. (2017)
43	2015	43 Male	Kidney transplant (dog variant)	China	Gong et al. (2017)
44	2015	17 Male	No known exposure	India	Manesh et al. (2018)
45	2015	Unknown	Dog	Brazil	Vargas et al. (2019)[b]
46	2015	Unknown	Cat	Brazil	Vargas et al. (2019)[b]
47	2016	10 Male	Dog	India	Agarwal (2017)
48	2016	37 Male	Vampire bat	Brazil	Lima et al. (2018)
49	2016	Unknown	Cat	Brazil	Vargas et al. (2019)[b]
50	2016	12 Male	Dog	Saudi Arabia	Dhayhi et al. (2019)
51	2017	65 Female	Dog (India)	USA (Virginia)	Murphy et al. (2019)
52	2017	Unknown	Bat	Brazil	Vargas et al. (2019)[b]
53	2018	6 Male	Bat	USA (Florida)	Sheridan (2018)

[a] Personal communication from Dr. Rita Medeiros, University of Para, Belem, Brazil.
[b] Additional information provided by Alexander Vargas, Ministerio da Saude, Brasilia, Brazil.
[c] Patient was initially in a vegetative state but died within 6 months while in hospice care (personal communication from Dr. Ya-Sung Yang, Tri-Service General Hospital, Taipei, Taiwan).

Updated from Jackson, A. C. (2011). Therapy in human rabies. In Research Advances in Rabies, Alan C. Jackson (Ed.), Advances in Virus Research, vol. 79, 2011, pp. 365–375; Copyright: Elsevier.

who likely did not have rabies, and others who received rabies vaccine prior to the onset of illness similar to many patients who survived without the protocol (Table 17.1). There have been 10 successes reported from India with critical care since 2015, but without other components of the protocol. Hence, critical care is probably the only effective component of the protocol and has been previously recommended for aggressive therapy of rabies patients (Jackson et al., 2003).

17.2 Future prospects for the aggressive management of rabies in humans

Aggressive care of a patient with rabies should include consideration of critical care, combination therapy, antiviral therapy, immunotherapy, and neuroprotective therapy. Each of components will be discussed separately.

17.2.1 Critical care unit

Supportive care in a critical (intensive) care unit is very important for an aggressive approach. This is essential for management of cardiac and respiratory complications of rabies, which are severe and may be fatal. Multiple organ failure is also a common development during the course of rabies. Complex medical problems are best managed by a team of specialists with broad expertise.

17.2.2 Combination therapy

Similar to the approach for cancer, chronic hepatitis C infection, and human immunodeficiency virus infection, the use of more than one therapeutic agent may also prove useful for rabies. This may not only include antiviral agents, but also agents in other categories (e.g., immunotherapy and/or neuroprotective therapy).

17.2.3 Antiviral therapy

Antiviral therapy for rabies has recently been reviewed in detail (Appolinario & Jackson, 2015). Penetration of the blood–brain and blood-spinal cord (Bartanusz, Jezova, Alajajian, & Digicaylioglu, 2011) barriers are important considerations for efficacy of a potential antiviral agent in rabies. In summary, there is no known antiviral agent with efficacy for rabies. Hence, no recommendation can be given for using a presently available antiviral drugs. Although there was some initial enthusiasm about favipiravir (T-705), two studies have shown promising results in vitro and yet discouraging results in animal models (Banyard et al., 2019; Yamada, Noguchi, Komeno, Furuta, & Nishizono, 2016). Favipiravir did not show efficacy for increasing survival time or mortality in a mouse model of rabies (Yamada et al., 2019). There was also no survival benefit in six humans treated with favipiravir in three countries (Willoughby & Epstein, 2019).

17.2.4 Immunotherapy

Therapy of rabies with antibodies is compromised by limitations in penetration through an intact blood–brain barrier. Hence, an effective approach may require combination with a method of "opening" the blood-brain barrier (e.g., osmotic approach with mannitol or another agent). Immunization of a patient with rabies vaccine may not be effective and there is concern that it actually may be detrimental. Inactivated rabies vaccines do not stimulate a cytotoxic T-cell response that is important for clearance of rabies virus infection. There are safety concerns with the use of live attenuated rabies vaccines, including reversion to virulence, and the limited therapeutic applications would inhibit future development of such a vaccine given the high efficacy of inactivated rabies vaccines for both pre-exposure and post-exposure rabies prophylaxis.

17.2.5 Neuroprotective therapy

Neuroprotective therapies favorably influence the underlying etiology or pathogenesis of a disease and forestall the onset of the disease or clinical decline (Shoulson, 1998). The development of neuroprotective therapies for acute neurologic disorders is still in its infancy; no effective agents are presently available for stroke or traumatic brain injury despite numerous research studies and clinical trials. There are no known effective neuroprotective agents for use in rabies.

The most effective "neuroprotective" therapy to date for an acute brain insult is therapeutic hypothermia (also called targeted temperature management), in which the body temperature is reduced by a variety of cooling methods in order to reduce neuronal injury and improve clinical outcomes. Efficacy has been established in Australian (Bernard et al., 2002) and European (The Hypothermia After Cardiac Arrest Study Group, 2002) studies for patients who remain unconscious after witnessed cardiac arrest due to ventricular fibrillation. Efficacy for hypothermia for traumatic brain injury has not yet been established (Christian, Zada, Sung, & Giannotta, 2008; Cooper et al., 2018), but continues to be an area of active investigation. Hypothermia decreases cerebral metabolism, production of reactive oxygen species, lipid peroxidation, and inflammatory response activity, which, at least in part, may explain its beneficial effects. There are generalized methods of inducing hypothermia and also regional methods that can be applied to the head and neck, which include use of a cooling helmet (Wang et al., 2004) and intranasal cooling (Busch et al., 2010; Castren et al., 2010). Intranasal cooling involves spraying an inert evaporative coolant via nasal prongs that rapidly evaporates after contact with the nasopharynx and it has the advantage that it reduces the temperature more rapidly. The regional methods are associated with less systemic adverse effects, and would also be expected to have a reduced effect on a natural or rabies vaccine-induced systemic immune response, which is important for viral clearance in rabies virus infection. Rabies virus replication is generally fairly efficient at lower-than-normal body temperatures (e.g., 33°C), particularly with infection of an epithelial cell line with a bat rabies virus variant (Morimoto et al., 1996). However, there may be reduced viral spread due to inhibitory effects of hypothermia on fast axonal transport (Bisby & Jones, 1978) and trans-synaptic spread as well as other beneficial and neuroprotective effects. Under natural conditions, hibernation of rabies vectors likely results in "suspension" of viral replication (Sulkin, Allen, Sims, Krutzsch, & Kim, 1960) and inhibition of viral spread by marked inhibition of axonal transport (Bisby & Jones, 1978) due to very low body temperatures (e.g., below 5 to 10°C). In contrast, therapeutic mild (34°C) or moderate (30°C) hypothermia maintained for periods of 24 to 48 h would be expected to be associated with much more modest, but potentially beneficial effects. Therapeutic brain hypothermia should be considered in the future as an adjunctive therapy for rabies to slow progression of the disease, reduce neuronal injury, and provide time for the development of an immune response (Jackson, 2019).

Initial studies on the noncompetitive N-methyl-D-aspartate (NMDA) antagonist ketamine suggested that it might be a promising agent for the treatment of rabies with some support in animal model studies (Lockhart et al., 1991). However, subsequent experimental work performed in primary neurons and in a mouse model of rabies failed to show efficacy of ketamine (Weli et al., 2006). Hence, empirical therapy of human rabies with ketamine is not recommended at this time.

17.2.6 Specific therapies to avoid

Corticosteroids are not recommended for the management of rabies. Immune-mediated injury is not thought to be important in the pathogenesis of rabies (see Chapter 8) and corticosteroids act to close the blood-brain barrier and reduce the entry of other therapeutic agents into the brain and spinal cord, which is detrimental. Minocycline is a broad-spectrum antimicrobial agent with anti-inflammatory, antiapoptotic, and antioxidant properties (Appolinario & Jackson, 2015). Empirical use of minocycline in rabies is strongly discouraged because harmful effects were observed in rabies virus infection of neonatal mice (Jackson, Scott, Owen, Weli, & Rossiter, 2007) and also in different animal models of neurodegenerative diseases (Jackson, 2012).

Recovery of patients with rabies has inspired physicians to aggressively manage patients with rabies in critical care units. In part, because of the mild neuropathological changes in rabies, the hope was that patients, even when previously unimmunized, could be maintained through the acute phase of their illness, and, if they could avoid medical complications, then perhaps they could clear the viral infection and recover. Overall, this approach has proved to be disappointing (Bhatt, Hattwick, Gerdsen, Emmons, & Johnson, 1974; Emmons et al., 1973; Gode, Raju, Jayalakshmi, Kaul, & Bhide, 1976; Lopez et al., 1975; Rubin, Sullivan, Summers, Gregg, & Sikes, 1970; Udwadia et al., 1989). However, the case from Milwaukee has provided optimism that aggressive therapy may become much more effective in the future, even in cases without previous administration of rabies vaccine. This case was the first documented survivor who did not receive rabies vaccine prior to onset of clinical rabies. As discussed in the accompanying editorial (Jackson, 2005), it is unknown if therapy with one or more specific agents given played an important role in the outcome of this case. The induction of coma per se has not been shown to be useful in the management of infectious diseases of the nervous system, and to date there is no evidence supporting this approach in rabies or other forms of viral encephalitis. Hence, this approach should not become routine for the management of rabies at this time. There is now increased doubt about the efficacy of ketamine therapy in rabies virus infection (see Section 7.2.5 above). Unlike other viral infections of the nervous system, including Sindbis virus encephalomyelitis (Darman et al., 2004; Nargi-Aizenman et al., 2004; Nargi-Aizenman & Griffin, 2001) and human immunodeficiency virus infection (Kaul & Lipton, 2007), there is not yet any established experimental evidence supporting excitotoxicity in rabies. Even where there is strong experimental evidence of excitotoxicity in animal models, multiple clinical trials in humans have shown a lack of efficacy of neuroprotective agents in stroke (Ginsberg, 2009). Hence, a strong neuroprotective effect of a therapy given to a single patient without a clear scientific rationale is highly unlikely to be responsible for a favorable outcome.

The presence of neutralizing antirabies virus antibodies early in a patient's clinical course was undoubtedly an important factor contributing to the favorable outcome. This occurs in less than 20% of all patients with rabies. The presence of neutralizing antirabies virus antibodies is a marker of an active adaptive immune response that is essential for viral clearance (see Chapter 11). Because most survivors of rabies had received one or more doses of rabies vaccine prior to the onset of their disease, this suggests that an early immune response is associated with a positive outcome. Diagnostic laboratory tests in rabies survivors are usually negative for rabies virus antigen and RNA in fluids and tissues (without testing of brain tissues). This may be because viral clearance was so effective that centrifugal spread of the infection to peripheral organ sites was reduced or rapid clearance occurred through immune-mediated mechanisms.

It will never be known whether the causative bat rabies virus variant in the Milwaukee case was, in some way, attenuated and different from previously isolated bat rabies virus variants because viral isolation was not successful.

An aggressive approach to therapy of rabies will require the full resources of a critical care unit and have a high risk of failure. Table 17.2 highlights "favorable" clinical factors for initiating an aggressive therapeutic approach: 1) therapy with dose(s) of rabies vaccine prior to the onset of illness, 2) young age, 3) healthy and immunocompetent individual, and 4) mild neurological disease at the time of initiation of therapy. The presence of neutralizing antirabies virus antibodies in serum and CSF is an important laboratory feature. Access to critical care facilities is very important and should be considered essential. Finally, upon embarking on an aggressive approach, one needs to accept the risk that a survivor may have severe neurological sequaelae, which has unfortunately been the outcome in most survivors of rabies to date. In the future, new approaches to treating human rabies need to be developed rather than repeating ineffective therapies. It remains highly doubtful that the Milwaukee Protocol will prove to be useful in the management of human rabies in light of the fact that at least 53 patients plus another six patients who were treated with favipiravir received similar therapeutic approaches with fatal outcomes (Table 17.3). Continued repetition of this therapy will likely impede progress in moving the development of new effective therapies for rabies forward. Finding an effective neuroprotective drug is highly unlikely with a "trial and error" approach, in light of the fact that so many clinical trials have failed to show efficacy for a neuroprotective drug for acute stroke with a much more rational approach based on experimental studies (Ginsberg, 2009).

Entirely new approaches need to be taken for the aggressive management of human rabies, which may combine a variety of different therapeutic approaches, including, for example, hypothermia and new antiviral and other therapeutic agents. More work is needed to identify new efficacious therapeutic agents and more basic research is needed to improve our understanding of basic mechanisms underlying rabies pathogenesis in humans and animals. Hopefully, in the future, with early clinical diagnosis of rabies good clinical outcomes can be achieved with the development of effective therapy.

17.3 Importance of palliation in the management of rabies

The vast majority of patients do not survive rabies and for most cases there is no attempt at aggressive therapy. For comfort of the patient, appropriate palliative measures are extremely important because in the absence of comfort measures, death from rabies can be a terrifying and unpleasant experience for the patient and their family. These measures have recently been reviewed (Jackson, 2019; Warrell, Warrell, & Tarantola, 2017). Palliative measures can be most effectively initiated in a hospital setting with liberal use of sedatives and analgesics, but hospitalization is not always feasible for a variety of reasons, including financial reasons. A quiet private room to prevent unnecessary stimulation is desirable. Pharmacologic restraints are strongly preferred over physical restraints. For sedation, benzodiazepines such as diazepam should be strongly considered and may be given intravenously, subcutaneously, or per rectum, whereas lorazepam and midazolam are alternatives that can be given intravenously or subcutaneously. The neuroleptic haloperidol should be considered for restlessness, agitation, hyperexcitability, delirium, hallucinations, and aggressive behavior (Marsden & Cabanban, 2006). For analgesia, morphine can be administered either intraveneously or subcutaneously. In hospital, a subcutaneous infusion can be given using a syringe driver that prevents the need for multiple injections (Thomas & Barclay, 2015). Two- or three-drug combinations can also be considered (e.g., morphine, haloperidol, and midazolam). Excessive salivary secretions can be managed with anticholinergics such as scopolamine and glycopyrrolate. Fever can be treated with sponging and antipyretics. Thirst can be alleviated using ice chips in the mouth. Some physicians advocate for hydration measures, including intravenous fluids, in rabies without citing supportive evidence (Warrell et al., 2017). However, a systematic review of the literature and studies evaluating the benefits of clinically assisted hydration in palliative care patients failed to show a clear benefit of hydration measures for either the length or quality of life (Good, Richard, Syrmis, Jenkins-Marsh, & Stephens, 2014); it is not clear that rabies is different from other fatal diseases.

References

Agarwal, A. K. (2017). The 'Milwaukee protocol' (MP) hope does not succeeds for rabies victim. *Medical Journal of Dr. D.Y. Patil University, 10*(2), 184–186.

Alvarez, L., Fajardo, R., Lopez, E., Pedroza, R., Hemachudha, T., Kamolvarin, N., ... Baer, G. M. (1994). Partial recovery from rabies in a nine-year-old boy. *The Pediatric Infectious Disease Journal, 13*, 1154–1155.

Apanga, P. A., Awoonor-Williams, J. K., Acheampong, M., & Adam, M. A. (2016). A presumptive case of human rabies: A rare survived case in rural Ghana. *Frontiers in Public Health, 4*(256), 256.

Appolinario, C. M., & Jackson, A. C. (2015). Antiviral therapy for human rabies. *Antiviral Therapy, 22*(1), 1–10.

Aramburo, A., Willoughby, R. E., Bollen, A. W., Glaser, C. A., Hsieh, C. J., Davis, S. L., ... Roy-Burman, A. (2011). Failure of the Milwaukee protocol in a child with rabies. *Clinical Infectious Diseases, 53*(9), 572–574.

Badillo, R., Mantilla, J. C., & Pradilla, G. (2009). Human rabies encephalitis by a vampire bat bite in an urban area of Colombia (Spanish). *Biomédica, 29*(2), 191–203.

Bagchi, S. (2005, July 4). Coma therapy. *The Telegraph*.

Banyard, A. C., Mansfield, K. L., Wu, G., Selden, D., Thorne, L., Birch, C., ... Fooks, A. R. (2019). Re-evaluating the effect of favipiravir treatment on rabies virus infection. *Vaccine 37*(33), 4686–4693.

Bartanusz, V., Jezova, D., Alajajian, B., & Digicaylioglu, M. (2011). The blood-spinal cord barrier: Morphology and clinical implications. *Annals of Neurology, 70*(2), 194–206.

Bernard, S. A., Gray, T. W., Buist, M. D., Jones, B. M., Silvester, W., Gutteridge, G., & Smith, K. (2002). Treatment of comatose survivors of out-of-hospital cardiac arrest with induced hypothermia. *New England Journal of Medicine, 346*(8), 557–563.

Bhatt, D. R., Hattwick, M. A. W., Gerdsen, R., Emmons, R. W., & Johnson, H. N. (1974). Human rabies: Diagnosis, complications, and management. *American Journal of Diseases of Children, 127,* 862–869.

Bisby, M. A., & Jones, D. L. (1978). Temperature sensitivity of axonal transport in hibernating and nonhibernating rodents. *Experimental Neurology, 61*(1), 74–83.

Blanton, J. D., Palmer, D., & Rupprecht, C. E. (2010). Rabies surveillance in the United States during 2009. *Journal of the American Veterinary Medical Association, 237*(6), 646–657.

Bokade, C. M., Gajimwar, V. S., Meshram, R. M., & Wathore, S. B. (2019). Survival of atypical rabies encephalitis. *Annals of Indian Academy of Neurology, 22*(3), 319–321.

Branswell, H. (2012, April 19). Testing suggests Toronto rabies case infected in Dominican Republic. *The Globe and Mail.*

Busch, H. J., Eichwede, F., Fodisch, M., Taccone, F. S., Wobker, G., Schwab, T., ... Janata, A. (2010). Safety and feasibility of nasopharyngeal evaporative cooling in the emergency department setting in survivors of cardiac arrest. *Resuscitation, 81*(8), 943–949.

Caicedo, Y., Paez, A., Kuzmin, I., Niezgoda, M., Orciari, L. A., Yager, P. A., ... Willoughby, R. E., Jr. (2015). Virology, immunology and pathology of human rabies during treatment. *The Pediatric Infectious Disease Journal, 34*(5), 520–528.

Castren, M., Nordberg, P., Svensson, L., Taccone, F., Vincent, J. L., Desruelles, D., ... Barbut, D. (2010). Intra-arrest transnasal evaporative cooling: A randomized, prehospital, multicenter study (PRINCE: Pre-ROSC IntraNasal cooling effectiveness). *Circulation, 122*(7), 729–736.

Chen, Y. G., Kan, L. P., Lee, C. H., Lin, S. H., Chu, D. M., Chang, F. Y., & Yang, Y. S. (2015). Symptomatic hypercalcemia in a rabies survivor underwent hemodialysis. *Hemodialysis International, 19*(2), 347–351.

Christenson, J. C., Holm, B. M., Lechlitner, S., Howell, J. F., Wenger, M., Roy-Burman, A., ... Rupprecht, C. E. (2007). Human rabies—Indiana and California, 2006. *Morbidity and Mortality Weekly Report, 56*(15), 361–365.

Christian, E., Zada, G., Sung, G., & Giannotta, S. L. (2008). A review of selective hypothermia in the management of traumatic brain injury. *Neurosurgical Focus, 25*(4), E9.

Cohen, R., Babushkin, F., Shipiro, M., Uda, M., Willoughby, R. E., & David, D. (2016). Paralytic rabies misdiagnosed as Guillain-Barre syndrome in a guest worker: A case report. *Journal of Neuroinfectious Diseases, 7*(1), 208.

Cooper, D. J., Nichol, A. D., Bailey, M., Bernard, S., Cameron, P. A., Pili-Floury, S., ... McArthur, C. (2018). Effect of early sustained prophylactic hypothermia on neurologic outcomes among patients with severe traumatic brain injury: The POLAR randomized clinical trial. *Journal of the American Medical Association, 320*(21), 2211–2220.

Damodar, T., Mani, R. S., & Prathyusha, P. V. (2019). Utility of rabies neutralizing antibody detection in cerebrospinal fluid and serum for ante-mortem diagnosis of human rabies. *PLoS Neglected Tropical Diseases, 13*(1)e0007128.

Darman, J., Backovic, S., Dike, S., Maragakis, N. J., Krishnan, C., Rothstein, J. D., ... Kerr, D. A. (2004). Viral-induced spinal motor neuron death is non-cell-autonomous and involves glutamate excitotoxicity. *Journal of Neuroscience, 24*(34), 7566–7575.

Dhayhi, N. S., Arishi, H. M., Al Ibrahim, A. Y., Khalaf Allah, M. B., Hawas, A. M., Alqasmi, H., ... Alali, A. (2019). First confirmed case of local human rabies in Saudi Arabia. *International Journal of Infectious Diseases, 87,* 117–118. https://doi.org/10.1016/j.ijid.2019.08.011.

de Souza, A., & Madhusudana, S. N. (2014). Survival from rabies encephalitis. *Journal of the Neurological Sciences, 339*(1–2), 8–14.

Drosten, C. (2007). Rabies—Germany (Hamburg) ex Morocco. ProMED-mail, 20070419.1287. Retrieved from http://www.promedmail.org.

Elsiesy, H., Hussain, I., Abaalkhail, F. A., Al-Hamoudi, W. K., Al Sebayel, M. I., Broering, D. C., et al. (2015). Rabies outbreak involving four transplant recipients in Kuwait and Saudi Arabia. Platform presentation at the XXVIth international meeting on research advances and rabies control in the Americas in Fort Collins, Colorado, USA on October 5, 2015.

Emmons, R. W., Leonard, L. L., DeGenaro, F., Jr., Protas, E. S., Bazeley, P. L., Giammona, S. T., & Sturckow, K. (1973). A case of human rabies with prolonged survival. *Intervirology, 1*(1), 60–72.

Francis, J. R., Nourse, C., Vaska, V. L., Calvert, S., Northill, J. A., McCall, B., & Mattke, A. C. (2014). Australian bat Lyssavirus in a child: The first reported case. *Pediatrics, 133*(4), e1063–e1067.

Galvez, S., Basque, M., Contreras, L., Merino, C., Ahumada, R., Jamett, J., et al. (2013). Survivor of rabies encephalitis in Chile. Platform presentation at the XXIVth international meeting on research advances and rabies control in the Americas in Toronto, Ontario, Canada on October 27, 2013.

Ginsberg, M. D. (2009). Current status of neuroprotection for cerebral ischemia: Synoptic overview. *Stroke, 40*(Suppl 3), S111–S114.

Gode, G. R., Raju, A. V., Jayalakshmi, T. S., Kaul, H. L., & Bhide, N. K. (1976). Intensive care in rabies therapy. Clinical observations. *Lancet, 2*, 6–8.

Gong, C., Li, X., Luo, M., Zhang, Z., Wang, Q., Wang, Q., … Wu, J. (2017). Laboratory investigation of the rabies transmission through organ transplantation in China (Letter). *The Journal of Infection, 74*(417), 427–431.

Good, P., Richard, R., Syrmis, W., Jenkins-Marsh, S., & Stephens, J. (2014). Medically assisted hydration for adult palliative care patients. *Cochrane Database of Systematic Reviews*, (4),CD006273.

Greer, D. M., Robbins, G. K., Lijewski, V., Gonzalez, R. G., & McGuone, D. (2013). Case records of the Massachusetts General Hospital: Case 1-2013: A 63-year-old man with paresthesias and difficulty swallowing. *New England Journal of Medicine, 368*(2), 172–180.

Hattwick, M. A. W., Weis, T. T., Stechschulte, C. J., Baer, G. M., & Gregg, M. B. (1972). Recovery from rabies: A case report. *Annals of Internal Medicine, 76*, 931–942.

Hemachudha, T., Sunsaneewitayakul, B., Desudchit, T., Suankratay, C., Sittipunt, C., Wacharapluesadee, S., … Jackson, A. C. (2006). Failure of therapeutic coma and ketamine for therapy of human rabies. *Journal of Neurovirology, 12*(5), 407–409.

Herriman, R. (2018, January 12). Rabies survivor: Milwaukee protocol saves Brazilian teen. In *Outbreak News Today*.

Holzmann-Pazgal, G., Wanger, A., Degaffe, G., Rose, C., Heresi, G., Amaya, R., … Rupprecht, C. E. (2010). Presumptive abortive human rabies—Texas, 2009. *Morbidity and Mortality Weekly Report, 59*(7), 185–190.

Houston Chronicle. (2006). *Rabies, human—USA (Texas)*. ProMED-mail, 20060513.1360. Retrieved from http://www.promedmail.org.

Hu, W. T., Willoughby, R. E., Jr., Dhonau, H., & Mack, K. J. (2007). Long-term follow-up after treatment of rabies by induction of coma (letter). *New England Journal of Medicine, 357*(9), 945–946.

Hunter, M., Johnson, N., Hedderwick, S., McCaughey, C., Lowry, K., McConville, J., … Fooks, A. R. (2010). Immunovirological correlates in human rabies treated with therapeutic coma. *Journal of Medical Virology, 82*(7), 1255–1265.

IAfrica.com. (2012). *Rabies—South Africa (03): (Kwazulu-Natal), human ex Mozambique*. ProMED-mail, 20120608.1160328. Retrieved from http://www.promedmail.org.

Jackson, A. C. (2005). Recovery from rabies (editorial). *New England Journal of Medicine, 352*(24), 2549–2550.

Jackson, A. C. (2012). Is minocycline useful for therapy of acute viral encephalitis? *Antiviral Research, 95*(3), 242–244.

Jackson, A. C. (2019). Treatment of rabies. In T. W. Post, M. S. Hirsch, M. S. Edwards, & J. Mitty (Eds.), *UpToDate*. Waltham, MA: Wolters Kluwer.

Jackson, A. C., & Garland, A. (2015). Fatal rabies case did not die "accidentally" and should not be considered a rabies survivor (letter). *The Pediatric Infectious Disease Journal, 34*(6), 677–678.

Jackson, A. C., Scott, C. A., Owen, J., Weli, S. C., & Rossiter, J. P. (2007). Therapy with minocycline aggravates experimental rabies in mice. *Journal of Virology, 81*(12), 6248–6253.

Jackson, A. C., Warrell, M. J., Rupprecht, C. E., Ertl, H. C. J., Dietzschold, B., O'Reilly, M., … Wilde, H. (2003). Management of rabies in humans. *Clinical Infectious Diseases, 36*(1), 60–63.

Javaid, W., Amzuta, I. G., Nat, A., Johnson, T., Grant, D., Rudd, R. J., … Lankau, E. W. (2012). Imported human rabies in a U.S. Army soldier—New York, 2011. *MMWR Morbidity and Mortality Weekly Report, 61*(17), 302–305.

Karahocagil, M. K., Akdeniz, H., Aylan, O., Sünnetçioglu, M., Ün, H., Yapici, K., & Baran, A. I. (2013). Complete recovery from clinical rabies: Case report. *Turkiye Klinikleri Journal of Medical Sciences, 33*(2), 547–552.

Karande, S., Muranjan, M., Mani, R. S., Anand, A. M., Amoghimath, R., Sankhe, S., … Madhusudana, S. N. (2015). Atypical rabies encephalitis in a six-year-old boy: Clinical, radiological and laboratory findings. *International Journal of Infectious Diseases, 36*, 1–3.

Kaul, M., & Lipton, S. A. (2007). Neuroinflammation and excitotoxicity in neurobiology of HIV-1 infection and AIDS: Targets for neuroprotection. In J. O. Malva, A. C. Rego, R. A. Cunha, & C. R. Oliveira (Eds.), *Interaction between neurons and glia in aging and disease* (pp. 281–308). New York: Springer Science.

Kumar, K. V., Ahmad, F. M., & Dutta, V. (2015). Pituitary cachexia after rabies encephalitis (letter). *Neurology India, 63*(2), 255–256.

Lima, F. M. G., Moura, F. B. P., Chaves, C. S., Maia, K. M., Lima, M. F. C., & Rorigues, V. C. (2018). Human rabies by Desmondus rotundus in Ceara/Brazil. Poster presentation at the XXIXth international meeting on research advances and rabies control in the Americas in Buenos Aires, Argentina on October 29, 2018.

Lockhart, B. P., Tordo, N., & Tsiang, H. (1992). Inhibition of rabies virus transcription in rat cortical neurons with the dissociative anesthetic ketamine. *Antimicrobial Agents and Chemotherapy, 36*, 1750–1755.

Lockhart, B. P., Tsiang, H., Ceccaldi, P. E., & Guillemer, S. (1991). Ketamine-mediated inhibition of rabies virus infection in vitro and in rat brain. *Antiviral Chemistry and Chemotherapy, 2*, 9–15.

Lopez, M., Neves, J., Moreira, E. C., Reis, R., Tafuri, W. L., Pittella, J. E., … Marra, U. D. (1975). Human rabies. i. Intensive treatment. *Revista do Instituto de Medicina Tropical de São Paulo, 17*(2), 103–110.

Luminos, M., Barboi, G., Draganescu, A., Streinu Cercel, A., Staniceanu, F., Jugulete, G., … Turcitu, M. A. (2011). Human rabies in a Romanian boy—an ante mortem case study. *Rabies Bulletin Europe, 35*(2), 5–10.

Madhusudana, S. N., Nagaraj, D., Uday, M., Ratnavalli, E., & Kumar, M. V. (2002). Partial recovery from rabies in a six-year-old girl (letter). *International Journal of Infectious Diseases, 6*(1), 85–86.

Maier, T., Schwarting, A., Mauer, D., Ross, R. S., Martens, A., Kliem, V., … Drosten, C. (2010). Management and outcomes after multiple corneal and solid organ transplantations from a donor infected with rabies virus. *Clinical Infectious Diseases, 50*(8), 1112–1119.

Manesh, A., Mani, R. S., Pichamuthu, K., Jagannati, M., Mathew, V., Karthik, R., … Varghese, G. M. (2018). Case report: Failure of therapeutic coma in rabies encephalitis. *American Journal of Tropical Medicine and Hygiene, 98*(1), 207–210.

Mani, R. S., Damodar, T., Divyashree, S., Domala, S., Gurung, B., Jadhav, V., … Tambi, P. (2019). Case report: Survival from rabies: Case series from India. *American Journal of Tropical Medicine and Hygiene, 100*(1), 165–169.

Manning, S. E., Rupprecht, C. E., Fishbein, D., Hanlon, C. A., Lumlertdacha, B., Guerra, M., … Hull, H. F. (2008). Human rabies prevention—United States, 2008: Recommendations of the advisory committee on immunization practices. *Morbidity and Mortality Weekly Report, 57*(RR-3), 1–28.

Manoj, S., Mukherjee, A., Johri, S., & Kumar, K. V. (2016). Recovery from rabies, a universally fatal disease. *Military Medical Research, 3*, 21–0089.

Marsden, S. C., & Cabanban, C. R. (2006). Rabies: A significant palliative care issue. *Progress Palliative Care, 14*(2), 62–67.

McDermid, R. C., Saxinger, L., Lee, B., Johnstone, J., Noel Gibney, R. T., Johnson, M., & Bagshaw, S. M. (2008). Human rabies encephalitis following bat exposure: Failure of therapeutic coma. *Canadian Medical Association Journal, 178*(5), 557–561.

Ministerio da Saude in Brazil (2008). *Rabies, human survival, bat—Brazil: (Pernambuco)*. ProMED-mail, 20081114.3599. Retrieved from http://www.promedmail.org.

Mohan, M. C., Sudeep, K. N., Pramod, V. K., Kumar, A., Prasanna, K. S., & Narayanan, P. V. (2014). Clinical rabies: Is cure possible? *International Journal of Research in Medical Sciences, 2*(4), 1735–1739.

Morimoto, K., Patel, M., Corisdeo, S., Hooper, D. C., Fu, Z. F., Rupprecht, C. E., … Dietzschold, B. (1996). Characterization of a unique variant of bat rabies virus responsible for newly emerging human cases in North America. *Proceedings of the National Academy of Sciences of the United States of America, 93*(11), 5653–5658.

Murphy, J., Sifri, C. D., Pruitt, R., Hornberger, M., Bonds, D., Blanton, J., … Wallace, R. M. (2019). Human rabies—Virginia, 2017. *MMWR Morbidity and Mortality Weekly Report, 67*(5152), 1410–1414.

Nargi-Aizenman, J. L., & Griffin, D. E. (2001). Sindbis virus-induced neuronal death is both necrotic and apoptotic and is ameliorated by N-methyl-D-aspartate receptor antagonists. *Journal of Virology, 75*(15), 7114–7121.

Nargi-Aizenman, J. L., Havert, M. B., Zhang, M., Irani, D. N., Rothstein, J. D., & Griffin, D. E. (2004). Glutamate receptor antagonists protect from virus-induced neural degeneration. *Annals of Neurology, 55*(4), 541–549.

Nat, A., Nat, A., Sharma, A., Pothineni, A., & Amzuta, I. (2014). Looking beyond the Milwaukee protocol (abstract). *Chest, 145*(3 Suppl), 117A.

NE 10. (2012). *Rabies, human—Brazil (03): (Ceara)*. ProMED-mail, 20120314.1070531. Retrieved from http://www.promedmail.org.

Netravathi, M., Udani, V., Mani, R. S., Gadad, V., Ashwini, M. A., Bhat, M., … Satishchandra, P. (2015). Unique clinical and imaging findings in a first ever documented PCR positive rabies survival patient: A case report. *Journal of Clinical Virology, 70*, 83–88.

Pathak, S., Horton, D. L., Lucas, S., Brown, D., Quaderi, S., Polhill, S., … Brown, M. (2014). Diagnosis, management and post-mortem findings of a human case of rabies imported into the United Kingdom from India: A case report. *Virology Journal, 11*(1), 63–69.

Porras, C., Barboza, J. J., Fuenzalida, E., Adaros, H. L., Oviedo, A. M., & Furst, J. (1976). Recovery from rabies in man. *Annals of Internal Medicine, 85*, 44–48.

Pue, H. L., Turabelidze, G., Patrick, S., Grim, A., Bell, C., Reese, V., … Robertson, K. (2009). Human rabies—Missouri, 2008. *Morbidity and Mortality Weekly Report, 58*(43), 1207–1209.

Ramya, M. (2014, February 22). Madras Christian College student, a dog lover, dies of rabies. In *The Times of India*.

Rao, A., Pimpalwar, Y., Mukherjee, A., & Yadu, N. (2017). Serial brain MRI findings in a rare survivor of rabies encephalitis. *Indian Journal of Radiology & Imaging, 27*(3), 286–289.

Rawat, A. K., & Rao, S. K. (2011). Survival of a rabies patient (letter). *Indian Pediatrics, 48*(7), 574.

Rubin, J., David, D., Willoughby, R. E., Jr., Rupprecht, C. E., Garcia, C., Guarda, D. C., … Stamler, A. (2009). Applying the Milwaukee protocol to treat canine rabies in Equatorial Guinea. *Scandinavian Journal of Infectious Diseases, 41*(5), 372–375.

Rubin, R. H., Sullivan, L., Summers, R., Gregg, M. B., & Sikes, R. K. (1970). A case of human rabies in Kansas: Epidemiologic, clinical, and laboratory considerations. *Journal of Infectious Diseases, 122*, 318–322.

Santos, A., Cale, E., Dacheux, L., Bourhy, H., Gouveia, J., & Vasconcelos, P. (2012). Fatal case of imported human rabies in Amadora, Portugal, August 2011. *Eurosurveillance, 17*(12), 20130.

Sheridan, K. (2018). *6-Year-old dies from rabies after experimental treatment, 'Milwaukee protocol,' fails. www.newsweek.com*.

Shoulson, I. (1998). Experimental therapeutics of neurodegenerative disorders: Unmet needs. *Science, 282*(5391), 1072–1074.

Sulkin, S. E., Allen, R., Sims, R., Krutzsch, P. H., & Kim, C. (1960). Studies on the pathogenesis of rabies in bats. II. Influence of environmental temperature. *Journal of Experimental Medicine, 112*, 595–617.

Tanner, C. (2018, January 11). Brazilian teen becomes fifth person cured of rabies. In *Daily Mail.com*.

Thakur, B. S. (2014, September 14). 2nd rabies survivor in country at P'kula hospital. In *Hindustan Times*.

The Hypothermia After Cardiac Arrest Study Group. (2002). Mild therapeutic hypothermia to improve the neurologic outcome after cardiac arrest. *New England Journal of Medicine, 346*(8), 549–556.

Thomas, T., & Barclay, S. (2015). Continuous subcutaneous infusion in palliative care: A review of current practice. *International Journal of Palliative Nursing, 21*(2), 60. 62–60, 64.

Tillotson, J. R., Axelrod, D., & Lyman, D. O. (1977a). Follow-up on rabies—New York. *Morbidity and Mortality Weekly Report, 26*, 249–250.

Tillotson, J. R., Axelrod, D., & Lyman, D. O. (1977b). Rabies in a laboratory worker - New York. *Morbidity and Mortality Weekly Report, 26*, 183–184.

Times Live. (2012). *Rabies—South Africa (02): (Kwazulu-Natal), human ex Mozambique*. ProMED-mail, 20120528.1147931. Retrieved from http://www.promedmail.org.

Turabelidze, G., Pue, H., Grim, A., & Patrick, S. (2009). First human rabies case in Missouri in 50 years causes death in outdoorsman. *Missouri Medicine, 106*(6), 417–419.

Udwadia, Z. F., Udwadia, F. E., Katrak, S. M., Dastur, D. K., Sekhar, M., Lall, A., … Sane, B. (1989). Human rabies: Clinical features, diagnosis, complications, and management. *Critical Care Medicine, 17*, 834–836.

van Thiel, P. P., de Bie, R. M., Eftimov, F., Tepaske, R., Zaaijer, H. L., van Doornum, G. J., … Kager, P. A. (2009). Fatal human rabies due to Duvenhage virus from a bat in Kenya: Failure of treatment with coma-induction, ketamine, and antiviral drugs. *PLoS Neglected Tropical Diseases, 3*(7), e428.

Vargas, A., Romano, A. P. M., & Merchan-Hamann, E. (2019). Human rabies in Brazil: a descriptive study, 2000–2017. *Epidemiologia e Serviços de Saúde, 28*(2). e2018275-49742019000200001.

Wallace, R. M., Bhavnani, D., Russell, J., Zaki, S., Muehlenbachs, A., Hayden-Pinneri, K., … Robinson, L. (2014). Rabies death attributed to exposure in Central America with symptom onset in a U.S. detention facility—Texas, 2013. *MMWR Morbidity and Mortality Weekly Report, 63*(20), 446–449.

Wang, H., Olivero, W., Lanzino, G., Elkins, W., Rose, J., Honings, D., … Wang, D. (2004). Rapid and selective cerebral hypothermia achieved using a cooling helmet. *Journal of Neurosurgery, 100*(2), 272–277.

Warrell, M. J., Warrell, D. A., & Tarantola, A. (2017). The imperative of palliation in the management of rabies encephalomyelitis. *Tropical Medicine and Infectious Disease, 2*(4), pii: E52. https://doi.org/10.3390/tropicalmed2040052.

Weli, S. C., Scott, C. A., Ward, C. A., & Jackson, A. C. (2006). Rabies virus infection of primary neuronal cultures and adult mice: Failure to demonstrate evidence of excitotoxicity. *Journal of Virology, 80*(20), 10270–10273.

Weyer, J., Msimang-Dermaux, V., Paweska, J. T., le Roux, K., Govender, P., Coertse, J., … Blumberg, L. H. (2016). A case of human survival of rabies, South Africa. *Southern African Journal of Infectious Diseases, 31*(2), 66–68.

Wiedeman, J., Plant, J., Glaser, C., Messenger, S., Wadford, D., Sheriff, H., ... Petersen, B. W. (2012). Recovery of a patient from clinical rabies—California, 2011. *Morbidity and Mortality Weekly Report, 61*(4), 61–65.

Willoughby, R. E., & Epstein, C. R. (2019). Favipiravir treatment of human rabies. Platform presentation at the XXXth International Meeting on Research Advances and Rabies Control in the Americas in Kansas City, Missouri, USA on October 28, 2019.

Willoughby, R. E., Jr., Tieves, K. S., Hoffman, G. M., Ghanayem, N. S., Amlie-Lefond, C. M., Schwabe, M. J., ... Rupprecht, C. E. (2005). Survival after treatment of rabies with induction of coma. *New England Journal of Medicine, 352*(24), 2508–2514.

World Health Organization. (2018). *WHO expert consultation on rabies: Third report (WHO technical report series; no. 1012)*. Geneva: World Health Organization.

Yamada, K., Noguchi, K., Komeno, T., Furuta, Y., & Nishizono, A. (2016). Efficacy of favipiravir (T-705) in rabies postexposure prophylaxis. *The Journal of Infectious Diseases, 213*(8), 1253–1261.

Yamada, K., Noguchi, K., Kimitsuki, K., Kaimori, R., Saito, N., Komeno, T., ... Nishizono, A. (2019). Reevaluation of the efficacy of favipiravir against rabies virus using in vivo imaging analysis. *Antiviral Research, 172*, 104641.

Zeiler, F. A., & Jackson, A. C. (2016). Critical appraisal of the Milwaukee protocol for rabies: This failed approach should be abandoned. *Canadian Journal of Neurological Sciences, 43*(1), 44–51.

CHAPTER 18

Dog rabies and its control

Darryn L. Knobel[a], Katie Hampson[b], Tiziana Lembo[b], Sarah Cleaveland[b], Alicia Davis[c]

[a]Department of Biomedical Sciences, Ross University School of Veterinary Medicine, Basseterre, St Kitts and Nevis [b]Institute of Biodiversity, Animal Health and Comparative Medicine, College of Medical, Veterinary and Life Sciences, University of Glasgow, Glasgow, United Kingdom [c]School of Social and Political Sciences/Institute of Health and Wellbeing, University of Glasgow, Glasgow, United Kingdom

18.1 Introduction

Domestic dogs are the major reservoir of rabies virus (RABV) throughout Africa and Asia. Rabid dogs are still responsible for the vast majority (>99%) of human deaths from rabies worldwide (World Health Organisation (WHO), 2018a). The control of the disease in domestic dogs thus has important implications for public health, particularly in Africa and Asia where canine rabies is endemic.

18.1.1 Historical perspectives and the current situation

In the 19th century, muzzling and dog movement restrictions were used to successfully eliminate dog rabies in parts of Europe. By the early 1900s, animal rabies vaccines were being developed in Japan, with the first mass dog vaccinations in 1921 (Umeno & Doi, 1921). Over the second half of the 20th century, widespread successes in eliminating dog rabies through concerted mass vaccination efforts were seen: dog rabies was eliminated from Western Europe (King, Fooks, Aubert, & Wandeler, 2004), Japan (Shimada, 1971); and from several countries in the Americas (Schneider et al., 2007). During periods of colonial occupation in Africa, rabies was successfully controlled in some countries through mass vaccinations, supported by strict implementation of tie-up orders (Shone, 1962), but rabies has since re-emerged as these measures lapsed. Vaccination campaigns and strictly enforced culling eliminated rabies from Malaysia in the 1960s (Tan, Abdul, Mohd, & Beran, 1972; Wells,

1954), and recent decades have seen improved control measures reduce rabies incidence in other parts of Asia (Hahn, 2009). However, recent resurgences have also occurred: in China, re-emergence has been attributed to a relaxation of control measures, use of ineffective local animal vaccines (Hu, Fooks, Zhang, Liu, & Zhang, 2008), and poor awareness (Wu, Franka, Svoboda, Pohl, & Rupprecht, 2009). Neighboring Nepal and Bhutan have also reported increases in incidence (Tenzin, Sharma, Dhand, Timsina, & Ward, 2010), and since the late 1990s rabies has spread to many previously rabies-free islands in Indonesia, including Flores in 1997 (Bingham, 2001; Windiyaningsih, Wilde, Meslin, Suroso, & Widarso, 2004), Maluku and Ambon in 2003, Bali in 2008, and Nias islands in 2010 (Lembo, Craig, Miles, Hampson, & Meslin, 2013).

Fig. 18.1 shows the current and historical distribution of canine rabies, illustrating successful elimination programs across a range of countries.

18.1.2 The burden of domestic dog-mediated rabies

More than 99% of all human deaths from rabies occur in Africa and Asia (WHO, 1999), where the disease is responsible for an estimated 59,000 human deaths and 3.7 million disability-adjusted life years (DALYs) annually (Hampson et al., 2015). These figures, estimated from models, are informed by data on bite injuries from suspect rabid dogs, and by probabilities of receiving postexposure prophylaxis (PEP) and of developing rabies in the absence of PEP. Models are needed because routine surveillance of rabies in many countries with endemic canine rabies remains weak, with human rabies deaths often not recorded in official records (Taylor, Hampson, Fahrion, Abela-Ridder, & Nel, 2017). Data from a range

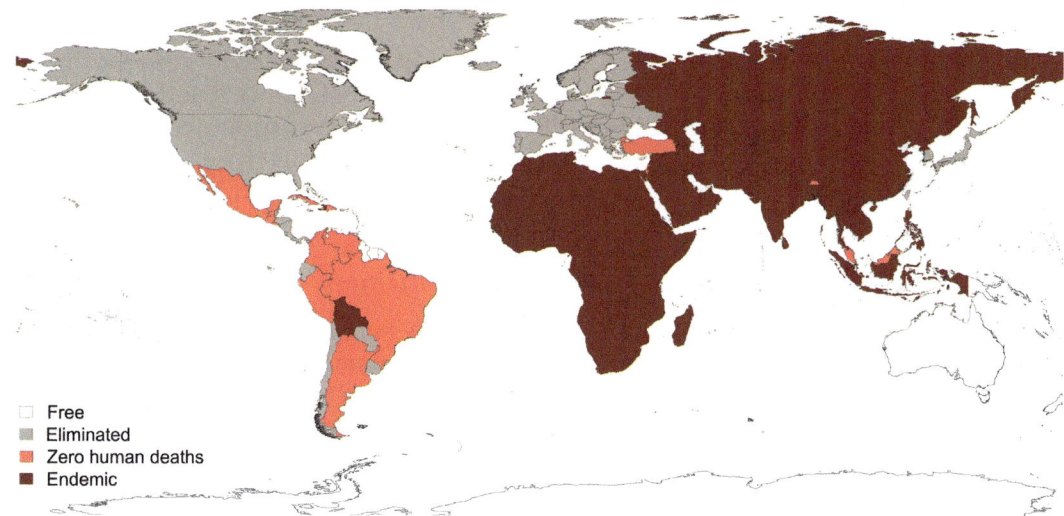

FIG. 18.1 Countries with rabies circulating endemically in domestic dog populations and ongoing human rabies deaths *(dark red)*, countries that report no human deaths from dog-mediated rabies but continue to have sporadic canine rabies cases or have geographically localized endemic foci *(pale red)* and countries that have been freed from canine rabies through concerted control efforts *(gray)* or have historically been rabies-free *(white)*. Data extracted from Schneider et al. (2011) and Hampson et al. (2015), and updated based on recent ProMED-mail reports (http://www.promedmail.org).

of sources, including contact tracing studies (Beyene, Mourits, Kidane, & Hogeveen, 2018; Hampson et al., 2008), enhanced surveillance (Hikufe et al., 2019; Rajeev et al., 2018; Wallace et al., 2015), a large multicentric study in India (Sudarshan et al., 2007), and country-specific decision tree models (Ly et al., 2009; Tenzin et al., 2011), consistently calculate a human rabies incidence between 1.5 and 5 deaths/100,000/year. In Africa, this death rate is approximately 100 times higher than indicated by official reports. Lost productivity due to premature death accounts for more than $4.7 billion dollars lost each year—over half of the economic burden of rabies (55% of $8.7 billion dollars).

A major economic cost of rabies is from the use of human rabies vaccines and immunoglobulins delivered as PEP to bite victims. In many low- and middle-income settings, bite patients are charged for PEP, which is costly and can be a major barrier to seeking PEP (Wentworth et al., 2019). Even where PEP is provided free of charge, access is often limited, with patients needing to travel long distances to centers where PEP is available (Rajeev et al., 2018; Tarantola et al., 2015), often resulting in delays and poor compliance with regimens. These obstacles are a major contributor to human rabies deaths. It is estimated that improved access to human rabies vaccines in the poorest countries could prevent half a million deaths between 2020 and 2035 (WHO Rabies Modelling Consortium, 2019). This approach would be highly cost effective at around $600 per death averted, and would increase the number of persons initiating PEP to around 45 million by 2035. However, it would not bring human rabies deaths to zero. Only in combination with scaled-up and sustained mass dog vaccination can human deaths from dog-mediated rabies be eliminated and transmission in dog populations interrupted. Over time, even with improved delivery of mass dog vaccination, costs may remain high due to continued use of PEP. In some countries in Asia and Latin America with very low or no circulation of rabies in dog populations, there is considerable overuse of rabies PEP (Amparo et al., 2018; Benavides et al., 2019; Rysava et al., 2018). In combination with poorly targeted surveillance, this presents a challenge for sustainability of rabies control. For these reasons, WHO is advocating the use of Integrated Bite Case Management (IBCM) (WHO, 2018a). This risk-based approach aims to differentiate bites from healthy animals from those from probable rabid animals, leading to more prudent use of PEP (Undurraga et al., 2017) and can increase case detection, which is needed for validation of freedom from dog-transmitted rabies (Hampson et al., 2016).

The human costs of dog rabies are borne mainly by countries in Africa and Asia, although even countries where the disease has been eliminated are impacted. For rabies-free countries, vaccine costs arise as a result of (a) preexposure vaccination of travelers, (b) the need for PEP in travelers returning with a recent history of exposure, and (c) inhabitants exposed to rabid dogs following new incursions or illegal importation. A study of PEP use in travelers returning home to Australia, New Zealand, and France indicated that only 11% of bitten travelers received both vaccine and rabies immunoglobulin at the time of exposure, and many only received appropriate PEP on return to clinics in their home country, which invariably resulted in delays and high risks (Gautret et al., 2008). In France, substantial costs are still incurred for PEP use, with peaks of demand linked with increased public awareness following cases of imported human rabies, as well as the illegal importation of rabid dogs from North Africa (Lardon et al., 2010). Although limited data are available for other components of the economic burden, costs associated with livestock rabies are also unlikely to be trivial (Anderson & Shwiff, 2015; Knobel et al., 2005; Lembo et al., 2010).

18.2 The influence of epidemiological and socioeconomic frameworks on dog rabies control

Domestic dogs are responsible for the maintenance of the rabies virus and transmission that leads to human rabies cases in many parts of the world, particularly Asia and Africa, including areas with abundant wildlife (Grover et al., 2018; Lembo et al., 2008). Prevention of the disease in humans should therefore include a strong focus on control in this reservoir population. Dog rabies control measures have the ultimate objective of decreasing the burden of human rabies by interrupting transmission in domestic dog populations. If sustained over time, such measures can lead to disease freedom. Combining epidemiological theory with broader methodological and theoretical approaches, particularly through interdisciplinary integration of social sciences, is critical to tackle public health challenges such as those posed by neglected tropical diseases (Reidpath, Allotey, & Pokhrel, 2011). However, in this setting, social science typically focuses on the economic implications and benefits of interventions and their effectiveness (for rabies, see e.g., Zinsstag et al., 2009; Fitzpatrick et al., 2014), or on broad socioeconomic contexts (Phua & Lee, 2005). When "ethnographic" research is implicated, it is often for case studies to provide these socioeconomic or sociopolitical contexts of the health and local community landscapes (Bardosh, Sambo, Sikana, Hampson, & Welburn, 2014; Parker & Harper, 2006). If theoretical concepts of the epidemiology of infectious diseases can help in the effective design and implementation of dog rabies control programs, so can theoretical concepts of medical anthropology and other social science disciplines. These latter areas explore not only intervention outcomes, but also the impacts of such interventions on everyday lives, the roles of health providers and recipients, and, critically, concepts that can be employed to ensure equitable health landscapes. A number of epidemiological concepts are reviewed in the first part of this section, with more detail provided in Chapter 20. In the second part, we argue not only for greater attention to be paid to the socioeconomic and cultural contexts in which rabies control interventions are enacted, but also to the methods of engagement with communities, the importance of sharing ownership and improving participation in health research and interventions, and the sociocultural environments.

18.2.1 Key epidemiological parameters

The basic reproductive number (R_0) of an infectious agent such as rabies virus is defined as the average number of secondary infections produced by an infected individual in an otherwise susceptible host population (Anderson & May, 1991). R_0 determines whether a pathogen can persist in such a population and is valuable for assessing control options. When R_0 is less than 1, on average each infectious individual infects less than one other individual, and the pathogen will die out in the population. In contrast, when R_0 exceeds 1, numbers of cases will on average rise over time, and an epidemic can result. R_0 is consistently estimated to be between 1 and 2 from rabies outbreaks in dog populations around the world (Hampson et al., 2009), which is relatively close to the extinction threshold of 1.

A closely related concept is that of the effective reproductive number (R_e), when transmission occurs in a population that is not entirely susceptible due to implemented control efforts.

For dog rabies, R_e is determined by the number of susceptible dogs bitten by each infected dog during its infectious period, and the probability that those bitten dogs go on to develop rabies (i.e., become infectious and show signs of disease themselves). The number of dogs bitten may depend upon genetic, behavioral, environmental, and anthropogenic factors, whereas the probability of developing rabies once bitten depends upon vaccination status (since there is no natural immunity to rabies), as well as other intrinsic factors, including viral dose, location of bite(s), and degree of tissue injury. The aim of control measures is to reduce transmission so that R_e is reduced to below the threshold of 1. For rabies, these control measures can therefore operate by reducing either the number of dogs bitten or the probability that bitten dogs develop rabies.

Historically, prior to the advent of effective vaccines, domestic dog rabies control measures focused on reducing the number of susceptible dogs bitten, through movement restrictions and culling rabid, bitten and "stray" dogs (Fooks, Roberts, Lynch, Hersteinsson, & Runolfsson, 2004; Muir & Roome, 2005; Tierkel, 1959; WHO, 1987). The low value of R_0 and cultural context of strict confinement and muzzling of dogs probably contributed to the success of these measures in some isolated locations, but few other successes were reported using only these measures. With the advent of effective animal vaccines (see Chapter 14), mass vaccination has become the mainstay of successful dog rabies control, and a more ethically and culturally acceptable control measure. The control of infectious diseases through mass vaccination is based on the concept of herd immunity, when the vaccination of a proportion of the population (or "herd") provides protection for individuals who are not vaccinated (Fine, 1993). That proportion of the population that needs to be vaccinated to achieve herd immunity and thus control disease depends on R_0. For rabies, low values of R_0 suggest that the critical vaccination coverage, P_{crit}, required to control disease should be roughly between 20% and 40%; that is, 20%–40% of the dog population should be immune at any point in time in order to prevent sustained outbreaks of rabies, although short chains of transmission can still occur in such partially vaccinated populations (Hampson et al., 2009). By comparison, some other infectious diseases that have been successfully controlled by mass vaccination (e.g., measles or rinderpest) have considerably higher values of R_0, and therefore require coverages well above 90%. These figures thus suggest that dog rabies is amenable to control by mass vaccination.

18.2.2 Epidemiological factors influencing the effectiveness of control efforts

Although only low levels of vaccination coverage are theoretically required to control dog rabies, in practice the achievement of this goal is hampered by factors responsible for declines in the proportion of the vaccinated population over time. Host demography plays a major role in influencing the long-term effectiveness of vaccination efforts, because coverage levels decline as vaccinated dogs die and new susceptible dogs are born (Fig. 18.2). The movement of dogs by people similarly affects coverage, particularly if large numbers of unvaccinated dogs are brought into an area. In most dog rabies-endemic areas, dog birth rates are high and therefore coverage levels decline rapidly (Cleaveland et al., 2018; Conan, Akerele, et al., 2015). If coverage falls below the target threshold of 20%–40%, rabies virus transmission can be

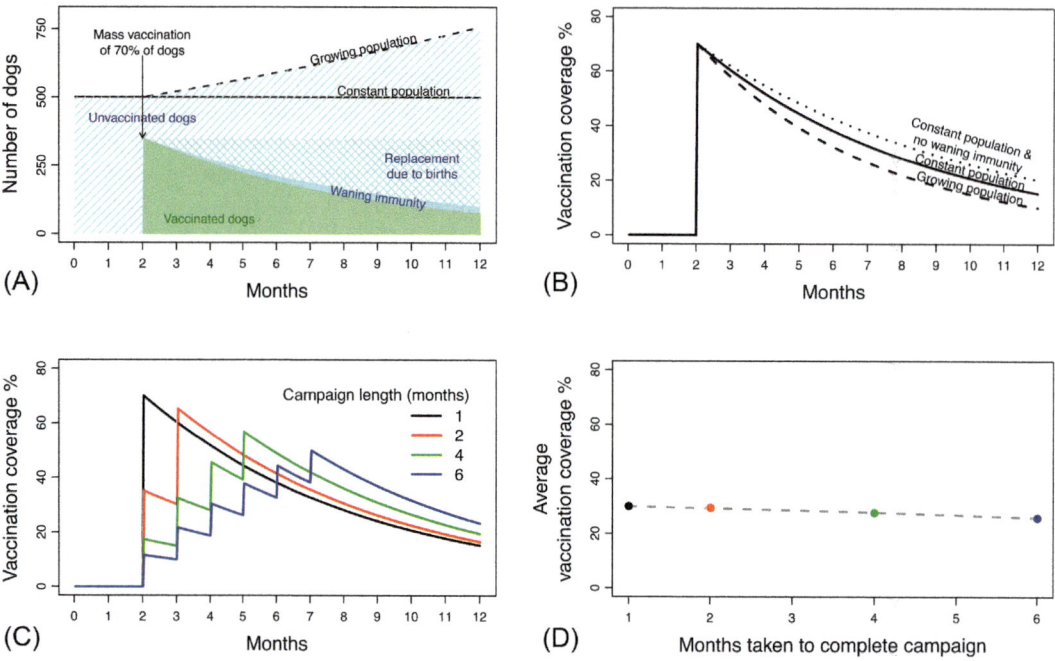

FIG. 18.2 Factors affecting vaccination coverage. (A) A population model showing the change in vaccinated individuals in a stable population *(solid line)* following a campaign implemented in month 2, whereby 70% of the population is vaccinated, and the corresponding change in a growing population *(dashed line)*. Vaccinated dog numbers *(green region)* decline due to deaths (crosshatched region) and waning immunity *(solid blue region)* while (B) coverage declines due to the replacement of the deaths of vaccinated dogs by susceptible puppies as well as the waning immunity. Deaths occur at the same rate in both vaccinated and unvaccinated animals; therefore, increases in dog lifespan only affect coverage by enabling dogs to reproduce for longer. However, if increased survival results in lower rates of reproduction then coverage may be stabilized, suggesting that research into the factors regulating dog populations is merited. Parameters used in the model are: birth rate = 1.5 year^{-1} in the constant population and 2 year^{-1} in the growing population, death rate = 1.5 year^{-1}, immunity wanes at a rate of 0.333 year^{-1}. (C) The effect of the speed of delivery of campaigns on vaccination coverage: faster but otherwise equivalent campaigns achieve a higher coverage than slower campaigns (see the campaigns implemented in 1, 2, 4, or 6 months—*black, red, green*, and *blue*, respectively) but coverage in the population is lower at the end of the year due to births and loss of immunity, while (D) average coverage over the entire year is only marginally affected by conducting a slower campaign.

sustained and outbreaks can persist. This phenomenon is particularly acute when mass dog vaccination is conducted as "pulsed" campaigns (rather than through the ongoing vaccination of new dogs born into the population), and is influenced by the interval between campaigns (Fig. 18.2).

The vaccination coverage that should be reached during any one campaign will therefore depend not only on the critical threshold coverage to be maintained (P_{crit}), but also on the dog population dynamics, duration of vaccine-induced immunity, and the intercampaign intervals because of the declines in coverage that occur over time. WHO- and OIE-recommended dog vaccines provide immunity lasting for 3 years. Empirically, consecutive annual vaccinations that achieve coverage of at least 70% of the dog population have repeatedly proven effective in eliminating endemic canine rabies (Coleman & Dye, 1996). On the contrary, campaigns that do not reach 60%–70% of the dog population can be effective, but are often not successful in controlling rabies in the long term, because of these declines in coverage in between campaigns (Hampson et al., 2009). A target of 70% coverage using high-quality vaccines would appear to be sufficient to maintain herd immunity even in populations with the highest demographic rates and should remain a universal target for programs that aim to eliminate rabies.

Unfortunately, the success of dog vaccination campaigns has sometimes been compromised by the quality of vaccines and the duration of vaccine-induced immunity. In general, the quality of dog vaccines has improved since the first dog vaccines were demonstrated to be effective (Umeno & Doi, 1921). Nerve tissue vaccines have been deployed at scale during mass dog vaccinations in many countries in Latin America (Belotto, Leanes, Schneider, Tamayo, & Correa, 2005), but these have largely now been replaced by safer cell culture vaccines. World Organisation for Animal Health (OIE) recommendations are designed to ensure use of only high-quality dog rabies vaccines (OIE, 2018). However, outbreaks have been reported from areas where there is suspicion that dog vaccines in use have not been inactivated properly. When highly efficacious vaccines are used, a negative relationship is observed between local vaccination coverage and local estimates of R_e; that is, the higher the vaccination coverage, the fewer secondary cases result from a primary infection. Estimates of R_e that are outliers to this relationship can be suggestive of poor quality or damaged vaccines (Hampson, De Balogh, & McGrane, 2019). Similarly, dog vaccines that require booster vaccination are sometimes procured because they are cheap, but in the long-term they are much less cost effective, especially in areas where free-roaming dogs need to be caught with nets. This is because dogs that have been caught before can be harder to catch again within a short space of time, and the cost of human resource and logistics required to deliver dog vaccination are generally much more than the cost of vaccine (Anderson et al., 2019). With the push for scaling up of mass dog vaccination activities, efforts therefore need to be made to support the procurement of high-quality vaccines and to prevent short-sighted planning that can have long-term negative consequences.

Other factors besides dog demography may impact the sustainability of control measures and prospects of dog rabies elimination. Declines in rabies incidence can occur relatively quickly given sufficient effort: two consecutive years of high coverage annual campaigns led to progressive declines in rabies incidence in the Americas (Schneider et al., 2007) and similar impacts have been observed in several African countries (Hikufe et al., 2019; Mpolya et al., 2017; Zinsstag et al., 2017). However, the period over which sustained coverage needs to be maintained to achieve elimination depends on a number of factors related to host population size and isolation, demography, spatial- and human-related factors, heterogeneities in coverage, as well as epidemic stochasticity (Ferguson et al., 2015; Selhorst et al., 2005; Townsend, Lembo, et al., 2013). In particular, the size, spatial distribution, and connectivity of host populations and interventions play a key role in disease dynamics and the efficacy of control measures. In contiguous landscapes, local transmission frequently introduces infections from dog populations with endemic rabies into neighboring areas, impeding any localized vaccination efforts (Beyer et al., 2011; Tenzin et al., 2010; Zinsstag et al., 2017). In more isolated or island populations, imported infections are less common, occurring mainly as a consequence of human-mediated transport of incubating or infectious dogs. Nonetheless, even in island populations, incursions can be frequent and a major impediment to achieving and sustaining freedom from disease (Tohma et al., 2016). The probability that imported infections spark new epidemics depends upon corresponding vaccination efforts (campaign frequency, spatial distribution, and coverage achieved) and demographically driven declines in coverage.

A further challenge to eliminating rabies is the need to maintain comprehensive vaccination coverage across populations. Gaps in coverage, particularly in high-density, large, and connected populations, can act as reservoirs of infection and help to maintain transmission between annual vaccination campaigns when coverage is waning (Ferguson et al., 2015; Townsend, Lembo, et al., 2013). Even though evidence suggests that rabies virus transmission does not depend on the density of dog populations (Hampson et al., 2009; Morters et al., 2013), epidemics are expected to last longer and more cases are expected to occur in larger (and higher density) populations. The time horizon to achieve elimination is therefore likely to be longer in larger and more connected populations. Careful planning of vaccination delivery to reach all communities, subsequent monitoring to ensure adequate coverage, and undertaking remedial vaccination where necessary are important components in the design of rabies elimination programs (see Section 18.3).

18.2.3 Beyond "effectiveness": Exploring social dimensions of rabies control

The critical interdependencies between sociobehavioral and cultural mechanisms and health are widely recognized (Anderson, 1998). Health interventions based on robust epidemiological evidence, but which lack understanding of broad social processes or theories, or of the contexts in which interventions take place, may not lead to the expected outcomes (McKinlay & Marceau, 2000; Power et al., 2004). These knowledge gaps are particularly large in health interventions in resource-limited settings where government-led, top-down delivery mechanisms tend to focus on reducing risks without engaging with the needs of those mostly affected (Long, 2001). Exploring how interventions are situated in broader social worlds is clearly an important first step toward understanding the landscapes that may enable optimal participation in health interventions. In impoverished populations, a myriad of other factors—ranging from infrastructural issues to health care access, availability and affordability—are likely to affect compliance. There is a need to understand and engage with such broader issues as well as the processes that may empower communities to build resilience against the disease threats they face. Yet, these aspects are often overlooked, despite this understanding being essential to the delivery of effective, equitable, and sustainable interventions (Reidpath et al., 2011). These factors are likely to influence individuals whose participation in vaccination campaigns will lead to the required coverage: those bringing their animals for vaccination, their communities more broadly, as well as those who implement such campaigns.

Further complexities arise from the idea that these phenomena not only operate at the level of the individual, whether receiver or provider of interventions, but they also affect the diverse contexts ("communities" or institutions) in which such individuals may be embedded, for example families, neighborhoods, peer groups, workplaces, etc. Such relationships, settings, and constraints may have important impacts on an individual's decisions around delivering interventions or seeking health care for themselves, or their family members or their animals. Furthermore, socioeconomic characteristics, shared values, social norms, or infrastructural constraints impact people at both individual and broader scales. For those implementing campaigns, in particular, it is important to consider that "communities" may not be so clear cut. They can include groups of people united by geography, kinship, religion, interests and ideas, and beyond. The influence of the broader environments people exist within is especially important in contexts where collective participation in an activity can critically contribute to a public good (Siegal, Siegal, & Bonnie, 2009).

Large-scale rabies vaccination programs exemplify this issue. Similar to other immunization programs, rabies control requires mass sustained participation in successive dog vaccination campaigns to achieve herd immunity, which generate health and welfare gains for a particular geographic community as a whole. If a small proportion of the population refrains from taking their dogs for vaccination, some benefit will still accrue from actions taken by others. However, a lack of sufficient participation will compromise vaccination rates to such an extent to threaten the public good and may lead to persistence of transmission and human deaths. The need to increase participation in mass dog vaccination campaigns in endemic areas is often cited (Bardosh et al., 2014; Lembo et al., 2010). However, key questions remain regarding the factors influencing individual contributions toward both the public good and individual benefit, particularly when these benefits are not clearly perceived. The design of programs that lead to sustained participation and interrupt rabies transmission needs to consider the salience of these interventions to the individual and to the broader communities and social environments of which they are part, and which they dually influence through their actions.

At the individual level, and from an economic frame of reference, it is important to consider how rabies interventions fit among competing daily tasks confronting those participating and ultimately influencing their behaviors, and how their salience could be increased in order to encourage participation. These considerations are particularly relevant given that the species involved in rabies transmission is not an economic commodity or valued within a household at the same level as other animal species such as livestock (Meltzer, 1995). In impoverished communities, rabies control itself is often neglected due to the multitude of other challenges in people's everyday lives (Bardosh et al., 2014). These situations are further exacerbated by the lack of prioritization given by agricultural sectors to the control of animal health problems unrelated to livestock species, despite their major implications for public health (Lembo et al., 2011). These perceptions (of health providers and recipients), combined with broader structural issues such as widespread national privatization policies or the move away from traditional public healthcare delivery in a number of rabies-endemic countries (Basu, Andrews, Kishore, Panjabi, & Stuckler, 2012; Price, 1988), will likely impact the willingness and ability of households to invest time and resources for rabies control. Framing rabies control interventions around concerted actions to protect a public good whose benefits are available to all, from recipients to policy makers, would be a critical step toward better recognition, creation, and preservation of collective behavior (Siegal et al., 2009). This necessitates first understanding these environments, then bringing communities and their health providers into the planning and implementation process for improved sustainability.

A question, which then arises, is who becomes the stakeholders in and of these efforts? Social interactions and cultural norms within any given community are an important component to understand in order to integrate new ideas and effect change at scales broader than the individual. Social interactions and interpersonal relationships, "social capital" (Coleman, 1990), and the hierarchies and roles individuals yield within communities mediate how interventions occur, and can shape community priorities and goals (Feierman, Kleinman, Stewart, Farmer, & Das, 2010). This is particularly important when considering interventions at community scales. Communal interventions that work toward shared goals have been examined in a range of contexts, such as in determining the level of health achievement of individuals and their communities (Berkman & Glass, 2000; House, Landis, & Umberson, 1988). These processes are central to the concept of community and individual empowerment (Wiggins, 2011), which builds on the Brazilian educator Paulo Freire's "empowerment education" theory and practice (Freire, 1973). For decades, this approach has been influential in addressing social, political, and economic contradictions (Cho, Paz y Puente, Louie, & Khokha, 2004), which arose from unequal distribution of power and resources within various countries. This philosophy has been increasingly applied to a public health context, whereby it is viewed as a multidimensional model comprising a range of interlinked levels and components. The levels include the individual, the organization and the community, and the components span from the most individual and internal (e.g., belief in one's capacity) to the most collective and external (e.g., to be able to use this capacity to tackle organizational and community needs) (Wallerstein, 2002; Wiggins, 2011). The use of empowerment education in a health context has been associated to positive outcomes at all these levels (for a review, see Wiggins, 2011), from the more individual enhancements in self-confidence, self-respect, self-esteem, and critical consciousness (which are considered essential steps toward collective action and social change) (Arenas-Monreal, Paulo-Maya, & López-González, 1999; Chang, Li, & Liu, 2004; Delp, Brown, & Domenzain, 2005; Minkler & Cox, 1980; Serrano-García, 1984), to the most collective and external levels such as a greater sense of community, participation in change, leadership, solidarity, and advocacy (Chang et al., 2004; Conner, Ling, Tuttle, & Brown-Tezera, 1999; Delp et al., 2005; Rivera, 2003; Valadaz Figueroa, Alfaro Alfaro, Guerra, Aldrete Rodriguez, & Mendoza Roaf, 2000). Empowerment education has also been linked to health-promoting behavioral changes particularly for building local capacity to address broad health challenges. For example, it has increased water-boiling practices in Honduras through training of rural women as health promoters (Minkler & Cox, 1980), led to dietary changes in rural Mexican communities by establishing health education groups (Valadaz Figueroa et al., 2000), and to reductions in exposure to pesticides among participants of empowerment education programs in Nicaragua (Weinger & Lyons, 1992). Such improvements have been reported also at the provider level. For instance, empowerment interventions targeting nurses in Taiwan led to enhancements in psychological empowerment, productivity, and innovations in their day-to-day practice (Chang, Liu, & Yen, 2008). Reframing rabies elimination from a collective point of view, whereby individuals act together to achieve and protect a public good, remains challenging. This is particularly true within settings where top down-heavy models of intervention delivery predominate (Long, 2001). Nevertheless, innumerable lessons can be learnt from instances across the globe where collective change has been applied in a number of contexts, such as development, conservation, education and health, through empowerment education approaches.

Sociological research suggests that societal characteristics can influence individual behavior, such as between those that are more "individualistic" or those that are more "collectivist." While other social science disciplines may not use such bounded frames of reference for social groups (Feierman et al., 2010), the sociocultural norms and rules that guide a member of a society do influence individual and collective behavior, beliefs, and norms. For example, it has been suggested that societies with a strong sense of solidarity, citizenry, and community, where emphasis is placed on group identities and cooperation in order to achieve group goals (Fjneman et al., 1996), may provide insights in successful resolutions of the collective action problem compared to more individualistic communities (Siegal et al., 2009). Such cultural and contextual factors are also central to how people respond to health interventions, and need to be considered from the design phase to carrying out the intervention. Issues ranging from breast cancer survival (Lim, Gonzalez, Wang-Letzkus, & Ashing-Giwa, 2009) to eating disorders (Orji & Mandryk, 2014) show that cultural differences between collectivist and individualist groups can influence the effectiveness of interventions. Tapping into how an individual is situated and views the collective units of which they are a part, while acknowledging the structural issues people face, can help inform what changes need to be made at multiple scales. This provides opportunities to support individuals in shifting from individual to collective empowerment, which can broaden health improvements.

Supporting communities and governments toward the ownership of their own health interventions can foster a culture of communal responsibility for public goods, and encourage a greater sense of community and voluntary commitment to program planning, development, and execution (see e.g., McFarlane & Fehir, 1994). This support can further the processes of empowerment. However, such an approach requires substantial time and effort that neither donor- nor government-led interventions, which typically require a quick-win solution or cost savings, allow for. The task is not impossible, especially if novel approaches that integrate lessons from social science are taken. For example, we can learn from traditional theories and models of volunteerism (Hustinx & Lammertyn, 2003). In cultural contexts with no tradition of volunteering, there may still be a role for voluntary engagement in collective activities. Collective actions that facilitate public good provisions, such as dog vaccination in rabies-endemic settings, can be based on culturally specific commitments to and norms within particular communities. Such an approach, which may involve time spent, no direct compensation, and indistinct benefits, may present challenges in areas where people rely on subsistence livelihoods and have competing primary needs, for instance across much of rural Africa. Nevertheless, understanding how individuals experience and contribute to their communities in these settings would provide valuable insights about the ways in which volunteering may be enacted. For instance, people contribute to a collective sense of identity and toward social cohesion in small rural communities when individuals give their time and efforts toward collective initiatives (e.g., village meetings to discuss specific health issues or development agendas). For example, in some settings in East Africa, local government bylaws are typically developed and enacted by communities themselves to formalize individual responsibilities and commitments to the communal whole. Additionally, there are often informal social norms, traditional rules, and local practices that guide social interactions as well as the rules of participation within a community. Simplified, social norms, if clearly understood and applied through volunteers in new ways, may positively impact the individual contributions to and sustainability of rabies intervention programs, thus benefiting the public good. However, programs that facilitate a sense of public good around rabies interventions may still face constraints due to the cost of delivery of vaccination, particularly in relation to the required human resources. Costs associated with supporting vaccinators and administrative staff (e.g., personnel responsible for registration of dogs and issuing vaccination certificates) have severely challenged budgets allocated to campaigns in some of the most affected countries (Hatch et al., 2016). If individuals are imparted with a sense of ownership, and thereby contribute to these programs through volunteering or through communally decided rules, they become not just passive recipients of interventions, but active participants in them. This could have cost-reducing benefits as an offshoot of the less tangible benefits accrued through empowerment. Nonetheless, high-quality vaccines must still be available for successful interventions and this remains the major barrier in many rabies endemic countries.

Another key consideration toward sustainable rabies control initiatives relates to identifying and engaging the critical gatekeepers within a community whose participation can lead to the required coverage, e.g., those that can influence others to participate or those most likely to bring their own animals for vaccination. This is essential in order to tailor methods to particular groups. Social sciences (e.g., anthropology or human geography) are critical in this regard because of their long history of working with and recognizing the importance of gatekeepers for sustained community engagement. While seeking out and interacting with gatekeepers of a community is time consuming, building trust and developing relationships has led to successes across a number of topical areas, from conservation initiatives (Davis & Goldman, 2017) to public health (Virhia, 2019). Additionally, targeting those who are most at risk of developing rabies, such as children, may require key gatekeepers to access these particularly vulnerable groups. Interactions with children may be necessary, since they are not only at primary risk, but they are also the ones most often responsible for taking care of dogs, including bringing them for vaccination. Therefore, it becomes important to include this particular group not only in the development of effective campaigns, but also in planning interventions aimed at raising awareness about rabies and its control. However, children—similar to adult groups—do not exist on their own, but are part of families and communities. Family and societal dynamics, and the nature of parent-child relationships in their specific cultural contexts, need to be understood when considering a specific target group such as children. Information delivered to children, particularly only when done through schools, may not necessarily translate into action (Power et al., 2004); thus, the most effective interventions should target children, their parents, and other key gatekeepers within a given community.

In conclusion, methods of intervention implementation and sustainability need to go beyond purely epidemiological factors, and include consideration of socioeconomic and cultural elements. Furthermore, individual and community empowerment efforts should be expanded to a rabies context, whereby the intervention is beneficial to entire communities, while also supporting individual benefits. Since rabies elimination is a public good, only defining individual choices, and economic decisions or program costs, is insufficient. We argue for paying greater attention to broader ideas beyond cost effectiveness and salience. Understanding how community needs, individual roles, social norms, and cultural influences interact is a critical step toward revealing how to improve collective contributions to a public good. Engaging with these ideas can facilitate individual and community empowerment while supporting sustainable intervention programs.

18.2.4 Dog population management

Dog population management (DPM) is a broad term encompassing interventions that usually target free-roaming domestic dogs, and aim to stabilize or reduce population size or density, adjust population structure, improve the overall health or alter the behavior of the dog population (Taylor, Wallace, et al., 2017). Interventions can include culling or sterilization programs, sometimes combined with rabies vaccinations (Reece & Chawla, 2006; Totton et al., 2010). There can be many benefits of DPM programs, including positive impacts on human behavior toward street dogs (Cleaveland, Hampson, Lembo, Townsend, & Lankester, 2014; Rowan, Lindenmayer, & Reece, 2014; Taylor, Wallace, et al., 2017), but we now restrict our discussion to issues related to rabies control. It is important to note that canine rabies can very effectively be controlled by vaccination programs alone.

18.2.4.1 Culling

Culling, or the widespread killing of hosts regardless of infection status, is often conducted on the assumption that, if host densities are sufficiently reduced, infectious diseases will be unable to invade, or persist (Morters et al., 2013). While the assumption that disease transmission in a population is dependent upon host density (i.e., that disease transmission rates increase with increasing host densities) is intuitively appealing, there is no evidence for density-dependent transmission of rabies in domestic dog populations (Hampson et al., 2009). The assumption that contact rate scales with host density ignores the complexity and heterogeneity of animal and human behavior. This is reflected in the evidence of the ineffectiveness of culling to control rabies. As an example, nearly 300,000 dogs were culled in Flores, Indonesia, over a 4-year period in response to a rabies outbreak on that island in 1997. Rabies was still endemic in 2004, even though the total dog population had been reduced by around half to approximately 400,000 dogs (Windiyaningsih et al., 2004). Similarly, culling of stray dogs and wildlife failed to control canine rabies in Israel, and it was not until mass vaccination of dogs was introduced that the disease was brought under control (Kaplan, Goor, & Tierkel, 1954). Mass dog culling also failed to control rabies in Dhaka, Bangladesh (Hossain et al., 2011; Tenzin, Ahmed, Debnath, Ahmed, & Yamage, 2015), whereas a subsequent focus on PEP and mass dog vaccination resulted in a consistent decrease in the annual incidence of rabies in that country (WHO, 2017). Historically, some culling efforts may have succeeded through strict authoritarian policies, by removing almost 100% of the dog population (Shone, 1962). However, there is no evidence that culling operations today are effective in controlling rabies, and enforcing such extreme levels of population reduction would now be logistically impractical and culturally unacceptable.

Two further points should be considered with regards to culling dogs to control rabies. Firstly, although culling programs may lead to the removal of infected animals, the low prevalence of canine rabies means that untargeted culling will remove far more healthy than sick dogs. Selective removal of suspect rabid dogs will usually still be carried out by communities, even in the absence of culling programs. Indeed, euthanasia of animals suspected to be rabid should be an essential part of rabies control programs (Medley et al., 2017). Secondly, although a variable degree of culling of free-roaming dogs—historically regarded as "strays"—has often been undertaken alongside mass vaccination programs, the vast majority of free-roaming dogs in most societies globally are owned (Butler & Bingham, 2000; Cleaveland & Dye, 1995; Gsell et al., 2012; Kaare et al., 2009; Kayali et al., 2003; Morters et al., 2014; Ratsitorahina et al., 2009; WHO, 1988; Windiyaningsih et al., 2004) and in reasonable health, and therefore culling these animals is ethically questionable. Culling can thus result in unintended negative consequences for the community, often provoking considerable upset, whereas such dogs can usually be vaccinated without any of these negative repercussions. Communities moving their dogs in response to the threat of culling, appears to have even led to more rapid dissemination of rabies in some situations (Townsend, Sumantra, et al., 2013).

18.2.4.2 Sterilization programs

Sterilization programs aim to reduce the birth rate in a population by permanently or temporarily removing the ability of individuals to reproduce. Currently, surgical sterilization through the removal of the ovaries and uterus in females (ovariohysterectomy) or testes in males (castration or orchidectomy) is the primary method to sterilize dogs, although there is ongoing research into safe, effective, and affordable chemical and immunological means (Rhodes, 2017).

Limited data suggest that sterilizations at the population level may stabilize or gradually reduce population size or density over time scales of several years (Reece & Chawla, 2006; Totton et al., 2010). However, as discussed in Section 18.2.4.1, there is no evidence that rabies transmission in domestic dog populations is density-dependent and that reductions in host density by sterilization (or culling) reduce the incidence of rabies. Although reductions in population density may plausibly reduce the number of dogs that require vaccination, timely reductions in density may be constrained by resources and population dynamics. Cost-effectiveness analyses of rabies control scenarios suggest that dog vaccination alone is more efficient than combined vaccination and sterilization (Anderson et al., 2019; Fitzpatrick et al., 2016).

Theory suggests that birth control could help to maintain levels of vaccination coverage, by reducing birth rates and increasing the longevity of sterilized dogs. Yet, empirical evidence for this is equivocal. Based on the limited data currently available (Totton et al., 2010), the effects of sterilizations on population structure and the maintenance of vaccination coverage are unclear. Where these programs are undertaken, there is no evidence that sterilization increases the life expectancy of dogs. Critically, there are no definitive studies providing evidence of how dog populations are regulated, nor what drives the demand for new animals. Given that most dogs, including roaming animals, are owned, the effect of sterilizations on herd immunity may be hampered by two possible responses to reductions in the local availability of puppies, coupled with an ongoing demand for dogs: (i) concomitant reductions in puppy mortality, which might result from reduced competition for food (although there is no empirical evidence that free-roaming dogs compete at the population level for food to survive), reductions in surplus puppies being dumped or improved care of puppies produced from any remaining unsterilized bitches, and (ii) an increase in puppies or dogs being sought from outside the population. Migration of people with their dogs may also affect population structure and vaccination coverage. It is unclear what effect these compensatory mechanisms have on vaccination coverage. If sterilizations precipitate an influx of dogs from external sources, this might result in an increase in the spread of rabies between populations or within very large populations. Arguably, a local supply of healthy puppies to meet demand might actually help to limit the spread of rabies. What is clear is that the influence of human factors on population structure, vaccination coverage, and the spread of rabies is currently poorly understood and warrant further research.

Sterilization to reduce the number of puppies and subadults in the population has been advocated on the basis that a proportionally higher incidence of rabies occurs in dogs under 12 months of age (Belcher, Wurapa, & Atuora, 1976; Beran, 1991; Kayali et al., 2003; Malaga, Lopez Nieto, & Gambirazio, 1979; Mitmoonpitak, Tepsumethanon, & Wilde, 1998) and an increased public health risk is often reported because of close contact with puppies (WHO, 1998; Widdowson, Morales, Chaves, & McGrane, 2002). This may be the effect of comparatively lower vaccination coverage in this subgroup (Belcher et al., 1976; Malaga et al., 1979; Widdowson et al., 2002) rather than a higher rate of contact (Mitmoonpitak et al., 1998; Mitmoonpitak, Wilde, & Tepsumethanon, 1997). Dogs younger than 12 weeks of age are not often vaccinated (Chomel et al., 1988; Mitmoonpitak et al., 1998) on the assumption that maternal antibody may prevent adequate immune responses (this issue is discussed in more detail in Section 18.3.2), although WHO (2018a) and OIE (2018) recommend vaccination of dogs younger than 12 weeks during mass dog vaccination campaigns. Rather than try to reduce the proportion of these age groups in the population through sterilizations, it is more efficient to simply vaccinate them as a means to maintain herd immunity.

Sterilization programs have only been shown to reduce the incidence of rabies when combined with rabies vaccinations (Reece & Chawla, 2006). However, as all sterilized dogs are also vaccinated, it is unclear to what extent sterilization itself impacts on rabies control. There are no data on the effect of sterilizations as a sole modality on disease incidence and currently no scientific evidence to support sterilizations as an essential component of rabies control programs. Rather, for the purposes of rabies control, available resources should be directed toward repeat vaccination campaigns (Anderson et al., 2019). For sterilization programs with the primary aim of addressing animal welfare concerns, animals should still be vaccinated at the time of sterilization to increase vaccination coverage. Legislation regulating the number of owned dogs as a means to control rabies (either to reduce recruitment or population densities) is also sometimes implemented (Lapiz et al., 2012), with differing degrees of enforcement. All these measures may have positive benefits if they increase awareness of rabies and responsible dog ownership, but direct impacts on reducing transmission are likely to be negligible if not carried out as part of a sustained, high-coverage vaccination program.

18.3 Practical aspects of dog rabies control

For infection to be eliminated from a given area, rabies control strategies must target the animal population mostly responsible for viral maintenance and transmission to humans (the reservoir host). Once rabies has been eliminated in the reservoir species, strategies need to be implemented to ensure that freedom from rabies is maintained. Regional and national targets for the elimination of dog-mediated rabies have now been set, with a global target to reach zero deaths from dog-mediated rabies by 2030 (Abela-Ridder et al., 2016). Several tools are available to support countries as they undertake rabies elimination programs, including the Stepwise Approach to Rabies Elimination (SARE) designed for use by countries to self-assess their national control programs, benchmark progress, and revise priorities along the complex pathway toward elimination (OIE, WHO, Food and Agriculture Organization of the United Nations (FAO), Global Alliance for Rabies Control (GARC), 2017). The SARE and other supporting materials are accessible from the Rabies Blueprint Platform (www. caninerabiesblueprint.org), a live document hosting up-to-date and comprehensive case studies, procedures, and protocols for rabies control and prevention (Lembo & Partners for Rabies Prevention, 2012).

18.3.1 Planning phase

The first steps in the SARE involve preparation and aim to (a) increase awareness at the national and community level to ensure commitment of politicians and policymakers (hence legislative and financial support) and intersectoral dialogue, as well as community participation in dog rabies control operations and responsible dog ownership; (b) establish local capacity for rabies surveillance, prevention and control, including training of relevant professionals; and (c) focus on initiating dog vaccination programs.

Emphasis has been placed by others on the need to conduct detailed surveys for planning purposes prior to the start of canine rabies control programs. While these surveys offer the opportunity to collect useful data on differences in dog demographics, relationships between dogs and people, levels of supervision and accessibility, and local community attitudes, they require specialized personnel, are costly to implement and may delay the implementation of control programs and divert resources from dog vaccination. The issue of inaccessibility of dogs to vaccination is often perceived as a critical impediment to successful dog rabies control, and is therefore viewed as an essential aspect to determine through preliminary surveys. Ecological studies from a range of rabies endemic settings have however demonstrated that, although most domestic dogs are allowed to roam freely, at least one household claims some degree of responsibility for them, and that most owners are willing to restrain their dogs for vaccination (Gsell et al., 2012; Kaare et al., 2009; Kayali et al., 2003; Kongkaew, Coleman, Pfeiffer, Antarasena, & Thiptara, 2004; Morters et al., 2014; Perry, 1993; Ratsitorahina et al., 2009; Robinson, Miranda, Miranda, & Childs, 1996; Suzuki et al., 2008; Wandeler, Matter, Kappeler, & Budde, 1993). Thus, in most circumstances, detailed surveys to determine these factors are not required.

Initial determination of the dog population size in a given community on the other hand is important in the preparatory and evaluation stages in order to determine campaign logistics (e.g., number of dogs to vaccinate to reach the recommended coverage, estimating supplies needed, etc.), and evaluate the intervention in terms of vaccination coverage. Complete dog counts are rarely available and, although techniques to estimate dog population sizes exist, they are not without problems. Rapid estimates can be made prior to program implementation (e.g., based on expert opinion or through simple household surveys to establish the mean number of owned dogs per household and dog-to-human ratios) and refined subsequently in combination with surveys for estimation of vaccination coverage. Household surveys potentially capture much of the owned population, but methods are also available to assess the number of free-roaming dogs for those localities where there is good evidence that a significant proportion of the dog population is unowned (WSPA, 2007). These include mark-resight methods (Meunier et al., 2019a; Tenzin, Ahmed, et al., 2015; Tenzin, McKenzie, et al., 2015; Tiwari et al., 2019; Tiwari, Vanak, O'Dea, Gogoi-Tiwari, & Robertson, 2018) or distance sampling methods (Meunier et al., 2019b). There are however several issues with all these approaches. Factors such as time of day or individual personnel may affect the reproducibility of estimates (Meunier et al., 2019a). Bias may be introduced by multiple extrapolations that may produce unrealistic estimates of dog population sizes. Counts obtained through these techniques are based on random selection of sampling units (e.g., households, sublocations, or routes), and care should be taken to attain adequate representation of the target population and not to extrapolate results beyond this population, due to heterogeneities in dog ecology and ownership patterns (Sambo et al., 2018). Similarly, estimates of owned dog populations are generally extrapolated from the total human population or number of households known through national population censuses. Human population censuses are generally carried out at wide time intervals with associated potential errors in dog size estimates resulting from population growth. In addition, the design, implementation, and analysis of data generated from these surveys are not trivial.

Mass dog vaccination campaigns have now been conducted, at least on a limited scale, in most countries around the world, and there is increasing knowledge on what approaches are most effective in different communities, as well as crude estimates of human:dog ratios across a range of geographical areas. This information is sufficient for mass dog vaccination campaigns to be conducted in pilot areas, which could then be evaluated to generate more realistic estimates of dog population sizes and used to extrapolate to nearby populations, to enable scaling up of vaccination activities. It is clear that obtaining accurate and precise estimates of dog populations at large scales is challenging, but postvaccination transects offer a useful tool for tailoring dog population estimates (Sambo et al., 2018). A lack of very accurate initial dog population estimates should therefore not be considered an obstacle to initiating vaccination campaigns. An additional option to consider would also be the inclusion of dogs within nationwide censuses of the human population, which are routinely carried out by census bureau offices.

18.3.2 Implementation of mass dog vaccination campaigns

The feasibility of rabies elimination through mass dog vaccination is now widely advocated, with a target set to achieve zero human deaths from dog-mediated rabies by 2030 (Abela-Ridder et al., 2016) (see Chapter 21). There are several approaches for mass dog vaccination, but—given the evidence of general accessibility of dogs for parenteral vaccination (Gibson et al., 2016; Kayali et al., 2003; Kongkaew et al., 2004; Lembo et al., 2010; Perry, 1993; Ratsitorahina et al., 2009; Robinson et al., 1996; Suzuki et al., 2008; Wandeler et al., 1993)—here we will focus on approaches that apply to free-roaming dogs that vary in accessibility and handleability. For highly accessible rural communities and dogs that are easy to handle, central-point vaccinations, consisting of mobile teams that set up temporary vaccination posts in central village locations, are generally the most cost-effective strategy (Kaare et al., 2009). This, along with continual vaccination at fixed vaccination posts in well-recognized sites within the community (e.g., private or government veterinary clinics), is also a suitable approach in urban settings (Gibson et al., 2016). In dispersed communities, or where dogs are difficult to handle, alternative delivery strategies may be required to achieve high coverage. In very remote communities, combined approaches using central point and house-to-house vaccination conducted by either mobile teams of permanent staff or trained local community-based animal health workers have proved very effective in achieving high coverage and eliminating rabies from given areas (Kaare et al., 2009; Lembo et al., 2010), although they require considerable investment in labor and capital and are operationally difficult (Kaare et al., 2009). Smartphone technology can be employed to assist with the coordination and supervision of field teams conducting rabies control activities (Gibson et al., 2018).

To ensure a high turnout among all communities in a population, vaccination campaigns need to be tailored to the local context. For instance, the timing of advertising is critical to reach the target population, as is the salience of the messages delivered. This may mean providing sufficient advance notice, but also reminding dog owners shortly before the campaign to tell others so as to encourage high turnout. Oftentimes small changes in the design of activities, such as the location and duration of vaccination points that take account of the local geography and community concerns, can greatly influence participation (Bardosh et al., 2014; Castillo-Neyra et al., 2017).

Trained dog-catchers with nets have been used to attain high coverage on Bali where dog densities are very high, and most dogs are not used to being restrained (Putra et al., 2013). The same technique has proven effective in communities in India (Gibson et al., 2015), and appears to be a feasible and effective approach for settings where a substantive proportion of dogs cannot easily be brought to vaccination stations. This capture-vaccinate-release (CVR) strategy is now being replicated in other settings, with training provided to dog catchers to ensure use of appropriate and humane methods (GARC, 2014).

Several candidate vaccines have been evaluated for oral rabies vaccination of dogs (Cliquet et al., 2018). Although oral rabies vaccination has not been used in any dog rabies elimination program, this strategy has the potential to supplement parenteral vaccination coverage in settings where a substantial proportion of dogs cannot be readily restrained (Darkaoui et al., 2014; Estrada, Vos, De Leon, & Muller, 2001; Gibson et al., 2019; Matter et al., 1998; WHO, 1998, 2007). A proof-of-principle study of oral bait handout (OBH) to access free-roaming dogs in a population in which only 30% of dogs could be handled for injection showed that fixed operational team costs were up to four times lower compared to CVR, although the cost per vaccine administered was similar due to the higher cost of oral vaccines (Gibson et al., 2019). The use of oral rabies vaccines could supplement parenteral vaccination, improving coverage in hard-to-reach populations, while also lowering operational costs, suggesting this could be a scalable approach. However, the method is not without its limitations. There is no universally attractive oral bait for dogs, and baits may therefore need to be developed and tested for local populations. Currently, all candidate vaccines for oral rabies vaccination of dogs are self-replicating agents (Cliquet et al., 2018), and delivery methods of oral vaccines may lead to wider environmental distribution than injectable vaccines. While oral rabies vaccination of wildlife has been used extensively with minimal human exposures (Mähl et al., 2014; Maki et al., 2017), the close association of dogs with people increases the likelihood of human exposure to oral rabies vaccines targeting dogs. Regulatory authorities will need to assess these aspects in decision making, to balance potential risks of oral vaccine-associated adverse events with the potential benefits of this method to enhance dog rabies control and elimination programs. An oral vaccination approach to supplement parenteral vaccination would likely be more socially acceptable and easier to repeat than CVR, which would help in sustaining high coverage.

An essential issue to be addressed in the design and implementation of dog rabies control programs is that of vaccination of young puppies. A lack of inclusion of puppies in vaccination efforts can result in insufficient coverage being attained (Kaare et al., 2009). Despite the widespread perception that vaccine-induced active immunity may be affected by the presence of maternally derived antibodies in young puppies, evidence from a range of settings demonstrates that vaccination of dogs younger than 3 months confers protection, independent of the health condition of dogs (Barrat et al., 2001; Chappuis, 1998; Morters et al., 2015). Puppies are an easy group to handle and therefore targeting puppies would be a simple approach that could easily enhance the coverage of existing strategies. Cost-effectiveness analysis also supports the inclusion of puppies in mass dog rabies vaccination campaigns, despite the high mortality rate in this age group (Anderson et al., 2019).

As discussed in Section 18.2.3, improving community participation in dog rabies control programs can be addressed through an understanding of the factors influencing participation, as well as awareness and appropriate campaign planning. Ensuring adequate engagement of local communities is undoubtedly a key step for successful dog rabies control (Bardosh et al., 2014; Castillo-Neyra et al., 2017). The input of other disciplines, dedicated to developing methods most compatible with local circumstances, to inform and influence individual and community decisions that enhance health (i.e., social sciences) is therefore increasingly recognized as essential in global rabies elimination strategies. To this effect, initiatives have been developed that have had considerable impacts worldwide (Costa, Briggs, & Dedmon, 2010; Costa, Briggs, Tumpey, Dedmon, & Coutts, 2009). A major factor that is likely to compromise community participation and that can be a difficult challenge to overcome is that of costs. Charging for vaccination is usually counterproductive in poorer communities without an effective regulatory culture, as turnout may be too low to achieve adequate vaccination coverage (Dürr et al., 2009; see also Section 18.3.5, Economics of dog vaccination for rabies control). Indeed, an approach of charging for vaccination could be counterproductive in terms of being both costly while only achieving very limited vaccination coverage (Jibat, Hogeveen, & Mourits, 2015).

Given that dog rabies control programs often operate under logistical and financial constraints, an important question relates to the design of vaccination campaigns that maximize the effectiveness of control efforts in terms of both reducing the occurrence of rabies, and the time and intensity of effort required to achieve this goal. The design of vaccine delivery in pulses is an important determinant of the effectiveness of vaccination campaigns (Nokes & Swinton, 1997), with factors such as pulse frequency and pulse size (number of vaccine doses) having a considerable influence on longer-term impacts of interventions (Beyer et al., 2012). When only limited vaccination resources are available, achieving sparse coverage throughout a large population will have only negligible impacts, whereas concentrating vaccination effort in a particular community, preferably one that is isolated, will generate greater gains (Beyer et al., 2012). A disadvantage of pulse vaccination campaigns is the need to attain higher coverage to account for declines in immunization coverage between campaigns. If puppies could be vaccinated without having to wait for annual campaigns, vaccination coverage could be sustained more effectively throughout the year (Cleaveland et al., 2018), with more rapid progress toward control and elimination. Approaches could be considered that allow dogs to be vaccinated by animal health professionals on a more continuous basis. A perceived constraint to this approach is lack of infrastructure to maintain vaccine cold chains continuously in remote areas. However, recent studies suggest that high-quality commercial vaccines have a high degree of thermotolerance (Lankester et al., 2016), which provides opportunities for exploring new models of vaccination that do not rely on annual pulse vaccination.

Although dog-associated rabies elimination should focus on eliminating infection at its source, the integration of dog rabies control (through mass dog vaccination) and human rabies prevention (through human prophylaxis) is required to avoid unnecessary human deaths when dog rabies is still prevalent. Efforts should be made toward correct and cost-effective utilization of costly human biologics by public health authorities and enhanced awareness about prevention behaviors among communities. Integrated Bite Case Management has the potential to address these concerns and enable progressive declines of canine rabies to also reduce demand for human biologics (see Sections 18.1.2 and 18.3.3).

18.3.3 Measuring the impact of mass dog vaccination campaigns

18.3.3.1 *Vaccination coverage*

Achieving high coverage is the most important aim of any vaccination program, since demographic processes cause coverage levels to rapidly decline and the logistic challenge of carrying out campaigns means that it takes time to achieve effective coverage (Fig. 18.2). It is therefore vital to reach as large a proportion of the population as possible: i.e., dogs of all ages, dogs that are free-roaming, and dogs in all communities.

To determine whether sufficient dogs have been vaccinated, coverage needs to be measured. The proportion of the total dog population that is vaccinated is the most direct estimate of coverage, but requires reliable denominator data on the dog population, which are often problematic to obtain (see Section 18.3.1). Coverage can also be estimated from postvaccination questionnaire surveys as the ratio of vaccinated to unvaccinated dogs in households, or from direct observation of marked (vaccinated) and unmarked (unvaccinated) dogs. If the latter method is used, identification of vaccinated dogs is required. In situations where permanent identification of dogs is constrained by a lack of resources, temporary forms of identification (e.g., colored tags or plastic collars) can be applied and are often very popular in local communities, providing motivation for owners to participate in vaccination campaigns (Kaare et al., 2009; Minyoo et al., 2015). In circumstances where dogs cannot easily be restrained to apply a collar, for example with OBH or CVR, animal-marking spray can be an effective alternative for short-term marking of dogs for postvaccination surveys (Conan, Kent, Koman, Konink, & Knobel, 2015). Postvaccination questionnaires are more expensive to conduct at scale and typically only sample a small proportion of the population, which can limit their utility, particularly in populations where there is considerable variability in dog ownership. In contrast to questionnaires, postvaccination mark-resight methods—also referred to as transects—can be conducted rapidly across all communities where vaccination has been conducted and dogs "marked" at vaccination. For this reason, transects typically provide more useful information, in terms of the levels of coverage achieved across large populations and are also able to identify specific gaps in coverage where remedial vaccination is necessary. However, it should also be noted that transects tend to overestimate coverage, and to miss puppies (which are typically not vaccinated), and therefore adjustments need to be applied to account for this (Sambo et al., 2017). Transects have increasingly been applied in a range of populations to estimate the coverage achieved during dog vaccination campaigns (Arief et al., 2017; Tenzin, Ahmed, et al., 2015; Tenzin, McKenzie, et al., 2015), and their analysis can be very effectively used to improve the delivery of future campaigns. Using mobile technology to record transect surveys allows assessments to be conducted in real-time and facilitates efficient campaign management (Gibson et al., 2015).

A common misperception is that serological surveillance is an alternative method for measuring coverage. The safety and efficacy of WHO-recommended cell-culture vaccines currently used for parenteral immunization of dogs are widely recognized, but serological studies can provide information about immunogenicity of alternative vaccines (e.g., locally produced vaccines/baits) and/or effectiveness of vaccine delivery (e.g., correct vaccine administration and maintenance of cold chain). However, there are several reasons why serology is unsuitable for estimating coverage. Firstly, gold standard serology assays are of limited availability in rabies-endemic areas because they require use of live rabies virus, are difficult to perform and are expensive (Rupprecht, Fooks, & Abela-Ridder, 2018). Alternative neutralization assays to assess seroconversion using lentiviral pseudotypes have been developed that remove the need for high containment laboratories and may be more suitable for a wider range of settings (Wright et al., 2008; Wright et al., 2009), but high-standard infrastructure for cell culture is still needed. Enzyme-linked immunosorbent assays (ELISA) are available and could be used to detect responses to vaccination, but these assays can have variable sensitivity and specificity (OIE, 2018). Secondly, quantification of serological responses to vaccination requires determining baseline and 4-week postvaccination antibody titers, making dog sampling strategies operationally cumbersome and very expensive. Thirdly, the minimum rabies virus neutralizing antibody titer of 0.5 IU per mL of serum considered satisfactory for the international transfer of vaccinated dogs should not be used to assess vaccine efficacy in dogs immunized in the context of mass campaigns in which lower values should be expected. As a result, there may be specific circumstances where serology might provide useful insight, but serosurveillance should not be considered as part of routine vaccination or monitoring activities.

18.3.3.2 Surveillance

To evaluate the effectiveness of intervention efforts, adequate surveillance measures need to be established so that the rabies situation can be determined at the start of the control program and impacts of intervention can be monitored through time. Surveillance is also essential to demonstrate freedom from disease and to detect new incursions that could compromise maintenance of freedom in areas where rabies has been eliminated.

International standards for rabies surveillance and recognition of freedom from rabies require laboratory confirmation of cases postmortem, hence effective systems for sample collection and submission, and prompt diagnosis (Rupprecht et al., 2018). However, limited capacity for sample collection and diagnosis is an enduring problem in many dog rabies-endemic countries. Therefore, laboratory-confirmed cases may not necessarily provide a good measure of incidence. In many countries, national statistics are still based on clinical signs and are largely incomplete (Dodet et al., 2008; Taylor & Knopf, 2015).

Over the past 40 years the direct fluorescent antibody technique (DFA) has been the global standard for rapid rabies diagnosis (Rupprecht et al., 2018), but in many parts of the world infrastructure and capacity for fluorescence microscopy for performing this test are still inadequate. Simplified techniques for sample collection, preservation, and diagnosis are now available, which have shown potential to increase in-country capabilities for rabies surveillance. A direct rapid immunohistochemical test (dRIT) based on light microscopy (Rupprecht et al., 2018) shows complete concordance with the DFA and performs well on samples preserved under field conditions (Dürr et al., 2008; Lembo et al., 2006; Tao et al., 2008). Other simple techniques have been described to successfully allow rapid rabies screening, including enzyme immunoassays (Vasanth, Madhusudana, Abhilash, Suja, & Muhamuda, 2004), dot blot enzyme immunoassays (Madhusudana, Paul, Abhilash, & Suja, 2004), and immunochromomatographic diagnostic tests (Kang et al., 2007; Nishizono et al., 2008). In particular, rapid diagnostic tests are very popular for obtaining an immediate diagnosis in the field, but should not be relied upon for decision-making regarding provisioning of PEP, given that in rabies-endemic areas PEP should be administered if there is any doubt about whether an animal is rabid. A major issue for rapid diagnostics is quality control. There has been a proliferation of lateral flow devices (also referred to as rapid immunochromomatographic diagnostic tests or lateral flow assays), but concerns remain over variability in the performance of these tests relative to gold-standard diagnostic tests (Eggerbauer et al., 2016). The use of lateral flow devices for the detection of rabies virus in saliva is not currently recommended, even if claimed by manufacturers (Rupprecht et al., 2018). Some commercially available kits appear to be as sensitive and specific as the gold standard when used on brain material collected postmortem, and can perform even better than the gold standard with fresh samples, which is important given the limited laboratory infrastructure and cold chain in many low- and middle-income countries (Léchenne et al., 2016). The cost of these devices is already relatively low; therefore, efforts to standardize and regulate quality, while also assuring low-cost production for widespread use in the field, could dramatically improve efforts to increase the confirmation of rabies in endemic countries.

Medical records of animal-bite injuries from suspect rabid animals can provide useful epidemiological information both in terms of rabies incidence and human exposures (Hampson et al., 2009) and for making inferences about spatial transmission dynamics and the efficacy of control measures (Beyer et al., 2011). However, the utility of medical records depends very much on the extent to which bite injuries are due to suspect rabid animals or healthy animals. In many endemic countries with very limited access to PEP, or where the cost of PEP to patients is very high, the proportion of bite patients bitten by probable rabid dogs can exceed 50% (Changalucha et al., 2018; Rajeev et al., 2018). In contrast, in countries that have improved access to PEP, the vast majority of bite victims who receive PEP are bitten by healthy animals (Amparo et al., 2018; Benavides et al., 2019; Rysava et al., 2018) and therefore bite records may not be indicative of trends in rabies incidence. Nonetheless, risk assessments conducted with bite patients can be very revealing about the likely rabies status of biting animals, and can inform PEP administration and guide surveillance activities. Risk assessments conducted as part of Integrated Bite Case Management (IBCM) can trigger investigations by veterinary personnel and have been shown to greatly increase the detection of rabies (Undurraga et al.,

2017) and have potential to sufficiently target case detection to enable verification of rabies freedom (Hampson et al., 2016). However, challenges to operationalizing IBCM remain, given the need for close intersectoral communication and sustained effort even when incidence is extremely low (Hampson et al., 2019). Mobile phone network access offers opportunities for real-time reporting/detection of cases and animal bite injuries (Mtema et al., 2016), and for supporting an IBCM approach. In general, channels of reporting and communication need to be improved, and increased deployment of field diagnostic tools and cheap and user-friendly means for reporting cases should be prioritized.

18.3.4 Maintaining rabies-free status

Once achieved, the economic and ethical incentives for maintaining freedom from rabies through sustained (potentially targeted) control (e.g., use of a cordon sanitaire) and effective surveillance are extremely strong, as demonstrated by the devastating effects of emerging epidemics in previously rabies-free areas (Tohma et al., 2016; Townsend, Sumantra, et al., 2013; Wera, Velthuis, Geong, & Hogeveen, 2013; Windiyaningsih et al., 2004). There is currently very little research available to advocate what kind of strategic approaches, such as targeted ring vaccination, would be most effective to eliminate residual foci in the final stages of an elimination program or for responding to new incursions. However, much can be learnt from countries that have achieved and maintained freedom from rabies.

After achieving rabies elimination, the island nations of Britain and Japan both suffered incursions, but were able to control these: swift dog muzzling and dog confinement contained an imported case to Britain in 1921 (Fooks et al., 2004), and mass vaccination freed Japan from rabies following the Second World War (Shimada, 1971). These nations have since maintained freedom from rabies through stringent quarantine systems, with Japan implementing mandatory dog registration and vaccination (Takahashi-Omoe, Omoe, & Okabe, 2008). While mandatory dog vaccination in Japan has not been strictly enforced and may not be necessary, sensitization about the risks of introduction would be useful (Kadowaki, Hampson, Tojinbara, Yamada, & Makita, 2018). In areas with much weaker surveillance capacity, continued mass dog vaccination should be a priority, while rabies circulates endemically in nearby areas (Castillo-Neyra et al., 2017; Pan American Health Organization (PAHO), 2015; Tohma et al., 2016).

Occasional introductions of dog rabies from North Africa to Europe have had significant economic ramifications (Lardon et al., 2010), but all have so far been contained without significant further spread. Similarly, despite periodic reintroductions, a vaccination "cordon-sanitaire" maintained at the Malaysia-Thailand border (Tan et al., 1972; Wells, 1954) had kept Malaysia effectively free from rabies with only one isolated and probably imported dog rabies case reported outside the buffer zone in 1996 (Hussin, 1997). However, in 2017 rabies was reported for the first time from the Malaysian State of Sarawak (ProMED-mail, 2017), likely having spread from West Kalimantan in Indonesia and threatening to emerge in Brunei. Generally, maintaining freedom from rabies across continental geopolitical boundaries has proven more difficult. Although rabies in North America has been largely controlled since the 1960s through mass vaccination campaigns followed by a long period of compulsory pet vaccination (Blanton, Hanlon, & Rupprecht, 2007; Held, Tierkel, & Steele, 1967; Korns &

Zeissig, 1948), the United States was only declared free of dog rabies in 2007 after concerted transboundary collaborations to prevent importations (Blanton et al., 2007). In southern Brazil, several states have relaxed mass dog vaccination campaigns as rabies has not circulated endemically for over two decades. However, the risk of rabies spread from neighboring Bolivia continues (Galhardo et al., 2019). For example, an outbreak occurred in the Brazilian state of Mato Grosso do Sul in late 2015, but was contained to the bordering municipalities through a strong mass dog vaccination response.

Geographic isolation, high levels of surveillance, local capacity and infrastructure for rapid mobilization, continued political commitment and intersectoral cooperation, and enforced legislation have been important factors in keeping countries rabies-free (Takahashi-Omoe et al., 2008). In contiguous landscapes, an assessment of the rabies situation in neighboring areas is of great importance, ideally followed by the establishment of rabies control and prevention efforts in these jurisdictions through liaison and collaborations between key stakeholders. The implementation of IBCM would be a means of strengthening surveillance in such areas, to identify incursions and enable rapid responses, as well as for reducing unnecessary use of PEP. Although contiguous international boundaries undoubtedly pose considerable challenges to maintaining freedom from infection, phylodynamic studies from North Africa indicate little mixing of viral sequences from Algeria, Tunisia, and Morocco, and evidence for only a few long-distance introductions, which are most likely the result of human-mediated spread (Talbi et al., 2010). The importance of human-mediated dispersal in North Africa gives grounds for cautious optimism that geopolitical boundaries may represent more of a barrier to canine rabies dispersal than may be expected from geographic features alone. Given that rabies spreads relatively slowly in comparison to some other directly transmitted pathogens and that its clinical signs are very characteristic, there is potential for rapid detection and response. With the scaling up of mass dog vaccination programs, there is definitely potential to progressively eliminate rabies and maintain freedom from disease using currently available tools, but strengthened surveillance will be critical to success.

18.3.5 Economics of dog vaccination for rabies control

To ensure that an area is continuously maintained free from rabies, careful consideration also needs to be given to building sustainability in established programs. Financial sustainability should address factors affecting the cost-effectiveness of different rabies control strategies, as well as operational issues associated with the design and implementation of dog vaccination campaigns. For prevention of human rabies deaths, rabies control strategies that incorporate mass vaccination of domestic dogs have the potential to be more cost effective than strategies relying on administration of human PEP alone, with dog vaccination typically becoming more cost effective within 5–6 years of the onset of mass dog vaccination campaigns (Bögel & Meslin, 1990; Léchenne et al., 2017; Zinsstag et al., 2009). These economic benefits can be explained by the high costs of PEP in comparison with delivery of dog vaccination (typically US$1.5–US$4/dog—Bögel & Meslin, 1990; Kaare et al., 2009; Kayali, Mindekem, Hutton, Ndoutamia, & Zinsstag, 2006), and the escalating costs of PEP over time as dog populations and rabies incidence continue to rise. However, in cost-effectiveness models, there is often an assumption of a linear relationship between dog rabies incidence, human exposure, and demand for human PEP, and it is clear that this relationship may vary in different settings. For example, the incidence of suspected rabid dog bites reported at local health facilities declined rapidly in rural Tanzania following the implementation of annual dog vaccination campaigns and decline in dog rabies incidence (Cleaveland, Kaare, Tiringa, Mlengeya, & Barrat, 2003; Lembo et al., 2010). In contrast, dog vaccination has been associated with increased demand for PEP in some parts of Asia (Hahn, 2009). Critical factors are likely to include education and awareness (of both bite victims and clinicians), availability and cost of human PEP, the size of national health budgets, and a society's tolerance of risk in clinical decision-making regarding administration of PEP. A further important factor relates to costs of different regimens of PEP. The latest WHO position paper promotes the use of intradermal vaccination administered over the course of 1 week (WHO, 2018b), which should reduce the costs of PEP administration (Hampson et al., 2018). Investment by Gavi, the Vaccine Alliance, has potential to dramatically improve PEP access (WHO Rabies Modelling Consortium, 2019). These initiatives to improve the cost-effectiveness of PEP should save many lives, but will not eliminate the disease. Implementation of rabies elimination demonstration projects in a range of settings is increasingly providing evidence of the cost effectiveness of mass dog vaccination, but also highlights the challenge of sustainability. It is clear that rabies incidence can be reduced to low levels through mass dog vaccination in just a few years through well-delivered dog vaccination campaigns, but achieving elimination is much harder and necessarily requires overcoming obstacles in those populations that are hardest to reach (Cleaveland & Hampson, 2017). However, interruption of transmission in the dog population would allow relaxing of dog vaccination efforts in large parts of the world and is therefore an attractive economic incentive. Once rabies transmission is interrupted, use of PEP could potentially also be drastically reduced.

18.3.6 Extended benefits of dog rabies control

Within the global health agenda, increasing attention is being paid to developing crosscutting approaches that enhance progress toward Sustainable Development Goals (SDGs) (https://sustainabledevelopment.un.org/), strengthen health systems, and accelerate health gains (Dye, 2018). Rabies provides an excellent example of how extended health benefits could be gained from investments in surveillance, control, and prevention of a locally relevant disease that affects disadvantaged communities.

Rabies is one of the 20 priority diseases within the WHO program on Neglected Tropical Diseases (NTDs) and progress toward dog rabies control contributes directly to a specific target on control and elimination of NTDs within SDG 3 (Good Health and Well Being). However, there is also potential for more extended benefits. As diseases of poverty, NTDs are increasingly considered an indicator of health inequality. This clearly applies to dog rabies, with human deaths from dog rabies occurring predominantly in impoverished communities that are poorly served by human and veterinary services. Substantial declines in human rabies deaths can be achieved through improved provision of PEP, particularly when directed to the most underserved communities (WHO Rabies Modelling Consortium, 2019), and this forms the basis of the decision by Gavi, the Vaccine Alliance to include human rabies vaccines for PEP within the new Gavi investment strategy (Gavi, 2018). Enhanced provision of PEP would undoubtedly contribute to the target of Universal Health Coverage, which remains a central pillar of SDG 3. However, until this can be achieved, it is inevitable that social, economic, and political factors will limit access of the poor to the quality health facilities that can provide effective PEP.

Mass dog vaccination offers an intervention that not only mitigates social and health inequalities, but can also contribute to several other SDG targets. Equity impacts of rabies can be mitigated because, as a One Health intervention targeted to the animal source, benefits are conveyed to all who are epidemiologically connected to the source of infection without regard to socioeconomic status—the benefit cannot be purchased or socially distorted to the detriment of the poor (Cleaveland et al., 2017).

Further health system benefits from dog rabies control arise through strengthening of core capacities for surveillance and response to other diseases, including emerging infectious diseases of concern to global health security. Effective control of dog rabies offers many opportunities for crosscutting benefits. The disease is highly visible, well recognized, and of local concern to most people in dog rabies-endemic populations. Mass dog vaccination can deliver appreciable and tangible benefits that can build trust between communities and health authorities, enhance community engagement and strengthen intersectoral collaboration, all factors that have been identified as constraints in responding to health crises from emerging zoonotic diseases (Cleaveland et al., 2017). Building core capacities and relationships to tackle endemic diseases of local concern and relevance will allow us to respond effectively to emerging disease outbreaks in times of crisis and address existing health inequalities (Halliday et al., 2017).

A further opportunity relates to the potential for dog rabies vaccination to provide a platform for integrated delivery of human and animal health services (Zinsstag, Schelling, Wyss, & Mahamat, 2005). Cost savings can be generated by using shared platforms for delivery of services in remote communities and can achieve higher coverages. This was demonstrated, for example, in pastoral communities in Tanzania where dog rabies vaccination was combined with mass anthelminthic treatment of people for the control of soil-transmitted helminths, generating both cost savings and a higher coverage of anthelminthic treatment that would have been achieved through existing school-based programs (Lankester et al., 2019). Integrated delivery of interventions targeting domestic dogs may also be possible, with options for addressing several important human and livestock health problems through a focus on dogs. For example, engaging with dog owners on rabies vaccination may provide opportunities for more effective delivery of deworming treatments to control human cystic echinococcosis, a further priority zoonoses within the WHO NTD program, as well as coenurosis, which is emerging as a disease of major concern to livestock production in East Africa (Hughes et al., 2019).

18.4 Conclusions

The declaration of the 2030 target to eliminate human rabies deaths due to dog-transmitted rabies has generated considerable momentum (Abela-Ridder et al., 2016) (see Chapter 21) and dog rabies control programs in many parts of the world are taking shape. The promise of a Gavi investment to improve access to rabies PEP has potential to accelerate progress, but alone will not enable the goal of zero human deaths to be achieved (WHO Rabies Modelling Consortium, 2019). While many of the key components of rabies control programs are described in this chapter, including the engagement and empowerment of individuals and communities within local contexts, it should also be recognized that rabies control is chronically underfunded. The limited budget to cover high-quality dog rabies vaccines is a crucial reason why many (if not most) dog rabies control initiatives are not undertaken or sustained. Other globally recognized public health concerns have overcome this challenge, through the recognition that they are a public good. For instance, childhood vaccines are provided for free within the Expanded Program on Immunization; many neglected tropical diseases are supported through the donation of drugs for mass administration; massive investments have led to the widespread provision of bed-nets and implementation of environmental management to control vector-borne diseases. But dog rabies vaccines are still not considered by the global health community as an essential commodity. Hence, no matter how much local enthusiasm exists for implementing and participating in rabies control, without animal vaccines only limited actions can take place. Lobbying of governments to purchase these vaccines may be able to advance this agenda, but the demands of large-scale rabies control are longer than political cycles and more widespread than local government units. National and regional governments in the Americas show that the large-scale control of rabies through dog vaccination can be achieved (Vigilato et al., 2013), but high-level commitment—including financing—is still a crucial requirement for the 2030 goal.

References

Abela-Ridder, B., Knopf, L., Martin, S., Taylor, L., Torres, G., & de Balogh, K. (2016). 2016: The beginning of the end of rabies? *Lancet Global Health*, 4(11), e780–e781.

Amparo, A. C. B., Jayme, S. I., Roces, M. C. R., Quizon, M. C. L., Villalon, E. E. S., III, Quiambao, B. P., et al. (2018). The evaluation of operating Animal Bite Treatment Centers in the Philippines from a health provider perspective. *PLoS One*, 13(7), e0199186. https://doi.org/10.1371/journal.pone.0199186.

Anderson, N. B. (1998). Levels of analysis in health science. A framework for integrating sociobehavioral and biomedical research. *Annals of the New York Academy of Sciences*, 840, 563–576.

Anderson, A., Kotzé, J., Shwiff, S. A., Hatch, B., Slootmaker, C., Conan, A., et al. (2019). A bioeconomic model for the optimization of local canine rabies control. *PLoS Neglected Tropical Diseases*, 13(5), e0007377. https://doi.org/10.1371/journal.pntd.0007377.

Anderson, R. M., & May, R. M. (1991). *Infectious diseases of humans: Dynamics and control*. Oxford: Oxford University Press.

Anderson, A., & Shwiff, S. A. (2015). The cost of canine rabies on four continents. *Transboundary and Emerging Diseases*, 62(4), 446–452. https://doi.org/10.1111/tbed.12168.

Arenas-Monreal, L., Paulo-Maya, A., & López-González, H. E. (1999). Educación popular y nutrición infantil: experiencia de trabajo con mujeres en una zona rural de México [Popular education and child nutrition: Experience with women in a rural area of Mexico]. *Revista de Saúde Pública*, 33(2), 113–121 (in Spanish).

Arief, R. A., Hampson, K., Jatikusumah, A., Widyastuti, M. D., Sunandar, Basri, C., et al. (2017). Determinants of vaccination coverage and consequences for rabies control in Bali, Indonesia. *Frontiers in Veterinary Science*, 3, 123. https://doi.org/10.3389/fvets.2016.00123.

Bardosh, K., Sambo, M., Sikana, L., Hampson, K., & Welburn, S. C. (2014). Eliminating rabies in Tanzania? Local understandings and responses to mass dog vaccination in Kilombero and Ulanga districts. *PLoS Neglected Tropical Diseases*, 8(6), e2935. https://doi.org/10.1371/journal.pntd.0002935.

Barrat, J., Blasco, E., Lambot, M., Cliquet, F., Brochier, B., Renders, C., … Aubert, M. F. A. (2001). Is it possible to vaccinate young canids against rabies and to protect them? In *Proceedings of the sixth Southern and Eastern African Rabies Group/World Health Organization meeting, Lilongwe, Malawi* (pp. 151–163). Lyon: Éditions Fondation Marcel Mérieux.

Basu, S., Andrews, J., Kishore, S., Panjabi, R., & Stuckler, D. (2012). Comparative performance of private and public healthcare systems in low- and middle-income countries: A systematic review. *PLoS Medicine, 9*(6), e1001244. https://doi.org/10.1371/journal.pmed.1001244.

Belcher, D. W., Wurapa, F. K., & Atuora, D. O. C. (1976). Endemic rabies in Ghana. *American Journal of Tropical Medicine and Hygiene, 25*, 724–729.

Belotto, A., Leanes, L. F., Schneider, M. C., Tamayo, H., & Correa, E. (2005). Overview of rabies in the Americas. *Virus Research, 111*(1), 5–12.

Benavides, J. A., Megid, J., Campos, A., Rocha, S., Vigilato, M., & Hampson, K. (2019). An evaluation of Brazil's surveillance and prophylaxis for canine rabies. *PLoS Neglected Tropical Diseases, 13*(8), e0007564. https://doi.org/10.1371/journal.pntd.0007564.

Beran, G. W. (1991). Urban rabies. In G. M. Baer (Ed.), *The natural history of rabies* (2nd ed., pp. 427–443). Boca Raton, FL: CRC Press.

Berkman, L. F., & Glass, T. (2000). Social integration, social networks, social support, and health. In L. F. Berkman & I. Kawachi (Eds.), *Social epidemiology* (pp. 137–173). London: Oxford University Press.

Beyene, T. J., Mourits, M., Kidane, A. H., & Hogeveen, H. (2018). Estimating the burden of rabies in Ethiopia by tracing dog bite victims. *PloS One, 13*(2), e0192313. https://doi.org/10.1371/journal.pone.0192313.

Beyer, H., Hampson, K., Cleaveland, S., Kaare, M., Lembo, T., & Haydon, D. T. (2011). Metapopulation dynamics of rabies and the efficacy of vaccination. *Proceedings of the Royal Society B—Biological Sciences, 278*(1715), 2182–2190. https://doi.org/10.1098/rspb.2010.2312.

Beyer, H. L., Hampson, K., Lembo, T., Cleaveland, S., Kaare, M., & Haydon, D. T. (2012). The implications of metapopulation dynamics on the design of vaccination campaigns. *Vaccine, 30*(6), 1014–1022. https://doi.org/10.1016/j.vaccine.2011.12.052.

Bingham, J. (2001). Rabies on Flores Island, Indonesia: Is eradication possible in the near future? In B. Dodet, F. -X. Meslin, & E. Heseltine (Eds.), *Proceedings of the fourth international symposium on rabies control in Asia, Hanoi, 5–9 March 2001* (pp. 148–155). Montrouge: John Libbey Eurotext.

Blanton, J. D., Hanlon, C. A., & Rupprecht, C. E. (2007). Rabies surveillance in the United States during 2006. *Journal of the American Veterinary Medical Association, 231*(4), 540–556.

Bögel, K., & Meslin, F.-X. (1990). Economics of human and canine rabies elimination – Guidelines for programme orientation. *Bulletin of the World Health Organization, 68*(3), 281–291.

Butler, J. R. A., & Bingham, J. (2000). Demography and dog-human relationships of the dog population in Zimbabwean communal lands. *The Veterinary Record, 147*, 442–446.

Castillo-Neyra, R., Brown, J., Borrini, K., Arevalo, C., Levy, M. Z., Buttenheim, A., et al. (2017). Barriers to dog rabies vaccination during an urban rabies outbreak: Qualitative findings from Arequipa, Peru. *PLoS Neglected Tropical Diseases, 11*(3), e0005460. https://doi.org/10.1371/journal.pntd.0005460.

Chang, L. C., Li, I. C., & Liu, C. H. (2004). A study of the empowerment process for cancer patients using Freire's dialogical interviewing. *Journal of Nursing Research, 12*(1), 41–49.

Chang, L. C., Liu, C. H., & Yen, E. H. (2008). Effects of an empowerment-based education program for public health nurses in Taiwan. *Journal of Clinical Nursing, 17*(20), 2782–2790.

Changalucha, J., Steenson, R., Grieve, E., Cleaveland, S., Lembo, T., Lushasi, K., et al. (2018). The need to improve access to rabies post-exposure vaccines: Lessons from Tanzania. *Vaccine*, S0264-410X(18)31243-X. https://doi.org/10.1016/j.vaccine.2018.08.086.

Chappuis, G. (1998). Neonatal immunity and immunisation in early age: Lessons from veterinary medicine. *Vaccine, 16*(14–15), 1468–1472.

Cho, E. H., Paz y Puente, F. A., Louie, M. C. Y., & Khokha, S. (2004). *BRIDGE: Building a race and immigration dialogue in the global economy – A popular education resource for immigrant and refugee community organizers*. Oakland, CA: National Network for Immigrant and Refugee Rights.

Chomel, B., Chappuis, G., Bullon, F., Cardenas, E., de Beublain, T. D., Lombard, M., & Giambruno, E. (1988). Mass vaccination campaign against rabies: Are dogs correctly protected? The Peruvian experience. *Reviews of Infectious Diseases, 10*, S697–S702.

Cleaveland, S., & Dye, C. (1995). Maintenance of a microparasite infecting several host species: Rabies in the Serengeti. *Parasitology, 111*, S33–S47.

Cleaveland, S., & Hampson, K. (2017). Rabies elimination research: Juxtaposing optimism, pragmatism and realism. *Proceedings of the Royal Society B—Biological Sciences, 284*(1869), 20171880. https://doi.org/10.1098/rspb.2017.1880.

Cleaveland, S., Hampson, K., Lembo, T., Townsend, S., & Lankester, F. (2014). Role of dog sterilisation and vaccination in rabies control programmes. *Veterinary Record, 175*(16), 409–410. https://doi.org/10.1136/vr.g6352.

Cleaveland, S., Kaare, M., Tiringa, P., Mlengeya, T., & Barrat, J. (2003). A dog rabies vaccination campaign in rural Africa: Impact on the incidence of dog rabies and human dog-bite injuries. *Vaccine*, *21*(17–18), 1965–1973.

Cleaveland, S., Sharp, J., Abela-Ridder, B., Allan, K. J., Buza, J., Crump, J. A., et al. (2017). One Health contributions towards more effective and equitable approaches to health in low- and middle-income countries. *Philosophical Transactions of the Royal Society of London. Series B, Biological Sciences*, *372*(1725), 20160168. https://doi.org/10.1098/rstb.2016.0168.

Cleaveland, S., Thumbi, S. M., Sambo, M., Lugelo, A., Lushasi, K., Hampson, K., & Lankester, F. J. (2018). Proof of concept of mass dog vaccination for the control and elimination of canine rabies. *Revue Scientifique et Technique de l'OIE*, *37*(2), 559–568.

Cliquet, F., Guiot, A. L., Aubert, M., Robardet, E., Rupprecht, C. E., & Meslin, F. X. (2018). Oral vaccination of dogs: A well-studied and undervalued tool for achieving human and dog rabies elimination. *Veterinary Research*, *49*(1), 61. https://doi.org/10.1186/s13567-018-0554-6.

Coleman, J. S. (1990). *Foundations of social theory*. Cambridge, MA: Harvard University Press.

Coleman, P. G., & Dye, C. (1996). Immunization coverage required to prevent outbreaks of dog rabies. *Vaccine*, *14*, 185–186.

Conan, A., Akerele, O., Simpson, G., Reininghaus, B., van Rooyen, J., & Knobel, D. (2015). Population dynamics of owned, free-roaming dogs: Implications for rabies control. *PLoS Neglected Tropical Diseases*, *9*(11), e0004177. https://doi.org/10.1371/journal.pntd.0004177.

Conan, A., Kent, A., Koman, K., Konink, S., & Knobel, D. (2015). Evaluation of methods for short-term marking of domestic dogs for rabies control. *Preventive Veterinary Medicine*, *121*(1–2), 179–182. https://doi.org/10.1016/j.prevetmed.2015.05.008.

Conner, A., Ling, C. G., Tuttle, J., & Brown-Tezera, B. B. (1999). Peer education project with persons who have experienced homelessness. *Public Health Nursing*, *16*(5), 367–373.

Costa, P., Briggs, D., & Dedmon, R. (2010). World Rabies Day (September 28, 2010): The continuing effort to 'make rabies history'. *Asian Biomedicine*, *4*, 671.

Costa, P., Briggs, D. J., Tumpey, A., Dedmon, R., & Coutts, J. (2009). World Rabies Day outreach to Asia: Empowering people through education. *Asian Biomedicine*, *3*, 451–457.

Darkaoui, S., Boué, F., Demerson, J. M., Fassi Fihri, O., Yahia, K. I., & Cliquet, F. (2014). First trials of oral vaccination with rabies SAG2 dog baits in Morocco. *Clinical and Experimental Vaccine Research*, *3*(2), 220–226. https://doi.org/10.7774/cevr.2014.3.2.220.

Davis, A., & Goldman, M. J. (2017). Beyond payments for ecosystem services: Considerations of trust, livelihoods and tenure security in community-based conservation projects. *Oryx*, *53*(3), 491–496. https://doi.org/10.1017/S0030605317000898.

Delp, L., Brown, M., & Domenzain, A. (2005). Fostering youth leadership to address workplace and community environmental health issues: A university-school-community partnership. *Health Promotion Practice*, *6*(3), 270–285.

Dodet, B., Adjogoua, E. V., Aguemon, A. R., Amadou, O. H., Atipo, A. L., Baba, B. A., et al. (2008). Fighting rabies in Africa: The Africa Rabies Expert Bureau (AfroREB). *Vaccine*, *26*(50), 6295–6298. https://doi.org/10.1016/j.vaccine.2008.04.087.

Dürr, S., Mindekem, R., Kaninga, Y., Moto, D. D., Meltzer, M. I., Vounatsou, P., & Zinsstag, J. (2009). Effectiveness of dog rabies vaccination programmes: Comparison of owner-charged and free vaccination campaigns. *Epidemiology and Infection*, *137*, 1558–1567. https://doi.org/10.1017/S0950268809002386.

Dürr, S., Naïssengar, S., Mindekem, R., Diguimbye, C., Niezgoda, M., Kuzmin, I., et al. (2008). Rabies diagnosis for developing countries. *PLoS Neglected Tropical Diseases*, *2*(3), e206. https://doi.org/10.1371/journal.pntd.0000206.

Dye, C. (2018). Expanded health systems for sustainable development. *Science*, *359*(6382), 1337–1339. https://doi.org/10.1126/science.aaq1081.

Eggerbauer, E., de Benedictis, P., Hoffmann, B., Mettenleiter, T. C., Schlottau, K., Ngoepe, E. C., et al. (2016). Evaluation of six commercially available rapid immunochromatographic tests for the diagnosis of rabies in brain material. *PLoS Neglected Tropical Diseases*, *10*(6), e0004776. https://doi.org/10.1371/journal.pntd.0004776.

Estrada, R., Vos, A., De Leon, R., & Muller, T. (2001). Field trial with oral vaccination of dogs against rabies in the Philippines. *BMC Infectious Diseases*, *1*, 23.

Feierman, S., Kleinman, A., Stewart, K., Farmer, P., & Das, V. (2010). Anthropology, knowledge-flows and global health. *Global Public Health*, *5*(2), 122–128.

Ferguson, E. A., Hampson, K., Cleaveland, S., Consunji, R., Deray, R., Friar, J., et al. (2015). Heterogeneity in the spread and control of infectious disease: Consequences for the elimination of canine rabies. *Scientific Reports*, *5*, 18232. https://doi.org/10.1038/srep18232.

Fine, P. E. M. (1993). Herd immunity: History, theory, practice. *Epidemiologic Reviews, 15*(2), 265–302. https://doi.org/10.1093/oxfordjournals.epirev.a036121.

Fitzpatrick, M. C., Hampson, K., Cleaveland, S., Lembo, T., Mzimbiri, I., Lankester, F., et al. (2014). Cost-effectiveness of canine vaccination to prevent human rabies in rural Tanzania. *Annals of Internal Medicine, 160*(2), 91–100. https://doi.org/10.7326/M13-0542.

Fitzpatrick, M. C., Shah, H. A., Pandey, A., Bilinski, A. M., Kakkar, M., Clark, A. D., et al. (2016). One Health approach to cost-effective rabies control in India. *Proceedings of the National Academy of Sciences of the United States of America, 113*(51), 14574–14581. https://doi.org/10.1073/pnas.1604975113.

Fjneman, Y. A., Willemsen, M. E., Poortinga, Y. H., Erelcin, F. G., Georgas, J., Hui, C. H., et al. (1996). Individualism-collectivism: An empirical study of a conceptual issue. *Journal of Cross-Cultural Psychology, 27*(4), 381–402.

Fooks, A. R., Roberts, D. H., Lynch, M., Hersteinsson, P., & Runolfsson, H. (2004). Rabies in the United Kingdom, Ireland and Iceland. In A. A. King, A. R. Fooks, M. Aubert, & A. I. Wandeler (Eds.), *Historical perspective of rabies in Europe and the Mediterranean Basin* (pp. 25–32). Paris: OIE (World Organisation for Animal Health).

Freire, P. (1973). *Education for critical consciousness*. New York: Continuum.

Galhardo, J. A., de Azevedo, C. S., Remonti, B. R., Gonçalves, V. M. N., Marques, N. T. A., Borges, L. O., & das Neves, D. A. (2019). Canine rabies in the Brazil-Bolivia border region from 2006 to 2014. *Annals of Global Health, 85*(1), 25. https://doi.org/10.5334/aogh.2334.

GARC (2014). *Mass dog vaccination in Bangladesh: The silent revolution towards rabies elimination*. https://rabiesalliance.org/resource/mass-dog-vaccination-bangladesh-silent-revolution-towards-rabies-elimination. (Accessed 23 July 2019).

Gautret, P., Shaw, M., Gazin, P., Soula, G., Delmont, J., Parola, P., et al. (2008). Rabies postexposure in returned injured travelers from France, Australia, and New Zealand: A retrospective study. *Journal of Travel Medicine, 15*(1), 25–30. https://doi.org/10.1111/j.1708-8305.2007.00164.x.

Gavi (2018). *Vaccine investment strategy*. https://www.gavi.org/about/strategy/vaccine-investment-strategy/. (Accessed 23 July 2019).

Gibson, A. D., Handel, I. G., Shervell, K., Roux, T., Mayer, D., Muyila, S., et al. (2016). The vaccination of 35,000 dogs in 20 working days using combined static point and door-to-door methods in Blantyre, Malawi. *PLoS Neglected Tropical Diseases, 10*(7), e0004824. https://doi.org/10.1371/journal.pntd.0004824.

Gibson, A. D., Mazeri, S., Lohr, F., Mayer, D., Burdon Bailey, J. L., Wallace, R. M., et al. (2018). One million dog vaccinations recorded on mHealth innovation used to direct teams in numerous rabies control campaigns. *PLoS One, 13*(7), e0200942. https://doi.org/10.1371/journal.pone.0200942.

Gibson, A. D., Ohal, P., Shervell, K., Handel, I. G., Bronsvoort, B. M., Mellanby, R. J., & Gamble, L. (2015). Vaccinate-assess-move method of mass canine rabies vaccination utilising mobile technology data collection in Ranchi, India. *BMC Infectious Diseases, 15*, 589. https://doi.org/10.1186/s12879-015-1320-2.

Gibson, A. D., Yale, G., Vos, A., Corfmat, J., Airikkala-Otter, I., King, A., et al. (2019). Oral bait handout as a method to access roaming dogs for rabies vaccination in Goa, India: A proof of principle study. *Vaccine: X, 1*, 100015. https://doi.org/10.1016/j.jvacx.2019.100015.

Grover, M., Bessell, P. R., Conan, A., Polak, P., Sabeta, C. T., Reininghaus, B., & Knobel, D. L. (2018). Spatiotemporal epidemiology of rabies at an interface between domestic dogs and wildlife in South Africa. *Scientific Reports, 8*(1), 10864. https://doi.org/10.1038/s41598-018-29045-x.

Gsell, A. S., Knobel, D. L., Cleaveland, S., Kazwala, R. R., Vounatsou, P., & Zinsstag, J. (2012). Domestic dog demographic structure and dynamics relevant to rabies control planning in urban areas in Africa: The case of Iringa, Tanzania. *BMC Veterinary Research, 8*, 236. https://doi.org/10.1186/1746-6148-8-236.

Hahn, N. T. H. (Ed.), (2009). *Proceedings of the 2nd international conference of Rabies in Asia (RIA) Foundation, 9–11 September, Hanoi, Viet Nam, 2009*. http://www.rabiesinasia.org/vietnam/riacon2009report.pdf. (Accessed 9 July 2019).

Halliday, J. E. B., Hampson, K., Hanley, N., Lembo, T., Sharp, J. P., Haydon, D. T., & Cleaveland, S. (2017). Driving improvements in emerging disease surveillance through locally-relevant capacity strengthening. *Science, 357*(6347), 146–148. https://doi.org/10.1126/science.aam8332.

Hampson, K., Abela-Ridder, B., Bharti, O., Knopf, L., Léchenne, M., Mindekem, R., et al. (2018). Modelling to inform prophylaxis regimens to prevent human rabies. *Vaccine*, S0264-410X(18)31519-6. https://doi.org/10.1016/j.vaccine.2018.11.010.

Hampson, K., Abela-Ridder, B., Brunker, K., Bucheli, S. T. M., Carvalho, M., Caldas, E., et al. (2016). Surveillance to establish elimination of transmission and freedom from dog-mediated rabies. *BioRxiv*. https://doi.org/10.1101/096883.

Hampson, K., Coudeville, L., Lembo, T., Sambo, M., Kieffer, A., Attlan, M., et al. (2015). Estimating the global burden of endemic canine rabies. *PLoS Neglected Tropical Diseases*, *9*(4), e0003709. https://doi.org/10.1371/journal.pntd.0003709.

Hampson, K., De Balogh, K., & McGrane, J. (2019). Lessons for rabies control and elimination programmes—A decade of experience from Bali, Indonesia. *Revue Scientifique et Technique de l'OIE*, *38*(1), 213–224. https://doi.org/10.20506/rst.38.1.2954.

Hampson, K., Dobson, A., Kaare, M., Dushoff, J., Magoto, M., Sindoya, E., & Cleaveland, S. (2008). Rabies exposures, post-exposure prophylaxis and deaths in a region of endemic canine rabies. *PLoS Neglected Tropical Diseases*, *2*(11), e339. https://doi.org/10.1371/journal.pntd.0000339.

Hampson, K., Dushoff, J., Cleaveland, S., Haydon, D. T., Kaare, M., Packer, C., et al. (2009). Transmission dynamics and prospects for the elimination of canine rabies. *PLoS Biology*, *7*, e1000053. https://doi.org/10.1371/journal.pbio.1000053.

Hatch, B., Anderson, A., Sambo, M., Maziku, M., Mchau, G., Mbunda, E., et al. (2016). Towards canine rabies elimination in south-eastern Tanzania: Assessment of health economic data. *Transboundary and Emerging Diseases*, *64*(3), 951–958. https://doi.org/10.1111/tbed.12463.

Held, J. R., Tierkel, E. S., & Steele, J. H. (1967). Rabies in man and animals in the United States, 1946–65. *Public Health Reports*, *82*, 1009–1018.

Hikufe, E. H., Freuling, C. M., Athingo, R., Shilongo, A., Ndevaetela, E. E., Helao, M., et al. (2019). Ecology and epidemiology of rabies in humans, domestic animals and wildlife in Namibia, 2011–2017. *PLoS Neglected Tropical Diseases*, *13*(4), e0007355. https://doi.org/10.1371/journal.pntd.0007355.

Hossain, M., Bulbul, T., Ahmed, K., Ahmed, Z., Salimuzzaman, M., Haque, M. S., et al. (2011). Five-year (January 2004–December 2008) surveillance on animal bite and rabies vaccine utilization in the Infectious Disease Hospital, Dhaka, Bangladesh. *Vaccine*, *29*(5), 1036–1040. https://doi.org/10.1016/j.vaccine.2010.11.052.

House, J. S., Landis, K. R., & Umberson, D. (1988). Social relationships and health. *Science*, *214*(4865), 540–545.

Hu, R. L., Fooks, A. R., Zhang, S. F., Liu, Y., & Zhang, F. (2008). Inferior rabies vaccine quality and low immunization coverage in dogs (*Canis familiaris*) in China. *Epidemiology and Infection*, *136*(11), 1556–1563. https://doi.org/10.1017/S0950268807000131.

Hughes, E. C., Kibona, T. K., de Glanville, W. A., Lankester, F., Davis, A., Carter, R. W., et al. (2019). *Taenia multiceps* coenurosis in Tanzania: A major and under-recognised disease problem in pastoral communities. *Veterinary Record*, *184*(6), 191. https://doi.org/10.1136/vr.105186.

Hussin, A. (1997). Malaysia: Veterinary aspects of rabies control and prevention. In B. Dodet & F.-X. Meslin (Eds.), *Rabies control in Asia* (pp. 167–170). Paris: Elsevier.

Hustinx, L., & Lammertyn, F. (2003). Collective and reflexive styles of volunteering: A sociological modernization perspective. *Voluntas: International Journal of Voluntary and Nonprofit Organizations*, *14*(2), 167–187.

Jibat, T., Hogeveen, H., & Mourits, M. C. (2015). Review on dog rabies vaccination coverage in Africa: A question of dog accessibility or cost recovery? *PLoS Neglected Tropical Diseases*, *9*(2), e0003447. https://doi.org/10.1371/journal.pntd.0003447.

Kaare, M., Lembo, T., Hampson, K., Ernest, E., Estes, A., Mentzel, C., & Cleaveland, S. (2009). Rabies control in rural Africa: Evaluating strategies for effective domestic dog vaccination. *Vaccine*, *27*(1), 152–160. https://doi.org/10.1016/j.vaccine.2008.09.054.

Kadowaki, H., Hampson, K., Tojinbara, K., Yamada, A., & Makita, K. (2018). The risk of rabies spread in Japan: A mathematical modelling assessment. *Epidemiology and Infection*, *146*(10), 1245–1252. https://doi.org/10.1017/S0950268818001267.

Kang, B., Oh, J., Lee, C., Park, B. K., Park, Y., Hong, K., et al. (2007). Evaluation of a rapid immunodiagnostic test kit for rabies virus. *Journal of Virological Methods*, *145*(1), 30–36.

Kaplan, M. M., Goor, Y., & Tierkel, E. S. (1954). A field demostration of rabies control using chicken-embryo vaccine in dogs. *Bulletin of the World Health Organization*, *10*(5), 743–752.

Kayali, U., Mindekem, R., Hutton, G., Ndoutamia, A. G., & Zinsstag, J. (2006). Cost-description of a pilot parenteral vaccination campaign against rabies in dogs in N'Djaména, Chad. *Tropical Medicine and International Health*, *11*, 1058–1065.

Kayali, U., Mindekem, R., Yemadji, N., Vounatsou, P., Kaninga, Y., Ndoutamia, A. G., & Zinsstag, J. (2003). Coverage of pilot parenteral vaccination campaign against canine rabies in N'Djamena, Chad. *Bulletin of the World Health Organization*, *81*, 739–744.

King, A. A., Fooks, A. R., Aubert, M., & Wandeler, A. I. (Eds.), (2004). *Historical perspective of rabies in Europe and the Mediterranean Basin*. Paris: OIE (World Organisation for Animal Health).

Knobel, D. L., Cleaveland, S., Coleman, P. G., Fèvre, E. M., Meltzer, M. I., Miranda, M. E. G., et al. (2005). Re-evaluating the burden of rabies in Africa and Asia. *Bulletin of the World Health Organization, 83*(5), 360–368.

Kongkaew, W., Coleman, P., Pfeiffer, D. U., Antarasena, C., & Thiptara, A. (2004). Vaccination coverage and epidemiological parameters of the owned-dog population in Thungsong District, Thailand. *Preventive Veterinary Medicine, 65*(1–2), 105–115.

Korns, R. F., & Zeissig, A. (1948). Dog, fox, and cattle rabies in New-York State – Evaluation of vaccination in dogs. *American Journal of Public Health, 38*, 50–65.

Lankester, F., Davis, A., Kinung'hi, S., Yoder, J., Bunga, C., Alkara, S., et al. (2019). An integrated health delivery platform, targeting soil-transmitted helminths (STH) and canine mediated human rabies, results in cost savings and increased breadth of treatment for STH in remote communities in Tanzania. *BMC Public Health, 19*(1), 1398. https://doi.org/10.1186/s12889-019-7737-6.

Lankester, F. J., Wouters, P. A. W. M., Czupryna, A., Palmer, G. H., Mzimbiri, I., Cleaveland, S., et al. (2016). Thermotolerance of an inactivated rabies vaccine for dogs. *Vaccine, 34*(46), 5504–5511. https://doi.org/10.1016/j.vaccine.2016.10.015.

Lapiz, S. M., Miranda, M. E., Garcia, R. G., Daguro, L. I., Paman, M. D., Madrinan, F. P., et al. (2012). Implementation of an intersectoral program to eliminate human and canine rabies: The Bohol Rabies Prevention and Elimination Project. *PLoS Neglected Tropical Diseases, 6*(12), e1891. https://doi.org/10.1371/journal.pntd.0001891.

Lardon, Z., Watier, L., Brunet, A., Bernède, C., Goudal, M., Dacheux, L., et al. (2010). Imported episodic rabies increases patient demand for and physician delivery of antirabies prophylaxis. *PLoS Neglected Tropical Diseases, 4*(6), e723. https://doi.org/10.1371/journal.pntd.0000723.

Léchenne, M., Mindekem, R., Madjadinan, S., Oussiguéré, A., Moto, D. D., Naissengar, K., & Zinsstag, J. (2017). The Importance of a participatory and integrated one health approach for rabies control: The case of N'Djaména, Chad. *Tropical Medicine and Infectious Disease, 2*(3), 43. https://doi.org/10.3390/tropicalmed2030043.

Léchenne, M., Naïssengar, K., Lepelletier, A., Alfaroukh, I. O., Bourhy, H., Zinsstag, J., & Dacheux, L. (2016). Validation of a rapid rabies diagnostic tool for field surveillance in developing countries. *PLoS Neglected Tropical Diseases, 10*(10), e0005010. https://doi.org/10.1371/journal.pntd.0005010.

Lembo, T., Attlan, M., Bourhy, H., Cleaveland, S., Costa, P., de Balogh, K., et al. (2011). Renewed global partnerships and re-designed roadmaps for rabies prevention and control. *Veterinary Medicine International, 2011*. Article ID 923149, 18 pp.

Lembo, T., Craig, P., Miles, M. A., Hampson, K., & Meslin, F. X. (2013). Zoonoses prevention, control, and elimination in dogs. In C. N. L. Macpherson, F. X. Meslin, & A. I. Wandeler (Eds.), *Dogs, zoonoses and public health* (2nd ed., pp. 205–258). Wallingford, Oxon, UK: CAB International.

Lembo, T., Hampson, K., Haydon, D. T., Craft, M., Dobson, A., Dushoff, J., et al. (2008). Exploring reservoir dynamics: A case study of rabies in the Serengeti ecosystem. *Journal of Applied Ecology, 45*(4), 1246–1257. https://doi.org/10.1111/j.1365-2664.2008.01468.x.

Lembo, T., Hampson, K., Kaare, M., Ernest, E., Knobel, D., Kazwala, R., et al. (2010). The feasibility of canine rabies elimination in Africa: Dispelling doubts with data. *PLoS Neglected Tropical Diseases, 4*(2), e626. https://doi.org/10.1371/journal.pntd.0000626.

Lembo, T., Niezgoda, M., Velasco-Villa, A., Cleaveland, S., Ernest, E., & Rupprecht, C. E. (2006). Evaluation of a direct, rapid immunohistochemical test for rabies diagnosis. *Emerging Infectious Diseases, 12*(2), 310–313. https://doi.org/10.3201/eid1202.050812.

Lembo, T., & Partners for Rabies Prevention. (2012). The blueprint for rabies prevention and control: A novel operational toolkit for rabies elimination. *PLoS Neglected Tropical Diseases, 6*(2), e1388. https://doi.org/10.1371/journal.pntd.0001388.

Lim, J., Gonzalez, P., Wang-Letzkus, M. F., & Ashing-Giwa, K. (2009). Understanding the cultural health belief model influencing health behaviors and health-related quality of life between Latina and Asian-American breast cancer survivors. *Supportive Care in Cancer, 17*, 1137. https://doi.org/10.1007/s00520-008-0547-5.

Long, N. (2001). *Development sociology: Actor perspectives*. London: Routledge.

Ly, S., Buchy, P., Heng, N. Y., Ong, S., Chhor, N., Bourhy, H., et al. (2009). Rabies situation in Cambodia. *PLoS Neglected Tropical Diseases, 3*(9), e511. https://doi.org/10.1371/journal.pntd.0000511.

Madhusudana, S. N., Paul, J. P., Abhilash, V. K., & Suja, M. S. (2004). Rapid diagnosis of rabies in humans and animals by a dot blot enzyme immunoassay. *International Journal of Infectious Diseases, 8*, 339–345.

Mähl, P., Cliquet, F., Guiot, A. L., Niin, E., Fournials, E., Saint-Jean, N., et al. (2014). Twenty year experience of the oral rabies vaccine SAG2 in wildlife: A global review. *Veterinary Research, 45*(1), 77. https://doi.org/10.1186/s13567-014-0077-8.

Maki, J., Guiot, A. L., Aubert, M., Brochier, B., Cliquet, F., Hanlon, C. A., et al. (2017). Oral vaccination of wildlife using a vaccinia-rabies-glycoprotein recombinant virus vaccine (RABORAL V-RG®): A global review. *Veterinary Research, 48*(1), 57. https://doi.org/10.1186/s13567-017-0459-9.

Malaga, H., Lopez Nieto, E., & Gambirazio, C. (1979). Canine rabies seasonality. *International Journal of Epidemiology, 8*(3), 243–245.

Matter, H. C., Schumacher, C. L., Kharmachi, H., Hammami, S., Tlatli, A., Jemli, J., et al. (1998). Field evaluation of two bait delivery systems for the oral immunization of dogs against rabies in Tunisia. *Vaccine, 16*(7), 657–665.

McFarlane, J., & Fehir, J. (1994). De madres a Madres: A community, primary health care program based on empowerment. *Health Education Quarterly, 21*(3), 381–394.

McKinlay, J. B., & Marceau, L. D. (2000). To boldly go .... *American Journal of Public Health, 90*(1), 25–33. https://doi.org/10.2105/ajph.90.1.25.

Medley, A. M., Millien, M. F., Blanton, J. D., Ma, X., Augustin, P., Crowdis, K., & Wallace, R. M. (2017). Retrospective cohort study to assess the risk of rabies in biting dogs, 2013–2015, Republic of Haiti. *Tropical Medicine and Infectious Disease, 2*(2), 14. https://doi.org/10.3390/tropicalmed2020014.

Meltzer, M. I. (1995). Livestock in Africa: The economics of ownership and production, and the potential for improvement. *Agriculture and Human Values, 12*(2), 4–18. https://doi.org/10.1007/BF02217292.

Meunier, N. V., Gibson, A. D., Corfmat, J., Mazeri, S., Handel, I. G., Bronsvoort, B. M. C., et al. (2019a). Reproducibility of the mark-resight method to assess vaccination coverage in free-roaming dogs. *Research in Veterinary Science, 123*, 305–310. https://doi.org/10.1016/j.rvsc.2019.02.009.

Meunier, N. V., Gibson, A. D., Corfmat, J., Mazeri, S., Handel, I. G., Gamble, L., et al. (2019b). A comparison of population estimation techniques for individually unidentifiable free-roaming dogs. *BMC Veterinary Research, 15*(1), 190. https://doi.org/10.1186/s12917-019-1938-1.

Minkler, M., & Cox, K. (1980). Creating critical consciousness in health: Applications of Freire's philosophy and methods to the health care setting. *International Journal of Health Services, 10*(2), 311–322.

Minyoo, A. B., Steinmetz, M., Czupryna, A., Bigambo, M., Mzimbiri, I., Powell, G., et al. (2015). Incentives increase participation in mass dog rabies vaccination clinics and methods of coverage estimation are assessed to be accurate. *PLoS Neglected Tropical Diseases, 9*(12), e0004221. https://doi.org/10.1371/journal.pntd.0004221.

Mitmoonpitak, C., Tepsumethanon, V., & Wilde, H. (1998). Rabies in Thailand. *Epidemiology and Infection, 120*(2), 165–169.

Mitmoonpitak, C., Wilde, H., & Tepsumethanon, V. (1997). Current status of animal rabies in Thailand. *Journal of Veterinary Medical Science, 59*, 457–460.

Morters, M. K., McKinley, T. J., Restif, O., Conlan, A. J., Cleaveland, S., Hampson, K., et al. (2014). The demography of free-roaming dog populations and applications to disease and population control. *Journal of Applied Ecology, 51*(4), 1096–1106. https://doi.org/10.1111/1365-2664.12279.

Morters, M. K., McNabb, S., Horton, D. L., Fooks, A. R., Schoeman, J. P., Whay, H. R., et al. (2015). Effective vaccination against rabies in puppies in rabies endemic regions. *Veterinary Record, 177*, 150. https://doi.org/10.1136/vr.102975.

Morters, M. K., Restif, O., Hampson, K., Cleaveland, S., Wood, J. L., & Conlan, A. J. (2013). Evidence-based control of canine rabies: A critical review of population density reduction. *Journal of Animal Ecology, 82*(1), 6–14. https://doi.org/10.1111/j.1365-2656.2012.02033.x.

Mpolya, E. A., Lembo, T., Lushasi, K., Mancy, R., Mbunda, E. M., Makungu, S., et al. (2017). Toward elimination of dog-mediated human rabies: Experiences from implementing a large-scale demonstration project in southern Tanzania. *Frontiers in Veterinary Science, 4*, 21. https://doi.org/10.3389/fvets.2017.00021.

Mtema, Z., Changalucha, J., Cleaveland, S., Elias, M., Ferguson, H. M., Halliday, J. E., et al. (2016). Mobile phones as surveillance tools: Implementing and evaluating a large-scale intersectoral surveillance system for rabies in Tanzania. *PLoS Medicine, 13*(4), e1002002. https://doi.org/10.1371/journal.pmed.1002002.

Muir, P., & Roome, A. (2005). Indigenous rabies in the UK. *Lancet, 365*(9478), 2175.

Nishizono, A., Khawplod, P., Ahmed, K., Goto, K., Shiota, S., Mifune, K., et al. (2008). A simple and rapid immunochromatographic test kit for rabies diagnosis. *Microbiology and Immunology*, *52*, 243–249. https://doi.org/10.1111/j.1348-0421.

Nokes, D. J., & Swinton, J. (1997). Vaccination in pulses: A strategy for global eradication of measles and polio? *Trends in Microbiology*, *5*, 14–19.

OIE. (2018). *Rabies (infection with rabies virus and other lyssaviruses). In Manual of diagnostic tests and vaccines for terrestrial animals* (pp. 578–612): OIE. Available from http://www.oie.int/fileadmin/Home/eng/Health_standards/tahm/3.01.17_RABIES.pdf.

OIE, WHO, FAO, & GARC. (2017). *The stepwise approach towards rabies elimination.* https://caninerabiesblueprint.org/A-stepwise-approach-to-planning. (Accessed 23 July 2019).

Orji, R., & Mandryk, R. L. (2014). Developing culturally relevant design guidelines for encouraging healthy eating behavior. *International Journal of Human-Computer Studies*, *72*(2), 207–223.

PAHO. (2015). *Epidemiological alert – Rabies.* https://www.paho.org/hq/dmdocuments/2015/2015-jun-12-cha-rabies-epi-alert.pdf. (Accessed 23 July 2019).

Parker, M., & Harper, I. (2006). The anthropology of public health. *Journal of Biosocial Science*, *38*(1), 1–5.

Perry, B. D. (1993). Dog ecology in eastern and southern Africa – Implications for rabies control. *Onderstepoort Journal of Veterinary Research*, *60*, 429–436.

Phua, K. L., & Lee, L. K. (2005). Meeting the challenge of epidemic infectious disease outbreaks: An agenda for research. *Journal of Public Health Policy*, *26*(1), 122–132.

Power, R., Langhaug, L., Nyamurera, T., Wilson, D., Bassett, M., & Cowan, F. (2004). Developing complex interventions for rigorous evaluation – A case study from rural Zimbabwe. *Health Education Research*, *19*(5), 570–575. https://doi.org/10.1093/her/cyg073.

Price, M. (1988). The consequences of health service privatisation for equality and equity in health care in South Africa. *Social Science and Medicine*, *27*(7), 703–716.

ProMED-mail. (2017). *Rabies – Malaysia: (Sarawak), human, dog bite susp. RFI.* ProMED-mail; 01 Jul: 20170701.5143911 http://www.promedmail.org. (Accessed 15 July 2019).

Putra, A., Hampson, K., Girardi, J., Hiby, E., Knobel, D., Mardiana, W., et al. (2013). Response to a rabies epidemic, Bali, Indonesia, 2008–2011. *Emerging Infectious Diseases*, *19*(4), 648–651. https://doi.org/10.3201/eid1904.120380.

Rajeev, M., Edosoa, G., Hanitriniaina, C., Adriamandimby, S. F., Guis, H., Ramiandrasoa, R., et al. (2018). Healthcare utilization, provisioning of post-exposure prophylaxis, and estimation of human rabies burden in Madagascar. *Vaccine*, S0264-410X(18)31520-2. https://doi.org/10.1016/j.vaccine.2018.11.011.

Ratsitorahina, M., Rasambainarivo, J. H., Raharimanana, S., Rakotonandrasana, H., Andriamiarisoa, M. P., Rakalomanana, F. A., & Richard, V. (2009). Dog ecology and demography in Antananarivo, 2007. *BMC Veterinary Research*, *5*, 21. https://doi.org/10.1186/1746-6148-5-21.

Reece, J. F., & Chawla, S. K. (2006). Control of rabies in Jaipur, India, by the sterilisation and vaccination of neighbourhood dogs. *The Veterinary Record*, *159*, 379–383.

Reidpath, D. D., Allotey, P., & Pokhrel, S. (2011). Social sciences research in neglected tropical diseases 2: A bibliographic analysis. *Health Research Policy and Systems*, *9*, 1. https://doi.org/10.1186/1478-4505-9-1.

Rhodes, L. (2017). New approaches to non-surgical sterilization for dogs and cats: Opportunities and challenges. *Reproduction in Domestic Animals*, *52*(Suppl. 2), 327–331. https://doi.org/10.1111/rda.12862.

Rivera, L. (2003). Changing women: An ethnographic study of homeless mothers and popular education. *Journal of Sociology and Social Welfare*, *30*(2), 31–51.

Robinson, L. E., Miranda, M. E., Miranda, N. L., & Childs, J. E. (1996). Evaluation of a canine rabies vaccination campaign and characterization of owned-dog populations in the Philippines. *Southeast Asian Journal of Tropical Medicine and Public Health*, *27*, 250–256.

Rowan, A. N., Lindenmayer, J. M., & Reece, J. F. (2014). Role of dog sterilisation and vaccination in rabies control programmes. *Veterinary Record*, *175*(16), 409. https://doi.org/10.1136/vr.g6351.

Rupprecht, C. E., Fooks, A. R., & Abela-Ridder, B. (Eds.), (2018). *Laboratory techniques in rabies* (5th ed.). Geneva: World Health Organization.

Rysava, K., Miranda, M. E., Zapatos, R., Lapiz, S., Rances, P., Miranda, L. M., et al. (2018). On the path to rabies elimination: The need for risk assessments to improve administration of post-exposure prophylaxis. *Vaccine*, S0264-410X(18)31627-X, https://doi.org/10.1016/j.vaccine.2018.11.066.

Sambo, M., Hampson, K., Changalucha, J., Cleaveland, S., Lembo, T., Lushasi, K., et al. (2018). Estimating the size of dog populations in Tanzania to inform rabies control. *Veterinary Sciences*, *5*(3), 77. https://doi.org/10.3390/vetsci5030077.

Sambo, M., Johnson, P. C., Hotopp, K., Changalucha, J., Cleaveland, S., Kazwala, R., et al. (2017). Comparing methods of assessing dog rabies vaccination coverage in rural and urban communities in Tanzania. *Frontiers in Veterinary Science, 4*, 33. https://doi.org/10.3389/fvets.2017.00033.

Schneider, M. C., Aguilera, X. P., Barbosa da Silva Junior, J., Ault, S. K., Najera, P., Martinez, J., et al. (2011). Elimination of neglected diseases in Latin America and the Caribbean: A mapping of selected diseases. *PLoS Neglected Tropical Diseases, 5*(2), e964. https://doi.org/10.1371/journal.pntd.0000964.

Schneider, M. C., Belotto, A., Adé, M. P., Hendrickx, S., Leanes, L. F., Rodrigues, M. J., et al. (2007). Current status of human rabies transmitted by dogs in Latin America. *Cadernos de Saúde Pública, 23*, 2049–2063.

Selhorst, T., Muller, T., Schwermer, H., Ziller, M., Schluter, H., et al. (2005). Use of an area index to retrospectively analyze the elimination of fox rabies in European countries. *Environmental Management, 35*, 292–302.

Serrano-García, I. (1984). The illusion of empowerment: Community development within a colonial context. In J. Rappaport, C. Swift, & R. Hess (Eds.), *Studies in empowerment: Steps toward understanding and action* (pp. 173–200). New York: Haworth Press.

Shimada, K. (1971). The last rabies outbreak in Japan. In Y. Nagano & F. M. Davenport (Eds.), *Rabies* (pp. 11–28). Baltimore, MA: University Park.

Shone, D. K. (1962). Rabies in southern Rhodesia: 1900 to 1961. *Journal of the South African Veterinary Medical Association, 33*, 567–580.

Siegal, G., Siegal, N., & Bonnie, R. J. (2009). An account of collective actions in public health. *American Journal of Public Health, 99*(9), 1583–1587. https://doi.org/10.2105/AJPH.2008.152629.

Sudarshan, M. K., Madhusudana, S. N., Mahendra, B. J., Rao, N. S. N., Ashwath Narayana, D. H., Abdul Rahman, S., et al. (2007). Assessing the burden of human rabies in India: Results of a national multi-center epidemiological survey. *International Journal of Infectious Diseases, 11*, 29–35.

Suzuki, K., Pereira, L. A., Frias, R., Lopez, R., Mutinelli, L. E., & Pons, E. R. (2008). Rabies-vaccination coverage and profiles of the owned-dog population in Santa Cruz de la Sierra, Bolivia. *Zoonoses and Public Health, 55*, 177–183. https://doi.org/10.1111/j.1863-2378.2008.01114.x.

Takahashi-Omoe, H., Omoe, K., & Okabe, N. (2008). Regulatory systems for prevention and control of rabies, Japan. *Emerging Infectious Diseases, 14*(9), 1368–1374. https://doi.org/10.3201/eid1409.070845.

Talbi, C., Lemey, P., Suchard, M. A., Abdelatif, E., Elharrak, M., Nourlil, J., et al. (2010). Phylodynamics and human-mediated dispersal of a zoonotic virus. *PLoS Pathogens, 6*(10), e1001166. https://doi.org/10.1371/journal.ppat.1001166.

Tan, D. S. K., Abdul, W. M. A., Mohd, N. K., & Beran, G. (1972). An outbreak of rabies in West Malaysia in 1970 with unusual laboratory observations. *Medical Journal of Malaysia, 27*, 107–112.

Tao, X. Y., Niezgoda, M., Du, J. L., Li, H., Wang, X. G., et al. (2008). The primary application of direct rapid immunohistochemical test to rabies diagnosis in China. *Zhonghua Shi Yan He Lin Chuang Bing Du Xue Za Zhi, 22*, 168–170 [in Chinese].

Tarantola, A., Ly, S., In, S., Ong, S., Peng, Y., Heng, N. Y., et al. (2015). Rabes vaccine and rabies immunoglobuin in Cambodia: Use and obstacles to use. *Journal of Travel Medicine, 22*(5), 348–352. https://doi.org/10.1111/jtm.12228.

Taylor, L. H., Hampson, K., Fahrion, A., Abela-Ridder, B., & Nel, L. H. (2017). Difficulties in estimating the human burden of canine rabies. *Acta Tropica, 165*, 133–140. https://doi.org/10.1016/j.actatropica.2015.12.007.

Taylor, L. H., & Knopf, L. (2015). Surveillance of human rabies by national authorities – A global survey. *Zoonoses and Public Health, 62*(7), 543–552. https://doi.org/10.1111/zph.12183.

Taylor, L. H., Wallace, R. M., Balaram, D., Lindenmayer, J. M., Eckery, D. C., Mutonono-Watkiss, B., et al. (2017). The role of dog population management in rabies elimination—a review of current approaches and future opportunities. *Frontiers in Veterinary Science, 4*, 109. https://doi.org/10.3389/fvets.2017.00109.

Tenzin, T., Ahmed, R., Debnath, N. C., Ahmed, G., & Yamage, M. (2015). Free-roaming dog population estimation and status of the dog population management and rabies control program in Dhaka City, Bangladesh. *PLoS Neglected Tropical Diseases, 9*(5), e0003784. https://doi.org/10.1371/journal.pntd.0003784.

Tenzin, Dhand, N. K., Gyeltshen, T., Firestone, S., Zangmo, C., Dema, C., et al. (2011). Dog bites in humans and estimating human rabies mortality in rabies endemic areas of Bhutan. *PLoS Neglected Tropical Diseases, 5*(11), e1391. https://doi.org/10.1371/journal.pntd.0001391.

Tenzin, T., McKenzie, J. S., Vanderstichel, R., Rai, B. D., Rinzin, K., Tshering, Y., et al. (2015). Comparison of mark-resight methods to estimate abundance and rabies vaccination coverage of free-roaming dogs in two urban areas of south Bhutan. *Preventive Veterinary Medicine, 118*(4), 436–448. https://doi.org/10.1016/j.prevetmed.2015.01.008.

Tenzin, Sharma, B., Dhand, N. K., Timsina, N., & Ward, M. P. (2010). Reemergence of rabies in Chhukha District, Bhutan, 2008. *Emerging Infectious Diseases, 16*(12), 1925–1930. https://doi.org/10.3201/eid1612.100958.

Tierkel, E. S. (1959). Rabies. In C. A. Brandly & E. L. Jungherr (Eds.), *Advances in veterinary science* (pp. 183–226). New York, NY: Academic Press.

Tiwari, H. K., Robertson, I. D., O'Dea, M., Gogoi-Tiwari, J., Panvalkar, P., Bajwa, R. S., & Vanak, A. T. (2019). Validation of Application SuperDuplicates (AS) enumeration tool for free-roaming dogs (FRD) in urban settings of Panchkula Municipal Corporation in North India. *Frontiers in Veterinary Science, 6*, 173. https://doi.org/10.3389/fvets.2019.00173.

Tiwari, H. K., Vanak, A. T., O'Dea, M., Gogoi-Tiwari, J., & Robertson, I. D. (2018). A comparative study of enumeration techniques for free-roaming dogs in Rural Baramati, District Pune, India. *Frontiers in Veterinary Science, 5*, 104. https://doi.org/10.3389/fvets.2018.00104.

Tohma, K., Saito, M., Demetria, C. S., Manalo, D. L., Quiambao, B. P., Kamigaki, T., et al. (2016). Molecular and mathematical modeling analyses of inter-island transmission of rabies into a previously rabies-free island in the Philippines. *Infection, Genetics and Evolution, 38*, 22–28. https://doi.org/10.1016/j.meegid.2015.12.001.

Totton, S. C., Wandeler, A. I., Zinsstag, J., Bauch, C. T., Ribble, C. S., et al. (2010). Stray dog population demographics in Jodhpur, India following a population control/rabies vaccination program. *Preventive Veterinary Medicine, 97*, 51–57. https://doi.org/10.1016/j.prevetmed.2010.07.009.

Townsend, S. E., Lembo, T., Cleaveland, S., Meslin, F. X., Miranda, M. E., Putra, A. A., et al. (2013). Surveillance guidelines for disease elimination: A case study of canine rabies. *Comparative Immunology, Microbiology and Infectious Diseases, 36*(3), 249–261. https://doi.org/10.1016/j.cimid.2012.10.008.

Townsend, S. E., Sumantra, I. P., Pudjiatmoko, Bagus, G. N., Brum, E., Cleaveland, S., et al. (2013). Designing programs for eliminating canine rabies from islands: Bali, Indonesia as a case study. *PLoS Neglected Tropical Diseases, 7*(8), e2372. https://doi.org/10.1371/journal.pntd.0002372.

Umeno, S., & Doi, Y. (1921). A study on the anti-rabic inoculation of dogs and the results of its practical application. *The Kitasato Archives of Experimental Medicine, 4*, 89–108.

Undurraga, E. A., Meltzer, M. I., Tran, C. H., Atkins, C. Y., Etheart, M. D., Millien, M. F., et al. (2017). Cost-effectiveness evaluation of a novel integrated bite case management program for the control of human rabies, Haiti 2014–2015. *American Journal of Tropical Medicine and Hygiene, 96*(6), 1307–1317. https://doi.org/10.4269/ajtmh.16-0785.

Valadaz Figueroa, I., Alfaro Alfaro, N., Guerra, J. F., Aldrete Rodriguez, G., & Mendoza Roaf, P. (2000). Una experiencia de educación popular en salud nutricional en dos comunidades del Estado de Jalisco, México [An experience with public education in nutritional health in two communities in Jalisco, Mexico]. *Cadernos de Saúde Pública, 16*(3), 823–829 [in Spanish].

Vasanth, J. P., Madhusudana, S. N., Abhilash, K. V., Suja, M. S., & Muhamuda, K. (2004). Development and evaluation of an enzyme immunoassay for rapid diagnosis of rabies in humans and animals. *Indian Journal of Pathology and Microbiology, 47*, 574–578.

Vigilato, M. A., Clavijo, A., Knobl, T., Silva, H. M., Cosivi, O., Schneider, M. C., et al. (2013). Progress towards eliminating canine rabies: Policies and perspectives from Latin America and the Caribbean. *Philosophical Transactions of the Royal Society of London. Series B, Biological Sciences, 368*(1623), 20120143. https://doi.org/10.1098/rstb.2012.0143.

Virhia, J. (2019). *Healthy animals, healthy people: Exploring health seeking strategies among livestock keepers in northern Tanzania.* PhD thesis: University of Glasgow.

Wallace, R. M., Reses, H., Franka, R., Dilius, P., Fenelon, N., Orciari, L., et al. (2015). Establishment of a high canine rabies burden in Haiti through the implementation of a novel surveillance program [corrected]. *PLoS Neglected Tropical Diseases, 9*(11), e0004245. https://doi.org/10.1371/journal.pntd.0004245.

Wallerstein, N. (2002). Empowerment to reduce health disparities. *Scandinavian Journal of Public Health, 30*(Suppl. 59), 72–77.

Wandeler, A. I., Matter, H. C., Kappeler, A., & Budde, A. (1993). The ecology of dogs and canine rabies: A selective review. *Revue Scientifique et Technique de l' Office International des Epizooties, 12*, 51–71.

Weinger, M., & Lyons, M. (1992). Problem-solving in the field: An action-oriented approach to farmworker education about pesticides. *American Journal of Industrial Medicine, 22*(5), 677–690.

Wells, C. W. (1954). The control of rabies in Malaya through compulsory mass vaccination of dogs. *Bulletin of the World Health Organization, 10*(5), 731–742.

Wentworth, D., Hampson, K., Thumbi, S. M., Mwatondo, A., Wambura, G., & Chng, N. R. (2019). A social justice perspective on access to human rabies vaccines. *Vaccine,* S0264-410X(19)30150-1. https://doi.org/10.1016/j.vaccine.2019.01.065..

Wera, E., Velthuis, A. G., Geong, M., & Hogeveen, H. (2013). Costs of rabies control: An economic calculation method applied to Flores Island. *PLoS One*, *8*(12), e83654. https://doi.org/10.1371/journal.pone.0083654.

WHO. (1987). *Guidelines for dog rabies control* [WHO/VPH/83.43 Rev. 1]. Geneva: WHO.

WHO. (1988). *Report of a WHO Consultation on dog ecology studies related to rabies control* [WHO/Rabies Research/88.25].Geneva: WHO.

WHO. (1998). *Field application of oral rabies vaccines for dogs* [WHO/EMC/ZDI/98.15]. Geneva: WHO.

WHO. (1999). *World survey of rabies No. 34 for the year 1998* [WHO/CDS/CSR/APH/99.6]. Geneva: WHO.

WHO. (2007). *Guidance for research on oral rabies vaccines and field application of oral vaccination of dogs against rabies*. Geneva: WHO.

WHO. (2017). *The Rabies Elimination Program of Bangladesh – A model for transformation and accomplishment*. https://www.who.int/neglected_diseases/news/Bangladesh-rabies-elimination-program/en/ (Accessed 23 July 2019).

WHO. (2018a). *WHO expert consultation on rabies, 3rd report:* WHO Technical Report Series, 1012. (pp. 1–9).

WHO. (2018b). Rabies vaccines: WHO position paper – April 2018. *Weekly Epidemiological Record*, *93*(16), 201–220.

WHO Rabies Modelling Consortium. (2019). The potential effect of improved provision of rabies post-exposure prophylaxis in Gavi-eligible countries: A modelling study. *Lancet Infectious Diseases*, *19*(1), 102–111. https://doi.org/10.1016/S1473-3099(18)30512-7.

Widdowson, M. A., Morales, G. J., Chaves, S., & McGrane, J. (2002). Epidemiology of urban canine rabies, Santa Cruz, Bolivia, 1972–1997. *Emerging Infectious Diseases*, *8*(5), 458–461. https://doi.org/10.3201/eid0805.010302.

Wiggins, N. (2011). Popular education for health promotion and community empowerment: A review of the literature. *Health Promotion International*, *27*(3), 356–371.

Windiyaningsih, C., Wilde, H., Meslin, F.-X., Suroso, T., & Widarso, H. S. (2004). The rabies epidemic on Flores Island, Indonesia (1998–2003). *Journal of the Medical Association of Thailand*, *87*, 1389–1393.

Wright, E., McNabb, S., Goddard, T., Horton, D. L., Lembo, T., Nel, L. H., et al. (2009). A robust lentiviral pseudotype neutralisation assay for in-field serosurveillance of rabies and lyssaviruses in Africa. *Vaccine*, *27*(51), 7178–7186. https://doi.org/10.1016/j.vaccine.2009.09.024.

Wright, E., Temperton, N. J., Marston, D. A., McElhinney, L. M., Fooks, A. R., & Weiss, R. A. (2008). Investigating antibody neutralization of lyssaviruses using lentiviral pseudotypes: A cross-species comparison. *Journal of General Virology*, *89*(Pt 9), 2204–2213. https://doi.org/10.1099/vir.0.2008/000349-0.

WSPA. (2007). *Surveying roaming dog populations – Guidelines on methodology*. Available from https://www.icamcoalition.org/download/surveying-roaming-dog-populations-guidelines-on-methodology/.

Wu, X. F., Franka, R., Svoboda, P., Pohl, J., & Rupprecht, C. E. (2009). Development of combined vaccines for rabies and immunocontraception. *Vaccine*, *27*(51), 7202–7209.

Zinsstag, J., Dürr, S., Penny, M. A., Mindekem, R., Roth, F., Gonzalez, S., et al. (2009). Transmission dynamics and economics of rabies control in dogs and humans in an African city. *Proceedings of the National Academy of Sciences of the United States of America*, *106*, 14996–15001. https://doi.org/10.1073/pnas.0904740106.

Zinsstag, J., Lechenne, M., Laager, M., Mindekem, R., Naissengar, S., Oussiguere, A., et al. (2017). Vaccination of dogs in an African city interrupts rabies transmission and reduces human exposure. *Science Translational Medicine*, *9*(421). eaaf6984. https://doi.org/10.1126/scitranslmed.aaf6984.

Zinsstag, J., Schelling, E., Wyss, K., & Mahamat, M. B. (2005). Potential of cooperation between human and animal health to strengthen health systems. *Lancet*, *366*(9503), 2142–2145.

CHAPTER 19

Rabies control in wild carnivores

Amy T. Gilbert[a], Richard B. Chipman[b]

[a]United States Department of Agriculture, Animal and Plant Health Inspection Service, Wildlife Services, National Wildlife Research Center, Fort Collins, CO, United States [b]United States Department of Agriculture, Animal and Plant Health Inspection Service, Wildlife Services, National Rabies Management Program, Concord, NH, United States

19.1 Introduction

Rabies virus (RABV) is the only lyssavirus that circulates naturally in wild carnivore hosts (Rupprecht, Turmelle, & Kuzmin, 2011). Circulation of RABV globally in domestic dogs (*Canis lupus familiaris*) is responsible for an excess of 59,000 human deaths annually (Hampson et al., 2015) and has resulted in host shifts of dog RABV to several wild carnivore species (Badrane & Tordo, 2001; Troupin et al., 2016). In comparison to the burden associated with domestic dogs, there have been fewer human deaths reported globally linked to wild carnivore RABV (Müller, Demetriou, et al., 2012; Velasco-Villa et al., 2017; Vora et al., 2013), yet there is a significant economic cost of wild carnivore RABV circulation due to spillover infections to other wildlife, domestic animals and livestock, which leads to human exposures and administration of postexposure treatment (PET) (Christian, Blanton, Auslander, & Rupprecht, 2009; Nunan et al., 2002; Pieracci et al., 2019).

Control efforts are implemented to protect public and agricultural health, or threatened and endangered wildlife, and reduce the burden and associated economic costs of RABV circulation. Control can be defined as a reduction in the incidence, prevalence, morbidity, or mortality of an infectious disease to a locally acceptable level. This can be contrasted with elimination, which is defined as the reduction to zero of the incidence of disease or infection within a defined geographical area. Control of RABV in wild carnivores is possible because the basic reproductive number associated with RABV circulation in reservoir populations is low, meaning that on average relatively few infections are successfully transmitted from a given rabid animal in a population of susceptible hosts. The level of vaccination needed however may be sensitive to enzootic versus epizootic RABV circulation dynamics, as well as control versus elimination management objectives. Despite well-documented cross-species transmission from sylvatic reservoirs, RABV perpetuation principally occurs among populations of conspecifics and control targeting the reservoir species generally leads to dramatic reduction or elimination of spillover infections in other species. Control of RABV in wildlife reservoir populations can be implemented to prevent virus establishment in a naïve population by use of a barrier (i.e., cordon sanitaire) technique, as a point-infection-control (PIC) response to focal outbreaks, or to eliminate virus circulation (Sterner, Meltzer, Shwiff, & Slate, 2009).

19.2 Historical aspects, milestones, and epizootiology

The RABV control measures employed with wildlife during the early 20th century involved destruction of populations, realized through activities, including targeted hunting, gassing of dens, and various forms of poisoning (Debbie, 1991; Lewis, 1975). These tactics were based on the premise that wildlife RABV transmission is a density-dependent phenomenon among carnivores, whereby reduction of a reservoir population to a lower density threshold could effectively eliminate RABV transmission (Anderson, Jackson, May, & Smith, 1981). The use of toxic baits and bait stations was shown to be much more economical and effective for reducing animal populations compared to trap-euthanize or hunting strategies (e.g., bounties) and likely paved the way for a transition to oral delivery of RABV vaccines. Population reduction alone was generally ineffective at eliminating RABV circulation in target host species inhabiting broad geographic areas (Aubert, 1994; Müller & Freuling, 2018), partly due to compensatory effects of the target host populations (Bögel, Moegle, Steck, Krocza, & Andral, 1981), and is no longer recommended as a standalone strategy for landscape-level RABV control in wild carnivore populations (WHO, 1992, 2018).

Modern strategies used and supported by the global community rely on vaccination of free-ranging target populations, through a parenteral approach using a trap-vaccinate-release (TVR) strategy or by an oral rabies vaccination (ORV) strategy using aerial or hand delivery of vaccine baits across broad landscapes, or a combination of tactics, to reduce the susceptible fraction of the population for control and elimination of RABV circulation (WHO, 1992, 2005, 2013, 2018). Theoretical advancements that led to the development of an oral vaccination strategy were predicated upon early experimental evidence of RABV infection by oral route (Baer, 1975; Correa-Giron, Allen, & Sulkin, 1970). These observations were transformative when applied in a preventive vaccination framework using attenuated laboratory strains of RABV, evaluated by oral route as a proof of concept during the early 1970s in North America (Baer, Abelseth, & Debbie, 1971; Black & Lawson, 1970; Debbie, Abelseth, & Baer, 1972) (Fig. 19.1). Black and Lawson (1973) were the first to experimentally demonstrate efficacy of oral rabies vaccines delivered by bait to red foxes (*Vulpes vulpes*). Following proof of concept and evidence of efficacy in field applications (Brochier et al., 1991; Hanlon et al., 1998; Johnston et al., 1988; Pastoret et al., 1988, 1987; Steck, Wandeler, Bichsel, Capt, Hafliger, et al., 1982; Steck, Wandeler, Bichsel, Capt, & Schneider, 1982; Wandeler, Capt, Kappeler, & Hauser, 1988) (Fig. 19.2), continued development and refinement of safe and potent vaccines and relative cost efficiency compared to other methods, ORV has gained popularity as a landscape-level management strategy for control of RABV in wild carnivores globally.

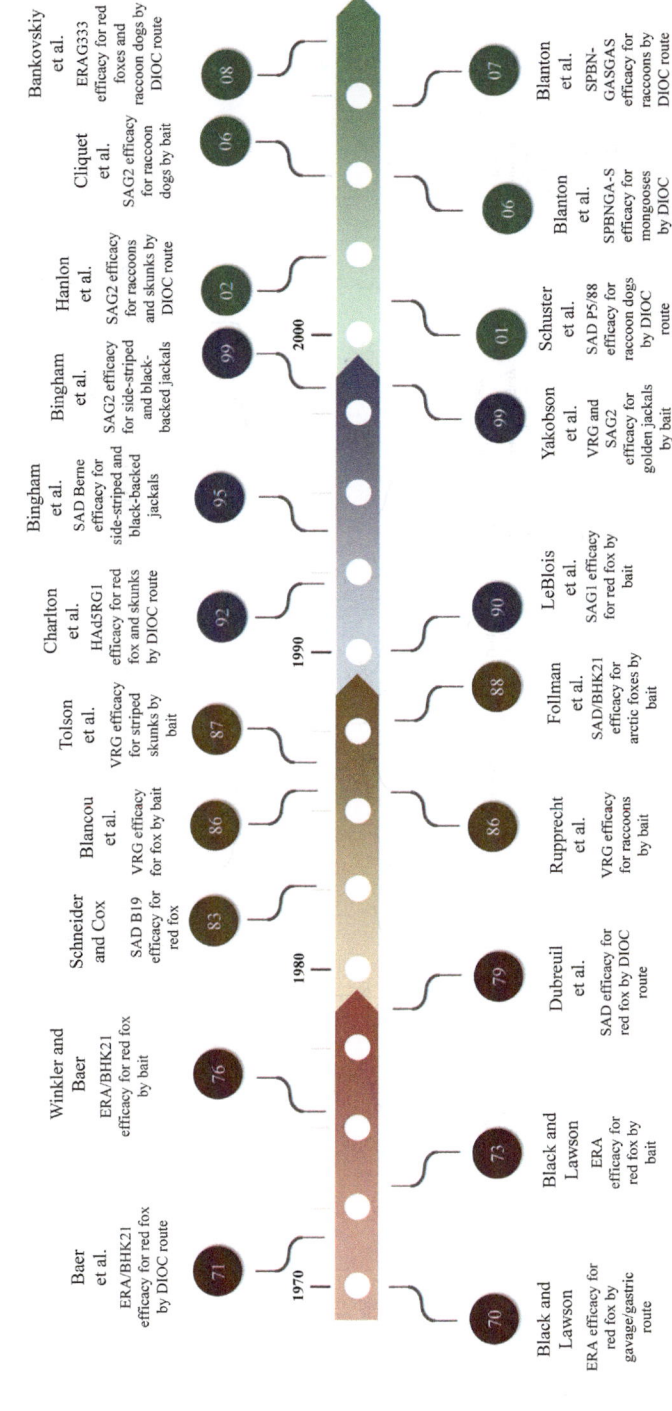

FIG. 19.1 Selected oral rabies vaccine research and development milestones during 1970–2010.

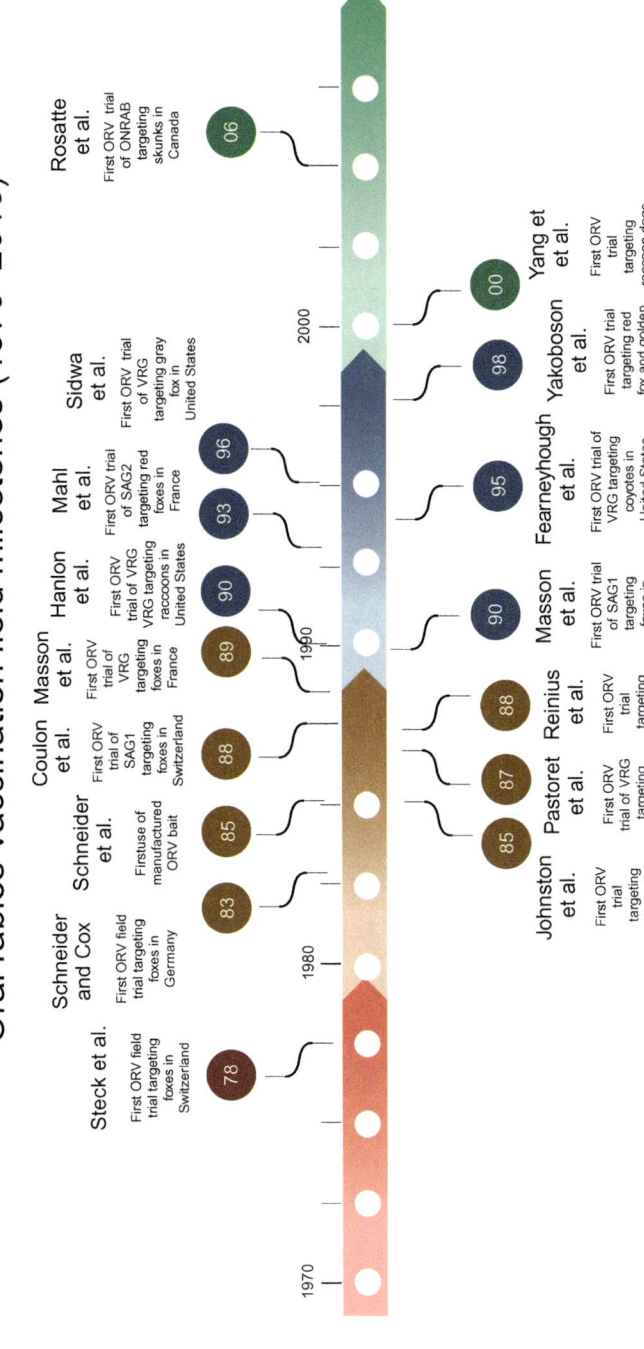

FIG. 19.2   Selected oral rabies vaccination field research and development milestones during 1970–2010.

19.2.1 Europe

The host shift of Cosmopolitan dog RABV to red foxes led to one of the most important wildlife disease epizootics documented in Europe (Artois, Delahay, Guberti, & Cheeseman, 2001). Although outbreaks of RABV among red foxes and other wildlife were described in Europe during the 19th century and earlier, a significant epizootic in red fox populations originated along the border of Poland and the former Soviet Union around 1939 and the wave-front of this epizootic spread at an average of 30–60 km/year west across Europe, infecting an estimated 25%–50% of the red fox population in its path and reaching France by 1968 (Macdonald, 1980; Steck & Wandeler, 1980; Wachendorfer & Frost, 1992; Winkler, 1975). Anthropogenic habitat modification prior to the Second World War likely increased red fox densities in certain areas of Europe, creating conditions favorable for the rapid spread and enzootic establishment of RABV (Bögel, 1992a). Natural barriers or areas with reduced rates of spread were associated with habitats harboring low densities of foxes (e.g., >2000 m above sea level; (Steck & Wandeler, 1980) and following the passage of the wave-front, most areas experienced a brief fadeout of cases for a few years, prior to enzootic re-establishment. The majority of cases were reported in red foxes but a lower fraction of spillover cases occurred in domestic animals, European badger (*Meles meles*), and Beech marten (*Martes foina*) (Steck & Wandeler, 1980; Wachendorfer & Frost, 1992). In northeastern Europe and Russia, infections were frequently reported among raccoon dogs (*Nyctereutes procyonoides*) which, although not a recognized reservoir in Europe, became a key target for RABV control efforts (EFSA, 2015; Müller, Freuling, et al., 2015; Reinius, 1992).

The first field application of ORV was conducted in Switzerland during 1978 to prevent the establishment of an advancing epizootic in Canton of Valais and an achieved ~60% herd immunity among the fox population was successful in halting the disease spread (Steck, Wandeler, Bichsel, Capt, Hafliger, et al., 1982; Steck, Wandeler, Bichsel, Capt, & Schneider, 1982). Pilot testing continued in Switzerland and was initiated in 1983 in Germany and 1984 in Italy, with limited testing up until 1986 (Bögel, Meslin, & Kaplan, 1992). Use of ORV expanded during 1986 and by the early 1990s, greater contiguous areas of ORV baiting were being connected across political borders (Stohr & Meslin, 1996). During this early period, concerns about the safety of the live attenuated RABV vaccines available (e.g., SAD Berne, SAD B19) motivated the development and testing of a recombinant ORV product on a limited scale in Belgium during 1987, which scaled up in 1989, and in France during 1989 (Brochier et al., 1991; Maki et al., 2017). Around the same time, a modified live attenuated RABV vaccine was selected from the SAD Berne strain (SAD Avirulent Gif, SAG1) to reduce pathogenicity and was tested in limited trials in Switzerland during 1988 and France in 1990 (Coulon et al., 1992; Mahl et al., 2014; Masson, Aubert, Barrat, & Vuillaume, 1996). First-generation live attenuated RABV vaccines, second-generation selected RABV vaccines, and third-generation recombinant RABV vaccines have all successfully been used in Europe for control and elimination of wild carnivore RABV (Mahl et al., 2014; Maki et al., 2017; Müller, Schroder, Wysocki, Mettenleiter, & Freuling, 2015; Vos et al., 1999).

Thirty European countries have reported the use of ORV to control wild carnivore RABV and 12 have self-declared freedom (Müller & Freuling, 2018; Müller, Schroder, et al., 2015). Despite many challenges and setbacks early on, stemming in part from a lack of crossborder cooperation regarding ORV activities and insufficient national long-term planning

(Stohr & Meslin, 1996), the cofinancing model of the EU introduced in 1989 was pivotal for enhancing multilateral and intersectoral coordination between Member States toward a common goal of eliminating RABV from red fox populations across Western and Central Europe (EC, 1989; Freuling et al., 2013; Müller, Batza, et al., 2012; Müller, Freuling, et al., 2015; Müller & Freuling, 2018). A financing model that included the Western Balkan region and specific States neighboring the EU (e.g., Kaliningrad) was also initiated in 2008 (Demetriou & Moynagh, 2011), further enhancing the network of Eastern European collaborating countries working toward fox RABV elimination.

As of 2017, a total of 22 Central and Eastern European countries were listed with active ORV programs, mostly following a north-south transect from Finland to Greece (Müller & Freuling, 2018). To the west of the Black Sea, the front lines of the dynamic cordon sanitaire are maintained by ORV in the Baltic States, Belarus, Poland, and Romania, ultimately to eliminate RABV from wildlife in EU member countries by 2020 (EC, 2017). An ORV program was developed in 2007 in Kaliningrad Oblast, a Russian territory bordered by the Baltic Sea, Lithuania and Poland, targeting red foxes and raccoon dogs, with elimination from this area reported since 2013. During the most recent 5 years (2014–18), Belarus, Moldova, Poland, Romania, and Ukraine reported more than 50 fox RABV cases (i.e., averaging more than 10 per year), with Belarus and Ukraine also reporting a notable proportion of RABV cases in raccoon dogs (WHO, 2019). Among these five countries with evidence of fox RABV circulation during 2014–18, only Poland and Romania are EU Member States.

19.2.2 North America

North America has the highest diversity of wild carnivore RABV reservoirs globally (Fehlner-Gardiner, 2018; Gilbert, 2018). Seven mainland and one invasive island species are recognized as distinct reservoirs, including red fox, coyote (*Canis latrans*), gray fox (*Urocyon cinereoargenteus*), arctic fox (*Vulpes lagopus*), raccoon (*Procyon lotor*), striped skunk (*Mephitis mephitis*), Eastern spotted skunk (*Spilogale putorius*), and small Indian mongoose (*Herpestes auropunctatus*), respectively. While infections and outbreaks among wild carnivores were reported as early as the 19th century in North America, the epidemiology of wildlife RABV was not fully appreciated until domestic dog RABV was controlled during the middle of the 20th century and RABV antigenic typing methods were developed and refined during the 1980s to facilitate discrimination of distinct RABV foci (Rupprecht, Glickman, Spencer, & Wiktor, 1987; Rupprecht, Smith, Fekadu, & Childs, 1995; Smith, Orciari, & Yager, 1995; Smith, Sumner, Roumillat, Baer, & Winkler, 1983; Tabel, Corner, Webster, & Casey, 1974; Wiktor & Koprowski, 1978, 1980). Host shifts of Cosmopolitan dog RABV were implicated in the establishment of red fox, coyote, gray fox, Eastern spotted skunk, and small Indian mongoose reservoirs in North America (Kuzmin et al., 2012; Troupin et al., 2016; Zieger et al., 2014), in contrast to the situation in Ontario, Canada, where a shift of the arctic fox RABV lineage to red foxes led to the establishment of red fox and striped skunk reservoirs. The circulation of RABV among striped skunks in prairie habitats of the north central United States and south central Canada represents a long-term enzootic focus (i.e., North Central skunk variant), which, along with distinct foci in California and northwestern states of Sinaloa, Durango, and Sonora, Mexico, is related to Cosmopolitan dog RABV (Davis, Nadin-Davis, Moore, &

Hanlon, 2013; Velasco-Villa et al., 2017). In contrast, RABV circulation among raccoons in the eastern United States, striped skunks in the south-central United States, and spotted skunks in north-central Mexico comprise foci historically related to bat rather than dog RABV (Davis et al., 2013; Kuzmin et al., 2012; Kuzmina et al., 2013; Velasco-Villa et al., 2017).

Rabies was first diagnosed among arctic foxes and gray wolves (*Canis lupus*) from the Northwest Territories of Canada around the middle of the 20th century (Plummer, 1947). A host shift of RABV from arctic to red foxes caused a significant epizootic among red foxes in Ontario during the 1950s, which remained enzootic in southern Ontario for over 40 years (Johnston & Beauregard, 1969; Nadin-Davis, Muldoon, & Wandeler, 2006b; Tabel et al., 1974). Spillover cases of arctic fox RABV were detected in striped skunks as early as 1957, with potential of a subsequent host shift from red foxes to striped skunks (Nadin-Davis et al., 2006b; Tabel et al., 1974). The North Central skunk RABV focus also spread north from the United States into Canada during the 1970s, despite intermittent population reduction control activities in the United States (Charlton, Webster, & Casey, 1991; Pybus, 1988). The circulation of the North Central skunk RABV has been routinely described in striped skunks from southern prairie habitats of Manitoba and Saskatchewan during the last decade, which were targeted for control in the past by population reduction methods (Pybus, 1988; Rosatte, Pybus, & Gunson, 1986). In Ontario, the red fox and striped skunk have been targeted for control since the 1950s and the elimination of arctic fox RABV from red fox populations in eastern Ontario was declared by the mid-1990s (MacInnes et al., 2001; Rosatte, Power, et al., 2007). Following arctic RABV elimination from foxes, raccoon RABV was detected in the province during 1999 and subsequently eliminated through multiyear coordinated ORV and PIC techniques with the last case detected in 2005 (Rosatte et al., 2001). Despite these successes, arctic fox RABV circulation persisted in limited foci among striped skunks (Nadin-Davis et al., 2006b), and a new ORV product intended for skunks was developed and distributed experimentally during 2006–07 and then operationally during 2008–10 at high densities (i.e., 300 baits/km^2) to eliminate arctic fox RABV from striped skunk populations (Rosatte, Donovan, Davies, et al., 2009; Rosatte et al., 2011). While Ontario remained free of arctic fox RABV during 2012–14, new cases were detected as part of enhanced surveillance following an incursion of raccoon RABV during 2015 (Middel, Fehlner-Gardiner, Pulham, & Buchanan, 2017). More recent cases in red foxes and striped skunks have been reported in the province with genetic relationship to previously circulating arctic fox RABVs (Lobo et al., 2018; Nadin-Davis & Fehlner-Gardiner, 2019; OMNRF, 2019). Cases of raccoon RABV were also detected in New Brunswick in 2000 and Quebec in 2006, yet these foci were considered eliminated by 2002 and 2009, respectively, by applying multifaceted strategies, including ORV and PIC (Stevenson, Goltz, & Masse, 2016). However, new cases of raccoon RABV were detected in New Brunswick during 2014–17 and Quebec in 2015. Following a single case in 2015, Quebec has not reported any additional cases of raccoon RABV during 2016–18 (Government of Quebec, 2019) and additional cases in New Brunswick have not been reported following a single case in 2017 (CFIA, 2019), yet the outbreak in southern Ontario has persisted. Current ORV activities in Canada occur in southeastern Ontario, targeting both arctic fox and raccoon RABV, and limited stretches of Quebec and New Brunswick along the US border targeting raccoons (Fig. 19.3).

During most of the 20th century, red foxes were a key reservoir of RABV in the eastern United States whereas striped skunks were the main reservoir in the central plains, respectively, and both species were independently targeted for localized population reduction to

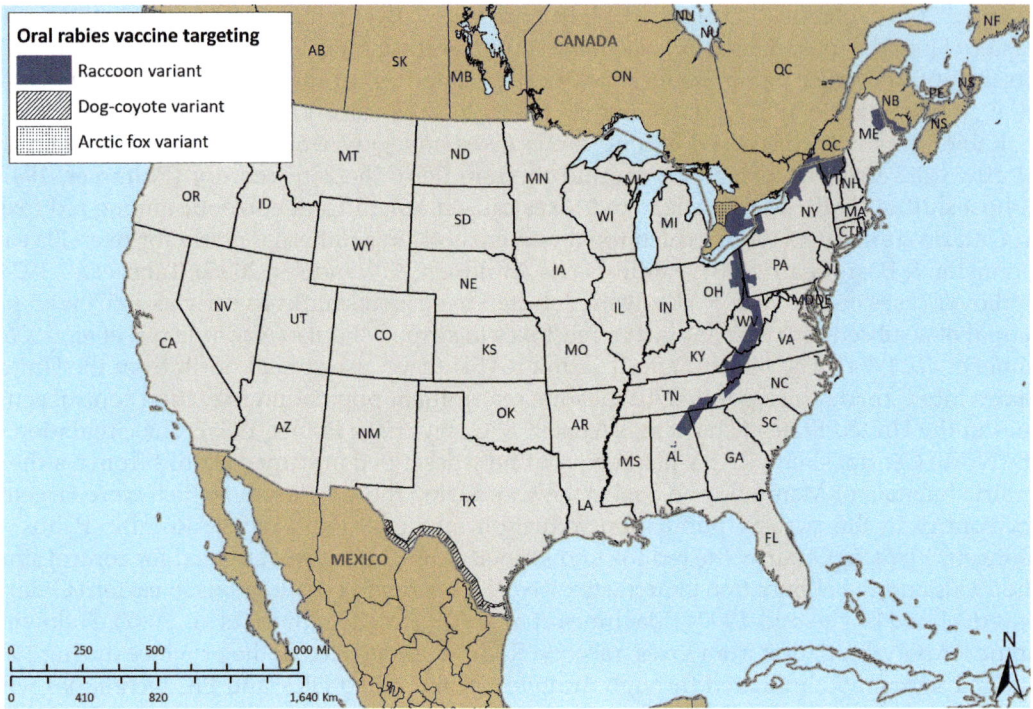

FIG. 19.3 Oral rabies vaccination zones targeting different rabies virus variants in North America, 2018.

control RABV (Linhart, 1960; Macdonald, 1980; Parks, 1968; Schnurrenberger, Beck, & Burson, 1964). In contrast, raccoons historically played a more minor role given that RABV circulation in this host was restricted to populations in Florida and neighboring southeastern states (McLean, 1975). A translocation of raccoons incubating RABV in the late 1970s for hunting purposes led to one of the most significant epizootics in North America and affected raccoon populations along the entire eastern seaboard of the United States, which reached southeastern Canada by the 1990s (Biek, Henderson, Waller, Rupprecht, & Real, 2007; Childs et al., 2000; Rupprecht & Smith, 1994). The first ORV trial targeting raccoons occurred on Parramore Island, Virginia (Hanlon et al., 1993, 1998). Based on compelling results and safety, subsequent field trials on the mainland occurred during 1991 in Pennsylvania, 1992 in New Jersey, 1994 in New York and Cape Cod, Massachusetts, and 1995 in Florida and New York (Hanlon & Rupprecht, 1998). The burden of human and animal rabies exposures in the United States continues to be greatest in areas where raccoon RABV is enzootic (Christian et al., 2009; Pieracci et al., 2019), and since the 1990s this species has been the primary target for RABV control (Elmore et al., 2017; Slate et al., 2005, 2009) (Fig. 19.4). A strategy of ORV and enhanced surveillance, with periodic contingency actions, has prevented appreciable westward spread of raccoon RABV in the United States, yet a low population antibody response in most control areas raised concerns about the ability to eliminate RABV circulation (Slate et al., 2009) and multiple incursions have been detected to the north (Trewby, Nadin-Davis, Real, & Biek, 2017). Striped skunks are the most frequent spillover host of raccoon RABV, raising concerns

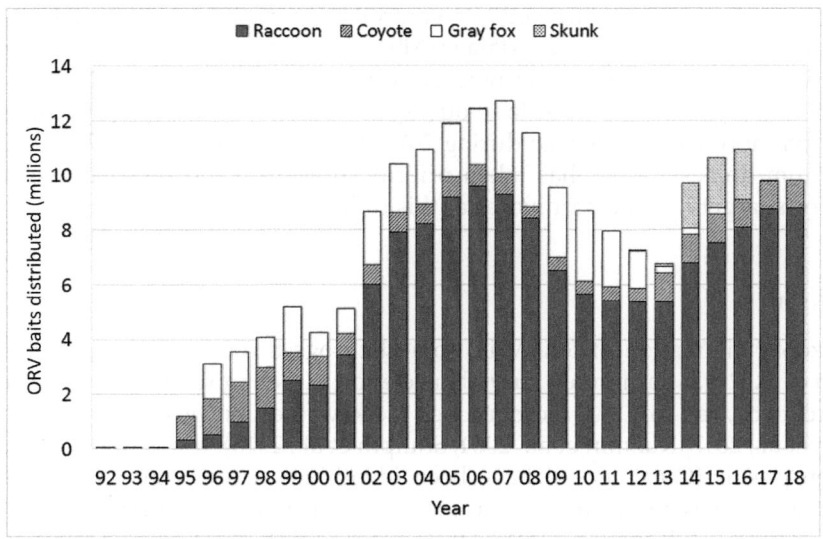

FIG. 19.4 Number of oral rabies vaccination (ORV) baits distributed in the United States targeting different rabies vector species, 1992–2018.

about the potential for a host shift in spillover hotspots (Guerra et al., 2003; Wallace et al., 2014). Experimental field trials since 2011 using a Canadian ORV product have resulted in antibody responses in the raccoon population that may enable progress on regional elimination in the United States (Davis, Kirby, et al., 2019; Davis, Nelson, et al., 2019; Gilbert et al., 2018; Moore et al., 2017; Pedersen et al., 2019; Slate et al., 2014). Control activities targeting other reservoir hosts have been more limited in contemporary time in the United States, with rare exceptions following multiple host shifts of big brown bat RABV to gray foxes and striped skunks during 2001–09 in Arizona (Kuzmin et al., 2012; Leslie et al., 2006) that were eliminated using ORV targeting gray foxes and TVR for skunks, as well as experimental ORV trials targeting striped skunks in Texas (Wohlers, Lankau, Oertli, & Maki, 2018).

The elimination of dog RABV circulation occurred twice in the United States. The first effort was achieved through mass parenteral vaccination of dogs, education, and effective stray animal control programs. The latter effort relied on ORV and other measures targeting coyote populations, which were the reservoir of a Mexican dog RABV along the United States/Mexico border area during the late 1980s (Clark et al., 1994). Coordinated ORV began during 1995 targeting coyotes and the dog-coyote RABV focus was eliminated from the United States by 2004 (Fearneyhough et al., 1998; Maki et al., 2017; Sidwa et al., 2005; Velasco-Villa et al., 2008). Also during the late 1980s, an independent cycle of dog-associated RABV was detected in gray foxes in Texas (i.e., Texas gray fox variant), and this species was also targeted as part of coordinated ORV starting in 1996 (Maki et al., 2017; Sidwa et al., 2005). The Texas gray fox RABV focus may be considered eliminated from the United States based on detection of the last case in 2013, although an independent gray fox RABV focus circulates in Arizona and western New Mexico (Ma et al., 2018). In contemporary time, a maintenance ORV zone approximately 30–65 km wide is baited annually along the Texas-Mexico border to prevent the reintroduction of dog-coyote RABV (Maki et al., 2017).

In Mexico, coyotes were targeted for control by population reduction techniques in the 1950s and 1960s (Cocozza & Malaga-Alba, 1962), yet a stable focus of dog-coyote RABV was reportedly circulating in northeastern Mexico during 1976–2002 (Velasco-Villa et al., 2005). Despite the apparent fade-out of this cycle, a case of dog-coyote RABV was reported from northeastern Mexico during 2011 (Velasco-Villa et al., 2017). The elimination of RABV from domestic dogs has taken priority in Mexico and there are no coordinated control efforts targeting wild carnivores at this time.

19.2.3 Asia

The enzootic circulation of RABV in domestic dogs throughout many Asian countries (Hampson et al., 2015) has limited surveillance effort targeting wildlife. It is well established that the arctic fox, red fox, and raccoon dogs are key reservoir hosts in the northern region (Deviatkin et al., 2017; Kuzmin et al., 2004; Shulpin et al., 2018). In the Middle East, red foxes and golden jackals (*Canis aureus*) have been identified as reservoirs in certain countries, whereas gray wolves more likely represent a spillover host or vector of RABV (Baghi, Alinezhad, Kuzmin, & Rupprecht, 2018; Horton et al., 2015). Israel began a biannual ORV program in 1998 targeting red foxes and golden jackals, with notable case reduction in the most frequently baited areas by 2004 (Yakobson et al., 2006). Recent reports indicate a re-emergence of RABV cases during 2017–18, which was subsequently brought under control with a relatively high density (i.e., 150 baits/km^2) bait application (Yakobson, King, Ingor, Markovich, & Goshen, 2019). The raccoon dog is the primary reservoir of RABV in Korea and ORV has been conducted since 2000 targeting this host, with reported elimination since 2014 (Yang, Cho, & Kim, 2018). In mainland China and Chinese Taipei, the Chinese ferret badger (*Melogale moschata*) has independently emerged as a sylvatic RABV reservoir. The first human RABV infection associated with ferret badger rabies was reported from China in 1997 and detection in ferret badgers from Chinese Taipei was reported during 2013 (Tu, Feng, & Wang, 2018; Zhao, Zhao, Liu, Jiang, & Yang, 2019). Several ORV products have been tested with ferret badgers in the laboratory and limited field settings (Hsu et al., 2017; Wallace et al., 2018). There appears to be a high diversity of sylvatic RABV reservoirs in Asia, yet coordinated control efforts targeting wildlife have been geographically limited to date.

19.2.4 Africa

Domestic dog RABV is enzootic throughout North and sub-Saharan Africa, yet several sylvatic reservoirs have been described (Bengoumi, Mansouri, Ghram, & Merot, 2018; Nel & Rupprecht, 2007; Sabeta & Ngoepe, 2018; Swanepoel et al., 1993). The black-backed jackal (*Canis mesomelas*) and bat-eared fox (*Otocyon megalotis*) are reservoirs of the Cosmopolitan (Africa-1) dog RABV cycle. Control activities targeting jackals during the middle of the 20th century focused on population reduction techniques, but did not eliminate RABV circulation (Swanepoel et al., 1993). The efficacy of oral rabies vaccines was reported for the black-backed and side-striped jackal (*Canis adustus*) (Bingham et al., 1995; Bingham, Schumacher, Hill, & Aubert, 1999) and bait uptake studies have been reported from Zimbabwe (Bingham et al., 1993). The Africa-3 RABV lineage represents an older host shift of dog RABV, which is endemic to Africa and circulates in the yellow mongoose (*Cynictis penicillata*) in South Africa and slender mongoose (*Galerella sanguinea*) in Zimbabwe (Nel et al., 2005). Control activities focused on population reduction were described in the early 20th century for the yellow mongoose, but these activities did not eliminate RABV circulation and were abandoned by the turn of the century (Swanepoel et al., 1993).

The African wild dog (*Lycaon pictus*) is an endangered carnivore whose populations continue to be threatened by RABV transmission from dogs and jackals in several African countries. Parenteral delivery of RABV vaccines to free-ranging African wild dogs can present logistical challenges and the feasibility of an ORV approach was explored. An oral rabies vaccine and associated baiting systems were evaluated in South Africa with captive and free-ranging African wild dogs, which indicated potential for high acceptance of chicken head baits and immunogenicity of SAG2 vaccine by bait delivery for adult and juvenile African wild dogs (Knobel & du Toit, 2003; Knobel, du Toit, & Bingham, 2002; Knobel, Liebenberg, & Du Toit, 2003).

The distribution of the Ethiopian wolf (*Canis simensis*), the world's rarest canid, is restricted to the highlands of Ethiopia and this species is also threatened by RABV transmission from domestic dogs. Parenteral immunization was used to control an outbreak of RABV in Ethiopian wolves during 2003–04 (Knobel et al., 2008; Randall et al., 2004), though an ORV approach has been developed and experimental field trials were recently reported (Randall et al., 2006; Sillero-Zubiri et al., 2016).

There is also frequent spillover of the Africa-1 lineage to the greater kudu (*Tragelaphus strepsiceros*) in Namibia, causing substantial economic losses due to its status as a major game species, though it remains controversial whether this ungulate is a RABV reservoir (Scott et al., 2013). Nevertheless, recent studies investigated oral delivery of rabies vaccines to the greater kudu, given that this host is largely inaccessible on the landscape and difficult to safely trap and immobilize for parenteral immunization (Hassel, Ortmann, et al., 2018; Hassel, Vos, et al., 2018).

19.2.5 Central America, South America, and the Caribbean

Elimination of domestic dog RABV circulation throughout much of Central and South America has led to a focal shift in attention to RABV circulation among common vampire bat (*Desmodus rotundus*) populations in the region (Schneider et al., 2009; Velasco-Villa et al., 2017). However, through enhanced surveillance and typing efforts, a limited number of wild carnivore reservoirs have also been described in the region. The crab-eating fox (*Cerdocyon thous*) and hoary fox (*Lycalopex vetulus*) have been recognized as reservoirs of Cosmopolitan dog RABV in Brazil (Carnieli et al., 2008; Cordeiro Rde et al., 2016; Favoretto et al., 2006). The public and agricultural health impact of sylvatic carnivore RABV circulation in this region may be geographically restricted in comparison to the situation in North America and control activities targeting wild carnivore reservoirs have not been reported in the region.

The small Indian mongoose was introduced during the late 19th century to several islands in the Caribbean to control rat damage to sugar cane plantations and is a recognized reservoir of Cosmopolitan dog RABV in Cuba, Grenada, Hispaniola, and Puerto Rico (Seetahal et al., 2018). Attempts to control RABV circulation in mongooses by population reduction tactics have been described, but these were unsuccessful in eliminating RABV (Everard & Everard, 1985, 1992). Advances have been made to establish the immunogenicity and efficacy of oral rabies vaccines for mongoose (Berentsen et al., 2020; Blanton et al., 2006; Ortmann et al., 2018; Vos et al., 2013) and placebo ORV field trials were recently reported from southwestern Puerto Rico (Berentsen, Chipman, et al., 2019). Experimental field trials using live vaccines are needed to develop and optimize an ORV strategy for this invasive tropical host.

19.3 Conceptual foundations and principles

Guidelines regarding harmonization of practices relating to wildlife rabies control were published by the World Health Organization (WHO) Expert Committee in 1992 (WHO, 1992). Guiding documents have also been available through the World Organization for Animal Health (OIE) Terrestrial Manual since 1989, with the most recent update in 2018 (OIE, 2018). Additional guidance on wildlife rabies control was published by the WHO Expert Consultation on Rabies, with the first edition published in 2005 and the most recent update in 2018 (WHO, 2005, 2013, 2018). Detailed guidelines for control of rabies in foxes among European countries were published in 2002 and updated in 2015 to include raccoon dogs (EC, 2002; EFSA, 2015). Moreover, a collection of documents describing a framework for the design and implementation of ORV programs to control fox rabies was adapted from an earlier blueprint for control of canine rabies and published in 2012 by the Partners for Rabies Prevention (Lembo & Partners for Rabies, 2012; Partners for Rabies, 2012).

The control of rabies in dogs and wildlife may require vaccination of approximately 70% of the susceptible population (Anderson et al., 1981; Coleman & Dye, 1996). One point elaborated by Anderson and others was that the vaccination coverage required in fox populations might vary positively with host density, whereby anywhere between 50% and 100% may be needed for RABV control and elimination. Several subsequent publications estimating vaccination coverage required for control and elimination of rabies in fox populations presented estimates within this range (Eisinger & Thulke, 2008; Thulke & Eisinger, 2008; Thulke, Eisinger, Selhorst, & Muller, 2008). Similarly for raccoons, theoretical estimates of vaccination coverage required for RABV control and elimination are reported to be in the range of 50%–100%, and appear to scale positively with raccoon densities (Coyne, Smith, & McAllister, 1989; Rees, Pond, Tinline, & Belanger, 2013; Reynolds, Hirsch, Gehrt, & Craft, 2015). Besides variation in host density across the landscape, reasons for possible discrepancies also relate to natural variation in contact and transmission probabilities, host movement, and epizootic or enzootic nature of RABV circulation in animal populations (McClure, Gilbert, Chipman, Rees, & Pepin, 2020; Newton et al., 2019; Rees et al., 2013; Smith & Harris, 1991; Tardy, Masse, Pelletier, & Fortin, 2018). While the 50%–100% vaccination coverage range appears consistent across theoretical studies involving various single host models and landscapes, the goals for control or elimination of RABV in one or more hosts should be context specific depending on the geographic area and relevant host ecology and RABV epizootiology.

Economics is a key element influencing multilateral commitment and cooperation for wildlife rabies control and elimination programs and serves as a guiding principle behind the development and refinement of products and strategies for specific hosts and geographic areas (e.g., Johnston et al., 1988; Selhorst, Thulke, & Müller, 2001; Sterner et al., 2009; Sterner & Smith, 2006). The costs of living with wildlife rabies are multifaceted, yet the greatest costs stem from case investigations and administration of PET for humans. The costs associated with ORV programs are also multifaceted, though the primary expenses are the vaccine baits and systems associated with delivery (e.g., personnel, flight operations, and fuel). Several reviews have been published on the economics of ORV targeting wildlife (Meltzer & Rupprecht, 1998; Shwiff, Elser, Ernst, Shwiff, & Anderson, 2018; Sterner et al., 2009; Sterner & Smith, 2006) and highlight how an economic analysis and forecasting of costs, as well as anticipated impacts, of a long-term rabies management program are critical during the strategic planning process.

19.4 Structure and operation of control programs

19.4.1 Communication and collaboration

Rabies prevention, control, and elimination in wild carnivore populations has proven successful yet challenging over the last four decades, as management at the landscape scale remains a complex enterprise requiring coordination, collaboration, and communication among diverse stakeholders (Müller & Freuling, 2018; Rupprecht & Slate, 2012; WHO, 1992, 2005, 2013, 2018). A multisector One Health approach with a long-term commitment to disease control and information sharing among policy makers, stakeholders, and the public is essential (Octaria et al., 2018; Vercauteren et al., 2014; WHO, 2005). Frequent communication, transparency, and information exchange promote a critical synergy dedicated to rabies prevention and control among government agencies, industry, nongovernment organizations, research institutions and the public, and leverages expertise in the public health, agriculture and wildlife management communities. Coordinated project planning, sustained funding and surveillance, strategic vaccine bait distribution and science-based program monitoring and adaptive management have been key elements of a successful approach in both Europe and North America for fox rabies elimination (Rosatte, 2011; Stohr & Meslin, 1996). Creating a committed, broad-based coalition of agencies and organizations provides a shared platform to encourage political and public support and timely reporting of technical and political milestones (Müller, Demetriou, et al., 2012). Increasing awareness of these important milestones and measurable successes across jurisdictions helps to underscore and promote advocacy and the need for political support and sustained funding to meet long-term rabies management objectives (Rupprecht & Slate, 2012).

Strategic planning for rabies management ensures a regional, national, and international strategy that establishes program priorities with short- and long-term goals and objectives and details a plan for evaluating success (Elmore et al., 2017; Slate et al., 2009, 2008, 2005; WHO, 2018). The framework for a national strategic plan targeting rabies in wildlife populations typically incorporates topics integral to proactive ORV research and management as well as regulatory compliance, surveillance and diagnostic laboratory support, as well as contingency action and communication planning (Rosatte, 2011; Slate et al., 2005). These plans can also function as a tool to increase transparency and manage expectations and serve as foundation for permitting and regulatory compliance through interjurisdictional collaboration at the municipal, county, state, federal, and international levels (MacInnes & LeBer, 2000;

Slate et al., 2009, 2005; WHO, 2018). Technical collaboration during the strategic and local project planning process facilitates the detailed development of surveillance objectives, standard criteria for evaluation of control programs, budgetary requirements, and defines roles and responsibilities of collaborating institutions (WHO, 1992, 2005). Regular communication among national and international partners reinforces crossborder cooperation and contributes to the alignment of management practices for coordinated ORV campaigns at a superregional level (Müller, Demetriou, et al., 2012; Slate et al., 2009, 2005).

The decision to cofinance fox rabies elimination in Europe to protect human and animal health has provided a framework for coordination, standardization, and program evaluation among EU Member and nonmember States (Müller & Freuling, 2018; WHO, 2018). Reporting of European ORV activities and surveillance data to a central database maintained by the WHO Collaborating Centre for Rabies Surveillance and Research in Germany since 1977 and sponsored by the WHO and OIE, published quarterly as the Rabies Bulletin Europe, has been an instrumental tool for timely communication and strategic planning relating to wildlife rabies control across Eurasia (Freuling, Kloss, Schroder, & Müller, 2006; Müller, Batza, et al., 2012). Detailed guidelines for ORV activities in EU Member States were prescribed in 2002, and updated in 2015 (EC, 2002; EFSA, 2015) and a rabies task force was created to monitor and assess disease elimination campaigns in Member States and progress toward meeting strategic goals (EC, 2017; Müller, Demetriou, et al., 2012). Laboratory harmonization of rabies diagnostics across the EU in support of wildlife ORV has been coordinated since 2008 by the Nancy Laboratory for Rabies and Wildlife in France (Cliquet et al., 2010, 2003; Robardet, Demerson, Andrieu, & Cliquet, 2012; Robardet, Picard-Meyer, Andrieu, Servat, & Cliquet, 2011; Wasniewski et al., 2016) and has been integral for standardization of surveillance and ORV monitoring initiatives across Eurasia.

In North America, the US compendium for animal rabies prevention and control and US state-based rabies task forces (Rupprecht & Slate, 2012; Slate et al., 2009) provide technical recommendations and oversight. A Canadian Rabies Committee was established in 2007 with membership among public health, agriculture, and wildlife agencies and this collaboration led to formalization of the first Canadian Rabies Management plan in 2009 (Rosatte, 2011). A US National Rabies Management Team was formed in 1997, comprising 10 subgroups to facilitate coordination of ORV programs across states and along the borders with Mexico and Canada (Slate et al., 2009), and these collaborations led to the development of the first US National Plan for Wildlife Rabies Management in 2008. A multilateral cooperative effort during 2006–08 to develop the North American Rabies Management Plan (NARMP) was an important breakthrough in rabies planning and communication that fostered a multisector collaboration and a multidisciplinary approach to the management of rabies in Canada, United States, Mexico and the Navajo Nation (Elmore et al., 2017; Rupprecht & Slate, 2012; Slate et al., 2009, 2008). The four pillars of collaboration outlined in the NARMP include information transfer, rabies prevention and control, surveillance and monitoring, and research (USDA, 2008). The collaboration under the NARMP has led to crossborder research, including ORV field trials (Fehlner-Gardiner et al., 2012; Mainguy, Fehlner-Gardiner, Slate, & Rudd, 2013), program evaluation (e.g., Davis, Nelson, et al., 2019) and strategic planning (e.g., Delphi workshops, Fig. 19.5).

The formation of national interagency committees, task forces, and strategic planning teams can enable technical working groups with diverse expertise and responsibilities to find consensus on original, practical, and effective solutions for management regardless of

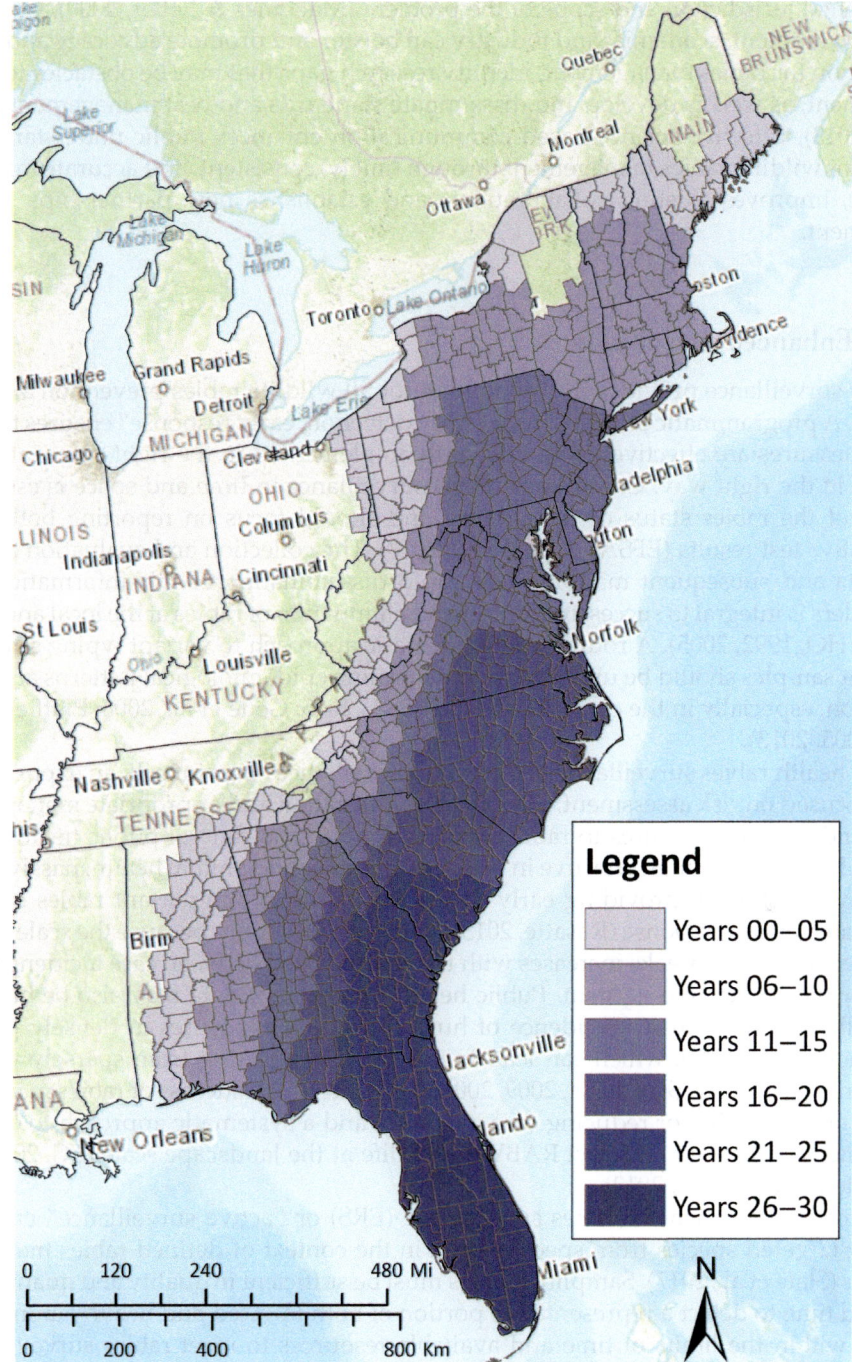

FIG. 19.5 Strategic management approach for achieving raccoon rabies elimination over a 30-year planning horizon showing oral rabies vaccination (ORV) zones in 5-year increments. ORV zones are based on a Delphi process at the county level.

departmental jurisdiction and scope of the problem (MacInnes & LeBer, 2000). Cooperation among government, academia, and industry can bolster and promote advocacy and resource mobilization for rabies management, identify research gaps that may be obstacles to effective management, as well as develop and disseminate standards and best management practices (WHO, 2013). Effective science-based communication enhances public understanding and support for wildlife rabies management through timely, consistent, and accurate information exchange, improves existing collaborations, and establishes new partnerships for rabies management.

19.4.2 Enhanced surveillance

Rabies surveillance provides the foundation for all wildlife rabies prevention and control activities. A programmatic philosophy of "early detection, early response" ensures that rabies control measures are effectively and efficiently implemented in the right place, at the right time and in the right way. Continuous rabies surveillance in time and space is essential regardless of the rabies status of the country and should focus on reporting both positive and negative test results (EFSA, 2015; WHO, 2013). The collection and evaluation of surveillance data and subsequent mapping and rapid dissemination of this information among stakeholders is integral to successful control and elimination of rabies at the local and regional levels (WHO, 1992, 2005). A routine and systematic approach to variant typing strategically important samples should be undertaken to determine epidemiological patterns and sources of infection, especially in the context of ORV (EFSA, 2015; Geue et al., 2008; Pfaff et al., 2018; WHO, 2005, 2013).

Public health rabies surveillance or "passive surveillance" is primarily an exposure-based system focused on risk assessment, diagnostic confirmation and appropriate management of human and animal exposures to rabid animals by state and federal public health agencies (Slate et al., 2009). Although effective in protecting human and animal health, passive surveillance may fall short in providing early detection of critically important rabies cases for a timely management response (Rosatte, 2013). Timing is important because the scale of control needed to address outbreaks increases with the time elapsed between case incidence and detection by the surveillance system. Public health surveillance data may also be biased geographically due to a greater incidence of human and pet exposures in densely populated areas (Slate et al., 2009), which can lead to inadequate reporting from sparsely populated areas (Kirby et al., 2017; Slate et al., 2009, 2008). Ultimately, the success of rabies management depends on preventing or reducing case incidence and a systematic approach to enhancing surveillance is necessary to detect RABV in wildlife at the landscape scale (EC, 2002; EFSA, 2015; Slate & Rupprecht, 2012).

The strategy of enhanced rabies surveillance (ERS) or "active surveillance" emphasizes sampling targeted species from specific areas in the context of defined rabies management objectives (Slate et al., 2017). Sampling efforts must be sufficient in quality and quantity across space and time to detect a representative portion of both infected and healthy animals while working within the limits of time and available resources to meet rabies surveillance and management goals (Davis, Kirby, et al., 2019; Rees, Bélanger, Lelièvre, Coté, & Lambert, 2011). Typically, surveillance is based on geographic and taxonomic risk, targeting suspect animals that have been involved with unprovoked bites, displayed clinical signs consistent with rabies or are found dead (WHO, 2013). The ability to detect rabies cases as part of enhanced surveillance efforts depends on public awareness of the disease and the ability of responsible authorities to collect and transport samples for diagnostic testing in a timely manner (WHO, 2018). Surveillance data from Quebec suggest that the highest proportion of rabies cases were found through citizen notification, and the initial confirmation of raccoon RABV cases in Quebec and Ontario was made possible by reports from the public (Rees et al., 2011; Rosatte et al., 2001). The current raccoon RABV epizootic in Ontario represents the largest outbreak of raccoon rabies in Canada and more than 96% of rabies cases were detected though ERS (Lobo et al., 2018). Although collection of apparently healthy animals not involved with a human or domestic animal exposure may not return a significant number of positive samples, in some situations a single confirmed case in a strategically important management area can necessitate extensive and expensive contingency actions (WHO, 2018). A recent analysis highlighted the strengths of using a multifaceted surveillance approach, where combining data from multiple active and passive sources enhanced the detection of raccoon RABV across space and time and revealed the strengths and weaknesses of individual methods (Davis, Kirby, et al., 2019). A sufficient level of surveillance should be maintained even in countries where animal rabies has been declared eliminated (WHO, 2013).

19.4.3 Management actions

Based on current tools and technology, rabies in wild carnivores can be controlled and successfully eliminated over considerable geographic areas as demonstrated by recent management milestones in Europe and North America (Elmore et al., 2017; Müller & Freuling, 2018; Rupprecht et al., 2008; Slate et al., 2009). The risk of RABV exposure from vector species can be reduced by applying basic wildlife damage management techniques, such as exclusion, habitat modification to reduce attractants and limiting artificial resource supplementation by humans (Slate, Owens, Connolly, & Simmons, 1992). Alternatives for wildlife rabies prevention and control at the landscape scale generally focus on a single or combination of strategies for managing the disease in wildlife populations, which may include local population reduction, TVR, and ORV (Rosatte, 2011; Rosatte, Power, MacInnes, & Campbell, 1992; Slate et al., 2008, 2005; WHO, 2013). An integrated program of two or more of these strategies may increase the likelihood of success and overall effectiveness depending on local program goals (Rosatte, Donovan, Allan, et al., 2009; Slate et al., 2008). Important questions remain regarding best practices for optimizing these tactics across reservoir species and habitats, as control and elimination has proven more difficult with raccoons and skunks in comparison to canid hosts (Hanlon, Childs, & Nettles, 1999; Rupprecht & Slate, 2012). The introduction of exotic species and challenges presented by highly developed landscapes may require a fundamental shift to more novel integrated management strategies, including habitat-based baiting coupled with focal population reduction of nonnative species in high-risk management zones (Müller, Freuling, et al., 2015; Slate et al., 2005).

Adaptive management is fundamental to rabies control, allowing program managers to take action in spite of uncertainty (Fontaine, 2011). This approach provides a feedback loop that focuses on "learning by doing" when a full understanding of the most appropriate management strategy may be lacking and the knowledge of downstream potential impacts in complex

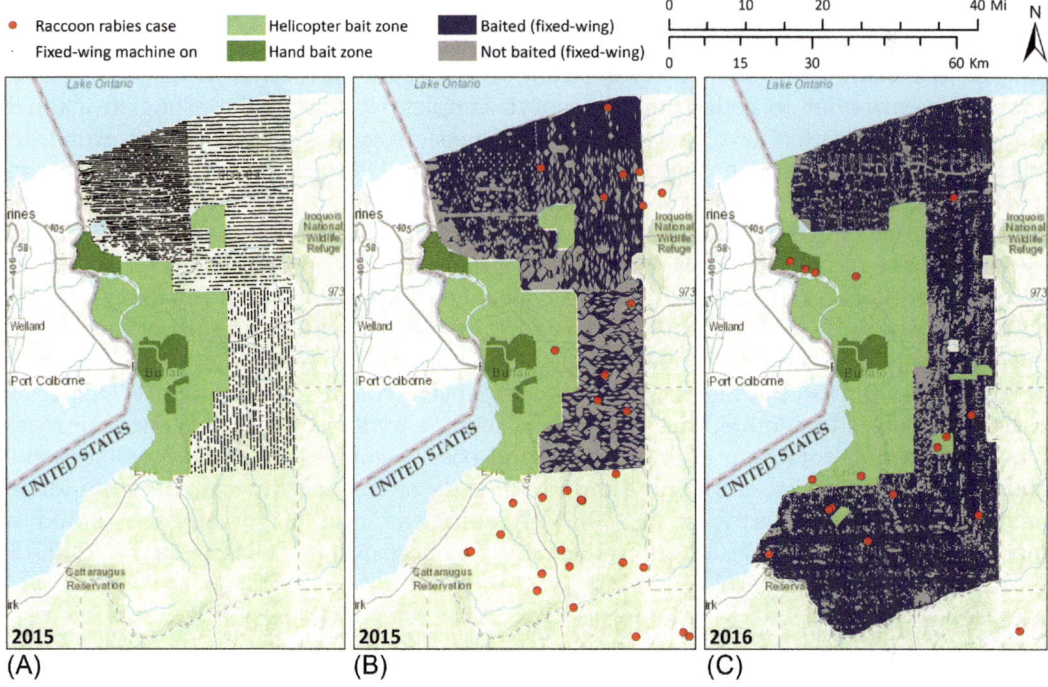

FIG. 19.6 Fixed-wing aircraft oral rabies vaccination (ORV) zones in the Buffalo, New York area, showing adaptive management over time: (A) 2015 ORV zone with fixed-wing bait machine dots and helicopter and hand-baited areas; (B) Delaunay triangulation analysis of 2015 fixed-wing aircraft ORV zone showing areas baited vs not baited relative to raccoon rabies variant cases; (C) Delaunay triangulation analysis of 2016 fixed-wing aircraft ORV zone showing areas baited vs not baited relative to raccoon rabies variant cases after expanding bait zone to the south, converting entire zone to a different vaccine, doubling bait density, decreasing flight line spacing, and enhancing navigator training to reduce bait machine off-time.

systems is incomplete (Fontaine, 2011). In the United States, one example involves the updating of preprogrammed ORV flight lines to maximize an even distribution of baits in target areas. An algorithm known as Delaunay Triangulation (Lee & Schachter, 1980) was used to triangulate points in space and time where the baiting machinery was turned on or off at the operator's discretion. This triangulation processing technique identified gaps on the landscape that were not adequately baited and the programmed flight lines were iteratively revised and updated to minimize baiting gaps on the landscape (Fig. 19.6). Adaptive management requires strong interagency collaboration, consensus building, and a well-designed communication plan to insure continued cooperation (Rosatte, Tinline, & Johnston, 2007).

19.4.3.1 Population reduction

Prior to the development of safe and effective oral rabies vaccines, reducing wildlife reservoir populations below a targeted level was the only method available for wildlife managers attempting to stop the spread and eliminate RABV at the landscape scale (Debbie, 1991; Fehlner-Gardiner, 2018; Lewis, 1975; Müller & Freuling, 2018). Globally, population reduction techniques for the management of mesocarnivores have included hunting, trapping, poisoning, gassing of dens, and bounties (Debbie, 1991; Lewis, 1975). These intensive, single-strategy management programs in Europe, North America, and Africa did not eliminate nor prevent the spread of wild carnivore RABV (Fehlner-Gardiner, 2018; Macdonald, 1980; Müller & Freuling, 2018; Slate & Rupprecht, 2012; Swanepoel et al., 1993; WHO, 1992, 2013).

Trapping with foothold traps was the most common method historically used for rabies control targeting foxes and skunks in North America (Debbie, 1991), often implemented by state or provincial agencies focusing on rabies management at a county level (Linhart, 1960). Large-scale trapping efforts were expensive and showed limited success in reducing RABV case incidence in New York (Debbie, 1991; Linhart, 1960). One advantage of trapping over the use of toxicants is greater host specificity and thus reduced impact for nontarget wildlife. Fox rabies control in Europe was reportedly enhanced once calcium cyanide was used for gassing fox dens in conjunction with more traditional population reduction techniques in some European countries. Other toxicants, including strychnine, thallium sulfate, and sodium fluoroacetate (1080), used in chicken baits and chicken eggs were the most common poisons used for wildlife control, although zinc phosphide bait was also used to control mongoose on the island of Grenada (Everard & Everard, 1992). Control efforts targeted coyotes, foxes, and other wild canids in Alberta, Canada, during the early 1950s in an aggressive population reduction campaign to combat the incidence and spread of fox RABV from the Northwest Territories (Ballantyne, 1956; Ballantyne & O'Donoghue, 1954). Lethal control efforts later targeted striped skunks to reduce the incidence of and limit the spread of skunk RABV in southern Alberta (Gunson, Dorward, & Schowalter, 1978; Pybus, 1988; Rosatte et al., 1986).

Despite some examples of short-term localized success in reducing rabies incidence, experts agree that the implementation of broad-scale population reduction as a single tactic is unlikely to achieve long-term management objectives of reducing the spread of or eliminating RABV from wildlife (Slate & Rupprecht, 2012). Increased movements and contacts of incubating or rabid target animals and demographic resilience of mesocarnivore populations makes this management approach unsuitable for most disease control situations (Bögel, Arata, Moegle, & Knorpp, 1974; Bögel et al., 1981; Macdonald, 1980; Slate & Rupprecht, 2012). Bögel et al. (1974) reported that a fox population reduced by 40%–50% quickly rebounded with in 3 years. Raccoon populations in Scarborough, Ontario, Canada, were reduced by 20% on an annual basis from 1987 to 1993, but were still able to maintain an average density of >11 raccoons/km^2 (Rosatte, 2000). One experiment removed raccoons from 30 different forest patches (3–12 ha in size) and then monitored immigration and recolonization dynamics for 3 years postremoval along with 25 separate control (i.e., no removal treatment) patches (Beasley, Olson, Beatty, Dharmarajan, & Rhodes, 2013). While only 40% of treatment patches had returned to preremoval densities after 3 years of monitoring, colonizing raccoons were mostly young males, suggestive that raccoon removal activities may enhance juvenile dispersal and the potential for RABV spread (McClure et al., 2020). Nevertheless, an integrated program similar to the PIC technique used in Canada for management of raccoon rabies involving ORV, TVR, and focused population reduction around an index case along with ERS may continue to play an important role in wildlife RABV prevention and control (WHO, 2005).

19.4.3.2 Trap-vaccinate-release

The TVR method involves live trapping and subsequent release of mesocarnivores at the site of capture after ear tagging and parenteral vaccination with an off-label use of veterinary vaccines (Fehlner-Gardiner, 2018; Sobey et al., 2010; WHO, 2013; Wohlers et al., 2018), which have been shown to induce nearly 100% seroconversion in raccoons and skunks by intramuscular injection (Rosatte, Howard, Campbell, & MacInnes, 1990; Rosatte et al., 1992; Sobey et al., 2010). The live capture and parenteral vaccination of skunks and raccoons has proven to be an effective rabies control strategy on a limited scale, often in urbanized and developed landscapes especially when conducted in conjunction with ERS and ORV as practical in an adaptive management framework (Elmore et al., 2017; Rosatte, 2011; Rosatte et al., 1992). The use of TVR for mesocarnivores in combination with robust domestic animal vaccination in treatment areas creates a buffer zone of vaccinated animals that reduces the susceptible fraction and limits RABV spread (Rosatte, 2000).

This vaccination strategy was originally conceived and implemented in Toronto, Ontario, Canada, for use in a more developed landscape and prior to the availability of an effective oral rabies vaccine for skunks (Rosatte et al., 1992). Early attempts using this strategy for red foxes in Europe and Canada were quickly abandoned due to logistical infeasibility and low success rates of live-trapping adequate numbers of foxes (Rosatte, 2011; WHO, 2005). However, TVR has been used successfully in North America to manage raccoon rabies as part of contingency and PIC operations (Elmore et al., 2017; Fehlner-Gardiner, 2018; Rosatte et al., 1992; Slate & Rupprecht, 2012; Sobey et al., 2010; WHO, 2013). The first major implementation of TVR occurred in Toronto, Canada, during 1987–1991, where 955 unique skunks and 2266 unique raccoons were captured and vaccinated in a 60 km^2 area (Rosatte et al., 1993, 1992), comprising an estimated 52%–72% of the skunk population and 61%–68% of the raccoon population over the 5-year period (Rosatte et al., 1992). From 1991 to 2008, more than 90,000 raccoons were live-trapped and vaccinated in Ontario, Canada, as part of TVR and PIC operations to address an incursion of raccoon rabies reported in 1999 (Sobey et al., 2010). A combined TVR and PIC strategy was also used to manage raccoon rabies in Quebec and New Brunswick, Canada (Sobey et al., 2010). However, TVR is labor intensive and expensive compared to ORV and remains best suited to focal areas of less than 2000 km^2 (Rosatte, 2013; Rosatte et al., 2001; Sterner et al., 2009). A coordinated TVR campaign was performed in three highly urbanized parks in Manhattan, New York City, United States, after an outbreak of raccoon rabies was reported (Slavinski et al., 2012), as management with ORV was considered impractical and population reduction was opposed by the public (Elmore et al., 2017). The rapidly mobilized TVR and ERS campaign controlled and eliminated raccoon RABV in Central Park (Slavinski et al., 2012). A recent study also highlighted beneficial effects of TVR along with ORV in reducing RABV occurrence in a high-risk rural area of the northeastern United States along the Canadian border, where over 17,000 raccoons were live-trapped and vaccinated during 2007–12 (Davis, Nelson, et al., 2019). In 2001, an outbreak involving 19 striped skunks infected with a big brown bat RABV was detected in Flagstaff Arizona and successfully controlled with a cooperative TVR program (Kuzmin et al., 2012; Leslie et al., 2006; Slate et al., 2009). TVR will continue to be a key management strategy for PIC and contingency actions in North America to control raccoon RABV, while more effective ORV products and strategies are refined to target striped skunks (Slate et al., 2009; Wohlers et al., 2018).

19.4.3.3 Oral rabies vaccination

The broad-scale distribution of oral rabies vaccines in Europe and North America as an intervention strategy for the control and elimination of RABV circulation in wild carnivores has proven to be successful for a diversity of species and habitats under varying ecological and epidemiological conditions (Elmore et al., 2017; Freuling et al., 2013; Müller, Demetriou, et al., 2012; Müller, Freuling, et al., 2015; Müller & Freuling, 2018; Rupprecht et al., 2008; Rupprecht & Slate, 2012; Sidwa et al., 2005; Slate et al., 2008, 2005; Vos, 2003). At its core, rabies management with ORV requires establishing a defined zone where a level of population immunity ("herd immunity") can be induced through vaccination to prevent, reduce, or eliminate RABV transmission among susceptible wild carnivore populations inhabiting rural and urban environments (Moore et al., 2017; Müller, Demetriou, et al., 2012; Müller, Freuling, et al., 2015; Slate & Rupprecht, 2012; WHO, 2005, 2013). The level of immunity achieved is dependent on vaccine effectiveness and stability, bait type and durability, bait density, baiting method, precise spatial and temporal vaccine distribution, and density of target species and nontarget bait competitors (Müller, Demetriou, et al., 2012; Selhorst et al., 2001; Slate & Rupprecht, 2012; WHO, 2005, 2013). This strategy was originally developed for managing geographic spread of rabies in red foxes and is now an essential tool used operationally in more than 30 countries across three continents targeting an increasing number of mesocarnivore species.

Most ORV operations are conducted by fixed-wing aircraft in rural areas; however, high elevation or urban-suburban habitats may require helicopter, hand baiting, and bait stations to safely and effectively distribute vaccine baits in favorable habitats (Boulanger, Bigler, Curtis, Lein, & Lembo, 2008; Elmore et al., 2017; Fearneyhough et al., 1998; Masson et al., 1996; Rosatte, 2011; Rosatte, Donovan, Allan, et al., 2009; Slate & Rupprecht, 2012; Slate et al., 2005; WHO, 2018). Landscape features such as rivers, lakes, mountains, major road systems, and metropolitan areas may all impact the spread of RABV and the effectiveness of ORV (Real & Biek, 2007; Real, Russell, Waller, Smith, & Childs, 2005; Slate et al., 2008). Adopting technologies, including satellite-based Global Positioning System (GPS), to improve accuracy and standardizing aircraft flight lines for aerial operations, and Geographic Information System (GIS) for bait distribution planning, managing, and analyzing distribution data have improved precision of management actions (Müller, Freuling, et al., 2012). These and other advances like automation of bait machinery have provided managers with the capability to distribute millions of baits across hundreds of thousands of square kilometers annually at prescribed bait densities ranging from 20 to over 300 baits/km^2 (Rosatte, 2011; Rosatte, Tinline, et al., 2007; Sidwa et al., 2005; Slate et al., 2005; WHO, 2013).

Despite these successes, ORV is an expensive management strategy with the majority of costs relating to project planning, surveillance, air time for bait distribution, fuel, ground baiting, and the purchase of vaccine baits (Müller, Schroder, et al., 2015; Slate et al., 2005). Since the late 1980s, the EU has cofinanced ORV programs for Member States and along some border areas with nonmember states (EC, 1989, 1990, 2009; Müller, Demetriou, et al., 2012; Müller & Freuling, 2018). This long-term, predictable funding source has allowed countries to standardize management approaches and collaboratively address local and regional impacts of wildlife rabies on human and animal health as a priority (Müller, Demetriou, et al., 2012; Müller & Freuling, 2018), yet these programs took between 5 and 26 years to meet management objectives with a median of 14 years (Freuling et al., 2013). Retrospective analysis of fox rabies management data from Germany showed that incomplete or inconsistent vaccination effort can increase the time to RABV elimination from red fox populations, highlighting the importance of consistent control effort over space and time to achieve management objectives (Baker, Matthiopoulos, Muller, Freuling, & Hampson, 2019). Additional research on economic costs and benefits of rabies control in combination with a better understanding of the social dimensions and business of ORV will promote efficiency and program sustainability (Rupprecht et al., 2008; Slate et al., 2009, 2005; Slate & Rupprecht, 2012; Sterner et al., 2009).

Oral vaccination of wildlife remains inherently complex from a technical, logistical, ecological, and economic perspective regardless of species (Rupprecht & Slate, 2012). Tailoring of programs and specific management strategies have been refined and standardized for several wild carnivore species (Table 19.1), but a single bait product or distribution strategy may not be successful across all target populations and areas. Targeting striped skunks for ORV given their role as the primary spillover host for both raccoon and arctic fox RABV in the Eastern United States and Canada has proven challenging and requires additional research on novel vaccine-baits and ORV approaches (Brown, Rosatte, Fehlner-Gardiner, Ellison, et al., 2014; Gilbert, Johnson, Walker, Beath, & VerCauteren, 2018; Slate & Rupprecht, 2012; Vos et al., 2017; WHO, 2018; Wohlers et al., 2018). The small Indian mongoose is a rabies reservoir on several islands in the Caribbean including Cuba, Grenada, Hispaniola, and Puerto Rico (Everard & Everard, 1992; Nadin-Davis et al., 2006; Nadin-Davis, Velez, Malaga, & Wandeler, 2008; Seetahal et al., 2018; Zieger et al., 2014) and can exist at very high densities (e.g., greater than 50 per hectare) in tropical habitats along with invasive and abundant nontarget bait competitors such as black rats (*Rattus rattus*) (Berentsen, Torres-Toledo, Rivera-Rodriguez, Davis, & Gilbert, 2018; Johnson, Berentsen, Ellis, & VerCauteren, 2016). Development of an effective ORV strategy for the small Indian mongoose could have global implications given the extensive geographic range of this host and may also be relevant for mongoose reservoirs described from southern Africa (Müller, Freuling, et al., 2015; Nel et al., 2005). The development of ORV baits and control strategies targeting RABV circulation by Chinese ferret badgers in tropical habitats of Chinese Taipei is ongoing yet has proven challenging given the elusive nature of this host and presence of multiple nontarget wild carnivore bait competitors in some habitats (Chang et al., 2016).

TABLE 19.1 General characteristics of oral rabies vaccination programs tailored to individual reservoir species.

Area	Target	Bait density	Frequency	Flight lines
Eurasia	Red fox, raccoon dog	20–30/km^2	2× annual	500 m
Israel	Red fox, golden jackal	15–150/km^2	2× annual	300 m
Canada	Red fox	20–150/km^2	1× annual	750–2000 m
Canada/USA	Striped skunk	300/km^2	1× annual	250 m
Canada/USA	Raccoon	37–300/km^2	1× annual	250–750 m
USA	Gray fox, coyote	20–40/km^2	1× annual	800 m

Even routine wildlife targets such as foxes and raccoon dogs in Europe, and raccoons, skunks, and foxes in North America pose additional health concerns in large metropolitan areas where an increasing number of RABV exposures have occurred (Broadfoot, Rosatte, & O'Leary, 2001; Slate et al., 2009; WHO, 1992). Ecological, environmental, and logistical challenges in developed landscapes include higher densities of reservoir hosts, habituation of target species to humans, smaller home ranges of reservoir hosts in developed and fragmented habitat, competing anthropogenic food sources, nontarget bait competition, and the inability to effectively and logistically bait all suitable habitat with current road-based distribution methods (Boulanger et al., 2008; Graser III, Gehrt, Hungerford, & Anchor, 2012; Riley, Hadidian, & Manski, 1998). Raccoons are ecological generalists and increased population densities (more than $100/km^2$) have been reported for some developed areas and suburban parks in North America and Europe (Graser III et al., 2012; Riley et al., 1998; Rosatte et al., 2010; Rupprecht & Slate, 2012; Slate et al., 2009, 2005). Raccoons have been introduced to much of mainland Europe and populations are expanding in many countries, providing a novel rabies threat and specialized management challenge particularly in urban habitats (Müller, Freuling, et al., 2015). Domestic dogs and cats are typically abundant in urban environments and may be susceptible to spillover infections or compete for baits (Lobo et al., 2018; Wallace et al., 2014). Other factors, which can impact ORV success, also include bait competition by nontarget wildlife (e.g., Virginia opossum [*Didelphis virginiana*] in North America and wild boar [*Sus scrofa*] in Europe), morbidity and mortality of target populations due to other diseases (e.g., canine distemper), parasites (e.g., toxoplasma), and urban-suburban sprawl.

The aerial method of bait delivery using low-flying aircraft was developed and refined by Canadian researchers and facilitated more cost-effective delivery across large rural landscapes (Johnston et al., 1988; Johnston & Voigt, 1982; MacInnes et al., 1992). Fixed-wing aerial vaccine distribution is primarily used for larger rural landscapes in Europe and North America whereas helicopter (rotary-wing) distribution is used to target high elevation habitats in Europe or smaller, more developed landscapes in the United States and Canada (Boulanger et al., 2008; Boyer, Canac-Marquis, Guerin, Mainguy, & Pelletier, 2011; EC, 2017; Elmore et al., 2017; Fehlner-Gardiner, 2018; Masson et al., 1996; Rosatte, 2011; Slate & Rupprecht, 2012; WHO, 2013) (Fig. 19.7). In Europe, early bait applications were often done by hand, requiring large crews of trained personnel typically utilizing roads or hunter networks for delivery (Kappeler, 1992), but the efficiency of operations improved substantially following a transition to fixed-wing aerial delivery with integration of a fully automated and quality controlled baiting system piloted in Germany starting in 1995, which has been subsequently adopted by many EU countries (Müller, Freuling, et al., 2012). Flight line spacing may be adjusted based on planned bait density and optimized for specific carnivores based on their home range behavior (Elmore et al., 2017; Slate & Rupprecht, 2012). Flight line spacing for skunks and raccoons in Canada has ranged from 250 to 750 m and, in the United States, from 125 to 750 m with a mode of 500 m (Elmore et al., 2017; Rosatte, 2011; Rosatte et al., 2008; Rosatte, Donovan, Davies, et al., 2009; Rosatte et al., 2011) (Table 19.1). Parallel flight line spacing targeting fox in Europe is generally 500 m in comparison to 800 m for coyotes and gray fox in Texas (EC, 2017; Fearneyhough et al., 1998; Sidwa et al., 2005). Distribution of oral rabies vaccine baits by fixed-wing aircraft is generally not feasible in urban areas and other methods, including helicopter baiting of large urban green spaces, river corridors, and other semideveloped areas in the urban environment, may be required (Lobo et al., 2018; Müller, Batza, et al., 2012).

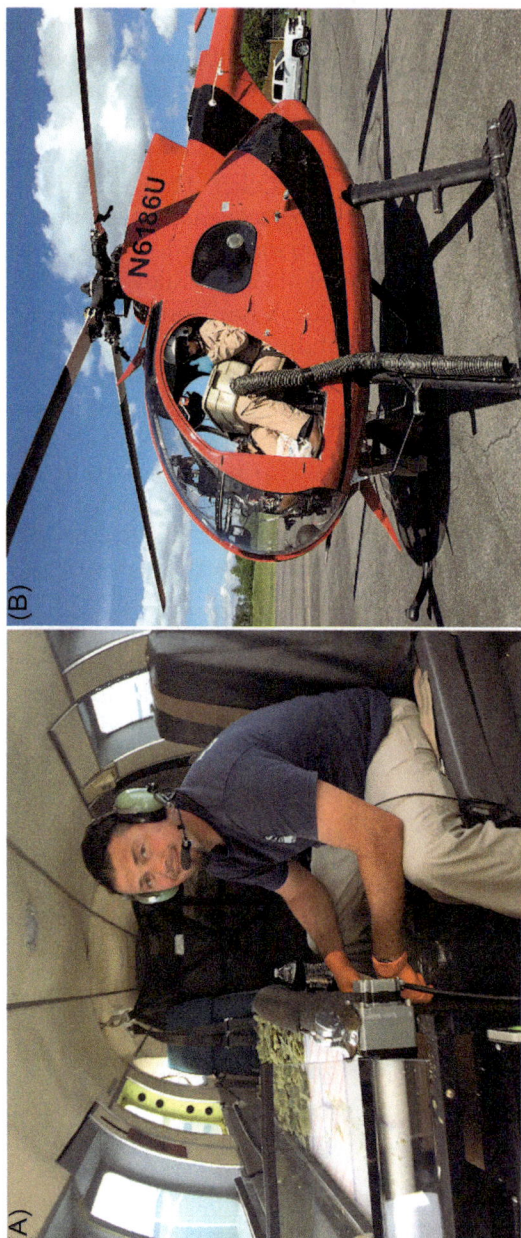

FIG. 19.7 Current machinery for distribution of oral rabies vaccine baits in the United States: (A) fixed-wing aircraft bait belt; (B) helicopter vacuum slide. *Photos by USDA.*

Hand baiting and to a lesser extent bait stations may be used in more highly developed habitats (Boulanger et al., 2008; EFSA, 2015; Fehlner-Gardiner, 2018; Roscoe et al., 1998; WHO, 2013). Hand baiting is an essential method for limited geographic scales and is typically conducted by walking or from vehicles appropriate to the terrain, including ATVs, trucks, and bicycles (Boulanger et al., 2008; Fehlner-Gardiner, 2018; Rosatte, Power, et al., 2007). It is now possible to monitor vaccine bait placement during hand-baiting operations using GPS technology and the USDA APHIS WS program began using Point of Interest GPS technology (Tsi GL-770, Transystem Inc., Hsinchu City, Chinese Taipei) during 2015. This push button technology allows the user to georeference baits distributed by hand in 1-second increments. During 2018, USDA recorded over 355,000 vaccine bait locations across seven states to improve the ORV planning process and implement adaptive management to control raccoon rabies in urban environments (Fig. 19.8). The use of bait stations also forms an important part of a comprehensive ORV strategy in urban-suburban landscapes (Bjorklund et al., 2017; Boulanger et al., 2008; Haley et al., 2017; Lobo et al., 2018; Slate & Rupprecht, 2012), yet additional research is needed to more effectively integrate this method operationally as a complementary strategy to hand baiting in suburban areas.

Regardless of vaccine distribution method, timing of ORV operations, frequency of baiting, and bait density must be carefully considered (Bachmann et al., 1990). Distribution of vaccine baits should target a bait density that results in approximately 70% of the target animal

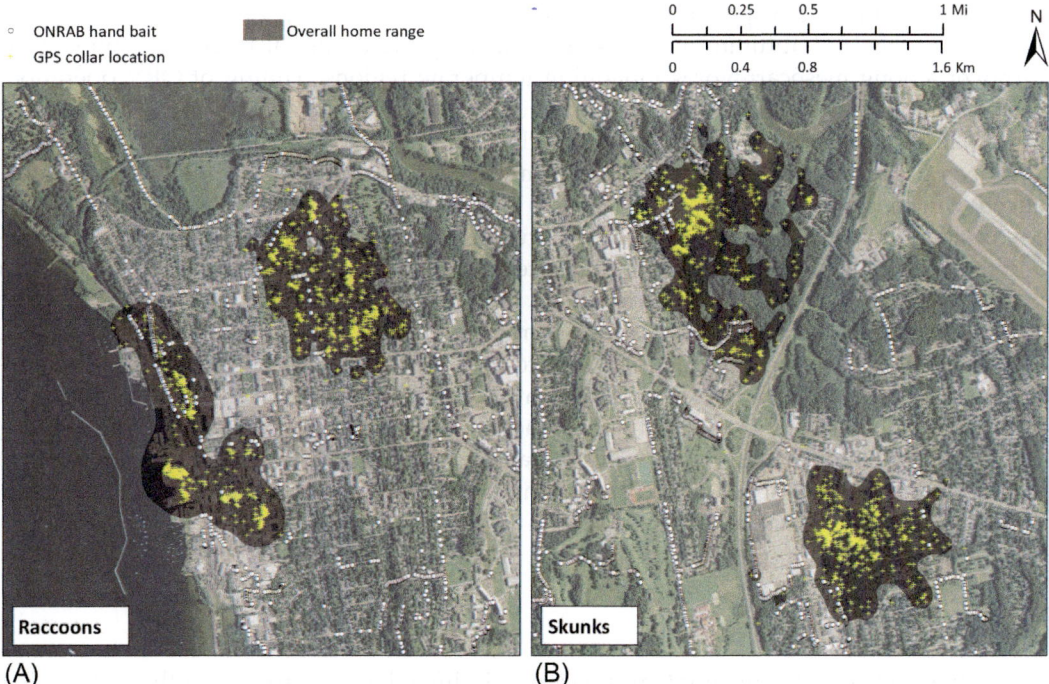

FIG. 19.8 Hand bait locations, GPS collar locations, and overall home ranges for (A) two raccoons and (B) two skunks in an urban area of Burlington, Vermont, 2016.

population consuming bait with an average of one bait consumed per individual to achieve ORV operational cost efficiency (Blackwell et al., 2004). Bait density is determined based on population abundance or density estimates of target species and should be adjusted based on the ecology of the target reservoir, abundance of nontarget competitors and land use features to optimize vaccine bait uptake (Elmore et al., 2017; Fehlner-Gardiner, 2018; Ramey, Blackwell, Gates, & Slemons, 2008). European ORV operations for red fox target a minimum bait density of 20 baits/km^2 with an increased density (25–30 baits/km^2) and narrower flight line spacing (300 m) for managing contingency actions (EC, 2017; EFSA, 2015; Freuling et al., 2013; Müller, Batza, et al., 2012). Typical bait densities for coyotes in Texas range from 19 to 27/km^2 in contrast to 27–39 baits/km^2 for gray fox in comparison to 20 baits/km^2 targeting red fox in Ontario (MacInnes et al., 2001; Sidwa et al., 2005; Slate & Rupprecht, 2012). Bait densities targeting raccoons and skunks typically range from 75 baits/km^2 in rural habitats to 150 baits/km^2 in suburban and urban habitats in North America (Sattler et al., 2009; Slate & Rupprecht, 2012). However, depending on the management activity, habitat, and population density, target bait densities for raccoons and skunks may range from 37 baits/km^2 in low-density, poor-quality habitat to 300 baits/km^2 in high-density, high-quality habitats in North America (Elmore et al., 2017).

Host densities are an integral component to conspecific contact rates, seasonal incidence of RABV, and timing of control (Reynolds et al., 2015). One consistent life history pattern among temperate dwelling mesocarnivores is seasonal synchronized parturition (Clayton, Duke-Sylvester, Gross, Lenhart, & Real, 2009; Duke-Sylvester, Bolzoni, & Real, 2011; Peel et al., 2014). Observations that juveniles comprise a significant fraction of the total population during summer and autumn months suggest a fairly rapid annual target host population turnover among mesocarnivores inhabiting temperate regions. Timing of ORV operations has been refined and standardized for foxes in Europe (spring baiting March-May and autumn baiting September-October) based on climatic conditions (WHO, 2013), though it has also been suggested that late autumn (November) or early winter (December) may be more effective in considering fox ecology in Europe (Vos, 2003). Rabies control programs targeting coyotes and gray foxes in Texas typically occur in January each year to avoid high temperatures that may negatively impact vaccine potency and to target adults and young of the year prior to seasonal dispersal (Sidwa et al., 2005). ORV campaigns for raccoons in Eastern North America generally occur once a year in late summer and early autumn when juveniles are able to forage for baits and to allow adequate time to undertake post-ORV serologic program monitoring (Elmore et al., 2017;Slate et al., 2009; Slate & Rupprecht, 2012), although a twice annual strategy may be employed for contingency actions (Sattler et al., 2009; Slate & Rupprecht, 2012). Regardless of the seasonal timing of baiting, ORV campaigns should continue for at least 2 years after the last confirmed case along with robust surveillance (Rosatte, Donovan, Allan, et al., 2009; WHO, 2013).

19.4.4 Contingency actions

Contingency actions are an integral part of wildlife rabies control in North America and proactive planning insures rapid response and control of outbreaks (Rosatte, 2013). However, this type of emergency management action is logistically complex to implement, labor intensive, and expensive compared to a standard ORV campaign (Slate et al., 2009). Data from Ontario suggest that contingency actions may be more than twice as expensive as ORV in similar habitats and therefore not sustainable for long-term management at the landscape scale (Rosatte et al., 2001; Slate et al., 2009). Since 1999, USDA APHIS WS has cooperatively implemented 15 contingency actions as part of management activities to control the spread of raccoon RABV. When a rabies case is documented beyond an existing ORV zone or in a strategically important rabies management area, a risk assessment is developed and consideration is given to implementation of a contingency action. This type of emergency intervention may include ERS targeting rabies reservoir and vector species in the vicinity of the index case (Slate et al., 2008). Contingency actions may also involve a more comprehensive management response involving focal TVR, biannual emergency ORV, and local population reduction (Slate et al., 2008). The striped skunk is a frequent spillover host of raccoon RABV and targeted populations show uniformly low seroconversion rates to standard ORV bait densities applied for fox or raccoon populations in North America (Elmore et al., 2017). Detection of high-risk raccoon RABV spread by skunks has resulted in the need for contingency actions in Ontario, Quebec, and New Brunswick to enhance ORV zones and meet rabies management objectives (Slate et al., 2009). The PIC approach involves population reduction followed by TVR and ORV around detected cases until they have been eliminated (Fehlner-Gardiner, 2018; Rosatte, Donovan, Allan, et al., 2009; Slate & Rupprecht, 2012). The reemergence of raccoon RABV in Hamilton, Ontario, Canada, in 2015 has resulted in the largest raccoon rabies epizootic in Canada and the contingency action response included coordination, public education, ERS, ORV distribution by hand, helicopter, fixed-wing aircraft, and bait stations as well as parenteral vaccination efforts targeting domestic animals (Lobo et al., 2018). Contingency actions will remain an integral component to ORV campaigns, but are more expensive than routine operations and are inherently limited as a management action in terms of treatment area and sustainability (Slate et al., 2009; Slate & Rupprecht, 2012).

19.4.5 Translocation

Translocation involves the capture and transport of wildlife from one location to another and subsequent release within their native range with an emphasis on resolving conflicts and reducing wildlife damage (Craven, Barnes, & Kania, 1998). Translocation remains an important technique used by wildlife biologists in many countries for conservation and management purposes involving a wide variety of species (Craven et al., 1998; Griffith, Scott, Carpenter, & Reed, 1989). Intentional relocation of animals has played a role in the introduction, reintroduction, and augmentation of populations of game species, furbearers, and rare and endangered wildlife and has been used as a tool to mitigate human wildlife conflicts (Chipman, Slate, Rupprecht, & Mendoza, 2008; Griffith et al., 1989). While unintentional translocation of wildlife by truck, train, and boat (often associated with the transport of solid waste) has been documented in Canada and the United States, the scale and scope of these types of accidental relocations is unknown but likely common (Chipman et al., 2008; Lobo et al., 2018; Rosatte, Donovan, Allan, Bruce, & Davies, 2007).

The capture, temporary possession and relocation of RABV reservoir species raises particular concerns regarding transmission risk to humans, pets, and wildlife as well as the potential to spread the disease from enzootic to naïve areas (Elmore et al., 2017; SEAFWA, 2016b; Slate et al., 2005). The source of the most recent 2015 epizootic of raccoon rabies in Southwestern Ontario likely resulted from a long-distance accidental translocation from southeastern New York (Lobo et al., 2018). Hundreds of thousands of individual animals are intentionally translocated each year in the United States, most often in an attempt to reduce and mitigate damage caused by common urban-suburban species (Craven et al., 1998). Interactions between humans and wildlife are increasing primarily in countries with rapidly developing urban-suburban landscapes where population densities of more common mesocarnivores like raccoons, skunks, gray fox, and red fox are greater due to the availability of anthropogenic food and den resources (Craven et al., 1998; Massei, Quy, Gurney, & Cowan, 2010). Translocation may be perceived as a humane, safe, and effective solution to nuisance wildlife and often community pressure rather than science or economics drives the use of this technique over alternative management strategies (Massei et al., 2010; Mengak, 2018). Alternative nonlethal management approaches for resolving conflicts with individual nuisance animals in residential areas include exclusion with physical barriers, repellents, habitat management to reduce attractants and modifying the behavior of people and wildlife (Slate et al., 1992).

Mesocarnivores have been observed to exhibit extensive postrelease exploratory and linear long-distance movements that may facilitate the spread of parasites and disease (Linnell, Aanes, Swenson, Odden, & Smith, 1997; Massei et al., 2010; Rosatte & MacInnes, 1989). Molecular epidemiological studies or "genetic tracking" has provided significant insight on raccoon populations targeted for management with ORV along the United States-Canadian border and provided new evidence on the source of long distance translocation events (Nadin-Davis et al., 2018; Nadin-Davis, Muldoon, & Wandeler, 2006a; Szanto, Nadin-Davis, Rosatte, & White, 2011; Trewby et al., 2017). Historically, the interstate shipment of raccoons in the southeastern United States to supply private hunting clubs was common practice (Nettles, Shaddock, Sikes, & Reyes, 1979). The translocation of infected raccoons from Florida to the mid-Atlantic states for this purpose resulted in the sizeable expansion of raccoon RABV in the United States (Massei et al., 2010; Slate et al., 2005). In Europe, there are several policies and regulations regarding nuisance wildlife and translocation, and the relocation of wildlife by the public or government agencies is uncommon. Most research now suggests that translocation should be curtailed as a method of resolving human-wildlife conflict because of the risk to animal welfare and potential for disease spread (Massei et al., 2010). Reducing or eliminating translocation of rabies reservoir species in proximity of ORV zones is essential to avoid undermining progress made in local and regional management (Slate et al., 2005), and euthanasia should be considered over translocation for rabies reservoir and vector species (Craven et al., 1998; SEAFWA, 2016a). Agencies with regulatory authority for movement of wildlife are encouraged to conduct risk assessments and reduce the importation, distribution and relocation of wildlife particularly among mesocarnivores in proximity to rabies management areas (Rosatte, 2013).

19.4.6 Program monitoring and evaluation

Measuring success of ORV management actions, or integrated contingency actions, requires well defined metrics and long-term, year-round monitoring (WHO, 2018). Globally, the gold standard metric for documenting the effectiveness of vaccination campaigns is based on prevention of cases in rabies free areas or demonstrated case reduction in areas with wild carnivore RABV circulation, but may also include monitoring of serologic responses to vaccination and detection of a biomarker as a measure of vaccine and bait uptake in the target population (Brochier et al., 1996; Müller, Batza, et al., 2012; Müller, Freuling, et al., 2015; Rosatte, Tinline, et al., 2007; WHO, 1992). Robust surveillance remains critical throughout all phases of multiyear ORV campaigns to ensure timely documentation and reporting of rabies cases or reemergence of cases in areas previously declared free of rabies and to adapt management strategies (Brochier et al., 1996; WHO, 1992, 2018). The EU Member States monitor ORV campaigns targeting wild carnivores by hunting bag collection of four animals per 100 km^2 distributed evenly across the landscape (EC, 2017). Many important questions remain regarding the intensity of spatiotemporal sampling required to sufficiently document case reduction and landscape-level elimination. A recent paper that analyzed decades of case data for raccoon RABV in the Eastern United States demonstrated that the intensity of sampling required is a function of the surveillance method or strategy (Davis, Kirby, et al., 2019).

Screening for RABV-specific binding (RVBA) or neutralizing antibodies (RVNA) in serum collected from wild carnivores in rabies management areas serves as an index to vaccine-induced population immunity and is the other primary metric besides surveillance for monitoring ORV campaigns in Europe and North America (EC, 2017; Müller, Freuling, et al., 2015; Rosatte, 2013; Rosatte, Tinline, et al., 2007). Inference may be enhanced if sampling occurs both pre- and postvaccine distribution (Brown et al., 2011; Fehlner-Gardiner, 2018; WHO, 2018), as naturally occurring RABV antibodies can be found in ORV naïve reservoir populations can complicate interpretation of management impact (Gilbert, Johnson, Nelson, Chipman, et al., 2018). Depending on the type of live attenuated or recombinant vaccine used for ORV, animal trapping for serologic sample collection should begin 2–8 weeks postbaiting (Brown et al., 2011; Fehlner-Gardiner, 2018; Moore et al., 2017; Sattler et al., 2009; WHO, 2018). Approved methods for quantitation of RABV-specific antibodies include the Rapid Fluorescent Focus Inhibition Test (RFFIT), Fluorescent Antibody Viral Neutralization (FAVN) test, or an enzyme-linked immunosorbent assay (ELISA) (OIE, 2018). For wildlife serology associated with ORV, the ELISA may have advantages over RFFIT and FAVN, including better reproducibility and a stronger correlate of protection against a lethal infection (Moore et al., 2017). Several ELISAs have been tested with wildlife sera (Cliquet et al., 2003; Cliquet, Sagne, Schereffer, & Aubert, 2000; Wasniewski et al., 2013), filter paper methods have been developed for the collection of blood samples (Wasniewski, Barrat, Combes, Guiot, & Cliquet, 2014), and interlaboratory test routines have enhanced laboratory standardization of ORV monitoring by serology (Wasniewski et al., 2016). An RVNA titer of 0.5 IU/mL is recognized as a specific and adequate immune response in humans to rabies vaccination by the WHO (1992). European and Canadian rabies management programs utilize a RVBA level equivalent to 0.5 IU/mL as evidence of adequate vaccine-induced seroconversion in animals (Cliquet, Aubert, & Sagne, 1998). In the United States, there is a lack of standardized serologic testing for monitoring ORV campaigns, including variable positive threshold cutoffs across laboratories (Wohlers et al., 2018). The USDA NRMP has historically used a level of 0.05–0.06 IU/mL as an index to population immunity for ORV program monitoring in the United States, based on an objective to maximize the assay sensitivity for detecting vaccinated animals as well as evidence that raccoons exposed to ORV under field conditions with RVNA titers less than 0.5 IU/mL may resist a lethal challenge (Blanton et al., 2018; Moore et al., 2017).

In addition to documenting the incidence of rabies and seroconversion rates in target populations, biomarkers incorporated into the oral vaccine bait matrix can provide an indirect measure of bait consumption (Rosatte et al., 2008; Slate & Rupprecht, 2012). Bait uptake data from Ohio, New Jersey, and Ontario suggest that 80%–100% of distributed vaccine baits were consumed within 1–2 weeks (Blackwell et al., 2004; Brown et al., 2011; Rosatte & Lawson, 2001; Roscoe et al., 1998). Bait uptake can be assessed by detection of a tetracycline biomarker in bone and tooth usually by sampling the second premolar from immobilized animals or cuspid (i.e., canine) teeth from euthanized or hunted animals (Algeo et al., 2013; Bachmann et al., 1990; Johnston et al., 1987; Robardet et al., 2012; Sidwa et al., 2005). The cuspid teeth generally have higher tetracycline deposition than the first and second premolar teeth, but removal of cuspid teeth is invasive and is more likely to negatively impact survival of the released animal (Algeo et al., 2013; Slate et al., 2009). Biomarker monitoring data for raccoons and skunks are variable and do not always exceed seroconversion rates in sampled animals (Fehlner-Gardiner, 2018; Slate et al., 2009).

Biomarkers can play an important role in evaluation of bait uptake, but sampling requires capture and immobilization of, as well as tooth extraction from, target carnivore hosts, increasing the risk of exposure and costs for program monitoring (Slate et al., 2009). Public health and environmental concerns over antibiotic resistance have been raised in a number of countries (e.g., Canada and Germany) over the continued use of tetracycline as a biomarker in vaccine baits (Müller, Batza, et al., 2012; Slate & Rupprecht, 2012). Germany was one of the first countries to discontinue the use of tetracycline as a biomarker in 1998 over food safety concerns related to the consumption of wild game meat harvested in ORV management zones (Müller, Batza, et al., 2012). Biomarkers requiring less invasive sampling methods and suitable for nonlethal monitoring of target species have been described, where RVNA and biomarker detection is possible from serum (Baer, Shaddock, Hayes, & Savarie, 1985; Berentsen, Sugihara, et al., 2019). However, there is a need for more effective monitoring and evaluation tools that reduce the need to capture and process animals and that simplify the real-time collection of data to inform science-based decision making for wildlife rabies management (e.g., Robardet, Rieder, Barrat, & Cliquet, 2019).

19.5 Bait and vaccine principles, research and developments

One of the key components of rabies control programs in Europe and North America has been the availability of safe and highly potent vaccines (Müller, Schroder, et al., 2015; Rupprecht et al., 2008). Collaboration with industry has facilitated the development and availability of cost-effective commercially produced vaccine baits. More than 10 different oral vaccines have been used for wildlife rabies management in Europe and 3 have been used in North America, which represents successful culmination of decades of research on rabies bait and vaccine development (Fehlner-Gardiner, 2018; Müller, Schroder, et al., 2015) (Table 19.2).

19.5.1 Vaccines

Four routes of vaccination were considered early in the development of control strategies for wildlife (Wandeler, 1991; Winkler, 1992). While parenteral route is the most efficient form

TABLE 19.2 A brief comparison of the main rabies vaccine bait products that have been utilized for oral rabies vaccination targeting wild carnivores.

Product name	Construct type	Construct virus	Strain	Licensing/use	Residual pathogenicity
LYSVULPEN	Attenuated	Rabies virus	Mix of SAD Bern and SAD B19-like	Europe (red fox, raccoon dog)	Some, but infrequent among target hosts
FUCHSORAL	Attenuated	Rabies virus	SAD B19	Europe (red fox, raccoon dog)	Some, but infrequent among target hosts
RABIGEN	Attenuated, selected	Rabies virus	SAG2 (SAD Bern)	Europe (red fox, raccoon dog)	None reported
RABITEC	Attenuated, recombinant	Rabies virus	SPBN-GAS-GAS (SAD B19 clone)	Europe (red fox, raccoon dog)	None reported
RABORAL V-RG	Recombinant	Vaccinia virus	ERA glycoprotein	Europe (red fox, raccoon dog); United States (raccoons, coyotes)	None reported
ONRAB	Recombinant	Human adenovirus type 5	ERA glycoprotein	Canada and United States[a] (striped skunks)	None reported

[a] *Experimental use in the United States.*

of delivery to animals, it is only a practical strategy for animals that are largely accessible on the landscape such as companion animals and livestock. Only for a short time was a modified "coyote-getter" device considered for automated parenteral injection of lyophilized vaccine to wild carnivores. Consequently, delivery of vaccines to wildlife had to be successful in the absence of trapping or handling animals directly. Fumigation of dens with vaccine aerosol instead of poisons was also considered for a short time, but the enhanced pathogenicity of first-generation attenuated RABVs by intranasal route for target animals and human safety concerns led to the discontinuation of research involving this method. Oral delivery of live attenuated vaccines was demonstrated to be relatively safe and effective, but the vaccine must make adequate contact with the host oropharyngeal tissues due to the destructive enzymatic activity and pH in the stomach. In particular, infection of the palatine tonsils may be critical for successful uptake in target animals (Baer, 1975; Orciari et al., 2001; Rupprecht, Hamir, Johnston, & Koprowski, 1988; Thomas et al., 1990; Vos et al., 2017, 1999). Although delivery of live attenuated RABV vaccines by intestinal route was reported to be highly immunogenic, the encapsulation for protection against passage through the stomach proved extremely challenging and this strategy was mostly abandoned (reviewed in Wandeler, 1991; Winkler, 1992). Delivery of vaccines by oral route also had the advantage of extensive prior research on attractive baits and delivery systems for target animals, which had developed as vehicles for lethal population control.

Oral rabies vaccines must be potent, innocuous to man, exhibit negligible excretion in the infected host and correspondingly low horizontal transmission risk, thermostable for several days at ambient temperatures, genetically stable concerning reversion to a virulent phenotype, free of contaminants, and relatively inexpensive to produce (Bögel, 1992b; Wandeler, 1991). All vaccines intended for use within an ORV framework are live virus vectors that infect the host and exhibit a limited degree of replication and expression of the RABV glycoprotein. The RABV glycoprotein is the primary protein that interacts with host cell and is responsible for the induction of RVNA (Dietzschold, Li, Faber, & Schnell, 2008). Live RABV vaccines also likely lead to expression of other RABV proteins besides the glycoprotein, which may enhance the protective effects of those vaccines (Dietzschold et al., 1987; Hooper et al., 1998).

The earliest rabies vaccines considered for use as part of control strategies targeting mass vaccination of wildlife were based on highly attenuated RABV derived from the Street-Alabama-Dufferin (SAD) strain, originally isolated from a rabid dog in Alabama in 1935. The SAD virus was passaged multiple times and later renamed Evelyn Rokitniki Abelseth (ERA) (Abelseth, 1964). The ERA strain grown on primary porcine kidney cells was marketed commercially as a parenteral domestic animal vaccine. Another ERA strain grown on baby hamster kidney (BHK)-21 cells (ERA-BHK21) was shipped to three European laboratories in 1972 where it was later renamed SAD Berne and became the progenitor for several subsequent oral live attenuated and recombinant RABV-based vaccines targeting wildlife. First-generation RABV vaccines for ORV were shown to have residual pathogenic effects in some target animals (especially skunks) and rodents (Artois et al., 1992; Cliquet et al., 2015; Fehlner-Gardiner et al., 2008; Hostnik, Picard-Meyer, Rihtaric, Toplak, & Cliquet, 2014; Müller et al., 2009; Pfaff et al., 2018; Rupprecht et al., 1990; Tolson, Charlton, Lawson, Campbell, & Stewart, 1988; Vos et al., 1999; Vuta et al., 2016), although arguably this did not preclude the utility of these vaccines in broad-scale elimination campaigns targeting red foxes (Geue et al., 2008; MacInnes et al., 2001; Müller, Schroder, et al., 2015). A meeting of experts was held in Tunis during 1983, which was influential in describing the antigenic structure of RABV and transferred key knowledge for bioengineering of safer next-generation RABV vaccines and to enhance detection and discrimination of vaccine-induced rabies cases associated with ORV in the field. Notably, the pathogenicity of the ERA strain could be significantly reduced by mutations at position 333 of the glycoprotein (Dietzschold et al., 1983), which became a key target for modification with second-generation vaccines, although RABV pathogenicity is likely a multigenic trait (Faber et al., 2004; Geue et al., 2008). Second-generation live RABV vaccines targeted changes at position 333, and include SAD stain mutants selected under monoclonal antibodies pressure, such as SAG1 and SAG2 (Lafay, Benejean, Tuffereau, Flamand, & Coulon, 1994; Le Blois et al., 1990). The SAG2 construct has been demonstrated to be safe and effective in the lab and field with a diverse array of wild carnivore species (Mahl et al., 2014). Third-generation RABV vaccines involve recombinants created by site-directed mutagenesis, such as the SPBN platform (Dietzschold et al., 2004; Faber et al., 2002) and ERA-g333 (Bankovskiy, Safonov, & Kurilchuk, 2008). Recombinant viruses that only express the ERA glycoprotein were also developed, including recombinant vaccinia virus rabies glycoprotein (Kieny et al., 1984) and recombinant human adenovirus type 5 rabies glycoprotein (Prevec, Campbell, Christie, Belbeck, & Graham, 1990). The recombinant vaccinia virus rabies glycoprotein construct has been successfully used in Europe and the United States targeting canids and raccoons (Maki et al., 2017). During the last decade, several oral rabies vaccine studies have been reported for diverse wild carnivores using third-generation vaccines (Brown, Rosatte, Fehlner-Gardiner, Bachmann, et al., 2014; Brown, Rosatte, Fehlner-Gardiner, Ellison, et al., 2014; Brown et al., 2012; Choi et al., 2015; Follmann, Ritter, Swor, Dunbar, & Hueffer, 2011; Freuling et al., 2017; Gilbert et al., 2018; Smith et al., 2017; Vos et al., 2017, 2013; Yang et al., 2016).

Other expression systems that were described, but did not advance to ORV field trial testing, include recombinant baculovirus (Prehaud, Takehara, Flamand, & Bishop, 1989), canine adenovirus type 2 (Hamir, Raju, & Rupprecht, 1992; Henderson et al., 2009; Sumner, Shaddock, Wu, & Baer, 1988; Tordo et al., 2008), and raccoonpox virus (Esposito, Knight, Shaddock, Novembre, & Baer, 1988). A canarypox RABV vaccine was developed for parenteral delivery to domestic animals (Taylor et al., 1995, 1991; Taylor & Paoletti, 1988; Taylor, Weinberg, Languet, Desmettre, & Paoletti, 1988), but was not evaluated in wild carnivores. Future delivery systems may benefit from improved vaccine preservation methods (Smith, Siirin, Wu, Hanlon, & Bronshtein, 2015), or consideration of controversial strategies such as transmissible rabies vaccines (Basinski, Nuismer, & Remien, 2019; Murphy, Redwood, & Jarvis, 2016).

19.5.2 Baits

Bait development targeting wild carnivores was initiated when population reduction techniques were being implemented for rabies control and where baits were vehicles for poisons and toxicants (Lewis, 1975; Linhart, Kappeler, & Windberg, 1993). Baits tended to be optimized to individual species given that the development of optimal multihost baits is challenging. Previous reviews have been published (Linhart, Kappeler, et al., 1993; Perry et al., 1988) in addition to individual case studies for certain target species (Farry, Henke, Anderson, & Fearneyhough, 1998; Knobel et al., 2002; Linhart et al., 1991; Linhart, King, et al., 1997; Linhart et al., 2002; Perry, Garner, Jenkins, McCloskey, & Johnston, 1989; Rosatte & Lawson, 2001; Steelman, Henke, & Moore, 2000).

The research and development of bait vehicles often begins with palatability testing in a captive environment, along with assessment of target and nontarget visitation and fate of placebo formats in a field setting, with the aid of a biomarker (e.g., Hadidian et al., 1989). The most common biomarker utilized has been tetracycline, although Rhodamine-B and iophenoxic acid have also been commonly used to mark wildlife (Ballesteros et al., 2013; Fisher, 1999). Candidate baits containing vaccine also must be subjected to efficacy testing with target hosts in captivity and vaccines must further undergo extensive safety testing for target and nontarget animals per regulatory requirements of the country or region and international guidelines (OIE, 2018). Effective and safe vaccines can then be evaluated in field studies targeting delivery to target host populations, also known as field trials (Brochier et al., 1988; Hanlon et al., 1998; Johnston et al., 1988; Pastoret, Boulanger, & Brochier, 1995; Slate et al., 2014). Upon demonstration of consistent and reliable marking and population immunity results from the operational use of a vaccine bait, biomarkers may eventually be discontinued. Early testing revealed potential virucidal effects of mixing live attenuated RABV vaccines directly with the bait matrix material, and all ORV bait products have maintained a separate container for liquid vaccine that is enclosed and protected by a bait matrix containing attractants and biomarker. Thermostability is an important guiding principle for development of vaccine baits for field use, given that distribution commonly targets spring and autumn seasons and considering that free-ranging target carnivores may require up to 2 weeks to locate and consume baits in the environment (Bachmann et al., 1990; Linhart, Kappeler, et al., 1993).

The earliest ORV bait vehicles for red foxes were chicken heads sourced from abattoirs and assembled with vaccine capsules by hand. Other early bait types evaluated for red foxes included commercial sausage (Winkler & Baer, 1976) and sponge baits coated with paraffin and tallow (Bachmann et al., 1990). The first manufactured bait comprised meal from slaughtered livestock was developed in Germany and used operationally starting in 1985, which was a major milestone that facilitated scaling-up of ORV activities. However, concerns about bovine spongiform encephalopathy and poultry diseases led to eventual phasing out of livestock and poultry-derived ingredients in bait formulations in Europe, and fish attractants or meal are now the most common bait ingredients used in production (Kappeler, 1992; Müller, Schroder, et al., 2015). The main ORV bait products used in contemporary time compromise a mix of first-, second-, and third-generation rabies vaccines (Table 19.2).

19.6 Conclusions

The control and elimination of RABV in wild carnivores has been demonstrated throughout broad landscapes in Europe and North America primarily using an ORV approach. National-level operations have also been successful in the control and elimination of RABV from wild carnivores in a number of Asian countries. While tools, methods, and guidelines in support of wildlife RABV elimination programs are available, lessons learned from successes and failures over the last 40 years illustrate the importance of long-term planning and funding, as well as the importance of intersectoral and multilateral partnerships, collaboration, and engagement toward a common goal. Timely and transparent communication of information among partners is critical and adaptive management frameworks are required to enhance and refine program operations targeting one or more target species. The final steps in the elimination process are cost and labor intensive, yet maintaining a high level of vigilance postelimination is also critical, particularly where RABV circulates in target host populations in neighboring areas and translocation of RABV reservoirs and vectors may occur.

Acknowledgments

This work was supported by the US Department of Agriculture, Wildlife Services National Rabies Management Program. The authors extend a special thanks to Dr. Ad Vos for providing constructive feedback that improved the chapter.

References

Abelseth, M. K. (1964). Propagation of rabies virus in pig kidney cell culture. *The Canadian Veterinary Journal*, 5(4), 84–87.

Algeo, T. P., Norhenberg, G., Hale, R., Montoney, A., Chipman, R. B., & Slate, D. (2013). Oral rabies vaccination variation in tetracycline biomarking among Ohio raccoons. *Journal of Wildlife Diseases*, 49(2), 332–337. https://doi.org/10.7589/2011-11-327.

Anderson, R. M., Jackson, H. C., May, R. M., & Smith, A. M. (1981). Population dynamics of fox rabies in Europe. *Nature, 289*(5800), 765–771.

Artois, M., Delahay, R., Guberti, V., & Cheeseman, C. (2001). Control of infectious diseases of wildlife in Europe. *Veterinary Journal, 162*(2), 141–152. https://doi.org/10.1053/tvjl.2001.0601.

Artois, M., Guittre, C., Thomas, I., Leblois, H., Brochier, B., & Barrat, J. (1992). Potential pathogenicity for rodents of vaccines intended for oral vaccination against rabies: A comparison. *Vaccine, 10*(8), 524–528.

Aubert, M. (1994). Control of rabies in foxes: What are the appropriate measures? *Veterinary Record, 134*, 55–59.

Bachmann, P., Bramwell, R. N., Fraser, S. J., Gilmore, D. A., Johnston, D. H., Lawson, K. F., … Voigt, D. R. (1990). Wild carnivore acceptance of baits for delivery of liquid rabies vaccine. *Journal of Wildlife Diseases, 26*(4), 486–501. https://doi.org/10.7589/0090-3558-26.4.486.

Badrane, H., & Tordo, N. (2001). Host switching in Lyssavirus history from the Chiroptera to the Carnivora orders. *Journal of Virology, 75*(17), 8096–8104.

Baer, G. M. (1975). Wildlife vaccination. In G. M. Baer (Ed.), *Vol. 1. The natural history of rabies* (pp. 261–266). New York, NY: Academic Press, Inc.

Baer, G. M., Abelseth, M. K., & Debbie, J. G. (1971). Oral vaccination of foxes against rabies. *American Journal of Epidemiology, 93*(6), 487–490.

Baer, G. M., Shaddock, J. H., Hayes, D. J., & Savarie, P. J. (1985). Iophenoxic acid as a serum marker in carnivores. *Journal of Wildlife Management, 49*(1), 49–51.

Baghi, H. B., Alinezhad, F., Kuzmin, I., & Rupprecht, C. (2018). A perspective on rabies in the Middle East—Beyond neglect. *Veterinary Sciences, 5*(3), 1–18.

Baker, L., Matthiopoulos, J., Muller, T., Freuling, C., & Hampson, K. (2019). Optimizing spatial and seasonal deployment of vaccination campaigns to eliminate wildlife rabies. *Philosophical Transactions of the Royal Society of London. Series B, Biological Sciences, 374*(1776), 20180280. https://doi.org/10.1098/rstb.2018.0280.

Ballantyne, E. E. (1956). Rabies control program in Alberta. *Canadian Journal of Comparative Medicine and Veterinary Science, 20*(1), 21–30.

Ballantyne, E. E., & O'Donoghue, J. G. (1954). Rabies control in Alberta. *Journal of the American Veterinary Medical Association, 125*(931), 316–326.

Ballesteros, C., Sage, M., Fisher, P., Massei, G., Mateo, R., de La Fuente, J., … Gortazar, C. (2013). Iophenoxic acid as a bait marker for wild mammals: Efficacy and safety considerations. *Mammal Review, 43*(2), 156–166.

Bankovskiy, D., Safonov, G., & Kurilchuk, Y. (2008). Immunogenicity of the ERA G 333 rabies virus strain in foxes and raccoon dogs. *Developmental Biology (Basel), 131*, 461–466.

Basinski, A. J., Nuismer, S. L., & Remien, C. H. (2019). A little goes a long way: Weak vaccine transmission facilitates oral vaccination campaigns against zoonotic pathogens. *PLoS Neglected Tropical Diseases, 13*(3), e0007251. https://doi.org/10.1371/journal.pntd.0007251.

Beasley, J. C., Olson, Z. H., Beatty, W. S., Dharmarajan, G., & Rhodes, O. E., Jr. (2013). Effects of culling on mesopredator population dynamics. *PLoS One, 8*(3), e58982. https://doi.org/10.1371/journal.pone.0058982.

Bengoumi, M., Mansouri, R., Ghram, B., & Merot, J. (2018). Rabies in North Africa and the Middle East: Current situation, strategies and outlook. *Revue Scientifique et Technique, 37*(2), 497–510. https://doi.org/10.20506/rst.37.2.2818.

Berentsen, A., Chipman, R., Nelson, K., Gruver, K., Boyd, F., Volker, S., … Gilbert, A. (2019). Placebo oral rabies vaccine uptake by small Indian mongooses (*Herpestes auropunctatus*) in southwestern Puerto Rico. *Journal of Wildlife Diseases*. https://doi.org/10.7589/2019-03-077 [in press].

Berentsen, A., Ellis, C., Johnson, S., Leinbach, I., Sugihara, R., & Gilbert, A. (2020). Immunogenicity of ontario rabies vaccine for small Indian mongooses (*Herpestes auropunctatus*). *Journal of Wildlife Diseases, 56*(1), 224–228.

Berentsen, A. R., Sugihara, R. T., Payne, C. G., Leinbach, I., Volker, S. F., Vos, A., … Gilbert, A. T. (2019). Analysis of iophenoxic acid analogues in small indian mongoose (*Herpestes auropunctatus*) sera for use as an oral rabies vaccination biological marker. *Journal of Visualized Experiments*, (147). https://doi.org/10.3791/59373.

Berentsen, A., Torres-Toledo, F., Rivera-Rodriguez, M., Davis, A., & Gilbert, A. (2018). Population density and home range estimates of black rat (*Rattus rattus*) populations in southwestern Puerto Rico. In: *Proceedings of the 28th Vertebrate Pest Conference*, pp. 299–303.

Biek, R., Henderson, J. C., Waller, L. A., Rupprecht, C. E., & Real, L. A. (2007). A high-resolution genetic signature of demographic and spatial expansion in epizootic rabies virus. *Proceedings of the National Academy of Sciences of the United States of America, 104*(19), 7993–7998.

Bingham, J., Kappeler, A., Hill, F. W., King, A. A., Perry, B. D., & Foggin, C. M. (1995). Efficacy of SAD (Berne) rabies vaccine given by the oral route in two species of jackal (*Canis mesomelas* and *Canis adustus*). *Journal of Wildlife Diseases, 31*(3), 416–419.

Bingham, J., Perry, B. D., King, A. A., Schumacher, C. L., Aubert, M., Kappeler, A., ... Aubert, A. (1993). Oral rabies vaccination of jackals: Progress in Zimbabwe. *The Onderstepoort Journal of Veterinary Research, 60*(4), 477–478.

Bingham, J., Schumacher, C. L., Hill, F. W., & Aubert, A. (1999). Efficacy of SAG-2 oral rabies vaccine in two species of jackal (*Canis adustus* and *Canis mesomelas*). *Vaccine, 17*(6), 551–558.

Bjorklund, B. M., Haley, B. S., Bevilacqua, R. J., Chandler, M. D., Duffiney, A. G., von Hone, K. W., ... Algeo, T. P. (2017). Progress towards bait station integration into oral rabies vaccination programs in the United States: Field trials in Massachusetts and Florida. *Tropical Medicine and Infectious Disease, 2*(3). https://doi.org/10.3390/tropicalmed2030040.

Black, J. G., & Lawson, K. F. (1970). Sylvatic rabies studies in the silver fox (*Vulpes vulpes*). Susceptibility and immune response. *Canadian Journal of Comparative Medicine, 34*(4), 309–311.

Black, J. G., & Lawson, K. F. (1973). Further studies of sylvatic rabies in the fox (*Vulpes vulpes*). Vaccination by the oral route. *The Canadian Veterinary Journal, 14*(9), 206–211.

Blackwell, B. F., Seamans, T. W., White, R. J., Patton, Z. J., Bush, R. M., & Cepek, J. D. (2004). Exposure time of oral rabies vaccine baits relative to baiting density and raccoon population density. *Journal of Wildlife Diseases, 40*(2), 222–229. https://doi.org/10.7589/0090-3558-40.2.222.

Blanton, J. D., Meadows, A., Murphy, S. M., Manangan, J., Hanlon, C. A., Faber, M. L., ... Rupprecht, C. E. (2006). Vaccination of small Asian mongoose (*Herpestes javanicus*) against rabies. *Journal of Wildlife Diseases, 42*(3), 663–666. https://doi.org/10.7589/0090-3558-42.3.663.

Blanton, J. D., Niezgoda, M., Hanlon, C. A., Swope, C. B., Suckow, J., Saidy, B., ... Slate, D. (2018). Evaluation of oral rabies vaccination: Protection against rabies in wild caught raccoons (*Procyon lotor*). *Journal of Wildlife Diseases, 54*(3), 520–527. https://doi.org/10.7589/2017-01-007.

Bögel, K. (1992a). Introduction. In K. Bögel, F. M. Meslin, & M. M. Kaplan (Eds.), *Wildlife rabies control* (pp. 3–6). Kent: Wells Medical Ltd.

Bögel, K. (1992b). Principles and obstacles to immunization of wildlife. In K. Bögel, F. M. Meslin, & M. M. Kaplan (Eds.), *Wildlife rabies control* (pp. 97–99). Kent: Wells Medical Ltd.

Bögel, K., Arata, A. A., Moegle, H., & Knorpp, F. (1974). Recovery of fox populations in rabies control 1. *Zentralblatt für Veterinärmedizin Reihe B, 21*(6), 401–412.

Bögel, K., Meslin, F. M., & Kaplan, M. M. (1992). *Wildlife rabies control*. Kent: Wells Medical Ltd.

Bögel, K., Moegle, H., Steck, F., Krocza, W., & Andral, L. (1981). Assessment of fox control in areas of wildlife rabies. *Bulletin of the World Health Organization, 59*(2), 269–279.

Boulanger, J. R., Bigler, L. L., Curtis, P. D., Lein, D. H., & Lembo, A. J., Jr. (2008). Comparison of suburban vaccine distribution strategies to control raccoon rabies. *Journal of Wildlife Diseases, 44*(4), 1014–1023.

Boyer, J. P., Canac-Marquis, P., Guerin, D., Mainguy, J., & Pelletier, F. (2011). Oral vaccination against raccoon rabies: Landscape heterogeneity and timing of distribution influence wildlife contact rates with the ONRAB vaccine bait. *Journal of Wildlife Diseases, 47*(3), 593–602. https://doi.org/10.7589/0090-3558-47.3.593.

Broadfoot, J. D., Rosatte, R. C., & O'Leary, D. T. (2001). Raccoon and skunk population models for urban disease control planning in Ontario, Canada. *Ecological Applications, 11*, 295–303.

Brochier, B., Aubert, M. F., Pastoret, P. P., Masson, E., Schon, J., Lombard, M., ... Desmettre, P. (1996). Field use of a vaccinia-rabies recombinant vaccine for the control of sylvatic rabies in Europe and North America. *Revue Scientifique et Technique, 15*(3), 947–970.

Brochier, B., Kieny, M. P., Costy, F., Coppens, P., Bauduin, B., Lecocq, J. P. ... (1991). Large-scale eradication of rabies using recombinant vaccinia-rabies vaccine. *Nature, 354*(6354), 520–522. https://doi.org/10.1038/354520a0.

Brochier, B., Thomas, I., Iokem, A., Ginter, A., Kalpers, J., Paquot, A., ... Pastoret, P. P. (1988). A field trial in Belgium to control fox rabies by oral immunisation. *Veterinary Record, 123*(24), 618–621.

Brown, L. J., Rosatte, R. C., Fehlner-Gardiner, C., Bachmann, P., Ellison, J. A., Jackson, F. R., ... Donovan, D. (2014). Oral vaccination and protection of red foxes (*Vulpes vulpes*) against rabies using ONRAB, an adenovirus-rabies recombinant vaccine. *Vaccine, 32*(8), 984–989. https://doi.org/10.1016/j.vaccine.2013.12.015.

Brown, L. J., Rosatte, R. C., Fehlner-Gardiner, C., Ellison, J. A., Jackson, F. R., Bachmann, P., ... Donovan, D. (2014). Oral vaccination and protection of striped skunks (*Mephitis mephitis*) against rabies using ONRAB[(R)]. *Vaccine, 32*(29), 3675–3679. https://doi.org/10.1016/j.vaccine.2014.04.029.

Brown, L. J., Rosatte, R. C., Fehlner-Gardiner, C., Knowles, M. K., Bachmann, P., Davies, J. C., ... Donovan, D. (2011). Immunogenicity and efficacy of two rabies vaccines in wild-caught, captive raccoons. *Journal of Wildlife Diseases, 47*(1), 182–194. https://doi.org/10.7589/0090-3558-47.1.182.

Brown, L. J., Rosatte, R. C., Fehlner-Gardiner, C., Taylor, J. S., Davies, J. C., & Donovan, D. (2012). Immune response and protection in raccoons (*Procyon lotor*) following consumption of baits containing ONRAB(R), a human adenovirus rabies glycoprotein recombinant vaccine. *Journal of Wildlife Diseases, 48*(4), 1010–1020. https://doi.org/10.7589/2012-01-023.

Canadian Food Inspection Agency (CFIA) (2019). *Rabies in Canada*. Retrieved from: http://www.inspection.gc.ca/animals/terrestrial-animals/diseases/reportable/rabies/rabies-in-canada/eng/1519159995664/1519159996478.

Carnieli, P., Jr., Fahl Wde, O., Castilho, J. G., de Novaes Oliveira, R., Macedo, C. I., Durymanova, E., … Kotait, I. (2008). Characterization of rabies virus isolated from canids and identification of the main wild canid host in Northeastern Brazil. *Virus Research, 131*(1), 33–46. https://doi.org/10.1016/j.virusres.2007.08.007.

Chang, S. S., Tsai, H. J., Chang, F. Y., Lee, T. S., Huang, K. C., Fang, K. Y., … Fei, C. Y. (2016). Government response to the discovery of a rabies virus reservoir species on a previously designated rabies-free island, Taiwan, 1999–2014. *Zoonoses and Public Health, 63*(5), 396–402. https://doi.org/10.1111/zph.12240.

Charlton, K. M., Webster, W. A., & Casey, G. A. (1991). Skunk rabies. In G. M. Baer (Ed.), *The natural history of rabies* (2nd ed., pp. 307–324). Boca Raton, FL: CRC Press, Inc.

Childs, J. E., Curns, A. T., Dey, M. E., Real, L. A., Feinstein, L., Bjornstad, O. N., & Krebs, J. W. (2000). Predicting the local dynamics of epizootic rabies among raccoons in the United States. *Proceedings of the National Academy of Sciences of the United States of America, 97*(25), 13666–13671. https://doi.org/10.1073/pnas.240326697.

Chipman, R., Slate, D., Rupprecht, C. E., & Mendoza, M. (2008). Downside risk of wildlife translocation. *Developments in Biologicals, 131*, 223–232.

Choi, J., Yang, D. K., Kim, H. H., Jo, H. Y., Choi, S. S., Kim, J. T., … Kim, H. W. (2015). Application of recombinant adenoviruses expressing glycoprotein or nucleoprotein of rabies virus to Korean raccoon dogs. *Clinical and Experimental Vaccine Research, 4*(2), 189–194. https://doi.org/10.7774/cevr.2015.4.2.189.

Christian, K. A., Blanton, J. D., Auslander, M., & Rupprecht, C. E. (2009). Epidemiology of rabies post-exposure prophylaxis—United States of America, 2006-2008. *Vaccine, 27*(51), 7156–7161. https://doi.org/10.1016/j.vaccine.2009.09.028.

Clark, K. A., Neill, S. U., Smith, J. S., Wilson, P. J., Whadford, V. W., & McKirahan, G. W. (1994). Epizootic canine rabies transmitted by coyotes in south Texas. *Journal of the American Veterinary Medical Association, 204*(4), 536–540.

Clayton, T., Duke-Sylvester, S., Gross, L. J., Lenhart, S., & Real, L. A. (2009). Optimal control of a rabies epidemic model with a birth pulse. *Journal of Biological Dynamics*. https://doi.org/10.1080/17513750902935216.

Cliquet, F., Aubert, M., & Sagne, L. (1998). Development of a fluorescent antibody virus neutralisation test (FAVN test) for the quantitation of rabies-neutralising antibody. *Journal of Immunological Methods, 212*(1), 79–87. https://doi.org/10.1016/S0022175997002123.

Cliquet, F., Freuling, C., Smreczak, M., Van der Poel, W. H. M., Horton, D., Fooks, A. R., … Müller, T. F. (2010). Development of harmonised schemes for monitoring and reporting of rabies in animals in the European Union. *EFSA Supporting Publications, 7*(67e), 1–60. https://doi.org/10.2903/sp.efsa.2010.EN-67.

Cliquet, F., Müller, T., Mutinelli, F., Geronutti, S., Brochier, B., Selhorst, T., … Aubert, M. (2003). Standardisation and establishment of a rabies ELISA test in European laboratories for assessing the efficacy of oral fox vaccination campaigns. *Vaccine, 21*(21–22), 2986–2993.

Cliquet, F., Picard-Meyer, E., Mojzis, M., Dirbakova, Z., Muizniece, Z., Jaceviciene, I., … Celer, V. (2015). In-depth characterization of live vaccines used in Europe for oral rabies vaccination of wildlife. *PLoS One, 10*(10), e0141537. https://doi.org/10.1371/journal.pone.0141537.

Cliquet, F., Sagne, L., Schereffer, J. L., & Aubert, M. F. (2000). ELISA test for rabies antibody titration in orally vaccinated foxes sampled in the fields. *Vaccine, 18*(28), 3272–3279. https://doi.org/10.1016/S0264-410X(00)00127-4.

Cocozza, J., & Malaga-Alba, A. (1962). Wildlife control project in Baja California. *Public Health Reports, 77*(2), 147–151.

Coleman, P. G., & Dye, C. (1996). Immunization coverage required to prevent outbreaks of dog rabies. *Vaccine, 14*(3), 185–186.

Cordeiro R. de, A., Duarte, N. F., Rolim, B. N., Soares Junior, F. A., Franco, I. C., Ferrer, L. L., … Sidrim, J. J. (2016). The importance of wild canids in the epidemiology of rabies in Northeast Brazil: A retrospective study. *Zoonoses and Public Health, 63*(6), 486–493. https://doi.org/10.1111/zph.12253.

Correa-Giron, E. P., Allen, R., & Sulkin, S. E. (1970). The infectivity and pathogenesis of rabiesvirus administered orally. *American Journal of Epidemiology, 91*(2), 203–215.

Coulon, P., Lafay, F., LeBlois, H., Tuffereau, C., Artois, M., Blancou, J., ... Flamand, A. (1992). The SAG: A new attenuated oral rabies vaccine. In K. Bogel, F. M. Meslin, & M. M. Kaplan (Eds.), *Wildlife rabies control* (pp. 105–111). Kent: Wells Medical Ltd.

Coyne, M. J., Smith, G., & McAllister, F. E. (1989). Mathematic model for the population biology of rabies in raccoons in the mid-Atlantic states. *American Journal of Veterinary Research, 50*(12), 2148–2154.

Craven, S., Barnes, T., & Kania, G. (1998). Toward a professional position on the translocation of problem wildlife. *Wildlife Society Bulletin, 26*(1), 171–177.

Davis, A. J., Kirby, J. D., Chipman, R. B., Nelson, K. M., Xifara, T., Webb, C. T., ... Pepin, K. M. (2019). Not all surveillance data are created equal—Prioritizing methods for esimation of rabies virus elimination from wildlife. *Journal of Applied Ecology, 56*, 2551–2561.

Davis, R., Nadin-Davis, S. A., Moore, M., & Hanlon, C. (2013). Genetic characterization and phylogenetic analysis of skunk-associated rabies viruses in North America with special emphasis on the central plains. *Virus Research, 174* (1–2), 27–36. https://doi.org/10.1016/j.virusres.2013.02.008.

Davis, A. J., Nelson, K. M., Kirby, J. D., Wallace, R., Ma, X., Pepin, K. M., ... Gilbert, A. T. (2019). Rabies surveillance identifies potential risk corridors and enables management evaluation. *Viruses, 11*(11), 1006.

Debbie, J. G. (1991). Rabies control of terrestrial wildlife by population reduction. In G. M. Baer (Ed.), *The natural history of rabies* (2nd ed., pp. 477–484). Boca Raton, FL: CRC Press, Inc.

Debbie, J. G., Abelseth, M. K., & Baer, G. M. (1972). The use of commercially available vaccines for the oral vaccination of foxes against rabies. *American Journal of Epidemiology, 96*(3), 231–235.

Demetriou, P., & Moynagh, J. (2011). The European Union strategy for external cooperation with neighbouring countries on rabies control. *Rabies Bulletin Europe, 35*(1), 5–7.

Deviatkin, A. A., Lukashev, A. N., Poleshchuk, E. M., Dedkov, V. G., Tkachev, S. E., Sidorov, G. N., ... Shipulin, G. A. (2017). The phylodynamics of the rabies virus in the Russian Federation. *PLoS One, 12*(2), e0171855. https://doi.org/10.1371/journal.pone.0171855.

Dietzschold, M. L., Faber, M., Mattis, J. A., Pak, K. Y., Schnell, M. J., & Dietzschold, B. (2004). In vitro growth and stability of recombinant rabies viruses designed for vaccination of wildlife. *Vaccine, 23*(4), 518–524. https://doi.org/10.1016/j.vaccine.2004.06.031.

Dietzschold, B., Li, J., Faber, M., & Schnell, M. (2008). Concepts in the pathogenesis of rabies. *Future Virology, 3*(5), 481–490. https://doi.org/10.2217/17460794.3.5.481.

Dietzschold, B., Wang, H. H., Rupprecht, C. E., Celis, E., Tollis, M., Ertl, H., ... Koprowski, H. (1987). Induction of protective immunity against rabies by immunization with rabies virus ribonucleoprotein. *Proceedings of the National Academy of Sciences of the United States of America, 84*(24), 9165–9169.

Dietzschold, B., Wunner, W. H., Wiktor, T. J., Lopes, A. D., Lafon, M., Smith, C. L., & Koprowski, H. (1983). Characterization of an antigenic determinant of the glycoprotein that correlates with pathogenicity of rabies virus. *Proceedings of the National Academy of Sciences of the United States of America, 80*(1), 70–74.

Duke-Sylvester, S. M., Bolzoni, L., & Real, L. A. (2011). Strong seasonality produces spatial asynchrony in the outbreak of infectious diseases. *Journal of the Royal Society Interface, 8*(59), 817–825. https://doi.org/10.1098/rsif.2010.0475.

Eisinger, D., & Thulke, H. H. (2008). Spatial pattern formation facilitates eradication of infectious diseases. *Journal of Applied Ecology, 45*(2), 415–423. https://doi.org/10.1111/j.1365-2664.2007.01439.x.

Elmore, S. A., Chipman, R. B., Slate, D., Huyvaert, K. P., VerCauteren, K. C., & Gilbert, A. T. (2017). Management and modeling approaches for controlling raccoon rabies: The road to elimination. *PLoS Neglected Tropical Diseases, 11* (3), e0005249. https://doi.org/10.1371/journal.pntd.0005249.

Esposito, J. J., Knight, J. C., Shaddock, J. H., Novembre, F. J., & Baer, G. M. (1988). Successful oral rabies vaccination of raccoons with raccoon poxvirus recombinants expressing rabies virus glycoprotein. *Virology, 165*(1), 313–316.

European Commission (EC), (1989). 89/455/EEC: Council Decision of 24 July 1989 introducing Community measures to set up pilot projects for the control of rabies with a view to its eradication or prevention. *Official Journal of the European Union, 32*(L223), 19–21.

European Commission (EC), (1990). 90/424/EEC: Council Decision of 26 June 1990 on expenditure in the veterinary field. *Official Journal of the European Union, 33*(L224), 19–27.

European Commission (EC), (2002). *The oral vaccination of foxes against rabies*. Brussels: European Commission.

European Commission (EC), (2009). 2009/470/EC: Council Decision of 25 May 2009 on expenditure in the veterinary field (codified version). *Official Journal of the European Union, 52*(L155), 30–45.

European Commission (EC), (2017). *Overview report: Rabies eradication in the EU*. [DG(SANTE) 2016-8980—MR] Luxembourg: Publications Office of the European Union.

Everard, C. O. R., & Everard, J. D. (1985). Mongoose rabies in Grenada. In P. J. Bacon (Ed.), *Population dynamics of rabies in wildlife* (pp. 43–69). London: Academic Press.

Everard, C. O., & Everard, J. D. (1992). Mongoose rabies in the Caribbean. *Annals of the New York Academy of Sciences, 653*, 356–366.

Faber, M., Pulmanausahakul, R., Hodawadekar, S. S., Spitsin, S., McGettigan, J. P., Schnell, M. J., & Dietzschold, B. (2002). Overexpression of the rabies virus glycoprotein results in enhancement of apoptosis and antiviral immune response. *Journal of Virology, 76*(7), 3374–3381.

Faber, M., Pulmanausahakul, R., Nagao, K., Prosniak, M., Rice, A. B., Koprowski, H., ... Dietzschold, B. (2004). Identification of viral genomic elements responsible for rabies virus neuroinvasiveness. *Proceedings of the National Academy of Sciences of the United States of America, 101*(46), 16328–16332. https://doi.org/10.1073/pnas.0407289101.

Farry, S. C., Henke, S. E., Anderson, A. M., & Fearneyhough, M. G. (1998). Responses of captive and free-ranging coyotes to simulated oral rabies vaccine baits. *Journal of Wildlife Diseases, 34*(1), 13–22.

Favoretto, S. R., de Mattos, C. C., de Morais, N. B., Carrieri, M. L., Rolim, B. N., Silva, L. M., ... de Mattos, C. A. (2006). Rabies virus maintained by dogs in humans and terrestrial wildlife, Ceara State, Brazil. *Emerging Infectious Diseases, 12*(12), 1978–1981.

Fearneyhough, M. G., Wilson, P. J., Clark, K. A., Smith, D. R., Johnston, D. H., Hicks, B. N., & Moore, G. M. (1998). Results of an oral rabies vaccination program for coyotes. *Journal of the American Veterinary Medical Association, 212*(4), 498–502.

Fehlner-Gardiner, C. (2018). Rabies control in North America—Past, present and future. *Revue Scientifique et Technique, 37*(2), 421–437. https://doi.org/10.20506/rst.37.2.2812.

Fehlner-Gardiner, C., Nadin-Davis, S., Armstrong, J., Muldoon, F., Bachmann, P., & Wandeler, A. (2008). ERA vaccine-derived cases of rabies in wildlife and domestic animals in Ontario, Canada, 1989-2004. *Journal of Wildlife Diseases, 44*(1), 71–85.

Fehlner-Gardiner, C., Rudd, R., Donovan, D., Slate, D., Kempf, L., & Badcock, J. (2012). Comparing ONRAB® AND RABORAL V-RG® oral rabies vaccine field performance in raccoons and striped skunks, New Brunswick, Canada, and Maine, USA. *Journal of Wildlife Diseases, 48*(1), 157–167.

Fisher, P. (1999). Review of using Rhodamine B as a marker for wildlife studies. *Wildlife Society Bulletin, 27*(2), 318–329.

Follmann, E., Ritter, D., Swor, R., Dunbar, M., & Hueffer, K. (2011). Preliminary evaluation of Raboral V-RG(R) oral rabies vaccine in Arctic foxes (*Vulpes lagopus*). *Journal of Wildlife Diseases, 47*(4), 1032–1035. https://doi.org/10.7589/0090-3558-47.4.1032.

Fontaine, J. J. (2011). Improving our legacy: Incorporation of adaptive management into state wildlife action plans. *Journal of Environmental Management, 92*(5), 1403–1408.

Freuling, C. M., Eggerbauer, E., Finke, S., Kaiser, C., Kaiser, C., Kretzschmar, A., ... Müller, T. (2017). Efficacy of the oral rabies virus vaccine strain SPBN GASGAS in foxes and raccoon dogs. *Vaccine*. https://doi.org/10.1016/j.vaccine.2017.09.093.

Freuling, C. M., Hampson, K., Selhorst, T., Schroder, R., Meslin, F. X., Mettenleiter, T. C., & Müller, T. (2013). The elimination of fox rabies from Europe: Determinants of success and lessons for the future. *Philosophical Transactions of the Royal Society of London Series B, Biological Sciences, 368*(1623), 20120142. https://doi.org/10.1098/rstb.2012.0142.

Freuling, C., Kloss, D., Schroder, C., & Müller, T. F. (2006). Rabies Bulletin Europe—New web-based rabies information system for Europe. *Rabies Bulletin Europe, 30*(1), 8–11.

Geue, L., Schares, S., Schnick, C., Kliemt, J., Beckert, A., Freuling, C., ... Müller, T. (2008). Genetic characterisation of attenuated SAD rabies virus strains used for oral vaccination of wildlife. *Vaccine, 26*(26), 3227–3235. https://doi.org/10.1016/j.vaccine.2008.04.007.

Gilbert, A. (2018). Rabies virus vectors and reservoir species. *Revue Scientifique et Technique, 37*(2), 371–384.

Gilbert, A. T., Johnson, S. R., Nelson, K. M., Chipman, R. B., VerCauteren, K. C., Algeo, T. P., ... Slate, D. (2018). Field trials of Ontario rabies vaccine bait in the Northeastern USA, 2012-14. *Journal of Wildlife Diseases*. https://doi.org/10.7589/2017-09-242.

Gilbert, A., Johnson, S., Walker, N., Beath, A., & VerCauteren, K. (2018). Flavor preference and efficacy of variable dose Ontario Rabies Vaccine Bait (ONRAB) delivery in striped skunks (*Mephitis mephitis*). *Journal of Wildlife Diseases, 54*(1), 122–132. https://doi.org/10.7589/2017-04-073.

Gilbert, A., Johnson, S., Walker, N., Wickham, C., Beath, A., & VerCauteren, K. (2018). Efficacy of Ontario Rabies Vaccine Baits (ONRAB) against rabies infection in raccoons. *Vaccine, 36*(32 Pt B), 4919–4926. https://doi.org/10.1016/j.vaccine.2018.06.052.

Government of Quebec (Quebec) (2019). Retrieved from: http://www.rageduratonlaveur.gouv.qc.ca/en/lutte/historique.asp#cas-rage.

Graser, W. H., III, Gehrt, S. D., Hungerford, L. L., & Anchor, C. (2012). Variation in demographic patterns and population structure of raccoons across an urban landscape. *Journal of Wildlife Management*, *76*, 976–986.

Griffith, B., Scott, J. M., Carpenter, J. W., & Reed, C. (1989). Translocation as a species conservation tool: Status and strategy. *Science*, *245*(4917), 477–480. https://doi.org/10.1126/science.245.4917.477.

Guerra, M. A., Curns, A. T., Rupprecht, C. E., Hanlon, C. A., Krebs, J. W., & Childs, J. E. (2003). Skunk and raccoon rabies in the eastern United States: Temporal and spatial analysis. *Emerging Infectious Diseases*, *9*(9), 1143–1150.

Gunson, J. R., Dorward, W. J., & Schowalter, D. B. (1978). An evaluation of rabies control in skunks in Alberta. *Canadian Veterinary Journal*, *19*(8), 214–220.

Hadidian, J., Jenkins, S. R., Johnston, D. H., Savarie, P. J., Nettles, V. F., Manski, D., & Baer, G. M. (1989). Acceptance of simulated oral rabies vaccine baits by urban raccoons. *Journal of Wildlife Diseases*, *25*(1), 1–9. https://doi.org/10.7589/0090-3558-25.1.1.

Haley, B. S., Algeo, T. P., Bjorklund, B., Duffiney, A. G., Hartin, R. E., Martin, A., ... Slate, D. (2017). Evaluation of bait station density for oral rabies vaccination of raccoons in urban and rural habitats in Florida. *Tropical Medicine and Infectious Disease*, *2*(3). https://doi.org/10.3390/tropicalmed2030041.

Hamir, A. N., Raju, N., & Rupprecht, C. E. (1992). Experimental oral administration of canine adenovirus (type 2) to raccoons (*Procyon lotor*). *Veterinary Pathology*, *29*, 509–513.

Hampson, K., Coudeville, L., Lembo, T., Sambo, M., Kieffer, A., Attlan, M., ... Global Alliance for Rabies Control Partners for Rabies Prevention. (2015). Estimating the global burden of endemic canine rabies. *PLoS Neglected Tropical Diseases*, *9*(4), e0003709. https://doi.org/10.1371/journal.pntd.0003709.

Hanlon, C. A., Buchanan, J. R., Nelson, E., Niu, H. S., Diehl, D., & Rupprecht, C. E. (1993). A vaccinia-vectored rabies vaccine field trial: Ante- and post-mortem biomarkers. *Revue Scientifique et Technique*, *12*(1), 99–107.

Hanlon, C. A., Childs, J. E., & Nettles, V. F. (1999). Recommendations of a National Working Group on Prevention and Control of Rabies in the United States. Article III: Rabies in wildlife. National Working Group on Rabies Prevention and Control. *Journal of the American Veterinary Medical Association*, *215*(11), 1612–1618.

Hanlon, C. A., Niezgoda, M., Hamir, A. N., Schumacher, C., Koprowski, H., & Rupprecht, C. E. (1998). First North American field release of a vaccinia-rabies glycoprotein recombinant virus. *Journal of Wildlife Diseases*, *34*(2), 228–239.

Hanlon, C. A., & Rupprecht, C. E. (1998). The reemergence of rabies. In W. M. Scheld, D. Armstrong, & J. M. Hughes (Eds.), *Emerging infections 1* (pp. 59–80). Washington, DC: American Society for Microbiology Press.

Hassel, R., Ortmann, S., Clausen, P., Jago, M., Bruwer, F., Lindeque, P., ... Müller, T. F. (2018). Baiting studies on oral vaccination of the greater kudu (*Tragelaphus strepsiceros*) against rabies. *European Journal of Wildlife Research*, *64*, 1–7.

Hassel, R., Vos, A., Clausen, P., Moore, S., van der Westhuizen, J., Khaiseb, S., ... Müller, T. (2018). Experimental screening studies on rabies virus transmission and oral rabies vaccination of the Greater Kudu (*Tragelaphus strepsiceros*). *Scientific Reports*, *8*(1), 16599. https://doi.org/10.1038/s41598-018-34985-5.

Henderson, H., Jackson, F., Bean, K., Panasuk, B., Niezgoda, M., Slate, D., ... Rupprecht, C. E. (2009). Oral immunization of raccoons and skunks with a canine adenovirus recombinant rabies vaccine. *Vaccine*, *27*(51), 7194–7197. https://doi.org/10.1016/j.vaccine.2009.09.030.

Hooper, D. C., Morimoto, K., Bette, M., Weihe, E., Koprowski, H., & Dietzschold, B. (1998). Collaboration of antibody and inflammation in clearance of rabies virus from the central nervous system. *Journal of Virology*, *72*(5), 3711–3719.

Horton, D. L., McElhinney, L. M., Freuling, C. M., Marston, D. A., Banyard, A. C., Goharrriz, H., ... Fooks, A. R. (2015). Complex epidemiology of a zoonotic disease in a culturally diverse region: Phylogeography of rabies virus in the Middle East. *PLoS Neglected Tropical Diseases*, *9*(3), e0003569. https://doi.org/10.1371/journal.pntd.0003569.

Hostnik, P., Picard-Meyer, E., Rihtaric, D., Toplak, I., & Cliquet, F. (2014). Vaccine-induced rabies in a red fox (*Vulpes vulpes*): Isolation of vaccine virus in brain tissue and salivary glands. *Journal of Wildlife Diseases*, *50*(2), 397–401. https://doi.org/10.7589/2013-07-183.

Hsu, A. P., Tseng, C. H., Barrat, J., Lee, S. H., Shih, Y. H., Wasniewski, M., ... Tsai, H. J. (2017). Safety, efficacy and immunogenicity evaluation of the SAG2 oral rabies vaccine in Formosan ferret badgers. *PLoS One*, *12*(10), e0184831. https://doi.org/10.1371/journal.pone.0184831.

Johnson, S. R., Berentsen, A. R., Ellis, C. K., & VerCauteren, K. C. (2016). Estimates of small Indian mongoose densities: Implications for rabies management. *The Journal of Wildlife Management*, *80*(1), 37–47.

Johnston, D. H., & Beauregard, M. (1969). Rabies epidemiology in Ontario. *Wildlife Disease*, *5*(3), 357–370.

Johnston, D. H., Joachim, D. G., Bachmann, P., Kardong, K. V., Stewart, R. E. A., Dix, L. M., ... Watt, I. D. (1987). Aging furbearers using tooth structure and biomarkers. In M. Novak, J. Baker, M. Obbard, & B. Mallock (Eds.), *Wild furbearer management and conservation in North America* (pp. 228–243). North Bay, ON: Ontario Trappers Association.

Johnston, D. H., & Voigt, D. R. (1982). A baiting system for the oral rabies vaccination of wild foxes and skunks. *Comparative Immunology, Microbiology and Infectious Diseases, 5*(1–3), 185–186.

Johnston, D. H., Voigt, D. R., MacInnes, C. D., Bachmann, P., Lawson, K. F., & Rupprecht, C. E. (1988). An aerial baiting system for the distribution of attenuated or recombinant rabies vaccines for foxes, raccoons, and skunks. *Reviews of Infectious Diseases, 10*(Suppl. 4), S660–S664.

Kappeler, A. (1992). Manual bait distribution in oral vaccination campaigns in Europe. In K. Bögel, F. M. Meslin, & M. M. Kaplan (Eds.), *Wildlife rabies control* (pp. 155–159). Kent: Wells Medical Ltd.

Kieny, M. P., Lathe, R., Drillien, R., Spehner, D., Skory, S., Schmitt, D., … Lecocq, J. P. (1984). Expression of rabies virus glycoprotein from a recombinant vaccinia virus. *Nature, 312*(5990), 163–166.

Kirby, J. D., Chipman, R. B., Nelson, K. M., Rupprecht, C. E., Blanton, J. D., Algeo, T. P., & Slate, D. (2017). Enhanced surveillance to support effective oral rabies vaccination of raccoons in the eastern United States. *Tropical Medicine and Infectious Disease, 2*(3), 34. https://doi.org/10.3390/tropicalmed2030034.

Knobel, D. L., & du Toit, J. T. (2003). The influence of pack social structure on oral rabies vaccination coverage in captive African wild dogs (*Lycaon pictus*). *Applied Animal Behaviour Science, 80*(1), 61–70.

Knobel, D. L., du Toit, J. T., & Bingham, J. (2002). Development of a bait and baiting system for delivery of oral rabies vaccine to free-ranging African wild dogs (*Lycaon pictus*). *Journal of Wildlife Diseases, 38*(2), 352–362. https://doi.org/10.7589/0090-3558-38.2.352.

Knobel, D. L., Fooks, A. R., Brookes, S. M., Randall, D. A., Williams, S. D., Argaw, K., … Laurenson, M. K. (2008). Trapping and vaccination of endangered Ethiopian wolves to control an outbreak of rabies. *Journal of Applied Ecology, 45*(1), 109–116. https://doi.org/10.1111/j.1365-2664.2007.01387.x.

Knobel, D. L., Liebenberg, A., & Du Toit, J. T. (2003). Seroconversion in captive African wild dogs (*Lycaon pictus*) following administration of a chicken head bait/SAG-2 oral rabies vaccine combination. *The Onderstepoort Journal of Veterinary Research, 70*(1), 73–77.

Kuzmin, I. V., Botvinkin, A. D., McElhinney, L. M., Smith, J. S., Orciari, L. A., Hughes, G. J., … Rupprecht, C. E. (2004). Molecular epidemiology of terrestrial rabies in the former Soviet Union. *Journal of Wildlife Diseases, 40*(4), 617–631.

Kuzmin, I. V., Shi, M., Orciari, L. A., Yager, P. A., Velasco-Villa, A., Kuzmina, N. A., … Rupprecht, C. E. (2012). Molecular inferences suggest multiple host shifts of rabies viruses from bats to mesocarnivores in Arizona during 2001-2009. *PLoS Pathogens, 8*(6), e1002786. https://doi.org/10.1371/journal.ppat.1002786.

Kuzmina, N. A., Lemey, P., Kuzmin, I. V., Mayes, B. C., Ellison, J. A., Orciari, L. A., … Rupprecht, C. E. (2013). The phylogeography and spatiotemporal spread of South-central skunk rabies virus. *PLoS One, 8*(12), e82348. https://doi.org/10.1371/journal.pone.0082348.

Lafay, F., Benejean, J., Tuffereau, C., Flamand, A., & Coulon, P. (1994). Vaccination against rabies: Construction and characterization of SAG2, a double avirulent derivative of SADBern. *Vaccine, 12*(4), 317–320.

Le Blois, H., Tuffereau, C., Blancou, J., Artois, M., Aubert, A., & Flamand, A. (1990). Oral immunization of foxes with avirulent rabies virus mutants. *Veterinary Microbiology, 23*(1–4), 259–266.

Lee, D. T., & Schachter, B. J. (1980). Two algorithms for constructing a Delaunay triangulation. *International Journal of Computer and Information Sciences, 9*(3), 219–242.

Lembo, T., & Partners for Rabies Prevention. (2012). The blueprint for rabies prevention and control: A novel operational toolkit for rabies elimination. *PLoS Neglected Tropical Diseases, 6*(2), e1388. https://doi.org/10.1371/journal.pntd.0001388.

Leslie, M. J., Messenger, S., Rohde, R. E., Smith, J., Cheshire, R., Hanlon, C., & Rupprecht, C. E. (2006). Bat-associated rabies virus in skunks. *Emerging Infectious Diseases, 12*(8), 1274–1277.

Lewis, J. C. (1975). Control of rabies among terrestrial wildlife by population reduction. In G. M. Baer (Ed.), *Vol. 1. The natural history of rabies* (pp. 243–259). New York, NY: Academic Press, Inc.

Linhart, S. B. (1960). Rabies in wildlife and control methods in New York State. *New York Fish and Game, 7*, 1–13.

Linhart, S. B., Blom, F. S., Dasch, G. J., Roberts, J. D., Engeman, R. M., Esposito, J. J., … Baer, G. M. (1991). Formulation and evaluation of baits for oral rabies vaccination of raccoons (*Procyon lotor*). *Journal of Wildlife Diseases, 27*(1), 21–33.

Linhart, S. B., Kappeler, A., & Windberg, L. A. (1993). *A review of baits and bait delivery systems for free-ranging carnivores and ungulates*. United States Department of Agriculture.

Linhart, S. B., King, R., Zamir, S., Naveh, U., Davidson, M., & Perl, S. (1997). Oral rabies vaccination of red foxes and golden jackals in Israel: Preliminary bait evaluation. *Revue Scientifique et Technique, 16*(3), 874–880.

Linhart, S. B., Wlodkowski, J. C., Kavanaugh, D. M., Motes-Kreimeyer, L., Montoney, A. J., Chipman, R. B., … Fearneyhough, M. G. (2002). A new flavor-coated sachet bait for delivering oral rabies vaccine to raccoons and coyotes. *Journal of Wildlife Diseases, 38*(2), 363–377. https://doi.org/10.7589/0090-3558-38.2.363.

Linnell, J. D. C., Aanes, R., Swenson, J. E., Odden, J., & Smith, M. E. (1997). Translocation of carnivores as a method for managing problem animals: A review. *Biodiversity and Conservation, 6,* 1245–1257.

Lobo, D., DeBenedet, C., Fehlner-Gardiner, C., Nadin-Davis, S., Anderson, M., Buchanan, T., ... Hopkins, J. (2018). Raccoon rabies outbreak in Hamilton, Ontario: A progress report. *Canada Communicable Disease Report, 44*(5), 116–121.

Ma, X., Monroe, B. P., Cleaton, J. M., Orciari, L. A., Li, Y., Kirby, J. D., ... Blanton, J. D. (2018). Rabies surveillance in the United States during 2017. *Journal of the American Veterinary Medical Association, 253*(12), 1555–1568. https://doi.org/10.2460/javma.253.12.1555.

Macdonald, D. W. (1980). *Rabies in wildlife: A biologists perspective (160 pp.).* New York: Oxford University Press.

MacInnes, C. D., Johnston, D. H., Bachmann, P., Pond, B. A., Fielding, C. A., Nunan, C. P., ... Tinline, R. L. (1992). Design considerations for large-scale aerial distribution of rabies vaccine-baits in Ontario. In K. Bögel, F. M. Meslin, & M. M. Kaplan (Eds.), *Wildlife Rabies Control* (pp. 160–167). Kent: Wells Medical Ltd.

MacInnes, C. D., & LeBer, C. A. (2000). Wildlife management agencies should participate in rabies control. *Wildlife Society Bulletin, 28,* 1156–1167.

MacInnes, C. D., Smith, S. M., Tinline, R. R., Ayers, N. R., Bachmann, P., Ball, D. G., ... Voigt, D. R. (2001). Elimination of rabies from red foxes in eastern Ontario. *Journal of Wildlife Diseases, 37*(1), 119–132.

Mahl, P., Cliquet, F., Guiot, A. L., Niin, E., Fournials, E., Saint-Jean, N., ... Gueguen, S. (2014). Twenty year experience of the oral rabies vaccine SAG2 in wildlife: A global review. *Veterinary Research, 45,* 77. https://doi.org/10.1186/s13567-014-0077-8.

Mainguy, J., Fehlner-Gardiner, C., Slate, D., & Rudd, R. J. (2013). Oral rabies vaccination in raccoons: Comparison of ONRAB® and RABORAL V-RG® vaccine-bait field performance in Quebec, Canada and Vermont, USA. *Journal of Wildlife Diseases, 49*(1), 190–193. https://doi.org/10.7589/2011-11-342.

Maki, J., Guiot, A. L., Aubert, M., Brochier, B., Cliquet, F., Hanlon, C. A., ... Lankau, E. W. (2017). Oral vaccination of wildlife using a vaccinia-rabies-glycoprotein recombinant virus vaccine (RABORAL V-RG®): A global review. *Veterinary Research, 48*(1), 57. https://doi.org/10.1186/s13567-017-0459-9.

Massei, G., Quy, R. J., Gurney, J., & Cowan, D. P. (2010). Can translocations be used to mitigate human-wildlife conflicts? *Wildlife Research, 37*(5), 428–439.

Masson, E., Aubert, M. F. A., Barrat, J., & Vuillaume, P. (1996). Comparison of the efficacy of the antirabies vaccines used for foxes in France. *Veterinary Research, 27,* 255–266.

McClure, K., Gilbert, A., Chipman, R., Rees, E., & Pepin, K. (2020). Variation in host home range size decreases rabies vaccination effectiveness by increasing the spatial spread of rabies virus. *The Journal of Animal Ecology.* https://doi.org/10.1111/1365-2656.13176, [in press].

McLean, R. G. (1975). Raccoon rabies. In G. M. Baer (Ed.), *Vol. II. The natural history of rabies* (pp. 53–77). New York, NY: Academic Press.

Meltzer, M. I., & Rupprecht, C. E. (1998). A review of the economics of the prevention and control of rabies. Part 2: Rabies in dogs, livestock and wildlife. *PharmacoEconomics, 14*(5), 481–498.

Mengak, M. T. (2018). *Wildlife translocation.* Fort Collins, CO: USDA Wildlife Services, National Wildlife Research Center.

Middel, K., Fehlner-Gardiner, C., Pulham, N., & Buchanan, T. (2017). Incorporating direct rapid immunohistochemical testing into large-scale wildlife rabies surveillance. *Tropical Medicine and Infectious Disease, 2*(3), 21.

Moore, S. M., Gilbert, A., Vos, A., Freuling, C., Ellis, C., Kliemt, J., & Müller, T. (2017). Rabies virus antibodies from oral vaccination as a correlate of protection against lethal infection in wildlife. *Tropical Medicine and Infectious Disease, 2*(31), 1–24.

Müller, T., Batza, H. J., Beckert, A., Bunzenthal, C., Cox, J. H., Freuling, C. M., ... Mettenleiter, T. C. (2009). Analysis of vaccine-virus-associated rabies cases in red foxes (*Vulpes vulpes*) after oral rabies vaccination campaigns in Germany and Austria. *Archives of Virology, 154*(7), 1081–1091. https://doi.org/10.1007/s00705-009-0408-7.

Müller, T., Batza, H. J., Freuling, C., Kliemt, A., Kliemt, J., Heuser, R., ... Mettenleiter, T. C. (2012). Elimination of terrestrial rabies in Germany using oral vaccination of foxes. *Berliner und Münchener Tierärztliche Wochenschrift, 125*(5–6), 178–190.

Müller, T. F., Demetriou, P., Moynagh, J., Cliquet, F., Fooks, A. R., Conraths, F. J., ... Freuling, C. (2012). *Rabies elimination in Europe: A success story.* Paper presented at the OIE global conference on rabies control: Towards sustainable prevention at the source, Incheon-Seoul, Republic of Korea.

Müller, T. F., & Freuling, C. M. (2018). Rabies control in Europe: An overview of past, current and future strategies. *Revue Scientifique et Technique, 37*(2), 409–419. https://doi.org/10.20506/rst.37.2.2811.

Müller, T., Freuling, C., Gschwendner, P., Holzhofer, E., Mürke, H., Rüdiger, H., ... Vos, A. (2012). SURVIS: A fully-automated aerial baiting system for the distribution of vaccine baits for wildlife. *Berliner und Münchener Tierärztliche Wochenschrift, 125*, 197–202.

Müller, T., Freuling, C. M., Wysocki, P., Roumiantzeff, M., Freney, J., Mettenleiter, T. C., & Vos, A. (2015). Terrestrial rabies control in the European Union: Historical achievements and challenges ahead. *Veterinary Journal, 203*(1), 10–17. https://doi.org/10.1016/j.tvjl.2014.10.026.

Müller, T. F., Schroder, R., Wysocki, P., Mettenleiter, T. C., & Freuling, C. M. (2015). Spatio-temporal use of oral rabies vaccines in fox rabies elimination programmes in Europe. *PLoS Neglected Tropical Diseases, 9*(8), e0003953. https://doi.org/10.1371/journal.pntd.0003953.

Murphy, A. A., Redwood, A. J., & Jarvis, M. A. (2016). Self-disseminating vaccines for emerging infectious diseases. *Expert Review of Vaccines, 15*(1), 31–39.

Nadin-Davis, S. A., & Fehlner-Gardiner, C. (2019). Origins of the arctic fox variant rabies viruses responsible for recent cases of the disease in southern Ontario. *PLoS Neglected Tropical Diseases, 13*(9), e0007699. https://doi.org/10.1371/journal.pntd.0007699.

Nadin-Davis, S. A., Fu, Q., Trewby, H., Biek, R., Johnson, R. H., & Real, L. (2018). Geography but not alternative host species explain the spread of raccoon rabies virus in Vermont. *Epidemiology and Infection, 146*(15), 1977–1986. https://doi.org/10.1017/S0950268818001759.

Nadin-Davis, S. A., Muldoon, F., & Wandeler, A. I. (2006a). A molecular epidemiological analysis of the incursion of the raccoon strain of rabies virus into Canada. *Epidemiology and Infection, 134*(3), 534–547. https://doi.org/10.1017/S0950268805005108.

Nadin-Davis, S. A., Muldoon, F., & Wandeler, A. I. (2006b). Persistence of genetic variants of the arctic fox strain of Rabies virus in southern Ontario. *Canadian Journal of Veterinary Research, 70*(1), 11–19.

Nadin-Davis, S. A., Torres, G., Ribas Mde, L., Guzman, M., De La Paz, R. C., Morales, M., & Wandeler, A. I. (2006). A molecular epidemiological study of rabies in Cuba. *Epidemiology and Infection, 134*(6), 1313–1324. https://doi.org/10.1017/S0950268806006297.

Nadin-Davis, S. A., Velez, J., Malaga, C., & Wandeler, A. I. (2008). A molecular epidemiological study of rabies in Puerto Rico. *Virus Research, 131*(1), 8–15. https://doi.org/10.1016/j.virusres.2007.08.002.

Nel, L. H., & Rupprecht, C. E. (2007). Emergence of lyssaviruses in the old world: The case of Africa. *Current Topics in Microbiology and Immunology, 315*, 161–193.

Nel, L. H., Sabeta, C. T., von Teichman, B., Jaftha, J. B., Rupprecht, C. E., & Bingham, J. (2005). Mongoose rabies in southern Africa: A re-evaluation based on molecular epidemiology. *Virus Research, 109*(2), 165–173. https://doi.org/10.1016/j.virusres.2004.12.003.

Nettles, V. F., Shaddock, J. H., Sikes, R. K., & Reyes, C. R. (1979). Rabies in translocated raccoons. *American Journal of Public Health, 69*(6), 601–602. https://doi.org/10.2105/ajph.69.6.601.

Newton, E. J., Pond, B. A., Tinline, R., Middel, K., Belanger, D., & Rees, E. E. (2019). Differential impacts of vaccination on wildlife disease spread during epizootic and enzootic phases. *Journal of Applied Ecology, 56*(3), 526–536. https://doi.org/10.1111/1365-2664.13339.

Nunan, C. P., Tinline, R. R., Honig, J. M., Ball, D. G., Hauschildt, P., & LeBer, C. A. (2002). Postexposure treatment and animal rabies, Ontario, 1958-2000. *Emerging Infectious Diseases, 8*(2), 214–217.

Octaria, R., Salyer, S. J., Blanton, J., Pieracci, E. G., Munyua, P., Millien, M., ... Wallace, R. M. (2018). From recognition to action: A strategic approach to foster sustainable collaborations for rabies elimination. *PLoS Neglected Tropical Diseases, 12*(10), e0006756. https://doi.org/10.1371/journal.pntd.0006756.

Ontario Ministry of Natural Resources and Forestry (OMNRF) (2019). *Rabies in wildlife*. Retrieved from: https://www.ontario.ca/page/rabies-wildlife.

Orciari, L. A., Niezgoda, M., Hanlon, C. A., Shaddock, J. H., Sanderlin, D. W., Yager, P. A., & Rupprecht, C. E. (2001). Rapid clearance of SAG-2 rabies virus from dogs after oral vaccination. *Vaccine, 19*(31), 4511–4518.

Ortmann, S., Vos, A., Kretzschmar, A., Walther, N., Kaiser, C., Freuling, C., ... Muller, T. (2018). Safety studies with the oral rabies virus vaccine strain SPBN GASGAS in the small Indian mongoose (*Herpestes auropunctatus*). *BMC Veterinary Research, 14*(1), 90. https://doi.org/10.1186/s12917-018-1417-0.

European Food Safety Authority Panel on Animal Health and Welfare (EFSA) (2015). Scientific opinion—Update on oral vaccination of foxes and raccoon dogs against rabies. *EFSA Journal, 13*(7), 1–70. https://doi.org/10.2903/j.efsa.2015.4164.

Parks, E. (1968). Control of rabies in wildlife in New York. *New York Fish and Game, 15*, 98–111.

Partners for Rabies Prevention. (2012). *Blueprint for fox rabies prevention and control*. Retrieved from: http://www.foxrabiesblueprint.org.

Pastoret, P. P., Boulanger, D., & Brochier, B. (1995). Field trials of a recombinant rabies vaccine. *Parasitology, 110* (Suppl), S37–S42.

Pastoret, P. P., Brochier, B., Languet, B., Thomas, I., Paquot, A., Bauduin, B., ... Desmettre, P. (1988). First field trial of fox vaccination against rabies using a vaccinia-rabies recombinant virus. *Veterinary Record, 123*(19), 481–483.

Pastoret, P. P., Frisch, R., Blancou, J., Wolff, F., Brochier, B., & Schneider, L. (1987). Campagne internationale de vaccination antirabique du renard par voie orale menée au grand-duché de Luxembourg, en Belgique et en France. *Annales de Médecine Vétérinaire, 131*, 441–447.

Pedersen, K., Gilbert, A. T., Nelson, K. M., Morgan, D. P., Davis, A. J., VerCauteren, K. C., ... Chipman, R. B. (2019). Raccoon (*Procyon lotor*) response to Ontario Rabies Vaccine Baits (ONRAB) in St. Lawrence County, New York, USA. *Journal of Wildlife Diseases, 55*(3), 645–653.

Peel, A. J., Pulliam, J. R., Luis, A. D., Plowright, R. K., O'Shea, T. J., Hayman, D. T., ... Restif, O. (2014). The effect of seasonal birth pulses on pathogen persistence in wild mammal populations. *Proceedings of the Biological Sciences, 281*(1786). https://doi.org/10.1098/rspb.2013.2962.

Perry, B. D., Garner, N., Jenkins, S. R., McCloskey, K., & Johnston, D. H. (1989). A study of techniques for the distribution of oral rabies vaccine to wild raccoon populations. *Journal of Wildlife Diseases, 25*(2), 206–217. https://doi.org/10.7589/0090-3558-25.2.206.

Perry, B. D., Johnston, D. H., Jenkins, S. R., Foggin, C. M., Garner, N., Brooks, R., & Bleakley, J. (1988). Studies on the delivery of oral rabies vaccines to wildlife and dog populations. *Acta Veterinaria Scandinavica. Supplementum, 84*, 303–305.

Pfaff, F., Muller, T., Freuling, C. M., Fehlner-Gardiner, C., Nadin-Davis, S., Robardet, E., ... Hoper, D. (2018). In-depth genome analyses of viruses from vaccine-derived rabies cases and corresponding live-attenuated oral rabies vaccines. *Vaccine*. https://doi.org/10.1016/j.vaccine.2018.01.083.

Pieracci, E. G., Pearson, C. M., Wallace, R. M., Blanton, J. D., Whitehouse, E. R., Ma, X., ... Olson, V. (2019). Vital signs: Trends in human rabies deaths and exposures-United States, 1938-2018. *MMWR Morbidity and Mortality Weekly Report, 68*(23), 524–528. https://doi.org/10.15585/mmwr.mm6823e1.

Plummer, P. J. G. (1947). Preliminary note on Arctic dog disease and its relationship to rabies. *Canadian Journal of Comparative Medicine, 11*, 154–160.

Prehaud, C., Takehara, K., Flamand, A., & Bishop, D. H. (1989). Immunogenic and protective properties of rabies virus glycoprotein expressed by baculovirus vectors. *Virology, 173*(2), 390–399.

Prevec, L., Campbell, J. B., Christie, B. S., Belbeck, L., & Graham, F. L. (1990). A recombinant human adenovirus vaccine against rabies. *The Journal of Infectious Diseases, 161*(1), 27–30.

Pybus, M. J. (1988). Rabies and rabies control in striped skunks (*Mephitis mephitis*) in three prairie regions of western North America. *Journal of Wildlife Diseases, 24*(3), 434–449. https://doi.org/10.7589/0090-3558-24.3.434.

Ramey, P. C., Blackwell, B. F., Gates, R. J., & Slemons, R. D. (2008). Oral rabies vaccination of a northern Ohio raccoon population: Relevance of population density and prebait serology. *Journal of Wildlife Diseases, 44*(3), 553–568. https://doi.org/10.7589/0090-3558-44.3.553.

Randall, D. A., Marino, J., Haydon, D., Sillero-Zubiri, C., Knobel, D. L., Tallents, L. A., ... Laurenson, M. K. (2006). An integrated disease management strategy for the control of rabies in Ethiopian wolves. *Biological Conservation, 131*, 151–162.

Randall, D. A., Williams, S. D., Kuzmin, I. V., Rupprecht, C. E., Tallents, L. A., Tefera, Z., ... Laurenson, M. K. (2004). Rabies in endangered Ethiopian wolves. *Emerging Infectious Diseases, 10*(12), 2214–2217.

Real, L. A., & Biek, R. (2007). Spatial dynamics and genetics of infectious diseases on heterogeneous landscapes. *Journal of the Royal Society Interface, 4*(16), 935–948.

Real, L. A., Russell, C., Waller, L., Smith, D., & Childs, J. (2005). Spatial dynamics and molecular ecology of North American rabies. *The Journal of Heredity, 96*(3), 253–260. https://doi.org/10.1093/jhered/esi031.

Rees, E. E., Bélanger, D., Lelièvre, F., Coté, N., & Lambert, L. (2011). Targeted surveillance of raccoon rabies in Québec, Canada. *The Journal of Wildlife Management, 75*(6), 1406–1416.

Rees, E. E., Pond, B. A., Tinline, R. R., & Belanger, D. (2013). Modelling the effect of landscape heterogeneity on the efficacy of vaccination for wildlife infectious disease control. *Journal of Applied Ecology, 50*(4), 881–891.

Reinius, S. (1992). Epidemiology of fox/raccoon dog rabies in Finland. In K. Bögel, F. M. Meslin, & M. M. Kaplan (Eds.), *Wildlife rabies control* (pp. 32–34). Kent: Wells Medical Ltd.

Reynolds, J. J., Hirsch, B. T., Gehrt, S. D., & Craft, M. E. (2015). Raccoon contact networks predict seasonal susceptibility to rabies outbreaks and limitations of vaccination. *The Journal of Animal Ecology, 84*(6), 1720–1731. https://doi.org/10.1111/1365-2656.12422.

Riley, S. P. D., Hadidian, J., & Manski, D. A. (1998). Population density, survival, and rabies in raccoons in an urban national park. *Canadian Journal of Zoology, 76*, 1153–1164.

Robardet, E., Demerson, J. M., Andrieu, S., & Cliquet, F. (2012). First European interlaboratory comparison of tetracycline and age determination with red fox teeth following oral rabies vaccination programs. *Journal of Wildlife Diseases, 48*(4), 858–868. https://doi.org/10.7589/2011-07-205.

Robardet, E., Picard-Meyer, E., Andrieu, S., Servat, A., & Cliquet, F. (2011). International interlaboratory trials on rabies diagnosis: An overview of results and variation in reference diagnosis techniques (fluorescent antibody test, rabies tissue culture infection test, mouse inoculation test) and molecular biology techniques. *Journal of Virological Methods, 177*(1), 15–25. https://doi.org/10.1016/j.jviromet.2011.06.004.

Robardet, E., Rieder, J., Barrat, J., & Cliquet, F. (2019). Reconsidering oral rabies vaccine bait uptake evaluation at population level: A simple, noninvasive, and ethical method by fecal survey using a physical biomarker. *Journal of Wildlife Diseases, 55*(1), 200–205. https://doi.org/10.7589/2018-02-045.

Rosatte, R. C. (2000). Management of raccoons (*Procyon lotor*) in Ontario, Canada: Do human interventions and disease have significant impacts on raccoon populations? *Mammalia, 64*, 369–390.

Rosatte, R. (2011). Evolution of wildlife rabies control tactics. In A. Jackson (Ed.), *Advances in virus research: Research advances in rabies* (pp. 397–419). London: Academic Press.

Rosatte, R. (2013). Rabies control in wild carnivores. In A. Jackson (Ed.), *Rabies* (pp. 617–670). (3rd ed.). London: Academic Press.

Rosatte, R., Allan, M., Bachmann, P., Sobey, K., Donovan, D., Davies, J. C., … Schumacher, C. (2008). Prevalence of tetracycline and rabies virus antibody in raccoons, skunks, and foxes following aerial distribution of V-RG baits to control raccoon rabies in Ontario, Canada. *Journal of Wildlife Diseases, 44*(4), 946–964. https://doi.org/10.7589/0090-3558-44.4.946.

Rosatte, R. C., Donovan, D., Allan, M., Bruce, L., Buchanan, T., Sobey, K., … Wandeler, A. (2009). The control of raccoon rabies in Ontario Canada: Proactive and reactive tactics, 1994-2007. *Journal of Wildlife Diseases, 45*(3), 772–784.

Rosatte, R., Donovan, D., Allan, M., Bruce, L., & Davies, C. (2007). Human-assisted movements of Raccoons, *Procyon lotor*, and Opossums, *Didelphis virginiana*, between the United States and Canada. *The Canadian Field Naturalist, 121*(2), 212–213.

Rosatte, R., Donovan, D., Allan, M., Howes, L. A., Silver, A., Bennett, K., … Radford, B. (2001). Emergency response to raccoon rabies introduction into Ontario. *Journal of Wildlife Diseases, 37*(2), 265–279.

Rosatte, R. C., Donovan, D., Davies, J. C., Allan, M., Bachmann, P., Stevenson, B., … Lawson, K. (2009). Aerial distribution of ONRAB[R] baits as a tactic to control rabies in raccoons and striped skunks in Ontario, Canada. *Journal of Wildlife Diseases, 45*(2), 363–374.

Rosatte, R. C., Donovan, D., Davies, J. C., Brown, L., Allan, M., von Zuben, V., … Fehlner-Gardiner, C. (2011). High-density baiting with ONRAB[R] rabies vaccine baits to control Arctic-variant rabies in striped skunks in Ontario, Canada. *Journal of Wildlife Diseases, 47*(2), 459–465. https://doi.org/10.7589/0090-3558-47.2.459.

Rosatte, R. C., Howard, D. R., Campbell, J. B., & MacInnes, C. D. (1990). Intramuscular vaccination of skunks and raccoons against rabies. *Journal of Wildlife Diseases, 26*(2), 225–230.

Rosatte, R. C., & Lawson, K. F. (2001). Acceptance of baits for delivery of oral rabies vaccine to raccoons. *Journal of Wildlife Diseases, 37*(4), 730–739.

Rosatte, R., & MacInnes, C. D. (1989). *Relocation of city raccoons.* Paper presented at the Proceedings of the 9th Great Plains Wildlife Damage Control Workshop, Fort Collins, CO.

Rosatte, R. C., MacInnes, C. D., Power, M. J., Johnston, D. H., Bachmann, P., Nunan, C. P., … Calder, L. (1993). Tactics for the control of wildlife rabies in Ontario (Canada). *Revue Scientifique et Technique, 12*(1), 95–98.

Rosatte, R. C., Power, M. J., Donovan, D., Davies, J. C., Allan, M., Bachmann, P., … Muldoon, F. (2007). Elimination of arctic variant rabies in red foxes, metropolitan Toronto. *Emerging Infectious Diseases, 13*(1), 25–27.

Rosatte, R. C., Power, M. J., MacInnes, C. D., & Campbell, J. B. (1992). Trap-vaccinate-release and oral vaccination for rabies control in urban skunks, raccoons and foxes. *Journal of Wildlife Diseases, 28*(4), 562–571. https://doi.org/10.7589/0090-3558-28.4.562.

Rosatte, R. C., Pybus, M. J., & Gunson, J. R. (1986). Population reduction as a factor in the control of skunk rabies in Alberta. *Journal of Wildlife Diseases, 22*(4), 459–467.

Rosatte, R., Ryckman, M., Ing, K., Proceviat, S., Allan, M., Bruce, L., … Davies, J. C. (2010). Density, movements, and survival of raccoons in Ontario, Canada: Implications for disease spread and management. *Journal of Mammalogy, 91*(1), 122–135.

Rosatte, R. C., Tinline, R. R., & Johnston, D. H. (2007). Rabies control in wild carnivores. In A. C. Jackson & W. H. Wunner (Eds.), *Rabies* (2nd ed., pp. 595–634). London: Academic Press.

Roscoe, D. E., Holste, W. C., Sorhage, F. E., Campbell, C., Niezgoda, M., Buchannan, R., … Rupprecht, C. E. (1998). Efficacy of an oral vaccinia-rabies glycoprotein recombinant vaccine in controlling epidemic raccoon rabies in New Jersey. *Journal of Wildlife Diseases, 34*(4), 752–763.

Rupprecht, C. E., Barrett, J., Briggs, D., Cliquet, F., Fooks, A. R., Lumlertdacha, B., … Wandeler, A. I. (2008). Can rabies be eradicated? *Developments in Biologicals, 131*, 95–121.

Rupprecht, C. E., Charlton, K. M., Artois, M., Casey, G. A., Webster, W. A., Campbell, J. B., … Schneider, L. G. (1990). Ineffectiveness and comparative pathogenicity of attenuated rabies virus vaccines for the striped skunk (*Mephitis mephitis*). *Journal of Wildlife Diseases, 26*(1), 99–102.

Rupprecht, C. E., Glickman, L. T., Spencer, P. A., & Wiktor, T. J. (1987). Epidemiology of rabies virus variants. Differentiation using monoclonal antibodies and discriminant analysis. *American Journal of Epidemiology, 126*(2), 298–309.

Rupprecht, C. E., Hamir, A. N., Johnston, D. H., & Koprowski, H. (1988). Efficacy of a vaccinia-rabies glycoprotein recombinant virus vaccine in raccoons (*Procyon lotor*). *Reviews of Infectious Diseases, 10*(Suppl. 4), S803–S809.

Rupprecht, C. E., & Slate, D. (2012). Rabies prevention and control: Advances and challenges. In R. G. Dietzgen & I. Kuzmin (Eds.), *Rhabdoviruses: Molecular taxonomy, evolution, genomics, ecology, host-vector interactions, cytopathology and control* (pp. 215–252). Norfolk: Caister Academic Press.

Rupprecht, C. E., & Smith, J. S. (1994). Raccoon rabies—The reemergence of an epizootic in a densely populated area. *Seminars in Virology, 5*(2), 155–164.

Rupprecht, C. E., Smith, J. S., Fekadu, M., & Childs, J. E. (1995). The ascension of wildlife rabies: A cause for public health concern or intervention? *Emerging Infectious Diseases, 1*(4), 107–114.

Rupprecht, C. E., Turmelle, A., & Kuzmin, I. V. (2011). A perspective on lyssavirus emergence and perpetuation. *Current Opinion in Virology, 1*(6), 662–670. https://doi.org/10.1016/j.coviro.2011.10.014.

Sabeta, C., & Ngoepe, E. C. (2018). Controlling dog rabies in Africa: Successes, failures and prospects for the future. *Revue Scientifique et Technique, 37*(2), 439–449. https://doi.org/10.20506/rst.37.2.2813.

Sattler, A. C., Krogwold, R. A., Wittum, T. E., Rupprecht, C. E., Algeo, T. P., Slate, D., … Slemons, R. D. (2009). Influence of oral rabies vaccine bait density on rabies seroprevalence in wild raccoons. *Vaccine, 27*(51), 7187–7193. https://doi.org/10.1016/j.vaccine.2009.09.035.

Schneider, M. C., Romijn, P. C., Uieda, W., Tamayo, H., da Silva, D. F., Belotto, A., … Leanes, L. F. (2009). Rabies transmitted by vampire bats to humans: An emerging zoonotic disease in Latin America? *Revista Panamericana de Salud Pública, 25*(3), 260–269.

Schnurrenberger, P. R., Beck, J. R., & Burson, F. (1964). Bat rabies. A discussion of problems existing in Ohio. *The Ohio State Medical Journal, 60*, 361–364.

Scott, T. P., Fischer, M., Khaiseb, S., Freuling, C., Hoper, D., Hoffmann, B., … Nel, L. H. (2013). Complete genome and molecular epidemiological data infer the maintenance of rabies among kudu (*Tragelaphus strepsiceros*) in Namibia. *PLoS One, 8*(3), e58739. https://doi.org/10.1371/journal.pone.0058739.

Seetahal, J. F. R., Vokaty, A., Vigilato, M. A. N., Carrington, C. V. F., Pradel, J., Louison, B., … Rupprecht, C. E. (2018). Rabies in the Caribbean: A situational analysis and historic review. *Tropical Medicine and Infectious Disease, 3*(3) https://doi.org/10.3390/tropicalmed3030089.

Selhorst, T., Thulke, H. H., & Müller, T. (2001). Cost-efficient vaccination of foxes (*Vulpes vulpes*) against rabies and the need for a new baiting strategy. *Preventive Veterinary Medicine, 51*(1–2), 95–109.

Shulpin, M. I., Nazarov, N. A., Chupin, S. A., Korennoy, F. I., Metlin, A. Y., & Mischenko, A. V. (2018). Rabies surveillance in the Russian Federation. *Revue Scientifique et Technique, 37*(2), 483–495. https://doi.org/10.20506/rst.37.2.2817.

Shwiff, S. A., Elser, J. L., Ernst, K. H., Shwiff, S. S., & Anderson, A. (2018). Cost-benefit analysis of controlling rabies: Placing economics at the heart of rabies control to focus political will. *Revue Scientifique et Technique (Office International des Epizooties), 37*(2), 681–689.

Sidwa, T. J., Wilson, P. J., Moore, G. M., Oertli, E. H., Hicks, B. N., Rohde, R. E., & Johnston, D. H. (2005). Evaluation of oral rabies vaccination programs for control of rabies epizootics in coyotes and gray foxes: 1995-2003. *Journal of the American Veterinary Medical Association, 227*(5), 785–792.

Sillero-Zubiri, C., Marino, J., Gordon, C. H., Bedin, E., Hussein, A., Regassa, F., … Fooks, A. R. (2016). Feasibility and efficacy of oral rabies vaccine SAG2 in endangered Ethiopian wolves. *Vaccine, 34*(40), 4792–4798. https://doi.org/10.1016/j.vaccine.2016.08.021.

Slate, D., Algeo, T. P., Nelson, K. M., Chipman, R. B., Donovan, D., Blanton, J. D., … Rupprecht, C. E. (2009). Oral rabies vaccination in North America: Opportunities, complexities, and challenges. *PLoS Neglected Tropical Diseases*, 3(12), e549. https://doi.org/10.1371/journal.pntd.0000549.

Slate, D., Chipman, R. B., Algeo, T. P., Mills, S. A., Nelson, K. M., Croson, C. K., … Rupprecht, C. E. (2014). Safety and immunogenicity of ontario rabies vaccine bait (ONRAB) in the first us field trial in raccoons (*Procyon lotor*). *Journal of Wildlife Diseases*, 50(3), 582–595. https://doi.org/10.7589/2013-08-207.

Slate, D., Kirby, J. D., Morgan, D. P., Algeo, T. P., Trimarchi, C. V., Nelson, K. M., … Chipman, R. B. (2017). Cost and relative value of road kill surveys for enhanced rabies surveillance in raccoon rabies management. *Tropical Medicine and Infectious Disease*. 2(2). https://doi.org/10.3390/tropicalmed2020013.

Slate, D., Owens, R., Connolly, G., & Simmons, G. (1992). *Decision making for wildlife damage management.* Paper presented at the 57th North American Wildlife and Natural Resources Conference, Charlotte, North Carolina.

Slate, D., & Rupprecht, C. E. (2012). Rabies management in wild carnivores. In R. E. Miller & M. E. Fowler (Eds.), *Fowler's zoo and wild animal medicine: Current therapy* (pp. 366–375). St. Louis, MO: Elsevier Sanders.

Slate, D., Rupprecht, C. E., Rooney, J. A., Donovan, D., Badcock, J., Messier, A., … Nelson, K. (2008). Attaining raccoon rabies management goals: History and challenges. *Developments in Biologicals*, 131, 439–447.

Slate, D., Rupprecht, C. E., Rooney, J. A., Donovan, D., Lein, D. H., & Chipman, R. B. (2005). Status of oral rabies vaccination in wild carnivores in the United States. *Virus Research*, 111(1), 68–76. https://doi.org/10.1016/j.virusres.2005.03.012.

Slavinski, S., Humberg, L., Lowney, M., Simon, R., Calvanese, N., Bregman, B., … Oleszko, W. (2012). Trap-vaccinate-release program to control raccoon rabies, New York, USA. *Emerging Infectious Diseases*, 18(7), 1170–1172. https://doi.org/10.3201/eid1807.111485.

Smith, G. C., & Harris, S. (1991). Rabies in urban foxes (*Vulpes vulpes*) in Britain: The use of a spatial stochastic simulation model to examine the pattern of spread and evaluate the efficacy of different control regimes. *Philosophical Transactions of the Royal Society of London Series B, Biological Sciences*, 334(1271), 459–479. https://doi.org/10.1098/rstb.1991.0127.

Smith, J., Orciari, L. A., & Yager, P. A. (1995). Molecular epidemiology of rabies in the United States. *Seminars in Virology*, 6, 387–400.

Smith, T. G., Siirin, M., Wu, X., Hanlon, C. A., & Bronshtein, V. (2015). Rabies vaccine preserved by vaporization is thermostable and immunogenic. *Vaccine*, 33(19), 2203–2206. https://doi.org/10.1016/j.vaccine.2015.03.025.

Smith, J. S., Sumner, J., Roumillat, F., Baer, G. M., & Winkler, W. G. (1983). *Epidemiological analysis of street rabies viruses from enzootic areas of the United States.* Paper presented at the Rabies in the Tropics, Tunis.

Smith, T. G., Wu, X., Ellison, J. A., Wadhwa, A., Franka, R., Langham, G. L., … Bronshtein, V. L. (2017). Assessment of the immunogenicity of rabies vaccine preserved by vaporization and delivered to the duodenal mucosa of gray foxes (*Urocyon cinereoargenteus*). *American Journal of Veterinary Research*, 78(6), 752–756. https://doi.org/10.2460/ajvr.78.6.752.

Sobey, K. G., Rosatte, R., Bachmann, P., Buchanan, T., Bruce, L., Donovan, D., … Wandeler, A. (2010). Field evaluation of an inactivated vaccine to control raccoon rabies in Ontario, Canada. *Journal of Wildlife Diseases*, 46(3), 818–831. https://doi.org/10.7589/0090-3558-46.3.818.

Southeastern Association of Fish and Wildlife Agencies (SEAFWA), (2016a). *Recommendations for state oversight of wildlife damage control agents.* Retrieved from: http://www.seafwa.org/Documents%20and%20Settings/46/Site%20Documents/Resources/Committee%20Resources/SEAFWA_WDCA%20recommendations_AS%20ADOPTED.pdf.

Southeastern Association of Fish and Wildlife Agencies (SEAFWA), (2016b). *Recommended best management practices to address possession, transportation and disposition of rabies vector species.* Retrieved from: http://www.seafwa.org/Documents%20and%20Settings/46/Site%20Documents/Resources/Committee%20Resources/SEAFWA_BMPs%20for%20Rabies%20Vector%20Species_AS%20ADOPTED.pdf.

Steck, F., & Wandeler, A. (1980). The epidemiology of fox rabies in Europe. *Epidemiologic Reviews*, 2, 71–96.

Steck, F., Wandeler, A., Bichsel, P., Capt, S., Hafliger, U., & Schneider, L. (1982). Oral immunization of foxes against rabies. Laboratory and field studies. *Comparative Immunology, Microbiology and Infectious Diseases*, 5(1–3), 165–171. https://doi.org/10.1016/0147-9571(82)90031-5.

Steck, F., Wandeler, A., Bichsel, P., Capt, S., & Schneider, L. (1982). Oral immunisation of foxes against rabies, a field study. *Zentralblatt für Veterinärmedizin. Reihe B*, 29, 372–396.

Steelman, H. G., Henke, S. E., & Moore, G. M. (2000). Bait delivery for oral rabies vaccine to gray foxes. *Journal of Wildlife Diseases*, 36(4), 744–751. https://doi.org/10.7589/0090-3558-36.4.744.

Sterner, R. T., Meltzer, M. I., Shwiff, S. A., & Slate, D. (2009). Tactics and economics of wildlife oral rabies vaccination, Canada and the United States. *Emerging Infectious Diseases*, *15*(8), 1176–1184. https://doi.org/10.3201/eid1508.081061.

Sterner, R. T., & Smith, G. C. (2006). Modelling wildlife rabies: Transmission, economics, conservation. *Biological Conservation*, *131*, 163–179.

Stevenson, B., Goltz, J., & Masse, A. (2016). Preparing for and responding to recent incursions of raccoon rabies variant into Canada. *Canada Communicable Disease Report*, *42*(6), 125–129.

Stohr, K., & Meslin, F. M. (1996). Progress and setbacks in the oral immunisation of foxes against rabies in Europe. *Veterinary Record*, *139*, 32–35.

Sumner, J. W., Shaddock, J. H., Wu, G. J., & Baer, G. M. (1988). Oral administration of an attenuated strain of canine adenovirus (type 2) to raccoons, foxes, skunk, and mongoose. *American Journal of Veterinary Research*, *49*, 169–171.

Swanepoel, R., Barnard, B. J., Meredith, C. D., Bishop, G. C., Bruckner, G. K., Foggin, C. M., & Hubschle, O. J. (1993). Rabies in southern Africa. *Onderstepoort Journal of Veterinary Research*, *60*(4), 325–346.

Szanto, A. G., Nadin-Davis, S. A., Rosatte, R. C., & White, B. N. (2011). Genetic tracking of the raccoon variant of rabies virus in eastern North America. *Epidemics*, *3*(2), 76–87. https://doi.org/10.1016/j.epidem.2011.02.002.

Tabel, H., Corner, A. H., Webster, W. A., & Casey, C. A. (1974). History and epizootiology of rabies in Canada. *The Canadian Veterinary Journal*, *15*(10), 271–281.

Tardy, O., Masse, A., Pelletier, F., & Fortin, D. (2018). Interplay between contact risk, conspecific density, and landscape connectivity: An individual-based modeling framework. *Ecological Modelling*, *373*, 25–38.

Taylor, J., Meignier, B., Tartaglia, J., Languet, B., VanderHoeven, J., Franchini, G., ... Paoletti, E. (1995). Biological and immunogenic properties of a canarypox-rabies recombinant, ALVAC-RG (vCP65) in non-avian species. *Vaccine*, *13*(6), 539–549. https://doi.org/10.1016/0264410X(94)00028L.

Taylor, J., & Paoletti, E. (1988). Fowlpox virus as a vector in non-avian species. *Vaccine*, *6*(6), 466–468.

Taylor, J., Trimarchi, C., Weinberg, R., Languet, B., Guillemin, F., Desmettre, P., & Paoletti, E. (1991). Efficacy studies on a canarypox-rabies recombinant virus. *Vaccine*, *9*(3), 190–193.

Taylor, J., Weinberg, R., Languet, B., Desmettre, P., & Paoletti, E. (1988). Recombinant fowlpox virus inducing protective immunity in non-avian species. *Vaccine*, *6*(6), 497–503.

Thomas, I., Brochier, B., Languet, B., Blancou, J., Peharpre, D., Kieny, M. P., ... Pastoret, P. P. (1990). Primary multiplication site of the vaccinia-rabies glycoprotein recombinant virus administered to foxes by the oral route. *The Journal of General Virology*, *71*(Pt 1), 37–42.

Thulke, H. -H., & Eisinger, D. (2008). The strength of 70%: Revision of a standard threshold of rabies control. *Developmental Biology (Basel)*, *131*, 291–298.

Thulke, H. -H., Eisinger, D., Selhorst, T., & Muller, T. F. (2008). Scenario-analysis evaluating emergency strategies after rabies re-introduction. *Developmental Biology (Basel)*, *131*, 265–272.

Tolson, N. D., Charlton, K. M., Lawson, K. F., Campbell, J. B., & Stewart, R. B. (1988). Studies of ERA/BHK-21 rabies vaccine in skunks and mice. *Canadian Journal of Veterinary Research*, *52*(1), 58–62.

Tordo, N., Foumier, A., Jallet, C., Szelechowski, M., Klonjkowski, B., & Eloit, M. (2008). Canine adenovirus based rabies vaccines. *Developmental Biology (Basel)*, *131*, 467–476.

Trewby, H., Nadin-Davis, S. A., Real, L. A., & Biek, R. (2017). Processes underlying rabies virus incursions across US-Canada Border as revealed by whole-genome phylogeography. *Emerging Infectious Diseases*, *23*(9), 1454–1461. https://doi.org/10.3201/eid2309.170325.

Troupin, C., Dacheux, L., Tanguy, M., Sabeta, C., Blanc, H., Bouchier, C., ... Bourhy, H. (2016). Large-scale phylogenomic analysis reveals the complex evolutionary history of rabies virus in multiple carnivore hosts. *PLoS Pathogens*, *12*(12), e1006041. https://doi.org/10.1371/journal.ppat.1006041.

Tu, C., Feng, Y., & Wang, Y. (2018). Animal rabies in the People's Republic of China. *Revue Scientifique et Technique*, *37*(2), 519–528. https://doi.org/10.20506/rst.37.2.2820.

United States Department of Agriculture (USDA), (2008). *North American rabies management plan*. Concord, NH: US Department of Agriculture. Retrieved from: https://www.aphis.usda.gov/wildlife_damage/oral_rabies/downloads/Final%20NARMP%209-30-2008%20(ENGLISH).pdf.

Velasco-Villa, A., Escobar, L. E., Sanchez, A., Shi, M., Streicker, D. G., Gallardo-Romero, N. F., ... Emerson, G. (2017). Successful strategies implemented towards the elimination of canine rabies in the Western Hemisphere. *Antiviral Research*, *143*, 1–12. https://doi.org/10.1016/j.antiviral.2017.03.023.

Velasco-Villa, A., Orciari, L. A., Souza, V., Juarez-Islas, V., Gomez-Sierra, M., Castillo, A., … Rupprecht, C. E. (2005). Molecular epizootiology of rabies associated with terrestrial carnivores in Mexico. *Virus Research, 111*(1), 13–27. https://doi.org/10.1016/j.virusres.2005.03.007.

Velasco-Villa, A., Reeder, S. A., Orciari, L. A., Yager, P. A., Franka, R., Blanton, J. D., … Rupprecht, C. E. (2008). Enzootic rabies elimination from dogs and reemergence in wild terrestrial carnivores, United States. *Emerging Infectious Diseases, 14*(12), 1849–1854.

Vercauteren, K., Deliberto, T., Shwiff, S., Ellis, C., Chipman, R. B., & Slate, D. (2014). *Rabies in North America: Need and call for a One Health approach*. Paper presented at the Fourteenth Wildlife Damage Management Conference, Nebraska City, NE.

Vora, N. M., Basavaraju, S. V., Feldman, K. A., Paddock, C. D., Orciari, L., Gitterman, S., … Transplant-Associated Rabies Virus Transmission Investigation Team. (2013). Raccoon rabies virus variant transmission through solid organ transplantation. *JAMA, 310*(4), 398–407. https://doi.org/10.1001/jama.2013.7986.

Vos, A. (2003). Oral vaccination against rabies and the behavioural ecology of the red fox (*Vulpes vulpes*). *Journal of Veterinary Medicine. B, Infectious Diseases and Veterinary Public Health, 50*(10), 477–483.

Vos, A., Freuling, C. M., Hundt, B., Kaiser, C., Nemitz, S., Neubert, A., … Muller, T. (2017). Oral vaccination of wildlife against rabies: Differences among host species in vaccine uptake efficiency. *Vaccine, 35*(32), 3938–3944. https://doi.org/10.1016/j.vaccine.2017.06.022.

Vos, A., Kretzschmar, A., Ortmann, S., Lojkic, I., Habla, C., Muller, T., … Schuster, P. (2013). Oral vaccination of captive small Indian mongoose (*Herpestes auropunctatus*) against rabies. *Journal of Wildlife Diseases, 49*(4), 1033–1036. https://doi.org/10.7589/2013-02-035.

Vos, A., Neubert, A., Aylan, O., Schuster, P., Pommerening, E., Muller, T., & Chivatsi, D. C. (1999). An update on safety studies of SAD B19 rabies virus vaccine in target and non-target species. *Epidemiology and Infection, 123*(1), 165–175.

Vuta, V., Picard-Meyer, E., Robardet, E., Barboi, G., Motiu, R., Barbuceanu, F., … Cliquet, F. (2016). Vaccine-induced rabies case in a cow (*Bos taurus*): Molecular characterisation of vaccine strain in brain tissue. *Vaccine, 34*(41), 5021–5025. https://doi.org/10.1016/j.vaccine.2016.08.013.

Wachendorfer, G., & Frost, J. (1992). Epidemiology of red fox rabies: A review. In K. Bögel, F. M. Meslin, & B. Kaplan (Eds.), *Wildlife rabies control* (pp. 19–31). Kent: Wells Medical Ltd.

Wallace, R. M., Gilbert, A., Slate, D., Chipman, R., Singh, A., Cassie, W., & Blanton, J. D. (2014). Right place, wrong species: A 20-year review of rabies virus cross species transmission among terrestrial mammals in the United States. *PLoS One, 9*(10), e107539. https://doi.org/10.1371/journal.pone.0107539.

Wallace, R. M., Lai, Y., Doty, J. B., Chen, C. C., Vora, N. M., Blanton, J. D., … Pei, K. J. C. (2018). Initial pen and field assessment of baits to use in oral rabies vaccination of Formosan ferret-badgers in response to the re-emergence of rabies in Taiwan. *PLoS One, 13*(1), e0189998. https://doi.org/10.1371/journal.pone.0189998.

Wandeler, A. (1991). Oral immunization of wildlife. In G. M. Baer (Ed.), *The natural history of rabies* (2nd ed., pp. 485–503). Boca Raton, FL: CRC Press, Inc.

Wandeler, A. I., Capt, S., Kappeler, A., & Hauser, R. (1988). Oral immunization of wildlife against rabies: Concept and first field experiments. *Reviews of Infectious Diseases, 10*(Suppl. 4), S649–S653.

Wasniewski, M., Almeida, I., Baur, A., Bedekovic, T., Boncea, D., Chaves, L. B., … Cliquet, F. (2016). First international collaborative study to evaluate rabies antibody detection method for use in monitoring the effectiveness of oral vaccination programmes in fox and raccoon dog in Europe. *Journal of Virological Methods, 238*, 77–85. https://doi.org/10.1016/j.jviromet.2016.10.006.

Wasniewski, M., Barrat, J., Combes, B., Guiot, A. L., & Cliquet, F. (2014). Use of filter paper blood samples for rabies antibody detection in foxes and raccoon dogs. *Journal of Virological Methods, 204*, 11–16. https://doi.org/10.1016/j.jviromet.2014.04.005.

Wasniewski, M., Guiot, A. L., Schereffer, J. L., Tribout, L., Mahar, K., & Cliquet, F. (2013). Evaluation of an ELISA to detect rabies antibodies in orally vaccinated foxes and raccoon dogs sampled in the field. *Journal of Virological Methods, 187*(2), 264–270. https://doi.org/10.1016/j.jviromet.2012.11.022.

World Health Organization (WHO), (1992). *WHO Expert Committee on Rabies: 8th report*. Geneva: WHO.

World Health Organization (WHO), (2005). *WHO Expert Consultation on Rabies, first report*. Geneva, Switzerland, Retrieved from: https://www.who.int/rabies/resources/who_trs_931/en/.

World Health Organization (WHO), (2013). *WHO Expert Consultation on Rabies, second report*. Geneva, Switzerland, Retrieved from: http://www.who.int/iris/handle/10665/85346.

World Health Organization (WHO), (2018). *WHO Expert Consultation on Rabies, third report.* Geneva, Switzerland, Retrieved from: https://www.who.int/rabies/resources/who_trs_1012/en/.

World Health Organization (WHO), (2019). Rabies surveillance. *Rabies Bulletin Europe,* Retrieved from: https://www.who-rabies-bulletin.org/site-page/queries.

Wiktor, T. J., & Koprowski, H. (1978). Monoclonal antibodies against rabies virus produced by somatic cell hybridization: Detection of antigenic variants. *Proceedings of the National Academy of Sciences of the United States of America, 75*(8), 3938–3942.

Wiktor, T. J., & Koprowski, H. (1980). Antigenic variants of rabies virus. *The Journal of Experimental Medicine, 152*(1), 99–112.

Winkler, W. G. (1975). Fox rabies. In G. M. Baer (Ed.), *Vol.1. The natural history of rabies* (pp. 3–22). New York, NY: Academic Press, Inc.

Winkler, W. G. (1992). A review of the development of the oral vaccination technique for immunizing wildlife against rabies. In K. Bögel, F. M. Meslin, & M. M. Kaplan (Eds.), *Wildlife rabies control* (pp. 82–96). Kent: Wells Medical Ltd.

Winkler, W. G., & Baer, G. M. (1976). Rabies immunization of red foxes (*Vulpes fulva*) with vaccine in sausage baits. *American Journal of Epidemiology, 103*(4), 408–415.

Wohlers, A., Lankau, E. W., Oertli, E. H., & Maki, J. (2018). Challenges to controlling rabies in skunk populations using oral rabies vaccination: A review. *Zoonoses and Public Health, 65*(4), 373–385. https://doi.org/10.1111/zph.12471.

World Organization for Animal Health (OIE) (2018). Rabies (infection with rabies virus and other lyssaviruses). In *Manual of diagnostic tests and vaccines for terrestrial animals* (pp. 578–612). World Animal Health Organization.

Yakobson, B. A., King, R., Amir, S., Devers, N., Sheichat, N., Rutenberg, D., … David, D. (2006). Rabies vaccination programme for red foxes (*Vulpes vulpes*) and golden jackals (*Canis aureus*) in Israel (1999-2004). *Developmental Biology (Basel), 125,* 133–140.

Yakobson, B., King, R., Ingor, M., Markovich, M. P., & Goshen, T. (2019). *Re-emergency of wildlife rabies in golden jackals (Canis aureus), Israel (2017-2018).* Paper presented at the 30th Annual Rabies in the Americas Conference, Kansas City, Kansas.

Yang, D. K., Cho, I. S., & Kim, H. H. (2018). Strategies for controlling dog-mediated human rabies in Asia: Using 'One Health' principles to assess control programmes for rabies. *Revue Scientifique et Technique, 37*(2), 473–481. https://doi.org/10.20506/rst.37.2.2816.

Yang, D. K., Kim, H. H., Choi, S. S., Kim, J. T., Lee, K. B., Lee, S. H., & Cho, I. S. (2016). Safety and immunogenicity of recombinant rabies virus (ERAGS) in mice and raccoon dogs. *Clinical and Experimental Vaccine Research, 5*(2), 159–168. https://doi.org/10.7774/cevr.2016.5.2.159.

Zhao, J. H., Zhao, L. F., Liu, F., Jiang, H. Y., & Yang, J. L. (2019). Ferret badger rabies in Zhejiang, Jiangxi and Taiwan, China. *Archives of Virology, 164*(2), 579–584. https://doi.org/10.1007/s00705-018-4082-5.

Zieger, U., Marston, D. A., Sharma, R., Chikweto, A., Tiwari, K., Sayyid, M., … Horton, D. L. (2014). The phylogeography of rabies in Grenada, West Indies, and implications for control. *PLoS Neglected Tropical Diseases, 8*(10), e3251. https://doi.org/10.1371/journal.pntd.0003251.

CHAPTER 20

Modeling canine rabies virus transmission dynamics

Malavika Rajeev[a], C. Jessica E. Metcalf[a], Katie Hampson[b]

[a]Department of Ecology and Evolutionary Biology, Princeton University, Princeton, NJ, United States [b]Institute of Biodiversity, Animal Health and Comparative Medicine, College of Medical, Veterinary and Life Sciences, University of Glasgow, Glasgow, United Kingdom

20.1 Introduction

Models of disease dynamics are a powerful tool in the arsenal of disease prevention and control efforts, and can be used to estimate key epidemiological parameters, establish targets for control, and guide policy (Heesterbeek et al., 2015). Modeling can also identify counterintuitive outcomes that emerge as interventions are implemented, and challenges in the endgame when disproportionate resources are necessary to reach the last mile of elimination (Klepac, Metcalf, McLean, & Hampson, 2013). In light of the global goal to eliminate human deaths due to dog-mediated rabies by 2030, models of rabies virus transmission have potential to inform control efforts as countries progress toward elimination.

20.1.1 History of modeling rabies virus transmission dynamics

Modeling rabies in domestic dog populations is a relatively nascent effort. In contrast, models of wildlife rabies guided early control efforts (Panjeti & Real, 2011). Elimination of fox rabies in Europe was kickstarted by modeling studies that demonstrated the feasibility of control (Anderson, Jackson, May, & Smith, 1981). Surveillance of rabies in wildlife systems in Europe and North America provided rich data sets to characterize dynamics, identifying the wave front of outbreaks to target control geographically (Murray, Stanley, & Brown, 1986), establishing that landscape features such as rivers act as barriers to disease dispersal (Smith, Lucey, Waller, Childs, & Real, 2002), and delineating how birth pulses shape seasonality in transmission (Duke-Sylvester, Bolzoni, & Real, 2011). This work provides a foundation for modeling canine rabies, but there are fundamental differences between wildlife and domestic dog systems. Human populations, behavior, and culture structure dog populations (Cleaveland, Beyer, et al., 2014). In addition, canine rabies persists in low- and middle-income countries where surveillance capacity is limited and representative disease data are lacking (Scott, Coetzer, Fahrion, & Nel, 2017). Beyond capturing core infection biology, models of canine rabies must also encompass human influences and be tractable to interpretation in data-sparse settings.

20.1.2 The modeling backbone for canine rabies

Rabies can be modeled in an **SEIV** framework, with **Susceptible**, **Exposed**, **Infectious**, and **Vaccinated** classes (Fig. 20.1). Dog demography governs the dynamics of the susceptible and vaccinated classes. The **Susceptible** population is replenished by births and depleted by mortality (both natural and disease-induced) and vaccination. The **Vaccinated** population is governed by the rate of vaccination, but depleted by natural mortality and waning of immunity generated by vaccines (most high quality vaccines are protective for at least 3 years, Lakshmanan et al., 2006). For canine rabies, evidence suggests that domestic dogs are the reservoir host even in areas with complex wild carnivore communities (Lembo et al., 2007, 2008). While other wildlife hosts may contribute to transmission, single-host models of rabies in the dog population are likely sufficient to understanding and predicting dynamics in most endemic areas (Cleaveland, Lankester, Townsend, Lembo, & Hampson, 2014).

Rabies virus is directly transmitted in the saliva of infectious animals, typically via bites. Transmission is on average low: most infectious cases do not result in forward transmission, and on average dogs infect one or two other dogs. However, there is also substantial heterogeneity in transmission, and some dogs are capable of biting upwards of 20 other dogs during their short infectious period (Hampson et al., 2009). The incubation period is about 21 days, but is highly variable. While most exposed dogs become infectious within 1 month, some infections manifest months after initial exposure (Foggin, 1988; Hampson et al., 2009; Hemachudha, Laothamatas, & Rupprecht, 2002). The infectious period, on the other hand, is predictably short, and infection results in death generally within 10 days of showing neurological signs of infection (Hampson et al., 2009; Tepsumethanon, Wilde, & Meslin, 2005). There is little evidence of subclinical infections.

Although transmission is mostly local (<1 km), rabies can cause erratic and unpredictable behavior, with infected dogs able to run more than 15 km, beyond the typical home range of most healthy dogs (Hampson et al., 2009). As a result, secondary cases often occur from disease-mediated incursions spread from neighboring populations (e.g., nearby populated settlements within the range of rabid dog movement). In addition, long-distance human-mediated incursions of incubating dogs can result in outbreaks being seeded from otherwise unconnected populations (Brunker, Hampson, Horton, & Biek, 2012).

20.1.3 How to model rabies virus transmission?

There has been considerable debate about how to model rabies virus transmission, which echoes a larger debate within the disease ecology community (Lloyd-Smith, Cross, et al., 2005). Theory indicates that for diseases with density-dependent transmission, i.e., when transmission scales with host density, there exists a threshold density below which the

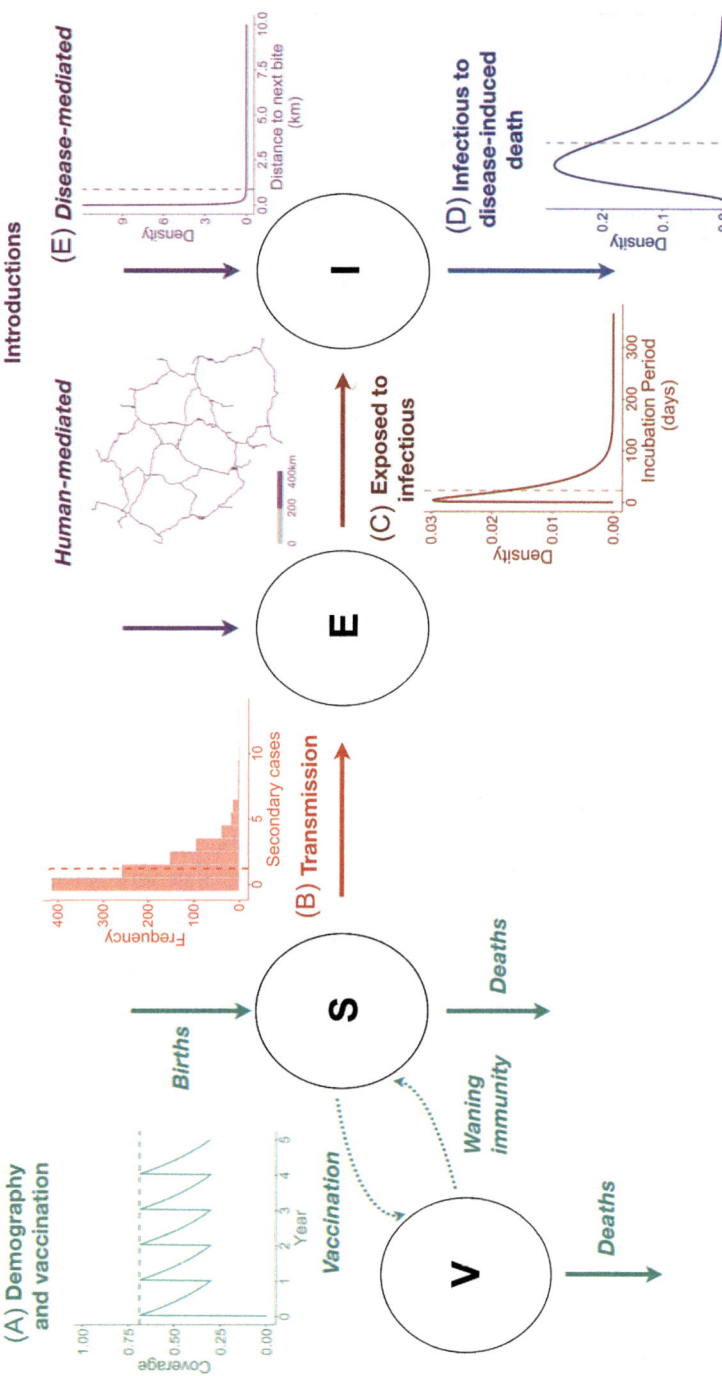

FIG. 20.1 The Susceptible-Exposed-Infectious-Vaccinated (SEIV) modeling framework for canine rabies: circles indicate epidemiological classes, arrows linking circles indicate how individuals can move between classes, insets describe underlying processes and influences. **(A) Host demography** (i.e., the balance between births and deaths) and vaccination govern the susceptible and vaccinated population dynamics. Following vaccination campaigns, vaccination coverage (y axis, inset) first increases (vertical jumps) then wanes over time (x axis) as vaccinated individuals die, susceptible individuals are born, or as immunity conferred by vaccination wanes (in this example, campaigns reach 70% of the population annually, but coverage wanes to approximately 35% before the next annual campaign). **(B) Transmission** is on average low, but highly heterogeneous. Inset shows number of secondary cases generated from a negative binomial distribution ($n = 1000$ draws, mean number of secondary cases = 1.2, red dashed line). **(C)** Individuals move from exposed to infectious on average after 22.3 days (inset, dashed line) but this is also highly variable with some infections occurring months to years after exposure. **(D) Disease-induced mortality** is complete, and the infectious period is short, on average 3.1 days (dashed line), with deaths due to infection occurring within 10 days. **(E) Introductions** from outside the population modeled may seed cases within. Introductions may result from **disease-mediated** movement of infectious dogs (sometimes upwards of 10 km; inset shows dispersal kernel, gamma distribution) and **human-mediated** movements of incubating dogs (potentially on the scale of 100s of km through movement along roads; the inset shows an example of a major road network in Tanzania). All parameters used and associated references are listed in Table 20.1.

TABLE 20.1 Key parameter values associated with underlying processes illustrated in Fig. 20.1.

Process	Distribution	Parameters	Value	Source	Inset
Birth rate	Exponential	Mean annual rate (dogs/yr)	0.5	Czupryna et al. (2016)	A
Death rate	Exponential	Mean annual rate (dogs/yr)	0.42	Czupryna et al. (2016)	A
Vaccine waning	Exponential	Mean annual rate (dogs/yr)	0.33	Lakshmanan et al. (2006)	A
Secondary cases (R_0)	Negative binomial	Mean secondary cases	1.2	Townsend, Lembo, et al. (2013); Townsend, Sumantra, et al. (2013)	B
		Dispersion parameter (k)	1.3		
Incubation period	Gamma, mean 22.3 days	Shape	1.15	Hampson et al. (2009)	C
		Rate	0.04		
Infectious period	Gamma, mean 3.1 days	Shape	2.9	Hampson et al. (2009)	D
		Rate	1.01		
Dispersal kernel	Gamma, mean 0.88 km	Shape	0.215	Townsend, Lembo, et al. (2013); Townsend, Sumantra, et al. (2013)	E
		Rate	0.245		

disease cannot persist (McCallum, Barlow, & Hone, 2001). However, there is no such threshold when transmission is frequency-dependent, i.e., transmission rates are independent of host density (Lloyd-Smith, Cross, et al., 2005).

For canine rabies, the basic reproductive number (R_o) or the average number of secondary cases resulting from a single infection in a completely susceptible population, is generally estimated as between 1 and 2 (Coleman & Dye, 1996; Hampson et al., 2009; Kurosawa et al., 2017; Townsend, Sumantra, et al., 2013). Such consistently low estimates of R_0 across a range of dog densities suggest that rabies virus transmission is largely frequency-dependent (Fitzpatrick et al., 2012; Hampson et al., 2009; Tian et al., 2018; Townsend, Sumantra, et al., 2013; Zinsstag et al., 2009). That is, rabid dogs have on average the same number of infectious contacts regardless of the density of dogs around them. As a result, reductions in population densities are not likely to be effective in eliminating rabies. In practice, although a common intervention and one predicated on assumptions of density-dependent transmission, indiscriminate culling of dogs does not curtail rabies transmission (Morters et al., 2012).

Despite evidence for frequency-dependent transmission, many modeling studies formulate rabies transmission as density-dependent (Fig. 20.2D). For a given R_0, this assumption of density-dependent transmission does not impact herd immunity thresholds; the critical proportion that needs to be vaccinated, p_c, is equal to $1 - 1/R_0$ regardless of the form of transmission (McCallum et al., 2001). However, density-dependent models predict reductions in transmission due to declining dog density (e.g., via culling or disease-induced mortality) that are unlikely to translate to the real world.

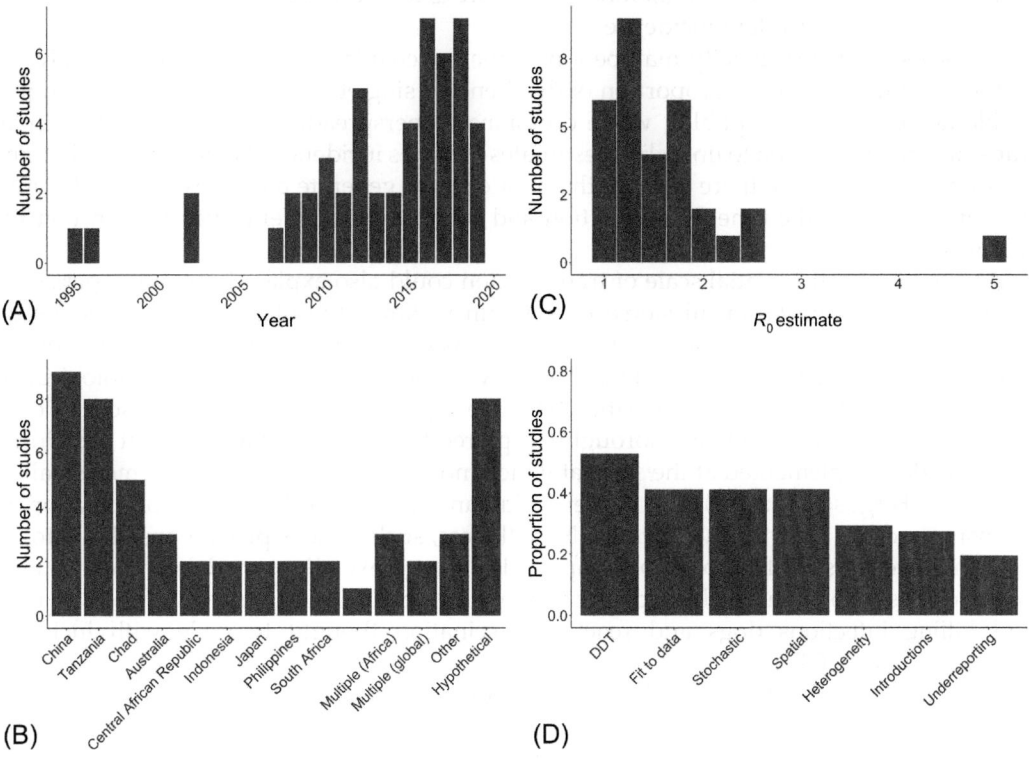

FIG. 20.2 Summary of studies with a dynamic model of canine rabies. A total of 51 studies were included. (A) Year of publication, with most studies published after 2006; (B) Countries where rabies dynamics were modeled: studies were concentrated in China, Tanzania, and Chad, but many also examined dynamics in hypothetical contexts, not specific to any geographic situation. (C) Estimates of R_0: most studies estimated R_0 below 2 (10 studies, with 31 estimates; estimates of R_e (the effective reproduction number which accounts for ongoing vaccination) and R_t (time-varying reproductive number) were excluded ($N=3$). (D) Key features of models ($N=51$): most assumed density-dependent transmission ($N=27$). Less than half were fit to data ($N=20$), stochastic ($N=20$), or spatially explicit ($N=19$). 15/51 studies incorporated individual heterogeneity in transmission and 14/51 introductions from outside the population modeled. Only 10 included an observation model in their analysis or accounted for underreporting in their inference. Full bibliography and metadata included in Supplementary Table S1.

Models with frequency-dependent transmission are also not entirely consistent with empirical observations. Frequency-dependent models that assume homogeneous mixing (i.e., equal contact probabilities between all individuals in a population, also referred to as "mass action") result in eventual population extinction for fatal pathogens like rabies (Keeling & Rohani, 2011). Only under very low transmission (1.01–1.02) and high population growth can rabies persist in models with frequency-dependent transmission. For models with density-dependent transmission, even with R_0 between 1.01 and 1.1, models of rabies show high annual incidence (Supplemental Fig. S1 at https://doi.org/10.1016/B978-0-12-818705-0.00020-0), which is at odds with empirical evidence. Where measured, rabies incidence is low (<2% annually) and consequently has little demographic impact on dog populations

(Hampson et al., 2016). Additional model structure is therefore necessary to explain how rabies can persist at such low incidence.

Transmission heterogeneity may be a potential mechanism to explain the relatively low incidence of rabies. A high proportion of dead-end or singleton transmissions result in negligible depletion of susceptibles, while occasional superspreaders may seed and maintain transmission. In addition to unrealistic estimates of rabies incidence, if heterogeneity in transmission is not captured, there is a risk that models may generate biased estimates of control indicators, such as the time to elimination and the threshold level of vaccination that this requires.

Accounting for the spatial scale of transmission could also explain how rabies persists at low incidence. As most transmission occurs within a 1-km radius of infected animals, susceptible depletion at such fine scales may limit transmission in a way that is not captured in mass action models (Ferrari, Perkins, Pomeroy, & Bjørnstad, 2011). Phenomenological approximations may offer a solution to this challenge (Aparicio & Pascual, 2007; Pascual, Roy, & Laneri, 2011), but have yet to be thoroughly explored for rabies. Spatially explicit individual-based models implemented at the scale at which most mixing occurs generate more realistic dynamics (Ferguson et al., 2015; Townsend, Sumantra, et al., 2013), but are computationally intensive and not analytically tractable. Nonetheless, such models provide insights into underlying mechanisms that could be simplified for more expedient models. Finally, human behavior has also been implicated in curtailing epidemics, with responses such as tying and killing infectious dogs and reactive vaccination thought to scale with incidence (Hampson et al., 2007).

There are limited data to disentangle these potential mechanisms, which could reconcile empirical observations with modeling results. Further work is necessary to ensure sufficient model realism to inform policy, but balancing realism and complexity is a key challenge for any modeling study (Grassly & Fraser, 2008). Building in realism requires additional parameterization and, often, additional assumptions. Robust epidemiological and biological data are therefore key to improving our understanding of how to model rabies transmission.

20.2 Existing modeling studies

Two systematic reviews of rabies models recently examined the effectiveness and cost effectiveness of control and prevention strategies. They concluded that estimates of R_0 are consistently below 2 and dog vaccination is an effective strategy, but vaccination coverage is critically influenced by dog demography (Rattanavipapong et al., 2018). Both mass dog vaccination and provisioning of PEP to bite patients are cost effective, in contrast to dog culling, which has rarely been identified as either economically feasible or effective (Anothaisintawee et al., 2018). Building off these reviews, we examined studies with a dynamic modeling component and synthesized insights generated and data used to inform them. We searched for papers that had the terms "rabies" AND ("domestic dog*" OR "canine") AND "model*" on PubMed and Scopus, including all English language papers published between January 1995 and July 2019 that incorporated a transmission model of rabies virus in domestic dogs. Of the 547 unique

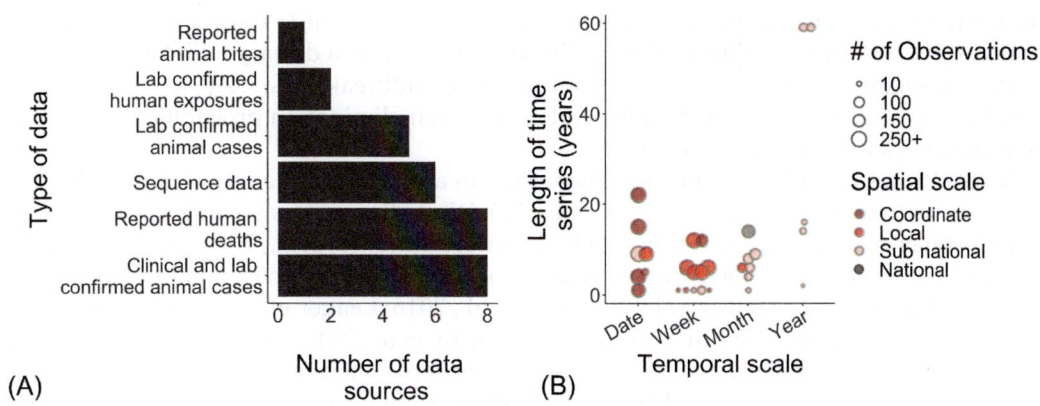

FIG. 20.3 **Rabies data reported in modeling studies** ($N=25$ studies reporting 30 unique data sources). (A) Type of data used. (B) The scales of temporal (x-axis) and spatial (colors) information available and the duration (y axis). The size of the points is proportional to the number of observations in each data set. Any rabies data that were reported in studies were included (even if not used for fitting purposes, only for qualitative comparison). If multiple data sets were used, they were included as separate data sources, and if the same data set was used in multiple studies it was only included once.

records retrieved, 51 papers fitted these inclusion criteria (Fig. 20.3, Supplementary Table S1 at https://doi.org/10.1016/B978-0-12-818705-0.00020-0).

20.2.1 Insights and limitations

Of studies that compared intervention strategies (generally: mass dog vaccination, human PEP provisioning, and dog population control including culling), the majority show that dog vaccination is most effective, and essential to achieve elimination. Despite the potential to maximize population-level immunity, synchronizing vaccination campaigns geographically had little impact on probability of elimination, at least for annual vaccination campaigns. In contrast, spatial heterogeneity in vaccination coverage had a greater impact, with even small contiguous coverage gaps reducing the probability of rabies being eliminated (Ferguson et al., 2015; Townsend, Sumantra, et al., 2013).

While the critical vaccination threshold (p_c or $1-1/R_o$) should theoretically be much lower than 70% for a disease with the low range of R_0 estimated for rabies (Fig. 20.2C), the coverage level recommended by WHO reflects an empirical consensus (Coleman & Dye, 1996; World Health Organization, 2013). Models show that due to high turnover in domestic dog populations, annual campaigns that reach at least 70% of the population are necessary to maintain coverage >20% throughout the year. Furthermore, heterogeneity in transmission and frequent introductions of rabies cases increase both the vaccination threshold necessary to interrupt transmission, and the probability of observing small outbreaks even when vaccination coverage is high (Hampson et al., 2009; Lloyd-Smith, Schreiber, Kopp, & Getz, 2005).

Most published models were deterministic (33/51) and did not incorporate heterogeneities in transmission (36/51, Fig. 20.2D). However, as R_0 for rabies appears to be low, the interaction between stochasticity and heterogeneity in transmission may influence disease dynamics. In general, for diseases with high transmissibility (i.e., measles), heterogeneities in transmission can often be ignored as these complexities have little impact on the emergent dynamics of infection (Keeling & Rohani, 2011). However, for a disease with lower transmission, heterogeneities may result in unpredictable outbreaks (Grassly & Fraser, 2008). Stochasticity is especially crucial in the endgame, when elimination probabilities and incursion dynamics depend on rare events.

Most studies model rabies virus transmission in a closed population, that is without introductions from neighboring areas (Fig. 20.2D). While this is a reasonable approach in island settings such as in Bali, Indonesia (Townsend, Sumantra, et al., 2013), recent modeling and phylogenetic work shows the importance of incursions in less isolated populations in sustaining rabies virus transmission (Bourhy et al., 2016; Laager et al., 2019) and that multiple strains co-circulate within a population (Bourhy et al., 2016; Zinsstag et al., 2017). Human behavior is also a key driver of transmission patterns, facilitating as well as dampening transmission (Brunker et al., 2015). Multiple studies have found signals of long-distance transmission beyond the range of disease-mediated dispersal, showing the role of human-mediated movement of incubating dogs (Brunker et al., 2015; Talbi et al., 2010; Tohma et al., 2016). Road networks have been identified as correlates of phylogenetic distance, indicating that human movement could shape the spatial structure of canine rabies virus (Brunker et al., 2015; Talbi et al., 2010; Tohma et al., 2016). There is also strong phylogenetic evidence that historical human-mediated long-distance movements underlie much of the contemporary global distribution of canine rabies (King, Fooks, Aubert, & Wandeler, 2004). This work emphasizes the need to understand how the size and connectivity of populations affects the persistence of disease. Metapopulation models, which model disease transmission in a set of connected sub-populations, have productively explored this historically important question for childhood infections such as measles (Bjørnstad & Grenfell, 2008). For canine rabies, while a few studies have used metapopulation models (Beyer et al. 2010; Beyer et al. 2012; Laager et al. 2019), defining the drivers of connectivity between populations remains an important challenge, one which may well define progress toward elimination.

A few studies look at how contact networks and movement behaviors could drive transmission (Hudson, Brookes, Ward, & Dürr, 2019; Laager et al., 2018; Wilson-Aggarwal et al., 2019). These studies simulated outbreaks on contacts networks constructed using data from healthy domestic dogs. They found that in general, targeting highly connected dogs or dogs with larger home ranges for vaccination results in a higher probability of disease elimination, but few predictors of connectivity of individuals emerged. Broadly, these results are consistent with previous work on transmission heterogeneity and could bring valuable benefits if it were possible to a priori identify and target high-risk animals. However, these traits are difficult to estimate in most endemic settings, where there are limited data on dog populations, let alone individual dog traits. Moreover, as rabies causes severe neurological symptoms, the validity of these findings depends on how representative data from healthy dogs are of movement and contact patterns of rabid dogs.

Dynamic models have been integrated with economic models to estimate cost effectiveness of interventions, demand for rabies PEP, and disease burden. Early cost-effectiveness models critically lacked data on the costs of PEP for those seeking care for nonrabid dog bites (Fitzpatrick et al., 2014; Hampson, Cleaveland, & Briggs, 2011; Zinsstag et al., 2009). Decision tree models have addressed these issues and provide a framework to integrate field data on rabies exposures, health-seeking, and access and adherence to PEP into estimates of burden

(Hampson et al., 2015; Knobel et al., 2005; WHO Rabies Modelling Consortium, 2019). These more recent studies demonstrate that PEP is still a very cost-effective intervention even when accounting for management of patients bitten by nonrabid animals and emphasize the potential value of administering rabies vaccine intradermally using the latest WHO recommended abridged regimens (Hampson et al., 2019; Tarantola et al., 2018). However, they also highlight two other critical points for policy. First, without strategies for more judicious use, costs of PEP will remain high and continue to rise even when dog rabies is controlled. Moreover, human rabies deaths will continue to occur and the target of zero deaths by 2030 cannot be achieved through PEP alone. A massive scaling up of dog vaccination is required in most endemic countries. Support for human rabies vaccines through Gavi, the Vaccine Alliance, is therefore a promising step toward the 2030 goal (WHO Rabies Modelling Consortium, 2019), but more investment and commitment is still needed.

20.3 The gap between models and data

Despite limited surveillance, few studies incorporated observation models into their analyses or conducted sensitivity analyses on how underreporting might bias their inferences (Fig. 20.2D). Developing models of the observation process and integrating them into dynamic models, often termed state-space modeling (Beyer et al., 2010; Beyer et al., 2012; Cori et al., 2018), is essential when fitting to incomplete data. But, these modeling frameworks can also guide surveillance strategies across the elimination timeline by estimating the minimum detection levels and time necessary to verify elimination (Townsend, Lembo, et al., 2013).

A major limitation of many existing modeling studies is a lack of data to inform conclusions, with less than 40% of models fit to data (Fig. 20.2D). For studies which did report incidence data, the scale and quality of the data also varied greatly. Human deaths reported at the national or regional level and numbers of clinical and laboratory confirmed animal cases were the most commonly used data (Fig. 20.3A). The number of observations and length of the time series varied greatly, from over 1000 observations at a fine spatiotemporal scale over a 15-year period to annual cases reported for only 2 years (Fig. 20.3B). Ultimately, integrating data on rabies incidence and dog populations into models of transmission is a critical step to moving modeling efforts forward. We now describe the various data sources that can be used to fit and inform models and associated challenges and solutions to collecting this data.

20.3.1 Bite data

Bite data (i.e., data on patients seeking care for animal bites) are often used as a proxy for rabies exposure incidence. However, these data often lack details on the status of the biting animal and are heavily skewed by who has access to care, both geographically and socioeconomically. Paradoxically, in settings where the direct cost of PEP is charged to patients, bite records may be more reflective of rabies exposures. People may be less likely to seek care when the perceived risk is low (i.e., fewer people seek care for provoked bites by known healthy and/or vaccinated animals) due to the associated costs (Changalucha et al., 2018; Hampson et al., 2008). In settings where PEP is provided for free and indiscriminately, a higher proportion of reported bites may be due to nonrabid animals (Rajeev et al., 2018;

Rysava et al., 2018; Wallace et al., 2015), and many Category 1 exposures, i.e., those for which PEP is not indicated (Rabies vaccines: WHO position paper, 2018), receive unnecessary PEP (Duong et al., 2016; Rajeev et al., 2018; Tenzin, Dhand, & Ward, 2011).

For data on bite patients to be more useful for modeling and surveillance purposes, supplementary information for each bite beyond the date reported and number of doses received is needed. Categorizing the type of exposure per the WHO categories can help to exclude Category 1 exposures. Reporting clinical signs and the outcome of the biting animal at each patient visit can identify probable rabies exposures and trigger field investigations and sample collection to improve surveillance. Finally, information on the geographical location where the patient was bitten, for example to the finest scale administrative unit identifiable, could be used to understand spatial patterns of transmission, estimate demand for PEP, and identify determinants of health-seeking behavior.

20.3.2 Laboratory-confirmed case data

Laboratory-confirmed case data are considered a gold standard due to the high sensitivity and specificity of diagnostic tests for rabies, but represent the tip of the iceberg in terms of true incidence (Scott et al., 2017; Townsend, Lembo, et al., 2013). Diagnostic confirmation of rabies cases is often lacking in many endemic settings due to limited laboratory and field capacity. Even with strong laboratory resources in country, collecting a brain sample from a suspected rabid animal or human case can be challenging. Lack of cold chain and accessibility to communities, limited veterinary capacity and training in euthanasia and sampling methods, and low reporting of suspected cases are all significant barriers to case confirmation. For humans, nuchal samples can be collected noninvasively (from nape of the neck) to confirm a rabies case ante-mortem (Dacheux et al., 2008). However, confirmation of a human case first requires a person to seek care, and rabies deaths are most common in populations with the least access to health care (Wentworth et al., 2019). For animal cases, field sample collection methods, like the straw method of sampling brain tissue that does not require the submission of the whole head, and alternative forms of sample storage and testing, such as rapid diagnostic tests and filter papers, have potential to address some of these challenges (Léchenne et al., 2016). While these alternative tests may not be appropriate for guiding patient treatment, they could greatly improve surveillance and understanding of rabies virus transmission if implemented more routinely.

Even with the gold-standard diagnostic test, using laboratory confirmation to guide administration of PEP in endemic settings may be impractical, due to delays in sampling and testing. Integrated bite case management (IBCM) (see Chapter 18) programs, which combine risk assessments, field investigations, animal observation/quarantine, and sampling of suspected cases, are a promising method of improving rabies surveillance and PEP provisioning. IBCM can increase both detection of and confirmation of clinically suspect animal cases and guide referrals for PEP, as well as limit further exposures by euthanizing rabid animals once detected (Undurraga et al., 2017). However, IBCM relies on coordination between human and animal health practitioners and resources to support clinical rabies diagnosis and field sample collection, which is still lacking in most low-income countries.

20.3.3 Sequence data

Sequence data can be used to make inferences about transmission processes, particularly when linked with epidemiological data (Bourhy et al., 2016; Brunker et al., 2015, 2018; Cori et al., 2018; Talbi et al., 2010; Zinsstag et al., 2017). Recent studies have demonstrated the added value of whole genome sequencing (WGS) for understanding finer-scale transmission dynamics of canine rabies (Brunker et al., 2015, 2018), but WGS has yet to be routinely generated for canine rabies. Sequencing capacity is even more limited than general laboratory capacity in rabies-endemic countries and exporting samples for sequencing is costly. Advances in portable, real-time sequencing could help to tackle these limitations in the field (ARTICnetwork, http://artic.network/index.html). Portable sequencers such as the MinION could support rapid generation and dissemination of sequence data. Methods to sequence from alternative sample types, such as rapid diagnostic tests and filter papers, could also help to overcome obstacles in field sample collection and transport (Léchenne et al., 2016). Bioinformatic pipelines and open sharing of sequences, such as those developed for other viral pathogens (Hadfield et al., 2018), could greatly facilitate our understanding of rabies dynamics at a regional and global scale. In general, low-cost, high-throughput sequencing methods should be developed to increase the timely availability of representative sequence data from endemic settings.

20.3.4 Dog population and vaccination data

Data on the dog population are necessary to further understand how the distribution, density, and connectivity of the host population drives transmission (Brunker et al., 2018). Estimates of vaccination coverage and other intervention efforts facilitate inference of the mechanisms driving transmission and the impact of interventions, helping to predict future outcomes given different control strategies (Zinsstag et al., 2017). In most endemic countries, limited systematic data are collected on dog populations. If integrated into more routine census or demographic surveys (i.e., the Demographic and Health Surveys, https://dhsprogram.com), questions on dog ownership and vaccination status at the household level could be a potential way to get data on dog populations where the majority of dogs are owned. However, if conducted as standalone surveys, these can be resource intensive and difficult to implement in a representative way, particularly in more rural/remote areas. Alternatively, integrating postvaccination coverage surveys into campaigns has been shown to be a cost-effective way to generate coverage and population estimates, and only requires temporary marking of vaccinated dogs (Gibson et al., 2015; Sambo et al., 2017; Tenzin et al., 2015). As spatial heterogeneity in coverage is likely a key factor driving the success of vaccination campaigns, such coverage estimates at the scale at which campaigns are implemented could be critical to understanding rabies persistence and elimination probabilities.

20.4 Conclusions

Modeling studies, in combination with decades of empirical evidence, have demonstrated that dog vaccination is the optimal intervention strategy for controlling canine rabies. As global momentum for implementing national rabies control programs grows, studies should move beyond comparing vaccination and other strategies in idealized populations toward linking models with field data to identify refinements to intervention strategies. To date, most work has focused on studying control efforts and identifying drivers of dynamics (often without using data), and studies of the impact of control have rarely been linked to analyses grounded in empirical data (i.e., studies that explained observed patterns or estimated key parameters, see Supplementary Fig. S2 at https://doi.org/10.1016/B978-0-12-818705-0.00020-0 for an overview of existing studies). Models should aim to integrate these questions and test specific vaccination strategies, such as ring vaccination or establishment of control corridors based on geographic barriers as implemented for wildlife rabies in Europe.

Key parameters to estimate from models and data include transmission heterogeneity (captured in the distribution of secondary cases), the dispersal kernel, and introduction rates (including how to differentiate ongoing local transmission from imported cases). Integrating models of surveillance into dynamic models can further establish surveillance requirements necessary to verify freedom from disease and inform policy decisions regarding the cessation and scaling back of control efforts. Importantly, models can predict how these requirements might change over the elimination timeline. Given the challenges in generating high-quality surveillance data for canine rabies, these models can also be used to account for underreporting and determine the minimum level of detection necessary for robust inference. Phylodynamic approaches, which combine both epidemiological and genetic data, are a promising avenue to tackle many of these questions. Critically, progress in this area will require strong surveillance systems and representative data from a range of populations.

Countries have made varying progress toward elimination, ranging from some that lack an operationalized national control policy and others in the end-game stages of elimination. Now, we are tasked with building flexible models that can capture rabies dynamics and the impacts of control across the elimination timeline. Identifying where and how implementation of control efforts needs improving and delivering such improvements will require a much closer collaboration between scientists, practitioners, and policymakers.

Data and code availability

All data and code used to generate figures and supplemental files, as well as the bibliography for the literature review are available online at https://github.com/mrajeev08/ModelingChapter.

References

Anderson, R. M., Jackson, H. C., May, R. M., & Smith, A. M. (1981). Population dynamics of fox rabies in Europe. *Nature, 289*(5800), 765–771. https://doi.org/10.1038/289765a0.

Anothaisintawee, T., Julienne Genuino, A., Thavorncharoensap, M., Youngkong, S., Rattanavipapong, W., Meeyai, A., et al. (2018). Cost-effectiveness modelling studies of all preventive measures against rabies: A systematic review. *Vaccine*. https://doi.org/10.1016/j.vaccine.2018.11.071.

Aparicio, J. P., & Pascual, M. (2007). Building epidemiological models from R_0: An implicit treatment of transmission in networks. *Proceedings of the Royal Society B: Biological Sciences, 274*(1609), 505–512. https://doi.org/10.1098/rspb.2006.0057.

Beyer, H. L., Hampson, K., Lembo, T., Cleaveland, S., Kaare, M., & Haydon, D. T. (2010). Metapopulation dynamics of rabies and the efficacy of vaccination. *Proceedings of the Royal Society B: Biological Sciences, 278*(1715), 2182–2190. https://doi.org/10.1098/rspb.2010.2312.

Beyer, H. L., Hampson, K., Lembo, T., Cleaveland, S., Kaare, M., & Haydon, D. T. (2012). The implications of metapopulation dynamics on the design of vaccination campaigns. *Vaccine, 30*(6), 1014–1022. https://doi.org/10.1016/j.vaccine.2011.12.052.

Bjørnstad, O. N., & Grenfell, B. T. (2008). Hazards, spatial transmission and timing of outbreaks in epidemic metapopulations. *Environmental and Ecological Statistics, 15*(3), 265–277. https://doi.org/10.1007/s10651-007-0059-3.

Bourhy, H., Nakouné, E., Hall, M., Nouvellet, P., Lepelletier, A., Talbi, C., et al. (2016). Revealing the micro-scale signature of endemic zoonotic disease transmission in an African urban setting. *PLoS Pathogens. 12*(4), e1005525–15. https://doi.org/10.1371/journal.ppat.1005525.

Brunker, K., Hampson, K., Horton, D. L., & Biek, R. (2012). Integrating the landscape epidemiology and genetics of RNA viruses: Rabies in domestic dogs as a model. *Parasitology, 139*(14), 1899–1913. https://doi.org/10.1017/S003118201200090X.

Brunker, K., Lemey, P., Marston, D. A., Fooks, A. R., Lugelo, A., Ngeleja, C., et al. (2018). Landscape attributes governing local transmission of an endemic zoonosis: Rabies virus in domestic dogs. *Molecular Ecology, 27*(3), 773–788. https://doi.org/10.1111/mec.14470.

Brunker, K., Marston, D. A., Horton, D. L., Cleaveland, S., Fooks, A. R., Kazwala, R., et al. (2015). Elucidating the phylodynamics of endemic rabies virus in eastern Africa using whole-genome sequencing. *Virus Evolution, 1*(1), vev011–11. https://doi.org/10.1093/ve/vev011.

Changalucha, J., Steenson, R., Grieve, E., Cleaveland, S., Lembo, T., Lushasi, K., et al. (2018). The need to improve access to rabies post-exposure vaccines: Lessons from Tanzania. *Vaccine.* https://doi.org/10.1016/j.vaccine.2018.08.086.

Cleaveland, S., Beyer, H., Hampson, K., Haydon, D., Lankester, F., Lembo, T., et al. (2014). The changing landscape of rabies epidemiology and control. *The Onderstepoort Journal of Veterinary Research, 81*(2), E1–E8. https://doi.org/10.4102/ojvr.v82i2.731.

Cleaveland, S., Lankester, F., Townsend, S., Lembo, T., & Hampson, K. (2014). Rabies control and elimination: A test case for one health. *The Veterinary Record, 175*(8), 188–193. https://doi.org/10.1136/vr.g4996.

Coleman, P. G., & Dye, C. (1996). Immunization coverage required to prevent outbreaks of dog rabies. *Vaccine.*

Cori, A., Nouvellet, P., Garske, T., Bourhy, H., Nakouné, E., & Jombart, T. (2018). A graph-based evidence synthesis approach to detecting outbreak clusters: An application to dog rabies. *PLoS Computational Biology, 14*(12)e1006554–22. https://doi.org/10.1371/journal.pcbi.1006554.

Czupryna, A. M., Brown, J. S., Bigambo, M. A., Whelan, C. J., Mehta, S. D., Santymire, R. M., et al. (2016). Ecology and demography of free-roaming domestic dogs in rural villages near Serengeti National Park in Tanzania. *PLoS One, 11*(11), e0167092. https://doi.org/10.1371/journal.pone.0167092.

Dacheux, L., Reynes, J.-M., Buchy, P., Sivuth, O., Diop, B. M., Rousset, D., et al. (2008). A reliable diagnosis of human rabies based on analysis of skin biopsy specimens. *Clinical Infectious Diseases, 47*(11), 1410–1417. https://doi.org/10.1086/592969.

Duke-Sylvester, S. M., Bolzoni, L., & Real, L. A. (2011). Strong seasonality produces spatial asynchrony in the outbreak of infectious diseases. *Journal of the Royal Society Interface, 8*(59), 817–825. https://doi.org/10.1098/rsif.2010.0475.

Duong, V., Tarantola, A., Ong, S., Mey, C., Choeung, R., Ly, S., et al. (2016). Laboratory diagnostics in dog-mediated rabies: An overview of performance and a proposed strategy for various settings. *International Journal of Infectious Diseases, 46*, 107–114. https://doi.org/10.1016/j.ijid.2016.03.016.

Ferguson, E. A., Hampson, K., Cleaveland, S., Consunji, R., Deray, R., Friar, J., et al. (2015). Heterogeneity in the spread and control of infectious disease: Consequences for the elimination of canine rabies. *Nature Publishing Group, 5*(1), 1–13. https://doi.org/10.1038/srep18232.

Ferrari, M. J., Perkins, S. E., Pomeroy, L. W., & Bjørnstad, O. N. (2011). Pathogens, social networks, and the paradox of transmission scaling. *Interdisciplinary Perspectives on Infectious Diseases, 2011*(2), 1–10. https://doi.org/10.1155/2011/267049.

Fitzpatrick, M. C., Hampson, K., Cleaveland, S., Meyers, L. A., Townsend, J. P., & Galvani, A. P. (2012). Potential for rabies control through dog vaccination in wildlife-abundant communities of Tanzania. *PLoS Neglected Tropical Diseases, 6*(8), e1796. https://doi.org/10.1371/journal.pntd.0001796.

Fitzpatrick, M. C., Hampson, K., Cleaveland, S., Mzimbiri, I., Lankester, F., Lembo, T., et al. (2014). Cost-effectiveness of canine vaccination to prevent human rabies in rural Tanzania. *Annals of Internal Medicine, 160*(2), 91–100. https://doi.org/10.7326/M13-0542.

Foggin, C. M. (1988). *Rabies and rabies-related viruses in Zimbabwe: Historical, virological and ecological aspects.* (unpublished doctoral thesis)University of Zimbabwe.

Gibson, A. D., Ohal, P., Shervell, K., Handel, I. G., Bronsvoort, B. M., Mellanby, R. J., et al. (2015). Vaccinate-assess-move method of mass canine rabies vaccination utilising mobile technology data collection in Ranchi, India. *BMC Infectious Diseases*, 1–10. https://doi.org/10.1186/s12879-015-1320-2.

Grassly, N. C., & Fraser, C. (2008). Mathematical models of infectious disease transmission. *Nature Reviews Microbiology, 180*, 1–12. https://doi.org/10.1038/nrmicro1845.

Hadfield, J., Megill, C., Bell, S. M., Huddleston, J., Potter, B., Callender, C., et al. (2018). Nextstrain: Real-time tracking of pathogen evolution. *Bioinformatics, 34*(23), 4121–4123. https://doi.org/10.1093/bioinformatics/bty407.

Hampson, K., Abela-Ridder, B., Bharti, O., Knopf, L., Léchenne, M., Mindekem, R., et al. (2019). Modelling to inform prophylaxis regimens to prevent human rabies. *Vaccine, 37*, A166–A173.

Hampson, K., Abela-Ridder, B., Brunker, K., Bucheli, S. T. M., Carvalho, M., Caldas, E., et al. (2016). Surveillance to establish elimination of transmission and freedom from dog-mediated rabies. *bioRxiv*, 096883. https://doi.org/10.1101/096883.

Hampson, K., Cleaveland, S., & Briggs, D. (2011). Evaluation of cost-effective strategies for rabies post-exposure vaccination in low-income countries. *PLoS Neglected Tropical Diseases, 5*(3), e982. https://doi.org/10.1371/journal.pntd.0000982.

Hampson, K., Coudeville, L., Lembo, T., Sambo, M., Kieffer, A., Attlan, M., et al. (2015). Estimating the global burden of endemic canine rabies. *PLoS Neglected Tropical Diseases, 9*(4), e0003709. https://doi.org/10.1371/journal.pntd.0003709.

Hampson, K., Dobson, A., Kaare, M., Dushoff, J., Magoto, M., Sindoya, E., et al. (2008). Rabies exposures, post-exposure prophylaxis and deaths in a region of endemic canine rabies. *PLoS Neglected Tropical Diseases, 2*(11), 1–9. https://doi.org/10.1371/journal.pntd.0000339.

Hampson, K., Dushoff, J., Bingham, J., Brückner, G., Ali, Y. H., & Dobson, A. (2007). Synchronous cycles of domestic dog rabies in sub-Saharan Africa and the impact of control efforts. *Proceedings of the National Academy of Sciences, 104*(18), 7717–7722. https://doi.org/10.1073/pnas.0609122104.

Hampson, K., Dushoff, J., Cleaveland, S., Haydon, D. T., Kaare, M., Packer, C., et al. (2009). Transmission dynamics and prospects for the elimination of canine rabies. *PLoS Biology, 7*(3), e53. https://doi.org/10.1371/journal.pbio.1000053.

Heesterbeek, H., Anderson, R. M., Andreasen, V., Bansal, S., De Angelis, D., Dye, C., et al. (2015). Modeling infectious disease dynamics in the complex landscape of global health. *Science, 347*(6227), aaa4339. https://doi.org/10.1126/science.aaa4339.

Hemachudha, T., Laothamatas, J., & Rupprecht, C. E. (2002). Human rabies: A disease of complex neuropathogenetic mechanisms and diagnostic challenges. *The Lancet Neurology, 1*(2), 101–109. https://doi.org/10.1016/S1474-4422(02)00041-8.

Hudson, E. G., Brookes, V. J., Ward, M. P., & Dürr, S. (2019). Using roaming behaviours of dogs to estimate contact rates: The predicted effect on rabies spread. *Epidemiology and Infection, 147*, e135. https://doi.org/10.1017/S0950268819000189.

Keeling, M. J., & Rohani, P. (2011). *Modeling infectious diseases in humans and animals.* Princeton University Press.

King, A. A., Fooks, A. R., Aubert, M., & Wandeler, A. I. (2004). *Historical perspective of rabies in Europe and the Mediterranean Basin.* Paris, France: OIE.

Klepac, P., Metcalf, C. J. E., McLean, A. R., & Hampson, K. (2013). Towards the endgame and beyond: Complexities and challenges for the elimination of infectious diseases. *Philosophical Transactions of the Royal Society B: Biological Sciences, 368*(1623), 20120137. https://doi.org/10.1098/rstb.2012.0137.

Knobel, D. L., Cleaveland, S., Coleman, P. G., Fèvre, E. M., Meltzer, M. I., Miranda, M. E. G., et al. (2005). Re-evaluating the burden of rabies in Africa and Asia. *Bulletin of the World Health Organization, 83*(5), 360–368.

Kurosawa, A., Tojinbara, K., Kadowaki, H., Hampson, K., Yamada, A., & Makita, K. (2017). The rise and fall of rabies in Japan: A quantitative history of rabies epidemics in Osaka prefecture, 1914–1933. *PLoS Neglected Tropical Diseases, 11*(3). e0005435–19. https://doi.org/10.1371/journal.pntd.0005435.

Laager, M., Léchenne, M., Naissengar, K., Mindekem, R., Oussiguere, A., Zinsstag, J., et al. (2019). A metapopulation model of dog rabies transmission in N'Djamena, Chad. *Journal of Theoretical Biology, 462*, 408–417. https://doi.org/10.1016/j.jtbi.2018.11.027.

Laager, M., Mbilo, C., Madaye, E. A., Naminou, A., Léchenne, M., Tschopp, A., et al. (2018). The importance of dog population contact network structures in rabies transmission. *PLoS Neglected Tropical Diseases, 12*(8), e0006680. https://doi.org/10.1371/journal.pntd.0006680.

Lakshmanan, N., Gore, T. C., Duncan, K. L., Coyne, M. J., Lum, M. A., & Sterner, F. J. (2006). Three-year rabies duration of immunity in dogs following vaccination with a core combination vaccine against canine distemper virus, canine adenovirus type-1, canine parvovirus, and rabies virus. *Veterinary Therapeutics: Research in Applied Veterinary Medicine, 7*(3), 223–231.

Léchenne, M., Naissengar, K., Lepelletier, A., Alfaroukh, I. O., Bourhy, H., Zinsstag, J., et al. (2016). Validation of a rapid rabies diagnostic tool for field surveillance in developing countries. *PLoS Neglected Tropical Diseases, 10*(10), e0005010–e0005016. https://doi.org/10.1371/journal.pntd.0005010.

Lembo, T., Hampson, K., Haydon, D. T., Craft, M., Dobson, A., Dushoff, J., et al. (2008). Exploring reservoir dynamics: A case study of rabies in the Serengeti ecosystem. *The Journal of Applied Ecology, 45*(4), 1246–1257. https://doi.org/10.1111/j.1365-2664.2008.01468.x.

Lembo, T., Haydon, D. T., Velasco-Villa, A., Rupprecht, C. E., Packer, C., Brandão, P. E., et al. (2007). Molecular epidemiology identifies only a single rabies virus variant circulating in complex carnivore communities of the Serengeti. *Proceedings of the Royal Society B: Biological Sciences, 274*(1622), 2123–2130. https://doi.org/10.1098/rspb.2007.0664.

Lloyd-Smith, J. O., Cross, P. C., Briggs, C. J., Daugherty, M., Getz, W. M., Latto, J., et al. (2005). Should we expect population thresholds for wildlife disease? *Trends in Ecology & Evolution, 20*(9), 511–519. https://doi.org/10.1016/j.tree.2005.07.004.

Lloyd-Smith, J. O., Schreiber, S. J., Kopp, P. E., & Getz, W. M. (2005). Superspreading and the effect of individual variation on disease emergence. *Nature, 438*(7066), 355–359. https://doi.org/10.1038/nature04153.

McCallum, H., Barlow, N., & Hone, J. (2001). How should pathogen transmission be modelled? *Trends in Ecology & Evolution, 16*(6), 295–300.

Mollentze, N., Nel, L. H., Townsend, S., le Roux, K., Hampson, K., Haydon, D. T., et al. (2014). A Bayesian approach for inferring the dynamics of partially observed endemic infectious diseases from space-time-genetic data. *Proceedings. Biological Sciences/the Royal Society, 281*(1782), 20133251. https://doi.org/10.1098/rspb.2013.3251.

Morters, M. K., Restif, O., Hampson, K., Cleaveland, S., Wood, J. L. N., & Conlan, A. J. K. (2012). Evidence-based control of canine rabies: A critical review of population density reduction. *Journal of Animal Ecology, 82*(1), 6–14. https://doi.org/10.1111/j.1365-2656.2012.02031.x.

Murray, J. D., Stanley, E. A., & Brown, D. L. (1986). On the spatial spread of rabies among foxes. *Proceedings of the Royal Society of London. Series B. Biological Sciences, 229*(1255), 111–150. https://doi.org/10.1098/rspb.1986.0078.

Panjeti, V. G., & Real, L. A. (2011). Mathematical models for rabies. *Advances in Virus Research, 79*, 377–395. https://doi.org/10.1016/B978-0-12-387040-7.00018-4.

Pascual, M., Roy, M., & Laneri, K. (2011). Simple models for complex systems: Exploiting the relationship between local and global densities. *Theoretical Ecology, 4*(2), 211–222. https://doi.org/10.1007/s12080-011-0116-2.

Rabies vaccines: WHO position paper. (2018). World Health Organization - Weekly Epidemiological Record. Retrieved from http://apps.who.int/iris/bitstream/handle/10665/272371/WER9316.pdf?ua=1.

Rajeev, M., Edosoa, G., Hanitriniaina, C., Andriamandimby, S. F., Guis, H., Ramiandrasoa, R., et al. (2018). Healthcare utilization, provisioning of post-exposure prophylaxis, and estimation of human rabies burden in Madagascar. *Vaccine*, 1–10. https://doi.org/10.1016/j.vaccine.2018.11.011.

Rattanavipapong, W., Thavorncharoensap, M., Youngkong, S., Genuino, A. J., Anothaisintawee, T., Chaikledkaew, U., et al. (2018). The impact of transmission dynamics of rabies control: Systematic review. *Vaccine*. https://doi.org/10.1016/j.vaccine.2018.11.035.

Rysava, K., Miranda, M. E., Zapatos, R., Lapiz, S., Rances, P., Miranda, L. M., et al. (2018). On the path to rabies elimination: The need for risk assessments to improve administration of post-exposure prophylaxis. *Vaccine*, 1–9. https://doi.org/10.1016/j.vaccine.2018.11.066.

Sambo, M., Johnson, P. C. D., Hotopp, K., Changalucha, J., Cleaveland, S., Kazwala, R., et al. (2017). Comparing Methods of Assessing Dog Rabies Vaccination Coverage in Rural and Urban Communities in Tanzania. *Frontiers in Veterinary Science, 4*(8), e0003709–e0003712. https://doi.org/10.3389/fvets.2017.00033.

Scott, T. P., Coetzer, A., Fahrion, A. S., & Nel, L. H. (2017). Addressing the disconnect between the estimated, reported, and true rabies data: The development of a regional African rabies bulletin. *Frontiers in Veterinary Science, 4*(4), 1–6. https://doi.org/10.3389/fvets.2017.00018.

Smith, D. L., Lucey, B., Waller, L. A., Childs, J. E., & Real, L. A. (2002). Predicting the spatial dynamics of rabies epidemics on heterogeneous landscapes. *Proceedings of the National Academy of Sciences*, *99*(6), 3668–3672. https://doi.org/10.1073/pnas.042400799.

Talbi, C., Lemey, P., Suchard, M. A., Abdelatif, E., Elharrak, M., Jalal, N., et al. (2010). Phylodynamics and human-mediated dispersal of a zoonotic virus. *PLoS Pathogens*, *6*(10), e1001166–10. https://doi.org/10.1371/journal.ppat.1001166.

Tarantola, A., Ly, S., Chan, M., In, S., Peng, Y., Hing, C., et al. (2018). Intradermal rabies post-exposure prophylaxis can be abridged with no measurable impact on clinical outcome in Cambodia, 2003–2014. *Vaccine*. https://doi.org/10.1016/j.vaccine.2018.10.054.

Tenzin, Dhand, N. K., & Ward, M. P. (2011). Human rabies post exposure prophylaxis in Bhutan, 2005–2008: Trends and risk factors. *Vaccine*, *29*(24), 4094–4101. https://doi.org/10.1016/j.vaccine.2011.03.106.

Tenzin, T., McKenzie, J. S., Vanderstichel, R., Rai, B. D., Rinzin, K., Tshering, Y., et al. (2015). Comparison of mark-resight methods to estimate abundance and rabies vaccination coverage of free-roaming dogs in two urban areas of South Bhutan. *Preventive Veterinary Medicine*, *118*(4), 436–448. https://doi.org/10.1016/j.prevetmed.2015.01.008.

Tepsumethanon, V., Wilde, H., & Meslin, F. X. (2005). Six criteria for rabies diagnosis in living dogs. *Journal of the Medical Association of Thailand = Chotmaihet Thangphaet*, *88*(3), 419–422.

Tian, H., Feng, Y., Vrancken, B., Cazelles, B., Tan, H., Gill, M. S., et al. (2018). Transmission dynamics of re-emerging rabies in domestic dogs of rural China. *PLoS Pathogens*, *14*(12), e1007392–18. https://doi.org/10.1371/journal.ppat.1007392.

Tohma, K., Saito, M., Demetria, C. S., Manalo, D. L., Quiambao, B. P., Kamigaki, T., et al. (2016). Molecular and mathematical modeling analyses of inter-island transmission of rabies into a previously rabies-free island in the Philippines. *Infection, Genetics and Evolution*, *38*(C), 22–28. https://doi.org/10.1016/j.meegid.2015.12.001.

Townsend, S. E., Lembo, T., Cleaveland, S., Meslin, F. X., Miranda, M. E., Putra, A. A. G., et al. (2013). Surveillance guidelines for disease elimination: A case study of canine rabies. *Comparative Immunology, Microbiology and Infectious Diseases*, *36*(3), 249–261. https://doi.org/10.1016/j.cimid.2012.10.008.

Townsend, S. E., Sumantra, I. P., Pudjiatmoko, Bagus, G. N., Brum, E., Cleaveland, S., et al. (2013). Designing programs for eliminating canine rabies from islands: Bali, Indonesia as a case study. *PLoS Neglected Tropical Diseases*, *7*(8), e2372. https://doi.org/10.1371/journal.pntd.0002372.

Undurraga, E. A., Meltzer, M. I., Tran, C. H., Atkins, C. Y., Etheart, M. D., Millien, M. F., et al. (2017). Cost-effectiveness evaluation of a novel integrated bite case management program for the control of human rabies, Haiti 2014–2015. *American Journal of Tropical Medicine and Hygiene*, *96*(6), 1307–1317. https://doi.org/10.4269/ajtmh.16-0785.

Wallace, R. M., Reses, H., Franka, R., Dilius, P., Fenelon, N., Orciari, L., et al. (2015). Establishment of a canine rabies burden in Haiti through the implementation of a novel surveillance program. *PLoS Neglected Tropical Diseases*, *9*(11), e0004245–15. https://doi.org/10.1371/journal.pntd.0004245.

Wentworth, D., Hampson, K., Thumbi, S. M., Mwatondo, A., Wambura, G., & Chng, N. R. (2019). A social justice perspective on access to human rabies vaccines. *Vaccine*, 1–3. https://doi.org/10.1016/j.vaccine.2019.01.065.

WHO Rabies Modelling Consortium. (2019). The potential impact of improved provision of rabies post-exposure prophylaxis in Gavi-eligible countries: A modelling study. *The Lancet Infectious Diseases*, *19*(1), 102–111. https://doi.org/10.1016/S1473-3099(18)30512-7.

Wilson-Aggarwal, J. K., Ozella, L., Tizzoni, M., Cattuto, C., Swan, G. J. F., Moundai, T., et al. (2019). High-resolution contact networks of free-ranging domestic dogs Canisfamiliaris and implications for transmission of infection. *PLoS Neglected Tropical Diseases*, *13*(7), e0007565. https://doi.org/10.1371/journal.pntd.0007565.

World Health Organization. (2013). *WHO expert consultation on rabies*. (Second report). World Health Organization Technical Report Series, (982), 1–139– Back cover.

Zinsstag, J., Durr, S., Penny, M. A., Mindekem, R., Roth, F., Gonzalez, S. M., et al. (2009). Transmission dynamics and economics of rabies control in dogs and humans in an African city. *Proceedings of the National Academy of Sciences*, *106*(35), 1–22. https://doi.org/10.1073/pnas.0904740106.

Zinsstag, J., Léchenne, M., Laager, M., Mindekem, R., Naïssengar, S., Oussiguere, A., et al. (2017). Vaccination of dogs in an African city interrupts rabies transmission and reduces human exposure. *Science Translational Medicine*, *9*(421), eaaf6984. https://doi.org/10.1126/scitranslmed.aaf6984.

CHAPTER 21

Strategies for the elimination of dog-mediated human rabies by 2030

Terrence P. Scott[a,b], Andre Coetzer[a,b], Louis H. Nel[a,b]

[a]Department of Biochemistry, Genetics and Microbiology, Faculty of Natural and Agricultural Sciences, University of Pretoria, Pretoria, Gauteng, South Africa [b]Global Alliance for Rabies Control, Manhattan, KS, United States

21.1 Introduction

Rabies elimination has been achieved in some regions and territories globally, but most countries remain rabies endemic. Toward changing this status, the last decade saw significant efforts by the international community to raise awareness about rabies at international, national, and community levels (Nel, Taylor, Balaram, & Doyle, 2015). One outcome has been that many governments of rabies-endemic countries have now listed rabies as a top five priority disease (Rist, Arriola, & Rubin, 2014; Salyer, Silver, Simone, & Behravesh, 2017). The next logical step for these governments and their partners would be to act upon their commitment toward the global goal of "Zero by 30," i.e., the achievement of zero dog-mediated human rabies deaths globally by the year 2030. This historical goal was agreed by consensus of the Global Conference (Geneva, December 10–11, 2015), where a framework that provides a coordinated approach and vision for the global elimination of dog-mediated human rabies was adopted (WHO, OIE, FAO, & GARC, 2018). With the intention to harmonize actions and provide practical and adaptable guidance for national and regional strategies, the next step to operationalize this framework was the formation of the United Against Rabies (UAR) collaboration and the subsequent output of the Global Strategic Plan (GSP) (Minghui, Stone, Semedo, & Nel, 2018). While the UAR global steering group consists of the United Nations Food and Agriculture Organization (UN FAO), World Organisation for Animal Health (OIE), World Health Organization (WHO), and the Global Alliance for Rabies Control (GARC), the key to the success of the GSP will be the engagement of the international rabies community, development partners, and individual national governments.

In this chapter, we will discuss the GSP and assess some of the tools and strategies that have been developed by the international rabies community. Furthermore, we will attempt to draw parallels with approaches that have been successful in other disease intervention and elimination/eradication efforts. With these examples, we justify the continued development, dissemination, and implementation of these tools for rabies as the most feasible means to achieve the global goal of "Zero by 30." In addition, we garner evidence to support the development of additional supporting tools that will be required to achieve elimination.

21.2 Sustainability requires government ownership and commitment

The sustainability of disease interventions is a key issue in global healthcare policy. As for many other diseases, the GSP for rabies elimination is country-centric, relying on the principle that governments would need to take ownership of their national rabies programs and drive the elimination strategy. Historical evidence is supportive of the intrinsic nature of this principle to ensuring sustainability (Bonita et al., 2013; Coker, Atun, & McKee, 2004; Epping-Jordan, Galea, Tukuitonga, & Beaglehole, 2005). A sustainable, government-driven approach has proven vital for the eradication of smallpox and rinderpest, and indeed also for the drastic reduction of polio and several noncommunicable diseases (Atun, Lennox-Chhugani, Drobniewski, Samyshkin, & Coker, 2004; Cochi, Hegg, Kaur, Pandak, & Jafari, 2016; FAO, 2012; Henderson, 2011; Hochman, 2009). This principle, however, does not imply that all nations will be able to undertake such major elimination programs without international support. Indeed, for smallpox, only wealthy countries with good infrastructure would have been able to eliminate the disease in the absence of global coordination. Such coordination only followed a global consensus to push toward disease eradication (Barrett, 2006). This drive facilitated successes in several additional countries, but a group of low-income countries remained endemic as the cost-benefit to their own national health system and current resources was not attractive (Barrett, 2006). Through cost-benefit analyses (for those countries that were maintaining disease elimination through mass vaccination) and international recognition of the benefit to achieving disease eradication as opposed to country-specific elimination, it was decreed by the Assistant Director General of the WHO that "Without a greatly intensified and well-coordinated effort, *and substantial additional resources* (emphasis added), global eradication was not a realistic goal in the foreseeable future." (World Health Assembly, 1966). Therefore, it became evident that the international community would need to support elimination efforts in those remaining countries to achieve global eradication. This notion still focused on sustainability, but with international support, finally resulting in a global program funded 70% from endemic-government contributions and 30% from international support (Barrett, 2006). As a further case in point, rinderpest regional eradication campaigns were initially disjointed and, when nearing eradication, were less effective than the initial interventions (Roeder, 2011). The obvious need for a globally coordinated strategy was then addressed by the establishment of the Global Rinderpest Eradication Campaign (GREP) (Rweyemamu & Cheneau, 1995). The GREP secretariat provided technical guidance, monitored and managed global progress, facilitated coordination, and provided practical assistance to select countries facing unique challenges that prevented success from the conventional approach (Roeder, 2011).

In the case of rabies, elimination has already been achieved in higher-income countries that includes the United Kingdom and Japan, Western Europe, and North America. In Latin America, a strong regional collaboration coordinated by PAHO, enabled dog rabies elimination in most of the member states (de Carvalho et al., 2018). As mentioned earlier, a global target for rabies elimination has recently been agreed and a coordinated strategy—the county-centric GSP—was developed by the UAR collaboration. These developments closely follow the events that eventually enabled smallpox and rinderpest eradication. In this analogy, a coordinated rabies strategy, based on government spending, regulation and public ownership—supported by the international community through a global goal and strategic plan as well as resources that may include financial contributions—would allow for global elimination. Critically, not a single rabies control program that was primarily or substantially implemented/driven by nongovernmental parties or international aid agencies has succeeded in sustained elimination of rabies in any sovereign country anywhere in the world (International Coordinating Group (ICG), 2014; Lembo et al., 2010; Vigilato, Cosivi, Knöbl, Clavijo, & Silva, 2013; World Health Organization, 2018). Indeed, such compromise of government ownership and responsibility is not only unsustainable, but evidence suggests that it could potentially lead to long-term damage to health care systems, comparable to the devastating impact of neoliberal policy on existing healthcare structures in low- and middle-income countries (LMICS) (Unger, De Paepe, Ghilbert, Soors, & Green, 2006).

It seems clear, therefore, that the key to rabies elimination lies therein that national governments initiate, control, and execute their own elimination strategies in line with a clear global plan and drive. While this approach is country-centric, critical support—such as that received through international aid—should be channeled through governmental structures subject to rigorous monitoring and evaluation toward continuous optimization of the elimination strategy (Coker et al., 2004). Such an approach should prevent programmatic collapse following the cessation of donor-supported projects and ensure sustainability through public ownership of the government's contribution toward reaching the global rabies elimination target.

21.2.1 Tools to support sustainable rabies elimination programs

21.2.1.1 Global disease elimination/eradication strategies

For rabies, the GSP in principle provides a sustainable, coordinated, and standardized approach to rabies elimination for each rabies-endemic country in the world (Minghui et al., 2018; WHO et al., 2018). Despite being government-centric, the GSP also considers those lessons learned from both smallpox (Barrett, 2006) and rinderpest (Roeder, 2011) whereby the need for international donor support will be critical in assisting LMICs that remain dog rabies-endemic in reaching the goal of "Zero by 30," followed by elimination of dog rabies globally. The GSP includes a detailed fund-raising structure and roll-out to support those countries currently lagging in their efforts, and to assist those countries that are nearing elimination (WHO et al., 2018).

In addition, the GSP seeks to provide technical guidance to countries through the host of rabies-specific tools that have been developed (Rabies Toolkit), as discussed later. From historical experience, it is not only the development of tools that is important—but also ensuring their effective uptake. For example, a thermostable vaccine that changed the course of rinderpest eradication programs took only 2 years to develop, yet social innovations to realize the true potential of this revolutionary advancement took more than a decade (Mariner et al., 2012).

21.2.1.2 Disease prioritization

For the GSP to succeed, governments need to be wholly supportive of the concept and the drive toward achieving rabies elimination. On face value, this seems logical and obvious, but what is required from governments to effectively implement and execute a global strategy such as the GSP?

Undoubtedly, strong leadership will be required, as demonstrated in many historical instances. As one example, New York City had great success in controlling tuberculosis in the 1980s and 1990s, with much of this attributed to successful programmatic implementation through strategic and social interventions and considerable financial support. Importantly, it was well recognized that these successes were dependent on strong political leadership, advocacy and political will, without which the program would have been far less effective (Coker et al., 2004). To gain similar political will in the case of rabies, the disease must be recognized as horrendous yet preventable; where elimination is possible and would represent a notable global public good.

In this regard, one tool that helps governments prioritize zoonotic diseases has been particularly useful. This tool, being One Health Zoonotic Disease Prioritization (OHZDP) workshops, has been developed and coordinated by the United States Centers for Disease Control and Prevention (US CDC). It was shown to be critical not only in driving disease prioritization, but in subsequently garnering the political will and support required to control and eliminate the prioritized diseases. Due to the considerable advocacy efforts for rabies over the last decade—including the establishment and celebration of World Rabies Day (Balaram, Taylor, Doyle, Davidson, & Nel, 2016), rabies has been listed among the top priority diseases in all of the rabies endemic countries investigated ($n = 24$) (Salyer et al., 2017; US CDC, 2019). For rabies, this indeed represents an important first step toward securing committed political support and resources.

21.2.1.3 Organization and planning for rabies elimination

Intervention strategies have been developed for various infectious diseases around the world. These strategies are based on the unique epidemiology of each and can therefore be quite different from one another. However, one common denominator in all disease intervention plans is the concept of organization (Bonita et al., 2013; Epping-Jordan et al., 2005). This refers to organization of human, financial, technical, logistical, operational, and other resources, and ultimately the coordination of all these resources into an effective operation. To achieve successful organization, a clear plan is required. For many diseases—especially those already eradicated or nearing eradication—a global plan has sufficed to assist national governments in developing strategies with direction and assistance from international coordination groups. This was the case with smallpox (WHO-taskforce), rinderpest (FAO-secretariat), and polio (WHO and partners) (Bonita et al., 2013; Cochi et al., 2016). In comparison to these diseases, rabies does however present a significant additional layer of complexity and challenges, given its zoonotic nature. Indeed, rabies control and elimination is ultimately dependent on a multisectoral approach that coalesces both animal and human health into a single coordinated program.

National rabies elimination strategies have been developed and endorsed for several LMICs. Such strategies are typically political documents vitally important for governance and advocacy purposes. Often these documents do not provide the required detail to effectively become a measured and focused strategy. Overarching objectives such as "To vaccinate at least 70% of the dog population annually for 5 years" might lack the necessary granular detail on how this is to be achieved and measured. Thus, although such national strategies serve an important purpose toward securing government support and endorsement, they might not necessarily serve as a defined workplan per se.

It is with this fact in mind that the principle of a "stepwise approach" to reducing various diseases had been developed (and successfully implemented) in a number of cases (Epping-Jordan et al., 2005). This principle provides national governments with clear, defined objectives around which their strategy can be developed, ensuring achievement of the global strategy and target to reduce human deaths associated with the disease (Bonita et al., 2013; Epping-Jordan et al., 2005). In evidence, the Progressive Control Pathway for Foot and Mouth Disease (PCP-FMD)—a standardized approach that also allows for the verification of elimination by international bodies such as the OIE—has helped to address the same challenges faced in FMD control (OIE & FAO, 2012).

To address these challenges and complexities for rabies, the Global Alliance for Rabies Control and the Partners for Rabies Prevention (Lembo et al., 2011; Nel et al., 2015; Taylor & Partners for Rabies Prevention, 2013) developed the Stepwise Approach toward Rabies Elimination (SARE) tool (Coetzer et al., 2016; Nel, 2018). The SARE tool has roots in these previous approaches and pathways, but added additional layers of detail, while maintaining the core elements of being simple yet comprehensive. As intended, the SARE is a dynamic instrument that evolves and improves over time. The latest version of the SARE tool consists of two components, viz. an assessment component and a component for the generation of a practical workplan. The SARE tool enables national governments to objectively assess the present-day rabies situation across the country and in terms of each of the seven core categories that are critical to effective rabies control and elimination. Using the concept of progressive stages within each category, every aspect of rabies control is carefully considered and addressed. The tool then highlights completed and pending activities in a clear summarized form, together with a score that is automatically calculated based on the responses and achieved activities. This interactive assessment enables objective planning and measurement of progress in the execution of a control plan. It is generally accepted that such a structured, systematic approach based on prioritized and sequential activities is likely to be most effective for the progression of disease control. This concept is well evidenced in the cases of other infectious and noncommunicable diseases (Bonita et al., 2013; Diall et al., 2017; Epping-Jordan et al., 2005; OIE & FAO, 2012). In addition to this monitoring and evaluation component, the SARE tool also drives the development of a practical workplan. With the use of this unique component, national authorities can prioritize the pending activities into a multiyear workplan that is actionable and measurable on an activity-by-activity basis.

While a well-considered strategy is an essential start, it is the implementation thereof that will determine its success. As a case in point, the control of tuberculosis in London in the early 21st century initially failed, despite a good understanding of what the strategy should be. In this case, the roles of each of the authorities responsible for the implementation of the action plan were not clearly defined and thus the failure to meaningfully progress the control of tuberculosis (Coker et al., 2004). In contrast, in New York several decades earlier, the roles of all stakeholders were clearly defined and enabled the rapid and efficient control of the disease (Coker et al., 2004). Thus, for rabies, the SARE-derived workplan includes the designation of a responsible authority for each activity, ensuring effective implementation. This tool therefore helps governments to develop a clear multiyear national plan based on the prioritization of pending activities and the definition of clear objectives, responsible authorities and deliverables—ensuring that rabies elimination strategies can be more time- and resource-efficient. By using this standardized tool, recognized by the international community, external interested stakeholders can easily identify key areas where they may be able to contribute in-kind, with expertise, or financially. Importantly, this tool will measure the impact of all actions and contributions, a concept which was one of the keys to success in rinderpest eradication (Roeder, 2011).

Organization and management of disease intervention campaigns is key to their success and ensures that financial investments are spent appropriately, gaining the greatest return on investment possible. For smallpox, India and China were faced with similar challenges in terms of the demographics and sheer scale of the challenge toward achieving elimination. However, with good planning and organization, China was able to eliminate smallpox without international assistance in the early 1960s, while India—a country with high incentive based on the strain of mass vaccination on the national health budget—lacked organization and eventually only achieved elimination by 1977 with large-scale support from the international community (Barrett, 2006). The SARE tool captures all the necessary elements for a successful rabies elimination strategy and allows for tangible progress to be demonstrated continually as the program advances in any or each of the seven key areas of the SARE. As a standardized approach, the SARE also enables national strategies to be evaluated thoroughly and on an equal footing, which should assist the global community of donors and stakeholders to determine where and how they may best contribute.

21.2.1.4 Financing

Despite the allocation of limited financial resources toward health budgets in most rabies-endemic countries, rabies elimination remains feasible through considered planning and strategic use of these financial resources (Wallace, Undurraga, Blanton, Cleaton, & Franka, 2017). It is commonly argued that the allocated national health budgets are insufficient to provide basic health services in most LMICs, let alone address additional burdens such as rabies elimination efforts. In 2016, LMICs had a mean per capita expenditure of USD $534.38 on health, with 4 of the 5 countries with the greatest rabies burden (39,572/58,990 estimated human deaths; 67%) having per capita expenditures below this (Table 21.1) (Hampson et al., 2015; World Bank Group, 2019). On face value, it thus appears that health budgets in many LMICs truly are insufficient to cope with the additional costs of rabies elimination, yet, with proper planning and budgeting, it is feasible.

TABLE 21.1 Comparison of the estimated number of human deaths to the annual per capita health spending of the five countries with the highest number of human rabies deaths.

Country	Estimated rabies burden	Per capita spent on health (USD $)
India	20,847	241.48
China	6002	761.49
Democratic Republic of Congo	5579	34.49
Myanmar	4552	291.09
Ethiopia	2771	69.52

Countries are ranked based on the human rabies death burden from highest to lowest, as estimated in the most recent global burden of dog rabies study (Hampson et al., 2015).

The budgeting of a rabies intervention plan presents several challenges, primarily due to the sheer expanse of variables and costs that need to be considered. Thus, budgets for rabies elimination may not be truly representative due to a lack of detailed fiscal evaluations of the required activities. These typically inaccurate budgets are infrequently funded as few stakeholders (government or other) have the capacity to pledge/provide the gross lump sum requested to achieve rabies elimination, without any supporting evidence.

The Global Dog Rabies Elimination Pathway (GDREP) tool, however, provides countries with a detailed cost breakdown of all activities required to implement a mass dog vaccination program, with both upper, lower, and median costs being available. This facilitates the generation of a clear, realistic, and comprehensive national budget (Wallace et al., 2017). The budgeting of the costs using the GDREP is broken into 3 distinct phases that consider the scaling of efforts and the premise that future funding will be more likely based on the successes shown through completion of the previous phase (Wallace et al., 2017). This phased approach disassembles the gross lump sum of a rabies elimination strategy budget into more manageable portions that can realistically be financed by national governments and supporting partners. Using the tool, rabies elimination strategies can make the case that funding is feasible—even in low-income countries. For example, the domestic general government expenditure on infectious and parasitic diseases for Ethiopia in 2016 was USD $463 million (World Health Organization, 2014). The GDREP estimates that the total amount required for rabies elimination via mass dog vaccination in Ethiopia is USD $17.5 million. Considering the phased approach provided by the tool, the first phase will require only USD 1.5 million over 3 years—a mere 0.1% of the annual infectious and parasitic disease budget. Therefore, using the GDREP tool, a concrete and well-considered case can be made for the adequate financing of a clear rabies elimination strategy in any rabies-endemic country, including those with below-average expenditure on health. Furthermore, by providing specific cost breakdowns and requirements, financial contributions can be easily monitored, and their impact assessed within the greater strategy—enabling donors to attribute impact-value to their investment. With steady and timely progression to the next phase, investors would receive returns on their investment and would thus be more likely to fund further activities (Meessen, Soucat, & Sekabaraga, 2011).

21.3 International guidelines/recommendations

International guidelines and recommendations provide essential information and serve two core functions: (1) To provide programs and governments with information regarding expert-recommended methods, strategies, and tools and (2) to provide information relating to the means in which to undertake intervention and elimination/eradication efforts, including criteria to declare freedom from a disease. Due to the zoonotic nature of rabies, both the WHO and the OIE developed guidelines and recommendations for national rabies strategies and the associated factors in driving these interventions. Despite a plethora of guidelines being available, there remain some gaps that will need to be addressed to drive momentum toward achieving global dog rabies elimination. Two of these that we believe will be critical to driving elimination would be legislative prevention of the purchase and use of substandard human and animal rabies vaccines (Taylor et al., 2019) and the international recognition of the declaration of freedom from rabies by individual countries.

21.3.1 Declaration of freedom

Presently, the OIE provides official recognition of freedom for six diseases, of which rabies is not one (OIE, 2019). The OIE does not grant official recognition for any other diseases, but upon meeting self-declaration criteria, the OIE may choose to publish this status claim. However, such self-declaration criteria are typically open ended and nonspecific to the disease. It could be argued that, although clear guidelines for the declaration of freedom from dog-mediated rabies exist, the current pathway toward official recognition of freedom is not particularly straightforward or rewarding.

Considering this, the case of rinderpest eradication might be particularly enlightening. Prior to 1993, no clear process for the eradication and accreditation of rinderpest freedom was available, hampering the ability of governments to declare freedom. However, in 1993, the FAO and OIE worked closely to develop a three-stage process as a clear guide with specific, objective, actionable criteria for the eradication and accreditation of rinderpest freedom. The role of the OIE was significant in monitoring the accreditation process, and finally in recognizing freedom in conjunction with an ad hoc group of Rinderpest experts (Roeder, 2011). Smallpox also demonstrated the importance of certified eradication to the global community, by providing confidence that the disease had in fact been eliminated from the identified region (Henderson, 1987). These examples demonstrate the need for similar such criteria and processes for rabies.

An international rabies expert group already exists in the Partners for Rabies Prevention (PRP), who meet on an annual basis. This group could serve as the body that could evaluate the accreditation of freedom for rabies in a country. Alternatively, following the example set by measles, regional verification commissions could be established to assess and officially recognize the declaration of freedom by member states (Perry et al., 2014). The impact of having official recognition of freedom would be a critical incentive for more governments and stakeholders to rapidly scale and accelerate efforts toward rabies elimination. As soon as one rabies endemic country becomes officially recognized as rabies free by an international authority, public and political pressure will likely mount with a resultant increase in political will and drive toward elimination in other territories.

21.4 Networks

The value and necessity for disease-focused networks and programs are well known and have been shown to be particularly critical in those cases where the ultimate goal is elimination and/or eradication, as was noted for smallpox, malaria, rinderpest, and polio (Cochi, Freeman, Guirguis, Jafari, & Aylward, 2014; Cotter et al., 2013; Taylor, Bhat, & Nanda, 1995). Networks are essential as it has been recognized that no single entity (being government, institution, or organization) has the capacity (financial or technical) to achieve global disease elimination (Mackenzie et al., 2014). For governments alone, close collaboration and coordination with national and international stakeholders is required to effectively address many of the complex political, social, environmental, and economic factors that are apparent during disease outbreaks and in endemic disease situations (Mackenzie et al., 2014). An example of international coordination is evident with the Global Outbreak Alert and Response Network (GOARN), which ensures that partner organizations work together to rapidly deliver technical support to affected countries, resulting in lessened outbreak-associated morbidities and mortalities (Mackenzie et al., 2014). Failure of TB programmatic implementation was driven by an environment whereby lessons were not drawn and shared, mistakes were inadvertently replicated and sustainability was not assured, originating as a result of different priorities of researchers and their failure to effectively communicate these experiences with others (Coker et al., 2004). Expert and coordinative networks are thus essential in their ability to share knowledge, discuss challenges, prevent duplication of activities, act as platforms for dissemination of new tools and ideas, and act as coordination structures driving a global strategy (Cochi et al., 2014; Coetzer, Scott, Amparo, Jayme, & Nel, 2018; Cotter et al., 2013; Scott, Coetzer, De Balogh, Wright, & Nel, 2015; Taylor et al., 1995).

For rabies, expert networks—such as the PRP and the UAR coalition—exist to fulfill many of these functions and to coordinate international rabies efforts. The UAR coalition is responsible for the development and oversight of the GSP (Minghui et al., 2018), while the PRP group is primarily responsible for tool development, global research (such as the global burden of dog rabies study), advocating for rabies, and fostering public-private partnerships (Hampson et al., 2015; Lembo et al., 2011; Nel et al., 2015; Taylor & Partners for Rabies Prevention, 2013). At the regional level, efforts were previously disjunctive, with different networks—often with differing approaches, priorities, and agendas—in overlapping regions working disharmoniously toward rabies elimination. This hindered the dissemination of tools and strategies developed by the PRP, and the duplication of efforts across the same region became apparent (Scott et al., 2015). In response, standardized rabies-dedicated networks have been created across most rabies-endemic regions. These networks are the Pan-African Rabies Control Network (PARACON) (Scott et al., 2015), the Asian Rabies Control Network (ARACON) (Coetzer et al., 2018), and the Middle East, Eastern Europe, Central Asia, and North Africa Rabies Control Network (MERACON). Given the regional context of these rabies control networks, those countries that are most lagging behind may be the primary beneficiaries as regional strategies for rabies elimination have to address the transboundary nature of the disease.

These rabies-focused networks, under the secretariat of GARC, act as a conduit between national authorities and expert groups for the dissemination of tools and global strategies. Member states are represented by designated persons from the human and animal health sectors and are supported by partners that include UAR representatives and international experts that constitute the PRP. In keeping with their unifying ambitions, these networks seek to engage with civil society, international experts, and the private sector, as well as regional political structures (EAC, SADC, ECOWAS, AU, SAARC, ASEAN)—something that was especially important in rinderpest eradication, specifically in terms of the acquisition of funding (Roeder, 2011). In addition to assisting countries in aligning their national strategies with the GSP through the SARE tool, these specialized rabies control networks also provide a much-needed platform for the dissemination of tools, the sharing of lessons learnt and the consolidated coordination of efforts across regions to ensure that rabies elimination is achieved in line with the global goal.

21.5 Monitoring and surveillance

The monitoring and evaluation of disease incidence is critical to the success of a disease elimination/eradication strategy—be it at the national or at the international level. This has been evident for all the currently eradicated diseases, as well as for those diseases striving toward eradication (De Gourville, Duintjer Tebbens, Sangrujee, Pallansch, & Thompson, 2006; Henderson, 1987; Roeder, 2011; WHO, 2012). A pertinent example of this is rinderpest. In 1994, there was no clear, discernible international surveillance, resulting in a lack of clarity regarding the current situation and the drive to eradication. Yet, after undertaking a series of surveillance studies, it was determined that rinderpest was in fact limited to only a handful of reservoirs, primarily in African and Asian communities (Roeder, 2011). This enabled the global strategy to be adapted to the new information and facilitated the achievement of disease eradication.

International reporting of disease incidence has been designed to guide global elimination/eradication efforts, to determine trade implications for disease outbreaks, to monitor and disseminate pertinent information pertaining to exceptional outbreaks, and to facilitate the monitoring process for declaration of freedom (Jebara, 2004). As discussed earlier, the drive toward global disease elimination/eradication is based on global political will that can only be obtained through good data. An international reporting and monitoring system is required to build this strong evidence base for continued support (Henderson, 1987). Rabies is not typically considered a disease of trade importance, yet international reporting remains essential to guide and monitor global (and national) strategies, as well as the potential evaluation of disease interventions with declaration of freedom in mind. In this regard, critical assessments such as a recent study of the global burden of dog rabies (Hampson et al., 2015) remain important base points for monitoring success and the advancement of national strategies International databases such as the WHO Global Health Observatory (GHO) and the OIE WAHIS provide the platforms for the collection and global dissemination of important monitoring and epidemiological information that can help drive the rabies elimination agenda (Jebara, 2004). However, surveillance that is capable of generating reliable and actionable disease information is more critical at the national level (Jebara, 2004). To this end, a national statutory requirement for the official reporting of all human and animal rabies cases will be most helpful in improving surveillance in individual countries. However, the disease is still not declared notifiable in a number of rabies-endemic countries. Indeed, vastly improved surveillance for rabies in most rabies-endemic countries is critically needed (Taylor & Knopf, 2015). Furthermore, surveillance should not only focus on current or endemic disease threats, but should also act as an early-warning system that can monitor potential introductions or reintroductions of diseases into the region (Jebara, 2004). For rabies, national surveillance systems in endemic countries are typically weak, with limited numbers of disease indicators built into a general public health/notifiable disease system. This contributes to the misinformation of stakeholders, hindering their capacity to generate clear, actionable, data-informed intervention plans (Nel, 2013). Being zoonotic, close collaboration and coordination between public human health and animal health sectors is required for any meaningful or efficient response; current national rabies surveillance systems often do not cater for such coordination and data sharing (Jebara, 2004; Scott, Coetzer, Fahrion, & Nel, 2017).

The Rabies Epidemiological Bulletin (REB)—developed by GARC—has been designed to address rabies surveillance challenges and assist countries in collecting the most pertinent information in an easy and efficient manner (Scott et al., 2017). It addresses critical challenges that are consistent across all communicable diseases such as a cumbersome reporting processes, the inability to analyze reports, the lack of a feedback mechanism, and a limited budget for maintaining the reporting system (Janati, Hosseiny, Gouya, Moradi, & Ghaderi, 2015). The comprehensive REB addresses data collection, analysis, has GIS features (mapping functionalities) and reliable feedback mechanisms, and works in a One Health capacity as a single platform for all stakeholders involved in rabies control. These attributes are critical for effective disease surveillance and data-driven decision-making processes (whether in response to an outbreak or as routine interventions) relating to the efficient and successful implementation of the national elimination strategy (Jebara, 2004).

The REB is unique in the sense that it is a comprehensive data system that caters for the needs of the country's government and is not solely a disease reporting platform. Despite the focus on the governmental capacity building and implementation, the REB does facilitate both regional and international collaboration. Regional and international "dashboards" that provide an overview of information for each member state that has provided permissions for their data to be shared publicly (Scott et al., 2017).

After the establishment of the national reporting component of the REB, it became evident that the relevant and most pertinent and effective data were not being collected at subnational levels. The resultant typically poor or incomplete national data affected national strategies, decisions, and hindered global efforts. High-resolution data and effective monitoring and evaluation at the community level has been demonstrated to be essential in driving effective control and eradication strategies, evidenced with the community animal health worker network and Participatory Disease Surveillance program for rinderpest (Roeder, 2011); the EPI program for smallpox and the use of community workers to monitor and deliver interventions (Cutts, Waldman, & Zoffman, 1993); and the use of community health workers in the

2014 West African Ebola outbreak (Crowe et al., 2015). Being a zoonotic disease, rabies requires coordination of monitoring and surveillance activities between community animal health workers, community health workers, veterinarians, medical professionals and laboratory professionals, among others, especially for integrated bite case management (IBCM) programs (Day, 2011; Pieracci et al., 2016). Thus, additional components of the REB—including IBCM components—have been developed to improve grassroots data collection within member states, while other IBCM-specific mobile apps and community reporting systems have been developed by global rabies stakeholders to assist in obtaining relevant, timely, and accurate community-level data to drive decision-making processes and improve understanding of the disease (Global Alliance for Rabies Control, 2019b; Mission Rabies, 2019).

21.5.1 Monitoring of vaccination events

For many disease intervention campaigns, mass vaccination is critical to either control and/or elimination. However, with weak information and data collection systems, timely data flow from the periphery (community, clinics, health workers, etc.) to the central system (typically at national level) is often lacking. Thus, national reporting and statistics rely on poor data to make decisions (Murray et al., 2003). A second key issue may be that community workers responsible for undertaking the activities (mass dog vaccination (MDV) in the case of rabies) may intentionally inflate figures and outputs as, in some cases, their salaries or remuneration may be based on an outcome-based/incentive-based mechanism (Murray et al., 2003). Furthermore, multiple disease intervention campaigns have demonstrated the inherent flaw in calculating vaccination coverage based on doses administered, as outcomes of greater than 100% coverage are often achieved (Alavian et al., 2009; Luman, Cairns, Perry, Dietz, & Gittelman, 2007; Zuber, Yameogo, Yameogo, & Otten, 2003). Without extensive and accurate campaign monitoring, data can also be drastically underreported, as evidenced during an Influenza mass vaccination campaign where 124 million doses of vaccine were distributed but only 81 million people reported receiving vaccine (Centers for Disease Control and Prevention, 2011). In addition, data associated with the vaccination of target groups may be missed due to poor data collection techniques, emphasizing the importance and need for efficient and effective real-time data collection tools capable of managing the large volumes of data collected during mass vaccination campaigns (McClung et al., 2018).

The organization and logistics surrounding an effective mass vaccination campaign are challenging and rely on accurate and timely data to (i) identify target areas; (ii) determine vaccination coverage; (iii) direct field teams to target areas; and (iv) identify and react to outbreaks. Well-organized, efficient vaccination campaigns require fewer financial and human resources and are able to achieve success in a shorter time than inefficient ones lacking good surveillance (Barrett, 2006). Similarly, if incorrectly implemented, mass vaccination can have a limited impact on disease control (irrespective of the resources utilized) as evidenced with India for smallpox and Egypt for avian influenza elimination and control efforts (Barrett, 2006; Peyre et al., 2009). For measles and rubella, reliable and contemporary data—especially vaccination coverage—were critical indicators considered by donors and funding agencies to determine success. Such indicators affected decisions for continued funding and provisioned more accurate vaccine forecasting (Burton et al., 2009; Fields, Dabbagh, Jain, & Sagar, 2013).

Mobile technology has helped address some of these challenges by streamlining data collection and analysis, while reducing costs (in comparison to paper-based techniques) and providing real-time surveillance data to decision makers (Mtema et al., 2016). The combination of reliable data with real-time analysis and outputs creates a more robust and efficient framework for effective surveillance. Even basic use of mobile technology (such as using the GPS for tracking) assisted Nigeria in identifying critically missed communities that previously hindered polio elimination, while also being able to track vaccinator performance—ensuring that inflated figures were less likely to be reported (Touray et al., 2016). In terms of monitoring rabies MDV events, various stakeholders have successfully utilized mobile technology and custom-developed technologies to address these inconsistencies and challenges (Gibson et al., 2018; Global Alliance for Rabies Control, 2019a; Humane Society International, 2018). Importantly, these technologies track the number of vaccines administered, along with the GPS location of each vaccinated animal and a time and date stamp (among other information in some cases), to effectively monitor the use of vaccine during a campaign. Using this data, vaccinator efficiency and data quality control can be determined—a development suggested for polio—and a more accurate vaccination coverage determined (Touray et al., 2016). As mentioned previously, one of the challenges in determining vaccination coverage using the "vaccines administered" approach is the determination of the population number (denominator), with one reason for above 100% coverage being the influx of vaccinated individuals from outside of the population area. Alternatively, inaccurate censuses are contributing factors. By using mobile technologies, improved dog census, postvaccination surveys, vaccination coverage, and geographic analyses can be obtained, facilitating more accurate and reliable coverage estimates and other associated indicators. These technologies also address commonly reported challenges relating to the validity of vaccination coverage for national health programs such as poor timeliness and completeness of reporting, poor data storage, insufficient analyses, and feedback mechanisms and confusion relating to the denominator for indicators (i.e., population numbers) (Murray et al., 2003).

After initial implementation of the MDV campaigns, campaign supervisors are able to use the maps and reports generated by such technologies to determine whether mop-up campaigns are required in any specific area, ensuring that an adequate coverage can be obtained (Gibson et al., 2018; Touray et al., 2016). With clear outputs and feedback mechanisms to community workers and a more efficient campaign, vaccinator, and surveillance fatigue—a common challenge faced in MDV campaigns—can also be reduced, ensuring sustainability for future maintenance or follow-up efforts (Averhoff et al., 2001; Conan et al., 2015; Hussain, McGarvey, Shahab, & Fruzzetti, 2012; Morice et al., 2003). Overall, it appears that digital and mobile health systems are important in improving campaign efficiency, thereby freeing resources (both financial and human) for distribution to other areas, improving surveillance and data-driven decision-making processes, and ensuring continued public and political will to improve national coverage and achieve elimination (Averhoff et al., 2001; Morice et al., 2003; Teng, Thomson, Lascher, Raymond, & Ivers, 2014; Vatsalan et al., 2010). Although this technology is not yet being widely implemented for rabies, the available evidence-base advocates its continued use and expansion to elimination efforts in all endemic countries. The use of such tools and technologies should be integral to all strategies for the achievement of the "Zero by 30" goal and the eventual elimination of dog-mediated human rabies.

21.6 Conclusions

Although the tools, strategies, and means to eliminate rabies exist, the effective delivery of such interventions in many endemic countries currently remains slow. Considerable international support (technical, advocative, and organizational) will thus be required to initiate change and momentum through global strategies such as the GSP (Goodman, Coleman, & Mills, 1999; Minghui et al., 2018). However, the will for such change—although typically initiated through such support—must be embedded within the national government and completely owned, directed, and driven by local stakeholders to ensure a sustainable, and therefore effective, elimination strategy (Coker et al., 2004).

We have demonstrated that rabies has all the necessary tools to successfully achieve "Zero by 30," with supportive historical evidence from other disease interventions. There should be a clear unequivocal message that dog-mediated human rabies elimination should not be hindered by concerns about wildlife vectors, as dogs are responsible for 99% of human rabies cases globally. While it might be argued that wildlife rabies could in future pose a threat to immunologically naïve dog populations, this could only be addressed meaningfully once the risk has been established in the face of dog rabies elimination. Of benefit to low-income dog rabies-endemic countries, the tools discussed facilitate the implementation of an efficient campaign that can be delivered with minimal resources and little additional burden to the health system. Therefore, the next steps will be for the continued education of stakeholders in the use of the tools—through regional rabies networks and other platforms. Continued efforts to prioritize rabies elimination should focus on rabies not only as a neglected zoonotic disease, but as a disease of which the elimination would unite the world as far as the One Health principle is concerned. Moving forward, political will, international interest, and public backing must be garnered to build international pressure and gather resources to facilitate efforts in the remaining rabies-endemic countries.

References

Alavian, S. M., Gooya, M. M., Hajarizadeh, B., Esteghamati, A. -R., Moeinzadeh, A. M., Haghazali, M., … Lankarani, K. B. (2009). Mass vaccination campaign against hepatitis B in adolescents in Iran: Estimating coverage using administrative data. *Hepatitis Monthly*, 9(3), 189–195.

Atun, R. A., Lennox-Chhugani, N., Drobniewski, F., Samyshkin, Y. A., & Coker, R. J. (2004). A framework and toolkit for capturing the communicable disease programmes within health systems. *European Journal of Public Health*, 14(3), 267–273. https://doi.org/10.1093/eurpub/14.3.267.

Averhoff, F., Deladisma, A., Shapiro, C. N., Bell, B. P., Simard, E. P., Margolis, H. S., … Kuter, B. (2001). Control of hepatitis A through routine vaccination of children. *Journal of the American Medical Association*, 286(23), 2968–2973. https://doi.org/10.1001/jama.286.23.2968.

Balaram, D., Taylor, L. H., Doyle, K. A. S., Davidson, E., & Nel, L. H. (2016). World Rabies Day—A decade of raising awareness. *Tropical Diseases, Travel Medicine and Vaccines*, 2(19), 1–9. https://doi.org/10.1186/s40794-016-0035-8.

Barrett, S. (2006). The smallpox eradication game. *Public Choice*, 130(1–2), 179–207. https://doi.org/10.1007/s11127-006-9079-z.

Bonita, R., Magnusson, R., Bovet, P., Zhao, D., Malta, D. C., Geneau, R., … Beaglehole, R. (2013). Country actions to meet UN commitments on non-communicable diseases: A stepwise approach. *The Lancet*, 381(9866), 575–584. https://doi.org/10.1016/S0140-6736(12)61993-X.

Burton, A., Monasch, R., Lautenbach, B., Gacic-Dobo, M., Neill, M., Karimov, R., … Birmingham, M. (2009). WHO and UNICEF estimates of national infant immunization coverage: Methods and processes. *Bulletin of the World Health Organization, 87*(7), 535–541. https://doi.org/10.2471/BLT.08.053819.

Centers for Disease Control and Prevention (2011). *Final estimates for 2009–10 seasonal influenza and influenza A (H1N1) 2009 monovalent vaccination coverage—United States, August 2009 through May, 2010.*

Cochi, S. L., Freeman, A., Guirguis, S., Jafari, H., & Aylward, B. (2014). Global polio eradication initiative: Lessons learned and legacy. *Journal of Infectious Diseases, 210*(Suppl. 1), S540–S546. https://doi.org/10.1093/infdis/jiu345.

Cochi, S. L., Hegg, L., Kaur, A., Pandak, C., & Jafari, H. (2016). The global polio eradication initiative: Progress, lessons learned, and polio legacy transition planning. *Health Affairs, 35*(2), 277–283. https://doi.org/10.1377/hlthaff.2015.1104.

Coetzer, A., Kidane, A. H., Bekele, M., Hundera, A. D., Pieracci, E. G., Shiferaw, M. L., … Nel, L. H. (2016). The SARE tool for rabies control: Current experience in Ethiopia. *Antiviral Research, 135*. https://doi.org/10.1016/j.antiviral.2016.09.011.

Coetzer, A., Scott, T. P., Amparo, A. C., Jayme, S., & Nel, L. H. (2018). Formation of the Asian Rabies Control Network (ARACON): A common approach towards a global good. *Antiviral Research, 157*, 134–139. https://doi.org/10.1016/j.antiviral.2018.07.018.

Coker, R., Atun, R., & McKee, M. (2004). Untangling Gordian knots: Improving tuberculosis control through the development of "programme theories". *International Journal of Health Planning and Management, 19*(3), 217–226. https://doi.org/10.1002/hpm.759.

Conan, A., Akerele, O., Simpson, G., Reininghaus, B., van Rooyen, J., & Knobel, D. (2015). Population dynamics of owned, free-roaming dogs: Implications for rabies control. *PLoS Neglected Tropical Diseases, 9*(11), e0004177. https://doi.org/10.1371/journal.pntd.0004177.

Cotter, C., Sturrock, H. J. W., Hsiang, M. S., Liu, J., Phillips, A. A., Hwang, J., … Feachem, R. G. A. (2013). The changing epidemiology of malaria elimination: New strategies for new challenges. *The Lancet, 382*(9895), 900–911. https://doi.org/10.1016/S0140-6736(13)60310-4.

Crowe, S., Hertz, D., Maenner, M., Ratnayake, R., Baker, P., Lash, R. R., … Centers for Disease Control and Prevention (CDC) (2015). A plan for community event-based surveillance to reduce Ebola transmission—Sierra Leone, 2014-2015. *MMWR. Morbidity and Mortality Weekly Report, 64*(3), 70–73. Retrieved from http://www.ncbi.nlm.nih.gov/pubmed/25632956%0Ahttp://www.pubmedcentral.nih.gov/articlerender.fcgi?artid=PMC4584562.

Cutts, F. T., Waldman, R. J., & Zoffman, H. M. D. (1993). Surveillance for the expanded programme on immunization. *Bulletin of the World Health Organization, 71*(5), 633–639.

Day, M. J. (2011). One health: The importance of companion animal vector-borne diseases. *Parasites & Vectors, 4*(1), 49. https://doi.org/10.1186/1756-3305-4-49.

de Carvalho, M., Vigilato, M. A. N., Pompei, J. A., Rocha, F., Vokaty, A., Molina-Flores, B., … Del Rio Vilas, V. J. (2018). Rabies in the Americas: 1998-2014. *PLoS Neglected Tropical Diseases, 12*(3), e0006271. https://doi.org/10.1371/journal.pntd.0006271.

De Gourville, E., Duintjer Tebbens, R. J., Sangrujee, N., Pallansch, M. A., & Thompson, K. M. (2006). Global surveillance and the value of information: The case of the global polio laboratory network. *Risk Analysis, 26*(6), 1557–1569. https://doi.org/10.1111/j.1539-6924.2006.00845.x.

Diall, O., Cecchi, G., Wanda, G., Argilés-Herrero, R., Vreysen, M. J. B., Cattoli, G., … Bouyer, J. (2017). Developing a progressive control pathway for African animal trypanosomosis. *Trends in Parasitology, 33*(7), 499–509. https://doi.org/10.1016/j.pt.2017.02.005.

Epping-Jordan, J. A. E., Galea, G., Tukuitonga, C., & Beaglehole, R. (2005). Preventing chronic diseases: Taking stepwise action. *Lancet, 366*(9497), 1667–1671. https://doi.org/10.1016/S0140-6736(05)67342-4.

FAO (2012). Lessons learned from the eradication of rinderpest for controlling other transboundary animal diseases. In: *Proceedings of the GREP symposium and high-level meeting*. FAO Animal Production and Health Proceedings: Rome. Retrieved from http://www.fao.org/3/a-i3042e.pdf.

Fields, R., Dabbagh, A., Jain, M., & Sagar, K. S. (2013). Moving forward with strengthening routine immunization delivery as part of measles and rubella elimination activities. *Vaccine, 31*(Suppl. 2), B115–B121. https://doi.org/10.1016/j.vaccine.2012.11.094.

Gibson, A. D., Mazeri, S., Lohr, F., Mayer, D., Burdon Bailey, J. L., Wallace, R. M., … Gamble, L. (2018). One million dog vaccinations recorded on mHealth innovation used to direct teams in numerous rabies control campaigns. *PLoS One, 13*(7), 1–19. https://doi.org/10.1371/journal.pone.0200942.

Global Alliance for Rabies Control (2019a). *GARC data logger (GDL)*. Retrieved July 19, 2019, from https://rabiesalliance.org/capacity-building/gdl.

Global Alliance for Rabies Control (2019b). *Rabies epidemiological bulletin (REB)*. Retrieved June 13, 2019, from https://rabiesalliance.org/capacity-building/reb.

Goodman, C. A., Coleman, P. G., & Mills, A. J. (1999). Cost-effectiveness of malaria control in sub-Saharan Africa. *Lancet, 354*(9176), 378–385. https://doi.org/10.1016/S0140-6736(99)02141-8.

Hampson, K., Coudeville, L., Lembo, T., Sambo, M., Kieffer, A., Attlan, M., … Dushoff, J. (2015). Estimating the global burden of endemic canine rabies. *PLoS Neglected Tropical Diseases, 9*(4), e0003709. https://doi.org/10.1371/journal.pntd.0003786.

Henderson, D. A. (1987). Principles and lessons from the smallpox eradication programme. *Bulletin of the World Health Organization, 65*(4), 535–546. https://doi.org/10.1016/j.eswa.2012.01.068.

Henderson, D. A. (2011). The eradication of smallpox—An overview of the past, present, and future. *Vaccine, 29* (Suppl. 4), D7. https://doi.org/10.1016/j.vaccine.2011.06.080.

Hochman, G. (2009). Priority, invisibility and eradication: The history of smallpox and the Brazilian public health agenda. *Medical History, 53*(2), 229–252. https://doi.org/10.1017/S002572730000020X.

Humane Society International (2018). *To mark World Rabies Day, animal charity uses game-changer smartphone app to vaccinate dogs in the Philippines, with global potential*. Retrieved December 12, 2018, from http://www.hsi.org/news/press_releases/2018/09/world-rabies-day-philippines-092618.html.

Hussain, R. S., McGarvey, S. T., Shahab, T., & Fruzzetti, L. M. (2012). Fatigue and fear with shifting polio eradication strategies in India: A study of social resistance to vaccination. *PLoS One, 7*(9), 1–8. https://doi.org/10.1371/journal.pone.0046274.

International Coordinating Group (ICG) (2014). *Report of the sixth meeting of the International Coordinating Group of the World Health Organization and the Bill & Melinda Gates Foundation project on eliminating human and dog rabies*. Durban.

Janati, A., Hosseiny, M., Gouya, M. M., Moradi, G., & Ghaderi, E. (2015). Communicable disease reporting systems in the world: A systematic review article. *Iranian Journal of Public Health, 44*(11), 1453–1465. Retrieved from http://www.ncbi.nlm.nih.gov/pubmed/26744702%0Ahttp://www.pubmedcentral.nih.gov/articlerender.fcgi?artid=PMC4703224.

Jebara, K. B. (2004). Surveillance, detection and response: Managing emerging diseases at national and international levels. *Revue Scientifique et Technique de l'OIE, 23*(2), 709–715.

Lembo, T., Attlan, M., Bourhy, H., Cleaveland, S., Costa, P., de Balogh, K., … Briggs, D. J. (2011). Renewed global partnerships and redesigned roadmaps for rabies prevention and control. *Veterinary Medicine International, 2011*, 923149.

Lembo, T., Hampson, K., Kaare, M. T., Ernest, E., Knobel, D. L., Kazwala, R. R., … Cleaveland, S. (2010). The feasibility of canine rabies elimination in Africa: Dispelling doubts with data. *PLoS Neglected Tropical Diseases, 4*(2), 1–9. https://doi.org/10.1371/journal.pntd.0000626.

Luman, E. T., Cairns, K. L., Perry, R., Dietz, V., & Gittelman, D. (2007). Use and abuse of rapid monitoring to assess coverage during mass vaccination campaigns. *Bulletin of the World Health Organization, 85*(September), 651. https://doi.org/10.2471/BLT.

Mackenzie, J. S., Drury, P., Arthur, R. R., Ryan, M. J., Grein, T., Slattery, R., … Bejtullahu, A. (2014). The global outbreak alert and response network. *Global Public Health, 9*(9), 1023–1039. https://doi.org/10.1080/17441692.2014.951870.

Mariner, J. C., House, J. A., Mebus, C. A., Sollod, A. E., Chibeu, D., Jones, B. A., … van 't Klooster, G. G. M. (2012). Rinderpest eradication: Appropriate technology and social innovations. *Science, 337*(6100), 1309–1312. https://doi.org/10.1126/science.1223805.

McClung, M. W., Gumm, S. A., Bisek, M. E., Miller, A. L., Knepper, B. C., & Davidson, A. J. (2018). Managing public health data: Mobile applications and mass vaccination campaigns. *Journal of the American Medical Informatics Association, 25*(4), 435–439. https://doi.org/10.1093/jamia/ocx136.

Meessen, B., Soucat, A., & Sekabaraga, C. (2011). Performance-based financing: Just a donor fad or a catalyst towards comprehensive health-care reform? *Bulletin of the World Health Organization, 89*(2), 153–156. https://doi.org/10.2471/BLT.10.077339.

Minghui, R., Stone, M., Semedo, M. H., & Nel, L. (2018). New global strategic plan to eliminate dog-mediated rabies by 2030. *The Lancet Global Health, 18*, 4–5. https://doi.org/10.1016/S2214-109X(18)30302-4.

Mission Rabies. (2019). *WVS data collection app*. Retrieved July 19, 2019, from http://www.missionrabies.com/app.

Morice, A., Carvajal, X., León, M., Machado, V., Badilla, X., Reef, S., … Castillo-Solórzano, C. (2003). Accelerated rubella control and congenital rubella syndrome prevention strengthen measles eradication: The Costa Rican experience. *The Journal of Infectious Diseases, 187*(s1), S158–S163. https://doi.org/10.1086/368053.

Mtema, Z., Changalucha, J., Cleaveland, S., Elias, M., Ferguson, H. M., Halliday, J. E. B., … Hampson, K. (2016). Mobile phones as surveillance tools: Implementing and evaluating a large-scale intersectoral surveillance system for rabies in Tanzania. *PLoS Medicine, 13*(4), e1002002. https://doi.org/10.1371/journal.pmed.1002002.

Murray, C. J. L., Shengelia, B., Gupta, N., Moussavi, S., Tandon, A., & Thieren, M. (2003). Validity of reported vaccination coverage in 45 countries. *Lancet, 362*(9389), 1022–1027. https://doi.org/10.1016/S0140-6736(03)14411-X.

Nel, L. H. (2013). Factors impacting the control of rabies. *Microbiology Spectrum, 1*(2), 1–12. https://doi.org/10.1128/microbiolspec.OH-0006-2012.f1.

Nel, L. H. (2018). The role of non-governmental organisations in controlling rabies: The Global Alliance for Rabies Control, Partners for Rabies Prevention and the Blueprint for Rabies Prevention and Control. *Revue Scientifique et Technique (International Office of Epizootics), 37*(2), 751–759. https://doi.org/10.20506/rst.37.2.2838.

Nel, L. H., Taylor, L. H., Balaram, D., & Doyle, K. A. S. (2015). Global partnerships are critical to advance the control of Neglected Zoonotic Diseases: The case of the Global Alliance for Rabies Control. *Acta Tropica*. https://doi.org/10.1016/j.actatropica.2015.10.014.

OIE. (2019). Terrestrial animal health code *General: Vol. 1* (28th ed.). Paris: World Organisation for Animal Health. Retrieved from http://www.oie.int/standard-setting/terrestrial-code/access-online/.

OIE, & FAO. (2012). *The global foot-and-mouth disease control strategy: Strengthening animal health systems through improved control of major diseases.* Retrieved July 25, 2019, from https://www.oie.int/doc/ged/D11886.PDF.

Perry, R. T., Gacic-Dobo, M., Dabbagh, A., Mulders, M. N., Strebel, P. M., Okwo-Bele, J.-M. M., … Centers for Disease Control and Prevention (CDC). (2014). Global control and regional elimination of measles, 2000-2012. *Morbidity and Mortality Weekly Report, 63*(5), 103–107. https://doi.org/mm6305a5 [pii].

Peyre, M., Samaha, H., Makonnen, Y. J., Saad, A., Abd-Elnabi, A., Galal, S., … Domenech, J. (2009). Avian influenza vaccination in Egypt: Limitations of the current strategy. *Journal of Molecular and Genetic Medicine: An International Journal of Biomedical Research, 3*(2), 198–204.

Pieracci, E. G., Schroeder, B., Mengistu, A., Melaku, A., Shiferaw, M., Blanton, J. D., & Wallace, R. (2016). Notes from the field : Assessment of health facilities for control of canine rabies—Gondar City, Amhara Region, Ethiopia, 2015. *MMWR. Morbidity and Mortality Weekly Report, 65*(17), 456–457. https://doi.org/10.15585/mmwr.mm6517a4.

Rist, C. L., Arriola, C. S., & Rubin, C. (2014). Prioritizing zoonoses: A proposed one health tool for collaborative decision-making. *PLoS One, 9*(10). https://doi.org/10.1371/journal.pone.0109986.

Roeder, P. L. (2011). Rinderpest: The end of cattle plague. *Preventative Veterinary Medicine, 102*(2), 98–106. https://doi.org/10.1016/j.prevetmed.2011.04.004.

Rweyemamu, M. M., & Cheneau, Y. (1995). Strategy for the global rinderpest eradication programme. *Veterinary Microbiology, 44*(2–4), 369–376. https://doi.org/10.1016/0378-1135(95)00030-E.

Salyer, S. J., Silver, R., Simone, K., & Behravesh, C. B. (2017). Prioritizing zoonoses for global health capacity building—Themes from One Health zoonotic disease workshops in 7 countries, 2014–2016. *Emerging Infectious Diseases, 23*(13), s57–s64. https://doi.org/10.3201/eid2313.170418.

Scott, T. P., Coetzer, A., De Balogh, K., Wright, N., & Nel, L. H. (2015). The Pan-African Rabies Control Network (PARACON): A unified approach to eliminating canine rabies in Africa. *Antiviral Research, 124*. https://doi.org/10.1016/j.antiviral.2015.10.002.

Scott, T. P., Coetzer, A., Fahrion, A. S., & Nel, L. H. (2017). Addressing the disconnect between the estimated, reported and true rabies data: The development of a regional African rabies bulletin. *Frontiers in Veterinary Science, 4*(18). https://doi.org/10.3389/fvets.2017.00018.

Taylor, E., Banyard, A. C., Bourhy, H., Cliquet, F., Ertl, H., Fehlner-Gardiner, C., … Fooks, A. R. (2019). Avoiding preventable deaths: The scourge of counterfeit rabies vaccines. *Vaccine, 37*, 2285–2287. https://doi.org/10.1016/j.vaccine.2019.03.037.

Taylor, W. P., Bhat, P. N., & Nanda, Y. P. (1995). The principles and practice of rinderpest eradication. *Veterinary Microbiology, 44*(2–4), 359–367. https://doi.org/10.1016/0378-1135(95)00029-A.

Taylor, L. H., & Knopf, L. (2015). Surveillance of human rabies by national authorities—A global survey. *Zoonoses and Public Health, 62*, 543–552. https://doi.org/10.1111/zph.12183.

Taylor, L., & Partners for Rabies Prevention (2013). Eliminating canine rabies: The role of public–private partnerships. *Antiviral Research, 98*(March), 314–318. https://doi.org/10.1016/j.antiviral.2013.03.002.

Teng, J. E., Thomson, D. R., Lascher, J. S., Raymond, M., & Ivers, L. C. (2014). Using mobile health (mHealth) and geospatial mapping technology in a mass campaign for reactive oral cholera vaccination in rural Haiti. *PLoS Neglected Tropical Diseases, 8*(7). https://doi.org/10.1371/journal.pntd.0003050.

Touray, K., Mkanda, P., Tegegn, S. G., Nsubuga, P., Erbeto, T. B., Banda, R., … Vaz, R. G. (2016). Tracking vaccination teams during polio campaigns in Northern Nigeria by use of geographic information system technology: 2013-2015. *Journal of Infectious Diseases, 213*(Suppl. 3), S67–S72. https://doi.org/10.1093/infdis/jiv493.

Unger, J. P., De Paepe, P., Ghilbert, P., Soors, W., & Green, A. (2006). Integrated care: A fresh perspective for international health policies in low and middle-income countries. *International Journal of Integrated Care, 18*(6), e15.

US CDC. (2019). *One health zoonotic disease prioritization. CDC 24/7*. Retrieved from https://www.cdc.gov/onehealth/global-activities/prioritization.html.

Vatsalan, D., Arunatileka, S., Chapman, K., Senaviratne, G., Sudahar, S., Wijetileka, D., & Wickramasinghe, Y. (2010). Mobile technologies for enhancing eHealth solutions in developing countries. In: *2nd international conference on eHealth, telemedicine, and social medicine, eTELEMED 2010, Includes MLMB 2010; BUSMMed 2010* (pp. 84–89). https://doi.org/10.1109/eTELEMED.2010.18.

Vigilato, M. A., Cosivi, O., Knöbl, T., Clavijo, A., & Silva, H. M. (2013). Rabies update for Latin America and the Caribbean. *Emerging Infectious Diseases, 19*(4), 678–679. https://doi.org/10.1098/rstb.2012.0143.

Wallace, R. M., Undurraga, E. A., Blanton, J. D., Cleaton, J., & Franka, R. (2017). Elimination of dog-mediated human rabies deaths by 2030: Needs assessment and alternatives for progress based on dog vaccination. *Frontiers in Veterinary Science, 4*(9). https://doi.org/10.3389/fvets.2017.00009.

WHO. (2012). *Global measles & rubella strategic plan 2012-2020*. Geneva, Switzerland. Retrieved from http://www.who.int/about/licensing/copyright_form/en/index.html.

WHO, OIE, FAO, & GARC. (2018). *Zero by 30: The Global Strategic Plan to end human deaths from dog-mediated rabies by 2030*. Geneva. Retrieved from https://rabiesalliance.org/resource/zero-30-global-strategic-plan-end-human-deaths-dog-mediated-rabies-2030.

World Bank Group. (2019). *World Bank data*.

World Health Assembly. (1966). Eighth meeting of the nineteenth World Health Assembly. *Official Records of the World Health Organization: Vol. 152*. Geneva. https://doi.org/10.1016/S0140-6736(57)91770-1

World Health Organization. (2014). *WHO global health expenditure database*. Retrieved from http://apps.who.int/nha/database/ViewData/Indicators/en.

World Health Organization. (2018). WHO expert consultation on rabies, third report. *WHO technical report series: Vol. 931*. Geneva: World Health Organization. Retrieved from http://apps.who.int/iris/handle/10665/272364.

Zuber, P. L. F., Yameogo, K. R., Yameogo, A., & Otten, M. W. (2003). Use of administrative data to estimate mass vaccination campaign coverage, Burkina Faso, 1999. *Journal of Infectious Diseases, 187*(Suppl. 1), 86–90.

CHAPTER 22

Future developments and challenges

Anthony R. Fooks[a], Alan C. Jackson[b]

[a]Director of a World Health Organization Communicable Disease Surveillance and Response Collaborating Centre for the Characterization of Rabies and Rabies-Related Viruses/Head of an OIE (World Organisation for Animal Health) Reference Laboratory for Rabies, The Animal and Plant Health Agency, Addlestone, Surrey, United Kingdom [b]Professor of Medicine (Neurology), University of Manitoba, Winnipeg, MB, Canada

22.1 Introduction

Rabies lyssavirus (RABV), the prototype virus species of the *Lyssavirus* genus along with other members of the genus in the family *Rhabdoviridae* of the order *Mononegavirales*, cause an acute progressive encephalitis (rabies) in mammals, being transmitted between susceptible individuals directly by bites, scratches, or contamination of mucous membranes with infected saliva (Amarasinghe et al., 2019). Rabies is a neglected viral zoonotic disease of antiquity that was first described in the 4th Century BC (see Chapter 1) and remains an important public health problem. In the 19th century, Louis Pasteur pioneered the development of the first rabies vaccine for human use that has been used after rabies exposures in order to prevent the disease. Rabies virus is principally transmitted by bites from infected animals, most frequently dogs, resulting in more than 99% of the human cases reported worldwide. Consequently, rabies remains an important threat to public health, and the disease is virtually always fatal once clinical symptoms develop. As rabies is entirely preventable through vaccination, it is unfortunate that rabies remains enzootic in many regions of the world, especially in resource-limited countries in Africa and Asia, where the disease burden is high (Hampson et al., 2015). In many underdeveloped countries with inadequate human health care and often an ineffective veterinary infrastructure, rabies remains a disease of the underrepresented and impoverished minorities (Fooks et al., 2017). The thousands of officially reported human deaths worldwide do not capture the true burden of this disease. Any deficiency in laboratory capabilities results in underreporting of the disease, which compounds a negative cycle of neglect, lack of advocacy and more importantly, placing rabies as a disease that has been controlled and consequently, no longer considered as a high-priority disease.

With only a clinical diagnosis of rabies in many countries and without laboratory confirmation, the disease in humans is often misdiagnosed (Mallawa et al., 2007). Long-term elimination strategies for reducing the burden of human rabies is by controlling rabies in dogs (Meslin & Briggs, 2013). This intervention strategy should be complemented with the control of rabies in wildlife, which remains a source of rabies reintroduction into the canine population (Horton et al., 2013). Control strategies in wildlife using oral vaccination programs have been demonstrated to be highly successful, especially in foxes and raccoons. Several regions of the world have demonstrated that rabies elimination is possible and there is a vision to eliminate human rabies transmitted by dogs in rabies-endemic countries by 2030 (Abela-Ridder et al., 2016) (Chapter 21). This final chapter will focus on future research efforts as well as strategies to reduce the human death toll of rabies.

22.2 Pathogenesis

Major advances in our understanding of the basic mechanisms involved in the pathogenesis of rabies have been slow. Unfortunately, research efforts have been limited in this area, probably at least in part, because human rabies can be effectively prevented and, hence, the disease is not considered an important public health problem in resource-rich countries. Further work is needed to gain an improved understanding of the basic mechanisms involved in how the nervous system infection in rabies produces clinical disease and death (see Chapter 9). Additionally, the process of how RABV infection can modulate neuronal function and death remains poorly understood. However, there have been some important recent insights into the basic mechanisms involved, at least in an animal model infected with a laboratory RABV strain. Neuronal process degeneration affecting dendrites and axons occurs as a result of oxidative stress (Jackson, Kammouni, Zherebitskaya, & Fernyhough, 2010) caused by mitochondrial dysfunction (Alandijany, Kammouni, Roy Chowdhury, Fernyhough, & Jackson, 2013). This dysfunction occurs as a result of interaction of the RABV phosphoprotein and mitochondrial Complex I, and amino acids positions at both 162 and 166 of the phosphoprotein are critical sites (Kammouni et al., 2015; Kammouni, Wood, & Jackson, 2017a, 2017b). The mechanisms underlying RABV evasion of the host immune system in the CNS, may provide an explanation for the reasons that in the absence of postexposure prophylaxis (PEP), rabies remains one of the few human infections with a 100% case-mortality rate, once clinical symptoms are observed. The exact mechanisms of immune evasion, however, remain unclear and improvements in our understanding of how RABV has evolved these immune evasive mechanisms to infect the CNS, may help identify new therapeutic targets (see Chapter 11).

22.3 Therapy of human rabies

A limited understanding of the basic mechanisms responsible for clinical disease and death in rabies (see Chapter 9) has made therapeutic efforts much more difficult. Further research is essential in good animal models in order to provide this critical information to assist in the development of therapeutic efforts to successfully move the field forward. There has been little recent progress in developing an effective therapy for human rabies (see Chapter 17).

Unfortunately, repetition of the ineffective Milwaukee Protocol (Willoughby Jr. et al., 2005) has proved to be an impediment in moving forward. The protocol included induction of therapeutic coma and concurrent administration of antiviral drugs. The original protocol has since undergone several modifications. There is no clear scientific rationale for the protocol and repeated failures and the absence of any well-documented success after the index survivor clearly indicate that new approaches are needed (Zeiler & Jackson, 2016). It has proved very difficult to get this message across to physicians who are faced with real cases and are not familiar with the complex treatment issues in rabies. No real progress will be made in the future until physicians are willing to take new approaches instead of repeating what is easy and has repeatedly failed in the past (see Chapter 17).

Many other viral infections of the CNS also do not have effective therapies, although there are notable exceptions such as the therapy of herpes simplex encephalitis with intravenous acyclovir. Where antiviral development has been attempted, in vitro antiviral data has rarely translated to in vivo successes. Unfortunately, this has also been the case with the broad-spectrum antiviral molecule favipiravir (6-fluoro-3-hydroxy-2-pyrazinecarboxamide), which has shown only negligible efficacy for rabies in rodent models (Banyard et al., 2019; Yamada, Noguchi, Komeno, Furuta, & Nishizono, 2016).

Palliative care is imperative even if critical care is not feasible due to medical, logistical, or financial reasons. Unfortunately, palliation in rabies has received little attention and guidelines for palliative care are generally lacking in endemic regions. Patients continue to be isolated and often physically restrained in dismal cubicles, partly as a result of concerns about the potential for aggression and transmission of infection to others. Appropriate comfort measures are extremely important for patients with rabies and a greater emphasis on this aspect of rabies care is badly needed (see Chapter 17).

22.4 Epidemiology

Asia and Africa continue to report the vast majority of the worldwide burden of rabies (see Chapters 4 and 18). About three billion people live in regions with endemic dog rabies and are at potential risk of a dog exposure. Impoverished people and particularly children carry a disproportional burden of the disease. The number of global human cases has been estimated at between 55,000 and 75,000 deaths per year. However, these are only very crude estimates and accurate numbers of cases are not available, including in India that has by far the greatest number of cases of human rabies of any country in the world.

Mesocarnivore species, including dogs, foxes, raccoon dogs, raccoons, and skunks, serve as rabies reservoirs. Other mammals, including humans, cattle, and equids, represent spillover (dead-end) hosts, caused by a cross-species transmission event without virus adaptation and transmission to other hosts. In contrast, RABV host-switching events, which involves onward virus transmission and host adaptation of the virus in a new species, including from bats to mesocarnivores, are reported to occur in nature with some regularity (see Chapter 3). With the advent of molecular tools, viral characterization methods have revealed the extensive diversity of the *Lyssavirus* genus. Phylogenetic analysis of rabies virus and the non-RABV lyssaviruses has revealed that the evolution of this species has likely evolved from a bat-associated progenitor (see Chapter 3). The generation of whole genome sequences (Marston et al., 2013) has provided new insights into the phylogeography of RABV and has identified factors that influence virus spread, information of value to control programs (see Chapter 5). Inadequate surveillance, however, of human and animal rabies in resource-limited countries without proper medical and veterinary infrastructures remains a widespread problem. This lack of available confirmatory laboratory diagnostic evaluations in many regions is contributory to a failure to report cases in animals or humans (see Chapter 12). The lack of accurate global surveillance data contributes to control efforts being underresourced because the full extent of the rabies problem is undocumented and not realized. Ongoing passive and active surveillance coupled with viral genetic sequencing is essential to further our knowledge of rabies spread and the intervention strategies required to control rabies, especially in developed countries.

22.5 Prevention of human rabies

Human rabies is not a major public health problem in developed countries, including North America and Europe, although a small number of sporadic cases continue to occur, typically due to unrecognized bat exposures (see Chapter 7) and "imported" cases from countries with endemic dog rabies. The Advisory Committee on Immunization Practices in the United States recommends initiation of rabies PEP in situations in which a bat exposure cannot be excluded, including, for example, a situation with an adult sleeping in a room in which a bat was present that subsequently escaped (Manning et al., 2008). However, recent evidence argues against this approach (De Serres et al., 2009), and changes in future recommendations are expected.

An awareness of intervention tools to prevent rabies is of paramount importance in preventing deaths from rabies (see Chapter 16). Where possible, rabies preimmunization (Pr-EP) is recommended for certain travelers to endemic regions although uptake of such advice may not always be followed. Regardless whether Pr-EP is administered or not, appropriate wound care and application of PEP is almost 100% effective at preventing deaths.

The high cost of rabies biologicals continues to be a major impediment to the more widespread use of both Pr-EP and rabies PEP measures. The development of vaccines for intramuscular administration from the early crude preparations of animal neuronal tissue through to the safer cell-culture-derived vaccines has increased the uptake of vaccine in both PrEP and PEP situations. WHO strongly recommends discontinuation of the production and use of rabies nerve-tissue vaccines (NTVs) because of their association with an unacceptable risk of neuroparalytic complications. In some countries however, particularly Algeria, Argentina, Bolivia, and Ethiopia, NTVs are still produced for human use. The use of concentrated, purified cell culture and embryonated egg-based vaccines (jointly referred to as CCEEVs) has proven to be safe and efficacious against rabies (WHO, 2018). Human rabies vaccines (cell culture-derived) are expensive with four doses required for PEP plus the cost of human rabies immune globulin for administration to one individual of at least US $1000, excluding associated medical costs. Clearly, costs of this magnitude are completely out of reach for most individuals in many countries of the world. New, low-cost rabies vaccines for human use are urgently needed (see Chapters 14 and 15).

Effective prevention of human rabies (see Chapter 16) will require that the disease receive appropriate recognition by the governments of countries in which it is an important problem. A comprehensive strategy is needed to combat the disease and rabies must gain the attention from government officials that it deserves. Control of canine rabies is fundamental and there needs to be cooperation by intergovernmental agencies and the necessary resources need to be allocated. The World Health Organization (WHO) should play a key role in efforts to combat rabies as a global problem. Without being on the "radar," rabies will not receive the attention or the resources that it deserves. The World Health Assembly is the decision-making body of the WHO. The assembly meets yearly and passes resolutions that are very influential on the activities of the WHO. As the Asian and African regions are overrepresented with their burden of human rabies, accounting for more than 95% of global rabies deaths, there has been a focus on passing WHO resolutions that meet a global need, and which can be converted into regional rabies control strategies. In 2018, one high-level meeting was held in Nepal on *Driving progress towards rabies elimination*. Representatives of 18 countries from the Asian region were hosted, with the aim to share knowledge and to identify a regional strategy to achieve rabies elimination in Asia. Similar meetings have also been held in Africa with a regional focus on the elimination of dog-mediated rabies. WHO and expert committees have agreed the critical factors of rabies control and elimination, including disease awareness, responsible dog ownership, mass dog vaccination, cross-sectoral collaboration, appropriate wound management, and access to postbite treatment (PEP). All WHO-recognized rabies human vaccines are inactivated, lyophilized, safe, and effective with three vaccines being recommended by the WHO. Critically, all licensed rabies vaccines have exemplary safety records with only very rare instances of adverse effects postvaccination being reported (Giesen, Gniel, & Malerczyk, 2015).

Because children may have unrecognized rabies exposures and they are at greater risk than adults and suffer a disproportionate burden of rabies (accounting for about half of all cases), children in countries with endemic canine rabies should receive Pr-EP immunization. The high association of rabies infection in children under 15 years of age has led to the proposal that Pr-EP should be considered as part of a pediatric regimen in high-risk regions (Fooks, Koraka, de Swart, Rupprecht, & Osterhaus, 2014). This would simplify the PEP measures required and would also likely provide some protection against unapparent and unreported exposures. One can safely assume that only a tiny minority of this highly susceptible population receives preexposure immunization, and economic factors are the main barrier. Further studies have attempted to shorten PEP regimens to 1 week making PEP more affordable and accessible (Wilde, Lumlertdacha, Meslin, Ghai, & Hemachudha, 2016). Postexposure tools available following suspected exposure to rabies include both vaccination and administration of RIG. The goal is to induce VNAs (see Chapter 13) that provide protection against RABV infection as part of a PEP regimen (Rupprecht, Nagarajan, & Ertl, 2016). The precise location and mechanisms by which VNAs inhibit viral replication are still unknown. While use of intradermal rabies vaccination schedules is expanding worldwide (Brown et al., 2008; Brown, Fooks, & Schweiger, 2011), WHO's requirement to use reconstituted vaccine within 6 hours limits this practice to larger clinics, which encounter several PEP cases daily. More research is needed evaluating vaccination schedules that would result in cost savings. Research outputs have focused on the development of novel vaccines that may reduce the number of clinic visits required through overexpression of the virus glycoprotein. A live-attenuated triple-glycoprotein RABV variant has proven to be effective as Pr-EP in murine and canine models. This attenuated virus vaccine was able to induce rabies neutralizing antibody in the CSF of mice following intrathecal administration and may be of future value in a PEP setting (Faber et al., 2009).

22.6 Control of animal rabies

Outbreaks and resurgences of rabies will continue to occur in unexpected places in the world (see Chapter 6). Resources will be mobilized when the problems become large, leading to subsequent improvements in local situations. However, in the absence of effective policies backed with appropriate funding, there will be an endless cycle of rabies outbreaks in different locations in the future. An important advance in the control of canine rabies was the development of a blueprint for rabies prevention and control (Lembo, 2012). This greatly facilitates organization of what needs to be done in controlling canine rabies (see Chapter 18).

There are many barriers to controlling canine rabies, including political, economic, religious, and cultural barriers. Some of the barriers result from lack of awareness of the burden and effects of rabies, lack of funding, lack of commitment from public health and veterinary services, and practical difficulties in bringing different authorities together to develop and implement an effective plan of action (Senior, 2012). Countries in Latin America largely reduced human deaths transmitted by dogs by controlling canine rabies with a regional elimination program, which included mass vaccination of dogs. This resulted in a reduction of human and canine cases by about 90% over a period of 20 years (Schneider et al., 2011). A highly effective oral rabies vaccine for dogs would be helpful in accessing stray dogs and would potentially facilitate vaccination of a difficult to reach population. Several candidate vaccines have been evaluated for oral rabies vaccination of dogs (Cliquet et al., 2018). An oral vaccination strategy would be used to complement a parenteral vaccination program, especially for dogs that are not easily accessible and cannot be readily vaccinated using a parenteral method (Darkaoui et al., 2014). In a recent study, a population of free-roaming dogs in which only 30% of dogs could be vaccinated parenterally showed that the costs of operationalizing the oral bait handout to access dogs were up to four times lower compared to a capture-vaccinate-release program, although the cost per vaccine administered was similar due to the higher cost of oral vaccines (Gibson et al., 2019). As the production costs of oral vaccines for use in free-roaming dogs are reduced, the likelihood of their use in a large vaccination campaign increases. Similarly, an inactivated rabies vaccine based on a highly attenuated triple RABV G protein variant of SAD (SPBAANGAS-GAS-GAS) might permit an antigen-sparing strategy with reduced production costs (Faber et al., 2009; Freuling et al., 2017). The effective control of wildlife rabies in some countries has been achieved using oral vaccination programs. Although the costs may be high, prevention of the development of endemic rabies in a region may result in a long-term financial benefit. Hence, decisions concerning management of wildlife rabies need to be proactive (see Chapter 19).

22.7 Health impact and economic burden of canine rabies

Globally, canine-mediated rabies that is responsible for the majority of human rabies deaths is estimated to cause a loss of over 3.7 million (95% CIs: 1.6–10.4 million) disability-adjusted life years (DALYs), a societal measure of health impact expressed as the number of years lost due to morbidity (resulting from adverse events of vaccination with nerve tissue vaccines) or mortality due to rabies (see Chapter 20). The annual global economic burden due to canine-mediated rabies is USD $8.6 billion (95% CIs: 2.9–21.5 billion); Lost productivity due to premature deaths (55%) comprises a major component, while direct costs of PEP (depending on the regimen and rabies biologicals used) and indirect costs of seeking PEP (travel, accommodation, multiple clinic visits, lost income, etc.) account for 20% and 15% of the economic burden, respectively (Hampson et al., 2015).

22.8 Conclusions

In the 21st century, plans are underway that will largely confine human rabies to history (see Chapter 21). The goal for elimination of rabies is in preventing the transmission of the virus between terrestrial animals, mainly dogs and humans. For this vision to be realized, a combination of improved diagnostic tools (see Chapter 12), cheaper and efficacious biologicals (vaccines, antivirals, and HRIG) as well as increased education and awareness of rabies in endemic countries is urgently required. One of the first steps in this process is to ensure that all regions have the capability to accurately diagnose the infection using a network of dedicated laboratories with validated diagnostic tools.

The WHO and OIE have clarified that rabies should be considered a notifiable disease. One challenge for governments is the timeliness for when rabies in animals and humans should be considered as a "reportable" or "notifiable" disease. Without the appropriate laboratory infrastructure, it is problematic for national governments to consider rabies as a notifiable disease.

The elimination of rabies from the principal reservoir, domestic dogs, relies on pragmatic actions that can be adopted where the veterinary support structures are in place to support a sustained approach to eliminate the virus in dogs and prevent the disease in humans (Fooks et al., 2017). The majority of human rabies cases would be eliminated if dog populations in endemic regions were controlled, dogs are licensed and vaccinated. While the actual burden of human rabies remains difficult to quantify, one factor that is evident is that rabies can be managed at a community level if human awareness of the disease increases and dog populations are controlled. Every bite should be considered a potential exposure. Often though, through a lack of awareness, the simplest actions required to follow PEP, including washing the wound and avoiding suturing wounds, are rarely implemented. Moreover, the biologicals required for provision of PEP present challenges; current vaccines require multiple inoculations and are often unavailable due to high costs and the lack of a sustained supply across endemic regions. As such the requirement for novel PEP vaccines and RIG alternatives exists. The use of a pediatric rabies vaccine as part of an Expanded Programme on

Immunization initiative would increase the chance of exposed children not developing rabies in the absence of immunoglobulin treatment. Optimization of vaccines to reduce production costs, number of doses, and noninjectable routes of vaccine administration may enable inclusion of rabies vaccines into existing childhood immunization schemes. Moreover, the menace from counterfeit and fake vaccines or unauthorized replicas of rabies vaccines for use in humans and animals also hampers control programs (Taylor et al., 2019).

Novel approaches are needed in order to overcome the scarcity of RIG, expensive production costs, and problems associated with blood products. Potential solutions to the availability and costs associated with RIG are urgently required.

The outlook for successful treatment of patients exhibiting clinical rabies remains poor. Despite promising in vitro proof-of-principle data, none of the antivirals tested have demonstrated sufficient in vivo efficacy in experimental models to warrant further investigation. An effective antiviral that could be used therapeutically in symptomatic human cases of rabies would have to cross the blood-brain-barrier and interfere with RABV replication, without affecting neuronal function and also without inhibiting detrimental host responses to RABV infection. Due to the overriding fact that RABV replicates in the CNS as an immune-privileged site and the variable stages of presentation, exposure history, viral species, and host immune response (see Chapter 11), it is envisaged that the development of antivirals for rabies will remain an urgent priority and yet pose a major challenge and success remains unlikely. It is only with a better understanding of the host factors and the pathogenesis of RABV that a safe, effective antiviral therapeutic protocol may be developed in the future. Other, novel therapies for rabies may be developed after we have an improved understanding in the basic mechanisms resulting in the disease with such a poor clinical outcome. It is essential that basic research work continues on rabies (Hooper, Roy, Kean, Phares, & Barkhouse, 2011; Israsena, Supavonwong, Ratanasetyuth, Khawplod, & Hemachudha, 2009).

As part of the Millennium Development Goals to reduce poverty and preventable childhood deaths from infectious diseases, especially in resource-limited regions of the world, dog-mediated rabies with a concomitant reduction in human cases of rabies has been targeted for global eradication by 2030 (Abela-Ridder et al., 2016). The eradication goal should promote global initiatives that will, hopefully, encourage pharmaceutical companies, international organizations, governments, charities and nongovernmental bodies to work together and eliminate rabies worldwide.

References

Abela-Ridder, B., Knopf, L., Martin, S., Taylor, L., Torres, G., & De Balogh, K. (2016). The beginning of the end of rabies? *The Lancet Global Health*, 4, e780–e781.

Alandijany, T., Kammouni, W., Roy Chowdhury, S. K., Fernyhough, P., & Jackson, A. C. (2013). Mitochondrial dysfunction in rabies virus infection of neurons. *Journal of Neurovirology*, 19(6), 537–549.

Amarasinghe, G. K., Ayllón, M. A., Bào, Y., Basler, C. F., Bavari, S., Blasdell, K. R., ... Kuhn, J. H. (2019). Taxonomy of the order Mononegavirales: Update 2019. *Archives of Virology*, 164(7), 1967–1980.

Banyard, A. C., Mansfield, K. L., Wu, G., Selden, D., Thorne, L., Birch, C., ... Fooks, A. R. (2019). Re-evaluating the effect of Favipiravir treatment on rabies virus infection. *Vaccine*, 37(33), 4686–4693. https://doi.org/10.1016/j.vaccine.2017.10.109.

Brown, D., Featherstone, J. J., Fooks, A. R., Gettner, S., Lloyd, E., & Schweiger, M. (2008). Intradermal pre-exposure rabies vaccine elicits long lasting immunity. *Vaccine*, 26, 3909–3912.

Brown, D., Fooks, A. R., & Schweiger, M. (2011). Using intradermal rabies vaccine to boost immunity in people with low rabies antibody levels. *Advances in Preventive Medicine, 2011*, 601789.

Cliquet, F., Guiot, A. L., Aubert, M., Robardet, E., Rupprecht, C. E., & Meslin, F. X. (2018). Oral vaccination of dogs: A well-studied and undervalued tool for achieving human and dog rabies elimination. *Veterinary Research, 49*(1), 61. https://doi.org/10.1186/s13567-018-0554-6.

Darkaoui, S., Boué, F., Demerson, J. M., Fassi Fihri, O., Yahia, K. I., & Cliquet, F. (2014). First trials of oral vaccination with rabies SAG2 dog baits in Morocco. *Clinical and Experimental Vaccine Research, 3*(2), 220–226. https://doi.org/10.7774/cevr.2014.3.2.220.

De Serres, G., Skowronski, D. M., Mimault, P., Ouakki, M., Maranda-Aubut, R., & Duval, B. (2009). Bats in the bedroom, bats in the belfry: Reanalysis of the rationale for rabies postexposure prophylaxis. *Clinical Infectious Diseases, 48*(11), 1493–1499.

Faber, M., Li, J., Kean, R. B., Hooper, D. C., Alugupalli, K. R., & Dietzschold, B. (2009). Effective preexposure and postexposure prophylaxis of rabies with a highly attenuated recombinant rabies virus. *Proceedings of the National Academy of Sciences of the United States of America, 106*, 11300–11305.

Fooks, A. R., Cliquet, F., Finke, S., Freuling, C., Hemachudha, T., Mani, R., … Banyard, A. (2017). Rabies. *Nature Reviews Disease Primers, 3*, 17091.

Fooks, A. R., Koraka, P., de Swart, R. L., Rupprecht, C. E., & Osterhaus, A. D. M. E. (2014). Development of a multivalent paediatric human vaccine for rabies virus in combination with Measles–Mumps–Rubella (MMR). *Vaccine*. pii: S0264-410X(14)00250-3. https://doi.org/10.1016/j.vaccine.2014.02.065.

Freuling, C. M., Eggerbauer, E., Finke, S., Kaiser, C., Kaiser, C., Kretzschmar, A., et al.Müller, T., (2017). Efficacy of the oral rabies virus vaccine strain SPBN GASGAS in foxes and raccoon dogs. *Vaccine*. https://doi.org/10.1016/j.vaccine.2017.09.093.

Gibson, A. D., Yale, G., Vos, A., Corfmat, J., Airikkala-Otter, I., King, A., et al.Mazeri, S., (2019). Oral bait handout as a method to access roaming dogs for rabies vaccination in Goa, India: A proof of principle study. *Vaccine: X, 1*, 100015. https://doi.org/10.1016/j.jvacx.2019.100015.

Giesen, A., Gniel, D., & Malerczyk, C. (2015). 30 years of rabies vaccination with Rabipur: A summary of clinical data and global experience. *Expert Review of Vaccines, 14*, 351–367.

Hampson, K., Coudeville, L., Lembo, T., Sambo, M., Kieffer, A., Attlan, M., … Global Alliance for Rabies Control Partners for Rabies Prevention. (2015). Estimating the global burden of endemic canine rabies. *PLoS Neglected Tropical Diseases, 9*, e0003709.

Hooper, D. C., Roy, A., Kean, R. B., Phares, T. W., & Barkhouse, D. A. (2011). Therapeutic immune clearance of rabies virus from the CNS. *Future Virology, 6*, 387–397.

Horton, D. L., Ismail, M. Z., Siryan, E. S., Wali, A. R. A., Dulla, H. E., Wise, E., … Fooks, A. R. (2013). Epidemiology of rabies in Iraq. *PLoS Neglected Tropical Diseases, 7*(2), e2075. https://doi.org/10.1371/journal.pntd.0002075.

Israsena, N., Supavonwong, P., Ratanasetyuth, N., Khawplod, P., & Hemachudha, T. (2009). Inhibition of rabies virus replication by multiple artificial microRNAs. *Antiviral Research, 84*, 76–83.

Jackson, A. C., Kammouni, W., Zherebitskaya, E., & Fernyhough, P. (2010). Role of oxidative stress in rabies virus infection of adult mouse dorsal root ganglion neurons. *Journal of Virology, 84*(9), 4697–4705.

Kammouni, W., Wood, H., & Jackson, A. C. (2017a). Lyssavirus phosphoproteins increase mitochondrial complex I activity and levels of reactive oxygen species. *Journal of Neurovirology, 23*(5), 756–762.

Kammouni, W., Wood, H., & Jackson, A. C. (2017b). Serine residues at positions 162 and 166 of the rabies virus phosphoprotein are critical for the induction of oxidative stress in rabies virus infection. *Journal of Neurovirology, 23*(3), 358–368.

Kammouni, W., Wood, H., Saleh, A., Appolinario, C. M., Fernyhough, P., & Jackson, A. C. (2015). Rabies virus phosphoprotein interacts with mitochondrial Complex I and induces mitochondrial dysfunction and oxidative stress. *Journal of Neurovirology, 21*(4), 370–382.

Lembo, T. (2012). The blueprint for rabies prevention and control: A novel operational toolkit for rabies elimination. *PLoS Neglected Tropical Diseases, 6*(2), e1388.

Mallawa, M., Fooks, A. R., Banda, D., Chikungwa, P., Mankhambo, L., Molyneux, E., … Solomon, T. (2007). Rabies encephalitis in a malaria-endemic area of Malawi, Africa. *Emerging Infectious Diseases, 13*(1), 136–139.

Manning, S. E., Rupprecht, C. E., Fishbein, D., Hanlon, C. A., Lumlertdacha, B., Guerra, M., … Hull, H. F. (2008). Human rabies prevention—United States, 2008: Recommendations of the Advisory Committee on Immunization Practices. *Morbidity and Mortality Weekly Report, 57*(RR-3), 1–28.

Marston, D. A., McElhinney, L. M., Ellis, R. J., Horton, D. L., Wise, E. L., Leech, S. L., ... Fooks, A. R. (2013). Next generation sequencing of viral RNA genomes. *BMC Genomics, 14*(1), 444.

Meslin, F. X., & Briggs, D. J. (2013). Eliminating canine rabies, the principal source of human infection: What will it take? *Antiviral Research, 98,* 291–296.

Rupprecht, C. E., Nagarajan, T., & Ertl, H. (2016). Current status and development of vaccines and other biologics for human rabies prevention. *Expert Review of Vaccines, 15,* 731–749.

Schneider, M. C., Aguilera, X. P., Barbosa da Silva, J. J., Ault, S. K., Najera, P., Martinez, J., ... Periago, M. R. (2011). Elimination of neglected diseases in Latin America and the Caribbean: A mapping of selected diseases. *PLoS Neglected Tropical Diseases, 5*(2), e964.

Senior, K. (2012). Global rabies elimination: Are we stepping up to the challenge? *Lancet Infectious Diseases, 12,* 366–367.

Taylor, E., Banyard, A. C., Bourhy, H., Cliquet, F., Ertl, H., Fehlner-Gardiner, C., ... Fooks, A. R. (2019). Avoiding preventable deaths: The scourge of counterfeit rabies vaccines. *Vaccine, 37*(17), 2285–2287.

WHO. (2018). *WHO Expert Consultation on Rabies, 3rd report.* Geneva: World Health Organisation. WHO Technical Report Series, No. 1012.

Wilde, H., Lumlertdacha, B., Meslin, F. X., Ghai, S., & Hemachudha, T. (2016). Worldwide rabies deaths prevention—A focus on the current inadequacies in postexposure prophylaxis of animal bite victims. *Vaccine, 34,* 187–189.

Willoughby, R. E., Jr., Tieves, K. S., Hoffman, G. M., Ghanayem, N. S., Amlie-Lefond, C. M., Schwabe, M. J., ... Rupprecht, C. E. (2005). Survival after treatment of rabies with induction of coma. *New England Journal of Medicine, 352*(24), 2508–2514.

Yamada, K., Noguchi, K., Komeno, T., Furuta, Y., & Nishizono, A. (2016). Efficacy of favipiravir (T-705) in rabies postexposure prophylaxis. *The Journal of Infectious Diseases, 213*(8), 1253–1261.

Zeiler, F. A., & Jackson, A. C. (2016). Critical appraisal of the Milwaukee protocol for rabies: This failed approach should be abandoned. *Canadian Journal of Neurological Sciences, 43*(1), 44–51.

Index

Note: Page numbers followed by *f* indicate figures and *t* indicate tables.

A

ABLV. *See* Australian bat lyssavirus (ABLV)
Abortive rabies, 259, 330
Active rabies surveillance, 125
Adaptive immune response
   $CD4^+$ T lymphocytes, 386–387
   collapsing response mediator protein 2 (CRMP2), 388–389
   neutralizing antibodies, 389
   in periphery, 387
   T-cell receptor (TCR), 386–387
   tumor-specific T-cell immunity, 389
Adenovirus-rabies glycoprotein recombinant vaccine baits (ONRAB), 209
African lyssaviruses
   classification, 243–244
   detection, 243–244
   Duvenhage virus (DUVV), 245–246
   Ikoma lyssavirus (IKOV), 246
   Lagos bat virus (LBV), 244–245, 253
   Mokola lyssavirus (MOKV), 246–247
   Shimoni bat lyssavirus (SHIBV), 246
AGP. *See* Antigenome promoter (AGP)
Amino acid sequence, 48
α-Amino-3-hydroxy-5-methyl-4-isoxazole propionate (AMPA), 312
γ-Amino-n-butyric acid (GABA), 325
Antigen-binding assays, 446–448
Antigenome promoter (AGP), 57
Anti-RABV glycoprotein monoclonal antibody, 446–448
Apoptosis
   cell death, 316–317
   cellular fragmentation, 315–316
   human T-lymphocyte cell line Jurkat, 316–317
   immunostaining, 317, 322*f*
   intracerebral inoculation, 318–320, 329*f*
   morphologic evidence, 320
   phagocytosis, 315–316
Aravan bat lyssavirus (ARAV), 241–242, 252–253
Arctic foxes *(Vulpes lagopus)*, 203–204
Asian bat lyssaviruses, 158–159
Australian bat lyssavirus (ABLV), 28, 156, 247, 252

B

Bat-eared foxes *(Otocyon megalotis)*, 204
Bat rabies, 23
   abortive infection, 258–259
   aerosol transmission, 253–254
   African lyssaviruses (*see* African lyssaviruses)
   Australian bat lyssavirus (ABLV), 247, 252
   Brazilian free-tailed bats, 248–249, 261
   clinical stages, 231
   disease control, 260
   distribution of, 232, 233*f*
   Eurasian lyssaviruses (*see* Eurasian lyssaviruses)
   excretion, 255–256
   frugivorous bats, 251
   Gannoruwa bat lyssavirus (GBLV), 248
   hematophagous bat, 250–251
   immunobiology, 259
   inoculation route and dose, 255
   insectivorous bats, 249
   laboratory studies, 256–257
   natural reservoir host, 249–250
   paralytic disorder, 231
   pathogenesis, 248
   pre-exposure prophylaxis, 262
   proactive local population reduction strategies, 260–261
   silver-haired bat RABV variant (SHBV), 250
   surveillance, 257–258
   Taiwan bat lyssavirus (TBLV), 248
   vaccine-mediated protection, 232
   vaccinia-rabies recombinant virus (VRG), 261
   Vampire bat rabies (*see* Vampire bat rabies)
   viral shedding and dissemination, 250
   virus:host interactions, 254–255
   virus neutralizing antibodies (VNA), 249–250
   in vivo vaccination-challenge experiments, 232
Biosafety Level 2 (BSL2) pathogen, 404
Blood-brain barrier (BBB), 315, 321
Bokeloh bat lyssavirus (BBLV), 156, 158, 243
Brain dysfunction
   acetylcholine, 324–326
   γ-amino-n-butyric acid (GABA), 325
   behavioral changes, 328–330

Brain dysfunction *(Continued)*
  electroencephalographic (EEG) recordings, 325–326
  excitotoxicity, 327
  ion channels, 326
  neuropeptide synthesis, 323–332
  nitric oxide (NO), 326–327
  oxidative stress, 328, 334*f*
  serotonin, 324–325
Branched-chain oligosaccharides, 47–48
Brazilian free-tailed bats, 233–234, 238, 248–249, 261, 277–280, 331–332
Budding process, 47

C
California skunk (CASK) variant, 166–167
Canadian Rabies Management plan, 618
Canidae, 199–200
Canine rabies virus transmission dynamics
  data and code, 666
  density-dependent transmission, 659–660
  disease prevention and control, 655
  frequency-dependent transmission, 658–660, 659*f*
  incubation period, 656
  landscape features, 655–656
  model, 656
    bite data, 663–664
    dog demography, 660–661
    dog population and vaccination data, 665
    insights and limitations, 661–663
    laboratory-confirmed case data, 664
    limitation, 663
    sequence data, 665
    surveillance, 663
    transmission model data, 660–661, 661*f*
  robust epidemiological and biological data, 660
  Susceptible, Exposed, Infectious, and Vaccinated (SEIV) framework, 656, 657*f*
  transmission heterogeneity, 660
Carnivora, 195, 199
Cats, 213
Cattle, 213–215
CCV techniques. *See* Cell-culture vaccine (CCV) techniques
Cell-culture vaccine (CCV) techniques, 5–6
Cellular immunity assays, 446
Cellular kinases, 47–48
Chiroptera, 88–89, 195, 199
Collapsing response mediator protein 2 (CRMP2), 388–389
Competitive ELISA (cELISA), 456–457
Corrective action/preventive action (CAPA System), 471–472
Coyotes, 205, 207
Crab-eating fox *(Cerdocyon thous)*, 204

CRMP2. *See* Collapsing response mediator protein 2 (CRMP2)
Cytosine phosphate-linked guanine (CpG), 494
Cytoskeleton proteins, 47–48

D
Democratic Republic of Congo (DRC), 13
Direct fluorescent antibody test (DFA)
  autolysis, 412
  fluorescein isothiocyanate (FITC), 411–413
  immunofluorescence test protocol, 413–414, 414*f*
  rabies virus antigen in cell culture (RTCIT), 411–412
Direct Rapid Immunohistochemical Test (DRIT), 414
DNA vaccines, 518–519
Dog-mediated human rabies elimination
  disease-focused networks, 679
  Global Strategic Plan (GSP), 671
  government ownership and commitment
    cost-benefit analysis, 672
    disease prioritization, 674
    financing, 676–677, 677*t*
    global disease elimination/eradication strategies, 673–674
    Global Rinderpest Eradication Campaign (GREP), 672
    international community, 673
    intervention strategies, 674–675
    low- and middle-income countries (LMICS), 673
    National rabies elimination strategies, 675
    organization and management, 676
    Stepwise Approach toward Rabies Elimination (SARE) tool, 675–676
  international guidelines/recommendations
    declaration of freedom, 678–679
    monitoring and evaluation, 680
    vaccination events, 682–683
  Rabies Epidemiological Bulletin (REB), 681
  rabies-focused networks, 680
  surveillance, 680–681
  Zero by 30, 671
Domestic dogs, 200–201
  animal rabies vaccines, 567–568
  control measures, 571
  disability-adjusted life years (DALYs), 568–569
  disease prevention, 570
  dog population management (DPM)
    culling, 579–580
    human behavior, 579
    sterilization programs, 580–581
  epidemiological factors, 571–574, 572*f*
  Integrated Bite Case Management (IBCM), 569
  mass dog vaccination campaigns
    capture-vaccinate-release (CVR) strategy, 584
    cost-effectiveness analysis, 585
    disadvantage, 585–586

economic benefits, 590–591
health system benefits, 592
Neglected Tropical Diseases (NTDs), 591–592
parenteral vaccination, 583–584
proof-of-principle study, 584–585
public health authorities, 586
rabies-free status, 589–590
surveillance, 587–589
Sustainable Development Goals (SDGs), 591
trained dog-catchers, 584
vaccination coverage, 586–587
planning phase, 582–583
populations, 568, 568f
postexposure prophylaxis (PEP), 568–569
rabies-free countries, 569
regional and national targets, 581–582
reproductive number, 570–571
risk-based approach, 569
social dimensions
communities and governments, 577–578
cultural norms, 575–576
empowerment education, 575–576
large-scale rabies vaccination programs, 575
parent-child relationships, 578
social capital, 575–576
sociobehavioral and cultural mechanisms, 574
socioeconomic characteristics, 574–575
socioeconomic/sociopolitical contexts, 570
Stepwise Approach to Rabies Elimination (SARE), 581–582
DRC. *See* Democratic Republic of Congo (DRC)
Duvenhage virus (DUVV), 109, 159, 245–246

E

EBLV. *See* European bat lyssavirus (EBLV)
ELISA. *See* Enzyme-linked immunosorbent assay (ELISA)
Encephalitic rabies
autonomic dysfunction, 282–283
bat-acquired rabies, 283–285
cardiac arrhythmias, 285
cardiopulmonary complications, 285
defensive reflexes, 283–285
hydrophobia, 283–285, 284f
incidence, 282–283
parasympathetic stimulation, 282–283
Endoplasmic reticulum (ER)-Golgi-plasma membrane secretion pathway, 53
Enhanced rabies surveillance (ERS), 620–621
Enzootic cycles, 115
Enzyme-linked immunosorbent assay (ELISA), 416, 454–456, 455f, 458–459, 469
Epidemiology. *See also* Molecular epidemiology
Africa, 108–109
Asia, 105–108
bats, 130–131
Caribbean islands, 112–113
community engagement, 103
disease transmission, 103
domestic dogs. (*see also* Domestic dogs)
household surveys, 132
postvaccination evaluations, 132–133
surveillance systems, 132
vaccinate-assess-move programs, 132–133
vaccination, 131–132
virus distribution, 104–105, 107f
Europe, 110
host-shift events, 104
intrareservoir transmission, 104
Middle East, 109–110
North America, 110–112, 111f
Oceania, 113
public health measures, 103, 129–130
rabies virus (RABV) transmission
atypical routes, 114
cross-species transmission, 116, 116f
enzootic cycles, 115
host shift events, 116–117
incubation period, 115
translocation events, 117
typical routes, 114
viral shedding, 115
South America, 113
spillover events, 104
surveillance
canine-rabies endemic countries, 118–119, 120t
human-animal-vector approach, 129
human rabies prevention, 123–124
implementation stages, 120
Integrated Bite Case Management (IBCM), 127–129
laboratory diagnosis, 127
monitoring and assessment, control measures, 124–125
point-of-care-testing, 127
postelimination, 126
proof of burden, 123
rabies assays and utility, 119, 121–122t
rabies elimination verification, 125–126
regional data sharing platforms, 118–119
risk matrix, 127, 128t
standards, 117–118
Word Health Organization, 118, 119t
World Organisation for Animal Health, 117–118
terrestrial carnivores, 131
wildlife species, 133
wildlife, virus distribution, 104–105, 105f
zoonotic pathogens, 104–105, 106t
Equine rabies immune globulin (ERIG), 533–534
Ethiopian wolf, 216

European bat lyssavirus (EBLV), 156–158
  antibodies, 157
  detection, 239
  European bat-1 lyssavirus (EBLV-1), 240–241, 251
  European bat-2 lyssavirus (EBLV-2), 241, 251–252
European Directorate for the Quality of Medicines and Health Care (EDQM), 454
Evolution
  cross-species transmission, 83
    adaptive evolution, 92
    baby hamster kidney (BHK) cells, 91–92
    deep sequencing study, 92–93
    fox-adapted strain, 92
    heterologous transfer, 93
    nonsynonymous substitutions, 93–94
    phenotypic changes, 92
    post-host shift adaptation, 92–93
    virulence-transmission tradeoff, 93
  genetics, 94–95
  lyssaviruses, 84
  macroevolutionary dynamics, 83, 84f
    bat and carnivore species, 87–88
    biological traits, 87–88
    host shifts, 89–91, 90f
    lyssavirus circulation, 88
    mesocarnivore families, 87–88
    New World bats, 88–89
    Old World bats, 88
    transmission cycles, 91
  mammalian species, 83
  microevolutionary dynamics
    experimental infections, 85–86
    high-throughput sequencing methods, 85
    homologous recombination, 86–87
    mathematical modeling, 85–86
    mongoose-associated RABV strains, 85–86
    phylogenetic studies, 85–86
    protein structures, 85
    rates of, 85–86, 86f
  molecular evolution, 83
  proofreading mechanism, 83

F

FAK. *See* Focal adhesion kinase (FAK)
Ferret badgers, 211–212
FFPE tissues. *See* Formalin-fixed paraffin-embedded (FFPE) tissues
Fixed-wing aircraft oral rabies vaccination, 621, 622f, 625–630
Fluorescein isothiocyanate (FITC), 411–413
Fluorescent antibody test (FAT), 360
Focal adhesion kinase (FAK), 50–51
Food and Drug Administration (FDA), 470–471
Formalin-fixed paraffin-embedded (FFPE) tissues, 415–416, 431
Foxes, 201–204, 202–203f
Fox-mediated rabies, 9–10

G

Gannoruwa bat lyssavirus (GBLV), 158–159, 248
GARC. *See* Global Alliance for Rabies Control (GARC)
Genetic vaccines, 518
Genome
  characteristic features, 56
  "high transcription" phenotype, 62
  leader-dependent transcription, 61–62
  Lyssavirus genes, 58
  N-vRNA replication, 60–61
  polymerase complex, 56
  primary structure, 57, 57f
  primary transcription, 61
  RNA synthesis, 56
  sequencing, 485
  transcription, 58–60
Global Alliance for Rabies Control (GARC), 401
Global Alliance Vaccine Initiative (GAVI), 528
Global canine rabies elimination
  Africa
    events, 11, 12–13t
    hydrophobia, 14
    low egg passage (LEP) vaccine, 15
    molecular insights, 15
    nonrabies lyssaviruses, 15
    public-private demonstration projects, 16
    regional translocations, 11–14
    wildlife involvement, 14
  Asia
    animal control, 6
    canine domestication, 2–4, 2–3t
    cell-culture vaccine (CCV) techniques, 5–6
    dog bites, 4
    duck embryos, 5–6
    human diploid cell vaccine (HDCV), 6
    incidence, 7
    mass dog vaccination, 7
    mesopotamians, 4
    Persian Avesta, 5
    postcolonial occurrence, 5
    primary chick embryo cell vaccine (PCECV), 6
    purified Vero cell rabies vaccine (PVRV), 6
    rabies virus (RABV) vaccine, 5–6
    Semple vaccine, 5–6
    viruses/slimy poisons, 4–5
    rabies virus (RABV) vaccine, 25–28, 26t
  bat rabies, 23
  Caribbean, 22–23

colonization impacts, 16
Europe
  coevolution, 7
  Eurasian cultures, 7–8
  European rabies elimination, 11
  human PEP post-Pasteur, 9
  (re)introduction, 10–11
  occurrence of, 8–9, 8–9t
  oral rabies vaccination (ORV), 10
  pre-Neolithic domestication, 7
  viral evolutionary data, 7
  wildlife-mediated rabies, 7
  wildlife rabies appreciation, 9–10
European colonization, 16–17
events, Americas, 17, 18–21t
hormone-based vaccine-induced contraception methods, 22–23
institution-initiated activities, 21
laboratory-based surveillance, 22
North America
  colonization impacts, 23–24
  madstones, 24, 24f
  mail-order human nerve tissue vaccine, 24, 25f
One World-One Health concept, 23
oral dog vaccination, 22–23
Pan American Health Organization (PAHO), 21–22
population immunity, 22–23
in South America, 16–17
"spillover" phenomenon, 17
Global elimination of human rabies via dogs (GEHRD), 1–2
Global Rinderpest Eradication Campaign (GREP), 672
Global Strategic Plan (GSP), 671
Glycoprotein (G protein), 53–55, 483–484, 509
Glycosylation, 54
Good manufacturing practice (GMP) quality standards, 445–446
Gray foxes (Urocyon cinereoargenteus), 204
Gray wolves, 215–216
Guillain-Barré syndrome, 291, 401–402

H

HDCV. See Human diploid cell vaccine (HDCV)
Heat/freeze-thaw cycles, 405
High-throughput sequencing (HTS) methods, 147–148
Hoary fox (Lycalopex vetulus), 204
Hormone-based vaccine-induced contraception methods, 22–23
Host adaptations, 116–117
Human and animal vaccines
  abbreviated vaccination regimens, 500
  allergies, 495
  canine rabies elimination, 500
  cell substrates, 487
  continuous cell lines, 489–490
  diploid cells, 488–489
  disease prevention, 481
  domestic animals, 498–499
  downstream processing, 486f, 491–492
  formulations, 493–494, 493t, 500
  friendly delivery systems, 500
  gelatin-containing vaccines, 496
  genome sequencing, 485
  glycoprotein (G protein), 483–484
  history of, 482
  hypersensitivity reactions, 495
  multiplicity of infection (MOI), 484–485
  neurological adverse events, 496
  nucleotide sequence analysis, 485
  oral rabies vaccination (ORV), 496–497
    monitoring, 499
  post-exposure prophylaxis (PEP), 481, 483, 484t
  potency testing, 494–495
  pre-exposure prophylaxis (PrEP), 481–483
  primary cells, 488
  production, 485–486, 487f, 490–491
  purification, 492
  tumor necrosis factor (TNF), 496
  vaccine strains, 499
  viral inactivation, 491
  wildlife immunization, 497–498
  zoonosis, 481
Human-animal-vector approach, 129
Human diploid cell vaccine (HDCV), 6, 537–538
Human rabies
  aggressive management
    antiviral therapy, 557
    combination therapy, 557
    corticosteroids, 559
    critical care unit, 557
    immunotherapy, 558
    neuroprotective agents, 559–560
    neuroprotective therapy, 558–559
    neutralizing antirabies virus antibodies, 560
    viral encephalitis, 559–560
  antirabies virus antibodies, 553–554
  autosterilization, 287–289
  brain biopsy, 551
  brain tissues, 289
  Brazilian free-tail bats, 277–280
  cerebrospinal fluid (CSF) analysis, 287–289
  computed tomographic (CT) studies, 287
  differential diagnosis, 290–291
  electroencephalogram, 287–289
  electrophysiologic evidence, 287–289
  encephalitic rabies

Human rabies (Continued)
    autonomic dysfunction, 282–283
    bat-acquired rabies, 283–285
    cardiac arrhythmias, 285
    cardiopulmonary complications, 285
    defensive reflexes, 283–285
    hydrophobia, 283–285, 284f
    incidence, 282–283
    parasympathetic stimulation, 282–283
  endotracheal intubation, 553–554
  factors, 547, 550t
  fatal neurologic illness, 277
  incidence, 277
  incubation period, 280–282
  infectious cycle, 277–280
  Lyssavirus species, 291–295, 292t
  magnetic resonance imaging (MRI) studies, 287
  Milwaukee protocol, 554–557, 554–556t
  neurologic deterioration, 551
  neutralizing antibodies, 551
  nonbite exposures, 277–280
  nonspecific prodromal symptoms, 282
  organ transplantation, 277–280, 279–280t
  palliation, 561
  paralytic rabies, 285–287
  preventative therapy, 548–550t
  purified chick embryo cell vaccine (PCECV), 552
  public health management (*see* Public health management)
  puncture wound, 280–282, 281f
  rabies immune globulin, 552
  rabies virus neutralizing antibodies, 553–554
  retro-orbital pain, 282
  reverse transcriptase polymerase chain reaction (RT-PCR), 289
  serum-neutralizing antibodies, 287–289
  skin biopsies, 287–289, 288f
  transmission, corneal transplantation, 277–280, 278t
  virus neutralizing antibodies (VNA), 280–282
Human rabies immune globulin (HRIG), 533–534
Hydrophobia, 283–285, 284f

I
IBCM. *See* Integrated Bite Case Management (IBCM)
Ideal oral rabies vaccination programs, 626, 626t
Ideal rabies vaccine, 520–521, 522t
Ikoma lyssavirus (IKOV), 109, 160, 246
Immunology
  adaptive immune response
    CD4$^+$ T lymphocytes, 386–387
    collapsing response mediator protein 2 (CRMP2), 388–389
    neutralizing antibodies, 389
    in periphery, 387
    T-cell receptor (TCR), 386–387
    tumor-specific T-cell immunity, 389
  challenge virus standard (CVS), 381
  CNS-mediated immune unresponsiveness, 390
  fatal encephalitis, 379
  immunological homeostasis, 379–381, 380f
  innate immune response, 379–381 (*see also* Innate immune response)
  interferon (IFN), 390–391
  virus replication, 379
Inactivated adjuvanted traditional rabies vaccines, 516
Inactivated genetically modified rabies vaccine, 516
Indirect fluorescent antibody (IFA), 454–455
Indirect immunoperoxidase virus neutralization (IPVN) technique, 452–454
Infection life cycle
  cell-to-cell spread (transport), 68–70
  early-phase events
    cell culture systems, 63–65, 64f
    internalization process, 65
    neural cell adhesion molecule (NCAM) CD56, 63–65
    reversible "fusion-inactive" conformation, 65
    target cell and penetration, 63–65
  events sequence, 62–63
  experimental cell culture systems, 62–63
  late phase, 67–68
  middle-phase events
    cytoplasmic inclusion bodies, 66–67
    N-encapsidation-dependent synthesis, 66
    genome and antigenome RNPs, 67
    protein synthesis, 66
    viral nucleocapsids, 66–67
    vRNA genome transcription, 66
  receptor-mediated endocytosis, 62–63
Innate immune response
  nervous system (NS), 384–386
  N protein, 382
  pathogen-associated molecular patterns (PAMPs), 381–382
  pattern recognition receptors (PRRs), 381–382
  in periphery, 382–383
  promyelocytic leukemia (PML) protein, 382
  toll-like receptors (TLRs), 381–382
Integrated Bite Case Management (IBCM), 127–129
Interferon regulatory factors (IRF-3), 51
International Alliance for Biological Standardization (IABS), 454
Irkut bat lyssavirus (IRKV), 242, 252–253
Irkut virus (IRKV), 158

J
Jackals, 205–207, 206f

K
Khujand bat lyssavirus (KHUV), 158–159, 242, 252–253
Kotalahti bat lyssavirus (KBLV), 243
Kudu, 215

L
Laboratory-based surveillance, 22
Laboratory diagnosis
  antemortem diagnosis, 434–436
  Biosafety Level 2 (BSL2) pathogen, 404
  biosafety practices, 404
  central nervous system (CNS) tissue, 402
  clinical signs, 402
  differential diagnosis, 401–402
  distribution and epidemiology, 403
  early detection, 401–402
  Fluorescent Antibody Test, 411
  Global Alliance for Rabies Control (GARC), 401
  Guillain-Barré syndrome, 401–402
  heat/freeze-thaw cycles, 405
  laboratory reporting practices, 409–410
  live virus detection, 433–434
  microbiological safety cabinets (MSCs), 405
  Negri bodies, 410–411
  neurological signs, 403
  nucleic-acid-based detection, 411
  optimal sampling, 408–409, 409t
  passive surveillance schemes, 403
  Personal Protection Equipment (PPE), 404
  pre- and postexposure prophylaxis, 401–402
  precautionary post exposure prophylaxis, 403
  quality assurance, 406–408
  robust diagnostic and surveillance program, 401
  viral antigen detection
    direct fluorescent antibody test (DFA) (*see* Direct fluorescent antibody test (DFA))
    Direct Rapid Immunohistochemical Test (DRIT), 414
    enzyme-linked immunosorbent assay (ELISA), 416
    formalin-fixed paraffin-embedded (FFPE) tissues, 415–416, 431
    lateral flow assay (LFA), 415
  viral RNA detection
    advantages and disadvantages, 416–417
    amplification methods, 432
    immunohistochemical methods, 430–431
    nucleic-acid sequence-based amplification (NASBA), 432
    recombinase polymerase amplification (RPA), 433
    reverse-transcription loop-mediated isothermal amplification (RT-LAMP) method, 432–433
    reverse transcription-polymerase chain reaction (RT-PCR) methods (*see* Reverse transcription-polymerase chain reaction (RT-PCR) methods)
    RNA extraction, 417–418
  in situ hybridization method, 430–431
  World Health Organization, 406
Lagos bat lyssavirus (LBV), 109, 159–160, 244–245, 253
Lateral flow assay (LFA), 415
Lateral Flow Devices (LFD), 127
LBV. *See* Lagos bat lyssavirus (LBV)
LEP vaccine. *See* Low egg passage (LEP) vaccine
LFD. *See* Lateral Flow Devices (LFD)
Limit of detection (LOD), 452–454, 463
Limit of quantitation (LLOQ), 463
Live vaccines, 515
Lleida bat lyssavirus (LLEBV), 156, 158, 243
Low-affinity p75 neurotrophin receptor, 309–310
Low egg passage (LEP) vaccine, 15
Lower limit of quantification (LLOQ), 452–454
Lyssavirus, 43, 143
  of Africa, 159–160
    Africa 2 lineage, 168
    Africa 3 lineage, 168
  American indigenous lineage, 171–174
  arctic-like 1 (AL1), 169
  Arctic lineage, 168–169
  Asian bat lyssaviruses, 158–159
  Asian lineage, 169–171, 170t
  Australian bat lyssavirus (ABLV), 156
  bat rabies, 231 (*see also* Bat rabies)
  circulation, 88
  cosmopolitan lineage, 165–168
  European bat lyssavirus (EBLV), 156–158
  evolution, 84
  functional significance, 155
  glycoprotein, 144–145
  Indian subcontinent, 169
  infections, terrestrial rabies (*see* Terrestrial rabies)
  Mononegavirales, 143–144
  nonrabies, 15
  phylogenetic analysis, 161–165, 161–164f
  population sequence heterogeneity, 143–144
  quasispecies, 143–144
  species, 153, 154–155t
  spill-over event, 144
  viral RNA polymerase, 144–145
  virus-host relationship, 144

M
Marmosets, 212
Marsupials, 217
Mass dog vaccination, 7
Matrix protein (M), 52–53
Maximum likelihood (ML) method, 149, 150–152f
Maximum parsimony (MP) method, 149, 150–152f
Mexican Baja California skunk (BCSK) variant, 166–167
Milwaukee protocol, 554–557, 554–556t

Mokola lyssavirus (MOKV), 109, 160, 246–247
Molecular epidemiology
  disease-causing agent, 143
  lyssavirus (*see* Lyssavirus)
  viral typing
    climate change, 177
    cost-effective tool, 145–146
    geopolitical boundaries, 174–175
    high-throughput sequencing (HTS) methods, 147–148
    host shift events, 176–177
    lyssaviruses, genetic characterization, 146
    mitochondrial genome, 175
    monoclonal antibodies (MAbs), 145
    phylogenetic analysis (*see* Phylogenetic analysis)
    risk assessments, 174
Mongooses, 210–211
Monoclonal antibodies (MAbs), 145
Mononegavirales (MNV), 43, 143–144
M protein-deleted rabies virus, 515–516
Multilayer cultivation systems, 490
Multiplicity of infection (MOI), 484–485

N

National Institute of Health (NIH) test, 454
National rabies elimination strategies, 675
National Rabies Prevention Program, 529
National Toxicology Program Interagency Center for the Evaluation of Alternative Toxicological Methods (NICEATM), 454
Negri bodies, 353–355, 356f, 410–411
Neighbor joining (NJ) tree, 149, 150–152f
Neural cell adhesion molecule (NCAM) CD56, 63–65
Neural cell adhesion molecule (NCAM) receptor, 308–309
Neuroinvasiveness, 313
Neuronophagia, 351–352
Neurotropism, 307
Nicotinic acetylcholine receptor (nAChR)
  glycoprotein, 307, 318f
  immunostaining, 308–309
  monoclonal antibodies, 307
  neuromuscular junction, 308–309
  opossums, 308
  postsynaptic cells, 308
  snake venom neurotoxins, 307
  viral antigen distribution, 307
Nonrabies lyssaviruses, 15
North American Rabies Management Plan (NARMP), 618
Nucleic acid amplification technologies, 146
Nucleic-acid-based detection, 411
Nucleic-acid sequence-based amplification (NASBA), 432
Nucleoprotein (N), 48–49, 509
Nucleotide sequencing technologies, 48

O

OIE. *See* World Organisation for Animal Health (OIE)
Oligosaccharyltransferase, 54
One World-One Health concept, 23
Opossums, 217
Oral dog vaccination, 22–23
Oral rabies vaccination (ORV), 10, 110, 172, 496–497, 606, 607–608f. *See also* Wild carnivores
  monitoring, 499

P

PAHO. *See* Pan American Health Organization (PAHO)
PAMPs. *See* Pathogen-associated molecular patterns (PAMPs)
Pan American Health Organization (PAHO), 21–22
Passive public health surveillance systems, 124
Passive surveillance systems, 124
Pathogen-associated molecular patterns (PAMPs), 381–382
Pathogenesis, 304f
  α-amino-3-hydroxy-5-methyl-4-isoxazole propionate (AMPA), 312
  apoptosis, 315–316
  blood-brain barrier (BBB), 315, 321
  brain dysfunction
    acetylcholine, 324–326
    γ-amino-n-butyric acid (GABA), 325
    behavioral changes, 328–330
    electroencephalographic (EEG) recordings, 325–326
    excitotoxicity, 327
    ion channels, 326
    neuropeptide synthesis, 323–332
    nitric oxide (NO), 326–327
    oxidative stress, 328, 334f
    serotonin, 324–325
  Brazilian free-tailed bats, 331–332
  central nervous system (CNS), 311–312
  challenge virus standard (CVS) strain, 305
  chronic rabies virus infection, 330–332
  colchicine, 310
  degeneration, neuronal processes, 316–320, 331f
  dynein light chain 8 (LC8), 310–311
  experimental models, 303
  fast axonal transport, 310–311
  incubation period, 303–305
  inflammatory changes, cats, 330–331, 334f
  innate immune responses, 315–323
  kainate receptors, 312
  low-affinity p75 neurotrophin receptor, 309–310
  mouse and hamster models, 311

neural cell adhesion molecule (NCAM) receptor, 308–309
neurectomy, sciatic nerve, 303–305
neurotropic virus, 303
neurotropism, 307
nicotinic acetylcholine receptor (nAChR) (*see* Nicotinic acetylcholine receptor (nAChR))
recovery, 330
reverse transcriptase-polymerase chain reaction (RT-PCR) amplification, 305
street rabies virus, 303–305
superficial and nonbite exposures, 306
time-dependent movement, 303–305
transneuronal tracer methods, 311
ultrastructural studies, 312
viral neurovirulence, 313, 319*f*
virus neutralization antibody (VNA), 332
Pathology
  cardinal pathological features, 347
  central nervous system (CNS)
    bullet-shaped or tubular virions, 355–357
    cell injury and cell death, 351–353
    clinical outcome, 348
    degeneration, neuronal processes, 358–360, 359*f*
    encephalitic rabies, 362
    fluorescent antibody test (FAT), 360, 363
    hematoxylin and eosin (HE), 354, 354*f*
    immunofluorescence, 360
    immunohistochemical distribution, 362–363
    immunoperoxidase staining, 360–361, 361*f*, 363
    inflammation, 349–351, 349*f*, 351*f*
    Mann methylene blue method, 354–355
    Negri bodies, 353–355, 356*f*
    neuronal and non-neuronal cell lines, 357–358
    polyclonal/monoclonal antiribonucleoprotein/nucleocapsid antibodies, 360–361
    rabies virus antigen (RVAg), 360–362, 364
    rabies virus genomic RNA, 361–362
    sequential immunofluorescence and infectivity titration studies, 362–363
  extraneural organs, 369–370, 370*f*
  inoculation site, 368–369
  macroscopic examination, 348
  neuron-to-neuron trans-synaptic viral dissemination, 347
  ocular pathology, 369
  peripheral nervous system (PNS)
    sensory and autonomic ganglia, 364–366, 365*f*
    spinal nerve roots and peripheral nerves, 366–368
    structural components, 364
Pattern recognition receptors (PRRs), 381–382
PCECV. *See* Primary chick embryo cell vaccine (PCECV)
Perivascular mononuclear inflammatory cell infiltrates, 349–350, 349*f*
Personal Protection Equipment (PPE), 404
Peruvian fox *(Lycalopex sechurae)*, 204
Pet Travel Scheme, 10–11
Phosphoprotein (P), 49–51
Phylogenetic analysis, 161–165, 161–164*f*
  Bayesian statistical methods, 153
  coalescent theory, 153
  data acquisition, 148
  maximum likelihood (ML) method, 149, 150–152*f*
  maximum parsimony (MP) method, 149, 150–152*f*
  neighbor joining (NJ) tree, 149, 150–152*f*
  nucleotide sequence data, 148
  nucleotide sequences, 153
  target sequences, 148
  taxon, 148
  tree reconstruction, 149
Phylogeography, 168
Phylogroup I, 43
Phylogroup II, 44
PML protein. *See* Promyelocytic leukemia (PML) protein
Polar fox. *See* Arctic foxes *(Vulpes lagopus)*
Polymerase chain reaction (PCR), 48
Postadaptation viral mutations, 116–117
Post-exposure prophylaxis (PEP), 481, 483, 484*t*
Postvaccinal encephalomyelitis, 291
Preadaptation viral adaptations, 116–117
Pre-exposure prophylaxis (PrEP), 481–483
Primary chick embryo cell vaccine (PCECV), 6
Probit analysis, 449–451
Procyonidae, 207
Promyelocytic leukemia (PML) protein, 382
Protein kinase C, 49–50
Protein vaccines, 517–518
PRRs. *See* Pattern recognition receptors (PRRs)
Public health management
  cell culture vaccines, 537
  economical PEP regimens, 539
  educational awareness, 528
  educational initiatives, 538
  exposure, 529–531, 530*t*
  Global Alliance Vaccine Initiative (GAVI), 528
  immunocompromised patients, 536–537
  integrated systematic approach, 528
  interchangeability and coadministration, 537–538
  multidiscipline strategies, 527–528
  National Rabies Prevention Program, 529
  postexposure prophylaxis (PEP), 528
  preexposure vaccination (PreP), 528
  pregnancy, 536
  travel medicine, 537
  vaccination protocols
    booster vaccinations, 534–536, 535*t*
    intradermally (ID), 531
    intramuscularly (IM), 531

Public health management *(Continued)*
    postexposure vaccination, 533–534
        preexposure vaccination, 531–532, 532t
    vaccine investment strategy (VIS), 528
Purified chick embryo cell vaccine (PCECV), 537–538, 552
Purified Vero cell rabies vaccine (PVRV), 6

Q
Quality management system (QMS), 445–446

R
Rabies Epidemiological Bulletin (REB), 681
Rabies immune globulin (RIG), 445–446, 533–534, 552
Rabies immunity measurement
    antigen-binding assays, 446–448
    anti-RABV glycoprotein monoclonal antibody, 446–448
    assay selection, 459–461
    assays, types, 448
    cellular immunity assays, 446
    effector functions, 446, 447f
    good manufacturing practice (GMP) quality standards, 445–446
    host factors, 445–446
    immune components, 448
    laboratory method parameters, 445–446
    mouse neutralization test (MNT), 445
    quality management system (QMS), 445–446
    quality results
        "fit for purpose" method, 463, 464t
        immunity, 463–465
        limit of detection (LOD), 463
        limit of quantitation (LLOQ), 463
        matrix effect, 462–463
        parameters, 462–463
        performance characteristics, 461–462, 465–466
        quality control practices, 466
        robustness evaluation, 462–463
        samples, 463, 464t
        serological titration assays, 466
        standard reference rabies immune globulin serum (SRIG), 465
    rabies immune globulin (RIG), 445–446
    RABV-neutralizing antibodies (RVNA), 446–448
    rapid fluorescent focus inhibition test (RFFIT), 445
    regulatory compliance
        corrective action/preventive action (CAPA System), 471–472
        development phases, 473–474
        drug development, 471–472, 472f
        Food and Drug Administration (FDA), 470–471
        fundamental needs, 473
        national/regional requirements, 470
        robust quality system, 470–472
        testing laboratory, 470–471, 471t
    serology, 446
        principle, 449
        serum neutralization (SN) assays (*see* Serum neutralization (SN) assays)
    vaccination, 466–470
Rabies vaccines, next generation
    clinical results, 512, 515t
    DNA vaccines, 518–519
    genetic vaccines, 518
    glycoprotein (G), 509
    ideal rabies vaccine, 520–521, 522t
    inactivated adjuvanted traditional rabies vaccines, 516
    inactivated genetically modified rabies vaccine, 516
    incidence and risk, 511
    intradermally (ID), 510
    intramuscularly (IM), 510
    live vaccines, 515
    matrix protein (M), 509
    M protein-deleted rabies virus, 515–516
    nucleoprotein (NP), 509
    peptide vaccines, 516–517
    phosphoprotein (P), 509
    polymerase (L), 509
    postexposure prophylaxis (PEP), 509
    preclinical results, 512, 513–514t
    protection, 511–512
    protein vaccines, 517–518
    rabies immunoglobulin (RIG), 509
    recombinant adenoviruses, 520
    recombinant poxviruses, 519
    RNA vaccines, 519
    viral vector vaccines, 519
    virus-neutralizing antibodies (VNAs), 509
Rabies virus (RABV)
    bacilliform, 44–46
    biologic and physicochemical features, 44
    canine rabies, health impact and economic burden, 695
    carbohydrate, 56
    composition, 44, 45f
    cone-shaped "defective" virions, 46–47
    control, 694
    cryo-electronmicroscopy, 46
    defective-interfering (DI) virions, 47
    electron density, 46
    epidemiology, 691–692
    evolution (*see* Evolution)
    genome (*see* Genome)
    life cycle (*see* Infection life cycle)
    Lyssavirus, 43
    Mononegavirales (MNV), 43
    oral vaccination programs, 689–690
    pathogenesis, 690

Phylogroup I, 43
Phylogroup II, 44
post-translational modifications, 44
prevention, 692–694
Rhabdoviridae, 43
RNA synthesis, 62
structure, 44, 45*f*
therapy of, 690–691
viral proteins, 56
  branched-chain oligosaccharides, 47–48
  cellular kinases, 47–48
  cytoskeleton proteins, 47–48
  glycoprotein (G), 53–55
  glycosylated protein, 47–48
  matrix protein (M), 52–53
  molecular chaperones, 47–48
  nucleoprotein (N), 48–49
  phosphoprotein (P), 49–51
  RNA polymerase activity, 47
  virion-associated RNA polymerase/large protein (L), 51–52
West Caucasian bat virus (WCBV), 44
Rabies virus antigen (RVAg), 360–362, 364
Rabies virus antigen in cell culture (RTCIT), 411–412
Rabies virus neutralizing antibodies, 553–554
Rabies virus (RABV) vaccine, 5–6, 25–28, 26*t*
RABV-neutralizing antibodies (RVNA), 446–448, 466–470
Raccoon *(Procyon lotor)*, 207–209, 208*f*
Raccoon dog *(Nyctereutes procyonoides)*, 205
Raccoon rabies variant (RRV), 171–172
Rapid fluorescent focus inhibition test (RFFIT), 445
  indirect immunoperoxidase virus neutralization (IPVN) technique, 452–454
  limit of detection (LOD), 452–454
  lower limit of quantification (LLOQ), 452–454
  procedure, 449–451, 450*f*
  results, 451–452, 451*t*
  virus neutralization testing method, 452–454
Reactive oxygen species (ROS), 50–51
Real-time reverse transcription-polymerase chain reaction (RT-PCR), 424*t*, 431
  application, 431–432
  assay selection, 429
  intercalating dyes, 427–428, 428*f*
  lyssavirus assays, 423
  Mock extraction control, 429–430
  5′ nuclease assay
    dual-labeled probe (DLP), 423–424
    emission wavelengths, 424
    reverse primer, 425–427
    TaqMan technology, 425, 426*f*
  positive and negative PCR controls, 430
  template integrity, 430

Recombinase polymerase amplification (RPA), 433
Red foxes *(Vulpes vulpes)*
  Asia, 201–202, 203*f*
  cross-species transmission events, 203
  geographic distributions, 201
  host-switching events, 203
  maintenance and epidemiology, 201–202
  in Middle East, 201, 202*f*
  rabies virus lineage, 201
  reservoir species, 202–203
  spatial and chronological detection, 202–203
Reed and Muench formula, 449–451
Reverse transcriptase-polymerase chain reaction (RT-PCR) amplification, 146, 305
Reverse transcription loop-mediated isothermal amplification (RT-LAMP), 146
Reverse transcription-polymerase chain reaction (RT-PCR) methods
  complementary DNA (cDNA) strand, 421
  deoxynucleotide triphosphates (dNTPs), 418
  DNA polymerase, 422
  extension phase, 422
  lyssavirus detection, 419
    real-time RT-PCRs (*see* Real-time reverse transcription-polymerase chain reaction (RT-PCR))
  primer design, 419–421, 420*t*
  principle of, 418
  product detection, 423
  sequence-specific primers, 421–422
  synthetic oligonucleotides/primers, 418
Rhabdoviridae, 43
RNA vaccines, 519
Rodents, 217
ROS. *See* Reactive oxygen species (ROS)
RRV. *See* Raccoon rabies variant (RRV)

S

Sanger sequencing methods, 146
SARE tool. *See* Stepwise Approach toward Rabies Elimination (SARE) tool
Semple vaccine, 5–6
Serum neutralization (SN) assays
  antibody kinetics, 458–459
  anti-IgG, 458–459, 459*t*
  anti-RABV virus antibody, 449–451
  blocking ELISA, 456–457
  competitive ELISA (cELISA), 456–457
  electrochemiluminescent (ECL) adaption, 457
  enzyme-linked immunosorbent assay (ELISA), 454–456, 455*f*, 458–459
  European Directorate for the Quality of Medicines and Health Care (EDQM), 454
  glycoprotein, 456–457

Serum neutralization (SN) assays *(Continued)*
   immunochromatographic assays, 457–458
   indirect fluorescent antibody (IFA), 454–455
   National Institute of Health (NIH) test, 454
   National Toxicology Program Interagency Center for the Evaluation of Alternative Toxicological Methods (NICEATM), 454
   nucleoprotein (N), 449–451
   Probit analysis, 449–451
   rapid fluorescent focus inhibition test (RFFIT)
      indirect immunoperoxidase virus neutralization (IPVN) technique, 452–454
      limit of detection (LOD), 452–454
      lower limit of quantification (LLOQ), 452–454
      procedure, 449–451, 450*f*
      results, 451–452, 451*t*
      virus neutralization testing method, 452–454
   Reed and Muench formula, 449–451
   Spearman-Karber formula, 449–451
   in vitro immunogenicity assay, 454
   World Health Organization (WHO), 449
Shimoni bat lyssavirus (SHIBV), 109, 160, 246
Silver-haired bat RABV variant (SHBV), 250
Skunks, 209–210
Snow fox. *See* Arctic foxes *(Vulpes lagopus)*
South central skunk (SCSK) variant, 171–172
Spearman-Karber formula, 449–451
"Spillover" phenomenon, 17
Standard reference rabies immune globulin serum (SRIG), 465
Steppe fox *(Vulpes corsac)*, 204
Stepwise Approach toward Rabies Elimination (SARE) tool, 675–676
Street-Alabama-Dufferin (SAD) strain, 636–637
Street rabies virus, 303–305
Susceptible, Exposed, Infectious, and Vaccinated (SEIV) framework, 656, 657*f*
Sylvatic (wildlife) rabies cycles, 108

T

Taiwan bat lyssavirus (TWBLV), 159, 248
TaqMan technology, 425, 426*f*
Terrestrial rabies
   acute progressive encephalomyelitis, 195
   Canidae, 199–200
   Carnivora, 195, 199
   Chiroptera, 195, 199
   clinical signs, 197–198
   coyotes, 205, 207
   disease distribution, 195
   domestic dogs, 200–201
   ferret badgers, 211–212
   foxes *(see* Foxes*)*
   incubation period, 196–197, 197*t*
   jackals, 205–207, 206*f*
   Lyssavirus infections *(see* Lyssavirus, infections*)*
   marmosets, 212
   mongooses, 210–211
   pathogenesis, 195–196
   raccoon *(Procyon lotor)*, 207–209, 208*f*
   raccoon dog *(Nyctereutes procyonoides)*, 205
   reservoir species, 198
   skunks, 209–210
   spillover (dead-end) hosts
      carnivorous animal species, 212
      cats, 213
      cattle, 213–215
      Ethiopian wolf, 216
      in Europe, 212, 213*f*
      fox epizootics, 216
      gray wolves, 215–216
      kudu, 215
      marsupials, 217
      opossums, 217
      rabid domestic dogs, 216
      rabid wolves, 216
      rodents, 217
      in United States, 212, 214*f*
Tetanus, 290–291
Toll-like receptors (TLRs), 381–382
Transneuronal tracer methods, 311
Trap-vaccinate-release (TVR) strategy, 606
Tumor necrosis factor (TNF), 496
TWBLV. *See* Taiwan bat lyssavirus (TWBLV)

V

Vaccinate-assess-move programs, 132–133
Vaccination protocols
   booster vaccinations, 534–536, 535*t*
   intradermally (ID), 531
   intramuscularly (IM), 531
   postexposure vaccination, 533–534
   preexposure vaccination, 531–532, 532*t*
Vaccine investment strategy (VIS), 528
Vaccinia-rabies glycoprotein recombinant vaccine baits (VRG), 209, 261
Vampire bat rabies
   in Americas, 237–239
   Brazilian free-tailed bats, 233–234
   epizootics, 234–236
   geographic range, reservoir, 234–236, 234–235*f*
   hematophagous bats, 234–236
   human infections, 236–237
   incidence, 237
   laboratory diagnoses, 233
   rabies-specific antibodies, 236–237

resource supplementation studies, 234–236
sex-biased host dispersal, 234–236
Viral mutations, 116–117
Viral neurovirulence, 313, 319f
Viral proteins
- branched-chain oligosaccharides, 47–48
- cellular kinases, 47–48
- cytoskeleton proteins, 47–48
- glycoprotein (G), 53–55
- glycosylated protein, 47–48
- matrix protein (M), 52–53
- molecular chaperones, 47–48
- nucleoprotein (N), 48–49
- phosphoprotein (P), 49–51
- RNA polymerase activity, 47
- virion-associated RNA polymerase/large protein (L), 51–52

Viral vector vaccines, 519
Virus-neutralizing antibodies (VNAs), 249–250, 332, 509
Vulpes, 201

W

West Caucasian bat lyssavirus (WCBV), 44, 156, 158, 242, 252–253
West Nile virus brain infection, 388
White fox. *See* Arctic foxes *(Vulpes lagopus)*
Wild carnivores
- Africa, 614–615
- Asia, 614
- bait delivery, 627
- bait development, 637–638
- bait products, 634, 635t
- Caribbean, 615–616
- Central America, 615–616
- communication and collaboration, 634
  - Canadian Rabies Management plan, 618
  - multisector One Health approach, 617
  - North American Rabies Management Plan (NARMP), 618
  - strategic planning, 617–618, 619f
- contingency actions, 630–631
- control and elimination, 616
- economics, 616–617
- enhanced rabies surveillance (ERS), 620–621
- Europe, 609–610
- hand baiting, 629, 629f
- host densities, 630
- ideal oral rabies vaccination programs, 626, 626t
- landscape-level management strategy, 606
- management actions
  - fixed-wing aircraft oral rabies vaccination, 621, 622f, 625–630
  - high-risk management zones, 621
  - population reduction, 622–623
  - trap-vaccinate-release, 624
- nontarget wildlife, 627
- North America, 610–614, 612–613f
- operational cost efficiency, 629–630
- program monitoring and evaluation, 632–634
- public and agricultural health, 605–606
- public health rabies surveillance, 620
- South America, 615–616
- Street-Alabama-Dufferin (SAD) strain, 636–637
- toxic baits and bait stations, 606
- translocation, 631–632
- trap-vaccinate-release (TVR) strategy, 606
- in United States, 627, 628f
- vaccination routes, 634–635
- vaccine distribution method, 629–630
- vaccine preservation methods, 637

Wildlife rabies surveillance systems, 108
World Health Organization (WHO), 6, 118, 119t, 406, 528
World Organisation for Animal Health (OIE), 123–125, 127, 407, 449, 498–499, 573, 671, 678